Fifth Interdisciplinary Scientific Forum with International Participation "New Materials and Promising Technologies"

IOP Conference Series: Materials Science and Engineering
Volume 848

Moscow, Russia
30 October - 1 November 2019

ISBN: 978-1-7138-4731-1
ISSN: 1757-8981

Printed from e-media with permission by:

Curran Associates, Inc.
57 Morehouse Lane
Red Hook, NY 12571

Some format issues inherent in the e-media version may also appear in this print version.

Institute of Physics
Dirac House, Temple Back
Bristol BS1 6BE UK

Phone: 44 1 17 929 7481
Fax: 44 1 17 920 0979

techtracking@iop.org

Additional copies of this publication are available from:

Curran Associates, Inc.
57 Morehouse Lane
Red Hook, NY 12571 USA
Phone: 845-758-0400
Fax: 845-758-2633
Email: curran@proceedings.com
Web: www.proceedings.com

TABLE OF CONTENTS

Preface

Peer review statement

A structure formation in Ti-Ni powder after high-energy mechanical treatment 1
E V Abdulmenova, S N Kulkov

Development of chitosan-hyaluronic acid nanoparticles and study of their physico-chemical properties for targeted delivery of anticancer drugs 5
N P Akentieva, A R Gizatullin, O Silvestre, O Savchuk, N I Shkondina, T P Prichodchenko, D V Mitschenko, A V Zhilenkov, P A Troshin, N A Sanina, N H Dremova, V I Torbov, S M Aldoshin, J B Nieder

Deoxidation Equilibrium of Silicon in the Nickel-Cobalt Melts 13
A A Aleksandrov, V Ya Dashevskii, A G Kanevskii

Minimal pore size reached upon sintering 20
M I Alymov, S I Averin, I V Saikov

Interaction of exogenous refractory nanoparticles of Al_2O_3 and $MgAl_2O_4$ with nonferrous metal impurities in iron melts 23
S N Anuchkin

In silico determination of substrate spectrum of lactonases, hydrolyzing various N-acyl homoserine lactones 33
A G Aslanli, E N Efremenko

High-temperature physical properties of the electrodes based on MAX-phase Ti-Al-C produced by SHS-extrusion method 42
O A Averichev, A M Stolin

Self-propagating high-temperature synthesis of ceramic composite material based on Al_2O_3 - SiO_2 - ZrB_2 49
A P Chizhikov, P M Bazhin, A M Stolin

TiB / 30 wt.% Ti layered composite material obtained by free SHS compression on a Ti6Al4V titanium alloy 56
P M Bazhin, A P Chizhikov, A S Konstantinov, A D Prokopets, E V Kostitsyna, A V Bolotskaya, A M Stolin, N Yu Khomenko

The impact of the lead on the physicochemical and optical properties of the fluorophosphate glasses doped by Nd3+ ions 62
O Bogdanov, E Kolobkova

Microstructural evolution of the AlMgMnZr alloy during severe plastic deformation 67
Yu Borisova, D Yuzbekova, A Kalinenko, A Mogucheva

Synthesis, photoluminescence and thermal properties of nanostructured organosilicon luminophore based on 2,2'-bithienyl and 4,7-diphenyl-2,1,3-benzothiadiazole 72
O V Borshchev, M S Skorotetcky, E A Svidchenko, N M Surin, S A Ponomarenko

Modification of barium aluminosilicate glass-ceramics with zirconium oxide additives 78
A S Chaynikova, V S Kovaleva, D A Zabelin, I O Belachenkov, S Y Modin

Refractive index of fluoride crystals doped with rare earth ions with low concentration 85
D S Chunaev, A Y Karasik, V A Konyushkin, A N Nakladov, P G Zverev

Functional properties and microstructure of TiNi alloy during multiple martensitic transformations 91
A A Churakova

Thermochemical grinding of diamond films in installation for chemical vapor deposition of diamond from the gas phase 97
S A Eremin, A M Kolesnikova, I A Leontiev, V N Anikin, O Yu Kudryashov, Y M Yashnov, M V Zhdanova

New method of obtaining heat-resistant biomorphic composites with carbide reinforcement......................... 107
A E Ershov, D V Prokhorov, T S Stroganova, I M Shmytko

Inhibitors based on new polysaccharides in technological processes of oil production. Efficiency of inhibition of scale formation and influence on stability of water-oil emulsion 117
A V Fakhreeva, A I Voloshin, A G Telin, L A Magadova, L V Spirikhin, V A Dokichev

Reactionary-solidified oxygen permeable membrane material based on cermet $Bi_{1.6}Er_{0.4}O_3$ – 26 wt % Ag – 4 wt % In 124
S V Fedorov, A S Lysenkov, I V Kulbakin

Effect of tantalum on short-term creep of a 12%Cr-3%Co-0.07%Ta martensitic steel for steam blades 129
A E Fedoseeva, I S Nikitin, A E Fedoseev, E S Tkachev, R O Kaibyshev

Spectrophotometric determination of epinephrine using new analytical systems based on label-free silver triangular nanoplates.. 137
A A Furletov, V D Zaytsev, V V Apyari, A V Garshev, P A Volkov, S G Dmitrienko

Silver nanocomposites based on copolymers of N,N-diallyl-N'-acylhydrazines with N-vinylpyrrolidone 145
M Gorbunova, T Batueva

Gold nanorods and their nanocomposites based on polyurethane foam for determination of catecholamines in biological fluids 152
M V Gorbunova, V V Apyari, A V Garshev, P A Volkov, V V Tolmacheva, S G Dmitrienko

Problems of aviation leaded gasoline application on aircraft 163
K I Gryadunov, M L Nemchikov, I S Melnikova, A N Timoshenko

Gradient structure formation in titanium alloys using thermohydrogen treatment technology 167
S V Skvortsova, O N Gvozdeva, A V Shalin, A S Stepushin

The study of model reactions occurring in the synthesis of adaptive multiblock polyurethanes..................... 175
M A Gorbunova, A M Imamutdinova, V P Lodygina, D V Anokhin, E R Badamshina

The application of copper-containing nanoparticles as a gas nitriding catalyst... 182
I Yu Isaeva, E A Eliseeva, I V Morenko, V A Aleksandrov, G Yu Ostaeva

The electrochemical synthesis and investigation of nanostructured Fe-Pt and Co-Pt systems 188
N Ivanova, A Lobanov, A Andyyakova, Yu Zakharov, A Popova, R Kolmykov

Application of optical methods in tests of new materials with gradient structure 196
I S Kamantsev, N A Drukarenko, A V Kuznetsov, S I Stepanov, M A Chesnokov

Technology of spark plasma sintering as an innovative solution of synthesis high-density of SiC ceramics 200
G D Kardashova

Oxide composite fibers with quasiplastic components and composites with brittle matrices on their basis 205
V M Kiiko

Rheological properties of Si_3N_4 and Si_3N_4 with sintering additive CaO-Al_2O_3 powders 213
K A Kim, A S Lysenkov, D D Titov, E A Gumennikova, M G Frolova, D O Lemeshev, S N Perevislov, Yu F Kargin

Characterization of spherical stainless steel powders prepared by electric arc spraying process 217
A A Kirsankin, T A Kalaida, M A Kaplan, M A Smirnov, A Yu Ivannikov, M A Sevostyanov

Electrochemical investigations of surfactants influence on copper ions cementation 222
E B Kolmachikhina, E A Ryzhkova, D V Starkova, V A Val'nev

Comparative analysis of the influence of urea additives on the properties of poly(vinyl alcohol) cryogels formed from the polymer solutions in water or dimethylsulfoxide media 229
O Yu Kolosova

The ratio of refractive index and molecular weight of copolymers of n-butyl acrylate and styrene obtained by the method of RAFT-radical polymerization 238
E V Vaganov, R I Komendant, E O Perepelitsina, V P Grachev, S A Kurochkin

Using of direct piezo-discharge in generation of plasma-activated liquid media 244
E M Konchekov, N G Gusein-Zade, L V Kolik, K V Artem'ev, A V Pulish

The effect of homogenizing annealing on the microstructure and microhardness of Ti-20Nb-(7.5-10)Ta alloys (at.%) 250
S V Konushkin, A Yu Ivannikov, E O Nasakina, K V Sergienko, L A Shatova, M A Sevostyanov

The investigation of the welded joints of the nitrogen containing cast austenitic steel, obtained by the manual arc welding 255
V S Kostina, M V Kostina, M G Sharapov, I O Samodurov, S O Muradyan

Molecular Docking simulation and Fluorescence lifetime characteristics of NIR Cyanine Dye Complexes with Albumin 261
A Kostyukov, M Mestergazi, A Shmykova, T Podrugina, V Pogonin, E Radchenko, V Palyulin, I Borissevitch, V Kuzmin

Effect of evaporating additive on sintering and properties ceramics based on magnesium aluminate spinel 272
E A Kozlovskaya, D S Kseneva, M O Senina, D O Lemeshev

Dicationic disiloxane liquids 278
V G Krasovskiy, O B Gorbatsrvich, L M Glukhov, E A Chernikova

Preparation and characterization of stable water soluble hybrid nanostructures of hydrophobic compounds by encapsulation into nanoparticles of amphiphilic N-vinylpyrrolidone copolymers of new generation 288
S V Kurmaz, D V Konev, V D Sen', V A Kurmaz, A V Kulikov

Determination of phase transition temperatures (melting, crystallization, rotator phases) of n-alkanes by the optical method 298
V N Kuryakov, D D Ivanova, A N Tkachenko, P N Sedenkov

Changes in electrochemical properties of a heavy tungsten alloy during its processing under the influence of DC current in ammonia-alkali solutions 305
O G Kuznetsova, A M Levin, M A Sevostyanov, O I Tsybin, A O Bolshikh

AC electrochemical oxidation of nickel and VNZh alloy in alkaline-ammonium solutions 311
O G Kuznetsova, A M Levin, M A Sevostyanov, O I Tsybin, A O Bolshikh

Study of the effect of crystallization conditions on the structure of polyurethane block copolymer based on poly-ε-caprolactone diol and isophoron diisocyanate 316
M A Gorbunova, V A Lelecova, D M Shukhardin, G V Malkov, D V Anokhin, E R Badamshina

Development of scientific and technological bases of processing of cast iron from thermal timing and ladle processing method of the resonant intermittent refining 322
D A Lubyanoi, A V Markidonov, V M Nevolin, D D Lubyanoi, I O Kambalin

Development of optimal formulations of natural alloyed cast irons for metals and engineering, and thermal timing subjected to secondary treatment by the method of resonance-intermittent refining 327
D A Lubyanoi, S Yu Shevchenko, D D Lubyanoi, A V Markidonov, E B Zvarych

Effect of synthesis parameters on the structure and properties of carbon particles formed from amorphous fullerites 332
I N Lukina, O P Chernogorova, E I Drozdova, E A Ekimov, M O Apostolova, D A Prokopenko, A V Soldatov, V Benavides

Fabrication and characterization of 1, 5, 10 at.% Ce:Y_2O_3 nanopowders 338
K E Lukyashin, A S Chepusov, V I Solomonov

21R-Sialon ceramics, obtained by hot pressing 343
A S Lysenkov, M O Stolbova, D D Titov, A I Plokhikh, K A Kim, D V Gridin, M G Frolova, N V Petrakova, K D Danilin, D O Lemeshev, Yu F Kargin

Application of mutagenic treatment of active silt for oxidation of cellulose nitrate 349
E A Saratovskikh, L V Avdeeva, V A Shcherbakova, A I Kazakov, R N Yarullin

Perfluorocarbon nanoemulsions containing fluorinated photosensitizer for photodynamic cancer therapy 359
A A Markova, E V Belyaeva, A S Radchenko, M V Dmitrieva, A L Sigan, A A Shtil', N D Chkanikov

Carbon-forming properties of oils influencing the service of engines and aircraft reducers 368
I S Melnikova, L V Kovba, D V Ratenko, V S Dvoyeglazov

Research of the intermetallics formation mechanism during the synthesis of functionally graded layered steel-aluminum compositions 375
R S Mikheev, I E Kalashnilov, L K Bolotova, A G Kolmakov

Effect of ions concentration in buffer solutions on the nucleation of hydroxyapatite surface on the octacalcium phosphate granules 382
P V Mikheeva, A Yu Teterina, I V Smirnov, A Yu Fedotov, V S Komlev

Study characteristics of failure multilayered steel materials in conditions at alternating symmetrical bending 388
A A Minakov, A I Plokhihk, D V Vlasova, K B Polikevich

Hardness and fracture-toughness of hot-pressed LaB_6-TiB_2 ceramics 395
D D Nesmelov, E S Novoselov, A S Lysenkov, S V Vikhman, S N Perevislov

Reactive hot-pressed LaB_6-W_2B_5 ceramics 401
D D Nesmelov, E S Novoselov, D D Titov, S S Ordan'yan, S N Perevislov

Copper and cerium co-substituted hydroxyapatite: powders synthesis and sintering........ 406
Y O Nikitina, N V Petrakova, A A Egorov, D D Titov, A A Ashmarin, S M Barinov, V S Komlev

Distribution of impurities in ceramics based on zirconium nitride obtained using the oxidative constructing approach........ 412
I A Kovalev, S V Shevtsov, K G Vorkachev, A A Ashmarin, A V Shokodko, A I Ogarkov, N A Ovsyannikov, G P Kochanov, A V Tenishev, V N Kazakova, D P Shornikov, E A Shokodko, S S Strelnikova, A S Chernyavskii, K A Solntsev

Hard radiation of reological explosion 419
I B Oparina, V V Shienok

Scandium extraction from multicomponent systems by crystallization of complex sulfates 423
L A Pasechnik, I S Medyankina, S P Yatsenko

Investigation of the influence LiF sintering additive on the properties of magnesium aluminate spinel ceramics 428
M O Senina, D O Lemeshev, M S Pedchenko

Thermal conductivity of SiC-B_4C materials obtained by reaction-sintering method 433
S N Perevislov, E S Motaylo, E S Novoselov, D D Nesmelov

Physical and mechanical properties of composite materials in the $MoSi_2$-SiC-TiB_2 system........ 439
S N Perevislov, M A Markov, E S Motailo, S V Vikhman, D D Titov

Sintering and physico-mechanical properties of materials based on silicon nitride nanoscale powders 444
S N Perevislov, O A Lukyanova, A S Lysenkov, K A Kim, A B Vysotin

Studying the properties of carbides in the system ZrC–HfC, TaC–ZrC and TaC–HfC........ 451
S N Perevislov, A B Vysotin, O Yu Shcherbakova

Structural changes after high-speed impact of tungsten powder with a steel target 456
E V Petrov, V S Trofimov, G R Saikova, V O Kopytskiy

Oxidation and biodegradation of polymeric composites based on polylactide: structure and properties........ 462
M V Podzorova, Yu V Tertyshnaya

Silver niobate doped lead-free perovskite KNN ceramics........ 468
E D Politova, G M Kaleva, N V Golubko, A V Mosunov, N V Sadovskaya, D A Kiselev, A M Kislyuk, T S Ilina, S Yu Stefanovich

Tunable parallel plate waveguide array based on VO_2 thin films........ 477
V I Polozov, S S Maklakov, S A Maklakov, V A Chistyaev, A A Politiko, K M Baskov, A D Mishin, D A Petrov, V N Kisel

The cobalt and nickel-containing polyethylene prepared by using metal-vapor synthesis........ 485
P V Pribytkov, A Yu Vasil'kov, A V Naumkin

Surface modification TRIP \ TWIP steels 495
D V Prosvirnin, M D Larionov, A G Kolmakov, A V Larionova, M E Pruckov, S V Pivovarchik

Metals Recovery from Molybdenite Concentrate by Electrooxidation and Leaching 499
S N Rasulova, V P Guro, E T Safarov, X F Adinaev

Precision human bone prototypes manufactured by 3D printing for training and experimental studies 507
I D Romanov, A D Romanov, N A Kulinchenko, R O Gorbatov, A A Klenin

Study of sintered aluminum materials with nanoparticles microadditions 512
L E Agureev, S V Savushkina, I N Laptev, B S Ivanov, A V Ivanov, V I Kostikov, Zh V Eremeeva

Research of thermoradiation properties and expansion of tantalum carbide at high temperatures 519
V N Senchenko, R S Belikov, A V Tenishev

Optimization potential of anaerobic biocatalytic processes using intracellular ATP concentration as the main criterion for decision making 525
O V Senko, O V Maslova, E N Efremenko

Elementoxanealumoxanes – modifiers of silicon carbide ceramic composites components 532
G I Shcherbakova, P A Storozhenko, T L Apukhtina, M Kh Blokhina, D V Zhigalov, E A Novokovskaya, M S Varfolomeev, D D Titov, A A Ashmarin

Study of high-frequency magnetic properties of Fe-Ti-B films obtained by magnetron sputtering 538
E N Sheftel, E V Harin, V A Tedzhetov, G Sh Usmanova, S Y Bobrovskii, K N Rozanov, P A Zezyulina, Ph V Kiryukhantsev-Korneev

Advantages of drawing and rolling metals with pulse current 544
O A Troitsky, V I Stashenko

Microscopic research of amorphous alloys AlFeNiLa exposed to magnetic pulse processing 552
A A Viryus, T P Kaminskaya, M N Shipko, N D Bakhteeva, V V Korovushkin, A G Savchenko, M A Stepovich, E S Savchenko, E V Todorova

Novel materials for high-resolution three-dimensional printing using surface-selective laser sintering 557
M Syachina, S Minaeva, E Krumins, S Howdle, V Popov

Mass-transfer kinetics, structure, and tribological properties of coatings deposited on steel in Ar or N_2+O_2 by electro-spark alloying using Cr_3C_2-NiAl electrodes 562
Ph V Kiryukhantsev-Korneev, A D Sytchenko, V A Gorshkov, E A Levashov

Reducing of holes effect on composite elements strength 567
E Kh Akhmedshin, A N Polilov, N A Tatus

Modeling of paths and energy losses of high-energy ions in single-layered and multilayered materials 571
D I Tishkevich, S S Grabchikov, E A Grabchikova, D S Vasin, S B Lastovskiy, A S Yakushevich, D A Vinnik, T I Zubar, I V Kalagin, S V Mitrofanov, D V Yakimchuk, A V Trukhanov

Rheological properties of $Y_3Al_5O_{12}$ powder obtained by preceramic organoyttroxanealumoxanes 581
D D Titov, E A Gumennikova, K D Danilin, N V Petrakova, A S Lysenkov, M G Frolova, G I Shcherbakova, E A Novokovskaya, D O Lemeshev, Yu F Kargin

Structure Of The $Al_{85}Ni_7Fe_4La_4$ Polycrystalline Alloy After Complex Impacts Of Deformation And Flash Lamp Annealing 587
N D Bakhteeva, E V Todorova, S V Kannykin

Effect of rotary swaging on structure and properties of low-carbon steel 592
A A Tokar, V A Lunev, A S Dolzhenko, O V Rybalchenko, M M Morozov, V S Yusupov, G V Rybalchenko, P D Odessky, S V Dobatkin

Influence of gallium oxide concentration on the properties of transparent ceramics based on magnesium aluminate spinel 602
A V Ulyanova, M O Senina, D O Lemeshev

Effect of sintering temperature on magnetic hysteresis characteristics of powder alloy Fe-30Cr-8Co (wt.%) 609
A S Ustyukhin, V A Zelensky, I M Milyaev, A B Ankudinov, V S Shustov

Phosphors based on Ce:(Pb,Gd)$_3$(Al,Ga)$_5$O$_{12}$ epitaxial films: synthesis, optical properties, application 615
D A Vasil'ev, D A Spassky, V G Plotnichenko, V V Voronov, A V Khakhalin, A M Galstyan, N V Vasil'eva

Magnetic properties of the Fe-24%Cr-15%Co-3%Mo-1.5%Ti hard magnetic alloy in anisotropic and isotropic states 622
T A Vompe, I M Milyaev

Preparation and researching the properties of organic aerogels based on epoxy resins 628
A E Voytik, G V Malkov, A T Kapasharov, E I Knerelman, A Yu Kostin

Ceria-based solid solutions for environmental application........ 637
I V Zagaynov, I V Shelepin, S V Fedorov, A A Klimashin, I V Kulbakin

Investigation of iron grains growth during solid-phase carbothermic reduction of red mud in the presence of iron metallic particles 641
A S Zakunov, P I Grudinsky, D V Zinoveev, A F Semenov, M O Panova, V G Dyubanov, A L Petelin

Study of active media based on microparticles of solid-state laser materials........ 652
O A Burdukova, V A Konyshkin, Yu L Kopylov, K V Lopukhin, V A Petukhov, Yu V Senatsky, P G Zverev

Ti-(15-25)Nb-5Ta Alloy Plate Hardness Research for Medical Applications........ 662
S V Konushkin, K V Sergienko, A S Baikin, A A Kolmakova, N V Berezina, A V Mikhailova, Y A Morozova, E P Balashov, A G Kolmakov, M A Sevostyanov

Corrosion resistance of nonnickel shape memory alloy 668
E O Nasakina, M A Sudarchikova, A M Tsareva, K V Sergiyenko, S V Konushkin, M A Kaplan, E P Balashov, M A Sevost'yanov

Research of mechanical properties of composite material based on titanium nickelide with a titanium surface layer depending on magnetron sputtering time........ 674
E O Nasakina, M A Sudarchikova, A M Tsareva, K V Sergiyenko, S V Konushkin, M A Kaplan, A S Baikin, L A Shatova, M A Sevost'yanov

Study of the effect of the catalyst on the mechanical properties of RTV grade siloxane 681
A M Tsareva, E O Nasakina, M A Sudarchikova, E P Balashov, A S Baikin, L A Shatova, M A Sevost'yanov

Study of the effect of the catalyst on the microstructure and film thickness of RTV siloxane........ 686
A M Tsareva, E O Nasakina, E P Balashov, A S Baikin, S V Konushkin, M A Sevost'yanov

The development of the high-strength corrosion resistant austenitic steels for the oil equipment shafts 692
S O Muradyan, O S Muradyan, M V Kostina, V S Kostina

Study of the processes of three-dimensional printing of caprolactone copolymers with methylphosphate groups 697
A O Mariyanats, A V Mironov, I E Nifant'ev, A V Shlyakhtin, V K Popov

Study of products influence of rice waste liquid-phase catalytic oxidation on growth and plant development 705
A V Khvatov, S M Sevostyanov, P A Sakharov, E A Bocharnikova, N Ph Deeva, B K Son, D Yu Aladin, S M Lomakin, Yu K Lukanina, A Yu Zherebker, A S Kononikhin, D V Demin

Synergetic flame retardant effect of bio-flame retardant based on oxidized wood in polyester's compositions 711
A V Khvatov, P A Sakharov, S M Lomakin, S D Varfolomeev, Yu K Lukanina, A A Minikh, S M Frolov

Numerical solution of stress-strain state during electromechanical treatment of plasma sprayed coatings 717
A Yu Ivannikov

Research and development of the inert gas atomization of the wire by means of arc spraying 724
A Yu Ivannikov, A A Kirsankin, T A Kalayda, M A Sevostyanov

Formation of $Si_3Al_3O_3N_5$ oxonitride from mixtures of xerogels and silicon and aluminum nitrides in the nitrogen atmosphere 730
S N Ivicheva, A A Klimashin, N A Ovsyannikov, A S Lysenkov, Yu F Kargin

Properties of composites SiC/SiC_f obtained by hot pressing of SHS of silicon carbide powder 734
M G Frolova, A S Lysenkov, D D Titov, E I Istomina, S N Perevislov, K A Kim, V V Zakorzhevsky, S V Fedorov, Yu F Kargin

Studies of the effectiveness of the chemical decomposition of polychlorinated biphenyls in alluvial meadow soils in situ under the influence of special physicochemical active compounds 738
D Yu Aladin, S M Sevostyanov, N Ph Deeva, D V Demin

Author Index

IOP Conf. Series: Materials Science and Engineering **848** (2020) 011001 doi:10.1088/1757-899X/848/1/011001

Introduction

From October 29, to November 1, 2019, the Fifth Interdisciplinary Youth Scientific Forum with international "New materials and promising technologies" was held in Moscow in the building of the Presidium of Russia Academy of Sciences, where topical and applied problems of creating new materials for the future and promising technologies for their processing were discussed. The forum was held for the fifth time. The Forum was organized by the Council of Young Scientists of the Russian Academy of Sciences and the Coordinating Council for Youth Affairs under the President of the Russian Federation. Co-organizers of the Forum were the Russian Academy of Sciences (RAS), The Russian Chemical Society DI. Mendeleev, Institute of Metallurgy and Materials Science A.A. Baykova and The Ministry of Science and Higher Education of the Russian Federation.

Thematic areas included in the program of the Forum, helped to gather in one place representatives of different interdisciplinary research groups in the area of interest which includes the creation of the fundamentals of materials and their application in various industries. The Forum program was expanded by discussing the problems of introducing scientific developments and interaction of scientists with consumers of scientific results, foundations and development institutions. Additionally, the role of international cooperation, public and educational organizations in the scientific process, and the strategy of popularizing scientific achievements were discussed at the Forum.

The scientific sections of the forum covered a wide range of topics on nanomaterials and nanotechnology, experimental methods for researching materials and structures, structural materials, functional materials, biomaterials and technologies, materials and technologies for green chemistry, additive technologies and promising processes in metallurgy.

Peer review statement

All papers published in this volume of *IOP Conference Series: Materials Science and Engineering* have been peer reviewed through processes administered by the proceedings Editors. Reviews were conducted by expert referees to the professional and scientific standards expected of a proceedings journal published by IOP Publishing.

IOP Conf. Series: Materials Science and Engineering **848** (2020) 012001 doi:10.1088/1757-899X/848/1/012001

A structure formation in Ti-Ni powder after high-energy mechanical treatment

E V Abdulmenova, S N Kulkov

[1]Institute of Strength Physics and Materials Science SB RAS, Tomsk, Russia
[2] National Research Tomsk Polytechnic University, Tomsk, Russia

E-mail: Ekaterina.V.Abdulmenova@yandex.ru

Abstract. In this work a structure formation in Ti-Ni powder after high-energy mechanical treatment in a planetary ball mill with an acceleration of 60g are presented. Mechanical treatment time of powder was 5-300 seconds. It was shown the powder in initial state consists of phase mixture TiN (B2, B19`), Ti_2Ni (E9$_3$) and $TiNi_3$ (D0$_{24}$). The lattice parameters of powders not significantly change at mechanical treatment and correspond in good agreement with literature data. The quasi-amorphous phase is formed after mechanical treatment and the relative content of phases change so that the relative content of quasi-amorphous phase increases steeply from 0 % to 45 % in first 10 second of treatment mechanical treatment and not significantly changes at more prolong milling. The full widths at half maximum of peaks not significantly change to 10 second mechanical treatment, but the more prolong milling leads to them increasing due to forming of dislocation in all phases, wherein the value of the dislocation density of Ti_2Ni phase is more than of TiNi and Ni_3Ti phases and corresponds to 10^{13} cm^{-2} at 60 second of mechanical treatment.

1. Introduction

Plastic deformation of metals leads to structural changes that it is reflected on properties of the material; the example could appear lattice stresses and generate stacking faults and dislocations [1]. However, the final state depends on many parameters, the example, chemical composition of materials, crystal structure, rate of deformation and other [2, 3].

Ti–Ni alloy of near-equiatomic composition is interesting material for investigation of plastic deformation of different phases existing in it and these phases - TiNi (B2, B19`), Ti_2Ni phase (E9$_3$) and $TiNi_3$ phase (D0$_{24}$) will obtain a new data for understanding deformation in multiphase material [4, 5]. Such studies possible to investigate only in the powdered state by using a high-energy mechanical activation [6]. Moreover a similar treatment often leads to amorphization [7-10], formation of new quasicrystalline phases [11], supersaturated solid solutions and nanostructures [12-14].

It is well known that in cold-worked metals for example in aluminium, the dislocation density may up to 10^{10} cm^{-2} [15], the dislocation density in equiatomic alloy $Ti_{50}Ni_{50}$ is 10^{11} cm^{-2} after 8 cycles of the equal channel angular pressing (ECAP). For copper powder after treatment in ball mill for 8 hour a dislocation density before and after milling changes from 10^{10} cm^{-2} to 10^{12} cm^{-2} [16].

However, the information on mutual influence of different phases in multiphase alloys of TiNi at mechanical treatment in the literature is limited. In this work it has been studied a structure formation in Ti-Ni powder after high-energy mechanical treatment.

IOP Conf. Series: Materials Science and Engineering **848** (2020) 012001 doi:10.1088/1757-899X/848/1/012001

2. Material and methods

Near equiatomic $Ti_{0.45}Ni_{0.55}$ powder was used as starting materials (Polema» plant), the average particle size of 11.1 μm with 7.5 μm of standard deviation.

A high-energy AGO-2 planetary ball mill with steel vial and steel balls were used for mechanical treatment. The rotation speed of planetary disk of 1820 rpm; acceleration during treatments was 60 g. Phase composition of the powder were characterized by X-ray diffraction (XRD) on a X-ray diffractometer using Cu K_{α} radiation. The coherently diffracting domain (CDD) of the powder was measured by X-ray line-broadening technique using Scherrer equation [17] on the first line of X-ray spectra. To evaluate of the relative content of crystalline phases and quasi-amorphous phase were used quantitative calculations of integral intensities. The total milling energy was calculated according to [11, 18]. The dislocation density was calculated according to [15] using the first line of X-ray spectra for all phases.

3. Results

The total energy of milling were calculated according to [11, 18] and corresponded 0.19 Joule for one second and has a linear dependence with milling time. On Fig. 1 a are shown XRD patterns of TiNi powders before and after mechanical treatment at 300 second and on Fig. 1 b – a relative content of crystalline and quasi-amorphous phases. As one can see (Fig. 1, a) the powder in initial state consists of a mixture of phases such as TiNi (B2 and 19`), Ti_2Ni ($E9_3$) and Ni_3Ti, vertical lines are indicated them the angular positions. In the process of mechanical treatment, widths of the peaks increase and after 300 s XRD pattern of powder looks like belongs to quasi-amorphous state of material. Mechanical treatment results in a non-monotonous change in the relative content of phases (Fig. 1, b), so that the relative content of crystalline phases decreases steeply from 100 % to 55 % after 10 seconds mechanical treatment and more slowly decreases from 55 % to 45 % after 300 seconds mechanical treatment, while the content of the quasi-amorphous phase increases in the same way. As one can see, the changes between phases stopped after first 10 sec of treatment, therefore, we can conclude that total energy is spend not only for formation quasi-amorphous phase but for particle

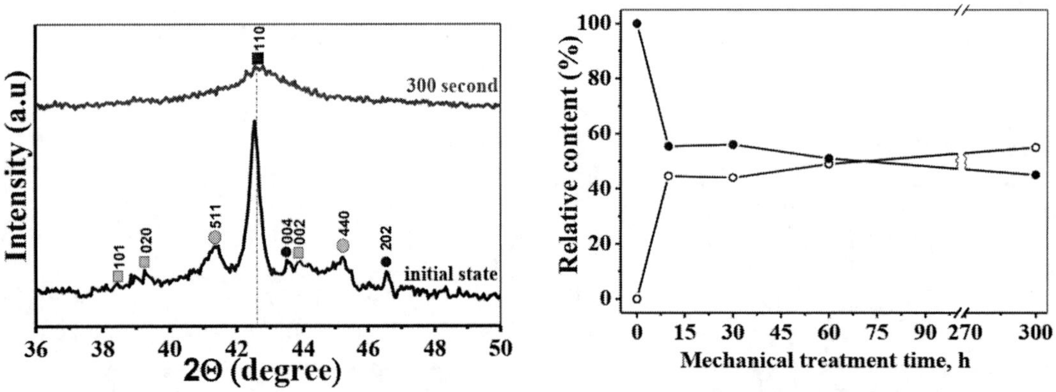

(a) (b)

Figure 1. XRD patterns of TiNi powders before and after mechanical treatment at 300 second (a): ■ – TiNi (B2); ■ – TiNi (B19`); ● – Ti_2Ni ($E9_3$); ● – Ni_3Ti ($D0_{24}$); The relative content (b) of crystallines phases —●— and —○— quasi-amorphous phase

aggregation, stiction of particles to the walls of vial, heating balls, powder, walls of vial and decreasing number of collisions between balls and powder/walls of vial, because the relative content of quasi-amorphous phase not significantly change.

The lattice parameters are in good agreement with the literature data [19-22]. The lattice parameters of TiNi (B2) phase decreases from 0.30074±0.00005 nm to 0.29968±0.00005 nm after

IOP Conf. Series: Materials Science and Engineering **848** (2020) 012001 doi:10.1088/1757-899X/848/1/012001

30 seconds mechanical treatment. The lattice parameter *a* of $TiNi_3$ phase not significantly changes and it is 0.509±0.005 nm, while the lattice parameters *c* and *c/a* decrease from 0.832±0.005 nm to 0,826±0.005 nm *(c)* and from 1.635±0.005 to 1.625±0.005 *(c/a)* after 30 seconds mechanical treatment. The lattice parameters of Ti_2Ni phase decreases from 1.13332±0.00005 nm to 1.12709±0.00005 nm after 60 seconds mechanical treatment.

The full widths at half maximum (FWHM) of peaks presented in Fig. 2 are obtained at different mechanical treatment time. One can see the FWHM changes insignificantly until 10 second of treatment and increasing of FWHM of peaks after this time can be directly related to strains produced by dislocations within these domains [23-25].

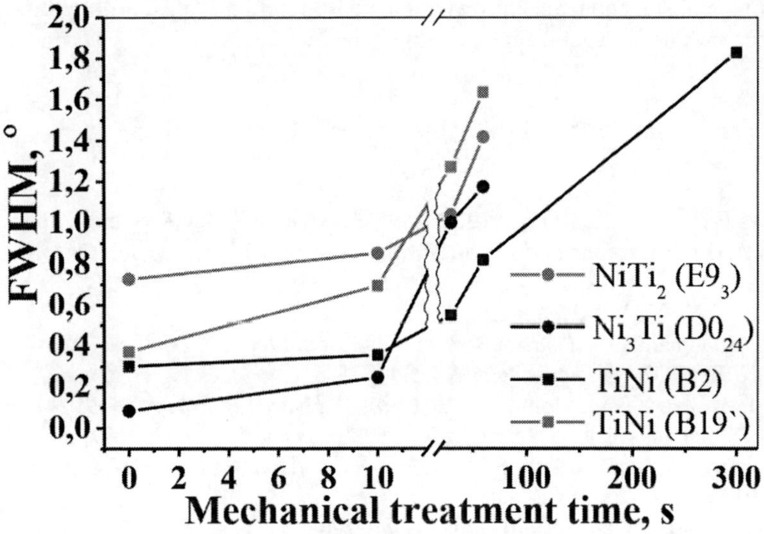

Figure 2. The dependence of the width (FWHM) of peaks on mechanical treatment time

The dislocation densities calculated like in [15] are shown in Table 1 for different phases and time of treatment. As one can see the dislocation density increases quicker in the Ni_3Ti phase than in the TiNi and Ti_2Ni phase up to 30 s of treatment, however value of the dislocation density at 60 second mechanical treatment for Ti_2Ni phase is more when 4 time higher than for TiNi phase and in ~17 time than for Ni_3Ti phase. These results well correlate with literature data [16] for Cu powder and dislocation density before and after ball milling changes from $\rho=2.3\cdot10^{10}$ cm^{-2} to $\rho=6.0\cdot10^{11}$ cm^{-2}, accordingly.

Table 1. The dislocation density of the samples.

Mechanical treatment time	**$\rho\cdot10^{11}$ (cm^{-2})**		
(s)	TiNi	Ni_3Ti	Ti_2Ni
ini	3.4	0.3	20.2
5	4.7	0.8	26.2
10	4.8	2.2	27.7
30	11.4	2.5	41.5
50	13.0	-	71.0
60	25.2	-	78.0
200	86.2	-	-
300	124.9	-	-

IOP Conf. Series: Materials Science and Engineering **848** (2020) 012001 doi:10.1088/1757-899X/848/1/012001

4. Conclusion

It has been shown that powder in initial state consists of TiN (B2, B19'), Ti_2Ni ($E9_3$) and $TiNi_3$ ($D0_{24}$) phases and the lattices parameters of this phases does not significantly change after mechanical treatment and are in good agreement with the literature data. It has been shown that relative content of these phases in crystalline state significantly decreases in a first 10 second of mechanical treatment and in system appeared quasi-amorphous phase, after this the phase content does not changes, this means that energy of mechanical activation introduced to a system is spend not only for formation quasi-amorphous phase but for particle aggregation, stiction of particles to the walls of vial and other for processes.

It has been shown that widths of peaks significantly increase after 10 second mechanical treatment due to increasing of dislocation densities which increase for a prolonged milling time and can reached up 10^{13} cm^{-2}, these value is in good agreement with the literature data.

Acknowledgement
The reported study was funded by RFBR, project number 19-38-90196.

References

[1] Chalmers B *Physical Metallurgy* 1962 (New York: John Wiley & Sons) p 468
[2] Cottrell A H *Dislocations and Plastic Flow in Crystals* 1958 (London:Oxford University Press) p 17.
[3] Mohamed F A 2014 *Int. Mater. Rev.* **59**, 394–416.
[4] Tang W, Sundman B, Sandström R et. al. 1999 *Acta Metall.* **47**, 3457–3468.
[5] Otsuka K, Ren X 2005 *Prog. Mater Sci.* **50**, 511–678.
[6] Gattia D M, Gizer G and Montone A 2014 *Int. J. Hydrogen Energ.* **39**, 9924–9930.
[7] Murty B S, Ranganathan S, Mohan M R 1992 *Muter. Sci. Engng. A* **149**, 231-240.
[8] Abdulmenova E V and Kulkov S N 2019 *IOP Conf. Series: Materials Science and Engineering* **696** 012018.
[9] Cocco G, Soletta I, Enzo S 1990 et al. *Colloa. Phvs*. **51**. C 4-181.
[10] Murty B S, Mohan M R, Ranganathan 1990 S. *Scripta metall. mater*. **24**, 1819.
[11] Burgio N, Iasonna A, Magini M 1991 *IL Nuovo Cimento D* **13(4)**, 459–476.
[12] Zadorozhnyi V Yu, Milovzorov G S., Skakov Yu A, 2008 *MiTOM* **8**, 46–52.
[13] SkakovYu A 2004 *MiTOM* **4**, 3–12.
[14] Schwarz R B, Koch C C 1986 *Appl. Phys. Lett.* **49(3)**, 146–148.
[15] Williamson G K, Smallman R E 1956 Philosophical Magazine 1(1), 34–46.
[16] Islamgaliyev R K, Nesterov K M, Valiyev R Z 2013 *Vestnik UGATU* **17** 81–89 (Russia)
[17] Scherrer P 1918 *Göttinger Nachrichten Gesellschaft* **2**, 98–101.
[18] Hodaei A, Ataie A, Mostafavi E 2015 *J. Alloys Compd.* **640**, 162–168.
[19] Laves F, Wallbaum H J 1939 *Kristallogr Z,* **101**, 78.
[20] Stuewe H.-P, Shimomura Y Z, 1960 *Metallkd* **51**, 180.
[21] Sinclair M G 1981*Acta Crystallogr.Sec. E* **37**, 1803.
[22] Mueller M H, Knott H W 1963 *Trans. Met. Soc. AIME* **227**, 674.
[23] Yao X, Wu C, Du A et al. 2007 *Chem. Soc.* **129** 15650 – 15654.
[24] Bratanich T I, Getman O I et al. 2006 *Powder Metall. Metal Ceram*. **45** 11 – 12
[25] Shan X, Payer J H and Jennings W D 2009 *Int. J. Hydrog. Energy* **34** 363-369.

IOP Conf. Series: Materials Science and Engineering **848** (2020) 012002 doi:10.1088/1757-899X/848/1/012002

Development of chitosan-hyaluronic acid nanoparticles and study of their physico-chemical properties for targeted delivery of anticancer drugs

N P Akentieva*[1,2], A R Gizatullin[1], O Silvestre[3], O Savchuk[3], N I Shkondina[1], T P Prichodchenko[1], D V Mitschenko[1], A V Zhilenkov[1], P A Troshin[1,4], N A Sanina[1,5], N H Dremova[1], V I Torbov[1], S M Aldoshin[1,5] and J B Nieder[3]

[1]Russia, Institute of Problems of Chemical Physics RAS, 142432, Russia, Chernogolovka, pr. Semenova, 1

[2]Karabuk University, 78050, Turkey, Kılavuzlar / Karabük, Kastamonu Yolu Demir

[3]International Iberian Nanotechnology Laboratory, Braga, Avda. Mestre José Veiga s/n, Portugal

[4]Skolcovo Institute of Science and Technology, 121205, Russia, Nobel st., 3, Moscow, Moscow region

[5]Lomonosov Moscow State University, 119991, Russia, Moscow, Lenin Hills st., 1

*E-mail: na_aken@icp.ac.ru

Abstract. Nanoparticles from chitosan and hyaluronic acid were obtained using ionotropic gelation technology. The size of the nanoparticles was investigated using electron microscopy and dynamic light scattering. Nanoparticles were obtained of an optimal size of ~ 100 nm. A physical association method has been developed of encapsulating nanoparticles with doxorubicin, a well-known antitumor drug, and dinitrosyl iron complex (donor of nitric oxide). The surface potential of nanoparticles was determined by dynamic light scattering. It was shown that HA-DOX: CS {6: 1} nanoparticles were stable and had a potential of -45.6 meV. The localization of nanoparticles in the cancer cells has been studied by confocal and FLIM microscopy. It was found that nanoparticles with doxorubicin are located in the cell near and inside the nucleus. It was shown that the encapsulation of DNIC in the composition of nanoparticles significantly increases the stability of DNIC, prolongs the formation and increases the yield of nitric oxide. A unique nano-system has been created for the delivery of anticancer drugs into the cell.

1. Introduction

Cancer is the second leading cause of death globally, and is responsible for an estimated 9.6 million deaths in 2018. Globally, about 1 in 6 deaths is due to cancer [1]. The relevance of developing nanoparticles for targeted delivery of anticancer drugs is determined by the high frequency of oncological diseases worldwide, the low survival rate of cancer patients and the expensive treatment of cancer. Currently, the main problem in oncology is the lack of specificity of the action of many antitumor drugs.

In this work, we used chitosan (CS) as a carrier in nanoparticles. Chitosan-aminosugar, a linear polysaccharide derivative whose macromolecules consist of randomly linked β-D-glucosamine units and N-acetyl-D-glucosamine [2]. Chitosan is a highly cationic, non-toxic, biocompatible and biodegradable compound [3]. Due to its cationic nature, chitosan is able to form insoluble complexes with the anionic polysaccharide-hyaluronic acid (hyaluronic acid, HA). Polysaccharide-coated nanoparticles are a new trend in drug delivery systems. Following this, we used a negatively charged polymer-HA as a delivery vector. HA is a non-sulfonated glucose aminoglycan that is part of the connective, epithelial and nervous tissues [4]. HA is a biocompatible, non-toxic and easily biodegradable molecule. HA can provide protection for the delivered drug and improve the solubility of hydrophobic drugs. In addition, since HA is a polyanionic polysaccharide, it has multiple charges for interaction with polycations [5]. HA is a physiological ligand of CD44 and RHAMM (receptor for hyaluronan-mediated mobility) receptors; therefore, it specifically interacts with CD44 and RHAMM receptors on the cell membrane, ensuring the efficiency of drug delivery to the cell [6]. HA-based nanoparticles are able to recognize CD44 and RHAMM-expressing cancer cells and, thus, specifically deliver particle-bound drugs through receptor-mediated endocytosis [7]. This cell targeting strategy allows you to deliver drugs directly to the cells, rather than concentrating in the kidneys, liver and then excreted from the body by the reticulo-endothelial system [8]. In addition, cellular enzymes (hyaluronidases) cleave HA inside the cell, thereby releasing the drug directly into the cell [9]. This strategy allows for specific and selective drug delivery.

The objective of this study was to develop nanoparticles based on chitosan-HA and to encapsulate doxorubicin (DOX) and dinitrosyl iron complex (DNIC) into nanoparticles to increase the stability and effectiveness of their action.

2. Materials and Methods

2.1. Materials

Commercial grade chitosan (M.w. = 100.000-300.000) manufactured by Acros organics (New Jersey, USA) was used. HA (sodium hyaluronate, low molecular weight <0.1 MDa) was obtained from My formula (Germany), HyClone Fetal Clone III Serum was bought from GE Healthcare (USA), doxorubizin (DOX) was purchased from Ebeve Pharma (Austria).

2.2. Fabrication of nanoparticles

Nanoparticles (CS-HA, HA-DOX: CS, HA-DNIC: CS) were obtained by ion gelation. A portion of chitosan (10 mg) was dissolved in 1 ml of HCl (0.01 M, pH = 2) and stirred for 12 hours. A portion of HA (40 mg) was dissolved in 1 ml of distilled water and stirred for 12 hours. Then, the obtained solutions of chitosan and HA were mixed in the different ratio, the pH of the solution was adjusted to pH = 6.8 with phosphate buffer (2 ml, 0.1 M, pH = 6.8) and stirred on a magnetic stirrer for 36 hours. Then the solution was dried by freeze drying and analyzed by electron microscopy. Chitosan and HA were also analyzed as control samples.

2.3. Analysis of nanoparticles by electron microscopy

The analysis of nanoparticles was carried out by electron microscopy. The samples of nanoparticles were prepared by vacuum deposition of carbon on an aluminum substrate. To apply carbon coating by an arc discharge method, a VUU-4 high-vacuum deposition device was used. Then, the nanoparticles were analyzed using a Supra-25 electron microscope (Zeiss, Germany), working distance WD = 2 mm, accelerating voltage EHT = 3.56 kV, magnification Mag = 100.02 KX, Signal detector A = InLens, vacuum in the system = 1.13e-0.006 mBar, vacuum in the cathode assembly = 8.49e-010 mbar.

2.4. Synthesis of mononuclear dinitrosyl iron complex with functional sulfur-containing ligands (DNIC)

Synthesis and identification of the DNIC compound was carried out according to the procedure [10]. As the tested compound used synthetic analog of mononuclear dinitrosyl iron complex with functional sulfur-containing ligands, namely with thiourea and its derivatives $[Fe(SC(NH_2)_2)_2(NO)_2]ClO_4Cl$

IOP Conf. Series: Materials Science and Engineering **848** (2020) 012002 doi:10.1088/1757-899X/848/1/012002

(Compound #6). DNIC secretes nitric oxide (NO) when dissolved in water solvents due to dissociation [10].

2.5. Analysis of size and zeta potential nanoparticles by dynamic light scattering (DLS)
Particle size distribution was studied using a DLS technique (Nanopartica SZ-100; HORIBA Ltd, Kyoto, Japan). The scattered light from the particles present in the sample was collected either at 90 or 173 degrees, which was automatically selected by the instrument as the optimum scattering angle based on sample concentration. The zeta potential, which is an indicator of dispersion and stability of the prepared nanoparticles, was measured, according to method [11].

2.6. Cell culture
Cell culture of HeLa cells (human cervix epitheloid carcinoma) are maintained in tissue culture flasks T-75 at 37^0C in 5% CO_2 atmosphere and culture with MEM cell medium supplemented with 10 % v/v HyClone Fetal Clone III Serum and 1 % v/v Penicillin/Streptomycin solution.

2.7. Analysis cells with confocal microscopy
The cells were imaged with an inverted laser scanning confocal microscope (LSCM) LSM780 (Zeiss, Germany), illuminated by a 488 nm laser and a fluorescence detection range from 550 nm to 700 nm. Fluorescence z-stack projections were obtained and analyzed with the ZEN2010 software [12].

2.8. Analysis of cells be Fluorescent Life Imaging Microscopy (FLIM)
FLIM was performed on an upright LSM 510 microscope (Carl Zeiss, Germany) with a 1.0 NA × 40 water-dipping objective using a 650-nm short-pass dichroic and 460±25 nm emission filter. Emission events were registered by an external detector attached to a commercial time-correlated single photon counting electronics module. Scanning was performed continuously for 4 min with a pixel dwell time of 1.6 μs.

2.9. Electrochemical determination of NO concentration, isolated from DNIC
An amperometric sensor electrode of the " amiNO-700 " system " inNO Nitric Oxide Measuring System " (Innovative Instruments, Inc., USA) was used to measure the concentration of NO generated from the DNIC. The NO concentration in the aqueous solution was fixed for ~ 500 seconds (in 0.2 second increments). A standard aqueous solution of $NaNO_2$ (100 μM) was used to calibrate the sensor electrode, to which was added a mixture of aqueous solutions of KI (0.12 M, 18 mL) and H_2SO_4 (1 M, 2 mL)[13].

3. Result and Discussion
Nanoparticles (CS-HA) were obtained using the method of physical association due to ion self-assembly of hyaluronic acid molecules (negatively charged) and chitosan molecules (positively charged) in nanogels. Nanoparticle's size was determined by electron microscopy. The obtained nanoparticles were water soluble due to the hydrophilicity of HA and chitosan. It was shown that the particles have a spherical morphology and the size is approximately equal to 100 nm as shown on Figure 1.

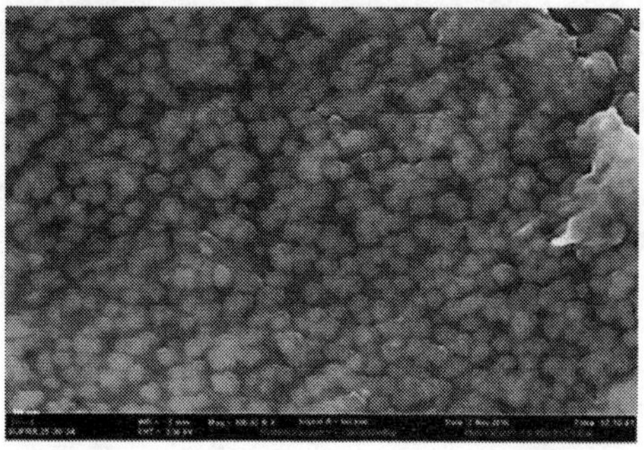

Figure 1. Images of nanoparticles (chitosan-HA) obtained by electron microscopy.

The nanoparticles were homogeneous in size and shape.

Thus, for the implementation of cell targeting, we have developed a technique for producing nanoparticles with a size of ~ 100 nm. This nanoparticle size is optimal for targeted delivery, since nanoparticles of this size can penetrate into the cells and they will not be removed by the reticulo-endothelial system from the bloodstream [8]. Therefore, these nanoparticles will carry out cell targeting.

Encapsulation of DOX in nanoparticles (HA-DOX: CS and HA-DNIC: CS) was performed by physical association. Using different ratios of chitosan and HA, a method was developed for producing nanoparticles of various sizes with encapsulated DOX. The size of the nanoparticles was determined by dynamic light scattering. The results showed that the nanoparticles have a size of 100 ~ 400 nm, the image of nanoparticles is shown on Figure 2. The optimal nanoparticle size (100 nm) was obtained with the ratio HA: CS = 6: 1.

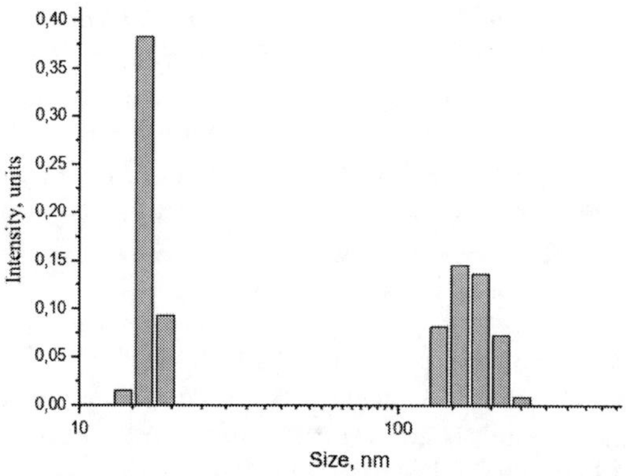

Figure 2. Determination of the size of nanoparticles (HA-DOX: CS) by dynamic light scattering.

The surface potential of nanoparticles was determined by dynamic light scattering. The data presented in Table 1. It was shown that HA-DOX: CS {6: 1} nanoparticles have a potential of -45.6 meV, which indicates that the nanoparticles are negatively charged and stable.

Table 1. Surface zeta potential of nanoparticles.

1	Nanoparticles (HA: CS {4: 1})	-1.74 mV
2	CS-DOX : HA {4:1}	-2.3 mV
3	HA-DOX : CS {6:1} stirred 10 min	-6.16 mV

4	HA-DOX : CS {12:1} stirred 30 min	-7.06 mV
5	HA-DOX : CS {12:1} stirred 30 мин + ultrasound	-8.1 mV
	After washing and ultrasound treatment	
6	Nanoparticles (HA:CS {4:1})	-47.22 mV
7	HA-DOX:CS {6:1} stirred 10 min	-45.6 mV
8	HA-DOX:CS {12:1} stirred 30 min	-39.4 mV

To improve the targeted efficiency of nanoparticles, we used the concept of active targeted delivery, which consists in using bio-molecular recognition of molecules on the surface of cells, which leads to a higher degree of specificity. In this case, we decorated nanoparticles with a biologically targeted molecule. As such a molecule, we used HA, which is the main component of the extra cellular matrix in the bone marrow, connective tissues [14]. This biopolymer regulates various cellular processes, including proliferation, differentiation, motility, invasiveness, cell adhesion, and gene expression in cancer cells [15]. It is HA that increases the possibility of targeted delivery and accelerates intracellular delivery through endocytosis, since all cells express endogenous receptors for this polymer (CD44 and RHAMM receptors) [15, 16]. The undoubted advantage of our nanoparticles is that we use natural biopolymer chitosan, which is a biocompatible and biodegradable material, as a nanostructured material [5].

The penetration of nanoparticles (HA-DOX: CS) into HeLa cancer cells was studied by confocal and FLIM microscopy using DOX, which has the property of auto-fluorescence. As can be seen from Figure 3, nanoparticles (HA-DOX: CS) are localized in the cytoplasm of HeLa cells (yellow fluorescence).

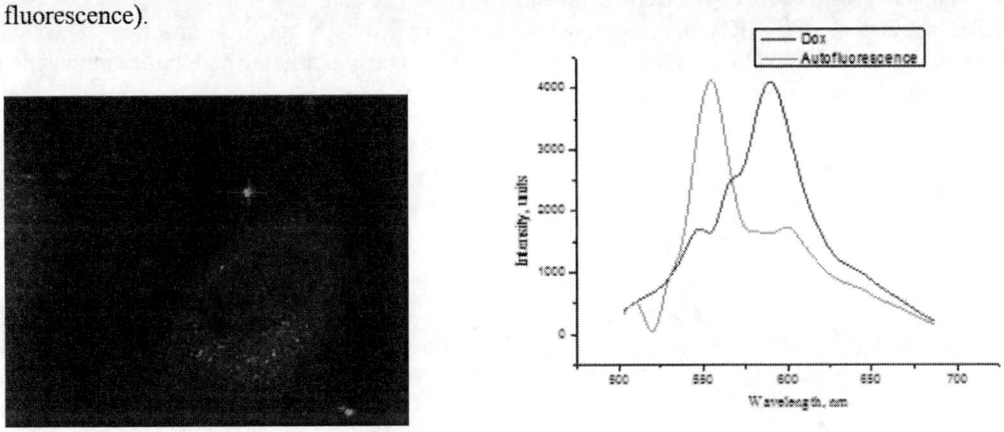

Figure 3. Confocal images of HeLa cells after incubation with nanoparticles (HA-DOX: CS) for 24 hours.

The results of the analysis of the localization of nanoparticles in living cells using FLIM microscopy showed that free DOX accumulates in the nucleus of HeLa cells after 24 hours of incubation (Figure 4, bottom panel). It was found that nanoparticles (HA-DOX: CS) after incubation with HeLa cells for 24 hours were localized around and inside the cell nucleus (Figure 4, upper panel).

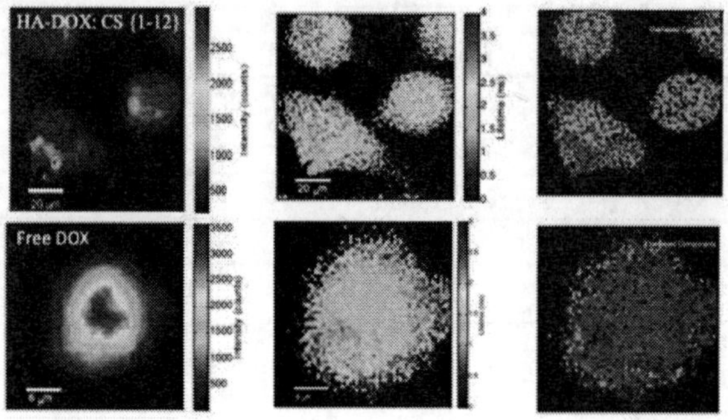

Figure 4. FLIM images of live HeLa cells incubated with nanoparticles (HA-DOX: CS) and with free DOX (control) for 24 hours.

Both methods confirmed the passage of nanoparticles through the cytoplasmic membrane of cell and showed the localization of nanoparticles in the cytoplasm, around and inside the cell nucleus after 24 hours of incubation.

At present time, in oncology, for the treatment of cancer deceases are used drugs that are donors of nitric oxide [17]. However, the main problem of these drugs is that these drugs are often unstable, nonspecific and have a number of side effects. Therefore, the development of nanoparticles for stabilization and targeted delivery of DNICs drugs is currently relevant.

The encapsulation of DNIC6 in nanoparticles was carried out by ion gelatinization. Then we investigated the release of NO from DNIC (compound #6), in aqueous solution and in the composition of nanoparticles (Figure 5).

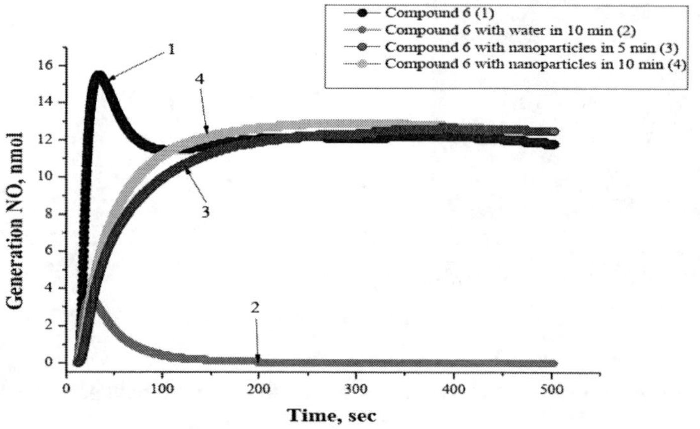

Figure 5. Generation of NO from the nanoparticle-DNIC complex (compound#6).

It was found that upon dissolution of DNIC in an aqueous solution, a sharp decrease in generation NO is observed, and after 100 seconds the level of generated NO drops to almost zero (Figure 5, curve 2). This indicated that the DNIC complex quickly decomposes in water and ceases to emit NO. Next, we analyzed the release of NO from the nanoparticle-DNIC complex (compound#6). It was shown that after adding DNIC to the nanoparticles, generation of 13 nmol NO was observed, and this level was maintained for a long time (> 500 seconds) (Figure 5, curve 3 and 4). Thus, the results showed that the encapsulation of DNIC in the composition of nanoparticles stabilizes these compounds and also prolongs and increases the amount of NO formed. Therefore, we first encapsulated DNIC in nanoparticles and increased the efficiency of the DNIC's action.

4. Conclusions
Thus, a universal nano-drug delivery system using nanoparticles (HA-DOX: CS) has been developed for target delivery anti-cancer drugs. The nanoparticles obtained by us are water-soluble, biocompatible and biodegradable, bio-permeable, capable of releasing the drug inside the cell due to the cleavage of HA with hyaluronidases. Such nanoparticles will undoubtedly be of interest to the pharmaceutical industry, since they are a combination of safe biomaterials, ideal for encapsulating cationic hydrophilic drugs, are able to provide intracellular drug delivery, can fulfill the necessary function of drug delivery for a long time without causing harmful effects to the human body, made simple and easy on a large scale nanotechnology, and finally are stable.

Acknowledgments
This work was financially supported by the Basic Research Program No. 1.42 P "Basic Research for Biomedical Technologies" and in accordance with the state assignment (registration No. 0089-2019-0014). Our gratitude to N. Zhelev, PhD, Professor at the University of Dundee (Scotland, UK) for the methodological assistance and discussion of the results. In the experiments, we used the equipment of the Scientific and Educational Centre of Medical University "Medical Chemistry" at the IPCP RAS, Chernogolovka.

5. References
[1] World Health Organization 2018 PRESS RELEASE **263** 1-3
[2] Shahidi F and Synowiecki J 1991 J. Agric. Food Chem. **39** 1527-32; https://doi.org/10.1021/jf00008a032
[3] Berger J, Reist M, Mayer J M, Felt O and Gurny R 2004 Eur. J. Pharm. Biopharm. **57** 35–52 ; doi:10.1016/s0939-6411(03)00160-7
[4] Almond A 2007 Cell Mol. Life Sci. **64** (13) 1591–96; doi:10.1007/s00018-007-7032-z
[5] Ossipov D A 2010 Expert Opin. Drug Deliv. **7** (6) 681-703; doi:10.1517/17425241003730399
[6] Day A J and Prestwich G D 2002 J. Biol. Chem. **277** 4585–88; DOI:10.1074/jbc.R100036200
[7] Aruffo A, Stamenkovic I, Melnick M, Underhill C B and Seed B 1990 Cell **61** (7) 1303–13; doi:10.1016/0092-8674(90)90694-a
[8] Choi K Y, Saravanakumar G, Park J H and Park K 2012 Colloids Surf. B. Biointerface 99 82–94; doi:10.1016/j.colsurfb.2011.10.029
[9] Kakizaki I, Ibori N, Kojima K, Yamaguchi M and Endo M 2010 FEBS J. **277** 1776 – 86; doi:10.1111/j.1742-4658.2010.07600.x
[10] Sanina N A, Aldoshin S M, Shmatko N Yu, Korchagin D V, Shilov G V, Knyaz'kina E A, Ovanesyan N S and Kulikov A V 2015 New J. Chem. **39** 1022-30; doi: 10.1039/C4NJ01693A
[11] Pasupuleti V R, Prasad T N, Shiekh R A, Balam S K, Narasimhulu G, Reddy C S, Ab Rahman I and Gan S H 2013 Int. J. Nanomedicine **8** 3355–64; doi: 10.2147/IJN.S49000
[12] Gonçalves Lopes R C F, Silvestre O F , Faria A R , C do Vale M L , Marques E F and Nieder J B 2019 Nanoscale **11** (13) 5932-41; doi: 10.1039/C8NR06346J
[13] Zhang X and Broderick M P 2000 Mod. Aspects Immunobiol. **1** (4) 160–5.
[14] Prè E D, Conti G and Sbarbati A 2016 Stem. Cell Rev. Rep. **12** (6) 664-81; doi:10.1007/s12015-016-9684-2
[15] Klarić M, Haller H, Brnčić Fischer A, Babarović E, Prijić A and Eminović S. 2019 Appl. Immunohistochem. Mol. Morphol. **27** (8) 606- 12; doi: 10.1097/PAI.0000000000000673
[16] Mylona E, Jones K A, Mills S T and Pavlath G K 2006 J. Cell Physiol. **209** (2) 314-21; doi:10.1002/jcp.20724

[17] Wu S C, Lu C Y, Chen Y L, Lo F C, Wang T Y, Chen Y J, Yuan S S, Liaw W F and Wang Y M 2016 Inorg. Chem. **55** (18) 9383-92; doi: 10.1021/acs.inorgchem.6b01562.

[18] Akentieva N P, Sanina N A, Gizatullin A R, Shmatko N Y, Goryachev N S, Shkondina N I, Prikhodchenko T R and Aldoshin S M 2017 Dokl. Biochem. Biophys. **477** (1) 389-93; doi:10.1134/S1607672917060126

Deoxidation Equilibrium of Silicon in the Nickel-Cobalt Melts

A A Aleksandrov, V Ya Dashevskii and A G Kanevskii
Baikov Institute of Metallurgy and Materials Science of the Russian Academy of Sciences, 119334, Moscow, Russia

E-mail: a.a.aleksandrov@gmail.com

Abstract. The oxygen solutions in the Ni-Co melts containing silicon have been thermodynamically analyzed. The activity coefficients at infinite dilution and the interaction parameters in the Ni-Co alloys at 1873 K are calculated. The oxygen solubility depending on the silicon content in the studied alloys are obtained. Silicon is characterized by the high affinity to oxygen in these alloys. The deoxidation ability of silicon decreases slightly with an increase in the cobalt content in melt. In the Ni-Co alloys with more than 20% of cobalt, at a silicon concentration of less than 0.5%, the deoxidation ability of silicon is virtually unchangeable.

1. Introduction

The Ni-Co alloys are widely used in modern technologies [1, 2]. Oxygen is a harmful impurity in these alloys; it occurs in this metal both in dissolved state and in form of nonmetallic inclusions. The metal making with the minimum oxygen concentration is one of the main problems of production of these alloys. Upon production of the Ni-Cr alloys, silicon is widely used as a deoxidizer and alloying element. For this reason, the effect of silicon on the oxygen solubility in these alloys is of great interest. The data published on thermodynamics of oxygen solutions in the liquid nickel and cobalt [3–6] allow one to estimate the effect of silicon on the oxygen solubility in the Ni-Co melts.

2. Thermodynamic Consideration

Based on the theory of quasiregular solutions, the method of recalculation of the interaction parameters of elements from one alloy base to another was developed [7]. As the most reliable, the reference books [8, 9] recommend the following values of the interaction parameters of silicon and oxygen in the iron melts: $e_{\mathrm{Si(Fe)}}^{\mathrm{Si}} = 0.103$; $e_{\mathrm{O(Fe)}}^{\mathrm{Si}} = -0.066$; $e_{\mathrm{Si(Fe)}}^{\mathrm{O}} = -0.119$. The data on similar interaction parameters in the cobalt melts are contradictory, namely: $e_{\mathrm{Si(Co)}}^{\mathrm{Si}}$ (0.197 [10]; 0.05 [4]; 0.139 [11]), $e_{\mathrm{O(Co)}}^{\mathrm{Si}}$ (–0.206 [10]; 0.069 [11]), $e_{\mathrm{Si(Co)}}^{\mathrm{O}}$ (–0.365 [10]; 0.118 [11]). The authors of [7] propose the method of estimation of the interaction parameters in cobalt using the recalculation of ones for iron [11]. The equations for the calculation of the interaction parameters of silicon and oxygen in cobalt are the following [7, 11]:

$$e_{\mathrm{O(Co)}}^{\mathrm{Si}} = 0.9911\frac{M_{\mathrm{Co}}}{M_{\mathrm{Fe}}}\left(e_{\mathrm{O(Fe)}}^{\mathrm{Si}} + e_{\mathrm{Co(Fe)}}^{\mathrm{Co}} - e_{\mathrm{O(Fe)}}^{\mathrm{Co}} - e_{\mathrm{Si(Fe)}}^{\mathrm{Co}}\right) + \frac{1}{230}\left(0.9911\frac{M_{\mathrm{Co}}}{M_{\mathrm{Fe}}} - 1\right)\left(\frac{M_{\mathrm{Co}}}{M_{\mathrm{Si}}} - 1\right),$$

$$e_{\mathrm{Si(Co)}}^{\mathrm{O}} = 0.9911\frac{M_{\mathrm{Co}}}{M_{\mathrm{Fe}}}\left(e_{\mathrm{Si(Fe)}}^{\mathrm{O}} + e_{\mathrm{Co(Fe)}}^{\mathrm{Co}} - e_{\mathrm{Si(Fe)}}^{\mathrm{Co}} - e_{\mathrm{O(Fe)}}^{\mathrm{Co}}\right) + \frac{1}{230}\left(0.9911\frac{M_{\mathrm{Co}}}{M_{\mathrm{Fe}}} - 1\right)\left(\frac{M_{\mathrm{Co}}}{M_{\mathrm{O}}} - 1\right),$$

$$e_{\mathrm{Si(Co)}}^{\mathrm{Si}} = 0.9911\frac{M_{\mathrm{Co}}}{M_{\mathrm{Fe}}}\left(e_{\mathrm{Si(Fe)}}^{\mathrm{Si}} + e_{\mathrm{Co(Fe)}}^{\mathrm{Co}} - 2e_{\mathrm{Si(Fe)}}^{\mathrm{Co}}\right) + \frac{1}{230}\left(0.9911\frac{M_{\mathrm{Co}}}{M_{\mathrm{Fe}}} - 1\right)\left(\frac{M_{\mathrm{Co}}}{M_{\mathrm{Si}}} - 1\right),$$

where M_i is the molecular mass of i component.

Taking into account that $e_{\mathrm{Co(Fe)}}^{\mathrm{Co}} = 0.00509$ [9], $e_{\mathrm{O(Fe)}}^{\mathrm{Co}} = 0.008$ [9], $e_{\mathrm{Si(Fe)}}^{\mathrm{Co}} = -0.0035$ [12], the calculation according to the above equations results in: $e_{\mathrm{O(Co)}}^{\mathrm{Si}} = -0.068$; $e_{\mathrm{Si(Co)}}^{\mathrm{O}} = -0.123$; $e_{\mathrm{Si(Co)}}^{\mathrm{Si}} = 0.121$. Table 1 shows the interaction parameters recommended for silicon and oxygen in the melts of iron, cobalt, and nickel.

Table 1. Interaction Parameters Recommended for Silicon and Oxygen in the Iron, Cobalt, and Nickel Melts at 1873 K.

Melt	$e_{\mathrm{Si}}^{\mathrm{Si}}$	$e_{\mathrm{O}}^{\mathrm{Si}}$	$e_{\mathrm{Si}}^{\mathrm{O}}$	Source
Fe	0.103	–0.066	–0.119	[8, 9]
Co	0.121	–0.068	–0.123	Present work
Ni	0.190	–0.065	–0.110	[5]

The interaction of silicon with oxygen in the Ni-Co melts can be described as

$SiO_2(s) = [Si] + 2[O]$, (1)

$$K_{(1)} = \frac{([\%\mathrm{Si}]\cdot f_{\mathrm{Si}})([\%\mathrm{O}]\cdot f_{\mathrm{O}})^2}{a_{\mathrm{SiO_2}}}.$$

Reaction (1) can be written as a sum of the reactions

$SiO_2(s) = Si(l) + O_2(g)$, (2)

$\Delta G^{\circ}_{(2)} = 949\,704 - 198.87T$, J/mole [10];

$Si(l) = [Si]_{1\%(\mathrm{Ni\text{-}Co})}$, (3)

$$\Delta G^{\circ}_{(3)} = RT\ln\left(\frac{\gamma^{\circ}_{\mathrm{Si(Ni\text{-}Co)}} M_{\mathrm{Ni\text{-}Co}}}{M_{\mathrm{Si}}\cdot 100}\right),$$

$O_2(g) = 2[O]_{1\%(\mathrm{Ni\text{-}Co})}$, (4)

$$\Delta G^{\circ}_{(4)} = 2RT\ln\left(\frac{\gamma^{\circ}_{\mathrm{O(Ni\text{-}Co)}} M_{\mathrm{Ni\text{-}Co}}}{M_{\mathrm{O}}\cdot 100}\right),$$

where $\gamma^{\circ}_{i(\mathrm{Ni\text{-}Co})}$ is the activity coefficient of component i in melt at infinite dilution. As a standard state for silicon and oxygen dissolved in the nickel-cobalt melt, 1% solution was chosen, which has the properties of ideal diluted solution.

The activity coefficients $\gamma^{\circ}_{i(\mathrm{Ni\text{-}Co})}$ for the Ni-Co melts were calculated by the following equation [13]:

$$\ln\gamma^{\circ}_{i\,(\mathrm{Ni-Co})} = X_{\mathrm{Ni}}\ln\gamma^{\circ}_{i\,(\mathrm{Ni})} + X_{\mathrm{Co}}\ln\gamma^{\circ}_{i\,(\mathrm{Co})} + X_{\mathrm{Ni}}X_{\mathrm{Co}}\Big[X_{\mathrm{Co}}\left(\ln\gamma^{\circ}_{i\,(\mathrm{Co})} - \ln\gamma^{\circ}_{i\,(\mathrm{Ni})} + \varepsilon^{\mathrm{Ni}}_{i\,(\mathrm{Co})}\right) +$$
$$+ X_{\mathrm{Ni}}\left(\ln\gamma^{\circ}_{i\,(\mathrm{Ni})} - \ln\gamma^{\circ}_{i\,(\mathrm{Co})} + \varepsilon^{\mathrm{Co}}_{i\,(\mathrm{Ni})}\right)\Big],$$

and the molecular mass $M_{\mathrm{Ni\text{-}Co}} = M_{\mathrm{Ni}}X_{\mathrm{Ni}} + M_{\mathrm{Co}}X_{\mathrm{Co}}$ [14].

Table 2 shows the equilibrium constants of reaction (1), the $\gamma^{\circ}_{\mathrm{Si}}$ and $\gamma^{\circ}_{\mathrm{O}}$ activity coefficients for nickel and cobalt, which were calculated for the nickel-cobalt melts at 1873 K. For the estimation of the $\gamma^{\circ}_{\mathrm{Si}}$ and $\gamma^{\circ}_{\mathrm{O}}$ values, the following interaction parameters were used: $\varepsilon^{\mathrm{Co}}_{\mathrm{Si(Ni)}} = 0.207$ [7]; $\varepsilon^{\mathrm{Ni}}_{\mathrm{Si(Co)}} = 0.575$ [11]; $\varepsilon^{\mathrm{Co}}_{\mathrm{O(Ni)}} = -1.159$ [6]; $\varepsilon^{\mathrm{Ni}}_{\mathrm{O(Co)}} = 0.164$ [6].

Table 2. Equilibrium Constants of Reaction (1), Activity Coefficients, and Interaction Parameters for the Ni-Co Melts at 1873 K.

Parameter	Co, %					
	0	20	40	60	80	100
$M_{\mathrm{Ni\text{-}Co}}$	58.69	58.738	58.787	58.836	58.884	58.933
X_{Ni}	1	0.801	0.601	0.401	0.201	0
X_{Co}	0	0.199	0.399	0.599	0.799	1
$\lg K_{(1)}$	−6.563	−6.446	−6.424	−6.448	−6.469	−6.435
$\gamma^{\circ}_{\mathrm{Si}}$	$9\cdot10^{-5}$ [5]	$1.1\cdot10^{-4}$	$1.5\cdot10^{-4}$	$2.1\cdot10^{-4}$	$2.8\cdot10^{-4}$	$2.9\cdot10^{-4}$ [6]
$\gamma^{\circ}_{\mathrm{O}}$	0.337 [3]	0.271	0.224	0.191	0.171	0.161 [4]
$e^{\mathrm{Si}}_{\mathrm{Si}}$	0.19 [5]	0.176	0.163	0.149	0.135	0.121
$e^{\mathrm{Si}}_{\mathrm{O}}$	−0.065 [5]	−0.0656	−0.0662	−0.0668	−0.0674	−0.068
$e^{\mathrm{O}}_{\mathrm{Si}}$	−0.110 [5]	−0.113	−0.115	−0.118	−0.120	−0.123
$e^{\mathrm{O}}_{\mathrm{O}}$	0 [3]	0	0	0	0	0 [4]

The equilibrium constant of reaction (1) depending on the alloy composition is given in figure 1. To be compared, this figure also shows similar data for the deoxidation of Ni-Co melts with chromium [14], manganese [15], titanium [16], and aluminum [17] at 1873 K. The equilibrium constants are estimated for the interaction of elements with only one oxygen atom dissolved in melt, which allows somebody to compare these dependences more conveniently. The equilibrium constants calculated for reaction (1) are in a good agreement with the data published [6].

The oxygen concentration being in equilibrium with the given silicon content in melt can be calculated by the equation

$$\lg[\%\mathrm{O}]_{\mathrm{Ni\text{-}Co}} = \frac{1}{2}\Big\{\lg K_{(1)} + \lg a_{\mathrm{SiO_2}} - \lg[\%\mathrm{Si}] - \Big[e^{\mathrm{Si}}_{\mathrm{Si(Ni\text{-}Co)}} + 2e^{\mathrm{Si}}_{\mathrm{O(Ni\text{-}Co)}}\Big][\%\mathrm{Si}] - \Big[2e^{\mathrm{O}}_{\mathrm{O(Ni\text{-}Co)}} + e^{\mathrm{O}}_{\mathrm{Si(Ni\text{-}Co)}}\Big][\%\mathrm{O}]\Big\}. \quad (5)$$

Oxide SiO_2 at 1873 K is solid (T_m = 1999 K [18]). Therefore, $a_{\mathrm{SiO_2}} = 1$. The [%O] value in the right-hand side of equation (5) can be expressed as the ratio $\left(K_{(1)}/[\%\mathrm{Si}]f_{\mathrm{Si}}f_{\mathrm{O}}^{2}\right)^{1/2}$. At [%O]→0, $f_{\mathrm{O}}\to1$. Due to the smallness of the [%O] value one can assume $\left(K_{(1)}/[\%\mathrm{Si}]f_{\mathrm{Si}}f_{\mathrm{O}}^{2}\right)^{1/2} \approx \left(K_{(1)}/[\%\mathrm{Si}]f_{\mathrm{Si}}\right)^{1/2}$. Such a change does not increase considerably the error of calculations [10]. Then equation (5) takes the form

$$\lg[\%\mathrm{O}]_{\mathrm{Ni\text{-}Co}} = \frac{1}{2}\Big\{\lg K_{(1)} - \lg[\%\mathrm{Si}] - \Big[e^{\mathrm{Si}}_{\mathrm{Si(Ni\text{-}Co)}} + 2e^{\mathrm{Si}}_{\mathrm{O(Ni\text{-}Co)}}\Big][\%\mathrm{Si}] - \Big[2e^{\mathrm{O}}_{\mathrm{O(Ni\text{-}Co)}} + e^{\mathrm{O}}_{\mathrm{Si(Ni\text{-}Co)}}\Big]\left(K_{(1)}/[\%\mathrm{Si}]f_{\mathrm{Si}}\right)^{1/2}\Big\}. \quad (6)$$

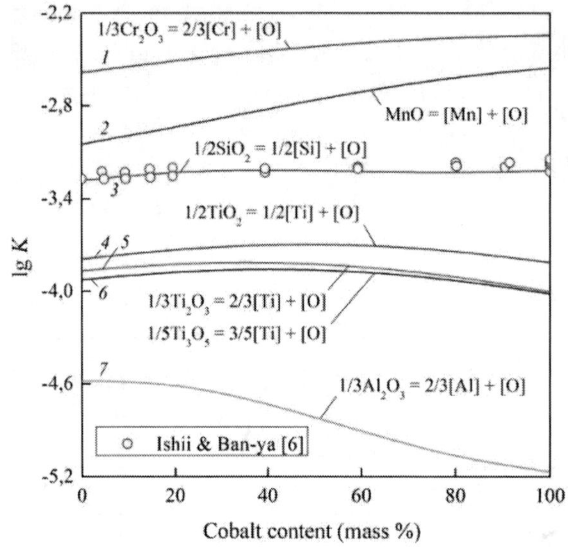

Figure 1. Dependence of the Equilibrium Constants of Deoxidation with Chromium (*1*), Manganese (*2*), Silicon (*3*), Titanium (*4–6*), and Aluminum (*7*) of the Ni-Co Melts on the Composition of Alloy Base at 1873 K.

The equilibrium oxygen content in the Ni-Co melts at different silicon concentrations were calculated by equation (6). Since the Ni-Co melts are close to ideal solutions [19], the interaction parameters used in these calculations (Table 2) were calculated by the equation $\varepsilon^{j}_{i(\text{Ni-Co})} = \varepsilon^{j}_{i(\text{Ni})} X_{\text{Ni}} + \varepsilon^{j}_{i(\text{Co})} X_{\text{Co}}$ [14].

The dependences of the oxygen concentration on the silicon content in melts are given below:

$$\lg[\%\text{O}]_{\text{Ni}} = -3.282 - \frac{1}{2}\lg[\%\text{Si}] - 0.030[\%\text{Si}] + \frac{0.288\cdot 10^{-4}}{[\%\text{Si}]^{1/2}}; \quad (6a)$$

$$\lg[\%\text{O}]_{\text{Ni-20\% Co}} = -3.223 - \frac{1}{2}\lg[\%\text{Si}] - 0.0225[\%\text{Si}] + \frac{0.337\cdot 10^{-4}}{[\%\text{Si}]^{1/2}}; \quad (6b)$$

$$\lg[\%\text{O}]_{\text{Ni-40\% Co}} = -3.212 - \frac{1}{2}\lg[\%\text{Si}] - 0.0151[\%\text{Si}] + \frac{0.354\cdot 10^{-4}}{[\%\text{Si}]^{1/2}}; \quad (6c)$$

$$\lg[\%\text{O}]_{\text{Ni-60\% Co}} = -3.224 - \frac{1}{2}\lg[\%\text{Si}] - 0.0076[\%\text{Si}] + \frac{0.352\cdot 10^{-4}}{[\%\text{Si}]^{1/2}}; \quad (6d)$$

$$\lg[\%\text{O}]_{\text{Ni-80\% Co}} = -3.234 - \frac{1}{2}\lg[\%\text{Si}] - 5\cdot 10^{-5}[\%\text{Si}] + \frac{0.351\cdot 10^{-4}}{[\%\text{Si}]^{1/2}}; \quad (6e)$$

$$\lg[\%\text{O}]_{\text{Co}} = -3.218 - \frac{1}{2}\lg[\%\text{Si}] + 0.0075[\%\text{Si}] + \frac{0.373\cdot 10^{-4}}{[\%\text{Si}]^{1/2}}. \quad (6f)$$

The oxygen concentrations depending on the silicon content in nickel and cobalt at 1873 K are given in figure 2. A good agreement between the obtained results and published data [5, 6, 20-22] characterizes an accuracy of calculation of the interaction parameters (Table 1).

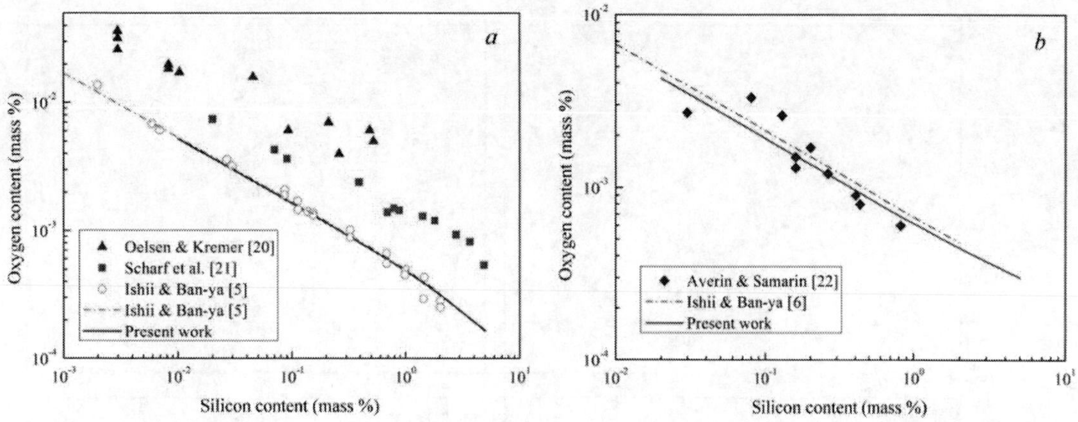

Figure 2. Oxygen concentration depending on the silicon content in nickel (*a*) and cobalt (*b*) at 1873 K.

Figure 3 shows the equilibrium oxygen concentrations depending on the silicon and cobalt contents in melt at 1873 K. As is seen, silicon has a considerable deoxidation ability in the Ni-Co melts. With an increase in the cobalt content, the deoxidation ability of silicon decreases slightly. In the Ni-Co melts with more than 20% Co, at a silicon content of less than 0.5%, the deoxidation ability of silicon is virtually constant.

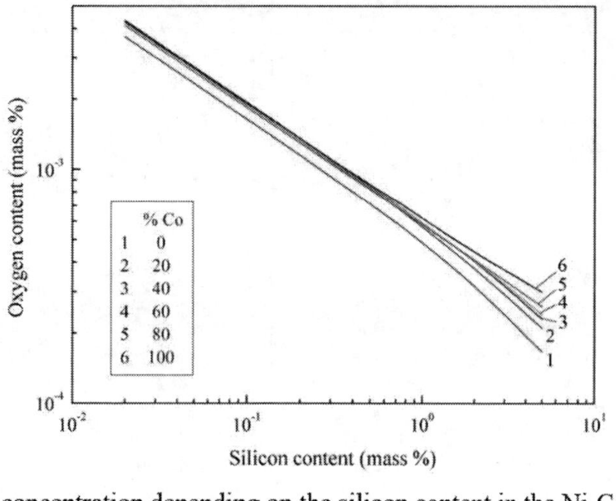

Figure 3. Oxygen concentration depending on the silicon content in the Ni-Co melts at 1873 K.

Figure 4 shows the dependences of oxygen concentration on the silicon content in the Ni-Co melts at 1873 K. Similar dependences for chromium [14], manganese [15], titanium [16], and aluminum [17] are also given in that figure. With an increase in the cobalt content, the deoxidation ability of aluminum rises, that of titanium remains virtually unchangeable, the deoxidation ability of silicon and chromium decreases inconsiderably, and the deoxidation ability of manganese diminishes substantially. Horizontal dotted lines demonstrate the oxygen solubility in the Ni-Co melts with the given compositions at 1873 K [23].

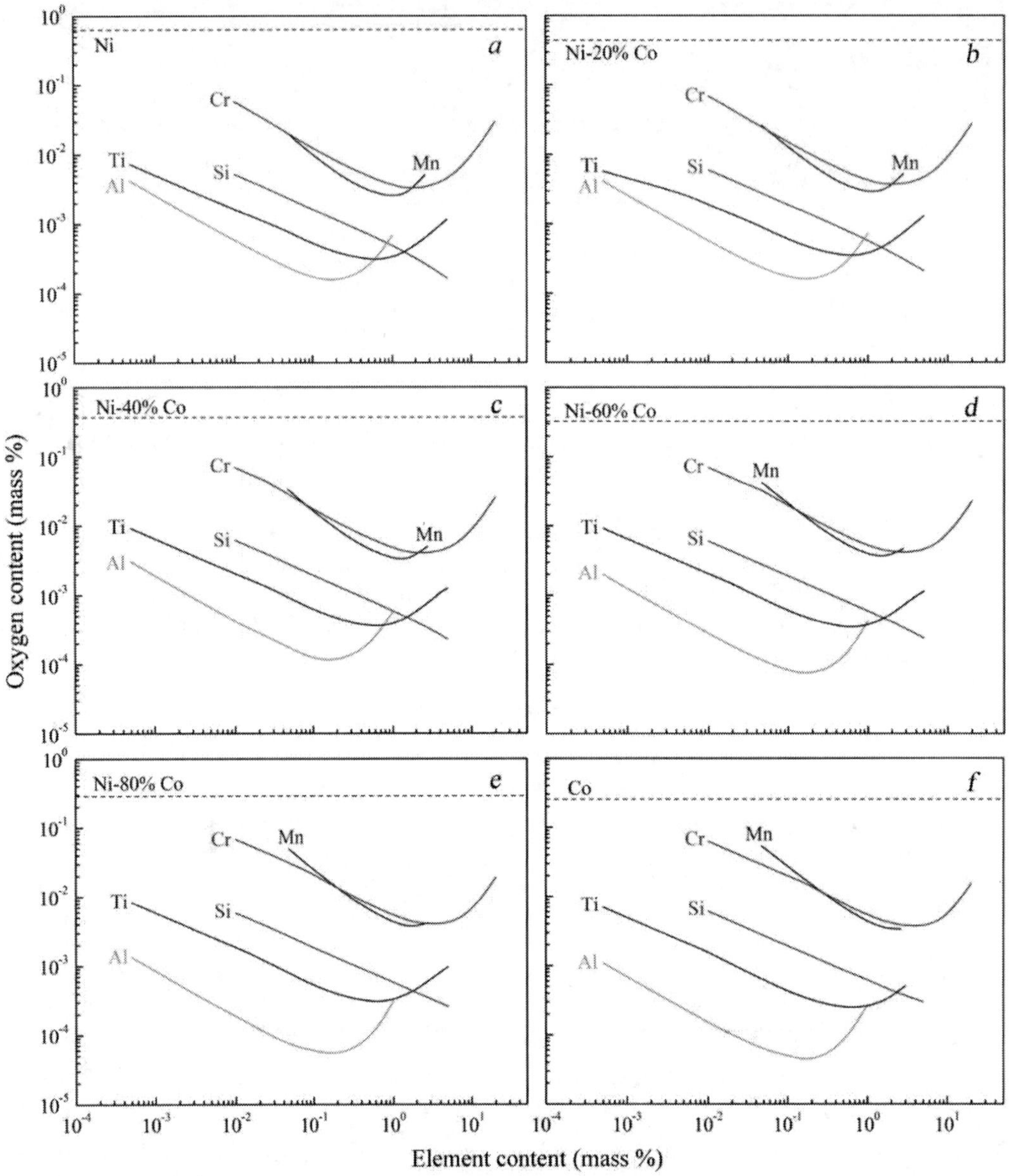

Figure 4. Oxygen concentration depending on the content of deoxidizers in the Ni-Co melts at 1873 K.

3. Conclusions

1. The interaction parameters of silicon and oxygen in the cobalt-based melts are determined. A good agreement between these parameters and the data published was exemplified by the oxygen solubility curves obtained for the nickel-cobalt melts.
2. The activity coefficients at infinite dilution and the interaction parameters in the Ni-Co melts are calculated. The concentration dependences of the oxygen solubility on the cobalt and silicon contents in these melts are obtained.
3. Silicon is characterized by the high affinity to oxygen in the nickel-cobalt melts. The deoxidation ability of silicon decreases inconsiderably with an increase in the cobalt content. In the Ni-Co

alloys with the more than 20% Co, at a silicon content of less 0.5%, the deoxidation ability of silicon remains virtually unchangeable.

Acknowledgments
This study was performed according to the RF Government Order No. 075-00746-19-00.

References

[1] Reed R C 2006 *The Superalloys. Fundamentals and Applications.* (Cambridge: University Press)
[2] Logunov A V and Shmotin Yu A 2013 *Sovremennye zharoprochnye nikelevye splavy dlya diskovykh gazovykh turbin [Modern heat-resistant nickel alloys for disks gas turbines].* (Moscow. Nauka i tekhnologii) (In Russ.)
[3] Sigworth G K., Elliott J F, Vaughn G and Geiger G H 1977 *Metallurgical Soc. CIM.* Annual Volume 104
[4] Sigworth G K and Elliott J F 1976 *Canadian Metallurgical Quarterly* **15**(2) 123
[5] Ishii F and Ban-ya S 1992 *ISIJ International* **32**(10) 1091
[6] Ishii F and Ban-ya S 1993 *ISIJ International* **33**(2) 245
[7] Belyanchikov L N 2009 *Elektrometallurgiya* 2 23 (In Russ.)
[8] *Steelmaking Data Sourcebook* 1988 (N.Y.-Tokyo: Gordon & Breach Science Publ.)
[9] Hino M and Ito K 2010 *Thermodynamic Data for Steelmaking* (Sendai: Tohoku University Press.)
[10] Kulikov I S 1975 *Raskislenie metallov [Deoxidation of Metals]* (Moscow: Metallurgiya Publ.) (In Russ.)
[11] Belyanchikov L N 2009 *Elektrometallurgiya* 4 16 (In Russ.)
[12] Snitko Yu P, Surovoy Yu N and Lyakishev N P 1983 *Doklady Akademii Nauk* **286**(5) 1154 (In Russ.)
[13] Frohberg M G and Wang M 1990 *Z. Metallkd* **81**(7) 513
[14] Aleksandrov A A and Dashevskii V Ya 2016 *Russian Metallurgy (Metally)* **2016**(7) 642
[15] Dashevskii V Ya, Aleksandrov A A and Leont'ev L I 2019 *Izvestiya. Ferrous Metallurgy* **62**(6) 475 (In Russ.)
[16] Aleksandrov A A and Dashevskii V Ya 2018 *Russian Metallurgy (Metally)* **2018**(11) 1081
[17] Aleksandrov A A, Dashevskii V Ya and Leont'ev L I 2017 *Russian Metallurgy (Metally)* **2017**(7) 590
[18] Lyakishev N P and Gasik M I 2005 *Fizikokhimiya i tekhnologiya electroferrosplavov [Physicochemistry and technology of electroferroalloys]* (Moscow: ELIZ) (In Russ.)
[19] Hultgren R, Desai P D, Hawkins D T, Gleiser M and Kelley K K 1973 *Selected Values of the Thermodynamic Properties of Binary Alloys.* (Ohio: Metals Park, Amer. Soc. Metals)
[20] Oelsen W and Kremer G 1936 *Mitt. Kaiser-Wilhelm-Inst. Eisenforsch.* Düsseldorf. **18** 90
[21] Scharf G, Averin V V, Polyakov A Yu and Samarin A M 1958 *Izvestiya VUZov. Chernaya Metallurgiya* **1**(11) 29 (In Russ.)
[22] Averin V V and Samarin A M 1963 *Transactions of the Baikov Institute of Metallurgy* **14** 50 (In Russ.)
[23] Dashevskii V Ya, Aleksandrov A A and Leont`ev L I 2015 *Steel in Translation* **45**(1) 42

IOP Conf. Series: Materials Science and Engineering **848** (2020) 012004 doi:10.1088/1757-899X/848/1/012004

Minimal pore size reached upon sintering

M I Alymov[1], S I Averin[2] and I V Saikov[1]

[1] Merzhanov Institute of Structural Macrokinetics and Materials Science Russian Academy of Sciences, 8 Academician Osipyan st., 142432 Chernogolovka, Moscow Region, Russia
[2] Baikov Institute of Metallurgy and Materials Science, Russian Academy of Sciences (IMET RAS), 49 Leninsky prospect, 119334 Moscow, Russia

Abstract. Analytical expression for minimal (critical) pore size reached upon sintering was derived. It has been shown that a pore size in a given material cannot be stable below some critical value characteristic of this material.

1. Introduction

As is known, thermomechanical processing of materials is accompanied by mass transfer caused by diffusion and drift of dislocations, which leads to a decrease in pore sizes and to an increase in material density. These processes can occur even in the absence of external pressure and interpore gas pressure [1]. The driving force of these processes is a decrease in the free energy of material due to a decrease in the free surface area under the action of surface tension forces. The application of external compressive pressure can be expected to result in further decrease in material porosity.

Shape of an individual pore, plastic deformation of material around a pore, and strain-induced distribution of vacancies have been analyzed in [2-3]. The latter factor (as well as strain magnitude) were found to play a key role in the kinetics of mass transfer responsible for pore healing. When the pore volume is less than the critical volume, the pore is energetically more profitable to heal, and when the pore volume is larger than the critical volume, the pore is energetically more favorable to grow [4, 5]. An expression for a minimal pore size in the absence of applied pressure was suggested in [1].

In this work, we aimed at determining a minimal pore size under applied pressure as in conditions of hot isostatic pressing (HIP).

2. Results and discussion

For a spherical pore with radius R and interpore pressure P_i under external pressure P_0, the problem can be formulated in terms of Lamé equation [6-7]. Accordingly, the distribution of radial σ_r and tangential σ_t strains around the pore can be described by Eqs. (1) and (2):

$$\sigma_r = -P_i\frac{R^3}{r^3} + P_0\frac{(R^3 - r^3)}{r^3} \quad (1)$$

$$\sigma_t = P_i\frac{R^3}{2r^3} - P_0\frac{(2r^3 + R^3)}{2r^3} \quad (2)$$

where r is running pore radius.

In the absence of applied pressure and at $P_i = 0$, pressure inside the pore is equal to the Laplace pressure (P_L):

$$P_L = \frac{2\alpha}{R} \quad (3)$$

IOP Conf. Series: Materials Science and Engineering **848** (2020) 012004 doi:10.1088/1757-899X/848/1/012004

where α is surface tension. This magnitude also contributes to the formation of deformation mode: In this case, the effect of the internal pressure P_i can be represented as the sum of the gas pressures inside the pore P_G and the Laplace pressure P_L: $P_i = P_G - P_L$.

The stresses are maximal at $r = R$:

$$\sigma_r = P_L - P_G \tag{4}$$

$$\sigma_t = -\frac{P_L - P_G}{2} - \frac{3P_0}{2} \tag{5}$$

Then the main normal strains can be represented in the form:

$$\sigma_1 = \sigma_r = P_L - P_G \tag{6}$$

$$\sigma_2 = \sigma_3 = \sigma_t = -\frac{P_L - P_G}{2} - \frac{3P_0}{2} \tag{7}$$

Plastic deformation around the pore gets started at some critical stresses.

There exist several explanations for material deformation when equivalent strains σ_{eq} exceed allowable stress σ. The theory of maximum-distortion-energy is based on the assumption of equal potential distortion energy accumulated to the moment of material creep upon both compound and simple stress. According to this theory for ductile materials [7], we have:

$$\sigma_{eq} = \left\{\frac{1}{2}[(\sigma_1 - \sigma_2)^2 + (\sigma_2 - \sigma_3)^2 + (\sigma_3 - \sigma_1)^2]\right\}^{1/2} \leq [\sigma] \tag{8}$$

Using (6–8), we obtain the expression for σ_{eq}:

$$\sigma_{eq} = \frac{3(P_L - P_G + P_0)}{2} \tag{9}$$

Let us now admit that $\sigma = \sigma_{0.2}$. If $\sigma_{eq} = \sigma_{0.2}$, the pore begins to deform until its healing. Hence, expressions (3) and (9) can be used to derive an analytical expression for critical pore size R_{cr}:

$$R_{cr} = \frac{6\alpha}{2\sigma_{0,2} + 3P_G - 3P_0} \tag{10}$$

The results of calculations for some metals at room temperature, $P_G = 0$ (pressing and sintering in vacuum) and $P_0 = 100$ MPa (isostatic pressing) are collected in Table 1. A decrease in $\sigma_{0.2}$ with increasing temperature favors an increase in R_{cr}, while the applied pressure decreases it. The increase in external pressure leads to the fact that large pores begin to collapse, that is, the critical pore size increases.

It follows that a minimal pore size in porous materials is restricted by a calculated value of R_{cr}.

IOP Conf. Series: Materials Science and Engineering **848** (2020) 012004 doi:10.1088/1757-899X/848/1/012004

Table 1. Properties [8–15] and critical pore radius for some metals at room temperature.

Material	α, N/m [7–11]	$\sigma_{0.2}$, MPa [7]	R_{cr}, nm ($P_0 = 0$)	R_{cr}, nm ($P_0 = 100$ MPa)
	a	b	*c*	*d*
Iron	1.90	170	34	285
Chromium	2.20	190	35	165
Titanium	1.70	300	17	34
	a	b	c	d
Cobalt (annealed)	1.97	300	20	39
Copper (cold-worked)	1.75	380	14	23
Molybdenum (annealed)	2.60	570	14	19
Nickel (deformed)	1.82	686	8	10
Tungsten	2.80	760	11	14

3. Conclusions

Plastic compression of pores occurs under the action of surface tension force and applied external pressure. By comparing main stresses with creep limit, we managed to derive the expression for a minimal pore size that can exist in a given material.

References

[1] Alymov M I, Markochev V M, Averin S I, Shustov V S 2017 Doklady Physics **62** (10) 447–9.
[2] Geguzin Ya E 1984 Phisics of Sintering (Moscow: Nauka Publ.) p 312.
[3] Kovalchenko M S 1980 Theoretical Fundamentals of Hot Pressure Working of Porous Materials (Kiev: Naukova Dumka Publ.) p 240.
[4] Cheremskoy P G, Slezov V V and Betekhtin V I 1990 Theory of Elasticity (Moscow: Energoatomizdat Publ.) p 376.
[5] McCartney L N 1977 Acta Metallurgica **25** 221–30.
[6] Timoshenko S P and Goodier J N 1970 Theory of Elasticity (NY: McGraw Hill) p 591.
[7] Lamé G 1852 Leçons Sur la Théorie Mathematique et L'elasticité des Corps Solides (Paris: Bachelier) p 384.
[8] Grigoriev I S, Meylikhov E Z 1991 Physical Quantities: Reference Book (Moscow: Energoatomizdat Publ.) p 1232.
[9] Lazarev S Yu 2003 Metal Working **2** 38–42.
[10] Yurov V M, Laurinas V Ch, Guchenko S A and Zavatskaya O N 2012 Modern High Technologies **3** 36–40.
[11] Missol W 1975 Energia Powierzchni Rozdzialu Faz w Metalach (Katowice: Wydaw "Śląsk").
[12] Oleshko V S, Pigovkin I S 2016 Science of Science **8** 131EVN316.
[13] Ustyukhin A S, Alymov M I and Milyaev I M 2014 Letters on Materials **4** (1) 59–61.
[14] Kablov E N, Orlov M P and Ospennikova O G 2012 Aviation Materials and Technologies **5** (57) 117–29.
[15] Kuznetsov V P, Lesnikov V P and Popov N A 2016 Structure and Properties of Singlecrystal Heat-resistant Nickel Alloys (Ekaterinburg: Ural Federal University Publ.) p 160.

IOP Conf. Series: Materials Science and Engineering **848** (2020) 012005 doi:10.1088/1757-899X/848/1/012005

Interaction of exogenous refractory nanoparticles of Al_2O_3 and $MgAl_2O_4$ with nonferrous metal impurities in iron melts

S N Anuchkin

A.A.Baikov Institute of Metallurgy and Materials Science, Russian Academy of Sciences (IMET RAS). Russia

E-mail: AnuchkinSN@yandex.ru

Abstract. The exogenous refractory nanoparticles is proposed for refining iron-based melts from nonferrous metal impurities. Investigated the heterophase interaction of Al_2O_3 and $MgAl_2O_4$ nanoparticles with Sb/Sn/Cu and showed that the degree of removal of nonferrous metal impurities was in the Fe-Sb system up to 30 rel.%, in the Fe-Sn system up to 29 rel. % and in the Fe-Cu system up to 23 rel. %. The existence of refractory nanoparticles in a liquid metal was proved by studying the structural properties of the melt. In the melts, after the introduction of the nanoparticles, inversion of the temperature coefficient of surface tension is observed, as well as multidirectional change depending on the nature of the nanoparticles and the holding time during interfacial interaction. Decompression and compression of the melts after the introduction of nanoparticles and the multidirectional nature of the dependences of properties indicate a possible change in the cluster structure of liquid metal.

1. Introduction

The development of the science of metallic nanomaterials [1], including various methods for producing refractory nanophases, allows implementing non-traditional methods of using nanopowders to develop more efficient and universal methods for improving properties of metal. One of the possible advanced methods is the use of exogenous refractory phase nanoparticles (RPNP) as reagents for refining iron-based melts from harmful impurities, and for inoculation in metal crystallization processes. Many modern studies deal with the interaction of ultrafine particles with iron-, nickel-, and aluminumbased melts during the solidification of commercial metallic parts. For example, in [2], the modification of steel such as Hatfield steel was studied with a fine powder of titanium carbonitride. The authors found that the modification of steel leads to an increase in the coefficient of abrasive wear resistance by 20-25%, impact-abrasive by 3-4 times, as well as grain refinement of austenite to 110-120 μm. It was noted that to ensure a high level of operational properties, it is necessary to introduce a modifier into the melt in an amount of 0.3-0.4% by weight of the alloy. In [3], the modification of low alloy steels was studied and the criteria for selecting nanopowder modifiers were identified: thermodynamic stability in the melt, accordance of the crystal lattices to the matrix metal, and proportionality to the critical radius of the austenite nucleus during crystallization. The authors justified the choice of Ti(CN) nanopowder with a size of less than 100 nm, and showed that the modification of 09G2 and 09G2S steels increases tensile strength by an average of 23%; yield strength - by 19%; δ - by 23%; ψ - by 6% and toughness by 39%. The alloying of 08Kh18N10T steel by nano- and microdispersed ZrO_2 particles during vacuum arc remelting was studied in work [4]. When studying the effect of the melting conditions and the alloying particle size on the ZrO_2 distribution in a metal ingot, the authors of showed that ZrO2 nanoparticles were more uniformly distributed in the metal volume than ZrO_2 microconglomerates did. When analyzing the mechanical properties, they found that the ultimate tensile strength and the yield strength increased as compared to the initial state when the ZrO_2 particle content increased to 0.3 wt.% and that these properties decreased at above 0.4 wt.% ZrO_2. The effect of nanopowders on the service characteristics of gray iron SCh20 was studied in [5]. Modifiers of two compositions were used in the work: I – obtained by plasma chemical technique in an iron matrix of a composition (α-Fe, TiC_xN_y, SiC) of size 30–70 nm; II - obtained by the SHS method, combined with the mechanochemical method, and consisting of a carbide phase (70-80% WC - 20-30% TiC) of size 30 - 35 nm with a metal protector zirconium. The use of nanomodifiers I and II composition in an amount of 0.002 wt.% (I) and 0.038 wt.% (II) increases Brinell hardness by 9.8 (I) and 13.1 (II) rel.%, tensile strength by 11.9 (I) and 19.8 (II) rel. %, relative wear resistance at 21 (I) and 69 (II) rel. %, corrosion resistance in hydrochloric acid at 12.3 (I) and 28 (II) rel. % compared with unmodified samples of cast iron. Studies of the effect of SiC nanoparticles on the structural and mechanical properties of nodular cast iron were carried out in [6]. It was experimentally shown that after the addition of nano-SiC, the spheroidization rate, spheroidization degree and graphitization of the graphite in casting structure are improved. It was noted that nano-SiC can promote the formation of ferrite, reduce pearlite and the content of carbide in the cast structure which is beneficial to improve the impact toughness of casting. In [7], the effect of modification with ultrafine powders of refractory metal oxides and cryolite on the structure and deformation behavior of SCh25 gray iron was investigated. The modifying composition (0.3 wt% of the mass of the melt) consisted of ultrafine powders TiO_2 and ZrO_2 and cryolite Na_3AlF_6. The authors found that after modification there is a significant change in the nature of distribution and size of lamellar graphite, as well as an increase in the dispersion of perlite (from 0.57 to 0.32 μm) and increase in tensile strength (by 90 MPa). At the same time, ductility of 0.17-0.18% and hardness of 207 HB ± 5% remain unchanged. This review supports the importance of using RPNP in iron-based melts for increasing the service properties of parts during solidification. However, the physicochemical laws of interaction of nanoparticles with metallic melt components were poorly considered in the reviewed works, and the obtained results are substantially different depending on an RPNP–melt system. The interaction of surfactants with RPNPs in a liquid metal was not considered, and the possible influence of surfactants on the nanoparticle redistribution in the melt volume was not analyzed.

The charge of modern steelmaking units contains significant amounts of nonferrous metal impurities (NMI), in particular copper, tin, antimony, arsenic, etc., and to remove them during metal refining is a complex physicochemical and technical problem, which has not been efficiently and unambiguously solved to date [8,9]. Effective and economically viable methods of metal refining from NMI are only being developed, but there are a large number of scientific papers devoted to this problem [10-13]. However, existing methods do not solve the problem of removal NMI and it remains quite relevant and requires a non-standard approach to solve it. It is note that NMI, as a rule, have surface-active properties in iron melts [14, 15], which suggests the use of these properties for refining melts from NMI.

In [16] have been proposed and experimentally confirmed the hypothesis about the interaction of nanoparticles of refractory phases with metal melt surfactants (SAS). The mechanism was schematically presented as follows: introduction exogenous RPNP (disperse phase) into a liquid metal (dispersion system) leads to redistribution of surfactants on the boundary particle-melt (lyophilic or lyophobic) and their adsorption on the surface of the nanoparticles. Taking into account the quasichemical theory of the structure of melts, structural compounds are formed - ensembles M + (RPNP-SAS) as result of absorption. Due to the surfactant gradient, the processes of moving the ensembles in the melt, their possible merging or degradation, and further removal under the influence of adsorption nature forces on the M-phase interface (ceramics, slag, gas) occur. As a result, one part of the ensemble is removed from the metal, i.e. the process of metal refining from harmful SAS is implemented. Another part of the ensembles will further influence the crystallization processes and mechanical characteristics of the metal. The purpose of this work is, firstly, the choice of RPNP for introducing into a liquid melt for refining it from NMI (Sb, Sn, Cu). Secondly, the study of heterophase interaction of various types of RPNP with NMI with surface-active properties in model iron melts. Thirdly, the study of the influence of RPNP on the structural properties of melts as proof of their presence in the metal and involvement in the processes of adsorption/desorption of surfactants (Sb, Sn, Cu).

2. Thermodynamic analysis

Based on the physicochemical analysis of the choice of RPNP, including the thermodynamics of dissociation reaction of refractory phase in the melt and information on the interfacial tension of the metal – refractory phase, the Al_2O_3 and $MgAl_2O_4$ nanoparticles were used in this work. Alumomagnesium spinel was chosen because of an attempt to reveal, first, the influence of a more complicated adsorbent structure and the related cavity content, the strength of cation–anion bonds, nonstoichiometry, and the defect content [17] on SAS adsorption and, second, the possibility of a change in the interfacial characteristics (interfacial tension, wetting contact angle) of the newly formed RPNP due to the dissociation of the nanospinel. The standard Gibbs energy of dissociation in iron was calculated according to the equations [18-20]:

$$Al_2O_3(s) = 2[Al]_{Fe} + 3[O]_{Fe}; \quad \Delta G_T^{\ 0} = 1209255 - 389{,}72 \cdot T$$

$$MgAl_2Os_4(s) = Al_2O_3(s) + MgO(s); \quad \Delta G_T^{\ 0} = 23600 + 5{,}91 \cdot T$$

$$MgAl_2O_4(s) = 2[Al]_{Fe} + [Mg]_{Fe} + 4[O]_{Fe}; \Delta G_T^{\ 0} = 1961455 - 622{,}21 \cdot T$$

Table 1. Gibbs energies for the dissociation of the compounds and their wettability with the iron melt.

Reaction	T_m(phase), K	$\Delta G^o_{1873\ K}$, J/mol	ln *Kp*	θ, deg (t, °C)
$Al_2O_3(s) = 2[Al]_{Fe} + 3[O]_{Fe}$	2327	479 309	-30,79	141 (1550) [21]
$MgAl_2O_4(s) = Al_2O_3(s) + MgO(s)$	2378	34 670	-2,23	107 (1560) [22]
$MgAl_2O_4(s) = 2[Al]_{Fe} + [Mg]_{Fe} + 4[O]_{Fe}$		796 056	-51,15	

The calculation results for 1873 K are presented in table 1. The data on the wettability of these compounds with iron melts were analyzed [21, 22], and the values of wetting angle θ are also given in table 1. An analysis of these data indicates that oxides can exist in the iron melt, being lyophobic to this melt.

The probability of removing Sn from the Fe-Sn(0.0561wt.%)-O(0.0059wt.%) model melt was considered, and the evaporation process was presented as [18.19]: $[Sn]_{Fe}$=Sn(g), ΔG_T^0=279700-58.32·T. Using the equation describing the temperature dependence of the equilibrium constant of the reaction, P_{Sn} was determined as $P_{Sn} = a_{Sn} \cdot \exp\{-\Delta G^0/RT\}$, where $a_{Sn} = f_{Sn}[Sn]$ and $\lg f_{Sn} = e_{Sn}^{Sn}[Sn] + e_{Sn}^{O}[O]$. The values of interaction parameters in the calculation were e^{Sn}_{Sn} = 0.0017 [23] and e^{O}_{Sn} = -0.11 [23]. The partial pressure P_{Sn} at 1873 K was $9{,}94 \cdot 10^{-2}$ Pa. Obviously, the amount of tin that passed from the melt to the gas phase under experimental conditions is negligibly small.

The possibility of removing Sb from Fe-Sb(0.0950wt.%)-O(0.0079 wt.%) model melts was considered, and the process was presented as a reaction [18.24]: $[Sb]_{Fe}$=Sb(g), ΔG_T^0=120995-14·T. Note that for the reaction $[Sb]_{Fe}$=Fe(l) the values of the reaction coefficients were obtained by extrapolating the data in [24] for [Sb] = 0.1 wt.%. Using the temperature dependence of the reaction equilibrium constant, P_{Sb} was determined as $P_{Sb} = a_{Sb} \cdot \exp\{-\Delta G^0/RT\}$, where $a_{Sb} = f_{Sb}[Sb]$ and $\lg f_{Sb} = e_{Sb}^{Sb}[Sb] + e_{Sb}^{O}[O]$. The value of e^{O}_{Sb} = -0.2 [23], and the value of e^{Sb}_{Sb} was taken equal to the value of e^{Sn}_{Sn} = 0.0017 [23]. The value of P_{Sb} at 1823 K was 17.34 Pa. Apparently, under the assumptions made in the calculations and in the experimental conditions, it indicates on possible evaporation of Sb from the melt, which was taken into account during special experiments.

The probability of a decrease Cu content due to evaporation process in the Fe-Cu(0.0445wt.%)-O(0.0078wt.%) system was considered, and the process was presented as a reaction [18,19]: $[Cu]_{Fe}$=Cu(g), ΔG_T^0=274363-69.36·T. Using the equation describing the temperature dependence of the equilibrium constant of the reaction, P_{Cu} was determined as $P_{Cu} = a_{Cu} \cdot \exp\{-\Delta G^0/RT\}$, where $a_{Cu} = f_{Cu}[Cu]$ and $\lg f_{Cu} = e_{Cu}^{Cu}[Cu] + e_{Cu}^{O}[O]$. In the calculation, the interaction parameters $e^O{}_{Cu}$ and $e^{Cu}{}_{Cu}$, were equal to -0.0645 [23] and -0.023 [23], respectively. The P_{Cu} value at 1873 K was 0.418 Pa, which indicates a low evaporation of Cu under the experimental conditions.

3. Methods and materials of research

The material analysis of the RPNP Al_2O_3, obtained by the plasmachemical method, showed that they were polydispersed and the particle size distribution was close to a logarithmic normal one. The dispersion composition obtained on a Mastersizer 2000 analyzer (Malvern) for a Al_2O_3 was characterized by the following data: d_{10} = 25 nm, d_{50} = 51 nm, and d_{90} = 83 nm, where d_x is the size below which the fraction contains x% particles. The specific surface was analyzed by the Brunauer–Emmett–Teller (BET) method on a Micrometrics TriStar 3000 analyzer and for Al_2O_3 it was 36.4 m^2/g. The composition of the RPNP Al_2O_3 was characterized by the presence of the following phases: δ (50–60%), θ (10–20%), and γ (10–20%). Spinel $MgAl_2O_4$ was obtained mechanochemically by treatment of Al_2O_3 and MgO RPNP in a Fritsch Pulverisette 7 planetary mill for 10 min followed by annealing in a TK.4-1400.1.F electric resistance furnace at 1000°C for 2 h. The X-ray diffraction pattern obtained on a Rigaku Ultima 4 instrument for the spinel after the treatment is 96.3 wt.% $MgAl_2O_4$, 1.9 wt.% Al_2O_3, and 1.8 wt.% MgO. The specific surface of $MgAl_2O_4$ is 24.55 m^2/g, and the average particle size $d_{av.}$ (BET) is 41 and 65 nm for Al_2O_3 and $MgAl_2O_4$, respectively.

A composite material was prepared from a mixture of the RPNP and iron powders to introduce RPNP into a liquid melt. An analysis of the iron powder (high-purity grade OSCh 6-2) showed the presence of 80% particles 15–25 μm in size. The composite material Fe(97.5wt.%)+RPNP(2.5wt.%) was prepared in a Fritsch Pulverisette 6 planetary mill for 1 h in an argon medium. After treatment, the mixture of powders was subjected to uniaxial pressing under a pressure of 35 MPa with degassing at 300 °C in argon and in vacuum followed by storage in vacuum at 25° C (for details, see [25]).

Model Fe-Sb, Fe-Sn and Fe-Cu alloys were melted in a vacuum resistance furnace with a graphite heater in a 46 ml corundum crucible at $P(He+10\%H_2)$=0.1 MPa. Heterophase interaction was investigated in a vacuum induction furnace (VIF), which has a special conical inductor that provides intensive mixing of the metal with RPNP both on the surface and inside the melt. The experiments were carried out at $P(He+10\%H_2)$=0.2 MPa (for details, see [16]). After the metal was melted completely, it was held for 2 min to achieve equilibrium with the gas phase. Then, the composite material (Fe+RPNP) was introduced into the liquid metal followed by the isothermal holding for 300–1200 s depending on the experimental conditions. The RPN content in the melt after the addition of the composite material was 0.12 wt.%. The contents of Sb, Sn, and Cu were determined using an Ultime 2 (Jobin Yvon) atomic emission spectrometer with an inductively coupled plasma with a sensitivity of 0.0005%, and the oxygen content was determined by reducing melting on a TC 436 (LECO) analyzer with a sensitivity of 0.0002%. Surface tension (σ) and melt density (ρ) were studied by the sessile drop method: a liquid metal drop was formed in a corundum conic cup (98% Al_2O_3 plus TiO_2 binder) in a vacuum resistance furnace with a graphite heater and a protective molybdenum tube. The system was pumped out to a residual pressure of $2 \cdot 10^{-3}$ Pa, and a sample was heated to 1000°C after the system was graduated against the melting points of pure metals. The processes of sample melting and drop formation and an equilibrium drop position were detected with a Nikon D70 digital camera upon holding for 600 s when the temperature was changed by 10–15°C above the previous temperature up to 1750°C at P_{Ar} = 0.1 MPa. The photographs were processed using the Drop software package (Laplace calculation) [26]. The theory and practice of the method were described in [27].

4. Results of heterophase interaction

IOP Conf. Series: Materials Science and Engineering **848** (2020) 012005 doi:10.1088/1757-899X/848/1/012005

More than 40 experiments with the following 8 types of alloys were conducted (wt.%): Fe-Sb(0.0950%); Fe-Sb(0.0950%)-Al_2O_3($d_{av.}$=41nm); Fe-Sn(0.0561%); Fe-Sn(0.0561%)-Al_2O_3(41nm); Fe-Sn(0.0561%)-$MgAl_2O_4$(65nm); Fe-Cu(0.0445%); Fe-Cu(0.0445%)-Al_2O_3(41nm) and Fe-Cu(0.0445%)-$MgAl_2O_4$(65nm), and the results were presented as the degree of removal of surfactants α = (([SAS]$_{initial}$-[SAS]$_{ending}$)/[SAS]$_{initial}$) ·100, rel. %. The experimental results of the heterophase interaction of the RPNP with surfactants using the dependences [SAS]=$f(\tau)$, as well as the SAS content in melts without RPNP (special experiments) are shown in figure 1.

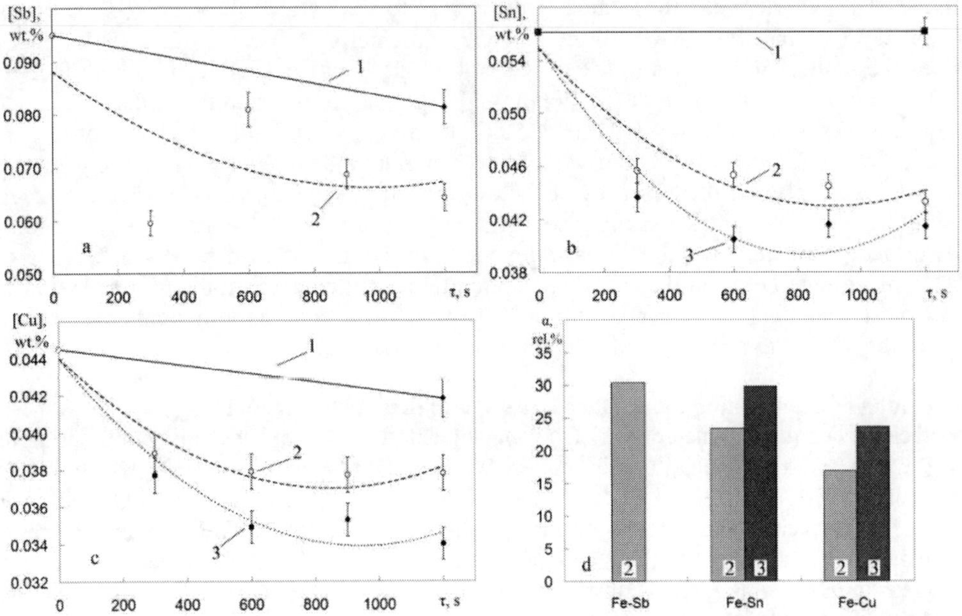

Figure 1. Dependence of SAS content on the duration of isothermal holding times in melts Fe-Sb (0.0950 wt.%) (a), Fe-Sn (0.0561 wt.%) (b) and Fe-Cu (0.0445 wt.%) (c) and extremum values of the degree of surfactant removal in these systems (d): 1 - Fe-SAS; 2 - Fe-SAS-Al_2O_3; 3 - Fe-SAS-$MgAl_2O_4$.

An analysis of the result showed that, firstly, the introduction of Al_2O_3 and $MgAl_2O_4$ RPNP at an isothermal holding time of 300-1200 s. led to a decrease in the content of SAS (Sb, Sn, and Cu), and the degree of removal was in the Fe-Sb system up to 30 rel.%, in the Fe-Sn system up to 29 rel. %, and in the Fe-Cu system up to 23 rel. %. Thermodynamic analysis of the removal processes and experimental data suggest that the recorded decrease is a result of heterophase interaction of RPNP with SAS and the removal of Fe+(RPNP-SAS) ensembles to the M-gas/ceramic interface. Secondly, data processing by the least squares method describing by polynomial function showed the presence of an extremum (see Figure 1) in the systems: Fe-Sb-Al_2O_3: [Sb]=0,0882-4,48·$10^{-5}\tau$+2,28·$10^{-8}\tau^2$ and R^2 = 0,64; Fe-Sn-Al_2O_3: [Sn]=0,0549-2,61·$10^{-5}\tau$+1,43·$10^{-8}\tau^2$ and R^2 = 0,94; Fe-Sn-$MgAl_2O_4$: [Sn]=0,0551-3,80·$10^{-5}\tau$+2,30·$10^{-8}\tau^2$ and R^2 = 0,96; Fe-Cu-Al_2O_3: [Cu]=0,0440–1,65·$10^{-5}\tau$+0,97·$10^{-8}\tau^2$ and R^2 = 0,94; Fe-Cu-$MgAl_2O_4$: [Cu]=0,0440–2,13·$10^{-5}\tau$+1,13·$10^{-8}\tau^2$ and R^2 = 0,95. The presence of extremum of the equations can serve as evidence of possible processes of surfactant desorption in the melt. Thirdly, a comparison of the dependences after the introduction of Al_2O_3 and $MgAl_2O_4$ RPNP shows that in all cases the introduction of $MgAl_2O_4$ leads to a higher degree removal of SAS. This indicates either the higher adsorption properties of spinel, which was indirectly noted in [17], or the dissociation of spinel nanoparticles into oxides with a surface energy different from the energy of individually introduced exogenous oxides. The extremum values of the functions [SAS]=$f(\tau)$ were determined, which indicate general dependencies of the adsorption/desorption mechanism of metal refining from SAS, and presented them as a histogram in figure 1d. The results showed that α depends on the surface-active properties of investigated nonferrous metal impurities, which probably indicates correlation of the processes of redistribution of nanoparticles and ensembles in a liquid metal with the surface-active properties of the studied impurity. To support the existence of an adsorption mechanism of the removal of RPNP from a metallic melt, we calculated the particle velocities under adsorption forces. Our calculation is based on the diffusion phoresis mechanism proposed and experimentally proved in [28]. This mechanism describes the motion of particles in the concentration gradient field induced by a surface tension gradient. Calculation is presented in detail in [29]. We found that the velocity induced by adsorption forces is equal to $\upsilon_a = 6{,}21\cdot10^{-5}$; $5{,}86\cdot10^{-5}$ and $4{,}43\cdot10^{-5}$ m/s for binary systems Fe-Sb, Fe-Sn and Fe-Cu, respectively. The gravitational component of the velocity that appears because of the difference between the densities of the disperse and dispersion phases was estimated, and the value $\upsilon_g = 2{,}01\cdot10^{-9}$ m/s. A comparison of the obtained rates shows that the determining role of the surface phenomena when nanoparticles move in the melt is evident at a difference of four orders of magnitude between the rates under the effect of the adsorption and gravitation forces. In this regard, the effect of the surface properties of impurities and hypothetical ensembles on the change in the structural properties of the melt was investigated.

5. The results of determining the surface tension and density of the melt

The surface tension (σ) and melt density (ρ) were studied after heterophase interaction during melting in a VIF with various isothermal holding times. We studied 7 types of materials: 1 - Fe, 2 - Fe-SAS (Sb, Sn, Cu), 3 - Fe-Sb-Al_2O_3, 4 - Fe-Sn-Al_2O_3, 5 - Fe-Sn-$MgAl_2O_4$, 6 - Fe-Cu-Al_2O_3 and 7 - Fe-Cu-$MgAl_2O_4$. The oxygen content in all systems was 0.0065-0.0082 wt.%. The experimental results are approximated by least squares method, and the results are presented in table 2 (see σ = f (T) and ρ = f (T), where σ in mN/m, ρ in g/cm³, T in K). Using the data on the melt densities for the experiments of all series, we estimated the degree of loosening when the temperature was increased by 100°C; to this end, we calculated parameter $\beta=\{(\rho^{1823}-\rho^{1923})/\rho^{1823}\}\cdot100$, rel %.

An analysis of the data of the first and second series of experiments, presented in figure 2a, shows that, firstly, the introduction of low concentrations of Sb, Sn, or Cu into the melt leads to a decrease σ at 1873 K (1647, 1439, 1534, and 1584 mN/m for Fe, Fe-Sb(0.0548 wt.%), Fe-Sn(0.0548 wt.%) and Fe-Cu(0.0445 wt.%), respectively). This confirms that even at low concentrations of antimony, tin and copper possess surface-active properties in metallic iron melts. Secondly, there is satisfactory agreement between the values of σ and ρ of pure iron and binary systems with published data (for details, see [14, 15]), which indicates the reliability of the results.

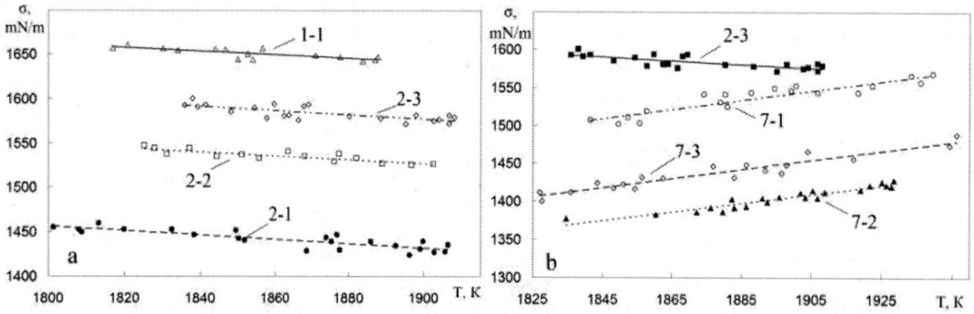

Figure 2. Temperature dependences of the surface tension of melts: 1-1 - Fe; 2-1 - Fe-Sb; 2-2 - Fe-Sn; 2-3 - Fe-Cu; (7-1 - 7-3) - Fe-Cu-MgAl2O4 (for details, see table 2).

Table 2. The measurement results of σ and ρ of the systems Fe, Fe-(Sb, Sn, Cu) and Fe-(Sb, Sn, Cu)-(Al_2O_3, $MgAl_2O_4$) depending on the type of RPNP and on the holding times in the VIF.

No.	System	τ^a, min	[SAS]·10^4, wt.%	σ = f(T)	R^2	σ^{1873K}, mN/m	ρ = f(T)	R^2	ρ^{1873K}, g/cm³	β^b, rel.%
1-1	Fe	20	-	2027-0,203T	0,79	1647	8,871-0,00102T	0,82	6,96	1,45
2-1	Fe-Sb	20	881	1905-0,249T	0,85	1439	7,683-0,00052T	0,78	6,71	0,77
2-2	Fe-Sn	20	548	1963-0,229T	0,81	1534	7,966-0,00057T	0,80	6,90	0,82
2-3	Fe-Cu	20	445	2037-0,242T	0,74	1584	7,990-0,00056T	0,93	6,94	0,80
3-1	Fe-Sb-Al_2O_3	5	590	2116-0,258T	0,73	1633	8,789-0,00079T	0,90	7,31	1,08
3-2		10	810	1821-0,119T	0,70	1598	7,740-0,00028T	0,87	7,22	0,39
3-3		20	640	1939-0,192T	0,75	1579	8,576-0,00080T	0,85	7,08	1,12
4-1	Fe-Sn-Al_2O_3	5	457	1069+0,290T	0,74	1612	8,071-0,00062T	0,84	6,91	0,89
4-2		10	453	1176+0,188T	0,72	1528	9,059-0,00114T	0,75	6,92	1,63
4-3		15	445	145,1+0,770T	0,92	1587	7,883-0,00051T	0,80	6,93	0,73
4-4		20	433	1834-0,195T	0,68	1469	8,332-0,00074T	0,89	6,95	1,06
5-1	Fe-Sn-$MgAl_2O_4$	5	436	87,9+0,779T	0,92	1547	8,519-0,00085T	0,90	6,93	1,22
5-2		10	405	665,5+0,502T	0,81	1606	7,901-0,00060T	0,83	6,78	0,88
5-3		15	416	-632,6+1,140T	0,88	1503	8,458-0,00085T	0,80	6,87	1,23
5-4		20	415	-634,7+1,156T	0,97	1530	8,684-0,00096T	0,88	6,89	1,38
6-1	Fe-Cu-Al_2O_3	5	389	554,2+0,521T	0,88	1530	8,558-0,00098T	0,86	6,72	1,45
6-2		10	379	485,8+0,565T	0,89	1545	8,409-0,00085T	0,87	6,82	1,24
6-3		20	378	234,1+0,719T	0,85	1581	8,481-0,00088T	0,85	6,83	1,28
7-1	Fe-Cu-$MgAl_2O_4$	5	377	373,1+0,616T	0,86	1527	7,714-0,00066T	0,88	6,48	1,01
7-2		10	349	312,4+0,576T	0,90	1391	7,615-0,00054T	0,76	6,59	0,83
7-3		20	340	293,0+0,610T	0,91	1435	7,599-0,00061T	0,85	6,46	0,94

[a] - holding times in the VIF; [b] - loosening of a melt.

An analysis of experiments with Fe-SAS-Al_2O_3 and Fe-SAS-$MgAl_2O_4$ melts after the introduction of RPNP showed, firstly, the fact of the inversion of $\partial\sigma/\partial T$ values in the metal, which, unlike the metal without RPNP, indicates a significant change in surface properties melts. Figure 2b shows the typical effect of the $MgAl_2O_4$ RPNP on the change in the surface tension of the Fe–Cu system. Secondly, there is a multidirectional change in the values of σ at 1873K: in the Fe-Sb system is observed an increase σ by 9.7–13.5 rel.%; in the Fe-Sn system – a change in values from -5.1 to +4.2 rel. %, and Fe-Cu – decrease in σ by 0.2–12.2 rel.%. This indicates different nature of the ensembles on the surface of melt depending on the surfactant.

To compare the behavior of Sb, Sn, Cu, and RPNP, the values of the temperature coefficient $\partial\sigma/\partial T$ on the exposure time in a VIF in the study of heterophase interaction are presented in Figure 3. Firstly, the multidirectional effect of surfactant-NMI on $\partial\sigma/\partial T$ values was found: unlike temperature dependences of the Fe-Sn-RPNP and Fe-Cu-RPNP systems, where the facts of temperature coefficient inversion are observed, this does not occur in the Fe-Sb-Al_2O_3. Secondly, a different rate of change in the values of $\partial\sigma/\partial T$ is recorded with an increase in the holding times in the VIF, depending on the nature of the RPNP. In the Fe–Sb system is observed a change the values of $\partial\sigma/\partial T$ from –0.258 to –0.192 after the introduction of Al_2O_3; in the Fe-Sn system – a decrease from +0.290 to -0.195 and an increase from 0.779 to 1.156 after entering Al_2O_3 and $MgAl_2O_4$ RPNP, respectively; in the Fe-Cu system – an increase from 0.521 to 0.719 and a slight change from 0.616 to 0.610 after the introduction of Al_2O_3 and $MgAl_2O_4$ RPNP, respectively. This may be due to the formation of complex ensembles (Fe_x+($RPNP_y$-SAS_z)) with long holding times that have different effects on the change in the structure of the surface layer.

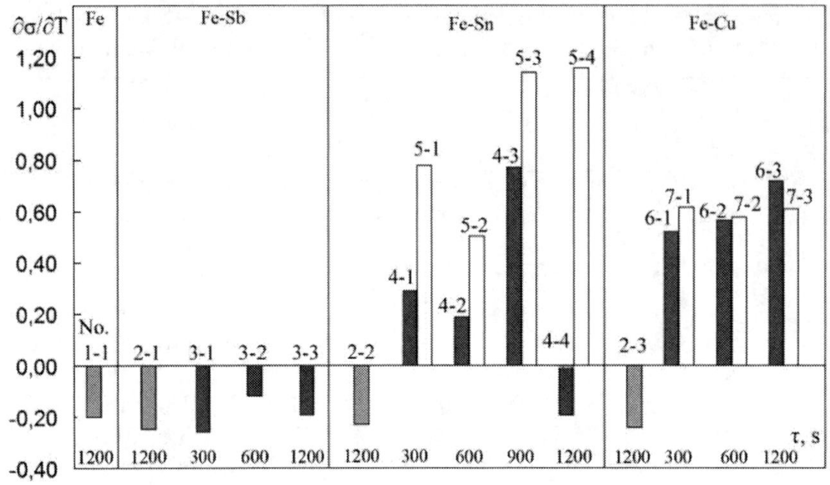

Figure 3. Dependence of $\partial\sigma/\partial T$ values on the nature of the RPNP and the holding times in the VIF (No. see table 2).

The multidirectional effect of RPNP is also observed in analysis of the melt density. Taking for comparison the arithmetic mean of ρ at 1873 K for experiments with different holding times in the VIF, it was determined that in the Fe-Sb system is observed an increase ρ of 7.3 rel. % after entering Al_2O_3. In Fe-Sn and Fe-Cu systems is observed a decrease ρ by 2.1 and 2.2 rel. % after entering Al_2O_3 and 3.1 and 6.2 rel. % after entering $MgAl_2O_4$, respectively. It can be assumed that, as a result of adsorption of impurity atoms on the surface of nanoparticles, there is a change in the cluster structure [30] both in the metal volume and at the nanoparticle – melt interface. Antimony, tin and copper, having different surface-active properties in iron and different bonding forces with iron atoms, as a result of the formation of ensembles, have different effects on the change of the metal cluster structure. Moreover, in Fe-Sn and Fe-Cu systems, higher values are observed in experiments after entering $MgAl_2O_4$, which indicates high adsorption properties of spinel compared to Al_2O_3, which is also observed in studies of the heterophase interaction of RPNP with SAS.

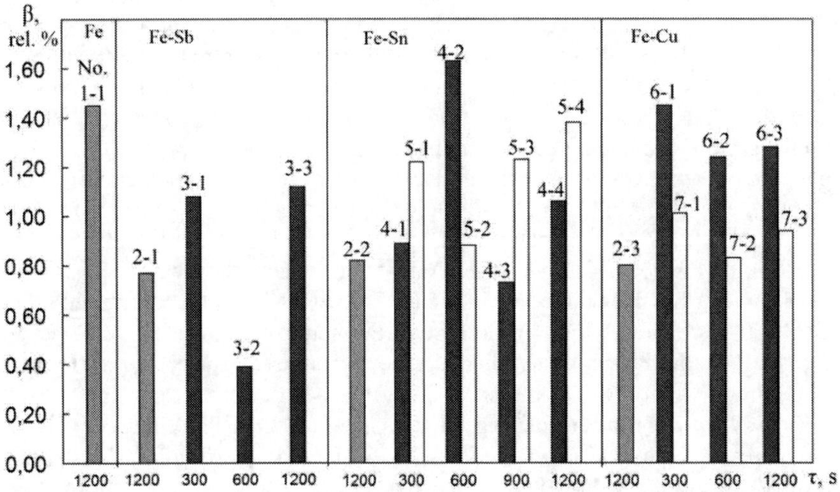

Figure 4. The dependence of the degree of loosening on the nature of the RPNP and the holding time in the VIF (No. see table 2).

The phenomenon of loosening (compression / decompression, see β in table 2 and in Figure 4) was examined. In figure 4 shows a histogram of changes in the degree of loosening of β depending on the nature of the RPNP and the holding time of the melt in the VIF. It can be seen that in the Fe-Sb-Al_2O_3 and Fe-Cu-Al_2O_3 systems, the dependence $\beta=f(\tau)$ is characterized by a minimum extremum, and in similar experiments in the Fe-Sn-Al_2O_3 system the dependence $\beta=f(\tau)$ has a maximum extremum. In all systems a minimal extremum is observed after entering $MgAl_2O_4$. Moreover, in most cases, an increase in β values (decompression) is observed. These dependences indicate different structures of the melts, which is possibly related to the different influence of the ensembles on the change in the cluster structure of the metal and requires further studies. Thus, due to the surface-active properties of surfactant-NMI in the iron melt as a result of its adsorption interaction with Al_2O_3 and $MgAl_2O_4$ RPNP, a significant change in the structural properties of liquid metal occurs, which affects on the behavior of ensembles in the melt and and their removal under the influence of adsorption nature forces on the M-phase interface.

6. Conclusions:

1. Based on the analysis of literature data, thermodynamic calculations on the dissociation of refractory compounds and the phenomena of wetting them with iron melts, were chose of refractory compounds Al_2O_3 and $MgAl_2O_4$ to study the interaction of these nanophases with Fe-Sb, Fe-Sn and Fe-Cu melts in reducing atmosphere.
2. The heterophase interaction of Al_2O_3 (41 nm) and $MgAl_2O_4$ (65 nm) nanoparticles with surfactant – antimony, tin and copper in Fe-Sb (0.0950 wt.%), Fe-Sn (0, 0561 wt.%) and Fe-Cu (0.0445 wt.%) model double melts was investigated. It was found that, depending on the nature of the nanoparticles and the holding time, the extremum values of the degree of surfactant removal were (rel.%): Fe-Sb-Al_2O_3 – 30,3 %; Fe-Sn-Al_2O_3– 23,4 %; Fe-Sn-$MgAl_2O_4$– 29,8 %; Fe-Cu-Al_2O_3 – 16,8 % and Fe-Cu-$MgAl_2O_4$– 23,6 %.
3. The surface tension and density of Fe-(Sb, Sn, Cu) and Fe-(Sb, Sn, Cu)-(Al_2O_3, $MgAl_2O_4$) melts were studied by the sessile drop method and showed that in systems after the introduction of nanoparticles, in most cases, an inversion of the temperature coefficient is observed, as well as its multidirectional change depending on the nature of the nanoparticles and the holding time in the VIF. This indicates the influence of ensembles (Fe+(RPNP-SAS)) on the surface properties of melts.Decompression of Fe-Sn and Fe-Cu melts and compression of Fe-Sn melts after nanoparticles are introduced and the multidirectional character of the loosening dependences $\beta=f(\tau)$ indicates a possible change in the cluster structure of liquid metal.

The work was carried out according to the state assignment number 075-00746-19-00

References

[1] Kolmakov A G, Barinov S M and Alymov M I 2012 Osnovy tekhnologii i primenenie nanomaterialov [Fundamentals of technology and the use of nanomaterials]. (Moscow: FIZMATLIT) 208 p. (In Russ.)

[2] Vdovin K N, Feoktistov N A, Gorlenko D A, Nikitenko O A and Khamidulina D D 2019 *Izvestiya. Ferrous Metallurgy* **62**(3) 188-94 (In Russ.)

[3] Bolshakov V I and Kalinin A V 2018 *Vestnik Kharkovskogo natsionalnogo avtomobilno-dorozhnogo universiteta* **82** 5-12 (In Russ.)

[4] Borts B V, Vanzha A F, Korotkova I M, Sytin V I and Tkachenko V I 2014 *Voprosy atomnoi nauki i tekhniki* **92**(4) 117–24 (In Russ.)

[5] Poluboyarov V A, Korotaeva Z A, Zhdanok A A, Kuznetsov V A and Samokhin A V 2015 *Izvestiya. Ferrous Metallurgy* **58**(8) 561-66. (In Russ.)

[6] Zhu R, Su Y and Qin X 2017 *IOP Conference Series: Materials Science and Engineering* **250** 012051

[7] Zykova A P, Lychagin D V, Chumaevsky A V, Kurzina I A and Novomeysky M Y 2014

Izvestiya. Ferrous Metallurgy **57**(11) 37-42 (In Russ.)
[8] Guzenkova A S, Ivanov S S, Isaev G A and Kudrin V A 2008 Proizvodstvo stali chistoi ot primesei tsvetnykh metallov [Manufacture of steel free from non-ferrous metal impurities]. (Moscow: MGVMI) 118 p. (In Russ.)
[9] Shakhpazov E H, Zaitsev A I and Mogutnov B M 2010 *Problemy chernoi metallurgii i materialovedeniia* **2010**(3) 5-12 (In Russ.)
[10] Warner N A 2004 *Metallurgical and materials transactions B* **35** 663-74
[11] Tohru M 2000 *Tetsu-to-Hagane* **86**(11) 741-47
[12] Labaj J, Oleksiak B and Siwiec G 2011 // Metalurgija **50**(4) 265-68
[13] Hu X, Yan Z, Jiang P, Zhu L, Chou K, Matsuura H and Tsukihashi F 2013 *ISIJ International* **53**(5) 920-22
[14] Nogi K, Chung W B, McLean A and Miller W A 1991 *Materials Transactions JIM.* **32**(2) 164-68
[15] Nizhenko V I and Floka L I 1981 Poverkhnostnoe natiazhenie zhidkikh metallov i splavov [Surface tension of liquid metals and alloys]. (Moscow: Metallurgiya Publ.) 208 p. (In Russ.)
[16] Anuchkin S N, Burtsev V T, Samokhin A V and Serov G V 2009 *Fizika i khimiia obrabotki materialov* **2009**(6) 78-85 (In Russ.)
[17] Khoroshavin L B 2009 Shpinelidnye nanoogneupory [Spineloid nano-refractories]. (Ekaterinburg: Ural Branch of the Russian Academy of Sciences) 600 p. (In Russ.)
[18] Turkdogan E T 1985 Fizicheskaia khimiia vysokotemperaturnykh protsessov [Physical chemistry of high-temperature processes]. (Moscow: Metallurgiya Publ.) 344 p. (In Russ.)
[19] Buzek Z 1979 Fundamental Thermodynamic Data on Metallurgical Reactions and Interactions of Elements in System Significant for Metallurgical Theory and Practice (Ostrava: Vyzkumny ustav hutnictvi zeleza) 110 p.
[20] Jacob K T, Jayadevan K P and Waseda Y 1998 *Journal of the American Ceramic Society* **81**(1) 209–12.
[21] Naiditsch Y B 1972 Kontaktnye iavleniia v metallicheskikh rasplavakh [Contact phenomena in metallic melts]. (Kiev: Naukova Dumka) 196 p. (In Russ.)
[22] Fukami N, Wakamatsu R, Shinozaki N and Wasai K 2009 *Materials Transactions* **50**(11) 2552-56
[23] *Steelmaking Data Sourcebook* 1988 (N.Y.-Tokyo: Gordon & Breach Science Publ.)
[24] Predel B 2013 *Landolt-Börnstein - Group IV Physical Chemistry* **12**C 67–69.
[25] Anuchkin S N, Gvozdkov I A, Samokhin A V, Serov G V and Burtsev V T 2011 *Fizika i khimiia obrabotki materialov* **2011**(2) 71-78 (In Russ.)
[26] Krylov A S, Vvedensky A V, Katsnelson A M and Tugovikov A F 1993 *J. Non-Crystalline Solids* **845** 156-58.
[27] Naidich Y V and Eremenko V I 1961 *Fizika metallov i metallovedenie* **11**(6) 883-88 (In Russ.)
[28] Minaev Y A 1984 Poverkhnostnye iavleniia v metallurgicheskikh protsessakh [Surface phenomena in metallurgical processes]. (Moscow: Metallurgiya Publ.) 152 p. (In Russ.)
[29] Anuchkin S N, Burtsev V T, Samokhin A V and Gvozdkov I A 2012 *Russian Metallurgy (Metally)* **2012**(3) 178-84
[30] Baum B A, Khasin G A, Tyagunov G V, Klimenkov E A, Bazin Y A, Kovalenko L V, Mikhailov V B and Raspopova G A 1984 Zhidkaia stal [Liquid steel]. (Moscow: Metallurgiya Publ.) 208 p. (In Russ.)

In silico determination of substrate spectrum of lactonases, hydrolyzing various N-acyl homoserine lactones

A G Aslanli[1] and E N Efremenko[1,2]

[1] Faculty of Chemistry, Lomonosov Moscow State University, Moscow 119991, Russia
[2] Correspondence: Leninskiye Gory, 1/3, Moscow 119991, Russia;
E-mail: elena_efremenko@list.ru

Abstract. The rapid growth in the number of resistant pathogenic bacteria has led to a decrease in the effectiveness of the existing antimicrobial agents. N-acyl homoserine lactones (N-AHLs) are the key molecules responsible for the formation of antibiotic resistance of gram-negative bacteria. The combination of various lactonases, capable of hydrolyzing a wide range of N-AHLs, with antibiotics, is one of the most appropriate ways to solve the problem of maintaining the effectiveness of the latter. The most interesting is the combination of lactonases with different substrate spectrum of action. In this study, using the molecular docking method, we investigated the substrate range of various lactonases in order to select enzymes suitable to combine with hexahistidine-tagged organophosphorus hydrolase (His_6-OPH), for which high lactonase activity against a number of N-AHLs and the possibility of complex formation with antibiotics have been shown previously. It was found that all the studied lactonases hydrolyze predominantly long chain N-AHLs, while, among all studied lactonases, the SsoPox enzyme from the class of phosphotriesterases-like lactonases was the best candidate for the development of combined enzyme preparations.

1. Introduction

Most of the bacteria (both gram-positive and gram-negative) are known to be resistant to various antibiotics. In recent years, the number of such resistant strains of pathogenic bacteria has increased significantly. This, in turn, led to a rapid loss in the effectiveness of the majority of antibiotics that are currently used in practice, and resulted in the need to increase their effective doses [1]. The study of the resistance mechanisms in pathogenic bacteria showed that in both gram-positive and gram-negative cells, one of the main mechanisms for the development of resistance is the Quorum Sensing (QS) formation. QS is the ability of pathogenic bacterial cells to interact with each other through signaling molecules within the stable highly concentrated populations [2]. Most gram-negative bacteria use various N-acyl homoserine lactones (N-AHLs) as inducers of QS development. Different bacteria are characterized by the synthesis of various N-AHLs, and the N-acyl chain length of N-AHLs can vary from C4 to C18. For example, such common bacterial pathogens as *Pseudomonas aeruginosa*, *Burkholderia cepacia*, *Yersinia enterocholitica* use C12-homoserine lactone (C12-HSL), C8-HSL and C6-HSL, respectively, as signaling molecules of QS [2,3].

To date, the most appropriate way to inhibit the QS mechanism is the destruction of the inducer molecules themselves. The destruction of these molecules leads to a loss of the ability of cells to transition into the QS state. This, in turn, increases their susceptibility to the action of a wide range of antimicrobial agents and eliminates the need to increase the applied dose of antibiotics [4]. Of greatest interest is the enzymatic hydrolysis of N-AHLs by lactonases - enzymes that break down the ester bond of lactones, thereby opening the lactone ring [4,5]. Currently, however, there are no known available and produced in significant quantities enzymes with lactonase activity that are effective against a wide range of N-AHLs, responsible for the formation of resistance of pathogenic bacterial cells. In this regard, there are no effective methods for the combined use of such enzymes with the antimicrobial agents in order to increase the effectiveness of existing antibiotics.

One of the enzymes which are exhibit lactonase activity is the hexahistidine-tagged organophosphorus hydrolase (His_6-OPH). It has been previously shown that, in addition to the ability to hydrolyze a wide range of organophosphorus compounds, His_6-OPH effectively catalyzes the hydrolysis of the lactone ring in molecules of various N-AHLs [6]. In addition, it was found that the enzyme is able to maintain a sufficiently high enzymatic activity and stability in the presence of a number of β-lactam antibiotics, as well as antimicrobial agents of different chemical structure [6,7,8]. However, the substrate spectrum of action of His_6-OPH is not as wide as the range of N-AHLs synthesized by gram-negative bacteria. Therefore, it is advisable to combine various enzymes capable of catalyzing the hydrolysis of a number of N-AHLs. The use of a rational combination of His_6-OPH and other lactonases for their subsequent combination with antibiotics should help to create new drugs with different focuses and a wide substrate spectrum of action.

So the purpose of this study was to select enzymes capable of hydrolyzing various N-AHLs, to study their substrate spectrum of action and the possibility of their combination with His_6-OPH, in order to obtain catalytically active drugs that provide antibiotics with increased effectiveness against resistant cells of gram-negative bacteria. Initially, an analysis of the literature data was carried out, and enzymes exhibiting lactonase activity were selected, and based on the use of computer simulation methods, their substrate spectrum of action was determined *in silico* for various N-AHLs with or without 3-oxo substitution.

2. Results and discussion

Based on an analysis of published data, 6 enzymes capable of hydrolyzing a range of N-AHLs were selected (Table 1). The choice of enzymes was based on their belonging to a different classes of lactonases: 4 enzymes (**AaL, AidC, AiiA** and **AiiB**) from the class of Metallo-β-lactamases, and 2 (**GKL** and **SsoPox**) from the class of Phosphotriesterases-like lactonases (PLLs), as well as on the differences in the structure of their active site.

Table 1. Characteristics of the enzymes hydrolyzing different N-AHLs.

Enzyme	Origin	Me in Active site	k_{eff}, ($10^3 s^{-1} \times M^{-1}$)	Substrate
AaL [9]	*Alicyclobacillus acidoterrestres*	Co^{2+}	1300	C4-HSL
AidC [10]	*Chryseobacterium sp.*	Zn^{2+}	1700	C7-HSL
AiiA [11]	*Bacillus thuringiensis*	Zn^{2+}	16	C6-HSL
AiiB [12]	*Agrobacterium tumefaciens*	Zn^{2+}	16	C6-HSL
GKL [13]	*Geobacillus kaustophilus*	Zn^{2+}; Fe^{2+}	0.430	C8-HSL
SsoPox [14]	*Sulfolobus solfataricus*	Co^{2+}; Fe^{2+}	36	3OC10-HSL

Using the molecular docking method the most probable interactions of various N-AHLs with or without 3-oxo substitution with the active sites of the selected enzymes were simulated at pH 7.5 (the closest pH to physiological conditions) and 10.5 (corresponding to the maximum enzymatic activity of His_6-OPH [15]) (Figures 1,2).

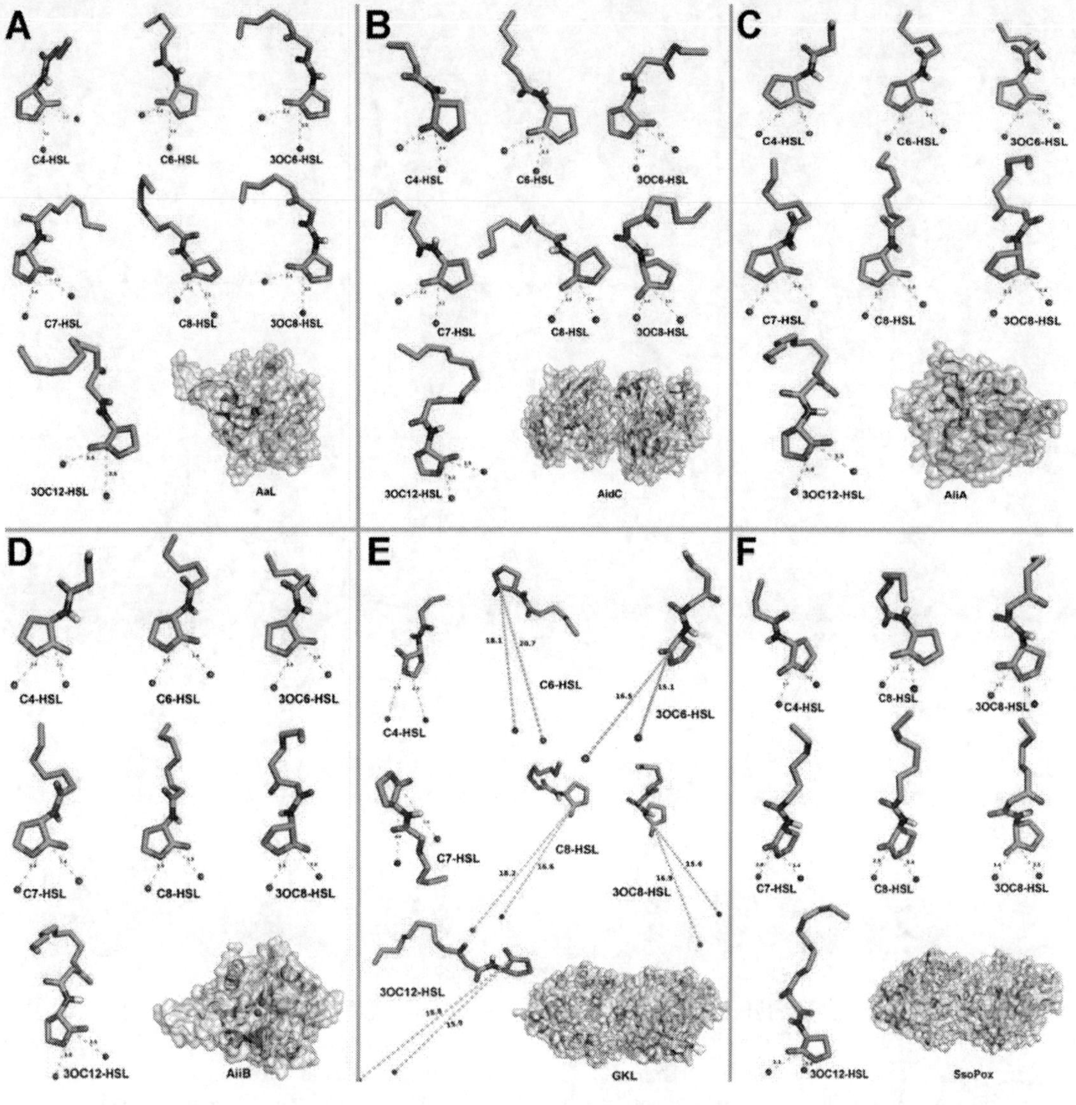

Figure 1. Possible interactions of N-AHLs with the active site of (A) Aal, (B) AidC, (C) AiiA, (D) AiiB, (E) GKL and (F) SsoPox enzymes at pH 7.5. N-AHLs are shown as green sticks. Ions in enzymes active site are colored differently: Co^{2+}, Zn^{2+} and Fe^{3+} ions are colored violet, blue and brown respectively. Interactions between lactone ring of N-AHLs and ions in the enzyme active site are indicated by yellow dashed lines.

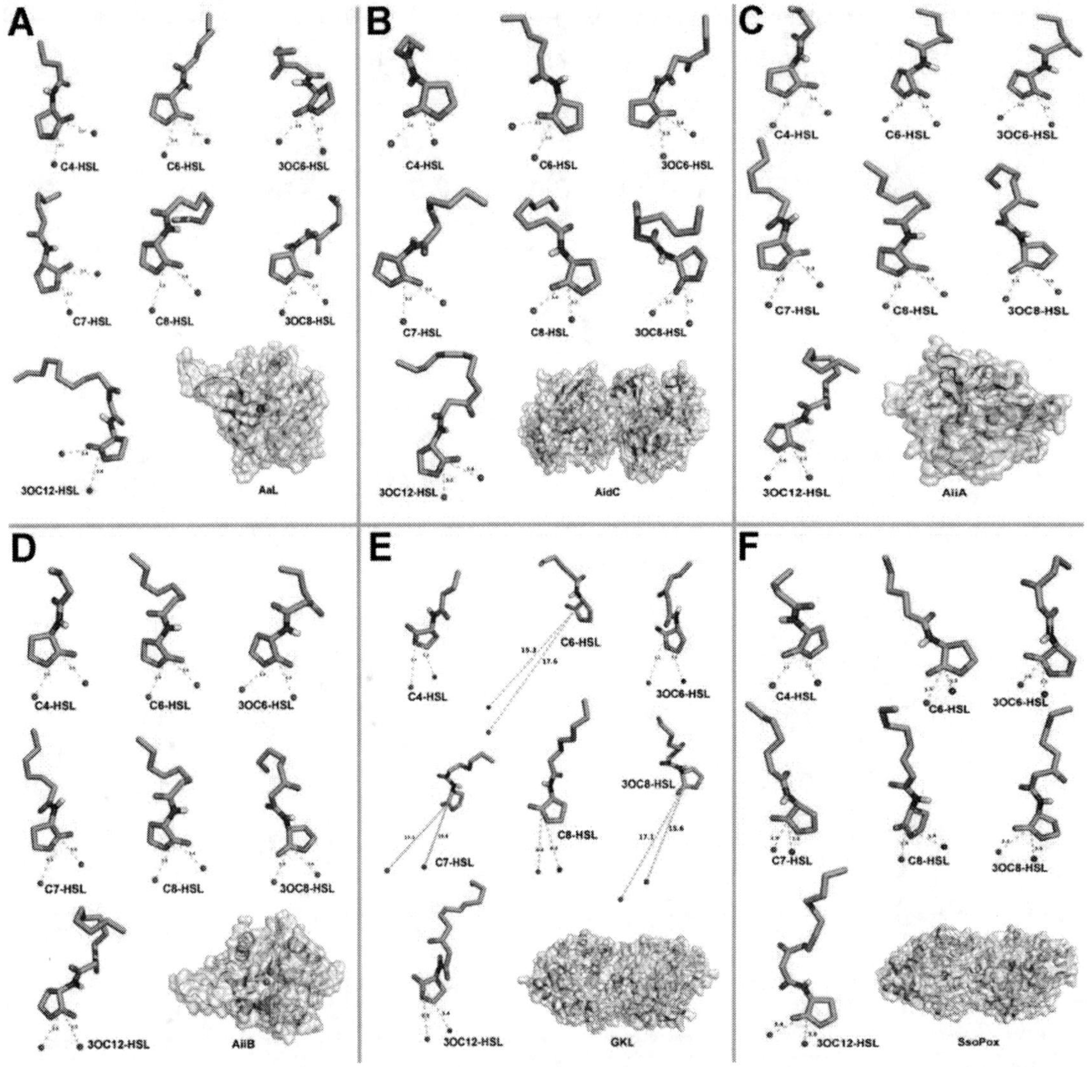

Figure 2. Possible interactions of N-AHLs with the active site of (A) Aal, (B) AidC, (C) AiiA, (D) AiiB, (E) GKL and (F) SsoPox enzymes at pH 10.5. N-AHLs are shown as green sticks. Ions in enzymes active site are colored differently: Co^{2+}, Zn^{2+} and Fe^{3+} ions are colored violet, blue and brown respectively. Interactions between lactone ring of N-AHLs and ions in the enzyme active site are indicated by yellow dashed lines.

The models of enzyme-substrate interaction, obtained from molecular docking, demonstrated that, regardless of the N-acyl chain length, both 3-oxo substituted and unsubstituted N-AHLs are located in the active site of the selected lactonases at a distance sufficient for efficient hydrolysis of their lactone ring. This suggests that the studied enzymes must be able to hydrolyze N-AHLs with different N-acyl chain length. The exception was the GKL enzyme, for which a significant remoteness of the molecules of the investigated substrates from his active sites was established and, therefore, the probability of effective hydrolysis was absent.

To more accurately assess the enzyme-substrate interactions, as well as to determine the effect of the substrate N-acyl chain length, in the obtained models, the affinity values of N-AHLs to the surface of the analyzed lactonases were calculated at both pH values (Table 2). The affinity was estimated as minimum of potential energy during electrostatic and hydrophobic interactions, and hydrogen bonding between ligand (N-AHL) and receptor (enzyme).

Table 2. Affinity of N-AHLs to the surfaces of various enzymes at different pH values

#	AMP	pH	Affinity (kJ·mol^{-1}) Mean	Median	Upper (Lower) Bounds	p value
AaL	C4	7.5	-20.9	-20.1±2.6	-18.9(-22.5)	0.296
		10.5	-21.6	-21.3±2.1	-19.7(-22.9)	
	C6	7.5	-22.8	-22.6±2.1	-20.9(-24.7)	0.400
		10.5	-22.0	-21.3±1.9	-20.5(-23.4)	
	3OC6	7.5	-20.8	-19.7±2.7	-19.2(-21.2)	0.006
		10.5	-22.9	-22.8±1.6	-21.3(-23.7)	
	C7	7.5	-22.9	-22.8±2.2	-20.9(-24.2)	0.873
		10.5	-22.8	-22.6±2.0	-21.0(-23.3)	
	C8	7.5	-22.9	-22.8±2.2	-21.0(-24.9)	0.714
		10.5	-22.6	-22.0±1.9	-21.3(-23.8)	
	3OC8	7.5	-23.5	-23.8±2.3	-21.0(-24.6)	0.751
		10.5	-23.7	-23.6±1.3	-22.6(-24.6)	
	3OC12	7.5	-24.9	-24.7±1.3	-23.9(-25.4)	0.815
		10.5	-24.9	-24.7±1.6	-23.8(-25.4)	
AidC	C4	7.5	-21.0	-21.1±0.9	-20.2(-21.3)	0.435
		10.5	-21.3	-21.1±0.8	-20.9(-21.6)	
	C6	7.5	-23.6	-23.4±1.3	-22.3(-24.2)	0.047
		10.5	-22.7	-22.4±1.3	-21.9(-23.0)	
	3OC6	7.5	-23.4	-23.0±1.1	-22.6(-24.0)	0.148
		10.5	-22.8	-22.6±0.9	-22.2(-23.3)	
	C7	7.5	-23.3	-23.4±1.3	-22.2(-23.8)	0.792
		10.5	-23.3	-23.2±1.4	-22.2(-24.4)	
	C8	7.5	-23.5	-23.0±1.2	-23.0(-23.4)	0.144
		10.5	-22.9	-22.4±1.5	-21.7(-24.0)	
	3OC8	7.5	-24.4	-24.5±1.0	-23.5(-25.1)	0.054
		10.5	-23.6	-23.4±0.5	-23.1(-24.2)	
	3OC12	7.5	-25.2	-25.1±1.2	-24.3(-25.7)	0.004
		10.5	-23.9	-23.6±1.0	-23.4(-24.3)	
AiiA	C4	7.5	-21.2	-20.7±1.6	-20.1(-22.8)	0.977
		10.5	-21.1	-20.5±1.6	-20.1(-22.5)	
	C6	7.5	-22.8	-22.4±1.9	-23.1(-24.0)	0.071
		10.5	-23.5	-23.2±0.8	-23.0(-24.2)	
	3OC6	7.5	-23.9	-23.6±1.6	-22.7(-24.6)	0.977
		10.5	-24.0	-23.8±1.5	-22.7(-24.8)	
	C7	7.5	-23.9	-22.8±2.4	-21.9(-25.9)	0.156
		10.5	-22.8	-22.0±1.8	-21.0(-24.9)	
	C8	7.5	-23.5	-23.0±1.8	-21.9(-25.0)	0.721
		10.5	-23.2	-23.4±1.5	-21.7(-24.6)	
	3OC8	7.5	-24.0	-23.8±1.0	-23.4(-24.7)	0.863
		10.5	-24.0	-23.8±1.0	-23.1(-24.7)	
	3OC12	7.5	-24.4	-23.8±2.2	-22.4(-26.3)	0.281
		10.5	-23.5	-23.2±1.5	-22.3(-25.0)	

AiiB	C4	7.5	-20.8	-21.1±1.7	-19.2(-22.1)	0.532
		10.5	-20.5	-20.5±1.5	-19.7(-21.7)	
	C6	7.5	-21.3	-21.3±1.8	-19..4(-22.6)	0.622
		10.5	-21.7	-21.5±1.2	-20.9(-22.1)	
	3OC6	7.5	-22.6	-22.8±1.3	-21.4(-23.8)	0.724
		10.5	-22.8	-22.8±1.0	-21.9(-23.3)	
	C7	7.5	-22.0	-22.0±0.9	-21.7(-22.9)	0.717
		10.5	-22.2	-22.2±1.5	-21.0(-23.6)	
	C8	7.5	-23.4	-23.4±1.1	-22.7(-23.8)	0.014
		10.5	-22.0	-22.0±1.6	-21.0(-22.9)	
	3OC8	7.5	-24.6	-24.7±1.0	-23.8(-25.4)	0.022
		10.5	-23.1	-22.6±1.6	-21.7(-24.7)	
	3OC12	7.5	-21.5	-21.5±2.0	-19.4(-23.3)	0.020
		10.5	-23.9	-23.4±2.0	-22.2(-25.9)	
GKL	C4	7.5	-20.4	-19.9±1.7	-19.4(-20.8)	0.025
		10.5	-21.8	-22.0±1.7	-20.2(-22.9)	
	C6	7.5	-21.5	-21.5±0.6	-21.0(-21.7)	0.157
		10.5	-22.7	-21.7±2.2	-21.3(-22.9)	
	3OC6	7.5	-21.0	-20.7±1.1	-20.5(-21.6)	0.412
		10.5	-21.3	-21.3±0.7	-20.9(-21.7)	
	C7	7.5	-23.5	-22.6±3.0	-21.0(-25.3)	0.200
		10.5	-21.8	-21.7±0.8	-21.0(-22.2)	
	C8	7.5	-21.7	-21.3±1.2	-20.9(-22.5)	0.011
		10.5	-22.9	-22.8±0.8	-22.2(-23.4)	
	3OC8	7.5	-23.6	-23.4±1.5	-22.6(-24.2)	0.431
		10.5	-23.1	-22.8±1.4	-21.9(-24.2)	
	3OC12	7.5	-26.3	-24.9±3.0	-23.9(-28.0)	0.485
		10.5	-26.5	-25.7±2.2	-24.7(-28.6)	
		10.5	-22.0	-21.5±2.6	-19.8(-23.0)	
SsoPox	C4	7.5	-22.7	-22.2±1.4	-21.7(-23.4)	0.075
		10.5	-23.4	-23.2±1.2	-22.7(-23.7)	
	C6	7.5	-24.7	-24.7±1.5	-23.4(-25.4)	0.306
		10.5	-25.1	-25.1±1.4	-23.9(-25.5)	
	3OC6	7.5	-26.6	-26.6±1.5	-25.2(-28.2)	0.158
		10.5	-25.5	-25.3±2.1	-23.8(-27.6)	
	C7	7.5	-25.5	-25.5±0.7	-25.1(-25.9)	0.103
		10.5	-24.6	-24.5±1.3	-23.5(-25.9)	
	C8	7.5	-26.8	-26.4±1.8	-25.6(-27.9)	0.005
		10.5	-24.7	-24.7±1.9	-23.1(-25.8)	
	3OC8	7.5	-23.2	-22.6±1.5	-22.3(-23.6)	<0.001
		10.5	-26.1	-25.9±1.3	-25.1(-27.2)	
	3OC12	7.5	-28.0	-27.8±2.4	-25.9(-30.3)	0.308
		10.5	-28.9	-29.1±2.1	-27.1(-30.5)	

As a result, it was found that with an increase in the N-acyl chain length of N-AHLs, their strength of interaction with the surface of enzymes increases (i.e., C6-HSL interacts more strongly than C4-HSL, etc. etc.). Moreover, for different lactonases, the affinity values in enzyme-substrate complexes were comparable. The strongest interaction was revealed in the case of the SsoPox enzyme, where the maximum value of the analyzed parameter was noted in the "SsoPox – 3OC12" model at pH 10.5 (**–28.9 kJ^{x}mol^{-1}**). Comparison of the data obtained for N-AHLs' pairs with or without 3-oxo substitution (C6-HSL vs 3OC6-HSL; C8-HSL vs 3OC8-HSL) showed that in most cases the presence of a substitution in the N-AHL molecule increased its strength of interaction with the surface of the enzyme.

Based on the results obtained from one-way analysis of variance (ANOVA) (Table 2, p value), it was found that, in almost all cases, a change in the pH of the medium did not have a statistically significant (**p> 0.05**) effect on the molecular affinity of N-AHLs to the surface of the enzyme. Only in 10 cases the pH variation turned out to be statistically significant for the enzyme-substrate interaction: (**AaL-3OC6** (p = 0.006); **AidC-C6** (p = 0.047); **AidC-3OC12** (p = 0.004); **AiiB-C8** (0.014) ; **AiiB-3OC8** (p = 0.022); **AiiB-3OC12** (p = 0.020); **GKL-C4** (p = 0.025); **GKL-C8** (p = 0.011); **SsoPox-C8** (p = 0.005) and **SsoPox-3OC8** (p <0.001)). It should be noted that the interaction of the AiiA enzyme with N-AHLs was pH independent.

Thus, it has been shown that all lactonases studied in this work, with the exception of GKL, are able to hydrolyze various N-AHLs, mainly long chain ones. Moreover, the best results (both in terms of accessibility of the active site of the enzyme and in terms of the affinity of N-AHLs to its surface) were obtained with the SsoPox enzyme belonging to the class of PLLs. Therefore, the combination of SsoPox with His_6-OPH is useful to obtain the most effective antimicrobial drugs with lactonase activity and a broader spectrum of action than each of the enzymes, introduced into this combination for its subsequent complexation with antibiotics, has.

3. Conclusions

This research enabled conclusion that among the studied lactonases, SsoPox from the PLLs class is the best candidate for the development of combined enzyme preparations with His_6-OPH. The use of molecular docking methods in further studies for the search and selection of enzymes capable of selectively hydrolyzing short-chain or long-chain N-AHLs with or without 3-oxo substitution molecules may allow the creation of combined antimicrobial agents with high efficiency and specificity of action.

4. Computational methods

The structures of N-AHLs were created using ChemBioDraw software (ver. 12.0, CambridgeSoft). After that, energy minimization using ChemBio3D with force field MM2 was applied. Finally, the structures in protein data bank (.PDB) format were converted to the PDBQT format (PDB format with partial charges and atom types) using AutoDockTools (as part of MGLTools ver. 1.5.6, available at http://mgltools.scripps.edu/) [16]. Atomic charges were calculated with the Gasteiger-Marsili method.

To calculate the surface charge distribution of enzymes at a certain pH, adaptive Poisson-Boltzmann solver (APBS) and PDB2PQR servers (ver. 1.4.2.1 and 2.1.1, respectively, available at http://www.poissonboltzmann.org/) with PARSE forcefield and default settings were used [17,18]. After that, structures in PQR format were converted to the PDBQT format using AutoDockTools [16].

The dockings of ligands to the enzyme surface were realized using AutoDockVina (ver. 1.1.2, available at http://vina.scripps.edu/) [19] on a desktop computer equipped with Intel Pentium Dual-Core CPU E5400 2.7 GHz and 3 GB of available memory. Briefly, the grid box was approximately centered on the center of mass of the enzyme. The size of the grid box was chosen so that any enzyme surface was within the box with an additional margin. Calculations were performed with default program options.

5. Acknowledgment

This work was financially supported by the Russian Science Foundation (Grant No.16-14-00061).

6. References

[1] Nathan, C. and Cars, O. (2014) 'Antibiotic Resistance — Problems, Progress, and Prospects', New England Journal of Medicine, 371(19), pp. 1761–1763. doi: 10.1056/NEJMp1408040.

[2] Antunes, L.C.M., Ferreira, R.B.R., Buckner, M.M.C. and Finlay, B.B. (2010) 'Quorum sensing in bacterial virulence', *Microbiology*. Microbiology Society, 156(8), pp. 2271–2282. doi:

10.1099/mic.0.038794-0.
[3] Parsek, M. R. and Greenberg, E. P. (2000) Acyl-homoserine lactone quorum sensing in Gram-negative bacteria: A signaling mechanism involved in associations with higher organisms. Proceedings of the National Academy of Sciences, 97(16), pp. 8789-8793. doi: 10.1073/pnas.97.16.8789
[4] Chen, F., Gao, Y., Chen, X., Yu, Z. and Li, X. (2013) 'Quorum Quenching Enzymes and Their Application in Degrading Signal Molecules to Block Quorum Sensing-Dependent Infection', International Journal of Molecular Sciences, 14(9), pp. 17477–17500. doi: 10.3390/ijms140917477.
[5] Kalia, V. C. (2013) 'Quorum sensing inhibitors: An overview', Biotechnology Advances, pp. 224–245. doi: 10.1016/j.biotechadv.2012.10.004.
[6] Aslanli, A., Lyagin, I. and Efremenko, E. (2018) 'Novel approach to quorum quenching: Rational design of antibacterials in combination with hexahistidine-tagged organophosphorus hydrolase', Biological Chemistry, 399(8). doi: 10.1515/hsz-2018-0162.
[7] Aslanli, A. and Efremenko, E. (2019) 'Simultaneous molecular docking of different ligands to His_6 -tagged organophosphorus hydrolase as an effective tool for assessing their effect on the enzyme', PeerJ, 7, p. e7684. doi: 10.7717/peerj.7684.
[8] Aslanli, A., Lyagin, I. and Efremenko, E. (2019) 'Charges' interaction in polyelectrolyte (nano)complexing of His_6-OPH with peptides: Unpredictable results due to imperfect or useless concept?', International Journal of Biological Macromolecules, 140. doi: 10.1016/j.ijbiomac.2019.08.137.
[9] Bergonzi, C. et al. (2018) 'Structural and Biochemical Characterization of AaL, a Quorum Quenching Lactonase with Unusual Kinetic Properties', Scientific Reports. Nature Publishing Group, 8(1). doi: 10.1038/s41598-018-28988-5.
[10] Mascarenhas, R. et al. (2015) 'Structural and Biochemical Characterization of AidC, a Quorum-Quenching Lactonase with Atypical Selectivity', Biochemistry, 54(28), pp. 4342–4353. doi: 10.1021/acs.biochem.5b00499.
[11] Kim, M. H. et al. (2005) 'The molecular structure and catalytic mechanism of a quorum-quenching N-acyl-L-homoserine lactone hydrolase', Proceedings of the National Academy of Sciences of the United States of America, 102(49), pp. 17606–17611. doi: 10.1073/pnas.0504996102.
[12] Liu, D. et al. (2007) 'Structure and Specificity of a Quorum-Quenching Lactonase (AiiB) from Agrobacterium tumefaciens', Biochemistry, 46(42), pp. 11789–11799. doi: 10.1021/bi7012849.
[13] Xue, B., Chow, J.Y., Baldansuren, A., Yap, L.L., Gan, Y.H., Dikanov, S.A., Robinson, R.C. and Yew, W.S. (2013) 'Structural Evidence of a Productive Active Site Architecture for an Evolved Quorum-Quenching GKL Lactonase', Biochemistry, 52(13), pp. 2359–2370. doi: 10.1021/bi4000904.
[14] Elias, M. et al. (2008) 'Structural Basis for Natural Lactonase and Promiscuous Phosphotriesterase Activities', Journal of Molecular Biology, 379(5), pp. 1017–1028. doi: 10.1016/j.jmb.2008.04.022.
[15] Votchitseva, Y.A., Efremenko, E.N., Aliev, T.K. and Varfolomeyev, S.D. (2006) Properties of hexahistidine-tagged organophosphate hydrolase. Biochemistry (Moscow). 71, pp. 167–72. doi: 10.1134/S0006297906020088.
[16] Morris, G.M., Huey, R., Lindstrom, W., Sanner, M.F., Belew, R.K., Goodsell, D.S. and Olson, A.J. (2009) 'AutoDock4 and AutoDockTools4: Automated docking with selective receptor flexibility', Journal of Computational Chemistry, 30(16), pp. 2785–2791. doi: 10.1002/jcc.21256.
[17] Baker, N.A., Sept, D., Joseph, S., Holst, M.J. and McCammon, J.A. (2001) 'Electrostatics of nanosystems: Application to microtubules and the ribosome', Proceedings of the National Academy of Sciences of the United States of America, 98(18), pp. 10037–10041. doi:

10.1073/pnas.181342398.
[18] Dolinsky, T.J., Czodrowski, P., Li, H., Nielsen, J.E., Jensen, J.H., Klebe, G. and Baker, N.A. (2007) 'PDB2PQR: Expanding and upgrading automated preparation of biomolecular structures for molecular simulations', Nucleic Acids Research, 35(SUPPL.2). doi: 10.1093/nar/gkm276.
[19] Trott, O. and Olson, A. J. (2010) 'AutoDock Vina: improving the speed and accuracy of docking with a new scoring function, efficient optimization and multithreading', Journal of Computational Chemistry, 31(2), pp. 461-464. doi: 10.1002/jcc.21334.

IOP Conf. Series: Materials Science and Engineering **848** (2020) 012007 doi:10.1088/1757-899X/848/1/012007

High-temperature physical properties of the electrodes based on MAX-phase Ti-Al-C produced by SHS-extrusion method

O A Averichev* and A M Stolin

Plastic Deformation Materials Laboratory, Merzhanov Institute of Structural Macrokinetics and Materials Science, Russian Academy of Sciences, Academician Osipyan, 8, Chernogolovka, Moscow Region, 142432, Russia

*E-mail: olegaverichev@gmail.com

Abstract. Physical properties of the Ti-Al-C compact rods with the content of the MAX phase have been studied. The rods were produced by energy-efficient SHS-extrusion method. It has been found that the oxidation mechanism is similar to nickel heat-resistant alloys, herewith the oxidation rate does not exceed 1 $g/m^2 \cdot h$ at 900 °C. The dependences of the corrosion resistance on stoichiometric parameters of green mixtures are reported. The electrical resistivity of the rods does not exceed 5-6 μOm * cm at room temperature with a slight increase in value to 18 μOhm*cm at 900 °C. These results can be useful for the potential applications of the rod as the high-temperature electrodes.

1. Introduction

Materials based on the MAX phase are thermodynamically stable layered ternary systems of the $M_{n+1}AXn$ (n= 1…3) phases with hexagonal close packing. Here M is a transition metal, A is an element of group "A" of the periodic table, X is carbon or nitrogen [1]. Among the variety of this class, ternary carbides such as Ti-Al-C, Ti-Si-C combine mechanical and physical properties most successfully. It can be said that such compounds are metals and ceramics at the same time [2]. MAX phase has high electrical and thermal conductivity, high elastic modulus, good machinability, on the one hand, and oxidation resistance, low density and thermal stability on the other one [3]. There are several methods to produce dense MAX-phase containing material, such as hot isostatic pressing (HIP), spark plasma sintering (SPS) or pulse discharge sintering [4-6]. Self-propagating high-temperature synthesis (SHS) is another highly effective method, which contributes the energy-efficient way to produce ready-to-use items in a fast mode [7,8]. SHS-extrusion technique allows to combine a combustion process with the high-temperature plastic deformation and gets long thin rods. As a result, dense compacts with different shapes and high purity of compounds can be obtained [9,10].

Ti-Al-C-based phases are one of the most appropriate candidates for this method, because of the high exothermic reaction between Ti and C during the SHS [11,12]. Long rods were produced by SHS-extrusion method previously. Sheng et.al. [13,14] fabricated a Ni-Al-Ti-C based composite rods by SHS. However, their effective length does not exceed 15 mm. In the case of aluminum electrolysis application, a desirable content of the MAX-phase in the rod is estimated at 80-90% wt. The MAX phases are very good conductors of electricity, therefore it is expected that they will not react with the cryolite–alumina melt. That is why so important to investigate the temperature dependences of the electrical resistance of such materials as well as the high-temperature oxidation behavior of the extruded items. [15].

IOP Conf. Series: Materials Science and Engineering **848** (2020) 012007 doi:10.1088/1757-899X/848/1/012007

In the current work, the rods of 10 mm in diameter and more than 100 mm in length were prepared from the Ti_2AlC or Ti_3AlC_2 main phases with the 96-98 % density of the compact. For the investigations, a choice of such materials was motivated by their relatively high heat resistance, chemical inertness to the cryolite–alumina melt and low electrical resistivity, thus giving hope for the application of these compounds as the aluminum electrolysis electrodes. For these rods, the dynamics of high-temperature oxidation, the dependences of electrical resistivity on temperature up to 900 °C as well as the evolution of microstructure with heating were studied.

2. Experimental procedures

Commercially available powders were used for the green mixtures preparation, shown on Table 1. Initial pellets were prepared with the special stoichiometric ratios to ensure the fullest formation of MAX-phases (Ti_2AlC and Ti_3AlC_2). Before the cold pressing, the raw powders were initially dried for 24 hours, than the pellets were dried again within 4 hours before the combustion processes.

Table 1. Initial powders parameters

Powder	Type	Purity, % wt.	Dimensions
Ti (titanium)	PTM	99,1	45
C (carbon)	PM-15T	99,1	1
Al (aluminium)	ASD-4	99,5	5

The stoichiometric ratios of the green mixtures were selected to ensure variability of aluminum content in the melt during the combustion. Some reports state that the exceeding of aluminum during the synthesis can improve MAX-phase content in the final product [16]. Table 2 shows the molar and mass ratios of green mixtures that were investigated in this paper.

Table 2. Stoichiometric and mass ratios of green mixtures

Molar ratio, %, Ti- Al - C	3 : 1 : 2	2 : 1,5 : 1	3 : 2,3 : 2
Mass ratio, %	74 : 14 : 12	65 : 27 : 8	63 : 27 : 10

Heat resistance tests of electrodes were carried out in special ceramic crucibles, which passed an oxidation atmosphere through and ensured the preservation of crumbling scale. The crucible with the samples was placed in an oven on ceramic supports. Before testing, the crucibles were calcined to constant weight. The quantitative characteristics of the heat resistance of the samples were determined by the weighing method based on analytical balance controlling the increasing sample's weight. Samples were weighed with an accuracy of 10^{-6} g after 2, 4, 9 and 16 hours of annealing at 900 °C. The true rate of weight increase of the samples (Vq) was calculated as the first time derivative of the specific mass loss of the oxidized sample. Let's use the equation (1) for calculating the true rate of the samples:

$$V_q = \frac{dq}{dt} = \frac{1}{S} \cdot \frac{dm}{dt}, \quad (1)$$

where dt is the oxidation time of the samples, dm is the mass change of the samples over the considering time, S is the surface area of the entire sample.

The electrical resistivity of the samples was measured by the four-point method, the scheme of such measurements is shown in Figure 1. The contacts were fixed in four places to the sample. An electric current I14 was excited through two contacts (1, 4), and the potential difference U23 was measured at the other two (2,3) contacts. From the measured values of the potential difference between points 2 and 3 and the current flowing through the points 1 and 4 the resistance was determined. The electrical resistivity was recalculated by equation (2):

$$\Omega = R * S/l, \tag{2}$$

where R is the measured resistance, S is the cross-sectional area of the sample, l is the distance between points 2 and 3.

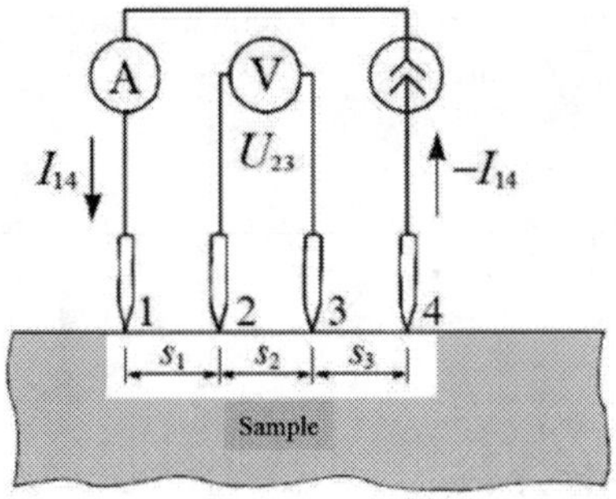

Figure. 1. The scheme of the electrical resistivity measurment of the extruded long rods.

The microstructure of the obtained items was studied by scanning electron microscopy (SEM) on a Zeiss Ultra Plus field emission scanning electron microscope equipped with an INCA 350 Oxford Instruments X-ray microanalysis attachment, which makes it possible to study the shape and size of the particles with a resolution of up to 2 nm.

3. Results and discussion

Figure 2 shows the dependences of the specific mass increase on the annealing time for the tested samples. This method also was used to determine the weight gain for a nickel heat-resistant EP741-NP alloy.

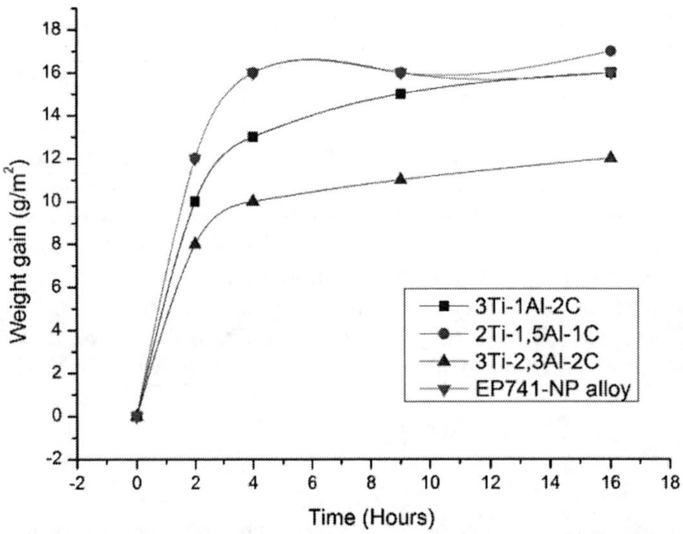

Figure 2. Dependences of specific weight gain on annealing time.

It was revealed that all the studied samples of stoichiometric ratios are possessing the characteristic oxidation curve dependences typical for the heat-resistant alloys. In the first stage of oxidation, the largest mass gain is caused due to the formation of oxide films on the surface, which is subsequently impede the penetration of oxygen into the samples. This oxidation stage is occurring during the first two hours. In order to evaluate the corrosion resistance of the samples, the specific rates of their oxidation

IOP Conf. Series: Materials Science and Engineering **848** (2020) 012007 doi:10.1088/1757-899X/848/1/012007

were calculated as the specific values of the mass gain given by the exposure time. Figure 3 shows the curves of growth rates (Vq) on the exposure time for each oxidized sample.

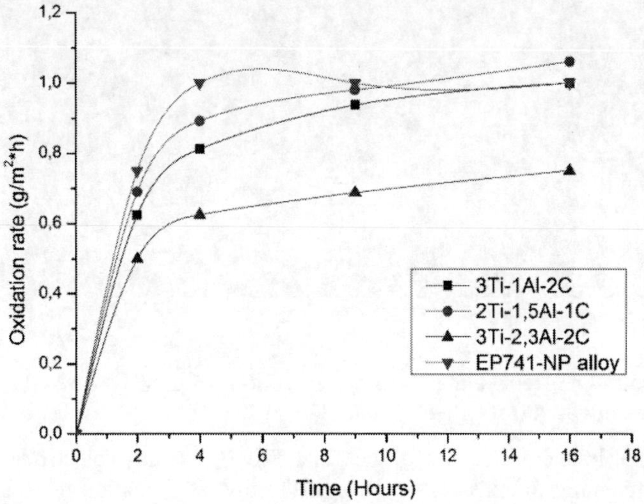

Figure 3. Oxidation rate of the extruded rods during 16 hours of annealing

As may see in Figure 3, after 4 hours of annealing, the oxidation rate of the samples after 4 hours of oxidation reaches a plateau, due to the oxide film formation at the samples surface. The obtained values of the oxidation rates for the studied samples were found to be close to that one for the EP741-NP alloy. Moreover, for 3Ti:2,3Al:2C and 3Ti:1Al:2C compositions, the oxidation rate was found to correspond to 0.65 g/m2 • h, which is by a factor of 1.5 less in comparison with that one for the EP741-NP alloy. The observed oxidation rate values mean that the nature of the oxidation process for all the samples in study is similar to that of corrosion resistant alloys. The rason of the samples mass gain by the oxidation process relates to the surface oxide film formation, which slows down the penetration of oxygen into the samples. The previously described oxidation rates (20 h of oxidation) for the SHS electrodes were found 0.84 g/m^2 h and 1.09 g/m^2 h for the 3Ti:1Al:2C and 3Ti:2,3Al:2C compositions, respectively. We assume that the following oxidation will not lead to a sharp increase of the sample's weight. Moreover, after recalculating, taking into account the oxidation time, the estimated oxidation rate at 100 h should decrease and may reach 0.5 g/m^2*h, which is on the level of the heat-resistant industrial alloys. For example, typical oxidation rate for the nickel-based alloys measured at 900 °C, 100 h duration of oxidation is in the range of 0.25-0.85 g/m^2*h, depending on composition. It is also a known fact that the intense oxidation occurs at the sites of defects. Therefore, to reduce the oxidation rate, it is necessary to further reduce the surface imperfection by grinding and polishing the surfaces of oxidizable samples. As an example, Figures 4a and 4b show a sample of 2Ti:1,5Al:1C composition before and after the high-temperature annealing.

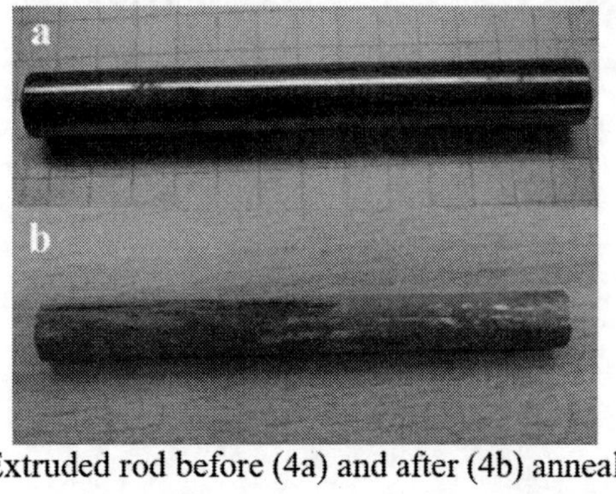

Figure 4. Extruded rod before (4a) and after (4b) annealing for 16 h.

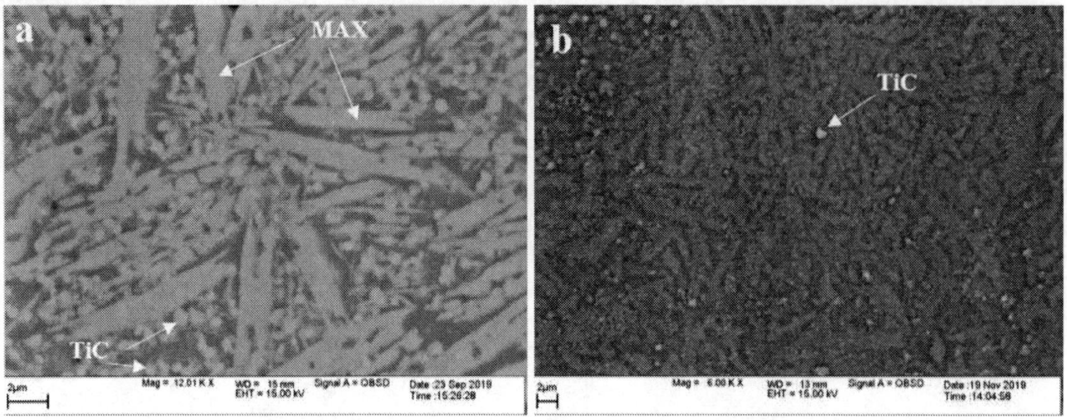

Figure 5. Backscattered electron images of cross-sections of the 2Ti-1,5Al-1C extruded rods before (a) and after (b) annealing at 900 °C after 16 hours.

The microstructure of the cross-section of the 2Ti:1,5Al:1C extruded electrode is presented in Figure 5. This sample is interesting to consider because of the lowest oxidation resistance among the entire series, so it was curious whether some morphological change occurs or not. The cross-section was taken from the central part of the rod. It can be noticed that the oxidation curve of this sample has almost full matching with that one for the heat-resistant EP741-NP alloy, which was shown in Figure 2. As may see in Figure 5a, the microstructure of the non-annealed electrode depicts a typical dendritic MAX phase grains reaching up to 10 µm in length and no more than 1 µm in width. The predominant phase corresponds to Ti_2AlC with Ti_3AlC_2 inclusion, by prior results. The orientation of dendrites is mainly elongate toward the direction of the heat sink, but it depends on the particular side of the rod. After the annealing at 900 °C, 16 hours, no severe morphological changes revealed, including a grain growth or mutual phase's diffusion. In this particular case, the presence of intermetallic compounds such as Ti_xAl_y (5-8% wt.) promotes increased mobility of the titanium carbide grains (10-15% wt.) during the heating, Figure 5b. We are considering that this occurs because of the diffusion mobility of the TiC grains at the temperatures above 882 °C. As a result, multiple micro-defects appear along grain boundaries, which leads to intensifying oxygen penetration into the sample. Based on the experimental high-temperature testing and subsequent calculation results, the following practical advice can be drawn; the decreasing of TiC grains in the SHS-extruded Ti-Al-C rods leads to enhancing of the high-temperature corrosion resistance.

The electrical resistivity of the samples with nominal compositions 3Ti:1Al:2C, 2Ti:1,5Al:1C and 3Ti:2,3Al:2C was measured in the temperature range of 25 – 900 °C. The samples were placed in a furnace, which was heated up to 900 °C with increments of 15 deg/min.

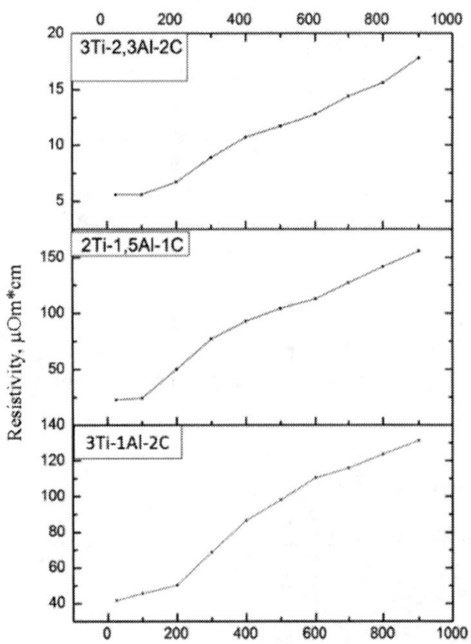

Figure 6. The temperature dependences of electrical resistivity of the samples

The measurement results are presented in Figure 6. As can be seen from the experimental graphs, for the 3Ti:2,3Al:1C sample, the electrical resistivity value is 5-6 μOhm*cm at room temperature. Whereas in the case of other systems, the value does not exceed 40 μOhm *cm. As the temperature increases to 900 °C, the electrical resistance increases to 18–20 μOhm*cm and 130-155 μOhm * cm for the 3Ti:2,3Al:2C and 2Ti:1,5Al:1C, 3Ti:2Al:1C samples, respectively. An order of magnitude difference between these values can be caused by the low content of TiC grains in the last two samples.

4. Conclusion

This work dedicated to high-temperature physical properties of the dense compact rods prepared by self-propagating high-temperature synthesis with subsequent extrusion from Ti, Al and C initial powders. The series of the ready-to-use samples with 3Ti:2Al:1C, 2Ti:1,5Al:1C and 3Ti:2,3Al:2C stoichiometric ratios were examined at 900 °C. It has been revealed that the visual dendritic structure of TiAlC MAX-phase do not undergo severe morphological changes under 900 °C of annealing, whereas its physical properties are changing slightly. The high-temperature oxidation mechanism for the studied SHS rods has been found to be similar to the nickel-based heat-resistant alloys, exhibiting the oxidation rate in the range of 0.65-1.05 g/m^2• h at 900 °C, depending on the carbides and intermetallic contents. For the samples in study, a monotonous increase of the electrical resistivity with temperature was observed. The reported results can be useful for the potential high-temperature application of such rods, especially as the electrodes for aluminum electrowinning process.

References.

[1]. Barsoum MW, El-Raghy T. The MAX phases: Unique new carbide and nitride materials, Am Sci 2001;89(4):334-343.
[2]. M. Radovic, M.W. Barsoum, MAX phases: bridging the gap between metals and ceramics, Am. Ceram. Soc. Bull. 92 (3) (2013) 20–27.
[3]. Z.M. Sun, Progress in research and development on MAX phases: a family of layered ternary compounds, Int. Mater. Rev. 56 (3) (2011) 143–146.
[4]. C. Yang, S. Jin, B. Liang, G. Liu, L. Duan, S. Jia, Synthesis of Ti_3AlC_2 by spark plasma sintering of mechanically milled 3Ti/xAl/2C powder mixtures, J. Alloy Compd. 472 (1) (2009) 79–83.
[5]. N.V. Tzenov, M.W. Barsoum, Synthesis and characterization of Ti_3AlC_2, J. Am. Ceram. Soc. 83 (4) (2000) 825–832.
[6]. Y. Zou, Z. Sun, S. Tada, H. Hashimoto, Synthesis reactions for Ti_3AlC_2 through pulse discharge sintering Ti/Al_4C_3/TiC powder mixture, Scr. Mater. 55 (9) (2006) 767–770.
[7]. Łopaciński M, Puszynski J, Lis J. Synthesis of ternary titanium aluminum carbides using self-propagating high-temperature synthesis technique, J. Am. Ceram. Soc. 84 (12) (2001) 3051–3053.
[8]. M. Akhlaghi, S.A. Tayebifard, E. Salahi, M. Shahedi Asl, G. Schmidt, Self-propagating high-temperature synthesis of Ti_3AlC_2 MAX phase from mechanically-activated Ti/Al/graphite powder mixture, Ceram. Int. (2018).
[9]. Y. Bai, X. He, Y. Li, C. Zhu, S. Zhang, Rapid synthesis of bulk Ti_2AlC by self-propagating high temperature combustion synthesis with a pseudo-hot isostatic pressing process, J. Mater. Res. 24 (8) (2011) 2528–2535.
[10]. A.M. Stolin, D. Vrel, S.N. Galyshev, A. Hendaoui, P.M. Bazhin, A.E. Sytschev, Hot forging of MAX compounds SHS-produced in the Ti-Al-C system, Int. J. Self- Propagating High-Temp. Synth. 18 (3) (2009) 194–199.
[11]. M.W. Barsoum, M. Radovic, Elastic and mechanical properties of the MAX phases, Annu. Rev. Mater. Res. 41 (2011) 195–227.
[12]. Kovalev, D.Y., Averichev, O.A., Luginina, M.A. et al. Phase Formation in the Ti–Al–C System during SHS. Russ. J. Non-ferrous Metals 60, 61–67 (2019).

[13]. Sheng, L. Y., et al. Investigation on NiAl–TiC–Al_2O_3 composite prepared by self-propagation high temperature synthesis with hot extrusion. Composites Part B: Engineering 45.1 (2013): 785-791.

[14]. Sheng, L. Y., et al. Microstructure evolution and mechanical properties of Ni_3Al/Al_2O_3 composite during self-propagation high-temperature synthesis and hot extrusion. Materials Science and Engineering: A 555 (2012): 131-138.

[15]. Wang, Dihua, and Wei Xiao. Inert anode development for high-temperature molten salts. Molten Salts Chemistry (2013) 171-186.

[16]. A. Hendaoui, D. Vrel, A. Amara, P. Langlois, M. Andasmas, M. Guerioune, Synthesis of high-purity polycrystalline MAX phases in Ti–Al–C system through mechanically activated self-propagating high-temperature synthesis, J. Eur. Ceram. Soc. 30 (4) (2010) 1049–1057.

IOP Conf. Series: Materials Science and Engineering **848** (2020) 012008 doi:10.1088/1757-899X/848/1/012008

Self-propagating high-temperature synthesis of ceramic composite material based on Al_2O_3 - SiO_2 - ZrB_2

A P Chizhikov, P M Bazhin and A M Stolin

Merzhanov Institute of Structural Macrokinetics and Materials Science Russian Academy of Sciences (ISMAN), Academician Osipyan str., 8, Chernogolovka, Moscow Region, 142432, Russia

e-male: chij@ism.ac.ru

Abstract. The present work is devoted to the production of a ceramic composite material in the mode of self-propagating high-temperature synthesis from a mixture of powder reagents: ZrO_2, Al, SiO_2, B, and Zr. The adiabatic combustion temperature was calculated for the stoichiometric composition of ZrO_2-Al-SiO_2 and for compositions with the addition of an energy additive from powders x (Zr + 2B), where x ϵ [10-30] wt. %. It was found that an increase in the fraction of Zr + 2B leads to the refinement of structural elements. An increase in its content to 30 wt. % leads to the formation of nanoscale structural components in the material. As a result of varying the initial composition, a ceramic composite material based on Al_2O_3-SiO_2-ZrB_2 was obtained containing an oxide matrix with particles of zirconium diboride distributed from 61 to 77 wt. %. It is shown that as a result of the SHS of the selected compositions, mullite is formed. Using SEM, EDS and XRD methods, we studied the structural features and phase composition of the materials obtained.

1. Introduction

Ceramic materials today are widely used in various industries [1-6]. One of the most common classes of ceramic materials is oxide ceramics. Oxide ceramic materials are widely used due to their high hardness, chemical inertness, ability to work at high temperatures and relative availability [7-10]. The most common type of oxide ceramic is aluminum oxide ceramic. Composite materials based on various oxides are also of great interest today. There are various types of such composites, for example, Al_2O_3-ZrO_2 [11, 12], Al_2O_3-SiO_2 [13, 14], Al_2O_3-SiC [15, 16], etc. The creation of composite materials makes it possible to compensate for some of the shortcomings of individual oxides, such as low ability of aluminum oxide to resist thermal shock. As a result, oxide ceramic composite materials are promising for various fields of application. For instance, refractory materials [17], cutting [18] ceramics, ceramic filters [19], etc.

Due to the great interest in the class of composite materials based on oxide ceramics, today there are a number of methods for the synthesis of such materials. In recent years, the sol-gel method [20], its variation Pechini method [21], hydrothermal synthesis method [22], solution combustion method [23], mechanochemical synthesis [24], and a number of other methods have become widespread. However, among all the methods used, it is possible to single out the method of self-propagating high-temperature synthesis (SHS), which is the process of an exothermic chemical reaction through a mixture of reagents with the formation of solid products [25, 26]. The reaction is initiated locally by rapidly heating a small portion of the sample. After local reaction initiation due to heat transfer, heating and initiation of a

IOP Conf. Series: Materials Science and Engineering **848** (2020) 012008 doi:10.1088/1757-899X/848/1/012008

chemical reaction in the adjacent layer of the starting components occurs. Thus, a chemical reaction localized in the layer spontaneously propagates through the sample in the form of a combustion wave. This mode of synthesis allows to avoid the need for external heating, provides refining of the starting reagents during the combustion process, and also allows to reduce the synthesis time to several seconds. Today, SHS has established itself as a promising method for producing a large number of materials, including composite oxide ceramics [27, 28].

This work is devoted to the preparation of a composite ceramic material based on Al_2O_3-SiO_2-ZrB_2 by the SHS method, as well as to the study of the influence of the initial composition on its phase composition and microstructure.

2. Materials and methods

In this work, the synthesis of the material was carried out in the mode of self-propagating high-temperature synthesis. Powders of zirconium oxide ZrO_2 (<50 μm, ≥ 98.5), aluminum (<5 μm, ≥ 99.5), silicon oxide SiO_2 (<1 μm, ≥ 99.3), boron (<10 μm, ≥ 99.5) and zirconium (<100 μm, ≥ 99.9). The listed powders were previously aged in an oven and then mixed in a ball mill for 4 hours. Then, cylindrical billets weighing 50 grams and a relative density of 0.65 were pressed from the resulting powder mixture. The SHS process was initiated by a tungsten helix. The proposed reaction scheme during SHS in the studied compounds:

$$ZrO_2 + Al + SiO_2 + B + x\,(Zr + 2B) \rightarrow Al_2O_3 + SiO_2 + ZrB_2 \quad (1)$$

where x ϵ [0, 30] wt. %.

The percentage of the starting powder reagents is given in table 1. The percentage of the phases of the synthesized materials was determined by the corundum number method.

Table 1. Percentage of starting reagents, wt. %.

№	ZrO_2	Al	SiO_2	B	Zr
1	55,8	16,3	18,1	9,8	-
2	50,2	14,7	16,3	9,7	8,1
3	44,6	13	14,5	11,6	16,2
4	39,1	11,4	12,7	12,6	24,3

3. Results and discussion

Firstly, the initial mixture was proposed, corresponding to composition No. 1 in table 1. The synthesis of the material in the selected mixture was planned to be carried out in the SHS mode due to the reduction of zirconium oxide by aluminum and the subsequent formation of zirconium diboride from reduced zirconium and boron. However, the reduction reaction of zirconium oxide with aluminum is weakly exothermic relative to other aluminothermic reactions; moreover, an inert component of silicon oxide is present in the system. The calculation of the adiabatic combustion temperature (table 2) showed that in such a system the SHS process cannot be initiated. The experiment also showed that in a system corresponding to composition 1, the SHS reaction could not be initiated without additional heating of the workpiece.

Table 2. Adiabatic combustion temperature of the studied compounds.

№	1	2	3	4
Tad, °C	1195	1866	1995	2054

To increase the thermal effect, it was decided to add to the composition 1 a mixture of powders x (Zr + 2B), where x ϵ [10-30] wt. % as an energy additive. The reaction of formation of zirconium diboride has a sufficiently high exotherm (ΔH^0_{298} = -76 kcal / mol, for comparison, ΔH^0_{298} (TiB_2) = -70 kcal / mol). As can be seen from table 2, the introduction of such an additive significantly increases the adiabatic combustion temperature and allows for SHS in this system. Thus, further results are presented for the starting mixtures corresponding to compositions 2-4 in table 1.

As a result of synthesis, multiphase products were obtained, the results of x-ray phase analysis (XRD) of which are shown in figure 1. As can be seen from the figure, the obtained materials consisted of four phases: aluminum oxide, silicon oxide, zirconium diboride and zirconium oxide. The percentage of these phases are shown in table 3.

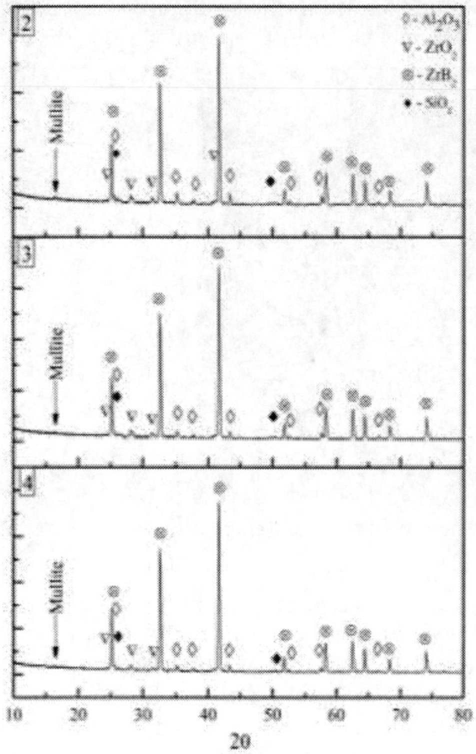

Figure 1. Results of XRD of synthesized materials.

Table 3. The percentage phase of the obtained materials, wt. %.

№	ZrB_2	Al_2O_3	SiO_2	ZrO_2
2	61	30	5	4
3	71	23	2	4
4	77	19	2	2

According to the obtained results, the main phase in the synthesized materials was zirconium diboride, while in its composition 4 its share reaches 77 wt. %, which indicates the inappropriateness of increasing the proportion of additives (Zr + 2B) to 40 wt. % in the original charge. An increase in the proportion of zirconium diboride leads to a decrease in oxide components in the resulting materials, while the proportion of aluminum oxide decreases from 30 to 19 wt. %. Figure 1 shows that the diffraction patterns contain weak peaks of silicon dioxide, the intensity of which decreases with increasing proportion of zirconium diboride. This is due, firstly, to the amorphous state of the initial silica powder, and, secondly, to the formation of mullite.

Confirm the presence of mullite in the resulting materials allows several factors. First, in the diffractograms in figure 1, in the region of the angle $2\theta = 16.3$ °, there is a confirming mullite peak, as well as a number of peaks of aluminum oxide and silicon oxide coincide with the peaks characteristic of mullite. It is difficult to quantify the fraction of mullite in products; most likely, this phase was in the process of nucleation, which did not manage to occur until the end due to the fleeting nature of the SHS process. Mullite is characterized by the formation of a lamellar structure. Figure 2 shows the results of SEM and EMF analysis of the material obtained on the basis of composition 2. In this figure, the formation of two phases is clearly visible, one of which has a pronounced plate structure (spectra S1 and S2) and consists of three elements: aluminum, silicon and oxygen. Thus, based on the foregoing, it can be concluded that mullite is formed in the synthesized materials. The presence of the described lamellar structure can also be seen in Figure 3b (spectrum S1) in the material based on composition 3.

Spectrum	B	O	Al	Si	Zr
S1		49.15	34.16	14.77	1.92
S2		58.00	21.00	17.46	3.54
S3	4.46	6.21	0.74	1.36	87.24
S4	4.13	6.02	1.78		88.07

Figure 2. Microstructure and elemental analysis of the material based on composition 2.

A general examination of the microstructure shown in figure 3 shows that the materials obtained had a composite structure. The dark regions corresponding to the oxide component were the matrix of the composite material. In figure 3a, the S1 spectrum corresponds to a large agglomerate of zirconium oxide particles, whose size was about 2-5 μm. This is unreacted zirconium oxide contained in the initial charge, the presence of which is also confirmed by the XRD results shown in figure 1. Table 3 indicates that the content in the obtained materials of the initial zirconium oxide was 2-4 wt. %. However, taking into account the error of the corundum number method used in calculating the percentage of phases, as well as the fact that no zirconium oxide deposits similar to composition 1 were found in the structure of materials based on compositions 3 and 4, we can say that with an increase in the proportion of diboride zirconium in materials decreases the amount of unreacted starting components. Thus, the completeness of the reaction of the studied compounds with increasing energy additives of zirconium diboride increases.

In addition to zirconium oxide agglomerates, the formation of a eutectic structure can be seen in figure 3a. Such a structure is characteristic of the Al_2O_3-ZrO_2 eutectic, which, most likely, was formed during SHS as a result of the interaction of unreacted initial zirconium oxide and the formed alumina.

As mentioned earlier, the oxide component was a matrix over the volume of which zirconium diboride particles were distributed, corresponding to the light regions in figure 3. When comparing the structures of the obtained materials, it can be noted that with an increase in the proportion of zirconium diboride in the materials, the particle size of ZrB_2 decreases. On the one hand, an increase in the combustion temperature of the system should promote particle growth. However, if we assume that the formation of ZrB_2 particles proceeds according to the diffusion mechanism as a result of the spreading of the zirconium melt along boron particles, then an increase in the content of zirconium diboride leads to an increase in the number of crystallization nuclei, the growth of which is hampered by the formation of the product layer. Further recrystallization is hindered due to the fleeting nature of SHS processes.

a

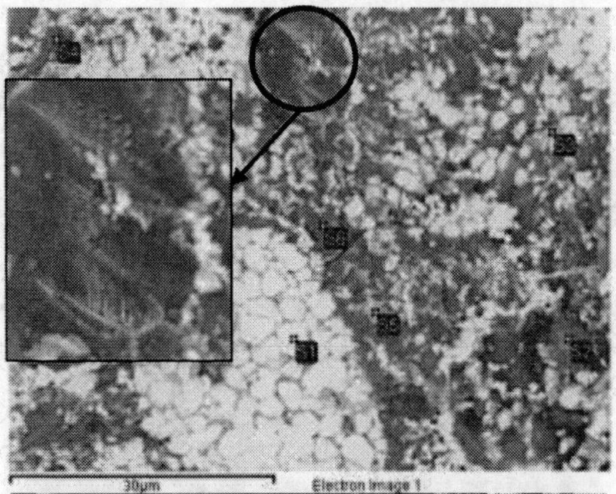

Spectrum	B	O	Al	Si	Z
S1		23.06			76.94
S2	10.56	23.13	11.44	47.41	7.46
S3	2.49	36.05	2.87	2.12	56.47
S4		58.10	14.09	20.05	7.74
S5		37.42	6.16	13.19	43.24
S6		47.76	23.69	19.52	9.03

b

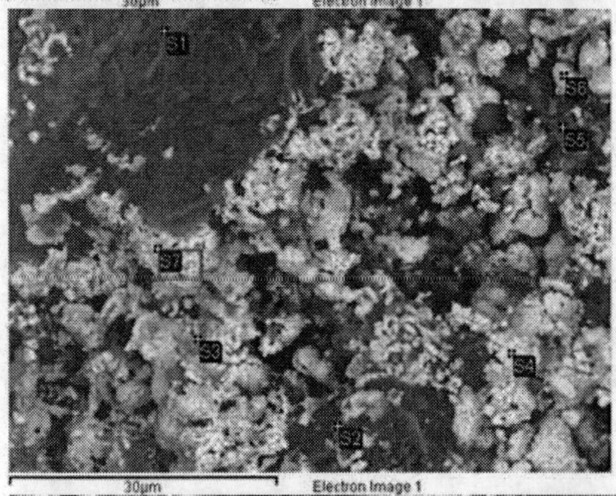

Spectrum	B	O	Al	Si	Zr
S1		60.49	16.87	20.45	2.19
S2		55.61	27.90	4.12	12.37
S3		4.30		30.30	65.40
S4	5.13	10.70		1.14	83.04
S5		53.81	30.17	12.88	3.14
S6	3.10	15.18	0.95	0.82	79.96
S7	2.96	8.80	0.95	3.25	84.04

c

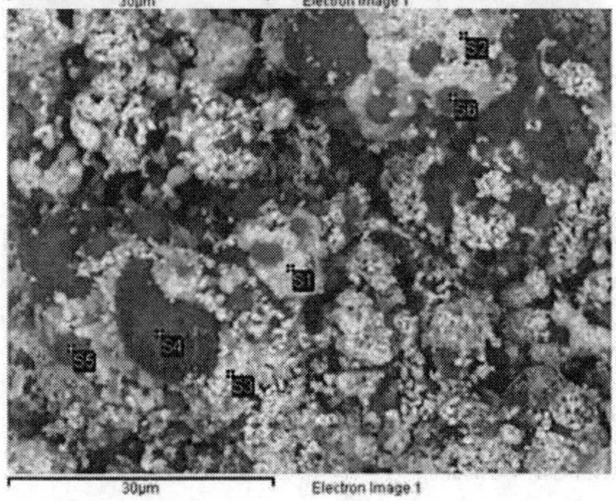

Spectrum	B	O	Al	Si	Zr
S1	4.69	11.00		0.58	83.73
S2	4.91	6.20	0.69	4.36	83.84
S3	3.98	7.69	0.60	1.12	86.60
S4		54.03	19.13	26.84	
S5		50.95	7.57	19.58	21.90
S6	4.59	32.09	3.15	9.02	51.16

Figure 3. SEM results of the materials obtained based on the compositions: a) No. 2 b) No. 3 c) No. 4.

In the structure of the material based on composition 2, zirconium diboride is mainly represented by particles of 1–3 μm in size, as shown in figure 4a. Smaller particles are also found, but their minority. When moving to a material based on composition 3, the thickness of the particles of zirconium diboride, which have a rectangular shape, becomes about 500 nm, as can be seen in figure 4b. When considering a material based on composition 4, it can be seen that the particles of zirconium diboride are agglomerates of rectangular grains, with the size of these grains not exceeding 100 nm. Thus, we can conclude that the increase in the proportion of energy additives (Zr + 2B) of the studied materials up to 30 wt. % leads to the formation of nanoscale grains of zirconium diboride.

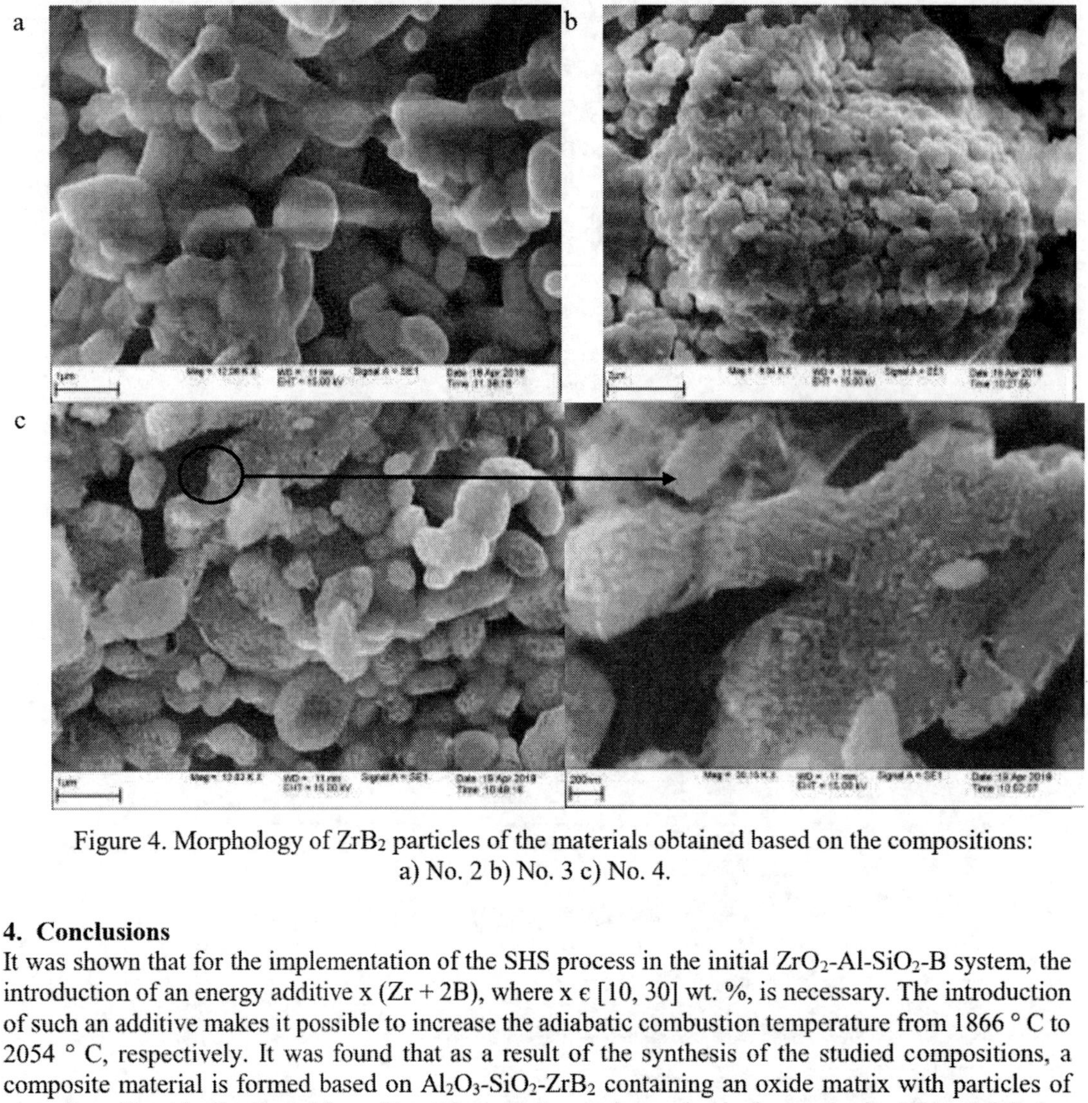

Figure 4. Morphology of ZrB_2 particles of the materials obtained based on the compositions: a) No. 2 b) No. 3 c) No. 4.

4. Conclusions

It was shown that for the implementation of the SHS process in the initial ZrO_2-Al-SiO_2-B system, the introduction of an energy additive x (Zr + 2B), where x ϵ [10, 30] wt. %, is necessary. The introduction of such an additive makes it possible to increase the adiabatic combustion temperature from 1866 ° C to 2054 ° C, respectively. It was found that as a result of the synthesis of the studied compositions, a composite material is formed based on Al_2O_3-SiO_2-ZrB_2 containing an oxide matrix with particles of zirconium diboride distributed from 61 to 77 wt. %. It was shown that in the synthesized material during the SHS of aluminum and silicon oxides, mullite is formed. It was shown that an increase in the fraction of the energy additive (Zr + 2B) leads to the refinement of structural elements, while an increase in the proportion of the energy additive in the materials under study is up to 30 wt. % leads to the formation of nanoscale grains of zirconium diboride.

References

[1] Li Z H, Liu S T, Song S G, Xu W L, Sun Y M and Dai Y Q 2019 *Compos. Commun.* **15** 168–78
[2] Katoh Y and Snead L L 2019 *J. Nucl.* **526** 151849
[3] Alizadeh-Osgouei M, Li Y C and Wen C E 2019 *Bioact. Mater.* **4** 22-36
[4] Vardanyan E L, Ramazanov K N, Abramov A N, Nazarov A Yu and Nagimov R Sh 2018 *JPCS* **1058** 012006
[5] Omanovic-Miklicanin E, Badnjevic A, Kazlagic A and Hajlovac M 2019 *Health and Technology* 9
[6] Zhang R Z and Reece M J 2019 *J. Mater. Chem.* A **7** 22148–62
[7] Tkalenko D M and Voronov V A 2019 *Glass Ceram.+* **76** 290–6

[8] Sevostyanov M A, Baikin A S, Shatova L A, Nasakina E O, Berezhnov A V, Gudkov S V, Sergienko K V, Konushkin S V, Baskakova M I, Kolmakov A G 2018 *Dokl. Chem.* **482** 204-6
[9] Panasyuk G P, Luchkov I V, Kozerozhets I V, Shabalin D G and Belan V N 2013 *Inorg. Mater.* **49** 899–903
[10] Abyzov A M 2019 *Refract. Ind. Ceram.+* **60** 24–32
[11] Boniecki M, Sadowski T, Gołębiewski P, Węglarz H, Piątkowska A, Romaniec M, Krzyżak K and Łosiewiczet K 2020 *Ceram. Int.* **46** 1033–39
[12] Liu X Y, Zou B, Xing H Y and Huang C Z 2020 *Ceram. Int.* **46** 937–44
[13] Luo W, Bao Z, Jiang W, Liu J, Feng G, Xu Y, Tang H and Wang T 2019 *Ceram. Int.* **45** 24750–6
[14] Sun Z, Chen X, Mao Y, Zhang L X and Feng J C 2020 *J. Eur. Ceram. Soc.* **40** 267–75
[15] Li Y C, Zhao H, Zhang H, Pan D, Feng Y, Li Y, Wang X and Guoet Y 2019 *Ceram. Int.* **45** 22723–30
[16] Akbari E, Kakroudi M G, Shahedifar V and Ghiasi H 2019 *Int. J. Appl. Ceram. Technol.* **00** 1–10
[17] Valenzuela-Gutierrez A, Lopez-Cuevas J, Gonzalez-Angeles A and Pilalua-Diaz N 2019 *Sn Applied Sciences* **7** 784
[18] Ge D L, Deng J X, Duan R, Li X M, Liu Y Y and Yue H Z 2019 *Ceram. Int.* **45** 24554–63
[19] Ramlow H, Ferreira R K M, Marangoni C and Machado R. A F 2019 *Int. J. Appl. Ceram. Technol.* **16** 2161–72
[20] Chen G, Cheng L, Wu H, Zhang S, Xu R, Zhang Q, Li Z and Fu C 2019 *Appl. Phys.* A **125** 848
[21] Ghorbani S, Razavi R S, Loghman-Estarki M R and Alhaji A 2017 *J. Clust. Sci.* **28** 1523–39
[22] Panasyuk G P, Kozerozhets I V, Voroshilov I L, Belan V N, Semenov E A and Luchkov I V 2015 *Russ. J. Phys. Chem.* A **89** 592–97
[23] Yu W J, Zheng Y T, Yu Y D and Su X Y 2019 *J. Eur. Ceram. Soc.* **39** 4313–21
[24] Myz A L, Shutilov R A, Batraev I C, Karagedov G R and Lyakhov N Z 2017 *Chemistry for Sustainable Development* **25** 287–92
[25] Merzhanov A G 2009 *Ind. Ceram.* **29** 196–9
[26] Sytschev A E and Merzhanov A G 2004 *Usp. Khim.* **73** 157–70
[27] Amosov A P 2017 *Russ. J. Non-ferr. Met.+* **58** 530–39
[28] Levashov E A, Mukasyan A S, Rogachev A S and Shtansky D V 2017 *Int. Mater. Rev.* **62** 203–39

IOP Conf. Series: Materials Science and Engineering **848** (2020) 012009 doi:10.1088/1757-899X/848/1/012009

TiB / 30 wt.% Ti layered composite material obtained by free SHS compression on a Ti6Al4V titanium alloy

P M Bazhin[1*], A P Chizhikov[1], A S Konstantinov[1], A D Prokopets[1], E V Kostitsyna[2], A V Bolotskaya[1], A M Stolin[1] and N Yu Khomenko[1]

[1]Merzhanov Institute of Structural Macrokinetics and Materials Science Academician Osipyan str., 8, Chernogolovka, Moscow Region, 142432, Russia
[2]National University of Science and Technology MISiS, Leninskii pr. 4, Moscow 119049, Russia

[*]E-mail: bazhin@ism.ac.ru

Abstract. In this work, a layered composite material based on TiB / 30 wt.% Ti on a Ti6Al4V titanium alloy was obtained by free SHS compression. Powders of titanium and amorphous black boron were used as initial components. The structural features, phase composition, porosity and distribution of microhardness along the height of the layered material were studied. It was shown that, as a result of synthesis and subsequent high-temperature deformation, the following phases were formed in the layered composite material: TiB, TiB_2, $Ti_{0.76}N$. It was found that at a compression pressure of 50 MPa a layered composite material with a porosity of up to 1-2% is formed. It was found that the delay time before applying pressure affects the size of the diffusion zone in the range of 10-50 microns. Due to the formation of hardening phases in the composite material on Ti6Al4V, the microhardness of its surface increased from 2.4 to 11 GPa.

1. Introduction

Titanium fame is widely used in the aviation industry due to its low density, high strength and hardness, corrosion resistance and elevated temperatures [1-4]. These include Ti6Al4V alloy, which is used for the manufacture of large-sized welded and prefabricated aircraft structures. Today, there are many different methods aimed at improving the physicomechanical properties of Ti6Al4V titanium alloys and obtaining composite materials based on them [5-8]. Each of the production methods has its own characteristics and areas of application. A lot of work has been devoted to the production of layered composite materials and multi-layered based on TiB or TiC on the titanium alloy Ti6Al4V [9-11].

A promising method for producing layered composite materials (Layered composite material or LCM) on titanium alloys is free SHS compression [12, 13]. This method combines in one technological stage the synthesis of the initial powder components and their subsequent high-temperature deformation. Due to the selection of the composition of the initial components, the location of macro layers, technological and structural parameters of the process, it is possible to vary the structure and physico-mechanical properties of the resulting composite materials over a wide range. Herewith, taking into account that the synthesis and preparation of a compact composite material takes tens of seconds, this confirms the great advantages of the method used.

IOP Conf. Series: Materials Science and Engineering **848** (2020) 012009 doi:10.1088/1757-899X/848/1/012009

This article is about obtaining a layered composite material based on TiB / 30 wt.% Ti on a Ti6Al4V titanium alloy by free SHS compression and studying its structure, phase composition, and microhardness.

2. Objects and methods of research

To obtain LCM, a model composition with an excess titanium content was taken from the calculation that, after synthesis, a TiB-30 wt.% Ti compound is formed. From pre-mixed powders, 81.6 wt.% Titanium (> 99.1%, 45 μm) and 18.4 wt.% Boron (amorphous black boron, 99%, 5 μm) were formed into cylindrical blanks with a diameter of 30 mm, a height of 15 mm, and a relative density of 0.55 -0.6 and placed them on a substrate of Ti6Al4V with dimensions of 30x30 mm and a thickness of 7 mm. After the initiation of combustion of the formed billet by a tungsten spiral, the passage of the combustion wave in the SHS mode, a delay time of 1 s, and a press plunger speed of 60 mm / s, the synthesized material was compressed with a force of 50 and 100 MPa. After compression and holding time under a pressure of 5 s, the sample was placed in a furnace for 2 h, followed by cooling at a temperature of 500 °C to relieve thermoelastic stresses. As shown by preliminary experiments at temperatures below 500 °C, due to thermoelastic stresses, cracking of the obtained samples occurs.

For the implementation of structural studies of synthesized powder materials the equipment of the Distributed Center for Collective Use of ISMAN was used: the high-resolution field-scanning electron microscope Carl Zeiss Ultraplus (Germany), the powder X-ray diffractometer ARL X'TRA, and other certified methods and techniques.

3. Discussion

As a result of the experimental work, by the method of free SHS compression, LCM of 30x30 mm in size with a ceramic material height based on titanium boride of 2 and 5 mm were obtained, depending on the applied pressure. It was found that at a compression pressure of 100 MPa, the porosity in the material is practically absent, however, in this case, main cracks are observed at the points of its interaction with the Ti6Al4V substrate. The destruction of the material occurs during cooling due to the loss of its ductile properties and increased external pressure. The study of the structure of the obtained SCMs was further conducted for a compression pressure of 50 MPa.

Based on the XRD results, it was found that during the chemical interaction of the initial components of boron and titanium, the main phase is formed - titanium monoboride, as well as the titanium diboride phase. Because of titanium was taken in excess, in the material it acted as a matrix. Due to experiments in the open air and high chemical activity of excess titanium, an additional Ti0.76N phase was formed (cubic crystal lattice, a = 4.235 Å), which, having increased hardness values (up to 20 GPa), increases the mechanical characteristics of the LCM surface.

The synthesis of the studied material proceeds with the melting of titanium (the combustion temperature of the selected composition was 1800 °C), which in the liquid state, under the action of gravity and external pressure, fills the pores and lower layers, thereby depleting or enriching the volumes in the material. According to the Ti-B phase diagram, the crystallization of titanium monoboride occurs by peritectic transformation from a Ti-B melt, which leads to the fact that initially TiB2 grains remain due to lack of free titanium in the material and crystallize in the form of hexagonal structures with sizes less than 1 μm (Figure 1). As you approach the substrate, the proportion of titanium increases, and the number of grains of diboride and titanium boride decreases.

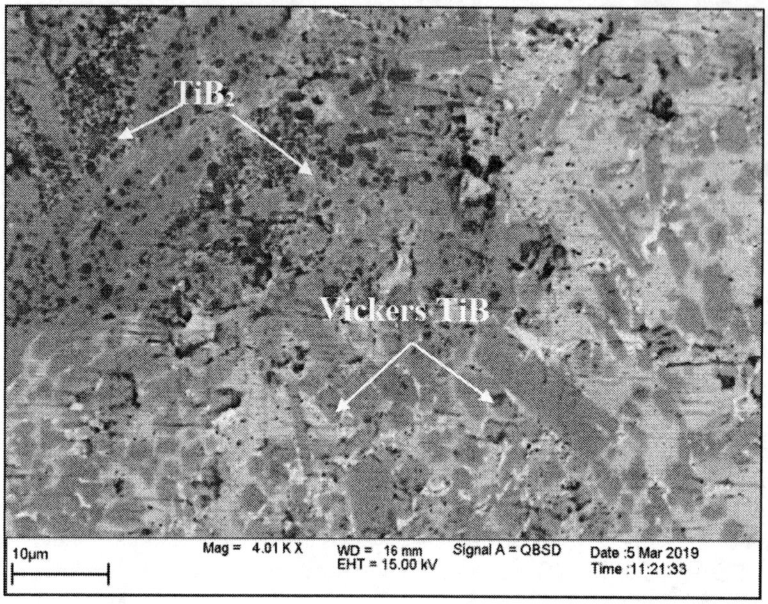

Figure 1. Results of SEM LCM

When compressed due to the contact of the surface of the synthesized material with the plunger of the press and the Ti6Al4V substrate, conductive heat loss occurs, which leads to faster cooling of the surface layers compared to the central part. This limits the residence time of the material at elevated temperatures and prevents grain growth. As a result of this cooling, finer titanium boride grains 1-10 μm in size are observed on the surface and next to the substrate in the obtained LCM, and as they approach the central part of the LCM, titanium boride grains grow to larger sizes up to 20 microns. In this case, the orientation of titanium boride grains along the direction of flow of the material during deformation is observed (Figure 2). The texture of the obtained material is also confirmed by the results of the XRD [13]. It was found that due to faster cooling of the material surface, the LCM surface has a porosity of up to 10-20%, which decreases as it approaches the central part to 1-2%. Due to the high-temperature shear deformation of the combustion products during SHS compression, the synthesized material moves into pores and defects, which reduces the porosity and defect formation of LCM.

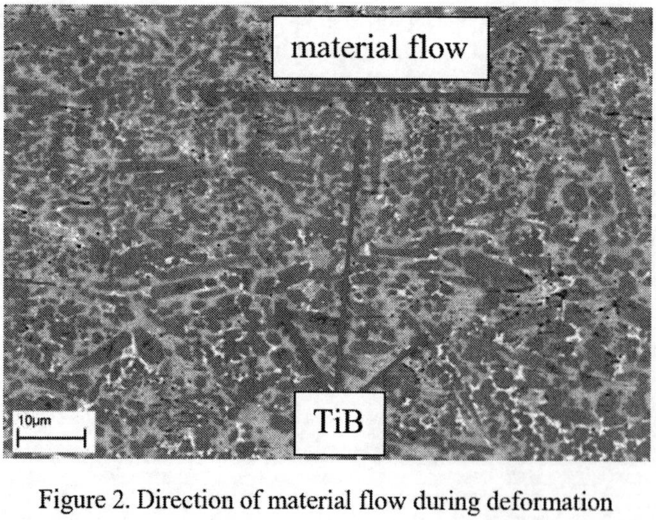

Figure 2. Direction of material flow during deformation

IOP Conf. Series: Materials Science and Engineering **848** (2020) 012009 doi:10.1088/1757-899X/848/1/012009

Under the experimental conditions, the surface layer of the Ti6Al4V substrate also melts, and when external pressure is applied after the combustion wave passes through the sample, the molten titanium and aluminum interact (mix) with the synthesized material of the ceramic composite at the interface. In this case, a diffusion zone of 10–50 μm is formed, the presence of which leads to an improvement in the adhesive strength between the layers. The main influence on the size of the diffusion zone is exerted by the delay time before applying pressure. With an increase in the delay time, the size of the diffusion zone decreases due to cooling and loss of plastic properties of the molten material. The diffusion of molten aluminum and vanadium (which is in the solid state) into the ceramic composite is confirmed by the results of SEM and EDS (Figure 3). It was found that aluminum and vanadium, due to diffusion processes and external pressure, mixed with the molten components of the synthesized material. Thus, aluminum and vanadium in LCM were detected up to 50 μm from the surface of the titanium substrate. EDS revealed traces of silicon in the LCM due to the preparation of samples for analysis.

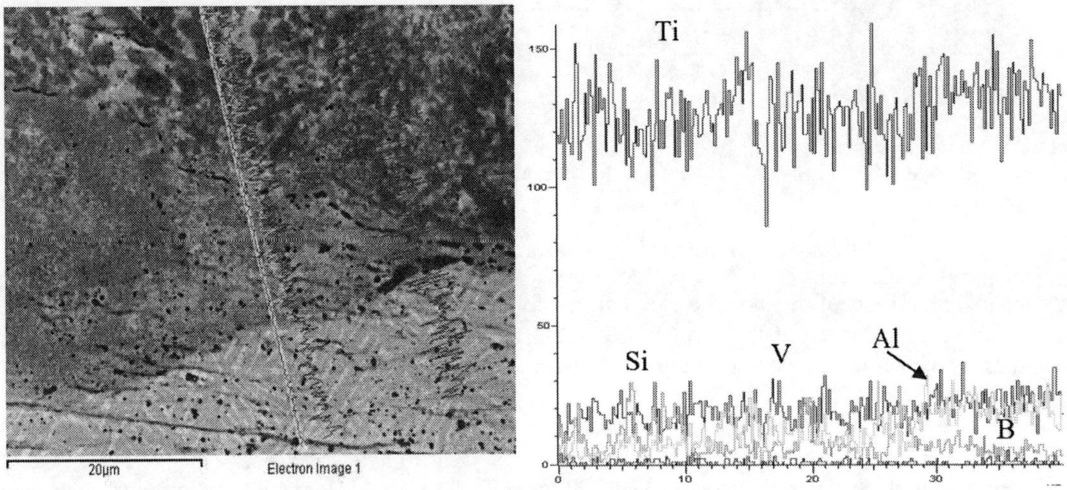

Figure 3. SEM and EDS a diffusion zone

Microhardness measurements along the LCM height were carried out on a PMT-3 microhardness meter with an indenter load of 100 g. Measurements showed that microhardness values decrease from 9.5 to 2.4 GPa when approaching the substrate (Figure 4). On the surface, increased microhardness values are due to the presence of a titanium nitride phase. The zones in which the titanium diboride phase is contained have even higher microhardness values up to 11 GPa. These zones are located approximately 1-2 mm from the surface of the LCM. As the distance from the surface of the LCM to a distance of more than 4 mm, a sharper decrease in the microhardness occurs due to a decrease in the amount of solid phase in the composite. At the interface between the ceramic composite and titanium Ti6Al4V substrate, a transition zone is observed, after which the microhardness of the substrate is 2.4 GPa and does not change further.

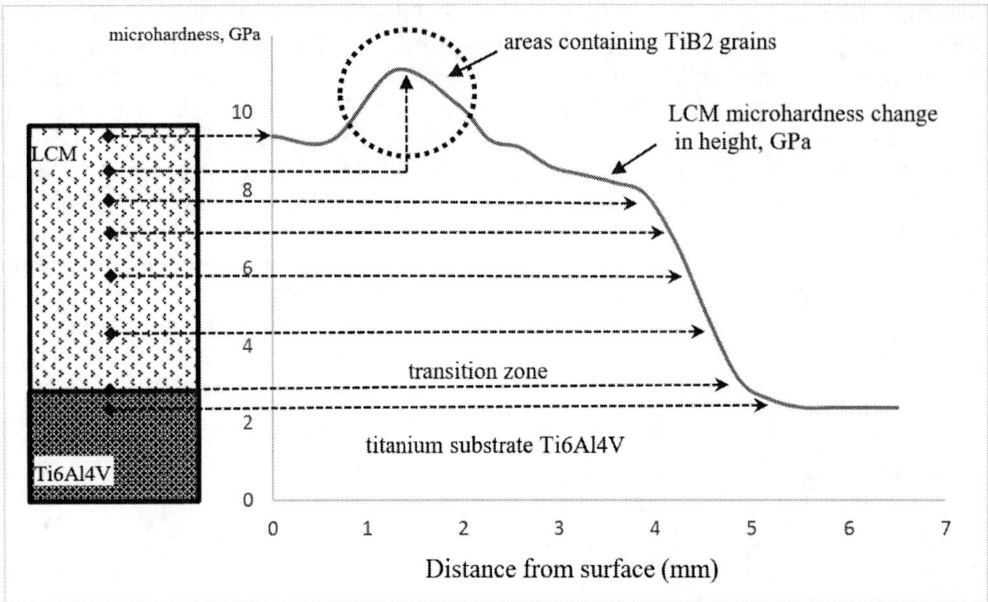

Figure 4. Microhardness measurement results for LCM height

Thus, on the basis of the work carried out, the formation of a layered composite material is shown due to the characteristics of the synthesis of the used composition of the initial titanium and boron reagents, as well as due to the high-temperature shear deformation realized under conditions of free SHS compression.

4. Conclusion

A layered composite material based on TiB / 30 wt.% Ti on a Ti6Al4V titanium alloy with a ceramic layer height of 2-5 mm, depending on the applied pressure, was obtained by free SHS compression. In the obtained material, zones enriched in titanium diboride and zones with the orientation of titanium boride whiskers along the direction of flow of the material are revealed.

It is shown that in the synthesis and subsequent high-temperature deformation of the material, depending on the conditions of the experiment, a diffusion zone of 10–50 μm is formed. The value of this zone is mainly influenced by the delay time before applying pressure. With an increase in the delay time, the size of the diffusion zone decreases due to cooling and loss of plastic properties of the molten material.

It has been established that due to the formation of titanium nitride and diboride phases in the material, the microhardness of the LCM surface increases to 11 GPa, which decreases to 2.4 GPa as it approaches the substrate. As a result, the microhardness of the surface of the titanium alloy increases by 4.5 times.

Acknowledgement
This work was supported by the Russian Science Foundation (project № 18-79-10254).

References

[1] Zhao P C, Li S X, Jia Y F, Zhang C C, Zhang X C and Tu S T 2018 Very high-cycle fatigu behavior of Ti-6Al-4V alloy under corrosive environment *Fatigue Fract. Eng. Mater. Struct.* **41** **4** 881-893 DOI: 10.1111/ffe.12735.
[2] Kumar S, Chattopadhyay K, Mahobia G S and et al. 2016 Hot corrosion behaviour of Ti-6Al-4V modified by ultrasonic shot peening *Mater. Des.* 110:196–206

[3] Boyer R R 2010 Attributes, Characteristics, and Applications of Titanium and Its Alloys *JOM JOM* **62 5** 21-24 DOI: 10.1007/s11837-010-0071-1.

[4] Gerov M V, Vladislavskaya E Y, Terent'ev V F, Prosvirnin D V, Kolmakov and Antonova O S 2016 Fatigue Strength of a Ti-6Al-4V Alloy Produced by Selective Laser Melting *Russ. Metall.* **10** 935-941 DOI: 10.1134/S0036029516100049

[5] Zhou F M, Zhang H X, Sun C X and Dai J 2019 Microstructure and wear properties of multi ceramics reinforced metal-matrix composite coatings on Ti-6Al-4V alloy fabricated by laser surface alloying *Surf. Eng.* **35 8** 683-691 DOI: 10.1080/02670844.2019.1570611

[6] Zhang M M, Xin L, Ding X Y, Zhu S L and Wang F H 2018 Effects Ti/TiAlN composite multilayer coatings on corrosion resistance of titanium alloy in solid NaCl-H2O-O-2 at 600 degrees C *J. Alloys Compd.* **734** 307-317 DOI: 10.1016/j.jallcom.2017.11.035

[7] Celik O N 2013 Microstructure and wear properties of WC particle reinforced composite coating on Ti6Al4V alloy produced by the plasma transferred arc method *Appl. Surf. Sci.* **274** 334-340 DOI: 10.1016/j.apsusc.2013.03.057

[8] Gu Y H, Chen L L, Yue W, Chen P, Chen F and Ning C Y 2016 Corrosion behavior and mechanism of MAO coated Ti6Al4V with a grain-fined surface layer *J. Alloys Compd.* **664** 770-6 DOI: 10.1016/j.jallcom.2015.12.108.

[9] Wei S L, Huang L J, Li X T, Jiao Y, Ren W and Geng L 2019 Network-Strengthened Ti-6Al-4V/(TiC+TiB) Composites: Powder Metallurgy Processing and Enhanced Tensile Properties at Elevated Temperatures *Metall. Mater. Trans. A* **50A 8** 3629-45 DOI: 10.1007/s11661-019-05244-7/

[10] Ivasishin O M, Markovsky P E, Savvakin D G, Stasiuk O O, Rad M N and Prikhodko S V 2019 Multi-layered structures of Ti-6Al-4V alloy and TiC and TiB composites on its base fabricated using blended elemental powder metallurgy *J. Mater. Process. Technol.* **269** 172-181 DOI: 10.1016/j.jmatprotec.2019.02.006

[11] Cai C, He S, Li L F, Teng Q, Song B, Yan C Z, Wei Q S and Shi Y S 2019 In-situ TiB/Ti-6Al-4V composites with a tailored architecture produced by hot isostatic pressing: Microstructure evolution, enhanced tensile properties and strengthening mechanisms *Composites Part B* **164** 546-558 DOI: 10.1016/j.compositesb.2019.01.080.

[12] Stolin A M, Bazhin P M, Konstantinov A S and Corresponding Member of the RAS Alymov M I 2018 Production of Large Compact Plates from Ceramic Powder Materials by Unconfined SHS Compaction *Dokl. Chem.* **480 2** 136–8 DOI: 10.1134/S0012500818060083.

[13] Bazhin P M, Stolin A M, Konstantinov A S, Chizhikov A P, Prokopets A D and Alymov M I 2019 tructural Features of Titanium Boride-Based Layered Composite Materials Produced by Free SHS Compression *Dokl. Chem* **488** 1 246–8. DOI: 10.1134/S0012500819090039.

IOP Conf. Series: Materials Science and Engineering **848** (2020) 012010 doi:10.1088/1757-899X/848/1/012010

The impact of the lead on the physicochemical and optical properties of the fluorophosphate glasses doped by Nd^{3+} ions

O Bogdanov[1] and E Kolobkova[2]
[1] Department of chemical technology of silicate materials, Saint-Petersburg State Institute of Technology, Saint-Petersburg, 190013, Russia
[2] The Faculty of Photonics and Optical Information Technology, ITMO University, Saint Petersburg 197101, Russia

Abstract. Fluorophosphate glasses of the composition 5Ba (PO3) 2- (38-x) AlF3-57RF2-xNdF3 were obtained, where R = Ba, Ca, Sr, Mg, Pb; x = 0.5, 3, 5 mol%. It was found that the introduction of lead fluoride reduces the glass transition and crystallization temperatures. It has been shown that for all systems containing MgF2, the formation of a uovite phase is characteristic. The UV-cutoff has shifted from 200 nm to 250 nm.

1. Introduction

Interest in fluorine-containing systems is due to the possibility of creating glasses based on them that have lower refractive indices and high values of the dispersion coefficient, increased relative dispersion in the blue part of the spectrum, special thermo-optical characteristics, and a wide range of spectral transparency [1]. Due to these properties, as well as a low level of phonon energy, fluorine-containing systems can be used in fiber optic and laser technology [2, 3] Fluorophosphate glasses have high crystallization ability, and are characterized by a small difference between the glass transition temperature and crystallization temperature [4, 5, 6] . Glasses of this system are characterized by surface crystallization [6], which negatively affects the drawing processes of the optical fiber. The introduction of lead fluoride into fluorophosphate glasses leads to an increase in the density ρ, refractive index nD, coefficient of linear thermal expansion α, and a decrease in the thermo-optical constant W. Lead is also an effective glass former, and its introduction into fluorophosphate glass is also expected to reduce the tendency of glass to crystallize. A study of the temperature dependences of fluorophosphate glass and its crystallization ability is necessary to determine the technological conditions for the production of glasses of optical quality and the subsequent drawing of the optical fiber from this material.

2. Materials and methods

Glass melting was carried out in an argon atmosphere at a temperature of 1050 ° C in a glass-carbon crucible according to the “crucible to crucible” scheme. This synthesis method was chosen in order to prevent fluoride from volatilization during melting in a furnace with silicon carbide heaters. Melting time was 45 minutes. Glass melt was produced by casting on a glassy carbon plate. After production, the glass was annealed in a furnace at temperatures close to Tg.

Table 1. Glass compositions (mol.%)

Compositions	Ba $(PO_3)_2$	BaF_2	AlF_3	CaF_2	MgF_2	SrF_2	PbF_2	NdF_3
Initial	5	10	38-x	18.5	10	18.5	0	x=0.5, 3, 5
B	5	0	38-x	18.5	10	18.5	10	x=0.5, 3, 5
S	5	10	38-x	18.5	10	8.5	10	x=0.5, 3, 5
M	5	10	38-x	18.5	0	18.5	10	x=0.5, 3, 5

The characteristic temperatures for determining the temperature-time regime of heat treatment of glasses were determined on the basis of data obtained by analyzing the differential scanning calorimetry (DSC) curve. Measurements and mathematical processing of data were carried out on a Nietzsche STA 449F1 Jupiter differential scanning calorimeter; the heating rate of the samples was 10 K / min.

3. Results and discussion

The DSC results are presented in Figure 1. The DSC of the original glass 1 (a) shows one complex peak. The introduction of lead due to the complete or partial replacement of one of the modifiers leads to the separation of crystallization peaks. Particular attention in this work is given to the composition of the phases formed during crystallization. For the original glass, the peak corresponds to the Usovite crystalline phase (MgCaSrBaAl2F14) [10,11].

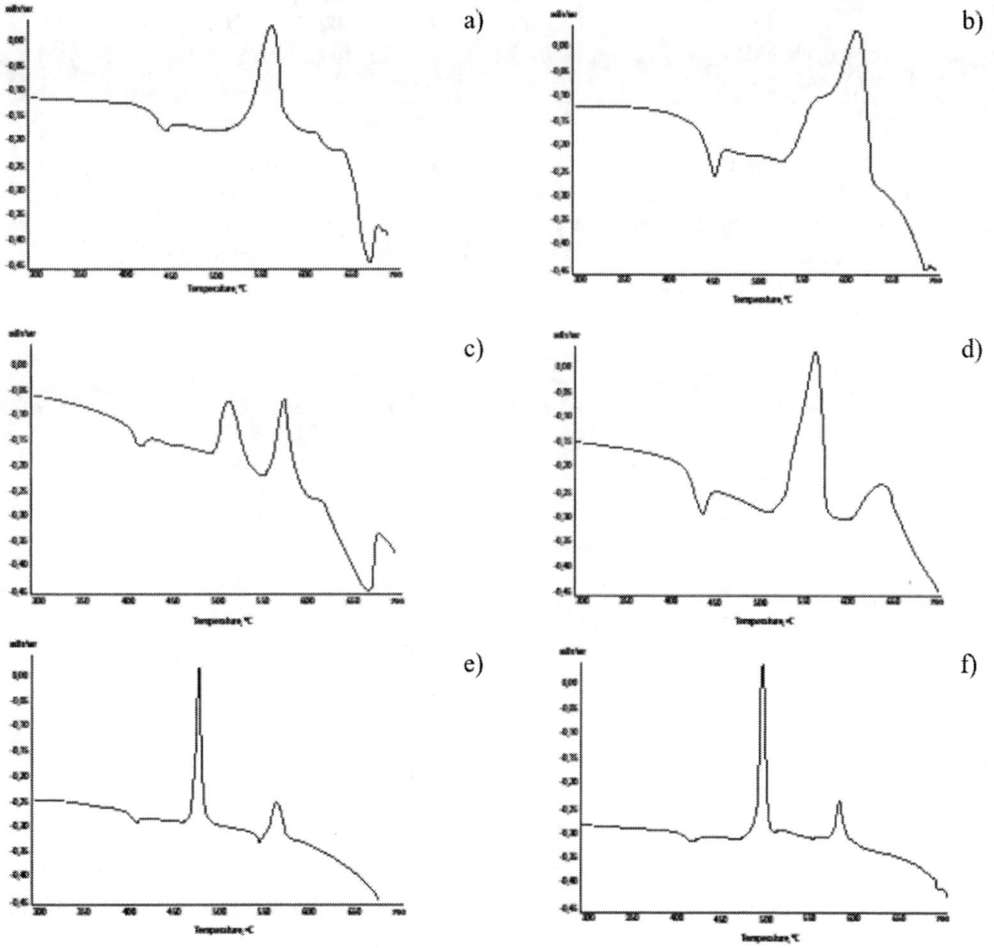

Figure 1. DSC curves for glasses: a), b) initial glass NdF_3 = 0.1 (a) 2 (b), mol.%; c) glass B, NdF_3 = 3 mol%; d) glass S, NdF_3 = 3 mol.% e), e) glass M, NdF_3 = 0.5 (d), 3 (e), mol.%.

Figure 1 shows that the introduction of lead leads to the formation of two peaks and a decrease in the glass transition temperature Tg. It should be noted that glasses B and S are characterized by a large difference in the glass transition temperatures and the onset of phase crystallization. On the contrary, the composition of M is characterized by a lower temperature of the onset of the formation of the first (low-temperature) peak. Also, the peaks of glasses B and S have a wider width than the peaks of glasses without magnesium fluoride (glass M). Characteristic temperatures are presented in table No. 2. Substitution of lead for strontium and barium in glass leads to a slight increase in the interval between the glass transition temperature and crystallization temperature. Overall, peak peaks of all compositions can be estimated as narrow, and, accordingly, it can be argued that volume crystallization is characteristic of all glasses. An increase in the concentration of neodymium leads to a decrease in the interval between the glass transition temperature and crystallization temperature.

Table 2. Glass transition temperatures T_g, crystallization temperatures T_{cr}

Glass compositions	T_g, °C	T_{cr}, °C	T_{cr1}-T_g, °C
Initial, 0.1% NdF_3	440	545	105
Initial, 2% NdF_3	445	535	90
B, 3% NdF_3	409	503, 563	94
S, 3% NdF_3	421	528, 602	107
M,0.5% NdF_3	412	485, 567	73
M, 3% NdF_3	405	486, 573	81

In order to study the crystallization process for all glasses, X-ray diffraction patterns were obtained (Figure 2). All glasses were crystallized in the temperature range corresponding to the first exothermic peak. An analysis of the obtained data shows that, for glasses B, S and the initial first crystalline phase the Usovit monoclinic crystal (map No. 010722129). Glass M is characterized by phase separation, which also has low symmetry, data on which are not yet available in the catalog.

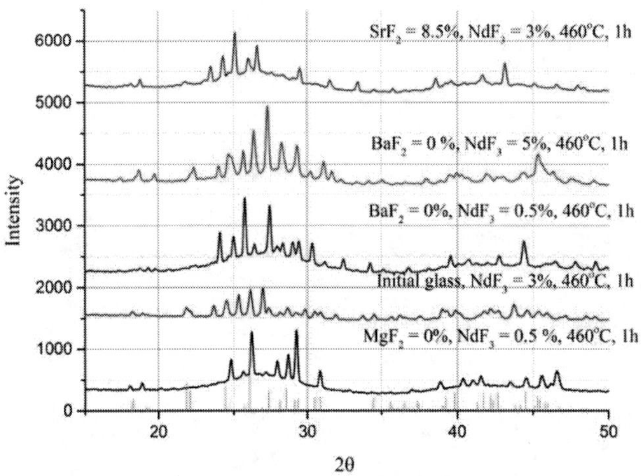

Figure 2. Comparison of diffraction patterns of crystallized source glass and B, M glasses at temperatures corresponding to low-temperature peaks. The bar diagram shows the phase of the usovit $Ba_2CaMgAl_2F_{14}$.

IOP Conf. Series: Materials Science and Engineering **848** (2020) 012010 doi:10.1088/1757-899X/848/1/012010

Optical characteristics were obtained for glasses of all compositions — the refractive index and absorption spectra. A logical increase in the density ρ, refractive index nD, and with the introduction of lead for all glasses was shown (Table 3).

Table 3. Refractive index and glass density

Glass compositions	n_D	ρ, г/см^3
Initial, 3 mol. % NdF_3	1.4535	3.725
B, 5 mol.% NdF_3	1.4800	4.178
S, 3 mol.% NdF_3	1.4826	4.257
M, 3 mol.% NdF_3	1.4842	4.310

The absorption was measured in the UV region and the near infrared region (3–7 μm). The measurement results are presented in Fig. 3. As you can see, the introduction of lead into the glass shifts the UV absorption limit of the glasses (Figure 3.a) to the long-wavelength region from 180 nm (initial glass) to 245 (glass B and S) and 250 nm (glass M). The infrared transmission limit for all compositions is determined by the composite frequencies of the natural vibrations of the phosphate groups (4.5 μm). In this part of the spectrum, one can also observe a shift of the absorption bands of phosphate groups by 60 nm to the region of large wavelengths.

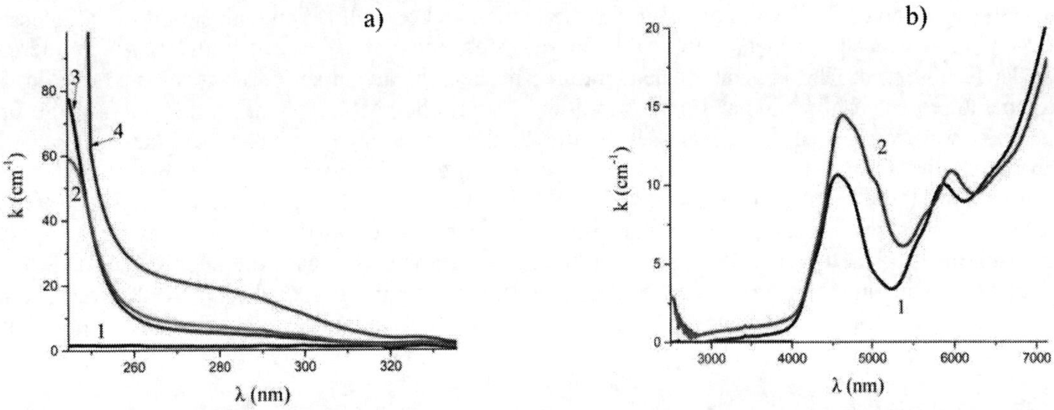

Figure 3. Absorption of glasses in the UV (a) and IR (b) region:
a - 1) Initial, 3 mol. % NdF_3; 2) B, 5 mol% NdF_3; 3) S, 3 mol% NdF_3; 4) M, 3 mol% NdF_3.
b - 1) initial, 3 mol. % NdF_3; 2) S, 3 mol% NdF_3.

The glass structure was analyzed based on the analysis of the Raman spectrum (Figure 4). In the spectra, vibration bands of P – O – P bridges are observed, the presence of which indicates the presence in glass structure of not only isolated tetrahedra (PO4), but also dimers (P2O7). Also a band corresponding to vibrations of the fluorine-aluminate network (580 cm-1) is observed.

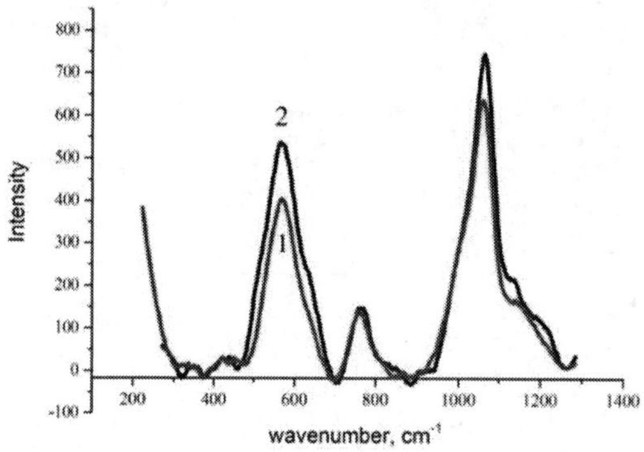

Figure 4. Raman spectra of light for two concentrations of neodymium: 1) 0.5 mol.% NdF_3; 2) 2 mol.% NdF_3.

4. Conclusion

The introduction of PbF2 leads to lower glass transition and crystallization temperatures in all glasses. In glasses with lead, a separation of DSC curves is observed and the formation of two exothermic peaks is observed. The content of neodymium fluoride in all glasses leads to an increase in the temperature of crystallization and glass transition. During the heat treatment in the region of the first peak, it was found that for all glasses B and S, the formation of a phase similar to usovite is characteristic. Glass M is characterized by the formation of a different phase, but also has low symmetry. The refractive index increases with the introduction of PbF2. The Fourier spectroscopy data did not show a change in the transparency of the glasses in the region from 2.5 to 4 μm. Measurement of the UV and IR absorption boundaries showed that with the addition of lead, these boundaries shift in the region of higher wavelengths. Despite the introduction of lead, these glasses still retain their high transparency in a wide spectral region, and can be considered as a promising photonic material.

Acknowledgements
This work was financially supported by Foundation for Assistance to Small Innovative Enterprises (Agreement 13486ГУ/2018).

References

[1] Haliev V.D., Karapetjan K.G., Bogdanov V.L., Nosyreva E.B., Janush O.V. 1990 Glass Physics and Chemistry.. T. 16.No 4. S. 529-533.
[2] Ebendorff-Heidepriem Heike, Tze-Cheung Foo, Roger C. Moore, Wenqi Zhang, Yahua Li, Tanya M. Monro, Alexander Hemming, and David G. Lancaster, 2008 , Opt. Lett. 33 (23), 2861;
[3] Xiushan Zhu, Xiushan Zhu and N. Peyghambarian, 2010. Advances in OptoElectronics Volume, Article, 23 pages;
[4] Gustavo Galleani , Yannick Ledemi, Elton Soares de Lima Filho, Steeve Morency, Gaelle Delaizir, Sebastien Chenu , Jean Rene Duclere, Younes Messaddeq, 2017, Optical Materials 64, pp.524 -532;
[5] Doris Ehrt, 2003, Current Opinion in Solid State and Materials Science 7 pp. 135–141;
[6] Burdaev P.A., Kolobkova E.V., Aseev V.A., Nikonorov N.V., Trofi mov A.O, 2015, Glass Physics and Chemistry.. T. 41. No 1. P. 132-136.

IOP Conf. Series: Materials Science and Engineering **848** (2020) 012011 doi:10.1088/1757-899X/848/1/012011

Microstructural evolution of the AlMgMnZr alloy during severe plastic deformation

Yu Borisova, D Yuzbekova, A Kalinenko and A Mogucheva
Belgorod State University, Belgorod, 308015, Russia

Corresponding author: borisovayuliya94@mail.ru

Abstract. The effect of severe plastic deformation (SPD) on microstructure of Al-5.4Mg-0.52Mn-0.1Zr alloy was studied. The methods of equal channel angular pressing (ECAP) and cold rolling were used to obtain fine-grained microstructure. The evolution of average misorientation angle and portion of HAB during SPD was determined. The lattice dislocation density was evaluated from EBSD data from the lattice curvature using the via Frank's equation.

Keywords – Aluminum alloys, Severe plastic deformation, Microstructure

1. Introduction

The development of various regimes of severe plastic deformation to obtain fine-grained microstructure in aluminum alloys has recently attracted more attention [1-2]. The formation of a fine-grained structure in these alloys is a convenient way to achieve a new level of physical and mechanical properties. Dynamic recrystallization is an important mechanism which make it possible to control the microstructure and mechanical properties of various materials [3-4]. The most significant advantage of dynamic recrystallization is that the necessary microstructure can be formed directly during deformation [3-5].

The process of dynamic recrystallization depends significantly on the temperature, strain and strain rate. In particular, large elongation ratios frequently used in the pressing technology, which lead to significant values of the strain. However, after reaching a certain critical degree of strain, it ceases to play the role of a control parameter, and then, the final properties of the alloy are determined by the temperature and strain rate [6]. Thus, the aim of the present study was to determine the evolution of the microstructure of the AlMgMnZr alloy various regimes of severe plastic deformation (SPD).

2. Material and methods

The studied Al-5.43%Mg-0.52%Mn-0.1%Zr wt.%. alloy was produced by continuous casting. The obtained billets were homogenized at 360°C for 6 h. Then the following treatments were used:

(a) To eliminate the disadvantages of the cast structure, an intermediate deformation was carried out: the rolling at room temperature with a total reduction of 70% and subsequent annealing at 400°C for 2 h followed by cooling in air. This material condition is denoted hereafter as coarse-grained (CG).

(b) The bars with cross section of 20×20 mm^2 and length of 100 mm were subjected to equal channel angular pressing (ECAP) via route BC with a 90° route after each pass. The ECAP processing was performed at 300°C to the true strain of ~12. This material condition is termed as fine-grained (FG).

(c) The samples of studied alloy after treatment (b) were subjected to addition cold-rolling at room temperature with a total reduction of 80%. This material condition is denoted as severely strained (SS).

The microstructural investigations and elemental analysis were performed by Quanta 600FEG scanning electron microscope (SEM) equipped with an electron back scatter diffraction (EBSD) pattern analyzer incorporating an orientation imaging microscopy (OIM) system. The transmission electron microscopy (TEM) observations were carried out in the bright field using JEOL JEM-2100 microscope equipped with an INCA energy dispersive x-ray spectrometer.

Due to the experimental error of the EBSD method, all low-angle boundaries with a misorientation of less than 2° were excluded from consideration using the grain expansion parameter in the TSL software. The misorientation of 15 ° was used as a criterion for distinguishing low- and high-angle boundaries (LAB and HAB, respectively).

3. Results and discussion

Representative microstructures of the studied AlMgMnZr alloy in various states is shown in figure 1.

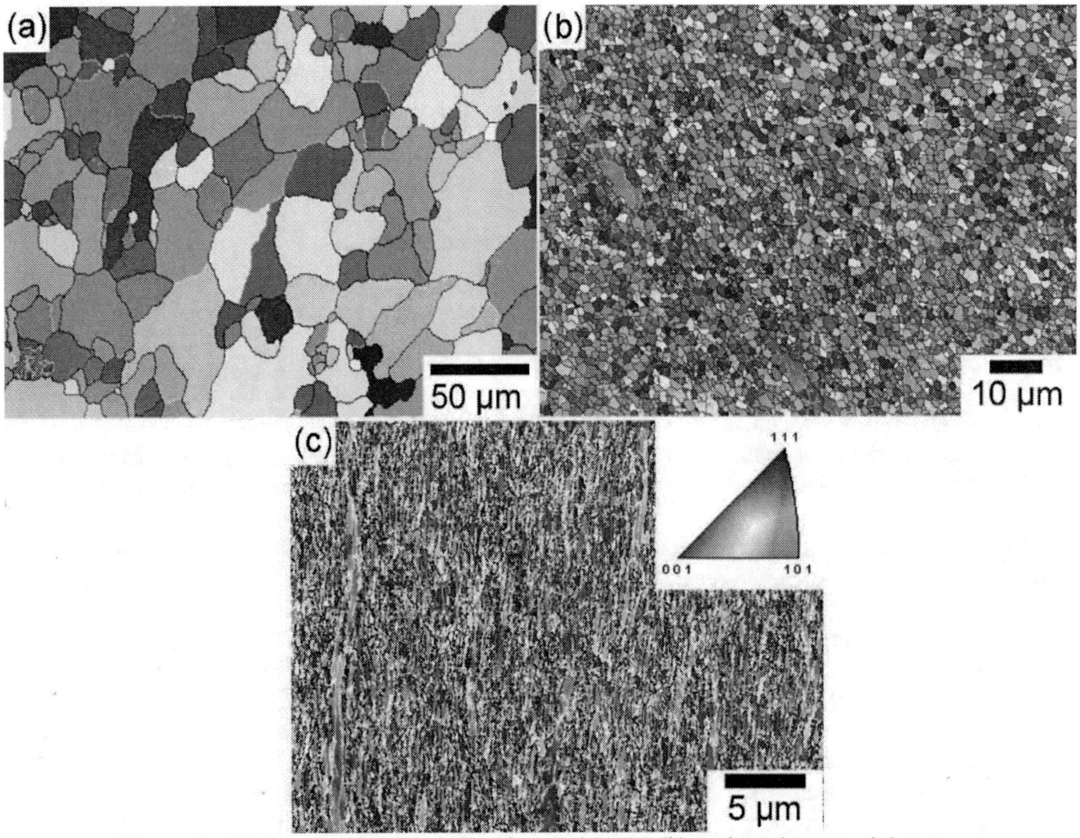

Figure 1. OIM maps of studied alloy in CG (a); FG (b) and SS (c) material states.

The main structural parameters are listed in Table 1. In the studied alloy recrystallization annealing at a temperature of 400°C, leading to the formation of the microstructure with heterogeneous grain size distribution (figures 1 (a) and 2 (a ')). Large and relatively equiaxed grains have an average size of 22.0 μm, and small grains located along the boundaries of large grains have an average size of 6.5

μm (figure 1 (a)). Uniformly distributed Al_6Mn particles with incoherent interfaces and average size of 35 nm were found inside the grains [7]. The specific fraction of HAB estimated using EBSD analysis is found to be ~ 93% and the average misorientation angle is 33 ° (figure 2 (a)). The dislocation density in the CG state is ~ 10^{13} m^{-2}.

Table 1. Microstructure; parameters of studied aluminum alloy in different material states.

State	CG	FG	SS
Grain size, μm	22	1.5	1.0
Fraction HAB's, %	93	62	43
Average Misorientation Angle, °	33	27	21
Size of Al_6Mn particles, nm	35	25	25

ECAP processing, leading to the refinement of recrystallized grains. The average grain size in this state is about 1.5 μm which is 15 times smaller as compared to the CG state (figures 1 (b) and 2 (b ')). The fraction of recrystallized grains is 90%. The mean size of Al_6Mn particles located inside the grains did not change remarkably during ECAP (table 1). The average misorientation angle and the portion of HAB in this state are decreased to 27 ° and 62% (figure 2 (b)), respectively.

Thc EBSD analysis of studied alloy in the SS material condition showed that the grains tends to acquire an elongated shape (figures 1 (c) and 2 (c')), the average grain size is ~ 1 μm. According to transmission electron microscopy, the average size of Al_6Mn particles was not affected by addition cold rolling (table 1). The average misorientation angle and the portion of HAB decreased to 21 ° and 43% (figure 2 (c)), respectively. Note that a gradual decrease in the average misorientation angle in the studied alloy is associated with a gradual increase in the fraction of LAB under deformation.

The lattice dislocation density was evaluated from EBSD data from the lattice curvature using the kernel average misorientation option [8] via Frank's equation:

$$\theta = 2\sin\frac{\theta}{2} = \frac{Nb}{h} \quad (1)$$

where θ is the misorientation angle created by a dislocation wall of height h consisting of N dislocations, and b is the Burgers vector. In EBSD analysis, distance h corresponds to the scanning interval. The dislocation density ρ is then given by the ratio of the dislocation number per surface area, Eq. (2):

$$\rho = \frac{N}{S} \quad (2)$$

where the surface area of the hexagon S (scanning geometry) is $S = \sqrt{3}h^2/2$. It follows that the dislocation density can be estimated from the following relationship:

$$\rho = \frac{2\theta}{\sqrt{3}hb} \quad (3)$$

The values used for dislocation density calculations are given in table 2.

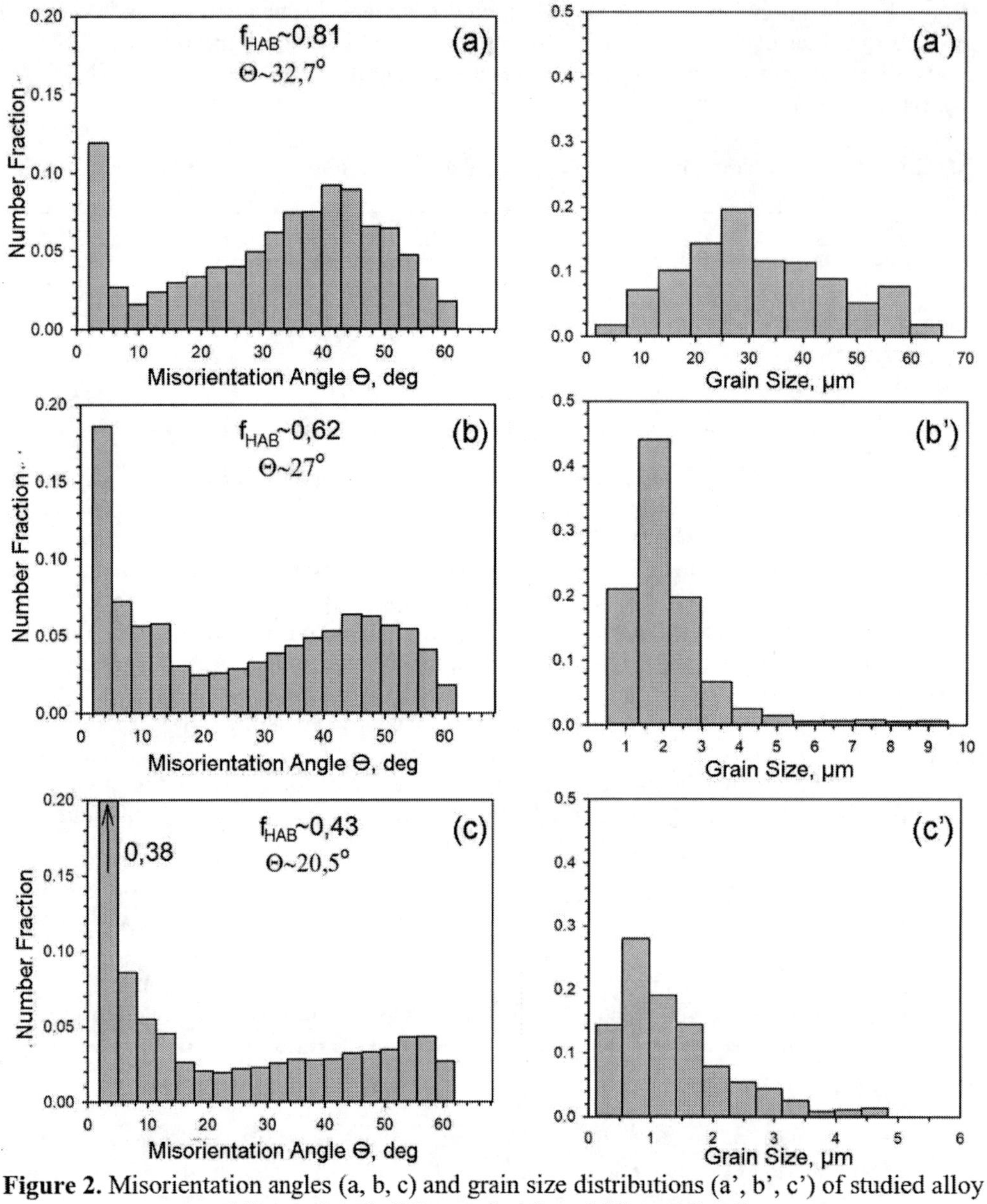

Figure 2. Misorientation angles (a, b, c) and grain size distributions (a', b', c') of studied alloy in CG state (a, a'); FG state (b, b') and SS state (c, c').

Table 2. The mean parameters for eqs. (1-3).

State	h, μm	Θ_{KAM}	b, nm	ρ, m^{-2}
FG	0.2	0.72830	0.286	1.4×10^{14}
SS	0.05	1.09021	0.286	8.8×10^{14}

It seen that calculated value of dislocation density in FG $(7.7\pm3.2)\times10^{13}m^{-2}$ and SS $(2.8\pm3.0)\times10^{15}$ m^{-2} material conditions is in good agreement with that measured using TEM images.

IOP Conf. Series: Materials Science and Engineering **848** (2020) 012011 doi:10.1088/1757-899X/848/1/012011

4. Summary and conclusions

The main conclusions derived based on the present study could be summarized as follows:

1) The average misorientation angle and portion of HAB are gradually decreased during grain refinement of studied aluminum alloy.

2) The mean size of Al_6Mn particles didn't change remarkably under severe plastic deformation and subsequent annealing.

3) The severe plastic deformation leading to significant increase of dislocation density from ~10^{13} m^{-2} in CG state to 8.8×10^{14} m^{-2} in SS state.

Acknowledgement

The study was financially supported by the Russian Science Foundation, Belgorod State University project No. 17-72-20239. The main results were obtained by using the equipment of the Joint Research Center, «Technology and Materials», Belgorod State University.

References

[1] Markushev M.V., Vinogradov A. 2006 *Severe Plastic Deformation: Towards Bulk Production of Nanostructured Mater.* 233.

[2] Dao M., Lu L., Asaro R. J., De Hosson J. T. M. and Ma E 2007 *Acta Mater.* **55** (12).4041.

[3] Sakai T., Belyakov A., Kaibyshev R., Miura H., & Jonas J. J 2014 *Progr. in mater. Sci.* **60** 130.

[4] Huang K., Logé R. E 2016 *Mater. & Design* **111** 548.

[5] Zhao P., Wang Y., Niezgoda S. R. 2018 *Int. J. Plast.* **100** 52.

[6] Aboulfadl H, Deges J, Choi ., Raabe D 2015 *Acta Mater.* **86** 34.

[7] Borisova Y., Yuzbekova D., Mogucheva A. 2019 IOP Conf. Series: Mater. Sci. and Eng. **597** (1) 012057.

[8] Zhilyaev A. P., Shakhova I., Morozova A., Belyakov A., and Kaibyshev R. 2016 *Mater. Sci. Eng.: A* **654** 131.

Synthesis, photoluminescence and thermal properties of nanostructured organosilicon luminophore based on 2,2'-bithienyl and 4,7-diphenyl-2,1,3-benzothiadiazole

O V Borshchev[1,a], M S Skorotetcky[1], E A Svidchenko[1], N M Surin[1] and S A Ponomarenko[1]
[1]Enikolopov Institute of Synthetic Polymeric Materials of Russian Academy of Sciences, Profsoyuznaya St. 70, Moscow 117393, Russia
[a] Corresponding author: borshchev@ispm.ru

Annotation. In this work, novel nanostructured organosilicon luminophore (NOL) containing four 2,2'-bithienyl donor units connected via silicon atoms to 4,7-diphenyl-2,1,3-benzothiadiazole acceptor unit was synthesized. Photophysical measurements of the NOL revealed the effect of solvents polarity on its photoluminescence quantum yield (PLQY). In THF solution the efficiency of photoluminescence was only 4%, whereas in toluene solution PLQY increased up to 76% without any changing the spectral characteristics. Under the normal conditions, the NOL synthesized is a viscous fluid, which contributes to its good solubility in various solvents. The NOL is thermostable up to 415 °C in argon atmosphere.

1. Introduction

Recently we suggested the concept of nanostructured organosilicon luminophores (NOLs) [1–3]. A NOL represents branched or dendritic organosilicon molecular antennae that is able to capture light quanta (photons) and nonradiatevely transfer electronic excitation energy from the periphery to its central unit, which leads to effective light absorption in a broad spectral range and emission in a narrower region with a longer wavelength [4–7]. In the NOL two types of organic chromophores are connected to each other via silicon atoms, which break the conjugation between them and fix them specifically in the space at 1-2 nm distance necessary for efficient Förster energy transfer [6,8]. NOLs possess several advantages: absorption in a wide spectral range; 5–10 times higher absorption cross-sections than those of the best low molar weight organic luminophores; very high photoluminescence quantum yield; emission in the defined region; short luminescence lifetime. For the NOL reported in this work, 2,2'-bithiophene (2T) was chosen as a donor, like in the most previously reported NOLs. While a novel acceptor based on 4,7-diphenyl-2,1,3-benzothiadiazole (PBtdP), which is known as a strongly fluorescent dye with large Stokes shift [9-12], was used in a NOL for the first time. For increasing stability and solubility of the NOL, each of the four donor units was end-capped with n-hexyl (Hex) groups.

2. Results and Discussion

2.1. Synthesis

The synthetic strategy, which was developed previously for bithiophenesilane dendrimers and branched organosilicon macromolecules [13-15] was extended here successfully for the new NOL. The synthesis was made via palladium-mediated cross-coupling reactions widely used for the synthesis of various conjugated oligomers and polymers [16]. The synthetic route to the NOL is summarized in Scheme 1. A key compound for the NOL synthesis, namely, bis(5'-hexyl-2,2'-bithien-5-yl)(methyl)[4-(4,4,5,5-tetramethyl-1,3,2-dioxaborolan-2-yl)phenyl]silane (**M1**), was prepared through a two-step scheme. First, bithiophenesilane monodendron (4-bromophenyl)[bis(5′-hexyl-2,2′-bithien-5-yl)]methylsilane **(Hex-2T)$_2$MeSiPhBr** was obtained by the reaction between (4-bromophenyl)(dichloro)methylsilane **BrPhSiMeCl$_2$** [7] and the lithium derivative of 5-hexyl-2,2′-bithiopene in 74% yield. Second, lithiation of **(Hex-2T)$_2$MeSiPhBr** followed by the treatment with 2-isopropoxy-4,4,5,5-tetramethyl-1,3,2-dioxaborolane (IPTMDOB) lead to pinacolatoborolane-functionalized monodendron **M1** in 89% yield.

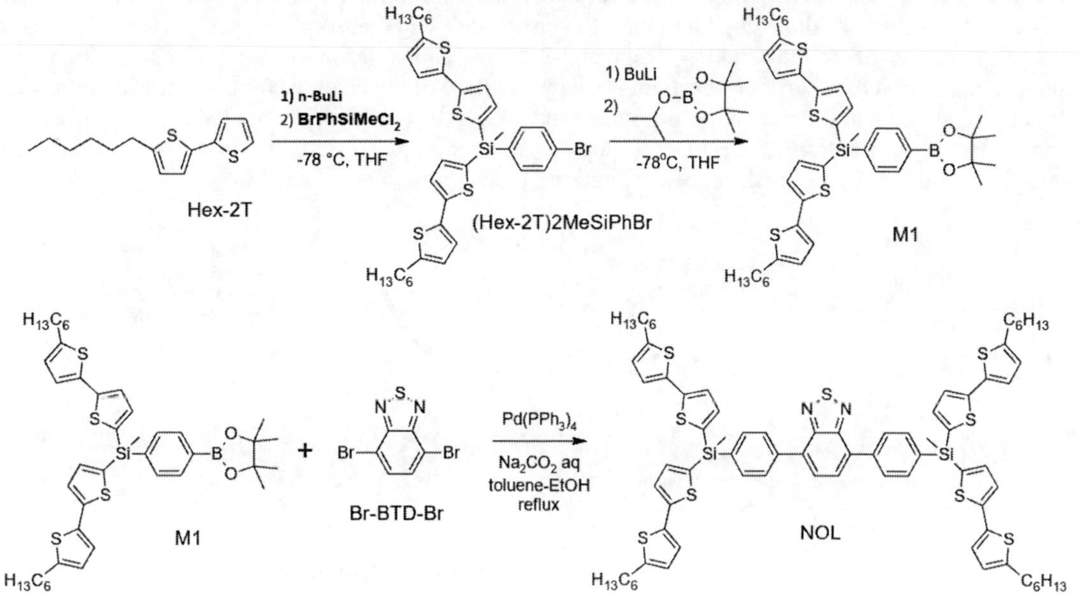

Scheme 1. The synthetic scheme of the NOL.

Finally, **M1** was reacted with 4,7-dibromo-2,1,3-benzothiadiazole under Suzuki conditions to give the desired 4,7-bis{4-[bis(5'-hexyl-2,2'-bithien-5-yl)(methyl)silyl]phenyl}-2,1,3-benzothiadiazole (**NOL**) in 81% reaction yield (according to GPC analysis) and 63% isolated yield. The reaction was carried out until the reaction mixture no longer changed according to GPC. The monitoring by GPC using an instrument equipped with a diode array detector, which allowed recording three-dimensional chromatograms in the retention time – wavelength – signal intensity coordinates, allows to optimize the reaction time that prevents the formation of by-products. Pure NOL was isolated by column chromatography. The chemical structure and purity of all precursors and final compounds were proved by a combination of TLC, GPC, ^{1}H-, ^{13}C- and ^{29}Si-NMR, MALDI-TOF MS and elemental analysis.

2.2. Optical properties of the NOL in dilute dolutions

The absorption and fluorescence properties of NOLs are determined by the combination of two types of non-conjugated chromophore fragments through silicon atoms [4]. With this combination, the absorption spectrum is an additive superposition of the absorption bands of each NOL fragments. The excited state of a NOL fragment with a larger band gap (donor) is deactivated by induction-resonance transfer of the electronic excitation energy to another NOL fragment with a smaller band gap (acceptor). If the rate constant of the induction-resonance transfer is much greater than the donor fluorescence rate constant, then only the acceptor fluorescence is observed [10].

The absorption spectra were recorded using a spectrophotometer (UV-2501PC, Shimadzu) in a standard 10 mm-thick photometric quartz cuvette filled with solutions at a concentration of 10^{-5} M. A scanning spectrofluorimeter ALS01M (Russia) with registration in a single photon counting mode at successive time intervals and automatic adjustment of the intensity of the emission measured was used for

registration of the photoluminescence spectra (PL) [17]. The PL excitation wavelengths were set to the absorption maxima. PL measurements were performed in 90° geometry at several optical densities in the range from 0.06 to 0.12 absorbance units in a 10 mm-thick quartz cuvette. The photoluminescence quantum yield (PLQY) was measured by comparing the integral PL intensity of 10^{-6} M diluted solutions of the NOL with the integral PL intensity of a standard as described elsewhere [18]. As the standard for measuring the PLQY, a diluted solution of 1,4-bis(5-phenyloxazol-2-yl)benzene (POPOP) in cyclohexane (PLQY = 1) was used.
The absorption and PL spectra of diluted solutions of the NOL in toluene and in THF are presented in Figure 1. As can be seen, the spectra in these solvents are almost identical. Deconvolution of the absorption spectrum of the NOL solution in toluene to individual absorption spectra corresponding to 2,2'-bithiophene 2T (donor) and 4,7-diphenyl-2,1,3-benzothiadiazole PBtdP (acceptor) fragments showed that the NOL absorption spectrum is a superposition of the bands of these fragments (Figure 2). When excited both in the acceptor absorption band and in the preferential donor absorption band , the same spectral distribution of the PL intensity is observed. Upon excitation of the acceptor, the PL quantum yield is 10% higher than upon excitation of the donor.

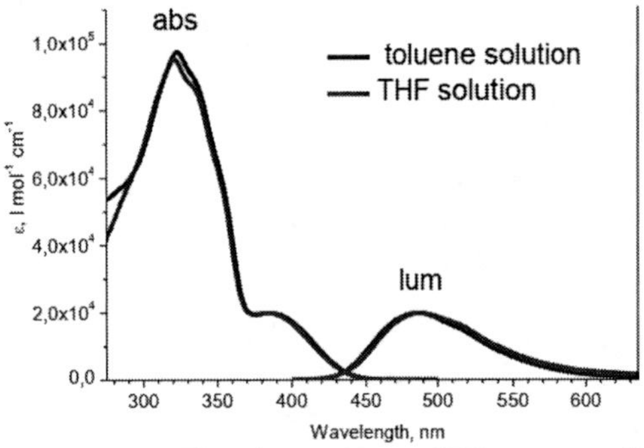

Figure 1. The absorption and PL spectra of solutions of the NOL in toluene and in THF.

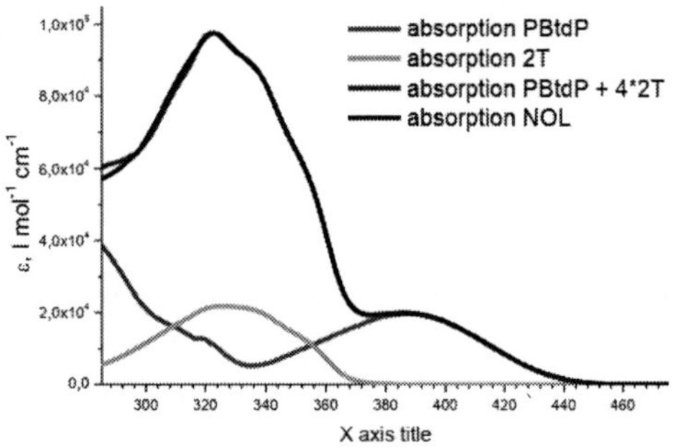

Figure 2. Deconvolution of the absorption spectrum of the NOL.

Thus, the NOL investigated is a typical representative of NOL family by its absorption-fluorescent characteristics [1]. However, with almost complete coincidence of the absorption and fluorescence spectra of NOL solutions in toluene and THF, the photoluminescence quantum yields for these two solutions are different. When excited in the absorption band of the acceptor in toluene solution PLQY = 76%, while in THF solution PLQY = 4%.
Figure 3 shows the dependence of the fluorescence spectra of NOL solutions in toluene on the concentration of THF (quencher). As can be seen, with the growth of THF concentration there is a rapid quenching of the NOL fluorescence without changing the shape of the fluorescence spectrum. There are various processes of dynamic and static quenching of the fluorescence molecules [19]. Dynamic quenching is caused by random collisions between the excited molecules and the quencher molecules. In this case, the excitation energy is transferred to the quencher. In the case of static quenching, PLQY decreases as a result of exposure to the unexcited luminophore molecules. Static quenching is usually associated with the formation of non-fluorescent complexes in the ground state. The kinetics of dynamic and static fluorescence quenching is characterized by the Stern-Volmer formula:

$$\frac{QY_0}{QY_T} = 1 + K \cdot [T],$$

where K – fluorescence quenching constant; $[T]$ – quenching concentration; QY_0 and QY_T – photoluminescence quantum yield in the absence and presence of a quenching agent, respectively.

The dynamic quenching constant $K = K_D$ is determined by the rate at which interacting molecules diffuse to each other. The static quenching constant is the equilibrium constant of the complexes formed

$K = K_{St}$. The concentration dependence of the processes of dynamic and static quenching, built in the coordinates $\frac{QY_0}{QY_T} - 1$ and $[T]$, can be described by a line passing through the origin.

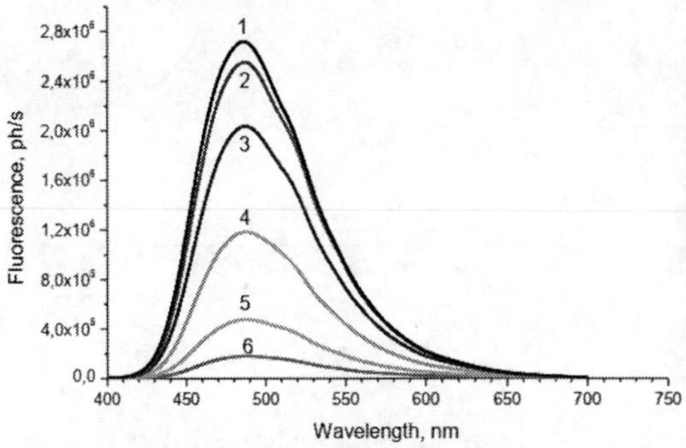

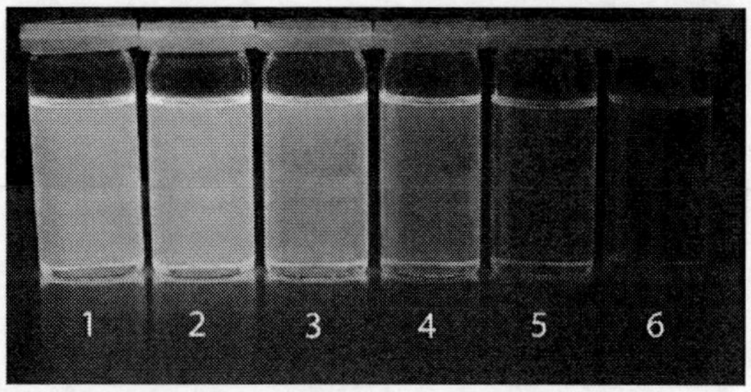

Figure3. PL spectra and emission images of solutions of the NOL in toluene-THF mixtures containing THF, % vol.: 0 (1); 20 (2); 40 (3); 60 (4); 80 (5); 100 (6).

Figure 4 shows the experimental dependence of photoluminescence quantum yield of the NOL on the concentration of THF in these coordinates. As one can see, the observed process of quenching NOL fluorescence by THF molecules cannot be characterized by the Stern-Vollmer formula. Therefore, it cannot be attributed to dynamic or static extinguishing processes. More detailed research is required to elucidate the mechanism of this phenomenon.

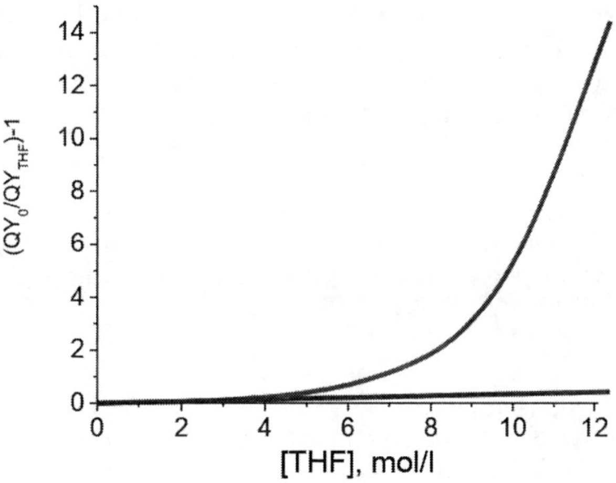

Figure 4. Experimental (red curve) and calculated by the Stern-Volmer equation (blue curve) concentration dependence of the PL quantum yield in the coordinates $\frac{QY_0}{QY_T} - 1$ and $[THF]$.

2.3. Thermal Behavior

Thermal properties of the NOL synthesized were investigated by thermogravimetric analysis (TGA) and differential scanning calorimetry (DSC). DSC curves show only a glass transition at -8 °C, so NOL is a viscous liquid under normal conditions (Figure 5).

IOP Conf. Series: Materials Science and Engineering **848** (2020) 012012 doi:10.1088/1757-899X/848/1/012012

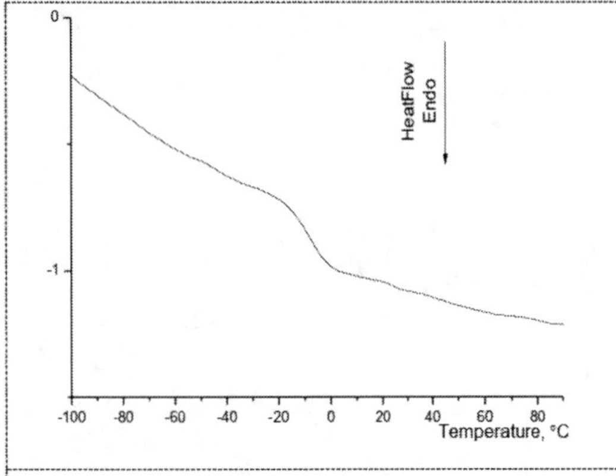

Figure 5. DSC heating curve of the NOL.

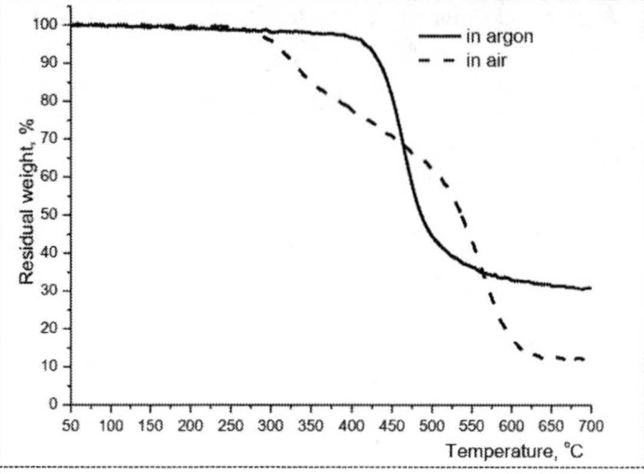

Figure 6. TGA curve of the NOL.

The thermal stability of the NOL was checked by the TGA method, which showed that it is highly thermostable – the 5% weight loss occurs only at 415 °C in argon (Figure 6). In the presence of oxygen their stability decreases - the destruction onset (5% weight loss) lies at 305 °C.

3. Conclusions

In summary, synthesis of a novel nanostructured organosilicon luminophore (NOL) containing four bithiophene fragments at the periphery and the 4,7-diphenyl-2,1,3-benzothiadiazole fragment at the center is described. The Suzuki cross-coupling reaction was controlled by GPC analysis, which allowed reducing the reaction time and preventing formation of by-products. The study of the optical properties of dilute solutions of the synthesized molecule showed that it is a typical representative of the NOL family by its absorption-fluorescent characteristics. The NOL obtained demonstrate a strong influence of the nature of solvents on the photoluminescence quantum yield. Further studies on to elucidate the mechanism of this phenomenon are currently in progress.

Acknowledgements

This work was supported by Russian Foundation for Basic Research (project #18-29-17006). NMR spectra registration and GPC analysis were performed with the financial support from Ministry of Science and Higher Education of the Russian Federation using the equipment of Collaborative Access Center “Center for Polymer Research” of ISPM RAS.

References

[1] Ponomarenko S.A., Surin N.M., Borshchev O.V., Luponosov Y.N., Akimov D.Y., Alexandrov I.S., Burenkov A.A., Kovalenko A.G., Stekhanov V.N., Kleymyuk E.A., Gritsenko O.T., Cherkaev G.V., Kechek’yan A.S., Serenko O.A., Muzafarov A.M. 2014 *Sci. Rep.* **4**, 6549

[2] Ponomarenko S.A., Surin N.M., Borshchev O.V., Skorotetcky M. S., Muzafarov A.M. 2015 *Proc. SPIE* **9545** 954509

[3] Borshchev O.V., Kleymyuk E.A., Surin N.M., Svidchenko E.A., Fedorov Yu.V., Dmitryakov P.V., Chvalun S.N., Ponomarenko S.A. 2017 *Organic Photonics and Photovoltaics* **5** 1-8

[4] Luponosov Y.N., Ponomarenko S.A., Surin N.M., Borshchev O.V., Shumilkina E.A., Muzafarov A.M. 2009 *Chem. Mater.* **21** 447

[5] Borshchev O.V., Ponomarenko S.A., Kleymyuk E.A., Luponosov Y.N., Surin N.M., Muzafarov A.M. 2010 *Russ. Chem. Bull.* **59** 797

[6] Starikova T.Y., Surin N.M., Borshchev O.V., Pisarev S.A., Svidchenko E.A., Fedorov Y.V., Ponomarenko S.A. 2016 *J. Mater.Chem. C* **4** 4699 – 4708

[7] Skorotetcky M.S., Borshchev O.V., Cherkaev G.V., Ponomarenko S.A. 2019 *Russian Journal of Organic Chemistry* **55** 25 – 41

IOP Conf. Series: Materials Science and Engineering **848** (2020) 012012 doi:10.1088/1757-899X/848/1/012012

[8] Ponomarenko S.A., Surin N.M., Borshchev O.V., Skorotetcky M. S., Borshchev O.V., Pisarev S.A., Svidchenko E.A., Fedorov Yu.V., Molins F., Brixner T. 2019 *J. Mater.Chem. C* **7** 14612-14624

[9] Kato S.-I., Matsumoto T., Shigeiwa M., Gorohmaru H., Maeda S., Ishi I.T. and Mataka S. 2006 *Chemistry – A European Journal* **12** 2303-2317

[10] Skorotetcky M.S., Krivtsova E.D., Borshchev O.V., Surin N.M., , Svidchenko E.A., Fedorov Yu.V., Pisarev S.A., Ponomarenko S.A. 2018 *Dyes and Pigments* **155** 284-291

[11] Raimundo J., Blanchard P., Brisset H., Akoudad S., Roncali J., 2000 *Chem. Commun.* 939-940

[12] Justin T.K., Lin J., Velusamy M., Tao Y.-T., Chuen C.-H. 2004 *Adv. Funct. Mater.* **14** 83-90

[13] Ponomarenko S.A., Muzafarov A.M., Borshchev O.V., Vodopyanov E.A., Demchenko N.V., Myakushev V.D. 2005 *Russian Chemical Bulletin, International Edition* **54** 684—690

[14] Borshchev O.V., Ponomarenko S.A., Surin N.M., Kaptyug M.M., Buzin M.I., Pleshkova A.P., Demchenko N.V., Myakushev V.V., Muzafarov A.M. 2007 *Organometallics* **26** 5165-5173

[15] Polinskaya M.S., Borshchev O.V., Luponosov Y.N., Surin N.M., Muzafarov A.M., Ponomarenko S.A. 2011 *Mendeleev Commun.* **21** 89-91

[16] Ananikov V.P. et al 2014 *Russ. Chem. Rev.* **83** 885

[17] Shumilkina E.A., Borshchev O.V., Ponomarenko S.A., Surin N.M., Pleshkova A.P., Muzafarov A. M. 2007 *Mendeleev Commun.* **17** 34

[18] Crosby G.A., Demas J.N. 1971 *J Phys Chem* **75** 991–1024

[19] Boaz H., Rollefson G. K. 1950 *J. Am. Chem. Soc.* **72** 8 3435-3443

Modification of barium aluminosilicate glass-ceramics with zirconium oxide additives

A S Chaynikova, V S Kovaleva, D A Zabelin, I O Belachenkov and S Y Modin

All-Russian Scientific Research Institute of Aviation Materials, 105275 Moscow, Russia

Corresponding author: anna-chanikova@mail.ru

Abstract. Barium aluminosilicate glass-ceramics modified with zirconium oxide (ZrO_2) in the presence and without of stabilizing yttrium oxide was synthesized by the sol-gel method. The influence of the concentration of zirconium oxide on gelation, phase formation and sintering of materials were investigated. It was found that modification with zirconium oxide leads to a reduction in the gelation time of the initial solutions. It was shown that the introduction of ZrO_2 in an amount of up to 10 wt.% leads to the intensification of sintering of glass-ceramics. However, with a further increase in the ZrO_2 content to 15 wt.%, it is necessary to increase the sintering temperature to 1600°C. But this is not permissible, since it leads to not a desired metastable phase – hexacylsian becoming predominant in all samples. Only glass-ceramics modified with 10 wt.% ZrO_2 contains only the monocelsian phase in the entire sintering temperature range. The use of the sol-gel method made it possible to synthesize glass-ceramics with a fine-crystalline structure and a uniform volume distribution of ZrO_2 grains.

1. Introduction

Barium aluminosilicate glass-ceramics is promising to use in aerospace technology as a structural material as a structural material that is operated at high temperatures and in aggressive environments [1-13]. This is because the crystalline phase dominant in this glass-ceramics – monocelsian ($BaAl_2Si_2O_8$), has a low density (3.39 g/cm3) and a high melting point – 1760°C [14]. However, low fracture toughness (critical stress intensity factor $K_{IC} < 2.5$ MPa · m1/2) significantly limits the use of these materials. A possible way to solve this problem is to modify barium aluminumsilicate glass-ceramics with various fillers, that is, the creation of glass-ceramic composite materials [15-17].

The most promising fillers are particles of refractory oxides, in particular zirconium oxide. The uniqueness of zirconia is its ability to perform a local martensitic phase transformation from the tetragonal phase to the monoclinic phase under the action of tensile stresses in the region of microconcentrators (in the region of microcracks, at the boundaries of particles of strengthening phases, etc.). During synthesis of materials it is necessary to ensure the stabilization of the high-temperature tetragonal form of ZrO_2, including with the use of stabilizing oxides (Y_2O_3, MgO, CeO). The local phase transition is accompanied by the development of shear and volume deformations that ensure stress relaxation and closure of the crack surface. As a result, microcracks become stable while maintaining the level of external load. This effect was called transformational reinforcement [18, 19].

Works on the use of ZrO_2 particles as a reinforcing fillers were carried out back in the 1980s on the example of aluminum oxide ceramics and magnesium aluminum silicate glass-ceramics (cordierite glass-ceramics). For example, it was shown in [18] that the introduction of 10-15 vol.% Y_2O_3-stabilized (2.2-2.5 mol% Y_2O_3) ZrO_2 made it possible to increase the fracture toughness of cordierite glass-ceramics by more than 50% and the change of the dielectric properties of material were insignificant. Reinforcement of aluminum oxide ceramics with particles of Y_2O_3 stabilized (3 mol% Y_2O_3) ZrO_2 in an amount of 15% by weight using the sol-gel method resulted in an increase in its critical stress intensity factor and bending strength up to 9.23 MPa·m$^{1/2}$ and 856 MPa, respectively, which is 2.4 and 1.3 times higher in comparison with the initial ceramic [19].

The most promising method for the synthesis of modified barium aluminumsilicate glass-ceramics is the sol-gel method, since it allows one to obtain materials with a fine crystalline structure and uniform distribution of modifying oxides in glass-ceramics, which provides their higher mechanical and thermal properties.

The goal of our study was sol-gel synthesis of barium aluminosilicate glass-ceramics modified with zirconium oxide (ZrO_2) in the presence and without of stabilizing yttrium oxide and study of the influence of the concentration of zirconium oxide on gelation, phase formation and sintering of materials.

2. Experimental

For the synthesis of sol-gel solutions, we used the composition of monocelsian ($BaO \cdot Al_2O_3 \cdot 2SiO_2$, BAS) with ZrO_2 additives in an amount of 5-15 wt.% in the presence and without stabilizing yttrium oxide (in an amount of 3 mol.% of the contents of ZrO_2). Solutions were prepared by sequentially adding components.

The gelation of solutions was studied by constructing kinetic curves of viscosity changes. The gelation time was determined visually by the meniscus deflection for 1 min. when tilting the glass with the gel at an angle of 45°. As the main methods of study were used differential scanning calorimetry and thermogravimetry (DSC/TG), X-ray phase analysis (XRD), and dilatometric method for determining the shrinkage during sintering. The analysis of the microstructure of materials was carried out by using scanning electron microscopy and energy dispersive X-ray spectroscopy. Apparent density was measured by using the Archimedes principle.

3. Results and discussion

A study of the kinetic curves of the change in the viscosity of solutions during gelation (Fig. 1) showed that two sections are observed on all curves. The first one, almost parallel to the abscissa axis, corresponds to a uniform course of hydrolysis of silicon alcoholates and condensation of the resulting monomers into dimers. The second, characterized by an intensive increase in viscosity, is associated with the polycondensation reaction with the formation of a three-dimensional network structure via ≡Si–O–Si≡ bonds.

It was found that the introduction of ZrO_2 in an amount of 5 wt.% Contributes to a significant reduction in the time of hydrolysis and condensation. This leads to a decrease in the total gelation time from 380 h to 23 h (Figure 1). However, with an increase in the content to 15 wt.%, This effect weakens. The presence of yttrium oxide has little effect on the kinetics of viscosity changes.

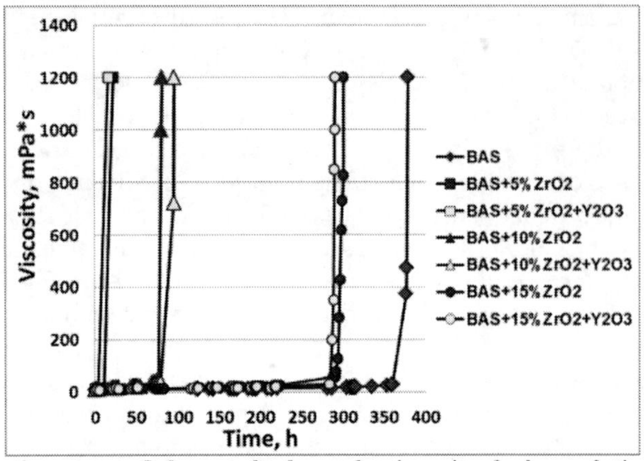

Figure 1. Kinetic curves of changes in dynamic viscosity during gelation of solutions.

After hydrolysis and polycondensation were completed, the gels were dried. Studying the dried gels by the DSC/TG method showed that in the temperature range 100-1000 ° C crystallization water and solvent are removed from them, and inorganic salts of the starting elements decompose as well. The occurring physicochemical transformations are accompanied by significant mass losses (about 45%), which stabilize only at a temperature of 1000°C. In the temperature range 1020-1450 ° C, gel crystallization is observed with the formation of monocelsian through successive transformations of barium silicates and hexacelsian. With the introduction of refractory zirconium oxide, the crystallization peaks in the thermograms of the samples become blurred, which makes their identification difficult.

According to the DSC/TG results, the dried gels were calcined and heat treated at temperatures of 1000°C and 1500°C, respectively. The resulting powders were ground in a planetary mill to a particle size of 3 μm.

The study of the curves of linear shrinkage of samples during heat treatment in the temperature range 25–1600 °C (Figure 2) showed that their sintering occurs due to the mechanism of viscous flow and is accompanied by shrinkage of 7–13% in the temperature range 1300–1600 °C. The maximum sintering speed is achieved in the range of 1500-1600 °C. It was found that the introduction of up to 10 wt.% ZrO_2 leads to an intensification of the sintering process of glass-ceramics. So, for example, the shrinkage achieved during sintering of barium aluminumsilicate glass-ceramics with 10 wt.% zirconium oxide increases by 1.7 times. However, a further increase in the concentration of zirconium oxide, characterized by a high melting point (2715°C), up to 15 wt.% leads to a shift in the maximum sintering speed to a higher temperature region.

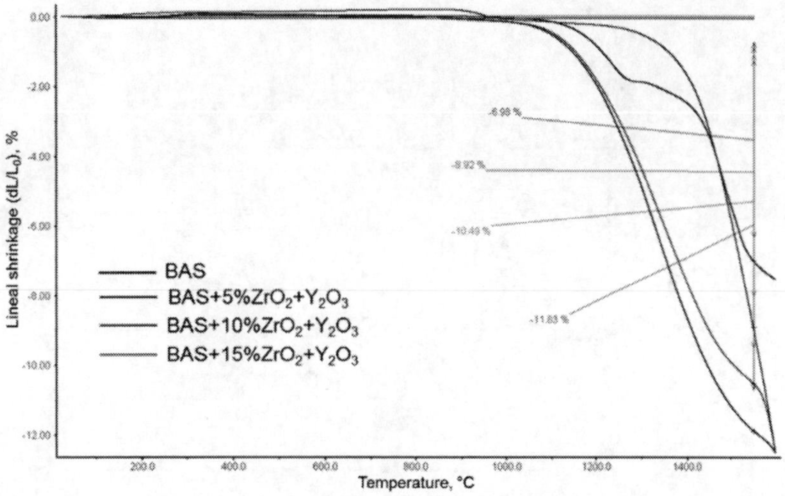

Figure 2. Linear shrinkage curves of samples containing zirconium and yttrium oxides depending on the concentration of the additive.

The results of studying the sintering of modified glass-ceramics by isothermal firing of samples in the temperature range 1500-1650 °C are consistent with the data of high-temperature dilatometry. Thus, it was shown that the introduction of ZrO_2 up to 10 wt.% helps to reduce the open porosity of barium aluminosilicate glass-ceramics. With the joint introduction of zirconium and yttrium oxides, this effect is enhanced. For samples containing up to 10 wt.% zirconium oxide additives in the presence and without yttrium oxide, the maximum density can be achieved by sintering at 1550°C. With an increase in the ZrO_2 content to 15 wt.%, it is necessary to increase the sintering temperature to 1600°C.

The study of sintered samples by XRD showed that, regardless of the concentration of ZrO_2 after firing at 1550°C, the main phase in the samples is monocelsian (m-$BaAl_2Si_2O_8$), which is consistent with the previously considered DSC results. However, in most samples there is also a metastable form – hexacylsian (h-$BaAl_2Si_2O_8$), which becomes predominant with an increase in firing temperature to 1600°C. According to published data, the hexacylsian phase is undesirable for high-temperature glass-ceramics, since it is characterized by a high thermal expansion coefficient (~ (8.0×10^{-6}) K^{-1}) and polymorphic transformation into an orthorhombic modification at a temperature of about 300°C, which is accompanied by a volume change of ~ 3% [13]. Thus, it was found that the sintering temperature of the modified barium aluminosidicate glass-ceramics should not exceed 1550°C. It was also revealed that glass-ceramics modified with 10 wt.% ZrO_2 is the most promising, since only the monocelsian is present in it at all sintering temperatures (Figure 3). It was established that, regardless of its concentration, zirconium oxide is present in glass-ceramics in monoclinic and tetragonal modifications.

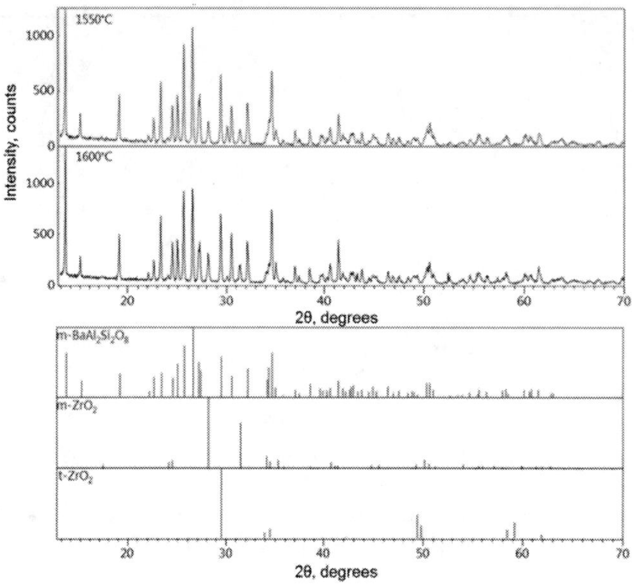

Figure 3. XRD results of samples containing 10 wt.% ZrO_2 after sintering at 1550°C and 1600°C (m – monoclinic, t – tetragonal).

A study by scanning electron microscopy (Figure 4) showed that all samples, regardless of ZrO2 concentration and sintering temperatures (1550°C or 1600°C), are characterized by a fine-crystalline structure with elongated celsian grains and zirconium grains. The grain sizes are at the level of micrometer units and increase slightly with increasing sintering temperature. All samples contain pores with sizes from 1 to 10 µm.

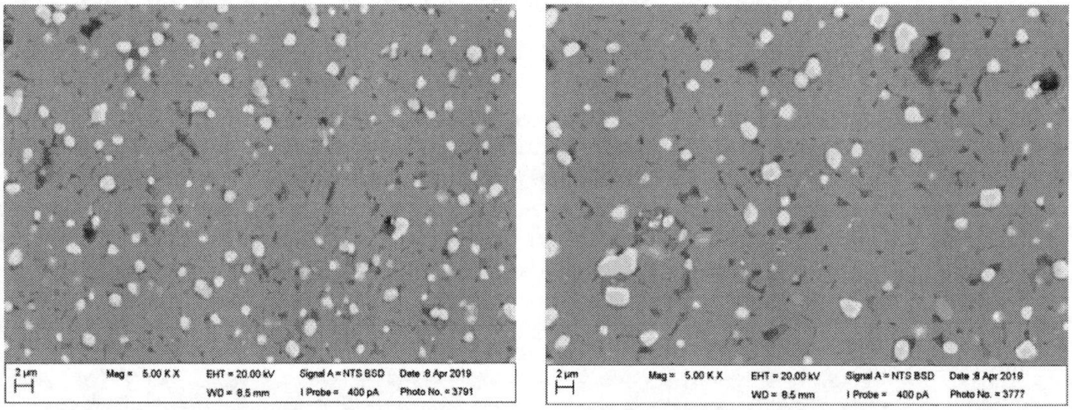

(a) Sample obtained by sintering at 1550°C (b) Sample obtained by sintering at 1550°C

Figure 4. Scanning electron microscopy images of glass-ceramic samples modified with 10 wt.% zirconium oxide obtained by sintering at 1550°C and 1600°C.

According to the results of the study of sintered samples by energy dispersive X-ray spectroscopy (Fig. 5), it was found that the use of the sol-gel method made it possible to create uniform distribution of zirconium oxide grains in the volume of glass-ceramics. It was shown that yttrium oxide is also distributed over the volume of materials; however, foci of its concentration near zirconium oxide grains are observed. This confirms the effect of Y_2O_3 as a stabilizer of the tetragonal form of ZrO_2 in glass-ceramics.

IOP Conf. Series: Materials Science and Engineering **848** (2020) 012013 doi:10.1088/1757-899X/848/1/012013

(b) Energy dispersive x-ray spectroscopy compositional map of the Zr (orange)

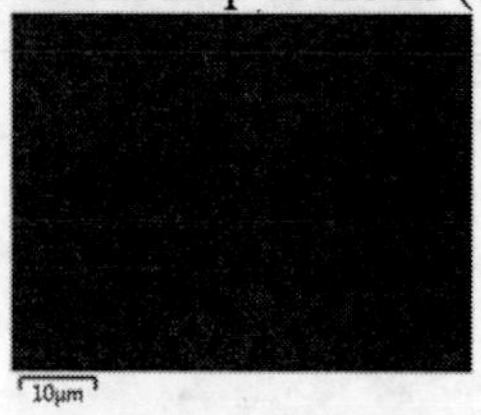

(c) Energy dispersive x-ray spectroscopy compositional map of the Y (red)

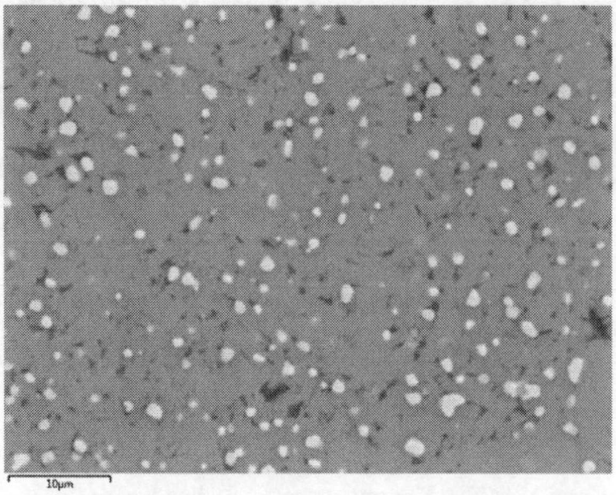

(a) Scanning electron microscopy image

Figure 5. The results of the study of polished specimens of glass-ceramics modified with 10 wt.% ZrO_2 in the presence of yttrium oxide, obtained by sintering at 1550°C, by scanning electron microscopy and energy dispersive X-ray spectroscopy.

4. Conclusions

Thus, the barium aluminum silicate glass-ceramics modified with zirconia additives in the presence and without stabilizing yttrium oxide were synthesized by the sol-gel method, and the influence of the ZrO_2 concentration on gelation, phase formation, and sintering of materials was studied.

It was found that the introduction of ZrO_2 reduces the gelation time of solutions. This is manifested to the greatest extent with the introduction of small amounts of ZrO_2 (5 wt.%). Crystallization of gels occurs in the temperature range 1020-1450 °C with the formation of monocelsian through successive transformations of barium silicates and hexacelsian. With the introduction of refractory zirconium oxide, the crystallization peaks in the thermograms of the samples become blurred, which makes their identification difficult.

It was shown that the introduction of ZrO_2 in an amount of up to 10 wt.% leads to an intensification of sintering of barium aluminumsilicate glass-ceramics. With the joint introduction of zirconium and yttrium oxides, this effect is enhanced. For samples containing up to 10 wt.% zirconium oxide additives in the presence and without yttrium oxide, the maximum density can be achieved by sintering at 1550°C. With an increase in the ZrO_2 content to 15 wt.%, it is necessary to increase the sintering temperature to 1600°C. However, according to the XRD results, this is not permissible, since it leads to the predominance of an undesirable hexacylsian phase in all samples. Glass-ceramics modified by the addition of 10 wt.% ZrO_2 is characterized by the presence of only the monocelsian phase in the entire temperature range of sintering.

It is shown that the synthesized samples are characterized by a fine-crystalline structure with grain sizes at the level of micrometer units. The use of the sol-gel method made it possible to ensure uniform distribution of zirconium oxide grains in the volume of glass-ceramics. It was established that yttrium oxide is predominantly concentrated near zirconium oxide grains. This confirms the effect of Y_2O_3 as a stabilizer of the tetragonal form of ZrO_2 in glass-ceramics.

Acknowledgements

The authors are grateful to Zhitnyuk S.V. and Zhuravleva P.L. (All-Russian Scientific Research Institute of Aviation Materials, Moscow, Russia) for scanning electron microscopy, energy dispersive X-ray spectroscopy and X-ray phase analysis of samples.

IOP Conf. Series: Materials Science and Engineering **848** (2020) 012013 doi:10.1088/1757-899X/848/1/012013

The study was carried out by a grant from the Russian Science Foundation (Project № 18-73-00325).

References

[1] Beall G.H. 2009 J. of the European Ceramic Society. **29** 1211–1219
[2] Bansal N.P. 1997 *J. of Materials Research*. **12(3)** 745–753
[3] Roether J.A., Boccaccini A.R. 2005 *Handbook of ceramic composites*. Boston: Kluwer Academic Publishers 485–511
[4] Ye F., Liu L., Wang Y., Zhou Y., Peng B., Meng Q. 2006 *Scripta Materialia*. **55** 911–914
[5] Liu L., Ye F., Zhou Y., Zhang Z. 2010 *Scripta Materialia*. **63** 166–169
[6] Academician Sarkisov P.D., Paleari A., Chainikova A.S., Orlova L.A., Popovich N.V., Filippov V.V. 2012 *Doklady Chemistry*. **446(2)** 220–222.
[7] Kablov E.N., Grashchenkov D.V., Isaeva N.V., Solntsev S.S., Sevast'yanov V.G. 2012 *Glass and Ceramics*. **69(3-4)** 109–112
[8] Kablov E.N., Grashchenkov D.V., Isaeva N.V., Solntsev S.S. 2010 *Ros. Khim. Zh.* **LIV(1)** 20–24
[9] Grashchenkov D.V., Vaganova M.L., Schegoleva N.E., Chainikova A.S., Lebedeva Yu.E. 2017 *Aviats. Mater. Tekhnol.* **S** 290–305 DOI: 10.18577/2071-9140-2017-0-S-290-305
[10] Chainikova A.S., Orlova L.A., Popovich N.V., Lebedeva Yu.E., Solntsev S.S. 2014 *Aviats. Mater. Tekhnol.* **3** 45–54 DOI: 10.18577/2071-9140-2014-0-3-45-54
[11] Chainikova A.S., Orlova L.A., Popovich N.V., Lebedeva Yu.E., Solntsev S.S. 2014 *Aviats. Mater. Tekhnol.* **S6** 52–58 DOI: 10.18577/2071-9140-2014-0-s6-52-58
[12] Sung Y.M., Kim S. 2000 *J. of Materials Science*. **35** 4293–4299
[13] Narottam P. Bansal. 2005 *Handbook of Ceramic Composites Edited by Narottam P. Bansal. Kluwer academic publisher: Norwell, MA.* 554
[14] Chainikova A.S., Voropaeva M.V., Alekseeva L.A., Orlova L.A., Samsonov V.I. 2014 *Aviats. Mater. Tekhnol.* **S6** 45–51 DOI: 10.18577/2071-9140-2014-0-s6-45-51.
[15] Kablov, E.N., Shchetanov, B.V., Ivakhnenko, Yu.A., Balinova, Yu.A., 2013 *Trudy VIAM*. **2** 5 (viam-works.ru)
[16] Ye F., Liu L., Zhang J., Meng Q. 2008 *Composites Science and Technology*. **68** 1073–1079
[17] Ye F., Liu L., Zhang J., Iwasa M., Su C.-L. 2005 *Composites Science and Technology*. **65** 2233–2239
[18] Hannink R.H.J., Kelly P.M., Muddle B.C. 2000 *J. Am. Ceram. Soc.* **83** 461–487
[19] Evans A.G., Cannon R.M. 1986 *Acta Metall.* **34(5)** 761–800

IOP Conf. Series: Materials Science and Engineering **848** (2020) 012014 doi:10.1088/1757-899X/848/1/012014

Refractive index of fluoride crystals doped with rare earth ions with low concentration

D S Chunaev*, **A Y Karasik, V A Konyushkin, A N Nakladov and P G Zverev**

A.M. Prokhorov General Physics Institute of RAS, 38 Vavilov str., Moscow, 119991, Russia

*Corresponding author: chunaev@lst.gpi.ru

Abstract. Refractive indices of SrF_2 crystals doped with low concentrations of holmium, erbium, and thulium ions are investigated. These crystals are attractive as active laser media for developing planar waveguides for near and mid-IR spectral regions. It is shown that, by varying the concentration of impurity rare-earth ions in SrF_2 crystals, it is possible to increase and control the refractive index of the medium. This allows one to develop planar waveguides with the core and shell of the same fluoride and control theirs numerical aperture and the number of excited modes. The design of single mode planar waveguide is calculated and proposed for the wavelength of 2 μm.

1. Introduction

Development of integral and fiber optics and optical computers requires compact solid state laser oscillators and amplifiers operating in a broad spectral range from near UV to mid IR. There are optical glass fibers doped with rare earth ions that satisfy these requirements. Due to their long length, they provide high gain and can control the radiation intensity. Meanwhile, the transparency range of glass fibers, which is 0.4÷1.8 μm, is narrower than that of fluoride materials (0.16÷11 μm). This considerably limits their possible applications in the near and mid-IR spectral region. The small core and low concentration of active dopants in fibers limit the efficiency of fiber amplifiers. Crystalline planar waveguides similar to bulk crystals can be doped with high concentration of impurities in the core area [1-5]. Planar waveguides have very small diffraction losses, which allows one to increase the length of interaction between the laser and pump radiation and thus to achieve higher amplification gain comparable with that in crystalline media. At the same time, planar waveguides conserve the linear polarization of the radiation and exhibit good thermal physical properties. In a simple planar waveguide, one can excite a large number of spectrally and spatially split channels which can increase the amplification efficiency via further coherent summation [6, 7].

During two decades, one observes a strong increase in the number of investigation on development, characterization, and application of planar waveguides [1-3]. Optimization of active media and cavity design allowed creating a great number of compact highly efficient laser sources. Planar waveguide lasers can operate in various spectral regions, in cw and pulsed temporal regimes with pulse durations up to tens of femtosecond. The research of planar waveguides is mostly restricted to oxide crystals which are usually produced by the molecular-beam epitaxial technique. It provides a very thin waveguide structure with dimensions of several microns. Recently, irradiation of crystals by femtosecond lasers was proposed as an efficient technique of planar waveguide production [4]. Such technique allows creating structures with various sizes and shapes. It was shown that highly effective waveguide lasers can also be prepared from oxide ceramics [5]. In contrast to oxide crystals, ceramics can be prepared with required shape, high concentration of doped ions, high distribution coefficient, and good physical and optical properties [8].

Since the invention of the first laser, optical materials based on fluorides doped with rare earth ions attracted great interest as solid state laser active materials [9, 10]. They have some advantages compared with oxides, specifically for developing of a wide range of lasers and amplifiers [11]. Fluoride crystals can be doped with high concentrations of rare earth ions up to 10^{21} cm^{-3} without considerable nonradiative quenching. They exhibit shorter phonon spectra than oxides, which allows one to reduce the probability of multiphonon relaxation processes, considerably diminish the nonradiative losses, and obtain laser radiation in mid-IR spectral region with high efficiency and low thermal losses [12, 13]. Fluoride materials have lower linear and nonlinear refractive indexes, which makes it possible to reduce the probability of various nonlinear processes, such as SRS, Brillouin scattering, etc., at high pump intensity [14]. In addition, fluorides have high mutual solubility in systems like $Me^{(1)}F_2$-$Me^{(2)}F_2$, as well as in $Me^{(1)}F_2$-$Me^{(2)}F_3$, which allows one to have different spectral properties of rare earth ions due to different local surrounding fields in more complex systems $Me^{(1)}F_2$-$Me^{(2)}F_2$-$Me^{(3)}F_3$ [15]. It is important to note that solid solutions $Me^{(1)}F_2$-$Me^{(2)}F_3$ keep a cubic structure for concentrations of $Me^{(2)}F_3$ up to 20%, which allows obtaining nanoceramic samples [16]. Fluoride nanoceramics has some advantages over similar single crystals due to its higher mechanical strength and fracture toughness [17]. Recently, four-component BaF_2–ZnF_2–CdF_2-YbF_3 fluoride nanoceramics was synthesized [18].

A larger relaxation lifetime of rare earth ions in fluorides allows them to accumulate a high population inversion, which simplifies the development of laser amplifiers excited by longer laser pulses. The thermal conductivity of fluoride materials is also rather high. It is necessary to note that, even at low concentrations, rare earth ions in fluorides create different optical centers with various luminescence properties [19, 20]. This proves that the local surrounding of rare earth ions can differ from center to center.

The refractive indices of nanostructured crystals composed of solid solutions $Me^{(1)}F_2$-$Me^{(2)}F_3$ (where $Me^{(1)}$ is Ca, Sr, Ba, Cd, or Pb ion and $Me^{(2)}$ is a rare-earth ion with the fraction from 10 to 50%) were recently investigated [21, 22]. It was shown that the refractive indices of two-component solid solutions linearly depend on the concentration of rare earth ions and can be calculated using a method of molecular refraction additivity for components with a rather high accuracy. The value can be determined by separating crystals into the components related to $Me^{(1)}F_2$, for which refractive indices are known, and $Me^{(2)}F_3$, for which the impact in refractive index linearly depends on the atomic number of $Me^{(2)}$ in the periodic table. Doping of the fluoride crystal with rare earth ions results in the change of the lattice constant due to various sizes of $Me^{(1)}$ and $Me^{(2)}$ ions. The authors of [21, 22] investigated considerable changes of the refractive index in highly doped solid solutions. In $Me^{(1)}F_2$-$Me^{(2)}F_3$ crystals with a low doping concentration, due to unequal ionic charges, rare earth ions can create double and triple optical centers, which are clearly registered by site-selective laser spectroscopy [23]. Thus, due irregular positions of $Me^{(2)}$ ions, the concentration dependence of the refractive index can be not linear.

The aim of this research was to investigate the concentration dependence of the refractive index of SrF_2 crystal doped with Tm, Ho, and Er ions with concentrations as low as $0.5 \div 5\%$ and to analyze the possibility of preparing single-mode planar waveguides composed of the shell of undoped SrF_2 crystal and the core of SrF_2 crystal doped with a low concentration of rare earth ions.

2. Investigation of the refractive index of SrF_2 crystal doped with Tm, Ho, and Er ions

In this paper, we present the results on measurement of the refractive indices of strontium fluoride crystals containing small concentrations of rare earth ions, which are usually used for laser-active media. Tm, Ho, and Er ions were chosen as they are active impurities used in lasers operating in the near-infrared spectral region $1.5 \div 3$ μm. The refractive indices were measured using an URL-2 refractometer, which allows one to measure refractive indices of liquids and solids at the Fraunhofer line D ($\lambda = 589$ nm) with an accuracy up to 0.0002 [24].

IOP Conf. Series: Materials Science and Engineering **848** (2020) 012014 doi:10.1088/1757-899X/848/1/012014

The high-quality SrF_2 crystals were grown by the Bridgman technique from the melt in a fluoride atmosphere of CF_4. The impurity ions were added in the form of rare-earth fluorides. The raw materials were prepared by appropriate mixture of initial fluoride powders with the purity of 99,99% and put into a graphite crucible heated up to the melting temperature of 1465°C. After complete melting, which was controlled visually, the crucible was pulled out with a speed of 5-6 mm per hour. Annealing of the crystal continued during 4 hours. Up to seven single crystal samples were obtained during one crystal growth process. For refractive index measurements, we made samples in the form of parallelepipeds with a length of 20 mm and transverse dimensions of 5 x 5 mm. All surfaces of the samples were polished.

Table 1. Refractive indices of SrF_2 crystals doped with Ho, Er, and Tm, their differences from that of undoped SrF_2, and ratio of the core size and the oscillation wavelength for a single-mode waveguide.

Material	Refractive index, *n* (589 nm)	Δn	d/λ
SrF_2	1.4380±0.0002	0	-
SrF_2+2 mol.%TmF_3	1,4416±0.0002	0.0036	4.9
SrF_2+4 mol.% TmF_3	1.4453±0.0002	0.0073	3.4
SrF_2+2 mol.% ErF_3	1,4411±0.0002	0.0031	5.3
SrF_2+4 mol.% ErF_3	1,4451±0.0002	0.0071	3.5
SrF_2+0.5 mol.% HoF_3	1,4389±0.0002	0.0009	10
SrF_2+1 mol.%Ho F_3	1,4401±0.0002	0.0021	6.4

The refractive index of the SrF_2 crystal at a wavelength of 589 nm was measured to be $n = 1.4380 \pm 0.0002$, which agrees well with the results published in [22]. Doping of the rare-earth fluoride as a second component in the SrF_2-$Me^{(2)}F_3$ crystal leads to an increase in the refractive index. The measured refractive indices for SrF_2 crystals with different molar concentrations of $Me^{(2)}F_3$ fluorides, where $Me^{(2)}$ stands for Ho, Er, Tm, are shown in table 1 and presented in figures 1-3. The dashed curves present the linear dependencies calculated according to the results described in [21]. One can see that the refractive index exhibits a rather linear behaviour in SrF_2 crystals with low concentrations of rare-earth impurities. However, the investigated samples have slightly lower values of refractive indexes than those calculated by the method of molecular refraction additivity for the components. This can be attributed to the presence of double and triple optical centers recently discovered in fluorides [18-20].

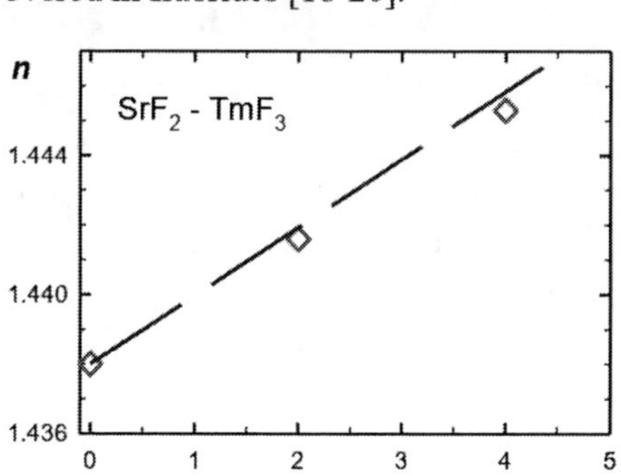

Figure 1. Dependence of the refractive index of SrF_2 – TmF_3 crystal on the molar concentration of TmF_3.

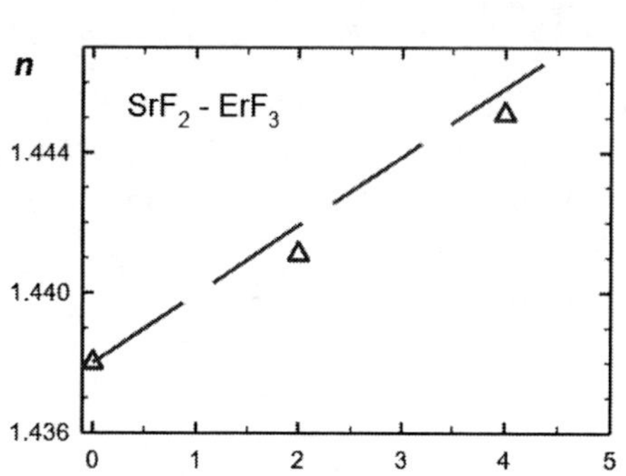

Figure 2. Dependence of the refractive index of SrF_2 – ErF_3 crystal on the molar concentration of ErF_3.

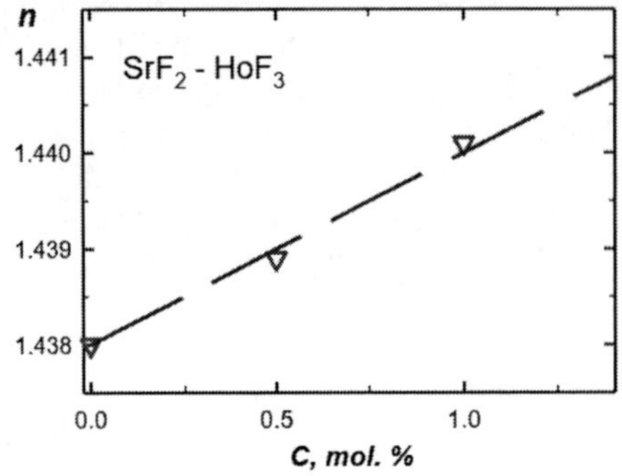

Figure 3. Dependence of the refractive index of SrF_2 – HoF_3 crystal on the molar concentration of HoF_3.

The obtained data show that the refractive index of SrF_2-$Me^{(2)}F_3$ crystal can be controlled by changing the $Me^{(2)}$ concentration, which makes it possible to develop planar waveguides with the core and reflective cladding made of one and the same fluoride. This opens the possibility of controlling the properties of planar waveguides, namely, numerical aperture and number of excited modes, which is demonstrated below.

3. Planar waveguide

The laser operation of planar waveguide based on a LaF_3:Nd^{3+} layer grown on a CaF_2 substrate by molecular-beam epitaxy was demonstrated for the first time in 1999 [25]. Recently, laser operation of a hot-pressed planar waveguide based on SrF_2 ceramics with a neodymium-doped core was described [26]. However, these researches used multimode waveguides. One of the key problems of waveguide development is to develop a single-mode planar waveguide with a maximum size of the core. This allows working with a high laser energy and a good divergence of the output beam. If we consider a one-dimensional case, the number of modes of a symmetrical planar waveguide is determined by parameter p, which is determined as [2]:

$$p = \frac{2d \cdot NA}{\lambda}, \tag{1}$$

where d is the waveguide thickness, λ is the operation wavelength, and NA is the numerical aperture. The NA of a waveguide with a simple step-index structure in air is

$$NA = \sqrt{n_2^2 - n_1^2}, \tag{2}$$

here n_2 is the refractive index of the core and n_1 is the refractive index of the waveguide shell. For a single-mode waveguide, parameter p should be $p \leq 1$. Thus, from (1) and (2) we find that the ratio of the maximal core size of a single-mode waveguide to the operation wavelength is determined by the difference of refractive indices of the core and the shell Δn as

$$\frac{d}{\lambda} = \frac{1}{2\sqrt{2n \cdot \Delta n}}, \tag{3}$$

Figure 4 presents the dependence of the ratio d/λ on Δn for a SrF_2 single-mode waveguide. This ratio calculated for investigated samples used as a core material is presented in table 1. It is clear that, the core dimension for a single-mode waveguide can be increased by using the core and shell made of materials with close values of refractive indices. In this paper, we propose to use SrF_2 crystal as a shell material and SrF_2-$Me^{(2)}F_3$ solid solution (where $Me^{(2)}$ stands for Tm, Ho, and Er ions with the low concentration of 0.5 ÷ 5%). These laser active ions oscillate in the wavelength region of 2-3 μm. As an example, we can consider a waveguide with the core made of a $SrF_2 - HoF_3$ crystal with a HoF_3 concentration of 0.5 mol %, which provides the difference between the refractive indices of 0.0009 and the ratio d/λ equals to 10. This gives the thickness of the core of a single-mode planar waveguide with a step-index structure operating at 2 μm to be about 20 μm. Due to the large thickness of the waveguide core, it is easier to fabricate it and it can amplify laser radiation to a higher output power due to a lower thermal load. It is necessary to note that further increase of the core size can be made by further reduction of the concentration of rare-earth fluorides and shifting the operation radiation to longer wavelengths. By varying the concentration of doping rare-earth ions in SrF_2 crystals, it is possible to control the refractive index of the medium and to develop single-mode planar waveguides with the core and the shell of the same fluoride.

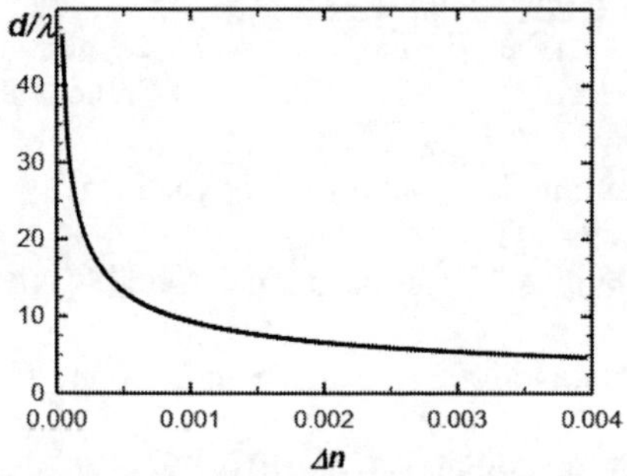

Figure 4. Dependence of the ratio of the core size of a SrF_2 single-mode waveguide to the operation wavelength on the difference between the refractive indices of the core and the shell.

4. Conclusion

The refractive indices of SrF_2 crystals doped with small concentrations of impurities of Ho, Er, and Tm ions were investigated. These crystals are well known as active media of lasers operating in the spectral region of 2-3 μm. It was shown that, by varying the concentration of impurity rare-earth ions in SrF_2 crystals, it is possible to precisely control the refractive index of the medium. The concentration dependences of the refractive index exhibit a linear behavior in SrF_2 crystals for all Ho, Er, and Tm impurities. However, they exhibit a slightly lower slope than those calculated by the method of molecular refraction additivity for the components for high doping levels. It was shown that these materials can be used for fabrication of planar waveguides with the core and shell of the same fluoride and for control of its numerical aperture and the number of excited modes. The design of SrF_2 single-mode planar waveguides with the core of the $SrF_2 - HoF_3$ crystal was calculated and proposed for a waveguide laser emitting at a wavelength of 2 μm.

References

[1] Mackenzie J I 2007 *IEEE J. Sel. Topics Quantum Electron.* **13** 626
[2] Grivas C 2011 *Progress Quantum Electronics* **35** 159
[3] Grivas C 2016 *Progress Quantum Electronics* **45** 3
[4] Chen F and V'azquez de Aldana J R 2014 *Laser Photonics Review* **8** 251
[5] Salamu G, Jipa F, Zamfirescu M and Pavel N 2014 *Optics Express* **22** 5177
[6] Thielen P A, Ho J G, Burchman D A, Goodno G D, Rothenberg J E, Wickham M G, Flores A, Lu C A, Pulford B, Robin C, Sanchez A D, Hult D and Rowland K B 2012 *Optics Letters* **37** 3741
[7] Pyrkov Yu N, Trikshev A I and Tsvetkov V B 2012 *Quantum Electronics* **42** 790
[8] Ikesue A 2002 *Optical Materials* **19** 183
[9] Batygov S K, Kulevskii L A, Prokhorov A M, Osiko V V, Savel'ev A D and Smirnov V V 1975 *Sov. J. Quantum Electronics* **4** 1469

IOP Conf. Series: Materials Science and Engineering **848** (2020) 012014 doi:10.1088/1757-899X/848/1/012014

[10] Moncorge R, Braud A, Camy P, Doualan J L 2013 *Handbook of solid-state lasers: materials, systems and applications*, eds B Denker and E Shklovsky (Amsterdam: Elsevier) p 28

[11] Lucca A, Jacquemet M, Druon F, Balembois F, Georges P, Camy P, Doualan J L and Moncorge R 2004 *Optics Letters* **29** 1879

[12] Basiev T T, Dergachev A Y, Kirpichenkova E O, Orlovskii Y V and Osiko V V 1987 *Sov. J. Quantum Electronics* **17** 1289

[13] Orlovskii Y V, Basiev T T, Pukhov K K, Vorobiev I N, Papashvili A G, Pelle F and Osiko V V 2001 *J. Luminescence* **94** 791

[14] Joubert M F and Moncorgé R 2003 *Optical Materials* **22** 95

[15] Basiev T T, Doroshenko M E, Fedorov P P, Konyushkin V A and Osiko V V 2007 *Quantum Electronics* **37** 934

[16] Fedorov P P, Osiko V V, Basiev T T, Orlovskii Yu V, Dukel'skii K V, Mironov I A, Demidenko V A and Smirnov A N 2007 *Russian Nanotechnologies* **2** 95

[17] Doroshenko M E, Demidenko A A, Fedorov P P, Garibin E A, Gusev P E, Jelinkova H, Konyshkin V A, Krutov M A, Kuznetsov S V, Osiko V V, Popov P A and Shulc J 2013 *Phys. Stat. Solidi C* **10** 952

[18] Alimov O K, Doroshenko M E, Pierpoint K A, Komandin G A, Nozdrin V S, Buchinskaya I I, Popov A I and Fedorov P P 2019 *Optical Materials* **94** 113

[19] Doroshenko M E, Pierpoint K A, Alimov O K, Papashvili A G, Konyushkin V A and Nakladov A N 2019 *J. Luminescence* **208** 475

[20] Doroshenko M E, Papashvili A G, Konyushkin V A, Nakladov N A, Martynova K A and Osiko V V 2018 *J. Luminescence* **199** 331

[21] Konstantinova A F, Krivandina E A, Karimov D N and Sobolev B P 2010 *Crystallography Reports* **55** 990

[22] Glushkova T M, Karimov D N, Krivandina E A, Zhmurova Z I and Sobolev B P 2009 *Crystallography Reports* **54** 603

[23] Alimov O K, Doroshenko M E, Konyushkin V A, Papashvili A G and Osiko V V 2016 *Quantum Electronics* **46** 68

[24] Karasik A Y, Konyushkin V A, Nakladov A N and Chunaev D S 2018 *Quantum Electronics*, **48** 854

[25] Daran E, Shepherd D P, Bhutta T and Serrano C 1999 *Electron. Lett.* **35** 398

[26] Konyushkin V A, Nakladov A N, Konyushkin D V, Doroshenko M E, Osiko V V and Karasik A Y 2013 *Quantum Electronics* **43** 60

IOP Conf. Series: Materials Science and Engineering **848** (2020) 012015 doi:10.1088/1757-899X/848/1/012015

Functional properties and microstructure of TiNi alloy during multiple martensitic transformations

A A Churakova[1,2*]

[1] Institute of Molecule and Crystal Physics - Subdivision of the Ufa Federal Research Center of the Russian Academy of Sciences, 151 pr. Oktyabrya, Ufa, 450075, Russia
[2] Ufa State Aviation Technical University, 12 K. Marx str., Ufa, 450008, Russia
* churakovaa_a@mail.ru

Abstract. The influence of multiple martensitic transformations on the functional properties and microstructure of the equiatomic TiNi alloy is considered. An analysis of mechanical tests showed that in the CG and UFG state in the $Ti_{50.0}Ni_{50.0}$ alloy as a result of thermal cycling with an increase in the number of cycles, a monotonic increase in the yield strength is caused by phase hardening. Moreover, in the UFG state, the yield strength increases by 185 MPa, which is 67.5 MPa more than the increase in σ_{YS} in the CG state. In this case, ductility remains at the level of 40%, which is somewhat lower than ductility in the CG state (~ 60%). The phase yield stress and the estimated reactive stress of the CG and UFG states as a result of the TC also increase, with σ_{reac} UFG reaching 830 MPa. The increase in the yield strength of the CG state in an overgrown alloy is 113 MPa, while in the UFG it is 155 MPa, in addition, an increase in the estimated reactive stress and the reversible deformation with an increase in the number of thermal cycling cycles.

1. Introduction

TiNi alloys with the shape-memory effect (SME) possess high functional characteristics, enhanced strength, ductility, corrosion resistance, biocompatibility, etc. [1-2]. They have a unique set of properties - high values of recoverable deformation and developed stresses, high corrosion resistance, due to which they are widely used in many industries and medicine [3-4].

At the same time, for many applications, especially in critical or small-sized products, the level of mechanical and functional properties possessed by TiNi alloys in the usual coarse-grained state is insufficient. Since physico-mechanical properties are structurally sensitive, methods of deformation-heat treatment are traditionally used to increase them, allowing to obtain various types of structures.

For these alloys, various metallurgical and metallurgical aspects are developed, new alloys and technologies for their processing are created, a systematic study of the structural and phase transformations in them, various properties, including the thermomechanical memory effects, unique in their temperature, deformation and power characteristics, is carried out.

In the development of products made of SMA (shape memory alloys), the following functional characteristics are most important: temperature interval of shape recovery, maximum reversible deformation, maximum fully reversible deformation, degree of shape recovery, similar characteristics of the reversible shape memory effect and superelasticity [1-10].

The heat treatment of this alloy requires high accuracy, as it greatly affects its properties. Heat treatment is the main means for finely controlling the temperature of the phase transformation. The duration and temperature of the heat treatment affect the deposition of various phases Ni-rich and thereby determine the amount of nickel deposited on the nickel-titanium crystal lattice; depleting the nickel matrix, heat treatment leads to an increase in the temperature of the phase transformation. The

IOP Conf. Series: Materials Science and Engineering **848** (2020) 012015 doi:10.1088/1757-899X/848/1/012015

combination of heat and cold treatment is the main means of regulating the properties of this alloy. An effective method of hardening of structural elements and increasing strength and functional characteristics is heat treatment and thermal cycling. Thermal cyclic treatment (TCT) is accompanied by multiple phase transformations during heating and cooling with optimal speeds.

Shape memory alloys have a unique ability to restore shape upon heating, and it is known that multiple heat exchanges through the temperature range of martensitic transformations lead to a change in their functional properties [1,2].

During thermal cycling in a free state (without stress), a decrease in the temperature of phase transitions and a change in their sequence are observed [11]. If external stress acted upon the sample during thermal cycling, along with a change in the characteristics of martensitic transformations, the parameters of the shape memory effects change, and one-sided irreversible deformation accumulates.

Thermal cycling can have a beneficial effect on the formation of the structure and properties of metallic materials and alloys. In this regard, thermal cycling is included as one of the operations in the production technology of materials. It is important to determine the effect of multiple martensitic transformations on the functional properties and microstructure of TiNi alloy.

2. Material and methods

As the study material it chose a two-component alloy of the TiNi system - an alloy of the equiatomic composition $Ti_{50}Ni_{50}$, which has the structure of B19′-monoclinic martensite at room temperature, the transformation temperatures are M_s = 63 °C, M_f = 40 °C, A_s = 94 °C, A_f = 110 °C. To form a solid solution the TiNi alloy were quenched from the homogeneity region (from 800 °C) in water. Thermal cycling on the samples in different initial states is performed via consecutive dipping of samples in liquid nitrogen (-196 °C) with subsequent heating to 150 °C on the electric heater, which respectively is lower and higher than the temperatures of direct and reverse martensitic transformations. Samples with a section of less than 1 mm are subjected to thermal cycling, which ensures quick heating and cooling of samples. The number of thermocycles “heating-cooling” varies from 0 to 100. The dwell time for heating and cooling is 5 minutes. The fine structure of the material was studied with a JEM-2100 transmission electron microscope (TEM) at an accelerating voltage of 200 kV. The foils for electron-microscopic studies were prepared on a TenuPol-5 twin jet electropolishing set according to the standard procedure with the help of the electrolyte 10%$HClO_4$ + 90%$CH_3(CH_2)_3OH$ (90% butanol). Tensile mechanical tests of small-sized flat samples with a gauge portion of 1*0.25*4 mm were conducted at room temperature with a strain rate of $1*10^{-3}$ s^{-1} on a Shimadzu AG-50kNXD machine.

3. Results and discussions

First of all, consider the structure of the alloy in the coarse-grained state after quenching. In the initial quenched state in the equiatomic alloy $Ti_{50.0}Ni_{50.0}$, large austenitic grains with a size of about 200±10 microns with clear boundaries are observed (figure 1, a). According to the results of optical metallography, a surface relief in the form of a set of parallel strips from the plates of the martensitic phase is detected inside them (figure 1, b).

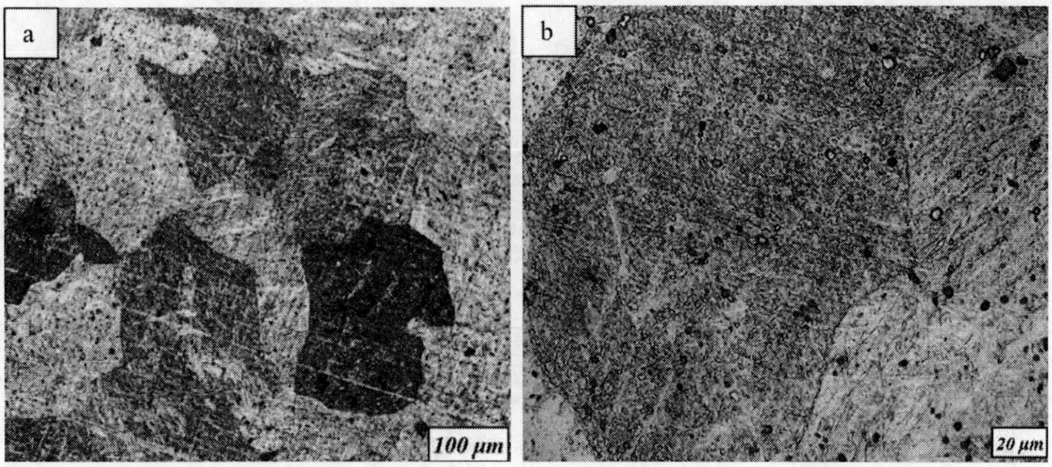

Figure 1. Microstructure of the alloy $Ti_{50}Ni_{50}$ in the quenched state (OM). Figure b shows a section with a large magnification and a martensitic relief.

The transmission electron microscopy analysis shows that the alloy has martensite microstructure at room temperature with mainly package morphology of paired twin plate crystals B19', which are twins $(\bar{1}11)$ and $(011)_{B19'}$ of the I type (on the whole it is typical of binary TiNi alloys). The width of twin martensite plates h_M is 85±15 nm, the boundaries between martensite crystals are clear and free of defects, Figure 2.

Figure 2. Typical TEM images of the microstructure of B19'-martensite of $Ti_{50}Ni_{50}$ alloy in the coarse-grained state.

Thermal cycling with consecutive increase of the number of cycles results in enhancement of the dislocation density and successive decrease of the martensite plate width. After thermal cycling in the temperature range MT B2 → B19' with a number of cycles equal to 20 in a microstructure consisting mainly of plate martensite, a complex dislocation substructure arises in some regions. With an increase in the number of thermal cycles from 20 to 50, developed dislocation clusters are observed in this state, forming the so-called "dislocation forest". Filming from various sections of the foil revealed the first composite nanotwins and stacking faults formed inside the martensite twins ($h\ (001)_{B19'} = 12 \pm 3$ nm). The width of the martensite plates decreases to $h_M = 65 \pm 10$ nm. In the state after thermal cycling with the maximum number of thermocycles, the microstructure is B19'-martensite with twins $(011)_{B19'}$ of I type 60±5 nm wide, inside which composite nanotwins $(001)_{B19'}$ several nanometers wide (~10±1 nm) and formed during thermal cycling are observed, Figure 3.

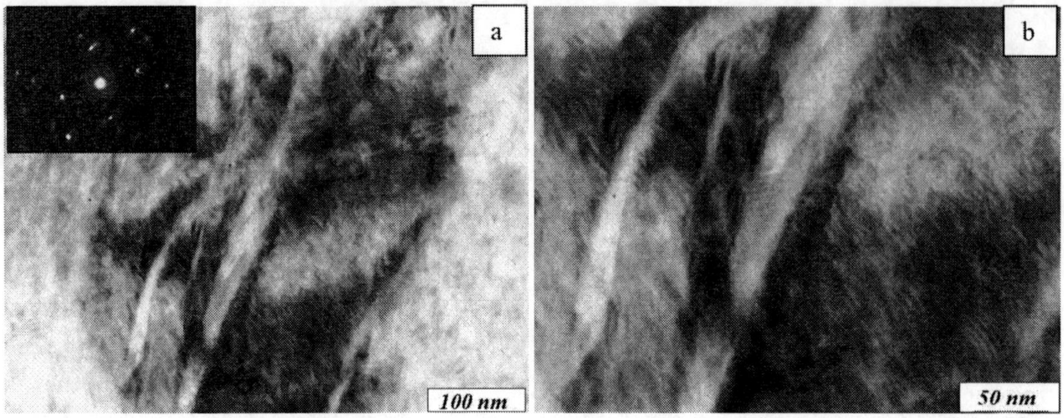

Figure 3. Typical microstructure of coarse-grained $Ti_{50}Ni_{50}$ alloy after 100 thermocycles (a), (b) displays composite nanotwins $(001)_{B19'}$

The ECAP n=8 at T=400°C leads to the formation of a complex martensitic structure in the material. The size of the initial austenitic grain in the form of martensite packets is about 600±30 nm, inside which there are twins with a plate width h_M=79±8 nm (Figure 4, a). Microdifraction has a close to annular distribution of reflexes, which indicates high-angle disorientation of grain boundaries in the deformed structure. In addition, there is an azimuthal blur of individual reflexes, indicating the presence of small-angle disorientation in the structure (Figure 4, b).

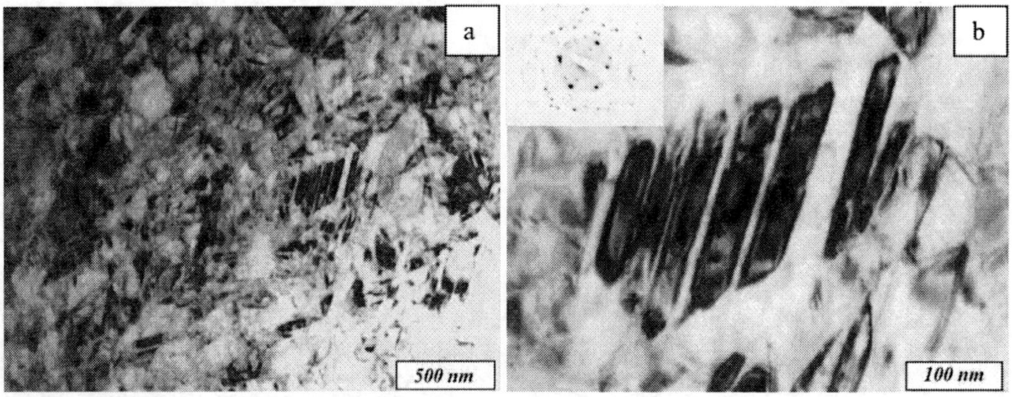

Figure 4. TEM image of microstructure of alloy $Ti_{50}Ni_{50}$ in UFG state: bright-field (a, b) images corresponding to UFG structure microdifraction (b).

Thermal cycling (n=20) leads to dislocations in martensitic plates. The width of the B19'-martensite twins is 70±5 nm. After 50 thermocycles, the density of dislocations and defects in the structure increases markedly, but it is difficult to make numerical estimates of changes in structural parameters from TEM images. Dislocation walls and thick extinction contours at the grain boundary also confirm that the accumulation of defects occurs in the alloy during multiple heat changes. Although this makes it difficult to quantify the structural elements, the average width of the measured plates is 55±9 nm. TC with the maximum number of thermal cycles (100 cycles) leads to the formation of a predominantly martensitic structure containing twins with a width of 40±7 nm (Figure 5, a), in addition, there is a noticeable increase in the density of dislocations in the martensite matrix (Figure 5, b). The increase in the individual grain allows to detect in the structure of composite nanowires $(001)_{B19'}$ (Figure 5, c). The width of the nanowires is about 7±4 nm, and the length is equal to the width of the martensite twins. Azimuthal diffusion of reflexes indicates high internal stresses.

IOP Conf. Series: Materials Science and Engineering **848** (2020) 012015 doi:10.1088/1757-899X/848/1/012015

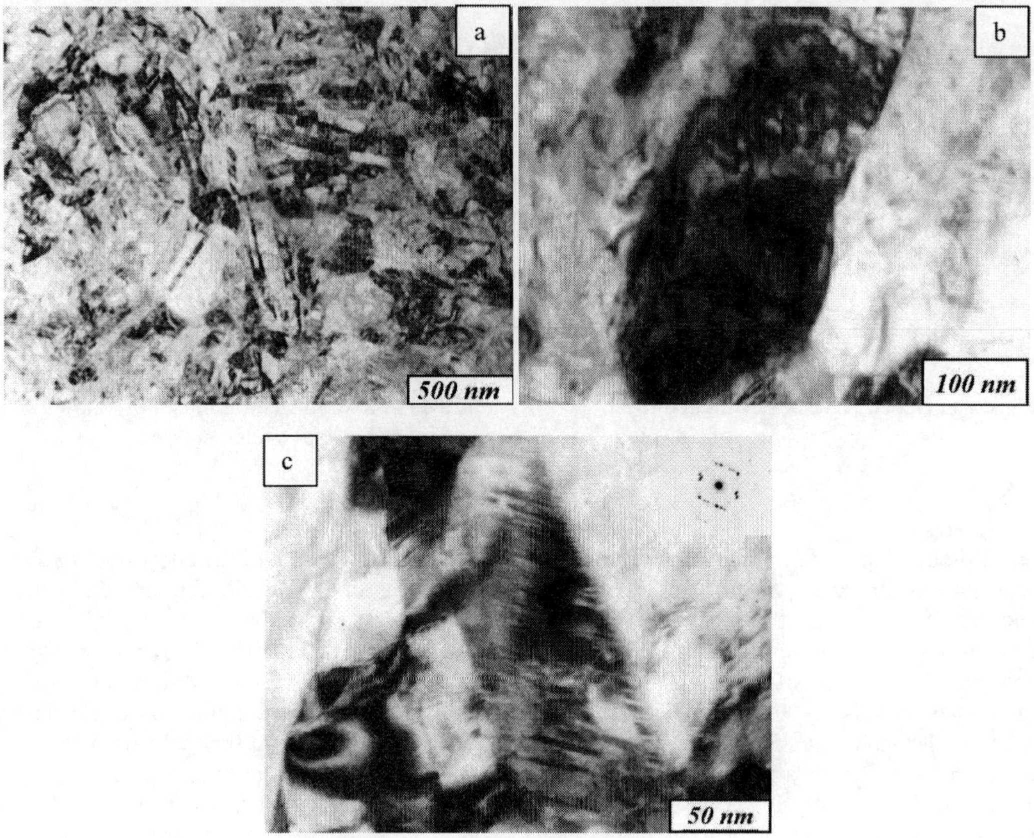

Figure 5. Typical TEM images of the microstructure of $Ti_{50}Ni_{50}$ alloy with UFG structure and maximum number (n=100) of cycles: bright-field (a,b,c) images: dislocation cluster (b), grain containing compound nanotwins (c)

In the initial coarse-grained state, the yield strength of the alloy is 430 MPa, with an increase in the number of thermal cycles, the yield strength of the alloy nonlinearly changes with the achievement of a maximum at 100 cycles and becomes equal to 547.5 MPa. The increase in the phase yield limit indicates that with the increase of thermal cycles, the phase transformation under load becomes more difficult, and a greater stress is necessary for its implementation, in this case it is caused by an increase in the density of defects accumulated at the TC. The estimated reactive stress, defined as the difference between the dislocation and phase yields, increases with the number of thermal cycles. Therefore, thermal cycling increases the functional properties of the alloy. The length of the phase flow plateau, which determines the possible resource of reversible deformation, increases from 5.1 to 5.45 % (figure 6, a). In the UFG state, the yield strength increases to 935 MPa. TC alloy in the UFG state with an increase in the number of thermal cycles leads to an increase in the yield strength to 1120 MPa at the maximum number of cycles. Thus, the yield strength increase in the UFG state is greater than in the CG (185 MPa - in the UFG, 117.5 MPa - CG). In the UFG state, there is an increase in the tensile strength by 110 MPa. The plasticity remains at an acceptable level of ~ 40 %, which, however, is slightly lower than the plasticity in the CG state (~ 60%). For the UFG state, along with the increase in the tensile strength and yield strength, a monotonic increase in the phase yield strength and the estimated reactive stress, up to 290 and 830 MPa, respectively, is also characteristic (figure 6, b).

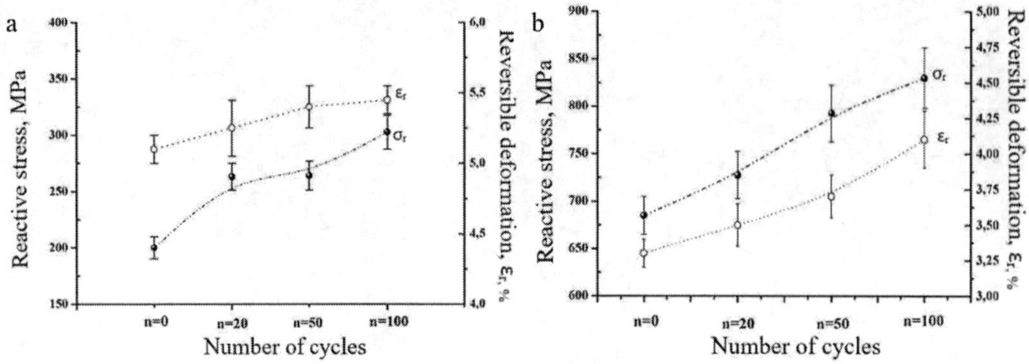

Figure 6. Dependences of estimated reactive stress and reversible deformation in CG (a) and UFG (b) states of $Ti_{50}Ni_{50}$ alloy on the number of thermal cycles

4. Conclusions

Under the influence of multiple martensitic transformation in the structure of the alloy $Ti_{50}Ni_{50}$ there is a decrease in the width of martensitic plates in both coarse-grained and ultrafine-grained states. In addition, it was found that at the maximum number of cycles in the martensite twins, composite nanowires $(001)_{B19'}$, several nanometers in size, are formed to reduce the level of internal stresses. Analysis of the results of mechanical tests showed that as a result of thermal cycling in the CG and UFG states there is an increase in the yield strength associated with the accumulation of defects in the alloy. At the same time, in the UFG state, there is a noticeable increase in the yield strength by 185 MPa, which is 67.5 MPa more than the increase in the CG state. The plasticity remains at the level of 40 %, which is slightly lower than the plasticity in the coarse-grained state (~60%). The ultrafine-grained state is characterized by a monotonic increase in the phase yield strength and estimated reactive stress to 290 and 830 MPa, respectively. The dependence of microhardness on the number of thermal cycles is similar to the behavior of the alloy under tension: with an increase in the number of cycles, an increase in the values of microhardness is observed.

Acknowledgments

This work was supported by a Grant of the Bashkortostan Republic of the Russian Federation for young scientists (No. 28 GR of 07.03.2019).

References

[1] V.N. Khachin, Titanium nickelide: structure and properties, Nauka, Moscow, Russia 1992 (in Russian).
[2] V. Brailovski, S. Prokoshkin, P. Terriaultet, F. Trochu, Shape memory alloys: fundamentals, modeling, applications, (ETS) Publ., Montreal, Canada 2003.
[3] K. Otsuka, Shape Memory Materials, Cambridge University Press, Cambridge, UK 1999.
[4] V.E. Gunther, G.Ts. Dambaev, P.G. Sysolyatin, et al. Medical materials and implants with shape memory (Ed.: V.E. Gunther), TSU, Tomsk, Russia 1998 (in Russian).
[5] S. Miyazaki, Y. Igo, K. Otsuka, 1986 *Acta metall.* **34** 2045.
[6] V.Y. Erofeev, L.A. Monasevich, V.A. Pavskaya, Y.I. Paskal, 1982 *Fiz. Met. Metalloved.* **53** 963 (in Russian).
[7] Yu.P. Mironov, P.G. Erokhin, S.N. Kul'kov, *Russian Physics Journal* 1997 **40** 210.
[8] Y. Furuya, Y.C. Park, 1992 *Nondestructive Testing and Evaluation* **8–9** 541.
[9] G.M. Lin, J.K.L. Lai, C.Y. Chung, 1995 *Scripta Metall. et Materialia* **32** 1865.
[10] S. Belyaev, N. Resnina, 2013 *International Journal of Materials Research* **104** 11.
[11] A.A. Churakova, D.V. Gunderov, 2015 *Acta Metallurgica Sinica (English Letters)* **28**(10) 1230.

IOP Conf. Series: Materials Science and Engineering **848** (2020) 012016 doi:10.1088/1757-899X/848/1/012016

Thermochemical grinding of diamond films in installation for chemical vapor deposition of diamond from the gas phase

S A Eremin[1], A M Kolesnikova[2], I A Leontiev[2], V N Anikin[1], O Yu Kudryashov[2], Y M Yashnov[2] and M V Zhdanova[1]

[1]The National Researching Technological University MISIS, Moscow, 6 Leninsky prospect

[2]TWINN, LLC, Moscow, 1 Feodosiaskaya, str. 30
Corresponding author: serega21_93@mail.ru

Abstract. The paper studies the process of thermochemical grinding of diamond polycrystalline films in atomic hydrogen, using microwave to generate a plasma discharge, in the temperature range 1000 – 1100 ^{0}C. It is established that the increase in temperature leads to an increase in the rate of diamond dissolution due to the formation of eutectic. The study of roughness showed that after thermochemical grinding roughness of the diamond film is about 0,8 μm. The speed of thermochemical grinding varies from 20 μm/h to 3 mm/h. Raman spectroscopy allowed us to establish that thermochemical grinding does not degrade the quality of the diamond film, and atomic hydrogen reduces the possible graphitization caused by heating.
It is suggested that the mechanism of gas bubbles in the process of thermochemical grinding associated with the formation of methane and its subsequent decomposition into a methyl radical and atomic hydrogen in the plasma discharge.

1. Introduction

Diamond is the hardest compound on Earth, its hardness on the Mohs scale is 10 [1], which creates difficulties in its processing. To give the necessary shape to the diamond, mechanical processing with free [2] or fixed abrasive [3], a laser for rough grinding and shaping [4], as well as thermochemical grinding of diamond based on the catalytic dissolution of carbon in iron [5] are used. In [6] proposed installation for thermochemical diamond cutting, grinding speed up to 0.3 mm per hour. One of the disadvantages is the inability to grind diamond polycrystalline films of large diameter, as when scaling the design of the installation will be complicated, due to the need to ensure the rotation of the heated part from 600 ^{0}C to 1300 ^{0}C in a hydrogen medium.

The method of thermochemical grinding has long been known, and in itself is not new. However, the use of the installation for the growth of diamond from the gas phase for thermochemical grinding can allow grinding diamond films up to 100 mm in diameter, which in turn is promising and will reduce the cost, for example, in the manufacture of diamond output windows for gyrotrons, where mechanical grinding is one of the expensive and long operations.

2. Methods and materials

For the experiments, a device for deposition of diamond from the gas phase in a microwave discharge was used. Hydrogen of high purity gases brand (99.9999%) was used as plasma-forming gas. For dissolve carbon steel grade ST2 was used. To wash the diamond after thermochemical grinding, chemical reagent solution (know-how) was used. The process of thermochemical grinding was carried out in the pressure range of 75-100 torus. A micrometer was used to measure the thickness of the diamond film. Structural studies of the samples were carried out using a scanning electron microscope of the ZEISS series EVO MA10, Raman spectra were obtained using a DXR Raman Microscope spectrometer.

The experimental design is shown in figure 1.

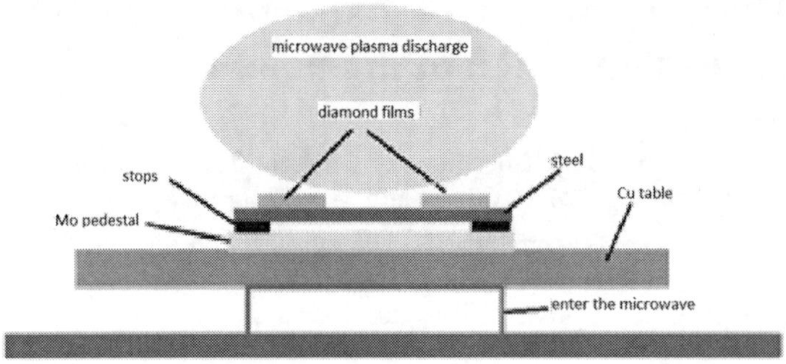

Figure 1. The scheme of the experiment on thermal grinding.

The samples of 10x10 mm in the amount of 4 pieces were placed on a steel plate with a size of 70x70x4 mm. The supports necessary for reducing the heat removal from the plate and the possibility of heating it to the temperatures of the active process of diamond graphitization and its subsequent dissolution were placed under the steel plate.

The plasma discharge heats the samples to activate the process of dissolution of carbon in iron, while the plasma discharge consisting entirely of atomic hydrogen should protect the diamond film from excessive graphitization in the area not in contact with the steel plate. The experiments were carried out at different sample temperatures. The sample temperature was controlled by chamber pressure and microwave power. Three experiments were carried out at 1000 °C, 1050 °C, 1100 °C.

3. Results and discussion

Figure 2 shows the appearance of the samples after the thermochemical grinding process.

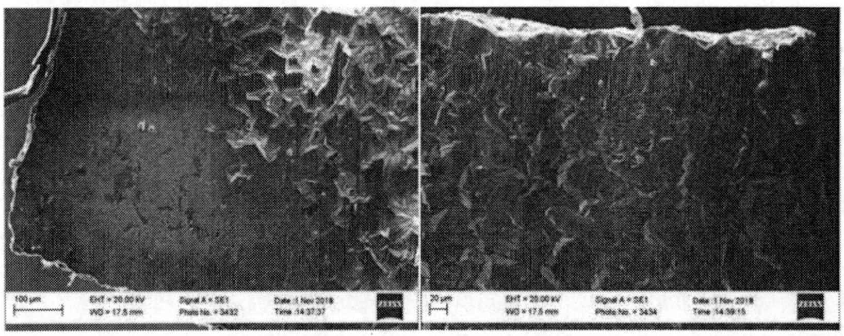

Figure 2. Appearance of the sample after the thermochemical grinding experiment.

In the process of iron dissolution, only the angular surface of the diamond film was involved. A gradual transition is visible (figure 2) from a smooth surface to a rough surface, represented by individual crystals with a large number of borders.

Measuring the thickness of the diamond film before and after the process of thermochemical, grinding allowed to set that the grinding speed at a temperature of 1000 °C is 20-30 μm /h. Thus, the temperature of 1000 °C is low for the active process of thermochemical grinding.

Figure 3 shows the appearance of diamond films before the thermal grinding process at 1050 °C.

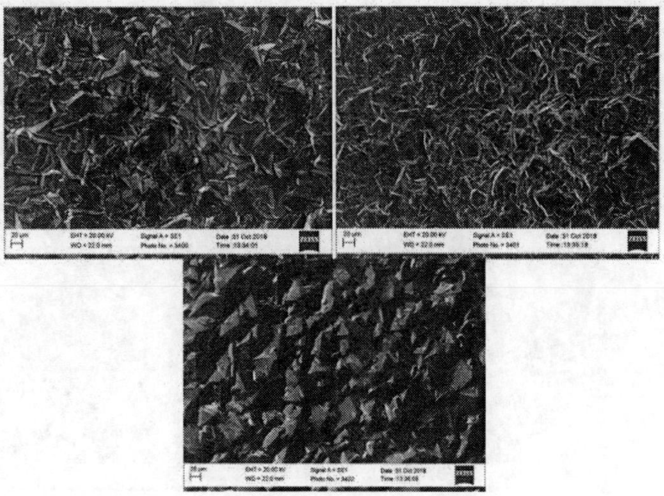

Figure 3. Appearance of original diamond films.

The structure of diamond films is rough, represented by crystals in the form of hexagons or pyramids.

Figure 4 shows the appearance of the samples after thermochemical grinding at 1050 °C.

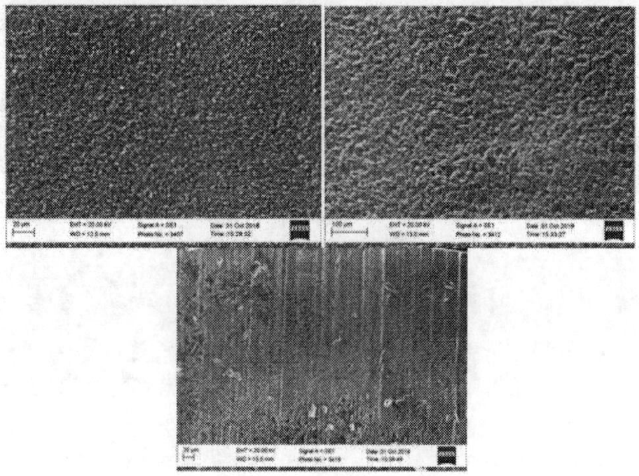

Figure 4. Appearance of samples after thermochemical grinding at 1050 °C.

The surface of all samples after thermochemical processing is almost the same; there are no crystals with faceting. The surface is smooth, represented by small craters.

Figure 5 shows the appearance of the samples before the experiment on thermal grinding at 1100 °C.

IOP Conf. Series: Materials Science and Engineering **848** (2020) 012016 doi:10.1088/1757-899X/848/1/012016

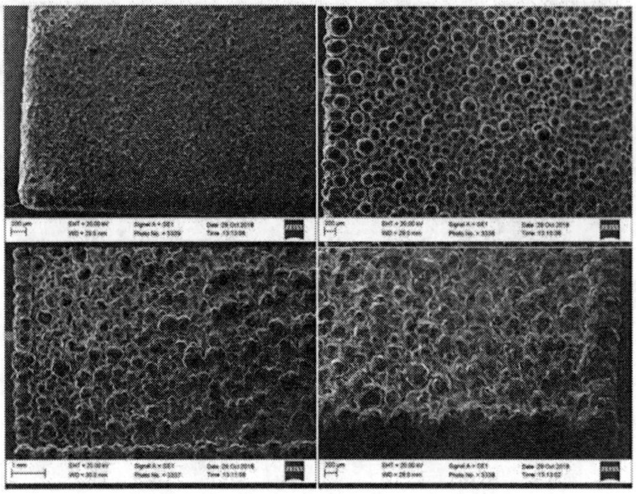

Figure 5. Appearance of the original diamond films.

Figure 6 shows the appearance of diamond films after the process of thermochemical grinding at 1100 °C.

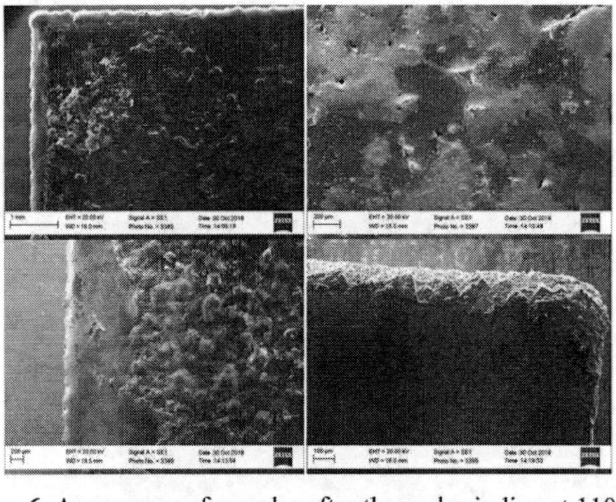

Figure 6. Appearance of samples after thermal grinding at 1100 °C.

After thermal grinding, globules and faceting are practically absent, the entire surface is smoothed to a certain level of roughness. It is worth to note that during the experiment, the samples floated on the surface of the steel plate, which indicates the formation of eutectic. Also under the surface of some diamond films formed bubbles of liquid metal lifting the samples. In the plasma discharge, flashes with a greenish tint were observed when the bubble burst. Figure 7 shows a diagram showing the process described.

IOP Conf. Series: Materials Science and Engineering **848** (2020) 012016 doi:10.1088/1757-899X/848/1/012016

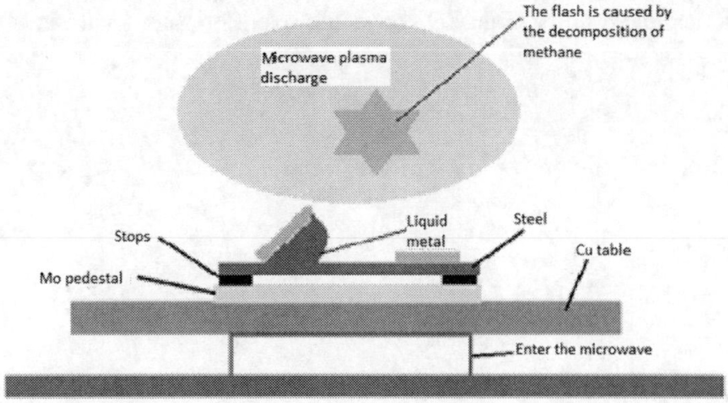

Figure 7. Scheme of methane formation and decomposition.

The formation of liquid eutectic as a result of metal saturation with carbon occurs at 1100 °C, since the pressure in the working chamber is 100 torus. As a result, atomic hydrogen begins to react with dissolved carbon, resulting in methane bubbles that escape from the liquid eutectic, fall into a plasma discharge, where methane decomposes into methyl radical or atomic hydrogen. However, this theory is hypothetical, since no spectra confirming the nature of this phenomenon has not been obtained.

In the process of thermochemical grinding, the penetration of iron into diamond is possible, as well as the graphitization of diamond. To investigate this assumption, the transverse kink of one of the experimental samples was studied. The result of the study is presented in figure 8.

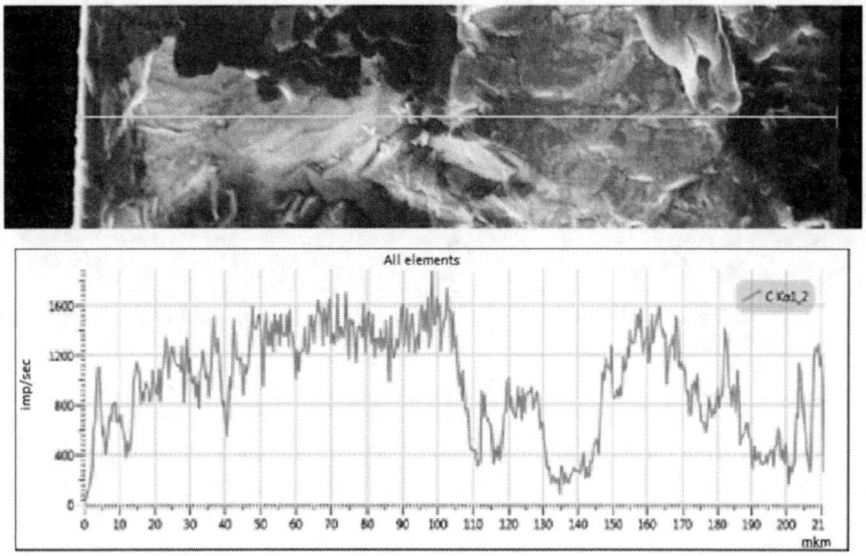

Figure 8. Elemental analysis of the transverse fracture of the diamond film.

Elemental analysis did not show any other elements besides carbon, which indicates that iron, did not penetrate deep into the sample.

After thermochemical grinding, the samples were washed in chemical reagent solution (know-how). The solution was brought to a boil, and then the samples were dipped in a boiling solution, and then washed with deionized water. Figure 9 shows the result of phase analysis of samples before and after washing.

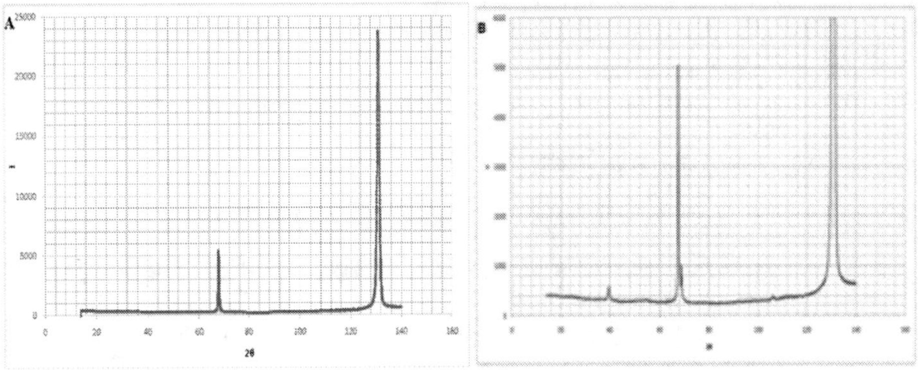

A. Sample after washing, **B.** Sample without washing

Figure 9. Diffractograms.

According to figure 9B, carbon, in the form of diamond and graphite, as well as iron, is present on the surface of the sample after thermochemical grinding. After washing the samples, not only iron is removed, but also graphite, as evidenced by the two peaks in figure 9A, which correspond only to diamond.

In the study, the roughness of the samples was studied, before and after thermochemical grinding. The initial roughness values ranged from 40-60 μm, table 1.

Table 1. Roughness Measurement Result.

Sample	Roughness, μm
Steel	1,10
Diamond films thermal grinding at a temperature of 1060 °C	
Sample №2	0,887
Sample№3	0,831
Sample№4	0,785
Diamond films thermal grinding at a temperature of 1100 °C	
Sample №1	0,548
Sample №2	1,26
Sample№4	0,522

The roughness results for the measured samples almost coincide and are equal to an average of 0.8 μm, which is significantly less than the initial roughness of the steel plate. Thus, thermochemical grinding can significantly reduce the level of surface roughness of the original diamond films, tables 2

Also, the thermal conductivity of samples after thermal grinding was measured. In order to calibrate the instrument, measurements were made on test samples with known thermal conductivity. Copper thickness of 1 mm – 315 W/m·K (table value 394 W/m·K).

Values of test samples:

- Copper thickness of 0,5 mm - 400 W/m·K (table value 394 W/m·K);
- PDF (polycrystalline diamond film) thickness of 0,4 mm growth side -713 W/m·K;
- PDF thickness of 0,4 mm reverse side – 615 W/m·K.

IOP Conf. Series: Materials Science and Engineering **848** (2020) 012016 doi:10.1088/1757-899X/848/1/012016

Table 2. Result of thermal conductivity measurement of diamond films after thermal grinding.

The temperature of experiment 1100 °C	The temperature of experiment 1050 °C
Sample №1:	Sample №1:
The growth side of the PDF – 737 W/m·K;	The growth side of the PDF – 362 W/m·K;
The reverse side of the PDF – 656 W/m·K;	The reverse side of the PDF – 616 W/m·K;
Sample №2:	Sample №2:
The growth side of the PDF – 975 W/m·K;	The growth side of the PDF – 817 W/m·K;
The reverse side of the PDF – 985 W/m·K;	The reverse side of the PDF – 915 W/m·K;
Sample №3:	Sample №3:
The growth side of the PDF – 235 W/m·K;	The growth side of the PDF – 53 W/m·K;
The reverse side of the PDF – 280 W/m·K;	The reverse side of the PDF – 174 W/m·K;
Sample №4	Sample №4
The growth side of the PDF – 448 W/m·K;	The growth side of the PDF – 577 W/m·K;
The reverse side of the PDF – 736 W/m·K;	The reverse side of the PDF – 895 W/m·K;

Based on the data in the table, the thermal conductivity of the thermally polished side decreased significantly, which is caused by contamination of the surface with iron, as well as its graphitization according to the data of X-ray diffraction analysis (figure 9B).

After washing the samples, the thermal conductivity was also measured. The results are presented in table 3.

Table 3. The result of measuring the thermal conductivity of diamond films after thermal grinding and washing.

The temperature of experiment 1100 °C	The temperature of experiment 1050 °C
Sample №1:	Sample №1:
The growth side of the PDF – 800 W/m·K;	The growth side of the PDF – 650 W/m·K;
The reverse side of the PDF – 690 W/m·K;	The reverse side of the PDF – 620 W/m·K;
Sample №2:	Sample №2:
The growth side of the PDF – 980 W/m·K;	The growth side of the PDF – 950 W/m·K;
The reverse side of the PDF – 918 W/m·K;	The reverse side of the PDF – 900 W/m·K;
Sample №3:	Sample №3:
The growth side of the PDF – 350 W/m·K;	The growth side of the PDF – 250 W/m·K;
The reverse side of the PDF – 270 W/m·K;	The reverse side of the PDF – 165 W/m·K;
Sample №4	Sample №4
The growth side of the PDF – 750 W/m·K;	The growth side of the PDF – 900 W/m·K;
The reverse side of the PDF – 700 W/m·K;	The reverse side of the PDF – 850 W/m·K;

Washing samples allowed to increase the thermal conductivity, while not affecting the thermal conductivity of the reverse side of the diamond polycrystalline film.

The thickness of samples after thermochemical grinding was measured in order to determine the average speed of thermochemical grinding. Figure 10 shows the dependence of the thermochemical grinding speed on the sample temperature.

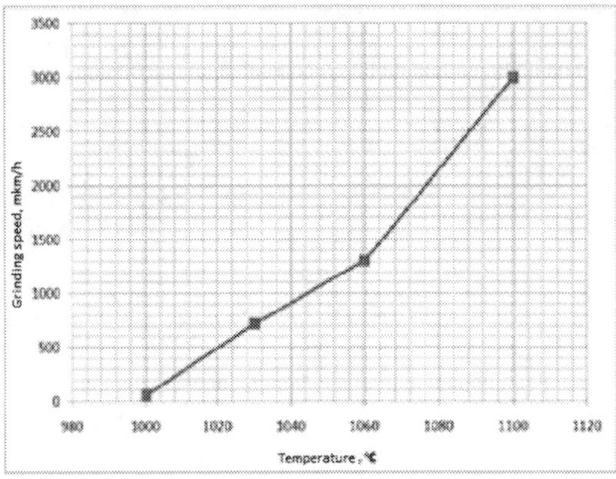

Figure 10. Graph dependence of the temperature of thermochemical grinding of diamond.

As a result, it was found that the grinding speed in the range from 1000 - 1100 ° C varies in the range of 20 μm/h to 3 mm/h. Such a large variation in temperature is caused by the formation of liquid eutectic, in which the process of graphitization and dissolution of carbon is very active.

Separately, a piece of white diamond was selected to assess the visual effect of the influence of thermal grinding on the quality of the diamond film. The appearance of the sample before the experiment is shown in figure 11.

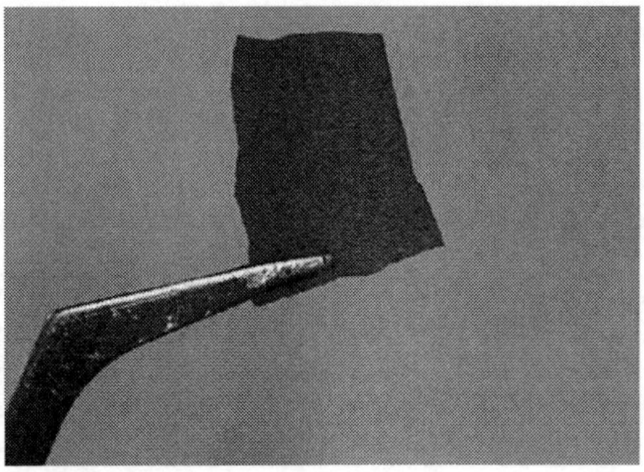

Figure 11. Appearance of the original white diamond film.

Figure 11 shows a diamond film directly subjected to thermochemical grinding. Also, it is worth noting that the diamond film had a bend.

After the process of thermochemical grinding and washing in chemical etchants, the diamond returned to its original transparency and color. The appearance of the sample after washing is shown in figure 12.

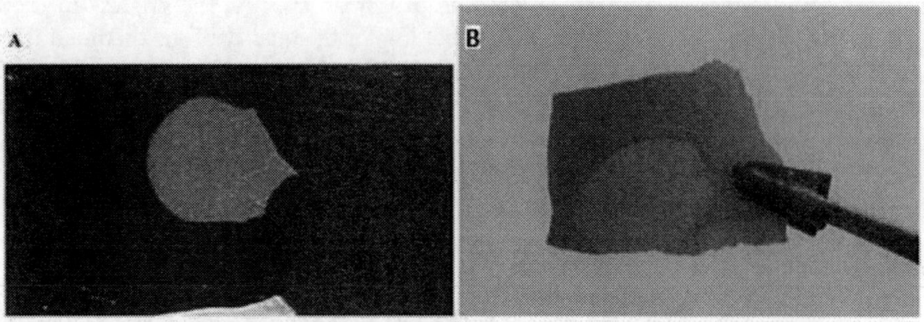

A diamond film before washing, B Diamond film after washing

Figure 12. Appearance of diamond film.

After thermochemical grinding, the area of the diamond in contact with the iron became completely black, but after washing it was possible to return the original appearance.

Raman spectroscopy of the area after washing and the area not involved in the grinding process was carried out. The spectra are shown in figure 13.

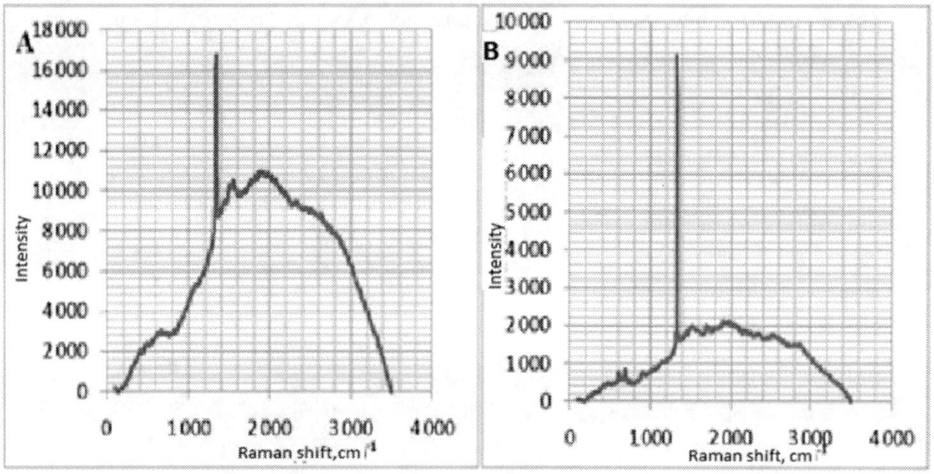

A Area without thermo-chemical polishing, B Area subjected to the thermochemical polishing

Figure 13. Raman spectra.

The area, which was not in direct contact with the steel, was only subjected to heat, which led to the release of a large number of defects on the surface of the diamond, which was expressed on the spectrum by a wide luminescent peak (figure12A). The area subjected to thermo-chemical grinding produces a smaller luminescence (figure12B). Perhaps this is due to the fact that the surface after thermochemical grinding is homogeneous. Thus, thermochemical grinding not only does not degrade the quality of the diamond film, but also allows to obtain uniform surface with uniform roughness and thickness.

4. Conclusion

In the process of studying the process of thermochemical grinding in a plant for the growth of diamond from the gas phase, it was found that in the temperature range 1000 - 1100 °C, the grinding

IOP Conf. Series: Materials Science and Engineering **848** (2020) 012016 doi:10.1088/1757-899X/848/1/012016

speed varies from 20 μm/h to 6 mm/h. According to Raman spectra, the process of thermochemical grinding does not lead to graphitization and does not impair the quality of the diamond film.

The method of thermochemical grinding is known for a long time, and in itself is not new. However, the use of the installation for the growth of diamond from the gas phase for grinding can allow grinding diamond films with a diameter of 1 mm to 100 mm, which in turn is promising and will reduce the cost, for example, in the manufacture of output windows for gyrotrons, where mechanical grinding is one of the expensive operations

References

[1] Sergey D 2017 Vickers Hardness of Diamond *Crystals* Vol **7(369)** p 1-13

[2] Burlakov V I 2017 Analysis of methods of processing parts with free abrasive *Technical science* p 132-137

[3] Avdeev D M, Avdeev D M, Miranchuk A I, Boyko A A, Alekseenko Y A 2000 Diamond grinding elements on a ceramic bond for fine diamond grinding of optical parts Vol **3** p 31-36

[4] Cardoso M S 2005 A valuation model for cut diamonds *Intl. Trans. in Op. Res* Vol **12** p 417-436

[5] Dongik J Yumkyum K, Minsoo S, Joonho L 2012 Kinetics of carbon dissolution of coke in molten iron *Metallurgical and Materials Transactions* Vol **43B** p1308-14

[6] Shamaev P P, Grigorieva A S, Botvin V V 2002 On thermochemical methods of diamond processing from new positions *Science and technology in Yakutia* Vol **1** p 27-29

IOP Conf. Series: Materials Science and Engineering **848** (2020) 012017 doi:10.1088/1757-899X/848/1/012017

New method of obtaining heat-resistant biomorphic composites with carbide reinforcement

A E Ershov[1, 2], D V Prokhorov[1], T S Stroganova[1] and I M Shmytko[1]
[1] Institute of Solid State Physics Russian Academy of Sciences, Chernogolovka, Moscow District, 2 Academician Ossipyan str., 142432, Russia.
[2] E-mail: ershov@issp.ac.ru

Abstract
In this work, an original method of obtaining biomorphic composites of the Me – MeC_x type, consisting of several stages, has been proposed. One of the steps is the Infiltration through Nonreactive Solvent (INS) method, which consists in converting biomorphic carbon preforms into porous microchannel MeC_x ceramics by transferring a carbide-forming element to carbon through a solvent that does not react with carbon. The interaction of biomorphic alder-based carbon templates with melts of Cu – Ti and Sn – Ti systems has been investigated. It has been shown that as a result of the interaction of biocarbon templates with Cu-10Ti and Cu-20Ti alloys, carbon walls turn into TiC grains of micron and nanometer sizes, which partially repeat the template microstructure. The mechanism of formation of such a structure has been proposed. It has been demonstrated that, upon the interaction with Sn – Ti system, TiC and Ti_2SnC layers are formed on the surface of the microchannels with good preservation of the initial template structure.

1. Introduction

The development of new heat-resistant materials is one of the main directions of modern materials science. Such materials are in demand in many industries. However, the most difficult problem is to increase the working temperatures and improve the efficiency of gas turbine engines, in particular by increasing the working temperatures of turbine blades. The main requirements for the materials used in their manufacture are high strength, creep resistance, and phase stability at temperatures above 1300 – 1400 °C, combined with good crack resistance at room temperature and low material density [1, 2].

There are three main types of heat-resistant materials, which are considered promising: composites with a metal matrix (MMCs), composites with a ceramic matrix (CMCs), and high-entropy alloys (HEAs). At the same time, blades made of MMCs are used in almost all existing engines. High characteristics of heat-resistant materials of this type are provided by their complex composite structure consisting of a viscous matrix based on a solid metal solution and a hardening phase, which is the elongated grains of intermetallic compounds, carbides, silicides, and other high-strength refractory compounds. At present, nickel alloys continue to be the most widespread material for the manufacture of blades. However, their operating temperatures are close to the theoretical limit, and their further increase is impossible [2]. Materials based on Nb – Si, Nb – Al, Mo – Si – B, and many other systems are being actively investigated [1 – 8].

There are many types of metal-matrix composites obtained by various methods. However, the method of directed crystallization by the lost-wax model is mainly used today for the manufacture of real products. This method allows manufacturing blades of a given shape, practically requiring no finishing. In this case, due to directed crystallization, the material structure contains uniformly distributed elongated inclusions of the strengthening phase in the metal matrix (eutectic structure), which ensure high operational characteristics of the blades [1, 2, 8]. With all the advantages of this technology, when changing the composition of melted alloys, additional work on determination of the optimal crystallization rates is required. The transition to other systems of elements causes search for new materials for the manufacture of crucibles that are resistant to melt effect, and due to high melting points and duration of the crystallization process, this is a separate difficult task. The lack of suitable crucibles is one of the main reasons limiting the preparation of promising heat-resistant composites of the Me – MeC_x type, where Me is a refractory metal (Nb, Hf, Ta, W, Ti, Zr, V, Cr), and MeC_x is its carbides. Hence, researchers are trying to use other methods of manufacturing such composites, in particular, layered composites obtained by diffusion welding, fiber-reinforced composites obtained by powder metallurgy, SHS, additive technologies, and others. Each of them, however, has its drawbacks, in particular, the presence of pores, non-uniform distribution of the hardening phase, low phase stability at high temperatures, high manufacturing costs, and many other factors that impede the widespread use of the materials obtained. In this regard, the search for new approaches to obtaining heat-resistant materials of this type is an urgent task.

2. Biomorphic Me – MeC_x composites. Method of obtaining.

This article proposes a new method of obtaining Me – MeC_x composites, consisting of several stages:

Stage 1 – obtaining a microchannel biomorphic carbon template.

Biomorphic carbon templates are obtained by pyrolysis of wood in an oxygen-free atmosphere to high temperatures [9, 10]. As a result, a porous carbon material is obtained that preserves the microchannel structure of wood. The specific parameters of this material (diameter and length of channels, total density, etc.) are determined by the choice of an initial species (figure 1). In addition to choosing a species, the structure of biomorphic materials can also be affected by prepressing of wood [11], which allows varying the characteristics of the materials obtained in an even wider range. An important feature of biomorphic materials is the ability to shape the final product at the stage of soft carbon preform, which enables minimizing the cost of subsequent processing.

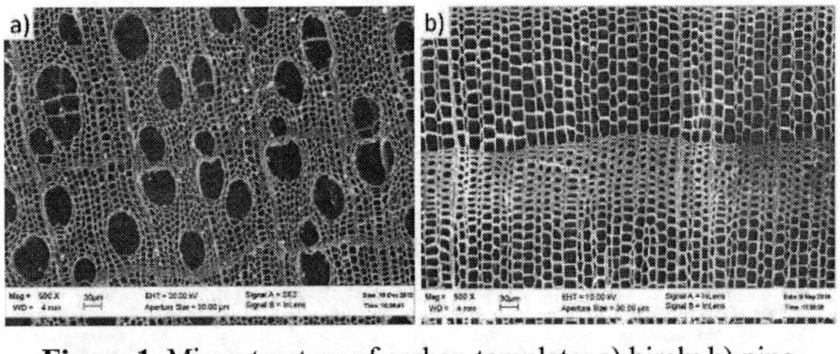

Figure 1. Microstructure of carbon template: a) birch; b) pine.

Stage 2 – obtaining a microchannel biomorphic MeC_x template.

The next task is to convert a biocarbon template into a microchannel template made of refractory metal carbide, and the treatment should not significantly disrupt the initial structure of the material. This problem can be solved by several methods, for example, CVD, PVD, sol-gel techniques [12, 13]. Such methods, however, have a number of disadvantages, in particular, low productivity, high requirements for equipment and reagents, and the difficulty of obtaining large-size samples.

As a result, in this paper, it is proposed to use the Infiltration through Nonreactive Solvent (INS) method, which needs to be described in more detail.

The main idea of this method coincides with the idea of the LMTA (liquid metal transfer agent) method and consists in using a melt that does not react with carbon as an agent for transporting a carbide-forming element to carbon [14]. The LMTA method is commonly used for coating carbon fibers that are placed in a bath with a melt, i.e., the process of carbide formation occurs along the entire surface of the fiber at a constant concentration of the carbide forming element.

Infiltration into a microchannel preform is noticeably different, which can be shown by the example of the interaction of a biocarbon preform with S+Me melt, where S is an inert to carbon solvent, Me is a carbide forming element. Upon contact of the preform with the melt, it will begin to penetrate into the microchannels, but Me will be spent on the formation of MeC_x compound. Since the solvent does not form compounds with carbon and does not wet it, the infiltration process will slow down and will be determined by the rate of Me diffusion from the melt to the infiltration front. Due to this, Me carbide will gradually grow on the carbon surface, which will accurately preserve the wood structure (figure 2).

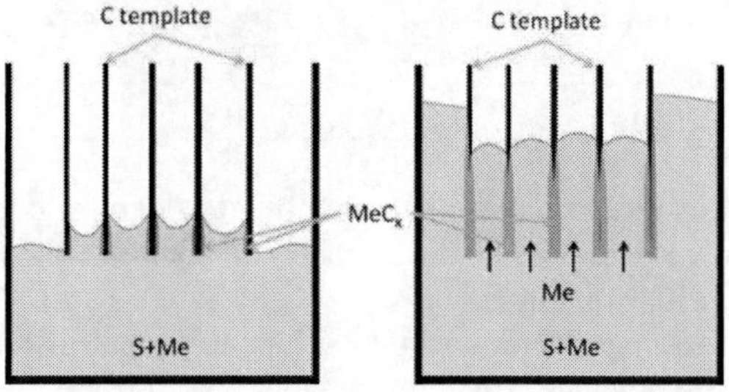

Figure 2. Diagram of biomorphic carbon template infiltration.

The main advantage of the method is that the process can be carried out at sufficiently low temperatures (800 – 1200 C). This greatly simplifies the selection of crucible materials, allows using long curing time and carrying out the process in inexpensive furnaces with metal or ceramic heaters, which removes restrictions on the size and shape of the resulting samples or details.

The second most important advantage is that the proposed scheme is universal, and in this way it is possible to obtain carbides of various (including refractory) elements for which suitable solvents exist.

The solvent must satisfy several requirements:

1. The solubility of carbon in the solvent melt should be small so that the microchannel structure of the biocarbon template is not destroyed.

2. The solvent cannot form stable carbides.

3. The solubility of the carbide-forming element in the melt should be sufficiently large at temperatures up to ~ 1100 – 1200 °C. This requirement is not critical. However, the higher the concentration of a carbide-forming element in the melt, the faster it will diffuse through the melt into the template. The solubility of refractory elements can be increased by increasing the melt temperature; however, the solubility of carbon also increases. In addition, with increasing temperature, the requirements for equipment increase.

When the melt interacts with carbon, the system becomes at least three-component; therefore, additional requirements arise.

4. The addition of a carbide-forming element to the solvent should not lead to a significant increase in the solubility of carbon in the melt.

5. The alloy of a solvent with a refractory element must not react with the carbide of this element

Based on paragraphs 1 – 2, one can propose several possible types of solvents:

Firstly, these are melts of some elements with a low melting point, such as Ag, Au, Bi, Cu, Ga, Ge, In, Pb, Sb, Sn, Zn. Taking into account paragraph 3, for almost all refractory metals, several possible solvents can be selected, which are presented in table 1, although there are difficulties with molybdenum and tantalum.

Secondly, these are melts of salts, in particular fluorides and chlorides, which weakly react with carbon. This method is a modification of the MSS (molten salt synthesis) process, where molten salts are usually used as a medium for the obtainment of complex oxides. It has already been used in a number of works, and biomorphic SiC and ZrC ceramics were obtained in this way [15, 16]. An additional advantage of using salts is that they dissolve in water, which simplifies the extraction of the melt from the microchannels after the process is carried out.

Table 1. Refractory metals solvents.

Metal	Solvent
Ti	Cu (70 % at 1100 °C), Sb, Sn
V	Ga (35 % at 1100 °C), Ge
Cr	Sb (45 % at 1100 °C), Ge, Ga
Zr	Cu (75 % at 1100 °C), Ag
Nb	In (40 % at 1100 °C), Sb
Mo	Ga, Ge (~5 % at 1100 °C)
Hf	Cu (45 % at 1100 °C)
Ta	Au (15 % at 1100 °C)

Probably, some multicomponent solvents can also be considered, for example, Ni – Pd, Cu – Sn, and other alloys. However, an increase in the number of components significantly complicates the process and analysis of the results obtained.

The proposed solvents satisfy paragraphs 1 – 3. The fulfillment of requirements 4 and 5 is much more difficult to predict in advance since an analysis of phase diagrams with three or more components that are often not built yet is required. Moreover, the situation is complicated by the fact that, firstly, the alloy changes its composition depending on the depth of penetration into the sample along the channel axis due to depletion of the refractory element on carbide formation. Secondly, carbides of refractory elements have a rather large region of homogeneity, and the ratio of an element to carbon can vary widely. All this can affect the processes of chemical interaction of carbide and a melt. As an example, we can cite the data from [17], where it was shown that the stability of TiC_x (where x vary from 0.5 to 1) in (Cu, Sn, Ag, Au) – Ti melts depends not only on titanium content in the melt, but also on the value of x.

Stage 3 – filling microchannels with metal and obtaining Me – MeC_x composites.

The main advantage of biomorphic carbide template is their microchannel structure. Provided that the carbide is wetted, the melt will quickly fill the preform under the action of capillary forces. This significantly reduces the duration of the process, lowers the requirements for crucible materials and equipment, and also allows using crucible-free methods.

As a result of the infiltration, a biomorphic composite will be obtained, where the hardening phase is a continuous carbide skeleton. In the view of mechanical characteristics, such a structure is not optimal, since cracks can propagate through the entire bulk of the sample. In this regard, it is necessary to divide the skeleton into separate fibers to obtain a classic fiber-reinforced MMC. To achieve this, deformation and/or heat treatment, electron-beam floating zone melting, or combinations of these methods can be used.

This concept is at an early stage of development, and many possible problems may arise during its implementation. However, there are several main advantages that make this method promising.

Advantages of the method of obtaining composites.

1. A wide range of possible Me – MeC_x systems.
2. The possibility of controlling the size and volume fraction of the hardening phase by selecting the source species.
3. Low requirements for crucible materials. The possibility of using crucible-free methods.
4. The possibility of manufacturing large-size products of a complex shape.
5. Minimum porosity of the material

3. Experimental.

To create composites according to the scheme above, it is necessary to obtain biomorphic carbide templates, for which purpose the interaction of biocarbon templates with Cu – Ti and Sn – Ti melts was studied in this work. The phase diagrams of these systems are shown in figure 3. Alder wood (Alnus incana) was used as an initial material. Large wood samples were pyrolyzed in the self-generated atmosphere of wood decomposition products in a semihermetic reactor, separated from the exterior atmosphere by a water seal (temperature of 900 °C, heating rate of 5 °C/h to 500 °C and 100 °C/h to 900 °C). The obtained samples were additionally annealed in vacuum at 1450 °C for 30 minutes to completely decompose the organic components. The density of the obtained preforms was measured. Then, samples for microscopic investigations (cleavages along the characteristic wood planes), as well as samples with sizes of ~ 6×6×23 mm, elongated in the direction of wood growth, were made from the preforms for the further examination.

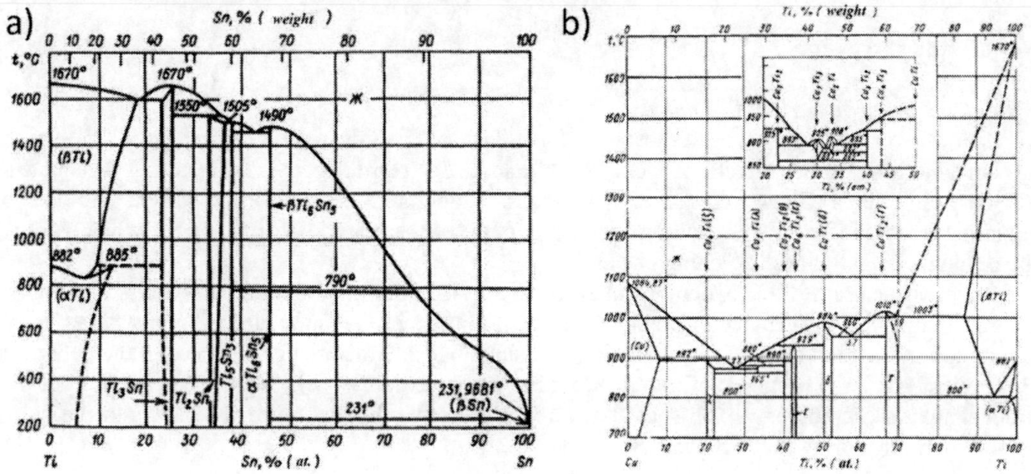

Figure 3. Phase diagrams: a) Sn – Ti; b) Cu – Ti. [18, 19]

Cu – Ti and Sn – Ti alloys were prepared by alloying from pure components (Ti – 99.9, Sn – 99.9, Cu – 99.8) in vacuum in a graphite crucible at 1300 and 1200 °C, respectively. The content of the elements varied from 10 to 70 at. % for Cu – Ti system and was 30 at. % for Sn – Ti system.

An alloy sample was placed on the bottom of a graphite crucible with an inner diameter of 10 mm to carry out the INS process. A biocarbon perform additionally loaded with a graphite weight with a mass of ~ 5 g was placed onto the sample. For short exposures (several hours), the process was carried out in a vacuum induction furnace, for long processes (several days) the process was carried out in evacuated and sealed quartz ampoules in a resistive furnace with a metal heater. The process parameters are described in more detail in Table 2. After the process was carried out, the samples were cut along the long side, polished, and examined using scanning electron microscopy (Tescan Vega II XMU, Carl Zeiss SUPRA 50 VP) and X-ray diffraction (Siemens D500, Cu K_α).

Table 2. Samples and experiments description.

Alloy	Titanium concentration (at. %)	Temperature (°C)	Duration of the INS process (h)
Cu-70Ti	70	1150	17
Cu-10Ti	10	1150	108
Cu-20Ti	20	1150	108
Cu-30Ti	30	1150	108
Sn-1150	30	1150	4
Sn-1000	30	1000	72
Sn-900	30	900	72
Sn-800	30	800	72

It should be noted that in a number of experiments, gas emission was observed during the interaction of biocarbon templates with alloys. This was particularly true for Cu – Ti system, which is probably related to the reduction of Cu_2O present in the melt by the carbon of the preform and crucible. Due to the fact that the crucible had an inner diameter of 10 mm, the resulting bubbles could not always leave the melt, and in some cases they combined into a large bubble, which separated the infiltrated sample from the main alloy bulk. This impeded the uniformity of infiltration, and made it difficult to determine the kinetics of the process. The characteristic infiltration rates for alloys with titanium content of 30 at. % or less were ~ 5 – 10 mm/day. However, in view of the above, the infiltration kinetics will not be considered in this paper in detail.

4. Results and discussion.

4.1 The structure of biocarbon template.
As a result of pyrolysis, alder samples lost ~ 73 % of the mass and decreased in size. Shrinkages in the axial, tangential and radial directions were 20, 35 and 27 %, respectively. The resulting density was 0.36 g/cm^3. Simple calculation can show that upon the complete conversion of carbon into TiC of stoichiometric composition, the resulting volume fraction of the material will be 36 %, which is close to the optimal content of the hardening phase.

The microstructure of bioC is demonstrated in figure 4. In the microstructure, vessels with a large (~ 50 μm) diameter, libriforms with a small (~ 10 μm) diameter, elongated along the growth axis, and rays directed in the radial direction can be distinguished. The characteristic thickness of the cell walls is 3 – 5 μm. A large number of pores are also present in the cell walls. Pores bind all the microchannels to each other, which should ensure an effective infiltration of the melt into the entire bulk of the sample during the INS process.

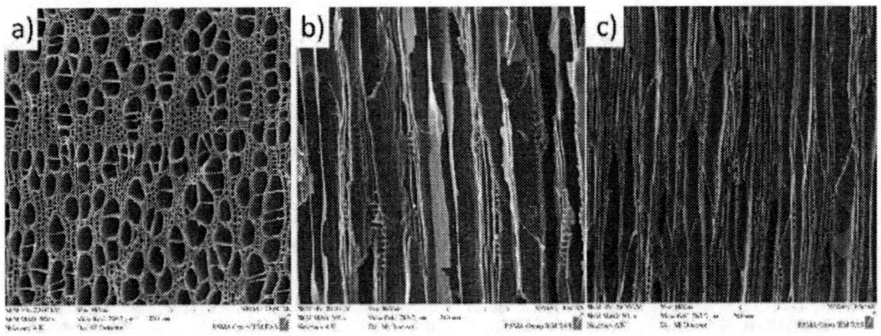

Figure 4. Microstructure of alder-based carbon template: a) axial; b) tangential; c) radial.

4.2 Cu – Ti system
The microstructure of the samples after the interaction with Cu – Ti alloys, shown in figure 5, depended significantly on titanium content. At a concentration of 70 at. %, an active reaction of the sample with the melt occurred, its shape was not preserved, and large pores with sizes up to several millimeters were present in the structure. An initial microchannel structure was preserved neither, and after cooling, the sample consisted of large grains of titanium carbide with sizes of ~ $10^0 – 10^1$ μm in a copper alloy matrix.

With a decrease in the titanium concentration to 20 at. % or lower, the samples retained a macroscopic directional structure close to the microchannel structure of wood. However, the solid walls were divided into separate well-faceted grains with sizes from hundreds of nanometers to several microns. The presence of grains of a submicron size with a high total surface energy is a rather unexpected fact, since holding times were several days at 1150 °C, i.e., the conditions for recrystallization processes seem to be suitable, especially taking into account the presence of a contact between the grains and the liquid phase. A similar structure was observed in the central part of the samples after the interaction with Cu-30Ti alloys; however, grains with sizes of ~ $10^0 - 10^1$ µm were mainly present in the outer layer bordering the melt.

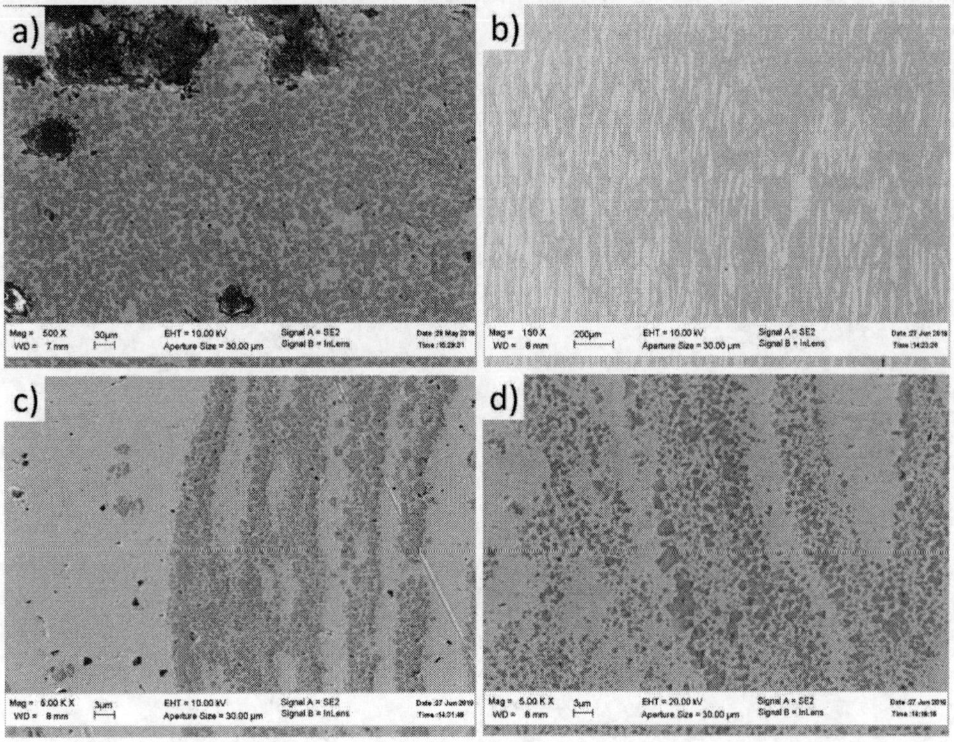

Figure 5. Microstructure of samples after reaction with Cu-Ti alloys: a) Cu-70Ti; b, c) Cu-10Ti; d) Cu-20Ti.

Probable explanation of the formation of such structures in this system is the features of thermodynamic equilibrium in Cu – Ti – C system.

The authors were unable to find a triple diagram of Cu – Ti – C system. However, it can be assumed that the solubility of carbon in Cu-70Ti melt is high, and the melt wets carbon well. Upon the contact of the melt with the biocarbon template, under the action of capillary forces, the microchannels is rapidly filled up, which is accompanied by dissolution of the cell walls. Due to high solubility of carbon in the melt, dissolution-crystallization of titanium carbide grains is actively taking place, which leads to an increase in their size.

The following model can be proposed to explain the formation of submicron TiC particles and maintain the texture. At low Ti concentrations in Cu, biocarbon does not dissolve in the melt, but turns into TiC, while the melt is depleted in titanium, and the infiltration front is shifted deeper into the sample. As the infiltration front moves forward, the regions being closer to the melt are enriched with titanium due to diffusion and, according to [17], TiC – Cu-Ti system is out of the local thermodynamic equilibrium. Stoichiometric titanium carbide begins to be depleted in carbon, and this process occurs by dissolving TiC in the melt, followed by precipitation of TiC_x, where $x < 1$. As a result, fragmentation of the walls occurs, and due to the fact that their initial thickness is small, the sizes of the formed grains have micron and submicron sizes. Considering that it is energetically more profitable for TiC_x grains to crystallize on a substrate, the fragmented walls remain connected for some time. Preservation of submicron grains during long exposures is probably related to the fact that after the formation of TiC_x with the minimum carbon content and as titanium content in the melt being in contact with it increases, the system again becomes thermodynamically stable, and the dissolution-crystallization processes significantly slow down. Nevertheless, this issue requires further study.

In the case of the interaction of biocarbon template with Cu-30Ti alloy, at an initial moment of time, the sample carbon actively dissolves in the melt. However, at a certain moment, the melt is saturated with carbon, and the process begins to occur similarly to alloys with low titanium content.

Thus, upon the interaction of bioC with melts of Cu – Ti system with titanium content of 10 to 70 at. % at 1150 °C, microchannel TiC template cannot be obtained. Nevertheless, the results obtained are quite promising from a practical viewpoint as a simple method of obtaining porous ceramics with nanoscopic grain size, as well as submicron-size TiC powders, which does not require complex equipment. Low rates of recrystallization of submicron TiC grains in this system are also of interest.

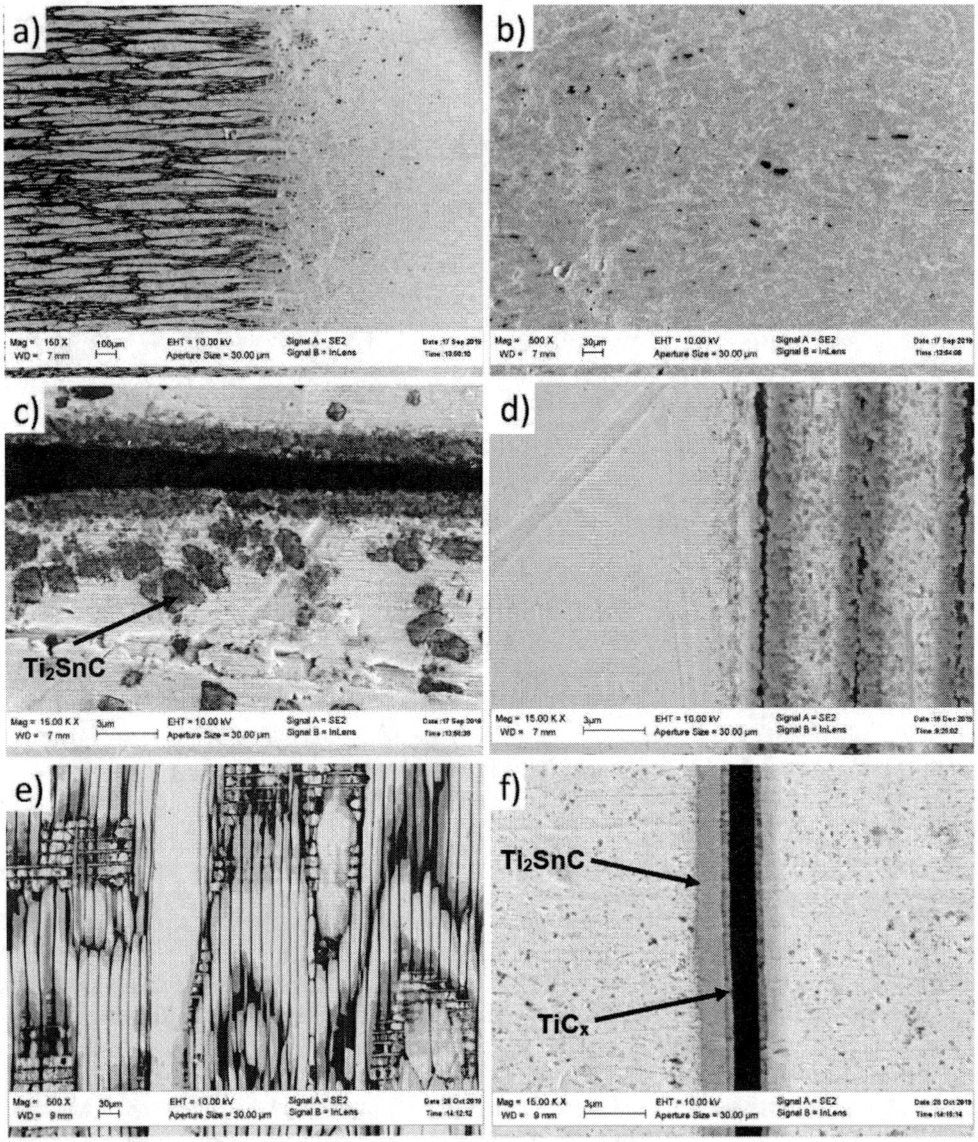

Figure 6. Microstructure of samples after reaction with Sn-Ti alloys: a) Sn-1150; b) Sn-1150, area near the melt/sample border; c) Sn-1150 front of infiltration area; d) Sn-1000 area near the melt/sample border; e, f) Sn-900.

4.3 Sn – Ti system

In Sn – Ti system, experiments were carried out at different temperatures (800 – 1150 °C) and at the same titanium concentration of 30 at. %. Due to the fact that the solubility limit of titanium at temperatures below ~ 980 °C is less than 30 at. %, in the experiments performed below this temperature, a part of titanium in the melt remained in the form of Ti_6Sn_5 compound.

At figure 6 it is seen that an initial structure of the wood was preserved much better than that in Cu – Ti system. At 1150 °C the cell walls are clearly visible in the region of the infiltration front, although their structure is rather incoherent, and in the channels there are also individual grains, which, according to the X-ray diffraction and EMPA analyzes, are Ti_2SnC compound. In addition, in the region directly bordering the melt, the cell walls were destroyed. Reducing the temperature to 1000 °C prevented the destruction of the cell walls in the region of a contact with the melt; however, the walls are also quite incoherent, and separate grains are present in the channels. At 800 °C, there are no individual grains in the microchannels; however, the carbon walls practically did not react with the melt.

The most interesting is the structure of materials obtained at 900 °C. It is clearly seen that carbon reacted quite actively with the melt, and the walls are a layered structure consisting of a central carbon part coated with a layer of titanium carbide and a dense Ti_2SnC layer at the boundary with the melt in the microchannels.

To explain the mechanisms of formation of such layers, one can use the data of [14], where TiC coatings on carbon fibers were obtained by the LMTA method at 1200 °C using Sn-Ti melt. It was shown that a TiC layer is formed directly on carbon: however, if Ti content in the melt exceeds 2.6 at. %, the Ti_2SnC phase is additionally formed at the boundary with the melt. The mechanism of formation of this phase consists in the diffusion of carbon through a layer of titanium carbide, its dissolution in the melt, followed by precipitation of grains and formation of a Ti_2SnC layer on the TiC surface.

This mechanism allows explaining the structure of the samples obtained at 1000 and 1150 °C, in particular, the presence of Ti_2SnC grains in the microchannels. The porosity of the Ti_2SnC layer can be related to the exposure time and the development of dissolution and recrystallization processes. A decrease in the temperature to 900 °C reduces the solubility of carbon in the melt, as evidenced by the absence of individual grains in the microchannels. In this regard, the mechanism of formation of the Ti_2SnC layer changes and occurs not due to the crystallization from the melt, but due to the interaction of diffusing carbon with tin at the boundary of TiC and later Ti_2SnC.

Thus, carrying out infiltration processes at 900 °C is the most promising, since it allows obtaining melt-resistant microchannel Ti_2SnC materials, which enables for long exposures and, as a result, removes restrictions on the volume of the parts obtained.

5. Conclusion.

In this work, an original method of obtaining biomorphic composites of the Me – MeC_x type, consisting of several stages, has been proposed. One of the steps is the INS method, which consists in converting biomorphic carbon preforms into porous microchannel MeC_x ceramics by transferring a carbide-forming element to carbon through a solvent that does not react with carbon. The interaction of alder-based biomorphic carbon templates with Cu-Ti and Sn-Ti systems, has been investigated. According to the research results, one can conclude that the proposed method of obtaining carbide templates can be used. However, the processes are quite complicated. This is due to a change in the chemical compositions of the alloy and carbide during the process, as well as the possibility of the formation of ternary compounds. These processes, in particular, have led to the formation of micron and nanometer TiC grains as a result of the interaction of biocarbon preforms with Cu-Ti alloys, as well as the formation of TiC and Ti_2SnC layers on the surface of microchannels while interacting with the Sn-Ti system..

Acknowledgments.
This work was supported by the RFBR project No. 19-03-01007 "Development of a new method of producing heat-resistant Me – MeC_x composites".

References.

[1] Zhao J C and Westbrook J H 2003 Ultrahigh-temperature materials for jet engines. *MRS bulletin* **28**, 622-630.

[2] Soboyejo W O and Srivatsan T S 2006 *Advanced structural materials: properties, design optimization, and applications*. CRC press.

[3] Svetlov I L, Karpov M I, Neuman A V and Stroganova T S 2018 Temperature Dependence of the Ultimate Strength of in situ Multicomponent Nb–Si–X (X= Ti, Hf, W, Cr, Al, Mo) Composites. *Russian Metallurgy (Metally)* **4** 348-353.

[4] Bewlay B P, Jackson M R, Subramanian P R and Zhao J C 2003 A review of very-high-temperature Nb-silicide-based composites. *Metallurgical and Materials Transactions A* **34** 2043-2052.

[5] Guan D L, Brooks C R and Liaw P K 2002 Microstructure and mechanical properties of as-cast and aged Nb–15 at.% Al–10 at.% Ti,–25 at.% Ti and–40 at.% Ti alloys. *Intermetallics* **10** 441-458.

[6] Jéhanno P, Heilmaier M, Saage H, Heyse H, Böning M, Kestler H and Schneibel J H 2006 Superplasticity of a multiphase refractory Mo–Si–B alloy. *Scripta Materialia* **55** 525-528.

[7] Korzhov V P, Ershov A E, Stroganova T S and Prokhorov D V 2016 Structure and mechanical properties of a multilayer carbide-hardened niobium composite material fabricated by diffusion welding. *Russian Metallurgy (Metally)* **4** 313-320.

[8] Svetlov I L, Kuzmina N A, Neiman A V, Ishadzhanova I V, Karpov M I, Stroganova T S, Korzhov V P and Vnukov V I 2015 Effect of the rate of solidification on the microstructure, phase composition, and strength of Nb/Nb_5Si_3 in-situ composites. *Bulletin of the Russian Academy of Sciences: Physics* **79** 1146-1150..

[9] Greil P, Lifka T and Kaindl A 1998 Biomorphic cellular silicon carbide ceramics from wood: I. Processing and microstructure. *Journal of the European Ceramic Society* **18** 1961-1973.

[10] Byrne C E and Nagle D C 1991 Carbonization of wood for advanced materials applications *Carbon* **35** 259–266.

[11] Ershov A E and Klassen N V 2015 Control of functional characteristics of biomorphous carbon matrices, SiC-Si-C composites, and SiC-C ceramics by prepressing wood *Inorganic Materials: Applied Research* **6** 79-90.

[12] Vogli E, Sieber H and Greil P 2002 Biomorphic SiC-ceramic prepared by Si-vapor phaseinfiltration of wood *Journal of the European Ceramic Society* **22** 2663-2668.

[13] Sun B, Fan T and Zhang D 2002 Porous TiC ceramics derived from wood template *Journal of Porous Materials* **9** 275-277.

[14] Vincent H, Vincent C, Mentzen B F, Pastor S and Bouix J 1998 Chemical interaction between carbon and titanium dissolved in liquid tin: crystal structure and reactivity of Ti_2SnC with Al. *Materials Science and Engineering: A* **256** 83-91.

[15] Kan X, Ding J, Yu C, Zhu H, Deng C and Li G 2017 Low-temperature fabrication of porous ZrC/C composite material from molten salts. *Ceramics International* **43** 6377-6384.

[16] Ding J, Deng C J, Yuan W J, Zhu H X and Li J 2013 Preparation of porous TiC/C ceramics using wooden template in molten salt media. *Advances in applied ceramics* **112** 131-135.

[17] Frage N, Froumin N and Dariel M P 2002 Wetting of TiC by non-reactive liquid metals. *Acta Materialia* **50** 237-245.

[18] *Phase diagrams of binary metallic systems Vol. 2* Ed. Lyakishev N P 1997 Moscow: Mashinostroenie 337

[19] *Phase diagrams of binary metallic systems Vol. 3 Book 2* Ed. Lyakishev N P 2000 Moscow: Mashinostroenie 330

IOP Conf. Series: Materials Science and Engineering **848** (2020) 012018 doi:10.1088/1757-899X/848/1/012018

Inhibitors based on new polysaccharides in technological processes of oil production. Efficiency of inhibition of scale formation and influence on stability of water-oil emulsion

A V Fakhreeva[1,2*], A I Voloshin[3], A G Telin[4], L A Magadova[5], L V Spirikhin[2] and V A Dokichev[2,3]

[1]Ufa State Aviation Technical University, Karl Marks st. 12, Ufa, Bashkortostan, 450008 Russia
[2]Ufa Institute of Chemistry, Russian Academy of Sciences, Oktyabrya pr. 71, Ufa, Bashkortostan, 450054 Russia
[3]OOO RN-BashNIPIneft', Sochinskaya st. 12, Ufa, Bashkortostan, 450103 Russia
[4]OOO Ufa-NTC, Aksakov st. 59, Ufa, Bashkortostan, 450076 Russia
[5]National University of Oil and Gas «Gubkin University», Leninskii pr. 65/1, Moscow, 119991 Russia
E-mail: alsu.allagulova@mail.ru

Abstract. Functionalized nitrogen-containing polysaccharides based on available low molecular weight sample of carboxymethylcellulose (CMC) with a molecular mass of 90 thousand and mono-, di- and tri- ethanolamine were obtained. The study of CMC salts with ethanolamine as inhibitors of salt deposition was carried out at 80 °C by capillary testing. It was shown that at the concentration of 10-30 mg/l these salts inhibit the processes of scale deposition of $CaCO_3$ and their efficiency is 53-80%. Rheological and sedimentation characteristics of artificially obtained water-oil (w/o) emulsions based on oil of the Priobskoye oilfield without and in the presence of low molecular weight sodium salt of carboxymethylcellulose (NaCMC) were studied. It was found that the NaCMC does not increase the viscosity and stability of W/O emulsions at concentrations of 50, 100, 200 mg/l.

1. Introduction

In the technological processes of development and operation of oil and gas fields, scale deposition in the bottom-hole zone of the formation, on downhole oilfield equipment and in the processes of oil preparation is one of the most common types of complications [1-4]. Scale deposition occurs as a result of crystallization of inorganic salts in the event that there is an excess of the actual concentration of scale-forming ions in mineralized water over the equilibrium concentration with changes in temperature and pressure.

The group of the most common scale deposits includes slightly soluble native minerals $CaCO_3$, $CaSO_4$, $BaSO_4$. Oilfield reagents based on phosphorus-containing compounds and organic polymers are widely used to prevent scale formation [4-7], the disadvantages of which include low thermal stability, high toxicity and relatively low efficiency in inhibiting sulfate deposits.

In recent decades, we can note intensive research in the field of search and development of new inhibitors of scale deposition based on polysaccharides [8], which is in good agreement with the general trend of "green" development of the processes of chemigation of modern industries, including oil and gas production.

Sodium salt of carboxymethylcellulose is one of the most accessible water-soluble polysaccharides, the physicochemical properties of which, as well as the availability, thermal stability, biodegradability in aerobic and anaerobic conditions determine the search and creation of new "green" oilfield reagents on its basis [9-11].

Without experimental data, it is difficult to assess the degree of influence of the polysaccharides on the properties of crude oil, the effectiveness of other oilfield reagents used in oil production. Therefore, the use of polysaccharides in oil production should be carried out only after studies of the impact on the production process and demulsification processes of w/o emulsions, which will prevent possible complications of the operation of oil preparation plants [12]. It is known that some polysaccharides can both stabilize and destroy w/o emulsion [13, 14].

This article presents the results of the synthesis of new mono-, di-, triethanolamine salts of CMC, their effectiveness in inhibiting the formation of calcium carbonate from supersaturated mineralized water. The results of studies of the effect of NaCMC, which is an effective inhibitor of scale deposition [15] on the viscosity and stability of the w/o emulsions of the Priobskoye oilfield, are considered.

2. Experimental part

NaCMC (1) produced by Aldrich with a molecular weight of 90 thousand was used in the work; the degree of substitution was 0.7.

The IR spectrum of polysaccharides was taken on the spectrometer "Nicolet iS10" on the prefix NPVO Diamant in the range of 400-4000 cm^{-1}.

Carboxymethylcellulose (H-CMC), (2). To 140 ml of 20% H_2SO_4 solution in 70% ethanol solution (pH = 2-2, 5) was added 5.0 g NaCMC and stirred on a magnetic stirrer, at room temperature for 20 minutes. H-CMC was filtered on a filter "blue ribbon", washed with 70% ethyl alcohol to a negative test for sulfate ions, dried in a vacuum. The yield of 4.51 g (97%).

Interaction of H-CMC with ethanolamine. To the suspension of 1 g H-CMC in 40 ml of distilled water in a three-neck flask of 100 ml, ethanolamine was added at the rate of 2 mol per carboxyl group and stirred for 5 hours at a temperature of 80 °C. the precipitated solid product was filtered and washed on a SCHOTT's funnel with isopropyl alcohol and dried in vacuum to a constant mass.

Monoethanolamine salt of carboxymethylcellulose (3). From 1 g H-CMC and 0.422 g (6.9 mmol) of monoethanolamine 1.034 g (3) was obtained. The yield 3 of 85%.

Diethanolamine salt of carboxymethylcellulose (4). From 1 g H-CMC and 0.727 g (6.9 mmol) diethanolamine 1.22 g (4) was obtained. The yield 4 of 89%.

Triethanolamine salt of carboxymethylcellulose (5). From 1 g of H-CMC and 1.033 g (6.9 mmol) triethanolamine, 1.414 g (5) was obtained. The yield 5 of 93%.

Elemental analysis data and IR spectra of CMC and ethanolamine salts of CMC 1-5 are presented in figure 1 and table 1.

Table 1. Elemental analysis of polysaccharides **1-5**.

polysaccharide	C, %	H, %	N, %
1	35.8, 35.9	4.9, 4.9	-
2	40.0, 40.0	6.2, 5.8	-
3	35.6, 35.5	6.7, 6.8	3.0, 2.9
4	38.2, 38.4	6.3, 6.0	2.8, 2.8
5	39.1, 39.1	6.1, 6.0	2.7, 2.6

The effect of NaCMC and its functional derivatives on the inhibition of $CaCO_3$ formation was determined at a temperature of 80 ^{0}C and a polysaccharide concentration of 10, 30 and 50 mg/l by a method based on blocking the steel capillary formed by calcium carbonate crystals. The reservoir water model (RWM) was pumped at a rate of 1 ml/min through a 2.5 m long stainless-steel capillary with an internal diameter of 1 mm and the dynamics of the pressure drop in the capillary caused by the deposition of $CaCO_3$ was recorded.

The effectiveness of inhibition of calcium carbonate formation was determined by the formula (1):

$$E = 100\times(\Delta P_1 - \Delta P_2)/\Delta P_1 \qquad (1)$$

where ΔP_1 is the pressure drop on the capillary when pumping RWM without polysaccharides; ΔP_2 is the pressure drop on the capillary when pumping RWM with polysaccharides.

To prepare emulsions, a sample of oil from the Priobskoye oilfield was used as a hydrocarbon phase. Anhydrous oil had the following physical and chemical characteristics: oil density 0.876 g/cm^3 at 20°C; water cut determined by the Dean-Stark method – 0%. Component composition of oil: paraffins – 3.6%, asphaltenes – 2.6%, resins – 15.2%. The formation water of the Priobskoye field had the following characteristics: mineralization – 13.2 g/l; density –1.01 g/cm^3; pH = 8.1; ionic composition: Ca^{2+} – 250; Mg^{2+} – 85.0; $Na^+ + K^+$ – 3048; Cl^- – 4425; HCO_3^- – 1658 mg/l.

The preparation of emulsions without polysaccharides was carried out by mixing the aqueous and hydrocarbon phases in a ratio of 1:1 at room temperature for 30 min using a mechanical stirrer at a speed of 1400 rpm. In a container put the oil in droplets and when mixing the injected aqueous phase for 10 min, after which the composition was stirred 20 min [12].

Water-oil emulsion in the presence of polysaccharides was obtained by dissolving in RWM of the Priobskoye oilfield the calculated amount of polysaccharide (H-CMC and arabinogalactan). The density of water samples was determined by a standard method using an aerometer.

Rheological characteristics (viscosity@shear rate) of artificial emulsions were determined on rotary viscometer RHEOTEST RN4.1, designed to measure the dynamic viscosity of the liquid and rheological studies. For the studied samples, a shear test was applied when the shear rate changed in the range from 1 to 300 c^{-1}, at which the dependence of the shear stress and viscosity on the shear rate was determined.

To characterize the process of separation of w/o emulsion in the absence and in the presence of polysaccharides, a complex system "Lumi Fuge Stability Analyser" was used to study dispersions on the basis of a multi-seat analytical centrifuge. The device allows you to quickly characterize any process with phase separation, calculate the rate of delamination depending on the centrifugal forces.

Micrographs of w/o emulsion were obtained using a microscope "Micromed 3" (LOMO) at a magnification of 400 times.

3. Results and discussion

For the synthesis of CMC salts with ethanolamines and CMC amides, the sodium salt of carboxymethylcellulose **1** was pre-converted to H-form **2** by treatment with a 20% H_2SO_4 solution in a 70% ethanol solution at room temperature.

The interaction of carboxymethylcellulose **1** with a 2-fold molar excess of ethanolamines is most successful in an aqueous solution at 80 ° C. Under these conditions, the only reaction products are monoethanolamine **3**, diethanolamine **4** and triethanolamine **5** carboxymethylcellulose salts formed with yields of 85, 89 and 93% respectively.

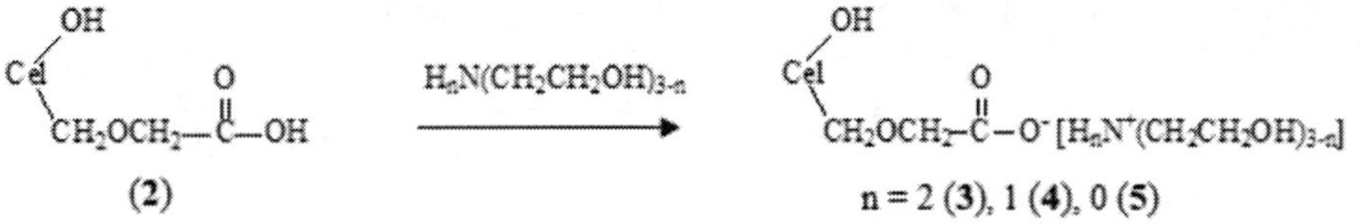

In the IR spectra of the obtained compounds, **3-5** valence oscillations of O-H groups are manifested in a wide frequency range of 3000-3600 cm^{-1}, absorption bands characteristic of carboxylation are observed in the region of 1577-1585 cm^{-1}, and for the ammonium group at ~ 2870 cm^{-1} (figure 1).

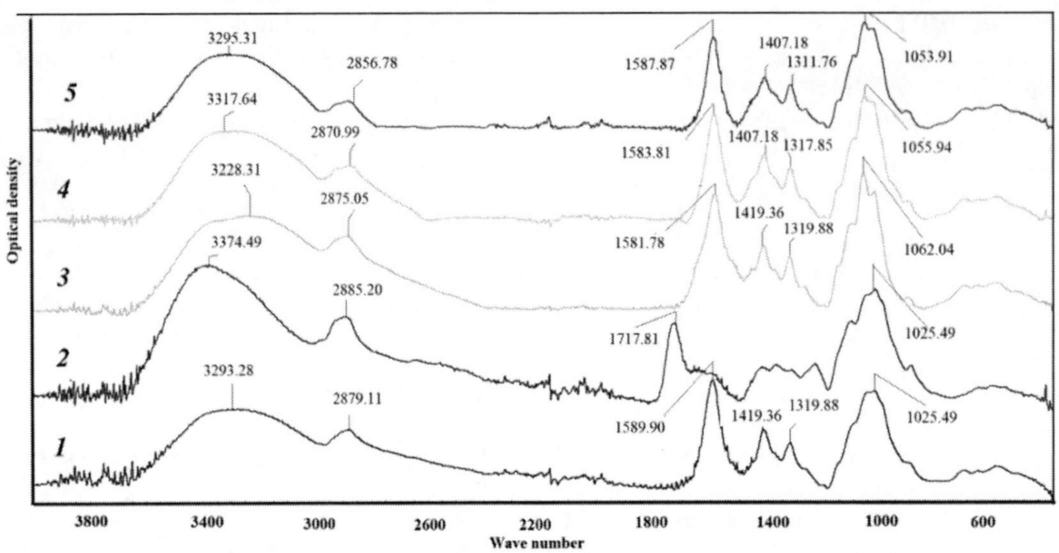

Figure 1. IR spectra of polysaccharides **1-5**. ν - wave number, cm^{-1}.

It is known that NaCMC can form sufficiently stable complexes with the Ca^{2+} ion [16,17]. Therefore, assuming that functionalized polysaccharides can prevent the precipitation of calcium salts from supersaturated aqueous solutions and inhibit scale deposition; studies were conducted on its effect on the crystallization process of $CaCO_3$.

The study of CMC salts with ethanolamines as inhibitors of salt deposition at a temperature of 80 °C showed that at a concentration of 10-30 mg/l, the reagents inhibit the processes of scale deposition of $CaCO_3$ and their efficiency is 53-80% (figure 2).

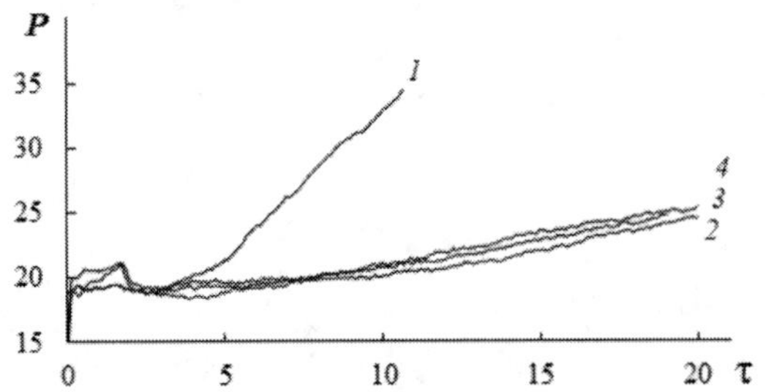

Figure 2. Effect of polysaccharide **3** on $CaCO_3$ salt deposition at concentrations 0 (1), 10 (2), 30 (3) and 50 (4) mg/l at 80 °C. *P* - pressure (kPa), *τ* - time (min).

The efficiency with respect to inhibition of calcium carbonate at a concentration of 10 mg/l is higher for ethanolamine salts of CMC than for NaCMC. At higher concentrations, this effect disappears. Considering the efficiency of inhibition may be noted that in the range from **3** to monoethanolamine diethanolamine and triethanolamine salt of CMC **5**, it decreases.

It is important that the reagents used in the production and preparation of oil are compatible with reservoir water and oil, and do not have a negative impact on w/o emulsions. In order to obtain data on the effect of polysaccharides on the stability of emulsion, we studied the rheological and sedimentation characteristics of artificially produced w/o emulsions based on oil from the Priobskoye field without and in the presence of **1**.

To solve the problem of the effect of polysaccharide on the properties of w/o emulsion, emulsions containing 50% of model water with and without polysaccharide were prepared. The concentration of NaCMC in the aqueous phase of the emulsion was 50, 100, 200 mg/l. The density of water-oil emulsions without polysaccharide was 0.940 g/cm^3, for the emulsion in the presence of NaCMC, the density practically did not change.

The rheological curve of oil in the Priobskoye field at a temperature of 20 °C deviates only slightly from Newton's law. The viscosity of oil at a shear rate of 2.8 s^{-1} is 34.6 mPa·s. The viscosity of w/o emulsion is 578 mPa·s and has increased more than 16 times compared to oil. The dependence of viscosity on shear rate is typical for a typical visco-plastic non-Newtonian liquid (figure 3). The viscosity of w/o emulsion in the presence of H-CMC at a concentration of 50 mg/l decreases by more than 30% at low shear rates. Moreover, the effect of reducing the viscosity of w/o emulsion is greatest at a concentration of 50 mg/l with increasing concentration; the effect of reducing the viscosity is less (figure 4). It can be assumed that with the increase in the concentration of polysaccharide, the properties stabilizing the w/o emulsion are manifested, in particular, in the formation of the armor layer at the phase separation. This effect of the polysaccharide is quite consistent with the known literary results [13,14].

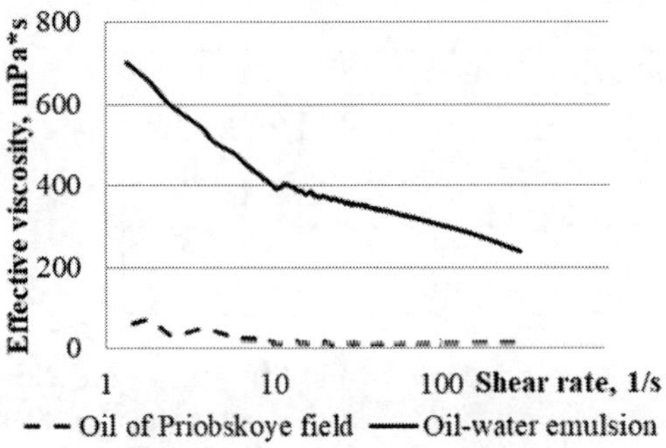

Figure 3. Dynamic viscosity as a function of shear rate of Priobskoye oil field and oil-water emulsion prepared without polysaccharide.

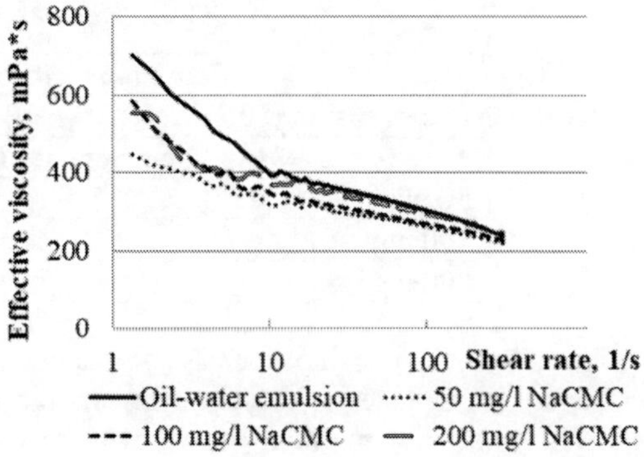

Figure 4. Dynamic viscosity as a function of shear rate for w/o emulsion without polysaccharide and in the presence of 50, 100, 200 mg/l NaCMC.

According to the results obtained from the transmission data (transmission profiles - "fingerprints"), the w/o emulsion of the Priobskoye oilfield is sufficiently stable, which is confirmed by the relatively low rate of emulsion delamination and the relatively low value of the Lumi Fuge instability index (figure 5). In the presence of polysaccharide, w/o emulsion is markedly destabilized, especially at a concentration of 100 mg/l (figure 5 and table 2).

The profiles of transmittance	**Kinetics of demulsifications**
w/o emulsion without polysaccharide	
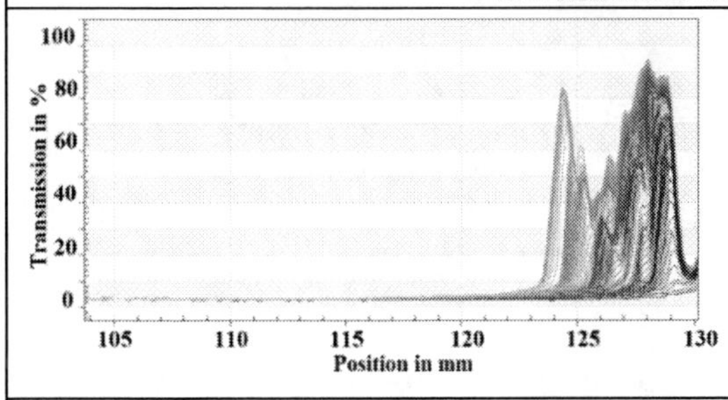	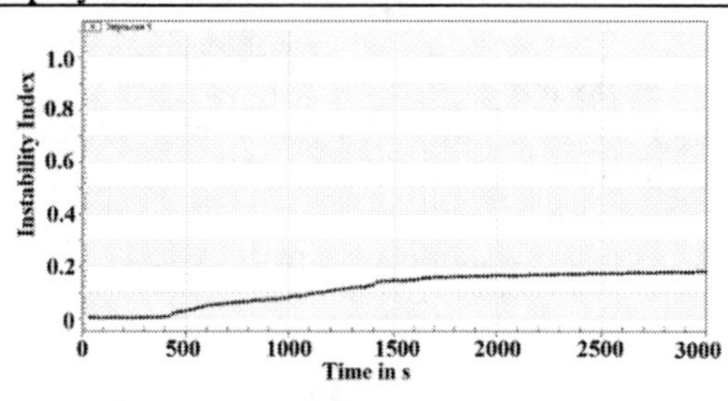
w/o emulsin in the presence of 100 mg/l NaCMC	

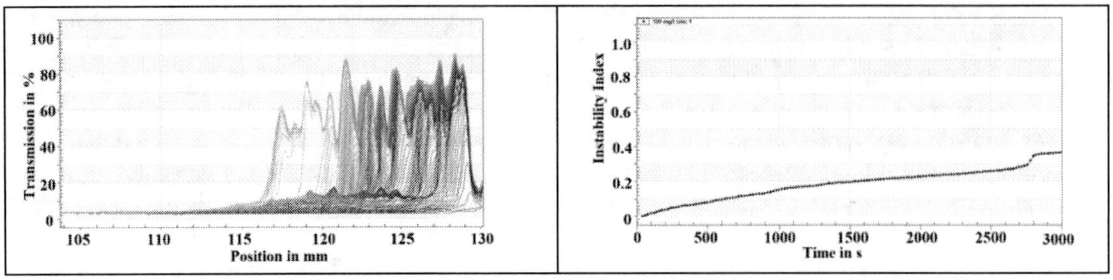

Figure 5. Transmission profiles and demulsification kinetics for w/o emulsion without polysaccharide and in the presence of 100 mg/l NaCMC.

Table 2. Instability Index Lumi Fuge

Emulsions without additives NaCMC		
w/o emulsion (50% water/50% oil)	0.18	0.21
Emulsions with additives NaCMC		
50 mg/l	0.21	0.17
100 mg/l	0.30	0.36
200 mg/l	0.17	0.17

Thus, NaCMC does not increase the stability of the w/o emulsion of the Priobskoye oilfield. For w/o emulsion samples without polysaccharide (figure 6a), as well as in the presence of 100 mg/l NaCMC (figure 6b), micrographs were obtained using at a magnification of 400 times. From the photo it can be seen that in the presence of NaCMC water globules have large sizes, causing less stability of the emulsion over time.

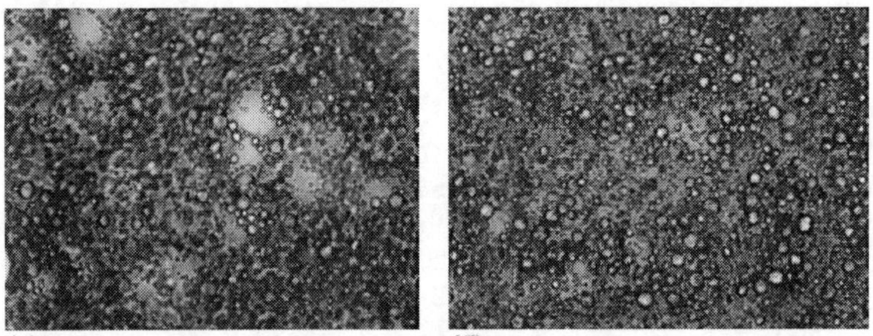

(a) (б)

Figure 6. Micrographs of w/o emulsion without polysaccharide (a), in the presence of 100 mg/l NaCMC (b).

4. Conclusion

On the basis of carboxymethylcellulose ethanolamine salts, new highly effective inhibitors of calcium carbonate deposits were obtained, which are superior in efficiency to NaCMC.

It was found that the sodium salt of carboxymethylcellulose does not increase the effective viscosity and stability of VNE at concentrations of 50, 100, 200 mg/l.

Acknowledgement

The research is performed under the scope of the State Order of the Ministry of Education and Science of the Russian Federation (№ 4.2703.2017/ПЧ) on subject №AAAA-A17-1170011910021-8.

References

[1] Crabtree M, Eslinger D, Fletcher P et al. 2002 *Oilfield Rev.* **11 N3** 30-45

IOP Conf. Series: Materials Science and Engineering **848** (2020) 012018 doi:10.1088/1757-899X/848/1/012018

[2] Olajire A A 2015 J. *Pet. Sci. Eng.* **135** 723-737
[3] *Pereyma A A* 2015 *Oil Industry* **N2** 84-87
[4] Kashchavtsev, V E, Mishchenko, I T 2004 *Soleobrazovanie pri dobyche nefti (Scaling in Oil Extraction),* Moscow: Orbita-M p 140
[5] Sawada K 1997 *Pure Appl. Chem.* **69 N5** 921-928
[6] Reddy M M 1977 *J. Cryst. Growth.* **41 N2** 287-295
[7] Ishmiyarov E R, Rakhimova N T, Latypova D R et al. 2015 *Russ. J. of Applied Chem.* **88 N7** 1174-1177
[8] Voloshin A I, Gusakov V N, Fakhreeva A V et al. 2018 *Neftepromislovoe delo (Oilfield engineering)* **N11** 60-72
[9] Fakhreeva A V, Gusakov V N, Voloshin A I et al. 2016 *Russ. J. Appl. Chem.* **89 N12** 1541-1545
[10] Dokichev V A, Fakhreeva A V, Voloshin A I et al. 2018 *Equipment and technologies for oil and gas complex* **5** 43-48
[11] Klemm D. Cellulose / Klemm D, Schmauder H 2002 *Biopolymers* **6** 275-319
[12] Graifer V I, Lazarev G A, Leontiev M I. 1973 *Neftepromislovoe delo (Oilfield engineering)* **N2** 21-23
[13] Hippmann S, Ahmed S S, Fröhlich P et al 2018 *Colloids and Surfaces A: Physicochemical and Engineering Aspects* **553** 71-79
[14] Martínez-Palou R, Mosqueira M de L, Zapata-Rendón B et al. 2011 *J. of Petroleum Science and Engineering* **75(3-4)** 274-282
[15] Fakhreeva A V, Voloshin A I, Telin A G et al *IOP Conference Series: Materials Science and Engineering* 2019 **525** 012050
[16] Zakharov N A, Ezhova Zh A, Koval' E M et al. 2005 *Inorg. Mater.* **41** 509-515
[17] Backfolka K, Lagergeb S, Rosenholmc J et al. 2002 *J. of Colloid and Interface Science* **248** 5–12.

IOP Conf. Series: Materials Science and Engineering **848** (2020) 012019 doi:10.1088/1757-899X/848/1/012019

Reactionary-solidified oxygen permeable membrane material based on cermet $Bi_{1.6}Er_{0.4}O_3$ – 26 wt % Ag – 4 wt % In

S V Fedorov[1], A S Lysenkov[1] and I V Kulbakin[1]

[1] Baikov Institute of Metallurgy & Materials Science (IMET RAS), 49, Leninsky Prospekt, Moscow, 119334, Russian Federation

E-mail: fedserv@rambler.ru

Abstract. Using hot uniaxially pressing in argon atmosphere at 800 °C and at a stress of 6 MPa, the high-density reactionary-solidified cermet material based on $Bi_{1.6}Er_{0.4}O_3$ - 26 wt % Ag - 4 wt % In with total porosity of less than 0.3 % was firstly synthesized. The relationship between phase composition, microstructure, oxygen permeability coefficient, and oxygen selectivity (vs air nitrogen) of this material before and after its long-term (100 hours) operation as an ion transport membrane was established. The dependences of oxygen transport through this cermet material at 650-800 °C were investigated: for example, at 800 °C and at a ratio of oxygen partial pressures on both sides of the membrane of about 20, the oxygen transfer regime changes from diffusion to kinetics while the characteristic thicknesses of membrane were of 0.11 cm. It was shown that at 800 °C the membrane material based on $Bi_{1.6}Er_{0.4}O_3$ - 26 wt % Ag - 4 wt % In had an oxygen permeability coefficient of $1.1 \cdot 10^{-8}$ mol/cm/s, while its selectivity for oxygen (over nitrogen) was about 100. The decrease in oxygen flux by 1.3-1.5 times during 100 hours of work at 700-800 °C with a slight change in oxygen selectivity was observed, thus the possible causes of this phenomenon were explained.

1. Introduction

Despite many years of researching, there is still no optimal medium-temperature (600-800 °C) membrane material for separation of oxygen from air, since the currently known single-phase (cobaltites, cobaltite-ferrites and manganites of alkaline-earth or rare-earth cations) or composite (cermets, perovskite-fluorite or spinel-fluorite composites) materials do not completely satisfy all the requirements for long-term stability or performance [1-4]. In some of them (perovskites and cermets), a high level of oxygen permeability has already been achieved, while, unlike perovskites, cermets have better thermal, chemical and mechanical stability. In addition, the indisputable advantage of cermet materials for ion-transport membranes compared to perovskite ones is a lower range of operating temperatures (from 500 °C), and also, when having both percolation of ion-conducting ceramic and electron-conducting metal nets, the much higher oxygen permeability is achieved [5-8]. However, due to the differences of ceramic and metallic components, there is a technological problem of creating high-density, highly oxygen-selective (vs air nitrogen) and oxygen-permeable cermets with percolative ceramic and metal components. This is mainly due to the formation of either scaly (i.e. incoherent, when silver is introduced into cermet in the form of metal with its subsequent pressing, resulting to the formation of a membrane with low oxygen permeability, but with high oxygen selectivity), or porous (i.e. non-gas tight, when silver is introduced in cermet in the form of Ag_2O, and the resulting membrane is highly oxygen-permeable, but poorly oxygen-selective) of the structure of the material. In this regard, the question of creating porous-free cermet materials with high selective

IOP Conf. Series: Materials Science and Engineering **848** (2020) 012019 doi:10.1088/1757-899X/848/1/012019

oxygen permeability is very relevant. This problem has been solved in our laboratory by creating the partly molten oxide composite membrane materials [9-11]. However, such membranes have a significant drawback that impedes their practical use - high creep, accompanied by deterioration in transport characteristics. To overcome this problem, in the present work, it is proposed to use a fusible component not at membrane operating temperatures, but in the process of preparation of the material. Metallic indium, known to have a melting point of about 160 °C and oxidation resistance up to 800 °C, will be taken as such a low-melting component [12]. It is assumed that liquid indium, which is a surface-active modifier of the silver particles [13], will provide compaction of the material and will prevent recrystallization of silver particles, and will also interact with silver above 330 °C to form a plastic α-Ag solid solution [12]. The choice of the ceramic component $Bi_{1.6}Er_{0.4}O_3$ is due to its high oxygen-ion conductivity at 500-800 °C [14].

In this regard, the main aim of this work is to develop a reactionary-solidified membrane material for oxygen separation based on cermet $Bi_{1.6}Er_{0.4}O_3$ - 26 wt % Ag - 4 wt % In. Its composition has been empirically selected based on the Ag-In phase diagram [12], as well as from considerations of further obtaining the maximum of electronic and ionic conductivity for the membrane material.

2. Experimental

The synthesis of samples was carried out in two stages. At the first stage, $Bi_{1.6}Er_{0.4}O_3$ powder was obtained by solid-phase synthesis from starting components of bismuth oxide (Bi_2O_3) and erbium oxide (Er_2O_3) purity of 99.99 %. The powder mixture of Bi_2O_3 and Er_2O_3 was ground in isopropyl alcohol for 30 minutes using a Pulverisette 5 planetary mill (Fritsch, Germany) at room temperature, then was dried an oven at 120 °C for 4 hours and was annealed at 800 °C for 20 hours with intermediate grinding. Silver oxide (I) was obtained by adding to the aqueous solution of silver nitrate (1 M) a twofold excess of potassium hydroxide (2.5 M) according to reaction: $2AgNO_3 + 2KOH$ (excess) $\rightarrow Ag_2O\downarrow + 2KNO_3 + H_2O$. The resulting black precipitate was filtered using a Buchner funnel, washed thoroughly and dried in an oven at 120 °C for 4 hours.

At the second stage, powder mixture of $Bi_{1.6}Er_{0.4}O_3$ - 26 wt % Ag - 4 wt % In composition was prepared, taking into account the reduction of Ag_2O to Ag, and then it was thoroughly mixed in a planetary mill for 1 h in isopropyl alcohol and was dried in oven at 120 °C for 4 hours. Then, the powder was placed in a corundum mold (diameter of 1.5 cm) and calcined in an HP20-3560-20 hot uniaxial pressing machine (Thermal Technology Inc., USA) at a temperature of 800 °C for 1 hour in an argon atmosphere at a pressure of 6 MPa. To relieve mechanical stresses, the obtained samples were calcined at 800 °C for 24 hours in a vacuum furnace in argon atmosphere. The obtained samples were in the form of tablets with a diameter of 1.5 cm and a height of 1.5-2.0 cm, from which the samples of 0.1-0.5 cm thick were made on a cutting machine (STRUERS, Denmark).

Phase composition of both oxide powders of $Bi_{1.6}Er_{0.4}O_3$, Ag_2O and cermet membranes was characterized by X-ray phase analysis on an XRD-6000 diffractometer (Shimadzu, Japan) using CuKα - radiation. The decoding of X-ray diffraction patterns was performed using ICDD PDF-2 database. Microstructure of polished cross section for the cermets before and after long-term tests, which had been prepared by STRUERS equipment, was investigated by Olympus GX-51 optical microscope. Open, closed and total porosity of the composite was measured by hydrostatic weighing.

The measurements of oxygen fluxes for cermet membranes (tablet of Ø = 1.5 cm, h = 0.1-0.4 cm) were performed by gas chromatography technique at 650 °C, 700 °C, 750 °C and 800 °C on Crystallux-4000M (Russia) gas chromatograph. To do this, the polished tablet was tightly connected to the end of the measuring cell, which was placed into a tube furnace and heated to the temperature of 650-800 °C. From the outside face, the tablet was in contact with air, and from the inner face it was connected to a gas chromatograph, while the sweep gas (helium) was blown through the measuring cell to remove the gas mixture penetrated through the sample. The selectivity of composite membrane was evaluated by the ratio of concentrations of oxygen over nitrogen in the analyzing gas mixture (consisted of oxygen, nitrogen, and helium). The rate of sweep gas flow was 5-20 ml/min. Oxygen fluxes, oxygen permeability coefficients, and characteristic thicknesses of membranes were measured according to methods, fully described earlier in our previous papers [9, 15].

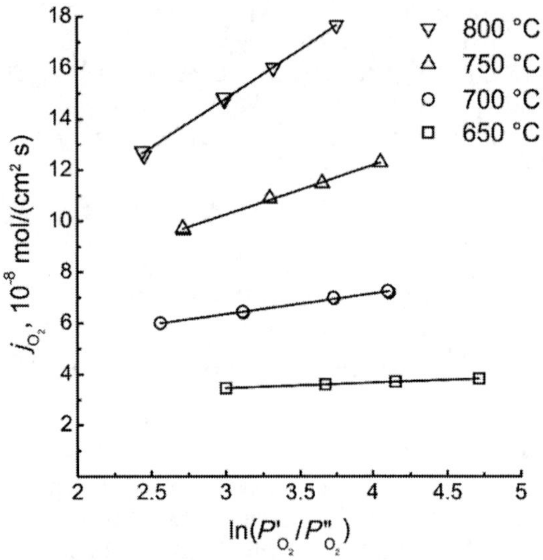

Figure 1a. Dependences of oxygen flux (j) for 0.1 cm thick cermet membrane on oxygen partial pressure difference (ln P'/P'') at 650-800 °C

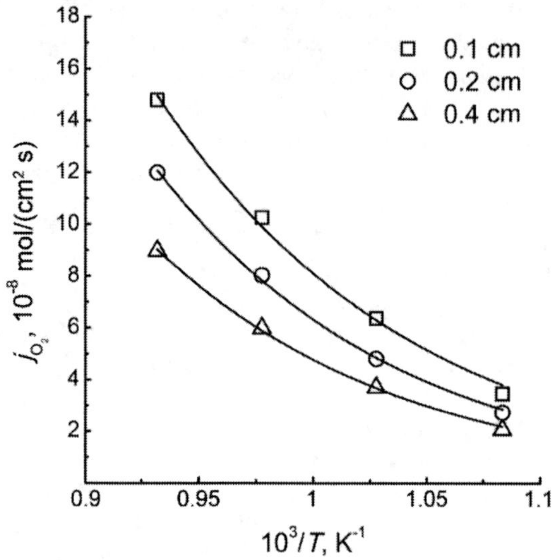

Figure 1b. Dependences of oxygen flux (j) for 0.1, 0.2, and 0.4 cm thick cermet membranes on inverse temperature (1000/T) at ln P'/P'' = 3

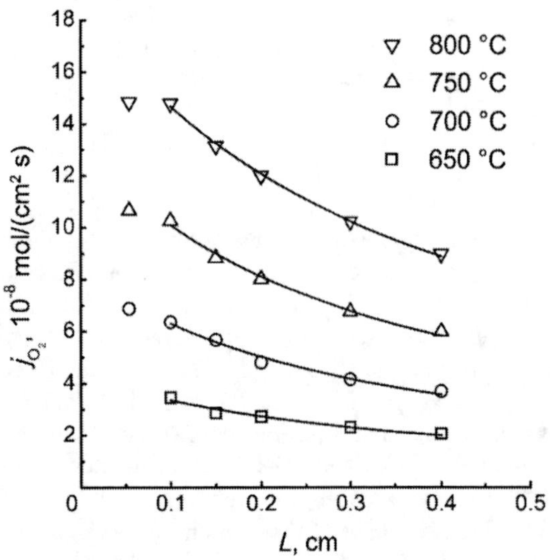

Figure 2a. Dependences of oxygen flux (j) for 0.1-0.4 cm thick cermet membranes on their thickness (L) at 650-800 °C when oxygen partial pressure difference ln P'/P'' = 3

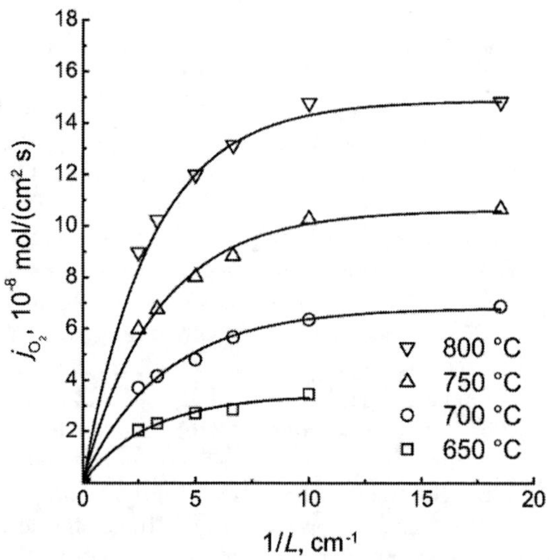

Figure 2b. Dependences of oxygen flux (j) for 0.1-0.4 cm thick cermet membranes on their inverse thickness (1/L) at 650-800 °C when oxygen partial pressure difference ln P'/P'' = 3

3. Results and discussion

Figure 1 illustrates the dependences of oxygen flux for 0.1 cm thick cermet membrane on oxygen partial pressure differences and on inverse temperature. The oxygen fluxes rise at an increase in the temperature and oxygen partial pressure differences, and evaluated activation energy of oxygen mass transfer is about 0.8 eV. To establish the regime of oxygen mass transfer through cermet membranes, oxygen permeation tests for some membranes with different thickness have been examined (Figure 2).The values of characteristic thicknesses of all the membranes at every temperatures, according to Burggraaf's model of mass transfer [16], were close to 0.11 cm. Such values are in good agreements with literature data [5] for the same membrane materials at the same temperatures, but without indium.

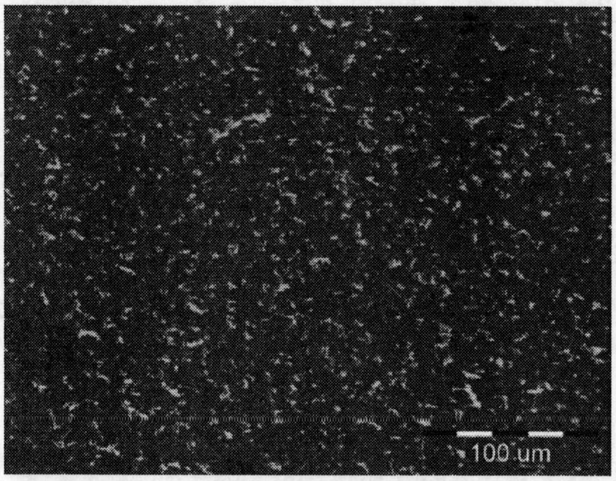

Figure 3a. Micrograph of polished cross section for as-sintered initial cermet membrane material of $Bi_{1.6}Er_{0.4}O_3$ - α-Ag

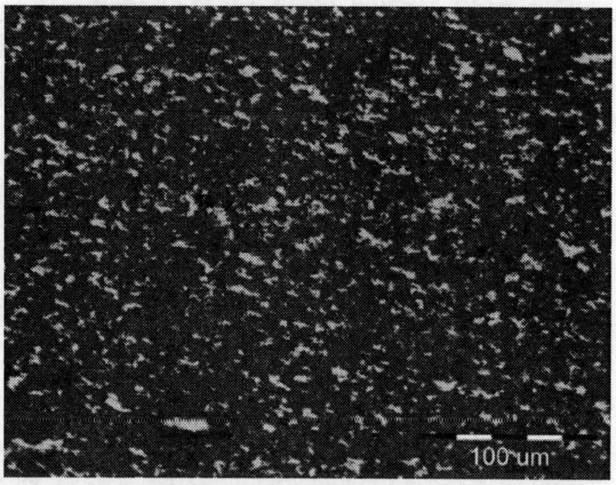

Figure 3b. Micrograph of polished cross section for the same cermet membrane material after gas chromatographic tests at 700 °C during 100 hours

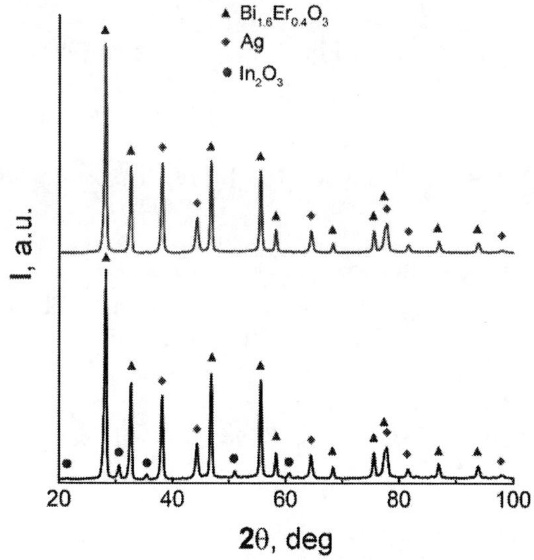

Figure 4. X-rays patterns of initial cermet membrane material (upper graph) and that of the same material after gas chromatographic tests at 700 °C during 100 hours (lower graph)

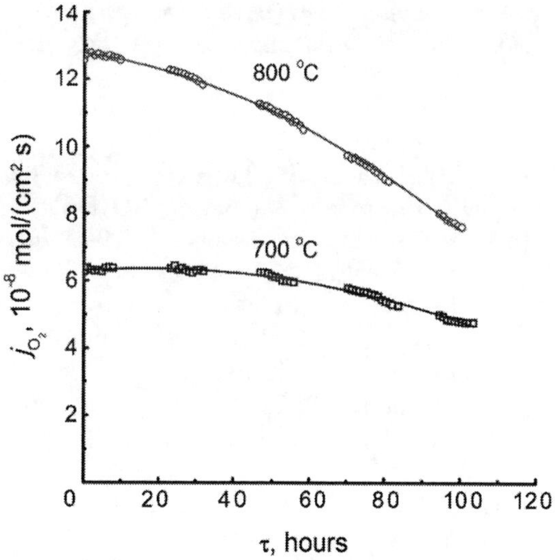

Figure 5. Time (τ) dependences of oxygen flux (j) for 0.1 cm thick cermet membrane at 700 °C and at 800 °C when oxygen partial pressure difference ln P'/P'' = 3

The calculated value of oxygen permeability coefficient was close to 1.1 • 10^{-8} mol/cm/s at 800 °C, which is also close to literature data [5], but in our case the content of expensive silver is 25 % lower. The estimated value of selectivity for oxygen (over nitrogen) was about 100 (the value of total porosity was less than 0.3 %), that is definitely very outstanding result for such cermet membrane material. But, the key factor for such membrane material with different components (metal and ceramics) is its long-term stability. Figures 3-5 indicate the problems of instability for membrane material, like agglomeration of silver (Figure 3) or formation of low conductive indium oxide (Figure 4), and, as the result, decreasing of oxygen fluxes during 100 hours of exploitation at 700 °C and at 800 °C by 1.3 and 1.5 times, accordingly. One more problem is the migration of oxygen-conductive silver [17-18] onto the opposite side of membrane, which is the main hindering factor, besides the cost of silver-containing cermets, overcoming their future application in oxygen membrane technology.

4. Conclusions

A new cermet membrane material based on $Bi_{1.6}Er_{0.4}O_3$ - 26 wt % Ag - 4 wt % In has been firstly synthesized and characterized. The idea of forming the reactionary-solidified oxygen-selective and oxygen-permeable membrane material is quiet workable and applicable. The evaluated selectivity of oxygen over nitrogen is outstanding, but in fact the instability of transport properties, phase composition, and microstructure of investigated cermet membrane material during long-term tests gives us no chance to the future application of such material in oxygen membrane technology.

Acknowledgements
This work was supported by the Russian Foundation for Basic Research (Project No. 17-08-01079-a) and was carried out according to the State Assignment of IMET RAS (No. 075-00947-20-00).

References

[1] Zhang C, Sunarso J and Liu S 2017 *Chemical Society Reviews* **46** 2941
[2] Zhu X and Yang W 2017 *Mixed conducting ceramic membranes: fundamentals, materials and applications* (Berlin: Springer) 367
[3] Sunarso J, Baumann S, Serra J M, Meulenberg W A, Liu S, Lin Y S and Diniz da Costa J C 2008 *Journal of Membrane Science* **320** 13
[4] Liu Y, Zhu X, Li M, Liu H, Cong Y and Yang W 2013 *Angewandte Chemie International Edition* **52** 3232
[5] Ten Elshof J E, Nguyen N Q, den Otter M W and Bouwmeester H J M 1997 *Journal of the Electrochemical Society* **144** 4361
[6] Kobayashi K and Tsunoda T 2004 *Solid State Ionics* **175** 405
[7] Kim J and Lin Y S 2000 *Journal of Membrane Science* **167** 123
[8] Ruiz-Trejo E, Boldrin P, Lubin A, Tariq F, Fearn S, Chater R, Cook S N, Atkinson A, Gruar R I, Tighe C J, Darr J and Brandon N P 2014 *Chemistry of Materials* **26** 3887
[9] Belousov V V, Kulbakin I V, Fedorov S V and Klimashin A A 2016 *ACS Applied Materials & Interfaces* **8** 22324
[10] Kulbakin I V, Fedorov S V and Belousov V V 2018 *Journal of the Electrochemical Society* **165** H861
[11] Belousov V V 2016 *Ionics* **22** 451
[12] Yatsenko S P 1987 *Indium. Properties and application* (Moscow: Nauka) 256
[13] Zadumkin S N and Pugachevich P P 1962 *Doklady of the USSR Academy of Sciences* **146** 1363
[14] Verkerk M J, Keizer K and Burggraaf A J 1980 *Journal of Applied Electrochemistry* **10** 81
[15] Belousov V V, Fedorov S V and Vorobiev A V 2011 *Journal of the Electrochemical Society* **158** B601
[16] Bouwmeester H J M, Kruidhof H and Burggraaf A J 1994 *Solid State Ionics* **72** 185
[17] Laubitz M J 1969 *Canadian Journal of Physics* **47** 2633
[18] Gryaznov V M, Gul'yanova S G and Serov Yu M 1989 *Russian Chemical Reviews* **58** 35

Effect of tantalum on short-term creep of a 12%Cr-3%Co-0.07%Ta martensitic steel for steam blades

A E Fedoseeva[1*], I S Nikitin[1], A E Fedoseev[1], E S Tkachev[1] and R O Kaibyshev[1]

[1]Belgorod National Research University, Pobeda 85 308015 Belgorod, Russia
[*]Corresponding author: fedoseeva@bsu.edu.ru

Abstract. 9-12%Cr martensitic steels are prospective materials for elements of boilers, tubes, pipes, heaters and steam blades for fossil power plants, which are able to work at ultra-supercritical parameters of steam. Decreasing N content and increasing B content together with increasing Cr content required the optimization of ferrite-stabilizing and austenite-stabilizing elements to avoid the formation of δ-ferrite at high temperatures. Addition of 3-4%Co, 0.01%B, 0.80%Cu and 0.07%Ta as austenite-stabilizing elements can compensate increasing Cr content up to 12% at decreasing N content to 0.003% and provides the δ-ferrite in the amount less than 10%. Two Co-containing 12%Cr steels were developed and investigated. The main difference between these steels is Ta addition; one of the 12%Cr steel studied contains 0.07%Ta, other 12%Cr steel is Ta-free one. Creep tests at 650°C and the applied stresses ranging 180-80 MPa with a step of 20 MPa for both Co-containing 12%Cr steels were carried out. Addition of Ta in the Co-containing 12%Cr steel provide the increment in the creep time to rupture at both high and low applied stresses. This increment in the creep properties due to Ta addition was explained by decrease in coarsening rate of Ta-rich MX carbonitrides during creep at 650°C.

1. Introduction

Nine to 12%Cr martensitic steels are prospective candidate as materials for elements of boilers, tubes, pipes, heaters and steam blades for new units of fossil power plants, which are able to work at ultra-supercritical parameters of steam [1-3]. High creep resistance of the 9-12%Cr martensitic steels is achieved by the formation of the tempered martensite lath structure (TMLS) after heat treatment consisted in normalizing and medium-temperature tempering [4-7]. The high-angle boundaries of prior austenite grains (PAGs), packets and blocks, as well as low-angle boundaries of martensitic laths are decorated by $M_{23}C_6$ carbides enriched by Cr, Fe and W [8,9], while fine MX carbonitrides play a role of the obstacles for rearrangements of free lattice dislocations into more stable configurations [10,11]. Main problem of 12%Cr steels with standard N content is the replacement of fine MX carbonitrides by coarse Z-phase during long-term thermal exposure that leads to appearance of creep strength breakdown and impede the prediction of long-term creep strength from short-term creep test data [12-15]. This replacement becomes possible due to the similar chemical compositions of MX carbonitrides and Z-phase (CrVN) particles [12]. In the steels with Cr content more than 9%, multiple acceleration of the Z-phase formation was fixed during creep [16]. Moreover, the mechanism of Z-phase nucleation are changed from nucleation of this phase on the MX/ferrite surface to direct transformation of face-centered cubic lattice of MX carbonitrides into tetragonal lattice of Z-phase. The last mechanism leads to the formation of coarse Z-phase particles with a mean size of 1 μm and full dissolution of fine MX carbonitrides; the steels loss the creep resistance after exposure of 10,000 h or more [16]. The addition of Ta content in the steel can provide the three-fold separation of MX carbonitrides on V-rich MX, Nb-rich MX and Ta-rich MX and slows down the replacement of MX carbonitrides by coarse Z-phase.

The recent investigations showed that development of microstructural design of 9-10%Cr steel by optimization of austenite-stabilizing and ferrite-stabilizing elements allowed significantly increasing the creep time to rupture at 650°C [17-20]. Only low N content together with high B content could provide excellent 100,000 h creep strength of 110 MPa at 650°C using Larson-Miller parameter [17]. Increasing Cr content up to 12% for increasing in oxidation resistance together with low N and high B contents were compensated by the increasing in the content of austenite-stabilizing elements such as Co to 4% and Cu to 1%. New alloying schema allowed obtaining the δ-Fe content less than 10% (Fig. 1a,b). The increase in Co and B contents provides the slowing down of diffusion-controlled processes such as dislocation climb, particle coarsening according to Ostwald ripening and others [21-23]. The addition of Cu allows forming the nanoparticles with a size of 3-5 nm after tempering, which are nucleation sites for the fine Laves phase particles that provides narrow size distributions of last phase and retards the coarsening of this phase during thermal exposure [24,25]. Complex influence of alloying elements must provide improved creep resistance of 12%Cr steels. There are a lot of studies focused on investigation of the effect of Co, Cu and B on improved creep resistance [17-25]. However, effect of Ta on creep properties and microstructural changes in 9-12%Cr steel with low N and high B contents is not clear yet. The aim of the present study is to investigate the effect of Ta on creep properties of Co-containing 12%Cr steels.

2. Experimental procedure

The Ta-containing 12%Cr steel denoted here as 12%Cr-3Co-0.11(Ta+Nb) and Ta-free 12%Cr steel denoted here as 12%Cr-4Co-0.07Nb were prepared by vacuum induction melting as 100 kg ingots in Steel Institute (Institut für Eisenhüttenkunde (IEHK)), Aachen, Germany. Square bars with a 11 mm × 11 mm cross-section were cast and hot forged by the Joint Research Center, «Technology and Materials», Belgorod National Research University, Belgorod, Russia. Chemical compositions of the Co-containing 12%Cr steels with low N and high B contents are listed in Table 1.

Table 1. Chemical compositions of the Co-containing 12%Cr steels with low N and high B contents.

	C	Cr	Co	Mo	W	V	Cu	Nb	B	N	Ta
12%Cr-4Co-0.07Nb	0.09	11.3	3.9	0.63	2.4	0.24	0.78	0.07	0.009	0.003	-
12%Cr-3Co-0.11(Ta+Nb)	0.11	11.4	3.0	0.62	2.5	0.23	0.76	0.04	0.01	0.003	**0.07**

Both 12%Cr-4Co-0.07Nb and 12%Cr-3Co-0.11(Ta+Nb) steels were solution treated at 1050°C and 1070°C for 1 h, cooled in air, respectively, and subsequently tempered at 770°C for 3 h. Different normalizing temperature was chosen to provide the same PAG size for both steels. Flat specimens with a gage length of 25 mm and a cross section of 7 mm × 3 mm and cylindrical specimen with gauge length of 60 mm and a 6 mm diameter were crept until rupture at 650°C under the applied stresses of 180-80 MPa with a step of 20 MPa. The structural characterization was carried out using a transmission electron microscope JEOL–2100 (TEM) with an INCA energy dispersive X-ray spectrometer (EDS) on ruptured creep specimens. Identification of the precipitates was performed based of combination of EDS composition measurements of the metallic elements. The TEM specimens were prepared by electropolishing at room temperature using a solution of 10 pct perchloric acid in glacial acetic acid with Struers «Tenupol-5» machine. The precipitates were identified from both the chemical analysis and the selected-area diffraction method on at least 50 particles on the each portion. The volume fractions of the precipitated phases were calculated using the Thermo-Calc software with the TCFE7 database. The following phases were selected independently for calculation: BCC, FCC, M23C6 carbide, Laves phase (Fe2(W,Mo) (C14).

3. Results and Discussion

3.1 Microstructures after heat treatment

Microstructures of the Co-containing 12%Cr steels studied after heat treatment obtained by optical metallography and TEM are represented in Figure 1. Microstructural parameters after heat treatment for both steels are summarized in Table 2.

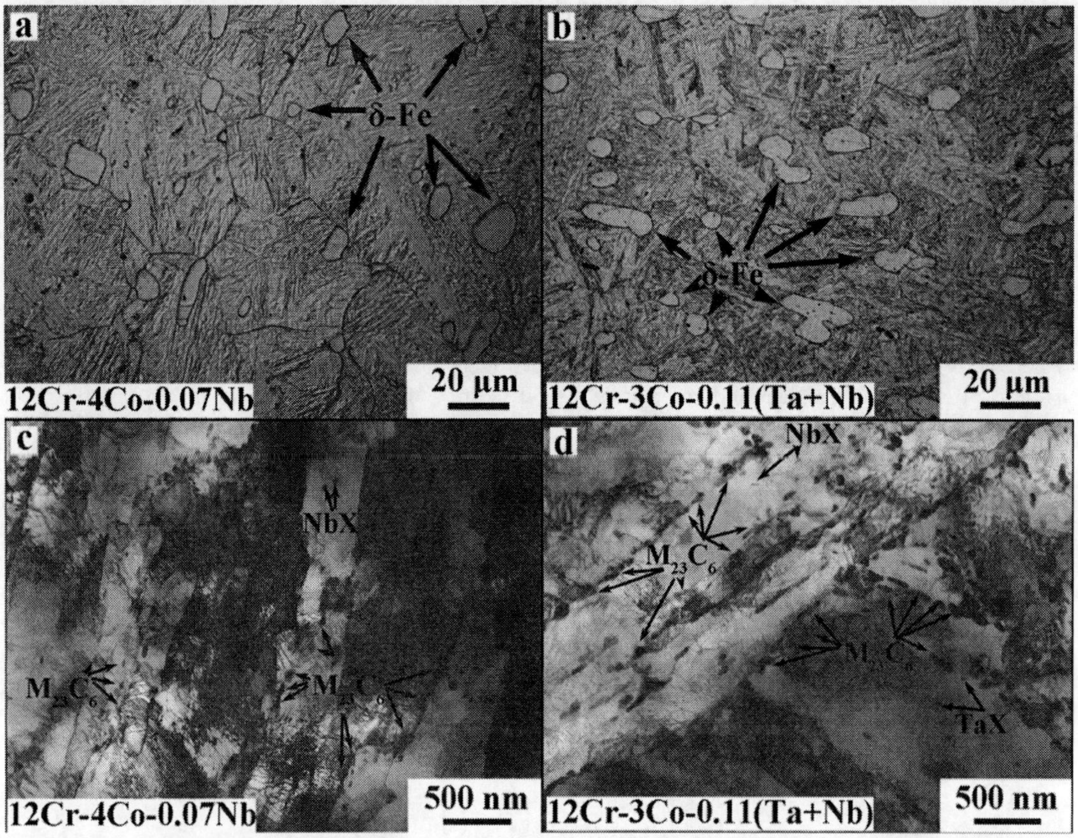

Figure 1. Microstructures of the 12%Cr-4Co-0.07Nb steel (a,c) and 12%Cr-3Co-0.11(Ta+Nb) steel studied (b,d) after normalizing at 1050-1070°C and tempering at 770°C obtained by optical metallography (a,b) and TEM of foils (c,d).

Table 2. Structural parameters of both 12%Cr steels studied after heat treatment.

	D_{PAG}, μm	Fraction of δ-Fe,%	d_{lath}, nm	Dislocation density, m^{-2}	Particle size, nm		Particle fraction, %	
					$M_{23}C_6$	MX	$M_{23}C_6$	MX
12Cr-4Co-0.07Nb	51±5	6±1	305±30	$(1.3\pm0.1)\times10^{14}$	61±5	29±3	1.59	0.076
12Cr-3Co-0.11(Ta+Nb)	48±5	9±2	290±30	$(2.0\pm0.1)\times10^{14}$	55±5	20±3	1.81	0.094

The PAG size was about 50 μm; fraction of δ was less than 10% for both steels after normalizing at 1050-1070°C (Figure 1a,b). After tempering at 770°C, the formation of tempered martensite lath structure (TMLS) was observed in both 12%Cr martensitic steels (Figure 1c,d). The transverse size of martensitic laths in both steels was nearly the same, whereas the dislocation density in the Ta-containing 12%Cr-3Co-0.11(Ta+Nb) steel is bigger than that in the Ta-free steel by a factor of 1.5 (Table 2). In both steels, Cr-rich $M_{23}C_6$ carbides decorated the high-angle PAG boundaries and low-angle boundaries of the martensitic laths (Figure 1c,d). The size of $M_{23}C_6$ carbides in both steels was similar, while the fraction of these particles in the 12%Cr-3Co-0.11(Ta+Nb) steel is higher than that in the Ta-free steel (Table 2). The main difference between tempered structures of Ta-containing and Ta-free steels was MX carbonitrides. Two-phase separation of MX carbonitrides on Nb-rich MX and (Ta,Nb)-rich MX particles in the 12%Cr-3Co-0.11(Ta+Nb) steel was observed, whereas only Nb-rich MX phase was revealed in the 12%Cr-4Co-0.07Nb steel after tempering at 770°C.

The chemical composition of Nb-rich MX particles in the 12%Cr-3Co-0.11(Ta+Nb) steel was 4 wt.%Ti, 18%V, 30%Cr, 10%Fe and 38%Nb. The chemical composition of (Ta,Nb)-rich MX particles in the 12%Cr-3Co-0.11(Ta+Nb) steel was 2 wt.%V, 3%Cr+Fe, 10%Nb and 85%Ta. The chemical composition of Nb-rich MX particles in the 12%Cr-4Co-0.07Nb steel was 15 wt.%V, 16%Cr, 5%Fe and 64%Nb. The size of MX carbonitrides in the Ta-containing 12%Cr-3Co-0.11(Ta+Nb) steel is less than that in the Ta-free steel by a factor of 1.5, the fraction of MX carbonitrides in Ta-containing steel is higher than that of Ta-free steel (Table 2).

3.2 Creep properties

Figure 2 shows the creep rupture data of the steels at a temperature of 650°C. Times to rupture after creep tests for both steels are summarized in Table 3. For the 12%Cr-4Co-0.07Nb steel, creep tests at 650°C under the applied stresses ranging of 180-80 MPa with a step of 20 MPa are finished now, whereas for the 12%Cr-3Co-0.11(Ta+Nb) steel, creep tests at 650°C under the applied stresses of 180, 160 and 120 MPa is finished and creep tests at 650°C under the applied stresses of 140, 100 and 80 MPa are in progress.

Both steels demonstrated no any breakdowns on curves "Applied stress vs. Time to rupture" and "Minimal creep rate vs. Applied stress" (Figure 2).

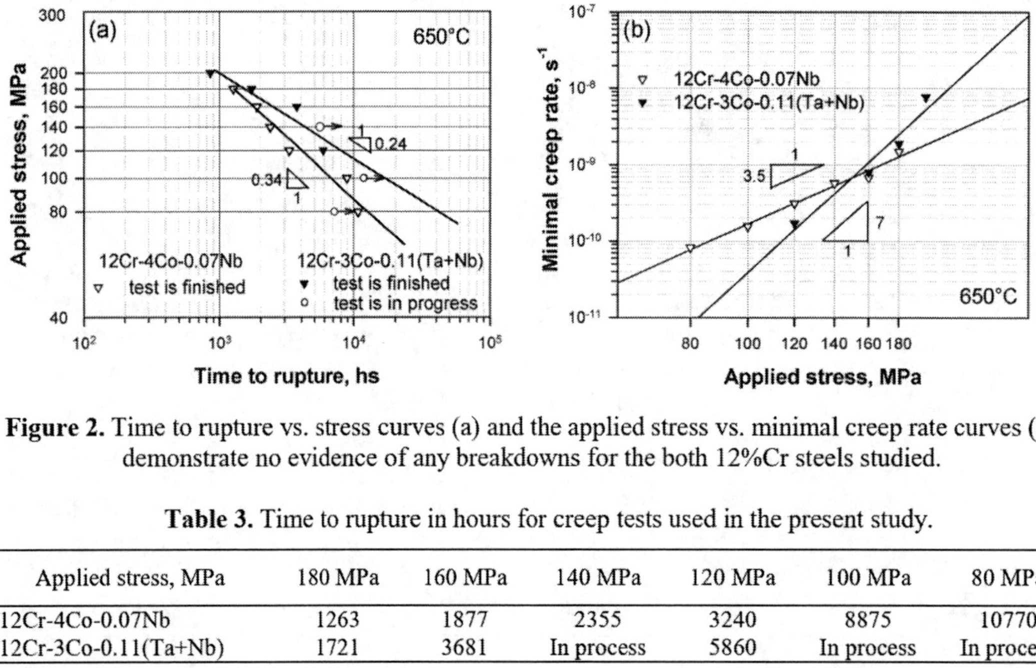

Figure 2. Time to rupture vs. stress curves (a) and the applied stress vs. minimal creep rate curves (b) demonstrate no evidence of any breakdowns for the both 12%Cr steels studied.

Table 3. Time to rupture in hours for creep tests used in the present study.

Applied stress, MPa	180 MPa	160 MPa	140 MPa	120 MPa	100 MPa	80 MPa
12Cr-4Co-0.07Nb	1263	1877	2355	3240	8875	10770
12Cr-3Co-0.11(Ta+Nb)	1721	3681	In process	5860	In process	In process

The 3%Co and 0.07%Ta additions significantly increased the creep strength under both the high and low applied stresses. Time to rupture in the Ta-containing steel was higher than that for Ta-free steel by a factor of 1.5-2.0 after at applied stresses of 180 MPa, 160 MPa and 120 MPa. Under the low applied stress (less 140 MPa), the effect of the Co and Ta additives on the creep strength did not tend to diminish in comparison with 9%Cr-3%Co steels, in which effect of Co decreased at the low applied stresses [6,20].

The experimental data obey a power law relationship throughout the whole range of the applied stress of the usual form [1,26,27]:

$$\dot{\varepsilon}_{min} = A \times \sigma^n \exp\left(\frac{-Q}{RT}\right), \quad (1),$$

where $\dot{\varepsilon}_{min}$ is the minimum creep rate, σ is the applied stress, Q is the activation energy for a plastic deformation, R is the gas constant, T is the absolute temperature, A is a constant, and n is the "apparent" stress exponent. These plots provided the best linear fit with a regression coefficient of 0.98 for n=3.5 for the 12%Cr-4Co-0.07Nb steel and for n=7 for the 12%Cr-3Co-0.11(Ta+Nb) steel the whole applied stress interval. This n value at all tested stress regimes remains constant for both steels. This clearly indicates that there is no creep strength breakdown during the steady-state creep of the Co-containing 12%Cr martensitic steels [1,28]. The steady-state creep of the Co-containing 12%Cr steels was controlled using the same process for short- and long-term regions at a creep rate ranging from 10^{-6} to 10^{-11} s^{-1}. However, the mechanisms controlled steady-state creep for Ta-containing and Ta-free steel are different.

3.3 Crept microstructures

After high applied stress of 180 MPa, the microstructures of both Co-containing 12%Cr steels are similar with tempered state, TMLS remained (Figure 3a,b). The transverse lath size increased from about 300 nm to about 500 nm, while dislocation density did not change for both steels (Table 4). W-rich Laves phase particles precipitated along PAG and lath boundaries. The average size of Laves phase was 140 nm for both steels(Table 4). $M_{23}C_6$ carbides increased in size up to 90 nm in both steels, while the fraction of these particles in the Ta-containing steel is higher than that of the Ta-free steel (Table 4). No evidence for Z-phase particles was revealed in both steels. Nb-rich MX carbonitrides in the 12%Cr-4Co-0.07Nb steel and (Ta,Nb)-rich MX carbonitrides in the 12%Cr-3Co-0.11(Ta+Nb) steel increased their sizes to 30-35 nm. The main difference between steels was the fractions of secondary phase particles at 650°C (Table 4).

After low applied stress of 120 MPa, the microstructures of both steels strongly evolved. TMLS partially transformed into subgrain structure with a mean size of subgrains of 0.70 μm (Figure 3c,d, Table 5). Dislocation density in both steels decreased down to 10^{13} m^{-2}. This is accompanied with the coarsening of grain boundary $M_{23}C_6$ and Laves phase particles (Table 5). (Ta,Nb)-rich MX carbonitrides in the 12%Cr-3Co-0.11(Ta+Nb) steel retained their size of about 35-37 nm, whereas NbX carbonitrides in the 12%Cr-4Co-0.07Nb steel increased their size up to 45-50 nm (Table 5).

The precipitation of the MX carbonitrides on dislocations indicates that their coarsening rate is controlled by pipe diffusion. In this case, the continuous growth of the MX particles in both steels (Figure 4) should obey the following relationship:

$$d^n - d_0^n = K_p \tau, \quad (2)$$

where d is subgrain/particle size at the time, d_0 is initial subgrain/particle size, K_p is constant of growth rate, τ is time, the coefficient n depends on the coarsening mechanism and is 5 that corresponds to pipe diffusion.

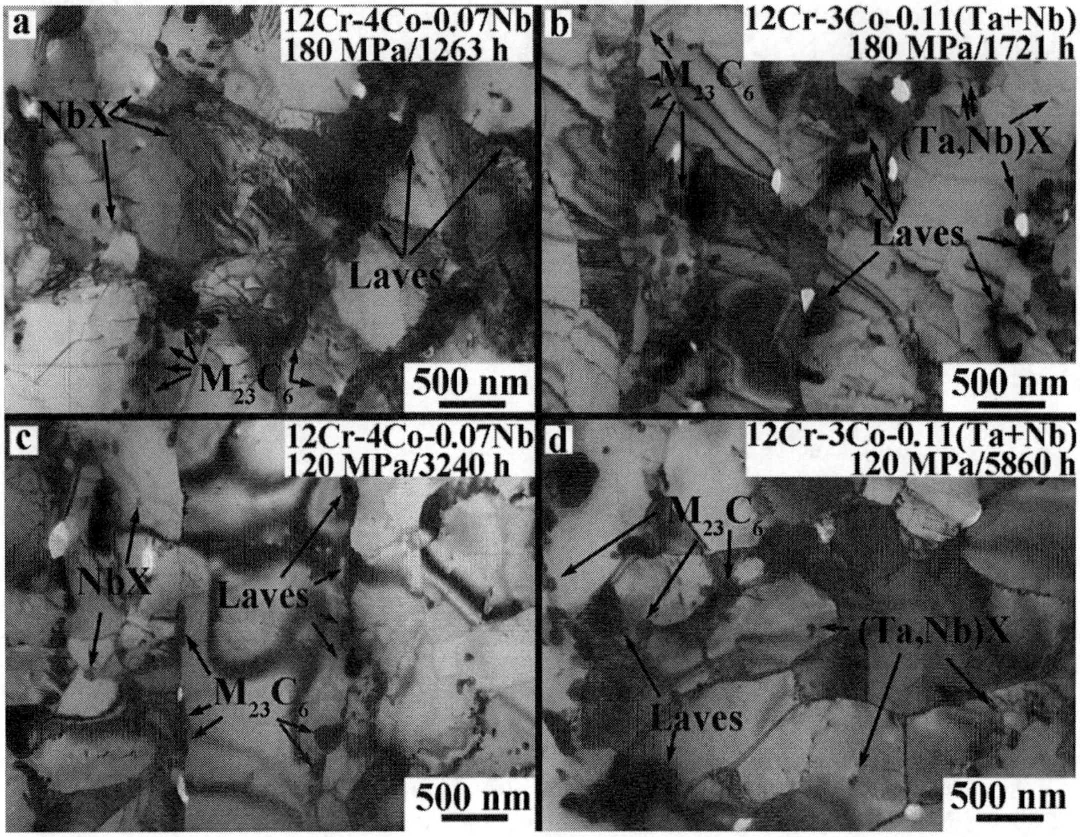

Figure 3. TEM images of microstructures of the 12%Cr-4Co-0.07Nb steel (a,c) and 12%Cr-3Co-0.11(Ta+Nb) steel studied (b,d) after creep tests at 650°C under the applied stresses of 180 MPa (a,b) and 120 MPa (c,d).

Table 4. Structural parameters of both 12%Cr steels studied after creep test at 650°C under an applied stress of 180 MPa.

	D_{PAG}, μm	Fraction of δ-Fe,%	d_{lath}, nm	Dislocation density, m^{-2}	Particle size, nm		Particle fraction, %	
					$M_{23}C_6$	MX	$M_{23}C_6$	MX
12Cr-4Co-0.07Nb	51±5	6±1	491±30	$(1.2±0.1)×10^{14}$	93±5	35±3	1.68	0.061
12Cr-3Co-0.11(Ta+Nb)	48±5	10±2	503±30	$(2.2±0.1)×10^{14}$	85±5	33±3	2.07	0.070

Table 5. Structural parameters of both 12%Cr steels studied after creep test at 650°C under an applied stress of 120 MPa.

	D_{PAG}, μm	Fraction of δ-Fe,%	d_{lath}, nm	Dislocation density, m^{-2}	Particle size, nm		Particle fraction, %	
					$M_{23}C_6$	MX	$M_{23}C_6$	MX
12Cr-4Co-0.07Nb	51±5	6±1	730±50	$(0.8±0.1)×10^{14}$	109±5	45±3	1.68	0.061
12Cr-3Co-0.11(Ta+Nb)	48±5	10±2	650±50	$(0.7±0.1)×10^{14}$	96±5	37±3	2.07	0.070

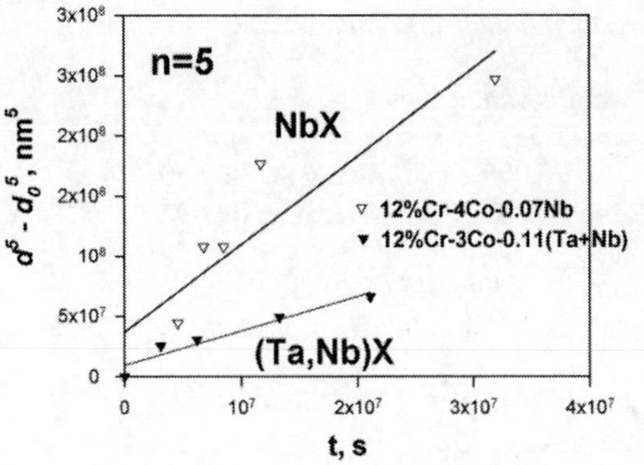

Figure 4. The dependence of MX size on creep time at 650°C in the gage section for n=5 corresponding to pipe diffusion. NbX corresponds to 12%Cr-4%Co-0.07%Nb steel and (Ta,Nb)X corresponds to 12%Cr-3%Co-0.11%(Ta+Nb) steel.

According to Figure 4, constant of growth rate for NbX carbonitrides in the 12%Cr-4Co-0.07Nb steel was 7.31×10^{-45} nm^{5}s^{-1} with correlation coefficient of 0.85 for all experimental points after creep tests for ~10 000 h, whereas constant of growth rate for (Ta,Nb)X carbonitrides in the 12%Cr-3Co-0.07Nb steel was 2.85×10^{-45} nm^{5}s^{-1} with correlation coefficient of 0.93 for all experimental points after creep tests for ~6 000 h. So, addition of Ta in the Co-containing 12%Cr martensitic steels decreased coarsening rate by 61%. Slowing down of the coarsening of (Ta,Nb)-rich MX carbonitrides stabilizes TMLS during creep and can be one of the reasons for improved creep resistance of the Ta-containing 12%Cr steels.

4. Conclusions

New developed 12%Cr-(3-4%Co) steel with low N and high B contents demonstrated no evidence for any breakdowns on curves of "Time to rupture vs. Applied stress" and "Applied stress vs. minimal creep rate" due to absence of Z-phase particles, which replace the fine MX carbonitrides during creep testing. The addition of Ta in the 12%Cr steel together with decreasing in Co and Nb contents increased the creep time to rupture at 650°C. During creep (Ta,Nb)X particles demonstrated high thermal stability to the coarsening; the coarsening rate of these particles in the Ta-containing 12%Cr steel was twice less than NbX particles in the Ta-free 12%Cr-3Co-0.07Nb steel. The thermal stability of (Ta,Nb)X carbonitrides is one of the reasons for the increment in creep properties at 650°C.

Acknowledgments

The study was financially supported by Ministry for Scientific Research and Higher Education in Russian Federation, President Grant for PhD – young scientists (grant number 075-15-2019-1165). The authors are grateful to the staff of the Joint Research Center, «Technology and Materials», Belgorod State University, for their assistance with instrumental analysis.

References

[1] Abe F, Kern T U, Viswanathan R 2008 *Creep resistant steels* (Cambridge: Woodhead Publishing in Materials) p 800
[2] Kern T U, Staubli M, Scarlin B 2002 *ISIJ International* **42** 1515
[3] Di Gianfrancesco A 2017 *Materials for Ultra-Supercritical and Advanced Ultra-Supercritical Power Plants* (Cambridge: Woodhead Publishing in Materials) p 900
[4] Abe F 2007 *Int. J Press. Vess.Piping* **84** 3
[5] Armaki H GH, Chen R P, Maruyama K, Igarashi M 2010 *Mater. Sci. Eng. A* **527** 6581
[6] Fedoseeva A, Dudova N, Kaibyshev R 2017 *Phys. Met. Metall.* **118** 591
[7] Mitsuhara M, Yamasaki Sh, Miake M, Nakashima H, Nishida M, Kusumoto J, Kanaya A 2016 *Phil. Mag. Lett.* **96** 76
[8] Kostka A, Tak K G, Hellmig R J, Estrin Y, Eggeler G 2007 *Acta Mater.* **55** 539
[9] Ghassemi-Armaki H, Chen R, Maruyama K, Igarashi M 2011 *Metall. Mater. Trans. A* **42** 3084

[10] Fedoseeva A, Dudova N, Kaibyshev R 2016 *Trans. Indian Inst. Met.* **69** 211
[11] Hald J 2016 *Trans. Indian Inst. Met.* **69** 183
[12] Danielsen H K, Di Nunzio P E, Hald J 2013 *Metall. Mater. Trans. A* **44A** 2445
[13] Strang A, Vodarek V 1996 *Mater. Sci. Technol.* **12** 552
[14] Suzuki K, Kumai S, Kushima H, Kimura K, Abe F 2000 *Tetsu-to-Hagane* **86** 550
[15] Suzuki K, Kumai S, Kushima H, Kimura K, Abe F 2003 *Tetsu-to-Hagane* **89** 691
[16] Danielsen H K 2016 *Mat. Sci. Technol.* **32** 126
[17] Dudova N, Mishnev R, Kaibyshev R 2019 *Mater. Sci. Eng. A* **766** 138353
[18] Fedoseeva A, Kozlov P, Dudko V, Skorobogatykh V, Shchenkova I, Kaibyshev R 2015 *Phys. Met. Metall.* **116** 1047
[19] Dudko V, Fedoseeva A, Belyakov A, Kaibyshev R 2015 *Phys. Met. Metall.* **116** 1165
[20] Fedoseeva A, Dudova N, Kaibyshev R, Belyakov A 2017 *Metals* **7** 573
[21] Abe F 2011 *Procedia Engineering* **10** 94
[22] Helis L, Toda Y, Hara T, Miyazaki H, Abe F 2009 *Mater. Sci. Eng. A* **510–511** 88
[23] Kipelova A, Odnobokova M, Belyakov A, Kaibyshev R 2013 *Metall. Mater. Trans. A* **44** 577
[24] Klueh R L 2005 *International Materials Reviews* **50** 287
[25] Futamura Yu, Tsuchiyama T, Takaki S 2001 *ISIJ International* **41** S106
[26] Cadek J 1994 *Creep in Metallic Materials* (Prague: Academia) p 372
[27] Kassner M E, Pérez-Prado M T 2004 *Fundamentals of creep in metals and alloys* 1st ed (New York: Elsevier Science) p 288
[28] Abe F 2015 *Metall. Mater. Trans. A* **46** 5610

IOP Conf. Series: Materials Science and Engineering **848** (2020) 012021 doi:10.1088/1757-899X/848/1/012021

Spectrophotometric determination of epinephrine using new analytical systems based on label-free silver triangular nanoplates

A A Furletov[1], V D Zaytsev[1], V V Apyari[1], A V Garshev[1,2], P A Volkov[3] and S G Dmitrienko[1]

[1] Lomonosov Moscow State University, Department of Chemistry, 1/3 Leninskie Gory, 119991 Moscow, Russia

[2] Lomonosov Moscow State University, Department of Materials Science, 1/73 Leninskie Gory, 119991 Moscow, Russia

[3] Scientific-Research Institute of Chemical Reagents and Special Purity Chemicals of National Research Center «Kurchatov Institute», 3 Bogorodskiy Val, 107076 Moscow, Russia

e-mail: aleksei_furletov@mail.ru

Abstract. Silver triangular nanoplates **(AgTNPs)** is a promising and still relatively poorly studied colorimetric probe for sensing various organic compounds. In particular, they undergo a change in their morphology when interacting with various catecholamines. This process is accompanied by a hypsochromic shift of the local surface plasmon resonance **(LSPR)** band of nanoparticles. The greatest spectral changes can be observed in the case of the interaction of AgTNPs with epinephrine which can be the basis for a sensitive method for its detection. It was found that the detection limit of epinephrine under the selected optimal conditions is equal to 3 μM, and the dynamic range is from 9 μM up to 50 μM. Selectivity of the proposed method for the epinephrine determination was evaluated as well. It was shown that the determination does not interfere with a 10-fold excess of vanillylmandelic acid and dopamine, and with a 1000-fold excess of common cations and anions. The proposed approach was successfully applied to the determination of epinephrine in a drug and a sample of artificial urine containing an epinephrine additive.

1. Introduction

It is well known that catecholamines such as dopamine, epinephrine and norepinephrine play a key role in the mechanisms of learning, memorization, emotions, sleep, as well as in psychomotor activity and neurohumoral regulation. The action of catecholamines is noticeable in stressful conditions when the adrenal medulla begins to intensively release epinephrine and norepinephrine to stimulate the central nervous system and increase the reaction rate for optimal response to the stimulus.

Epinephrine, also known as adrenaline, is the most important direct-acting hormone and medication. In addition to participating in homeostasis, epinephrine increases the strength and heart rate, innervates the vessels of the digestive system, bronchi and brain, as well as muscles. The action of epinephrine on the body is extensive and affects most of the major organs. Epinephrine narrows the vessels leading to the skin and digestive tract, and dilates the vessels supplying the muscles and brain, increases the strength and heart rate, and also lowers the sensitivity threshold of the senses in a stressful situation. A change in the concentration of epinephrine in biological fluids (blood, urine, cerebrospinal fluid) can serve as a reliable indicator of a violation of homeostasis [1–3].

According to existing literature data, mainly HPLC/MS and various electrochemical methods are used to determine epinephrine. In chromatographic methods, high selectivity is achieved with respect to the components of the matrix. However, chromatographic methods often require an unreasonably large amount of various organic solvents. It is also worth noting the high cost of chromatographic equipment and the low speed of a single analysis. Electrochemical methods, despite their wide distribution and low cost of analytical equipment, are highly susceptible to matrix effects and often require complex preparation of electrodes [4].

In this way, the development of inexpensive and available analytical techniques that allow rapid screening of a large number of samples using simple methods or visual detection is one of the priority fields of modern analytical chemistry. From this point of view, the attention of many researchers is attracted to silver nanoparticles [5]. A large number of scientific papers are devoted to synthesis and application of the spherical silver nanoparticles as well as materials based on them [6–20]. Significantly fewer papers are devoted to the anisotropic silver nanoparticles, such as nanocubes [21], nanowires [22], nanodiscs [23], nanorods [24] and triangular nanoplates [25–30]. The unique optical properties of silver triangular nanoplates **(AgTNPs)** arise from the local surface plasmon resonance **(LSPR)** phenomenon resulting in an intense absorption band in the visible spectral region. High molar extinction coefficients of AgTNPs, as well as the ability to change LSPR band in the presence of substances that cause changes in the morphology of nanoparticles and in the state of their surface, allow us for considering AgTNPs as a distinct colorimetric reagent.

In this study, new analytical systems based on label-free silver triangular nanoplates were assessed to outline their possibilities as a colorimetric probe for spectrophotometric determination of epinephrine.

2. Materials and methods of research

2.1. Reagents and instruments

The following reagents were used: *L*-epinephrine (*Acros Organics*, 99%), silver nitrate (*PZTsM-Vtormet*, analytical grade), sodium citrate (*Sigma*, ≥99.5%), poly(N-vinyl-2-pyrrolidone) (*Acros Organics*, M.W. 58000 g mol^{-1}, 99%), hydrogen peroxide (*Sigma-Aldrich*, 30 wt. % in H_2O, ACS), sodium borohydride (*Acros Organics*, 99%), sodium hydroxide (*Reachim*, pure grade) and acetic acid (*IREA 2000*, pure grade). Working solutions of these substances were prepared by dissolving their weighed portions or aliquots and dilution in deionized water obtained with the Millipore Simplicity water purification system (*Merck Millipore*, USA).

Absorption spectra of AgTNPs in the wavelength range 400–1100 nm were recorded in standard quartz cuvettes with an optical path length of 1 cm using the SF-103 spectrophotometer (*Aquilon*, Russia). Stirring of solutions was carried out using the Ekros PE-6100 magnetic mixer (*Ekros*, Russia). pH was measured using the Ekspert 001 pH-meter (*Ekoniks Ekspert*, Russia).

Chromatographic determination of epinephrine in samples was carried out using the Tsvet-Yauza high-performance liquid chromatograph (*Khimavtomatika*, Russia) with an amperometric detector (E = 1.2 V). The Luna C18 chromatographic column (*Phenomenex*, USA) was used as a stationary phase. A mixture consisting of degassed acetonitrile (50% vol.) and degassed deionized water (50% vol.) was used as a mobile phase. Flow rate of the mobile phase was equal to 0.4 mL min^{-1}. The injected volume of samples was equal to 20 μL.

TEM images of AgTNPs were recorded using the Libra 200 transmission electron microscope (*Carl Zeiss*, Germany) with a thermal field emission cathode at the accelerating voltage of 200 kV, the limit of information in a bright field transmission microscopy registration mode is better than 0.1 nm. Dispersions of the AgTNPs samples in hexane/heptane were deposited onto a copper grid support with a formvar film covered by amorphous carbon Formvar®/Carbon Reinforced Copper Grids 3440C-MB

(*SPI*, USA). Measurements of ζ-potentials of AgTNPs were carried out by the dynamic light scattering method using the Zetasizer Nano ZS (*Malvern Instruments*, UK) with non-invasive backscattering optics NIBS.

2.2. Preparation of label-free silver triangular nanoplates

Preparation of AgTNPs was performed as described by G.S. Metraux and C.A. Mirkin [30] with minor modifications. Glassware used was pre-washed with freshly prepared nitrohydrochloric acid, thoroughly rinsed with distilled water and dried. A 0.5 mL aliquot of 0.01 mol L^{-1} silver nitrate aqueous solution was diluted with 4.3 mL of deionized water. Then 2.3 mL of 1% sodium citrate aqueous solution, 0.6 mL of 20 g L^{-1} poly(N-vinyl-2-pyrrolidone) aqueous solution and 1.2 mL of 3% hydrogen peroxide aqueous solution were successively added under vigorous stirring. Then 1.0 mL of freshly prepared 35.0 mmol L^{-1} sodium borohydride aqueous solution was dropwisely added to the solution under stirring. Mixture got pale yellow-green color, which half an hour later abruptly changed to intense emerald-green and then to blue-violet. The stirring was stopped. The as-prepared AgTNPs colloidal solution was stored at room temperature. The calculated final concentration of AgTNPs in the solution was 56 μg mL^{-1} (0.52 mmol L^{-1} in terms of silver atoms). A TEM image of synthesized AgTNPs is present in Fig. 1a.

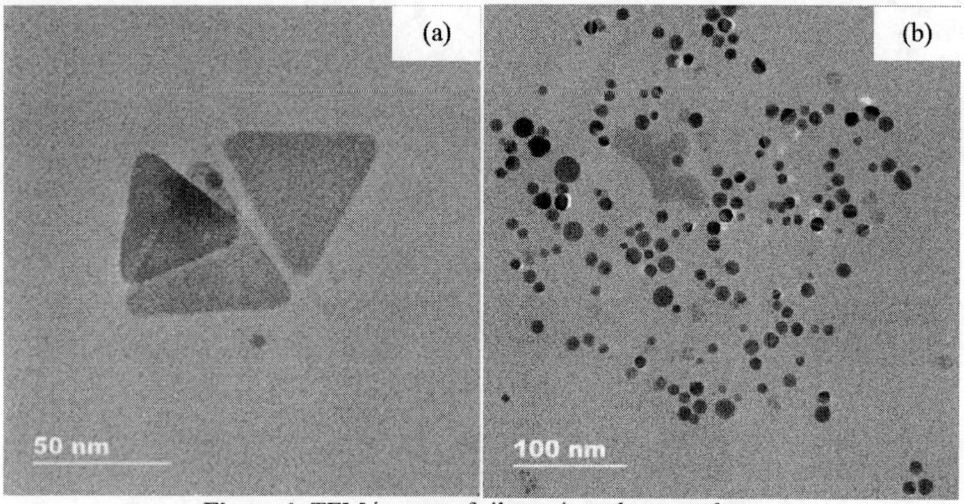

Figure 1. TEM images of silver triangular nanoplates before (a) and after (b) interaction with *L*-epinephrine.

2.3. Sample preparation

The medicinal form of the «Epinephrine» drug (solution for infusions) was 200 times diluted with deionized water. For the determination, an aliquot portion of this solution was taken.

Sorption pre-concentration of epinephrine from a sample of artificial urine containing epinephrine supplement was carried out in a dynamic mode. A cartridge filled with 20 mg of hypercrosslinked polystyrene was pre-conditioned with 3 mL of acetonitrile and washed with 10 mL of deionized water. Then 30 mL of artificial urine containing epinephrine was passed through the cartridge at an average rate of 1 mL min^{-1}. After that, the cartridge was washed with 10 mL of deionized water. Epinephrine was eluted with 3.00 mL of 0.1 M acetic acid. For the determination, an aliquot portion of this solution was taken.

3. Results and discussion

3.1. Interaction of silver triangular nanoplates with epinephrine

The study of the interaction of AgTNPs with epinephrine was carried out according to the following procedure. First of all, 1.55 mL of AgTNPs solution was added into polypropylene test-tubes. After that, certain amounts of *L*-epinephrine solution were injected. The final volume was adjusted to 5.0 mL with an acetate buffer solution (CH_3COOH/CH_3COONa) with pH 4.5. After 15 minutes, UV-Vis absorption spectra of nanoparticles were recorded.

One can see that there were a large number of silver nanodiscs and silver nanospheres presented on a TEM image of nanoparticles after the interaction (Fig. 1b). We propose that a change in the morphology of AgTNPs when interacting with epinephrine takes place. The value of the electrokinetic potential of AgTNPs after interaction with epinephrine changed from –27 mV to –6 mV which indicates a significant loss of nanoparticles stability.

It was also found that strong hypsochromic shift of the LSPR band of AgTNPs after interaction with *L*-epinephrine can be observed in the recorded absorption spectra (Fig. 2). The shift value of AgTNPs LSPR band maximum ($\Delta\lambda$) monotonically increases with increasing concentration of the analyte so that it can be considered as the analytical signal. To achieve maximum sensitivity of the AgTNPs-based analytical systems for epinephrine, it was necessary to study the effect of various factors and select the optimal values of experimental parameters that provide us with the largest value of $\Delta\lambda$.

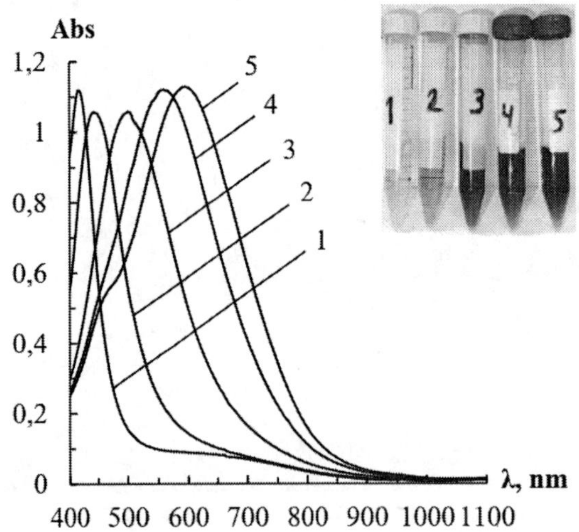

Figure 2. UV-Vis absorption spectra and photo of silver triangular nanoplates after interaction with *L*-epinephrine. *c*(epinephrine) = 40 μM (1), 30 μM (2), 20 μM (3), 10 μM (4), 0 (5); *c*(AgTNPs) = 0.13 mM Ag; pH 4.5; *t* = 15 min.

3.1.1. Optimization of AgTNPs concentration. The effect of AgTNPs concentration upon the analytical signal was studied. It was found that the sensitivity of epinephrine determination increases with increasing AgTNPs concentration in the range 0–0.32 mM Ag. At the same time, a measurement error of the analytical signal increases sympathetically. For further experiments, AgTNPs concentration was chosen equal to 0.16 mM Ag which corresponds to the middle of the above-mentioned concentration range.

3.1.2. Optimization of pH value. To study the effect of pH upon the analytical signal, a certain amount of *L*-epinephrine was added into the test-tubes, 1.55 mL of AgTNPs was further injected, and the solution volume was adjusted to 5.0 mL with aqueous solutions of 0.1 M CH_3COOH and 0.1 M NaOH in various ratios. After 15 minutes, UV-Vis absorption spectra of nanoparticles were recorded.

At pH < 4, the shape of nanoparticles changes from triangular nanoplates to nanodisks or nanospheres [31]. This fact was indirectly evidenced by the shift of the LSPR band maximum to shorter wavelengths. For this reason, the above-mentioned pH range was not studied in detail. The dependence of the analytical signal on the pH > 4 values is shown in Fig. 3. We assume that reactivity of AgTNPs increases with a decrease of pH value. This contributes to the occurrence of an interaction between AgTNPs and epinephrine. One can see that the maximum shift of the LSPR band can be observed at pH 4.3. Nonetheless, this value is very close to the stability boundary of colloidal AgTNPs solution. Moreover, the reproducibility of the analytical signal at pH 4.3 was not rather good. Therefore, pH 4.5 was chosen for further experiments. This pH provides us with a sufficiently high analytical signal and, at the same time, good reproducibility of the results.

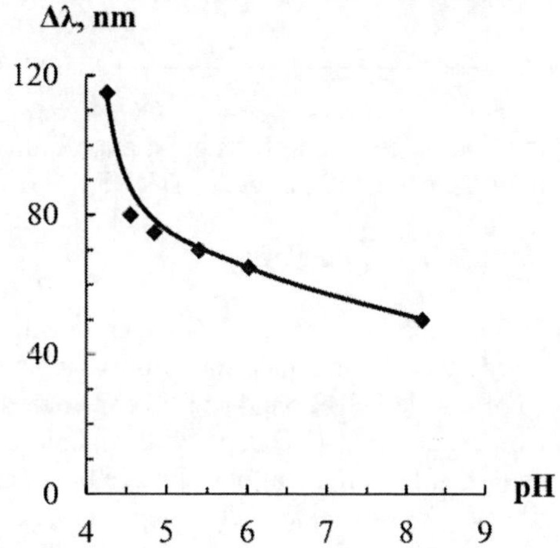

Figure 3. The shift value of AgTNPs LSPR band maximum depending on pH in solutions containing 20 μM epinephrine. c(AgTNPs) = 0.16 mM Ag, t = 15 min.

3.1.3. Optimization of the interaction time. The effect of the interaction time upon the analytical signal was studied. A certain amount of epinephrine was introduced into polypropylene test-tubes, and 1.55 mL of AgTNPs solution was added. The final volume was adjusted to 5.0 mL with an acetate buffer solution with pH 4.5. After 1, 2.5, 5, 10, 15, 20, 25 and 30 minutes, UV-Vis absorption spectra of nanoparticles were recorded.

The dependence of the analytical signal on the interaction time is shown in Fig. 4. One can see that during the first minute after the addition of AgTNPs a fast shift of the LSPR band is observed, and during the next 14 minutes, its slow shift continues. There is an almost linear time shift of the nanoparticles LSPR band during this time period. We assume that such behaviour of the analytical system is connected with the fact that the angles of AgTNPs have the largest excess surface energy and lose faceting in the first place. Further slow changes are likely to occur with the participation of silver atoms located on the edges of AgTNPs. Starting from 15 minutes, the LSPR band shift is terminated. Therefore, this time value was used in further experiments.

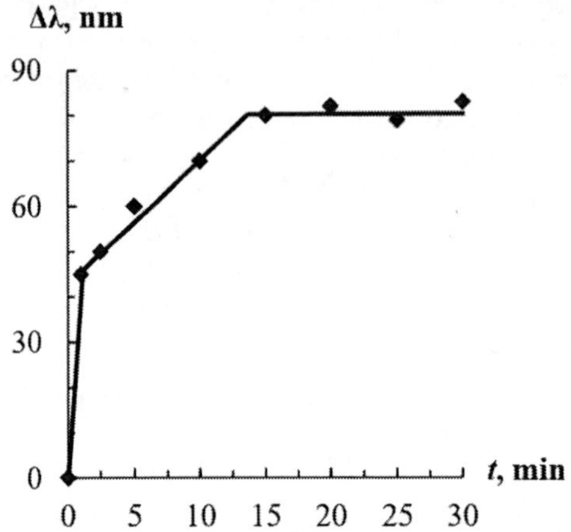

Figure 4. The shift value of AgTNPs LSPR band maximum depending on time in solutions containing 20 μM epinephrine. *c*(AgTNPs) = 0.16 mM Ag, pH 4.5.

3.2. Spectrophotometric determination of epinephrine

3.2.1. Analytical performance. The observed spectral changes during the interaction with AgTNPs can be used as a basis for spectrophotometric determination of epinephrine. As shown above, the hypsochromic shift value (Δλ, nm) of the LSPR band can be considered as the analytical signal. Limit of detection (LOD) and limit of quantitation (LOQ) were estimated as $3s_0/k$ and $9s_0/k$ respectively; here s_0 is the standard deviation of LSPR band maximum for the blank experiment and k is the slope of a calibration curve for epinephrine: Δλ, nm = $k \times c$(epinephrine), μM. It was found that LOD for epinephrine under the selected optimal conditions is equal to 3 μM, and the dynamic determination range is from 9 μM up to 50 μM. The calculated RSD value for 30 μM epinephrine is equal to 4%.

3.2.2. Selectivity studies. Selectivity of the proposed method for epinephrine determination was evaluated. It was found that the determination is not interfered by a 10-fold excess of vanillylmandelic acid and dopamine as well as with a 4000-fold excess of CH_3COO^-, Na^+ and K^+, and 1000-fold excess of NO_3^-, Mg^{2+}, Ca^{2+}, Al^{3+} and Pb^{2+}.

3.2.3. Application. To show applicability of the proposed method to analysis of real samples, the determination of epinephrine in the «Epinephrine» drug (solution for infusions) and a sample of artificial urine containing an epinephrine additive was performed after appropriate dilution of the samples. To confirm the accuracy of the epinephrine determination the studied samples were also analyzed by an independent method which was the reversed-phase high-performance liquid chromatography with amperometric detection (E = 1.2 V). The results are shown in Table 1. In all cases, the content of epinephrine found using the proposed method coincided with the content obtained by the independent method and the data declared by the manufacturers which proves good accuracy of the determination using AgTNPs. It should be stressed that the analysis of these samples could be done with a minimum sample preparation.

4. Conclusion

As shown in the present article, label-free silver triangular nanoplates are a promising colorimetric probe for spectrophotometric determination of epinephrine. It has been proposed that a change in morphology of AgTNPs when interacting with epinephrine takes place. It was shown that the hypsochromic shift value of the local surface plasmon resonance band of nanoparticles can be considered as the analytical signal. The detection limit of epinephrine under the selected optimal conditions is equal to 3 μM, and the dynamic determination range is from 9 μM up to 50 μM. The determination of epinephrine is not interfered by a 10-fold excess of vanillylmandelic acid and dopamine, and 1000-fold excess of common cations and anions. Other advantages of the approach are simplicity, rapidity, good analytical performance, availability of the equipment and ease of the test-method implementation.

Acknowledgement
This study was financially supported by the Russian Science Foundation [grant number 18-73-10001]. Some studies were performed using instrumentation provided according to the M.V. Lomonosov Moscow State University Program of Development.

Table 1. Determination of *L*-epinephrine in samples ($n = 3$, $P = 0.95$).

Sample	Declared epinephrine content	Found epinephrine content	
		Present method	HPLC
«Epinephrine» drug (solution for infusions)	1 g L^{-1}	(1.11 ± 0.05) g L^{-1}	(1.09 ± 0.01) g L^{-1}
Model solutions prepared on the basis of artificial urine[a] with addition of 20 μM epinephrine	—	(19 ± 2) μM	(20.3 ± 0.2) μM

[a] Composition of artificial urine: 22 g L^{-1} urea, 5.2 g L^{-1} sodium chloride, 1.5 g L^{-1} creatinine, 1.08 g L^{-1} sodium dihydrogen phosphate dihydrate, 1 g L^{-1} sodium azide, 0.97 g L^{-1} sodium hydrogen phosphate monohydrate, 10 mg L^{-1} tetrazine.

References

[1] Djelić N, Radaković M, Borozan S, Dimirijević-Srećković V, Pajović N, Vejnović B, Borozan N, Bankoglu E, Stopper H and Stanimirović Z 2019 Oxidative stress and DNA damage in peripheral blood mononuclear cells from normal, obese, prediabetic and diabetic persons exposed to adrenaline in vitro *Mutat. Res. Genet. Toxicol. Environ. Mutagen.* **843** pp 81–9

[2] Whiting M and Doogue M 2009 Advances in biochemical screening for pheochromocytoma using biogenic amines *Clin. Biochem. Rev.* **30** pp 3 – 17

[3] Y-Hassan S 2019 Plasma epinephrine level and its causal link to Takotsubo syndrome revisited: critical review with a diverse conclusion *Cardiovasc. Revasc. Med.* **20(10)** pp 907–14

[4] Tsunoda M 2006 Recent advances in methods for the analysis of catecholamines and their metabolites *Anal. Bioanal. Chem.* **386(3)** pp 506–14

[5] Gorbunova M, Gutorova S, Berseneva D, Apyari V, Zaitsev V, Dmitrienko S and Zolotov Y 2019 Spectroscopic methods for determination of catecholamines: A mini-review *Appl. Spectrosc. Rev.* **54(8)** pp 631–52

[6] Terenteva E, Apyari V, Kochuk E, Dmitrienko S and Zolotov Y 2017 Use of silver nanoparticles in spectrophotometry *J. Anal. Chem.* **72** pp 1138–54

[7] Siripattanakul-Ratpukdi S and Fürhacker M 2014 Review: issues of silver nanoparticles in engineered environmental treatment systems *Water. Air. Soil. Pollut.* **225** pp 1939–57

[8] Fan M, Andrade G and Brolo A 2011 A review on the fabrication of substrates for surface enhanced Raman spectroscopy and their applications in analytical chemistry *Anal. Chim. Acta.* **693** pp 7–25

[9] Oliveira E, Núñez C, Santos H, Fernández-Lodeiro J, Fernández-Lodeiro A, Capelo J and Lodeiro C 2015 Revisiting the use of gold and silver functionalised nanoparticles as

colorimetric and fluorometric chemosensors for metal ions *Sensor. Actuat.* B. **212** pp 297–328
[10] Beyene H, Werkneh A, Bezabh H and Ambaye T 2017 Synthesis paradigm and applications of silver nanoparticles (AgNPs), a review *Sustain. Mater. Technol.* **13** pp 18–23
[11] Ravindran A, Chandran P and Khan S 2013 Biofunctionalized silver nanoparticles: advances and prospects *Colloids. Surf.* B. **105** pp 342–52
[12] Tran Q, Nguyen V and Le A 2013 Silver nanoparticles: synthesis, properties, toxicology, applications and perspectives *Adv. Nat. Sci. Nanosci. Nanotechnol.* **4** p 033001
[13] Shenashen M, El-Safty S and Elshehy E 2014 Synthesis, morphological control, and properties of silver nanoparticles in potential applications *Part. Part. Syst. Charact.* **31** pp 293–316
[14] Wei L, Lu J, Xu H, Patel A, Chen Z and Chen G 2015 Silver nanoparticles: synthesis, properties, and therapeutic applications *Drug. Discov. Today.* **20** pp 595–601
[15] Vilela D, González M and Escarpa A 2012 Sensing colorimetric approaches based on gold and silver nanoparticles aggregation: chemical creativity behind the assay. A review *Anal. Chim. Acta.* **751** pp 24–43
[16] Liang A, Liu Q, Wen G and Jiang Z 2012 The surface-plasmon-resonance effect of nanogold/silver and its analytical applications *Trends. Anal. Chem.* **37** pp 32–47
[17] Haider A and Kang I 2015 Preparation of silver nanoparticles and their industrial and biomedical applications: a comprehensive review *Adv. Mater. Sci. Eng.* **2015** p 165257
[18] Rycenga M, Cobley C, Zeng J, Li W, Moran C, Zhang Q, Qin D and Xia Y 2011 Controlling the synthesis and assembly of silver nanostructures for plasmonic applications *Chem. Rev.* **111** pp 3669–712
[19] Cobley C, Skrabalak S, Campbell D and Xia Y 2009 Shape-controlled synthesis of silver nanoparticles for plasmonic and sensing applications *Plasmonics.* **4** pp 171–9.
[20] Li M, Cushing S and Wu N 2015 Plasmon-enhanced optical sensors: a review *Analyst.* **140** pp 386–406
[21] Personick M, Langille M, Zhang J, Wu J, Li S and Mirkin C 2013 Plasmon mediated synthesis of silver cubes with unusual twinning structures using short wavelength excitation *Small.* **9** pp 1947–53
[22] Han S and Lee J 2012 Synthesis of length-controlled polyvalent silver nanowire-DNA conjugates for sensitive and selective detection of DNA targets *Langmuir.* **28** pp 828–32
[23] Kim B and Lee J 2015 One-pot photochemical synthesis of silver nanodisks using a conventional metal-halide lamp *Mater. Chem. Phys.* **149** pp 678–85
[24] Zhang J, Langille M and Mirkin C 2011 Synthesis of silver nanorods by low energy excitation of spherical plasmonic seeds *Nano. Lett.* **11** pp 2495–8
[25] Zhang Q, Li N, Goebl J, Lu Z and Yin Y 2011 A systematic study of the synthesis of silver nanoplates: is citrate a «magic» reagent? *J. Am. Chem. Soc.* **133** pp 18931–9
[26] Tang B, Xu S, Hou X, Li J, Sun L, Xu W and Wang X 2013 Shape evolution of silver nanoplates through heating and photoinduction *ACS. Appl. Mater. Interfaces.* **5** pp 646–53
[27] Furletov A, Apyari V, Garshev A, Volkov P, Tolmacheva V and Dmitrienko S 2018 Sorption of triangular silver nanoplates on polyurethane foam *Rus. J. Phys. Chem.* A. **92** pp 357–60
[28] Furletov A, Apyari V, Garshev A, Dmitrienko S and Zolotov Y 2017 Triangular silver nanoplates as a spectrophotometric reagent for the determination of mercury(II) *J. Anal. Chem.* **72** pp 1203–7
[29] Apyari V, Gorbunova M, Shevchenko A, Furletov A, Volkov P, Garshev A, Dmitrienko S and Zolotov Y 2018 Towards highly selective detection using metal nanoparticles: a case of silver triangular nanoplates and chlorine *Talanta.* **176** pp 406–11
[30] Metraux G and Mirkin C 2005 Rapid thermal synthesis of silver nanoprisms with chemically tailorable thickness *Adv. Mater.* **17** pp 412–5
[31] Millstone J, Hurst S, Metraux G, Cutler J and Mirkin C 2009 Colloidal gold and silver triangular nanoprisms *Small.* **5** pp 646–64

IOP Conf. Series: Materials Science and Engineering **848** (2020) 012022 doi:10.1088/1757-899X/848/1/012022

Silver nanocomposites based on copolymers of N,N-diallyl-N'-acylhydrazines with N-vinylpyrrolidone

M Gorbunova and T Batueva

Institute of Technical Chemistry of Ural Branch of RAS, 614013, Korolev str., 3 Perm, Russia

E-mail: mngorb@yandex.ru

Abstract. New nanocomposites based on Ag and copolymers of *N*-vinylpyrrolidone with N,N-diallyl-N'-acetylhydrazine [poly(VP-DAAH)], N,N-diallyl-N'-butanoylhydrazine [poly(VP-DABH)] and N,N-diallyl-N'-benzoylhydrazine [poly(VP-DABEH)] have been developed. The average silver particle size ranged from 10 to 20 nm for poly(VP-DAAH), from 13 to 21 nm for poly(VP-DABEH) and from 30 to 42 nm for poly(VP-DABH), with the corresponding UV-vis absorption peak position at 400-410 nm. New nanocomposites based on copolymers of N,N-diallyl-N'-acylhydrazines have cytotoxic activity against rhabdomyosarcoma RD and melanoma MS. This effect is comparable to the activity of camptothecin, an alkaloid with high antitumor activity.

1. Introduction

High-molecular compounds and their metal-containing nanocomposites are characterized by unique properties and are promising for medicine, optoelectronics, nanophotonics, and catalysis [1-4]. Silver nanoparticles have antimicrobial activity, which allows them to be used to obtain antibacterial drugs [3]. Silver nanoparticles are used as anti-cancer therapeutic agents for the treatment of leukemia, breast cancer, hepatocellular carcinoma, lung cancer, skin cancer, and oral cancer [5-9].

Recently, among the rapidly developing methods for producing nanomaterials, the greatest attention is paid to methods for producing composite materials based on organic polymer matrices and metal nanoparticles. The polymer matrix acts as a nanostabilizing component, preventing the aggregation of metal nanoparticles. Moreover, the polymer nature has a significant effect on the particle stability [10]. Polymers of both synthetic (polyethylene glycol, polyvinylpyrrolidone, polyethylene, etc.) and natural (cellulose, chitosan, arabinogalactan, etc.) origin are studied. The highly stable hydrosols with isolated silver nanoparticles stabilized by poly-N-vinylpyrrolidone were obtained by borohydride method [11]. Hydrophilic poly-N-vinylpyrrolidone is also used to stabilize sols and simultaneously restore silver ions to a molecular metallic state in water-alcohol solutions in the preparation of poviargol [12]. The nanocomposites based on silver nanoparticles and homo- and copolymers of 1-vinyl-1,2,4-triazole were obtained [13, 14] which are promising for the development of new water-soluble antiseptics and bactericidal preparations. We obtained biocidal silver nanocomposites based on copolymers of 2,2-diallyl-1,1,3,3-tetraethylguanidinium chloride with N-vinylpyrrolidone and vinyl acetate [15,16].

Our work is focused on the incorporation of silver nanoparticles into the matrices of *N*-vinylpyrrolidone copolymers with N,N-diallyl-N'-acylhydrazines (DAH). Thorough structural and optical characterization of nanocomposites was performed using UV- spectroscopy and scanning electron microscopy. Antimicrobial efficiency of new nanocomposites was tested against both Gram-positive and Gram-negative bacteria. Cytotoxic activity of nanocomposites towards lung carcinoma, rhabdomyosarcoma and melanoma was determined by the MTT method.

2. Experimental

2.1. Materials

N-vinylpyrrolidone (VP) was dried by potassium hydroxide and purified by vacuum distillation. A fraction with T_b= 97 °C/13 mm Hg, n_D^{20}= 1.5117 was used.

N,N-diallyl-N'-acylhydrazines (DAH) were obtained as a result of the interaction of the corresponding hydrazide and allyl chloride by the technique as described in our previous work [17].

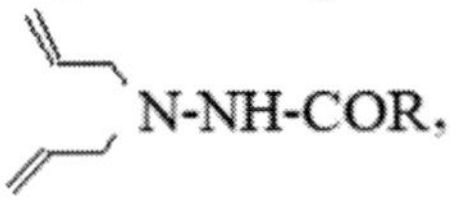

R= CH_3 (DAAH), C_3H_7 (DABH), C_6H_5 (DABEH)

The fractions of DAAH with T_b=120°C/5 mmHg, n_{20}^D =1.4771, DABH with T_b=145°C/8 mmHg, n_{20}^D =1.4740, DABEH with T_{mel}=108°C (aceton:water=1:1) were used in our experiments.

All the other chemicals were obtained from commercial suppliers. The characteristics of applied initiator (2,2'-azobisisobutyronitrile (AIBN)) and solvents (DMSO, acetone, diethyl ether, methanol) conformed to the reference data after purification by conventional methods.

2.2. Copolymerization

Copolymerization of VP with DAH was conducted in bulk in the presence of $3.0 \cdot 10^{-2}$ mol L^{-1} AIBN. Copolymers were precipitated and purified by three-fold reprecipitation by diethyl ether from methanol solution. The purified copolymers were dried under vacuum at 50 °C until constant weight was achieved. The copolymer composition was calculated from the elemental analysis data.

2.3. Synthesis of nanocomposites

Synthesis of silver nanocomposites was conducted as follows. Poly(VP-DAAH), poly(VP-DABH) or poly(VP-DABEH) (10^{-2} mol) were dissolved in water or methanol. Then $AgNO_3$ (5 x 10^{-4} mol of 1% aqueous solution) was added and the reactive mixture was stirred for one hour at room temperature. Then $NaBH_4$ (10^{-3} mol, 0.038 g) was added dropwise with the constant intensive stirring and the reactive solution was stirred for ten hours at room temperature. Nanocomposites were separated by dialysis. The purified nanocomposites were dried under vacuum at 50 °C until constant weight was achieved.

2.4. Instruments

Fourier transform infrared spectra (FT-IR) were recorded using a IFS 66/S Bruker spectrometer at a resolution of 4 cm^{-1}. Samples were coated on KBr plates and dried. The measurements were performed in the range from 400 to 4000 cm^{-1} at room temperature.

The ^1H and ^{13}C NMR spectra were recorded on a Bruker Avance II spectrometer operating at 400 and 100 MHz, respectively, using a broad-band proton decoupling and in a JMOD (J-modulated) mode. DMSO-d_6 was used as a solvent; tetramethylsilane was used as an internal standard.

The optical properties of the nanocomposites were measured using a CF-2000 spectrophotometer in a wavelength range of 200–600 nm.

Concentration of Ag in aqueous solutions was determined with using of atomic-absorption spectrometer iCE 3500 («Thermo Fisher Scientific»,USA).

Samples of nanocomposites were studied by means of an Evex Mini-SEM HR-3000 microscope.

2.5. Microbiological test

Microbiological tests were performed by serial two-fold dilution of preparations in LB medium. Test cultures were *Staphylococcus aureus,* ATCC 25923; *Staphylococcus epidermidis* 33; *Micrococcus luteus*, NCIMB 196; *Escherichia coli,* ATCC 25922; *Bacillus subtilis* ATCC 6633; *Klebsiella pneumonia,* ATCC 700 603. Microbial loads were 10^6 cells in 10 µl in LB medium. The measurements were made after incubation of bacteria for 20 h at 37°C. Experiments were repeated three times.

2.6. Cytotoxicity test
Cytotoxicity of compounds was performed as follows. Cell lines of human lung carcinoma (A549), human rhabdomyosarcoma (RD TE32) and human melanoma (MS) were obtained from the Research Institute of Experimental Tumor Diagnostics and Therapy, N.N. Blokhin Cancer Research Center, Russian Academy of Medical Sciences (Moscow). Cells were kept in DMEM medium (for A549 and RD) and in RPMI 1640 medium (for MS) supplemented with 10% fetal bovine serum, 2 mM L-glutamine and 1% gentamicin at 37°C in the Isotemp Barnstead CO_2 incubator. The 50% cell growth inhibitory concentration (IC_{50}) of the synthesized compounds was determined by the MTT method. The optical density was measured at 544 nm using a FLUOstar Optima microplate reader. Concentrations (IC_{50}) were calculated according to the dose-dependent inhibition curves. All experiments were performed for three times and the data were presented as means ± standard deviation. To test the significance of observed differences between the study groups, Student's t-test was applied. A value of $p< 0.05$ was considered to be statistically significant.

3. Results and discussion
Copolymers of N-vinylpyrrolidone with N,N-diallyl-N'-acetylhydrazine (poly(VP-DAAH)), N,N-diallyl-N'-butanoylhydrazine (poly(VP-DABH)), N,N-diallyl-N'-benzoylhydrazine (poly(VP-DABEH)) were obtained by radical polymerization in the presence of AIBN.

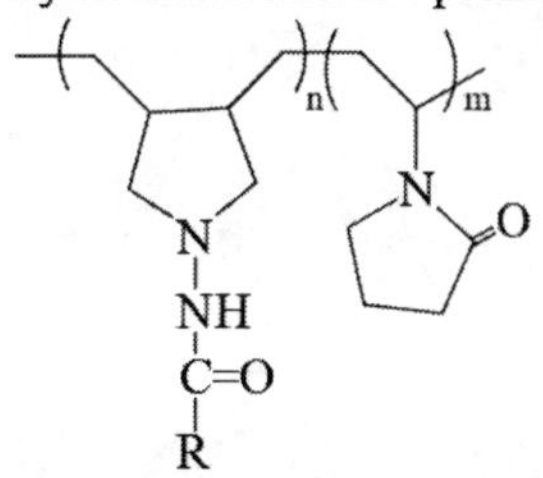

$R = CH_3, C_2H_5, C_3H_7, C_6H_5$

The IR spectra of copolymers contain several characteristic bands that are useful to the structural characterization of the polymers. We observe characteristic aliphatic methylene stretches between 3100 and 2850 cm^{-1} and the broadened carbonyl band at 1668 cm^{-1} VP. This broadened band is the result of spectral overlap between the acylhydrazine C=O stretch and the pyrrolidone C=O stretch.

By ^{13}C NMR it was found that N,N-diallyl-N'-acylhydrazines copolymerize with VP, both double bonds participating with formation of five-membered pyrrolidine structures in a cyclolinear polymer chain. Analysis of ^{13}C NMR spectra of copolymers indicates that block fragments of a vinyl monomer with inclusion of DAH units predominate in the copolymer structures. The DAH units consist of a mixture of cis-, trans-stereoisomers with a significant predominance of cis-heterocycles.

The copolymers obtained are light powders. They are soluble in methanol, chloroform, DMSO, dimethylformamide. The poly(VP-DAAH) is also soluble in water.

The intrinsic viscosity of poly(VP-DAAH) (80:20 mol%) is 0.11 dl/g (water, 25°C). Molecular weigth of poly(VP-DAAH) (80:20 mol%), calculated by the Levy and Frank equation for PVP ($[\eta] = 6.76\, M^{0.55} \cdot 10^{-4}$[18]), is equal to 10500.

We used our novel VP copolymers as stabilizing agents in the synthesis of silver nanoparticles. The synthesis of silver nanocomposites was carried out by the reduction of silver nitrate with sodium

IOP Conf. Series: Materials Science and Engineering **848** (2020) 012022 doi:10.1088/1757-899X/848/1/012022

borohydride in an aqueous (DAAH) or alcohol (DABH and DABEH) copolymer solution. Reaction proceeds via formation of the stable dark brown sols, from which soluble nanocomposites were separated by dialysis. The silver content depends on the copolymer used and is from 4 to 19%. The ratio of silver nitrate, reducing agent and copolymer also affects the concentration of silver in nanocomposites.

In the IR spectra of the obtained nanocomposites, except for the signals of the initial copolymers, there are no other signals, this fact indicates the constancy of copolymer structure. Nevertheless, in the IR spectra of nanocomposites based on DAH copolymers, one can observe an increase in the intensity of the absorption bands of NH group stretching vibrations at 1553 cm^{-1} and CN group stretching vibrations at 1282 cm^{-1}. We can notice a decrease in the intensity of the absorption band of stretching vibrations of carboxyl groups at 1668 cm^{-1}. These changes indicate the involvement of O and N atoms of VP and DAH in the interaction with silver nanoparticles.

It should be noted that due to the coordination binding of silver ions with the functional groups of macromolecules, a favorable microenvironment is created for the reduction of these ions [19]. Consequently, the obtained silver nanoparticles are incorporated into the macromolecules and are retained therein mainly through numerous specific electrostatic or covalent bonds between the functional groups of the copolymers and silver atoms on the surfaces of the nanoparticles. In addition, physical adsorption (dipole interactions, van der Waals forces) also makes a certain contribution, but the effect of the latter is much smaller than the effect of chemisorption.

In the UV spectra of nanocomposite solutions, a characteristic plasmon absorption band with a maximum in the region of 400–410 nm is observed (Fig. 1). The shape of the absorption spectra and the half-width of the maximum indicate the formation of sols with isolated silver nanoparticles [20].

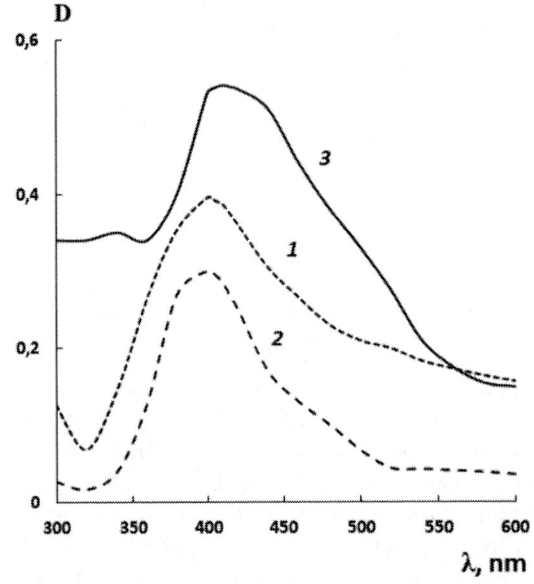

Figure 1. UV extinction spectra of nanocomposite solution: *1* – poly(VP-DAAH), water, C=7.5·10^{-3} mol/l; *2* – poly(VP-DABH), methanol, C=10^{-3} mol/l; 3 – poly(VP-DABEH), methanol, C=5·10^{-4} mol/l.

The results of SEM microscopy confirm the preparation of nanocomposites with a uniformly narrow distribution of silver nanoparticles in a polymer matrix. SEM micrographs and the size distribution of nanoparticles are shown in Fig. 2. It was established that the nanoparticles of spherical and elliptical shapes are formed.

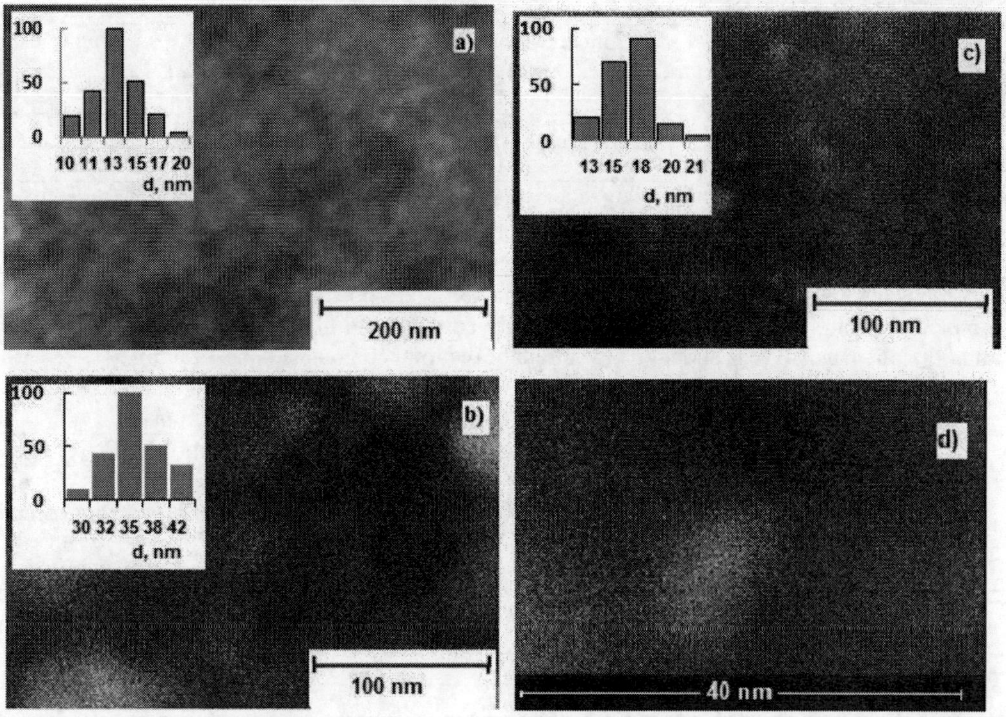

Figure 2. SEM images of silver nanoparticles and particle size distribution histograms of AgNCs based on: poly(VP-DAAH) (a,d), poly(VP-DABH) (b), poly(VP-DABEH) (c).

The influence of molecular weight of copolymer, the copolymer concentration and Ag^+ concentration on the size of Ag nanoparticles was investigated (Table 1, 2, 3). It was determined that with increasing molecular weight of copolymer, the copolymer concentration and Ag^+ concentration the average particle size decreased.

Table 1. The influence of M_n on the size of Ag nanoparticles.

Copolymer	M_n	Size of Ag nanoparticles, nm
poly(VP-DAAH) C(copolymer) = 2.0 wt. % C(Ag^+) = 0.05 mol/L	13000	16
	15900	13
	21000	12
poly(VP-DABEH) C(copolymer) = 2.0 wt. % C(Ag^+) = 0.05 mol/L	15700	20
	17500	18
	28000	17

Table 2. The influence of the copolymer concentration on the size of Ag nanoparticles.

Copolymer	Concentration, wt. %	Size of Ag nanoparticles, nm
poly(VP-DAAH) Mn = 15900 C(Ag+) = 0.05 mol/L	1.0	14
	2.0	13
poly(VP-DABEH) Mn = 17500 C(Ag+) = 0.05 mol/L	1.0	22
	2.0	18

IOP Conf. Series: Materials Science and Engineering **848** (2020) 012022 doi:10.1088/1757-899X/848/1/012022

Table 3. The influence of the Ag+ concentration on the size of Ag nanoparticles.

Copolymer	Concentration Ag+, mol/L	Size of Ag nanoparticles, nm
poly(VP-DAAH) M_n = 15900 C(copolymer) = 2.0 wt. %	0.025	15
	0.050	13
poly(VP-DABEH) M_n = 17500 C(copolymer) = 2.0 wt. %	0.025	22
	0.050	18

Studies of antimicrobial activity of nanocomposites obtained showed (Table 4) that our nanocomposites have a moderate activity both against Gram positive and Gram negative microflora. The biocide effect of new nanocomposites is higher compared to initial copolymers. It should be kept in mind that the nanosilver is about one twentieth of the total composite concentration.

Table 4. Antimicrobial activity of copolymers and their nanocomposites

Test cultures	Minimal bacteriostatic concentration (MBsC), μg/ml			
	poly(VP-DAAH)	Silver nanocomposite of poly(VP-DAAH)	poly(VP-DABEH)	Silver nanocomposite of poly(VP-DABEH)
Staphylococcus aureus ATCC 25923	>1000	125	500	125
Staphylococcus epidermidis 33	>1000	250	500	62.5
Micrococcus luteus NCIMB 196	>1000	250	250	62.5
Bacillus subtilis ATCC 6633	>1000	500	500	125
Escherichia coli ATCC 25922	>1000	500	500	125
Klebsiella pneumonia ATCC 700 603	>1000	500	>1000	62.5

New nanocomposites based on DAH copolymers have cytotoxic activity against rhabdomyosarcoma RD and melanoma MS cell lines (Table 5). Nanocomposites at a concentration of 0.36 and 2.59 μM, for poly(VP-DAAH) and poly(VP-DABEH) respectively, inhibit 50% of MS melanoma cells. The death of 50% RD rhabdomyosarcoma cells is observed at nanocomposites concentrations of 10.14 and 11.28 μM, for poly(VP-DAAH) and poly(VP-DABEH), respectively. The cytotoxic activity of the new nanocomposites is comparable to the activity of camptothecin, an alkaloid with high antitumor activity.

Table 5. Cytotoxic activity of AgNCs based on poly(VP-DAAH) and poly(VP-DABEH).

Culture	IC_{50}, μM				
	Camptothecin	Poly(VP-DAAH)	Silver nanocomposite of poly(VP-DAAH)	Poly(VP-DABEH)	Silver nanocomposite of poly(VP-DABEH)
Rhabdomyosarcoma RD	1.72±0.37	28.16±1.34	10.14±0.01	32.18±2.00	11.28±0.26
Bronchial carcinoma A549	1.31±0.03	no effect	85.17±3.59	no effect	85.49±0.084
Melanoma MS	0.77±0.34	72.34±2.86	0.36±0.00	65.48±1.44	2.59±0.42

The discovered biological activity of DAH-based nanocomposites makes them promising for the development of new medicines and biocides.

4. Conclusion

Thus, new water-soluble nanocomposites based on Ag and copolymers of *N*-vinylpyrrolidone with N,N-diallyl-N'-acylhydrazines have been developed. These nanocomposites have cytotoxic activity against rhabdomyosarcoma RD and melanoma MS. Following these results, the newly developed nanocomposite scaffold may be considered for the development of new drugs for the complex treatment of malignant neoplasms.

Acknowledgment

The study was funded by Russian Foundation for Basic Research and Government of the Perm Region according to the research project № 19-43-590019 r_a.

Analytical, spectroscopic, and biological studies were performed at the "Research of materials and substances" collective Center of PFRC UB RAS.

The authors gratefully acknowledge Dr. Darya Eroshenko, ITCh UB RAS, for providing cytotoxic experiments and Dr. Larisa Lemkina, IEGM UB RAS, for microbiological tests. We also wish to thank Dr. Dmitriy Kisel'kov, ITCh UB RAS, for his generous advices for SEM analysis.

References

[1] Pomogailo AD, Rozenberg AS, Uflyand IE 2000 *Metal nanoparticles in polymers* (Moscow: Chemistry)
[2] Rao CNR, Muller A, Cheetham AK 2004 *The chemistry of nanomaterials* (Weinheim: Wiley-VCH Verlag GmbH & Co. KGaA)
[3] Blagitko EM, Burmistrov VA, Kolesnikov AP, Mikhaylov YuI, Rodionov PP 2004 *Silver in medicine* (Novosibirsk: Nauka-center)
[4] Krutyakov YuA, Kudrinskii AA, Olenin AYu, Lisichkin GV 2008 *Uspekhi khimii* **77** 242
[5] Ostad SN, Dehnad S, Nazari ZE, Fini ST, Mokhtari N, Shakibaie M, Shahverdi AR 2010 *Avicenna J Med Biotechnol* **2** 187
[6] Wei L, Lu J, Xu H, Patel A, Chen ZS, Chen G 2015 *Drug Discov Today* doi: 10.1016/j.drudis.2014.11.014.
[7] Goodman AM, Cao Y, Urban C, Neumann O, Ayala-Orozco C, Knight MW, Joshi A, Nordlander P, Halas NJ 2014 *ACS Nano* **8** 3222
[8] Braun GB, Friman T, Pang H-B, Pallaoro A, Hurtado de Mendoza T, Willmore AM, Kotamraju VR, Mann AP, She ZG, Sugahara KN, Reich NO, Teesalu T, Ruoslahti E 2014 *Nature Materials* **13** 904
[9] Ong C, Lim JZ, Ng C-T, Li JJ, Yung LY, Bay BH 2013 *Curr Med Chem* **20** 772
[10] Vegera AV, Zimon AD 2006 *Rus J Appl Chem* **79(9)** 1419
[11] Saifullina IR, Chiganova GA, Karpov SV, Slabco VV 2006 *Rus J Appl Chem* **79(10)** 1660
[12] Kopeikin VV, Panarin EF 2001 *Dokl Akad Nauk, Ser Khim* **380(4)** 497
[13] Myachina GF, Korzhova SA, Ermakova TG, Sukhov BG, Trofimov BA 2008 *Dokl Akad Nauk, Ser Khim* **420(3)** 344
[14] Myachina GF, Korzhova SA, Ermakova TG, Kon'kova TV, Pozdnyakov AS, Sukhov BG, Trofimov BA 2009 *Dokl Akad Nauk, Ser Khim* **427(6)** 790
[15] Gorbunova M, Lemkina L 2014 *J Nanopart Research* **16** 2566
[16] Gorbunova MN, Lemkina LM 2016 *J Biomed Mat Res* **104** 630
[17] Gusev VYu, Vorob'eva AI, Radushev AV, Kolesov SV, Muslukhov RR, Tolstikov AG 2003 *Dokl Akad Nauk, Ser Khim* **12** 2603
[18] Levy GB, Frank HP 1955 *J Polym Sci* **17** 247
[19] Gorbunova M, Lemkina L, Eroshenko D., Gileva K 2019 *Polym Adv Technol* **30** 336
[20] Karpov SV, Slabko VV 2003 *Optical and photophysical properties of fractally structured metal sols* (Novosibirsk: SB RAS)

IOP Conf. Series: Materials Science and Engineering **848** (2020) 012023 doi:10.1088/1757-899X/848/1/012023

Gold nanorods and their nanocomposites based on polyurethane foam for determination of catecholamines in biological fluids

M V Gorbunova[1], V V Apyari[1], A V Garshev[2], P A Volkov[3], V V Tolmacheva[1,4] and S G Dmitrienko[1]

[1] Department of Chemistry, Lomonosov Moscow State University, Leninskie gory, 1/3, Moscow 119991, Russia
[2] Faculty of Materials Science, Lomonosov Moscow State University, Leninskie gory, 1/3, Moscow 119991, Russia
[3] Scientific-Research Institute of Chemical Reagents and Special Purity Chemicals of National Research Center "Kurchatov Institute", Bogorodsky Val, 3, Moscow 107076, Russia
[4] Kurnakov Institute of General and Inorganic Chemistry, Russian Academy of Sciences, Moscow, 119991 Russia

masha13_1992@mail.ru

Abstract. Capability of gold nanorods and their nanocomposites based on polyurethane foam as spectrophotometric reagents for catecholamines determination was examined. Comparison of the analytical properties of the systems based on interaction of nanoparticles of different morphology with various reagents in the presence of catecholamines was carried out, gold nanorods-based system is characterized by the highest sensitivity. Applicability of gold nanorods and their nanocomposites for catecholamines determination in biological fluids (urine, serum) was proved by HPLC analysis. Catecholamines preconcentration by dynamic sorption on hypercrosslinked polystyrene and further elution with 6M acetic acid could serve to improve the sensitivity of the analysis.

1. Introduction

Nanoparticles (NPs) are widely used in chemical analysis for determination of various compounds. Special optical properties of NPs contribute a lot to their use as peculiar spectrophotometric reagents [1–5]. Correlation of the NPs surface plasmon resonance (SPR) spectrum with their state serves to determine compounds that can directly or indirectly change NPs characteristics. Non-spherical nanoparticles, e.g. gold nanorods (AuNRs) are of special interest as far as their SPR-spectrum is characterized by the presence of two SPR-bands in the visible spectral region – an analytical signal in this case could be based not only on the change of the SPR-maxima intensity but also on the mutual arrangement of maxima and the ratio of their intensities. Most of the analytical techniques are based on aggregation or AuNRs morphology changes, nevertheless, approaches based on formation of nanoparticles of "core-shell" type are also promising. Thus, an approach to catecholamines determination based on formation of silver layer on the AuNRs surface in the presence of catecholamines that leads to a hypsochromic shift of SPR-maxima was reported [6, 7].

Along with NPs themselves their composites are of certain interest as far as immobilization of nanoparticles in solid matrices could stabilize them and contribute to their ease of use and analytical characteristics [8–11]. A promising material for such nanocomposites is polyurethane foam (PUF) that is characterized by sustainability, ease of use, low cost and absence of its own coloring [12–14].

An important modern task of analytical chemistry is determination of various biologically active compounds, catecholamines (CAs) among them. Such scientific interest to catecholamines, especially to natural epinephrine, norepinephrine and dopamine, is caused by their important vital functions – regulation of heart and psychomotor activity, processes of nervous system [15]. Timely detection of catecholamines abnormal concentrations serves to predict several diseases or to begin treatment at the illness initial stage. Use of nanoparticles could contribute to development of rapid and sensitive techniques of catecholamines determination.

2. Experimental

2.1. Materials

Hydrogen tetrachloroaurate (chemically pure) and silver nitrate (pure) were used for AuNRs synthesis. Sodium borohydride (analytical grade) and ascorbic acid (chemically pure) were used as reducing agents, cetyltrimetylammonium bromide (CTAB, analytical grade) was used as a stabilizer of AuNRs. Epinephrine (E), norepinephrine hydrochloride (NE) and dopamine hydrochloride (DA) (all of analytical grade) were used as subjects of the study. The substances stock solutions were prepared by dissolving their weighed portions in deionized water. Epinephrine was dissolved in 0.01 mol L^{-1} HCl. The stock solutions were frozen and kept in a refrigerator. Working 0.2 mmol L^{-1} solutions were prepared from the stock solutions immediately before use. Glycine buffer (pH 9.5) was prepared by addition of 74.8 mL of 0.1 mol L^{-1} glycine (7.51 g of glycine and 5.85 g of NaCl in 1 L) to 25.2 mL of 0.1 mol L^{-1} NaOH. Copper (II) chloride (chemically pure) was used for the comparative studies of various nanoparticles analytical characteristics. Phosphoric acid (H_3PO_4), octane-1-sulfonic acid sodium salt, acetonitrile (all of analytical grade) were used for preparation of chromatographic eluent. Acetic acid (analytical grade) was used for catecholamines solid-phase extraction.

Commercially available polyether-type polyurethane foam (PUF) was used as a matrix for nanocomposites. Tablets of PUF of 16 mm in diameter (20 ± 2 mg) were cut with a metal punch from an industrial sheet of the polymer. To remove impurities, the PUF tablets were placed in acetone and shaken for 10 minutes. The procedure was repeated twice, after which the tablets were dried under stream of air.

The extraction of catecholamines from biological fluids was carried out with the use of a microcolumn filled with hypercrosslinked polystyrene (Diapak P-3 cartridges, ZAO "BioChemMak ST), the particle size of the sorbent was 50 – 100 µm, the pore diameter was 10 – 1000 Å.

2.2. Instruments

Absorption spectra of solutions were recorded by SF-103 spectrophotometer (Akvilon, Russia). Diffuse reflection in the visible region was recorded by Eye-One Pro mini-spectrophotometer (X-Rite, USA).

Chromatographic determination of catecholamines was performed using a Tsvett Yauza chromatograph (NPO Khimavtomatika, Russia) with an amperometric detector (E = +0.8 V). The chromatographic column "Eclipse XDB-C18" (Agilent) and guard column Security Guard C18 were used. Eluent contained acetonitrile and 0.1% H_3PO_4 in water (10:90, v/v) with addition of sodium 1-octanesulfonate (0.3 mmol L^{-1}). The flow rate was 0.4 mL min^{-1}. The eluent was degassed in a Bransonic 1510R-DTH ultrasonic bath (USA). Sample volume was 20 µL, injection was performed with a loop.

A concentrating microcolumn (l = 6 cm, d = 10 mm) and a vacuum unit for solid-phase extraction M6 (Manifold, Russia) were used for catecholamines preconcentration.

Deionized water was obtained from the purification system Millipore Simplicity (Millipore, Germany). A mechanical shaker and a magnetic stirrer were also used.

2.3. Synthesis of AuNRs and their nanocomposites based on polyurethane foam

AuNRs were synthesized according to the method previously reported by Nikoobakht and El-Sayed [16] with some modifications. A detailed scheme of the synthesis is described in [7]. The obtained nanoparticles are characterized by the presence of two SPR-bands in the visible spectral region, corresponding to transverse and longitudinal plasmon oscillations. TEM-image and typical absorption spectrum of the obtained AuNRs are presented in figure 1.

The nanocomposite material based on polyurethane foam and AuNRs was obtained by sorption modification of polyurethane foam [13]. Samples of the nanocomposite were modified with silver nitrate, this procedure is described in detail in [14]. SEM-images and typical diffuse reflectance spectra of the obtained nanocomposites before and after modifying are presented in figure 2.

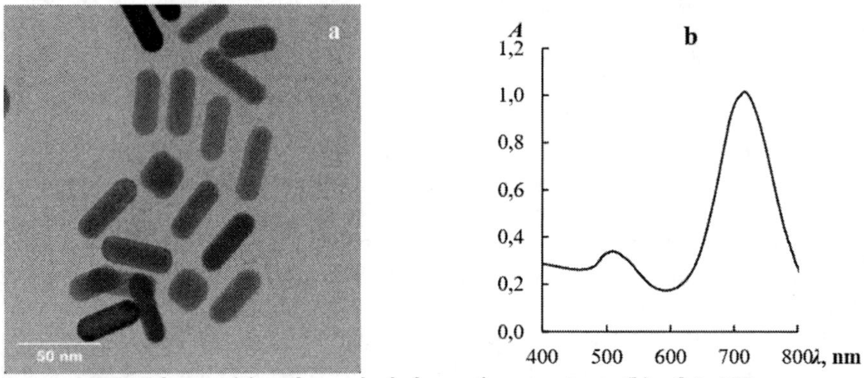

Figure 1. TEM-image (a) and a typical absorption spectrum (b) of AuNRs.

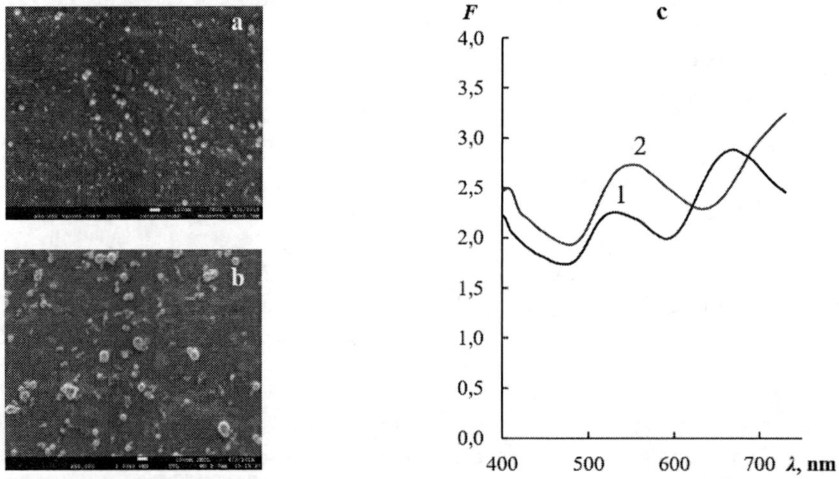

Figure 2. SEM-images (a, b) and typical diffuse reflectance spectra (c) of the nanocomposites before (a, c1) and after (b, c2) modifying with silver nitrate.

2.4. Synthesis of spherical gold nanoparticles and their nanocomposites

Synthesis of spherical nanoparticles was performed similarly to AuNRs synthesis, the difference was the absence of silver nitrate that initiates formation of long nanoparticles. Previously the seeds were synthesized by adding 5.00 mL of 0.50 mmol L^{-1} $HAuCl_4$, 5.00 mL of 0.20 mol L^{-1} CTAB aqueous solutions to a 250 mL round-bottom flask with stirring. Then, 0.60 mL of 0.010 mol L^{-1} $NaBH_4$ solution cooled to 0–4 ° C was added dropwise. To synthesize spherical nanoparticles 2.25 mL of deionized water, 45 mL of 0.20 mol L^{-1} CTAB, 45 mL of 1.0 mmol·L^{-1} $HAuCl_4$ were added to a 250 mL round bottom flask with stirring. After this, 0.56 mL of 0.10 mol·L^{-1} ascorbic acid was added to the solution dropwise, then 0.10 mL of the pre-synthesized seed solution was added. The resulting solution was kept for 2 weeks, then centrifuged for 30 min at 8000 rpm, the supernatant was removed by decantation, NPs were washed with deionized water by dispersing the precipitate in it. Separation of the supernatant by centrifugation and redispersion of NPs in deionized water was repeated twice.

For the synthesis of nanocomposites, 2.0 mL of non-redispersed in deionized water spherical NPs were placed in a test-tube, 2 mmol of NaCl was added, the total volume was 5.0 mL. Then, a PUF tablet was added, the tube was shaken on a mechanical shaker for 30 minutes. The tablet was removed, dried between sheets of filter paper and placed into 5.0 mL of 1 mmol·L^{-1} $AgNO_3$ solution. Modifying with silver nitrate was carried out for 10 min with stirring on a shaker. A tablet of the nanocomposite was removed from the solution and dried with filter paper.

2.5. General procedure

To study the interaction of colloidal solutions of AuNRs with catecholamines, 0.3 mL of AuNRs, 20 μL of 0.01 mol L^{-1} $AgNO_3$, 0.1 mL of glycine buffer solution and 0–0.5 mL of 25 μmol L^{-1} catecholamines solution were added into a test-tube. The total volume of all components was 2.5 mL. The absorption spectra of the obtained mixtures were measured after 3 minutes. The position of the long-wavelength maximum was considered as the point at which the first derivative of the function A (λ) is zero.

Interaction of the nanocomposite modified with silver nitrate with catecholamines was studied by placing the nanocomposite into 5 mL of solution containing 1–100 μmol L^{-1} catecholamines and 1 mL of glycine buffer solution. It was shaken on a mechanical shaker for 30 minutes. Then the nanocomposite was removed and dried between sheets of filter paper. The diffuse reflection coefficients (*R*) of the obtained nanocomposite were measured in a range of 400–730 nm with the 10 nm increment. Then, the values of the Kubelka-Munk function (*F*) were calculated from the formula: $F = \frac{(1-R)^2}{2R}$, here *R* is the diffuse reflection coefficient. The plotted spectra *F* versus λ were used to estimate changes in state of AuNRs in the phase of polyurethane foam. The exact position of the maximum was determined by calculation an arithmetic mean of two λ values of inflection points of the spectrum (zero-points of the F(λ) second derivative function).

To carry out solid-phase extraction of catecholamines in the dynamic mode, a concentrating microcolumn (l = 6 cm, d = 10 mm) packed with 0.02 g of hypercrosslinked polystyrene and a vacuum unit for solid-phase extraction M6 (Manifold, Russia) were used. Before use, the column was conditioned with 1 mL of acetonitrile and 10 mL of water. The rate of passing the solution through the column was 1.0 mL/min. Before desorption, the column was washed with 10 mL of water. Desorption of catecholamines was performed by 3 mL of 6 mol L^{-1} acetic acid.

3. Results and discussion

3.1. Interaction of catecholamines with $AgNO_3$ in AuNRs colloid solutions and nanocomposites

As it has been described before [7, 14, 17], the interaction of catecholamines with silver ions in the presence of AuNRs or the nanocomposite is accompanied by a hypsochromic shift of the SPR bands and an increase of their intensity. These effects are visually detected as change of color of the AuNRs solution from pale purple to bright green; the nanocomposites color changes from pale purple to grey. The shift of the maxima (Δλ) can be considered as a characteristic of completeness of the interaction.

The scheme of the interaction includes oxidation of the catechol group by silver ions with formation of the ortho-quinonoid structures and subsequent deposition of metallic silver onto the AuNRs surface.

Dependencies of the shift of maxima on concentration of individual epinephrine, norepinephrine, dopamine and their equimolar mixture are represented in figure 3. Linear parts of these dependencies can be used as calibration graphs for determination of catecholamines. Some analytical characteristics of the determination are presented in table 1. Determination of catecholamines with the use of colloidal solutions of AuNRs is characterized by higher sensitivity in comparison to the nanocomposites. Nevertheless, analysis with the use of nanocomposites is characterized by larger linear range and higher selectivity in the presence of inorganic ions that could be ascribed to the matrix effects.

IOP Conf. Series: Materials Science and Engineering **848** (2020) 012023 doi:10.1088/1757-899X/848/1/012023

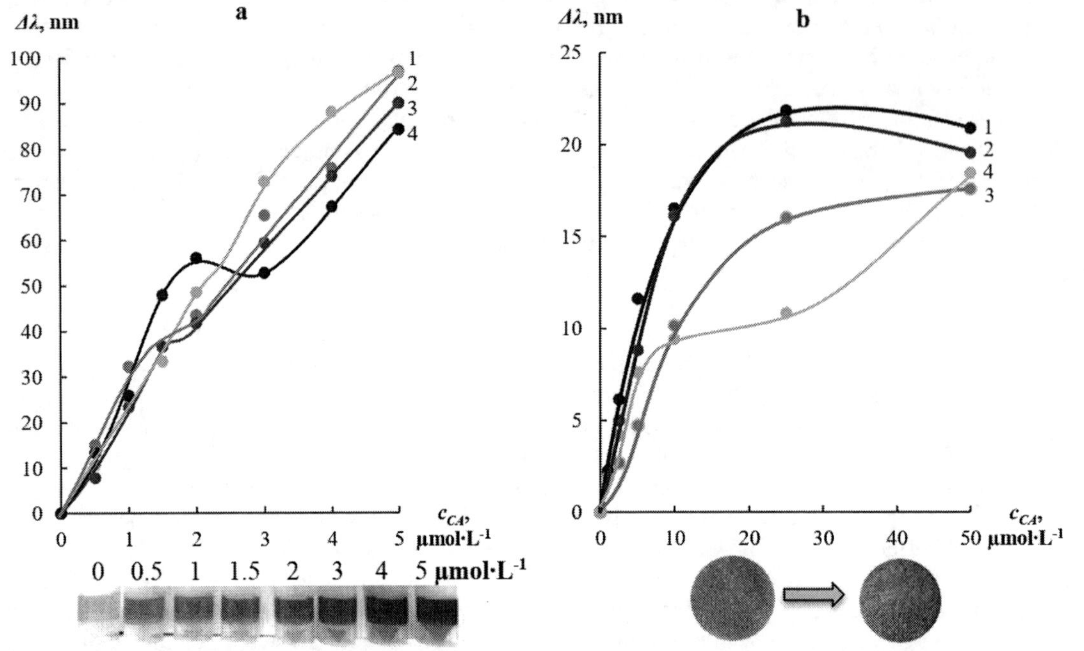

Figure 3. Dependencies of the shift of maxima of AuNRs solutions (a) and nanocomposites (b) on concentration of individual epinephrine (1), dopamine (2), norepinephrine (3) and their equimolar mixture (4) and photos of the obtained solutions and composites.

Table 1. Analytical characteristics of the catecholamines determination (n = 3, P = 0.95).

		dopamine	norepinephrine	epinephrine	equimolar mixture
With the use of AuNRs colloidal solution	LOD, μmol·L^{-1}	0.08	0.1	0.1	0.1
	Linear range, μmol·L^{-1}	0.2 – 5.0	0.3 – 5.0	0.3 – 5.0	0.3 – 5.0
With the use of nanocomposites	LOD, μmol·L^{-1}	0.3	0.3	0.4	0.3
	Linear range, μmol·L^{-1}	0.9 – 25	1.0 – 25	1.1 – 25	0.9 – 25

3.2. Comparison of the analytical characteristics of the catecholamines determination using various nanosystems

Epinephrine was used to study the interaction of catecholamines with silver nitrate, chloroauric acid and copper (II) chloride as oxidizing agents in the presence of colloid solutions of AuNRs and spherical gold NPs, spectra of the obtained solutions are presented in figure 4. In case of AuNRs and silver nitrate, as it was described above, hypsochromic shift of both maxima is observed (figure 4 a). In case of the interaction of catecholamines with chloroauric acid in the presence of AuNRs, a bathochromic shift of the SPR maxima of AuNRs is observed, which is presumably caused by the growth of AuNRs in the longitudinal direction as a result of the predominant reduction of chloroauric acid at the end faces of AuNRs (figure 4 b). If copper (II) chloride is used as an oxidizing agent, no significant changes in the absorption spectra of AuNRs occur (figure 4 c). Varying the amount of CA (1–4 μmol L^{-1}) and copper chloride (8–160 μmol L^{-1}) does not lead to any spectral changes. It is likely that under the given conditions, the formal potential of CA oxidation is insufficient for the reduction of Cu (II); in addition, the formation of complex compounds of Cu (II) with CA is possible, that causes decrease of the reduction potential of copper.

Along with AuNRs, spherical NPs were studied as spectrophotometric reagents for the determination of catecholamines. As can be seen from figure 4 d, as a result of the interaction, a hypsochromic shift of the SPR-band of the nanoparticles occurs, this is caused by formation of a silver shell on the nanoparticles surface. The obtained dependence of Δλ on epinephrine concentration is shown in figure 5. It is seen that in the case of the interaction of catecholamines with $AgNO_3$ in the presence of NPs, lower Δλ values are observed than in the presence of AuNRs (figure 3 a, table 2). The same interaction has been studied under heating. To do this, the reaction mixtures were kept in a water bath at 70°C for 25 min, then cooled for 5 min in ice water, and the absorption spectra were recorded. Heating contributes to an increase of the analytical signal, but the Δλ values do not exceed the values obtained in the system with AuNRs, the color change of the solutions is less contrast (figure 5 a). The determination range is also narrower because of the impossibility of fixing the exact position of the SPR maximum at high catecholamine contents.

Comparison of the catecholamines interaction with the nanocomposites based on AuNRs and spherical NPs has been also carried out. Figure 5 b shows the dependence of a hypsochromic shift of the SPR maximum of spherical NP-based nanocomposites on the epinephrine concentration. The initial section of this curve was used as a calibration graph. Table 3 summarizes analytical characteristics of the epinephrine determination using nanocomposites based on spherical NPs and AuNRs. According to the data, the sensitivity of the catecholamines determination using nanocomposites based on spherical nanoparticles is noticeably worse than the sensitivity of the technique based on AuNRs nanocomposites. This fact is associated with poor reproducibility of the SPR maximum position in the control experiment.

All these data prove that the most applicable nanosystem for catecholamines determination is that based on AuNRs/AuNRs-nanocomposites and silver nitrate, these systems should be used for the catecholamines determination in real samples.

Table 2. Characteristics of the epinephrine determination with the use of various nanoanalytical systems (n = 3, P = 0.95).

Characteristic	**Analytical system**			
	AuNRs, $AgNO_3$	**AuNRs, $HAuCl_4$**	**AuNRs, $CuCl_2$**	**Spherical NPs, $AgNO_3$**
Δλ in the presence of 2 μmol L^{-1} epinephrine	43 74*	-9.6	0	6.5 16
LOD, μmol L^{-1}	0.1 0.1*	0.1	Determination is impossible	0.2 0.1*
Linear range, μmol L^{-1}	0.3 – 5.0 0.4 – 5.0*	0.4 – 4.0		0.6 – 5.0 0.4 – 3.0*

* Determination was carried out at 70°C

Table 3. Analytical characteristics of the epinephrine determination using nanocomposites based on AuNRs and spherical NPs (n = 5).

Characteristic	Nanocomposites are based on	
	AuNRs	Spherical NP
LOD, μmol·L^{-1}	0.4	3
Linear range, μmol·L^{-1}	1.1 – 25.0	–

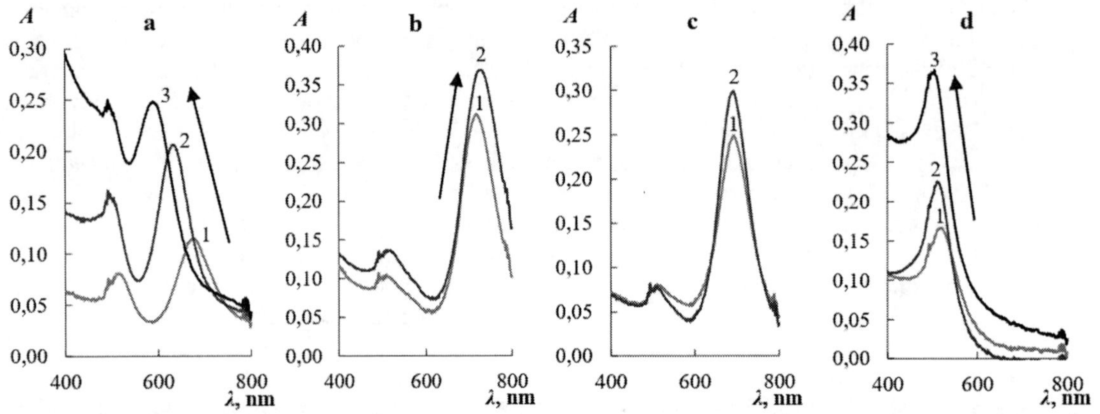

Figure 4. Absorption spectra of AuNRs (a–c) and spherical NPs (d) in the presence of 80 μmol L^{-1} $AgNO_3$ (a, d), $HAuCl_4$ (b), $CuCl_2$ (c) and 0 (1) / 2 (2) μmol L^{-1} epinephrine registered after the reaction carried out at 25°C (1, 2) and 70°C (3).

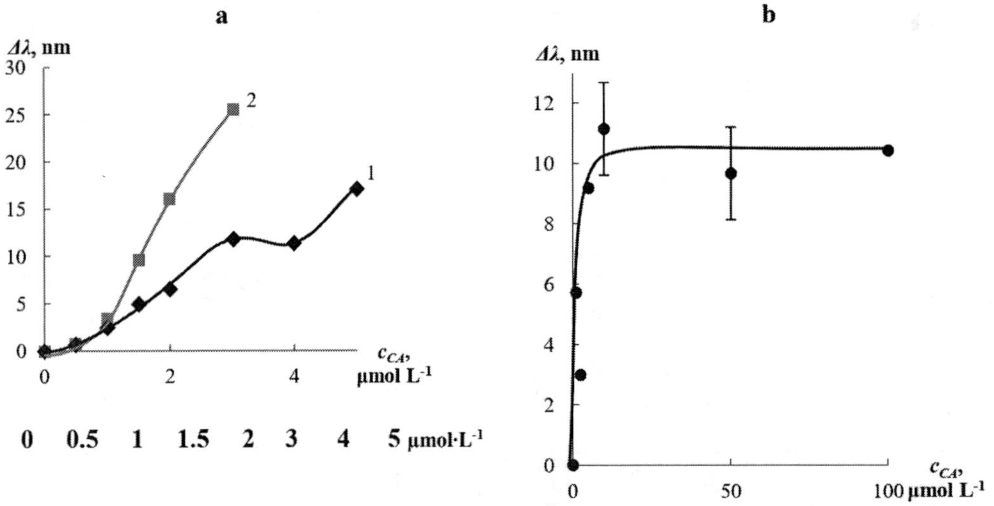

Figure 5. Dependencies of the SPR maximum shift of colloid solutions (a) and nanocomposites (b) of spherical NPs on epinephrine concentration after interaction at 25°C (1) and 70°C (2); and photos of the solutions obtained at 70°C (c_{CA} = 0–5 μmol L^{-1} (a), c_{CA} = 0–100 μmol L^{-1} (b), c_{AgNO_3} = 80 μmol L^{-1} (a), glycine buffer solution, pH 9).

3.3. Catecholamines determination in biological fluids

Previously, it was reported that the use of AuNRs is appropriate to total catecholamines determination, as far as the sensitivity coefficients in the calibration graph equations differ slightly [17]. Determination of total catecholamines is relevant for diagnosis of such diseases as Alzheimer's disease, Addison's disease, pheochromocytoma – illnesses that cause change of the one of the catecholamines content or simultaneous increase/decrease of several catecholamines content [18–20]. In this work applicability of AuNRs colloid solutions and their nanocomposites for catecholamines determination in biological fluids (urine, serum) was examined.

3.3.1. Determination of the total catecholamine content without preconcentration

Urine samples were analyzed using the standard addition method by introducing urine samples and catecholamines mixture (epinephrine : norepinephrine : dopamine = 1:7:32) to the reaction mixture, the additives of catecholamines in case of AuNRs solutions were 0/1/2/3 μmol L^{-1}, in case of nanocomposites – 0/5/15 μmol L^{-1}.

It was found out that introduction of 1.0 mL of urine into the reaction mixture containing AuNRs solution leads to aggregation of nanorods, determination of catecholamines in these conditions is impossible. Introduction of 0.25 mL of urine into the reaction mixture does not violate the stability of the colloidal system; registration of the analytical signal is possible. It was found that 3 min is not enough for the reaction: the Δλ values of some solutions were negative, in other solutions the Δλ values were significantly lower than the corresponding data for aqueous solutions in the absence of urine. The maximum Δλ values were reached after 1.5 h after mixing the reagents.

Nanorods immobilized on PUF are characterized by greater stability in comparison to the colloidal system: introduction of 3 mL of urine into the reaction mixture does not cause AuNRs aggregation, CAs determination is possible. Good agreement between the data obtained for the same urine sample using both colloidal solutions and using the nanocomposites, as well as the results of chromatographic analysis (table 4), indicates good accuracy of the proposed methods.

As can be seen from the results of HPLC analysis, the ratio of catecholamines contents in the studied samples does not correspond to the average catecholamines ratio in human daily urine (1:7:32) [15]. This can be explained by that the samples were collected in separate short time intervals. Since it was found that the concentration ratio of catecholamines in the samples was not 1:7:32, in further experiments equimolar catecholamine mixtures were used as additives.

Table 4. Results of total catecholamines determination (μmol L^{-1}) in urine samples with the use of AuNRs colloid solutions, nanocomposites and HPLC

Analysis method	Sample 1		Sample 2	
With the use of AuNRs colloid solutions			14.0 ± 0.7* (RSD 6%)	
With the use AuNRs nanocomposites	10.2 ± 0.9 (RSD 9%) 9.9 ± 0.4* (RSD 5%)		14.5 ± 0.9 (RSD 5%) 14 ± 1* (RSD 9%)	
HPLC	10 ± 3 (RSD 20%)	E: 7 ± 3 NE: 1.4 ± 0.2 DA: 0.9 ± 0.7	14 ± 5* (RSD 20%)	E: 10 ± 2 NE: 1 ± 1 DA: 3 ± 4

* The analyzed samples were diluted with deionized water

To evaluate the possibility of catecholamines determination in blood serum the spiked samples were used. To do this, 0.10 mL of serum was introduced into the reaction mixture in the case of using AuNRs solutions and 0.25 mL of serum in case of nanocomposites. Preliminarily, the dependences of Δλ on the total concentration of catecholamines in the system were measured, the initial sections of which were used as calibration plots. According to the results presented in table 5, the developed techniques are applicable for determination of catecholamines in the reaction mixture containing blood serum at the levels of 0.5 and 5 μmol L^{-1} in case of AuNRs colloidal solutions or their nanocomposites, respectively. However, the sensitivity of the proposed methods in some cases is insufficient for analysis at a clinically significant levels, therefore the development of methods for pre-concentration of catecholamines is required.

Table 5. Results of total catecholamines determination in blood serum with the use of AuNRs colloid solutions and their nanocomposites (n = 3, P = 0.95).

Analysis method	Added, μmol L^{-1}	Found, μmol L^{-1}	RSD, %
With the use of AuNRs colloid solutions	0.5	0.5 ± 0.1	8
With the use AuNRs nanocomposites	5	4.8 ± 0.3	2

3.3.2. Determination of the total catecholamines content after their solid-phase extraction with the use of hypercrosslinked polysterene

As can be seen from the data presented in tables 4, 5, the developed methods allow reliable determination of CAs if their total content is in the region of 0.5–15 μmol L^{-1}. Since in some cases the content of CAs in biological fluids is lower than this value, it is important to study the possibility of increasing the sensitivity of the developed determination techniques by pre-concentration of analytes. A promising option is solid-phase extraction (SPE) of CAs with the use of hypercrosslinked polystyrene, followed by their elution with a small volume of a solvent. It is well-known [21–23] that this sorbent is universal in relation to various organic compounds and, due to its microporous structure, allows to separate these compounds from the matrix high molecular weight substances. Therefore, this sorbent can be used to reduce the matrix effect.

To analyze urine samples, they were 10 times or 1.2 times diluted and 0, 0.5 and 1 μmol L^{-1} or 0, 5 and 10 μmol L^{-1} CAs equimolar mixtures were added. Total volumes of the samples with the additives were 10.0 mL and 15.0 mL respectively. The mixtures were passed through a microcolumn packed with hypercrosslinked polystyrene. Then, CAs were eluted by 3 mL of 6 mol L^{-1} acetic acid. Before the CAs determination with the use of AuNRs, solution of 10 mol L^{-1} NaOH was added to the eluates in order to provide pH 8. In the first case, the determination by the standard addition method is possible using colloidal solutions of AuNRs and HPLC. The data obtained are presented in table 6. Since the content of CAs in the eluates without additives was below LOD of the procedure using nanocomposites, analysis by this method was not applicable. The second variant of CAs preconcentration (1.2 times dilution of urine samples) turned out to be appropriate for the analysis with the use of nanocomposites. This is confirmed by a good agreement with the results of the analysis using AuNRs solutions and HPLC (table 6). The LODs for total CAs for these samples in the case of AuNRs solutions are below the normal CA content. In the case of nanocomposites, they correspond to the lowest values of the normal content.

Use of the spiked samples proved that catecholamines determination in blood serum after their SPE on hypercrosslinked polystyrene is also possible. The solutions contained 1.0 mL of serum and various additives of the equimolar mixture of CAs. SPE technique was the similar to analysis of urine samples. The data obtained are presented in table 7. Since the content of CAs in the blood was below LOD of the developed technique, this experiment shows only the potential possibility of using AuNRs to determine CAs in blood serum.

SPE of catecholamines from biological fluids allows to reduce the influence of external components of the matrix, which is especially evident in the case of HPLC determination. Also it increases sensitivity of the determination due to preconcentration, as was demonstrated by the analysis of urine samples.

Table 6. Results of total catecholamines determination (μmol L^{-1}) in urine samples after solid-phase extraction with the use of hypercrosslinked polystyrene

Analysis method	Sample 1		Sample 2	
With the use of AuNRs colloid solutions	1.50 ± 0.05[a] (RSD 1%)		3.0 ± 0.1[a] (RSD 2%)	
With the use of AuNRs nanocomposites	–		3 ± 1[b] (RSD 20%)	
HPLC	1.5 ± 0.4[a] (RSD 10%)	E: 0.83 ± 0.07 NE: 0.1 ± 0.3 DA: 0.6 ± 0.2	3.7 ±0.7[a] (RSD 8%)	E: 0.7 ± 0.5 NE: 0.06 ± 0.04 DA:2.9 ± 0.5

Composition of the solutions:
[a] 1.0 mL of urine, 0/0.5/1 μmol L^{-1} additive of catecholamines equimolar mixture, V = 10.0 mL
[b] 12.5 mL of urine, 0/5/10 μmol L^{-1} additive of catecholamines equimolar mixture, V = 15.0 mL

Table 7. Results of total catecholamines determination (μmol L^{-1}) in serum after solid-phase extraction with the use of hypercrosslinked polystyrene (n = 3, P = 0.95)

Analysis method	Added, μmol L^{-1}	Found, μmol L^{-1}	RSD, %
With the use of AuNRs colloid solutions	0.5	0.5 ± 0.1	8
With the use of AuNRs nanocomposites	2.5	2.1 ± 0.3	6

4. Conclusions

Gold nanorods colloid solutions and their nanocomposites proved to be appropriate spectrophotometric reagents for catecholamines determination. Determination is based on measuring hypsochromic shift of the SPR-maxima as a result of "core-shell" nanorods formation caused by covering AuNRs surface by silver shell, because of silver reduction from silver nitrate by catecholamines. Comparison of this system with other nanoanalytical systems has showed that use of AuNRs and silver nitrate is characterized by the best analytical characteristics. The developed techniques for catecholamines determination with the use of AuNRs solutions and the nanocomposites could be applied to analysis of urine and serum samples. As an approach to enhance the analysis sensitivity, solid-phase extraction of catecholamines with the use of a microcolumn packed with hypercrosslinked polystyrene could be applied.

Acknowledgments
Some studies were performed using instrumentation provided according to the M.V. Lomonosov Moscow State University Program of Development.

References

[1] Teerasong S, Sonsa-Ard T, Vimolkanjana C, Choengchan N, Chompoosor A, Nacapricha D 2013 Colorimetric sensor using silver nanoparticles for determination of hydrogen peroxide based on a flow injection system *J. Nanoelectron. Optoe.* **8** 446–9

[2] Teerasong S, Sani M, Numsawat P, Martchoo R, Chompoosor A, Nacapricha D 2015 A silver nanoparticle thin film modified glass substrate as a colourimetric sensor for hydrogen peroxide *J. Exp. Nanosci.* **10** 1327–35

[3] Trisaranakul W, Chompoosor A, Maneeprakorn W, Nacapricha D, Choengchan N, Teerasong S 2016 A simple and rapid method based on anti-aggregation of silver nanoparticles for detection of poly(diallyldimethylammonium chloride) in tap water *Anal. Sci.* **32** 769–73

[4] Jinnarak A, Teerasong S 2016 A novel colorimetric method for detection of gamma-aminobutyric acid based on silver nanoparticles *Sens. Actuat. B* **229** 315–20

[5] Teerasong S, Jinnarak A, Chaneam S, Wilairat P, Nacapricha D 2017 Poly(vinyl alcohol) capped silver nanoparticles for antioxidant assay based on seed-mediated nanoparticle growth *Talanta* **170** 193–98

[6] Liu J M, Wang X X, Cui M L, Lin L P, Jiang S L, Jiao L and Zhang L H 2013 A promising non-aggregation colorimetric sensor of AuNRs–Ag^+ for determination of dopamine *Sens. Actuat. B* **176** 97–102

[7] Gorbunova M V, Apyari V V, Dmitrienko S G and Garshev A V 2016 Formation of core-shell Au@Ag nanorods induced by catecholamines: a comparative study and an analytical application *Anal. Chim. Acta* **936** 185–94

[8] Pavlovic M, Nafradi M, Rouster P, Murath S, Szilagyi I 2019 Highly stable enzyme-mimicking nanocomposite of antioxidant activity *J. Colloid. Interf. Sci.* **543** 174–82

[9] Rouster P, Pavlovic M, Szilagyi I 2018 Immobilization of Superoxide Dismutase on Polyelectrolyte-Functionalized Titania Nanosheets *Chembiochem* **19** 404–10

[10] Shtykov S N, Rusanova T Yu 2008 Nanomaterials and nanotechnologies in chemical and biochemical sensors: Capabilities and applications *Russ. J. Gen. Chem.* **78** 2521–31

[11] Markina N E, Markin A V, Zakharevich A M, Gorin D A, Rusanova T Yu, Goryacheva I Yu 2016 Multifunctional silver nanoparticle-doped silica for solid-phase extraction and surface-enhanced Raman scattering detection *J. Nanopart. Res.* **18** DOI: 10.1007/s11051-016-3661-1

[12] Apyari V V, Arkhipova V V, Gorbunova M V, Volkov P A, Isachenko A I, Dmitrienko S G and Zolotov Yu A 2016 Towards the development of solid-state platform optical sensors: aggregation of gold nanoparticles on polyurethane foam *Talanta* **161** 780–88

[13] Gorbunova M V, Matveeva M A, Apyari V V, Garshev A V, Volkov P A, Dmitrienko S G and Zolotov Yu A 2017 Sorption of gold nanorods on polyurethane foam as a way to obtain a nanocomposite material with a surface plasmon resonance for chemical analysis purposes

Nanotechnologies in Russia **12** 185–92

[14] Gorbunova M V, Apyari V V, Zolotov I I, Dmitrienko S G, Garshev A V, Volkov P A and Bochenkov V E 2019 A new nanocomposite optical sensor based on polyurethane foam and gold nanorods for solid-phase spectroscopic determination of catecholamines *Gold. Bull.* DOI 10.1007/s13404-019-00267-9

[15] Gardner D G, Shoback D 2011 *Greenspan's Basic & Clinical Endocrinology* 9th edition. McGraw-Hill Education. Medical: New York 897

[16] Nikoobakht B and El-Sayed M A 2003 Preparation and growth mechanism of gold nanorods (NRs) using seed-mediated growth method *Chem. Mater.* **15** 1957–62

[17] Gorbunova M V, Shlenova A O, Klimenko R V, Gutorova S V, Apyari V V, Dmitrienko S G 2019 Gold nanorods and a nanocomposite material based on them: analytical possibilities for spectrophotometric determination of total catecholamines *IOP Conf. Series: Materials Science and Engineering* **525** DOI 10.1088/1757-899X/525/1/012012

[18] Lv C, Li Q, Liu X, He B, Sui Z, Xu H, Yin Y, Liu R, Bi K 2015 Determination of catecholamines and their metabolites in rat urine by ultra-performance liquid chromatography–tandem mass spectrometry for the study of identifying potential markers for Alzheimer's disease *J. Mass Spectrom.* **50** 354–63

[19] Bornstein S R, Breidert M, Ehrhart-Bornstein M, Kloos B, Scherbaum W A 1995 Plasma catecholamines in patients with Addison's disease *Clin. Endocrinol.* **42** 215–8

[20] Berends A M A, Eisenhofer G, Fishbein L, van der Horst-Schrivers A N A, Kema I P, Links T P, Lenders J W M, Kerstens M N 2019 Intricacies of the molecular machinery of catecholamine biosynthesis and secretion by chromaffin cells of the normal adrenal medulla and in pheochromocytoma and paraganglioma *Cancers* **11** DOI 10.3390/cancers11081121

[21] Dmitrienko S G, Tikhomirova T I, Apyari V V, Tolmacheva V V, Kochuk E V, Zolotov Yu A 2018 Application of Hypercrosslinked Polystyrenes to the Preconcentration and Separation of Organic Compounds and Ions of Elements *J. Analyt. Chem.* **73** 1053–63

[22] Tolmacheva V V, Yarykin D I, Serdiuk O N, Apyari V V, Dmitrienko S G, Zolotov Y A 2018 Adsorption of catecholamines from their aqueous solutions on hypercrosslinked polystyrene *React. Funct. Polym.* **131** 56–63

[23] Tolmacheva V V, Yarykin D I, Gorbunova M V, Apyari V V, Dmitrienko S G, Zolotov Yu A Preconcentration of catecholamins on hypercrosslinked polystyrene and their determination by high-performance liquid chromatography 2019 *J. Analyt. Chem.* **74** 1057–63

Problems of aviation leaded gasoline application on aircraft

K I Gryadunov[1], M L Nemchikov[1], I S Melnikova[2], A N Timoshenko[1]

[1]Russia Moscow state technical university of civil aviation, Moscow, Russia
[2]Russia State research institute of civil aviation, Moscow, Russia

E-mail: k.gryadunov@mstuca.aero

Abstract. Special aviation gasoline with specified operational properties is used for aircraft piston engines. The level of operational properties indicators is specified by the special additives introduction. One of these additives is an antiknock additive – tetraethyl lead. The influence of tetraethyl lead contained in aviation gasoline on the oil systems of aircraft piston engines operation are shown in the article. The insignificant content of TEL decomposition products in aviation oils sharply worsens their pumpability and leads to complete fine oil filters clogging in a few minutes of engine operation, which is negative from the point of view of flight safety.

1. Introduction

Modern trends in the development of civil aviation (CA) show a stable increase in the small aircraft fleet. These aircrafts are equipped with piston engines operated on aviation gasoline, which has raised standards. The main aviation gasoline brands for CA at the moment are the Russian B-91/115 and American 100LL. A number of additives are added to these gasolines to set the required operational properties. As an anti-knock additive, lead-containing additives are used: mainly this is tetraethyl lead (TEL).

Tetraethyl lead ($Pb(C_2H_5)_4$) used in the form of ethyl liquid is a mortal danger poison with a pronounced cumulative effect, causes disorders of the Central nervous system, has an allergic effect, belongs to the 1st class of danger. TEL penetrates into the body even through the skin. The maximum permissible concentration of TEL vapors is 0.005 mg/m^3 [1-3]. TEL is extremely unsafe for the environment, and therefore its use in motor gasoline in most countries of the world is prohibited.

However, in aviation, this additive is a necessary component for a number of reasons. First, it is an effective anti-knock additive that provides the necessary level of gasoline rating on a rich mixture. Secondly, the oxides of lead formed during the combustion of gasoline have a positive effect on the anti-wear properties of fuels (in relation to the parts of the piston-cylinder-unit).

The influence of TEL on the gasoline stability is ambiguous: in some cases, TEL is a catalyst for the oxidation of hydrocarbons, and in others is an inhibitor [4].

TEL has the negative impact on carbon forming properties, TEL oxidation products with air oxygen form insoluble precipitates in gasoline, which can cause clogging of fuel filters, combustion products of TEL have a negative impact on catalytic converters, etc. [5].

The content of TEL in aviation gasoline is shown in Table 1 [1, 2].

For an objective comparison of the Table 1 data, it's necessary to reduce to the TEL concentration values. Taking the aviation fuel 100LL density at 750 кг/м^3 with 20 °C, we obtain gPb/kg = 0,75.

IOP Conf. Series: Materials Science and Engineering **848** (2020) 012024 doi:10.1088/1757-899X/848/1/012024

Table 1. The content of TEL in aviation gasoline.

TEL concentration	100LL	B-91/115
mlTEL/dm^3, no more	0,53	-
gPb/dm^3, no more	0,56	-
gPb/kg, no more	-	2,5

The letters in the marking of aviation gasoline 100LL means low lead (low lead content), in some sense, it corresponds true (the difference of content of TEL compared to the petrol B-91/115 is more than 3 times), but we can talk that the negative impact of these gasoline brands on the environment as a result of the TEL toxicity, and the aircraft engines operation are practically identical (as we'll see later).

As is known, the peculiarity of piston engines is the following: the parts of the cylinder-piston group (CPG) work in conditions of high irregular thermal stresses and shock loads, while the piston lower part works in the liquid friction conditions, and the upper part works in the boundary friction conditions. During operation the piston is deformed and at certain moments takes the ellipse form, elongated along the axis of the piston pin. It conditions a guaranteed (and often increased) clearance between the piston and the cylinder. Thus, the used gasoline and its combustion products can get into the oil, especially in the case of increased CPG parts wear [6, 7]. Accordingly, the TEL gets in the oil, as well as the products of its decomposition after aviation gasoline combustion in the combustion chamber.

2. Method

To assess the composition of metallic impurities in aviation oils of piston engines, an x-ray fluorescence spectral analysis method was used, implemented at the Prisma ADC installation (Figure 1). ADC Prizma allows to conduct quantitative and qualitative analysis of oil samples for the metal presence in them with high reliability and is characterized by sample preparation simplicity.

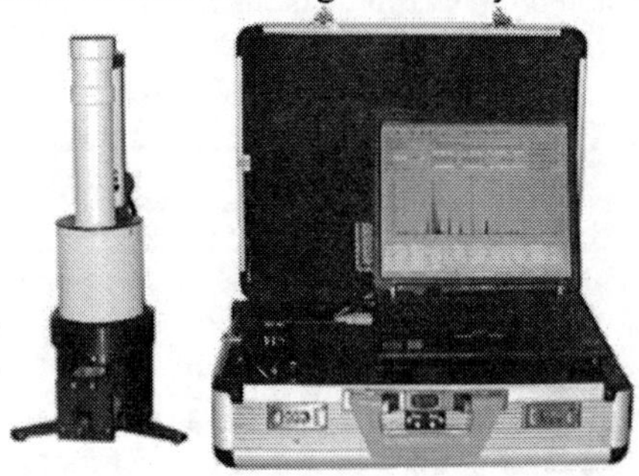

- ability to determine 74 the elements from Ca to Am;
- measured concentrations range: from 0,1 to 250 g/t;
- concentration measurements relative error limits: from 5 to 20 % subject to the concentration.

Figure 1. General view of ADC Prisma and its main parameters.

3. Results and discussion

X-ray fluorescence spectral analyses of aviation oil samples from piston engines showed a significant content of lead in them (Table 2). Source of lead in the oil is a TEL of aviation gasoline. At the same time, there is no significant difference in this parameter depending on the used aviation gasoline brand.

Table 1 shows that the presence of lead in the samples can be significant even when the engine is running on fresh oil. This can be explained by the fact that the engine is not washed when changing the oil. Any dependence of Pb presence in samples of working oil depending on oil operating time is also not found.

The presence of a small amount of lead in the oil samples impairs the pumpability of the oil and its filterability. In the presence of lead, the oil blackens and forms clots.

Oil fine filters of piston engines which meshes size is from 10 to 25 microns depending on a type of the applied filter are clogged up at work on such oil almost instantly (Figure 2,3), and contaminated oil continues to be supplied to the engine through the bypass valves bypassing the filters, that negatively affects the flight safety.

It is also worth noting that TEL in the form of ethyl fluid dissolved in the fuel, does not cause problems with the fuel pumpability through the fuel filters, the cell size which is from 10 to 25 μm depending on the applied filter, and especially not to cause their clogging. Clogging of filters is caused only by TEL oxidation or decomposition products after its combustion. In this case, these are decomposition products-lead dioxide (PbO_2) as part of other combustion products and oxidation of aviation gasoline, which is inevitable when using aviation gasoline with TEL.

Table 2. Lead content in oil samples from piston engines with different oil operating time.

Sample №	Engine operating time ASH-62IR on mineral oil MS-20, hour	Fe, g/t	Pb, g/t	Cu, g/t	Zn, g/t	Cr, g/t	Ag, g/t	Zr, g/t
1	1,5	1,76	8,19	1,7	0,31	-	-	-
2	100	1,29	8,81	0,67	-	0,21	2,62	-
3	2	1,24	1,91	0,82	-	-	-	-
4	2	2,02	3,09	0,72	-	-	-	0,67
5	100	1,73	2,67	0,81	0,23	0,25	-	-

In the design of PD oil systems, multi-stage oil scavenging can be used, which consists in the fact that the oil before entering the fine filter is pre-cleaned in a centrifuge. The oil centrifugation before the filter in these circumstances, obviously, has some positive effect, because a significant amount of fairly large particles (from 50 microns) for a certain period of time settles on the walls of the centrifuge. However, practice shows that centrifugation does not prevent oil clogging, and when they are clogged, it cannot be an effective system for cleaning oil from contamination instead of fine oil filters.

These facts also point to the oil toxicity used in aircraft piston engines, respectively, to its disposal and to safety should be applied the appropriate requirements.

The ways to solve this problem can be different, including:

1. Replacement of anti-knock additive to aviation gasoline.
2. Changing the regulations for oil change, fine filters replacement (cleaning) and centrifuge cleaning (if available), washing of the piston engine.
3. Changing the design of filter elements or oil filtration system.

Any of these ways will require significant time-labor and economic costs, but is necessary to ensure the flight safety of aircrafts equipped with piston engines, and environmental friendliness of these engines.

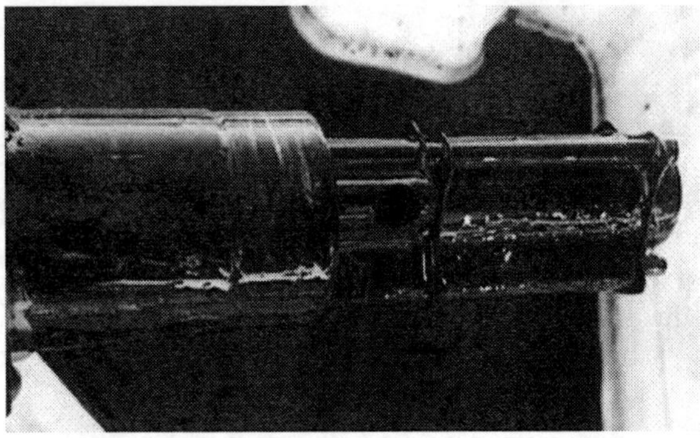

Figure 2. The filter MFM-25 after 1,5 hours of engine operation.

Figure 3. Centrifuge TCM-25 after 100 hours of engine operation.

References

[1] GOST R 55493-2013 *Aviation gasoline Avgas* 100LL 2014.07.11
[2] GOST 1012-2013 *Aviation gasolines* 2015.01.01
[3] Tetraethyl. Physical and chemical properties and toxicity. https://helpiks.org/9-10286.html, 2017.03.29
[4] Eiginson A S 1959 *Processing of sulfur crude oil* (Leningrad: Gostoptekhizdat) p 279
[5] Litvinov A A 1987 *Fundamentals of the use of fuels and lubricants in civil aviation* (Moscow: Transport) p 308
[6] Uglov B A 1992 *ASH-62IR Aircraft Engine: Tutorial* (Samara: Samara State Aerospace University) p 112
[7] Gryadunov K I, Nemchikov M L, Kozlov A N, Melnikova I S 2019 *Civil aviation high technologies* **3** 35

IOP Conf. Series: Materials Science and Engineering **848** (2020) 012025 doi:10.1088/1757-899X/848/1/012025

Gradient structure formation in titanium alloys using thermohydrogen treatment technology

S V Skvortsova[1], O N Gvozdeva[1], A V Shalin[1] and A S Stepushin[1]
[1]MAI, 125993, Volokolamskoe Shosse, 4, Moscow, Russia

Abstract. This paper shows the possibility of creating a "linear gradient" structure, which varies from finely dispersed globular grains to coarse lamellar ones over the cross section of a 13-mm semi-finished product of VT6 alloy, using the thermohydrogen treatment method. The formation of the oxide layer and the depth of oxygen penetration were studied depending on the heating temperature and duration of treatment. The oxide stability was assessed during subsequent annealing in vacuum. It was established that the only oxide film which provides protection from hydrogen penetration during hydrogenous annealing at 800 ° C, was formed during heat treatment at 900°C for 4 hours.

1. Introduction

Armor weight reduction while maintaining a high level of its working properties has been a major field of research in the area of vehicles anti-impact protection [1, 2]. This is promising field of study due to the fact that the most popular protection class is V and higher [3]. Also, regardless of the protection type, it should prevent penetration of the damaging elements into the vehicle cabin, while preserving the possibility of safe movement of this armored car on public roads [4]. Therefore, the specific characteristics of the material used in this area are important [5]. One of the ways to solve the existing problem is usage of titanium alloys, which have already proven themselves as the good individual armor protection material [6, 7]. It is known [3, 5, 8, 9] that in order to increase the dynamic resistance of protection, it is necessary for the used material to combine a hard surface layers with a ductile bulk, which ensures greater absorption of impact energy. This could be achieved by creating multilayer "sandwich-type" structures in various materials (Figure 1 a) or using layered composite materials (Figure 1 b) [10-12]. There are also promising methods that aimed at creating gradient structures in metallic alloys, which consists of gradual transition from one type of structure to another between the surface and the center of material (Figure 1 c). For example, the authors of selected works [13–16] created gradient structures using intense plastic deformation or local heating with high-energy heat sources.

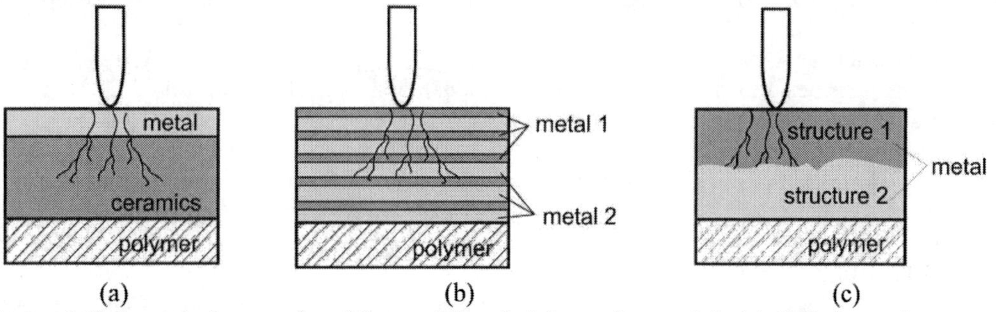

Figure 1. Schematic image of multilayered "sandwich-type" materials (a) [10], layered composite materials (b) [12] and gradient structure materials (c) [14].

IOP Conf. Series: Materials Science and Engineering **848** (2020) 012025 doi:10.1088/1757-899X/848/1/012025

A thermohydrogen treatment, based on reversible hydrogen addition, can be used to create gradient structures in titanium alloys [16–18]. Due to the differentce of diffusion mobility between the hydrogen atoms and the main alloying elements, thermal hydrogen treatment can be used not only for volumetric structure transformation (Figure. 2 a, b) [17], but also for its surface transformation (Fig. 2 c) [18]. Moreover, the mechanical properties of the material are largely determined by the depth of the transformed zone, as well as by the morphology, and the size of the structural components [18].

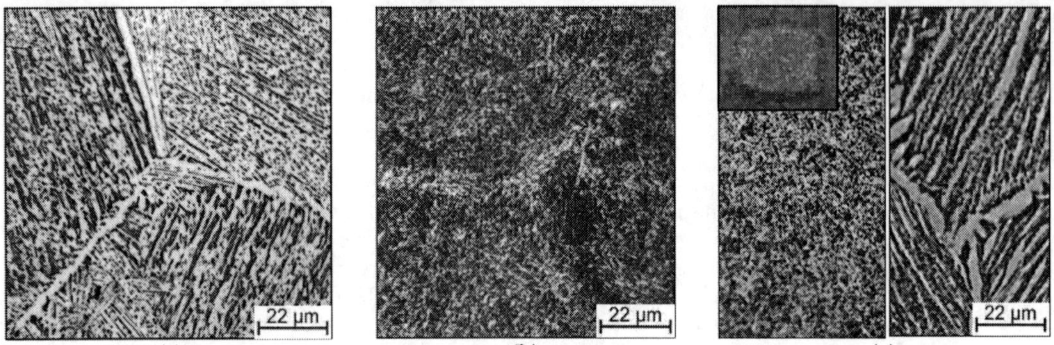

(a) (b) (c)

Figure 2. Transformation of VT6 titanium alloy volumetrical lamellar-like structure (a) to volumentrical dispersed globular structure (b) and «mixed» gradient structure (c) using thermohydrogen treatment [19]

However, when it comes to armored materials, it is necessary to create a «linear» gradient structure (Fig. 3), i.e. when the structure changes occur from one surface to the opposite, and not simultaneously from all surfaces to the center (Figure 2 c) [18].

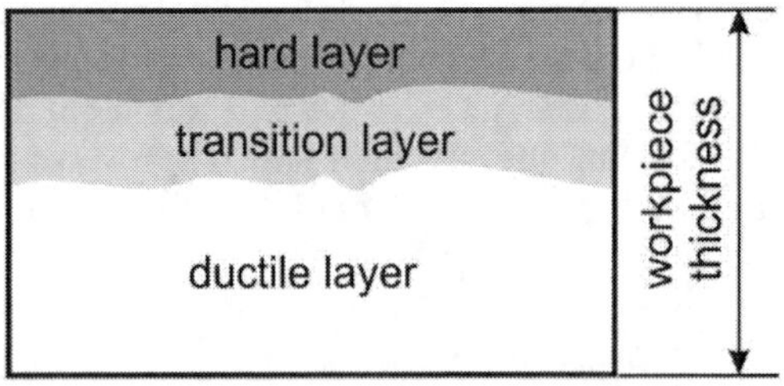

Figure 3. Schematic image of "linear" gradient structure.

Thus, it is necessary to create such conditions when the transformation of the structure due to hydrogen addition will occur only in one direction from the surface. Therefore, the aim of this work was to study the capability of providing a "one-way" hydrogenation in VT6 titanium alloy samples.

2. Materials and procedures

Investigations were performed for the 13 mm thick industrial manufactured hot-rolled plates made of VT6 titanium alloy.

Heat treatment was performed using SNOL-2.2,5.1,8/10-I3. Addition of hydrogen up to 0,4 and 0,6 wt. % was performed in pure molecular hydrogen environment of Sieverts device at β and (α+β)-area temperatures.

Quality of oxide film was visually tested. Hardness was measured according to Rockwell method using the BUEHLER Macromet 5100T device.

Microstructures was investigated in a light-field mode in air atmosphere using a Cals Zeiss Axio-Observer.A1m graphic microscope at x1000 magnification.

The value of added hydrogen during hydrogen annealing was determined by weight gain with an accuracy of 0.0001 g. The amount of residual hydrogen was determined by the spectral method on an ISP-51 spectrograph device with an electronic analytical attachment MORS-1/2048 / PCI.

IOP Conf. Series: Materials Science and Engineering **848** (2020) 012025 doi:10.1088/1757-899X/848/1/012025

3. Results and discussion

It is known [17] that a structure with a lamellar morphology of particles has a higher impact strength. The initial structure of the VT6 alloy plate was of a mixed type: it consisted of small primary α-phase lamellar grains and and more dispersed α+β-mix, located between them (Figure 4 a). To create an even (α+β)-structure in the semi-finished product, all samples were annealed in the β-area temperatures, followed by slow cooling to normal temperature. The resulting structure is uniform in cross section and is represented by the initial 80-100 μm β-grains, surrounded by α-border and 3-5 μm thick lamellar α-grains, located in the volume of β-grains. Hardness of the samples after the annealing was 31-33 HRC (Figure 4 b).

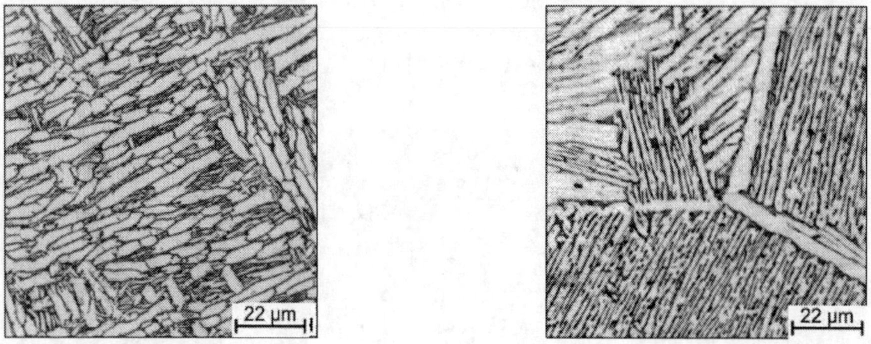

(a) (b)

Figure 4. Microstructure of VT6 alloy plate in initial (a) and annealed (b) state.

To determine the optimal value of added hydrogen, which is necessary to completely transform the lamellar structure and to ensure a significant increase in hardness, volumetric hydrogen treatment was carried up to a concentration of 0.4% and 0.6% at temperatures of 800°C (β-area) and 700°C (α+β-area). After the completion of hydrogen adsorption, the samples were kept under equilibrium pressure of hydrogen for 3 hours to ensure its uniform volume distribution. The residual hydrogen content in the samples did not exceed 0.006 wt.%.

Studies have shown that if hydrogen annealing up to 0.4 wt.% is carried out at a temperature of the β-area with subsequent cooling at 1K/s speed to room temperature, a structure consisting of β-phase and α″-martensite (Figure 5 a) is formed. Increasing the hydrogen percentage up to 0.6 leads to the formation of an almost single-phase structure of β-particles (Figure 5 b). Decreasing the temperature of thermohydrogen treatment to 700°C keeps the (α+β)-structure retained in the alloy, characterized by an increased volume fraction of β-phase in comparison with an alloy in its initial state (without hydrogen) (Figure 5c).

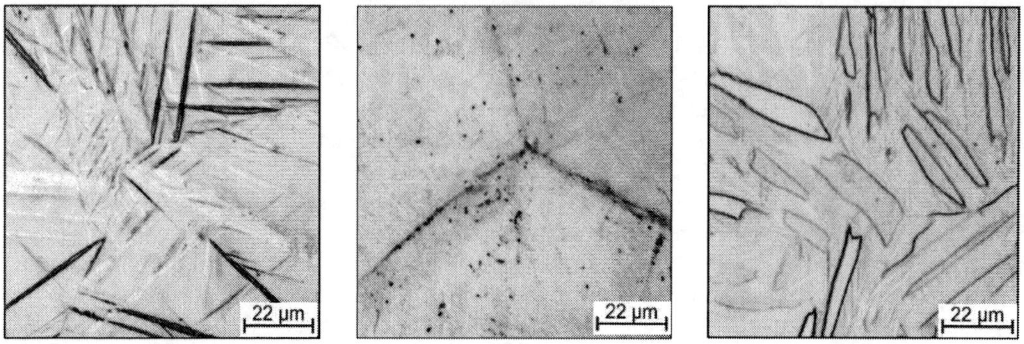

(a) (b) (c)

Figure 5. Microstructure of VT6 alloy samples after volumetric hydrogen addition up to 0.4 wt.% (a) and 0.6 wt.% (b) at β-area temperature and up to 0.4 wt. % (c) at (α+β)-area temperature.

During the subsequent gas removal process, the structure is transformed due to the development of β→α-transformation. Following investigations have shown that in the process of vacuum annealing at 625°C, the processes of nucleation of α-phase particles prevail over the processes of their growth. Thus, the structure of samples preliminarily hydrogenated up to 0.4 wt.% at a β-region temperature is represented by thin lamellar α-phase, which inherits the shape of martensite (Figure 6 a). The structure of samples hydrogenated to 0.6% after vacuum annealing is represented by dispersed particles of the α-phase (Figure 6 b). Low-temperature vacuum annealing of the samples after hydrogenation in the (α+β)-area leads to the formation of a bimodal structure of the α-phase, consisting of large lamellar grains of the primary α-phase, preserved in the structure during hydrogen addition, and dispersed particles of the secondary α-phase, formed during gas removal process (Figure 6 c). The maximum hardness of 42 HRC provided by a structure with a dispersed α-phase, while thin lamellar morphology structure provides slightly lower hardness (40-41HRC). Test samples with a bimodal structure have the lowest hardness of 37-39HRC.

(a) (b) (c)

Figure 6. Microstructure of preliminary hydrogenated up to 0.4 wt.% (a) and 0.6 wt.% (b) at β-area temperature and up to 0.4 wt. % (c) at α+β-area temperature VT6 alloy samples after subsequent low-temperature vacuum annealing.

Further investigations showed that it is possible to change the morphology and size of the structural components of the VT6 alloy over a wide range by changing the temperature and percentage parameters of thermohydrogen treatment.

Since the main alloying elements diffusion in titanium alloys proceeds much slower than the hydrogen diffusion [19], there is a possibility to control the depth of hydrogen penetration and the completeness of β→α-transformation by changing the kinetic parameters of the thermohydrogen treatment. It was investigated that to obtain the maximum hardness after vacuum annealing, the alloy structure must contain β-phase particles only after the preliminary thermohydrogen treatment. Hence, the alloy should contain at least 0.6 wt.% of hydrogen. It was calculated that to provide at least 0.6 wt.% of hydrogen in the surface layers, the volumetric amount of introduced hydrogen should be around 0.3-0.4 wt. %.

During surface «linear» gradient structure creation the transformation of the structure due to hydrogen addition must be occured only in one direction from the surface. In this case the main task is to protect all sides of semi-finished product from hydrogen penetration but one. It is known that a thin oxide film formed on the surface of titanium prevents the penetration of hydrogen, but it dissolves when heated in vacuum [19]. Therefore, at the next stage of this study, the kinetics of dissolution of oxide films was studied during processing in vacuum. The samples were subjected to oxidation in an air atmosphere in the temperature range from 600 to 900°C for 1-4 hours. Visual inspection of the samples after oxidation at various temperatures for 4 hours showed that a dense oxide film forms on their surface (Fig. 7). The thickness of which, together with the alpha layer, decreases from 150 µm after treatment at 900 ° C to 20 µm at 600°C (Table 1). Reducing the exposure time from 4 hours to 1 hour at heating temperatures of 800-900° C also leads to the formation of oxide films firmly adhered to the surface. In this case oxide film thickness almost half as thin (Table 1).

IOP Conf. Series: Materials Science and Engineering **848** (2020) 012025 doi:10.1088/1757-899X/848/1/012025

(a) (b) (c)

Figure 7. VT6 alloy samples exterior after air atmosphere oxidation at 900°C (a) and 600°C (b) for 4 hours, as well as at 900°C for 1 hour (c).

Table.1 Oxide film thickness after air atmosphere oxidation depending on temperature and exposure time.

Temperature	900°C	900°C	900°C	800°C	700°C	600°C
Duration	4 hours	2 hours	1 hour	4 hours	4 hours	4 hours
Thickness, μm	130	120	70	100	20	20
After 800°C, 1 hour vacuum annealing .						
Thickness, μm	250	170	100	120	100	100

The resistance of the oxide films on the samples surface during subsequent processing in vacuum was investigated. Since the maximum temperature of the thermohydrogen treatment is 800°C, the same temperature of vacuum annealing was chosen to evaluate the stability of the oxide.

It was found that vacuum annealing at 800°C allows to completely preserve the oxide film formed during 900°C isothermal exposure for 4 hours (Figure 8 a), while oxide films formed during 800 and 700°C isothermal treatment are partially dissolves during the same method of processing in vacuum. Oxide films formed during 600 °C isothermal exposure are completely dissolving in vacuum (Figure 8 b). Exposure time decreasing also leads to a partial dissolution of the oxide film (Figure 8 c). In the process of vacuum annealing, oxygen diffuses deeper into the sample. Thus, the thickest and the most stable oxide layer is observed in samples which was preliminary annealed at 900°C for 4 hours (Table 1).

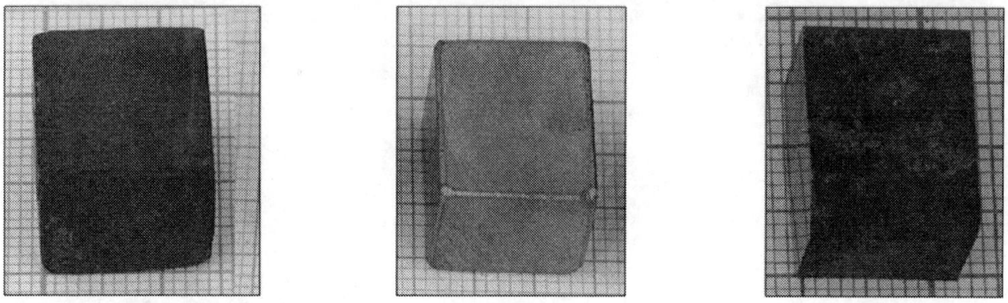

(a) (b) (c)

Figure 8. VT6 alloy samples exterior after air atmosphere oxidation at 900°C (a) and 600°C (b) during 4 hour exposure and at 900°C during 1 hour exposure (c), with subsequent 1-hour long high-temperature vacuum annealing.

IOP Conf. Series: Materials Science and Engineering **848** (2020) 012025 doi:10.1088/1757-899X/848/1/012025

At the final stage, the "protective" properties of the oxide film formed at a temperature of 900°C for 4 hours during thermal hydrogen treatment were investigated. Oxide film from one side of the sample was mechanically removed for further comparison of oxide protective ability. Thermohydrogen treatment up to 0.4% of hydrogen was carried out at a temperature of 900°C. In order to exclude a distribution of hydrogen over the cross-section of the samples, their cooling was carried out immediately after the completion of the hydrogen absorption process. To conduct a comparative analysis of thermal hydrogen treatment, samples not protected by an oxide film were also subjected.

Studies have shown that during hydrogen annealing, almost all of the hydrogen is concentrated in the surface layer of the sample; therefore, the structure is not transformed in the entire volume of sample, but only in the surface layer. The structure of the transformed layer is represented by α″-martensite and β-phase (Figure 9 a). At a distance of 4000 μm from the surface, the structure no longer differs from its initial state (before hydrogen adsorption) (Figure 9 b). Investigation of structures from the other three sides («isolated» by the oxide film) showed that already at a distance of 100 μm from the surface it is represented by α- and β-phases, which indicates a significantly lower hydrogen values (Figure 9 c).

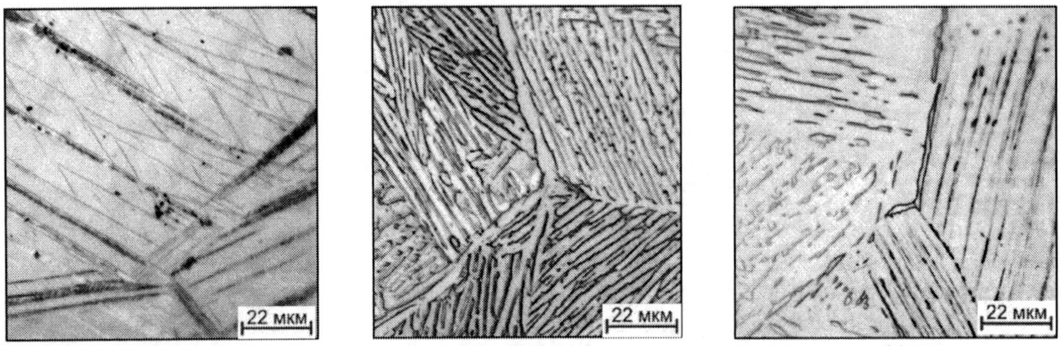

(a) (b) (c)

Figure 9. Microstructure of VT6 alloy samples after 900°C oxidation and thermohydrogen treatment up to 0.4% of hydrogen at the β-phase area temperatures 1500 μm (a) and 4000 μm (b) from the oxide-free surface and 100μm (c) from surface with the oxide film.

Microstructure analysis of a oxide-free sample showed that the depth of hydrogen penetration during the treatment from all sides is approximately the same and averages 2000 μm (Figure 10).

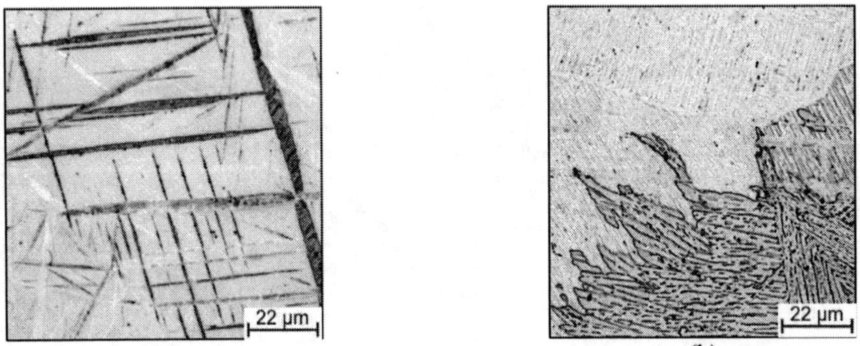

(a) (b)

Figure 10. Microstructure of VT6 alloy oxide-free samples after thermohydrogen treatment at the depth of 1000 (a) μm and 2500 (b) μm from surface.

Subsequent low-temperature vacuum annealing makes it possible to completely transform the lamellar structure in the surface layers oxide-free side to the dispersed one, while maintaining it practically unchanged from the other three sides, which was «protected by oxide film» (Figure 11). The hardness of the 2800 μm transformed layer with a finely dispersed (α+β)-grains is about 39 HRC (Figure 11 a). Dispersion degree of the structural components is decreasing with the distance from surface (Figure 11 b). Structure became almost identical to initial (before hydrogen addition) state at a depth of 4000 μm (Figure 11 c). The hardness of such layers is about 34 HRC.

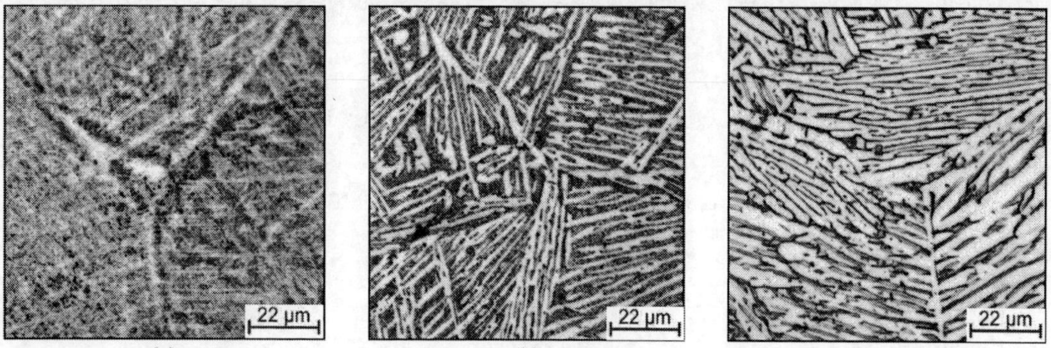

(a) (b) (c)

Figure 11. Microstructure of VT6 alloy samples after oxidation at 900°C for 4 hours and thermohydrogen treatment at a depth of 1500 μm (a), 2800 μm (b) and 4000 μm (c) from oxide-free surface direction.

4. Conclusions

It was investigated that 250 μm oxide film formed during high-temperature treatment can be used to «protect» particular side of semi-finished or finished product from hydrogen penetration during thermohydrogen treatment. This can also be used to create the «linear» gradient structure, which changes from dispersed globular to coarse lamellar one over the cross-section of such products.

References

[1] Alexeev M O, Chistyakov E N and Kupryunin D G 2017 *Defense Industry* **130** pp 50-56.
[2] Kupryunin D G, Gavze A L and Chusov S Yu 2017 *Defense Equipment Issues* **121** pp 114-121.
[3] GOST R 50963-96 2003 (Moscow: Government Standards Publisher) p 30.
[4] Rossikhin K G and Selifonov V V 2010 *"Automobile and tractor engineering in Russia: development priorities and training" Conference, Dedicated to the 145th Anniversary of MSTU "MAMI"*, pp 269-277.
[5] Grigoryan V A, Kobylkin I F, Marinin V M and Chistyakov E N 2008 *Materials and protective structures for local and individual armor protection* (Moscow: Radiosoft publishing) p 406.
[6] Mylnikov V V, Abrosimov A A, Romanov I D and Romanov A D 2014 *Technical Science* **9** pp 143 – 147
[7] Gavze A L, Chusov S Yu, Yankov V P Tetyukhin V V, Zhloba A V, Berestov A V and Bondaryuk N N 2013 *Titan* **1** pp 46 - 48
[8] Kovalenko V.V. 2009 *The physical basis for the formation and evolution of gradient structural phase states in steels and alloys* (Novokuznetsk: Poligraphist) p 557.
[9] Anastasiadi G P and Silnikov M V 2004 *The performance of armor materials* (St. Petersburg: Asterion) p 624.
[10] Khovrich M V, Saharov S A and Kamaev E A 2007 *Russian Patent №2 296 288*
[11] Kryukov D B and Kozlov D V 2016 *Penza State University Bulletin* **2 (14)** pp 103–108.
[12] Agarkov G A and Bliznik M G 2018 *Russian Patent № 2 647 963.*
[13] Golenkov V A, Radchenko S Yu, Dorofeev O V and Dorokhov D O 2009 *Fundamental and applied problems of engineering and technology* **3 (275)** pp 42-46.

IOP Conf. Series: Materials Science and Engineering **848** (2020) 012025 doi:10.1088/1757-899X/848/1/012025

[14] Raab G I, Kodirov I S, Alyoshin G N, Raab A G and Cenev N K 2019 *Magnitogorsk State Technical University Bulletin* **V 17 I 1** pp 64-75.
[15] Volodin V L, Khasanov O L, Volodin T V, Gromov V E and Konovalov S V 2011 *The formation of structural phase states of metal alloys during magnetic pulse processing* (Novokuznetsk: Inter-Kuzbass) p 211.
[16] Mamonov A M, Safaryan A I, Agarkova E O, Zhilyakova M A 2018 *Metal Science and Heat Treatment* **V 60 I 1-2** pp 80-88.
[17] Ilyin A A, Skvortsova S V, Spektor V S, Kudelina I M and Oreshko E I 2011 *Titan* **1(31)** pp 26-29
[18] lyin A A, Skvortsova S V, Spektor V S, Kudelina I M and Mamontova N A 2011 *Technology of Light Alloys* **2** pp 37-41.
[19] Ilyin A A, Kolachev B A, Nosov V K and Mamonov A M 2002 *Hydrogen technology of titanium alloys* (Moscow: Moscow Institute of Steel and Alloys State Technological University) p 392.

IOP Conf. Series: Materials Science and Engineering **848** (2020) 012026 doi:10.1088/1757-899X/848/1/012026

The study of model reactions occurring in the synthesis of adaptive multiblock polyurethanes

M A Gorbunova [1], A M Imamutdinova [2], V P Lodygina [1], D V Anokhin [1] and E R Badamshina [1]

[1] Institute of Problems of Chemical Physics of RAS, 142432, Moscow region, Russia
[2] Lomonosov Moscow State University, Moscow, Russia

[2]imamutdinovaarina@rambler.ru

Abstract. In the paper a study of formation process of linear polyurethane on the basis of 4,4'-diphenylmethane diisocyanate, poly(1,4-butylene adipate) diol and 1,4-butanediol is presented. The kinetics of the first stage of macrodiisocyanate formation reaction was studied by IR spectroscopy. Experimental and calculated values of the rate constant of the first reaction stage were obtained. The influence of the solvent on the polyurethane structure was defined. The conditions to fabricate polymer with a maximum crystallinity degree have been determined. Important results for the synthesis of the shape memory multiblock polyurethane thermoplastic elastomer have been obtained.

1.Introduction

A detailed study of the properties and synthesis methods of biodegradable and biocompatible polymers with shape memory effect (SME), one of which is a thermoplastic polyurethane block-copolymer (TPU), consisting of soft, crystallizable polyester blocks, and hard diol-urethane segments, is currently being conducted. Such polymers are used in medicine for drug-release systems, for manufacture of prostheses, implants, also there is the widespread application in textile, automobiles and aircrafts industry [1]. A wide range of materials based on these polymers applications is associated with their specific structure, the possibility of varying mechanical properties by changing the ratio of reagents during synthesis [2].

It is important to note that properties of TPU with shape memory effect (SME) depend on the method and conditions of synthesis, stoichiometry of functional groups of reagents, molecular and weight parameters, which are defined by a kinetics of reaction that occurs during the formation of these polymers.

Analysis of literature on the synthesis of TPU based on 4,4'-diphenylmethane diisocyanate (MDI), which, in fact, is one of the most popular diisocyanates used in the production of polyurethanes, it should be noted that, mainly, results of the research of properties and structure of TPU based on it are presented, and there is practically no information on the kinetics of synthesis.

In the literature about the kinetics of reaction, there are studies of stoichiometric reaction of MDI with low molecular weight diols in chloroform [3], the formation of macrodiisocyanate (MDC) with polypropylene oxide carried out in THF [4] and polypropylene glycol carried out in bulk at 60°C [4] with an excess of NCO-groups is discussed.

As assumed, the reaction has second-order kinetics and two reaction rate constants were presented in accordance with the first stage of MDC formation and the second final stage after adding a chain extension. It is important to note that the synthesis of multiblock thermoplastic polyurethanes with SME is usually carried out in a methylene chloride solution. However, this reaction was not described in details.

In this paper, the formation process of linear polyurethane based on hydroxyl-containing poly(1,4-butylene adipate) diol (PBA), MDI and a chain extender 1,4-butanediol (BD) in the presence of a methylene chloride solvent and a tin dibutyltin dilaurate catalyst, as a model reaction that occurs during synthesis of TPU with SME has been studied.

2.Experimental part

2.1. Materials

Poly(1,4-butylene adipate) diol (PBA), M_n ~ 2000 Da, by Aldrich, was used after pre-drying at 60°C on a rotary evaporator. The content of hydroxyl groups determined by the chemical method [5] was 1.76%

4,4'-diphenylmethane diisocyanate (MDI) by Alfa Aesar and dibutyltin dilaurate catalyst (DBTDL), were used without prior purification.

1,4-butanediol (1,4-BD), methylene chloride (purity □99.9%) and THF (purity □ 99.9%) were used preliminarily purified according to the standard method [6], equals to 99.9% of the theoretical one.

2.2. Fourier Transform Infrared Absorption Spectroscopy (FTIR)

IR spectra for the identification of the reagents and the synthesized TPU were recorded on an Alpha Bruker spectrometer under the following conditions: measurement range 4000 cm^{-1} - 360 cm^{-1}; measurement step 2 cm^{-1}; the number of scans per spectrum 56; samples were obtained in a film from a methylene chloride solution on KBr glass.

The kinetics of the catalytic reaction of urethane formation was studied by IR spectroscopy by the total consumption of NCO-groups and the accumulation of urethane groups. At certain intervals after the start of the reaction of PBA with MDI taken in the ratio [NCO] / [OH] = 2, samples of the reaction mixture were taken. Then, the IR spectra of solutions of the reaction mixture in methylene chloride were recorded on a spectrometer.

The current concentration of isocyanate groups was determined by changing the intensity of the absorption peak corresponding to the antisymmetric vibration of the NCO-groups of MDI (υ_{NCO} = 2275-2260 cm^{-1}) and the stretching vibration of the urethane group carbonyl ($\upsilon_{C=O}$ = 1720 cm^{-1}).

To obtain the molar extinction coefficient of MDI, MDI solutions in various concentrations of methylene chloride were prepared, after which a calibration dependence of the optical density of the analytical band on the concentration of isocyanate groups was constructed. The molar extinction coefficient of MDI was determined by equation (1).

$$\varepsilon = \frac{\Delta D}{\Delta l \cdot C} \qquad (1)$$

where: ε is the molar extinction coefficient of MDI
ΔD is the optical density
Δl is the cuvette thickness, cm
C is the concentration of the solution, g-eq /l.

The effective reaction rate constant was determined by equation (2), based on equal reactivity of both of the isocyanate groups of MDI with a twofold excess of them relative to the hydroxyl groups of PBA:

$$ln\frac{[OH]_0+[OH]}{2[OH]} = k \cdot \Delta\tau \qquad (2)$$

where: k is the reaction rate constant, l/g-eq·s
$\Delta\tau$ is the change in the reaction time, s

[OH] - the current concentration of hydroxyl groups of PBA during the reaction, g-eq/l
$[OH]_o$ is the concentration of hydroxyl groups at a time t = 0.

2.3. Theoretical calculation of the kinetics of the studied reactions of urethane formation
The experimental values of the effective rate constant of the first stage of the MDC formation were confirmed by solving direct kinetic equations by means of standard Kinet program using numerical Gear method [7]. For the calculation, the following assumptions were made: the reaction mixture is homogeneous throughout the urethane formation process, and the reactivity of hydroxyl and isocyanate groups does not change during the reaction.

For the studied PBA-MDI system with equal reactivity of the functional groups, a system of equations describing the interaction of the functional groups is written. As a result of the calculation, the value of the kinetic constant is obtained, which characterizes the interaction between the functional groups of PBA and MDI. The proof of the obtained value of the rate constant was made by comparison of the calculated and experimental kinetic curve, the deviation between which does not exceed 1-2%.

The calculation of the rate constant was carried out according to the following parallel-sequential reaction scheme (Scheme 1):

Scheme 1. Scheme of parallel-sequential reactions for calculating the rate constant of the interaction of NCO-groups of MDI with OH groups of PBA

$I + OH \xrightarrow{k} U_1 + I_2$

$I + OH \xrightarrow{k} U_2 + I_1$

$I_1 + OH \xrightarrow{k} U_1$

$I_2 + OH \xrightarrow{k} U_2$

where: I is the concentration of MDI at the initial time
OH is the concentration of OH groups of PBA at the initial time
U_1 is the concentration of urethane groups formed by OH-groups of PBA and NCO-group MDI in position 4-
U_2 is the concentration of urethane groups formed by OH-groups of PBA and NCO-group MDI in position 4′-
I_1 - concentration of NCO-groups of MDI in position 4-
I_2 - concentration of NCO-groups of MDI in position 4′-

2.4. DSC method
The structure of TPU was investigated by DSC using DSC 30 Mettler Toledo calorimeter in dynamic mode. The sample (10.0 ± 0.2 mg) were placed in an aluminium pan, rapidly cooled down to -20°C and further heated at rate 10°C/min in a nitrogen atmosphere to 70°. The degree of crystallinity (CD) of TPU was determined by the formula (3):

$$CD = \frac{\Delta H_m}{\Delta H_{m(100\%)} \cdot \omega} \quad (3)$$

where: ΔH_m - total melting enthalpy measured from DSC curves;
ω - mass fraction of PBA in the polymer;
$\Delta H_{m(100\%)} = 95$ J/g - equilibrium melting enthalpy of PBA at its 100% crystallinity.

2.5 Synthesis of polyurethane
Linear polyurethane block copolymer was obtained by a two-stage method through the stage of formation of macrodiisocyanate in methylene chloride solution (or THF) at a room temperature in the presence of DBTDL catalyst from PBA oligoether diol, MDI and BD as chain extender (figure 1). At the first stage, MDI interacted with PBA at $[NCO]_{MDI} / [OH]_{PBA} \approx 2$), then the chain extension 1,4-BD

was added to the stoichiometric ratio of functional groups ($[NCO]_{MDI}$ / $[OH]_{PBA + BD}$ = 1). Upon reaching the degree of transformation in the groups of NCO ~ 98%, the reaction mass was poured into a flat form of Teflon and dried at 40°C for a day to a constant weight.

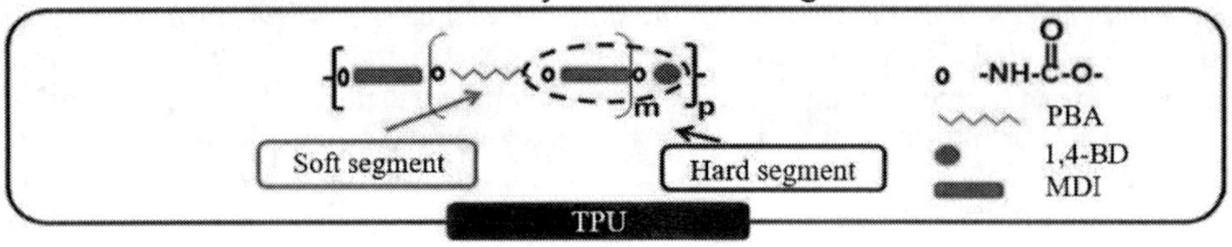

Figure 1. TPU scheme.

3.Results and discussion

The synthesis of TPU based on MDI, according to published data, is usually carried out in DMSO, DMAA or DMF [8]. However, the disadvantage of these solvents for the synthesis of the polymer is their high boiling point, which complicate removing solvents from the polymer matrix, as well as the difficulties in the study of kinetics by IR spectroscopy due to strong absorption in this wave range. To avoid above-mentionned problems, in the presented study the formation of TPU was carried out in methylene chloride and THF. For identification and purity of the starting compounds, their IR spectra were recorded for a film from a methylene chloride solution (figure 2).

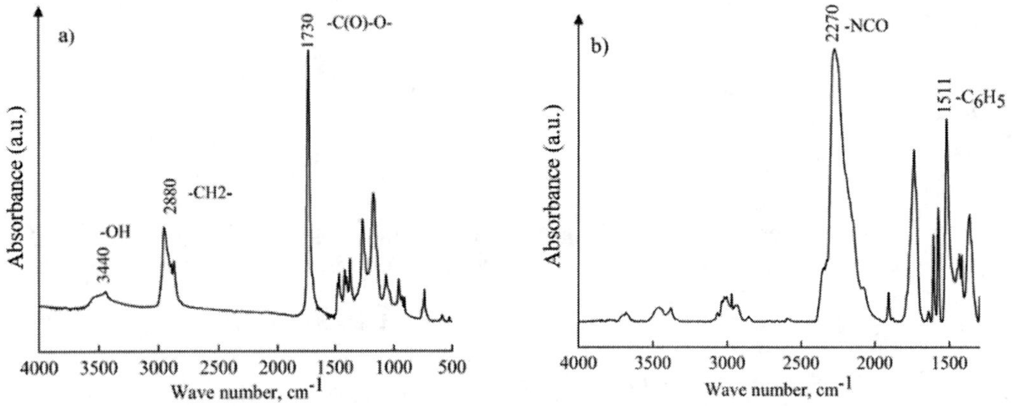

Figure 2. IR spectra of films of PBA (a) and MDI (b) in methylene chloride solution.

In the IR spectrum of PBA (figure 2a), absorption bands of hydroxyl groups (3670-3580 cm^{-1}) and carbonyl C = O (1730 cm^{-1}) which belong to the functional groups of diols based on esters, are observed, which confirms the chemical composition of polyether diol used in the work. On the IR spectrum of MDI (figure 2b), characteristic bands of isocyanates are observed in the region of absorption of stretching vibrations of NCO groups at 2270 cm^{-1}, a characteristic absorption band at 1511 cm^{-1} corresponding to stretching vibrations of the aromatic ring is also visible, which confirms the presence of the C_6H_5- group in the MDI isocyanate. Thus, from figure 2, we can conclude that IR spectra correspond to the proposed structures.

Figure 3 illustrates the change in the intensity of the absorption band of the NCO group of MDI in the solutions of various concentrations in methylene chloride (figure 3a). Using this dependence, a calibration plot of the optical density of the analytical band on the concentration of isocyanate groups was calculated (figure 3b). As a result, the value of the molar extinction coefficient for MDI in methylene chloride was calculated, which amounts to 872 l/g-eq·cm.

The kinetics of the catalytic reaction of the first stage of the synthesis - PBA with MDI in the presence of a dibutyltin dilaurate catalyst with a ratio of NCO/OH = 2 was studied by changing the intensity of the absorption band corresponding to the antisymmetric stretching vibrations of the NCO groups of MDI (figure 4a). The concentration of the catalyst was 0,003 mol/l; the initial concentration of MDI isocyanate groups was 0.86 g-eq/l.

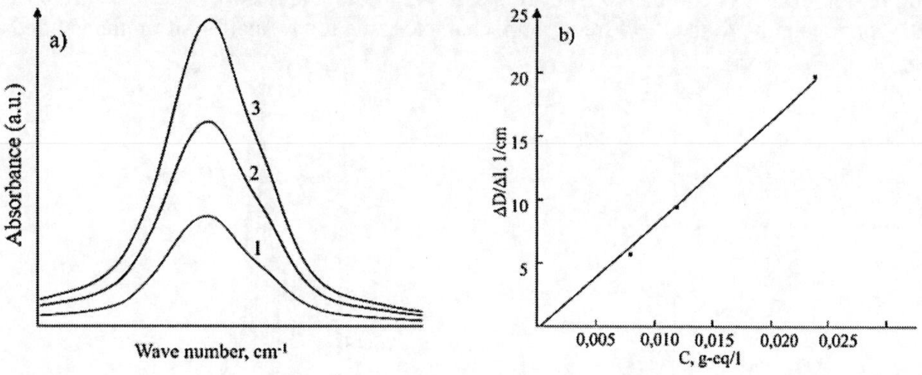

Figure 3. IR spectra of MDI solutions in methylene chloride at 0.006 (1), 0.015 (2), 0.029 (3) g-eq/l in the absorption region of NCO groups, Δl = 0.078 cm (a) and typical calibration dependence (b).

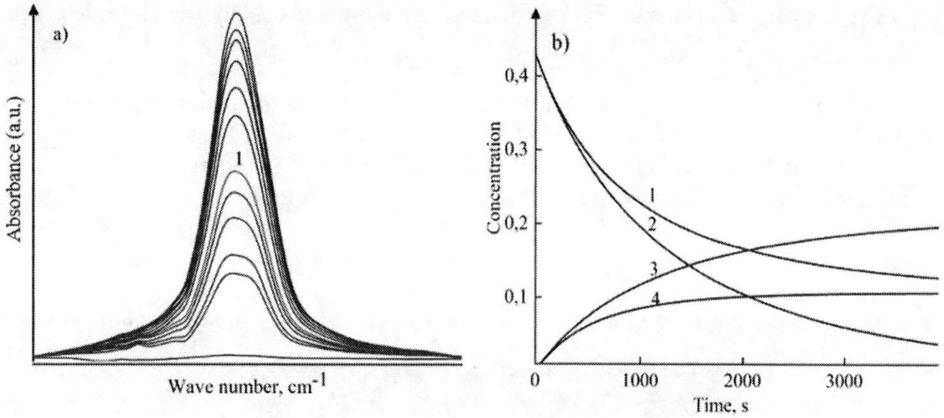

Figure 4. IR spectra of the reaction media, showing the change in the concentration of the NCO group in the process of urethane formation, where 1 is the end of the first stage of urethane formation (a) and the dependence of the change in the concentration of MDI, mol/l (1), the concentration of OH-groups of PBA, g-eq/l (2), the concentration of MDC, mol/l (3) and the concentration of urethane groups, g-eq/l (4) of the reaction time of urethane formation (b).

It was found that the catalytic reaction was almost completed in 60 minutes after the beginning of the synthesis. The conversion of OH groups of PBA was 98%. Based on the results obtained, the kinetic constant of urethane formation, calculated by the formula (1), was estimated to be $1.09\cdot10^{-3}$ l/g-eq·s. The calculation by the numerical method using the Kinet program confirmed the validity of this value for the kinetic constant and made it possible to determine the structural composition of the obtained prepolymer (figure 4b).

The curves presented in figure 4b make it possible to estimate the concentration of reagents and the structural composition of the reaction mass at any time during synthesis. After the OH-groups of PBA completely reacted, a 1,4-BD (chain extender) was added to the prepolymer to the stoichiometric ratio of the reacting groups.

As it can be seen from the IR spectra (figure 5), the absorption band corresponding to the stretching vibration of the NCO group of MDI in the region of 2270 cm^{-1} and the appearance of peaks in the absorption region of hydrogen-bonded NH groups at 3420 cm^{-1} and absorption bands are observed at 1780 cm^{-1}, corresponding to the C=O free group, characterizing the formation of urethane bonds.

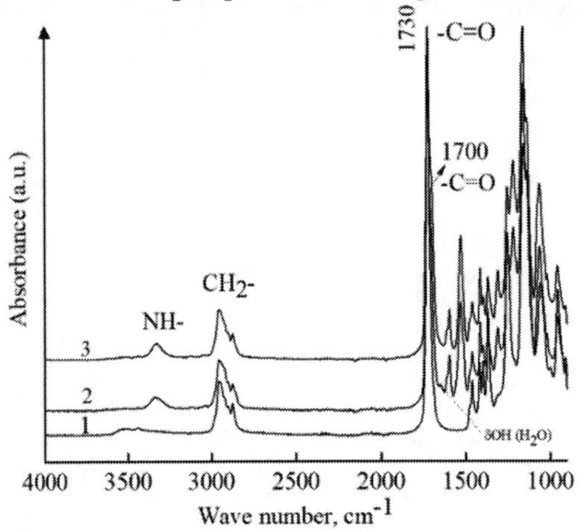

Figure 5. IR spectra of PBA (1) and TPU synthesized in a solution of methylene chloride (2) and THF (3).

The IR spectrum of TPU obtained in a THF solution shows a small shoulder in the region of free OH groups of water, which is probably due to the high hydrophilicity of the solvent. The presence of moisture in the sample will affect the formation of the structure of TPU and its properties.

The obtained kinetic results were used in the synthesis of TPU carried out in methylene chloride and THF.

The synthesized polyurethane is a phase-separated linear block copolymer consisting of thermodynamically incompatible soft block of PBA and a hard block based on MDI and BD. The polymer is opaque under normal conditions, which indicates its heterogeneous structure (figure 6).

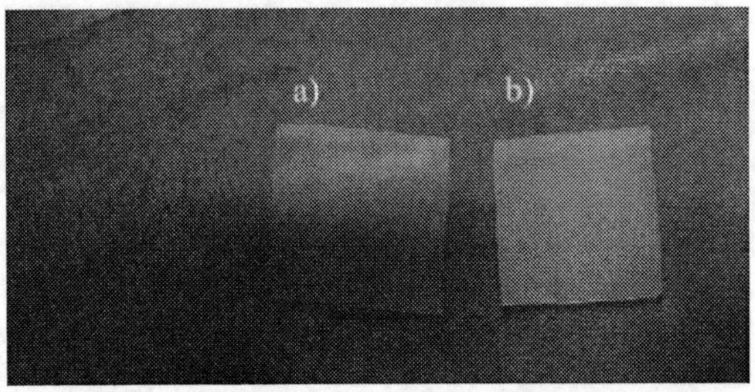

Figure 6. TPU synthesized in methylene chloride solution (a) and in THF (b).

The results of the DSC analysis of polymers, depending on the type of solvent in which the polymers were synthesized, showed differences in their structure (figure 7). In the DSC thermogram of samples immediately after synthesis, multiple melting peaks of crystals of the soft polyester block are observed. Such behavior is typical for which melting of imperfect crystals of different thicknesse [9]. It was established that TPU synthesized in methylene chloride solution shows higher degree of crystallinity (33.8%) in comparison with TPU synthesized in THF (23.05%). It is important to note that in the synthesis of multiblock TPUs with SME, phase separation and crystallization processes will determine thermal and mechanical parameters of adaptive materials.

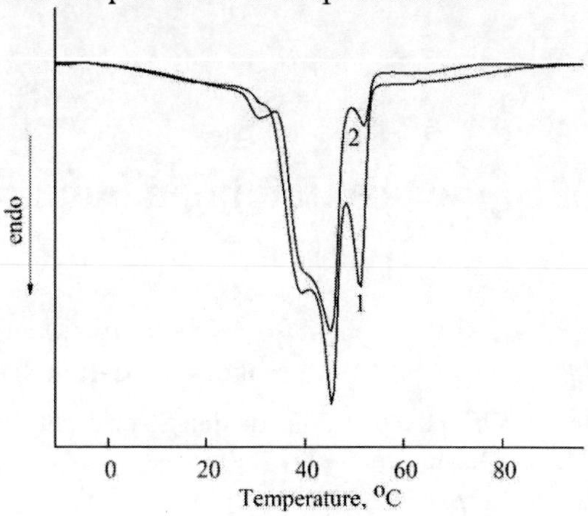

Figure 7. DSC heating thermograms of TPU synthesized in methylene chloride (1) and THF (2)

4.Conclusions

The formation of linear polyurethane based on PBA, MDI, and 1,4-BD in the presence of a catalyst in different solvents was studied. Based on the kinetics data, the optimal conditions for the synthesis of TPU were proposed. Using a combination of IR and DSC methods, the influence of a solvent on the formation of the structure of TPU was shown. A polymer synthesis strategy that allows to obtain TPU without side reactions with a maximum degree of crystallinity has been developed. The results will help for the synthesis of thermoplastic polyurethanes with SME.

Acknowledgements
The authors acknowledge the Russian Foundation for Basic Research for financial support (contract № 19-53-15016). The work was done on the theme of the state task, № 0089-2019-0012, registration № AAA-A19-119032690060-9.

References

[1] Yilmaz F 2017 *Aspects of Polyurethanes* pp 61-65
[2] Kang S M, Lee S J, Kim B K 2012 *Express Polymer Letters* **6(1)** 63–69
[3] Zaverkina M A, Lodygina V P, Komratova V V, Stovbun E V and Badamshina E R 2006 *Polymer Science A* **48(4)** 382–387
[4] Polo M L, Spontón M E, Jaramillo F, Estenoz D A and Meira G R 2017 *Journal of Applied Polymer Science* **135(4)** 45747
[5] Gafurova M P, Lodygina V P, Grigoreva V A, Tchernyi G I, Komratova V V and Baturin S M 1982 *Russ. J. Vysokomolek.Soed.* **24** 858-862
[6] Weissberger A, Proskauer E, Riddick J and Toops E 1955 Organic Solvents. Physical Properties
and Methods of Purification Wiley, New York
[7] Khairiah B H, Wong C S, Maisara S B R, Liow C H, Norhafiza Y B and Nor R A 2016 *Solid State Sci. Technol.* **18(2)** 2010 1-8
[8] Abramenkov AV, Ageev E P, Atyaksheva L F et al., ed. Lunina V V and Ageeva E P 2012 *Pr. Phys. Chem.: Kin. Cat. Electrochem.* pp 70-102
[9] Anokhin D V., Gorbunova M A, Estrin Ya I, Komratovaa V V, Badamshinaa E R 2016 *Phys. Chem. Chem. Phys.*, **18(46)** 31769-31776

IOP Conf. Series: Materials Science and Engineering **848** (2020) 012027 doi:10.1088/1757-899X/848/1/012027

The application of copper-containing nanoparticles as a gas nitriding catalyst

I Yu Isaeva, E A Eliseeva, I V Morenko, V A Aleksandrov and G Yu Ostaeva

Moscow Automobile and Road Construction State Technical University (MADI), Leningradskii pr. 64, Moscow 125319, Russia

E-mail: irina-razumova-xim@yandex.ru

Abstract. The results of the study of the possibilities of using copper-containing nanoparticles obtained by pseudomatrix synthesis as a catalyst for cyclic gas nitriding of steel products were presented in this paper. It was shown by the example of steel 08Yu (corresponding to the European standard of steel DC04/DC03 EN 10139) that the preliminary application of nanocomposite-catalyst including copper-containing nanoparticles and macromolecules of poly-(N-vinylpyrrolidone) polymer pseudomatrix to the product significantly intensified the chemical heat treatment process. It was found that under identical treatment process conditions the thickness of the diffusion layer for a sample with a pre-applied nanocomposite was 90 μm, and without such pretreatment — 42 μm. It is noted that the achievement of this effect is due to the catalyst synthesis peculiarities, which allows to obtain a copper sol with a narrow size distribution of particles resistant to aggregation.

1. Introduction

The currently proposed intensification methods of the steel treatment process have a few disadvantages: delayed nitrogen diffusion during the process of saturation through the surface layer leading to an increase in the duration of the process, the formation of the insufficient layer thickness of metal-ceramics on the surface of products, the non-equilibrium structures formation with high microhardness [1, 2]. However, the proposed methods are characterized by high cost and strict requirements for the application and manufacture.

In this matter, a study aimed at improving the steel surface process hardening productivity, due to the intensification of the diffusion nitriding process is relevant.

Various options have been proposed as catalysts for the nitriding process, for example: solids, liquids and gases [3, 4]. It was found that accelerators do not affect the hardness and depth of the layer, since they only change the composition of the gas phase and do not change the value of the nitrogen diffusion coefficient.

The use of copper as a nitriding catalyst was also described in the literature [5]. The thickness unevenness of the copper deposited by the galvanic method, especially on products profile complex is main disadvantages. This leads to lose of film catalytic properties and does not allow to receive a monolithic zone of metal ceramics in the diffusive nitrided layer.

The obtaining of copper nanoparticles by reducing Cu^{2+} ions in aqueous solutions of polymers used as regulators of the size resulting nanoparticles, as well as the possibility of controlling the size of synthesized copper-containing nanoparticles was investigated in our early works [6-8]. The sol nanocomposite obtained and used by us as a catalyst for nitriding allows to eliminate the above disadvantages, due to the nanoscale range and optimally selected components. This article focuses on the preparation and evaluation of the influence of copper nanoparticles on the diffusion processes of nitriding.

2. Experimental

Copper(II) sulfate pentahydrate (reagent grade, Reakhim, Russia), tert-butylamine borane (TBAB) (98% Aviabor, Russia), poly(N-vinylpyrrolidone (PVP) (Fluka, Mn = 3.6×10^5) and hydrochloric acid (reagent grade, Reakhimn Russia) were used without further purification. Solutions were prepared in twice distilled water. Low-alloy steel 08Yu was used 08Yu (GOST – state standard of the Russian Federation - 9045-93, corresponding to the European standard of steel DC04/DC03 EN 10139) as an object of chemical and heat treatment.

The sizes and morphology of metal nanoparticles were studied using transmission electron microscopy (TEM). Studies were carried out on LEO 912 AB OMEGA microscope ("Carl Zeiss", Germany).

When preparing samples, a drop of sol was applied on a copper grid covered by formvar and the excess liquid was removed by filter paper. The sample was dried at room temperature (drying time did not exceed 5-10 minutes).

The morphology of the samples after heating was examined using scanning electron microscopy (SEM). SEM images were measured on a Versa 3D DualBeam scanning electron/focused ion beam microscope (FEI, USA) equipped with energy dispersive X-ray spectroscopy system (EDAX, USA) in high vacuum mode using a backscattered electron detector.

Processing of micrographs was carried out by software Gatan Digital Micrograph (Gatan, USA) and ImageJ (NIH, USA).

The durometric analysis of the microhardness distribution over the nitrided layer thickness of samples was carried out on a microhardness tester PMT-3 (Lomo, Russia).

Metallographic studies were carried out using a light microscope "AXIOVERT 25CA"(Carl Zeiss, Germany). Marble reagent (4 g $CuSO_4$ + 20 ml HCl + 20 ml H_2O) was used to identify the structure.

At the first stage of the experiment, the reduction reaction of copper ions TBAB at room temperature in air was used to obtain copper-containing nanoparticles. TBAB (0.01 – 0.03 mol/l) was added to an aqueous solution containing copper salt (0.01 mol/l) and PVP (0.01 mol/l).

At the second stage of the experiment, the resulting composite system was applied to the surface of 08Yu steel product. The steel sample was maintained in the sol of copper-containing nanoparticles for 2-3 minutes at room temperature. Further, the steel product was dried at room temperature with air flow.

At the third stage, the steel sample was exposed to chemical heat treatment. At the beginning, it was placed in a sealed container, heated in an atmosphere of ammonia and carbon dioxide with the volume ratio of 1:1 to the temperature of 575 °C. Then, the nitriding was performed under isothermal exposure at the same temperature. At the same time, the cyclic change of the gas mixture was carried out: in the first stage of the cycle, a mixture of ammonia and water vapor was used, in the second stage — a mixture of ammonia and carbon dioxide. The amount of ammonia during saturation was kept constant and equal the volume ratio of 1:1 to the volume of the other saturating atmosphere component in the first and second stages.

After 3 hours from the beginning of the nitriding process, the sample was cooled with the furnace in an ammonia atmosphere to a temperature of about 200 °C and unloaded from the furnace reactor.

The chemical heat treatment of steel 08Yu sample was carried out by described method (the third stage of the experiment) without application of nanocomposite to estimate the effectiveness of the catalytic system.

3. Results and discussion

The formation of copper sol is occurred as a result of copper ions reduction by TBAB in an aqueous solution of $CuSO_4$ and poly(N-vinylpyrrolidone). The color of the reaction mixture changes during the reduction process: without the reducing agent – blue, after addition of TBAB becomes yellow, then

IOP Conf. Series: Materials Science and Engineering **848** (2020) 012027 doi:10.1088/1757-899X/848/1/012027

yellow-orange, and then lilac. The color change of the reaction mixture indicates the formation of Cu^{2+} ion reduction products.

The obtained nanocomposite is a complex of copper-containing nanoparticles with the polymer pseudomatrix macromolecules, which is PVP. Polymer macromolecules shield particles, preventing their aggregation. Figure 1 depicts the TEM image of the obtained sol of nanocomposite.

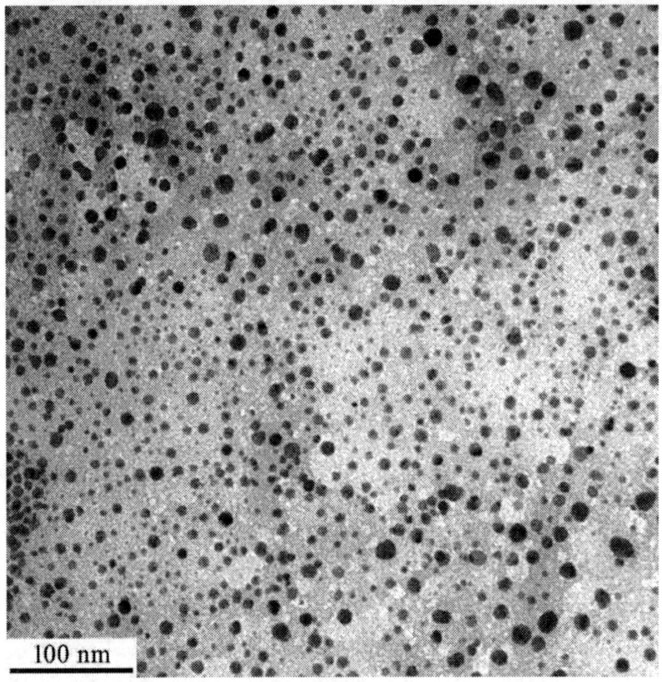

Figure 1. TEM image of particles in the Cu sol obtained by the reduction of copper ions in solution of PVP.

The theory of pseudomatrix (pseudotemplate) synthesis of similar sols which is based on the concepts of the cooperative character of noncovalent interactions of macromolecules with the nanoparticles surface makes it possible to effectively control the dimensional characteristics of nanoparticles in the processes of synthesis of sols [9,10] The particle numerical size distribution histogram is presented on figure 2. The particle size distribution is narrow. The main fraction of sol particles diameters is in the range of 4-12 nm.

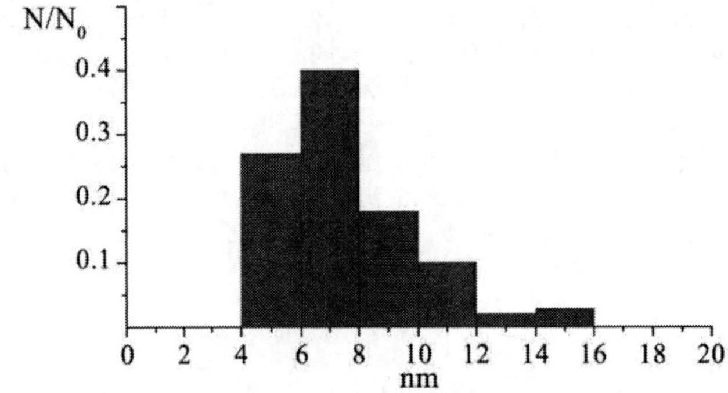

Figure 2. Histogram of numerical size distribution of particles in the Cu sol obtained by the reduction of copper ions in solution of PVP.

IOP Conf. Series: Materials Science and Engineering **848** (2020) 012027 doi:10.1088/1757-899X/848/1/012027

Upon the exposure of nanocomposite coated steel products at the temperature of 575 °C (in a flowing atmosphere of ammonia and water vapor mixture) metallic copper is oxidized to CuO, which then acts as a catalyst for ammonia dissociation. The formation of copper oxides from copper-containing nanoparticles is provided by the supply of water vapor at the first stage of the nitriding cycle. At the second stage of the cycle, copper oxides are reduced to pure copper by hydrogen formed as a result of ammonia dissociation.

A diffusion layer thickness of 90 μm (figure 3) was obtained as a result of chemical heat treatment of the nanocomposite coated steel 08Yu sample. Under the standard method of nitriding under similar conditions, the layer thickness was 42 μm (figure 4). The microhardness levels in both cases were the same.

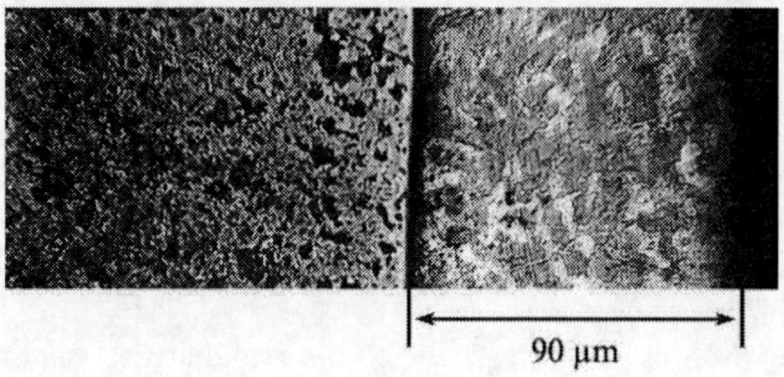

Figure 3. Microstructure of the steel 08Yu sample pre-coated by nanocomposite after chemical heat treatment.

Figure 4. Microstructure of the steel 08Yu sample after chemical heat treatment without application of copper-containing nanocomposite.

Studies have shown that copper-containing nanoparticles deposited on the surface as part of the composite coating exhibit high catalytic activity, which allows to obtain an additional effect in the reaction of ammonia dissociation with the generation of nitrogen ions, and also lead to a significant increase in the number of active centers through which nitrogen penetrates into the steel. This is facilitated by the fact that the particles do not aggregate when heated, but are distributed over the surface at a distance from each other, thereby accelerating the diffusion of nitrogen ions into steel products in high concentration. Figure 5 depicts the micrograph of the sample after heating up to 1000 °C.

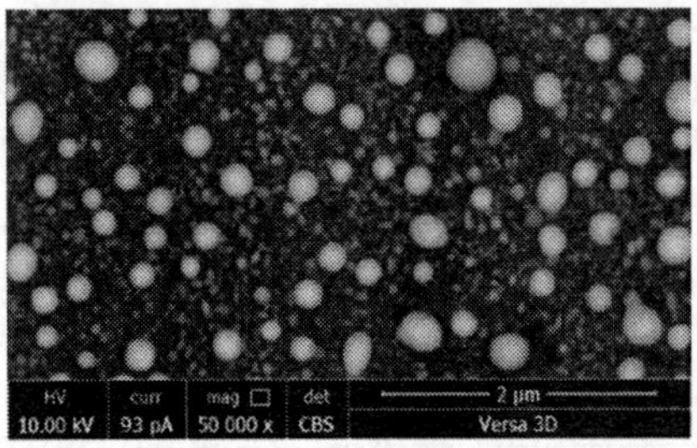

Figure 5. SEM image of particles in the sample after heating up to 1000 °C.

4. Conclusion

The results of a study confirming the possibility of using copper-containing nanoparticles to improve the efficiency of the chemical heat treatment process of steel were described in the article. It was found that the Cu-containing nanoparticles used as a nitriding catalyst are able to influence the rate of ammonia dissociation reaction. The use of copper-containing nanoparticles as a catalyst for gas nitriding allows to intensify the process, as well as to obtain a uniform hardened layer on the steel surface with a monolithic zone of cermets.

The efficiency of the resulting catalyst is due to the peculiarities of synthesis. Using of polymer protector PVP provides a narrow particle size distribution in the range of 4-12 nm and the absence of their aggregation at the initial stage (during synthesis) and subsequent (during heating).

Acknowledgment

This work was supported by the Ministry of Education and Science of the Russian Federation (project no. 16.11777.2018/11.12.) within the framework of a design part of the state assignment in the field of scientific research.

References

[1] Lakhtin Yu M, Kogan Ya D, Shpis G and Boehmer Z 1991 *Theory and technology of nitriding* (Moscow: Metallurgy)

[2] Chudina O V, Alexandrov V A and Brezhnev A A 2009 Promising technologies for creating diffusion coatings on steels with high mechanical characteristics *Bulletin of Kharkov national automobile and highway university* **46** 72

[3] Muravitsky Yu P 1987 Mar. 19 Method of nitriding steel products *Soviet Union patent* USSR1541303 (C23C 8/24)

[4] Seidenberg K I 1938 Apr. 29 Method of accelerated nitriding with catalyst *Soviet Union patent* USSR57851 (C23C16/28)

[5] Alexandrov V A, Petrova L G, Brezhnev A A and Demin P E 2015 Dec. 24 Method of cyclic gas nitriding of parts made of constructional alloy steels *Russian Federation patent* RU2614292 (S23S 28/04, S23S 8/26, S23S 8/34)

[6] Papisov I M, Isaeva I Yu, Ostaeva G Yu, Eliseeva E A, Papisova A I and Kozlovskii V F 2015 Formation of copper(I) oxide nanoparticles at an early stage of copper sol synthesis in the presence of poly(vinylpyrrolidone) *Colloid J.* **77** 780

[7] Ostaeva G Yu, Isaeva I Yu, Grushina V V, Stuzhuk A N and Odinokova I V 2018 Influence of the Molecular mass of poly(N-vinylpyrrolidone) on formation of Cu_2O nanoparticles during reduction of divalent copper ions with tert-butylamine borane in polymer solution *Polymer Science B* **60** 455

[8] Ostaeva G Yu, Isaeva I Yu, Morenko I V, Eliseeva E A and Litmanovich A A 2019 Synthesis and redox interconversions of copper-containing nanoparticles stabilized by poly(N-vinylpyrrolidone) *Polymer Science B* **61** 254

[9] Litmanovich A A and Papisov I M 1997 Preparation of nanocomposites in the processes

IOP Conf. Series: Materials Science and Engineering **848** (2020) 012027 doi:10.1088/1757-899X/848/1/012027

controlled by macromolecular pseudomatrices *Vysokomol. Soedin. B* **39** 323
[10] Papisov I M and Litmanovich A A 1999 On recognition phenomena in polymer-minute particle interactions and pseudo-matrix processes *Colloids and Surfaces* 1999 **151** 399

IOP Conf. Series: Materials Science and Engineering **848** (2020) 012028 doi:10.1088/1757-899X/848/1/012028

The electrochemical synthesis and investigation of nanostructured Fe-Pt and Co-Pt systems

N Ivanova[1], A Lobanov[1], A Andyyakova, Yu Zakharov[1,2], A Popova [2] and R Kolmykov[2]

[1] Institute of Fundamental Sciences, Kemerovo State University, Kemerovo 650000, Krasnaya, 6, Russia
[2] The Federal Research Center of Coal and Coal Chemistry SB RAS, Kemerovo 650000, Russia
E-mail: sayganta@mail.ru

Abstract. By means cyclic and anodic linear sweep/stripping voltammetry methods the electrochemical behavior of bimetallic electrolytic nanoalloys and nanostructured powders Fe–Pt, Co–Pt was investigated in acidic (chloride, sulfate, tartrate) and ammonia buffer media with the glassy carbon electrode. The conditions (supporting electrolytes, concentrations of metal ions, voltammetric parameters) of the electrochemical synthesis and voltammetric characterization of the binary systems were established.

1.Introduction

Unique magnetic performance is typical for nanostructured bimetallic Co-Pt and Fe-Pt systems due to the existence of high-coercivity phases with perpendicular anisotropy (*e.g.*, $L1_0$) in their structure. This determines a variety of application fields for such systems as thin film coatings, powders, and core-shell systems. Functional properties of bimetallic systems platinum-transition metal are employed when developing magnetic recording means, microelectromechanical systems, and also agents of directional action in cancer diagnosis and therapy [1, 2]. Furthermore, apparent catalytic properties determine the use of the specified systems in chemical and electrochemical catalysis, and also in developing sensor devices to determine bioactive organic compounds [3-5].

The main problem for the synthesis of Co-Pt and Fe-Pt systems is producing phases free of impurities of metal hydroxides and precursor components. Therefore, the task of developing electrolytic deposition methods that may allow producing pure synthesis products under soft conditions is relevant in this respect. Various structures, such as thin films, powders, nanoparticles, and nanofibers, may be produced *via* these techniques [6-8]. One of the evident advantages for electrodeposition is an opportunity to investigate the phase and elemental composition of the resulting systems through *in situ* voltammetric methods after selecting appropriate experimental conditions [8 – 10]. The information regarding electrodeposition conditions for nanostructured Co-Pt and Fe-Pt systems is scarce. The known data presented in literary sources deals with the use of electrolytes containing chloride and sulphate ions, organic additives (for example, sulphamate and citrate ions), and metal electrodes [10–15]. An opportunity to perform the electrochemical characterization of nanostructured Co-Pt and Fe-Pt systems both in the form of electrolytic films and powders has not been previously explored. Due to the above, the objective of this work is the exploration of electrodeposition processes and the search for electrochemical conditions for the nanostructured Co-Pt and Fe-Pt systems characterization.

2.Experimental

The CV and ASLV electrochemical measurements were performed in a three-electrode cell using potentiostat/galvanostat Parstat 4000. The working and counter electrodes were the glassy carbon cylinders (l = 15 mm, d = 2 mm) or the glassy carbon (Sigradur) plate (l = 25 mm, w = 5 mm) and the

IOP Conf. Series: Materials Science and Engineering **848** (2020) 012028 doi:10.1088/1757-899X/848/1/012028

reference electrode – Ag/AgCl in 0.1 M KCl. Before experiment working and counter electrode surfaces were polished down to polishing cloths, etching in 5 M H_2SO_4, thoroughly washed with pure water and transferred to the electrochemical cell. Chemicals used were $FeSO_4$, $Fe_2(SO_4)_3$, $CoCl_2$, H_2PtCl_6, supporting electrolytes: 0.1 M NH_4Cl + 0.1 M NH_4OH (pH 9.8), 0.1 M Na_2SO_4 + HCl (pH 2.0); 0.1 M KCl + HCl (pH 1.7); 0.1 M NaCl; 0.2 M $C_4H_6O_6$ + 0.2M NaOH; all of analytical grade. All solutions were prepared with double-distilled water. pH measurements were carried out using pH-meter Expert 001.

The electrodeposition of the nanoalloys Fe–Pt, Co–Pt was carried out in the potentiostatic mode from the background electrolytes with various ratios of the metal ions concentrations ($1 \cdot 10^{-5} - 1 \cdot 10^{-3}$ M) at −1.6 V, time of the deposition was 2 – 10 min. During this process the solution stirred with magnetic stirrer. Anodic oxidation was carried out in the potentiodynamic mode, scan rate – 100 mV/s.

The nanostructured Fe–Pt, Co–Pt powders synthesized according to the procedure of chemical reduction given in [16-18] were preliminary cleaned by water or etching agents and then were immobilized on the surface of the working glassy carbon electrode.

The X-ray diffraction (XRD) data were recorded by diffractometer Bruker D8 Advance A25 [18, 19]. The chemical composition of the powders was carried out by the method of inductively coupled plasma optical emission spectrometry (ICP-OES) using an iCAP-6500 DUO spectrometer (USA). The SEM images of the NPs were obtained by a Jeol JSM 6390 390 scanning electron microscope.

3.Result and Discussion

The research on the electrochemical behaviour of platinum, iron, and cobalt has been performed through cyclic (CV) and anodic linear stripping voltammetry (ALSV). Moreover, conditions for the formation of electrolytic deposits of metals and binary systems on a glassy carbon substrate have been experimentally determined. Various electrolytes, such as solutions of potassium chloride and sulphate (pH = 1.5) with additives of an organic complexing agent, *i.e.* tartaric acid, and also an ammonia buffer solution for cobalt and the cobalt – platinum system were scrutinized considering chemical properties of the metals.

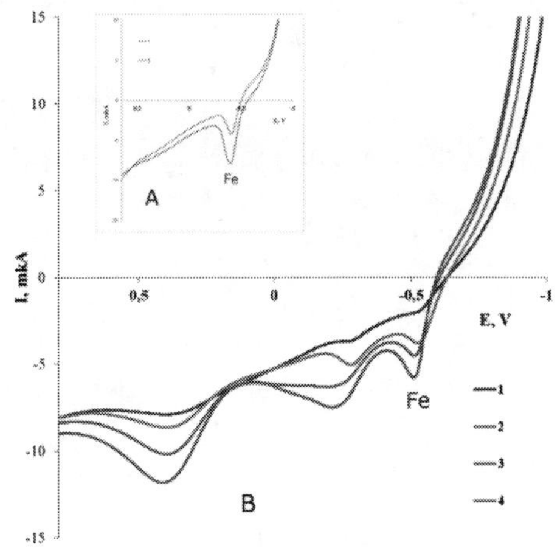

Figure 1. Anodic voltammetric curves acquired in different electrolytes, 0,1 M NaOH + H_2Tartr, pH 3, 0.5 M NaCl. c(Fe^{3+}), 10^{-4} M: **1** – 2 ; **2** – 5; **3** – 10; **4** – 16.

The criterium for efficient cathode deposition is the presence of reproducible oxidation peaks of metals in the anode voltammetric curve and the dependence of the peak value on the ion content of the corresponding metal in the electrolyser. As it is known, cobalt and iron form slightly soluble hydroxides, which should be considered when an electrolyte is selected. Solubility constants of hydroxides are as follows: $Fe(OH)_2 - 8\cdot10^{-16}$, $Fe(OH)_3 - 6{,}3\cdot10^{-38}$; $Co(OH)_2 - 2\cdot10^{-16}$ [20].

Hydrogen evolution process occurs in parallel in the cathodic region upon metal electrodeposition. Due to this, there are the local alcalinization of the near-electrode layer and even the formation of cobalt and iron oxide-hydroxide compounds on the electrode surface. These phenomena are especially clearly apparent for iron due to the extremely low solubility constant of $Fe(OH)_3$. In the anodic stage, that results in the appearance of a series of additional peaks in the potential range between +0,2 and +0,7 B (Fig.1).

The peaks correspond to iron oxidation processes to yield iron (II) oxide-hydroxide phases followed by transformation of the latter to iron (III) oxide or iron (III) oxo-hydroxide [21]. There is a qualitatively similar oxidation pattern after cobalt deposition. Parameters of these electropositive peaks and the fact itself of their formation allow determining the degree of purity of the resulting metal deposit and, on the other side, identifying the oxide-hydroxide phases by their response in the anodic process upon voltammetric research of nanostructured powders.

As determined by means of ASLV and CV the electrodeposition process of nanostructured iron deposits free of impurities of oxygeneous compounds proceeds in the electrolytes listed in Table 1, except for tartrate. As can be seen, by varying the nature of the background electrolyte and pH, one may change the potential range for iron oxidation by a value between 30 and 120 mV, which is crucial upon the phase analysis via the voltammetric method. Herewith, there have been acquired linear relationships of current peaks vs iron ion concentration in all media, except for 0.1 M KCl + HCl. There is the highest deposition rate in 0.5 M NaCl solution. Potassium chloride-based electrolytes with a pH of not higher than 2, and also ammonium buffer solution may be used for cobalt electrodeposition, as unlike iron ions, cobalt (II) species form stable complex compounds with ammonia. As can be seen in the table, the potentials of cobalt peaks are significantly different when using chloride and ammonia media. Cathodic peaks for iron and cobalt ions have not been recorded, as the former are shielded by background currents.

Table 1. Parameters of iron and cobalt peaks in various background electrolytes

Background electrolyte, ion	pH	E_{peak}, mV	Linearity formula, $c\cdot10^4$ mol/L	Range $c(M^{n+})$, 10^{-4} mol/L
0.1 M KCl + HCl, Fe^{3+}	1.7	–430	–	2,6 – 9,3
0.1 M Na_2SO_4 + HCl, Fe^{3+}	2.0	–480	I = 0.39c + 1.4	1.7 – 16
0.5 M NaCl, Fe^{3+}	5.5	–405	I = 1.48c – 4.1	13 – 33
0.2 M $C_4H_6O_6$ + 0.2M NaOH, Fe^{3+}	3.1	–530[a]	I = 0.35c + 2.0	1.7 – 16
0.5 M NaCl, Fe^{2+}	5.5	–405	I = 0.92c + 3.0	1.1 – 6.7
0.1 M KCl + HCl, Co^{2+}	1.7	–150	I = 0.95c + 28	0.02 – 3
0.1 M NH_4Cl+NH_4OH, Co^{2+}	9.8	–450	I = 2.90c – 4.0	2 – 30

*Parameters are presented for the peak of iron oxidation.

Electrodeposition of binary systems was carried out in a mode of the preliminary modification of the electrode surface with adatoms of a less precious component. The technique is considered in detail in [9, 10]. Experiments were carried out at various concentrations of iron (II/III), or cobalt (II) ions and chloroplatinic acid. As can be seen in voltammograms shown in Fig. 2 for electrolyte 0.1 M Na_2SO_4 + HCl (pH 2.0), upon the co-deposition of iron and platinum, the rate of iron deposition is sharply reduced and, as a consequence, the anodic peak current of iron is decreased; in the region E_{an} = – 190 mV, the hydrogen desorption peak appears from the electrolytic platinum surface. Herewith, there is the corresponding peak of hydrogen sorption at a potential of –520 mV on the cathodic branch. There is no anodic oxidation of platinum under experimental conditions, however, the presence of electrolytic platinum is indirectly proven by hydrogen sorption/desorption peaks. There are similar regularities for all acidic electrolytes explored.

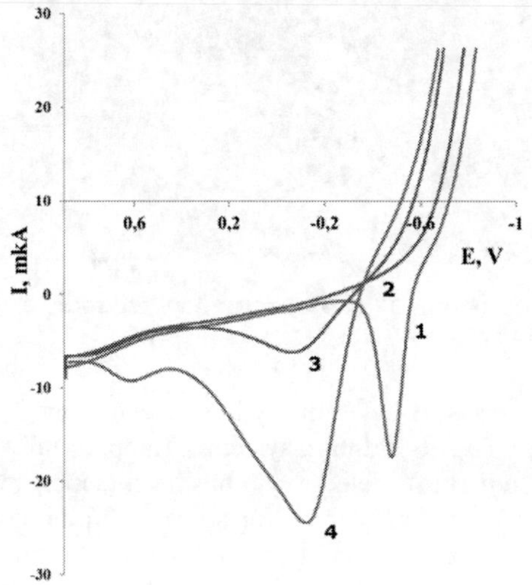

Figure 2. Anodic voltammetric curves for iron and iron-platinum system in 0.1 M Na_2SO_4 + HCl, c(Fe^{2+}), 10^{-4} M: **1** – 7; c($[PtCl_6]^{2-}$), 10^{-7} M: **2** – 4; **3** – 8; **4** – 28; **5** – 68.

As determined using quantity of electricity for the cathodic/anodic peaks, at least a hundredfold excess of iron (II/III) ions over chloroplatinate species is required to deposit binary iron-platinum phases. In this concentration region a subsequent increase in chloroplatinate ion content results in a shift of iron anodic peak in the positive direction. That is explained by the more energy consumption which requires for iron selective oxidation from the binary Fe-Pt alloy.

Electrode processes for the iron-platinum system can be presented by the following scheme:

Cathodic deposition

- $Fe^{3+} + 3e^- \rightarrow Fe$
- $Fe^{3+} + [PtCl_6]^{2-} + 7e^- \rightarrow Fe_xPt_y + 6Cl^-$
- $[PtCl_6]^{2-} + 4e^- \rightarrow Pt + 6Cl^-$

Hydrogen adsorption peak:

- $Pt + H^+ + e \rightarrow Pt(H_{ads})$

Anodic oxidation

Iron oxidation peak:

- $Fe \rightarrow Fe^{3+} + 3e^-$

Hydrogen desorption peak:

- $Pt(H_{ads}) \rightarrow Pt + H^+ + e$

As demonstrated by scanning electron microscopical investigations of iron-platinum nanostructured alloys deposited on glassy carbon plates from a chloride electrolyte (Figure 3), this system is characterized by island-type growth. Moreover, areas with dendritic and fibrous structure are formed. Cyclic voltammetry data show, the resulting nanoalloys have an average formula of $Fe_{20}Pt_{80}$. However, there is no iron in the samples according to the EDX, therefore its shielding with platinum metal occurs, which is explained by the lateral growth of the deposit. The observed features

IOP Conf. Series: Materials Science and Engineering **848** (2020) 012028 doi:10.1088/1757-899X/848/1/012028

of the morphology of the nanostructured deposit are due to the hydrogen evolution process in parallel. As a consequence, the electrode surface is partially blocked and the rate of the reduction process of metal ions is decreased.

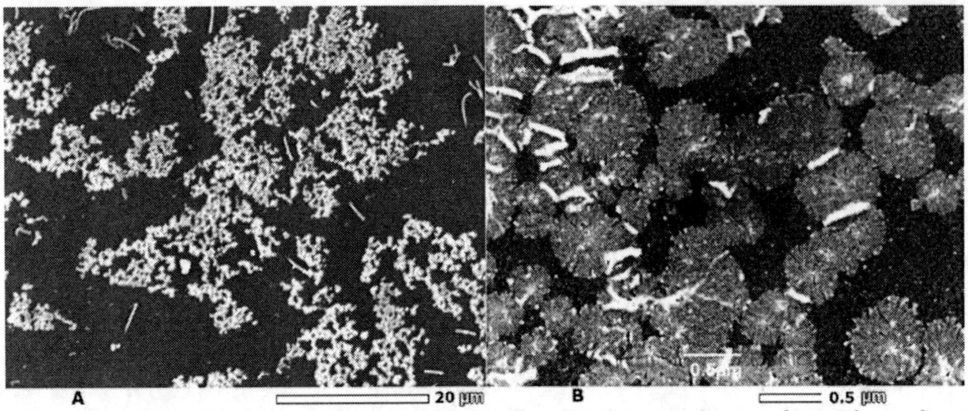

Figure 3. SEM images for the nanostructured Fe-Pt alloy on glassy carbon surface. Electrodeposition in 0.1 M KCl + HCl electrolyte; c([$PtCl_6$]$^{2-}$) = $5.7 \cdot 10^{-5}$ M c(Fe^{2+})=$7.7 \cdot 10^{-3}$ M.

Conditions that allow eliminating the effect of processes of catalytic hydrogen evolution in the presence of platinum have been determined for the binary cobalt-platinum system. An opportunity for the electrolytic synthesis of Co-Pt thin films in an ammonia buffer electrolyte has been demonstrated through stripping and cyclic voltammetry. It is possible to explicitly trace for a change in the cobalt content in the alloy under these conditions when varying deposition conditions, as there is an oxidation peak for this metal in voltammetric curves upon any platinum content. As can be seen in Fig. 2, the combined progression of reduction processes for ammonia complexes of cobalt and chloroplatinate ions in general leads to a reduction in the cobalt deposition rate. However, it has been noted for small Co^{2+} ion concentrations (till 10^{-5} mol/L) the electrodeposition of platinum leads to some increase in cobalt current apparently due to at the initial step due to facilitating the nucleation process (Fig. 4).

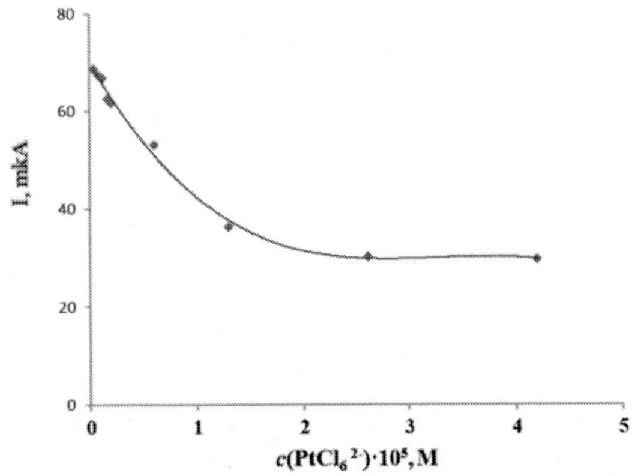

Figure.4. Cobalt peak current vs chloroplatinate ion concentration in the cobalt-platinum binary system, c(Co^{2+}) = $2 \cdot 10^{-4}$ M; 0,1 M NH_4OH + 0,1 M NH_4Cl.

As illustrated by research performed in the ammonia buffer medium, there is a shift of the cobalt peak towards more positive potentials in the presence of platinum due to the formation of the nanostructured cobalt-platinum alloy on the electrode surface (Figure 5 b). The existence of solid solutions and intermetallic compounds (IMC), such as CoPt and $CoPt_3$, recorded for micro-alloys and appropriate nanostructured powders [16, 21] is likely in the case considered. However, IMC phases do not show themselves voltammetrically under experimental conditions.

There are no hydrogen adsorption/desorption peaks in the ammonia buffer, therefore in order to assess the platinum content in the nanostructured system, a method for transferring the electrode into an acid chloride electrolyte has been envisaged. The main experimental steps involve selective cobalt oxidation in the platinum alloy in the course of potential cycling and chemical etching in an HCl solution. Afterwards, the platinum electrode is placed in a chloride electrolyte of a certain acidity and a series of CV curves are recorded until the stabilization of adsorption/desorption peaks in the cathodic and anodic stages of the process. As determined, the optimum parameters for these peaks may be recorded in an electrolyte pH range of 2.2–2.8.

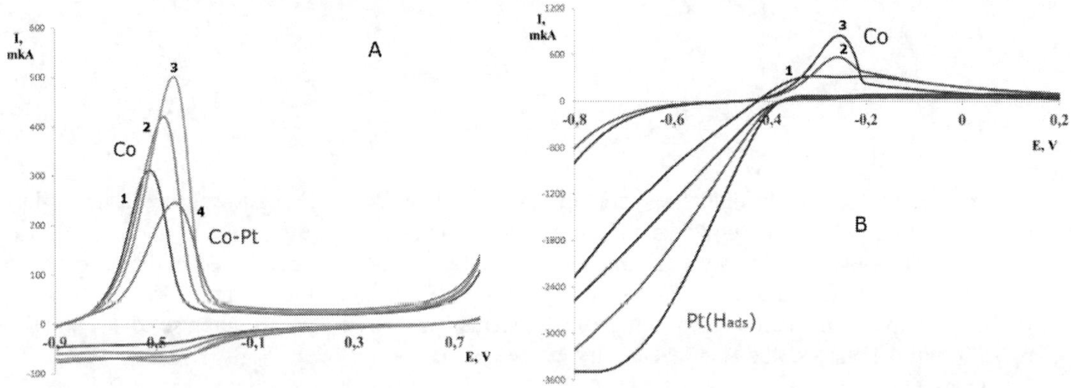

Figure 5. CV curves for cobalt–platinum system in ammonia buffer (A) and chloride (B) medium; A - $c(Co^{2+}) = 7\cdot10^{-5}$ M (*1*), $c(PtCl_6^{2-})$, 10^{-6} M: 0.4 (*2*); 1 (*3*); 7(*4*); B – $c(PtCl_6^{2-}) = 2.3\cdot10^{-5}$ M (*1*); $c(Co^{2+})$, 10^{-4} M: 3.1 (*2*), 6.4 (*3*). Anodic branch on the top.

In chloride medium (0.1 M KCl + HCl), as with the phenomena observed in the iron-platinum system, voltammograms are complicated with hydrogen adsorption/desorption peaks (Fig.5 B). Furthermore, phase cobalt peak becomes apparent upon a significant excess (in more than 50 times) of cobalt ions over chloroplatinate ones in the electrolyser. Hence, there is also a significant decrease in a deposition rate of cobalt in the presence of platinum in this instance.

The performed experiments have formed the basis for developing the voltammetric technique of the electrochemical response investigations of nanostructured Fe-Pt powders and suspensions specifically attached to the surface of glassy carbon. The fundamental concept of the technique is the sequential performance of the steps of gradient anodic oxidation and cathodic reduction in the mode of stripping and cyclic voltammetry. As a consequence, metal and oxide-hydroxide phase peaks are recorded in the anode process and those of metals in the form of oxides and hydroxides. The latter peaks are registered after cathode reduction. Figure 6 reports cyclic voltammetry curves for samples of Fe-Pt of diverse spatial arrangement.

Hydrogen adsorption/desorption peaks are recorded in a potential range of –0.5 ÷ –0.2 V. They indicate the presence of the phase of pure platinum or platinum in the binary Fe-Pt phase (it is impossible to differentiate these phases at this step of research) and a series of oxygen-containing iron compounds formed upon chemical synthesis and storage of the samples (a potential range of + 0.3 ÷ +0.6 V in the anodic scan). Current peaks correlate with absolute metal concentrations; there is no changing potentials and the total picture of oxidation/reduction when varying an Fe/Pt ratio. The appearance of the anodic iron peak after carrying out the electrodeposition step indicates that iron in the sample is found as a part of the binary phase.

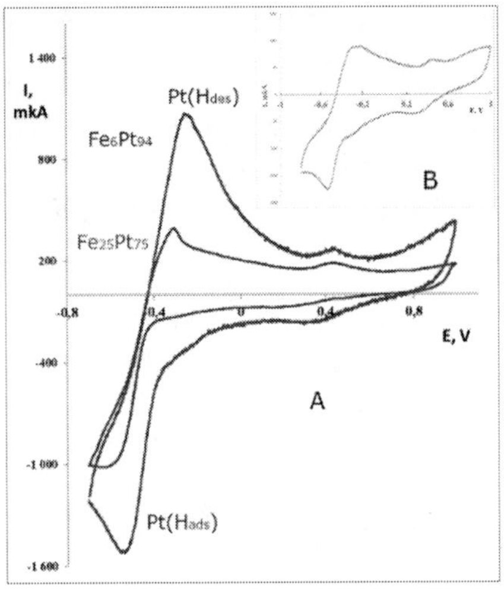

Figure.6. Cyclic voltammograms for the nanostructured powders (A) and suspension (B) Fe-Pt attached to glassy carbon, 0,1 M KCl + HCl. Anodic branch on the top.

Electrochemical research outcomes are in agreement with XRD data regarding the phase composition of nanostructured powders and atomic emission analysis results. The phase is comprised of solid solutions of varying composition and its content indicates the correspondence of metal concentrations to the formula determined upon synthesis [16]. Thus, the voltammetric method may be used to explore nanostructured bimetallic systems, in particular, to analyse the elemental composition of the phases, and also their purity extent.

4.Conclusion

The electrochemical response of nanostructured bimetallic systems Fe–Pt and Co–Pt by means voltammetric methods (CV and ASLV) can be used in the investigation of their physico-chemical characteristics such as phase and elemental composition, spatial phase arrangement, stability to the oxidation, etc. It is worth indicating that the electrolytic synthesis of nanostructured iron-platinum systems free of oxo-hydroxide inclusions is likely over glass-carbon substrates when using acid chloride, sulphate media with pH ≤ 2, whereas cobalt-platinum – acid chloride and ammonia buffer media. In order to perform the voltammetric research on nanostructured Me-Pt systems, it is required to consider catalytic activities of the specified metals in hydrogen emission processes and an opportunity to form oxygen-containing analogues by virtue of changing the pH of the medium.

Acknowlegments

The authors like to thank Sozinov S.A. for the assistance with SEM analysis.

References

[1] Sternik M, Couet S, Łażewski J, Jochym P T, Parlinski K, Vantomme A, Temst K and Piekarz P 2015 *J. Alloys Compd.* **651** 528–536
[2] Fan T–E, Liu T– , Zheng J–W, Shao G–F, Wen Y–H 2016 *J. Alloys Compd.* **685** 1008-15
[3] Cai S, Qi C, Li Y, Han Q, Yang R and Wang C 2016 *J. Mater. Chem. B*, **4** 1869-77
[4] Mei H, Wu W, Yu B, Wu H, Wang S, Xia Q 2016 *Sens. Actuator B-Chem.* **223** 68-75
[5] Termpornvithit C, Chewasatn N and Hunsom M 2012 *J. Appl. Electrochem.* **42** 169-178
[6] Arshad M S, Pečko D, Šturm S, Escrig J, Komelj M, McGuiness P J, Kobe S and Žužek Rožman K 2014 *IEEE Trans. Magn.* **50** 2302904
[7] Srivastav A K and Shekhar R 2015 *J. Nanopart. Res.* 17:14
[8] Ivanova N 2017 *Electrochem. Commun.* **80** 48-54
[9] Ivanova N, Pugatchev V, Nevostruev V and Kolpakova N 2002 *J. Solid State Chem.* **6** 540–544

[10] Ivanova N, Lobanov A, Emel'yanova V, Vanicheva A and Zakharov Yu A 2018 *Chem. Sustain. Develop.* **26** 635-639
[11] Fardi-ilkhchy A, Nasirpouri F, Vázquez M and Palmero E 2017 *Prot. Met. Phys. Chem. Surf.* **53** 57-67
[12] Yanai T, Honda J, Hamamura R, Omagari Yu, Furutani S, Yamada H, Fujita N, Morimura T, Nakano M and Fukunaga H 2019 *J. Electron. Mater.* **48** 1412-15
[13] Tabakovic I, Qiu J – M and Dragos O 2016 *J. Electrochem. Soc.* **163** D287-D294
[14] Cavallotti P, Bestetti M and Franz S 2003 *Electrochim. Acta* **48** 3013-20
[15] Oniku O, Qi B and Arnold D 2014 *J. Appl. Phys.* **115** 17E521
[16] Pugachev V, Zaharov Yu and Popova A 2017 *Butlerov Commun.* **51** 155-160
[17] Zaharov Y, Pugachev V, Datiy K, Popova A, Valnyukova A and Bogomyakov A 2015 *Key Eng. Mater.* **670** 49-54
[18] [Popova A 2012 *J. Phys. Conf. Ser.* **345** 012030
[19] ICDD, PDF-2 2011 (Database), ed. by Dr. Surya Kalakkodu, International Centre for Diffraction Data, Newtown Square, PA, USA.
[20] McKenzie K, Asogan D and Marken F 2002 *Electrochem. Commun.* **4** 820–824

Application of optical methods in tests of new materials with gradient structure

I S Kamantsev[1], N A Drukarenko[1], A V Kuznetsov[1], S I Stepanov[2] and M A Chesnokov[3]

[1] Institute of Engineering Science UB of RAS, 34 Komsomolskaya street, Ekaterinburg, Russia,
[2] Ural Federal University named after the first President of Russia B.N.Yeltsin, 18, Mira street, Ekaterinburg, Russia
[3] JSC «Uralturbo», 17 Artinskaya street, Ekaterinburg, Russia

ks@imach.uran.ru

Abstract. The main idea of the article is to demonstrate the possibility of applying the laser dynamic speckle interferometry technique to control the processes of fatigue failure of specimens with a gradient structure. A distinctive feature of the described technique is the possibility of practical application on high-frequency machines as an experimental non-contact method for recording surface changes in the studying objects with a complex structure. Besides, the proposed technique can be used as an indicator characterizing irreversible processes during fatigue failure and for analyzing the fatigue crack growth kinetics.

1. Introduction

At present finding the fracture precursors of structural elements as well as monitoring existing fracture centers under conditions of multi-cycle fatigue loading is an urgent not yet completely solved task. Among the existing methods for observing the fracture process, such as the use of contact measuring instruments (strain gauges, foils, piezoelectric elements) and non-contact based on measurements of magnetic and thermal properties optical methods deserve special attention. One of the most common and widely used optical control methods is the Digital Image Correlation method (DIC) [1]. At the same time tests not of standard samples, but of structural elements, including, among other things, materials of a new generation (for example, composites or products obtained by additive technologies) are finding increasing application. However, the use of DIC in the analysis of the fracture kinetics of materials with complex geometry under high-frequency fatigue loading is difficult due to the shallow depth of field, which is due to the wide field of view, relatively low sensitivity and implementation features of the algorithm for constructing strain distribution diagrams, which does not allow controlling the object when exposed to small loads. The methods of electronic speckle interferometry and holographic interferometry which are successfully used in non-destructive testing of materials and products to detect defects that cannot be detected by visual examination are devoid of these disadvantages [2].

IOP Conf. Series: Materials Science and Engineering **848** (2020) 012029 doi:10.1088/1757-899X/848/1/012029

Holographic interferometry allows you to solve many problems, such as defect detection, strain analysis, object shape control, etc. A variety of methods of holographic interferometry is based on the principle of comparing two wavefronts, one of them or both being recorded and restored using the holographic method [3]. Speckle interferometry, as well as holographic interferometry, allows you to measure displacements (static and dynamic) and to research the relief of an optically rough surface with sensitivity on the order of the light wavelength. And with the use of computer technology the removal and processing of images no longer takes a lot of time which can significantly expand the scope of this method. Besides, coherence methods use a small diaphragm size to increase the depth of field which is especially important when examining objects with a complex structure.

2. Materials and methods

In the present work, the results of researching the fatigue crack growth kinetics, the evolution of the size and shape of the plastic deformation zone during its nucleation and propagation are presented. Results are obtained using a technique based on recording changes in speckle images of the testing objects surface - laser dynamic speckle interferometry which was successfully used earlier to research processes of plastic deformation and fracture of metals under conditions of quasistatic loading [4, 5].

As objects of research metal products were selected that have a significant gradient in the structure or distribution of properties over the cross-section — a gas turbine engine guide blade, a titanium alloy Ti-6Al-4V with a cellular structure, obtained on a 3D printer (EOS 280), and a layered metal composite material consisting of alternating layers of low-carbon steels 09G2S and EP678. The speckle images of the surfaces of the researched objects and the distribution of the correlation coefficients of their fragments recorded directly during cyclic loading are shown in Fig. 1-3.

Fatigue loading was performed on an MIKROTRON RUMUL high-frequency resonance testing machine using three-point bending (for layered composite), cantilever bending (turbine blade) and uniaxial compression-compression (Ti-6Al-4V alloy) schemes. The loading frequency ranged from 70 to 110 Hz.

The research of the features of fracture and localization of the fatigue crack nucleation region, the evolution of the size and shape of the plastic deformation zone during its nucleation and propagation, was carried out using laser dynamic speckle interferometry techniques. It is based on finding the correlation of two frames of speckle images recorded at different stages of cyclic loading. Speckle images of the object were recorded using an optical device consisting of a laser module with a wavelength of 650 nm, a power of 40 mW, and a USB video camera VIDEOSKAN-415M [6].

One of the advantages of the technique is the ability to work in real-time, so it is possible to obtain pictures of averaging speckles over the time range of the exposure time of the camera. For each type of loading the optimal conditions for fixing images and the parameters for constructing correlation patterns (the step between the compared images, the minimum and maximum values of the correlation coefficient, the size of the cells into which the image is divided during calculations) were selected. All calculations are based on a modern hardware and software system that allows you to compare information with data obtained directly from the testing machine [7].

3. Results and discussion

According to the results of fatigue tests of gas turbine engines blades made of VT3-1 titanium alloy, it is shown that the dynamic speckle interferometry method allows us to unambiguously record the initial stage of formation of the fatigue fracture center on the blades near the fillet (Figure 1, b), which was also confirmed by a change in the resonance loading frequencies. Subsequent loading led to the growth of a fatigue crack and upon reaching 5×106 cycles, the length of the fatigue crack was 2 mm. According to the results of speckle image processing, the growth trajectory of the fatigue crack was established as well as the shape and size of the localized zone of plastic deformation at the crack tip.

IOP Conf. Series: Materials Science and Engineering **848** (2020) 012029 doi:10.1088/1757-899X/848/1/012029

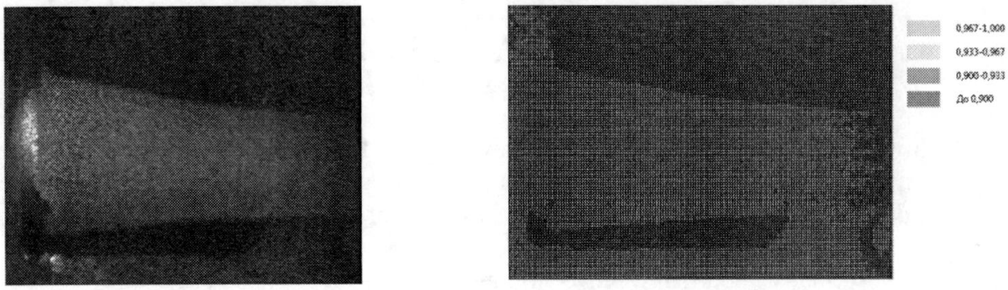

a b

Figure 1. Typical speckle image (a) and distribution of correlation coefficients for speckle image fragments (b) of a gas turbine engine blade.

Samples of a titanium alloy with a cellular structure, that is, with a periodic arrangement of pores of a given geometry, obtained by layer-by-layer fusion of the powder, have a difference in the relative surface height of up to 2 mm. Besides, the complex geometric structure is supplemented by production features that can cause anisotropy of the material properties, and therefore, the fracture can occur randomly in different areas, which greatly complicates the analysis of the fracture kinetics. Based on the results of the tests it was shown the possibility of processing images of a similar class of objects under cyclic loading and the program settings for processing speckle images have been selected (Fig. 2).

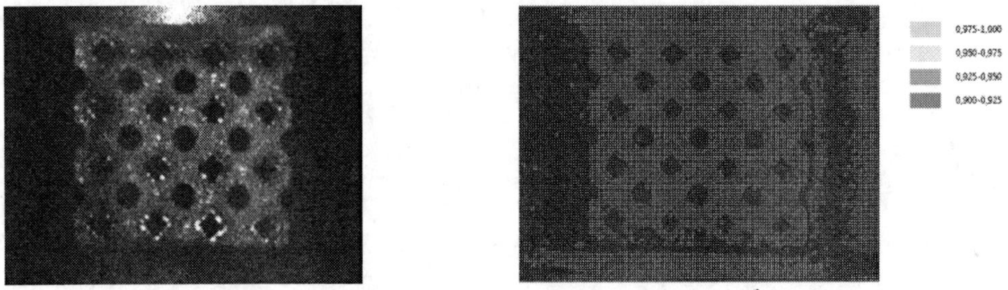

a b

Figure 2. Typical speckle image (a) and distribution of correlation coefficients of speckle image fragments (b) of a titanium alloy sample Ti-6Al-4V.

Using the example of a layered metal composite material of the steel-steel system, the features of the intersections of a layer connecting boundary by a fatigue crack were revealed and the dominant mechanisms of fatigue crack retardation were analyzed [8].

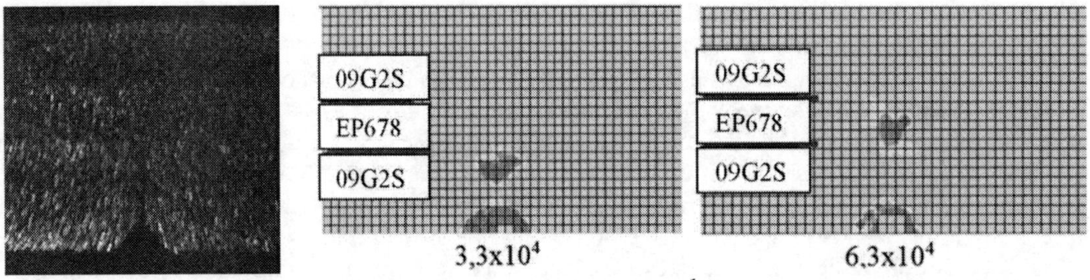

a b

Figure 3. Typical speckle image (a) and distribution of correlation coefficients of speckle image fragments (b) of a sample from a layered composite of the steel-steel system.

4. Conclusion

Thus, on the example of fatigue tests of a different class of objects with a gradient structure the possibilities of using the laser dynamic speckle interferometry technique to control fracture processes are shown. It should be emphasized that the developed speckle interferometric technique allows us to determine in situ such an important parameter of fracture mechanics as the size of the localized plastic deformation zone at the crack tip. This parameter is unambiguously related to the stress intensity coefficient characterizing the ability of the material to brittle fracture in the vicinity of the fracture focus. Moreover, since the stress intensity factor is invariant to the geometry of the controlled object, its value determined on standard samples can also be used to assess the level of accumulated damage to structural elements from the same materials as standard samples. A distinctive feature is the possibility of practical application on high-frequency machines as an experimental non-contact method for recording changes in the objects under research with a complex structure.

The proposed technique can be used as an indicator characterizing irreversible processes during fatigue failure and for analyzing the fatigue crack growth kinetics. The obtained experimental results indicate the possibility of reliable registration of the fatigue failure formation stages and crack propagation during cyclic loading. The registration of various stages of material fatigue degradation is fundamentally important for determining the limits of applicability of the approaches of damage mechanics and fracture mechanics which are used in assessing the durability of objects.

Acknowledgments

The authors are grateful to Dr.Eng.Sc. Vladimirov A.P. for assistance in preparing the publication material.

The work was performed using the equipment of the collective use center «Plastometriya» with partial financial support from the RFBR grant №18-38-00339_mol_a regarding the study of the behavior of laminated materials under cyclic loading and state assignment topic №AAAA-A18-118020790148-1 regarding the creation of methods for detecting prefracture zones in metal composite materials and structural elements.

References

[1] Vildeman V E 2012 *Experimental studies of the materials properties under complex thermomechanical processes* (Fizmatlit, Moscow) 204 (in Russian)

[2] Razumovskiy I A 2007 *Interference optical methods of deformable solid mechanics* (MGTU N.E. Baumana) 240 (in Russian)

[3] Vest Ch 1982 *Holographic interferometry* (Mir, Moscow) 504 (in Russian)

[4] Vladimirov A P 2004 *Dynamic speckle-interferometry of deformable objects* (UrO RAN, Ekaterinburg) 241 (in Russian)

[5] Vladimirov A P 2015 *Diagnostics, resource and mechanics of materials and structures* **6** 27-57 (in Russian)

[6] Vladimirov A P, Kamantsev I S, Veselova V E, Gorkunov E S and Gladkovskii S V 2016 *Technical physics* **61** (4) 563-568

[7] Kamantsev I S, Vladimirov A P and Drukarenko N A Svidetel'stvo o gosudarstvennoy registratsii programmi dlya EVM #2017613195 ot 13.03.2017

[8] Kamantsev I S, Kuznetsov A V and Drukarenko N A 2019 *XXX International Conference of Young Scientists and Students "Topical Problems of Mechanical Engineering"* (Moscow, Ras. Acad. Sci.) 94-97 (in Russian)

IOP Conf. Series: Materials Science and Engineering **848** (2020) 012030 doi:10.1088/1757-899X/848/1/012030

Technology of spark plasma sintering as an innovative solution of synthesis high-density of SiC ceramics

G D Kardashova[1, 2]

[1]Department of Engineering Physics, Dagestan State University, 43-a Gadzhiev Street, Makhachkala, the Republic of Dagestan, 367000, Russian Federation
[2]National Research Nuclear University MEPhI (Moscow Engineering Physics Institute), Kashirskoe shosse, 31, Moscow, 115409, Russian Federation

E-mail: gulya-ka11@yandex.ru

Abstract. The main advantages of spark plasma sintering (SPS) method compared to the classical hot pressing method for obtaining high-density ceramic based on silicon carbide are presented. Shows the perspective of obtaining a wide range of materials using this method. The possibilities of technology of spark plasma sintering of multicomponent ceramics based on silicon carbide of different composition have been studied. Established optimal SPS modes for sintering multicomponent ceramics of composition SiC (75%) - 22% (AlN) -Y_2O_3 (3%) (1800 °C/50MPa/15 min) provide formation of 100% dense composite, which allows to recommend these modes for making light and high-density ceramic material of specified composition with predicted properties.

The most widely used technology in the ceramic industry in recent years has been the method of spark plasma sintering due to the short duration of the process and energy efficiency. Using pulsed electric current and the so-called "spark plasma effect" ("plasma spark effect") is achieved very fast heating of the workpiece (up to 1000 ° C / min), so the sintering process is usually very short (several minutes). This allows you to suppress grain growth and obtain an equilibrium state, which opens up opportunities for creating new materials with previously unattainable compositions. In this work, we applied this method of spark plasma sintering (Spark Plasma Sintering, SPS), which is also known as the “Field Assisted Sintering Technology (FAST)”, as an innovative solution for the synthesis of high-density ceramic based on silicon carbide.

It is known that sintering of ceramics based on silicon carbide with aluminum nitride by the classical technology of hot pressing at temperatures below 1900 ° C is undesirable due to the low diffusion rate of AlN in SiC at these temperatures. Sintering at temperatures above 2250 ° C leads to the decomposition of aluminum nitride. The optimal sintering time is 2-6 hours. With a sintering time of less than 2 hours, the formation of SiC-AlN solid solutions in the grain boundary region does not occur, which leads to a decrease in the strength and crack resistance of ceramics. With an increase in sintering time of more than 6 hours, diffusion processes between the backfill and sintered ceramics leads to embrittlement of the latter. The solution to these problems is seen in the application of the method of spark plasma sintering for this system.

The studied samples of SiC-AlN ceramics were obtained under various SPS-sintering modes, with the introduction of an additional sintering additive of yttrium oxide (Y_2O_3), which is a promoter of the sintering process, into the initial mixture of powders.

The objective of the research was to optimize the compositions of the mixtures of the starting components and the SPS modes to achieve the complex of the highest physical and mechanical properties (density, thermal conductivity, microhardness). The effect of the concentration of activating additives, temperature, and holding time in the SPS-sintering mode on the density of ceramics based on silicon carbide is estimated.

Based on the experience of many decades and the successful application of the classical technology of hot pressing, it was established [1] that the optimum sintering temperature of SiC-AlN-based ceramics, depending on the AlN content, is 2050-2300 K for 1 hour in an N2 or Ar medium and a pressure of 10- 50 MPa. In early studies, the activating effect of yttrium oxide during sintering of ceramics based on SiC-AlN was studied [2-4]. When the content of yttrium oxide in the initial mixture is less than 2 wt.%, The final ceramic density does not exceed 90% TP, and an increase in the content of Y2O3 above 8 wt.% Leads to an increase in the oxidizability of the material due to the high diffusion mobility of yttrium ions at temperatures exceeding 1300 ° C. In addition, yttrium oxide is the most expensive component of the charge and an increase in its content leads to an increase in the cost of the material. We used this technique for SPS sintering of the compositions of the SiC-AlN system. Yttrium oxide was initially introduced into AlN powder, and then the corresponding mixtures with SiC were prepared. The starting components were mixed in the ratios shown in table 1.

Table 1. Composition and particle size of the powder (dispersion).

	Composition Components		
	SiC	AlN	Y_2O_3
Weight % component.	75	25	0
	75	22	3
	75	18	7
Dispersion, microns	14.22	5.69	7.72

Granulometric analysis of the powders was carried out on a Fritsch Analysette 22 laser particle analyzer (Germany). The measuring range of the analyzer is 80 to 2000 nm. A fluid cell was used. Below are the results of the analysis of the particle size of the powder mixtures prepared for sintering (Figures 1 and 2).

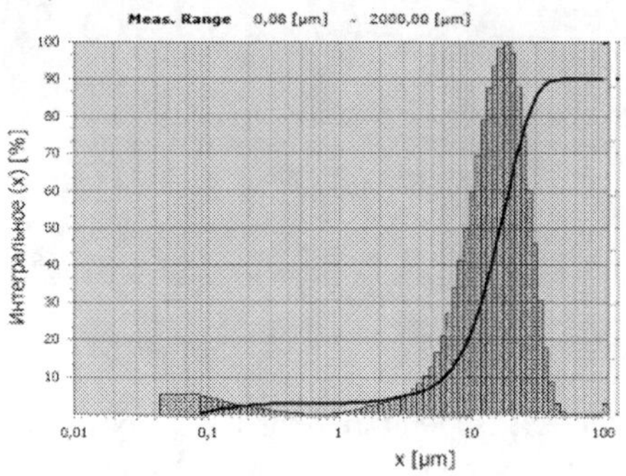

Figure 1. Particle size measurements using a laser analyzer for composition SiC (75%) - AlN (25%)

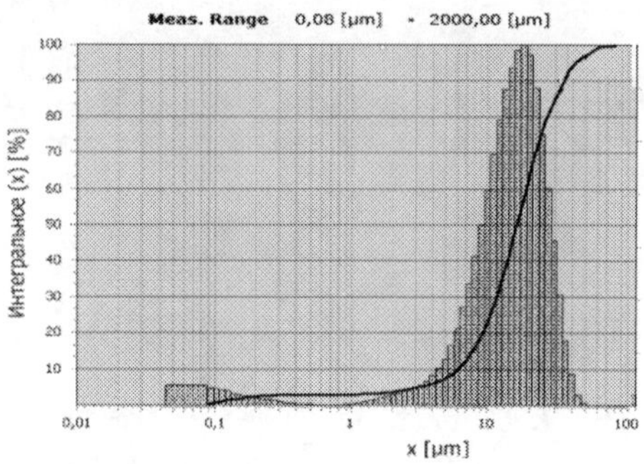

Figure 2. Particle size measurements using a laser analyzer for composition SiC (75%) - AlN (18%) - Y_2O_3 (7%)

Compositions after thorough mixing mainly contain a fraction of 8-14 microns.

The mixture of powders was molded and sintered by passing electric current through the billet in molds with an inner diameter of 24 mm made of isostatic graphite I-3 (area 4.52 cm^2), using an additional layer of graphite paper between the matrix and powder filling and between graphite punches and powder to prevent the reaction between the powder and the tool material. Download - 10g. Temperature control was carried out by an optical pyrometer. Sintering was carried out by applying constant pressure to the powder throughout the entire time. The samples were cooled freely in the chamber.

Using the well-known and previously described technological methods [5, 6], experimental batches of ceramic samples of various compositions were obtained (Figure 3). Figure 3 clearly shows how the color changes, and therefore the density of the obtained samples, depending on the concentration of the starting components and sintering conditions (temperature, time).

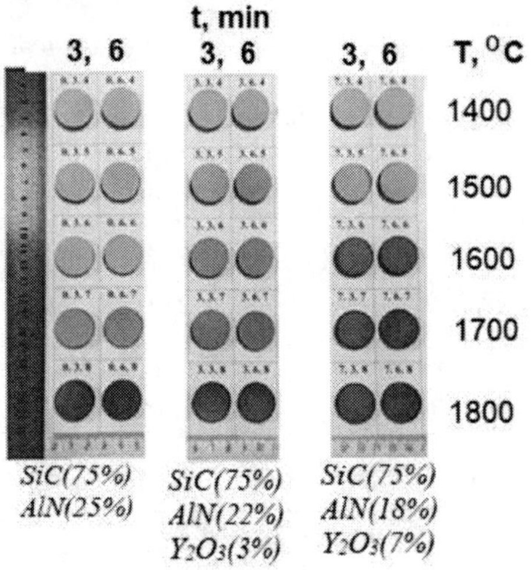

Figure 3. Ceramic samples obtained by the SPS method of various compositions under different sintering conditions.

The resulting samples were removed from the mold, cleaned and tested for density. It was found that the addition of yttrium oxide significantly increases the density of SiC-AlN-based ceramics. SPS sintering at a temperature of 1800 ° C of finely dispersed mixtures of SiC and AlN with the addition of yttrium oxide already at a content of 18 and 22% AlN provides a relative density of 0.95 and 0.98, respectively, which is significantly higher than the density value for ceramics obtained by hot pressing. (Figures 4 and 5).

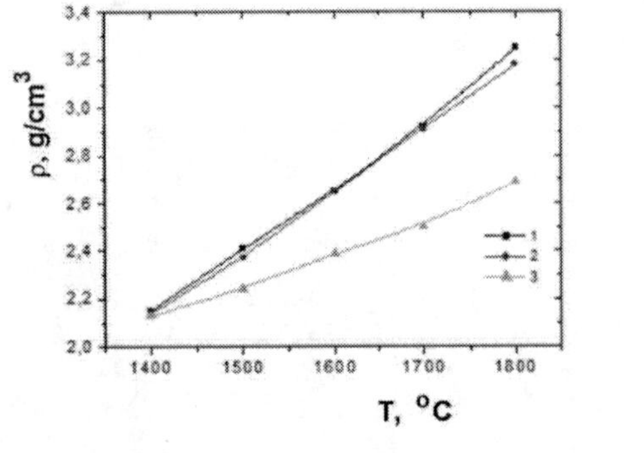

Figure 4. Temperature dependence of density of ceramic samples of various composition 1.SiC(75%)-AlN(18%)-Y_2O_3(7%), 2.SiC(75%)-AlN(22%)-Y_2O_3(3%), 3.SiC(75%)-AlN(25%). Sintering time - 3min.

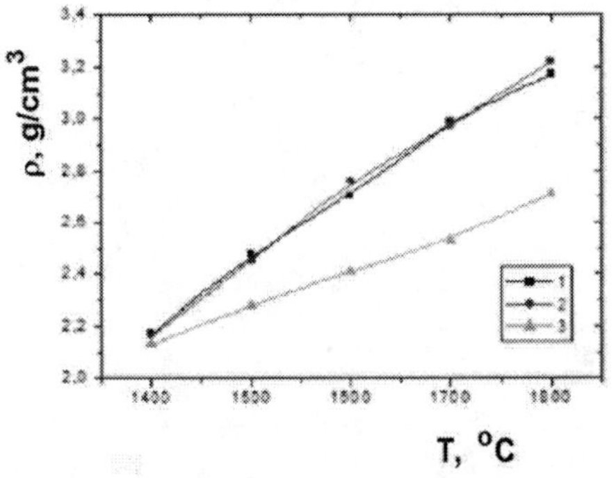

Figure 5. Temperature dependence of density of ceramic samples of various composition 1.SiC(75%)-AlN(18%)-Y_2O_3(7%), 2.SiC(75%)-AlN(22%)-Y_2O_3(3%), 3.SiC(75%)-AlN(25%). Sintering time - 6min.

During the experiment, it was found that upon sintering of samples with the addition of 7% yttrium oxide at 1800 ° C in the mold, the formation of a liquid phase is observed, which fills the space between the particles of silicon carbide until a dense material is formed. This is confirmed by the microstructure photographs shown in Figures 6 and 7, obtained on a field emission SEM with a nanolithography unit, Raith150TWO (Germany).

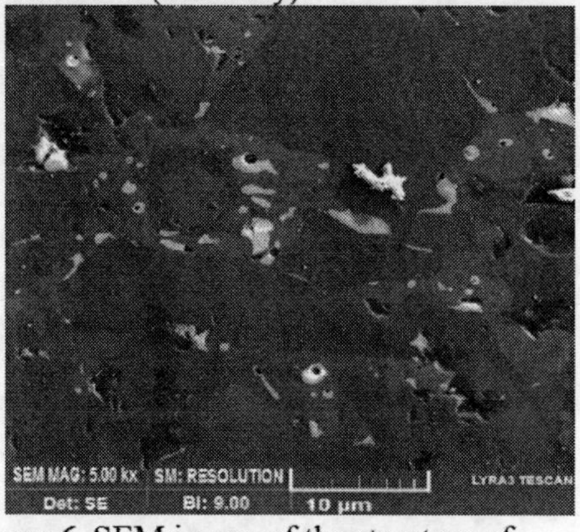

Figure 6. SEM image of the structure of ceramic SiC(75%)-AlN(18%)-Y_2O_3(7%) obtained by the SPS method.

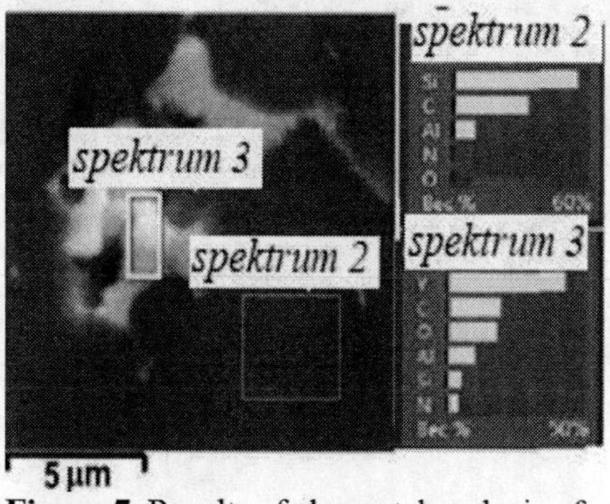

Figure 7. Results of elemental analysis of ceramic SiC(75%)-AlN(18%)-Y_2O_3(7%) obtained by the SPS method.

An increase in the concentration of activating additive leads to an increase in the evaporation of oxides and an increase in their interaction with silicon carbide, which reduces the density of the SiC material. At a high content of oxides (≥ 7% wt.), They fill the pore space, and the excess liquid melt is squeezed out of the sintered sample, corrodes the protective graphite foil and interacts with the die matrix. The oxide melt adheres to the graphite form and destroys the sample.

Thus, for SiC-AlN ceramics, the optimal additive is Y_2O_3 - 7 wt.% At a sintering temperature of 1800 ° C. Therefore, further studies were conducted on the effect of the exposure time on the sintering mode at a maximum sintering temperature of 1800 ° C on the density of the obtained samples for only two compositions: 0% and 3% yttrium oxide. Research results for density are shown in Figure 8.

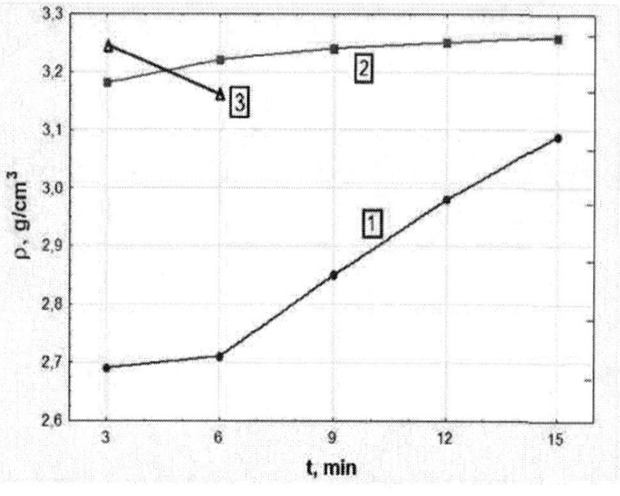

Figure 8. The dependence of the density of ceramic samples on the sintering time in the regime at T = 1800 °C: 1.SiC(75%)-AlN(25%). 2.SiC(75%)-AlN(22%)-Y_2O_3(3%). 3.SiC(75%)-AlN(18%)-Y_2O_3(7%).

It was found that the highest density value of ~ 95.9% of the theoretical value for the SiC (75%) - AlN (25%) composition is achieved by SPS sintering at a temperature of T = 1800 0C for a time t = 15 minutes. And with the addition of 3% yttrium oxide at the same technological parameters, 100% of the theoretical density is achieved.

The presented results show that a high-density silicon carbide based ceramic with perfect microstructure and high (up to 100%) density can be obtained using the plasma-micron method from a micron fraction powder, which opens up prospects for the widespread use of the SPS method for the economical production of high-quality ceramic products from silicon carbide of a given composition with predictable properties.

The research results are useful for optimizing the manufacturing process of structural and functional high-density ceramics. Using the available data, you can find the optimal parameters of the sintering process to obtain certain properties. The dependences presented in the work will help to evaluate the influence of plasma-spark sintering parameters on the structure and properties of the obtained ceramic materials. Studies of these promising results are ongoing, and experiments are being conducted with finely divided raw materials.

This work was supported by the Ministry of Education and Science of the Russian Federation, project No. 11.1957.2017 / 4.6.

References

[1] G D Kardashova and Sh Sh Shabanov 2019 *IOP Conf. Ser.: Mater. Sci. Eng.* 525 012020

[2] Eom, JH., Seo, YK., Kim, YW. et al. *Met. Mater. Int.* (2015) 21: 525

[3] Kim, Yong-Hyeon & Kim, Young-Wook & Lim, kwang-young & Lee, Seoung-Jae. (2018). Mechanical and thermal properties of silicon carbide ceramics with yttria–scandia–magnesia. *Journal of the European Ceramic Society*. V.39, p 144-149

[4] Kardashova G D 2004 Sintering processes and electrophysical properties of ceramics based on silicon carbide with activating additives (Makhachkala: PhD dissertation)

[5] Tokita M. Spark plasma sintering (SPS) method, systems and applications. *In: Handbook of Advanced Ceramics. Chapter* 11.2.3. 2013. p 1149–1177.

[6] Bilalov B.A., Kardashova G.D., Magomedova E.K., Akhmedov R.R. Features of the technology for the formation of ceramics based on silicon carbide by electropulse sintering. *Fundamental Problems of Radioelectronic Instrumentation* 2010 v.10. p 176-178

IOP Conf. Series: Materials Science and Engineering **848** (2020) 012031 doi:10.1088/1757-899X/848/1/012031

Oxide composite fibers with quasiplastic components and composites with brittle matrices on their basis

V M Kiiko

Institute of Solid State Physics of the Russian Academy of Sciences, Chernogolovka Moscow district, 142432 Russia

E-mail: kiiko@issp.ac.ru

Abstract. By the internal crystallization method oxide composite fibers containing high-strength sapphire and hexaaluminates of calcium, lanthanum or barium are obtained. The presence of "weak" planes in the layered structure of hexaaluminates provides cracks inhibition. The dependences of the ultimate strain (strength) of the fibers of these compositions on the length are obtained. The positive effect of such fibers was evaluated on composites with brittle matrices of recrystallized molybdenum and nickel-aluminum intermetallic. The fracture toughness of composite materials with such fibers significantly exceeds the fracture toughness of composites with sapphire fibers. The composites were tested in the temperature range of 20–1400 °C and showed non-brittle behavior under load, and their strength and fracture toughness correspond to the requirements for structural materials of this kind. The structure of fibers and composites was also studied by electron microscopy.

1. Introduction

This work is a continuation and development of researches on the fabrication of non-brittle fibrous composite materials from brittle components, some examples of which are from different years [1–6]. These works are based on the idea, apparently, first clearly articulated in [7]. It consists in the organization of an inhomogeneous non-plastic medium containing relatively weak interfaces on the crack propagation path. Under the influence of the stress concentration arising in the zone of the tip of the crack and before it, the material delaminates along "weak" interfaces. These delaminations lead to a blunting of the crack, a decrease in stress concentration, and ultimately to a braking of the crack that destroys the material. The problem of obtaining non-brittle materials from brittle components is particularly acute due to the need to develop high-temperature materials, where sufficiently ductile metals can no longer be used due to high operating temperatures, and more refractory but brittle compounds (intermetallic compounds and various ceramics) begin to be used. In the case of fibrous composites this idea is usually realized through special coatings on fibers that provide a relatively weak fiber – matrix interface [1–4]. As a rule, these coatings have low creep resistance, complicate and increase the cost of fabrication composite materials. The basis of the present work is composite fibers [5, 6], which contain both high-strength components and layered components containing relatively weak planes in their structure, which provide the necessary fracture toughness according to the mechanism described above. Their high strength combined with the ability to provide sufficient crack resistance of high-temperature composite materials has led to the need to expand the range of compositions of such fibers, obtaining and researching composites based on them. This paper gives examples of the results of the fabrication and testing of fibers of three different compositions containing sapphire Al_2O_3 and hexaaluminates of calcium $CaAl_{12}O_{19}$, lanthanum $LaAl_{11}O_{18}$, and

IOP Conf. Series: Materials Science and Engineering **848** (2020) 012031 doi:10.1088/1757-899X/848/1/012031

barium $BaAl_{12}O_{19}$. Fibers and composites with a recrystallized brittle molybdenum matrix were obtained by the internal crystallization method [8]. The difficult crystallization conditions for the melts of mixtures of the initial oxides in complex molybdenum channels determined by the method not only complicate the task of controlling the process of obtaining fibers of the required composition, but also make the fibers even more composite in composition, expanding the set of their characteristics, including useful ones. The fibers extracted from the molybdenum matrix were tested for strength and also used as a reinforcing component in a composite with an intermetallic nickel-aluminum matrix. Such composites were obtained by hot pressings a prepregs containing unidirectionally arranged fibers and nickel-aluminum powder.

All fibers are united by the presence of high-strength components: aluminum oxide (sapphire) and hexaaluminates, which are designed to provide, respectively, sufficient strength and non-brittle, “quasiplastic” behavior of composite structures that do not contain plastic deformable components.

2. Methods of fabrication and research

In the work mainly two technological methods were used. To obtain composites with oxide composite fibers and a molybdenum matrix, as well as to obtain fibers for use in matrices other than molybdenum, as already mentioned, the internal crystallization method was used [8]. The method in brief consists in the fabrication of a molybdenum carcass with hollow axial channels, the volume fraction of which in the carcass of the present work was approximately 0.4 of the total volume. The channels serve as shapers for the fibers, a characteristic cross-sections of which are shown in Figures 1b, 2b, 3b, 4b, 5b. In addition, a batch is prepared from a given mixture of oxide powders. The powders are weighed, mixed in ethanol, dried and pressed into tablets. Then the tablets are melted in a molybdenum crucible in a special high-temperature vacuum installation. In the work we used, as already mentioned, mixtures of oxides Al_2O_3 and CaO, Al_2O_3 and La_2O_3, Al_2O_3 and BaO. Then the molybdenum carcass, in which the longitudinal axis of the channels occupy a vertical position, is lowered endwise into the oxide melt. The melt fills the channels due to capillary forces. Further, the carcass with filled channels rises at a certain speed into the cold zone, during which the fibers crystallize. The resulting fiber composite specimen can be used for further researches, or fibers can be extracted from it for use, for example, in other matrices. In the process of obtaining such composite specimens they are heated to a temperature of about 2000 °C and then cooled. As a result, molybdenum recrystallization occurs, it is embrittled at the same time, and composite specimens with a brittle matrix and oxide fibers are obtained. Naturally, the structure and properties of the fibers are determined by both the initial composition of the oxides and technological regimes, the essential parameter of which is the crystallization rate of the fibers, which is determined by the rate of drawing of the specimen into the cold zone. The initial weight composition of the oxides to obtain the fibers and the drawing rate V of the specimen during crystallization of the fibers are given in the captions to Figures. The fibers were separated from the molybdenum matrix by etching it with a mixture of acids. To obtain fiber composites with the fibers and matrices from powders the hot pressing method was used. Powder of matrix material was poured into a graphite mold interspersed with layer-by-layer laying with a given step of unidirectionally arranged fibers. Thus obtained preform was subjected to hot pressing in vacuum at high temperature. The flat a specimen was then cut into rods of rectangular cross section with a longitudinal axis along the fibers suitable for mechanical testing.

Composite specimens were tested under three-point bending of rectangular cross-section rods; moreover, in the case of fracture toughness tests, specimens with a side notch were used to determine the critical stress intensity factor K* [9]. Strength values were calculated using well-known strength of materials formulas. The values of the effective surface fracture energy were determined as $g^* = (1/2S)\int_x P(x)dx$, where P is the load on the specimen, x is the displacement of the point of application of the load to the specimen (flexure of the specimen), S is the cross-sectional area of the specimen. Specimens were tested in the range of 20÷1400 °C; testing at high temperatures were carried out in a vacuum chamber.

IOP Conf. Series: Materials Science and Engineering **848** (2020) 012031 doi:10.1088/1757-899X/848/1/012031

Single fibers were tested at room temperature according to the procedure [10], which briefly consists in the following. The fiber is fixed to the metal foil substrate with rubber glue and wound sequentially on a series of rigid cylinders of decreasing diameters. At each step the number of breaks is calculated and the average fiber length between them is determined, the ultimate (maximum) deformation in the fiber is also calculated at the same time and the correspondence is established: ultimate strain – fiber length. Since the materials of the oxide fibers have an almost linear strain curve, fracture strength can be determined by multiplying the Young's modulus by the ultimate stress value.

The microstructure of composites and fibers was studied using the digital scanning electron microscopes CamScan MV230 and Tescan VEGA-II equipped with W-cathodes and detectors of secondary and reflected electrons and energy dispersive X-ray microanalyzer. This allowed not only to observe the features of structures, but also to determine their composition by chemical compounds. Analysis by chemical elements and constitution diagrams of the corresponding oxide systems allow us to do this in the first approximation.

3. Results and discussions

The results of fabrication and testing of fibers based on sapphire – calcium hexaaluminate, sapphire – hexaaluminate lanthanum, sapphire – hexaaluminate barium are given as examples. Figures 1a, 2a, 3a show the dependences of ultimate strain (fracture strain) on the length of fibers of various batches. The decreasing dependences in semilogarithmic coordinates have a linear shape, which indicates their power dependence, which is typical for objects of this type. Considering that the elastic modulus of fiber materials can be estimated at 300–400 GPa, the strength of fibers over a length of about 1 mm can be more than 1000 MPa, which is quite sufficient for using these fibers as reinforcing components of structural materials.

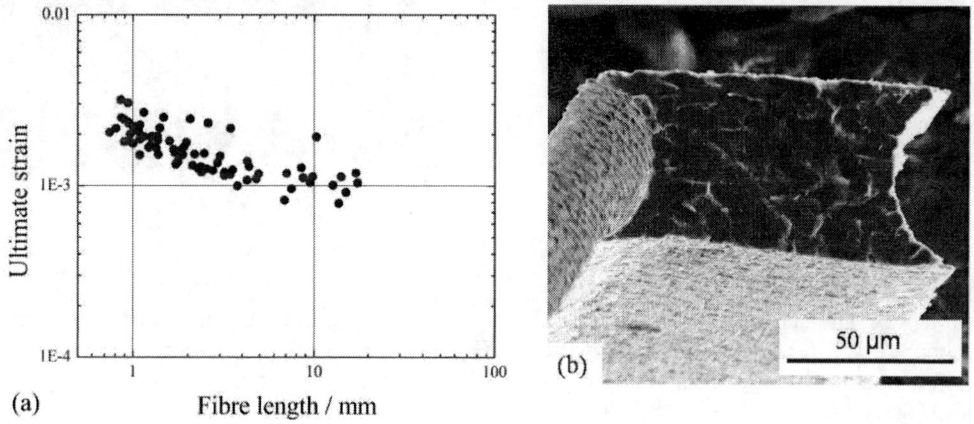

Figure 1. The dependence of the ultimate strain on the length of the fibers of the batch of initial composition Al_2O_3 – 4.2 wt.% CaO, V = 250 mm/min (a); the fracture surface of the tested fiber (b): black areas – Al_2O_3, light areas – calcium containing compounds including $CaAl_{12}O_{19}$.

In addition to these compounds, the fibers can also contain other complex oxides containing aluminum, calcium [5], lanthanum, barium presented in the corresponding constitution diagrams of the oxide systems. The appearance of these compounds is primarily caused by the difficult crystallization conditions of the initial melts in the channels of the molybdenum carcass. However, the analysis of this multifaceted process is beyond the scope of this work and should be the object of future research.

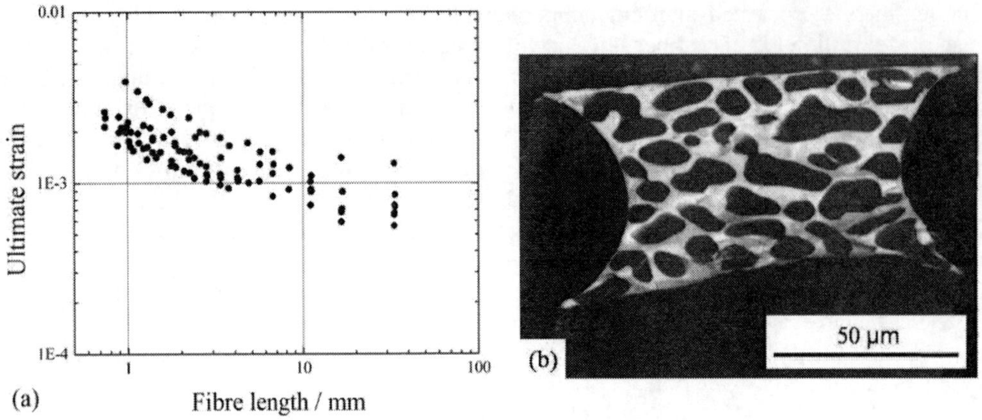

Figure 2. The dependence of the ultimate strain on the length of the fibers of the batch of initial composition equivalent to $LaAl_{11}O_{18}$, V = 50 mm/min (a); the fracture surface of the tested fiber (b): black areas – Al_2O_3, light areas – lanthanum containing compounds including $LaAl_{11}O_{18}$.

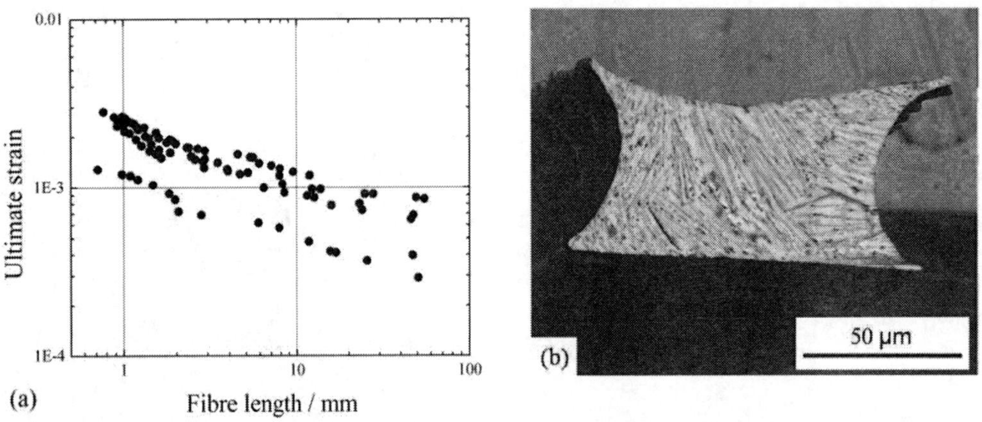

Figure 3. The dependence of the ultimate strain on the length of the fibers of the batch of initial composition equivalent to $BaAl_{12}O_{19}$, V = 800 mm/min (a); the fracture surface of the tested fiber (b): black areas – Al_2O_3, light areas – barium containing compounds including $BaAl_{12}O_{19}$.

Thus, with the initial compositions of oxide mixtures equivalent to lanthanum or barium hexaaluminates, ordered composite structures with other components are also formed as a result of crystallization. Fibers contain a high-strength component sapphire (black in images of fiber fracture surfaces) and complex oxides (gray) containing hexaaluminates, which should provide the necessary fracture toughness of composite structures, Figures 1b, 2b, 3b. Complex fracture surfaces of fibers and composites (Figures 1b, 2b, 3b, 4b, 5b) confirm this.

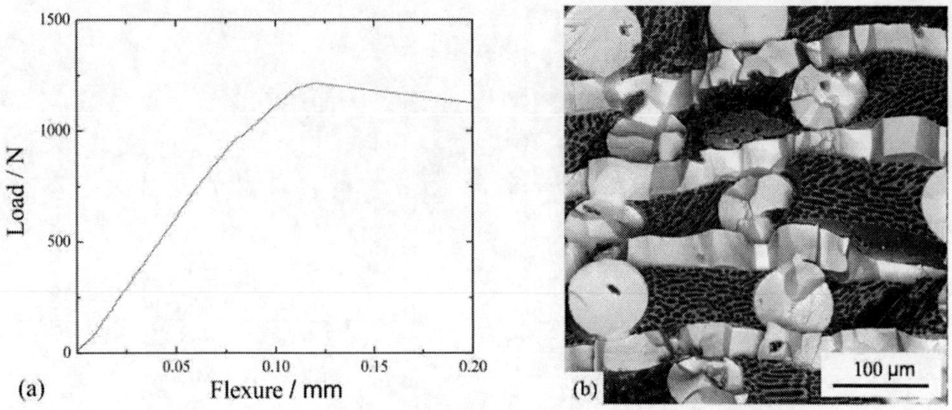

Figure 4. The flexure dependence on the load of one of the composite specimen (Al_2O_3 – $LaAl_{11}O_{18}$) / Mo, V = 250 mm/min with a notch for fracture toughness testing, specimen size ≈ 5×15×60 mm (a); the fracture surface of the tested specimen (b): Mo–matrix has white colour, fibres have black-gray colour.

Figure 4 shows the testing results of one of the specimens obtained by internal crystallization with composite oxide fibers containing sapphire and lanthanum hexaaluminate and a brittle molybdenum matrix from recrystallized molybdenum. “Quasiplasticity” at the macroscopic level is manifested in the nonlinearity of the deformation curve of the specimen during bending testing (Figure 4a). Debondings and shifts along the fiber-matrix interfaces, pulling out the fibers from the matrix, fracture of the fibers along weak layers are visible on the relief of the fracture surface of the specimen (Figure 4b). These microfracture processes lead to non-brittle fracture of the composite material as a whole.

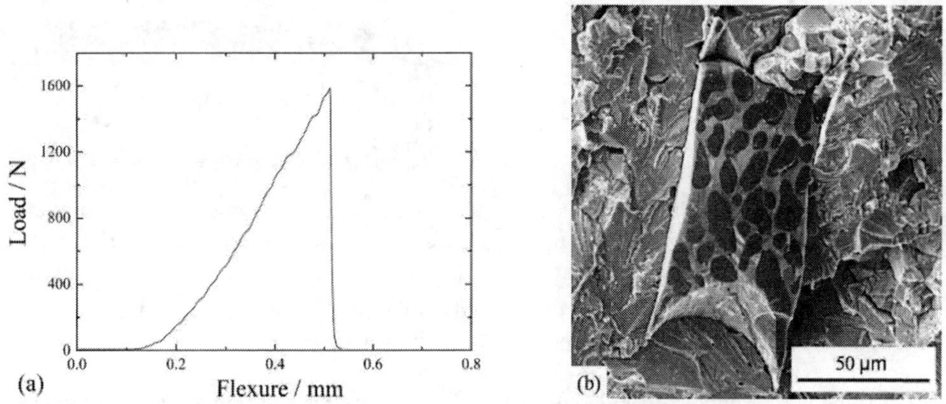

Figure 5. The flexure dependence on the load of one of the composite specimen (Al_2O_3 – $LaAl_{11}O_{18}$) / NiAl tested on strength, specimen size ≈ 4×4×20 mm (a); the area of a fracture surface of the tested specimen (b): (Al_2O_3 – $LaAl_{11}O_{18}$)-fibre in NiAl–matrix.

Similar results were obtained on a composite material with the same fibers and a brittle intermetallic nickel-aluminum matrix (Figure 5) made by hot pressing in the regime of 1400 °C – 30 min – 30 MPa.

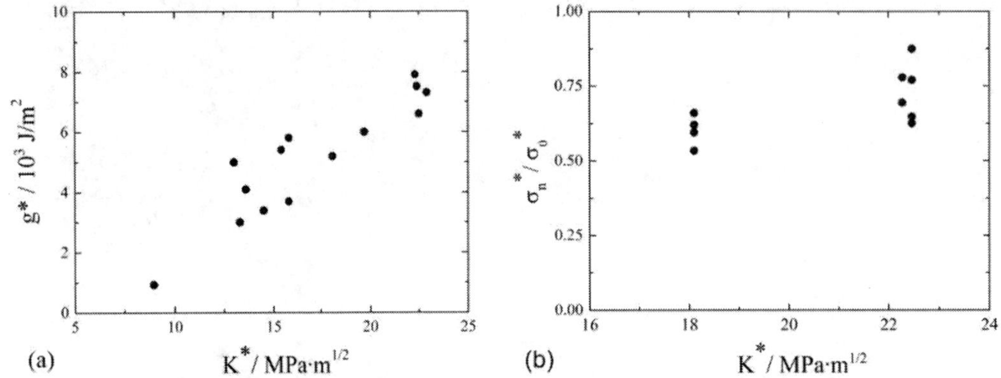

Figure 6. Correlations between critical stress intensity factors K* and effective fracture surface energy g* of composite specimens with a brittle molybdenum matrix and oxide fibers of different composition of the Al_2O_3 – CaO system (a) and the sensitivity of the specimens to notches σ_n*/σ_0* (b), σ_n* is the strength of the specimens with a notch, σ_0* is the strength of the specimens without a notch.

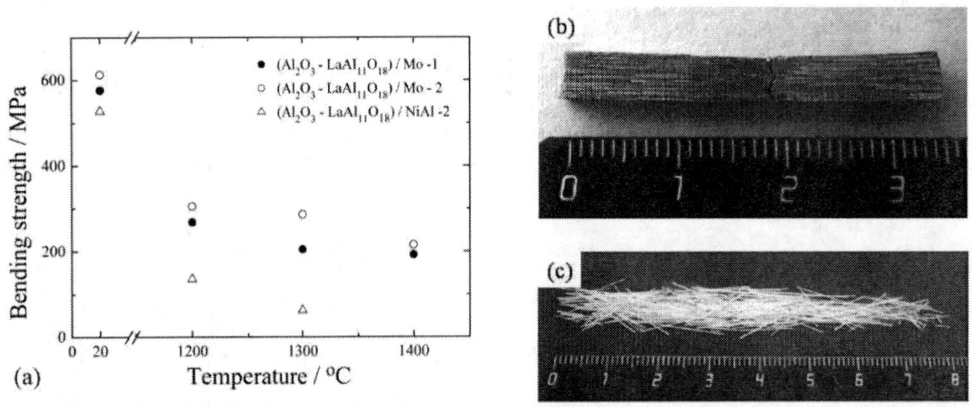

Figure 7. The temperature dependences of the strength of composite specimens with oxide fibers and a molybdenum or nickel-aluminium matrix (a), the compositions of specimens are indicated in the graph field (details in the text); a specimen with a molybdenum matrix after testing at a temperature of 1400 °C (b); fibers used for the making of specimens with a nickel-aluminum matrix (c). The numbers on the ruler scale indicate centimeters.

For composites with composite oxide fibers with initial variations of mixtures of oxides Al_2O_3 – CaO included from 100% Al_2O_3 to Al_2O_3 – 8.0 wt.% CaO containing sapphire and calcium hexaaluminate, and a brittle molybdenum matrix from recrystallized molybdenum, a series of testings were carried out to determine the critical stress intensity factor K*, effective surface energy g* and notch strength sensitivity σ_n*/σ_0* (Figure 6). In addition, a specimen of molybdenum carcass without fibres was tested, which showed K* = 8 MPa·m$^{1/2}$. And for the specimen with sapphire fibers the value K* = 13 MPa·m$^{1/2}$ was obtained. An almost linear correlation was established between the K* and g* values (Figure 6a), which indicates the reliability of the results and the correctness of approaches to their estimations. Strength values σ_0* were obtained on specimens without notches. And the values of σ_n* were obtained on specimens with a side notch reaching the middle of the cross section of the specimen, while calculating the strength value the remaining half of the section was taken into account. The values of σ_n*/σ_0* can reach values close to unity, which indicates the almost complete insensitivity of the specimens to notches. And the values K* and σ_n^*/σ_0^* correlate quite naturally with each other (Figure 6b).

In addition, tests were conducted on composite specimens over a wide temperature range. Specimens with a molybdenum and nickel-aluminum matrix were tested. The specimens used fibers with initial composition equivalent to $LaAl_{11}O_{18}$, and crystallized at different drawing rates V = 50 mm/min (designation "1" on the graph), V = 250 mm/min (designation "2" on the graph), Figure 7a. Figure 7b shows one of the specimens with molybdenum matrix after the testing, visible trajectory of the crack is nonlinear, which indicates the non-brittle nature of the fracture of the composite material. Figure 7c shows fibers of type "2" used for the manufacture of specimens with a nickel-aluminum matrix.

Thus, the test results of specimens of composite materials showed that their characteristics fully satisfy the requirements for structural materials.

4. Conclusion

Using the internal crystallization method from the initial oxides of aluminum, calcium, lanthanum and barium, fibers of three different compositions containing sapphire and hexaaluminates of calcium, lanthanum or barium were obtained. As a result of testing, the dependences of the ultimate strain (strength) of the fibers on the length were obtained. The strength of the fibers was quite sufficient to use them as a reinforcing component of composite materials.

Composites with the mentioned fibers and brittle matrices of recrystallized molybdenum and nickel-aluminum intermetallic were made and tested in the temperature range of 20 ÷ 1400 °C. The results of testing for strength and fracture toughness showed that the level of the obtained values fully corresponds to the requirements for structural materials of this kind.

The shapes of the deformation curves during testing and observations of the fracture structure of specimens of composite materials containing non-plastic components are showing the non-brittle "quasiplastic" nature of their fracture.

These results on composite materials were achieved mainly due to the presence of oxide composite fibers containing components of sapphire and hexaaluminates which provide the necessary strength and fracture toughness respectively.

Acknowledgements

The work was performed in accordance with State assignment and with the financial support of the Russian Foundation for Fundamental Research (Project No. 17-03-01136). The author also thanks S.T. Mileiko, V.A. Chumichev, S.A. Abashkin, A.A. Kolchin, D.G. Pizhenin, A.N. Nekrasov for help in carrying out the work.

References

[1] Atkins A G 1975 Intermitted bonding for high toughness/high strength composites *J. Mater. Sci.* **10** 819–32

[2] Cinibulk M K 2000 Hexaluminates as a cleavable fiber-matrix interphase: synthesis, texture development, and phase compatibility *J. Europ. Ceram. Soc.* **20** 569–82

[3] Kolchin A A, Kiiko V M, Sarkissyan N S, Mileiko S T 2001 Oxide/oxide composites with fibres produced by using internal crystallization method *Compos. Sci. and Technol.* **61** 1079-72

[4] Popova N A, Tolstun A N 2007 A monazite imterphase in oxide/oxide composites with ICM-fibres *Sci. to Industry* **2** 58–61 (in Russian)

[5] Mileiko S, Kolchin A, Kiiko V, Novokhatskaya N 2017 Composite oxide fibres grown by internal crystallisation method *Fibers* **5** 48

[6] Kiiko V M, Kolchin A A, Mileiko S T, Novokhatskaya N I 2019 Composite oxide fibers and brittle matrix composites based on them *IOP Conf. Series: Materials Science and Engineering* **525** 012021

IOP Conf. Series: Materials Science and Engineering **848** (2020) 012031 doi:10.1088/1757-899X/848/1/012031

[7] Cook J, Gordon J E 1964 A mechanism for the control of crack propagation in all-brittle systems *Proc. of Roy. Soc.* A **282** 508–20
[8] Mileiko S T, Kazmin V I 1992 Crystallization of fibres inside a matrix: a new way of fabrication of composites *J. Mater. Sci.* **27** 2165–72
[9] Broek D 1974 *Elementary Engineering Fracture Mechanics* Leyden 74
[10] Kiiko V M, Mileiko S T 1999 Evaluation of room temperature strength of oxide fibres produced by the internal crystallization method *Compos. Sci. and Technol.* **59** 1977–81

IOP Conf. Series: Materials Science and Engineering **848** (2020) 012032 doi:10.1088/1757-899X/848/1/012032

Rheological properties of Si_3N_4 and Si_3N_4 with sintering additive CaO-Al_2O_3 powders

K A Kim[1], A S Lysenkov[1], D D Titov[1], E A Gumennikova[1,2], M G Frolova[1], D O Lemeshev[2], S N Perevislov[3] and Yu F Kargin[1]

[1]A.A. Baikov Institute of Metallurgy and Materials Science, Russian Academy of Sciences,
Moscow, Russia
[2]Mendeleev University of Chemical Technology of Russia, Moscow, Russia
[3]Institute of Silicate Chemistry of Russian Academy of Sciences, Saint Petersburg, Russia

e-mail: const552@gmail.com

Abstract
In this paper were considered the rheological characteristics of Si_3N_4 powders and the mixture of Si_3N_4 powders with sintering additive based on CaO-Al_2O_3. The dependence of optimal density on the applied stress during pressing was established. The powders compressibility coefficients were studied. It was shown that upon addition of a sintering additive CaO-Al_2O_3 the compressibility coefficient decreases and the compressibility modulus increases.

1. Introduction
Modern ceramic materials are in great demand because they have a wide variety of types and properties. In particular, structural ceramic materials based on silicon nitride Si_3N_4 are in demand, due to their physical and mechanical characteristics. Since ceramics are obtained from the initial ceramic powders, which are pressed previously in green body, the properties of ceramics directly depend on the accompanying processes of the powder particles deformation, the density of the green body and rheological properties.

The dilatometry and the sintering kinetic analyses of ceramics are very demanding on the quality of a green body, which should have a similar or identical density. The study of powders rheology, in turn, allows establishing the relationship between the applied pressure and the density of powder samples [1]. The rheology properties of the powder allow setting the relationship between applied stress σ and deformation of powder particles ε.

2. Experimental part
In this study, the powders based on silicon nitride Si_3N_4 (Figure 1A) and Si_3N_4 with 10 wt.% CaO-Al_2O_3 (sintering additive) mixture (Figure 2) were studied.

The Si_3N_4 powder was crystallized properly with fibrous particles structure, the average size was from 1,5-2 µm in a length and up to 200 µm in a width. Sintering additive CaO-Al_2O_3 was obtained by solid-phase synthesis and previously used in study [2-3]. The sintering additive particles had sizes from 100 up to 300 nm and represented as agglomerates with dimensions up to 3 µm (Figure 1B).

The samples prepared as follows: 0,15 g for each sample was pressed by INSTRON, which registered measures of applied pressure and powder deformation. Max pressing pressure was 200 MPa for each test, and the total quantity of tests amounted to 3 for both samples. The total height of the samples was

IOP Conf. Series: Materials Science and Engineering **848** (2020) 012032 doi:10.1088/1757-899X/848/1/012032

4,45 ± 0,2 for pure Si_3N_4 and 4,25 ± 0,2 for the mixture Si_3N_4 with 10 wt.% CaO-Al_2O_3. At least 3 measurements were carried out for each sample, then the values were averaged (the average value of the standard deviation of deformation in the entire interval of the experiment was σ_1= 0,26% and σ_2 =1,28% for Si_3N_4 powder and Si_3N_4 with 10 wt.% CaO-Al_2O_3, respectively).

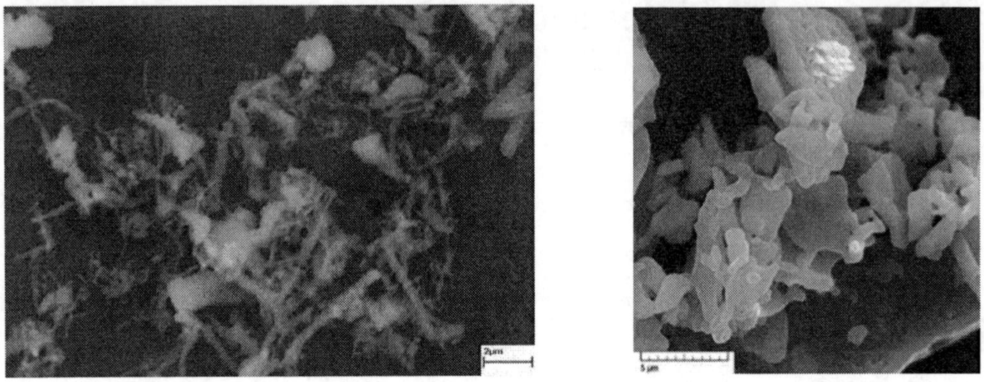

(A) (B)

Figure 1. SEM microstructure of powders: A – Si_3N_4 «Plasmoterm», B – sintering additive CaO-Al_2O_3

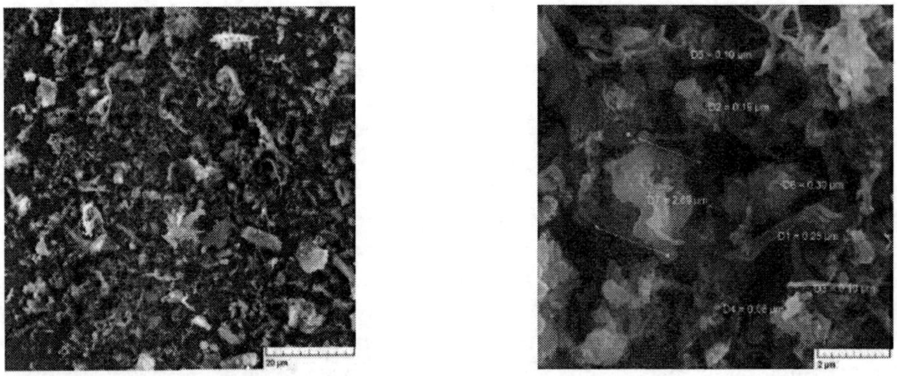

Figure 2. SEM microstructure of a mixture Si_3N_4 with 10 wt.% CaO-Al_2O_3 powders

3. Discussion of results

As a result of the measurements, stress-strain curves were constructed for pure silicon nitride and the mixture of silicon nitride with a sintering additive (Figure 3). The stress-strain curves represent a transition from linear to exponential increase, however, to study the compressibility modulus (*G*), compressibility coefficient (k_{strain}) and final strain (ε*), it is the linear section that corresponds to the first stage of deformation (Figure 4). At the first stage, an initial increase in stresses occurs and is accompanied by compaction of the structure due to the sliding of particles and filling of voids with smaller powder particles [4-5]. Further stages correspond to the exponential growth of deformation and density due to sliding, deformation, and growth of particle contacts.

Considering the fact that Si_3N_4 and CaO-Al_2O_3 powders have a completely different structure, this implies a difference in the values of deformation for pure Si_3N_4 and the mixture of Si_3N_4 with 10 wt. % CaO-Al_2O_3 powders. It is noticeable the Si_3N_4 powder had the highest strain value, and, consequently, the compression coefficient (Figure 3).

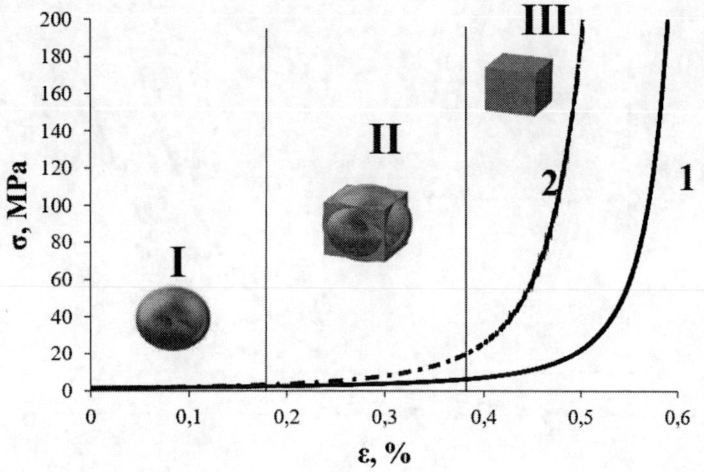

Figure 3. Stress-deformation curves of: 1 – Si_3N_4, 2 – Si_3N_4 + 10 wt. % $CaO-Al_2O_3$

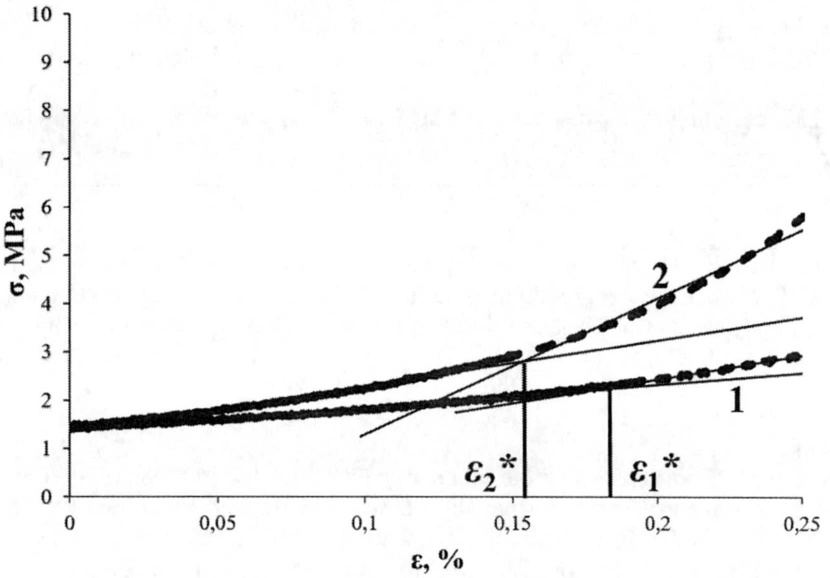

Figure 4. Linear sections of rheological stress-strain curves: 1 – Si_3N_4 «Plasmoterm», 2 – Si_3N_4 «Plasmoterm» with 10 wt.% $CaO-Al_2O_3$

The dependence of the relative density of the powders on the stress was established. It was found that samples from a mixture of Si_3N_4 with 10 wt. % $CaO-Al_2O_3$ had the highest density at the same stress value (Fig. 5). This dependence allows us to determine the optimal stress range, which is 60-120 MPa for the mixture and 100-200 MPa for pure Si_3N_4 to obtain a similar green body density (45%).

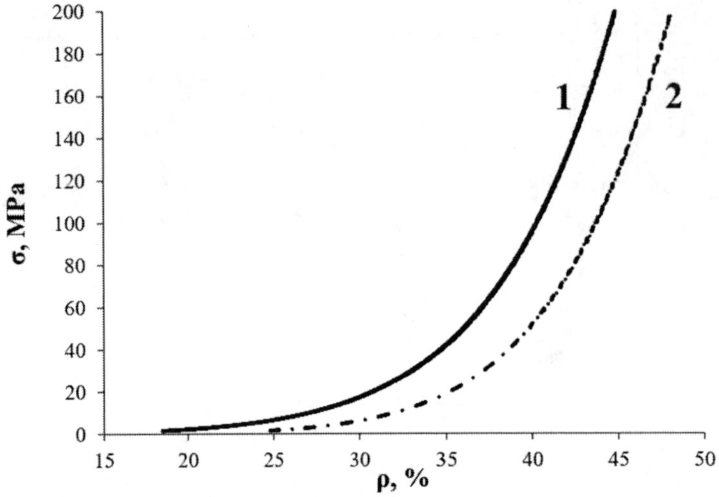

Figure 5. Dependence curves of the relative density of the powders on the stress: 1 – Si_3N_4, 2 – Si_3N_4 with 10 wt.% CaO-Al_2O_3

From the linear sections of the curves, the modulus of compressibility (*G*), compressibility coefficient (k_{strain}), and final strain values of the linear sections (ε^*) were determined (Tab. 1).

Table 1. The rheological characteristics of Si_3N_4 and Si_3N_4 + 10 wt.% CaO-Al_2O_3 powders

Powders	*G*, MPa	ε^*, %	k_{strain}, 10^{-3} Pa^{-1}
Si_3N_4	4,81	17,56	7,7
Si_3N_4 + 10 wt.% CaO-Al_2O_3	8,78	15,12	5,4

The sintering additive CaO-Al_2O_3 (10 wt.%) into Si_3N_4 powder leads to an increase in the elastic modulus and a decrease in the compressibility coefficient. It is associated with the structural features of the powders. Si_3N_4 powder particles have a fibrous structure and have a certain elasticity, which affects their compressibility. When granules of CaO-Al_2O_3 powder are added, voids are filled with spherical particles due to this fact the samples have the highest relative density.

4. Conclusions

- The rheological characteristics of the compressibility modulus (*G*), compressibility coefficient (k_{strain}), and final deformation (ε^*) were studied for Si3N4 powders and a mixture of Si_3N_4 + 10 wt. % CaO-Al_2O_3. It was found the sintering additive CaO-Al_2O_3 in an amount of 10 wt.% lead to decreases of the compressibility coefficient from $7{,}7 \cdot 10^{-3}$ to $5{,}4 \cdot 10^{-3}$ Pa^{-1} and the compressibility modulus increases from 4,81 to 8,78 MPa.
- The dependence of the powders relative density on the applied stress was established. The optimal stress range is 60-120 MPa for a mixture of Si_3N_4 + 10 wt.% CaO-Al_2O_3 and 100-200 MPa for pure Si_3N_4 to obtain the dense green body (40-45%).

Acknowledgments
The work was financially supported by the Grant «UMNIK» No. 13685GU/2018.

References

[1] Stolin A M, Bazhin P M, Pugachev D V 2009 *Russian Journal of Non-Ferrous Metals* **50** 290
[2] Lysenkov A S *et al.* 2018 *Journal of Physics: IOP Conf. Ser.* **1134** 012036
[3] Kargin Y F *et al.* 2010 *Inorganic Materials* **46** 799
[4] Kempen D, Piccolroaz A, Bigoni D 2019 *International Journal of Mechanical Sciences* **156** 146
[5] D D Titov et al. 2019 *IOP Conf. Ser.: Mater. Sci. Eng.* **525** 012077

IOP Conf. Series: Materials Science and Engineering **848** (2020) 012033 doi:10.1088/1757-899X/848/1/012033

Characterization of spherical stainless steel powders prepared by electric arc spraying process

A A Kirsankin[1], T A Kalaida[1], M A Kaplan[1], M A Smirnov[1], A Yu Ivannikov[1] and M A Sevostyanov[1,2]

[1]Baikov Institute of Metallurgy and Materials Science RAS;
[2]State Scientific Institution "All-Russian Research Institute of Phytopathology", Russian Agricultural Academy",st. Institute, own 5, Bolshie Vyazemy, Odintsovo distr., Moscow reg., 143050, Russia

E-mail: kirsankin@mail.ru

Abstract. Spherical powders were obtained by electric arc spraying of stainless-steel wire. The effect of the electrical power during the arc spraying process was investigated. The fractional composition of the obtained powder was determined by laser diffractometry. Scanning Auger microprobe was used to study surface morphology of the powder particles. The obtained spherical stainless-steel powders will be used to fabricate products by additive manufacturing methods.

1. Introduction

Stainless steel, as a class of ferrous alloys, is widely well-known for corrosion- and creep resistance, and for high-temperature applications. Due to the demand increasing of powder metallurgy components made from stainless steel in a variety of applications, including aerospace, automotive, chemical processing and biomedical field, this material became of great interest in the research domain. To obtain spherical metal powders, various methods can be used: gas atomization, plasma atomization, and plasma rotating electrode process. In these methods, the material of the tungsten cathode, the copper anode or the copper water-cooled crucible can get into the sprayed powder [1]. In this work, the properties of spherical stainless-steel powders prepared by the electric arc spraying process were studied. The main advantage of this method is the ability to obtain low impurity powder. The electric arc wire spraying method allows to obtain from 5 to 30 kg of powder per hour. In this work, the stainless-steel composition Fe-18%Cr-10%Ni-1%Ti (AISI 321) was used. Stainless steel powders are used to manufacture complex shapes products by different additive methods: selective laser melting, direct metal laser sintering or electron-beam melting [2]. The properties of the end product are determined by both characteristics of the metal powder and the type of the 3D printing process [3].

The aim of this work is to determine the properties of stainless steel Fe-18% Cr-10% Ni-1% Ti obtained by electric arc spraying. The wire was sprayed with an argon jet by means of the arc spraying process in the air.

2. Methods and materials

A commercial wire of Fe-18%Cr-10%Ni-1%Ti was chosen for spraying and atomization. This wire has a diameter of 1.6 mm. The chemical composition of steel is given in Table 1.

Table 1. The chemical composition of steel.

Cr, %	Ni, %	Cu, %	Mn, %	S, %	P, %	C, %	Si, %	Ti, %	Fe, %
18	10	<0.3	<2	<0.02	<0.035	<0.12	<0.8	1	Balance

The spraying equipment was the twin-wire arc spray system C-16-CT(M) (GMBh "Thermal-Spray-Tec", Moscow, Russia) with the arc spray gun. The maximum power of the arc spray gun is 29 kW. The arc spray gun had a cylindrical nozzle with a diameter 4.5 mm. The argon was used as the spray gas with the average flow range of 1 m^3/min. The spray process was in the cylindrical pipe in the air. The current was changed during the spraying process. The principal scheme of arc spraying process is given in Figure 1.

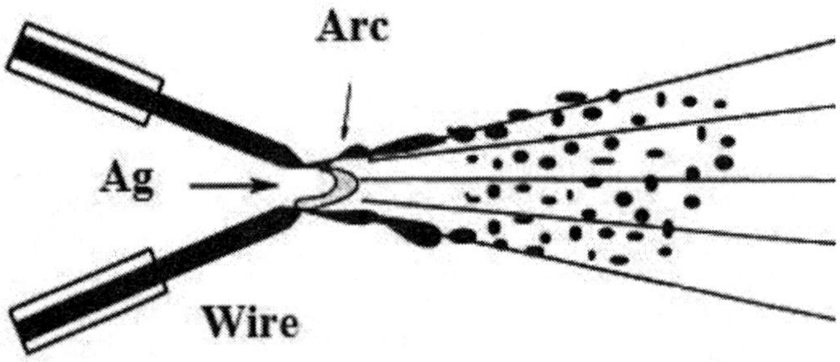

Figure 1. Principal scheme of arc spraying process.

The particle size distributions of powders were determined using an Analysette 22 NanoTec laser diffraction particle size analyzer from Fritsch. The particle size analyzer measuring cell contains the sample particles prepared by the dispersion unit which are irradiated with laser light. By changing the spacing between the detector and measuring cell, a different angle range of the scattered light is detected. The particle size distribution is calculated from this data.

The Leco TC-600 is a software-controlled instrument that measures both nitrogen and oxygen in a wide variety of metals, refractories, and other inorganic materials. The inert gas fusion principle is employed. A weighed sample, placed in a high-purity graphite crucible, is fused under a flowing helium gas stream at temperatures sufficient to release oxygen, nitrogen, and hydrogen. The oxygen in the sample combines with the carbon from the crucible forming primarily carbon monoxide (CO). In some instances, depending upon sample type and furnace temperature, some oxygen can be released directly as carbon dioxide (CO). The nitrogen present in the sample releases as molecular 2 nitrogen, and any hydrogen present is released as hydrogen gas.

Scanning Auger microprobe JEOL JAMP-9500F was used to study the surface morphology and to carry out the fractographic investigation of the specimens after the static tests. Powder particles were placed on double-sided carbon tape.

3. Results and discussion

The particle size distribution of spherical powder is shown in Figure 2. As can be seen, the particle size of the powder is in the range from 20 to 125 microns. Moreover, more than 60% of the powder particles have a size of 60 ± 10 μm.

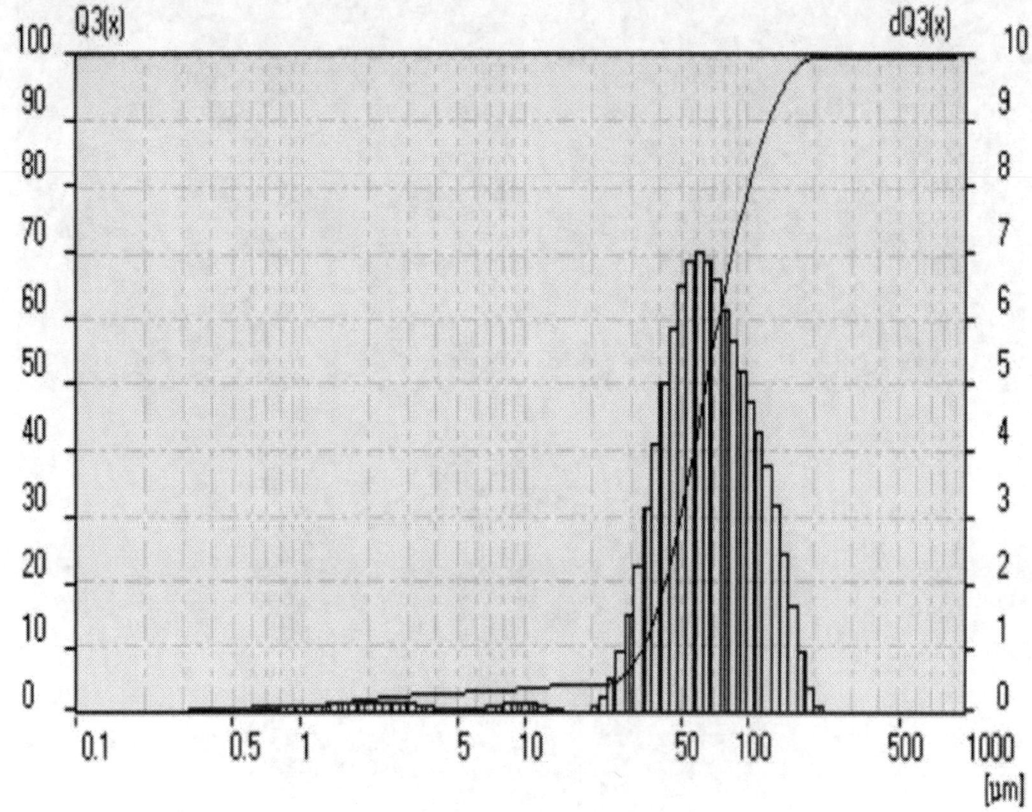

Figure 2. Particle size distribution of the Fe-18%Cr-10%Ni-1%Ti powder.

The determination of the content of gas-forming impurities in the initial wire and the obtained samples showed that during the spraying process, the powder particles are actively oxidized. So the oxygen content in the powder samples is more than 5%, while in the original wire the oxygen content was less than 0.005%. The nitrogen content in the powder samples and wire was 0.09% and 0.01%, respectively. The amount of carbon in the powder slightly decreased compared to its content in the wire - 0.05% and 0.07%. Since the process of spraying the wire took place in the air, the resulting particles were strongly oxidized. The carbon decreasing in powder particles compared to the wire is also due to oxidizing processes.

The study of the powder morphology was carried out using a JEOL JAMP-9500F scanning Auger microprobe. Images presented in figure 3 show that most particles have a spherical or near-spherical shape. However, there are some defects on the particle surfaces such as pores, holes, and cracks. The holes on the surface were formed due to the impacts between particles. There are no satellites on the surface of particles.

Analysis of this data showed that spherical particles can be obtained by the arc spray process. Particle size distribution complies with the requirements of the additive methods. To reduce particle oxidation, the atomization process must be carried out in an inert atmosphere.

a) b) c)

Figure 3. Images of Fe-18%Cr-10%Ni-1%Ti powder with typical spherical and near-spherical morphologies. a) – the scale bar represents 100 μm and the image is roughly 450 μm wide; b) – the scale bar represents 400 μm and the image is roughly 1800 μm wide; c) – the scale bar represents 200 μm and the image is roughly 900 μm wide.

4. Conclusion

The powder obtained by the electric arc spraying process from the Fe-18%Cr-10%Ni-1%Ti wire has a spherical and near-spherical shape. The spherical powder was formed during the solidification of liquid steel drops. The majority of particles has a size distribution from 20 to 60 μm. Therefore the particle size distribution complies with the requirements of the additive methods. Increased oxygen content impact on the morphology of the surface particles powder due to the mix of the argon and oxygen from the air.

Acknowledgments: The reported study was funded by RFBR, project number 19-38-90238.

IOP Conf. Series: Materials Science and Engineering **848** (2020) 012033 doi:10.1088/1757-899X/848/1/012033

References

[1] Liu Y et al 2015 Investigation into spatter behavior during selective laser melting of AISI 316L stainless steel powder *Materials & Design.* **87** pp 797-806
[2] Kaplan M A et al 2019 Properties of spherical stainless steel powders *IOP Conf. Ser. Materials Science and Engineering.* **525** (1) 012075
[3] Smirnov M A et al 2019 Investigation of the properties of heat-resistant spherical powders *IOP Conf. Ser. Materials Science and Engineering.* **525** (1) 012076

IOP Conf. Series: Materials Science and Engineering **848** (2020) 012034 doi:10.1088/1757-899X/848/1/012034

Electrochemical investigations of surfactants influence on copper ions cementation

E B Kolmachikhina[1*], E A Ryzhkova[1], D V Starkova[1] and V A Val'nev[1]

[1] Ural Federal University, 17, Mira St., Ekaterinburg, 620002, Russia
*Corresponding author: e.b.khazieva@urfu.ru

Abstract. The paper presents the results of experiments aimed to determine the effect of surfactants on the cementation rate and activation energy. The study was carried out on model solutions with the addition of surfactants. The results of the study showed that sodium lignosulfonate in a mixture with sodium dodecyl sulfate and sodium dodecylbenzene sulfonate increases the cementation rate over the entire range of studied concentrations. It is noted that the addition of individual surfactants leads to an increase in activation energy. Activation energy increases according to following order: sodium dodecylbenzene sulfonate → sodium dodecyl sulfate → sodium lignosulfonate.

1. Introduction

Cementation by zinc dust is widespread method for zinc solutions purification from ions of cobalt, copper and nickel. According to numerous investigations, organic impurities (incl. surfactants) have negative influence on cementation and zinc electrowinning results. Main cause of this phenomena is surfactants adsorption on zinc and copper particles [1,2]. Surfactant are used at pressure leaching of zinc sulfide concentrates to eliminate sulfur-sulfide pellets formation. Remaining surfactants in solution after leaching complicate further purification from ferrous and ferric ions, cementation purification from copper, cadmium, nickel and cobalt ions. Also surfactants could affect on results of zinc electrowinning.

Balkhash zinc plant (Kazakhstan) practices demonstrated that sodium lignosulfonate (SL) usage raises zinc extraction at leaching, but complicates cementation of solutions later on [3]. Presence of SL claims to enhance of zinc dust consumption at purification stages. SL significantly decreases degree of purification from cobalt and nickel ions [4].

Work results [5] confirm that presence of lignosulfonates degradation product leads to cementation velocity decreasing, zinc consumption increasing for copper ions reduction.

It was established [4, 6], that sodium dodecylsulfate (anionic surfactant), cetyl trimethylammonium bromide (cathionic surfactant) and Triton X-100 (nonionic surfactant) has negative influence of 32-94 % on copper cementation effectiveness.

Cementation purification in presence of organic reagents is complicated due to several reasons: decreasing of hydrogen overvoltage, organic impurities adsorption, stable complex compounds formation with metals ions and organic components.

The aim of this study is surfactants (SL, sodium dodecylsulfate (SDS), sodium dodecyl benzenesulfonate (SDBS) influence investigation on results of copper ions cementation.

2. Methods

Electrochemical investigations were carried out for cementation process tracking through current intensity movement during experiment.

IOP Conf. Series: Materials Science and Engineering **848** (2020) 012034 doi:10.1088/1757-899X/848/1/012034

For electrochemical investigations a model sulfate solution of 0.5 g·l^{-1} Cu^{2+} was used. Concentrations of SL, SDS and SDBS and their mixtures were varied from 50 to 400 mg·l^{-1}. Experiments temperature was fixed at 25, 45, 65 °C. Mixing rate was 500 rpm.

Tests were carried out on electrochemical workstation «ZIVE SP2». Work electrode (cathode) was made from zinc rod with diameter 0.4 cm, subsidiary electrode – from lead thin plate, reference electrode – silver chloride electrode.

Experiments were conducted under constant potential (0 V vs SCE). Solutions after experiments were analyzed on copper concentration by atomic absorption spectrometry (Analytik Jena, NovAA 300).

Cementation velocity was estimated through copper concentration (1) and current changing (Faraday's laws of electrolysis).

$$W = \frac{(C_0 - C_t) \cdot V}{t \cdot S}, \qquad (1)$$

W – specific cementation velocity, g·cm^{-2}·s^{-1};
V – solution volume, l;
C_o – initial copper concentration, g·l^{-1};
C_t – copper concentration at time t, g·l^{-1};
t – test duration, s;
S – disc area, cm^2.

3. Experiments results and discussions

Using STATISTICA software obtained results were presented as three-dimensional graph characteristic of cementation velocity from temperature and surfactants concentrations (figures 1-3).

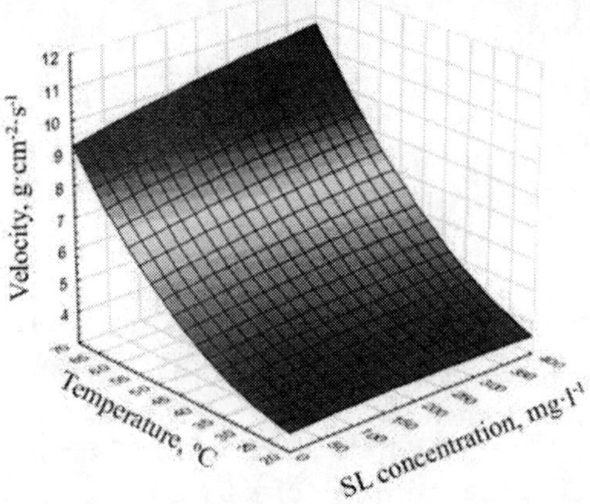

Figure 1. Dependence of copper cementation velocity from temperature and SL concentration

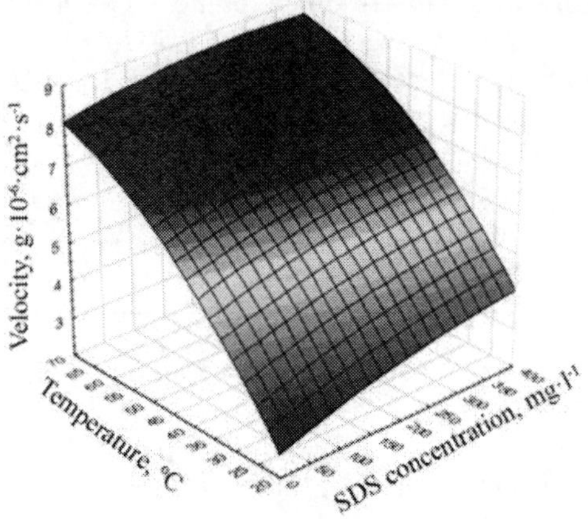

Figure 2. Dependence of copper cementation velocity from temperature and SDS concentration

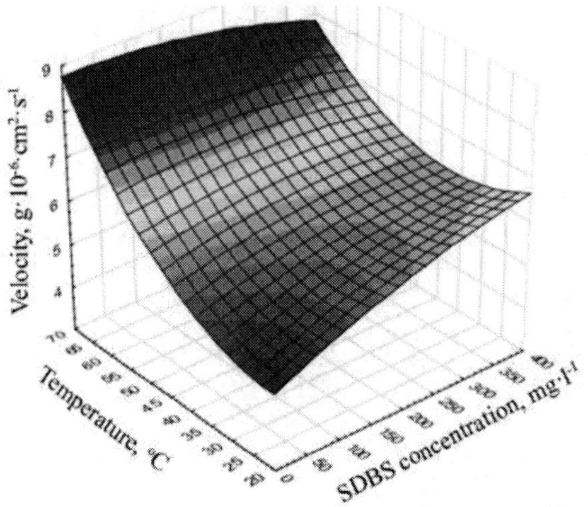

Figure 3. Dependence of copper cementation velocity from temperature and SDBS concentration

Obtained results testify that at 25 °C copper cementation velocity decreases from $4.06\cdot10^{-6}$, $g\cdot cm^{-2}\cdot s^{-1}$ without surfactants to ~$3.7\cdot10^{-6}$ $g\cdot cm^{-2}\cdot s^{-1}$ at all SL concentrations. However, at elevated temperature process rate increases from $5.24\cdot10^{-6}$ $g\cdot cm^{-2}\cdot s^{-1}$ to $6.01\cdot10^{-6}$ $g\cdot cm^{-2}\cdot s^{-1}$ at 45 °C and from $8.40\cdot10^{-6}$ $g\cdot cm^{-2}\cdot s^{-1}$ to $9.45\cdot10^{-6}$ $g\cdot cm^{-2}\cdot s^{-1}$ at 65 °C in absence and presence of 400 $mg\cdot l^{-1}$ SL, respectively.

SDS and SDBS usage at 25 °C and 45 °C leads to cementation velocity increasing. At SDS addition velocity increases at 25 °C and 45 °C on 21.8 % and on 31.6 %, respectively; at SDBS rate increases on 52.6 % and 24.6 % at 25 °C and 45 °C, respectively. However, SDBS at 65 °C leads to decreasing of copper cementation rate.

Also surfactant mixes influence was investigated, surfactants concentrations were in range from 50 to 400 $mg\cdot l^{-1}$ at 65 °C (figures 4, 5).

IOP Conf. Series: Materials Science and Engineering **848** (2020) 012034 doi:10.1088/1757-899X/848/1/012034

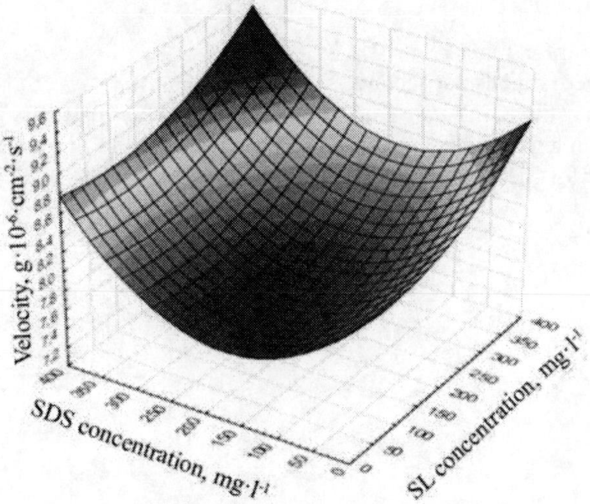

Figure 4. Cementation velocity dependence on LS and SDS concentrations

LS in mixes with SDS leads to cementation rate increasing at all range of studied concentrations. In some cases, the rates were lower than at surfactants absence ($8.40 \cdot 10^{-6}$ $g \cdot cm^{-2} \cdot s^{-1}$). Maximum rates were observed in presence of mixes of LS 50 $mg \cdot l^{-1}$+SDS 400 $mg \cdot l^{-1}$ ($8.64 \cdot 10^{-6}$ $g \cdot cm^{-2} \cdot s^{-1}$), LS 200 $mg \cdot l^{-1}$+SDS 400 $mg \cdot l^{-1}$ ($8.87 \cdot 10^{-6}$ $g \cdot cm^{-2} \cdot s^{-1}$), LS 400 $mg \cdot l^{-1}$+SDS 200 $mg \cdot l^{-1}$ ($8.92 \cdot 10^{-6}$ $g \cdot cm^{-2} \cdot s^{-1}$), LS 400 $mg \cdot l^{-1}$+SDS 400 $mg \cdot l^{-1}$ ($9.42 \cdot 10^{-6}$ $g \cdot cm^{-2} \cdot s^{-1}$) at 65 °C.

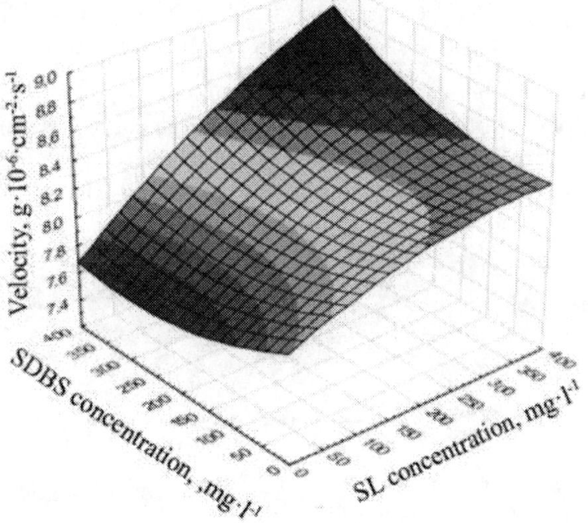

Figure 5. Cementation rate dependence on SL and SDBS concentration

It was established that SDBS with SL increases copper reduction rate at surfactants concentration enhancing. At low SL concentrations (50 $mg \cdot l^{-1}$) cementation rate was lower than at surfactants absence. At the same time, in case of SL usage (200, 400 $mg \cdot l^{-1}$) it was marked a slight rate increasing up to 4 %.

Experimental rate constant was estimated by followed equation (2):

$$K_e = \frac{W_{av}}{S \cdot C_t \cdot n^{0.5}}, \quad (2)$$

Ke – experimental rate constant;

Wav – average cementation rate, $mole \cdot cm^{-2} \cdot s^{-1}$;

IOP Conf. Series: Materials Science and Engineering **848** (2020) 012034 doi:10.1088/1757-899X/848/1/012034

S – disk area, cm^2;

Ct – copper concentration, $mole \cdot l^{-1}$;

n – disk rotation speed, rounds per second.

Energy of activation estimations were carried out by Arrhenius equation in logarithmical form (lgKe – 1/T). At figure 6 dependence of - lgKe on reciprocal temperature in SL presence is shown. analogous dependences were built for SDS, SDBS and their mixes with SL.

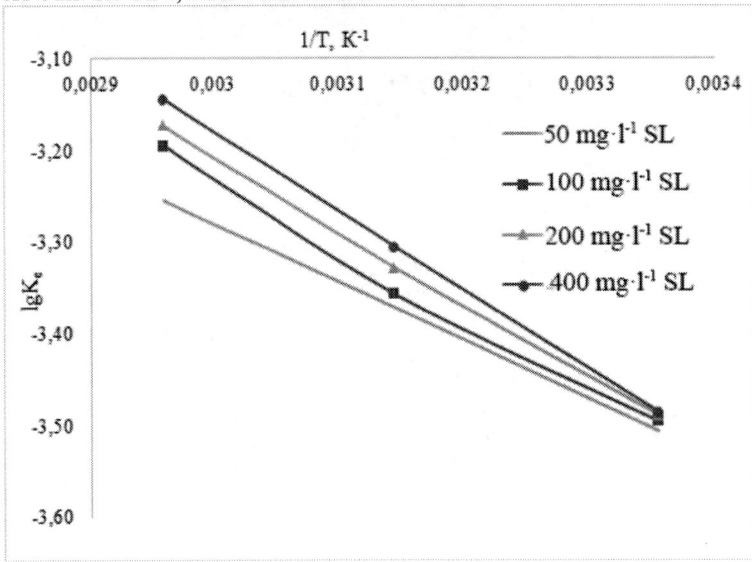

Figure 6. Dependence of - lgK_e on 1/T in SL presence

Addition of surfactants led to activation energy enhancing in the order SDBS→SDS→SL (figure 7). After achievement of 200 $mg \cdot l^{-1}$ of any surfactants an activation energy remains at a plateau, that could be related with reagents adsorption monolayer formation. Such effect was observed at surfactants mixes addition. Maximum activation energy was at SL addition (16.49 $kJ \cdot mole^{-1}$), at SDS usage – 11.23 $kJ \cdot mole^{-1}$, at SDBS – 10.84 $kJ \cdot mole^{-1}$.

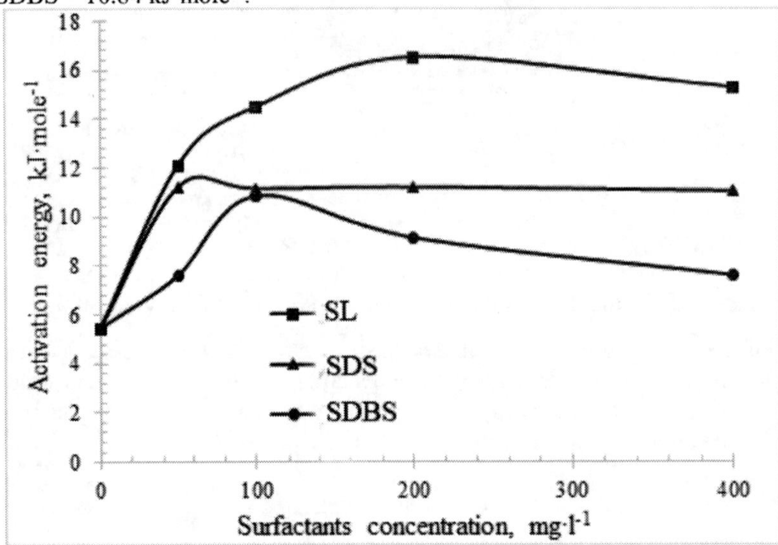

Figure 7. Activation energy dependence on surfactants concentration

Such influence was noted with all used mixtures: the higher SL concentration, the higher activation energy. Comparably with SL addition while using mixes remarkable activation energy increasing was observed: SL 50 mg·l^{-1}+SDS 200 mg·l^{-1} on 34.5 %, SL 200 mg·l^{-1}+SDS 200 mg·l^{-1} on 24.2 %, SL 400 mg·l^{-1}+SDS 200 mg·l^{-1} on 38 % (figure 8). While using of SL and SDBS mixes it was shown that there is an additive effect of reagents on energy activation growth (figure 9).

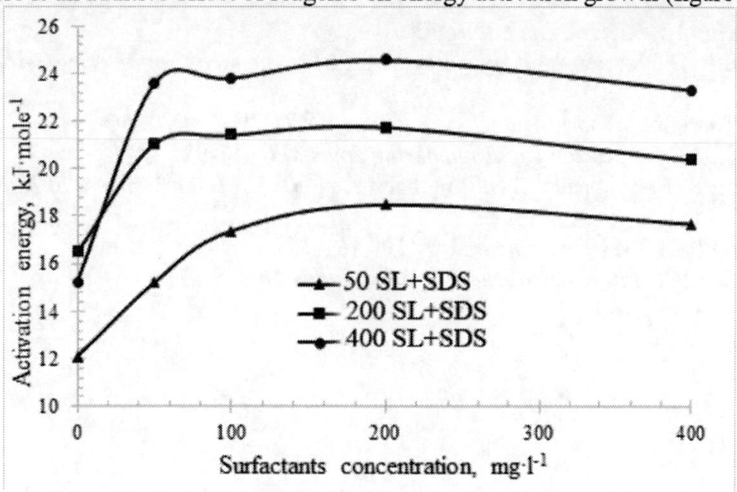

Figure 8. Activation energy dependence on concentration of SL and SDS mixes

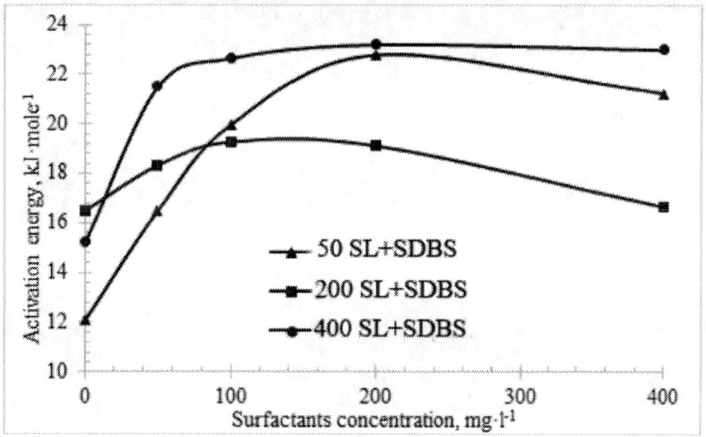

Figure 9. Energy activation energy on concentrations of SL and SDBS mixes

Energy activation increasing at SL usage could be associated with participating of SL function groups in donor-acceptor interactions with metal ions that leads to formation of complexes with Cu^{2+}, Co^{2+}, Zn^{2+}, Fe^{2+}, Fe^{3+} by means of sulfonate groups and phenolic hydroxyl groups [7]. At complexes formation activation energy increases due to stronger chemical bonding and it needs more energy for their breaking.

4. Conclusion

1. Surfactants addition (SDS, SDBS, SL) at 45 °C don't lead to copper cementation velocity decreasing;
2. Mixes of SL+SDS, SL+SDBS increases of cementation velocity;
3. Surfactant addition increases activation energy: SDBS→SDS→SL;
4. Surfactants concentration increasing up to 200 mg·l^{-1} leads to energy activation enhancing.

IOP Conf. Series: Materials Science and Engineering **848** (2020) 012034 doi:10.1088/1757-899X/848/1/012034

Acknowledgment
Leaching experiments and grade size analysis were conducted at the expense of the Russian Science Foundation grant (project No. 18-19-00186).

References
[1] Karavasteva M 1997 *Hydrometallurgy* **3** 91-98
[2] Demeev B B, Dauletbay A, Nauryzbaiev M K 2016 *Chemical Engineering Transactions* **47** 211-216
[3] Piskunov V M 2006 *Sbornik nauchnykh trudov VNIITCVETMETa* 56-58
[4] Drweesh M A 2004 *Alexandria Engineering Journal* **6** 917-925
[5] Khazieva E B, Naboichenko S S, Bolatbaev K N 2015 *Russian Journal of Non-Ferrous Metals* **56** 142-145
[6] Abdel-Aziz M H 2011 *Hydrometallurgy* **109** 161-167
[7] Khvan A M 1990 *Chemistry of Natural Compounds* **26** 575-577

IOP Conf. Series: Materials Science and Engineering **848** (2020) 012035 doi:10.1088/1757-899X/848/1/012035

Comparative analysis of the influence of urea additives on the properties of poly(vinyl alcohol) cryogels formed from the polymer solutions in water or dimethylsulfoxide media

O Yu Kolosova

A.N. Nesmeyanov Institute of Organoelement Compounds of Russian Academy of Sciences, Russian Federation

Abstract. Various poly(vinyl alcohol)-based gels are known for a long time and they are related to the gel materials of significant scientific and applied interests. Among different poly(vinyl alcohol) (PVA) hydrogels those formed via the so-called cryotropic gelation approach, i.e. PVA cryogels, attract special attention owing to, firstly, simplicity of their preparation and, secondly, very good properties of such matrices. The aim of this work was to compare the physicomechanical properties of PVA cryogels formed from the urea-containing aqueous and dimethylsolfoxide solutions of the polymer. It was shown that not only the solution's nature influences greatly on the properties of the resultant cryogels, but also the low-molecule additives in the initial polymer solutions lead to the opposite effect in cryogels characteristics. The results of our investigation can expend the area of PVA cryogels usage in many fields.

1. Introduction.

Poly(vinyl alcohol) (PVA) cryogels are the macroporous heterophase gels formed as a result of freezing of concentrated solutions of this polymer, their further incubation in a frozen state and subsequent thawing [1]. In this case, a prerequisite for the formation of such heterogeneous gels is the crystallization (freezing) of the bulk of the low molecular weight liquid (solvent). The physicochemical properties and porous structure of PVA cryogels depend on many factors, such as the characteristics of poly(vinyl alcohol) itself (its molecular weight, chain tacticity, the amount of unsaponificated O-acyl groups), polymer concentration, cryogenic treatment regimes, the number of freeze-thaw cycles [2]. Most often, aqueous media are used as initial solvents in the preparation of initial PVA solutions subjected to further cryogenic treatment. However, the literature describes PVA cryogels formed from polymer solutions in organic solvents, in particular in dimethylsulfoxide (DMSO) [3].

The physicomechanical as well as thermophysical properties of PVA cryogels formed from the DMSO solutions differ noticeably from the properties of PVA cryogels obtained in aqueous media. In particular, PVA cryogels formed in DMSO medium have lower rigidity and heat endurance than the samples formed from aqueous polymer solutions [4]. Since DMSO is a thermodynamically better solvent for poly(vinyl alcohol) than water, that is, the affinity of the polymer for DMSO is higher, this reduces the gelation efficiency due to competition between polymer-solvent and polymer-polymer interactions.

In addition, it is known that the introduction of low molecular weight substances into the initial PVA solution has an effect on cryotropic gelation and the properties of the resulting cryogels. For example, the effect of chaotropic and cosmotropic additives on the thermophysical and rheological properties of PVA cryogels formed aqueous media has been studied [5, 6], but only a few cases are known about the effect of low molecular weight additives on samples formed from the DMSO solutions of PVA [7]. Thus, the aim of this work was a comparative analysis of the effect of solvents and urea additives (Figure 1) on the properties of PVA cryogels formed from a polymer solution in water and DMSO containing urea in various concentrations.

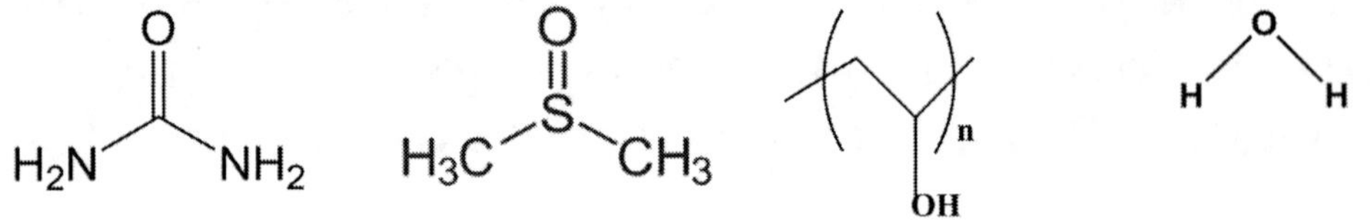

1 2 3 4

Figure 1. Chemical structures of: 1- urea, 2-dimethylsulfoxide, 3- poly(vinyl alcohol), 4 –water.

2. Materials and Methods

2.1 Materials

The following compounds were used: poly(vinyl alcohol) (molecular weight of ca. 86 kDa, the deacetylation degree of 100%; Acros Organics, USA), dimethyl sulfoxide (>99.8%; Komponent-Reaktiv, Russian Federation), urea (ultra pure grade; Sigma, USA), the Congo red dye (Aldrich, USA), as well as gelatine (photo quality), phenol (pure for analysis) and glycerol (pure for analysis) (all from Reakhim Co., Russian Federation).

2.2 Preparation of Feed PVA Solutions

The dissolution of PVA in DMSO or in water was performed as described elsewhere [8]. Briefly, a known amount of dry polymer was dispersed in a calculated volume of DMSO to reach a PVA concentration of 100 g/L. The mixture was incubated for 18 h at room temperature for swelling of the polymer, and then, the system was heated for 1 h on a boiling water bath under stirring until the completion of PVA dissolution. When preparing the feed solutions that should contain urea the required amounts of additive were dissolved in the polymer solution at room temperature. The final solutions were sonicated for 20 min at room temperature in an UNITRA ultrasonic bath (Unitra, Poland) to remove air bubbles.

2.3 Preparation of PVA Cryogels

The samples of PVA cryogels were prepared analogously to the earlier described techniques [8]. In these experiments, samples for physicomechanical tests were formed in sectional duralumin molds (inner diameter 15 mm, height 10 mm). To determine T_f values, cryogels were formed in transparent polyethylene test tubes (inner diameter 10 mm), 3-mL portions of the polymer solution were poured, and a stainless steel ball (diameter 3.5 mm, weight 0.275 ± 0.005 g) was placed on the bottom of each tube. The containers and the tubes were put into the chamber of an FP 45 HP precision programmable cryostat (Julabo, Germany), and the samples were frozen and incubated at a preset temperature for 12 h. Then, the temperature was raised to 20°C at a rate of 0.03 °C/min controlled by the cryostat microprocessor.

2.4 Characterization of PVA cryogels samples

The compression Young's modulus (E) and the fusion temperature (T_f) of PVACG samples were evaluated in accordance with the protocols described elsewhere [8]. The E and T_f values were measured for three samples; the samples were examined in three to five independent experiments. The results obtained were averaged.

In brief, the E modulus of the PVACG samples was determined from the linear portion of the stress–strain dependence that was found using a TA-Plus automatic texture analyzer (Lloyd Instruments, UK) at a loading rate of 0.2 mm/min. The tests were performed until reaching a 30% deformation. The measurements were accomplished either for PVA cryogels prepared in water medium, either for the PVACGs prepared in DMSO and also for the samples in which the organic

IOP Conf. Series: Materials Science and Engineering **848** (2020) 012035 doi:10.1088/1757-899X/848/1/012035

liquid phase (that is, DMSO and the dissolved solutes) was replaced by pure water. In the latter case, the cryogel samples were incubated for seven days at room temperature in glass beakers that each contained deionized water (100-fold excess relative to the volume of the PVACG sample); the water was replaced with a fresh portion every day. The fusion temperatures of the PVACG samples were measured by placing the tightly corked polyethylene tube containing cryogel with the stainless steel ball at the bottom upside down into the water bath. The bath temperature was increased at a rate of 0.4 °C/min. The gel fusion point was determined as the temperature when the ball fell down onto the stopper of the test tube after passing through the fused gel.

2.5 Macroporous morphology of PVACG samples

The macroporous morphology of PVA cryogels was examined using an Eclipse 55i (Nikon, Japan) optical microscope equipped with digital photocamera as it was described in the previous study [9]. DMSO in the respective PVACGs has been replaced for pure water prior to the thin sections preparation. Such sections of the 10 μm thickness were cut in the direction orthogonal to the axis of cylindrical samples using a cryomicrotome SM-1900 (Leica, Germany). Each section was placed on the microscope glass, which was then immersed into 1% aqueous solution of Congo red for staining for 10 seconds and then rinsed with pure water. The excess of liquid was removed with filter paper. Then the section was poured with one drop of "fixing medium" (solution of 1 g of gelatin in 12 mL of 50% aqueous glycerol and 0.2 g of phenol as a bactereostatic agent) and sealed with a cover glass. Prior to studies, the samples were stored at 4°C in a closed vessel.

3. Discussion

3.1 Features of the influence of urea on the properties of PVA cryogels in an aqueous medium.
Urea is a classical chaotropic agent; in an aqueous medium it inhibits intermolecular hydrogen bonding of the OH groups of the polymer, which leads to a decrease in the stiffness and heat endurance of the resulting cryogels. With an increase in the urea concentration in the initial polymer solution, the elastic moduli (E) and the melting temperature (T_f) of the resulting PVA cryogels decrease.

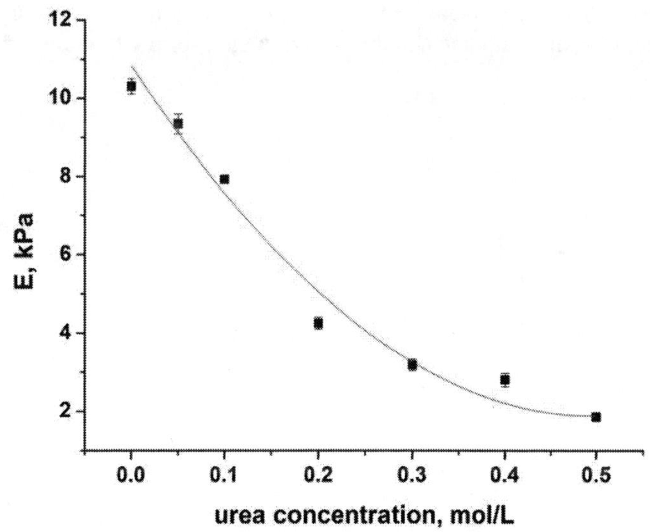

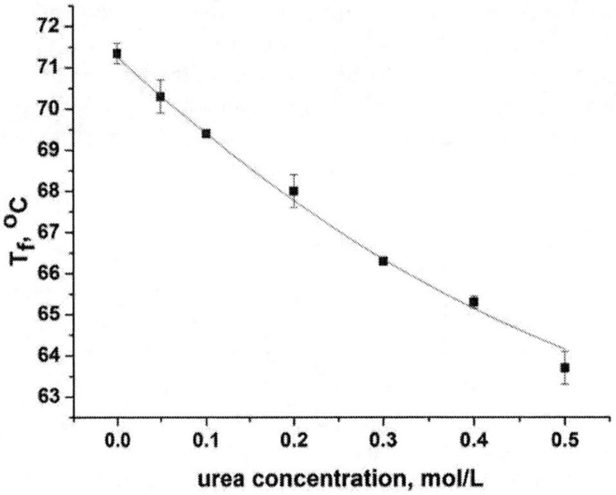

Figure 2. Dependences of the elastic modulus E and the fusion temperature T_f of PVA cryogels formed in the aqueous medium at -20°C on the urea concentration in the initial polymer solution.

3.2 Features of the effect of urea on the properties of PVA cryogels in dimethylsulfoxide.

Since the crystallization temperature of DMSO is T_0 = 18.4°C, describing the cryogenic treatment mode of such systems, it is more convenient to use not the absolute value of the temperature, but the difference value $\Delta T = T_i - T_0$, where T_0 and T_i are, respectively, the crystallization temperature of the pure solvent and the temperature of a specific experiment (expressed in the Celsius degrees). In our case, the cryogenic treatment temperature was -21.6°C, thus ΔT = -40°. PVA cryogels were formed under the following conditions: a concentration of PVA of 10 g/dL, a freezing temperature of -21.6°C, a freezing time was 12 h, a heating rate upon the system thawing was 0.03°C/min.

When studying the effect of urea on the properties of PVA cryogels formed in its presence, we conducted a series of experiments varying the concentration of a low molecular weight agent; moreover, it should be noted that the range of working concentrations of urea in dimethylsulfoxide was significantly higher than in an aqueous medium, and ranged from 1 to 4 mol/L.

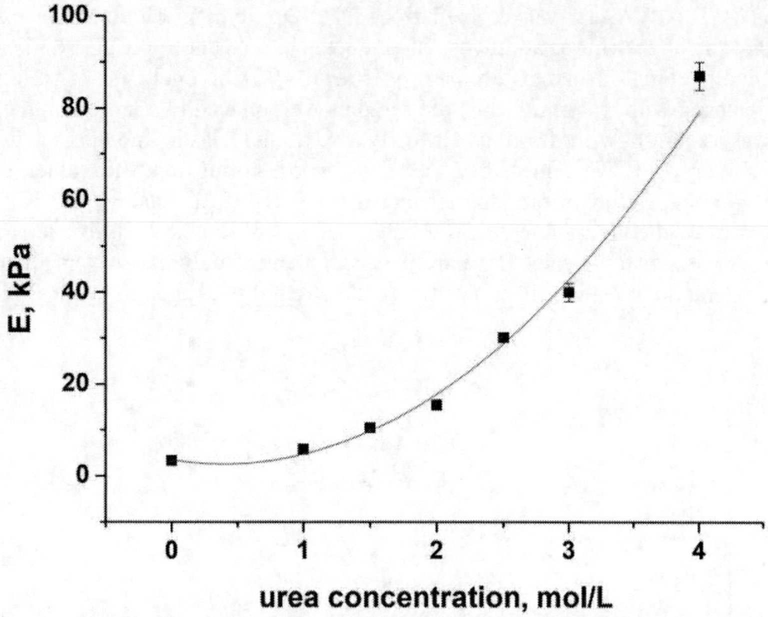

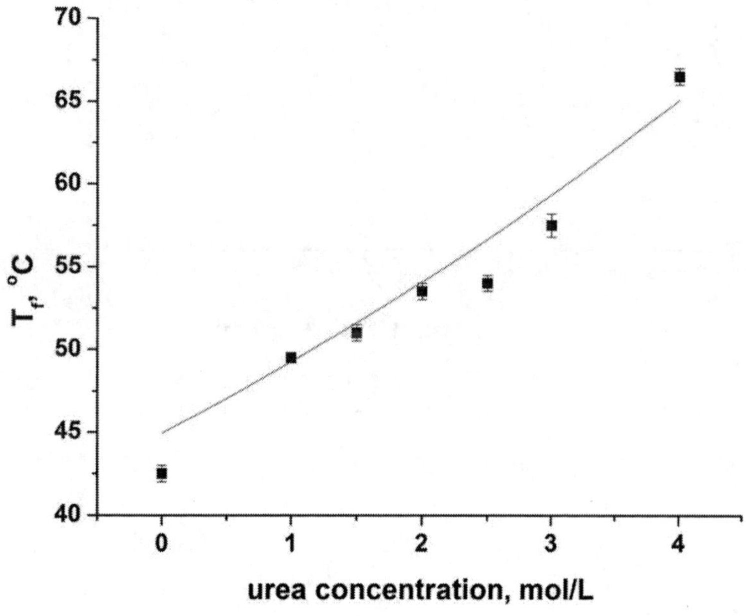

Figure 3. Dependences of the elastic modulus E and the fusion temperature T_f of PVA cryogels formed in DMSO at -21.6°C on the urea concentration in the initial polymer solution.

As follows from the diagrams in Figure 3, in the case of PVA cryogels formed from DMSO polymer solutions, with an increase in the urea concentration in the samples, in contrast to the effects in the aqueous medium (Fig. 2), the values of both the fusion temperature and the elastic modulus increase. So, when the urea content in the sample was 4 mol/L, the melting temperature increased by 20°C compared with the PVA cryogel formed without additives, and the elastic modulus E increased by 26 times. That is, for PVA cryogels formed from DMSO polymer solutions, we discovered the anomalous nature of the effect of urea additives when it acted as a cosmotropic agent.

In order to study the effect of solvents on the properties of PVA cryogels, the organic solvent from the samples was replaced with water. For this, the samples were placed in a large volume of water for a week; water was replaced with fresh water daily. Next, the elastic modulus and the melting temperature of PVA cryogels were measured. So, it is worth noting that the values of the elastic moduli and the fusion temperature of the samples increased to 210±5 kPa and 83±0.5 °C, respectively. Thus, the replacement of the thermodynamically better solvent of dimethyl sulfoxide (for poly(vinyl alcohol)) with the worst - water - leads to a partial replacement of polymer-solvent interactions with polymer-polymer interactions, which, in turn, affects the strength and thermophysical characteristics of PVA cryogels.

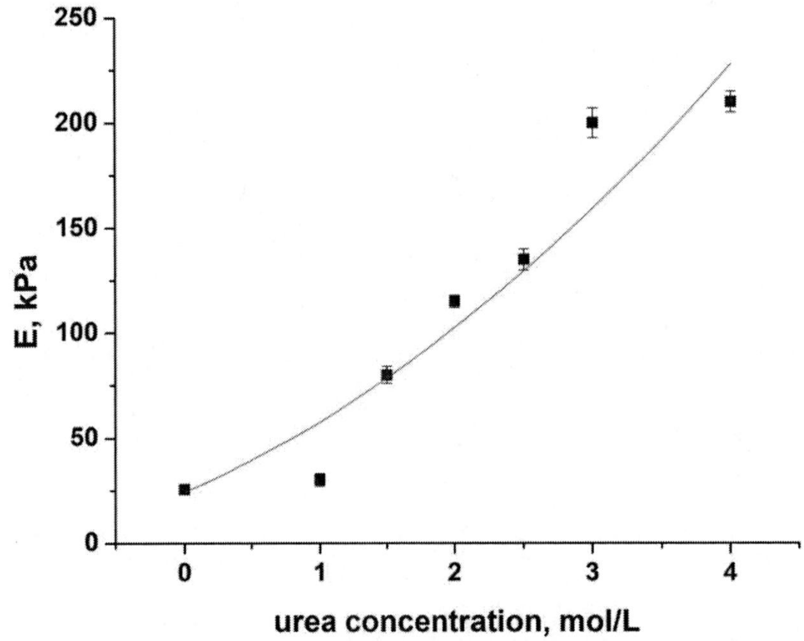

IOP Conf. Series: Materials Science and Engineering **848** (2020) 012035 doi:10.1088/1757-899X/848/1/012035

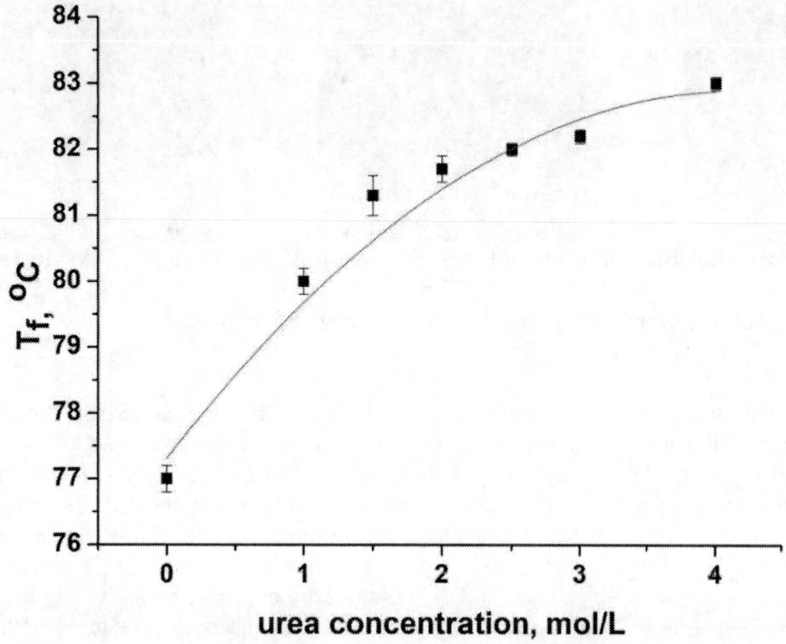

Figure 4. Dependences of elastic modulus E and fusion temperature T_f of PVA cryogels formed in DMSO at -21.6°C on urea concentration after replacing DMSO with water.

3.3 Peculiarities of the morphology of PVA cryogels formed in aqueous and organic media.

Figures 5 and 6 show microphotographs of thin sections of PVA cryogels equiconcentrated with respect to the polymer and formed in aqueous and organic media both without urea and in its presence. The urea concentration was 0.1 or 0.3 mol/L for aqueous solutions of poly(vinyl alcohol), or 1 and 3 mol/L for a solution of PVA in dimethyl sulfoxide. Microphotographs of thin sections perpendicular to the heat transfer gradient during freezing of the samples show a pattern of alternation of dark (gel phase stained with Congo red) and light (macropores) structural elements of various shapes and sizes.

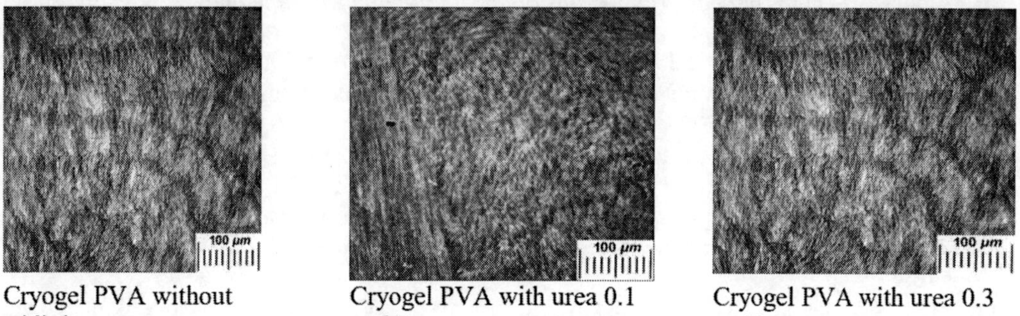

Cryogel PVA without addittives	Cryogel PVA with urea 0.1 mol/L	Cryogel PVA with urea 0.3 mol/L

Figure 5. Microphotographs of PVA cryogels formed in an aqueous medium.

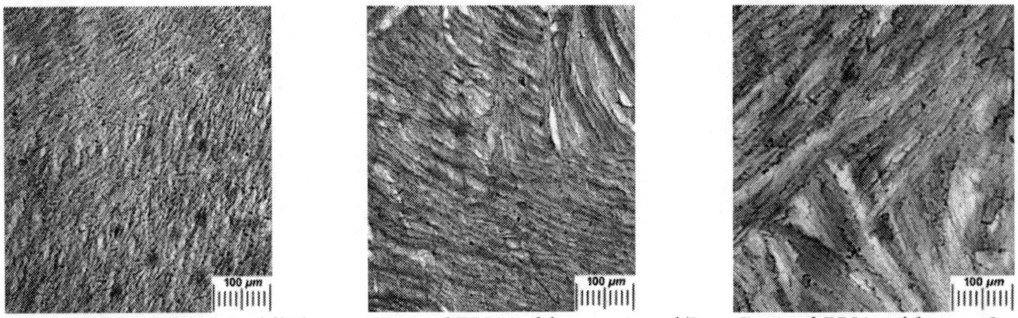

Cryogel PVA without addittives   Cryogel PVA with urea 1 mol/L   Cryogel PVA with urea 3 mol/L

Figure 6. Microphotographs of PVA cryogels formed in an organic medium.

4. Conclusion

Thus, the strong influence of dimethylsulfoxide in the presence of urea on the properties of the obtained PVA cryogels can be explained by hydrogen bonding between urea and dimethylsulfoxide molecules, which makes the latter a thermodynamically poorer solvent for poly(vinyl alcohol).

A study of the influence of the nature of the solvent on the physicochemical characteristics of PVA cryogels formed from polymer solutions in DMSO with urea additives of various concentrations showed that heterophase cryogels with higher rigidity and heat endurance than cryogels of DMSO – water composition are formed. Since various PVA cryogels are currently being widely used, primarily as materials of biotechnological and biomedical applications, the discovered features of the properties of PVA cryogels formed in an organic medium can serve as the basis for the development of such gel materials, but with new characteristics useful for practice.

5. Acknowledgments.

This work was supported by the Ministry of Science and Higher Education of the Russian Federation.

6. References.

[1] Lozinsky V I 1998 *Polymeric cryogels as a new family of macroporous and supermacroporous materials for biotechnological purposes* Russ Chem Bull vol. 57 (Springer-Verlag GmbH Heidelberg, Germany) p. 1015

[2] Hassan C N and Peppas N A 2000 *Structure and applications ofpoly(vinyl alcohol) hydrogels produced by conventional crosslinking or by freezing/thawing methods* Adv. Polym. Sci. (Springer) vol.153. P.37

[3] Horii F, Masuda K and Kaji H 1997 *CP/MAS 13C NMR spectra of frozen solutions of poly(vinyl alcohol) with different tacticities* Macromolecules (Acs Publications) vol 30 P 2519

[4] Trieu H and Qutubuddin S 1995 *Poly(vinyl alcohol) hydrogels: effects of processing parameters on structure and properties.* Polymer (Elsevier) vol. 36. p. 2531

[5] Kolosova O Yu, Kurochkin I N, Kurochkin I I and Lozinsky V I 2018 *Cryostructuring of polymeric systems. 48. Influence of organic chaotropes and kosmotropes on the cryotropic gel-formation of aqueous poly (vinyl alcohol) solutions.* Europ. Polym. J. (Elsevier) vol. 102 p. 169

[6] Lozinsky V I, Sakhno N G, Damshkaln L G, Bakeeva I V, Zubov V P, Kurochkin I N and Kurochkin I I *Study of cryostructuring of polymer systems: 31. Effect of additives of alkali metal chlorides on physicochemical properties and morphology of poly(vinyl alcohol) cryogels* Coll. J. vol. 73 (Pleiades Publishing, Ltd (Road Town, United Kingdom) p. 234

[7] Lozinsky V I, Leonova I M, Ivanov R I and Bakeeva I V 2017 *A study of cryostructuring of polymer systems. 46. Physicochemical properties and microstructure of poly(vinyl alcohol) cryo-gels formed from polymer solutions in mixtures of dimethyl sulfoxide with low-*

IOP Conf. Series: Materials Science and Engineering **848** (2020) 012035 doi:10.1088/1757-899X/848/1/012035

molecular-mass alcohols Coll J vol 79 (Pleiades Publishing, Ltd Road Town, United Kingdom) p. 788

[8] Lozinsky V I, Domotenko L V, Zubov A L and Simenel I A 1996 *Study of cryostructuration of polymer systems. XII. Poly(vinyl alcohol) cryogels: Influence of low-molecular electrolytes* JAPS vol 61 (John Wiley & Sons Inc. United States) p. 1991

[9] Masri C, Chagnon G and Favier D 2017 *Influence of processing parameters on the macroscopic mechanical behavior of PVA hydrogels* Mater. Sci. Eng., C. (Elsevier) vol. *75* p. 769

IOP Conf. Series: Materials Science and Engineering **848** (2020) 012036 doi:10.1088/1757-899X/848/1/012036

The ratio of refractive index and molecular weight of copolymers of n-butyl acrylate and styrene obtained by the method of RAFT-radical polymerization

E V Vaganov, R I Komendant, E O Perepelitsina, V P Grachev and S A Kurochkin

Institute of Problems of Chemical Physics, RAS, Academician Semenov avenue 1, Chernogolovka, Moscow region, 142432 Russian Federation

E-mail vaganov@icp.ac.ru, komendant@icp.ac.ru, oligo@icp.ac.ru

Abstract. The kinetic and molecular weight characteristics of the copolymers of n-butyl acrylate and styrene (95: 5 wt.%) obtained by the method of radical polymerization with reversible chain transfer were studied. The dependence of the refractive index on the molecular weight of the copolymers of n-butyl acrylate and styrene is shown.

1. Introduction

A promising method for controlling the processes of macromolecule formation is radical polymerization proceeding by the mechanism reversible addition fragmentation chain transfer (RAFT). The main difference between RAFT polymerization and conventional radical polymerization is the suppression of the quadratic death of macroradicals due to their reversible interaction with special substances – RAFT agents, compounds with di- or tri- tiocarbonate groups [1-10]. As a result of the addition of a macroradical in the double bond C=S, radical intermediates are formed, the fragmentation of which leads to a reactivation of the macroradicals, which further participate in the chain propagation reaction. The use of these processes makes it possible to effectively control copolymerization to obtain compositionally uniform copolymers with given molecular weights and a narrow molecular weight distribution from monomers that differ in their activity [10–16], which is difficult to achieve with traditional synthetic methods.

Light refractive index (n_D) is one of the most important characteristics of any transparent material, including polymeric, used for the manufacture of various optical elements. At constant temperature and pressure, the value of the refractive index depends on the composition and chemical nature of the polymer.

The purpose of this work is to establish the ratio of refractive index and molecular weight of butyl acrylate and styrene copolymers obtained by radical RAFT copolymerization. These elastomers of n-butyl acrylate and styrene can be used as promising high-impact modifying additives in the production of polymethyl methacrylate glasses [17]. In order to maintain the optical transparency of plexiglass, it is necessary that the elastomeric additive be compositionally uniform and have a refractive index equal to that of polymethyl methacrylate (PMMA) [18].

2. Materials and Methods

As the main monomer used n-butyl acrylate (BA) (Aldrich ≥ 99%). Styrene (ST) was used as a comonomer that regulates the refractive index of an elastomer, which was purified from hydroquinone with a 10% aqueous NaOH solution, washed with distilled water until neutral, dried over calcined $CaCl_2$, distilled and thawed in vacuum (<0.01 mm. hg).

Dibenzyl trithiocarbonate (BTC) synthesized by a known method [19, 20] was used as an agent for reversible transfer of the radical polymerization chain. 2,2′-azo-bis-isobutyronitrile (AIBN), which was purified from impurities by recrystallization from ethanol, was used as an initiator.

The radical copolymerization of BA and ST was carried out in a toluene solution in sealed glass ampoules. Oxygen dissolved in the mixture was removed in a vacuum unit with repeated repetition of the freeze-thaw cycle in liquid nitrogen to a residual pressure of 0.01 mm hg. The calorimetric method in isothermal mode (T = 80°C) on a Calve DAK-1-1 microcalorimeter recorded the heat release rate, from where the reduced copolymerization rate was determined from the known polymerization heats of BA and ST (83.3 and 73.8 kJ/mol, respectively [21]) and built kinetic curves. Companion ampoules in which ~ 5 g of copolymers were synthesized were placed in an air thermostat (T=80°C), and polymerization was carried out to the maximum conversion.

The molecular mass characteristics of the synthesized copolymers were determined by size-exclusion chromatography (SEC) on a Waters Alliance GPCV 2000 chromatograph equipped with a differential refractometer (*RI*) and a multi-angle laser light scattering detector (*MALLS*).

The refractive index (n_D) of the copolymers was determined using an IRF-22 laboratory refractometer (wavelength 589.3 nm) at T = 20°C.

3. Results and discussion

We synthesized copolymers of BA and ST (95:5 wt.%) in a toluene solution by radical polymerization in reversible addition fragmentation chain transfer (RAFT). By varying the initial concentration of the RAFT agent calculated from the law of "living" chains, the molecular weight of the copolymers was varied within the range of *MM* = 5000-200000. The conditions for the synthesis of copolymers are presented in table 1.

Table 1. - Initial conditions for the radical RAFT copolymerization of BA and ST. $[AIBN]_0$=0.03 mol/l, T = 80 ° C.

№	M_{cal}, g/mol	[BA], mol/l	[ST], mol/l	[BTC], mol/l
1	5000			0.078
2	10000			0.039
3	50000	3.27	0.21	0.008
4	100000			0.004
5	200000			0.002

The kinetic curves (Figure 1) show that the initial polymerization rate in all cases is almost the same value, which is due to the same initial concentration of the initiator. However, an increase in the reduced polymerization rate is observed, the intensity of which depends on the initial concentration of the RAFT agent. The lower the concentration of BTC, the stronger the growth of W/[M]. The maximum values of the reduced polymerization rate also depend on the concentration of BTC, which is due to the partial death of growth radicals by reaction with the intermediate [22–26], which leads to a slowdown in the polymerization reaction. On the whole, an increase in the polymerization rate is not associated with the effect of the auto-acceleration of polymerization at a high degree of conversion, due to the diffusion-controlled reaction of quadratic chain termination, since the reaction proceeds in a 50% toluene solution, but is associated with a change in the current ratio of BA and ST concentration since their copolymerization constants (r_1 = 0.2, r_2 = 0.8) [21] contribute to the formation of an alternating copolymer with the predominant incorporation of styrene units into the composition of the copolymer.

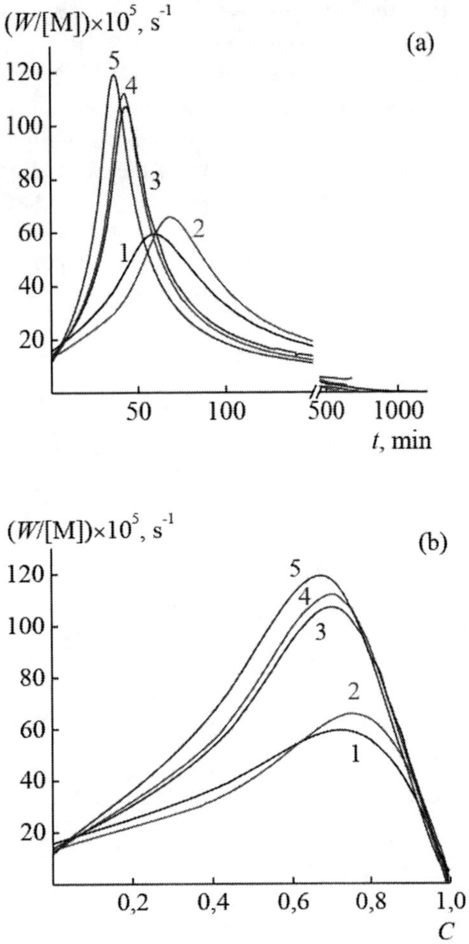

Figure 1. Kinetic curves of the radical RAFT copolymerization of BA and ST in the coordinates "reduced polymerization rate *W*/[M] - time *t*" (a), "reduced polymerization rate *W*/[M] - double bond conversion *C*" (b). Conditions according to table 1.

Therefore, when styrene is deficient, its conversion exceeds the conversion of n-butyl acrylate and the monomer ratio increases in favor of n-butyl acrylate, whose homopolymerization constant is more than an order of magnitude higher than styrene homopolymerization constant (k_{pBA} = 5800 l/mol s, k_{pST} = 300 l/mol s) [21].

The use of RAFT polymerization makes it possible to obtain compositionally uniform copolymers as compared to classical radical polymerization, in which compositionally nonuniform copolymers are formed due to the difference in copolymerization constants [21]. In compositionally nonuniform copolymers, some of the polymer chains that make up the polymer product can have different refractive indices that affect the optical properties of the polymer product [27, 28].

SEC analysis of the samples showed that the molecular weight distribution curves of BA and ST copolymers are unimodal in nature and have a narrow distribution. It can be seen (Figure 2) that the chromatographic curves shift toward higher molecular weights during the transition from experiment 1 to experiment 5 (Table 1).

It is seen (see Table 2) that the absolute molecular masses of the copolymers in experiments 1 and 2 coincide with the set values and the polydispersity M_w/M_n is 1.2, which indicates the implementation of "living" chains mode under the given conditions of copolymerization. With a decrease in the concentration of the RAFT agent (BTC), the polydispersity parameter increases, which is due to the occurrence of the mixed "living" mode and conventional radical polymerization. This is also confirmed by the fact that the experimental molecular weights of the copolymers in experiments 3-5 strongly differ from the calculated molecular weights. The main reason is the high concentration of AIBN initiator, which did not change from experiment to experiment and was chosen for polymerization under the conditions of experiment 1 at the highest concentration of BTC.

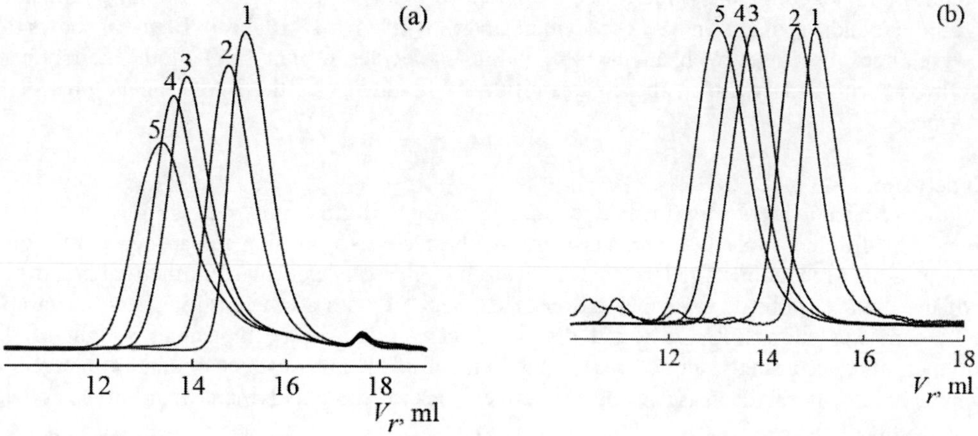

Figure 2. SEC curves of BA and ST copolymers synthesized under the conditions of experiments in Table 1. detected by a refractometric detector (RI) (a) and a multi-angle laser light scattering detector (MALLS) (b).

Table 2. - Molecular mass characteristics of BA and ST copolymers synthesized under the experimental conditions in table 1.

№	*RI*			*RI+MALLS*		
	M_n	M_W	*PDI*	M_n	M_W	*PDI*
1	3200	5400	1.7	5400	6500	1.2
2	4100	9200	2.2	9400	10900	1.2
3	5900	21200	3.6	16700	24700	1.5
4	6000	24700	4.1	19400	30400	1.6
5	5800	29600	5.1	22300	36800	1.7

IR spectra (Figure 3) of BA and ST copolymers were recorded in the regime of impaired total internal reflection. The IR spectra show characteristic absorption bands of the functional groups of poly(n-butyl acrylate) in the region of 1730 cm^{-1}, corresponding to stretching vibrations of the C=O bond and polystyrene in the region of 700 cm^{-1}, corresponding to vibrations of C-H bonds. IR spectra with different molecular weights are of a similar nature, which indicates the inclusion in the composition of the copolymers of units of both monomers and a qualitatively identical composition of BA and ST copolymers of different molecular weights.

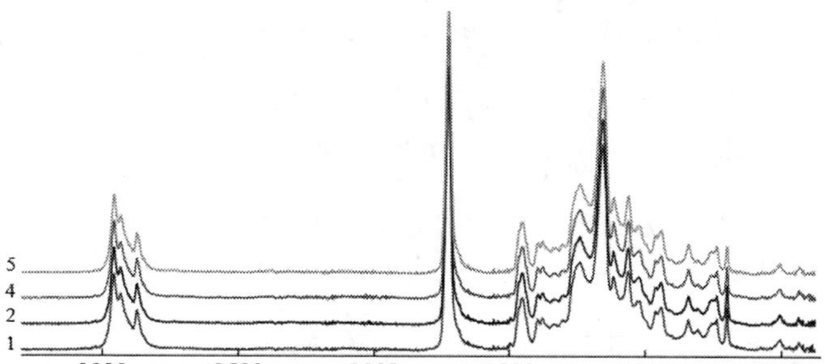

Figure 3. IR spectra of BA and ST copolymers synthesized under the experimental conditions in Table 1.

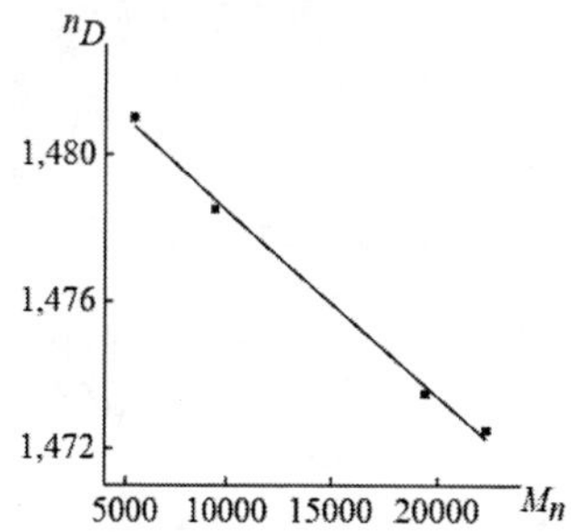

Figure 4. The dependence of the refractive index n_D on the molecular weight M_n of the BA and ST copolymer.

For BA and ST copolymers, the refractive index (n_D) was determined on an Abbe refractometer. The refractive indices of BA and ST copolymers are $n_D = 1.481 \div 1.472$. From Figure 4 shows that the refractive index decreases with an increase in the molecular weight (M_n) of the copolymer and, accordingly, with a decrease in the concentration of fragments of the RAFT agent present in the copolymer.

4. Conclusion

Using the RAFT polymerization method, compositionally uniform copolymers of n-butyl acrylate and styrene with different molecular weights were synthesized as promising impact-resistant modifying additives in the production of polymethyl methacrylate glasses. A decrease in the reduced maximum rate of the "living" radical copolymerization of BA and ST with an increase in the concentration of RAFT agent was shown. It was found that with an increase in the molecular weight of the M_n copolymer, and, accordingly, an decrease in the number of dibenzyl trithiocarbonate fragments in the polymer chain, a linear decrease in the refractive index of the copolymers of n-butyl acrylate and styrene occurs.

5. Acknowledgements

The work has been performed using the equipment of the Multi-User Analytical Center of IPCP RAS. This work was carried out as part of the state assignment of the IPCP RAS (state registration number AAAA-A19-119041090087-4) with partial financial support of the comprehensive program KP19-271 (P13) (state registration number AAAA-A18-118111390034-6).

References

1. Chernikova E V and Sivtsov E V 2017 *Polymer Science* B **59** 92-123
2. Barner-Kowollik, Christopher, ed. *Handbook of RAFT polymerization*. John Wiley & Sons, 2008.
3. Benaglia M, Chiefari J, Chong Y K, Moad G, Rizzardon E and Thang S H 2009 *J. Am. Chem. Soc.* **131** 6914–6915.
4. Gregory A and Stenzel M H 2012 *Progress in Polymer Science* **37** 38– 105.
5. Chiefari J, Mayadunne R T, Moad C L, Moad G, Rizzardo E, Postma A, Skidmore M A and Thang S H 2003 *Macromolecules* V **36** 2273-2288.
6. Matyjaszewski, Krzysztof, and Thomas P. Davis. *Handbook of radical polymerization*. John Wiley & Sons, 2003.
7. Yakimansky A V 2005 *Polymer Science* V **47** 1-8.
8. Matyjaszewski, Krzysztof, and Axel HE Müller, eds. *Controlled and living polymerizations: from mechanisms to applications*. John Wiley & Sons, 2009.
9. Leary J B, Calitz F M, Kenzie J M, Tonge M P, Sanderson R D and Klumperman B 2005 *Macromolecules* V **38** 3151-3159.
10. Schilli C, Lanzendorfer M G and Muller A E 2002 *Macromolecules* V **35** 6819-6823.
11. Hadjichristidisa N, Iatroua H, Pitsikalisa M and Mays J 2006 *Progress Polymer Science* **31** 1068–1132.
12. Moad G, Rizzardo E and Thang S H 2009 *Aust. J. Chem.* V **62** 1402-1411.
13. Gruendling T, Weidner S, Falkenhagen J and Barner-Kowollik C 2010 *Polym. Chem.* V **1** 599-616.
14. Xue X, Zhang W, Cheng Z, Zhu J and Zhu X 2008 J. *Polym. Sci.* V **46** 257-288.
15. Gody G, Maschmeyer T, Zetterlund P B and Perrier S 2014 *Macromolecules* V **47** 3451-3460.
16. Chernikova E V, Serkhacheva N S, Smirnov O I, Prokopov N I, Plutalova A V, Lysenko E A and Kozhunova E Y 2016 *Polymer Science* B **58** 629-639.
17. Schultes, Klaus, Suefke, T., Mueller, R., Arndt, T., Michler, G., and Laatsch, J."PMMA moulding compounds with improved impact resistance." U.S. Patent No. 7,605,193. 20 Oct. 2009.

18. Numrich, U., Wicker, M., Rhein, T., & Suefke, T. "Color-stable, weather-, and impact-resistant molding compositions based on polymethylmethacrylate, and process for production thereof." U.S. Patent No. 5,726,245. 10 Mar. 1998.
19. Chernikova E V, Yerpugova P S, Garina E S and Golubev V B 2007 *Polymer Science* A **49** 208-221.
20. Le T P, Moad G, Rizzardo E, Thang S H, Pat. WO98/01478 USA
21. Araujo, Patricia LB, Katia AS Aquino, and Elmo S. Araujo. "Effects of gamma irradiation on PMMA/polyaniline nanofibre composites." *International Journal of Low Radiation* 4.2 (2007): 149-160.
22. Moad G, Rizzardo E and Thang S H 2008 *Accounts of chemical research* V **41** 1133-1142.
23. Kurochkin S A and Grachev V P 2015 *Polymer Science* C **57** 20-31.
24. Korolev G and Berezin M Prepr. Book of 2nd IUPAC-sponsored international symposium on Free radical polymerization: kinetics and mechanisms; Santa Margherita Ligure, Italy. – (1996).: 315-318.
25. Chaduc I, D'Agosto F, Lansalot M and Charleux B 2012 *Macromolecules* V **45** 1241-1247.
26. Chernikova E V, Plutalova A V, Mineeva K O, Nasimova I R, Kozhunova E Y, Bol'shakova A V, Tolkachev A V, Serkhacheva N S, Zaitsev S D, Prokopov N I and Zezin A B 2015 *Polymer Science* B **57** 547-559.
27. Li G, Shi L, An Y, Zhang W and Ma R 2006 *Polymer* V **47** 4581-4587.
28. Chernikova E V, Yulusov V V, Mineeva K O, Garina E S and Sivtsov E V 2012 *Polymer Science* B **54** 349-360.

IOP Conf. Series: Materials Science and Engineering **848** (2020) 012037 doi:10.1088/1757-899X/848/1/012037

Using of direct piezo-discharge in generation of plasma-activated liquid media

E M Konchekov[1], N G Gusein-zade[1], L V Kolik[1], K V Artem'ev[1], A V Pulish[2]

[1] Prokhorov General Physics Institute of the Russian Academy of Sciences, Moscow, 119991 — Russia
[2] Pirogov Russian National Research Medical University, Moscow, 117997 — Russia

Abstract. This article shows the possibility of plasma-activated liquid media generation using the least energy-consuming source of low-temperature plasma, based on the use of a piezoelectric transformer. The experiments determined the possibility of producing reactive oxygen and nitrogen species in deionized water and culture media. The concentrations of H_2O_2 and NO_x were estimated depending on the energy deposition.

1. Introduction

Plasma treatment of liquids using cold atmospheric plasma (CAP) has recently attracted much attention due to a wide range of applications, such as water purification, sterilization of surfaces and liquids, increasing seed germination, stimulating plant growth, regenerating wounds, treating socially significant diseases (in particular, oncological) [1–9]. As a rule, for plasma activation of liquids, low-temperature nonequilibrium atmospheric-pressure plasma jets and dielectric barrier discharge (DBD) are used [4, 8–11]. However, both in the case of a plasma jet and DBD, the rate of production of reactive oxygen and nitrogen species (RONS), which play a key role in bactericidal and therapeutic effects, is low [4, 8-14]. In addition, in most cases, plasma is generated using a flow of inert gases (argon, helium), and the cross section of the plasma jet and the region of its interaction with the liquid are limited to several millimeters [3, 6, 13–15]. The low efficiency of RONS production, the need for gas flow and the limited processing volume make this approach to the creation of plasma-activated media (PAM) expensive for use in medicine and agriculture. The most effective and economical method for the production of PAM is the use of high-density low-temperature plasma created by direct discharge in air [2, 12, 16–18]. Studies [19–21] show that plasma-treated liquids can retain bactericidal properties for a sufficiently long time. To describe the processes that occur when a CAP acts on various objects and liquids, it is necessary to understand well what chemicals and in what proportions are formed in the medium itself, how does the processing conditions affect the effectiveness of the result.

2. Experimental technique

In fig. 1 shows a scheme of an experimental setup for plasma treatment of liquids. It includes a prototype of a source of CAP, created at the GPI RAS. The creation of CAP in this device is carried out using a piezoceramic transformer [22], whereby it is possible to create plasma both in ambient air (direct piezoelectric discharge) and using an inert gas flow of atmospheric pressure (plasma jet). The distance to the processing object can vary in the first case from 0.5 to 3 mm, and in the second from 1 to 15 mm. Deionized water was treated in glass vessels; the liquid volume was 10 ml (Figure 2). Cultural media were treated in six-well plates with a volume of 5 ml in each of the wells (Figure 3). The duration of CAP exposure to liquid media varied from 30 seconds to 10 minutes with an energy deposition of 80

J/min in the case of direct piezoelectric discharge and 40 J/min in the case of a plasma jet. The power consumed by the cold plasma generator did not exceed 10 watts. The objects of study in this work were liquids (deionized water, DMEM culture medium) treated with direct piezoelectric discharge.

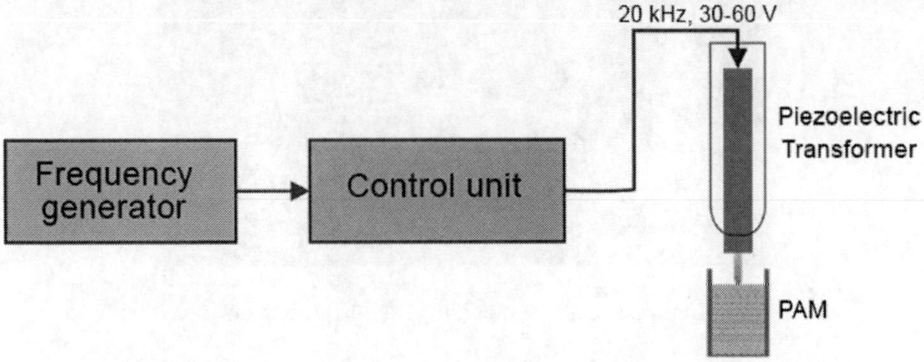

Figure 1. PAM generation setup.

a) b)

Figure 2. Deionized water treatment with direct piezo-discharge (a) and with plasma-jet (b)

Chemical reactions that occur in the gas phase and on the surface of a liquid when it is treated with CAP lead to the appearance of long-lived RNS and ROS in the liquid (the generic term RONS is often used) — NO_2–, NO_3–, H_2O_2, etc. These compounds determine the bactericidal and therapeutic properties of PAM, and their concentration affects the properties of the liquid, such as the redox and pH. In this work, redox was measured using an Expert 001-1 multielectrode meter (Econix), pH using an I-500 pH meter (Aquilon). Hydrogen peroxide concentration was measured using the enhanced chemiluminescence method in the luminol-paraiodophenol-peroxidase system (the measurement procedure can be found in detail in [23–25]). The production of nitrite/nitrate ions NO_x was evaluated by measuring the ultraviolet absorbance spectra (range 190 ... 280 nm) of the treated medium (the procedure was described in [26]). These measurements were carried out at various doses of plasma treatment, i.e. depending on the operating time, while the power of the source was kept stable, and the distance to the object remained constant.

Figure 3. Culture medium (DMEM) treatment.

3. Experimental results

Figure 4 shows the results of redox and pH measurements of plasma-activated water obtained by treating deionized water with a CAP (direct piezoelectric discharge). The power of the plasma source remained unchanged, and the energy deposition varied with an increase in the processing time. The measurements were carried out immediately after treatment and after 24 hours. The graphs show that redox increases linearly with increasing exposure time, and the pH decreases to a value of 3.7. Both of these parameters change slightly after 24 hours, which may indicate that the concentration of long-lived RONS is almost unchanged.

Hydrogen peroxide H_2O_2, formed in PAM, is involved in damage to the cell membrane of microbe and in the weakening of resistance, so it is important to evaluate its concentration. The H_2O_2 concentration increases according to a law close to linear in the processing time interval from 30 seconds to 3 minutes (Figure 5).

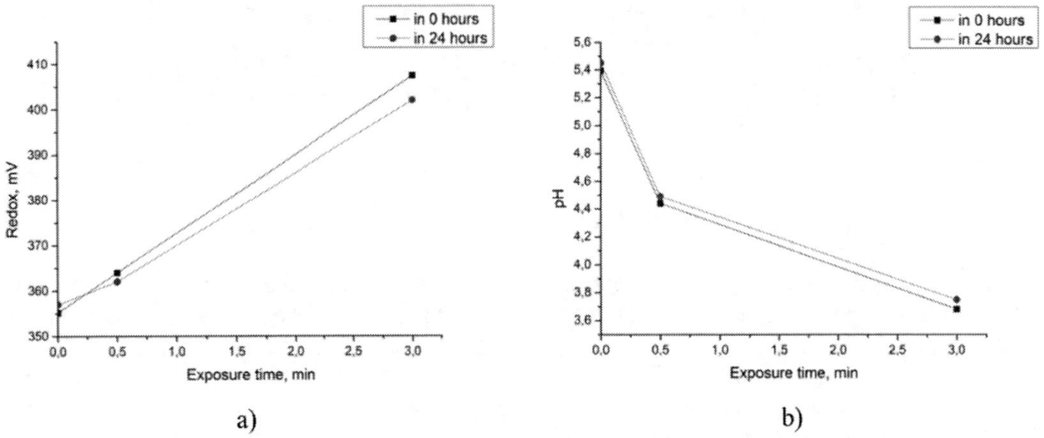

Figure 4. Dependence of redox potential (a) and pH (b) on the exposure time. Measurements were taken immediately after processing and after 24 hours.

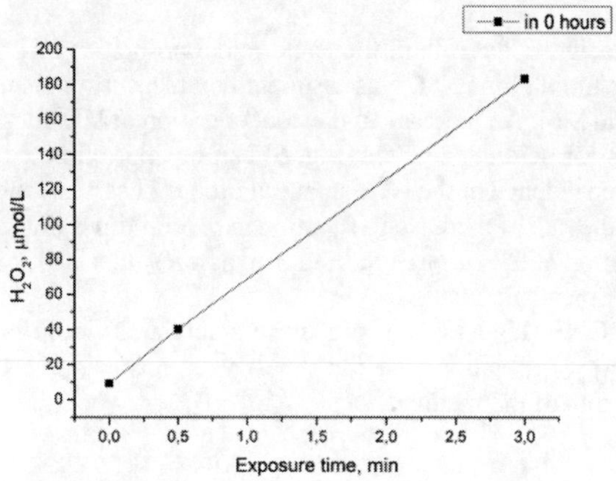

Figure 5. Dependence of H_2O_2 concentration in PAW on the exposure time immediately after processing.

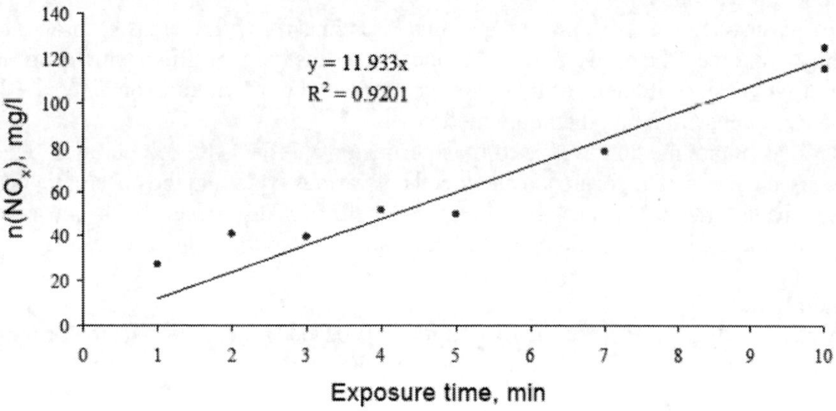

Figure 6. Dependence of NO_x concentration in PAW on the exposure time immediately after processing.

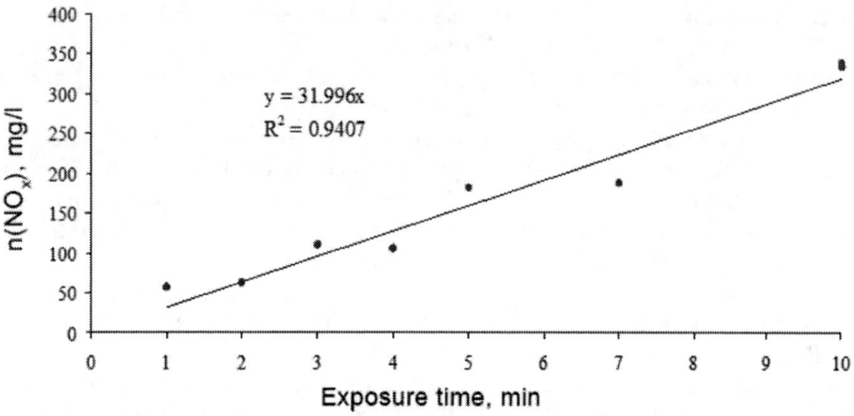

Figure 7. Dependence of NO_x concentration in culture medium (DMEM) on the exposure time immediately after processing.

Other RONS that play a significant role in the bactericidal properties of PAM are NO_2- and NO_3-. The accumulation of nitrite/nitrate ions NO_x was estimated by measuring the absorbance spectra of UV radiation in the treated liquid [26]. An increase in the concentration of NOx depending on the processing time (from 1 to 10 minutes) of deionized water (Fig. 6) can be approximated with high accuracy by a linear function. A similar dependence of the NO_x accumulation is observed when processing the DMEM culture medium, however, the rate of increase in function is several times higher. This is due to the more complex chemical composition of the culture medium, and as a result, a large number of reactions taking place on the surface of the object and inside it.

The concentration of RONS (H_2O_2, NO_x) in a plasma-activated liquid is determined by the total processing time (total energy deposition), so the result was not sensitive to interruptions (up to 5 minutes) during the processing of the medium.

4. Conclusion

In this paper, we studied the characteristics of deionized water and the DMEM culture medium after treatment with CAP generated by a direct piezoelectric discharge in ambient air. The higher energy deposition of direct piezoelectric discharge when exposed to a liquid medium compared with a plasma jet, and as a result, faster RONS production, allows us to consider this method of PAM generation as more efficient and economical.

Redox of plasma-activated deionized water increases linearly with increasing processing time and varies slightly over time. The pH decreases with increasing energy deposition, which correlates with the production of NO_x. Concentrations of hydrogen peroxide and nitrite/nitrate ions increase linearly with increasing energy deposition into the liquid medium.

Plasma-activated culture medium was used in experiments on effects on cell cultures. The first results showed the effectiveness of exposure to cancer cells of various types and the differentiated cytotoxicity of plasma-activated media created by direct piezoelectric discharge. Such an energy-efficient technology can find application both in biomedical applications and in agriculture.

Acknowledgments

The work was partially supported by the RFBR grant 19-02-00378 and the research program "Physical Methods in Agriculture and Ecology" (No. 0024-2019-0004).

References

[1] Chen Z Lin L Cheng X Gjika E Keidar M 2016 Effects of cold atmospheric plasma generated in deionized water in cell cancer therapy *Plasma Process. Polym.* **13** 1151–1156

[2] Gamaleev V Iwata N Oh J S Hiramatsu M Ito M 2019 Development of an Ambient Air Flow Rotating Arc Jet for Low-Temperature Treatment *IEEE Access* **7** 93100–93107

[3] Iwata N Gamaleev V Hashizume H Oh J Ohta T Ishikawa K Hori M Ito M 2019 Simultaneous achievement of antimicrobial property and plant growth promotion using plasma-activated benzoic compound solution *Plasma Process Polym* e1900023.

[4] Zhang J Kwon T Kim S Jeong D 2018 Plasma farming: Non-Thermal dielectric barrier discharge plasma technology for improving the growth of soybean sprouts and chickens *Plasma* **1** 285–296

[5] Kucerova K Henselova M Slovakova L Hensel K 2019 Effects of plasma activated water on wheat: Germination, growth parameters, photosynthetic pigments, soluble protein content, and antioxidant enzymes activity *Plasma Process. Polym.* **16** 1800131

[6] Li Y Wang J Zhang J Fang J 2017 In vitro studies of the antimicrobial effect of non-thermal plasma-activated water as a novel mouthwash *Eur. J. Oral Sci.* **125** 463–470

[7] Shaw P Kumar N Kwak H S Park J H Uhm H S Bogaerts A Choi E H Attri P 2018 Bacterial inactivation by plasma treated water enhanced by reactive nitrogen species *Sci. Rep.* **8** 11268

[8] Ito M Oh J S Ohta T Shiratani M Hori M 2018 Current status and future prospects of agricultural applications using atmospheric-pressure plasma technologies *Plasma Process. Polym.* **15** e1700073

[9] Judee F Simon S Bailly C Dufour T 2018 Plasma-activation of tap water using DBD for agronomy applications: Identification and quantification of long lifetime chemical species and production/consumption mechanisms *Water Res* **133** 47–59

[10] Zhang Z Shen J Cheng C Xu Z Xia W 2018 Generation of reactive species in atmospheric pressure dielectric barrier discharge with liquid water *Plasma Sci. Technol.* **20** 44009

[11 Tang Q Jiang W Cheng Y Lin S Lim T.M Xiong J 2011 Generation of reactive species by gas-phase dielectric barrier discharges *Ind. Eng. Chem. Res.* **50** 9839–9846

[12] Zhou R Zhou R Prasad K Fang Z Speight R Bazaka K Ostrikov K 2018 Cold atmospheric plasma activated water as a prospective disinfectant: The crucial role of peroxynitrite *Green Chem.* **20** 5276–5284

[13] Chauvin J Judee F Yousfi M Vicendo P Merbahi N 2017 Analysis of reactive oxygen and nitrogen species generated in three liquid media by low temperature helium plasma jet *Sci. Rep.* **7** 4562

[14] Ogawa K Oh J Gaur N Hong S Kurita H Mizuno A Hatta A Short R D Ito M Szili E J 2019 Modulating the concentrations of reactive oxygen and nitrogen species and oxygen in water with helium and argon gas and plasma jets *Jpn. J. Appl. Phys.* **58** SAAB01

[15] Kim D Y Kim S J Joh HM Chung T H 2018 Characterization of an atmospheric pressure plasma jet array and its application to cancer cell treatment using plasma activated medium *Phys. Plasmas* **25** 073505

[16] Schmidt M Hahn V Altrock B Gerling T Gerber I C Weltmann K von Woedtke T 2019 Plasma-Activation of Larger Liquid Volumes by an Inductively-Limited Discharge for Antimicrobial Purposes *Appl. Sci.* **9** 2150.

[17] Gharagozalian M Dorranian D Ghoranneviss M 2017 Water treatment by the AC gliding arc air plasma *J. Theor. Appl. Phys.* **11** 171–180

[18] Starek A Pawłat J Chudzik B Kwiatkowski M Terebun P Sagan A Andrejko D 2019 Evaluation of selected microbial and physicochemical parameters of fresh tomato juice after cold atmospheric pressure plasma treatment during refrigerated storage *Sci. Rep.* **9** 8407

[19] Duan J Lu X He G 2017 The selective effect of plasma activated medium in an in vitro co-culture of liver cancer and normal cell *J. Appl. Phys.* **121** 013302

[20] Tanaka H Ishikawa K Mizuno M Toyokuni S Kajiyama H Kikkawa F Metelmann H Hori M 2017 State of the art in medical applications using non-thermal atmospheric pressure plasma *Reviews of Modern Plasma Physics* **1(1)** 89

[21] Kurake N Tanaka H Ishikawa K Takeda K Hashizume H Nakamura K Kajiyama H Kondo T Kikkawa F Mizuno M Hori M 2017 Effects of OH and NO radicals in the aqueous phase on H2O2 and NO2 – generated in plasma activated medium *J. Phys. D: Appl. Phys.* **50(15)** 155202

[22] Johnson M J Boris D R Petrova T B Walton S G 2018 Characterization of a Compact, Low-Cost Atmospheric-Pressure Plasma Jet Driven by a Piezoelectric Transformer *IEEE Transactions on Plasma Science* **47(1)** 434–444

[23] Simakin A V Astashev M E Baimler I V Uvarov O V Voronov V V Vedunova M V Sevost'yanov M A Belosludtsev K N Gudkov S V 2019 The Effect of Gold Nanoparticle Concentration and Laser Fluence on the Laser-Induced Water Decomposition *J. Phys. Chem. B.* **123** 1869–1880

[24] Gudkov S V Garmash S A Shtarkman I N Chernikov A V Karp O E Bruskov V I 2010 Long-lived protein radicals induced by X-ray irradiation are the source of reactive oxygen species in aqueous medium *Dokl. Biochem. Biophys.* **430** 1–4

[25] Bruskov V I Karp O E Garmash S A Shtarkman I N Chernikov A V Gudkov S V 2012 Prolongation of oxidative stress by long-lived reactive protein species induced by X-ray radiation and their genotoxic action *Free Radic. Res.* **46** 1280–1290

[26] Oh J Szili E J Gaur N Hong S Furuta H Kurita H Mizuno A Hatta A Short R D 2016 How to assess the plasma delivery of RONS into tissue fluid and tissue. *J. Phys. D: Appl. Phys.* **49(30)** 304005

The effect of homogenizing annealing on the microstructure and microhardness of Ti-20Nb-(7.5-10)Ta alloys (at.%)

S V Konushkin[1], A Yu Ivannikov[1], E O Nasakina[1], K V Sergienko[1], L A Shatova[2] and M A Sevostyanov[1]

[1]Baikov Institute of Metallurgy and Material Sciences, 49 Leninsky Avenue, Moscow, Russia;
[2]Voronezh State Technical University, st. 20-letiya Oktyabrya, 84/4, Voronezh, Russia.

venev.55@mail.ru

Abstract. The microstructure and microhardness of Ti-20Nb-7.5Ta and Ti-20Nb-10Ta alloys (at. %) after smelting and after homogenizing annealing at 900°C and 950°C for 12 hours were studied. It was shown that, after smelting, the alloys have a dendritic structure, and after annealing, they have a polycrystalline structure with equiaxed grains. It is also shown that annealing leads to an increase in the microhardness of alloys.

1. Introduction

In modern medicine, the task of manufacturing complex-loaded orthopedic implants with high strength and biocompatibility is an urgent problem. At present, Ti-based alloys are being investigated and actively used in endoprosthetics. Alloying materials, along with metals already used, such as Al, V, Mo, etc., include metals with high biocompatibility with the human body: Ta, Nb, Zr, etc. Alloys of the Ti – Nb – Ta system have high biocompatibility, low Young's modulus, and can exhibit shape memory and superelasticity effects [1, 2]. These features of these alloys make them promising candidates as materials for implants.

After smelting the alloy ingot, it is necessary to remove the possible thermal stress and ensure the homogeneity of the distribution of alloying elements in the entire volume before further processing of the ingot and the manufacture of the endoprosthesis. For this, homogenizing annealing is used. During homogenizing annealing, the chemical composition is leveled, the alloy recrystallizes, and the mechanical properties change [1-11].

This paper presents the results of a study of the microstructure and microhardness of Ti-20Nb-7.5 and Ti-20Nb-10Ta alloys (at. %) after smelting and homogenizing annealing at 900°C and 950°C for 12 hours.

2. Materials and methods

For the study, the following alloys were selected: Ti-20Nb-7.5Ta and Ti-20Nb-10Ta since these alloys belong to the range of Ti-Nb-Ta alloys exhibiting the shape memory effect or the effect of superelasticity [1]:

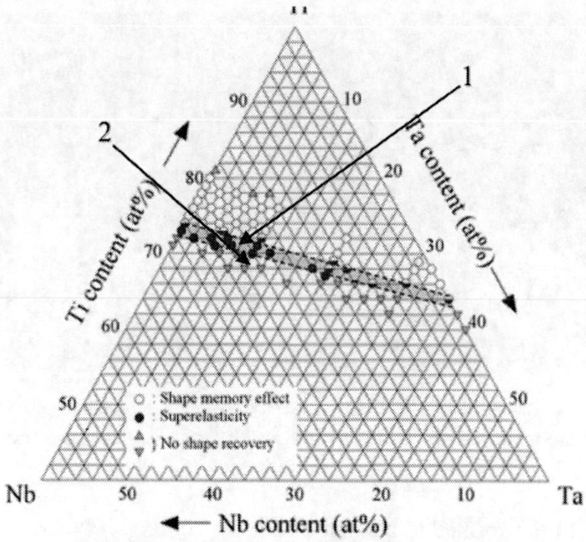

Figure 1. Diagram of Ti-Nb-Ta alloys [1]
1 - Ti-20Nb-7,5Ta alloy; 2 - Ti-20Nb-10Ta alloy.

The alloy ingots of the selected compositions were smelted by the vacuum-arc remelting method in a water-cooled copper crystallizer in an argon medium. The ingots had the shape of "boats" 20 mm long, 10-12 mm wide, 8-10 mm high, weighing 15 g. After smelting, the ingots were cut into pieces 4-5 mm thick across the length and annealed in vacuum at a temperature of 900-950°C for 12 hours at a pressure of $3*10^{-5}$ mm Hg. The temperature range was chosen on the basis of published data based on their task to align the chemical composition along the cross section of the ingot, while not allowing excessive grain growth [3].

Samples were prepared for metallographic studies by sequential grinding after pressing on a Piatto diamond disk with a grain size of P120 for 3 minutes, P320 for 5 minutes, P800 for 5 minutes, on an Aka-Allegran-3 fine grinding diamond disk with DiaMaxx Poly suspension with sizes of diamond particles 6 and 3 microns, on Akasel DARAN woven acetate with a DiaMaxx Poly suspension with a diamond particle size of 1 μm; on Akasel CHEMAL foamed neoprene with an aluminum oxide suspension with a particle size of 50 nm. The microstructure was revealed by etching in a solution of hydrofluoric and nitric acids with distilled water in a volume ratio of HF: $3H_2SO_4$: $16H_2O$. Sections were wiped with a solution for 30-60 s, washed with water and ethyl alcohol. The study was performed using a Neophot 2 light microscope manufactured by Carl Zeiss Jena. After etching and analysis of the microstructure, the microhardness of the samples was investigated. To study the microhardness, the Vickers method was used. The load was 300 g, exposure 15 sec.

3. Results and discussion

3.1 The study of the microstructure

As can be seen from Figure 2, the microstructure of the ingots after smelting is grains of oppositely directed dendrites. The axes of dendrites with increased content of refractory elements (Ta and Nb) (light areas of the microstructure - Figure 2a), the more etched inter-axis dark zones (Figure 2a) with larger content of Ti, and the etched grain boundaries of the dendrites (Figure 2b) are visible in the photographs.

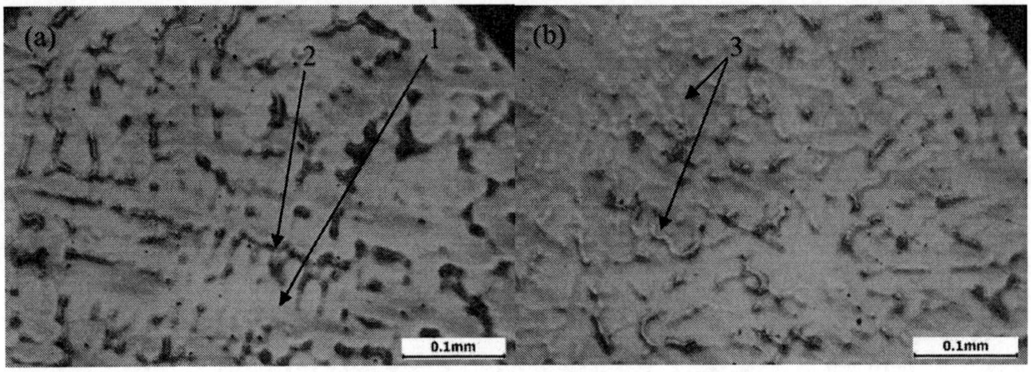

Figure 2. The microstructure of the ingot Ti-20Nb-7.5Ta (a) and Ti-20Nb-10Ta (b) after smelting 1 - dendrite, 2 - the area between the "branches" of the dendrite, 3 - the grain boundaries of the dendrite.

The dendritic structure is formed as a result of uneven cooling of the molten ingot due to more rapid cooling of the lower part of the ingot in contact with the copper mold compared to the upper part. Initially, at the point of contact of the ingot with the copper tray “branches” containing more refractory elements (Ta and Nb) crystallize and begin to grow along the temperature gradient. Between the “branches” there remain areas containing less refractory elements (Ti), which solidify later. The resulting dendritic structure is a defect that is corrected by homogenizing annealing.

The structure of the ingots of the selected alloys after annealing at 900 ° C for 12 hours is shown in Figure 3a, b.

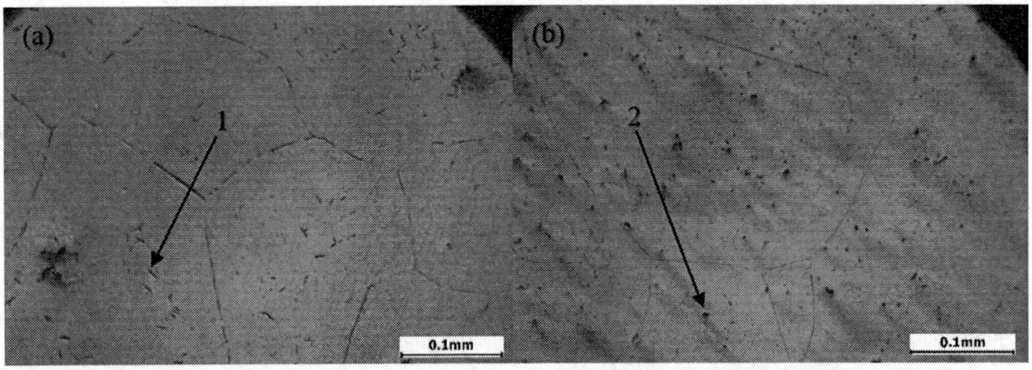

Figure 3. The microstructure of the ingot Ti-20Nb-7.5Ta (a) and Ti-20Nb-10Ta (b) after annealing at a temperature of 900 ° C for 12 hours, 1 - needle discharge, 2 - point discharge.

As can be seen from the photographs of the microstructure presented in Figure 3, grains recrystallized and dendrites dissolved. Samples consist of rather large equiaxed grains with sizes of 0.2-0.4 mm with boundaries of 120 °.

The structure of the ingots of the selected alloys after annealing at 900 ° C for 12 hours is shown in Figure 4.

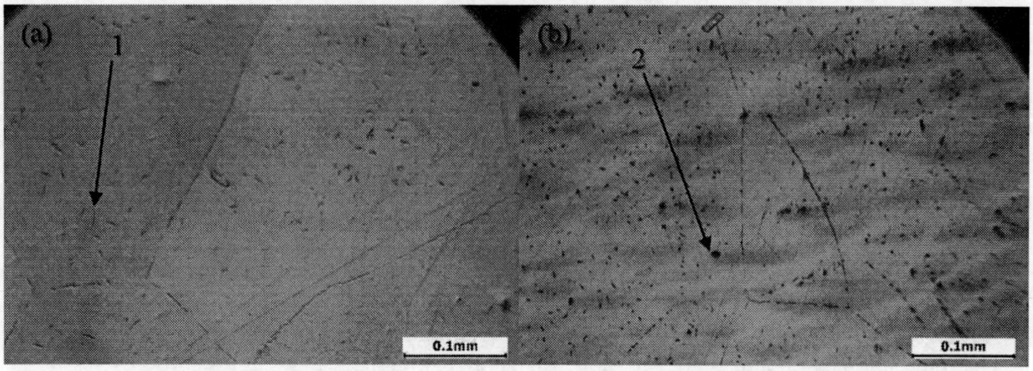

Figure 4. The microstructure of the ingot Ti-20Nb-7.5Ta (a) and Ti-20Nb-10Ta (b) after annealing at a temperature of 950 ° C for 12 hours: 1 - needle discharge, 2 - point discharge.

As can be seen from the photographs of the microstructure presented in Figure 4, there was a further growth of grains to sizes of 0.5-0.7 mm. The boundaries remained equal to 120°.

Micrographs of the samples show the separation of needle-shaped (Ti-20Nb-7.5Ta alloy, Figure 3a, 4a) and point (Ti-20Nb-10Ta alloy, Figure 3b, 4b) structural elements inside the main grains. Perhaps this is the α-phase, precipitated as a result of decay after annealing and slow cooling to room temperature of the main β phase, which is characteristic of high alloy titanium alloys. To confirm or refute this statement, an XRF of melted and annealed samples will be carried out in the future.

3.2 Microstructure study

The alloy samples microhardness test results are presented in Figure 5.

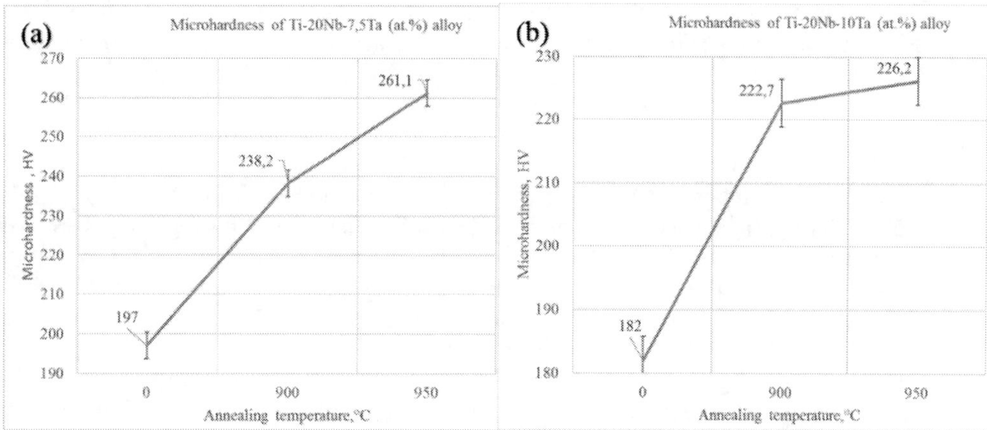

Figure 5. Microhardness of Ti-20Nb-7.5Ta (a) and Ti-20Nb-10Ta (b) alloys samples after smelting and homogenizing annealing at temperatures of 900°C and 950°C, respectively.

As can be seen from the research results, the microhardness of the alloys increases significantly after annealing for both alloys. The microhardness of the Ti-20Nb-7.5Ta alloy after annealing at 900°C increases by 20.9%, and after 950 ° C - by 32.5%. The microhardness of the Ti-20Nb-10Ta alloy after annealing at 900°C increases by 22.3%, and after 950°C - by 24.3%. We can note a general decrease in the effect of annealing on the microhardness of the Ti-20Nb-10Ta alloy compared to the Ti-20Nb-7.5Ta alloy. After smelting (by 8%) and after annealing (by 7% during annealing at 900°C and by 14% during annealing at 950°C), this also indicates the plasticizing effect of Ta on these alloys. In general, an increase in the microhardness of all samples after annealing, possibly indicates the release of additional phases, which is partially confirmed by the microstructural studies described in the previous paragraph. The expected decrease in the microhardness of the alloys due to the removal of thermal stresses arising during the cooling of the ingot after smelting was not found [4].

4. Conclusions

1.After smelting using vacuum-arc remelting, Ti-20Nb-(7.5-10)Ta (at. %) alloys have a dendritic structure.
2. Homogenizing annealing at 900°C and 950°C leads to the dissolution of dendrites and recrystallization of the alloy.
3. Homogenizing annealing leads to the release of additional phases in the grains.
4. The microhardness of alloys after homogenizing annealing increases significantly.
5. The plasticizing effect of Ta on the microhardness of Ti-20Nb-(7.5-10)Ta alloys is shown.

Acknowledgments
The work was supported by the Ministry of Education and Science of Russia (grant identifier RFMEFI60419X0242).

References

[1] Kim, H.Y.; Fu, J.; Tobe, H.; Kim, J.I.; Miyazaki, S. Crystal structure, transformation strain, and superelastic property of Ti–Nb–Zr and Ti–Nb–Ta alloys. Shape Mem. Superelast. 2015, 1, 107–116
[2] Tang, X., Ahmed, T. & Rack, H.J. Phase transformations in Ti-Nb-Ta and Ti-Nb-Ta-Zr alloys. Journal of Materials Science 35, 1805–1811 (2000). https://doi.org/10.1023/A:1004792922155
[3] Konopatskii A.S. Zhukova Y.S.,Dubinskii S.M., Korobkova A.A., Filonov M.R., Prokoshkin, S.D., Microstructure of Superplastic Alloys Based on Ti–Nb for Medical Purposes, Metallurgist, Volume 60, Issue 1-2, 1 May 2016, Pages 223-228, DOI: 10.1007/s11015-016-0277-8.
[4] Verhoeven, J.D. Fundamentals of Physical Metallurgy, Wiley, New York, 1975, p. 326
[5] Sevostyanov M A, Baikin A S, Shatova L A, Nasakina E O, Berezhnov A V, Gudkov S V, et al. 2018 Biocompatibility of the Ti81Nb13Ta3Zr3 Alloy Doklady Chemistry 482(1) 204–206.
[6] Nasakina E O, Sudarchikova M A, Sergienko K V, Konushkin S V, Sevost'yanov M A 2019 Ion Release and Surface Characterization of Nanostructured Nitinol during Long-Term Testing Nanomaterials 9(11) 1569
[7] Sergienko K V, Sevost'yanov M A, Konushkin S V, Nasakina E O, Baikin A S, Shatova L A, Kolmakov A G 2018 Investigation of static properties of medical alloys Ti-(20-30)Nb-(10-13)Ta-5Zr *IOP Conf Ser: Mater Sci Eng* **347** 012049
[8] Konushkin S V, Baskakova M I, Leonov A V, Nasakina E O, Sudarchikova M A, Kolmakova A A, Bespamiatnova A, Sergiyenko K V, Leonova Yu O, Sevostyanov M A 2019 The structure of the alloy Ti - (20-30) Nb - 5Zr after smelting and homogenizing annealing *IOP Conf Ser: Mater Sci Eng* **525** 012060
[9] Davis J R. 2003 *Handbook of materials for medical devices* (ASM Int.) 350 p
[10] Petrini L, Migliavacca F 2011 Biomedical Applications of Shape Memory Alloys Journal of Metallurgy 2011 1-15
[11] Duerig T W, Melton K N, Wayman C M, Stockel D 1990 Engineering aspects of shape–memory alloys (Oxford: Butterworth Heinemann Ltd) pp 181 – 194

The investigation of the welded joints of the nitrogen containing cast austenitic steel, obtained by the manual arc welding

V S Kostina[1], M V Kostina[2], M G Sharapov[3], I O Samodurov[4] and S O Muradyan[5]

[1] Junior researcher, A.A.Baikov Institute of Metallurgy and Materials Science, Russian Academy of Sciences, Moscow, Russia
[2] Head of laboratory, A.A.Baikov Institute of Metallurgy and Materials Science, Russian Academy of Sciences, Moscow, Russia
[3] Researcher, Federal State Unitary Enterprise «Central Research Institute of Structural Materials «Prometey» named by I.V. Gorynin of National Research Centre «Kurchatov Institute»», Saint-Petersburg, Russia
[4] Researcher, Federal State Unitary Enterprise «Central Research Institute of Structural Materials «Prometey» named by I.V. Gorynin of National Research Centre «Kurchatov Institute»», Saint-Petersburg, Russia
[5] Researcher, A.A.Baikov Institute of Metallurgy and Materials Science, Russian Academy of Sciences, Moscow, Russia

E-mail: vskostina@yandex.ru

Abstract. A literature analysis showed, that in the world there are not the welding additives, which is contain in own composition more than 0,37% mass. N. The investigations of the welding joints of the austenitic cast steel 05Kh21AG15N8MFL with the nitrogen containing 0,57% mass with using the available welding additives was provided. The technology of TIG welding with using two industrial welding wires Sv-10Kh20N18M3AFS (0,24% N) and Sv-25Kh25N16AG7 (0,10% N) was developed. It was obtain the defect-free welding joints, having a high rate of the mechanical properties, resistance to the pitting corrosion in 3,5% NaCl, containing up to 0,4% ferrite.

1. Introduction

It is known, that the austenitic stainless steels with the high nitrogen equilibrium concentration (~0,5% mass.), have a high rate of the physical-mechanical properties, corrosion resistance and cold resistance [1-8]. Using them as a constructional material often not complete without the welding. So, it is important to get the defect-free welding joints. However, the welding thus steels is considered as a difficult technological operation, in operating which can grow up the hot cracks, lack of fusion, segregations of the second phases in the zone of thermal effect, pores, nitrogen loses [9].

Developed in IMET RAN the cast steel of austenitic class 05Kh21AG15N8MFL with 0,57% of nitrogen containing is the high strength, corrosion- and cold resistant material to manufacturing the difficult form construction elements [10]. It is overtop applying in the Russia the cast austenitic steels as 12Kh18N9TL, 10Kh18N11BL, 12Kh18N12M3TL at the limit of liquidity in ~2 times, at percussive viscosity in ~ 4,5 times. It is not subjected to the cold brittleness. At cooling to -70°C it has the percussive viscosity values ~2,3 MJ/m^2 [11-13].

In a literature there are the works about the welding the similar in the chemical compound nitrogen containing hot deforming steels 05Kh22AG16N8M and 05Kh22AG15N8M2F [14]. Welding joints was obtained by the manual arc welding under flux (SAW) with using as the welding additive of a such metal, cutting to the strips (so-called «strips»). Welding joints has a high rate of the properties ($\sigma_{0,2}$ = 450 - 465 MPa, σ_e = 800 - 820 MPa), however the weld wire with a such chemical compound do not producing, thus using the such welding additive costly, in main, by the showed possibility to obtain the welded joints of a given steel without the cracks and the gas pores, with a good rate of the properties.

Due to this the purpose of the investigation consist of the possibility obtaining quality high strength weld joints from the steel 05Kh21AG15N8MFL with available industrial welding wires applying.

In present time in industry are absent the additive materials, which have the similar compound with the nitrogen steels, in which strength rate is providing by the nitrogen concentration near 0,5%, which confirmed by our literature analysis and authors data [15-17]. Producing in the world additive materials contain in their compound no more than 0,37% of a nitrogen. Literature data analysis by the welding of the austenitic corrosion resistance steels with the high nitrogen concentration, analysis the proposed producer welding materials, also showed that, as rule, using weld additives is low carbon, contain 20-27% Cr and 2,5-9,5% Mo; the high nickel concentration in their compound (17-25% in steels and 32-60% in alloys) is required to providing the austenitic structure and the high crack resistance. The nitrogen concentration in the welding additives – a balance between the concentrations Cr and Mo (promoting the nitrogen solution, increasing the corrosion resistance and providing the solid solution strengthening) and between the concentration of obstructing to the nitrogen solving Ni. When choosing the material of the welding additives for testing in present work at obtaining welding joints of the steel 05Kh21AG15N8MFL with 0,5% N the next condition was formulated:

- the welding wire material must be austenitic;
- the mechanical properties rate must be the such or higher that the welded base metal to providing the high mechanical properties rate;
- a wire material must providing the corrosion resistance rate no low the base metal.

2. Materials and methods

Leaning on the aforesaid and known investigations experiments, the welding of the cast steel 05Kh21AG15N8MFL thickness of 22 mm was providing with using welding wires: Sv-10Kh20N18M3AFS and Sv-25Kh25N16AG7. The model chemical composition BM the welding wires metal showed in the table 1.

Table 1. The chemical composition of the base metal and additive materials.

Model	Chemical composition, % mass (Fe – and impurities – another)									
	N	Si	Cr	Ni	Mn	Mo	V	Nb	C	S	P
									No more		
05Kh21AG15N8M2F	0,55-0,6	0,1-0,2	21-22	7,8-9	15-16	1-2	0,3	-	0,04	0,008	0,012
Sv-10Kh20N18M3AFS	0,2-0,25	0,7-0,1	19,5-20	16-17	3-3,5	1-2	0,8	-	0,12	0,008	0,015
Sv-25Kh23N16AG7	0,1-0,15	0,1-0,2	24,7-26	15-16	6,5-7	0,1-0,3	0,7	-	0,25	0,006	0,012

In this investigation was used the manual arc welding by non-melting electrode in the argon environ (TIG), as widely used and the available method. To obtain the welding joints with the high rate of properties, it was developed the welding technology, the mode of them showed in the table 2. Welding was provided at low currency and every welded roll was cooled by the welding wire to the room temperature. Thus, obtained welding joint was schematically looked, as showed at the figure 1.

Table 2. The mode of the manual arc welding in argon of the cast austenitic steel 05Kh21AG15N8M2FL under a developed technology.

Base metal / wire	Welding mode	Edges cutting
05Kh21AG15N8M2F / 10Kh20N18M3AFS	I=70-100 A, U=8-10 V,	V-looked
05Kh21AG15N8M2F / 25Kh25N16AG7	welding velocity 3 м/c	

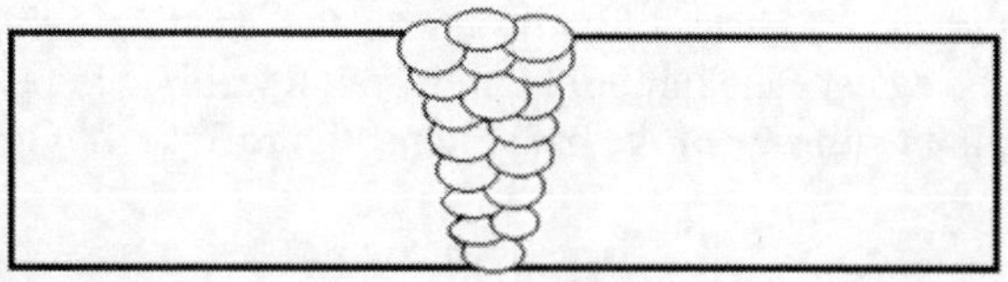

Figure 1. Schematic picture of obtained welding joints.

The microstructure of welded joints was examined on an Olympus light microscope and a Kyky 2800 scanning electron microscope with an Oxford Instruments X-ray Microscope Analysis Unit (MRSA). The nitrogen content in samples from different zones of welded joints was determined on the Eltra ONH-2000 gas analyzer ("Venta Lab"). Ferritometry was provided using IMP-2M ferritometer. Tensile tests were carried out according to GOST 1497-84 on Instron 3382 equipment; (test speed 1 mm / min). Impact bending tests were carried out according to GOST 9454-78 on Amsler RKP-450 (copra energy - 300 J). The microhardness of the various structural components of the welded joints (fusion line, weld, base metal, and heat-affected zone of weld) was determined according to GOST 9450-76 on Volpert 402MVD hardness meter (load 50g for 10 s). Corrosion tests on pitting corrosion – 3,5% NaCl at 20°C providing by GOST 9.912-89. A velocity of the potential change consist of 0,2 mV/s.

3. Results and discussions

Obtained under the developed mode the welding joints turned as qualitative, without the pores and cracks, under the radiographic and visual control data, and also by the microstructure investigations. In all WJ the microstructure of BM consist of the large austenitic cast grains, with the size 500-700 mkm, WJ have the characteristic cell-dendritic structure with the grains size 4-20 mkm (figure 2).

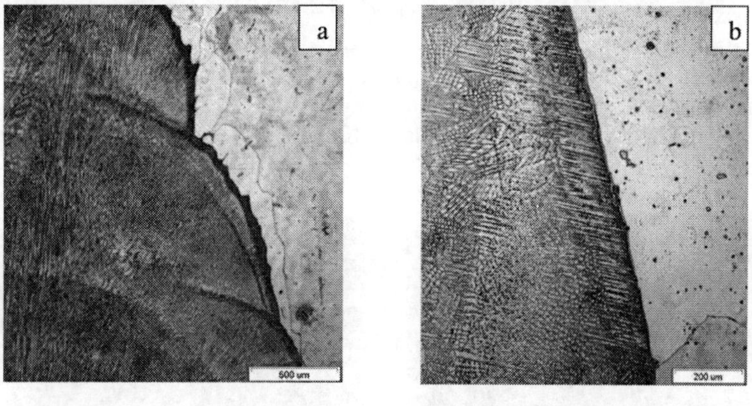

Figure 2. Microstructure of the different zones welding joints:
a) 05Kh21AG15N8M2F/10Kh20N18M3AFC; b) 05Kh21AG15N8M2F/25Kh25N16AG7.

The chemical composition investigation of the metal joint showed, that the chosen welding mode is not to provide the manganese loses more than 0,3-0,7%. Also, it was established, that in the welding process practically not happened the nitrogen loses in the weld joints. Including, the nitrogen containing in the metal of the wires / in the welding joint to make:

1) 05Kh21AG15N8M2F / 10Kh20N18M3AFS: 0,24% / 0,22% N;
2) 05Kh21AG15N8M2F / 25Kh25N16AG7: 0,10% / 0,11% N.

A condition to obtain the austenitic structure of the welding joints was fulfill. The microstructure investigation of the base metal and the wire metal showed, that the influence of the thermal cycle of the multi pass welding are not provide to the δ-ferrite formation in the metal wire. The δ-ferrite containing in the welding wire (0,00%), then in the base metal (~0,40%).

The strength characteristics WJ was investigated on the bursting cylindrical samples, where the welding joint located in the center. The limit of liquidity of all welding joints higher than the base metal. It is due to more smaller the grain size of the metal joint (figure 3).

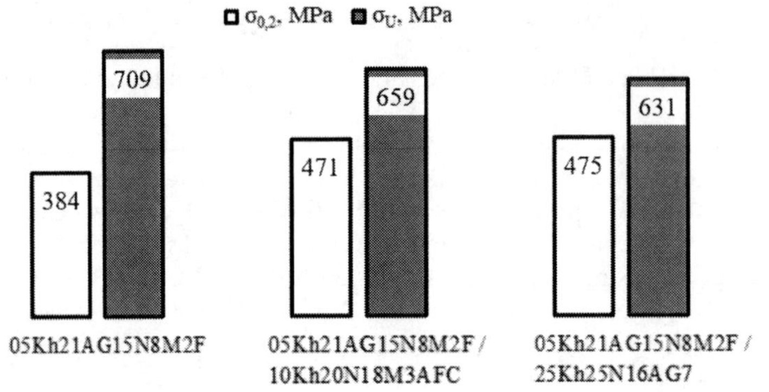

Figure 3. The mechanical properties of the base metal and it welding joints.

The thermal cycle of welding influence provide to the percussion viscosity declining of all the weld joints at +20 and at -70°C, with comparison to the base metal (KCV^{+20} = 331 J/cm^2):

1) 05Kh21AG15N8M2F / 10Kh20N18M3AFS: KCU^{+20} (90 J/cm^2) and KCV^{-70} (31 J/cm^2);

2) 05Kh21AG15N8M2F / 25Kh25N16AG7: KCU^{+20} (111 J/cm^2) and KCV^{-70} (51 J/cm^2).

However obtained indicators are settled to the criteria values of the percussion viscosity of the pipeline equipment, working under the pressure [18].

Fractographic analysis of the welded joints fractures after the tests on the distension and percussive bend showed, that in the fractures of all the samples prevails the viscous component. On the figure 4 are pictured the fractures of the samples of the welding joints after the distension tests.

All the welding joints have shown the resistance to the corrosive environment influence at the tests on the pitting corrosion due to the formation of the protective passivating film.

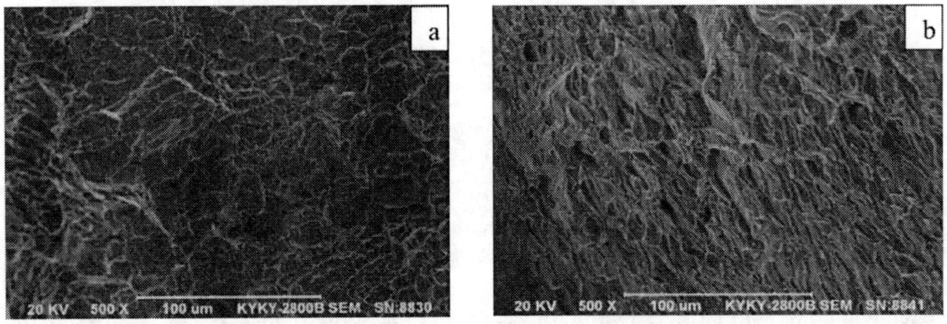

Figure 4. Fractography of the fractures of the welding joints after the tests on the distension: a) 05Kh21AG15N8M2F / 10Kh20N18M3AFS; b) 05Kh21AG15N8M2F / 25Kh25N16AG7.

4. Conclusions

1. It was developed the technology of TIG welding, allowing to obtain the non-defect welding joints of the austenitic cast steel 05Kh21AG15N8MFL with the nitrogen containing of 0,57% mass. with using the two welding wires: Sv-10Kh20N18M3AFS (0,24% N) and Sv-25Kh23N16AG7 (0,10% N).

2. Using mode of the welding allow to obtain the austenitic structure in all zones of the welding joints and practically does not lead to the changes of the nitrogen quantity in the metal joint.

3. The liquidity limit of the welding joints higher than the base metal. The strength limit is slightly lower (no more 11%).

4. The welding joints percussive viscosity after the welding thermal cycle influence has decreasing, however the obtained values has fit to the requirements of the criteria values of the percussive viscosity of the pipeline equipment.

5. The welding joints resistance to the pitting corrosion in 3,5% NaCl solution at 20°C was studied. It is shown that they not inclined to the corrosion in present environment due to the defensive passivating film formation on the surface.

5. Acknowledgements

The work was carried out according to the state assignment No. 075-00746-19-00.

References

[1] Berns H 1996 *ISIJ Int* Manufacture and application of high nitrogen steels 36 (7) pp 909-904

[2] Simmons J W 1996 *Materials Science and Engineering Ser A* Overview: high-nitrogen alloying of stainless steels Vol 207 pp 159-169

[3] Foct J June/August2003 *Sadhana* Future developments and applications of nitrogen-bearing steels and stainless steels Vol 28 parts 3-4 pp731–737

[4] Speidel M O 2006 *Mat-wiss. u. Werkstoiftech* Nitrogen Containing Austenitic Stainless Steels 37 № 10 pp 875-880

[5] Lo K H, Shek C H and Lai J K L 2009 *Materials Science and Engineering* Recent developments in stainless steels R 65 pp 39–104

[6] Berns H 2013 High Interstitial Stainless Austenitic Steels, Engineering Materials 10.1007/978-3-642-33701-7 Springer-Verlag Berlin Heidelberg

[7] Bannych O A, Blinov V M and Kostina M V 1998 *The institute of metallurgy and material science by Baycov A A 60 years: CSP* Moscow Eliz pp 192-201

[8] Kostina M V, Bannych O A, Blinov V M and Dymov A V 2001 *Material science* Nitrogen alloying chromium corrosion resistance steels of new generation №2 Moscow pp 35-44

[9] Harzenmoser M 2004 *Materials and Manufacturing Processes* Welding of High Nitrogen Steels Vol 19 Issue 1 pp 75-86

[10] Bannych O A, Blinov V M, Blinov Y V, Kostina M V, Muradyan S O and Rigina L G 2012 *Patent 2445397* High strength cast non-magnetic corrosion resistance steel and the product produced from it

[11] Muradyan S O 2016 *Candidate dissertation* Structure and properties of the cast corrosion resistance steel Moscow p 132

[12] Kostina M V, Bannych O A, Blinov V M, Muradyan S O and Chadyev M S 2011 *Procurement in engeneering* The development of new cast high corrosion resistance and high strength austenitic steel alloying by nitrogen. Part 3. The structure and mechanical properties of new cast high nitrogen corrosion resistance Cr-Mn-Ni-Mo-N steel №9 Moscow pp 39-45

[13] Kostina M V, Muradyan S O, Chadyev M S and Korneev A A 2011 *Metals* Investigation of heat treatment on the structure, phase compound and mechanical properties new cast high nitrogen corrosion resistance Cr-Mn-Ni-Mo-N steel №5 Moscow pp 33-48

[14] Bannych O A, Blinov V M, Kostina M V, Blinov Y V and Zvereva T N 2007 *Metals* Investigation of weldability high nitrogen corrosion resistant austenitic steels type Kh22AG16N8M №4 Moscow pp 51-67

[15] Madeleine du Toit June 2002 *Journal of Materials Engineering and Performance* Filler Metal Selection for Welding a High Nitrogen Stainless Steel Vol 11 Issue 3 pp 306–312

[16] Raffi Mohammed, Madhusudhan Reddy G and Srinivasa Rao K 2016 *Trans Indian Inst* Met Effect of Filler Wire Composition on Microstructure and Pitting Corrosion of Nickel Free High Nitrogen Stainless Steel GTA Welds 69(10) pp 1919–1927

[17] Bishokov R V, Baryshnikov A P, Gezha V V and Melnikov P V 2014 *Questions of material science* Welding materials and technologies the high strength steels welding №2 (78), pp 128-137

[18] *PNAE G-7-002-86* The computation norms to the strength of equipment and pipelines of atomic energy settings

IOP Conf. Series: Materials Science and Engineering **848** (2020) 012040 doi:10.1088/1757-899X/848/1/012040

Molecular Docking simulation and Fluorescence lifetime characteristics of NIR Cyanine Dye Complexes with Albumin

A Kostyukov,[*a] M Mestergazi,[a,b] A Shmykova,[b] T Podrugina,[b] V Pogonin,[a,c] E Radchenko,[b] V Palyulin,[b] I Borissevitch[d,e], V Kuzmin[a]

a. Emanuel Institute of Biochemical Physics RAS, Kosygin st. 4, Moscow,119334 Russia. E-mail: aak@sky.chph.ras.ru

b. Department of Chemistry, Lomonosov Moscow State University, Lenin Hills 1, build. 3, Moscow, 119991 Russia

c. Vernadsky Institute of Geochemistry and Analytical Chemistry of RAS, Kosygin st. 19, Moscow, 119991 Russia

d. Departamento de Física, Faculdade de Filosofia, Ciências e Letras de Ribeirão Preto, Universidade de São Paulo, Ribeirão Preto, SP 14040-901, Brazil

e. Instituto de Física, Universidade Federal de Goiás, Caixa Postal 131, Goiânia, GO 74001-970, Brazil.

Abstract The high sensitivity of modern techniques for lifetime detection opens new significant opportunities for investigation of various processes in the biological media. Here, we report on the novel sub-class of cyanine dyes. We examined the effect of electronic coupling on the optical properties of the complex of the dye with the human serum albumin. TCSPC measurements of fluorescence decay provided evidence that the studied dye generates more than two kinds of complexes with albumin. It was established that the effect of "frozen structure" in complex changes the rate of photoisomerization way of the dye's excited state degradation. Furthermore, by means of molecular docking, the experimentally proposed mechanism of complex formation and the existence of several binding sites were confirmed.

1. Introduction

In recent years the application of cyanine dyes in medicine has tangibly grown due to their good ability to accumulate in tissues and cells, low toxicity and satisfactory fluorescence quantum yields. Nowadays, they are widely used in bioimaging [1], as photosensitizers in photodynamic therapy [2] or as conjugates for selective delivery of drugs in order to improve the bioavailability of the latter [3]. Particularly, utilization of cyanine dyes as fluorescent contrast agents in molecular biology is centered on their ability to accumulate in tissues [4], and now it is a promising technique of bioimaging for visualization of blood vessels and tumors. At present, there are only two cyanine dyes approved by FDA for medical application – fluoresceine and NIR dye – cardio green (ICG). Both of them have a low range of intersystem crossing, which makes them safe, as there is no photosensitizing during irradiation. ICG long polymethine chains contribute to its absorption in NIR, which enables ICG to penetrate the tissues. A crucial role for the biological usage of dyes is devoted to the complex formation with blood proteins, such as albumin and lipoproteins, which ensure the delivery of a dye to the cell target [5]. In [6] it was shown that the interaction of dyes with biomacromolecules leads to an increase of the fluorescence intensity of complex, rigidity and stability of dye structure.

In this paper, we present a new biscarbocyanine dye containing two chromophore systems, which further on will be referred to as "BCD". The molecular docking experiment was also carried out in order to estimate the possible interactions between the dye and the protein and to elicit any special interactions. The obtained data may serve as a basis for further investigation of intercellular accumulation and distribution for fluorescence lifetime imaging and confocal autocorrelation microscopy.

2. Experimental Details

All of the chemicals used in the current study were purchased from commercial vendors and used as received without further purification unless otherwise noted. Human serum albumin (HSA) (Sigma, US) was used. Experiments in the water phase (Millipore, 18 MOhm) were carried out in phosphate-buffered saline (PBS) at pH 7.4 at room temperature.

2.1 The binding constant of dye

The binding constant was determined by fluorescence spectroscopy. The interaction of protein and dye corresponds with a bathochromic shift and an increase of fluorescence. The binding constant Kb was calculated from the binding curves according to the equation (1):

$$\theta = (K_b[HSA])/(1 + K_b[HSA]) \quad (1)$$

where $\theta = (F - F_0)/(F_\infty - F_0)$ is the fraction of the dye bound with the protein and F_0, F_∞ and F are the fluorescence intensity at [HSA] = 0 and at the complete and intermediate binding of the dye by protein, respectively[7].

Steady state spectroscopy

UV-visible absorbance spectra were recorded using the Shimadzu UV-3101PC model in EtOH and PBS buffer solutions with or without human serum albumin HSA (Sigma, USA). Fluorescence spectra were obtained using the Panorama fluorescence spectrometer.

2.2 TCSPC experiments

Lifetimes of singlet excited state registration were obtained by TCSPC UV-IR spectrofluorometer Fluotime 300 Picoquant. Dye dissolved in EtOH or PBS with albumin and irradiated by 640 nm (LDH-D Picoquant) 70 ps laser 40 MHz rep. rate with vertically polarized beam. Emission detected with a magic angle polarizer to prevent undesired polarization effect. Fluorescence registered in wavelength interval from 750 to 800 nm and obtained data were analyzed using FluoFit software (PicoQuant).

Characteristic times and the contribution of individual components to the detected fluorescence were calculated using a multi-exponential model with deconvolution based on internal response function (IRF) by the following formula:

$$I(t) = \int_{-\infty}^{t} IRF(t')\sum_i A_i \exp((-t - t')/\tau_i)dt' \quad (2)$$

where i is the number of components, A is the amplitude that determines the contribution of each component, and τ is the decay time of the components.

The values of A_i and τ_i can be used to determine the fractional contribution (Q_i) of each decay time to the steady-state intensity. This value can be expressed as:

$$Q_i = A_i\,\tau_i / \Sigma_j A_j\,\tau_j\,, \quad (3)$$

where j is the total number of measured components.

The terms $A_i\,\tau_i$ are proportional to the area under the decay curve for each decay time. For a mixture of fluorophores, the values of Q_i represent the fractional intensity of each fluorophore group at each observation wavelength [8].

The fitting quality of the decays was controlled by the least-squares method (χ^2), weighted residuals, and an autocorrelation function [8]. The residual plot and autocorrelation function are of particular advantage, in that they show where the lack of fitting occurs. The weighted residuals should be below four standard deviations and randomly fluctuate around zero with no obvious trend [8], [9]. The values of the autocorrelation function should be small and randomly distributed around zero for times >0.

To resolve the components of a mixture with different lifetimes, the data sets were subjected to global analysis [8–10] with FluoFit software (FluoFit Manual, PicoQuant, 2017). The fluorescence decay kinetic curves were registered sequentially in increments of 10 nm in the selected range of wavelengths using a single fluorescence excitation wavelength. For each data set, three fluorescence spectra were

IOP Conf. Series: Materials Science and Engineering **848** (2020) 012040 doi:10.1088/1757-899X/848/1/012040

constructed. Each point in the spectrum corresponded to the integral intensity of one of the three components (τ_1, τ_2 or τ_3) obtained from a single kinetic curve at taken wavelength. The calculation of χ^2 extended over several data sets.

2.3 Molecular Docking

The molecular docking processes of the BCD molecules into the HSA structure were simulated using AutoDock Vina 1.1.2 software [11]. The ligand molecules were analyzed in the ionized form. The atomic charges were calculated using the Gasteiger-Hückel method and geometric parameters were optimized using the Tripos force field in the Sybyl-X 2.1 software [12]. For the analysis and visualization of the results, the UCSF Chimera 1.13.1 [13] and Sybyl-X 2.1 [12] software was used.

3. Results and discussion.

3.1 Optical properties.

Polymethine dyes are a special class of organic compounds: they have the unique property of sensitizing light in visible and NIR range [14]. Polymethine compounds are characterized by relatively narrow intense main absorption bands, indicating a relatively small change in the bond length during excitation [15]. The study of the complexation of cyanine dyes with biomacromolecules is currently in the spotlight, as these dyes are used to detect nucleic acid molecules, and induce sensitized photochemical reactions in chemical and biological systems [16]. A special place among cyanine dyes is occupied by dyes whose molecules contain two conjugated chromophores, which exhibit two effects: dipole-dipole interaction and electron tunneling through the central heterocycle [17]. The result of these effects is the splitting of excited levels, which leads to a strong bathochromic shift in the absorption spectrum and makes these compounds promising photosensitizers [18]. Recently, biscarbocyanine dyes have been investigated as potential photosensitizers for photodynamic therapy, and high photocytotoxicity of these dyes was established on the example of cancer cells (melanoma). This discovery opens up new opportunities for the use of biscarbocyanines, the absorption spectrum of which is strongly shifted to the near-infrared region, as sensitizers in photodynamic therapy in oncology [19]. It was found that the localization of cyanine dyes in cancer cells occurs in the mitochondria, which determines their high photocytotoxicity [20]. An important role in the delivery of photosensitizers to cells is played by the process of complexation of dyes with human serum albumin and further photochemical reactions of excited dyes with proteins belonging to various cell organelles. In this paper, the spectral and kinetic characteristics of excited states of the complexes of biscarbocyanine dye 1,4-dimethyl-2,6-bis-((1E,3E)-3-(1,1,3-trimethyl-1,3-dihydro-2H-benzo[e]indol-2-ylidene)prop-1-ene-1-yl)-pyridin-1-ium iodide (BCD) and albumin were investigated.

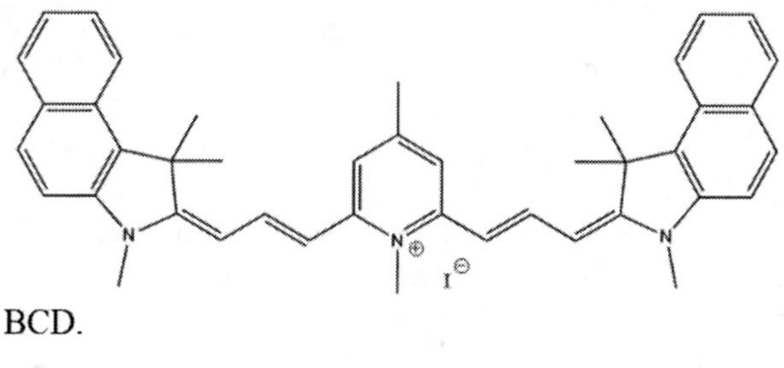

Scheme.1 Structure of BCD.

The absorption spectrum of BCD in the presence of HSA in PBS is shown in figure 1. According to our studies, the BCD absorption peak is near 660 nm, while a fluorescence peak is near 700 nm. This indicates that there is only a double chromophore system in this dye, and the effect of Davydov splitting is observed.

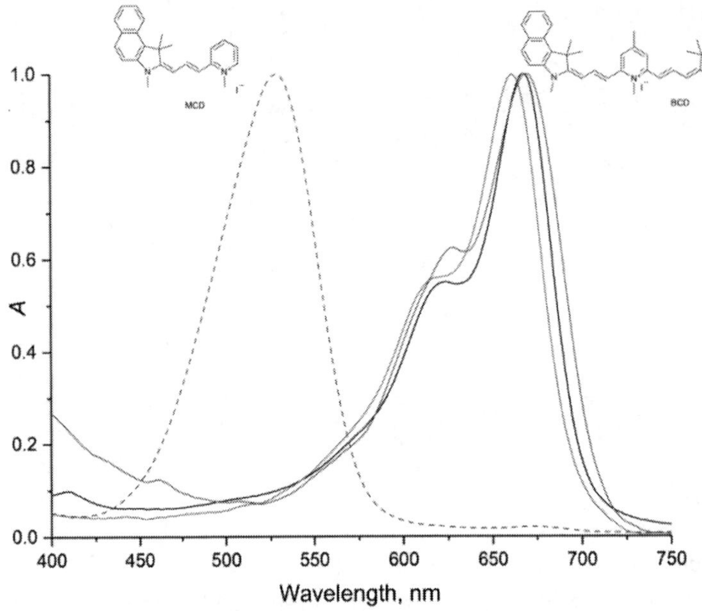

Figure 1. Normalized absorbion spectra of BCD in ethanol (λmax 668 nm) (black solid), PBS solution (λmax 661 nm) (red) and in the complex with HSA 10^{-4}M (λmax 669 nm) (blue). monochromophoric analog of BCD – (λmax 528 nm) (dashed line).

The addition of HSA to the dye in PBS solution shifts the absorption and the emission spectra by 10-20 nm to a longer wavelength range. Also, the intensity of fluorescence increased almost ten-fold. The increase of fluorescence while adding HSA indicates that internal conversion proceeds to the ground state from the excited state through the fluorescence pathway more effectively. It is due to the growth of dye's structure rigidity in complex with protein. The increase of protein concentration results in the reorganization of the complex. The binding constant can be assessed either from the changes in the absorption spectra on adding HSA or from the fluorescence spectra. The determination of a binding constant from the absorption spectra is less accurate mainly because of the aggregation of a dye in PBS. The value of K_b with HSA estimated ~ 10^5 M^{-1}.

3.2 TCSPC analysis of fluorescence kinetics.

Cyanine dyes are generally positively-charged molecules if no special substituents are added. Their structure consists of $2p_z$ conjugated orbitals with alternated and delocalized electron density. This makes the interaction of cyanine dyes with biomacromolecules quite complicated. [21] The contribution of Coulomb interaction is insignificant, and the contribution of the entropy factor to binding is indeterminate because albumin has different binding sites. The lowest value of fluorescence lifetime characterized by component τ_1 is constant for all concentrations of protein. It is equal to a characteristic fluorescence lifetime of unbound BCD molecules. In the presence of HSA at high concentrations of protein all BCD molecules are likely to be associated with HSA because of a high value of K_d[6]. Two components of the fluorescence decay of BCD associated with HSA correspond with two types of BCD-HSA complexes[6]. Complex formation characterized by lifetimes of τ_2 and τ_3 is associated with different types of complexes. The contribution of these components varies depending on the concentration of HSA. (Table 1).

Table 1 Fluorescence lifetimes τ_i ns of the mixtures of BCD with different HSA concentrations, the relative amplitudes of the components in % are given in parentheses.

690 nm			
10^{-6}	0.20 (96)	1.26 (4)	–
10^{-5}	0.19 (40)	0.92 (22)	1.68 (38)
10^{-4}	0.22 (13)	0.96 (36)	1.65 (51)

The lifetime value of the long-living component increases sufficiently with the increase of protein concentration. The interchange of amplitudes of components τ_2 and τ_3 is associated with migration of BCD between binding sites and the alteration of complexes' types (Figure 2-4). We attribute a change of fluorescence wavelength from 660 nm to 690 nm to this process.

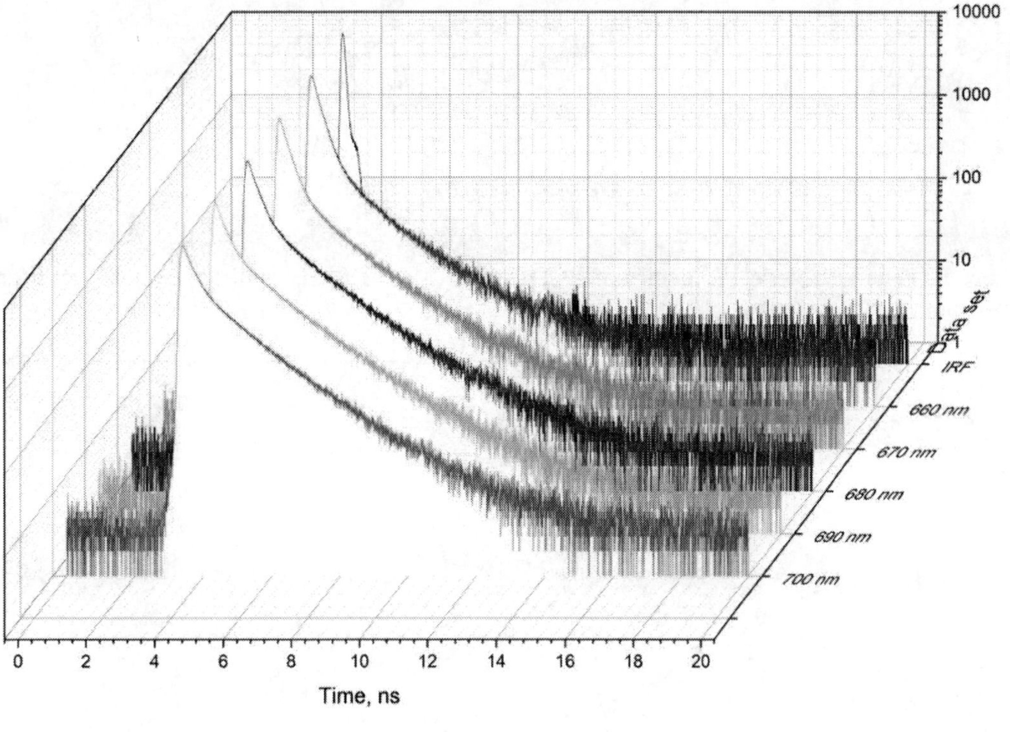

Figure 2. Fluorescence decay kinetics of BCD [1×10^{-7} M] + HSA [1×10^{-6} M] complex at different wavelengths, excitation 640 nm 70ps pulse.

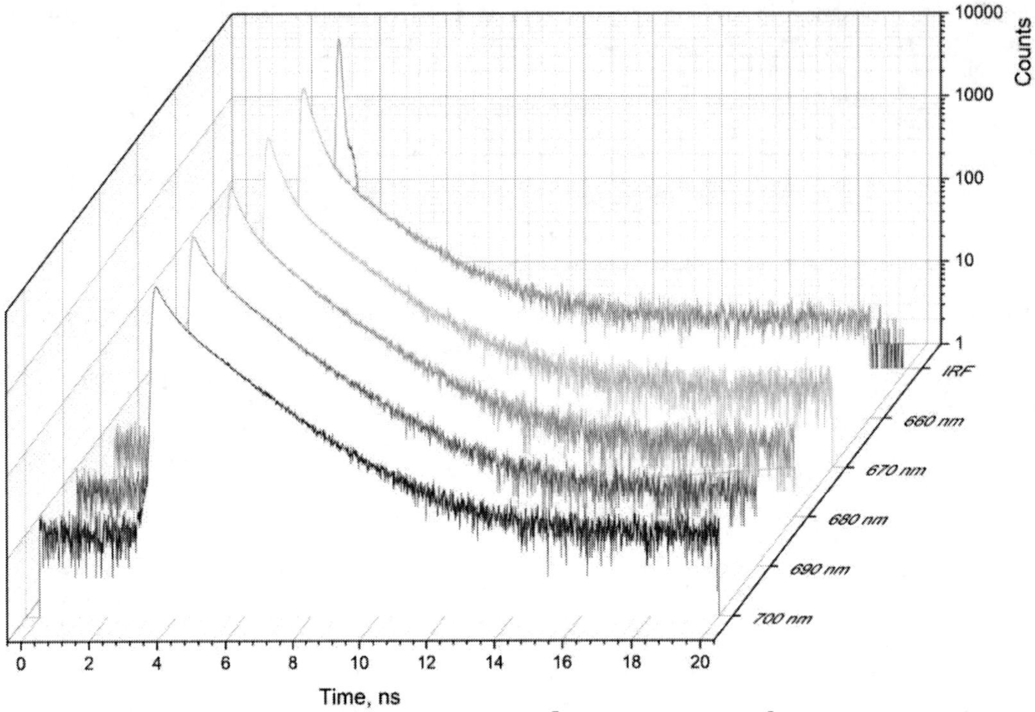

Figure 3. Fluorescence decay kinetics of BCD [1×10^{-7} M] + HSA [1×10^{-5} M] complex at different wavelengths, excitation 640 nm 70ps pulse.

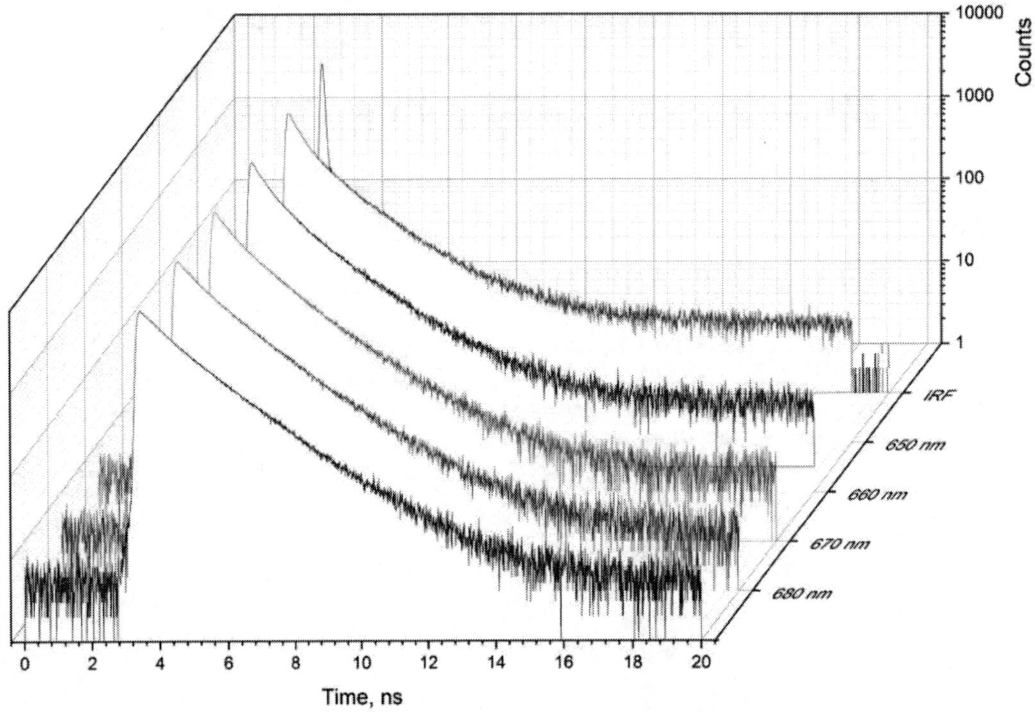

Figure 4. Fluorescence decay kinetics of BCD [1×10^{-7} M] + HSA [1×10^{-4} M] complex at different wavelengths, excitation 640 nm 70ps pulse.

IOP Conf. Series: Materials Science and Engineering **848** (2020) 012040 doi:10.1088/1757-899X/848/1/012040

The location of the BCD molecule in binding sites is stabilized and refers to the longest fluorescence lifetime. The interaction of BCD and albumin increase the rigidity of dye structure, and the electrostatic π-stacking with aromatic amino acid can change the rate of intersystem crossing to increase the quantum yield of the triplet state. The distance between BCD and tryptophan or tyrosine residues has great value to the possibility of the electron transfer between triplet state of a dye and the aromatic ring of amino acid[22]. Electron phototransfer leads to the formation of a reduced form of BCD, which when interacting with an oxygen molecule forms an intermediate – superoxide anion-radical, which further leads to the formation of highly reactive intermediates that destroy cell structures as a result of necrosis and apoptosis.

3.3 Molecular docking

The possible interactions between the BCD molecule and HSA were analyzed using a molecular docking method. All calculations were performed using AutoDock Vina 1.1.2 software [11]. The crystal structure of the human serum albumin was obtained from PDB ID (4L9Q). Among several proposed binding modes two most stable docking models should be noted. In both cases the BCD molecule is located in a wide, moderately polar pocket between the IIA and IIIA albumin subdomains (Figure 5). One substituted indoline fragment is oriented inside deep into the pocket while the other fragment is oriented towards the outer protein surface.

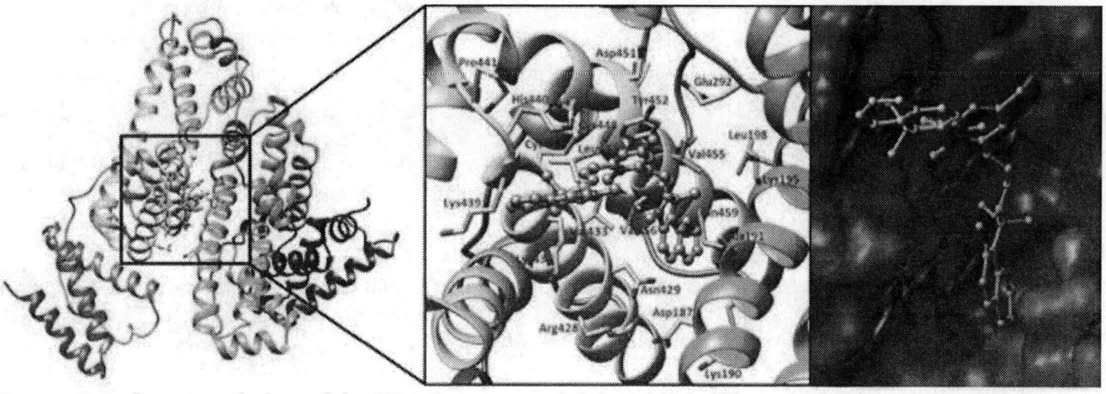

Figure 5. Left – general view of the HSA structure and the position of the polymethine dye binding site. Intermediate – detailed structure of the polymethine dye binding site. The ligand molecule is represented by beige ball-and-stick model. The amino acid residues located within 4 Å from it are shown as light-blue stick models. Right – view of the dye binding pocket in the HSA structure. The protein Connolly surface is colored according to the molecular lipophilic potential from brown (hydrophobic) to blue (hydrophilic) regions. The ligand molecule is represented by a grey ball-and-stick model.

Typically for the albumin ligands, the binding has mostly non-specific nature primarily based on hydrophobic interactions. As far as the second mode is concerned, there is an extra contribution to the binding from weak noncovalent π-π stacking interaction between the Tyr452 residue and the central pyridine ring (fig. 6). In order to estimate the distance between two aromatic rings that interact through the π-π stacking, an approach to calculate the distance between two rings' centroids (Figure 6) was suggested. The obtained value equals 4.241Å. Though, it is questionable if such π-π stacking distance value is acceptable. However, authors of [23] suggest that the values of 4.1 – 4.4 Å for the distances between two aromatic rings in parallel-displaced π-π stacking interaction counted for centroids are reasonable. Consequently, the obtained value satisfies the range of possible displaced π-π stacking distances as a preliminary estimation. However, it definitely requires further revision as a better approach to calculate this parameter might be devised.

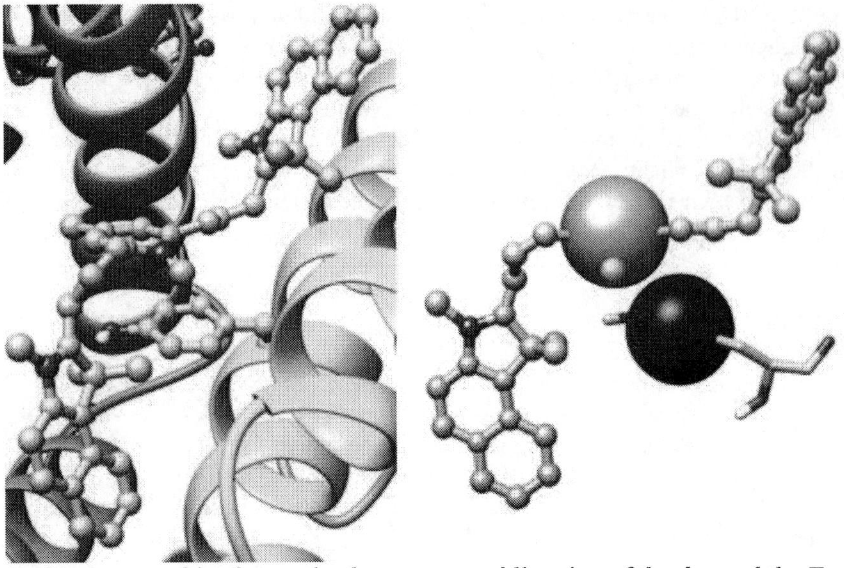

Figure 6. Left - $\pi - \pi$ stacking interaction between a pyridine ring of the dye and the Tyr452 residue. Right – centroids of Tyr452 benzene ring and the BCD pyridine ring to calculate the distance between the two aromatic rings. The ligand molecule is represented by a grey ball-and-stick model, the Tyr452 residue is represented by a blue ball-and-stick model.

The obtained data from the AutoDock shows that these two modes have slight differences in the binding energy of 0.3 kcal/mol. Consequently, it can be inferred that both binding modes are almost equally possible and, presumably, can interchange. Therefore, the dye molecule may insignificantly alter its conformation within the binding site skipping from one binding mode to another. Moreover, an overlay of two lowest energy conformers in their corresponding conformations in the binding site suggests the need of minor rotation of certain structural moieties for the molecule to transform from one state into another (Figure 7.). The supposition about such energetically low transitions does not conflict with the mechanism concluded from the spectral studies.

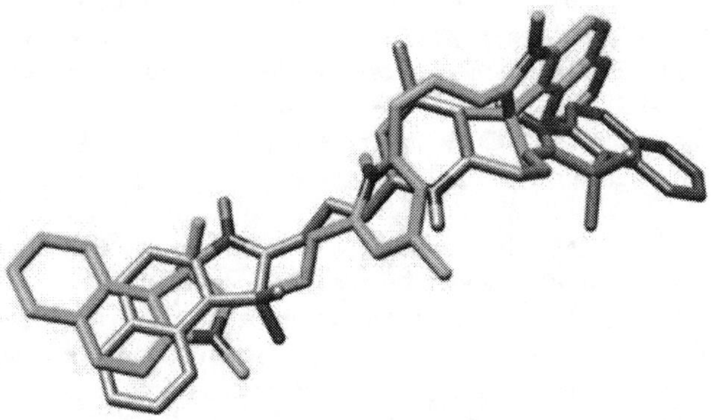

Figure 7. Overlay of two binding modes' structures. The conformers' molecules are represented by a grey and cyan stick models, and nitrogen atoms are colored in dark blue, respectively.

Ultimately, this binding site is the most probable one. However, it should be noted that binding in other albumin sites is also possible. Thus, the analysis of molecular docking results allows proposing a viable mechanism of the complex formation between the BCD molecule and HSA, which correlates well to the mechanism concluded from the spectral and fluorescent experiments.

4. Conclusion

In conclusion, a new bis-carbocyanine dye with an integrated heterocyclic fragment in the polymethine chain was synthesized, the complexation of the dye with HSA was studied using spectrophotometry, spectrofluorimetry, TCPSC, and molecular docking. The complexation of BCD and albumin changes the absorption spectrum and increases the fluorescence intensity. The experimentally determined value of the binding constant (~10^5 M^{-1}) indicates a high affinity of the dye to albumin. Analysis of fluorescence decays allows us to suggest the existence of at least three types of complexes between BCD and HSA. BCD-albumin complexes form as a result of rearrangements of the initially formed complex by re-distribution between HSA binding sites. A more rigid complex with a longer fluorescence lifetime demonstrates an increase in fluorescence and therefore a decrease in internal conversion by *cis-trans*-isomerization. The study of the binding site and conclusions about the complex's nature have been confirmed by molecular docking. The results of this study serve as a basis for further studies of the cellular accumulation and distribution of BCD-HSA using fluorescent imaging and confocal microscopy.

Conflicts of interest

The authors declare no conflict of interest.

Acknowledgements

Alexey A. Kostyukov and Mikhail G. Mestergazi acknowledge the financial support provided by the Russian Foundation for Basic Research, project no. 18-33-01112. Spectral measurements were performed in the Shared Research Facilities of IBCP RAS "New Materials and Technologies".

References

[1] Shershov V E, Lapa S A, Kuznetsova V E, Spitsyn M A, Guseinov T O, Polyakov S A, Stomahin A A, Zasedatelev A S и Chudinov A V 2017 Comparative Study of Novel Fluorescent Cyanine Nucleotides: Hybridization Analysis of Labeled PCR Products Using a Biochip *J. Fluoresc.* **27** 2001–16

[2] Murakami L S, Ferreira L P, Santos J S, da Silva R S, Nomizo A, Kuzmin V A и Borissevitch I E 2015 Photocytotoxicity of a cyanine dye with two chromophores toward melanoma and normal cells *Biochim Biophys Acta* **1850** 1150–7

[3] Guan Y, Zhang Y, Xiao L, Li J, Wang J, Chordia M D, Liu Z-Q, Chung L W K, Yue W и Pan D 2016 Improving therapeutic potential of farnesylthiosalicylic acid: Tumor specific delivery via conjugation with heptamethine cyanine dye *Mol. Pharm.* **14** 1–13

[4] Kovar J, Chen J, Draney D R, Olive D M, Volcheck W M, Xu X, Lugade A G и Narayanan N 2009 Optical fluorescent imaging

[5] Kostyukov A A, Pozdnyakova N V, Shevelev A B, Radchenko A S, Golovina G V, Klimovich O N, Shtil' A A, Codognato D C K, Gonçalves P J, Pavanelli A L S, Ferreira L P, Amado A M, Borisevich Y E и Kuzmin V A 2017 Complexes of alpha-fetoprotein and serum albumin with biscarbocyanine dye *High Energy Chem.* **51**

[6] Kuzmin V A, Nekipelova T D, Podrugina T A, Golovina G V, Kostyukov A A, Temnov V V, Doroshenko I A, Radchenko E V, Palyulin V A и Zefirov N S 2016 Complex formation of albumin with tricarbocyanine dyes containing phosphonate groups *Photochem. Photobiol. Sci.* **15** 1377–84

[7] Phillips D 1997 *Chemical Mechanisms in Photodynamic Therapy with Phthalocyanines* (Science and Technology Letters)

[8] Lakowicz J 2006 *Principles of Fluorescence Spectroscopy* т 1

[9] Lakowicz J R 1999 Instrumentation for fluorescence spectroscopy *Principles of fluorescence spectroscopy* (Springer) сс 25–61

[10] Nairat M, Konar A, Kaniecki M, Lozovoy V V и Dantus M 2015 Investigating the role of human serum albumin protein pocket on the excited state dynamics of indocyanine green using shaped femtosecond laser pulses *Phys Chem Chem Phys* **17** 5872–7

[11] Trott O и Olson A J 2010 AutoDock Vina: improving the speed and accuracy of docking with a new scoring function, efficient optimization, and multithreading *J Comput Chem* **31** 455–61

[12] Sybyl-X 2.1 St. Louis, Mo. 2013. и Sybyl-X 2.1 St. Louis, Mo., 2013. www.certara.com C L P No Title

[13] Pettersen E F, Goddard T D, Huang C C, Couch G S, Greenblatt D M, Meng E C и Ferrin T E 2004 UCSF Chimera—a visualization system for exploratory research and analysis *J. Comput. Chem.* **25** 1605–12

[14] Kuzmin V A, Golovina G V, Kostyukov A A, Nekipelova T D, Podrugina T A, Kriveleva A S, Matveeva E D, Proskurnina M V и Zefirov N S 2016 Spectral and kinetic parameters of photoexcited complexes of albumin and indotricarbocyanine dye with phosphonate substituents *High Energy Chem.* **50**

[15] Kuzmin V A, Podrugina T A, Nekipelova T D, Doroshenko I A, Proskurnina M V, Golovina G V, Radchenko E V, Kostyukov A A, Temnov V V, Matveeva E D, Palyulin V A и Zefirov N S 2016 New phosphonate-substituted tricarbocyanines and their interaction with bovine serum albumin *Dokl. Chem.* **470** 264–7

[16] Yin J, Kwon Y, Kim D, Lee D, Kim G, Hu Y, Ryu J H и Yoon J 2014 Cyanine-based fluorescent probe for highly selective detection of glutathione in cell cultures and live mouse tissues *J Am Chem Soc* **136** 5351–8

[17] Silva É R, Pavanelli A L S, Mostaço L B, Schaberle F A, Galembeck S E, Gonçalves P J, Costa e Silva R, Ferreira L P, Nekipelova T D, Kostyukov A A, Radchenko A S, Shtil A A, Kuzmin V A и Borissevitch I E 2017 Phototransformation of cyanine dye with two chromophores. Effects of oxygen and dye concentration *J. Photochem. Photobiol. A Chem.* **349** 42–8

[18] Kostyukov A A, Nekipelova T D, Radchenko A S, Golovina G V, Klimovich O N, Shtil' A A, Codognato D C K, Gonçalves P J, Pavanelli A L S, Ferreira L P, Amado A M, Borisevich Y E и Kuzmin V A 2017 Triplet states of the complexes of biscarbocyanine dye with albumin *High Energy Chem.* **51** 148–50

[19] Podrugina T A, Pavlova A S, Doroshenko I A, Kuzmin V A, Kostyukov A A и Shtil' A A 2018 Synthesis and photophysical properties of conformationally fixed tricarbocyanines with phosphonate groups *Russ. Chem. Bull.* **67**

[20] Radchenko A S, Kostyukov A A, Markova A A, Shtil A A, Nekipelova T D, Borissevitch I E и Kuzmin V A 2019 Photoactivated biscarbocyanine dye with two conjugated chromophores: complexes with albumin, photochemical and phototoxic properties *Photochem. Photobiol. Sci.* **18** 2461–8

[21] Shi C, Wu J B и Pan D 2016 Review on near-infrared heptamethine cyanine dyes as theranostic agents for tumor imaging, targeting, and photodynamic therapy *J Biomed Opt* **21** 50901

[22] Kostyukov A A, Nekipelova T D, Borissevitch I E и Kuzmin V A 2019 Interaction of the Triplet State of Biscarbocyanine Dye with a Nitroxyl Radical *High Energy Chem.*

IOP Conf. Series: Materials Science and Engineering **848** (2020) 012040 doi:10.1088/1757-899X/848/1/012040

53 87–8

[23] Avasthi K, Shukla L, Kant R и Ravikumar K 2014 Folded conformations due to arene interactions in dissymmetric and symmetric butylidene-linker models based on pyrazolo [3, 4-d] pyrimidine, purine and 7-deazapurine *Acta Crystallogr. Sect. C Struct. Chem.* **70** 555–61

IOP Conf. Series: Materials Science and Engineering **848** (2020) 012041 doi:10.1088/1757-899X/848/1/012041

Effect of evaporating additive on sintering and properties ceramics based on magnesium aluminate spinel

E A Kozlovskaya, D S Kseneva, M O Senina and D O Lemeshev

D Mendeleev University of Chemical Technology of Russia, Moscow, Russia
kozlovskaya.elen163@gmail.com

Annotation: Prospects of appliance of transparent ceramic based on magnesium aluminate spinel are considered. Sintering additive that makes it possible to synthesise transparent ceramic material from spinel was selected. Influence of evaporating additive B_2O_3 on properties ceramics based on magnesium aluminate spinel is shown.

1. Introduction

In the modern world, a promising development direction of the ceramic industry is the production of transparent ceramics used in the defense industry as visors for helmets and armor window of civil and military transport. Currently, technologies for producing transparent armor based on magnesium aluminate spinel (MAS) are actively used, which has a unique combination of properties, such as chemical inertness to aggressive media, thermal stability, low density, high mechanical performance, as well as excellent optical properties. The stable cubic structure of $MgAl_2O_4$ provides light transmission up to 92 % in the visible spectrum and isotropy of optical characteristics [1]. Due to the above properties, as well as the availability of raw materials MAS is one of the most promising materials for the manufacture of transparent ceramics with improved performance.

One of the most important stages of obtaining ceramics from MAS is the synthesis of precursor powder. To obtain materials with a high degree of transparency, it is necessary to achieve perfection of the technological process through the use of special methods of qualitative preparation of the charge, called chemical. The most promising of them should be considered two methods: co-precipitation from aqueous solutions, since it does not require the use of special equipment and high temperatures, and is characterized by a relatively short reaction time; the method of thermal synthesis, which is the easiest to use, does not require special equipment and is reproducible, because its course is not affected by any external parameters, whether the atmospheric pressure or ambient temperature.

It is known that a prerequisite for the creation of transparent ceramics is the complete absence of porosity in it. This is achieved by using expensive and time-consuming methods, such as spark plasma sintering (SPS), hot pressing (HP) and hot isostatic pressing (HIP). Numerous studies conducted over the years [2, 3, 4] have established the effect of evaporating boron oxide additive on the sintering process of spinel ceramics by the method of HIP and SPS at a temperature of 1300 °C. The results of the experiment show that the introduction of B_2O_3 additive in an amount of 0.15 wt. % significantly reduces porosity and increases the light transmission of the material in the visible spectrum up to 80 % at HIP and up to 70 % at SPS. Further, in [5] the effect of B_2O_3 additive in the amount of 0-10 wt. % on the properties of ceramics obtained by sintering in a vacuum at 1700 °C. As a result of the experiment, it was revealed that in order to obtain high light transmission and low porosity, it is necessary to introduce no more than 7 wt. % boron oxide.

The aim of this study is to investigate the effect of the concentration of evaporating additives on the sintering process of sample from magnesium aluminate spinel, the precursor of which was obtained by chemical co-precipitation ant thermal synthesis.

2. Materials and Methods

The method of reverse chemical co-precipitation from aqueous solutions is carried out by spraying a saturated solution of magnesium and aluminum nitrates taken in a stoichiometric ratio (in terms of oxides: MgO – 28.33 wt. %, Al_2O_3 – 71.67 wt. %, taking into account losses by calcination) in excess of a solution-precipitator as which the aqueous solution of ammonia is used.

The starting materials for thermal synthesis are aluminum hydroxide $Al(OH)_3$, obtained from aluminum nitrate by precipitation, and the basic magnesium carbonate $mMgCO_3 \cdot Mg(OH)_2 \cdot nH_2O$ in a ratio of 71.67 and 28.33 wt. %, respectively, in terms of oxides.

Differential scanning calorimetry of the precursors obtained by chemical precipitation was carried out. The results are shown in Figure 1. DSC mixtures of magnesium carbonate and aluminum hydroxide have a similar character.

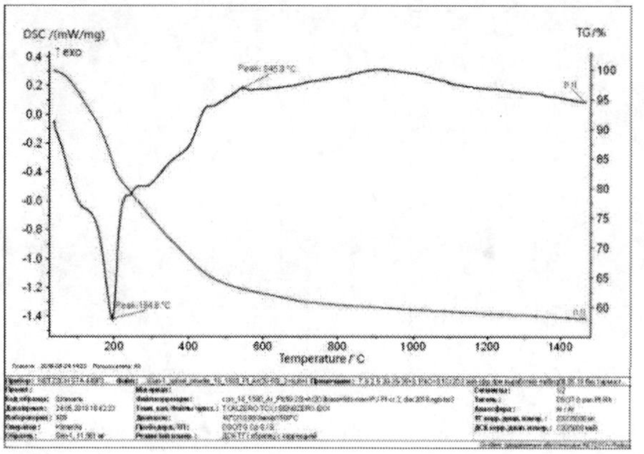

Figure 1. Results of DSC of the precursor spinel obtained by chemical precipitation.

On the basis of the results of differential scanning calorimetry, the temperature of the end of the crystallization of spinel equal to 1150 °C. For the purpose of more complete passage of the spinel formation process, a temperature of 1200 °C, was chosen.

To reduce porosity, B_2O_3 was selected as an additive in the concentration range from 5.0 to 6.0 wt. %, which is inserted into spinel powder by wet method. Mixing was carried oud for 15 minutes in a planetary mill. Molding of semi-finished products was carried out by semi-dry pressing at a pressing of 100 MPa with paraffin and a solution of carbon tetrachloride as a temporary technological bundle. Removal of bundle was carried out at a temperature of 1100 °C, and firing of samples – at a temperature of 1750 °C in a vacuum with a holding time of 3 hours.

3. Results and discussion

The choice of synthesis method and firing temperature is confirmed by the data of X-ray phase analysis (Figure 2, 3), which show the presence of a single phase of spinel.

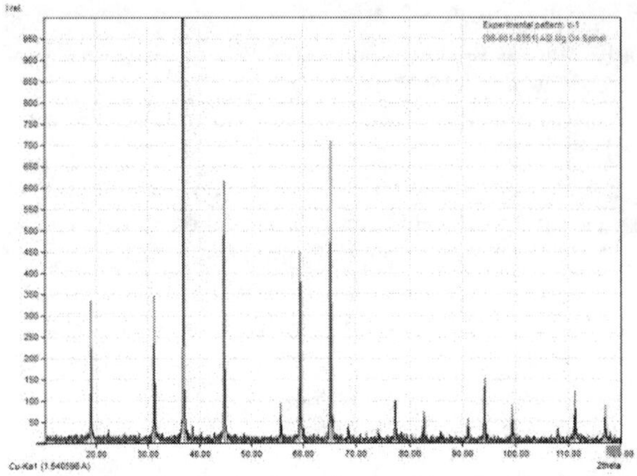

Figure 2. Radiograph of spinel synthesized by chemical precipitation.

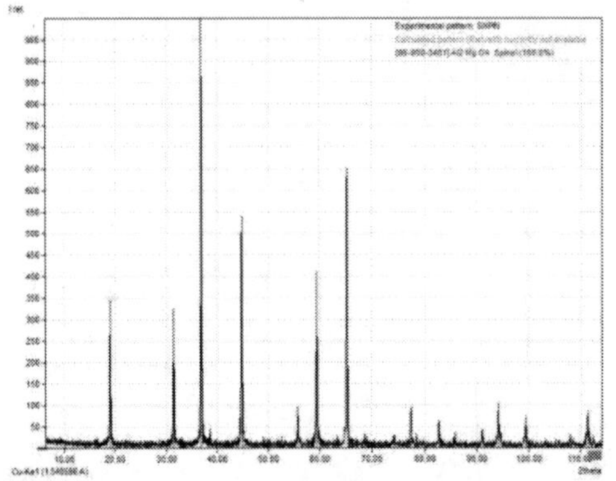

Figure 3. Radiograph of spinel synthesized by thermal synthesis.

The choice of temperature of removal of bundle is confirmed by results of the x-ray phase analysis of a material (Figure 4, 5) after firing at a temperature of 1100 °C. Radiographs shows the presence of a single spinel phase and the absence of boron compounds, which indicates the complete evaporation of boron oxide as a result of high-temperature treatment.

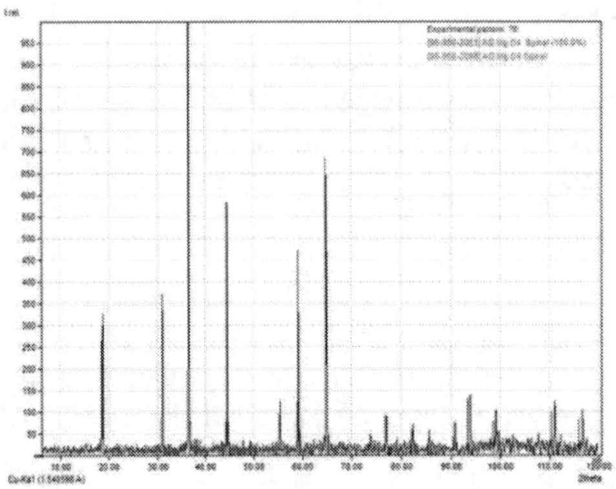

Figure 4. Radiograph of spinel synthesized by chemical precipitation after removal of bundle.

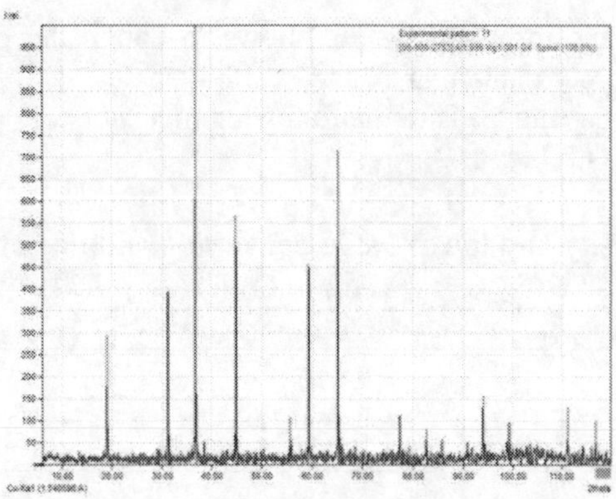

Figure 5. Radiograph of spinel synthesized by thermal synthesis after removal of bundle.

Photos of spinel microstructure (Figure 5, 6) synthesized both by precipitation and by thermolysis, show that the powder is represented by a polyfractive composition with a predominant content of isometric particles of 5-20 micrometers in size, as well as a small number of agglomerates with an average size up to 50-60 micrometers, which can affect the formation of pores in the material.

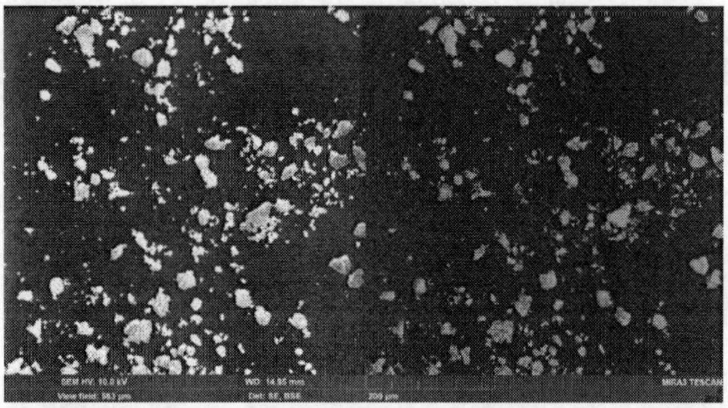

Figure 6. Photos of the microstructure of spinel synthesized by chemical precipitation.

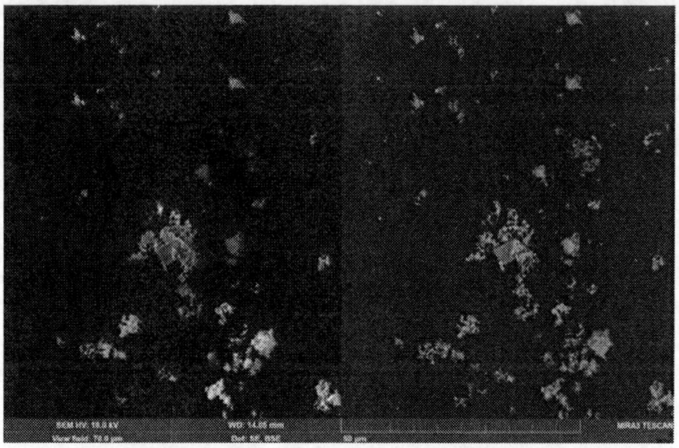

Figure 7. Photos of the microstructure of spinel synthesized by thermal synthesis.

It should be noted that this form of crystals allows to obtain the densest ceramics due to the formation of contacts of the type "plane-plane" and "knife-plane".

The analysis of the obtained data leads to the following indicators of open porosity and the average density of the burned samples (Tables 1, 2).

Table 1. Open porosity and the average density of samples obtained by chemical precipitation.

The concentration of additive, %	Open porosity, %	Average density, g/cm^3
5.0	4.7	3.32
5.5	4.5	3.37
6.0	5.3	3.25

Table 2. Open porosity and the average density of samples obtained by thermal synthesis.

The concentration of additive, %	Open porosity, %	Average density, g/cm^3
5.0	5.3	3.23
5.5	9.6	3.18
6.0	15.0	3.01

On the basis of experimental data, it can be concluded that the best properties have samples with a concentration of 5.5 wt. % additive obtained by chemical co-precipitation and samples with an additive concentration of 5.0 wt. % obtained by the method of thermal synthesis. To clarify the more optimal conditions for obtaining ceramics on the basis of MAS, a comparative analysis of the data obtained (Figure 7).

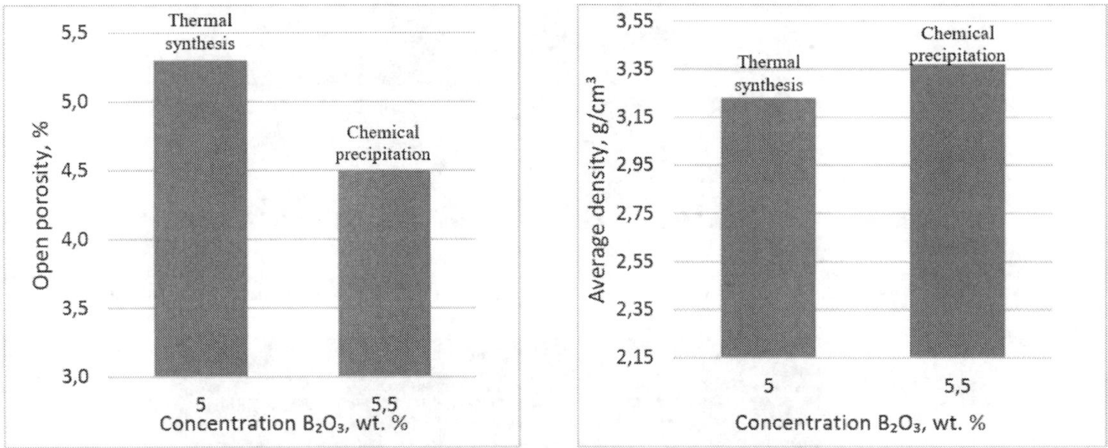

Figure 8. Properties of annealed samples.

4. Conclusion

As the outcome of the research it is revealed that the ceramics from magnesium aluminate spinel received by chemical co-precipitation with addition of boron oxide in quantity of 5.5 wt. % has the best properties. When included into the powder MAS sintering additives of this concentration, the open porosity becomes quite low – 4.5 %, and the average density – high – 3.37 g/cm^3. This is explained by the fact that the additive acts first by mechanism of liquid-phase sintering at a temperature of 490 °C, which is confirmed by the data of x-ray phase analysis presented in figures 4 and 5. In this case, boron oxide does not affect the composition and properties of the sample. Then sintering passes through a solid-phase mechanism.

Acknowledgments
The reported study was funded by RFBR according to the research project № 18-33-00507.

References
[1] Pavlyukova L T, Lukin E S, Popova N A 2013 Proceedings of "The VII project contest of young scientists" p 28
[2] Kong L B [et al] 2015Topics in Mining, Metallurgy and Materials Engineering p.29
[3] Tsukuma K 2006 Journal of the Ceramic Society of Japan 114 p 802
[4] Sarıdaş S, Göller G, Yücel O, Şahin F 2016 18th International Metallurgy & Materials Congress p 145
[5] M. O. Senina [et al]. Modern methods and technologies of materials creation and processing: proceedings of the XIII International scientific and technical conference 2018

IOP Conf. Series: Materials Science and Engineering **848** (2020) 012042 doi:10.1088/1757-899X/848/1/012042

Dicationic disiloxane liquids

V G Krasovskiy[1], O B Gorbatsrvich[2], L M Glukhov[1] and E A Chernikova[1]

[1]N. D. Zelinsky Institute of Organic Chemistry, Russian Academy of Sciences, 47 Leninsky prosp., 119991 Moscow, Russian Federation. e-mail: miyusha@mail.ru
[2]N. S. Enikolopov Institute of Synthetic Polymeric Materials, Russian Academy of Sciences, 70 ul. Profsoyuznaya, 117393 Moscow, Russian Federation

Abstract: Bis(trifluoromethylsulfonyl)imide dicationic ionic liquids containing disiloxane linkers between imidazole cations were synthesized. Their thermal stabilities, melting points, viscosities, and volatilities in vacuo were measured. The possibility of tuning the properties of dicationic ILs by introducing substituents of different nature into the cation structure is shown.

1. Introduction

Polar ionic liquids with very low volatility, low viscosity and wide operating temperature range are used in various fields of science and technology. Ionic liquids are used as polar solvents in organic synthesis [1,2], in the synthesis of metal nanoparticles [3], in electrochemistry [4,5] as electrolytes in lithium and solar batteries, in separation processes as liquid phases in capillary columns gas-liquid chromatography [6] and as extracting agents [7]. New application areas for ionic liquids are constantly emerging. The possibilities of their application as depolymerizing reagents in the processing of plastic waste [8,9], as the media for storing carbon dioxide [10], drying agents for industrial gases [11,12] and air conditioning systems [13,14] are also investigated. Ionic liquids can be used as working fluids in compressors to compress oxygen [15]. There are works on the use of ionic liquids as high-energy matrices [16]. In recent years, many papers have been published on the possibility of using ionic liquids as heat transfer fluids for heat transport in solar panels [17,18] and refrigeration plants [18-20]. There are some works on the estimation of possibility of use of ILs as heat transfer agents under conditions of dynamic vacuum [21-23].

The ionic interactions in ILs causes their low volatility and contribute to an increase in viscosity. The presence in the structure of an IL of polar groups and other heteroatoms generally reduces their thermal stability. To obtain an IL with low volatility, low viscosity and high thermal stability it is necessary to solve the optimization problem on the interrelated three parameters "viscosity-volatility-thermal stability". Generally, the most thermally stable ILs are imidazolium liquids with bis(trifluoromethylsulfonyl)imide anion (Tf_2N^-). This anion also allows obtaining ILs with low viscosities. Monocationic imidazolium ionic liquids with Tf_2N^- anion, without polar substituents in the cation structure, are characterized by a volatility value of 20-30 mg/h·cm^2 in a vacuum of ~0.013 Pa [24]. The presence in the cation structure of hydroxyl groups capable of participating in intermolecular interactions leads to a decrease in the volatility of ILs to ~10 mg/h·cm^2, but significantly increases their viscosity [24]. Introduction of a second ion pair into an IL structure (dicationic ILs) significantly reduces its volatility (up to ~0.1 mg/h·cm^2), but often leads to an increase in the melting point of the IL above room temperature and/or to a significant increase in viscosity [25,26]. The rotational energy barriers of Si-O and C-C bonds are 0.8 and 11.3 kJ/mol, respectively, and ones for Si-CH_3 and C-CH_3 bonds are 6.7 and 15.1 kJ/mol [27,28]. Use of siloxane structural fragments characterized by greater mobility as a linker in dicationic ILs will lead to a decrease in the viscosity and the tendency to crystallization of the ILs [29,30]. The aim of this work is the synthesis of room-temperature dicationic imidazolium ionic liquids (RTDIL) with disiloxane linker with low viscosity, low volatility and high thermal stability.

2. Experimental

2.1. Materials and methods

The melting points and glass transition temperatures of ILs were determined via differential scanning calorimetry (DSC) using a DSC-822e instrument (Mettler-Toledo, Switzerland). The measurements were performed in the temperature range from −100 to 100°C at a sample heating rate of 10°C/min in an argon atmosphere. Thermogravimetric analysis was performed on a Derivatograph-C instrument (MOM, Hungary) in an argon atmosphere at a heating rate of 10°C/min (sample mass, ~20 mg). Kinematic viscosity was measured using an Ostwald viscometer. The viscometer was calibrated at 25°C using ethylene glycol (Aldrich, 99.8%, water content < 0.01%) as the reference liquid. The viscosity of ethylene glycol was also checked against bidistilled water in another viscometer. The evaporation of ILs in vacuum was studied using a McBain quartz spring balance. Each sample (~0.2 g) was placed in a ceramic cup attached to the movable end of the spring of the balance. The surface area of the liquid layer was about 1.7 cm^2. The tube casings of the spirals with samples were placed in a thermostatted aluminum block. Spring elongation was determined from the change in the positions of reference marks using a KM-8 cathetometer with an accuracy of ±0.02 mm. The spring had a sensitivity of 0.3709 mm/mg. The setup was evacuated with a diffusion pump. Before measurements the samples were dried to constant weight (for ~15 h) in vacuum ~0.013 Pa at 100°C. ^1H and ^{13}C NMR spectra were recorded on a Bruker AM300 spectrometer (300.13 and 75.47 MHz, respectively) in DMSO-d_6. IR ATR spectra were measured on a Nicolet iS50 IR spectrometer (built-in ATR attachment, with diamond crystal) with resolution of 4 cm^{-1}.

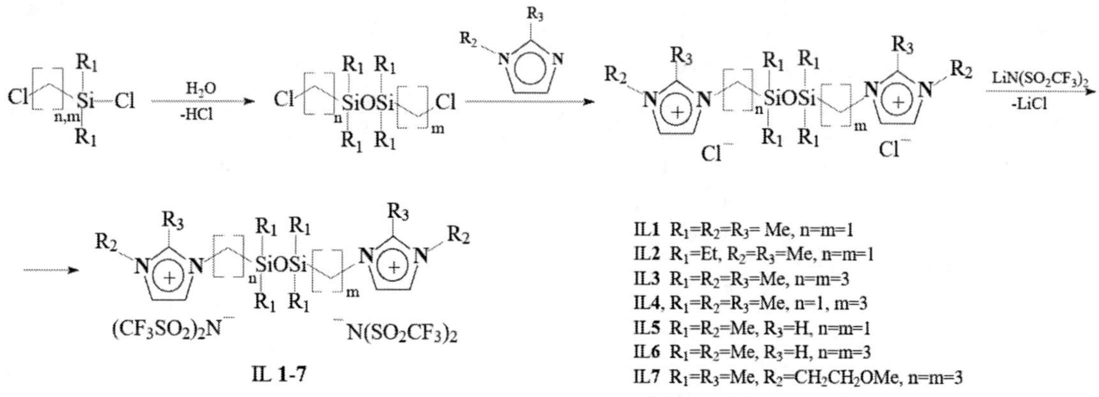

Scheme 1. Synthesis of disiloxane dicationic IL **1-7**.

Dimethyl(chloromethyl)chlorosilane (98%), dimethyl(3-chloropropyl)chlorosilane (97%) and diethyl(chloromethyl)chlorosilane (97%) were purchased from ABCR. 1-Methylimidazole (99%), 1,2-dimethylimidazole (98%), 1-(2-methoxyethyl)-2-methylimidazole (97%), lithium bis(trifluoromethylsulfonyl)imide (99%) were purchased from Acros and Sigma-Aldrich. Before addition of chloroalkylsiloxanes, the solutions of imidazoles were dried by azeotropic distillation of acetonitrile. All organic solvents used in the synthesis were previously dried over CaH_2 and distilled.

2.2. Methods for the synthesis of ionic liquids.

Synthesis of the ionic liquids was carried out in three stages [23,31]. At the first step, a disiloxane linker with chloralkyl end groups was obtained by hydrolytic condensation of the corresponding chlorosilane in water. At the second stage, alkylimidazole was quaternized with equimolar amount of siloxane linker in acetonitrile, and in the third stage Cl^- anion was exchanged for Tf_2N^- anion with lithium bis(trifluoromethylsulfonyl)imide (10% excess) in acetonitrile (Scheme 1). Lithium salts were removed by washing with water.1,1,3,3-Tetramethyl-1-chloromethyl-3-(3-chloropropyl)disiloxane used as asymmetric linker was synthesized (Scheme 2) by the disproportion reaction from symmetric disiloxanes 1,1,3,3-tetramethyl-1,3-di(chloromethyl)disiloxane and 1,1,3,3-tetramethyl-1,3-bis(3-chloropropyl) disiloxane without solvent in the presence of cation exchange resin Purolite ct175/2429 [32], and the reaction mixture was fractionally distilled (target product T_b 72-73°C/133 Pa). The yield of asymmetric disiloxane was 96%.

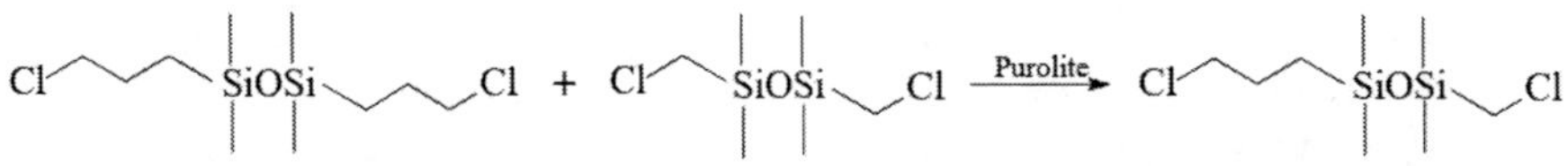

Scheme 2. Synthesis of the asymmetric disiloxane linker.

The structures of the synthesized ionic liquids were confirmed by IR and NMR spectroscopy and elemental analysis. All chloride salts synthesized in the second stage are soluble in water; the final ILs with Tf_2N^- anion are hydrophobic.

*2.2.1. 1,1,3,3-Tetramethyl-1,3-di([1,2-dimethylimidazolium-3-yl]methyl)disiloxane bis(trifluoromethylsulfonyl)imide (**1**):*

First stage of the synthesis (Scheme 1).

Dimethyl(chloromethyl)chlorosilane (100 mL, 108.6 g) was added dropwise to 300 mL of distilled water with stirring on a magnetic stirrer. Then the reaction mixture was stirred for 1 hour, followed by addition of 200 mL of hexane. The acidic water layer was removed, and the organic layer was washed with water until neutral pH. Hexane was removed by distillation, and then in vacuum at room temperature. Yield of 1,1,3,3-tetramethyl-1,3-di(chloromethyl)disiloxane was 85.6 g (97.5 %).

Second stage of the synthesis.

1,1,3,3-Tetramethyl-1,3-di(chloromethyl)disiloxane (3.15 g, 0.0136 mol) was added to a solution of 2.61 g (0.0272 mol) of 1,2-dimethylimidazole in 6 mL of acetonitrile under argon atmosphere. The reaction mixture was stirred at reflux temperature for 72 hours. After cooling, the product was collected by filtration and dried in a vacuum. Yield 5.07 g, 88 %.

Third stage of the synthesis.

To 5 g (0.0118 mol) of chloride salt a solution of 7.45 g (0.0260 mol) of lithium bis(trifluoromethylsulfonyl)imide ($LiNTf_2$) in 15 mL of acetonitrile was added under argon atmosphere. The reaction mixture was stirred at room temperature for 3 hours. The precipitate of lithium chloride was filtered off, acetonitrile was removed from the filtrate by rotary evaporation, and the residue was dissolved in 50 mL of dichloromethane. The solution was repeatedly washed with water to remove lithium chloride until no chloride can be detected in wash waters with silver nitrate. The target ionic liquid was dried by azeotropic distillation of dry dichloromethane. Tf_2N-IL **1** was obtained with yield of 10.02 g (93%). ^{1}H NMR (DMSO-d_6, 300.13 MHz), δ, ppm: 0.16, s (12H, CH_3-Si), 2.52, s (6H, CH_3-C), 3.75, s (6H, CH_3-N), 3.82, s (4H, CH_2-N), 7.41, m (2H, CH=), 7.62, m (2H, CH =). ^{13}C NMR (DMSO-d_6, 75.47 MHz), δ, ppm: 0.48, 9.74, 35.26, 40.21, 113.45, 11.70, 121.83, 122, 07, 122.81, 126.37, 143.97. Gross formula: $C_{20}H_{32}N_6O_9F_{12}S_4Si_2$, molecular weight: 912.92. Calculated, %: C 26.31, H 3.53, N 9.21, F 24.97, S 14.05, Si 6.15. Found, %: C 26.23, H 3.63, N 9.16, F 24.94, S 13.91, Si 6.26.

*2.2.2. 1,1,3,3-Tetraethyl-1,3-bis([1,2-dimethylimidazolium-3-yl]methyl)disiloxane bis(trifluoromethylsulfonyl)imide (**2**):*

IL **2** was synthesized a method similar to the method of obtaining IL **1**. Yield 90%. ^{1}H NMR (DMSO-d_6, 300.13 MHz), δ, ppm: 0.63 m (8H, $\underline{CH_2}CH_3$), 0.86 m (12H, $CH_2\underline{CH_3}$), 2.53 s (6H, C-CH_3), 3.74 s (6H, N-CH_3), 3.87 m (4H, N-CH_2), 7.38 m (2H, =CH), 7.62 m (2H, =CH). ^{13}C NMR (DMSO-d_6, 75.47 MHz), δ, ppm: 5.54, 6.25, 9.68, 35.25, 37.80, 113.60, 117.80, 121.90, 121.96, 122.94, 126.40, 144.04. Gross formula: $C_{24}H_{40}N_6O_9F_{12}S_4Si_2$, molecular weight: 969.03. Calculated, %: C 29.75, H 4.16, N 8.67, F 23.53, S 13.23, Si 5.80. Found, %: C 29.71, H 4.22, N 8.63, F 23.51, S 13.27, Si 5.85.

2.2.3. 1,1,3,3-Tetramethyl-1,3-bis(3-[1,2-dimethylimidazolium-3-yl]propyl)disiloxane bis(trifluoromethylsulfonyl)imide (3):

IL **3** was synthesized a method similar to the method of obtaining IL **1**. Yield 95%. ^{1}H NMR spectrum (DMSO-d_6, 300.13 MHz), δ, ppm: 0.04, s (12H, CH_3-Si), 0.49, m (4H, $CH_2CH_2CH_2Si$), 1.67, m (4H, $CH_2CH_2CH_2Si$), 2.55, s (6H, CH_3-C), 3.74, s (6H, CH_3-N), 4.06, m (4H, $NCH_2CH_2CH_2Si$), 7.61, m (4H, CH=). ^{13}C NMR spectrum (DMSO-d_6, 75.47 MHz), δ, ppm: 0.33, 9.43, 13.50, 23.90, 34.97, 50.51, 113.52, 117.79, 121.16, 122.05, 122.70, 126.32, 144.54. Gross formula: $C_{24}H_{40}F_{12}N_6O_9S_4Si_2$, molecular weight: 969.03. Calculated, %: C 29.75, H 4.16, N 8.67, F 23.53, S 13.23, Si 5.80. Found, %: C 29.66, H 4.27, N 8.60, F 23.50, S 13.19, Si 5.87.

*2.2.4. 1,1,3,3-tetramethyl-1-([1,2-dimethylimidazulium]-methyl)-3-(3-[1,2-dimethylimidazulium-3-yl]propyl))disiloxane bis(trifluoromethylsulfonyl)imide (**4**):*

IL **4** was synthesized a method similar to the method described for IL **1**. Yield 89%. ^{1}H NMR (DMSO-d_6, 300.13 MHz), δ, ppm: 0.05, s (6H, CH_3-Si), 0.17, s (6H, CH_3-Si), 0.49, m (2H, $CH_2CH_2\underline{CH_2}Si$), 1.67, m (2H, $CH_2\underline{CH_2}CH_2Si$), 2.53, s (3H, CH_3-C), 2.58, s (3H, CH_3-C), 3.76, s (6H, CH_3-N), 3.80, s (2H, CH_2-N), 4.07, m (2H, $N\underline{CH_2}CH_2CH_2Si$), 7.41, m (1H, CH=), 7.62, m (3H, CH=). ^{13}C NMR (DMSO-d_6, 75.47 MHz), δ, ppm: -0.51, 0.09, 9.45, 9.61, 14.31, 23.75, 34.99, 35.13, 40.45, 50.58, 113.55, 117.81, 121.18, 121.81, 122.08, 122.64, 122.74, 126.34, 143.86, 144.58. Gross formula: $C_{22}H_{36}N_6O_9F_{12}S_4Si_2$, molecular weight: 940.98. Calculated, %: C 28.08, H 3.86, N 8.93, F 24.23, S 13.63; Si 5.97. Found, %: C 27.96, H 3.91, N 8.85, F 24.43, S 13.33, Si 6.04.

IOP Conf. Series: Materials Science and Engineering **848** (2020) 012042 doi:10.1088/1757-899X/848/1/012042

*2.2.5. 1,1,3,3-Tetramethyl-1,3-bis([1-methylimidazolium-3-yl]methyl)disiloxane bis(trifluoromethylsulfonyl)imide (**5**):*
IL **5** was synthesized by a method similar to the method described for IL **1**. Yield 88%. ^{1}H NMR (DMSO-d$_6$, 300.13 MHz), δ, ppm: 0.16 s (12H, CH_3-Si), 3.87 s (6H, CH_3-N), 3.89 s (4H, CH_2-N), 7.52 m (2H, =CH), 7.69 m (2H, =CH), 8.90 s (2H, NCHN). ^{13}C NMR (DMSO-d$_6$, 75.47 MHz), δ, ppm: 0.96, 36.16, 41.37, 113.42, 117.81, 122.07, 123.54, 124.15, 126.31, 136.25. Gross formula: $C_{18}H_{28}N_6O_9F_{12}S_4Si_2$, molecular weight: 884.87. Calculated, %: C 24.43, H 3.20, N 9.50, F 25.76, S 14.49, Si 6.35. Found, %: C 24.33, H 3.35, N 9.46, F 25.73, S 13.51, Si 6.45.

*2.2.6. 1,1,3,3-Tetramethyl-1,3-bis(3-[1-methylimidazolium-3-yl]propyl)disiloxane bis(trifluoromethylsulfonyl)imide (**6**):*
IL **6** was synthesized by a method similar to the method described for IL **1**. Yield 96%. ^{1}H NMR (DMSO-d$_6$, 300.13 MHz), δ, ppm: 0.06, s (12H, CH_3-Si), 0.45, m (4H, $CH_2CH_2\underline{CH_2}Si$), 1.78, m (4H, $CH_2\underline{CH_2}CH_2Si$), 3.86, s (6H, CH_3-N), 4.13, m (4H, $N\underline{CH_2}CH_2CH_2Si$), 7.69, m (4CH=), 7.73, (4CH=), 9.08, s (2H, NCHN). ^{13}C NMR ((DMSO-d$_6$, 75.47 MHz), δ, ppm: 0.47, 14.53, 24.27, 36.13, 51.88, 113.54, 117.80, 122.07, 122.59, 124.05, 126.33, 136.94. Gross formula: $C_{22}H_{36}F_{12}N_6O_9S_4Si_2$, molecular weight: 940.98. Calculated, %: C 28.08, H 3.86, N 8.93, F 24.23, S 13.63, Si 5.97. Found, % C 27.99, H 3.95, N 8.90, F 24.20, S 13.51, Si 6.07.

2.2.7. 1,1,3,3-Tetramethyl-1,3-bis(3-[1-(2-methoxyethyl)-2-methylimidazolium-3-yl]propyl)disiloxane bis(trifluoromethylsulfonyl)imide (7):
IL **7** was synthesized by a method similar to the method described for IL **1**. Yield 94%. ^{1}H NMR (DMSO-d$_6$, 300.13 MHz), δ, ppm: 0.05, s (12H, CH_3-Si), 0.49, m (4H, $CH_2CH_2CH_2Si$), 1.71, m (4H, $CH_2CH_2CH_2Si$), 2.61, s (6H, CH_3-C), 3.25, s (6H, CH_3-N), 3.66, m (4H, CH_2-O), 4.09, m (4H, $NCH_2CH_2CH_2Si$), 4.32, m (4H, CH_2-N) , 7.65, m (4H, CH=). ^{13}C NMR (DMSO-d$_6$, 75.47 MHz), δ, ppm: 0.42, 9.76, 14.51, 23.85, 47.91, 50.51, 58.59, 70.29, 113.55, 117.81, 121.55, 122.08, 126.34, 144.70. Gross formula: $C_{28}H_{48}F_{12}N_6O_{11}S_4Si_2$, molecular weight: 1057.14. Calculated, %: C 31.81, H 4.58, N 7.95, F 21.57, S 12.13, Si 5.31. Found, %: C 28.67, H 4.66, N 7.88, F 21.62, S 12.02, Si 5.40.

3. Results and Discussion

3.1. Aggregate state of IL at RT
IL **1** (Table 1) obtained from 1,2-dimethylimidazole, despite the presence of a bulky fluorine-containing Tf_2N-anion, is a solid with T_m = 69°C. At the first stage of the study, an attempt was made to reduce T_m by introducing more bulky substituents at the silicon atom of the linker, i.e., to increase the steric effect. Replacement methyl for ethyl substituents at the silicon atom (IL **2**, Table 1) changes the melting point by only 1°C, and the replacement of the methylene spacer between imidazole and the silicon atom with propylene (IL **3**) reduces the T_m by 10° (Table 1). From the obtained data it is concluded that the increase in the steric effect with an increase in the size of substituents in the silicon atom in the linker has a weak effect on the crystallization ability of IL. Thus, increasing the steric effect in the linker and using an asymmetric linker does not result in RTDIL.

Table 1. Melting point/glass transition temperature, decomposition temperature and volatility of disiloxane ILs.

№ IL	Structure of IL (Scheme 1)	T_m,°C	T_d, °C	Volatility, mg/h·cm^2 *in vacuo* ~0.013Pa at 150°C/190°C/220°C
1	R_1=R_2=R_3=Me, *n*=*m*=1	69	437	0.00/0.03/0.04
2	R_1=Et, R_2=R_3=Me, *n*=*m*=1	68	432	0.02/0.05/0.07
3	R_1=R_2=R_3=Me, *n*=*m*=3	59	419	0.01/0.02/0.17
4	R_1=R_2=R_3=Me, *n*=1, *m*=3	42	420	0.02/0.11/0.40
5	R_1=R_2=Me, R_3=H, *n*=*m*=1	-52*	435	0.02/0.08/0.22
6	R_1=R_2=Me, R_3=H, *n*=*m*=3	-58*	402	0.03/0.33/0.95
7	R_1=R_3=Me, *n*=*m*=3, R_2=$CH_2CH_2OCH_3$	-48*	416	0.02/0.22/0.94

*t_g

It is known that the presence of polar groups in the cation reduces the tendency of IL to crystallization [24]. The use of 1-methylimidazole having a rather mobile hydrogen atom in the position 2, instead of 1,2-dimethylimidazole, results in ILs **5** and **6**, which are liquid at room temperature (RTDIL). The absence of crystallization for ILs **5** and **6** (Tg = -52 and -58°C, respectively, table 1) is explained by the interaction of the mobile hydrogen atom in imidazole and Tf_2N-anion, which prevents the crystallization of an IL [33]. The introduction of an oxygen-containing polar substituent - 2-methoxyethyl group into the imidazole cation also prevents the crystallization of IL **7** (Tg = -48°C, table 1). However, the use of hydroxyl substituents in the cation (in particular, 2-hydroxyethyl groups) leads to protolysis of the linker with methylene spacer during quaternization reaction [31].

Thus, the presence of polar groups in the cation, capable of additional intermolecular interaction with the formation of hydrogen bonds, prevents the crystallization of dicationic disiloxane liquids. It is interesting to assess the effect of such substituents on other properties of IL: thermal stability, viscosity and volatility.

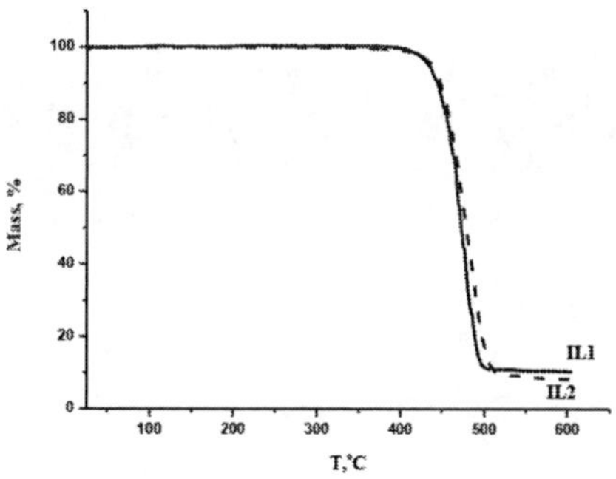

Figure 1. TGA curve IL **1** и **2**.

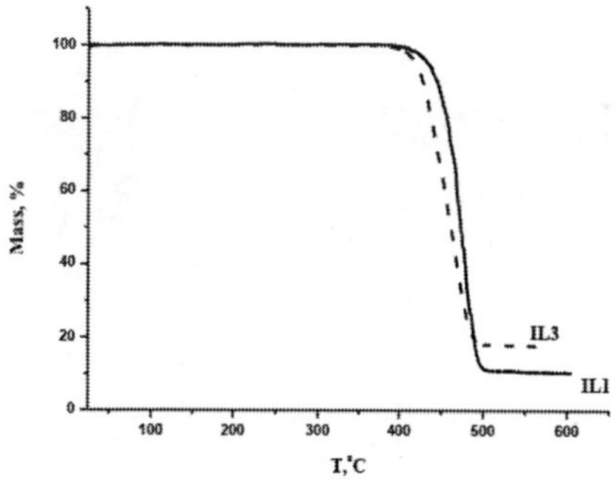

Figure 2. TGA curve IL **1** и **3**.

3.2. Thermal stability of ILs

The use of 1,2-dimethylimidazole moiety together with Tf_2N^- anion are expected to provide high thermal stability of the resulting ILs. For IL **1-4**, with different substituents at the silicon atom, the thermal destruction temperature (T_d) varies within 437-419°C. T_d of an IL decreases when the length of the substituents at the silicon atom is increased, and the presence of a propyl group in the linker has a greater influence on thermal stability (Figure 1 and 2).

The presence of a mobile hydrogen atom in the position 2 of imidazole has an unusual effect on the thermal stability of ILs. Use of 1-methylimidazole instead of 1,2-dimethylimidazole in the synthesis of ILs with a methylene spacer (ILs 1 and 5, Fig.3) practically does not affect their thermal stability (T_d 437 and 435°C, respectively), but an increase in the spacer length in such ionic liquids leads to a significant decrease in thermal stability – from 437 to 402°C (IL **1**, **5** and **6**, table.1, Fig. 3).

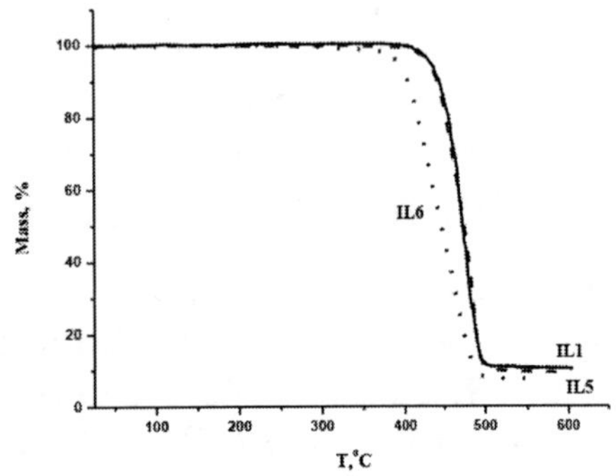

Figure 3. TGA curve IL **1**, **5** и **6**.

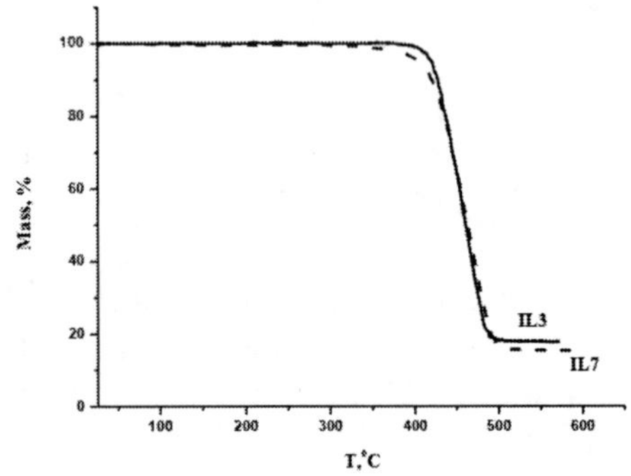

Figure 4. TGA curve IL **3**, и 7.

The presence of 2-methoxyethyl group in the cation changes the thermal stability of an IL to a much lesser extent – from 419 to 416°C (IL **3** and **7**, table 1, Fig. 4), but the combined effect of spacer elongation and the presence of a polar group reduces the thermal stability of an IL by 21°C (ILs **1** and **7**).

Thus, of the three RTDIL synthesized, only IL **5**, obtained from 1-methylimidazole and having a methylene spacer, has almost the same thermal stability as IL **1**. Elongation of the spacer in a disiloxane linker leads to a decrease in the thermal stability of the IL by 18°C. Combining a propyl spacer with an imidazole with unsubstituted position 2 causes a further decrease in thermal stability (IL **6**).

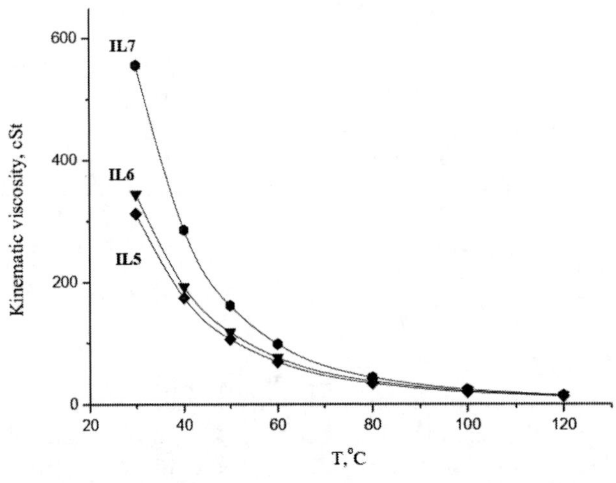

Figure 5. Dependence of the kinematic viscosity of IL **5-7** on temperature.

3.3. Viscosity of ILs

The kinematic viscosity values for RTDIL **5-7** are presented in Table 3. Ionic liquid **5** (Table 2), characterized by the highest T_d, has the lowest viscosity at 30°C among the obtained IL - 312 cSt. The increase in viscosity to 344 cSt for IL **6** is explained by the presence of a propyl spacer in its structure. The high viscosity value for IL **7** is most likely due to the additional intermolecular interactions of the oxygen-containing substituent in its cation.

The dependence of kinematic viscosity on temperature for ILs **5-7** (Fig. 5) is well approximated (R2>0.99) by the Vogel–Tamman–Fulcher (VTF) equation for melts [34]:

$$\ln\nu = \frac{a}{T - T_0} + b$$

The equation parameters are shown in Table 2.

Table 2. Viscosity of disiloxane RTDIL **5-7**.

№ IL	Viscosity, cSt (30°C)	Constants of the VTF equation		
		a	*-b*	T_0
5	312	798	1.402	191.5
6	344	901.3	1.681	183.3
7	555	890.9	1.784	193.3

3.4. *Volatility of ILs*

McBain quartz spring balance was used to measure the dynamic characteristics of volatility. Measurements were carried out at temperatures of 150, 190 and 220°C in vacuum of ~ 0.013 Pa. The synthesized dicationic ionic liquids practically do not evaporate at 150°C: the values of the mass loss rate of the IL samples are in the range of 0-0.03 mg/h·cm^2. When the temperature in the experiment increases to 190 and 220°C, the upper limits of the volatility range increase to 0.33 and 0.95 mg/h·cm^2, respectively.

4. Conclusions

In result of the above studies, dicationic imidazolium ionic liquids with disiloxane linkers were prepared. The steric effect of the substituents at the silicon atom practically does not affect the T_m of the ILs due to short length of the linker, and only the presence of polar groups in the cation structure significantly reduces the tendency of ILs to crystallization. Introducing polar groups into the cation leads to some decrease in thermal stability and a significant increase in the volatility of the resulting ILs in a vacuum.

5. Referens

[1] Zhao W, Sun J 2018Chem Rev 118 10349–10392
[2] Itoh T 2017Chem Rev 117 10567-10607
[3] Wegner S, Janiak C 2017Top Curr Chem (Z) 375 65-72
[4] Azov V A, Egorova K S, Seitkalieva M M, Kashin A S, Ananikov V P 2018Chem Soc Rev 47 1250—1284
[5] Watanabe M, Thomas M L, Zhang S, Ueno K, Yasuda T, Dokko K 2017Chem Rev 1177190–7239

[6] [6] Brown L, Earle M J, Gılea M A, Plechkova N V, Seddon K R 2017Top Curr Chem (Z)375 74-81
[7] Ventura S P M, e Silva F A, Quental M V, Mondal D, Freire M G, Coutinho J A P 2017Chem Rev 117 6984–7052
[8] Kamimura A, Shiramatsu Y, Kawamoto T 2019Green Energy & Environment 4 166–170
[9] Dutta T, Isern N G, Sun J, Wang E, Hull S, Cort J R, Simmons B A, Singh S 2017ACS Sustainable Chem Eng 5 11 10116-10127
[10] Zeng S, Zhang X, Bai L, Zhang X, Wang H, Wang J, Bao D, Li M, Liu X, Zhang S 2017Chem Rev 117 9625–9673
[11] Heym F, Haber J, Korth W, Etzold B J M, Jess A 2010Chem Eng Technol 33 1625–1634
[12] Krannich M, Heym F, Jess A 2016J Chem Eng Data 61 1162-1176
[13] Luo Y, Shao S, Qin F, Tian C, Yang H 2012Solar Energy 86 2718–2724
[14] Watanabe H, Komura T, Matsumoto R, Ito K, Nakayama H, Nokami T, Itoh T 2019Green Energy & Environment 4 139-145
[15] Predel T, Schlücker E 2009Chem Eng Technol 32 1183–1188
[16] Zhang Q, Shreeve J M 2014Chem Rev 114 10527–10574
[17] Van Valkenburg M, Vaughn R L, Williams M, Wilkes J S 2005Thermochimica Acta 425, 181–188
[18] Paul T C, Morshed A K M M, Fox E B, Khan J A 2017Appl Therm Eng 110 1–9
[19] Kühn A, Seiler M, Radspieler M, Kotenko O. Ionic liquids as new absorbents for absorption chillers and heat pumps. Chapter 6. Research on Working Pairs and Apparatus Technology in Book "Thermally driven heat pumps for heating and cooling". Ed. A. Kühn – Berlin: Universitätsverlag der TU Berlin, 2013, pp. 215-221
[20] Becker T M, Wang M, Kabra A, Jamali S H, Ramdin M, Dubbeldam D, Ferreira C A I, Vlugt T J H 2018Ind Eng Chem Res 57 5442 – 5452
[21] Bondareva N V, Glukhov L M, Koroteev A A, Krasovskii V G, Kustov L M, Nagel' Yu A, Safronov A A, Filatov N I, Chernikova E A, 2015Izv Ross Akad Nauk Energ 4 130–142
[22] Chernikova E A, Glukhov L M, Krasovskiy V G, Kustov L M, Koroteev A A, 2014Russ Chem Bull Int Ed 63 2702-2707
[23] Glukhov L M, Krasovskiy V G, Chernikova E A, Kapustin G I, Kustov L M, Koroteev A A . 2015Russ J Phys Chem A 89 2204-2209
[24] Krasovskiy V G, Chernikova E A, Glukhov L M, Kapustin G I, Koroteev A A 2018Russ J Phys Chem A 92 2379-2385
[25] Sun Y-X, Wang Y-Y, Shen B-B, Zhang B-X, Hu X-M S 2018Royal Soc Open Sci 5181230
[26] Mandai T, Imanari M, Nishikawa K 2012Chem Phys Let 543 72–75
[27] Stark F O, Falendar J R, Wright A P. Comprehensive Organometallic Chemistry V2 Pergamon Press New York 1982 1020 p
[28] Liebau F. Structural Chemistry of Silicates: Structure, Bonding and Classification Springer-Verlag New York 1985 347 p
[29] Chernikova E A, Glukhov L M, Krasovskiy V G, Kustov L M, Koroteev A A 2014Russ Chem Bull 63 2702-2706
[30] Niedermeyer H, Ab Rani M A, Lickiss P D, Hallett J P, Welton T, White A J P, Hunt P A 2010Phys Chem Chem Phys 12 2018–2029

[31] Krasovskiy V G, Kapustin G I, Glukhov L M, Gorbatsevich O B, Chernikova E A, Koroteev A A, Kustov L M 2020 Mendeleev Commun 30 Is. 1.
[32] Krasovskiy V G, Glukhov L M, Chernikova E A, Kapustin G I, Gorbatsevich O B, Koroteev A A, Kustov L M 2017Russ Chem Bull Int Ed 66 1269-1277
[33] Kempter V, Kirchner B 2010J of Mol Struct 972 22–34
[34] Solution Chemistry Research Progress, ed. D. V. Bostrelli, Nova Science, New York, 2008. 185 p.

Preparation and characterization of stable water soluble hybrid nanostructures of hydrophobic compounds by encapsulation into nanoparticles of amphiphilic *N*-vinylpyrrolidone copolymers of new generation

S V Kurmaz [1], D V Konev[1], V D Sen'[1], V A Kurmaz[1, *] and A V Kulikov[1,2]

[1]Institute of Problems of Chemical Physics, Russian Academy of Sciences, 1 prosp. akad. Semenova, 142432 Chernogolovka, Moscow Region, Russian Federation.
[2]Lomonosov Moscow State University, Department of Fundamental Physical and Chemical Engineering, 1 Leninskie Gory, 119991 Moscow, Russian Federation

*E-mail: kurmaz@icp.ac.ru

Abstract. The conditions for obtaining stable aqueous solutions of prospective pharmaceutical: a lipophilic aminonitroxyl platinum complex, namely, e-amine-d-(4-amino-2,2,6,6-tetramethylpiperidine 1-oxyl)-a,fbis(octanoato)-b,c-dichloroplatinum(IV) (PNC), and a dye zinc tetraphenylporphyrinate (ZnTPP) by means of their solubilization with polymer particles of N-vinylpyrrolidone with dimethacrylate. Aqueous solutions of PNC and ZnTPP encapsulated into polymer particles were investigated by dynamic light scattering, and the temperature dependence of nanostructure sizes was studied. According to transmission electron microscopy, the hybrid nanostructures are core—shell particles, in which the Pt^{IV} complex forms small clusters (~4 nm). It was established by ESR and voltammetric measurements that the free radical fragment of the aminonitroxyl Pt^{IV} complex kept its mobility and the same ESR and redox characteristics as the 4-acetylamino-2,2,6,6-tetramethylpiperidine 1-oxyl radical. The obtained results indicate that biocompatible copolymers of N-vinylpyrrolidone are of interests as carriers and delivery tools for ZnTPP and organic Pt^{IV} complexes. Hybrid structures of given dimensions and properties can be created on their basis necessary for biomedical applications.

1. Introduction

Poor solubility of many organic and inorganic compounds in aqueous media produces many problems in their practical use and scientific investigations. In part, efficiency of many pharmaceuticals increases sufficiently at lipofilicity growth. However, their water solubility is decreased considerably in the same time that may retard seriously their targeting delivery. Meanwhile, up to 70% of the synthesized active pharmaceutical ingredients are sparingly soluble in water [1]. Many approaches can be used to solve this problem, partially, chemical modification (e.g., covalent grafting of hydrophilic groups to a substrate, coordination / coordinating binding) or encapsulation in water-soluble matrix, partially polymer one [2-9]. This immobilization is important for practical biomedical applications as a kind of target delivery of lipophilic drugs into a living cell.

Synthetic amphiphilic polymers with a complex architecture can be promising objects for these purposes. However, the difficulty of synthesizing dendrimers and highly branched polymers with a regular structure prevents their widespread use for these purposes [10]. Therefore, polymer nanostructure materials based on less regular amphiphilic copolymers with a branched structure were developed. They are characterized by rich architectural diversity and a wide range of physicochemical properties.

These materials include amphiphilic copolymers of N-vinylpyrrolidone with triethylene glycol dimethacrylate (VP-TEGDM), synthesized by radical copolymerization in toluene [11]. The advantages of these copolymers include the simplicity of their preparation in combination with wide possibilities to control their composition, molecular weight, content of branches, dimensions of nanoparticles, as well as their low cytotoxicity for cells. These copolymers have proven themselves as potential agents for the delivery of various hydrophobic compounds of pharmaceutical activity, namely, fullerene C_{60} [12-14], and also have been tested as NO-carrying systems of binuclear iron tetranitrosyl complexes with antitumor activity [15].

Interest toward branched VP-copolymers is due to modern advances in the use of linear polymers consisting of a hydrophilic polyvinylpyrrolidone (PVP) fragment of various molecular weights and a hydrophobic alkyl fragment of various lengths as carriers of hydrophobic biologically active (antitumor) substances [16, 17]. They exhibited low toxicity and high safety *in vitro* and *in vivo* conditions, and also had the ability to efficiently deliver the active agent into the cell and increase the bioavailability of hydrophobic drugs.

In this connection, the goal of the study is to obtain water-soluble forms of *e*-amine-*d*-(4-amino-2,2,6,6-tetramethylpiperidine-1-oxyl)-*a*,*f*-bis(octanoato)-*b*,*c*-dichloroplatinum(IV) (PNC, **1**) of antitumor activity, and 5, 10, 15, 20-Tetraphenyl-21*H*,23*H*, porphine zinc (ZnTPP, **3**) as a prospective diagnostics remedy based on VP-TEGDM copolymers and their characterization. PNC choose to study was determined with its efficacy against cisplatin-resistant tumor cell lines *in vitro* and *in vivo* [18]. The stable radical 4-acetylamino-2,2,6,6-tetramethylpiperidine 1-oxyl (**2**) was used for a comparative electrochemical and ESR identification and study of the radical fragment with a free valence in PNC. Metalloporphyrins are of interest for various radiological and magnetic resonance diagnostic methods due to their high optical absorption and fluorescence in red area of spectrum, relatively high quantum yields of the triplet state, and fluorescence, along with the paramagnetism [19, 20, 21]. Biomedical application of water-insoluble ZnTPP is seriously limited, and to increase it is an important task. Figure 1 presents chemical structures of **1-3**.

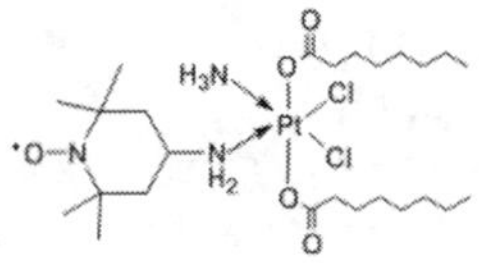
1
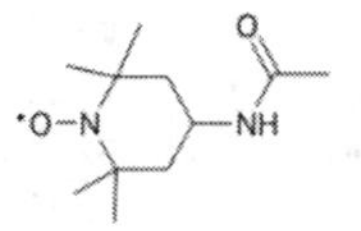
2
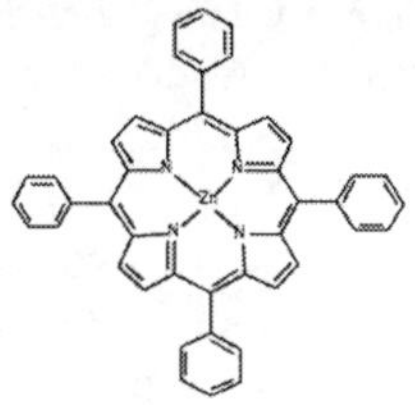
3

Figure 1. Chemical structures of **1-3**.

2. Materials and Methods

2.1 Materials, reagents and preparation of hybrid nanostructures

The copolymers VP-TEGDM were prepared via radical copolymerization [19] in toluene at a molar ratio of the reagents, VP : TEGDM, of 100 : 5, initiated by 2,2'-azo-bis-isobutyronitrile. F1 and F2 were fractions of VP-TEGDM copolymer, and F3 was non-fractionated copolymer. After the polymerization of active dimethacrylates, a polymer structure is formed containing side double bonds, to which the growing PVP chains are predominantly added. As a result, a 3D-structure is formed with a core consisting of methacrylate fragments and a shell formed by polymer chains of VP units. F1 and F2 differing in composition and ratio of VP and TEGDM units: copolymer F1 consists mainly of VP units, while F2 contains four times as many TEGDM units. NMR data evidenced a difference in monomer composition of copolymers F1 and F2 (Figure 2a). The absolute average molecular mass M_w of copolymer F2 enriched with TEGDM (196.0 kDa) was sufficiently higher with respect to F1 (26.0 kDa) due to a globular nature of its branched macromolecules, and for F3 M_w = 56.4 kDa.

The compound PNC (M = 740.68) appears as yellow crystals, m.p. 169—170 °C (decomp.) (from an ethyl acetate—hexane mixture). The complex structure is consistent with the results of elemental analysis, UV, IR spectroscopy, ESR, and electrospray ionization mass spectrometry [18].

The aqueous solutions were prepared based on bidistillate, DMSO (Lab-Scan, Poland) was purified by freezing, propan-2-ol (*i*-PrOH) (reagent grade Khimmed, Russia) was distillated. MeCN (Lab-

IOP Conf. Series: Materials Science and Engineering **848** (2020) 012043 doi:10.1088/1757-899X/848/1/012043

Scan, Poland), components of buffer solutions (reagent grade Khimmed, Russia), $TBAPF_6$ and radical **2** (all from Aldrich, USA) were used without additional purification. The water content in DMSO (less than 0.05%) was controlled by IR spectroscopy according to the previously described procedure [22].

Encapsulation of PNC into polymer particles was carried out according to the following procedure. Solutions of F1 and F2 in *i*-PrOH (3.5 and 3.7 mg ml^{-1}, respectively) and of the complex in DMSO (8.2 mg mL^{-1}) were prepared. Various volumes of the Pt^{IV} complex solution were added dropwise to the alcohol solution of the copolymer (4 ml) with continuous magnetic stirring. Calculating on the copolymer, solution of F1 contained 2, 3, 4, 5, 9, and 18 wt.% of Pt^{IV} complex, and in the solution of copolymer F2 these values were equal to 2, 3, 4, and 18 wt.% respectively. Organic solvents were then removed in air and in vacuo. An aqueous phosphate buffer solutions (PBS) with pH 6.8—7.0 (8 ml, 137 mmol l^{-1} of NaCl, 2.68 mmol l^{-1} of KCl, 4.29 mmol l^{-1} of Na_2HPO_4, 1.47 mmol l^{-1} of KH_2PO_4) was added to the dry residues to obtain solutions of encapsulated into polymer particles complexes PNC_nF1 (the complex content n = 2—5, 9, 18%) and PNC_nF2 (n = 2—4, 18%). Solutions of PNC_nF1 were slightly opalescent and had a bluish tint typical of micellar solutions and good stability, if n = 2 or 3 but where n = 4, 9, and 18 they were opalescent noticeably.

Hybrid macromolecular structures of the ZnTPPF3 were prepared using solutions of copolymer F3 with different concentrations (from 0.007 to 1 mg/ml) in *i*-PrOH and a solution of ZnTPP in freshly distilled toluene (0.07 mg/ml). The content of ZnTPP per the copolymer varied from 25 to 0.19%, and the next stages of the procedure were like for PNC_nF structures. Optical transparence of ZnTPP macromolecular structures depended on the dye content in a solution: it decreased at ZnTPP content increasing as a result of opalescence appearance, and a solution color was changed from yellowish to lilac one.

2.2 GPC-analysis of VP-TEGDM copolymers, absorbance and IR-spectroscopy, TEM

The absolute molecular weights M_w of copolymers were measured using a Waters GPCV 2000 liquid chromatograph (USA) (two PS-gel columns, 5 μm, MIXED-C, 300 × 7.5 mm) equipped with a refractometric detector and a WYATT DAWN HELEOS II light scattering detector (λ = 658 nm) in N-methylpyrrolidone in the presence 1% LiCl to prevent macromolecules aggregation.

The absorption spectra of the solutions in *i*-PrOH and buffer solutions were measured on Specord M40 (Germany) and PerkinElmer EZ-210 spectrophotometers in 1 cm cells. IR spectra of the initial PNC, copolymer powders, and PNCF after drying in vacuo were recorded in the disturbed total internal reflection (FTIR) regime, using a Bruker a Fourier-transform IR spectrometer (Bruker, Germany) in the range of 400—4000 cm^{-1}, the number of scans was 16.

The aqueous solution of polymer particles loaded with PNC was deposited on the substrate, a solvent was quickly evaporated, and images of hybrid nanostructures were obtained by Leo 912 AB equipment.

2.3 Dynamic light scattering

Solutions of the copolymer and hybrid structures in PBS were studied by dynamic light scattering (DLS) at a detection angle of 90° using a Photocor Compact installation (Photocor Instruments Inc., USA) equipped with a diode laser operating at a wavelength of 654 nm. The aqueous solutions were filtered through a microporous filter (0.45 mm). Before measurements, the vials with solutions were thermostated for ~20 min. The experimental data were processed using DynaLS v. 2.8.3 software by means of regularization (distribution analysis). The values of hydrodynamic radii R_h of particles was calculated using Einstein–Stokes equation $D = kT/6\pi\eta R$, where D is the coefficient of diffusion, k is the Boltzmann constant, T is the absolute temperature, and η is the viscosity of the medium in which hybrid nanostructures were suspended. The experimental results were presented in the form of distribution of the intensity of scattered light I by the sizes of the particles (hydrodynamic radii R_h).

2.4 ESR-spectroscopy

ESR spectra of aqueous solutions of encapsulated PNC were recorded on an Elexsys II E500 instrument (Bruker, Germany) at room temperature. The concentration and the g-factor were determined using the Xepr software package. A solution of nitroxyl in DMSO with a known concentration, $CuSO_4 \cdot 5H_2O$ powder, and a sample of diphenylpycrylhydrazyl (DPPH) with a *g*-factor of 2.0036 were used to verify these procedures. ESR spectra of the initial complex in DMSO and in a water—DMSO mixture ($4.4 \cdot 10^{-5}$ mol l^{-1}) and of radical **2** in DMSO (10^{-5}—10^{-3} mol L^{-1}) were recorded for comparison.

2.5 Cyclic, differential pulse and square-wave voltammetry

A cycle of electrochemical studies of the depolarisers **1-3** was carried out *via* cyclic (CVA), differential pulse (DPVA) and square-wave voltammetry (SWVA) in aqueous neutral PBS (pH 6.86), non-aqueous solvents (DMSO, MeCN) and in their admixtures with H_2O. It were used an Autolab/PGSTAT302N universal high-speed potentiostat/galvanostat (ECOCHEMIE, Netherlands) and an IPC-ProL potentiostat (IPCE of RAS, Russia), and a three-electrode quartz or glass cells; the volume of the working solution was ~10 cm^3. The range of sweep rates *v* in CVA was 0.02 - 5 V s^{-1}; in SWVA, the potential step was 5 mV, the amplitude was 20 mV, and the frequency was 25 Hz. The working electrode were Au, Pt (Menica, Poland) and glassy carbon (GC) (HTW Segradur-G, Germany) disc electrodes soldered into glass with a diameter of ~1 mm^2, and in ZnTPP case an Au wire electrode (visible area is ~0.1 cm^2) and a carbon paper (Sigraset 39AA, Germany) as well. The auxiliary electrode was always a Pt wire, and the reference electrode was a silver/silver chloride electrode (Ag/AgCl) in aqueous solutions. The measurements in non-aqueous solutions were carried out relative to the Ag/Ag^+-reference electrode (10 mmol l^{-1} of $AgNO_3$, 100 mmol l^{-1} of Bu_4NPF_6 in the used solvent, *i.e.*, MeCN or DMSO), separated from the working compartment of the cell by a double porous glass diaphragm. Its potential was +300 mV vs Ag/AgCl. Before the start of the experiment, the electrodes were polished with a diamond suspension (particle diameter ~1 mm), then cleaned by ultrasonic treatment in the medium in which the studies were carried out (water or a respective non-aqueous solvent). The experimental procedure is described in more detail in [20, 23, 24].

3. Results and Discussion

3.1 Composition and physical chemical properties of hybrid nanostructures of PNC

Solubilization of PNC by polymer particles results in the formation of the "guest—host" type nanostructures, in which the "guest" molecules retained through non-covalent interactions (van der Waals, hydrophobic, as well as hydrogen bonding). Chemical analysis of solid PNCF indicated the presence of Pt and Cl in them. Thus the content of Pt and Cl in PNC_5F1 (from an *i*-PrOH—DMSO mixture) was ~1.3 and 0.6%, which is in good agreement with the calculated values and indirectly indicated that the encapsulated complex is stable.

IR spectra of powders of copolymers PNCF1 and PNCF2 containing 2—3% of the Pt^{IV} complex are practically the same. Due to the low Pt^{IV} content, its characteristic absorption bands do not appear in the spectrum. Absorption bands characteristic of PNC are clearly visible in the IR spectrum of the insoluble product isolated from an aqueous solution of $PNC_{18}F1$ (Figure 2b). The IR spectrum in the region of 1800—1500 cm^{-1} has a complex pattern; absorption bands with a maximum at 1725 cm^{-1} can be distinguished (stretching vibrations of the C=O bond in the TEGDM units); a shoulder at 1681 cm^{-1}; a maximum at 1652 cm^{-1}, which is characteristic of stretching vibrations of hydrogen bonded C=O groups of VP units. The absorption maxima at 1629 cm^{-1} and 1576 cm^{-1} should be attributed to the aminonitroxyl complex. The absorption band at 1580 cm^{-1} is apparently related to the bending vibrations of the N—H bond. It is shifted toward lower frequencies, from 1597 to 1580 cm^{-1}, as compared to the band in the spectrum of the initial complex. It can be assumed that a hydrogen bond is formed between copolymer F1 and the complex involving its NH_2 groups. The band belonging to C=O groups of the carboxylate ligand is also shifted toward lower frequencies from 1638 to 1630 cm^{-1}, which can also be related to its bonding to the copolymer.

According to TEM (Figure 3a), PNC_3F1 and PNC_4F2 are a core—shell-type particles. The contrast core can be ~30 nm with a total size of ~70 nm. PNC_4F2 particles are less than 50 nm. Dark inclusions of small aggregates, apparently consisting of platinum complex molecules, can be distinguished within particles of both types.

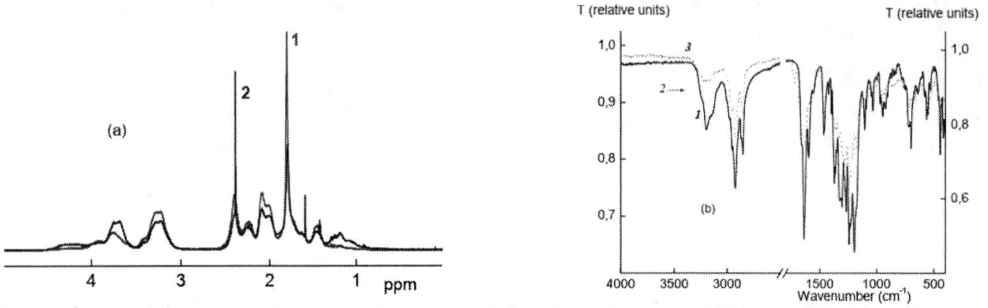

Figure 2. ^{1}H NMR spectra of the copolymers F1 (*1*) and F2 (*2*) recorded in $CDCl_3$ (a) and IR-spectra of PNC (*1*), copolymer F1 (*2*), and insoluble in water fraction of $PNC_{18}F1$ (*3*) (b).

Aqueous solutions of PNC_3F1 and PNC_3F2 were studied by DLS. The intensity of light scattering increased almost linearly with temperature (Figure 3b), while the sizes of scattering centers decreased. In the temperature range of 28—40 °C, the principal contribution to light scattering was made by particles with R_h ~50 nm. The same linear dependence I(T) was observed for F2 polymer particles. In this case, the size of the predominant type of particles did not change and was lower than that of the polymer particles encapsulated with the Pt^{IV} complex. Thus, the behavior of hybrid structures in a PBS was determined by the polymer component. The changing temperature of the aqueous solution of the encapsulated complex PNC_3F2 also was accompanied by aggregation, which was reversible, as opposed to the case of PNC_3F1.

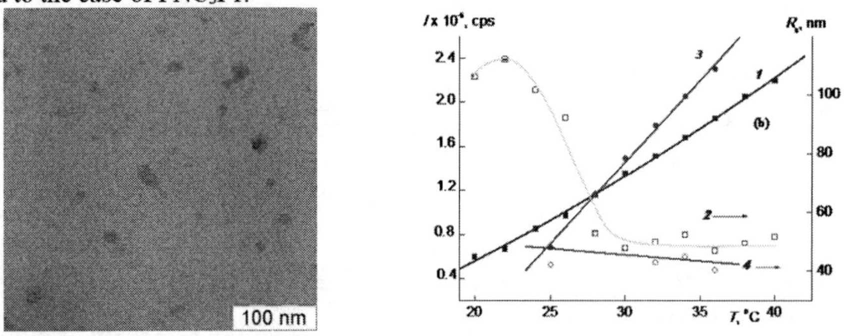

Figure 3. TEM image of the nanostructures of PNC_4F2 (a) temperature dependences of light scattering intensity (*1*) and hydrodynamic radius (*2*) for an aqueous solution of PNC_3F2. For comparison, the dotted lines show similar dependences for F2 polymer particles in an aqueous PBS (*3, 4*) (b).

The ESR spectra of aqueous dispersions of the encapsulated complex (Figure 4) contained three lines of approximately the same intensity, which is characteristic of its free form in a water—DMSO mixture, and DMSO. Apparently, the same local environment, namely water molecules, is the reason for the absence of considerable differences in the ESR spectra of the free Pt^{IV} complex and of the one encapsulated into polymer particles of various types. It was demonstrated by recording of ESR spectra of aqueous dispersions of PNC_3F1 and PNC_3F2, the free PNC complex in DMSO and in a water—DMSO mixture, and spectra of radical **2** in DMSO and MeCN, respectively (Figure 4) for comparison that the encapsulated aminonitroxyl fragment of radical nature retained its structure and mobility in polymer particles (Table 1). Measurements performed after four months showed that aqueous solutions of PNC encapsulated into polymer nanoparticles remained stable during storage, and the concentrations of radicals in aqueous solutions and the ESR spectra parameters did not undergo considerable changes.

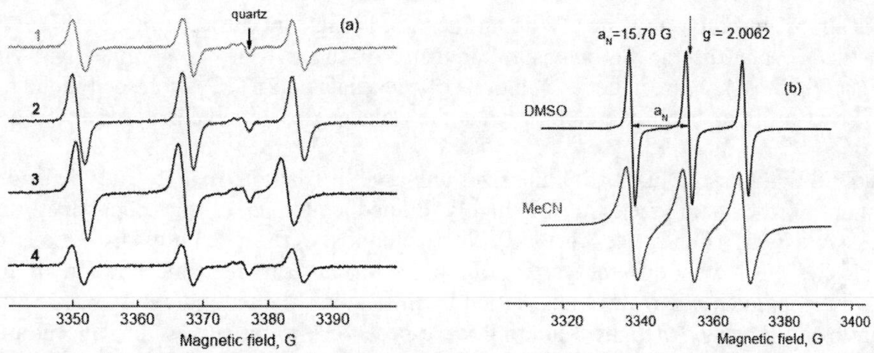

Figure 4. (a) - ESR spectra of aqueous solutions of PNC_3F1 (*1*) and PNC_3F2 (*2*), and of the free PNC in DMSO (*3*) and in a water—DMSO (*4*) mixture. The experimental conditions and numbers of spectra respect to that of in Table 1. (b) - ESR spectra of radical **2** (1 mmol l^{-1}) in DMSO and MeCN.

Table 1. Concentrations of nitroxyls and radical **2**, and parameters of ESR spectra of aqueous PBS of PNC_3F1 (sample *1*), PNC_3F2 (sample *2*), the free Pt^{IV} complex in DMSO (sample *3*) and in a water—DMSO mixture (sample *4*), as well as the radical **2** in DMSO (sample *5*).

Sample	$C \cdot 10^5$, mol L^{-1} (±15%)	*g*- factor (± 0.0002)	a_N, G (±0.1)	A_0/A_{-1} (±0.02)
1	4/3.8[a]	2.0054	16.8	1.16
2	6.5/7.5[a]	2.0054	16.8	1.14
3	6.3	2.0058	15.8	1.41
4	3.0	2.0054	16.8	1.19
5	100	2.0062	15.7	1.13

[a]The numerator shows the values for freshly prepared aqueous dispersions of the complex encapsulated into polymer nanoparticles, the denominator indicates those values after four months of storage.

It were sufficiently similar also the potentials of anodic and cathodic peaks for reversible one-electron oxidation/reduction or radical **2** being in good agreement with literature [25], and that of the radical moiety of PNC (Figure 5). The difference in several tenth of mV were apparently due to the influence of central Pt atom and the ligand environment in PNC, and with substituents in *para*-positions of radicals.

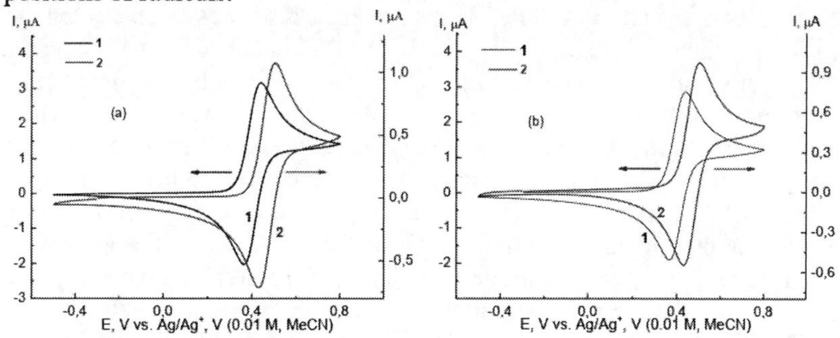

Figure 5. Anodic branches of CVA curves of the radical **2** (*1*) and the Pt^{IV} complex (*2*) on Pt (a) and GC (b) electrodes in MeCN + 0.1 mol L^{-1} of $TBAPF_6$. *v* is 100 mV s^{-1}.

3.2. Composition and physical chemical properties of hybrid nanostructures of ZnTPPF3

Water soluble ZnTPP forms were prepared by its solubilization with a copolymer F3 micellar particles [19, 23]. In electron absorption spectra of PBS of hybrid nanostructures, absorption in the region of 350–700 nm is associated only with the water-soluble form of ZnTPPF3, since the solution of copolymer F3 is optically transparent in the range of wavelengths above 250 nm, and the

IOP Conf. Series: Materials Science and Engineering **848** (2020) 012043 doi:10.1088/1757-899X/848/1/012043

hydrophobic dye is almost insoluble in water. With an increase in the concentration of ZnTPP in the hybrid structures, the *A* value of the Soret absorption band of the dye grows nonlinearly. With excitation using light with $\lambda = 422$ nm, aqueous solutions of encapsulated ZnTPP fluoresced in the red spectral region, and two emission bands were observed at the wavelengths with the maxima at 603 and 643 nm.

It follows from DLS that the structure of F3, the sizes and size distribution of aggregates depended on the concentration of the copolymer. In the highly diluted solutions of the copolymer, the distribution curves were wide (Figure 6a, curve *1*). The solutions were apparently dispersions of monomolecular micelles [26] with different sizes. Upon an increase in the concentration of the copolymer in the solutions, intense narrow peaks caused by the stable polymer aggregates making the major contribution to the intensity of light scattering were observed in the curves. In concentrated solutions, the distribution of the intensity of light scattering according to the sizes of the particles was virtually unimodal (Figure 6b), and the R_h of ZnTPPF3 nanoparticles was ~106 nm at the peak maximum. In PBS, hybrid nanostructures respond to the change in the temperature. Upon an increase in temperature, the intensity of light scattering grows sharply, and R_h drops intermittently to reach 50–60 nm in the 30–60°C range of temperatures.

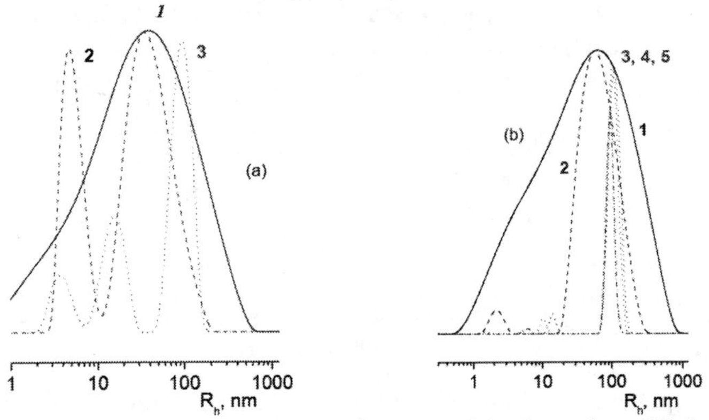

Figure 6. Curves of the distribution of the intensity of light scattering by the sizes R_h in solutions containing 0.4 (*1*), 1.2 (*2*) и 3.5 (*3*) mg ml^{-1} F3 (a), and the nanoparticles of ZnTPPF3 (*1*) 0.007, (2) 0.07, (3) 0.3, (4) 0.7, and (5) 0.91 mg ml^{-1}. Aqueous PBS. 20°C.

The aqueous PBS of high concentration ZnTPPF3 hybrid nanostructures were taken for electrochemical investigations. The anodic branch of CVA curve of PBS contained a small oxidation signal at $E = +1.22 \pm 0.02$ V (Figure. 7a) It was absent on the CVA curves of the background, and in the saturated solutions (suspensions) of the insoluble dye in water. According to the literature [27-30] and our ones, it corresponded roughly to the second oxidation–reduction peak in the anode branch of ZnTPP in nonaqueous solutions, and was located at the potentials which are 100–150 mV more positive than the oxidation (decomposition) peak of water on an Au electrode with the formation of gold oxide in the background solutions. In the cathodic branch of the CVA curve, the peak of its reduction was retained at about the same potentials (around 0.4–0.5 V, the insert on Figure 7a) as in the solutions of the background. Since the total concentration of ZnTPP in the solution was low, this peak was adsorption and not diffusion in nature. Polarization curves of ZnTPP in non-aqueous solutions (e.g., in MeCN, Figure 7b) contained two pairs of reversible cathodic and anodic diffusion peaks. We proposed a model of electrochemical transformations of nanoparticles ZnTPPF3 [20] considered for non-adiabatic electron transfer probability. For nanoparticles "core-shell" structure, the probability to realize of the mechanism have to depended first of all on their hydrodynamic radius R_h. Because of the maximum length of electron tunneling generally does not exceed several tens of ångströms [31], this mechanism may be realized only in individual cases where R_h values are equal or less of this length. It is a consequence of non-unimodal distribution of particles on R_h (Figure 6).

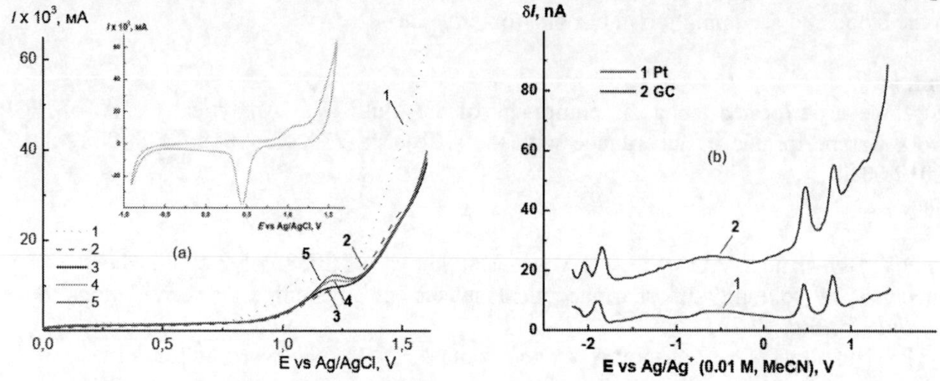

Figure 7. (a) - Anodic branches of CVA curves of the background (aqueous PBS): (*1*); (*2–5*) PBS containing ZnTPPF3. The times of potentiostating at $E = 0.2$ V are 2 (*2, 3,* and *5*) and 30 s (*4*). Insert - CVA curve of Au-electrode in aqueous PBS. v is 100 mV/s. (b) – DPVA-curves of of Pt- and GC-electrodes in MeCN + 0.1 mol L^{-1} of $TBAPF_6$. v is 100 mV s^{-1}.

A dynamics of polymer micelles destruction in MeCN and H_2O – MeCN mixtures (10-50%) was also studied spectroscopically (Figure 8a) on the Soret absorption band at λ = 425 nm, and electrochemically (Figure 8b), on its redox reactions. The polymer micelles ZnTPPF3 were destroyed sufficiently fast, for several seconds. The Soret band was absent in an aqueous PBS but was sufficiently large in MeCN and to increase with the non-aqueous solvent content increasing in the admixture proportionally. It was practically absent in 90% PBS, was getting visible in 70% ones and clearly visible at 50%. The band near 325 nm was changed oppositely. An uniform increasing of the absorbance after his band pointed on a colloid character of PBS with hybrid nanostructures of ZnTPP.

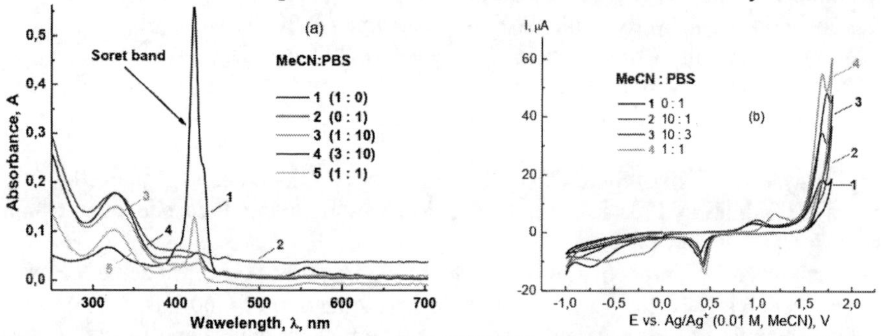

Figure 8. (a) – Absorbance spectra of ZnTPPF3 in MeCN, PBS and MeCN-PBS mixtures. (b) - CVA curves of ZnTPPF3 on Au-electrode in MeCN, PBS and MeCN-PBS mixtures. v is 100 mV/s. 1st scan.

Note that such a fast destruction we observed earlier for nanostructures of Pt^{IV} complex in DMSO that it followed from a comparison of CVA curves of polymer particles loaded with the complex in an aqueous PBS and DMSO [23, 24].

4. Summary

Thus, it was shown for a lipophilic aminonitroxyl Pt^{IV}-complex and zinc tetraphenyl porphyrinate dye that water-soluble forms of functional materials for various biomedical applications can be obtained through the solubilization of poorly and insoluble component by amphiphilic branched copolymers of *N*-vinylpyrrolidone with triethylene glycol dimethacrylate of different monomer composition and the topology presented in solutions as individual macromolecules and their aggregates (nanoscale polymer particles). As a result, new possibilities are opened to control not only their solubility but also bioavailability, prolonged action, distribution in the body, duration of circulation in the blood and accumulation in tumors, toxicity, etc.

Acknowledgment
The work has been performed using the equipment of the Multi-User Analytical Center of IPCP RAS. This work was performed in accordance with the state task (№ 0089-2019-0008, 0089-2019-0012, 0089-2019-0014).

References
[1] Yudina D V, Blynskaja E V, Alekseev K V and Marakhova A I 2018 Polymers for the enhancement of solubility of pharmaceutical substances in solid drug formulations *Russ. Chem. Bull. Int. Ed.* **67** 787
[2] Oberoi H S, Nukolova N V, Kabanov A V and Bronic T K 2013 Nanocarriers for delivery of platinum anticancer drugs *Adv. Drug Delivery Revs.* **65** 1667
[3] Torchilin V P 2001 Structure and design of polymeric surfactant-based drug delivery systems *J. Control. Release* **73** 137
[4] Duncan R 2003 The dawning era of polymer therapeutics *Nat. Rev. Drug Discov.* **2** 347
[5] Peer D, Karp J M, Hong S, Farokhzad O C, Margalit R and Langer R. 2007 Nanocarriers as an emerging platform for cancer therapy *Nat. Nanotechnol.* **2** 751
[6] Rieter W J, Pott K M, Taylor K M L and Lin Wenbin 2008 Nanoscale Coordination Polymers for Platinum-Based Anticancer Drug Delivery *J. Am. Chem. Soc.* **130** 11584
[7] Xiao H H, Song H Q, Zhang Y, Qi R G, Wang R, Xie Z G, Huang Y B, Li Y X, Wu Y and Jing X B 2012 The use of polymeric platinum(IV) prodrugs to deliver multinuclear platinum(II) drugs with reduced systemic toxicity and enhanced antitumor efficacy *Biomaterials* **33** 8657
[8] Kang R H, Kwon J Y, Kim Y and Lee S M 2017 Cisplatin-Mediated Formation of Polyampholytic Chitosan Nanoparticles with Attenuated Viscosity and pH-Sensitive Drug Release *Langmuir* **33** 9091
[9] Shirakura T, Kelson T J, Ray A, Malyarenko A E and Kopelman R 2014 Hydrogel Nanoparticles with Thermally Controlled Drug Release *ACS Macro Lett.* **3** 602
[10] Smeets N M B 2013 Amphiphilic hyperbranched polymers from the copolymerization of a vinyl and divinyl monomer: The potential of catalytic chain transfer polymerization *Eur. Polym. J.* **49** 2528
[11] Kurmaz S V and Pyryaev A N 2010 Synthesis of N-vinyl-2-pyrrolidone-Based Branched Copolymers via Crosslinking Free-Radical Copolymerization in the Presence of a Chain-Transfer Agent *Polym. Sci. Ser. B* **52** 1
[12] Kurmaz S V, Obraztsova N A, Perepelitsina E O, Anokhin D V, Shilov G V, Kabachkov E N, Torbov V I and Dremova N N 2014 Synthesis and characterization of C60-based composites of amphiphilic N-vinylpyrrolidone/triethylene glycol dimethacrylate copolymers *Polym. Compos.* **35** 1362
[13] Kurmaz S V, Obraztsova N A and Kabachkov E N 2016 The features of the formation of the hybrid nanostructures of C60 fullerene and amphiphilic copolymer of N-vinylpyrrolidone with (di)methacrylates in isopropyl alcohol and its mixtures with water *Colloid Polymer Sci.* **294** 2087
[14] Kurmaz S V, Obraztsova N A Perepelitsina E O, Shilov G V, Anokhin D V and Pechnikova E V 2015 New hybrid macromolecular structures of C60 fullerene–amphiphilic copolymers of N-vinylpyrrolidone and triethylene glycol dimethacrylate *Materials Today Commun.* **4** 130
[15] Kurmaz S V, Rudneva T N and Sanina N A 2018 New nitric oxide-carrier systems based on an amphiphilic copolymer of N-vinylpyrrolidone with triethylene glycol dimethacrylate Mendeleev Commun. **28** 73
[16] Tsatsakis A, Stratidakis A K, Goryachaya A V, Tzatzarakis M N, Stivaktakis P D, Docea A O,

Berdiaki A, Nikitovic D, Velonia K, Shtilman M I, Rizos A K and Kuskov A N 2019 In vitro blood compatibility and in vitro cytotoxicity of amphiphilic poly-N-vinylpyrrolidone nanoparticles Food Chem. Toxicol. **127**, 42

[17] Luss A L, Kulikov P P, Romme S B, Andersen C L, Pennisi C P, Docea A O, Kuskov A N, Velonia K, Mezhuev Y O, Shtilman M I, Tsatsakis A M and Gurevich L 2018 Nanosized carriers based on amphiphilic poly-N-vinyl-2-pyrrolidone for intranuclear drug delivery Nanomedicine **13** 703

[18] Cetraz M, Sen V, Schoch S, Streule K, Golubev V, Hartwig A, and Köberle B 2017 Platinum(IV)-nitroxyl complexes as possible candidates to circumvent cisplatin resistance in RT112 bladder cancer cells *Arch. Toxicol.* **91** 785

[19] Kurmaz S V, Obraztsova N A, Balakina, A A and Terent′ev A A 2016 Preparation of the amphiphilic copolymer of N-vinylpyrrolidone with triethylene glycol dimethacrylate nanoparticles and the study of their properties in vitro *Russ. Chem. Bull., Int. Ed.* **65** 2097

[20] Kurmaz S V, Gak V Yu, Kurmaz V A and Konev D V 2018 Preparation and Properties of Hybrid Nanostructures of Zinc Tetraphenylporphyrinate and an Amphiphilic Copolymer of N-Vinylpyrrolidone in a Neutral Aqueous Buffer Solution *Russ. J. Phys. Chem. A* **92** 329

[21] Aksenova N A, Savko M A, Uryupina O Y, Roldugin V I, Timashev P S, Kuz'min P G, Shafeev G A and Solov'eva A B 2017 Effect of the preparation method of silver and gold nanoparticles on the photosensitizing properties of tetraphenylporphyrin-amphiphilic polymer-nanoparticle systems *Russ. J. Phys. Chem. A* **91** 124

[22] Krivenko A G, Kotkin A S and Kurmaz V A 2005 Thermodynamic and Kinetic Characteristics of Intermediates of Electrode Reactions: A Comparative Investigation of a Number of Alkylaryl and Alkyl Halide Radicals by the Laser Photoemission Methods *Russ. J. Electrochem.* **41** 137

[23] Kurmaz S V, Sen′ V D, Kulikov AV, Konev D V, Kurmaz V A, Balakina A A and Terent′ev A A 2019 Polymer nanoparticles of N-vinylpyrrolidone loaded with an organic aminonitroxyl platinum(IV) complex. Characterization and investigation of their in vitro cytotoxicity *Russ. Chem. Bull. (Int. Ed.)* **68** 1769

[24] Kurmaz S V, Fadeeva N V, Fedorov B S, Kozub G I, Emel'yanova N S, Kurmaz V A, Manzhos R A, Balakina A A and Terent′ev A A 2020 New hybrid materials of platinum(IV) organic complex based on polymer nanoparticles of N-vinylpyrrolidone and (di)methacrylates with antitumor activity *Mendeleev Commun.* **30** (submitted)

[25] Sen' V D, Tikhonov I V, Borodin L I, Pliss E M, Golubev V A, Syroeshkin M A and Rusakov A I 2015 Kinetics and thermodynamics of reversible disproportionation–comproportionation in redox triad oxoammonium cations – nitroxyl radicals – hydroxylamines *J. Phys. Org. Chem.* **28** 17

[26] Zhou Y and Yan D 2009 Supramolecular self-assembly of amphiphilic hyperbranched polymers at all scales and dimensions: progress, characteristics and perspectives *Chem. Commun.* 1172

[27] Kuzmin S M, Chulovskaya S A and Parfenyuk V I 2018 Structures and properties of porphyrin-based film materials part I. The films obtained via vapor-assisted methods *Adv Colloid Interface Sci.* **253** 23

[28] Zhang J, Abe T and Kaneko, M 1998 Charge transfer of tetraphenylporphyrin zinc complexes incorporated in a poly(2-vinylpyridine) matrix *J. Porphyrins and Phthalocyanines* **2** 93

[29] Yoshimoto S, Tsutsumi E, Suto K, Honda Y and Itaya K 2005 Molecular assemblies and redox reactions of zinc(II) tetraphenylporphyrin and zinc(II) phthalocyanine on Au(111) single crystal surface at electrochemical interface *Chemical Physics* **319** SI 147

[30] Fan Yanru, Huang Yu, Wang Shixia, Jiang Yuan, Shan Duoliang and Lu Xiaoquan 2016 Investigation of Interfacial Consecutive Electron Transfer and Redox Behaviors of Zinc-tetraarylporphyrins *Electrochim. Acta* **190** 419

[31] Gavaghan D J and Feldberg S W 2000 Extended electron transfer and the Frumkin correction *J. Electroanalyt. Chem.* **491** 103

IOP Conf. Series: Materials Science and Engineering **848** (2020) 012044 doi:10.1088/1757-899X/848/1/012044

Determination of phase transition temperatures (melting, crystallization, rotator phases) of *n*-alkanes by the optical method

V N Kuryakov[1, 2], D D Ivanova[3], A N Tkachenko[4] and P N Sedenkov[5]

[1]Oil and Gas Research Institute of the Russian Academy of Sciences, 3, Gubkin Street Moscow, 119991, Russian Federation
[2]Gubkin University, Department of Physical and Colloid Chemistry, 65/1 Leninsky prospect, Moscow, 119991, Russian Federation
[3]D. I. Mendeleev University of Chemical Technology of Russia, 9, Miusskaya Street, Moscow, 125047, Russia
[4]Department of Landscape Geochemistry and Soil Geography, Faculty of Geography, Lomonosov Moscow State University, Leninskiye Gory, Moscow, 119899, Russia
[5]Moscow Pedagogical State University, Institute of Biology and Chemistry, 6/2 Kibalchich Street, Moscow, 129164, Russia

vladimir.kuryakov@gmail.com

Abstract. The temperatures of phase transitions (melting, crystallization, rotator phases) were determined for a number of individual *n*-alkanes from $C_{19}H_{40}$ (*n*-nonadecane) to $C_{28}H_{58}$ (*n*-octacosane) by the optical method, a modified droplet technique. For this research, *n*-alkanes were presented as an emulsion in water without the addition of surfactants. Emulsions were prepared by ultrasonic dispersion of a small amount of *n*-alkane in water at a temperature above the melting point of the investigated *n*-alkane. The hydrodynamic radius of the dispersions measured by the dynamic light scattering (DLS) method was about 100 nm and did not change within a wide temperature range within the measurement accuracy. The temperatures of melting, crystallization, and rotator phase transitions of pure *n*-alkanes in the form of emulsion were determined from temperature-dependent measurements of the light scattering by the dispersions. The supercooling values for studied *n*-alkanes were also determined.

1. Introduction

An emulsion is a mixture of two liquids that do not mix under normal conditions. Emulsions belong to the class of colloidal systems in which both the dispersed phase and the dispersion medium are liquids. The use of emulsions is diverse: in the agricultural industry they are used as carriers for insecticides, fungicides, and pesticides, in the cosmetic industry – emulsions are used for hair and skin care, in pharmaceuticals they are used to increase the efficiency of delivery of active ingredients in the human body, in the food industry fat emulsions, such as milk, mayonnaise. Currently, one of the promising objects of research is paraffin emulsions. They are interesting in that they are emulsions at temperatures above the melting point and are sols (solid particles in liquid) at temperatures below the melting point. This property determines the scope of their application. A paraffin emulsion in water can be used as a phase-change material (PCM) [1–3].

When the paraffin in the emulsion crystallizes, the stored energy is released in the form of latent heat of phase transition. Conversely, when the paraffin in the emulsion melts, energy is absorbed. This property of PCM materials can be used to store thermal energy, provide thermal barriers or thermal insulation. PCM materials are used, for example, in the transportation of chilled food and drinks, the transport of medicines and vaccines, in conditioning, the accumulation of solar energy.

Conventional methods for studying the phase transitions of *n*-alkanes are adiabatic and differential scanning calorimetry [4, 5]. Furthermore, X-ray diffraction methods [6, 7], the Dynamic mechanical spectroscopy (DMS) method [8], and the recently proposed atomic force microscopy approach [9] are used to study the rotator phases of *n*-alkanes.

In this work, we propose a new method for determining the temperatures of phase transitions of *n*-alkanes, based on dynamic and static light scattering technique – a powerful experimental methods for studying microheterogeneity and phase equilibria in various fluid samples including microcapsules [10], emulsions [11], protein solutions [12], near-critical ternary systems [13], and near-critical polymer solutions [14]. This methods feature the study of *n*-alkane emulsion in water. Such a technique is, in fact, a well-known droplet technique with modifications [15, 16], and has been already used to study phase transitions of pure *n*-alkanes and their mixtures [17–20].

2. Experimental section

Emulsions were prepared from *n*-alkanes ($C_{19}H_{40}$, $C_{20}H_{42}$, $C_{21}H_{44}$, $C_{23}H_{48}$, $C_{24}H_{50}$, $C_{25}H_{52}$, $C_{26}H_{54}$ и $C_{28}H_{58}$) purchased from Acros Organics with a purity of 98-99% and medical double-distilled water for injection (Solopharm, Russia). Using ultrasonic dispersion of the *n*-alkane-water mixture, at a temperature above the melting point of the used *n*-alkane, we prepared emulsion samples with a weight concentration of *n*-alkane in water of 0.01%. Surfactants were not used in the preparation of emulsions. Sartorius BP 301S analytical balance (0.1 mg) was used for sample component weighing. Ultrasonic dispersion of the water/*n*-alkane mixture was carried out for 30 seconds on an ultrasonic disperser UZDN-A (Russia) with an operating frequency of 22 kHz and a power of 75 watts. Measurements by dynamic light scattering showed that the characteristic hydrodynamic radius of the particles in the resulting dispersions was about 100 nm. Dynamic and static light scattering measurements were performed using Photocor Compact-Z equipment (Russia). The measurements were carried out at a scattering angle of 90 degrees. For measurements by dynamic and static light scattering, the samples were diluted 100 times with double-distilled water. The heating and cooling during the experiment were carried out at a speed of about 6 K/h and the temperature step was 0.1 K. The obtained samples, both in an emulsion state and in a sol state, did not coalesce. More details on the sample preparation procedure can be found in the articles [18-20].

3. Results and discussion

For all the emulsions studied, the temperature dependences of the intensity of the scattered light during heating and cooling were measured. At least two heating-cooling cycles were measured for each sample. Figure 1 shows a typical temperature dependence of the intensity of the scattered light on the dispersion of *n*-alkane in water.

The light scattering intensity, in this case, is determined by the size of the emulsion droplets, their concentration, and the difference in the refractive indices of the *n*-alkane phase and water. Phase transitions of melting and crystallization of *n*-alkane in the emulsion lead to a change in the refractive index of the *n*-alkane phase, which, in turn, leads to a change in the measured intensity of the scattered light. For example, for *n*-alkane $C_{19}H_{40}$ ($T_{Melt} \sim 31$ °C), the refractive index at 20 °C (solid phase) and at a temperature of 40 °C (liquid phase) is 1.4409 [21] and 1.43095 [22], respectively. The concentration of particles during the experiment does not change. The contribution to the change in the intensity of the scattered light from the change in particle size associated with different densities of the liquid and solid phases is about 2-3%.

For *n*-alkane emulsions $C_{20}H_{42}$ and $C_{28}H_{58}$, temperature dependences of the intensity of the scattered light were measured at different rates. Figure 2 shows the temperature dependences of the intensity of the scattered light during heating and cooling of the emulsions of these two *n*-alkanes. From the results obtained, it can be concluded that the measurement results of the temperature dependences of the scattered light intensity at different heating rates are well reproducible.

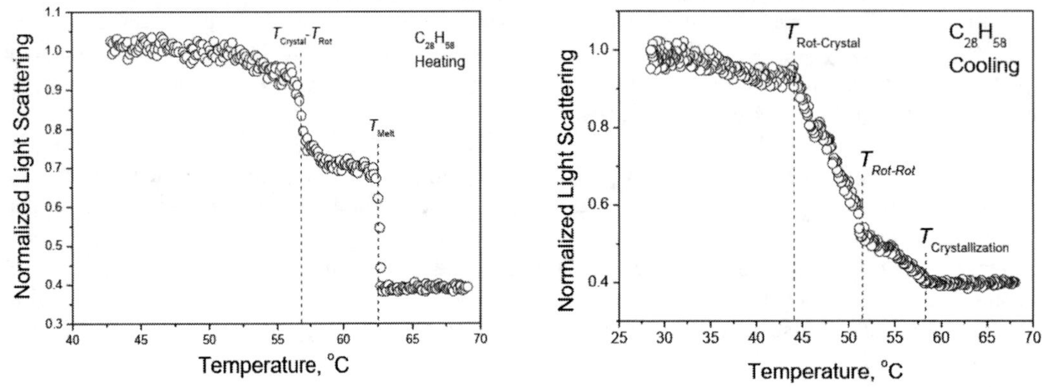

Figure 1. Temperature dependence of the scattered light intensity during heating and cooling for the dispersion of *n*-alkane $C_{28}H_{58}$ in water.

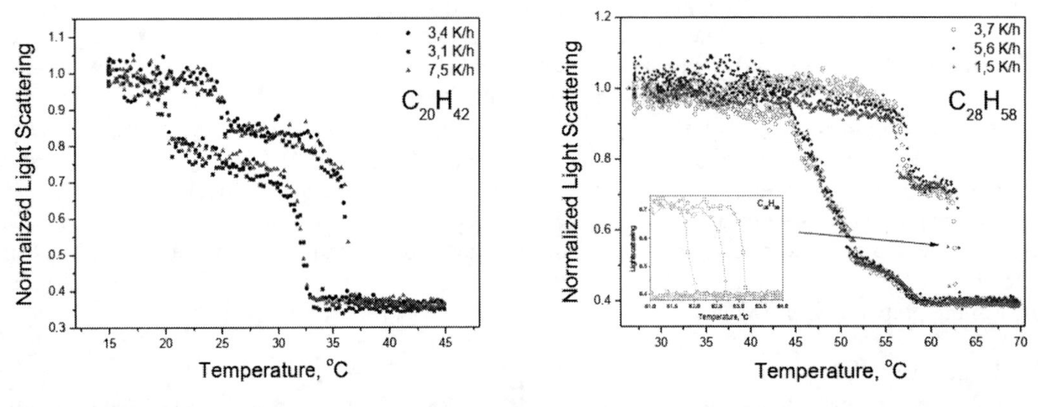

Figure 2. Temperature dependence of the intensity of the scattered light during heating and cooling for the dispersion of *n*-alkane $C_{20}H_{42}$ and $C_{28}H_{58}$ in water at different rates of temperature change.

Thus, from measurements of the scattered light intensity during heating and cooling of emulsions of individual *n*-alkanes in water, it is possible to determine the temperature of phase transitions (melting, crystallization, rotator phases) of *n*-alkanes. In this work, the phase transition temperature was determined as the maximum of the peak of the derivative of the intensity of the scattered light with respect to temperature.

Figure 3 shows the melting and crystallization temperatures of the studied *n*-alkanes measured by the optical method. The melting temperatures measured by the optical method are in good agreement with the melting temperatures of bulk samples measured in [23] by scanning calorimetry. The

IOP Conf. Series: Materials Science and Engineering **848** (2020) 012044 doi:10.1088/1757-899X/848/1/012044

difference between the crystallization temperatures measured by the optical method and the crystallization temperatures of emulsified *n*-alkanes samples in [23] is probably due to the use of surfactants for emulsification of *n*-alkanes in [23]. Surfactants can lead to a significant undercooling of the *n*-alkane drops in the emulsion.

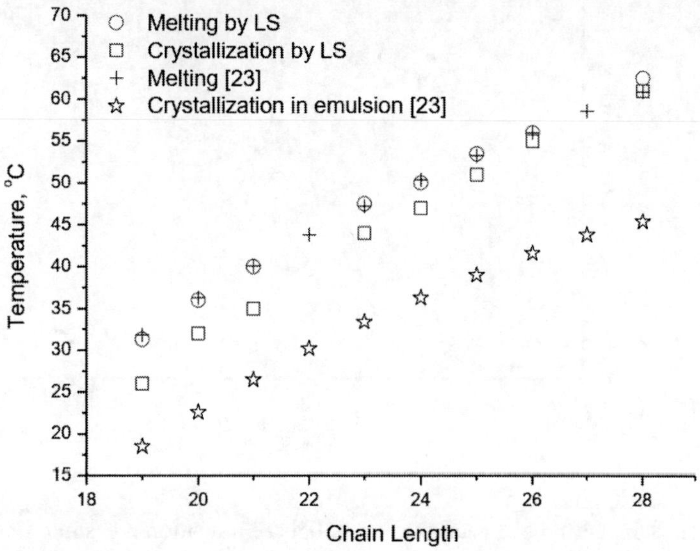

Figure 3. Melting and crystallization temperatures of *n*-alkanes measured in this work (red circles and blue squares) and data from Sirota E.B. work [23] (crosses and stars).

Figure 4 and Figure 5 show the temperatures of phase transitions of the crystal-crystal type (rotator phases), measured by the optical method, during heating and cooling of the samples under study.

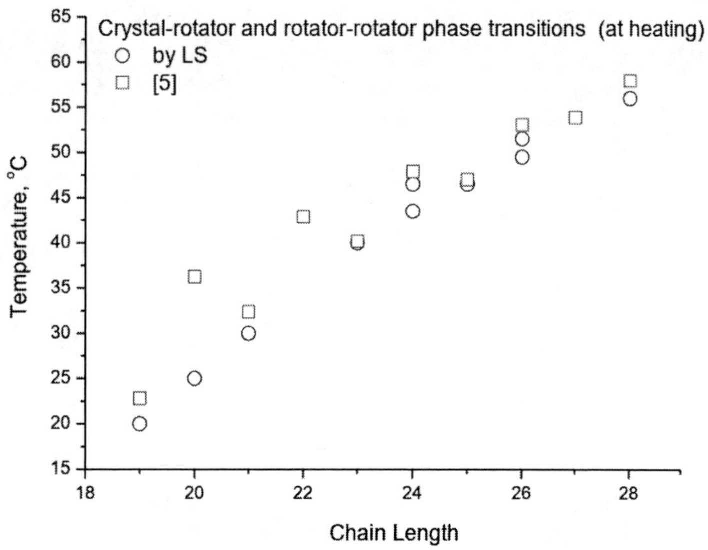

Figure 4. The temperatures of crystal to rotator phase transition measured in this work (black circles) and data from Sirota E.B. work [5] (red squares).

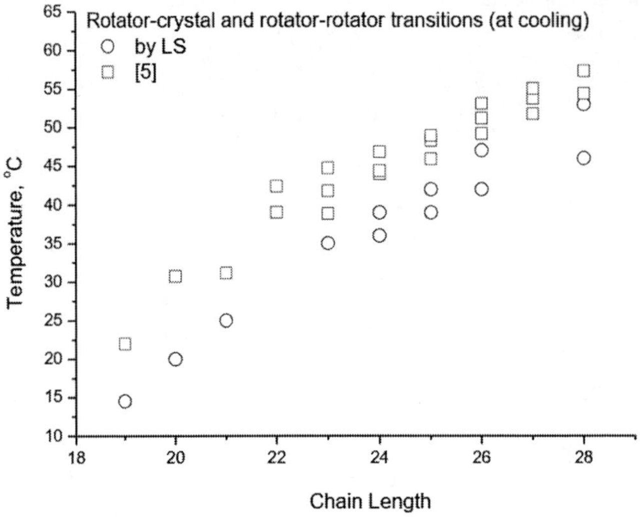

Figure 5. The temperatures of rotator to crystal phase transition measured in this work (black circles) and data from Sirota E.B. work [5] (red squares).

Figure 6 shows the dependence of the magnitude of undercooling (T_{Melt}-$T_{Crystal}$) on the number of carbon atoms in the *n*-alkane molecule.

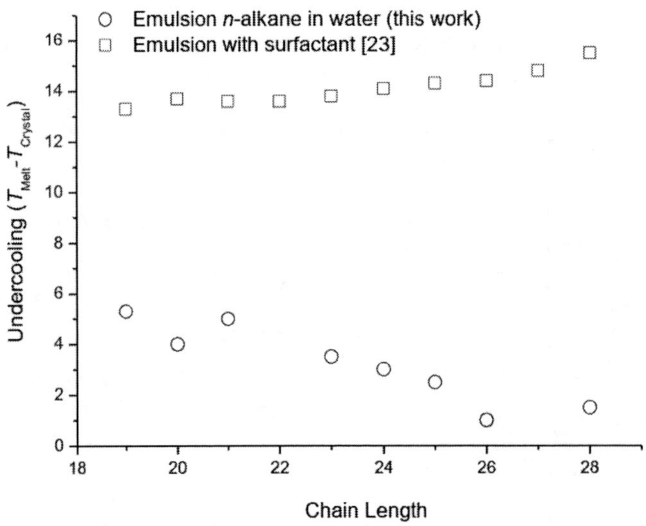

Figure 6. Dependence of the magnitude of undercooling (T_{Melt}-$T_{Crystal}$) on the number of carbon atoms in the *n*-alkane molecule, this work (black circles) and data from Sirota E.B. work [23] (red squares).

Figure 6 shows that for the studied emulsion samples, the undercooling value is significantly less than the undercooling value for *n*-alkane emulsions prepared using surfactants in [23].

Table 1 presents the temperatures of phase transitions measured by the optical method.

Table 1. Phase transition temperatures.

Chain length	$T_{Crystal\text{-}Rot}$, °C	$T_{Rot\text{-}Rot}$, °C at heating	T_{Melt}, °C	$T_{Crystallization}$, °C	$T_{Rot\text{-}Rot}$, °C at cooling	$T_{Rot\text{-}Crystal}$, °C	$T_{Melt}-T_{Crystallization}$, °C
19	20		31.5	26		14.5	5.5
20	25		36	32		20	4
21	30		40	35		25	5
23	40		47.5	44		35	3.5
24	43.5	46.5	50	47	36	39	3
25	46.5		53.5	51	39	42	2.5
26	49.5	51.5	56	55	42	47	1
28	56		62.5	61	46	53	1.5

4. Conclusion

Optical methods were used to determine the temperatures of phase transitions (melting, crystallization, rotator phases) of *n*-alkanes. It is shown that the method of dynamic and static light scattering can measure not only the size of paraffin emulsions but also determines the temperature of the phase transitions of *n*-alkanes. The proposed methodology allows studies of the influence of various factors on the phase transition temperatures of *n*-alkanes in emulsions, such as impurities to *n*-alkanes, emulsion droplet size, surfactants, pressure, and impurities in a dispersion medium.

Acknowledgment

The reported study was funded by RFBR, project number 20-08-00984.

References

[1] Du K, Calautit J, Wang Z, Wu Y and Liu H 2018 A review of the applications of phase change materials in cooling, heating and power generation in different temperature ranges *Appl. Energy* **220** 242–73

[2] Kahwaji S, Johnson M B, Kheirabadi A C, Groulx D and White M A 2018 A comprehensive study of properties of paraffin phase change materials for solar thermal energy storage and thermal management applications *Energy* **162** 1169–82

[3] Bukhalkin D D, Semenov A P, Novikov A A, Mendgaziev R I, Stoporev A S, Gushchin P A and Shchukin D G 2019 Phase change materials in the energy sector: current state of research and application prospects *Chemistry and Technology of Fuels and Oils* **55** 685–97

[4] Voronov V P 2000 Surface precrystallization of normal C24 alkane in porous glass *J. Exp. Theor. Phys.* **91** 144–9

[5] Sirota E B and Singer D M 1994 Phase transitions among the rotator phases of the normal alkanes *J. Chem. Phys.* **101** 10873–82

[6] Sirota E B, King H E, Singer D M and Shao H H 1993 Rotator phases of the normal alkanes: an X-ray scattering study *J. Chem. Phys.* **98** 5809–24

[7] Cholakova D and Denkov N 2019 Rotator phases in alkane systems: In bulk, surface layers and micro/nano-confinements *Adv. Colloid Interface Sci.* **269** 7–42

[8] Craievich A F, Denicolo I and Doucet J 1984 Molecular motion and conformational defects in odd-numbered paraffins *Phys. Rev. B* **30** 4782–7

[9] Jung N, Yun M and Jeon S 2012 Phase transitions between the rotator phases of paraffin investigated using silicon microcantilevers *J. Chem. Phys.* **136** 104903

[10] Trushina D B, Bukreeva T V, Borodina T N, Belova D D, Belyakov S and Antipina M N 2018

Heat-driven size reduction of biodegradable polyelectrolyte multilayer hollow capsules assembled on $CaCO_3$ template *Colloids Surf. B.* **170** 312–21

[11] Cabaleiro D, Hamze S, Agresti F, Estellé P, Barison S, Fedele L and Bobbo S 2019 Dynamic Viscosity, Surface Tension and Wetting Behavior Studies of Paraffin–in–Water Nano–Emulsions *Energies* **12** 3334

[12] Nikfarjam S, Ghorbani M, Adhikari S, Karlsson A J, Jouravleva E V, Woehl T J, & Anisimov M A 2019 Irreversible Nature of Mesoscopic Aggregates in Lysozyme Solutions *Colloid Journal* **81** 546–54.

[13] Novikov A A, Semenov A P, Monje-Galvan V, Kuryakov V N, Klauda J B and Anisimov M A 2017 Dual action of hydrotropes at the water/oil interface *J Phys Chem C* **121** 16423–31

[14] Zheng X, Anisimov M A, Sengers J V and He M 2017 Mesoscopic Diffusion of Poly (ethylene oxide) in Pure and Mixed Solvents *J Phys Chem B* **122** 3454–64

[15] Turnbull D 1949 The Subcooling of Liquid Metals *J. Appl. Phys.* **20** 817–817

[16] Vonnegut B 1948 Variation with temperature of the nucleation rate of supercooled liquid tin and water drops *J. Colloid Sci.* **3** 563–9

[17] Kuryakov V N and Ivanova D D 2019 Determination of melting point of *n*-alkanes by means of light scattering technique *J. Phys. Conf. Ser.* **1385** 12045

[18] Kuryakov V N and Ivanova D D 2019 Crystallization Behavior of Pure *n*-Alkane (*n*-Nonadecane) in a form of Nanoemulsion *Int. J. Nanosci.* **18** 1940032

[19] Kuryakov V N, De Sanctis Lucentini P G and Ivanova D D 2018 Tricosane (C23H48) and Octacosane (C28H58) mixture phase transition insight via Light scattering techniques *IOP Conference Series: Materials Science and Engineering* **347** 012034

[20] Kuryakov V N, Ivanova D D, Novikov A A, Ivanov E V, Gushchin P A, Semenov A P, Yusupova T N, & Shchukin D G 2019 The study of phase transitions in *n*-tricosane/bitumen aqueous dispersions by the optical method *Energy & Fuels (in Press)* DOI:10.1021/acs.energyfuels.9b03566

[21] Haynes W M 2014 CRC Handbook of Chemistry and Physics, 95th Edition (Florida: CRC Press)

[22] Lechner M D 1996 Refractive Indices of Inorganic, Organometallic, and Organononmetallic Liquids, and Binary Liquid Mixtures vol 38A (Berlin/Heidelberg: Springer-Verlag)

[23] Kraack H, Sirota E B and Deutsch M 2000 Measurements of homogeneous nucleation in normal-alkanes *J. Chem. Phys.* **112** 6873–85

Changes in electrochemical properties of a heavy tungsten alloy during its processing under the influence of DC current in ammonia-alkali solutions

O G Kuznetsova, A M Levin, M A Sevostyanov, O I Tsybin and A O Bolshikh

A.A. Baikov Institute of Metallurgy and Materials Science of Russian Academy of Sciences, 49 Leninsky pr., Moscow, Russian Federation

Abstract. The method of linear voltammetry in the potentiodynamic mode was used to study the change in the anode behavior of a heavy tungsten VNZh alloy (in wt.%: 78.8 W; 15.2 Ni; 6.0 Fe), as well as the phase composition and morphology of its surface during its electrochemical dissolution under the influence of DC current. It was determined, that the phase composition of VNZh alloy surface is a solid solution of iron-nickel after tungsten dissolution from the surface of the alloy. It has been established, that in ammonia-alkali electrolytes the process of tungsten passing into solution does not stop after dissolution of the metal from the alloy surface, but continues both due to diffusion of the metal from the depth of the sample and the pits formation in the processed alloy.

1. Introduction

Nowadays, a lot of research is being carried out, related to the use of various electrochemical ways for the tungsten alloys processing [1-6]. Electrolysis of fused salts is effective method, but extremely energy-consuming and quite complicated in implementation of a process [7, 8]. It is worth noting the hydrometallurgical method for processing heavy alloys, which does not require preliminary grinding of the alloys and consists in electrolysis in alkaline-chlorine solutions [9], however, the presence of chloride ions in the electrolyte will significantly complicate the further preparation of pure tungsten salts. The using of alkaline electrolyte is most optimal in ecological terms, however, it is accompanied by passivation of the soluble alloy during the process and a significant inhibition of its dissolution [10]. It is known, that the using of ammonia-alkali electrolytes can increase the oxidability of heavy tungsten alloys components [11, 12]. This work is devoted to the study of changes in the surface properties of heavy tungsten alloys, processed under the influence of DC current in alkali-ammonia electrolytes in order to increase their oxidation rate.

2. Experimental part

The studies were carried out by linear voltammetry in the potentiodynamic mode, using the universal IPC-Pro potentiostat. A heavy tungsten alloy of VNZh type (wt.%: 78.8 W; 15.2 Ni; 6.0 Fe) was applied as a working electrode. The measurements were carried out relative to a saturated chlorine-silver reference electrode with a platinum auxiliary electrode. A mixed solution of sodium hydroxide (2 M) and ammonium hydroxide (0÷6 M) was used as the electrolyte. The potential sweep speed was 3 mV/s and a temperature was 20°C. The samples were preliminarily cleaned and washed with a solution of hydrochloric acid (4 M) and distilled water.

The change in the surface composition of the VNZh alloy during hydro electrochemical processing was determined using X-ray spectral analysis (ISM-6380LV equipped with an Energy 250 analyser and X-ray diffractometry (ARL X'TRA). The morphology of alloy surfaces were examined with TESCAN VEGA II SBU scanning electron microscope (SEM).

Figure 1 shows the anode polarization curves of the fresh VNZh alloy in alkaline-ammonium-electrolytes. It is seen, that the dissolution rate of the alloy in the studied solutions is high due to the predominant tungsten passing into solution. The introduction of ammonium hydroxide into a solution of sodium hydroxide (2 M) to a concentration of 6 M does not significantly affect the value of the maximum current density of the alloy and only shifts this value to the region of more electropositive potentials.

Figure 2 shows the anodic polarization curves of the partially processed VNZh alloy, where tungsten was partially dissolved from its surface in ammonia-alkali electrolytes. Tungsten dissolution from the surface of the VNZh alloy was carried out according to [11], under the influence of DC current in the galvanostatic mode at a current density of 50 mA/cm^2 and the composition of the electrolyte NaOH 2M and NH_4OH 4M for 1 and 9 hours for partial and complete dissolution of tungsten from the alloy surface, respectively. It is seen, that with partially processed VNZh alloy for 1 hour, the dissolution rate of the alloy decreases. In this case, the introduction of ammonium hydroxide into the electrolyte still does not have a positive effect on the process.

Figure 3 shows the polarization curves of the completely processed VNZh alloy, where tungsten was completely dissolved from its surface for 9 hours, in a solution of sodium hydroxide (2 M) with the addition of ammonium hydroxide (2 ÷ 6 M). The polarization curves have two maxima: the first in the potential region - 0.3 ÷ - 0.2 V, the second - in the range of 0.1 ÷ 0.2 V. It can be seen, that in this case the introduction of ammonium hydroxide in the electrolyte leads to almost double increase the oxidation rate of the alloy. It is important to note, that neither for the fresh VNZh alloy, nor for its individual components (tungsten, nickel, iron), such phenomena were observed [11]. Nickel is the only alloy component, whose anodic oxidation is activated, when ammonium hydroxide is introduced into the electrolyte.

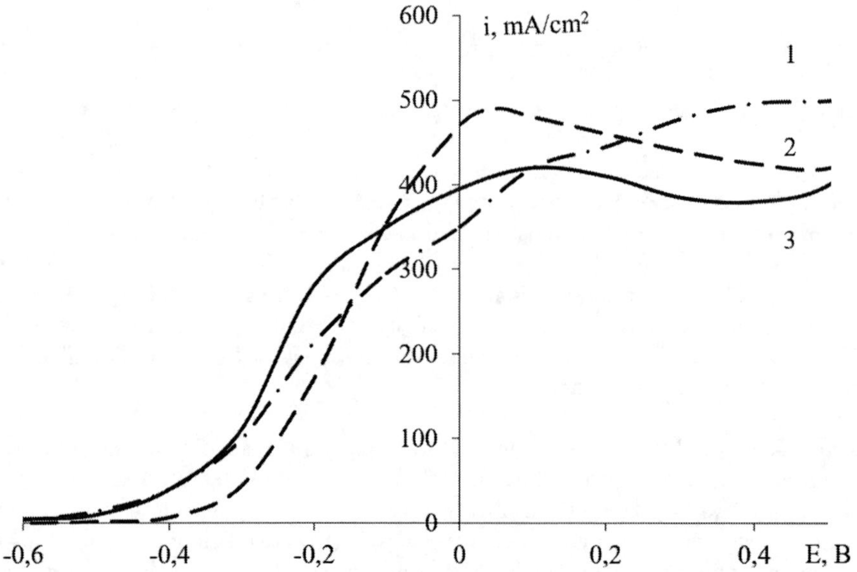

Figure 1. Anodic polarization curves of fresh VNZh alloy in a solution of 2M NaOH + NH_4OH. The concentration of NH_4OH, M: 1 - 6, 2 - 0, 3 - 2.

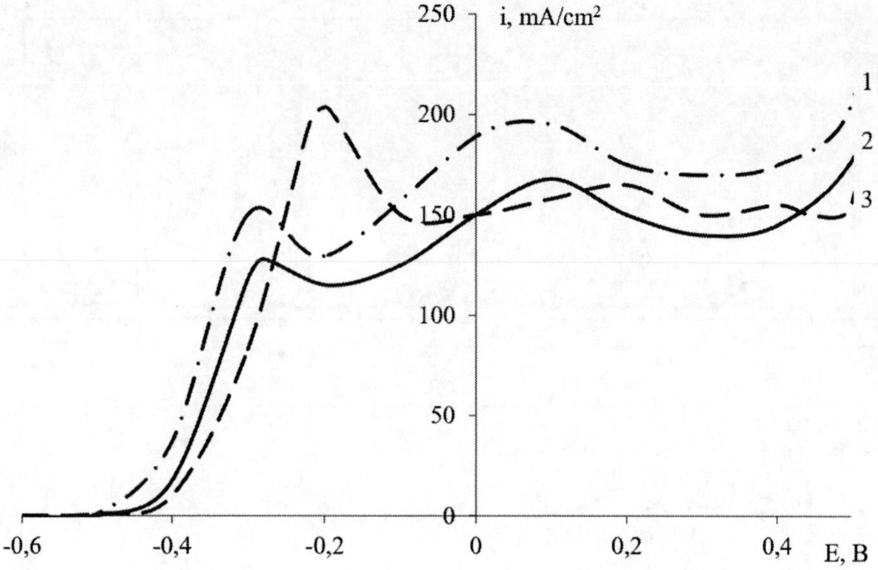

Figure 2. Anodic polarization curves of a partially processed VNZh alloy in a solution of NaOH + NH_4OH. The concentration of NH_4OH, M: 1 - 6, 2 - 2, 3 - 0.

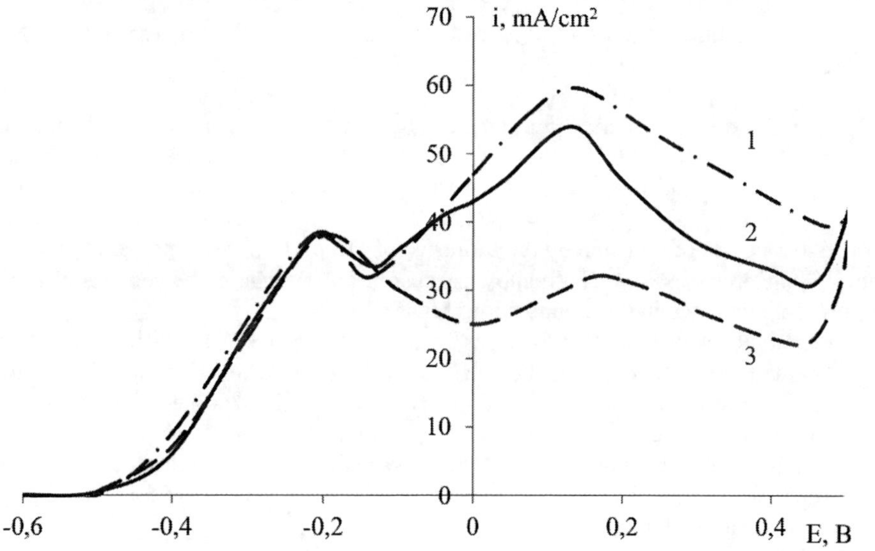

Figure 3. Anodic polarization curves of a completely processed VNZh alloy in a solution of 2M NaOH + NH_4OH. The concentration of NH_4OH, M: 1 - 6, 2 - 2, 3 - 0.

Thus, the complete dissolution of tungsten from the surface of the VNZh alloy significantly changes its electrochemical properties due to a change in the composition of the alloy surface, which is related with a change in the chemical and phase composition of the dissolved sample. The results of X-ray diffractometry of the fresh and completely processed VNZh alloys are shown in Figure 4.

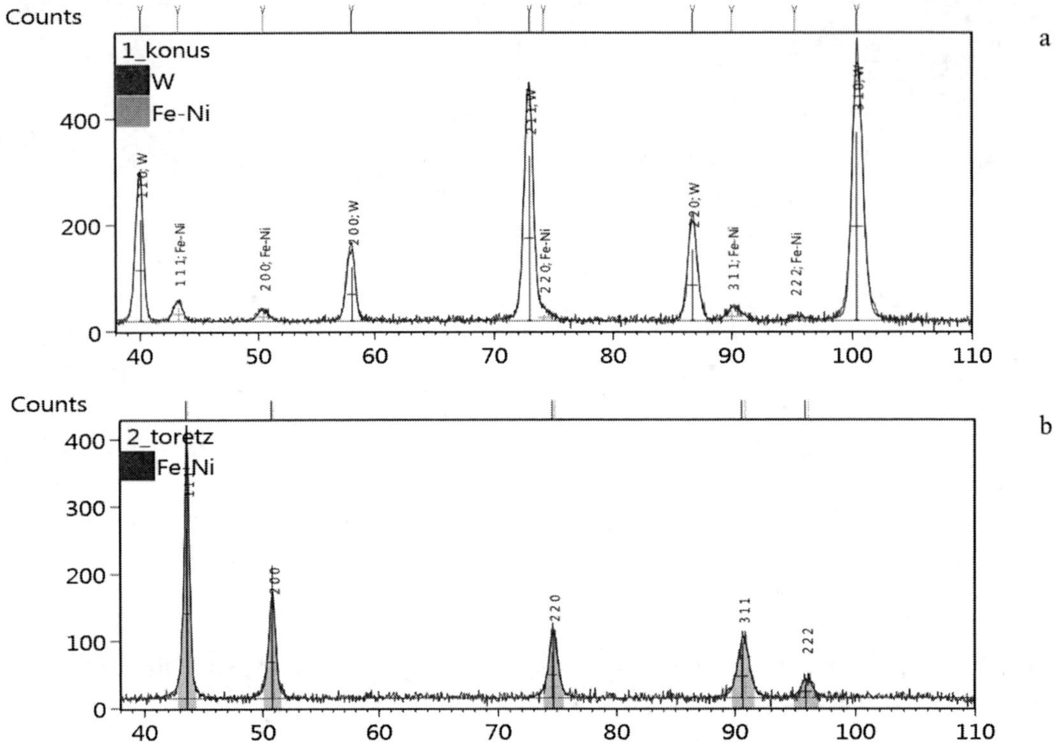

Figure 4. X-ray diffraction patterns of the fresh (a) and completely processed (b) VNZh alloy surfaces.

According to X-ray diffractometry (Figure 4), the fresh VNZh alloy is two-phase and consists of tungsten and iron-nickel solid solution and it is observed the predominance of tungsten with iron-nickel solid solution. It should be noted, that neither solid solutions of tungsten in ferronickel, nor ferronickel in tungsten were found in the fresh VNZh alloy (Figure 4a). However, after the electrochemical dissolution of tungsten by the influence of DC current in ammonia-alkali solution, the surface of the completely processed VNZh alloy has become single-phase and consists of a iron-nickel solid solution with no tungsten in the composition (Figure 4b).

In order to identify the processes occurring on the iron-nickel surface of VNZh alloy after complete dissolution of tungsten, we studied the anodic behavior of the ferronickel alloy (which is similar in the ratio of iron and nickel in the investigated VNZh alloy) in ammonia-alkaline solutions. It was found, that the ferronickel alloy behaves passively in the sodium hydroxide solution until the oxygen evolution potential is reached at a potential of ~0.55 V. When ammonium hydroxide is introduced into the electrolyte (up to a concentration of 6 M), the region of alloy oxidation appears on the polarization curves in the potential range 0.6÷0.8 V.

To understand the reasons for the difference in the anode behavior of completely processed VNZh alloy with the iron-nickel surface (Figure 3) and the ferronickel alloy, the morphology of the surface of the fresh and completely processed VNZh alloy surfaces was studied (Figure 5). It raises questions the reason for the presence of anode peaks on the polarization curves of the completely processed VNZh alloy in the potential range of -0.3 ÷ -0.2 , corresponding to tungsten oxidation. It can be seen, that as a result of the electrochemical processing of the VNZh alloy under the influence of DC current, pits are formed on the alloy surface (Figure 5b), in the depth of which the processes of tungsten oxidation can continue. This is reflected in the electrochemical properties of the processed heavy alloy.

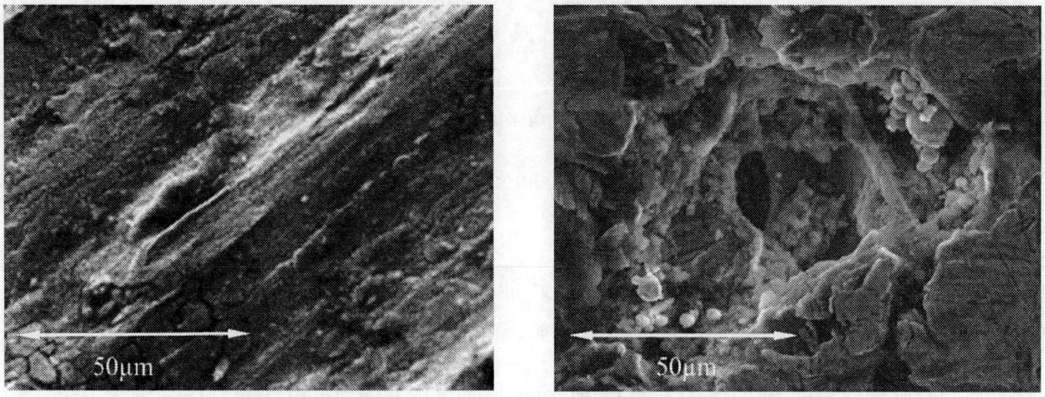

a b

Figure 5. Morphology of the fresh (a) and completely processed (b) VNZh alloy surfaces.

3. Conclusions

Using the method of linear voltammetry in the potentiodynamic mode, we studied the anode behavior of a heavy tungsten-containing alloy, such as VNZh alloy, in sodium hydroxide solutions (2 M) with ammonium hydroxide (0 ÷ 6 M). It has been established, that the VNZh alloy is intensively soluble in ammonia-alkali solutions and its electrochemical properties are determined by the behavior of the tungsten, which is the main alloy component.

It was determined, that the electrochemical properties of the VNZh alloy substantially change during its electrochemical dissolution in ammonia-alkali electrolytes, which is due to a decrease in the tungsten content in its composition and enrichment with a solid solution of iron-nickel.

It was found, that the introduction of ammonium hydroxide in the composition of an alkaline electrolyte has a positive effect on the oxidation of the VNZh alloy after the main amount of tungsten is passed into the solution during alloy processing.

It was revealed, that during the VNZh alloy processing under the influence of DC current in ammonia-alkali electrolyte, tungsten dissolution continues even after it is completely dissolved from the alloy surface due to the diffusion of metal from the depth of the sample and the formation of pits in the processed alloy.

4. Acknowledgements

The work was carried out according to the state ask №007-00129-18-00.

References

[1] Lin Luo, Lui Kein, Atsushi Shibayama, Wantai Yen 2004 Recovery of tungsten and vanadium from tungsten alloy scrap *Hydrometallurgy* **72** 1 1-8

[2] Berezina S L, Goryacheva V N, Dvulichanskaya N N, Ermolaeva V I 2017 Anodic dissolution of a tungsten-zirconium alloy in an alkaline electrolyte *Bulletin of MSTU N.E. BAUMAN* **2** 96-104

[3] Shemi A, Magumise A, Ndlovu S, Sacks N 2018 Recycling of tungsten carbide scrap metal: A review of recycling methods and future prospects *Mineral Engineering* **122** 195-205

[4] Kuznetsova O G, Levin A M., Sevastyanov M A and Bolshikh A O 2018 The improvement of rhenium recovery technology from W-Re alloys *IOP Conf Ser: Mater Sci Eng* 1134

[5] Kovalenko V 2017 Selective anodic treatment or W(WC)-based superalloy scrap *Eastern-European Journal of Enterprise Technologies* **85** 1/5 53-58

[6] Luo L, Miazaki T, Shibayama A, Yen W, Fujita T 2003 A novel process for recovery of tungsten and vanadium from a leach solution of tungsten alloy scrap *Minerals Engineering* **16** 665-670

IOP Conf. Series: Materials Science and Engineering **848** (2020) 012045 doi:10.1088/1757-899X/848/1/012045

[7] Longgang Ye, Zhen Ouyang, Yongming Chen, Shufen Liu 2020 Recovery of rhenium from tungsten-rhenium wire by alkali fusion in KOH-K2CO3 binary molten salt electrolysis *Int Journal of Refractory Metals and Hard Materiels* **87** 1-7
[8] Ming Li, Xiaoli Xi, Zuiren Nie, Liwen Ma 2019 Recovery of tungsten from WC-Co hard metal scraps using molten salts electrolysis *Journal of Material Research and Technology* **8** 1 1440-1450
[9] Parshutin V V 2008 Corrosive and electrochemical behavior of pseudo-alloys based on tungsten and their components *Electronic Processing of Materials* **6** 27-45.
[10] Hairunnisha S, Sendil G K, Prabhakar Rethinaraj J, Srinivasan G N, Adaikkalam P, Kulandaisamy S 2007 Studies on the preparation of pure ammonium paratungstate from tungsten alloy scrap *Hydrometallurgy* **85** 67-71
[11] Kuznetsova O G, Levin A M, Sevast'yanov M A, Tsybin O I, Bolshikh A O 2019 Electrochemical oxidation of a heavy tungsten-containing VNZh-type alloy and its components in ammonia–alkali electrolytes *Russian Metallurgy (Metally)* **5** 507–510
[12] Kuznetsova O G, Levin A M , Sevastyanov M A and Bolshikh A O 2019 Electrochemical recycling of nickel-cobalt-containing tungsten alloys *IOP Conf Ser: Mater Sci Eng* **525** 012088

IOP Conf. Series: Materials Science and Engineering **848** (2020) 012046 doi:10.1088/1757-899X/848/1/012046

AC electrochemical oxidation of nickel and VNZh alloy in alkaline-ammonium solutions

O G Kuznetsova, A M Levin, M A Sevostyanov, O I Tsybin and A O Bolshikh

A.A. Baikov Institute of Metallurgy and Materials Science of Russian Academy of Sciences, 49 Leninsky pr., Moscow, Russian Federation

Abstract. It was studied the electrochemical oxidation of nickel in an electrolyte of the composition NaOH 2M + NH_4OH 4M by using linear and cyclic voltammetry. A significant increase in the process rate has been revealed due to the cathodic polarization of the electrode under study. It was shown the possibility of heavy tungsten alloy processing, such as VNZh-type alloy (in wt%: 78.8 W, 15.2 Ni, 6.0 Fe), under the influence of alternating electric current of industrial frequency in alkaline-ammonium solutions.

1. Introduction

Alkaline solutions are often used for electrochemical processing tungsten alloy waste [1-4] and, in particular, for nickel-containing heavy alloys, based on tungsten [5-9]. The anodic polarization of nickel in these electrolytes is accompanied by the formation of a passivating oxide film with various oxidation levels, as well as oxygen evolution [10-12]. The formation of nickel oxides during the electrochemical dissolution of nickel-containing heavy tungsten alloys under the influence of direct current prevents the passing of tungsten into solution and significantly reduces the rate of alloys oxidation [5, 8]. According to the data of [8, 9], the introduction of ammonium hydroxide (4M) into the composition of the alkaline electrolyte significantly accelerates the oxidation of nickel and cobalt without significantly affecting the behavior of iron. It is known [10-16], that oxides of the iron group metals can be removed from the electrode surface due to their cathodic reduction and mechanical detachment by hydrogen, released during cathodic polarization. In this regard, the aim of this work is to:

- investigate the effect of cathodic polarization of nickel in the hydrogen evolution region on the process of its subsequent electrochemical oxidation in alkaline ammonia solutions;
- study the possibility of using alternating electric current for nickel-containing heavy tungsten alloys processing in alkaline-ammonium electrolytes on the example of VNZh alloy (tungsten-nickel-iron).

2.Experimental part

Using the methods of cyclic voltammetry (CVA) and linear voltammetry in the potentiodynamic mode, the anodic behavior of nickel in a solution, containing 2M sodium hydroxide and 4M ammonium hydroxide, was studied [8]. The studies were carried out with the IPC-Pro universal potentiostat with the use of grade N0 nickel, an auxiliary glassy-carbon electrode, and a saturated chlorine-silver reference electrode.

Preliminarily, the samples were stripped and washed with distilled water. The polarization of the electrode was carried out from the steady state potential of the nickel electrode (E_{st}) to the oxygen evolution potential (+0.9 V) and in the opposite direction to the potential of -1.5 V. In this case, the potential sweep speed in the range of -1.0 ÷ +0.9 V was 25 mV/s, while in the range of potentials -1,5 ÷ -1.0 V (in the hydrogen evolution region) the polarization rate was set in the range from 1 to 25 mV/s. Immediately, after the end of the CVA cycle, the nickel electrode was polarized from E_{st} up to a potential value of +0.9 V with a potential sweep speed of 25 mV/s. Studies were carried out at the temperature of 20°C. The amount of electricity, that passed during the cathode period of CVA (Q_{cat}), was calculated as the area under the corresponding dependence $I_{cat} = f\ (\tau_{cat})$, Where I_{cat} is the cathode current strength (A), τ_{cat} - cathodic polarization time (h).

Electrochemical oxidation of a heavy nickel-containing tungsten alloy (VNZh alloy, wt.,%: 78.8 W, 15.2 Ni, 6.0 Fe) under the influence of alternating current (ac) of industrial frequency (50 Hz) was carried out in a solution, containing 2M sodium hydroxide and 4M ammonium hydroxide, in a thermostatic cell at a temperature of 25°C by using two electrodes from the investigated alloy with a fixed surface area. The oxidation rate of VNZh alloy was determined based on the decrease in the electrodes mass during the experiment. The current efficiency for the alloy was determined in terms of tungsten for the ionization of tungsten at an oxidation level of +6. Distilled water and reagent grade reagents were used for the preparation of solutions.

Figure 1 shows the anodic polarization curves of nickel in the solution of NaOH 2M + NH_4OH 4M

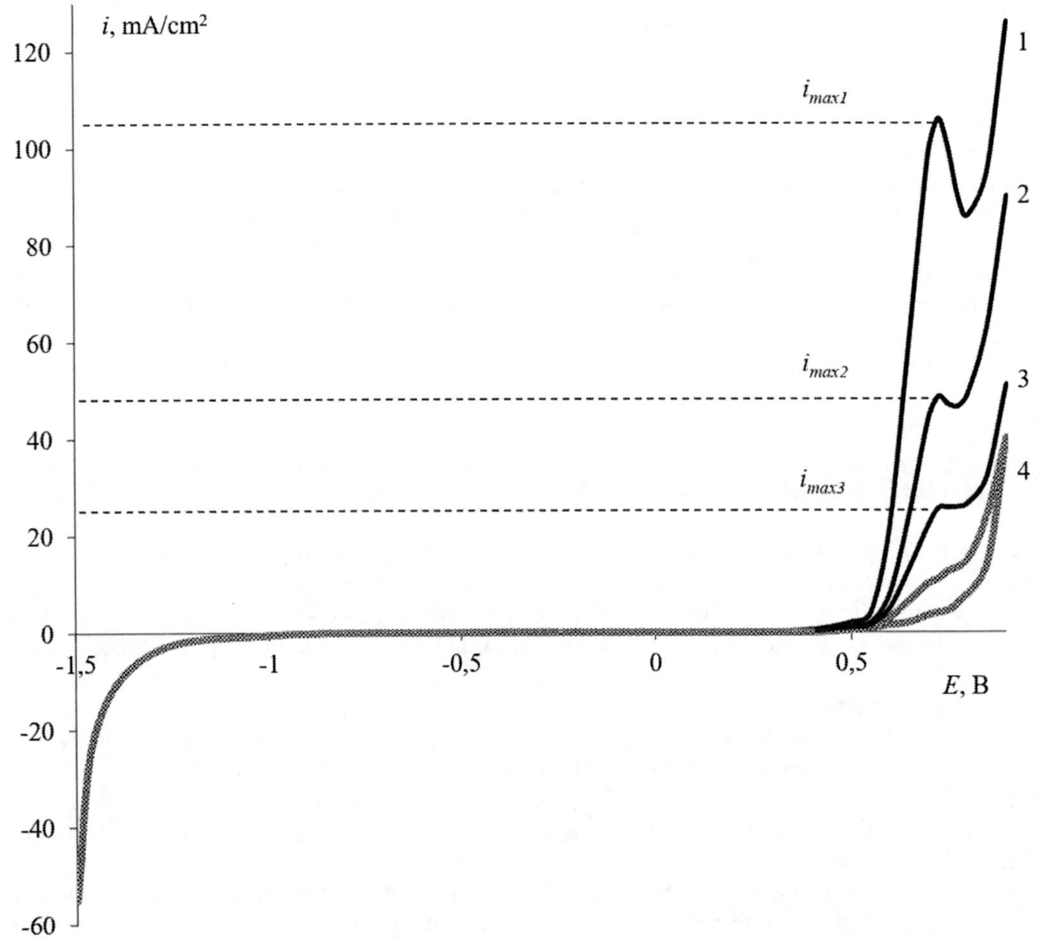

Figure 1. Anodic polarization of nickel in a NaOH 2M + NH_4OH 4M solution (curves 1-3) after its CVA (curve 4) at the potential sweep speed in the potential range of -1.5 ÷ -1.0 V, mV/s: 1 - 1 , 2 - 5, 3 - 25. The potential sweep speed in the range of potentials -1.0 ÷ +0.9 V is 25 mV/s.

IOP Conf. Series: Materials Science and Engineering **848** (2020) 012046 doi:10.1088/1757-899X/848/1/012046

(curves 1–3), each curve was taken after a CVA cycle (curve 4) with a different potential sweep speed (from 1 to 25 mV/s) in the hydrogen evolution region (-1.5 ÷ -1.0 V).

It can be seen, that a decrease in the potential sweep speed in the cathode region of CVA (-1.5 ÷ -1.0 V) from 25 to 1 mV/s leads to a four-fold increase in the maximum anode current density (i_{max}) from 26 to 106 mA/cm². It is important to note, that similar studies in a NaOH 2M solution, without ammonium hydroxide, showed the absence of anode effects of nickel, regardless of the potential sweep speed in the cathode region of CVA.

The effect of the amount of electricity, transmitted during the CVA cathode cycle, on the value of the maximum anode current density of nickel in the solution of NaOH 2M + NH_4OH 4M is presented in the table.

Table 1. Effect of the amount of electricity, transmitted during the CVA cathode cycle, on the value of the maximum anode current density of nickel in the solution of NaOH 2M + NH_4OH 4M.

Q_{cat}, mA·h	0	0.06	0.12	0.38	1.12	10
i_{max}, mA/cm²	12	26	34	47	106	200

It can be seen, that the maximum anode current density of nickel is only 12 mA/cm² in the absence of cathodic polarization. Cathodic polarization of the nickel electrode at Q_{cat} equal to 1.12mA·h leads to a tenfold acceleration of the nickel electrode oxidation to i_{max}= 106 mA/cm². Further increase in Q_{cat} is impractical, since at Q_{cat}= 10 mA·h the i_{max} only doubles (up to 200 mA/cm²).

Thus, the obtained data show, that the introduction of ammonium ions into the alkaline electrolyte significantly activates the oxidation processes on the nickel electrode, when using the current mode, which includes both the anode and cathode components. In this case, the most technologically simple current mode with different polarity is a sinusoidal alternating electric current of industrial frequency. This can be used to intensify the oxidation processes of VNZh-type alloys during their electrochemical processing in alkaline electrolytes.

Figure 2 shows the dependences of the oxidation rate of VNZh alloy and its current efficiency on the ac current density in the solution of NaOH 2M + NH_4OH 4M. It is seen, that an increase in current density leads to a substantial growth in the rate of the alloy with a simultaneous increase in current efficiency. Studies have established, that at a current density of 7 A/cm², the alloy oxidation rate is ~°1250 mg/cm²·h, and the current efficiency reaches 30%.

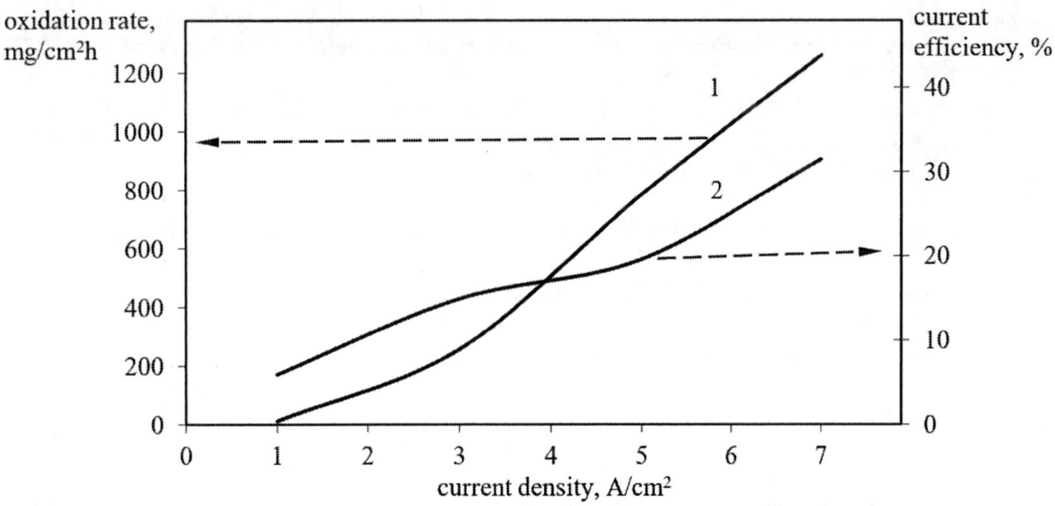

Figure 2. Dependences of oxidation rate (1) of VNZh alloy and its current efficiency (2) on the ac current density.

IOP Conf. Series: Materials Science and Engineering **848** (2020) 012046 doi:10.1088/1757-899X/848/1/012046

3. Conclusions

The positive effect of the cathodic component of the current on the process of nickel anodic oxidation in an alkaline-ammonium electrolyte has been established. It consists in the increasing the oxidation rate of nickel. It has been shown, that the electrochemical oxidation of a nickel-containing tungsten heavy alloy of the VNZh-type under the influence of alternating current of industrial frequency reaches ~1250 $mg/cm^2 \cdot h$ for current efficiency of ~ 30% in a solution of NaOH 2M + NH_4OH 4M.

Acknowledgments

The work was carried out according to the state ask №007-00129-18-00.

References

[1] Berezina S L, Goryacheva V N, Dvulichanskaya N N, Ermolaeva V I 2017 Anodic dissolution of a tungsten-zirconium alloy in an alkaline electrolyte *Bulletin of MSTU N.E. BAUMAN* **2** 96-104

[2] Kuznetsova O G, Levin A M ., Sevastyanov M A and Bolshikh A O 2018 The improvement of rhenium recovery technology from W-Re alloys *IOP Conf Ser: Mater Sci Eng* 1134

[3] Palant AA., Levin A M, Palant L A 2013 The method of electrochemical processing of metal tungsten-copper waste alloys RU 2479652*Bulletin***113**.

[4] Luo L, Miazaki T., Shibayama A, Yen W, Fujita T 2003 A novel process for recovery of tungsten and vanadium from a leach solution of tungsten alloy scrap *Minerals Engineering* **16** 665-670

[5] Hairunnisha S, Sendil GK, Prabhakar Rethinaraj J, Srinivasan GN, Adaikkalam P, Kulandaisamy S 2007 Studies on the preparation of pure ammonium paratungstate from tungsten alloy scrap *Hydrometallurgy* **85** 67-71

[6] Parshutin V V 2008 Corrosive and electrochemical behavior of pseudo-alloys based on tungsten and their components *Electronic Processing of Materials* **6** 27-45.

[7] Guriev V R 2001. *Research and development of technology for processing waste products from the production of refractory and heavy non-ferrous metals using electrochemical methods* Ext abst PhD disser Vladikavkaz

[8] Kuznetsova O G, Levin A M., Sevast'yanov M A, Tsybin O I, Bolshikh A O 2019 Electrochemical oxidation of a heavy tungsten-containing VNZh-type alloy and its components in ammonia–alkali electrolytes *Russian Metallurgy (Metally)* **5**, 507–510

[9] Kuznetsova O G, Levin A M ., Sevastyanov M A and Bolshikh A O 2019 Electrochemical recycling of nickel-cobalt-containing tungsten alloys *IOP Conf Ser: Mater Sci Eng* **525** 012088

[10] Obraztsova E Yu, Klimnik A B 2017 Investigation of the patterns of formation and physicochemical properties of nickel oxide powders synthesized by the method of nickel electrochemical dispersion *Vestnik TSTU* 23 104-110

[11] Klimnik A B, Nikiforova E Yu 2013 Electrochemical behavior of nickel and its oxides in concentrated solutions of sodium hydroxide *Electrochemistry* **12** 1251-1255.

[12] Klimnik AB,OstozhkovaE.Yu2012 *Electrochemical synthesis of nanodispersed powders of metal oxides* (Tambov: FSBEI HPE "TSTU") 144

[13] Korobochkin V V 2004 *The process of obtaining nanodispersed oxides using electrochemical oxidation of metals under the influence of alternating current methods* Ext abst PhD disser Tomsk

[14] Popova E V, Ustyugov A M 2016 Patterns of electrochemical oxidation by alternating current of metallic nickel and aluminum *Int Sci Conf "Chemistry and chemical technologies in the 21st century"*Tomsk Polytech Inst116-117

[15] Pchelnikov AP 2006 *Electrochemistry and corrosion of hydrogenated nickel, copper and copper-nickel alloys in aqueous electrolytes* Ext abst PhD disser Moscow 44

[16] Demyan V V, Bespalova Zh I, Fesenko L I 2017 To the question of the mechanism of nickel oxidation in aqueous alkali solutions during polarization by an alternating asymmetric current *Conf "Fund.and appl. questions of electrochem. and chemical-cat. deposition of metals and alloys"* FGBUN Inst of Physchem and Electrochem RAS 24

IOP Conf. Series: Materials Science and Engineering **848** (2020) 012047 doi:10.1088/1757-899X/848/1/012047

Study of the effect of crystallization conditions on the structure of polyurethane block copolymer based on poly-ε-caprolactone diol and isophoron diisocyanate

M A Gorbunova [1], V A Lelecova [2], D M Shukhardin [2], G V Malkov [1], D V Anokhin [1,2] and E R Badamshina [1]

[1]The Institute of Problems of Chemical Physics of the Russian Academy of Sciences, Academician Semenov avenue 1, Chernogolovka, Moscow region, 142432 Russian Federation

[2]Lomonosov Moscow State University, Faculty of fundamental physical and chemical engineering, Leninskiye Gory 1b.51, Moscow, 119991 Russian Federation

[1]zenyatharr@gmail.com

Abstract. In the paper a study of isothermal crystallization kinetics of polyurethane on the basis of poly-ε-caprolactone diol, isopforon diisocyanate and 1,4-butanediol is presented. The process of isothermal crystallization in the temperatures range from -15 to 30°C was studied by DSC method. Analysis of total melting enthalpy and degree of crystallinity of the polyurethane allows to obtain material with one or two melting points of polyester block at desired temperature. These results will help in design of new thermoplastic elastomers with one- or two-stage shape memory effect.

1. Introduction

Shape memory polymers (SMPs) attract a lot of attention of researchers because of their ability to obtain a temporary shape and restore their initial shape under the influence of external factors. In most cases, polymers with SME are thermosensitive, i.e. they recover to their original shape after local heating. Among polymers with SME, particular interest is given to semi-crystalline multi-block thermoplastic polyurethanes (TPU) with thermodynamically incompatible soft and rigid blocks, which lead to microphase separation. Rigid blocks are built from urethane groups forming a network of hydrogen bonds, providing physical cross-links. Soft blocks are crystallizing hydroxyl-containing polyesters and the crystals play the role of additional units of the physical network, providing high strength of the material. Therefore, a rigid block is responsible for the strength of the material, while a soft block is responsible for its elasticity and ability to take a temporary shape.

For applications, it is important to control switching temperature of heat-sensitive TPUs, i.e. the temperature at which the material changes its temporary shape. For TPU with crystallizing soft blocks the switching temperature is the melting point of polyester crystal phase [1]. Nowadays one of the most popular oligoether diols used as a soft block is biodegradable poly-ε-caprolactone (PCL), which has a melting point close to physiological environment temperature, which allows to use such TPU in medicine [2]. It is known that crystallization of PCL significantly changes the structure and properties of TPU. The size of crystals and the growth rate depend on the crystallization conditions and determine switching temperature. In such way, control of thermal properties of PCL-based TPU, particularly, switching temperature can be done by adjustment of crystallization conditions in soft

IOP Conf. Series: Materials Science and Engineering **848** (2020) 012047 doi:10.1088/1757-899X/848/1/012047

block domains without changing the polymer composition. The choice of optimal crystallization conditions allow to fabricate materials with desired properties which is important for practical applications.

In this paper, the structure of a model block copolymer based on PCL, isophoron diisocyanate (IPDI) and 1,4-butanediol (1,4-BD) has been studied by the DSC method depending on the crystallization conditions. The presented results are aimed at the development of multi-block urethane and urethanurea copolymers with programmable shape memory effect for medical applications.

2. Experimental part

2.1. Materials

Poly-ε-caprolactone diol (PCL), M_n~ 2000 Da, by Aldrich, was used after pre-drying at 60°C on a rotary evaporator. The content of hydroxyl groups determined by the chemical method [3] was 1.9%.

Isophoron diisocyanate (IPDI), 1,4-butanediol (1,4-BD) by Aldrich company, methylene chloride (purity ~ 99.9%) were used preliminarily purified according to the standard method [4]. The content of isocyanate groups in IPDI and hydroxyl groups of 1,4-BD, determined by the chemical method [4], equals to 99.9% of the theoretical one.

Dibutyltin dilaurate catalyst by Aldrich was used without additional purification.

2.2. Synthesis of polyurethane

Linear polyurethane block copolymer was obtained by a two-stage method through the stage of formation of macrodiisocyanate in methylene chloride solution in the presence of dibutyltin dilaurate catalyst. At the first stage, IPDI interacted with PCL at $[NCO]_{IPDI}/[OH]_{PCL}$ = 2, then the chain extension 1.4-BD was added to the stoichiometric ratio of functional groups. Upon reaching the degree of transformation in the groups of NCO ~ 98%, the reaction mass was poured into a flat form of Teflon and dried at 40°C for a day to a constant weight.

2.3. Methods

Polyurethane crystallization kinetics was investigated by DSC method in isothermal conditions using DSC 30 Mettler Toledo calorimeter. The samples (10±0,2 mg) were placed in an aluminium pan, rapidly cooled down to -20°C and further heated at rate to 10°C/min in a nitrogen atmosphere to 70°C. After 2 minutes at 70°C, the samples were cooled down to isothermal crystallization temperature (-15 to +30°C). After holding for period of time from 15 to 960 minutes at given temperature, the samples were heated up to 70°C to determine the melting enthalpy of crystal phase.

The degree of crystallinity (X) was determined by the formula (1):

$$X = \frac{\Delta H_m}{\Delta H_m^{100\%} \cdot \omega(PCL)} \quad (1)$$

where ΔH_m – total melting enthalpy measured from DSC curves, $\Delta H^{100\%}{}_m$ =243 J/g - equilibrium melting enthalpy of PCL, taking into account that crystallinity of Poly-ε-caprolactone diol determined by X-ray diffraction is about 40%.

Microscopic images of thin films with thickness of 10-30 microns were obtained with optical microscope Carl Zeiss AxioScope A1 POL in polarized light with 100x objective. The films were applied to cover glasses made of THF solution, heated up to 70°C, then crystallized at room temperature and at -10°C during twenty-four hours. To heat the films, a Linkam LTS420 heating stage with a working temperature range from -196°C to 420°C was used. Images were exported to the PC using 5 MP CMOS camera.

3. Results and discussion

Synthetic polyurethane is a block copolymer consisting of a soft block of PCL and a rigid block formed by IPDI and chain extensioner 1,4-BD (Figure 1).

IOP Conf. Series: Materials Science and Engineering **848** (2020) 012047 doi:10.1088/1757-899X/848/1/012047

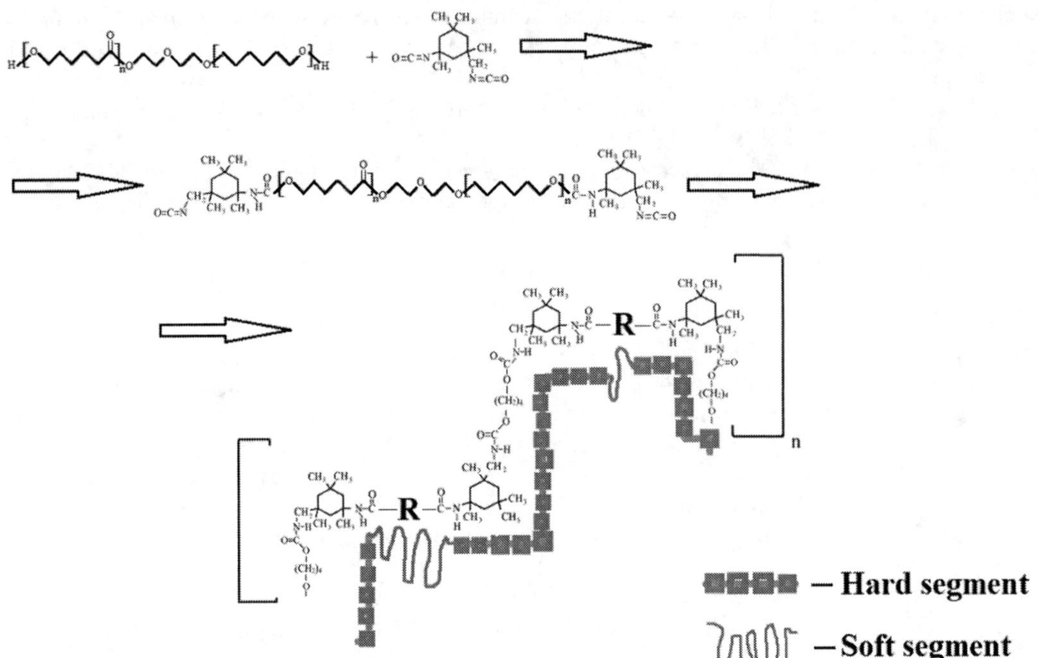

Figure 1. Scheme of reaction of two-stage synthesis of polyurethane based on IPDI and PCL using 1,4-BD as a chain extender.

Figure 2 shows DSC-thermograms of the studied samples heated from crystallization temperature from -15°C to 25°C. At the crystallization temperature -15°C two endothermic melting peaks are observed during heating, that can be associated with the melting of crystals of different perfection – less rgular $T_{melt}(1)$ and more regular $T_{melt}(2)$, respectively. With the increase of the crystallization temperature, the intensity of endothermic melting peaks changes due to the difference in the nucleation rate. With the increase of crystallization time up to 960 minutes the area of peaks gradually increases with a shift of the first peak to higher temperatures (Table 1). This is probably related to the improvement of the PCL crystal quality during heating.

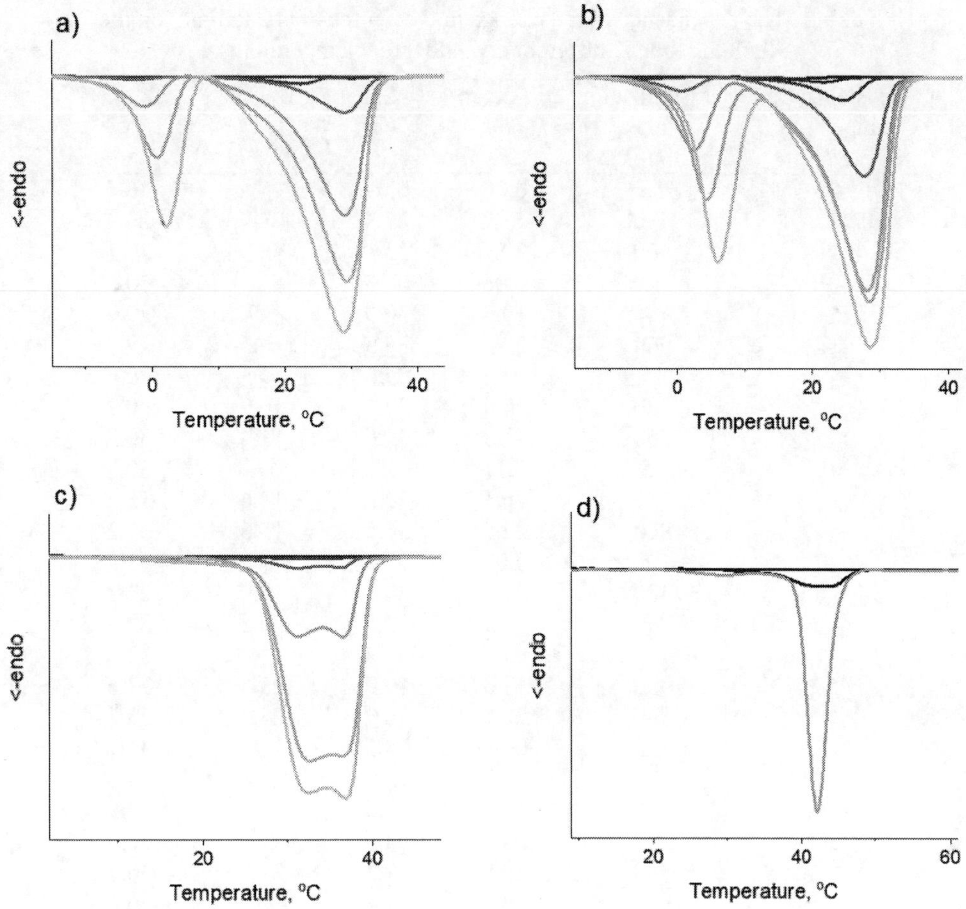

Figure 2. DSC heating thermograms of polyurethane crystallized at -15°C (a), 5°C (b), 20°C (c) and 25°C (d) for 960 minutes.

The increase of the crystallization temperature to 25°C results in decrease of the nucleation and crystallization rate and formation of perfect thick crystals due to homogeneous nucleation. This is represented on the DSC-thermogram as single endothermic melting peak. On the DSC curves of sample crystallized at higher temperature the endothermic peaks do not appear indicating the absence of crystallization because of very slow nucleation rate.

Table 1 shows the degree of crystallinity of TPU depending on crystallization temperature and time, calculated from area of endothermic melting peaks at certain temperature intervals using equation (1). Table 1 shows that the maximum degree of crystallinity of the TPU is obtained after crystallization at 5°C.

The influence of crystallization conditions on the TPU structure is confirmed by the results of optical microscopy (Figure 3).

IOP Conf. Series: Materials Science and Engineering **848** (2020) 012047 doi:10.1088/1757-899X/848/1/012047

Table 1. The dependence of the change in the degree of crystallinity on the holding time at different crystallization temperatures.

Crystallization temperatures *(˚C)*	Crystallization time *(min)*	X *(%)*	-ΔH *(J/g)*	$T_{melt}(1)$ *(˚C)*	$T_{melt}(2)$ *(˚C)*
-15	15	0.06	0.09	-	30.4
	480	17.1	27.3	0.6	29.3
	960	20.7	33.1	2.3	29.1
-10	15	0.06	0.10	-	30.5
	480	19.3	30.9	5.7	29.9
	960	22.2	35.5	7.2	29.7
-5	15	0.06	0.06	-	29.6
	480	20.4	32.6	10.2	30.2
	960	21.3	34.0	11.4	30.2
0	15	0.11	0.11	-	30.5
	480	16.3	32.8	14.6	15.5
	960	22.8	36.5	15.5	31.2
5	15	0.07	0.10	-	31.2
	480	21.5	34.3	19.0	32.3
	960	24.9	39.8	20.2	32.5
10	15	0.07	0.12	-	32.3
	480	12.5	19.9	23.4	33.6
	960	22.5	35.9	24.2	33.8
15	15	0.04	0.03	-	31.1
	480	8.5	13.6	27.8	34.8
	960	18.1	28.8	30.0	34.9
20	15	0.02	0.03	-	36.9
	480	10.8	17.2	32.4	36.7
	960	10.8	17.4	32.8	36.6
25	15	-	-	-	-
	480	0.17	0.27	-	37.1
	960	0.99	1.6	-	42.0

The microphotographs show the change in the size of the crystalline domains depending on the crystallization conditions. At the image of TPU crystallized at 25°C one can observe the presence of "Maltese cross" type texture with the size of objects of about 20-50 microns. These objects represent radially growing spherulites which are typical for crystallizing block copolymers [5]. In contrast, crystallization at -10°C results in formation of smaller spherulites of 10-20 microns. The obtained results indicate changes in nucleation rate which affects the total degree of crystallinity of the polymer and crystals quality.

IOP Conf. Series: Materials Science and Engineering **848** (2020) 012047 doi:10.1088/1757-899X/848/1/012047

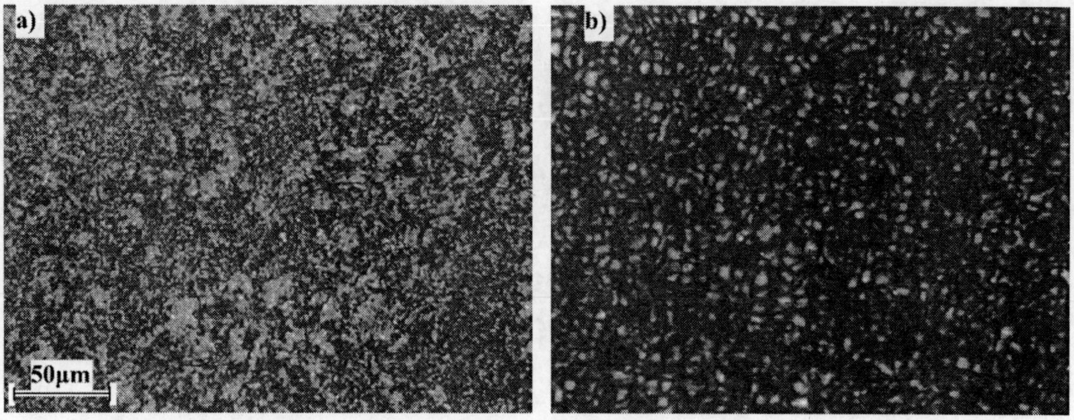

Figure 3. Images of the polyurethane based on PCL and IPDI crystallized at 25 (a) and -10°C (b).

4. Conclusions

Using combination of DSC and optical microscopy the effect of crystallization conditions on thermal behavior of PCL-based TPU was studied. It was shown that increasing of crystallization temperature leads to decrease of nucleation and growth rate. In the result, total degree of crystallinity decreases with improvement of crystal quality. In addition, variation of crystallization conditions allows varying switching temperature of the model linear polyurethane block copolymer from 30 to 42 °C for the same sample. This result is very important for fabrication of heat-sensitive TPU with controllable shape memory effect for particular medical application.

Acknowledgements

The authors acknowledge the Ministry of Science and Higher Education of Russian Federation for financial support (contract № 14.587.21.0052, identifier RFMEFI58718X0052). The work was done on the theme of the state task, № 0089-2019-0012, registration № AAA-A19-119032690060-9.

References

[1] Mucha M, Tylman M and Mucha J 2015 *Polimery* **60** 686

[2] Muthukumar M Reiter G Strobl G R 2007 *Progress in understanding of polymer crystallization* (Berlin: Springer-Verlag Berlin Heidelberg) **714** pp 1-18

[3] Gafurova M P, Lodygina V P, Grigor'yeva V A, Chernyi G I, Komratova V V and Baturin S M 1982 *J. Polym. Sci. USSR* **24** 964

[4] Weissberger A, Proskauer E, Riddick J and Toops E., 1955 *Organic solvents. Physical properties and methods of purification* (New York, London: Intersci. Publ. Wiley) p 254

[5] Wunderlich B, 1976 *Macromolecular physics - crystal nucleation, growth, annealing* (New York: Academic Press) p 212

IOP Conf. Series: Materials Science and Engineering **848** (2020) 012048 doi:10.1088/1757-899X/848/1/012048

Development of scientific and technological bases of processing of cast iron from thermal timing and ladle processing method of the resonant intermittent refining

D A Lubyanoi[1], A V Markidonov[2], V M Nevolin[3], D D Lubyanoi[4] and I O Kambalin[5]

[1] Department of Technology and Integrated Mechanization of Mining, T.F. Gorbachev Kuzbass State Technical University, Branch KuzSTU in Prokopyevsk, 19A Nogradskaya str., Prokopyevsk 653039 Russia

[2] Department of Informatics and Computer Engineering, Novokusnetsk Branch-Institute of Kemerovo State University, 23 Tsiolkovskii str., Novokuznetsk 654041 Russia

[3] Technology Recycling LLC, 29 Proizvodstvennaya str., Novokuznetsk 654004 Russia

[4] Kuznetsk Casting LLC, 17 Kuibyshev str., Novokuznetsk 654034 Russia

[5] Lomonosov Moscow State University, 1 Leninskie Gory, Moscow 119991 Russia

e-mail: lubjanoy@yandex.ru

Abstract. The paper considers the application of the technology of resonant pulsating refining to improve the quality of cast iron products. This technology is used in the production of molds from pig iron and induction cast iron. It is characterized by comparative simplicity and easily fits into existing production. The effectiveness of this technology is at the level of the best world indicators.

1. Introduction

Currently, the processing of metallurgical waste (magnetized kings of cast iron and steel slag dumps, recycling scrap pallets and molds, etc.) is relevant. These types of waste are characterized by an increased content of sulfur, phosphorus, an unstable structure of cast iron products obtained from a heterogeneous charge, and an increased gas content of the metal. The need for technologies to remove these impurities and gases becomes relevant. Semisynthetic iron smelting technology was developed, including the processing of both cast iron and steel scrap and waste. This technology was distinguished by the production of castings with a stable microstructure and mechanical properties of cast iron and operational performance of products from this cast iron. Such high performance of the products is mainly due to the application of the thermal-melt processing modes in this technology. The methods for determining the optimal temperature regimes were based on the approaches developed in the schools of Samarina A.M., Bauma B.A., Kudrina V.A. and Elansky G.N.

To obtain high quality products, the following tasks were solved: 1. A comprehensive technology has been developed that allows the processing of a heterogeneous metal charge; 2. A method has been developed and introduced to reduce the sulfur content in acidic induction furnaces; 3. A method has been developed to neutralize the harmful effects of phosphorus in gray cast iron; 4. Stabilized the structure and properties of cast iron induction melting by the optimal modes of thermal treatment of

IOP Conf. Series: Materials Science and Engineering **848** (2020) 012048 doi:10.1088/1757-899X/848/1/012048

the melt; 5. The gas content of the metal is stabilized before pouring the products by the method of resonantly pulsating metal refining; 6. A mathematical model is constructed for conducting numerical experiments to study the effect of phosphorus on the thermodynamic properties of iron.

In the iron foundry, the problem of increasing the sulfur content in pig iron scrap [1] is becoming more acute every year, reaching a level of 0.17%, as the practice of plants shows. According to the data of [1, 2] over the past 200 years, the sulfur content in cast irons has increased by almost 10 times, which is associated with the use of coke in cast iron for melting. Sulfur removal is a rather complex metallurgical task, the solution of which is associated with considerable costs of energy carriers and reagents, which can lead to higher prices for products, and often makes products unprofitable. All this requires the development and implementation of new resource-saving technologies [3-5], one of which is metal desulfurization in induction furnaces. This is all the more relevant in connection with the appearance in the industry of induction melting plants with wide technological capabilities for producing alloys of ferrous and non-ferrous metals [6-8].

2. Technology description

We have proposed, successfully tested and implemented at a number of enterprises the technology of refining cast iron melt from sulfur with basic slags by the "extraction" mechanism, which was successfully implemented in acid induction furnaces. The essence of technology is as follows. When the crucible of the furnace [9-11] is filled with (60 ... 70)%, the single-circuit metal circulation already exists in the furnace bath, which when the metal is heated to a temperature of 1420° C intensively crushes the slag cover and draws the slag particles deep into the metal. Since the slag particle is drawn deep into the metal, it is shielded by liquid metal from the lining of the furnace. This technique allows the use of basic slags that will not interact with the acid lining. The essence of extraction refining [12] as applied to the desulfurization process is as follows: particles of calcareous slag are drawn by electromagnetic fluxes deep into the induction furnace. The surface layer of slag particles is deoxidized by carbon metal and silicon. As a result, the solubility of sulfur increases dramatically, and the surface layer of slag particles absorbs sulfur from the metal. Further, metal flows carry sulfur-rich slag particles to the surface in contact with the atmosphere of the furnace, where the surface layer of slag particles is oxidized by atmospheric oxygen. As a result, the solubility of sulfur in the surface layer decreases sharply, and, being released from the slag, it is oxidized by oxygen to SO_2 and is removed into the atmosphere of the furnace. This process is repeated many times. The sulfur removal mechanism in induction furnaces with acid lining was investigated in the conditions of the foundry of ZSMK OJSC. The experiments were carried out on IChT-10M furnaces when casting iron for casting pallets and heaters. When slagging, it is advisable to use lumpy lime instead of fines. During processing, the pieces of lime do not have time to completely dissolve and do not form a homogeneous slag melt, which, when aged during the loading of slag, may fall on the lining and corrode it. As experiments have shown, such slag does not form during the work with lump lime, which explains the fact that this method of metal desulfurization practically did not affect the stability of the furnace lining. Analysis of slag samples taken during and after desulfurization showed that during processing by this method, slag is usually heterogeneous, consisting of separate small (1 ... 4) mm pieces of lime with a slag surface. Despite this, they provide a high degree of desulfurization.

3. Results and discussion

In order to increase the effectiveness of this technology, statistical methods evaluated the influence of various technological parameters on the rate and degree of metal desulfurization in acidic induction furnaces of industrial frequency. We compare the performance and kinetic parameters of the IChT-10M furnace with industrial (50 Hz) frequency and the average (250 Hz) frequency of the MFTGe6.000 furnace. An analysis of the dependences of the experimental treatment showed that the main influence on the degree of desulfurization of Cdes of cast iron is exerted by the basicity of the slag and the degree of filling of the Cf crucible (Fig. 1). This is explained by the fact that with decreasing crucible filling, the specific mixing power increases, and this leads to an increase in the circulation speed of the slag-metal emulsion, which ensures an increase in the degree of metal desulfurization.

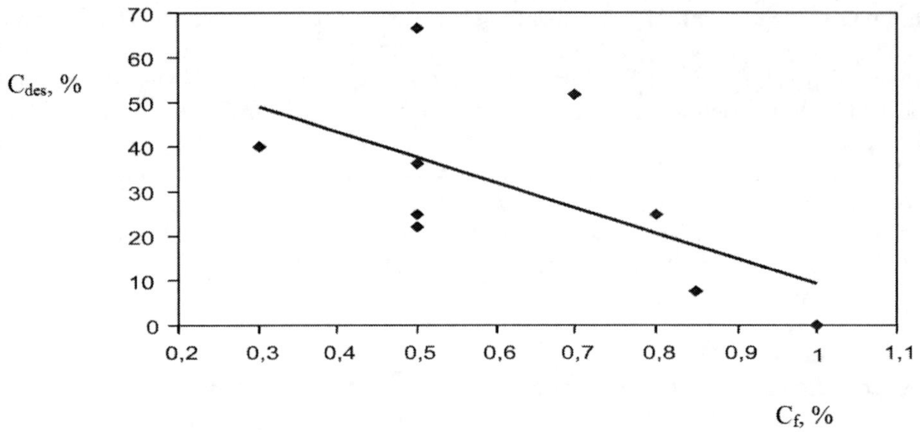

Figure 1. Dependence of the degree of desulfurization on the degree of filling of the crucible on the IChT-10M furnace.

The influence of various technological parameters on the rate of desulfurization was also evaluated. It was found that the rate of desulfurization is most affected by the degree of filling of the crucible and the initial sulfur content in the melt. As you know, medium-frequency furnaces have a higher specific power, for example, an industrial frequency induction furnace IChT-10M has a specific power when the crucible is completely filled (200 ... 300) kW/tn, when the crucible is not completely filled (400 ... 700) kW/tn, the medium-frequency furnace MFTGe6.000 has a specific power (800 ... 900) kW/tn. Therefore, it is of interest to evaluate the effect of mixing power on the kinetic parameters of metallurgical reactions, in particular desulfurization. The joint results obtained in the induction furnaces IChT-10M and MFTGe6.000 are presented in (Fig. 2).

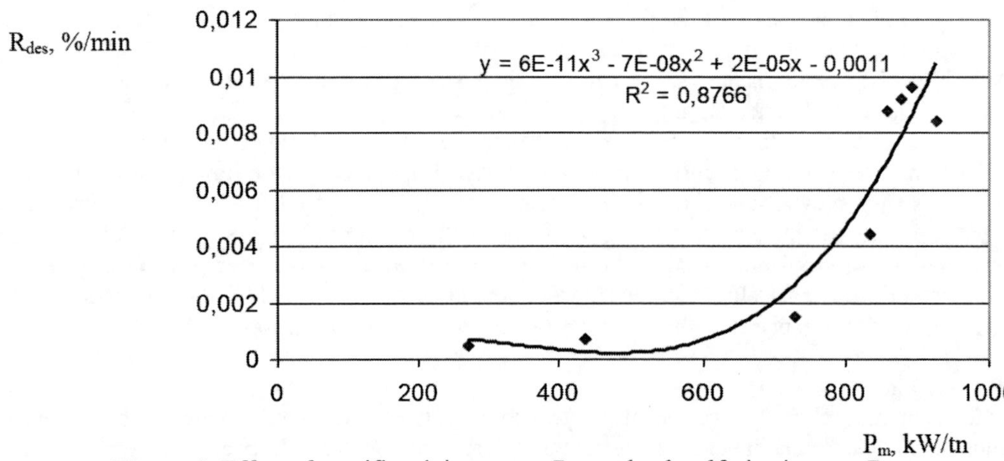

Figure 2. Effect of specific mixing power P_m on the desulfurization rate R_{des}.

An analysis of these results showed that the rate of desulfurization of cast iron with an increase in the mixing power from (200 ... 300) kW/tn to (850 ... 900) kW/tn increases by an order of magnitude from (0.0005 ... 0.00072) %/ min. up to (0.0084 ... 0.0096) %/min. At the same time, there was a significant difference in the change in the rate of sulfur removal with different filling of the crucible in induction furnaces of industrial and medium frequency. So, in induction furnaces of industrial frequency with the full filling of the crucible, sulfur removal practically does not occur, since inert cold slag floats on the surface of the crucible and does not react with the metal. When the crucible filling is reduced to 80%, the mixing intensity increases significantly, the slag heats up and sulfur is removed according to the classical mechanism during the transition of sulfur from metal to slag floating on the metal surface, while the desulfurization rate is insignificant and does not exceed 0.0005 %/min. When filling the crucible (60 ... 70)% or less, the rate of desulfurization of cast iron changes significantly. This is due to a change in the mechanism of sulfur removal during the transition from classical to extraction metal refining. In this case, the sulfur removal rate increases significantly and reaches 50% 0.00072 %/min when filling the crucible, and 0.0015 %/min at 30%. To assess the development of processes by this mechanism, a study was conducted on the content of SO2 in the atmosphere of the furnace and gas exhaust paths using the universal gas analyzer UG-2. As shown by the balance melts and the study of the gas content in the exhaust gases from the IChT-10M induction furnace, this mechanism removes from 16 to 80% of the removed sulfur, while the specific SO2 content in the exhaust gases increases by almost 3 times. A higher desulfurization value by this mechanism corresponds to a lower filling of the crucible. Under such conditions, slag is not an accumulator of removed sulfur, but an intermediate phase through which sulfur is transported to the gas phase. The use of pig iron desulfurization technology can significantly reduce the sulfur content in pig iron from 0.09-0.17% to an acceptable level of 0.015-0.040% in practice. This allows, as shown by operational tests, to increase the operational resistance of heat-resistant castings (slag for slag formed in the production of aluminum) by 2 times. One-circulation circulation is also advisable to use in the process of smelting cast iron, for example, when alloying cast iron with oxides, such as manganese and vanadium. Innovative technologies in waste processing are a reserve that, with relatively small investments, allows you to create new jobs and reduce the negative impact on the environment. In recent years, EVRAZ ZSMK JSC has developed [13, 14] and passed technical testing and technologies of resonant-pulsating refining of metal (RPR) have been introduced.

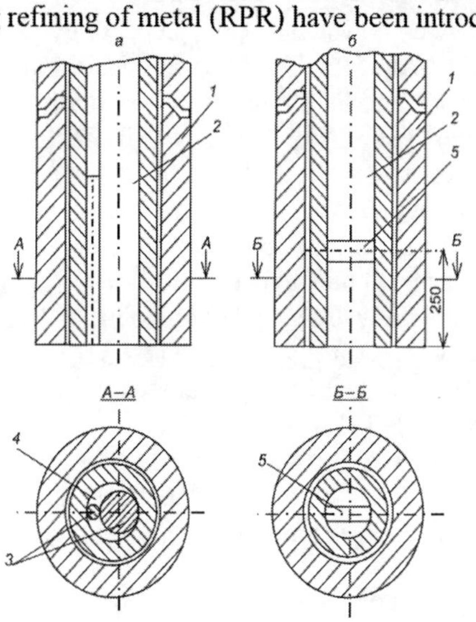

Figure 3. Diagram of tuyeres with a sickle-shaped nozzle (a) and a nozzle with a cylindrical pulsator (b): 1 - refractory pipe; 2 - steel pipe for gas supply; 3 - tuyere tip with a rod; 4 - crescent gap; 5 - cylindrical pulsator.

Physical blowing of a metal with a tuyere with a gas-dynamic pulsator, as shown by physical modeling, is distinguished by the special nature of the outflow of gas into the liquid. It was noted that the depth of jet penetration is much less than when blowing with a sickle lance (high-speed blowing), but more than when blowing through cylindrical nozzles. This technology allows you to simply control the processing of the melt, introducing the required flow rate of inert gas into the lance, intensifying the processing of the melt. In addition to the low-frequency region, the melt is subjected to processing by a whole spectrum of vibrations from low-frequency to high-frequency [13, 14]. The experimental results showed that melt processing is performed in the range from 1 to 23000 Hz. This technology has been successfully implemented on cast iron of blast furnace and induction melting of metal. It allowed to significantly reduce the gas content of the metal, increase the density of cast iron castings from 6890-6900 to 7000-7200 kg/m^3, the mechanical properties of cast iron and the performance of its products. The high efficiency of RPR was demonstrated in the processing of blast furnace pig iron for the production of molds and slag bowls [15].

4. Conclusion

This technology has significantly improved the strength properties of the metal, to neutralize the harmful effects of phosphorus due to changes in the microstructure of cast iron. The operational durability of cast iron products subjected to RPR is at the level of the best domestic and foreign indicators.

References

[1] Zhurilo A G and Zhurilo D Yu 2014 Chernyye metally **1** 23-27
[2] Girshovich N G 1978 Cast Iron Handbook (Lviv: Engineering Press) p 757
[3] Tsepelev V S, Selyanin I F, Lubyanoi D A, Baum B A and V'yukhin V V 1995 Steel in Translation **5** 42-45
[4] Lubyanoi D A 2007 Liteyshchik Rossii **3** 30-32
[5] Lubyanoi D A and Lubyanaya S V 2013 Problemy chernoy metallurgii I materialovedeniya **4** 36-39
[6] Luzgin V I, Petrov A Yu and Rachkov S A 2005 Liteynoye proizvodstvo **5** 23-27
[7] Luzgin V I, Petrov A Yu and Rachkov S A 2006 Liteyshchik Rossii **10** 24-26
[8] Luzgin V I, Petrov A Yu and Faerman L I 2006 Chernyye metally **7** 14-25
[9] Lubyanoi D A, Solov'yev S A, Yapparov R T and Starchenko V G 1998 Bulletin "Izobreteniya. Poleznyye modeli" **10** 3
[10] Sofroshenkov A F, Lubyanoi D A and Sinyavskiy I A 2005 Izvestiya vysshikh uchebnykh zavedeniy. Chernaya metallurgiya **6** 45-47
[11] Lubyanoi D A 2006 Liteyshchik Rossii **12** 22-25
[12] Yavoyskiy V.I., Yavoyskiy A.V 1987 Scientific Basis of Modern Steel Production Processes (Moscow: Metallurgy Press) p 184
[13] Lubyanoy D A, Mamedov R O and Perekhodov V G 2018 Problemy chernoy metallurgii i materialovedeniya **2** 36-39
[14] Lubyanoi D A, Kustov B A and Novikov N I 2004 The effectiveness of modern ways to improve the quality of cast iron products and the development of metallurgical and engineering enterprises ina competitive environment (Novosibirsk: Publishing House of IEOPP SB RAS) p 131
[15] Andreev V V, Lubyanoi D A, Samsonov Y N, Kaminskaya I A and Lubyanaya S V 2014 Metallurgist **58** 5-6 492-495

IOP Conf. Series: Materials Science and Engineering **848** (2020) 012049 doi:10.1088/1757-899X/848/1/012049

Development of optimal formulations of natural alloyed cast irons for metals and engineering,and thermal timing subjected to secondary treatment by the method of resonance-intermittent refining

D A Lubyanoi[1,2], S Yu Shevchenko[3], D D Lubyanoi[4], A V Markidonov[5] and E B Zvarych[2]

[1] JSC EVRAZ Consolidated West Siberian Metallurgical Plant, 16 Kosmicheskoye rte., Novokuznetsk 654042 Russia
[2] Department of Technology and Integrated Mechanization of Mining, T.F. Gorbachev Kuzbass State Technical University, Branch KuzSTU in Prokopyevsk, 19A Nogradskaya str., Prokopyevsk 653039 Russia
[3] Bauman Moscow State Technical University, 5 2nd Baumanskaya str., Moscow 105005 Russia
[4] Kuznetsk Casting LLC, 17 Kuibyshev str., Novokuznetsk 654034 Russia
[5] Department of Informatics and Computer Engineering, Novokusnetsk Branch-Institute of Kemerovo State University, 23 Tsiolkovskii str., Novokuznetsk 654041 Russia
e-mail: lubjanoy@yandex.ru

Abstract. This article presents the results of the introduction of resonant-pulsed refining technology in production, which made it possible to change the microstructure of blast-furnace irons and significantly increase the operational stability of molds and pallets.

1. Introduction

In [1], the preparation of the melt for obtaining products from heat-resistant cast irons was considered, as noted above, the molten iron was purified from sulfur, stabilized at a certain level in terms of oxygen, hydrogen, and nitrogen. Thermo-time treatment within the specified melt limits allowed melt to be obtained with the cast iron structure most favorable for a particular grade of cast iron [2]. Purging cast iron with argon or nitrogen made it possible to refine cast iron from gases and inclusions of mellow graphite, to grind graphite inclusions to the required sizes to increase the strength properties of cast iron (table 1) without additives of additional alloying additives.

Table 1. Change in the mechanical properties of blast furnace iron.

Type of processing	Tensile strength (MPa)
Nitrogen purge by resonant pulsed refining method	130..170
No purge	91..105

IOP Conf. Series: Materials Science and Engineering **848** (2020) 012049 doi:10.1088/1757-899X/848/1/012049

2. Technology description

Titanium and vanadium, which fall into cast iron with charge materials, microalloy cast iron by the mechanism [3, 4], and can neutralize the harmful effect of phosphorus in heat-resistant cast irons as well. All this made it possible to proceed to the development of the chemical composition of economically alloyed cast irons for castings with enhanced operational properties.

The greatest effect of this technology is achieved when nitrogen is treated with natural blast furnace titanium and vanadium cast iron. With this technology, endogenous carbonitrides of titanium and vanadium are formed in the metal volume at the nanoscale, which serve as substrates for the formation of fine graphite. These inclusions are shown in Figure 1. Metal processing by this technology allowed, due to a significant change in the mechanical properties of the cast iron base, to reduce the amount of alloying additives and cast castings from economically alloyed cast iron. The application of this technology has significantly improved the microstructure of cast iron products, increase and stabilize the resistance of replaceable steel casting equipment from blast iron.

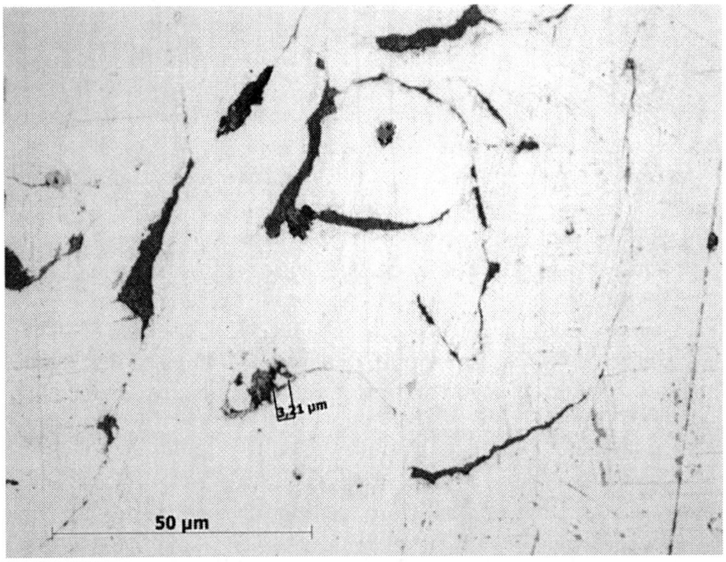

Figure 1. Carbonitride support initiates graphite formation.

3. Results and discussion

In the shop of cast iron molds, continuous, deep-flow molds and molds weighing up to 42 tons are currently being cast for casting forging ingots (Figure 2). To increase the resistance of through molds that fail due to bottom erosion, it is necessary to increase the strength of cast iron by alloying with manganese and modified by ferrosilicon; to increase the resistance of through molds to cracks, it is necessary to increase the heat resistance of cast iron of molds, while reducing the amount of interdendritic graphite in the cast iron structure and eliminating cementite in the structure cast iron, for which it is necessary to reduce the alloying of cast iron with manganese. Molds, both through and deaf, are cast from one ladle with the same chemical composition (table 2).

Figure 2. Large mold for forging ingots of power engineering from blast furnace pig iron weighing 42 tons.

Table 2. Chemical composition of cast iron for casting molds and slag bowls.

C, %	Si, %	Mn, %	P, %	S, %	Cr, %	Ti, V, %
			No more			
Not regulated	0.6..1.3 (molds) 0.9..1.3 (slag bowls)	0.45..1	0.3	0.05	0.1	0.03..0.20 (optimum 0.06..0.12)

Therefore, when choosing the chemical composition of cast iron, they are guided by what type of molds prevails when casting molds and the economic effect of saving ferroalloys when preparing cast iron for casting molds. Due to the fact that through mold production at JSC "EVRAZ ZSMK", through molds are predominantly cast, as well as taking into account the savings of ferrosilicon and silicomanganese, molds are tried to be cast from cast iron by the silicon and manganese content mainly with the lower values of these elements indicated in Table 2.

Molds of different types can be cast from one bucket, therefore, the chemical composition of cast iron depends on the prevailing type. To improve the microstructure of cast iron, nitrogen purging is used by the method of resonant pulsating refining [4]. The essence of such refining consists in applying vibrations to the metal by pulsating blasting, in the spectrum of which there is a low-frequency component, which coincides with the natural frequency of the metal oscillations in the bucket. The use of this relatively simple technology, as shown by many years of experience in its practical use, can significantly improve the quality of casting. These works allowed to significantly improve the quality of products, to develop compositions of economically alloyed cast irons (table 2), which provide a significant reduction in the consumption of ferroalloys. After operation, cast iron molds are broken in combat pits, and the resulting mold battle is used as a charge in induction furnaces.

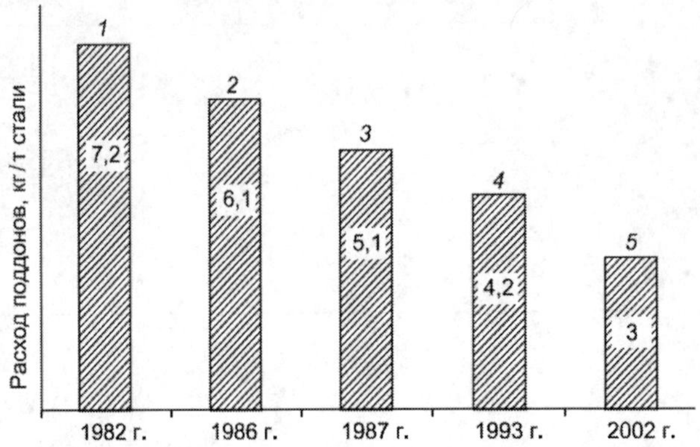

Figure 3. Dynamics of reducing the consumption of pallets for through molds at JSC "EVRAZ ZSMK": 1 - cupola (ferritic iron); 2 - induction furnace, remelting of foundry and pig iron, without thermal treatment (ferritic-pearlite cast iron); 3 - induction furnace, with thermal treatment (perlite-ferritic cast iron); 4 - induction furnace, with thermal treatment, purging with argon (pearlitic low manganese cast iron); 5 - induction furnace, with thermal treatment, economically alloyed cast iron (structure - perlite).

Induction melting pig iron is used to obtain pallets, molds and covers for intermediate ladles of continuous casting machines. So, depending on the chemical composition and structure of cast iron, the operational durability of pallets of through molds is increased and their consumption is reduced (Fig. 3), the operational durability of the covers of intermediate ladles of casting continuous casting machines made of economically alloyed cast iron is increased, and the resistance of ingot molds for forging ingots is increased. The chemical composition of cast iron products is shown in table 3.

Table 3. The chemical composition of cast iron for the production of heat-resistant castings at JSC "EVRAZ ZSMK".

Alloy group	Grade (or purpose) of the alloy	C %	Si %	Mn %	Cr %	Ti %	V %	S, % no more	P, %
Gray heat resistant cast iron	Pallets for molds and for casting ferroalloys	3.6.. 3.8	1.8.. 2.0	0.4..0.8		0.03.. 0.05	0.03.. 0.05		before 0.27
				0.4..0.75		0.06.. 0.15	0.06.. 0.15		
	Economically alloyed cast iron for pallets, ladle covers of continuous casting machines	3.85.. 4.0	0.9.. 1.05	0.4.. 0.7	no more 0.15	0.02.. 0.05	0.02.. 0.05	0.05	0.14.. 0.27
		3.85.. 4.0	1.05.. 1.3	0.4.. 0.6		0..0.15	0..0.15		before 0.27

After the introduction of steel frames installed in the form for the pallet, the durability of the pallets increased, and in 2019 their consumption reached 2.7 kg per ton of steel. Thus, the introduction of scientific methods allowed to reduce the consumption of pallets by 2.66 times.

Due to the fact that through-casting molds are mainly cast at the plant, as well as taking into account the saving of ferroalloys, molds for forging ingots and similar equipment are tried to be made of cast iron with a low content of manganese (0.45 - 0.8 %) and silicon (0.6 - 1.3 %), while carbon is not regulated, ≤ 0.3 % P, ≤ 0.05 % S, ≤ 0.1 % Cr, the optimal content of Ti and V is 0.06 - 0.12 %. In this regard, to obtain molds of maximum resistance, the content of titanium and vanadium was limited to 0.06-0.12 %.

A decrease in the content of titanium and vanadium below this limit adversely affects the resistance of molds to heat. The average operational stability of the blacksmith molds in this treatment reaches 50 bulk. The resistance of through molds at the plant reached 90 bulk (in 2007-2016 it was at the lowest level in the CIS of 11.98-12.05 kg per ton of steel) - this is the lowest consumption of molds in the industry [5]. In fact, the enterprise has reached the technological limit on the resistance of molds for gray cast iron. However, the use of cast iron molds with a given titanium and vanadium content is not limited to this. After operation, the molds are broken up in the pile shop and the battle is used as a charge in the induction furnaces of the foundry.

4. Conclusion

The developed technologies [1-5] allowed the plant to master the production of almost all types of blacksmith molds weighing up to 42 tons, allowing casting forge ingots for power engineering weighing up to 70 tons. The developed technologies made it possible to almost completely change the microstructure of blast-iron irons, to significantly increase the operational stability of molds, pallets, and their total consumption was reduced to the level of the best world indicators [5].

References

1. Lubyanoi D A, Kustov B A and Novikov N I 2004 The effectiveness of modern ways to improve the quality of cast iron products and the development of metallurgical and engineering enterprises ina competitive environment (Novosibirsk: Publishing House of IEOPP SB RAS) p 131
2. Tsepelev V S, Selyanin I F, Lubyanoi D A, Baum B A and V'yukhin V V 1995 Steel in Translation **5** 42-45
3. Lubyanoi D A, Sofroshenkov A F and Sinyavskiy I A 1999 Steel in Translation 29. **10** 86-90.
4. Andreev V V, Lubyanoi D A, Samsonov Y N, Kaminskaya I A and Lubyanaya S V 2014 Metallurgist **58** 5-6 492-495
5. Lubyanoy D A, Mamedov R O and Perekhodov V G 2018 Problemy chernoy metallurgii i materialovedeniya **2** 36-39

IOP Conf. Series: Materials Science and Engineering **848** (2020) 012050 doi:10.1088/1757-899X/848/1/012050

Effect of synthesis parameters on the structure and properties of carbon particles formed from amorphous fullerites

I N Lukina[1], O P Chernogorova[1], E I Drozdova[1], E A Ekimov[2], M O Apostolova[3], D A Prokopenko[4], A V Soldatov[5,6] and V Benavides[5]

[1] A.A. Baikov Institute of Metallurgy and Materials Science, Russian Academy of Sciences, Leninskii pr., 49, 119334, Moscow, Russia
[2] Institute for High Pressure Physics, Russian Academy of Sciences, Moscow Troitsk 108840, Russia
[3] Bauman Moscow State Technical University, ul. Baumanskaya 2-ya, 5, 105005 Moscow, Russia
[4] Moscow Aviation Institute (National Research University), Volokolamskoe shosse, 4, 125993, Moscow, Russia
[5] Department of Engineering Sciences and Mathematics, Luleå University of Technology, SE-97187 Luleå, Sweden
[6] Department of Physics, Harvard University, Cambridge, MA 02138, USA

lukina.i.n@yandex.ru

Abstract. The effect of high-pressure synthesis temperature on the structure and indentation characteristics of the superelastic hard carbon formed from amorphous fullerites and on the tribological properties of the Co-based composite materials (CM) reinforced by the particles of such carbon has been studied by Raman spectroscopy, high-resolution transmission electron microscopy (HRTEM), indentation measurements, and tribological tests. It is shown that ball milling (for 48 h) of C_{60} fullerite crystals results in the amorphization of the product of fullerite transformation upon their high-pressure treatment at temperatures above the stability limit of fullerene molecule (~800°C). An increase in synthesis temperature at 8 GPa from 800°C to 1200°C leads to a gradual graphitization of the structure of amorphous fullerite derived carbon. This decreases its hardness and indentation modulus from 32 to 18 GPa and from 256 to 95 GPa, respectively, and increases the elastic recovery (from 80% to 86%). The best tribological characteristics of the CM are attained at the maximum particle hardness, which is realized in the CM synthesized at 800°C. When the synthesis temperature is elevated to 1200°C, the friction coefficient and wear rate of the CM increase, but they remain substantially lower than those of the matrix cobalt.

1. Introduction

The products of high-pressure high-temperature fullerene decomposition are characterized by a unique combination of high hardness and superelasticity. For this reason, they are advantageous, in particular, for reinforcing metal matrix for the development of a new class of composite materials with a low friction coefficient of and high wear resistance [1]. The structure of bulk transformation products of C_{60} crystals at moderate pressure (3–8 GPa) was studied in [2–5]. It was shown that, at temperatures above the upper temperature limit of the stability of fullerene molecule, their collapse under pressure leads to the formation of hard graphite-like phases consisting of curved graphene planes. The structure of a disordered graphite (DG) phase consisting of nanoscale clusters containing 5-10 atomic layers (graphene planes) is characterized by long-range correlations [4], since the layers of the DG structure are parallel to the close-packed (111) planes of the parent fcc C_{60} crystals. Such “hard” carbon exhibits an extremely high elastic recovery upon stress relief [3] due to the ability of curved graphene layers to be compressed without shear and then to expand back to the initial structure. The collapse of amorphous fullerite C_{60} upon heating under pressure at a relatively low synthesis temperature (800°C) results in the formation of highly disordered carbon phase consisting of randomly arranged packets of 2-3 short graphene layers [6]. Such carbon was shown to be substantially harder but less elastic than the carbon formed from crystalline fullerites at the same synthesis conditions. The pressure-temperature diagram of the transformations of amorphous C_{60} fullerite has been constructed in [7]. In the present work, we studied the effect of high-pressure synthesis temperature on the structure and properties of carbon particles obtained from amorphous fullerites as well as on the tribological characteristics of metal-matrix composite materials reinforced by the carbon particles.

2. Experimental

The cobalt-based CM samples of 5 mm in diameter and 3-4 mm high were prepared from mixtures of cobalt powder and 10 wt % C_{60}. The initial fullerites were processed in a Frisch ball mill in an agate mortar with agate balls at low milling rates in a helium atmosphere. The powder mixtures were sintered at a pressure of 8 GPa in a toroid-type high-pressure chamber (toroid-15). The powder pellets pre-compressed at a pressure of less than 1 GPa were placed inside the tubular heater in the central part in order to reduce temperature gradients in the sample upon subsequent sintering. Under pressure, the samples were heated at a fixed rate of up to 100 K/min to a predetermined temperature by passing current through the heater. The sintering duration at fixed pressure-temperature parameters was 45 s. The temperature in the high-pressure cell was controlled by chromel-alumel thermocouples. The thermocouple junctions were placed on the outer wall of the graphite heater at the middle of its height. The structure of carbon phase was identified by the Raman spectra recorded with a WITec high-resolution CRM 200 spectrometer equipped with a confocal attachment. The size of the laser spot at the focus was about 1 μm. The fine structure of the superelastic hard carbon was studied by high-resolution transmission electron microscopy (TEM/STEM TITAN 80-30) at an accelerating voltage of 300 kV. The samples for transmission electron microscopy were prepared by ion beam cutting at Helios NanoLab (FEI, USA). Registration and processing of the obtained images was carried out with Digital Micrograph programs. The indentation properties of the high-pressure carbon phase were measured with Shimadzu dynamic ultra micro hardness tester DUH-211/DUH-211S. The tribological and abrasive wear tests were performed with a UMT-3MO (CETR) Universal Tester.

3. Results and discussion

The powders of the initial crystalline C_{60} fullerite are substantially nonuniform in morphology and particle size. Along with crystallites of 1-2 μm in size, there are nanosize crystallites, the total fraction of which is quite small (Figure 1a). After ball milling, the crystallite size decreases to 100-500 nm, and the grains smaller than 100 nm in size are also present (Figure 1b). The XRD spectra of the ball-milled fullerites exhibit substantial broadening of the C_{60} dffraction peaks and the presence of amorphous structure constituent, as was shown in [8].

The high-pressure treatment of the powder mixtures of cobalt and fullerites at temperatures above the limit of fullerene molecule stability leads to the synthesis of cobalt-base composite materials reinforced by superelastic hard carbon particles. Amorphization of the initial fullerites upon long-time ball milling breaks the packing order of fullerenes and reduces the probability of the formation of regularly oriented graphene plane packets upon heating under pressure [5, 9]. With increasing synthesis temperature from 800°C to 1200°C, the fullerites tend to transform into more stable graphite-like phase, which is exhibited by a series of Raman spectra (Figure 2).

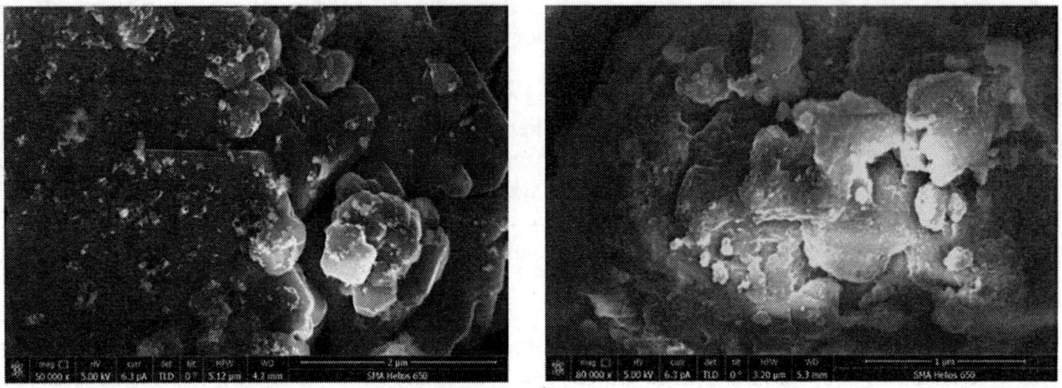

a b

Figure 1. C_{60} fullerites: (a) initial and (b) after ball milling for 48 hours

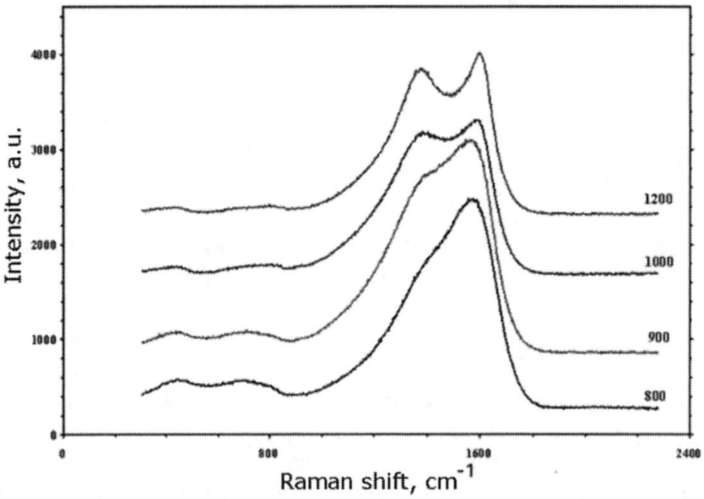

Figure 2. Raman spectra of the carbon particles formed from amorphous C_{60} at a pressure of 8 GPa at different synthesis temperatures.

The structure of the carbon particles formed at a relatively low synthesis temperature (800°C) generally corresponds to the amorphous state (Figure 3a) and contains short packets of 2-3 graphene layers arranged randomly in the amorphous matrix. The Fourier transform of such image is represented by a wide ring typical of amorphous structure (insert in Figure 3a). The structure of the carbon particles formed from amorphous fullerites at a high synthesis temperature (1200°C) consists of more extended curved graphene layers (Figure 3b) than those observed in the structure formed at 800°C. However, unlike graphene nanoclusters obtained from fullerite crystals [5], such packets are randomly oriented, as is clearly shown by the Fourier transform represented by a ring without pronounced preferred directions (insert in Figure 3b) This ring is much narrower than that of the phase synthesized from ball-milled fullerites at 800°C. Thus, the tendency to the formation of graphite-like structure from amorphous fullerite increases with synthesis temperature elevation, which increases the thermodynamic stimulus to achieve the equilibrium state (graphite).

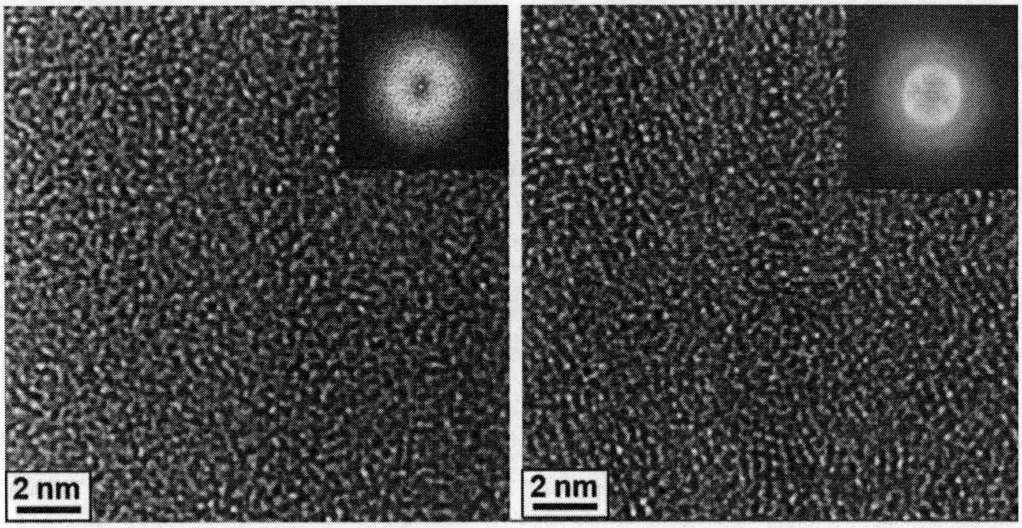

a b

Figure 3. HRTEM images of the carbon particles formed from amorphous C_{60} fullerites at (a) 800°C and (b) 1200°C at a pressure of 8 GPa.

The ball milling of C_{60} fullerites for 48 h leads to a significant increase in hardness (18 to 32 GPa) and elastic modulus (from 104 to 256 GPa) of their transformation product at a pressure of 8 GPa at 800°C and decreases its elastic recovery (from 94 to 80%) [6]. This is explained by the fact that the amorphization of the parent fullerites decreases the probability of the formation of graphene plane packets upon heating under pressure and, therefore, the fullerene fragments upon collapse of amorphous (ball-milled) fullerites under optimum synthesis conditions (high pressure and the temperature just above the limit of the stability of fullerene molecule) form more dense spatial network of strong covalent bonds. As the temperature of the synthesis of carbon particles from amorphized fullerites increases from 800°C to 1200°C, the formation of a more equilibrium graphite-like phase is accompanied by a decrease in hardness and indentation modulus of the synthesized carbon particles with a slight increase in elastic recovery (Table 1). Thus, the sintering temperature dependence of the indentation characteristics reflects the structure evolution of the fullrene-derived carbon particles: the hardness decreases and the elasticity increases with increasing degree of graphitization. As is seen from Table 1, the tribological properties of the CM spatially reinforced by 10 wt % of superelastic hard carbon particles substantially depend on the hardness of the carbon phase. The best combination of such properties, i.e., the lowest friction coefficient and wear rate, is exhibited by the CM synthesized at 800°C. With increasing synthesis temperature, the tribological properties of the CM are deteriorated, although, compared with the matrix cobalt (characterized by μ = 0.41 and wear rate = 6.30 mg/m), they remain very attractive in the entire range of synthesis regimes used. Note that, in our case, the ratio of hardness to indentation modulus increases from 0.13 to 0.19 with increasing synthesis temperature. Similar results were obtained in [10] for nanocarbon phases derived from C_{60} fullerenes at different synthesis conditions. Although the *H/E* ratio is commonly recognized to better predict the wear resistance of the material than hardness only [11], the data given in Table 1 show that, in the range of properties under consideration, a decrease in hardness is accompanied by an increase in the H/E ratio, but, at the same time, the best tribological characteristics are exhibited by the CM reinforced by the hardest particles.

Table 1. Indentation characteristics (indentation hardness H_{IT}, indentation modulus E_{IT}, and elastic recovery $\eta_{IT} = W_{elast}/W_{total}$, where W_{elast} and W_{total} are the elastic and total indentation works, respectively) of the carbon particles derived from ball-milled C_{60} fullerites and the tribological properties (friction coefficient μ and wear rate) of the composite materials reinforced by such particles.

Synthesis temperature (°C)	Properties of carbon phase			Properties of CM	
	H_{IT} (GPa)	E_{IT} (GPa)	η_{IT} (%)	μ	Wear rate (mg/m)
800	32	256	80	0.13	0. 07
900	23	140	84	0.18	0.13
1000	22	127	85	0.22	0.26
1200	18	95	86	0.23	1.50

4. Conclusion

Ball milling (for 48 h) of C_{60} fullerite crystals results in their substantial refinement and in the amorphization of the product of fullerite transformation upon their high-pressure treatment at temperatures above the temperature limit of the stability of fullerene molecule (~800°C). An increase in the high-pressure (8 GPa) synthesis temperature from 800°C to 1200°C leads to a gradual graphitization of the structure of amorphous fullerite derived carbon. This is expressed by the isolation of the D peak in the Raman spectrum and in an increase in the number and extension of graphene layers in HRTEM structure images. The graphitization of the structure of the carbon phase with increasing temperature of high-pressure synthesis decreases its hardness and indentation modulus from 32 to 18 GPa and from 256 to 95 GPa, respectively, and slightly increases elastic recovery (from 80% to 86%). The best tribological characteristics of the Co-based CM reinforced by the carbon particles correspond to the maximum particle hardness, which is realized in the CM synthesized at 800°C. When the synthesis temperature is increased to 1200°C, the friction coefficient and wear rate of the CM increase (from 0.13 to 0.23 and from 0.07 to 1.50 mg/m, respectively), but, at the same time, they remain substantially lower than those of the matrix cobalt.

Acknowledgment
The work was carried out within the Governmental task 075-00947-20-00 (14P).

References

[1] Chernogorova O, Drozdova E, Ovchinnikova I, Soldatov A and Ekimov E 2012 Structure and properties of superelastic hard carbon phase created in fullerene-metal composites by high temperature-high pressure treatment. *J. Appl. Phys.* **111** 112601-1–5
[2] Tat'yanin E , Lyapin A, Mukhamadiarov V, Brazhkin V and Vasiliev A 2005 Mechanism of formation of the superhard disordered graphite-like phase from fullerite C_{60} under pressure *J. Phys.-Condens. Mat.* **17** 249 - 56
[3] Wood R, Lewis M, West G, Bennington S, Cain M and Kitamura N 2000 Transmission electron microscopy, electron diffraction and hardness studies of high-pressure and high temperature treated C_{60} *J. Phys.-Condens. Mat.* **12** 10411 - 21
[4] Dzwilewski A, Talyzin A, Bromiley G, Dub S, and Dubrovinsky L Characterization of phases synthesized close to the boundary of C_{60} collapse at high temperature high pressure conditions 2007 *Diam. Relat. Mater* . **16** 1550–56
[5] Chernogorova O, Potapova I, Drozdova E, Sirotinkin V, Soldatov A, Vasiliev A and Ekimov E

2014 Structure and physical properties of nanoclustered graphene synthesized from C_{60} fullerene under high pressure and high temperature *Appl. Phys. Lett* . **104** 043110-1-4
[6] Chernogorova O, Drozdova E, Ushakova I, Bulychev S, Ekimov E, Benavides V and Soldatov A 2016 Indentation behavior of superelastic hard carbon *Philos. Mag.* **32–34** 3451–60
[7] Borisova P, Blanter M, Brazhkin V, Somenkov V and Filonenko V 2015 Phase transformations in amorphous fullerite C_{60} under high pressure and high temperature *J. Phys. Chem. Solids* **83** 104–8
[8] Sirotinkin V and Agafonov S 2014 X-Ray diffraction study of the microstructure of C_{60} fullerite powder after grinding in a planetary ball mill *Inorg. Mater.* **50** 464–8
[9] Lepoittevin C, A´lvarez-Murga M, Marques L, Mezouar M and Hodeau J 2013 Structural characterization of corrugated anisotropic graphene-based carbons obtained from the collapse of 2D C_{60} polymers *Carbon* **52** 278-87
[10] Brazhkin V, Solozhenko V, Bugakov V, Dub S, Kurakevych O, Kondrin M and Lyapin A 2007 Bulk nanostructured carbon phases prepared from C_{60}: approaching the 'ideal' hardness *J. Phys.-Condens. Mat.* **19** 236209-16
[11] Leyland A and Matthews A 2000 On the significance of the *H*/*E* ratio in wear control: a nanocomposite coating approach to optimised tribological behaviour *Wear* **246** 1–11

IOP Conf. Series: Materials Science and Engineering **848** (2020) 012051 doi:10.1088/1757-899X/848/1/012051

Fabrication and characterization of 1, 5, 10 at.% Ce:Y_2O_3 nanopowders

K E Lukyashin[1], A S Chepusov[1] and V I Solomonov[1,2]

[1] Institute of Electrophysics UrB RAS, Yekaterinburg, Russia
[2] Ural Federal University, Yekaterinburg, Russia
E-mail: kostya@iep.uran.ru

Abstract. Nanopowders (particle size of about 7-14 nm) of yttrium oxide doped with cerium ions with concentrations of 1, 5, 10 at.% have been synthesized by laser ablation of the target. For the first time in the world the cerium concentration of 10 at.% in the yttrium oxide lattice has been obtained. Their morphology, structure, and scintillation properties have been investigated. No significant cathodoluminescence of the activator in the Ce^{3+} state has been detected in the powders, and X-ray luminescence was completely absent. Presumably, the cerium ion is in the nonradiative state of Ce^{4+}.

1. Introduction

Ceramic materials research continues to evolve around the world. Moreover, one of the main advantages of ceramics is the possibility of doping a large number of active centers [1] with concentrations up to 30 at.% [2], which is not achievable in single crystals, where the threshold does not exceed 1 at.%.

One of the applications of ceramics is their use as scintillators. Yttrium aluminum garnet doped with cerium ions is a well-known and well-established scintillation material. Previously, this material was already synthesized and examined by the authors. So, for example, we have already demonstrated the relative light yield surpassing its analogues for single crystals and making up 32% of one of the CsI-Tl reference scintillator (light output 20800 photon/MeV) at the cerium concentration of 1 at.%, in our paper on the synthesis and characterization of the optical-luminescent properties of Ce^{3+}:YAG ceramics with different Ce concentrations (0.1 - 5 at.%) [3]. In addition, an increase in the speed of the scintillator was recorded (from 110 to 26 ns) with an increase in the concentration of cerium activating ions (from 0.1 to 5 at.%). It can be assumed that a further increase in the concentration of cerium ions, for example, up to 10 at.%, will lead to even greater speed. The authors believe that this is possible, and this is due to the concentration quenching effect.

Due to its unique scintillation properties, Ce:YAG is used as an X-ray and soft gamma-ray detector [4, 5]; it is characterized by high efficiency, mechanical and thermal stability. At the same time, there is a problem of correct registration of short (ns) X-ray radiation pulses and measurements of dose loads [6], which (pulsed radiation) are increasingly used in science and technology as more effective (Decreasing the dose load on people or inanimate objects by tens of times, reducing by times the mass and size parameters of X-ray sources). It is planned to solve this problem by synthesizing a new ceramic material based on yttrium aluminum garnet.

At the first stage of research, it is necessary to obtain and characterize nanopowders of the corresponding composition, which are needed in the synthesis of Ce:YAG ceramics. For a

IOP Conf. Series: Materials Science and Engineering **848** (2020) 012051 doi:10.1088/1757-899X/848/1/012051

comprehensive study and construction of the corresponding physical and mathematical model in the future, three compositions of 1, 5, 10 at.% Ce:Y_2O_3 nanopowders were chosen.

Thus, the aim of this work is to synthesize and characterize Y_2O_3 nanopowders doped with cerium ions with 1, 5, 10 at.% contents of the latter for the synthesis of new high-speed ceramic Ce:YAG scintillators for detecting nanosecond X-ray radiation pulses.

2. Experimental

Yttrium oxide nanopowders doped with cerium ions with concentrations of 1, 5, 10 at.% were synthesized by laser ablation of a solid porous target [7]. The cerium concentration of 10 at.% in the yttrium oxide lattice was obtained for the first time in the world. This method makes it possible to distribute the dopant evenly in the main material still at the synthesis stage, whereby it is immediately incorporated into the lattice of the main material. In this case, cerium ions replace yttrium ions.

The synthesized nanopowders were studied using a JEOL JEM 2100 (JEOL Ltd., Japan) transmission electron microscope (TEM). The specific surface area of powders (BET analysis) was determined by an automated gas adsorption analyzer (TriStar 3000, Micromeritics Instrument Corporation, USA). The structures of the sintered nanopowders were determined by X-ray diffraction (XRD) on copper radiation (CuKα, λ = 1.542Å) with the graphite monochromator on a secondary beam (D8 Discover GADDS, Bruker AXS, Germany). The processing of the XRD results was carried out via the full-height analysis program named TOPAS 3.

The pulsed cathodoluminescence (PCL) and X-ray luminescence (XRL) of powders were measured both immediately after their laser synthesis and after their transformation in the main cubic phase at 900°C for 3 hours in an atmospheric furnace. In addition, compacts fabricated from the obtained nanopowders by pressing at 200 MPa into disks (Ø14×3 mm) were examined.

The luminescent properties in the range 220–850 nm were measured by pulsed cathodoluminescent spectroscopy method (PCL) using our own KLAVI setup (IEP, Russia) with the maximum electron energy in the beam of 180 keV, pulse duration of 2 ns, pulse repetition rate of 1 Hz, and electron current density of 120 A/cm^2 [8]. The PCL spectrum was recorded in air at room temperature of the samples using a multichannel photodetector based on a CCD array combined with a diffraction spectrograph. The pulse X-ray luminescence (XRL) spectra were measured using an experimental generator of nanosecond X-ray radiation pulses consisting of two main parts: a high-voltage generator and a pulsed explosive-emission tube (IEP, Russia). The signal was measured using an experimental setup based on MicroFC 60035 photodiodes and a Tektronix TDS648 oscilloscope. Unique output characteristics of the used high-voltage generator (voltage amplitude up to 115 kV, pulse duration at half maximum up to 50 ns) require a special X-ray tube for operation as a source of pulsed X-ray radiation. The axially-symmetrical explosive-emission tube and a blade metal cathode were used in this case. The measurements were carried out with the X-ray radiation exposure of 1 s at the pulse repetition rate of 2 kHz.

3. The Results and Discussion

The morphological properties of the synthesized nanopowders were examined by scanning electron microscopy methods. For example, Figure 1 shows an image of the synthesized 1 at.% Ce:Y_2O_3 nanopowder and the size distribution function of nanoparticles taken after counting 4000 particles. It can be seen that the particles have an almost perfect spherical shape (Figure 1a). From the particle size distribution (Figure 1b), we observe that the peak falls at 11 nm, and the average particle size is 7-14 nm. The distribution lies in a very narrow range. Similar data were obtained for the other two powder compositions.

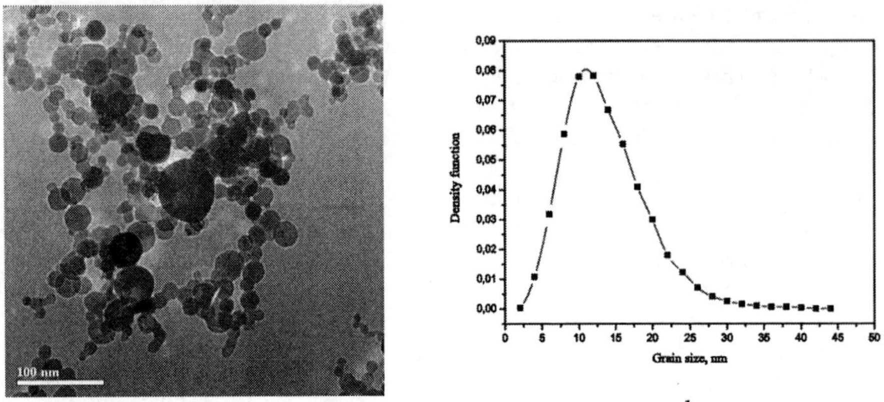

a b

Figure 1. Transmission electron microscopy (TEM) images of 1 at.%Ce:Y_2O_3 nanopowders produced by the laser ablation method (a). Size distribution of the nanopowders (b).

The structure of the powders was confirmed using X-ray diffraction analysis (XRD), the results of which are presented in Table 1. All powders have been obtained in a metastable monoclinic phase. The powder with the cerium content of 10 at.% stands out against the overall picture; it has been found that it contains a second phase (cubic cerium oxide CeO_2) with the content of about 3 wt.%. The presence of the second phase indicates that the cerium oxide has been unable to be incorporated into the lattice of the yttrium oxide in full, but has remained in the initial state partly. In addition, it is worth noting that, according to the same XRD results, the coherent scattering region (CSR) of particles decreases with the cerium concentration increase.

Table 1. XRD results.

Sample	mon – Y_2O_3 (main phase)	
	SCR, nm	Lattice constants, Å
1 at.%Ce:Y_2O_3	24±2	a = 13.921±0.007 b = 3.496±0.002 c = 8.614±0.005 β=100.19°
5 at.%Ce:Y_2O_3	22±2	a = 13.91±0.02 b = 3.502±0.005 c = 8.62±0.02 β=100.28°
10 at.%Ce:Y_2O_3*	20±3	a = 13.910±0.008 b = 3.510±0.002 c = 8.639±0.005 β=100.41°

* The sample with 10 at.% of Ce:Y_2O_3 contains the second phase: CeO_2 (Cerianite, cubic, content ≈3 wt.%), SCR = 29±3 nm, a = 5.327±0.005 Å

The results of the analysis of the specific surface area of particles (BET) show (Table 2) that the specific surface area increases with the increase in the concentration of the cerium activating ion, that is, the particle size decreases. This is also evidenced by the SCR data given in Table 1. This fact indicates that, when being incorporated into, cerium inhibits the growth of nanoparticles and distorts the yttrium oxide lattice, because their ionic radii are different: Y^{3+} (0.122 nm), Ce^{4+} (0.128 nm), Ce^{3+} (0.148 nm).

Table 2. Characteristics of synthesized nanoparticles.

	1 at.% Ce:Y_2O_3	5 at.% Ce:Y_2O_3	10 at.% Ce:Y_2O_3
Specific surface area (BET analysis), m^2/g	50.5152	51.01128	52.7950
Diameter of particles calculated upon BET data, nm	21.83	21.62	20.89
SCR (XRD), nm	24±2	22±2	20±3

IOP Conf. Series: Materials Science and Engineering **848** (2020) 012051 doi:10.1088/1757-899X/848/1/012051

Studies have shown that the PCL spectra of the powders and compacts are qualitatively similar to each other (Figure 2).

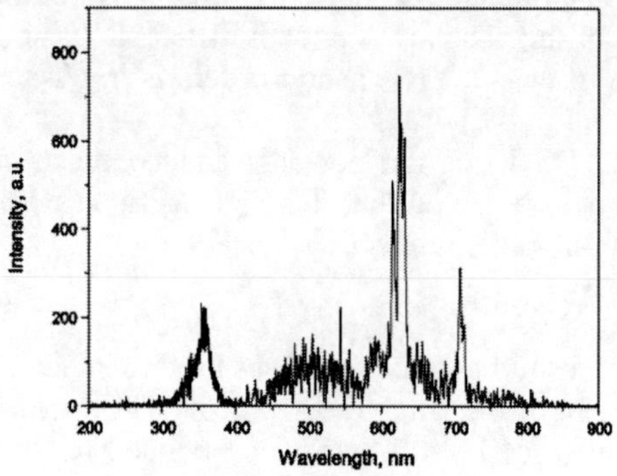

a

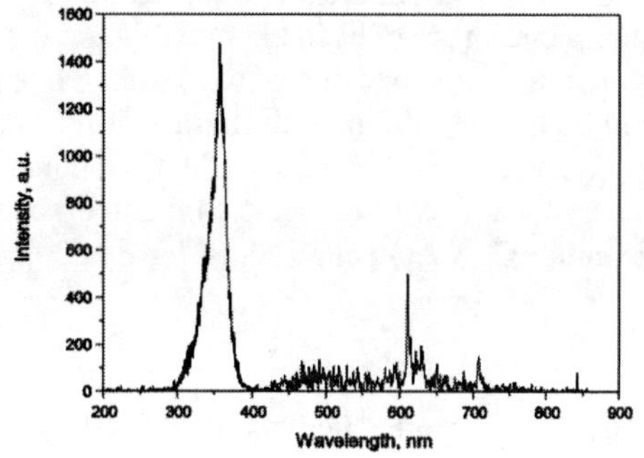

b

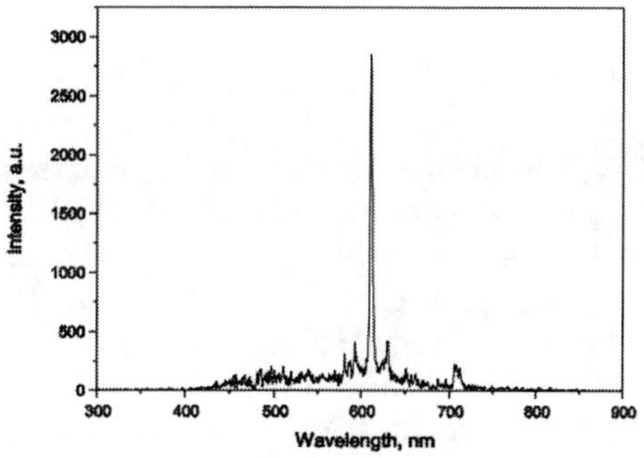

c

Figure 2. PCL spectra of compacts: monoclinic phase 1 at.% $Ce:Y_2O_3$ (a), cubic phase 1 at.% $Ce:Y_2O_3$ (b), cubic phase10 at.% $Ce:Y_2O_3$ (c).

In the spectral range 400–850 nm, the luminescence of the intrinsic surface center (bound radical Y-O) in the form of a system of narrow bands and a broad band with a center of about 500 nm has been observed in all spectra [8–10]. The band with a maximum at 350 nm, shown in Figure 2a,b, is emitted by self-trapped excitons (STEs), which are O^{-2} hole centers associated with the Y^{3+} vacancy [10]. The luminescence of these STEs has been observed only in the nanopowders with the cerium content of 1 at.%. Moreover, in the nanopowders of the cubic phase (Figure 2b), their luminescence intensity increases by almost an order of magnitude, compared with the monoclinic phase (Figure 2a). However, with an increase in the cerium content, this band disappears from the PCL spectrum (Figure 2c), i.e. cerium helps to eliminate this defect. However, the luminescence bands of Ce^{3+} ions are absent in the PCL spectra of the nanopowders with different cerium contents. This situation may be due to the fact that cerium in the Ce:Y_2O_3 nanopowders is in the tetravalent state Ce^{4+}, which is inactive in luminescence.

Pulsed X-ray luminescence (XRL) of the powders and compacts of the monoclinic and cubic phases was not detected. It should be noted that the XRL is much weaker than the PCL, due to the smaller cross section for radiation-matter interaction.

4. Conclusions

The method of laser ablation of a solid porous target was used to synthesize Y_2O_3 nanopowders doped with cerium ions with the content of the latter 1, 5, 10 at.%. The yttrium oxide nanopowder with the cerium (Ce) activating ion content of 10 at.% has been obtained for the first time in the world. The obtained nanopowders are in a metastable monoclinic phase, have an almost perfect spherical shape and an average particle size of 7-14 nm. The second phase was detected in the powder with 10 at.% of Ce:Y_2O_3: the undissolved cerium oxide CeO_2 in a cubic modification with a content of ≈3 wt.%. In all powders, the pulsed X-ray luminescence has not been excited, but the pulsed cathodoluminescence has been well excited. Its spectra for free and compacted powders have been similar. The intrinsic center luminescence bands of Y-O and STE type dominate in them, but no luminescence bands of the Ce^{3+} ions have been detected. Apparently, the cerium ion in Ce:Y_2O_3 nanopowders is in the tetravalent state, which is inactive in luminescence. Nevertheless, the obtained nanopowders can be used for the synthesis of new high-speed ceramic Ce:YAG scintillators for detecting nanosecond X-ray radiation pulses.

Acknowledgments

The reported study was funded by RFBR, project number 19-08-00117.

The authors gratefully acknowledge E.V. Tihonov and Dr. A.I. Medvedev from Institute of Electrophysics UrB RAS for laser synthesis of nanopowders and XRD analysis, respectively.

References

[1] Tokurakawa M., Takaichi K., Shirakawa A. et al 2007 *Appl.Phys.Lett.* **90** (5) 071101_1-071101_3

[2] Mourona J., Dujardin C., Tillement O. et al 2008 J. *Alloys and Comp.* **464** 407-411

[3] Osipov V., Ishchenko A., Shitov V. et al 2017 *Opt.Mat.* **71** 98-102

[4] Kobayoshi M., Shinkawa T., Sato T. et al 1994 *Nucl.Instr.Meth.* **A337** 355

[5] Fyodorov A., Pavlenco V., Korzhik M. et al 1996 *Radiat.Meas.* **26** (2) 215

[6] Miloichikova I. et al 2014 *IOP Conf.Ser.: Mat.Sci. and Eng.* **66** (012031) 4

[7] Osipov V., Kotov Yu. et al 2006 *Las. Phys.* **16** 116-125

[8] Osipov V., Solomonov V., Spirina A. 2011 *J.Opt.Technol.* **78** 408-412

[9] Osipov V., Rasuleva A., Solomonov V. 2008 *J.Opt. Spectrosc.(Russia)* **105** (4) 591-597

[10] Solomonov V., Osipov V., Shitov V., Lukyashin K., Bubnova A. 2020 J. *Opt. Spectrosc. (Russia)* **128** (1) 5-9

21R-Sialon ceramics, obtained by hot pressing

A S Lysenkov[1], M O Stolbova[1,2], D D Titov[1], A I Plokhikh[2], K A Kim[1], D V Gridin[3], M G Frolova[1], N V Petrakova[1], K D Danilin[1,4], D O Lemeshev[4] and Yu F Kargin[1]

[1]A.A. Baykov Institute of Metallurgy & Material Science RAS, Moscow, Russia
[2]Bauman Moscow State Technical University, Moscow, Russian Federation
[3]Plazmoterm, Moscow, Russia
[4]Mendeleev University of chemical technology of Russia, Russia, Moscow
E-mail: toxa55@bk.ru

Abstract. The formation, densification behaviour and mechanical properties of 21R polytype were studied. Self-propagating high-temperature synthesis obtained the starting powder of 21R-sialon. Sealing was carried out by hot pressing in the temperature range 1650-1800°C. With increasing temperature, there is an increase in density and mechanical properties.

1. Introduction

Silicon nitride is one of the most promising high-temperature materials. Silicon nitride and materials based on it have a high operating temperature up to 1400-1750°C, wear resistance, chemical inertness, high (especially silicon nitride) strength and crack resistance [1, 2]. Therefore, the appearance of these materials made it possible to raise the issue of replacing metal products with ceramic ones for operation in extreme thermomechanical conditions, for example, in the hot zone of gas turbine and piston engines. The replacement of carbide and other types of cutting tools with a more efficient ceramic tool continues, ceramics favourably differs from them, for example, in the field of processing of cast iron or superalloys. Currently, research is being conducted to use nitride and silicon carbide products in virtually all industries in developed countries. They have already found wide application in aerospace, metallurgy, chemical, electronic and other industries [3, 4].

There are several Sialon types: α; β; X; O1; H; R [1-8]. Sialons are Si-Al-O-N phase and related systems. They consist of tetrahedra (Si, Al)(O, N)$_4$. Double compounds are distributed in the edges of the tetrahedron. All of existed combinations that contain four elements are distributed in the plane of Si_3N_4-SiO_2-Al_2O3-AlN. It is possible to separate the areal to two isosceles triangles, they consist of two systems Si_3N_4-Al_2O_3-SiO_2 and Si_3N_4-Al_2O_3-AlN [9-11].

R-sialons - one of the types of sialons - are poorly studied materials, and are practically not mentioned in the literature. There are three modifications of R-Sialons: 15R ($Si_{6-x}Al_{4+x}O_xN_{12-x}$), 21R ($Si_{6-x}Al_{8+x}O_xN_{16-x}$) and 27R ($Si_{6-x}Al_{10+x}O_xN_{20-x}$) [12]. Literary sources describe very scarce information on the synthesis of each polytype and the results of studying the properties of these materials are practically absent. Basically, R-Sialons are mentioned as side phases in the synthesis of α- and β-sialons [13-15].

In the last two years, a number of articles by the Indian scientist Mita Biswas from the Central Research Institute of Glass and Ceramics of India have been published. Mita Biswas actively studies the properties of R-Sialons and states in its works that this material is underestimated, as it has high mechanical properties that are almost identical to α- and β-sialons [16]. Ceramics based on the R-modification of sialon mainly contain elongated or prismatic grains - this contributes to the emergence of the so-called self-reinforcing effect, which makes it possible to use sialon ceramics to create wear-resistant products. The researcher also notes the lack of generalized information on R-sialons [16-18].

The work aimed to obtain, by hot pressing, 21R-Sialon ceramics based on powders synthesised by the method of self-propagating high-temperature synthesis (SHS). SHS technology is based on the initiation of an exothermic chemical reaction of silicon with nitrogen and the further spontaneous propagation of the combustion front in powder mixtures with the formation of silicon nitride. The significant thermal effect of the SHS reaction of the interaction of nitrogen with silicon makes it possible to obtain powders with high yield without supplying heat from external sources [19].

2. Materials and Method

Commercial 21R-Sialon powders obtained by the SHS method (PLATINA LLC) were used as the initial ones. The 21R-Sialon powder is represented by well-crystallized particles with an average of 0.1-2 microns (Figure 1). The specific surface of the powder is 12 m^2/g. The phase composition of the powders is shown in Figure 2.

The powder was subjected to mechanical activation in a planetary mill for 60 min in isopropyl alcohol. Balls made of zirconium dioxide with a diameter of 5 mm were used as grinding bodies. The dried mixture was forced through a sieve № 0063.

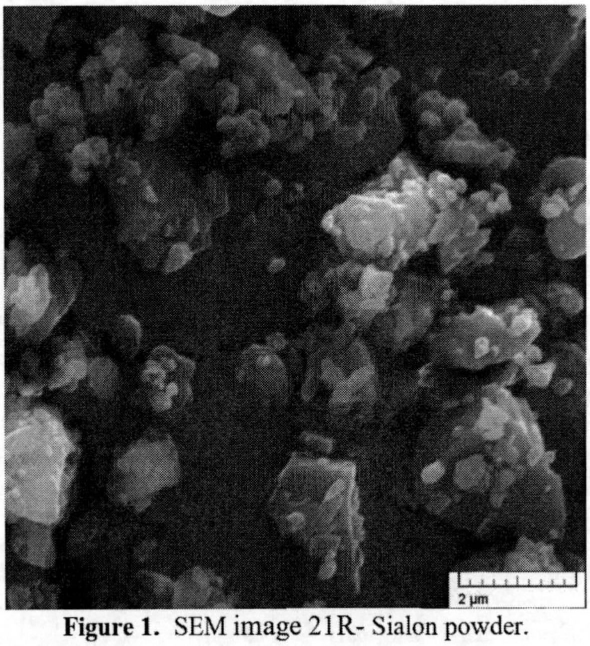

Figure 1. SEM image 21R- Sialon powder.

Before firing, dilatometric studies of the studied powder were performed. Figure 3 shows the curve of continuous shrinkage of 21R-Sialon in the temperature range 20-1800°C. The sintering temperature of the 21R-Sialon is 1689°C. Based on this temperature, subsequent hot pressed in the temperature range 1650-1800°C in increments of 50°C. Before firing samples were prepared in the form of disks (25 mm diameter) charge by the semi-dry pressing method from the received.

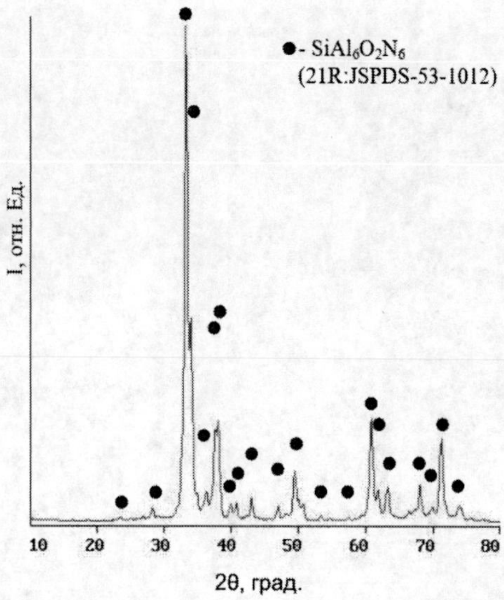

Figure 2. XRD patterns 21R- Sialon powder

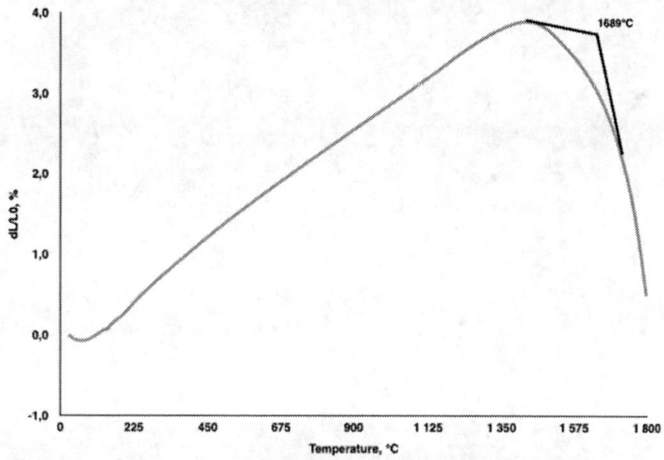

Figure 3. The curve of continuous shrinkage of 21R-Sialon in the temperature range of 20-1800°C.

For dense ceramics used hot pressing furnace Thermal Technology Inc. high temperature experts, model HP20-3560-20. To identify the phase and chemical composition were used X-ray diffraction analysis was used (XRD 6000 SHIMADZU diffractometer, CuKα radiation, λ = 1.5406 Å, scanning speed 2θ = 2 deg/min). The phase composition of the samples was identified using the PDF-2 database, JCPDS-ICDD (Set 1-2002). Morphology and structural features of the samples were studied by scanning electron microscopy (electron microscope NVision 40, Carl Zeiss). Density and open porosity were measured by the Archimedes method using water. The flexural strength three bending test were studied by Instron 5581. The microhardness test was studied by Micro-hardness Tester 401/402 MVD.

3. Results and discussion

The structure and ceramic properties of these samples were studied after sintering. With an increase in the sintering temperature, 21R-Sialon crystals grow (Figure 4). If at a firing temperature of 1650°C there is a large number of crystals of 1-2 μm with single crystals of 5 μm (Figure 4a), then at a firing temperature of 1800°C the size of crystals is 5-6 μm (Figure 4d).

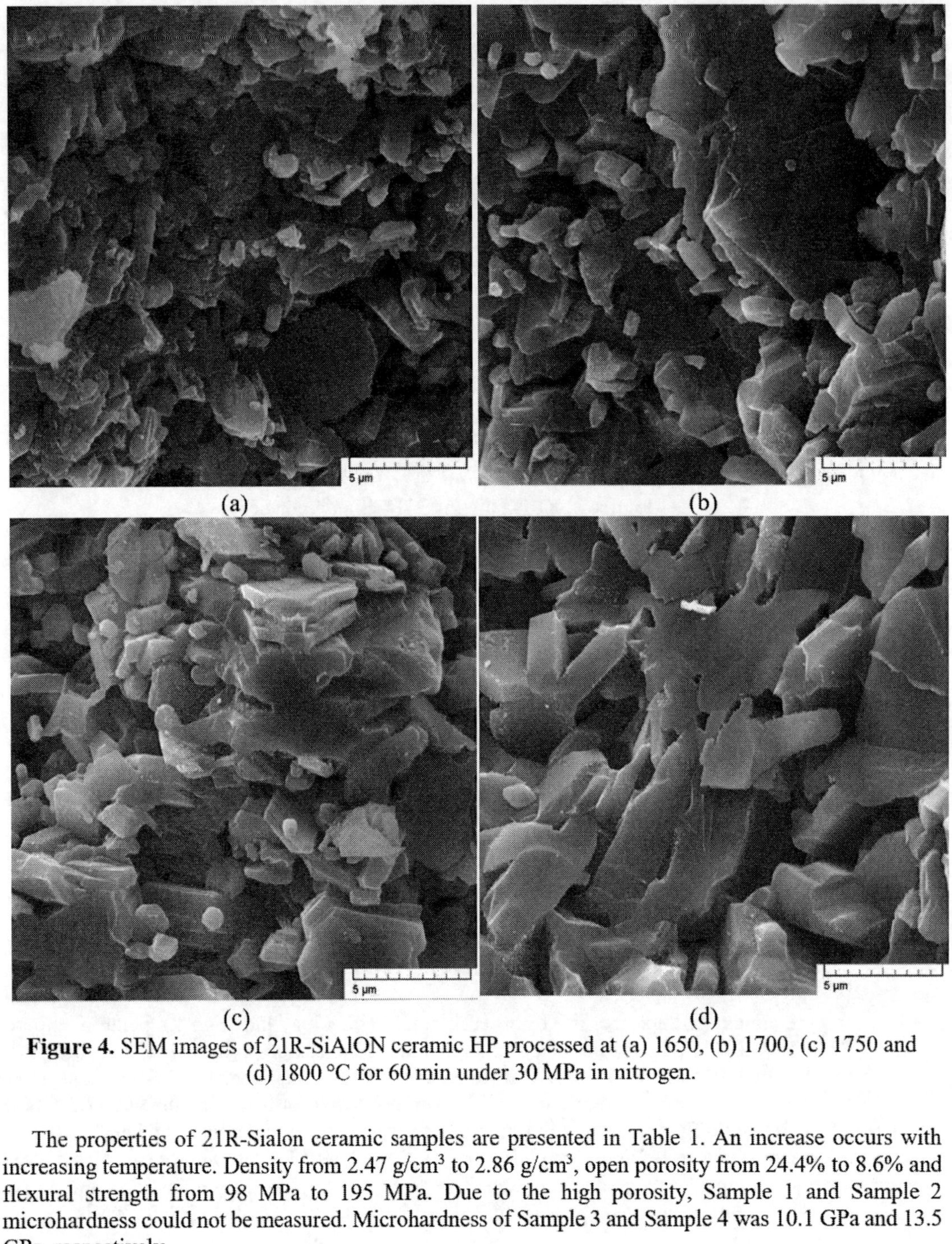

Figure 4. SEM images of 21R-SiAlON ceramic HP processed at (a) 1650, (b) 1700, (c) 1750 and (d) 1800 °C for 60 min under 30 MPa in nitrogen.

The properties of 21R-Sialon ceramic samples are presented in Table 1. An increase occurs with increasing temperature. Density from 2.47 g/cm^3 to 2.86 g/cm^3, open porosity from 24.4% to 8.6% and flexural strength from 98 MPa to 195 MPa. Due to the high porosity, Sample 1 and Sample 2 microhardness could not be measured. Microhardness of Sample 3 and Sample 4 was 10.1 GPa and 13.5 GPa, respectively.

The maximum density of the samples is not large and amounts to 2.86 g/cm^3, but it is higher than in [7], where under the same sintering conditions a density of 2.605 g/cm^3 was achieved. To maximize compaction of this material, it is necessary to either raise the calcination temperature or use sintering additives in a small amount such as Sm_2O_3, Nd_2O_3, and YAG [7].

Table 1. The sintering parameters, density, open porosity, flexural strength and microhardness of the 21R-SiAlON ceramics.

Sample	Sintering parameters (°C/min)	Density (g/cm^3)	Open porosity (%)	Flexural strength (MPa)	Microhardness (GPa)
1	1650/60	2.47	24.4	98±15	-
2	1700/60	2.61	18.1	115±17	-
3	1750/60	2.68	14.2	151±17	10.1±1.2
4	1800/60	2.86	8.6	195±18	11.5±1.1

4. Conclusion

As a result, 21R-Sialon ceramics were obtained by hot pressing in the temperature range 1650-1800°C. The phase composition, strength, hardness and microstructure of the obtained ceramic samples were studied. The mechanical properties of 21R vary with firing temperatures. They are in the range of 2.47-2.86 g/cm^3 for dence, 24.4-8.6% for open porosity, 98-195 MPa flexural strength for and 10.1-11.5 GPa for microhardness. To maximize the compaction of this material, it is necessary either to raise the firing temperature or to use sintering additives.

Acknowledgments

The reported study was funded by RFBR project No 20-03-00455. The methodical part of the work (X-ray, XRD, SEM, flexural strength, density) was carried out according to the state task № 075-00947-20-00.

References

[1] Izhevskiy V A 2000 *Journal of the European Ceramic Society* **13** 2275
[2] Rosenflanz A 1999 *Current Opinion in solid state and materials science* **5** 453
[3] Li Q, Yang Z, Zhong J, Sun Y, Jia D, Duan X and Zhou Y 2019 *Journal of the European Ceramic Society* **4** 934
[4] Kargin Y F, Lysenkov A S, Ivicheva S N, Zakharov A I, Popova N A and Solntsev K A 2010 *Inorganic Materials* **7** 799
[5] Li H X, Sun W Y, Yan D S 1995 *Journal of the European Ceramic Society* **7** 697
[6] Jack K H 1976 *Journal of materials science* **6** 1135
[7] Wang P L, Sun W Y, Yan D S 2000 *Journal of the European Ceramic Society* **1** 23
[8] S Ivicheva *et al* 2019 *IOP Conf. Ser.: Mater. Sci. Eng.* **525** 012084
[9] Kovziridze Z, Nijaradze N, Tabatadze G, Cheishvili T, Mshvildadze M, Mestvirishvili Z and Daraxvelidze N 2017 *Journal of Electronics Cooling and Thermal Control* **4** 103
[10] Ekström T, Nygren 1992 *Journal of the American Ceramic Society* **2** 259
[11] Jack K H 1978 *Materials Research Bulletin* **12** 1327
[12] Huang Z, Wu L 2018 *Phase Equilibria Diagrams of High-Temperature Non-oxide Ceramics* (Springer: Singapore) p 181
[13] Falk L K, Shen Z J and Ekström T 1997 *Journal of the European Ceramic Society* **9** 1099
[14] Bandyopadhyay S 1997 *Journal of the European Ceramic Society* **7** 929
[15] Calloch P, Brown I W, MacKenzie K J, Hanna J V and Rees G J 2016 *Ceramics International* **2** 2330-2338
[16] Biswas M, Sarkar S and Bandyopadhyay S 2018 *Ceramics International* 15 18703-18710
[17] Biswas M, Sarkar S and Bandyopadhyay S 2019 *Metallurgical and Materials Transactions A* **5** 2381-2390
[18] Biswas M, Sarkar S and Bandyopadhyay S 2019 *Materials Chemistry and Physics* **222** 75-80

IOP Conf. Series: Materials Science and Engineering **848** (2020) 012052 doi:10.1088/1757-899X/848/1/012052

[19] Amosov A P, Borovinskaya I P and Merzhanov A G 2007 *Poroshkovaya tekhnologiya samorasprostranyayushchegosya vysokotemperaturnogo sinteza materialov* (Moscow: Mashinostroyeniye-1) 471 p

IOP Conf. Series: Materials Science and Engineering **848** (2020) 012053 doi:10.1088/1757-899X/848/1/012053

Application of mutagenic treatment of active silt for oxidation of cellulose nitrate

E A Saratovskikh[1], L V Avdeeva[1], V A Shcherbakova[2], A I Kazakov[1] and R N Yarullin[3]

[1]Institute of Problems of Chemical Physics, Russian Academy of Sciences, Chernogolovka, Moscow Region, Russia
[2]G. K. Skryabin Institute of Biochemistry and Physiology of Microorganisms, Russian Academy of Sciences, Pushchino, Moscow Region, Russia
[3]Kazan (Near-Volga) Federal University, Kazan, Russia

E-mail: easar@icp.ac.ru

Abstract. The oxidation of native cellulose nitrate (13.38% N) and cellulose nitrate (CN) treated with UV radiation and ozone (UV+ozone) by a biocenosis of microorganisms of active silt (AS) from sewage disposal plants and by a combination of AS with sulfate-reducing bacteria *Desulfovibrio desulfuricans* BKM B-1388 and microscopic fungi *Fusarium solani* BKM F-819 was studied. The use of the preliminary treatment significantly increased the degree of decomposition of CN during its subsequent biodegradation by a symbiosis of AS microorganisms with *D. desulfuricans* and *F. solani*. The application of the mutagenic treatment of AS with nitrosomethylurea (NMU) allows the biocenosis of AS microorganisms with CN acting as a pollutant to retain high oxidation ability with good sedimentation properties for 65 days of incubation. The degree of decomposition was 24.36%. Substantial parameters were achieved during the incubation period of treated CN (CN_{treat}) within the first 5 days, which is promising from the point of view of practical use.

1. Introduction

The Materials characterized by high energy capacity, such as aromatic compounds, nitroethers, and nitroamines, are often serve as components of polymer composite materials. The scale of industrial use of cellulose di- and trinitrates with the nitrogen content higher than 10% (which are highly explosive) is being permanently increasing. Considerable amounts of sewage with wastes containing ecologically dangerous substances are formed during the industrial production of polymers of this kind. Waste waters are decontaminated at sewage disposal plants, the complete stage of operation in which is the biological treatment in aerotanks on active silt (AS). Active silt represents a complex biocenosis of many microorganisms, mainly bacteria, in which the cells exist in the medium of soluble or poorly soluble extracellular polymeric aggregates: polysaccharides, proteins, and ribonucleic and deoxyribonucleic acids.

The development of methods and technological approaches aimed at accelerating the decomposition of the compounds in the sewage of CN manufacturing would solve environmental problems related to CN accumulation and would diminish the negative load on the environment. Biological methods of utilization are most promising and environmentally friendly [1,2].

There are variants of application of chemical mutagenesis developed by Rapoport [3] increasing the oxidation ability of AS from sewage disposal plants toward poorly degradable or nearly non-degradable contaminating chemical substances [3-5]. After the mutagenesis, the enzymatic activity of AS microorganisms increases by more than two times [6]. For example, the use of the genetic method [7,8] made it possible to accelerate the destruction of benzene and synthetic grease substitutes when nitrosomethylurea (NMU) is applied as a mutagen.

We studied the biological decomposition of CN by sulfate-reducing bacteria *Desulfovibrio desulfuricans* [9] and mycelial fungus *Fusarium solani* [10,11] and also considered the influence on the process of the preliminary treatment of CN with UV radiation and ozone (CN_{treat}).

The purpose of this work is the evaluation of the possibility of application of AS from sewage disposal plants subjected to the mutagenic treatment with NMU and in combination of AS with fungi *F. solani* and bacteria *D. desulfuricans* during the decontamination of sewage from industrial CN production and the choice of the most appropriate variant of microbiological destruction for the industrial use.

2. Experimental

2.1. Materials

The following reagents and solvents purchased from Sigma (USA), BioRad (USA), and Reakhim (Russia) were used for the preparation of cultural nutrient media and buffer systems and for carrying out physicochemical studies: KCl, NH_4Cl, $CaCl_2x2H_2O$, K_2HPO_4, $MgCl_2x6H_2O$, $NaNO_3$, sodium lactate, sodium salicylate, NaOH, glucose, tetrahydrofuran (THF), and acetone.

The CN samples kindly presented by the Kazan Gunpowder Plant (trade mark M/l 26 432-01, nitrogen content 13.38%, weight of the elementary unit of the macromolecule 284.4, degree of polymerization 1140, molecular weight 324 216, technical conditions GOST R 50461-92) were used.

2.2. Microorganisms and methods of study

Mycelial fungus *Fusarium solani* BKM F-819 and sulfate-reducing bacterium *Desulfovibrio desulfuricans* (strain BKM B-1388) obtained from the All-Russia Collection of Microorganisms (G. K. Skryabin Institute of Biochemistry and Physiology of Microorganisms of the Russian Academy of Sciences, Pushchino Research Center of Biological Investigation of the Russian Academy of Sciences) were used.

Active silt (AS) taken from the aerotank at the sewage installations of the Chernogolovka Research Center of the Russian Academy of Sciences (Chernogolovka, Moscow Region) was used as a consortium of microorganisms. The AS had a brown-grayish color with slightly boggy smell without predominating smell of chemical impurities.

Mutagenic treatment of AS. *N*-Methyl-*N*-nitrosourea (NMU, reagent grade, Sigma (USA)) in a concentration of 0.07% was used as a mutagenic agent. The treatment of AS was made in 4 days of contact of the AS with the studied CN. The treatment was conducted for 18 h [3]. On the 29th day from the experiment onset, the AS was repeatedly treated with NMU also for 18 h. The studies were carried out in parallel for both the initial and treated AS.

Growing of biomass was carried out according to the earlier described method [10].

Medium for experiments on biodegradation of CN. According to the described method [10], AS was diluted with the medium and CN, closed with a cap (in the case without oxygen purging), and left to stay in a warm place for 1 h.

Biological destruction of CN by the cultures of bacteria and fungi was carried out according to the published procedure [11]. The variants of the study of the biological degradation of CN using the microorganisms are presented in Table 1.

Destruction of CN by the mixed culture of fungi and bacteria was carried out according to the published procedure [11].

Oxidation of CN by AS. Experiments on the study of the biological oxidation of CN by the microbial consortium of AS (Table 1) were conducted in 3-L laboratory glass reactors under natural solar irradiation with air purging. The reactors contained the nutrient medium (1000 mL), AS (200 mL), and CN (10 g/L). In the first series of experiments, the reactors (reactors 1, 3, and 5) were loaded with the initial CN, whereas CN_{treat} was loaded in the second series (reactors 2, 4, and 6).

In the variants providing the study of CN oxidation by AS, bacteria, and fungi, the cultures of bacterium *D. desulfuricans* and fungus *F. solani* were added to a solution of AS in the nutrient medium as described above (reactors 3, 4, and 6). Samples for analysis were taken on the 5th, 16th, 36th, and 65th day from the experiment onset in a volume of 50 mL after mechanical stirring of the solution.

Treatment of CN with ultraviolet and ozone was carried out according to the described procedure [10].

Isolation of CN from solution was conducted according to the described method [10].

Determination of nitrogen and other elements in CN was carried out according to the described procedure [10].

Determination of nitrates and nitrites was carried out according to the described procedure [10].

Determination of the heat release rate and heat in the thermal decomposition of the isolated CN samples was carried out according to the described procedure [12].

Table 1. Variants of the study of the biological degradation of CN

Reactor no.	Conditions	Volume of medium, mL	Volume of seeded bacteria, mL	Number of seeded fungi, shoals	Volume of AS, mL	Concentration of CN, g/L
Control	AS	1000	-	-	200	-
1	AS+CN	500	-	-	100	5.0
2	AS+CN_{treat}	1000	-	-	200	10.0
3	AS+CN+ bacteria+fungi	1000	50	3x2	200	10.0
4	AS+CN_{treat}+bacteria +fungi	1000	50	3x2	200	10.0
5	AS+NMU+CN	1000	-	-	200	10.0
6	AS+NMU+CN_{treat}+ bacteria+fungi	1000	50	3x2	200	10.0

3. Results and discussion

It is seen from the data in Table 1 that the studies were carried out in parallel: (1) on CN and CN_{treat} and (2) on AS and AS treated with NMU. In addition, the possibility to create a symbiosis of AS with the mixed culture of bacteria *D. desulfuricans* and mycelial fungi *F. solani* was considered.

3.1. Degree of biological oxidation of CN

Biodestruction was evaluated by a change in the characteristic parameters: pH, viscosity, MWD, contents of C, H, and N in CN, the formation of nitrates and nitrites in the solution, and changes in the value of heat release of the samples.

Table 2. Contents of carbon and hydrogen in CN and nitrate and nitrite ions in the microorganism growth medium and in the AS after biodegradation

Reactor no.	**Incubation time, day**	**pH**	**Content**			
			C, wt.%	**H, wt.%**	**NO_3^-, μg/mL**	**NO_2^-, μg/mL**
Control	0	6.5				
	5	7.04				
	16	7.15			2.13	0.15
	38	7.45			1.62	0.44
	65	7.68			2.63	0.35
1	0	6.5				
	5		31.23	3.570	4.30	not found
	16	8.39	27.07	3.264	3.64	not found
	38	5.98	26.46	3.135	2.02	12.48
	65	5.73	25.70	2.794	16.03	not found
2	0	5.0				
	5		29.44	3.744	12.90	not found
	16	8.15	24.92	2.606	7.58	not found
	38	6.99	25.80	3.038	2.02	0.18
	65	6.40	25.16	2.721	17.50	44.6
3	0	6.0				
	5		29.92	3.810	3.80	not found
	16	8.60	26.01	2.993	8.08	not found
	38	6.84	25.91	2.759	3.03	0.15
	65	6.34	25.00	2.865	6.62	23.4
4	0	4.50				
	5		32.52	4.333	10.86	not found
	16	8.21	27.59	2.970	16.90	0.57
	38	7.00	27.55	3.396	7.37	0.99
	65	5.41	26.58	3.213	4.04	2.52
5	0					
	5		28.54	3.550		
	16		25.01	2.621		
	38		25.55	2.976		
	65		25.96	2.845		
6	0					
	5		29.65	3.683		
	16		26.49	2.902		
	38		25.91	3.477		
	65		26.09	2.895		

As can be seen from the data in Table 2, sulfur was observed in none of the studied CN samples taken from reactors 1–6.

The initial CN in water is characterized by pH 6.0. The content of NO_3^- found in it ranged from 5.94 to 6.24 μg/mL, and no NO_2^- was observed.

The pH value increases from 6.5 to 7.68 within 65 days of incubation in the control reactor containing AS only. The concentrations of nitrates and nitrites in this reactor are insignificant: $[NO_3^-]_{max} = 2.63$ μg/mL and $[NO_2^-]_{max} = 0.44$ μg/mL.

In reactors 1–4, pH increases from neutral and weakly acidic to 8.15–8.60 from the experiment onset to the 16th day, whereas on the 65th day pH decreases to 5.41–6.40. During biodegradation, the amount of nitrate ions in the AS growth medium in reactor 1 on the 38th day insignificantly differs from that in the control reactor and only on the 65th day increases considerably: to 16.03 μg/mL. The situation is principally different in reactor 2 containing CN_{treat} on AS: after 5 days of storage, the content of NO_3^- is 12.9 μg/mL, then smoothly decreases almost to the values for the control reactor, and increases again to 17.5 μg/mL only on the 65th day (Table 2).

In reactor 4 (AS+CN_{treat}+bacteria (b)+fungi (f)), the content of nitrate ions observed on the 5th day of incubation is significant (10.86 μg/mL), increases more to the 16th day (up to 16.9 μg/mL), and begins to decreases on the 38th day.

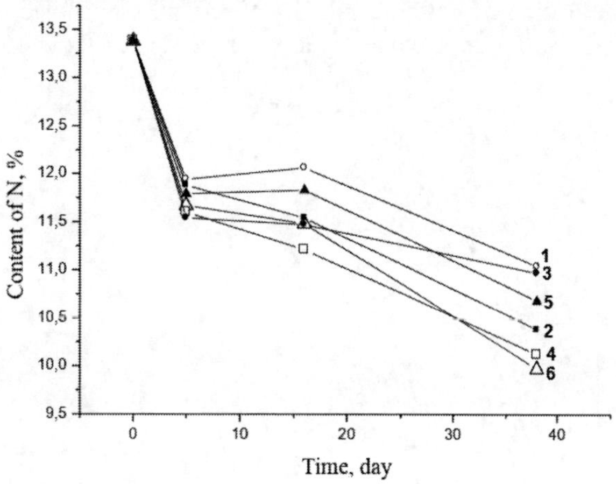

Figure 1. Change in the nitrogen content in the CN samples stored on AS.
Reactor 1: AS+CN; reactor 2: AS+CN_{treat}; reactor 3: AS+CN+b+f; reactor 4: AS+CN_{treat}+b+f; reactor 5: AS+NMU+CN; reactor 6: AS+NMU+CN_{treat}+b+f.

Figure 1 demonstrates that the residence of CN on AS results in a decrease in the nitrogen content in time. The preliminary UV+ozone treatment of CN increases the rate of CN decomposition on AS. For example, after 38 days of incubation in reactor 2, the nitrogen content is considerably lower (10.38%) than that in reactor 1 (10.97%). Similarly, reactor 4 contains a lower amount of nitrogen than reactor 3, whereas the nitrogen content in reactor 6 is lower than that in reactor 5.

The treatment of AS with the mutagen (NMU) exerts an additional positive effect. To the 38th day of incubation, the nitrogen contents in reactor 5 (10.12%) and the more so in reactor 6 (9.97%) are significantly lower than those in reactor 1 (10.97%) and reactor 2 (10.38%).

The introduction of bacteria and fungi into AS, i.e., an attempt to create a symbiosis of microorganisms, does not significantly improve the decomposition of CN. The nitrogen content in reactor 3 was stable and higher within the whole observation period (from 5 to 38 days) than that in reactor 1 containing AS only.

It follows from the results presented in Fig. 1 that the low nitrogen content in the samples is achieved already to the 16th day in reactor 4 in which bacteria and fungi were added to AS and CN_{treat} was used. The nitrogen content in this reactor decreases to 10.68% on the 38th day of exposure. The lowest nitrogen content is achieved within 38 days in a similar variant of treatment (CN_{treat}+b+f) but on the AS treated with NMU, i.e., in reactor 6, where the content of N was 10.12%, the maximum nitrogen loss was 3.26%, and, correspondingly, the degree of decomposition with respect to nitrogen was 24.36%.

The sum of all effects is observed in reactor 6: AS+NMU+b+f on CN_{treat}. However, the decrease in the nitrogen content is this reactor differs insignificantly from that in reactor 4. Therefore, it seems doubtful and unreasonable to introduce an additional technological procedure into the real technological process.

3.2. Heat release rates and heats for the thermal decomposition of CN

The kinetics of thermal decomposition of CN subjected to different variants of microbiological treatment in order to decrease the degree of nitration was studied. We found in our previous study [12] that the total heat of the heat release of the initial CN was about 4080 J/g_{CN} on the average. The heat of the heat release of CN_{treat} is somewhat lower being about 3560 J/g_{CN}. This can indirectly indicate that nitro groups are eliminated upon the UV-irradiation of CN.

The results of studying the kinetics of thermal decomposition of CN and CN_{treat} upon the storage on AS are presented in Figs. 2–5 along with the curves for the initial CN and CN_{treat} samples shown for comparison. The decomposition of both the initial CN and all CN subjected to the microbiological treatment and then isolated from an acetone solution proceeds via the autocatalytic reaction law as described earlier [12].

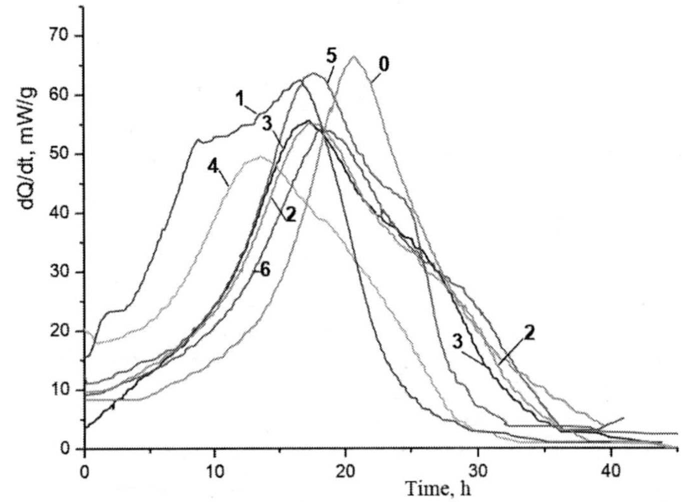

Figure 2. Time dependences of the heat release rate during the thermal decomposition of the CN samples at 139.8°C isolated after 38 days of the microbiological treatment of CN: 0, CN_{init}; *1*, reactor 1 (AS+CN); *2*, reactor 2 (AS+CN_{treat}); *3*, reactor 3 (AS+CN+b+f); *4*, reactor 4 (AS+CN_{treat}+b+f); *5*, reactor 5 (AS+NMU+CN); *6*, reactor 6 (AS+NMU+CN_{treat}+b+f).

It is seen from Fig. 2 that after 38 days of different variants of biological treatment the heat release rate in all reactors is lower than that of the initial CN (curve 0). The highest heat release rate is retained in reactor 5 (AS+NMU+CN) and is close to that in reactor 1 (AS+CN).

It is seen that the time of achieving the maximum heat release rate somewhat decreases when using various types of biological treatment as compared to reactor 1 (AS+CN). All variants are rather close. The lowest heat release rate was established in reactor 4 (AS+CN_{treat}+b+f). In this case, the time of achieving maximum is also minimal (12 h). For other variants of the treatment, the time of achieving maximum ranges from 16 to 18 h. A comparison of reactors 5 and 6 in which AS was subjected to the mutagenic treatment shows that within the indicated time (38 days of incubation of CN on AS) the performed mutagenic treatment of AS does not accelerate the decomposition of CN. It is seen that the preliminary treatment of CN induces the decrease in the heat release rate to a higher extent: pairs 1 and 2; 3 and 4; and 5 and 6. A comparison of reactor 4 (AS+b+f) and reactor 6 with the preliminary mutagen treatment (AS+NMU+b+f) also does not show a decrease in the rate on the treated AS.

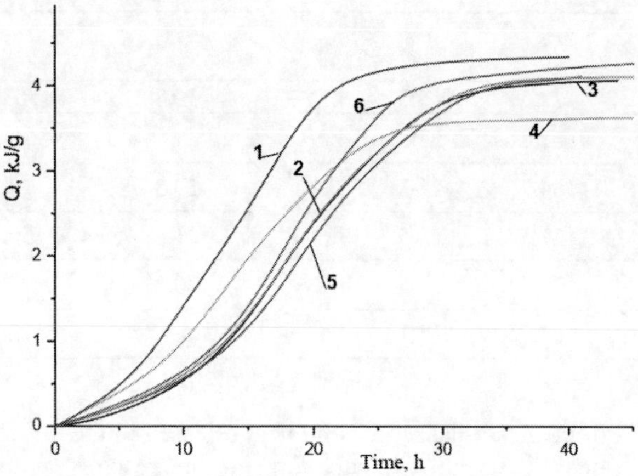

Figure 3. Time dependences of the heat released to the given moment upon the thermal decomposition at 139.8°C of the CN samples isolated after 38 days of the microbiological treatment of CN (numbers at the curves are variants of the microbiological treatment of CN): *1*, reactor 1 (AS+CN); *2*, reactor 2 (AS+CN_{treat}); *3*, reactor 3 (AS+CN+b+f); *4*, reactor 4 (AS+CN_{treat}+b+f); *5*, reactor 5 (AS+NMU+CN); *6*, reactor 6 (AS+NMU+CN_{treat}+b+f).

The time dependences of the heat released to the given moment in the thermal decomposition reaction are presented in Fig. 3. Reactor 4 (AS+CN_{treat}+b+f) can be distinguished by a minimum amount of the isolated heat and a minimum time of achieving the maximum heat release rate, which corresponds to the data in Fig. 2.

The initial $(dQ/dt)_{t=0}$ and maximum $(dQ/dt)_{max}$ heat release rates, times of achieving maximum rates t_{max}, and total heats of decomposition Q_{tot} for the studied CN samples are presented in Table 3. The weak dependences of the initial heat release rate, maximum heat release rate, time of achieving it, and total heat of the process on the microbiological treatment conditions show that the nitrogen content in the CN fraction dissolved in acetone differs slightly from the nitrogen amount in the initial CN. Under the conditions of microbiological processes with nitrogen consumption during the living activity of microorganisms that are present in the initial AS and seeded bacteria and fungi, the balance with respect to nitrogen should be calculated taking into account the amount of nitrogen consumed in the growth of population of all types of microorganisms, nitrogen transited to the mother liquor in the form of nitrite and nitrate anions and sugar nitrates, and nitrogen in the CN fraction that was not dissolved in acetone and remained in AS.

Of the whole array of microbiological conditions of the CN treatment, the conditions in reactor 4 (AS+CN_{treat}+b+f) result in a lower heat of thermal decomposition and, hence, in a lower content of nitrogen in the CN dissolved in acetone after 38 days of incubation.

Table 3. Kinetic parameters of the thermal decomposition of CN after 38 and 65 days of biological degradation.

Reactor no.	after 38 days of biological degradation				after 65 days of biological degradation			
	$(dQ/dt)_{t=0}$, mW/g	$(dQ/dt)_{max}$, mW/g	t_{max}, h	Q_{tot}, kJ/g	$(dQ/dt)_{t=0}$, mW/g	$(dQ/dt)_{max}$, mW/g	t_{max}, h	Q_{tot}, kJ/g
initial	9.6	77.2	20.8	4.1				
initial	9.4	72.2	18.3	4.1				

UV+ozone irradiation	21.5	65.5	14	3.8				
1	18.1	72.6	16.7	4.3	19.8	103.3	9.3	4.6
2	11.2	65.2	17.3	4.1	7.1	74.6	13	4.1
3	4.3	64.7	17.3	4.1	19.3	74.7	5.5	4.5
4	23.7	57.5	13.5	3.6	21.6	59.8	9.5	3.8
5	10.6	73.7	17.5	4.3	20.8	77.9	15.6	4.2
6	14.5	62.7	18.1	4.1	6.6	72.6	13.5	3.8

The time dependences of the heat release rate and heat released to the given moment in the thermal decomposition of CN after 65 days of incubation in the reactors with different variants of microbiological treatment are presented in Figs. 4 and 5. As can be seen from Fig. 4, the maximum rate of heat release still remains in reactor 1 (AS+CN). The high rate of heat release is also observed in reactor 3 (AS+CN+b+f). However, the amount of the released heat in reactor 3 is substantially lower and almost the same as in reactors 2, 5, and 6. Reactor 4 (AS+CN_{treat}+b+f) can be distinguished in which the heat release rate and the amount of released heat are minimum.

The initial rate of heat release is 14–15 cal/(h g) on the average, and its deviations toward overestimation are observed for the variants of microbiological treatment in reactor 4. The times of achieving the maximum rate of heat release during CN decomposition after the microbiological treatment for 65 days are shorter than the corresponding times during the treatment for 38 days. The shortest time of achieving the maximum rate of heat release during CN decomposition after the microbiological treatment for 65 days was observed for reactor 3.

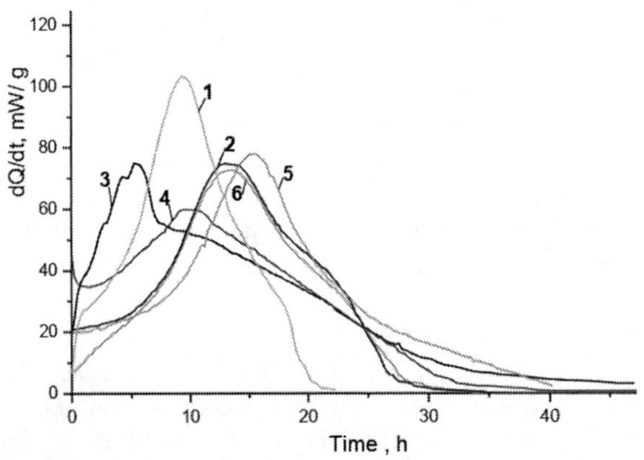

Figure 4. Time dependences of the heat release rate during thermal decomposition at 139.8°C of the CN samples treated for 65 days in the microbiological reactor: 0, CN_{init}; *1*, reactor 1 (AS+CN); *2*, reactor 2 (AS+CN_{treat}); *3*, reactor 3 (AS+CN+b+f); *4*, reactor 4 (AS+CN_{treat}+b+f); *5*, reactor 5 (AS+NMU+CN); *6*, reactor 6 (AS+NMU+CN_{treat}+b+f).

The heat of CN decomposition for all variants of treatment and both treatment times is about 1000 cal/g on the average. It is seen from the data in Table 3 that the amount of the heat released during CN decomposition after 65 days of incubation under various conditions differs weakly from that released in the specific variant of biological treatment after 38 days of incubation, i.e., an additional time of the microbiological process does not result in a significant decomposition of CN. The data in Figs. 2 and 4 and Table 3 show that a comparison of reactors 5 and 6 in which AS was subjected to the mutagenic treatment revealed weak dependences in both the initial rates of heat release, maximum heat release rates, times of achieving it, and total heat of the process on the time of the microbiological process.

The observed parameters of the heat release rates and total heats of the process allow us to conclude that the preliminary UV treatment affects substantially the decrease in the nitrogen content in the CN fraction subjected to the microcalorimetric study: the subsequent biological treatment of these samples makes it possible to decrease the nitrogen content compared to the initial CN. Of the whole array of microbiological conditions of the CN treatment for 38 days, the conditions in reactor 4 (CN_{treat}+b+f) still result in a lower heat of thermal decomposition and, hence, a lower nitrogen content in the CN subjected to the microbiological treatment.

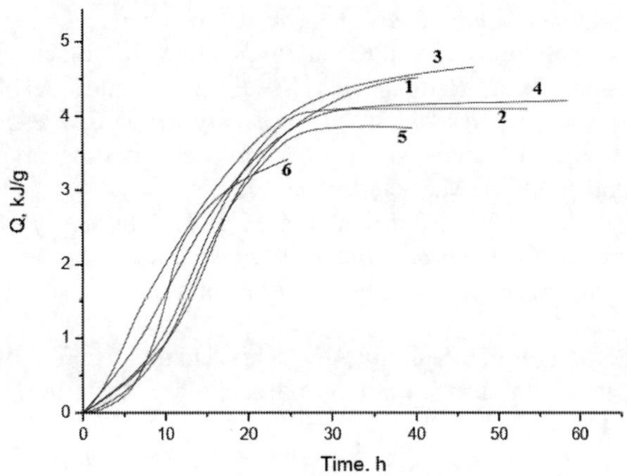

Figure 5. Time dependences of the heat released to the given moment upon thermal decomposition at 139.8°C for the CN samples treated for 65 days in the microbiological reactor (numbers at the curves indicate the variants of the microbiological treatment of CN): 0, CN_{init}; *1*, reactor 1 (AI+CN); *2*, reactor 2 (AS+CN_{treat}); *3*, reactor 3 (AS+CN+b+f); *4*, reactor 4 (AS+ CN_{treat}+b+f); *5*, reactor 5 (AS+NMU+CN); *6*, reactor 6 (AS+NMU+CN_{treat}+b+f).

4. Conclusion

It is found in the present study that the biological destructors (microorganisms) decompose CN. The degree of biodegradation of CN by the studied microorganisms was 24.36% within a period of up to 38 days of residence under the action of various microorganisms, potential biological destructors.
Among particular types of microorganisms, the most substantial results on the CN decomposition were achieved for the oxidation of CN by mycelial fungus *Fusarium solani* BKM F-819.
The use of the preliminary UV+ozone treatment significantly increases the degree of decomposition of CN.

A symbiosis of microorganisms *Desulfovibrio desulfuricans* 1388 and *Fusarium solani* BKM F-819 and a biocenosis of AS microorganisms are efficient for biodegradation.
It is shown that the application of the mutagenic treatment of AS with NMU allows the biocenosis of AS microorganisms to retain the high oxidation ability with good sedimentation properties within up to 65 days of incubation with the highly toxic and poorly degradable contaminator CN.

Acknowledgments
This work was carried out in the framework of the state task (state registration no. AAAA-A19-119101500098-3), theme 0089-2019-0005.

References

[1] Gallo B, Allen A, Bagalawis R L, Woodbury C, Yang A, Austin P et al 1993 Microbial degradation of nitrocellulose. *Nitrocellulose fines separation and treatment workshop proceedings*. Champaign, IL: US Army Construction Engineering Research Laboratory

[2] White G F, Snape J R 1993 Microbial cleavage of nitrate esters; defusing the environment. *J. Gen. Microbiol.* **139** 1947 – 1957

[3] Rapoport I A 1981 *The use of chemical mutagens in protecting the environment from pollution and in agricultural practice.* (M.: Science) p 267

[4] Lisichkin G V, Chernov N N 2003 *Man and his environment* (M.: World) p 460

[5] Zhmur N S 2003 *Technological and biochemical processes for wastewater treatment at facilities with aeration tanks.* (M.: AQUAROS) p 512

[6] Silantieva N S, Zamelin V I, Fedosova N F 1981 *Intensification of the biological wastewater treatment using chemical mutagens*. In: The use of chemical mugagens in protecting the environment from pollution and in agricultural practice. (M.: Science) 49 - 52

[7] Lebed E S, Zagusnina N A, Bartoshevich Yu E, Bezborodov A M 1981 *The use of the mutagenic effect of UV rays and N-nitrosomethylbiuret in the selection of Geotrichum candidum*. In: The use of chemical mugagens in protecting the environment from pollution and in agricultural practice. (M.: Science) 71 - 73

[8] Vanifatyeva M T, Filippov N A, Sergienko Z M 1981 *The use of chemical mutagenesis in wastewater treatment for the production of synthetic fat substitutes*. In: The use of chemical mugagens in protecting the environment from pollution and in agricultural practice. (M.: Science) 41 - 42

[9] Saratovskikh E A, Kazakov A I, Kulikov A V, Khryachkov V A, Yarullin R N 2016 The effect of UV treatment on biodegradation of nitrocellulose bacteria *D. desulfuricans Rus. J. General Chem.* **4** 106-115.

[10] Khryachkov V A, Saratovskikh E A, Yarullin R N, Kulikov A V 2017 Effect of the *D. desulfuricans* bacterium and UV radiation on nitrocellulose oxidation *Russ. J. Phys. Chem. B.* **11 (4)** 697–703 DOI: 10.1134/S1990793117040169

[11] Saratovskikh E A, Shcherbakova V A, Yarullin RN 2018 Nitrocellulose Degradation by the Fungus *Fusarium solani Appl. Biochem. and Microbiol.* **54 (1)** 45-52 DOI: 10.1134/S0003683818010106

[12] Saratovskikh E A, Avdeeva L V, Yarullin R N, Kazakov A I 2018 Evaluation of denitration of nitrocellulose by microbiological treatment for industrial waste effluents using Calorimetry analysis *J. Therm. Anal. Calorim.* **134 (2)** 653-664 DOI: 10.1007/s10973-018-7658-0

[13] Losev I P, Trostyanskaya E B 1960 *Chemistry of synthetic polymers*. (M.: Goskhimizdat) p 560

IOP Conf. Series: Materials Science and Engineering **848** (2020) 012054 doi:10.1088/1757-899X/848/1/012054

Perfluorocarbon nanoemulsions containing fluorinated photosensitizer for photodynamic cancer therapy

A A Markova[1], E V Belyaeva[1], A S Radchenko[2], M V Dmitrieva[3], A L Sigan[1], A A Shtil'[3] and N D Chkanikov[1]

[1] A.N.Nesmeyanov Institute of Organoelement Compounds, Russian Academy of Sciences, Moscow, Russian Federation
[2] Emanuel Institute of Biochemical Physics, Russian Academy of Sciences, Moscow, Russian Federation
[3] Blokhin National Medical Research Center of Oncology, Moscow, Russian Federation

alenmark25@gmail.com

Abstract. The composition of the photosensitizer and fluorocarbons with high oxygen capacity allows to increase the effectiveness of photodynamic therapy, which is shown on tumor cells. To achieve the solubility of photosensitizers in fluorocarbons,a number of fluorinated porphyrins were synthesized and their spectral characteristics were obtained. Fluorocarbon emulsions without a photosensitizer are inert for tumor cells of human colon adenocarcinoma. Fluorocarbon nanoemulsion with leader fluorinated porphyrin with fluoroaliphatic substituents accumulates in tumor cells and causes photoinduced necrosis due to peroxidation of membrane lipids. Perfluorocarbon nanoemulsions in combination with a dissolved fluorine phase photosensitizer are promising compositions for the photodynamic therapy of cancer.

1. Introduction

The promising development direction of photodynamic therapy (PDT) is using of a photosensitizer (PS) with fluorocarbon nanoemulsions in combination [1]. Fluorocarbon emulsions with high oxygen capacity increase the efficiency of PDT. In hypoxia conditions they create a fundamental possibility of using PDT method to initiate the death of tumor cells. Introduction of PS into the fluorine phase of the emulsion and modification of PS to give them solubility in fluorocarbons are demanded. The additional introduction of such fragments besides the solubility, can also contribute to a higher resistance to oxidation by reactive oxygen species. Development of such emulsions requires solving the following tasks: synthesis of PS (porphyrins) soluble in fluorocarbons, investigation of spectral characteristics and cytotoxicity, developing the composition of stable fluorocarbon emulsions, and investigation of the emulsions photodynamic activity *in vitro*.

2. Photosensitizes soluble in fluorocarbons

2.1. Synthesis

Fluorine-containing porphyrins were obtained according to the previously report [2]. from unsubstituted pyrrole and perfluoroalkyl-containing aldehydes. Perfluoroalkyl-substituted benzaldehydes were prepared from commercially available pentafluorobenzaldehyde (PFBA) and polyfluorinated alcohols with different aliphatic chain length. Nucleophilic substitution of fluorine in the *para*-position of PFBA was carried out using a soft base - potassium fluoride in dimethylformamide (DMF). The pyrrole-aldehyde condensation was carried out according to the Lindsey method in methylene chloride under with boron trifluoride etherate as a Lewis acid and a dehydrating agent. Then unstable porphyrinogen (without isolation) was oxidized by 2,3-dichloro-5,6-dicyano-1,4-benzoquinone (DDQ) (Scheme 1).

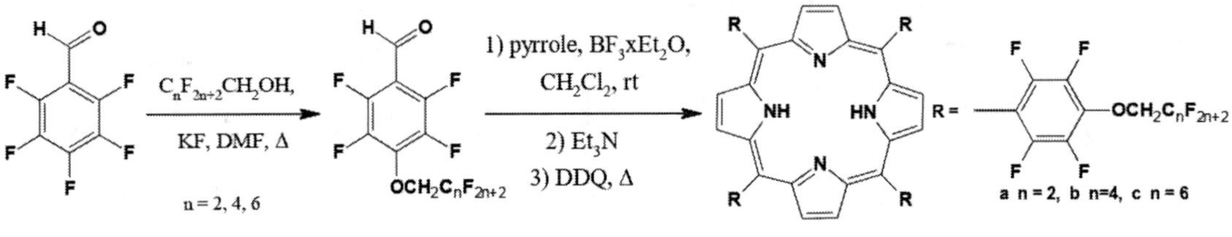

Scheme 1. Synthesis of porphyrins **2a-c** from perfluoroalkyl-substituted benzaldehydes **1a-c**.

Synthetic strategy for obtaining porphyrins from aliphatic polyfluorinated aldehydes (Scheme 3) has several differences. In the first step pyrrole-aldehyde condensation proceeds without acid, the product (oxyalkylated pyrrole) was isolated by crystallization from the reaction mixture and was used further without purification. The tetramerisation of oxyalkylated pyrroles **3a,b** was carried out in the presence of a strong acid (para-toluenesulfonic acid) in toluene with azeotropic water distillation. The oxidation of porphyrinogen, as in the previous method, was carried out in situ by DDQ.

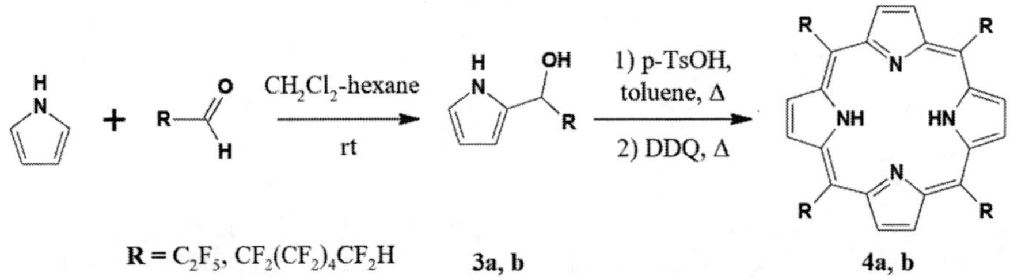

Scheme 3. Synthesis of porphyrins **4a,b** from polyfluoroalifatic aldehydes.

Porphyrins obtained by these methods were purified by silica gel column chromatography. Synthesis procedures and characterization of all compounds are given in Supplementary materials.

2.2. Solubility

The introduction of polyfluoroaliphatic substituents into the porphyrin structure was necessary first of all to give them sufficient solubility in fluorocarbons, in particular in perfluorodecalin, the main component of our fluorocarbon emulsions. Saturated solutions of porphyrins were prepared by the following steps: heating to complete dissolution of porphyrin, keeping at room temperature for a day. The final concentration of porphyrins was determined by spectrophotometry using a series of dilutions to obtain more correct results. The obtained porphyrins showed good solubility in perfluorodecaline -

1 mM, and the solubility values were practically independent of the length of the perfluorocarbon fragment.

2.3. Optical characteristics

Absorption spectra of porphyrins with perfluoroaliphatic substituents were classic for this type of compounds, i.e. the spectra of all compounds consist of an intensive Soret band of about 400 nm and four Q-bands of lower intensity (Figure 1A), all data are shown in Table 2. The extinction coefficient of the fourth Q-band at 650 nm for porphyrins with aliphatic substituents in the *meso*-position several times higher than aromatic analogues, which is generally characteristic of non-fluorinated compounds (Figure 1B).

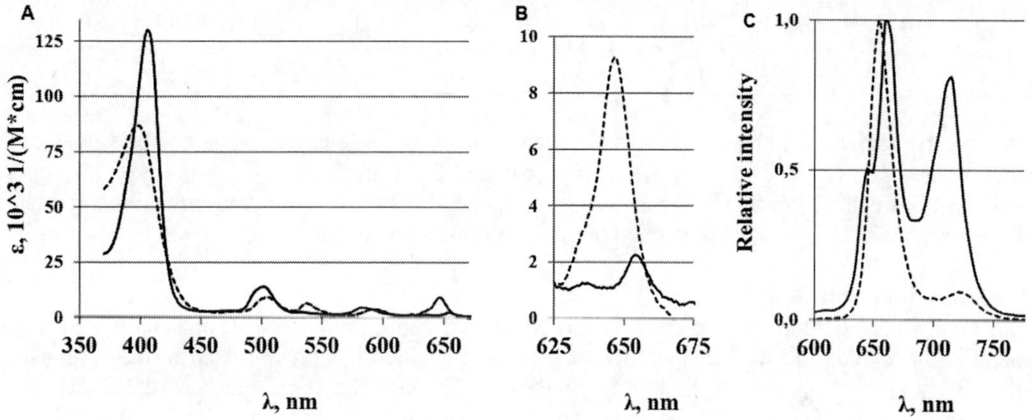

Figure 1. Spectra of porphyrins **2c** (solid line) and **4b** (dashed line). A: Absorption spectra in visible region; B: Absorption maxima in "red" region; C: Fluorescence spectra.

The fluorescence spectra of porphyrins have two maxima (Figure 1C), their intensities for aromatic derivatives are close, while for aliphatic long-wave maximums are noticeably lower. The fluorescence quantum yield for all PSs is low, which is explained by the predominant formation of an excited triplet state, rather than a singlet, which is necessary for the fluorescence emission.

2.4. Reactive oxygen species generation

Reactive oxygen species (ROS) are formed during energy transfer from the excited triplet state of a PS to an unexcited oxygen molecule. It is believed that in nonpolar media (nonpolar organic solvents, perfluorocarbons), singlet oxygen significantly prevails over other types of ROS, so they can be neglected in this case [3]. The efficiency of the singlet oxygen formation by the PS can be quantified by measuring its phosphorescence at 1275 nm. The quantum efficiency of this process is calculated relative to the well-known standard with which measurements were made under the same conditions. The values of the quantum efficiency of the singlet oxygen generation were calculated according to formula (1) relative to the literature standard *meso*-tetrakis(pentafluorophenyl)porphyrin [3], the obtained values are presented in Table 1.

$$\varphi = \varphi_{st} \times \frac{a_{\exp}}{a_{st}}, \tag{1}$$

where φ is the quantum yield, *a* is the slope of the I-D graph (I is the emission intensity, D is the optical density).

Table 1. Optical characteristics of porphyrins **2a-c** and **4a,b**, the singlet oxygen generation efficiency.

№	Absorption λ, нм (ε, 10³ М⁻¹*см⁻¹)					Fluorescence λ, нм		φ	Singlet oxygen generation φ
	Soret	Qy(1,0)	Qy(0,0)	Qx(1,0)	Qx(0,0)	Qx(0,0)	Qx(0,1)		
2a	407	503 *(22.5)*	533 *(2.2)*	582 *(6.4)*	653 *(1.0)*	660	715	0.016	
2b	407	502 *(17.9)*	532 *(1.9)*	582 *(5.0)*	653 *(1.7)*	660	715	0.017	
2c	406 *(137.2)*	501 *(13.9)*	*	583 *(4.7)*	654 *(2.3)*	660	715	0.016	0.83
4a	408 *(55,3)*	512 *(5,1)*	546 *(4,4)*	595 *(2,2)*	649 *(5,7)*	655	720	0.012	0.70
4b	410	514	547	598	650	655	720	0.011	0.72

All spectra registered in benzene at the same conditions as the selected standard. The values of the singlet oxygen generation efficiency allow us to conclude that the fluorine-containing porphyrins are promising PS for PDT. Porphyrins **2c** and **4b** were selected for further study. Fluorocarbon emulsions were prepared with them and were studied in vitro for photodynamic activity.

3. Fluorocarbon emulsions

Direct emulsions perfluorocarbons-in-water are stabilized using pluronics [4], phospholipids [5] and some others [6]. We used the proxanol-268 as emulsifier, which has proven itself in the composition of the "Perftoran" (medical acceptable fluorocarbon emulsion). The composition of the fluorocarbon phase was also taken by analogy with Perftoran - perfluorodecalin (PFD) - / perfluoromethylcyclohexylpiperidine (PMCP) = 2: 1. An ultrasonic dispersion apparatus Sonicator W-375 (Heat systemultrasonics) for emulsification was used. Our experience has shown that to obtain nanoscale emulsions, it is necessary to use a glass reactor cooled in ice-bath (sufficient heat removal and the possibility of visual control of the process). The dispersion conditions were selected by varying the power and exposure time, the particle size was controlled by dynamic light scattering, the obtained data are shown in Table 2.

Table 2. Test conditions of preparation of fluorocarbon emulsion.

Entry	Full time, min	Ultrasound mode (time, min/power, %)	Average particle size, nm
1	0.5	0.5/50	384±5
2	2	1/50 + 1/50	219±6
3	3	3/80	204±10
4	4	2/50 + 2/50	205±2
5	6	3/50 + 3/50	216±16
6	6	3/50 + 3/80	212±7
7	6	3/80 + 3/80	224±11
8	6	2/50 + 2/50 + 2/80	232±16
9	6	2/50 + 2/80 + 2/80	229±8

As a result of preliminary experiments, optimal conditions for preparing emulsions were found. The average particle size was about 205-220 nm, which is optimal for us (grey lines in Table 2). Subsequently, for the preparation of “single” (without PS) and loaded (with PS) fluorocarbon emulsions, we used the following dispersion conditions: two times 3 min at 50% and 80% power with and cool 2-5 min between actions. The content of PS in the fluorocarbon phase of the emulsion was chosen based on their solubility. As shown in Section 2.2., the solubility of all porphyrins in PFD was

IOP Conf. Series: Materials Science and Engineering **848** (2020) 012054 doi:10.1088/1757-899X/848/1/012054

about 1 mM. To prevent possible precipitation, because the PFD / PMCP mixture is used, we took half of this value, i.e. 0.5 mM in the fluorocarbon phase and 0.033 mM in the emulsion. Thus, the ratio of the emulsion components was taken as follows: porphyrin - 0.033 mM, a mixture of fluorocarbons - 12.2 wt. %, proxanol-268 - 2.5 wt. %, NaCl - 0.7 wt. % and water. The average particle size of the emulsion samples obtained under optimized conditions was 210-220 nm.

4. Cytotoxicity of porphyrins in emulsion on tumor cells *in vitro*

A photosensitizer (PS) as a potential agent for the photodynamic treatment of tumors should be low cytotoxic in the dark and cause the death of tumor cells due to its photoactivation. We investigated the dark and photoinduced cytotoxicity of fluorinated porphyrins using the MTT assay [2]. As PS, we chose compounds **2c** and **4b** containing fluoroaryl and fluoroalkyl substituents, respectively. Photosensitizers were used in a "free" form and in "emulsion" form. "Free PS" is a solution of PS in DMF (10 mM), diluted serially with DMEM culture medium to final concentrations when introduced into cells. Emulsions with **2c** and **4b** contained 0.033 mM of the corresponding PS and were introduced into the cells until the final concentrations of the PS were reached. Compounds were introduced into the cells of human colon adenocarcinoma HCT116 in the range of photosensitizer concentrations of 5-20 μM. In this experiments, the final concentration of DMF solvent did not exceed 0.5%, which allows us to neglect the toxic effect of this solvent for free PS.

Dark cytotoxicity after 72 hours of incubation of cells with **2c** free, **4b** free, **2c** emulsion, **4b** emulsion was insignificant (half maximal inhibitory concentrations IC_{50} were not achieved in the concentration range up to 20 μM and 72h incubation). This means that dark cytotoxicity of fluorporphyrins is very low (less than 10%, curves not shown). An emulsion without a PS ("Emulsion w/o PS") is also non-toxic in the dark, it contributed to an increase in cell growth up to 10%.

The study of photoinduced toxicity of fluorporphyrins **2c** and **4b** (free form and emulsion form) included the following steps. Porphyrins were accumulated in HCT116 cells during 24 h of incubation, laser irradiated, incubated for 24 h for death, the survival rate was determined by the MTT test (data on cell survival in relation to the number of untreated cells was taken as 100%). As is known, absorption in the red region is important for potential photosensitizers, since this light has the greatest ability to penetrate tissues. For the photoexcitation of porphyrins, a 633 nm laser was used, a dose of 33 J / cm^2. Upon photoexcitation of fluoroporphyrins **2c** and **4b**, photoinduced cell death was present for both free and emulsion forms (Figure 2).

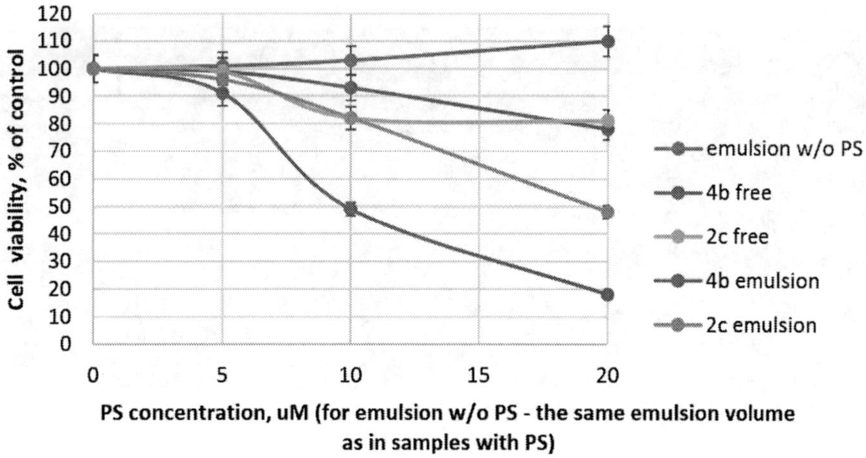

Figure 2. Photoinduced cytotoxicity of fluorporphyrins in DMF and emulsions on HCT116 cells.

In the free form, fluoroporphyrins **2c** and **4b** cause small (up to 30%) cell death due to photoinduced cytotoxicity. They are photoactive, but the small cell death for the free form can be explained by their increased hydrophobicity, since they have fluorine substituents. An increase in the efficiency of the photo-damaging effect of fluoroporphyrins was achieved due to their inclusion in the composition of fluorocarbon emulsions. The photoinduced cytotoxicity of emulsions with porphyrins is greater than in free form due to the solubility of PS and oxygen in fluorocarbons. Emulsion **4b** is more cytotoxic than emulsion **2c**. This correlates with spectrometry data: porphyrins with fluoroalkyl substituents in comparison with fluoroaryl analogues have a large absorption in the red region for the presented series. Thus, the formulation of fluoroporphyrins in emulsions contributes to an increase in their photoinduced cytotoxicity compared to free fluoroporphyrins.

Fluoroporphyrin **4b** is characterized by a large difference between the photoinduced cytotoxicity in the emulsion compared to the free form than by fluoroporphyrin **2c**. Thus, the advantage of introducing fluoroporphyrins into the composition of fluorocarbon emulsions with the aim of increasing their photoinduced cytotoxicity has been shown. The data obtained made it possible to choose emulsion **4b** as a leader composition for characterization the mechanism of cell death.

5. Emulsion-accumulated cells die according to the mechanism of photoinduced necrosis

It is known that photosensitizers after photoexcitation generate reactive oxygen species, which leads to damage to cells, in particular, their lipid membranes. Along with the necrotic effect, PS can also cause other types of cell death. For fluorinated porphyrins **2c** and **4b**, dark cytotoxicity after 72h of incubation with cells is minimal; therefore, we characterized photoinduced necrosis as the main mechanism of cell death. The excitation of PS by such a dose of light, which is capable of causing necrosis, is interesting for initiating cell death of various tissue origin. The fast and irreversible destruction of plasma membranes, characteristic of necrosis, allows the use of the PDT method on a wide range of tumor cells, including cell subline resistant to proapoptotic chemotherapy drugs. In the case of initiation of photonecrosis, the main events of the mechanism are the accumulation of a PS in the cell (in this case, emulsions), its photoexcitation with the generation of reactive oxygen species, and lethal damage to membranes, in particular, cytoplasmic membrane.

The distribution of the leader composition in the cells was visualized after 24 h incubation HCT116 cells with **4b** emulsion (final PS concentration 10 μM) by confocal laser scanning fluorescence microscopy microscopy (Figure 3). The **4b** emulsion enters the cells and remains into it for 24 hours; it accumulates not only in the cytoplasmic membrane, but also in the cytoplasm. The emulsion does not entres the nucleus.

Figure 3. Cytoplasmic distribution of the **4b** emulsion. Green - Hoechst Ex405Em415-485, purple - porphyrin **4b** Ex405Em600-700.

Lipid peroxidation was visualized by confocal microscopy and shown on **4b** emulsion. HCT116 cells were incubated with a emulsion for 24 h (final PS concentration 10 μM), after which they were stained with bodipy ceramide, a marker of lipid peroxidation. The reduced and oxidized form of the marker has different characteristics of excitation and fluorescence, which allows us to separate the signals in microscopy (Figure 4).

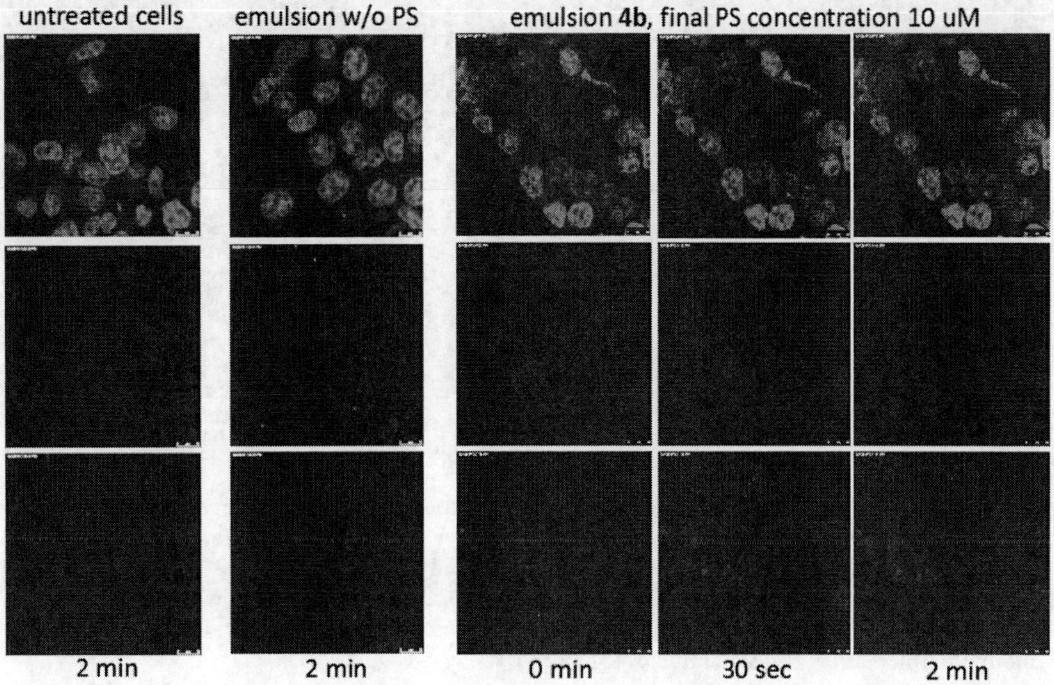

Figure 4. Lipid peroxidation after photoexcitation of fluoroporphyrin **4b** in an emulsion. Green - Hoechst Ex405Em415-485, blue - BODIPY581/591C11 Ex543Em560-650, red - BODIPY581/591C11 Ex488Em500-580.

Untreated control cells and cells incubated with an emulsion without a photosensitizer do not cause lipid peroxidation. Fluoroporphyrin in the emulsion causes an increase in the ratio of the oxidized form of the lipid peroxidation marker to the reduced one. This effect develops within 2 minutes after photoexcitation. This corresponds to lipid peroxidation of the cytoplasmic membrane and membrane organelles in accordance with the intracellular distribution of the PS. The oxidation of membranes, especially cytoplasmic, is lethal for the cell and is an event preceding necrotic death due to photoexcitation.

The cell membranes photodamage resulting from the excitation of a PS is visualized through the penetration of propidium iodide (PI) into the cell, and such membrane permeability is irreversible and is a marker of cell necrosis [7]. Emulsion **4b** (final concentration of a 10 μM PS, 24h) accumulates in HCT116 cells and, after illumination, after illumination, lipid peroxidation leads to cell membrane damage. Propidium iodide penetrates into such damaged cells and intercalates in the nuclei - this necrotic effect is visualized in fluorescence microscopy (Figure 5). Fluoroporphyrin **2b** is not detected by a fluorescence microscope, which does not interfere with the signal of PI.

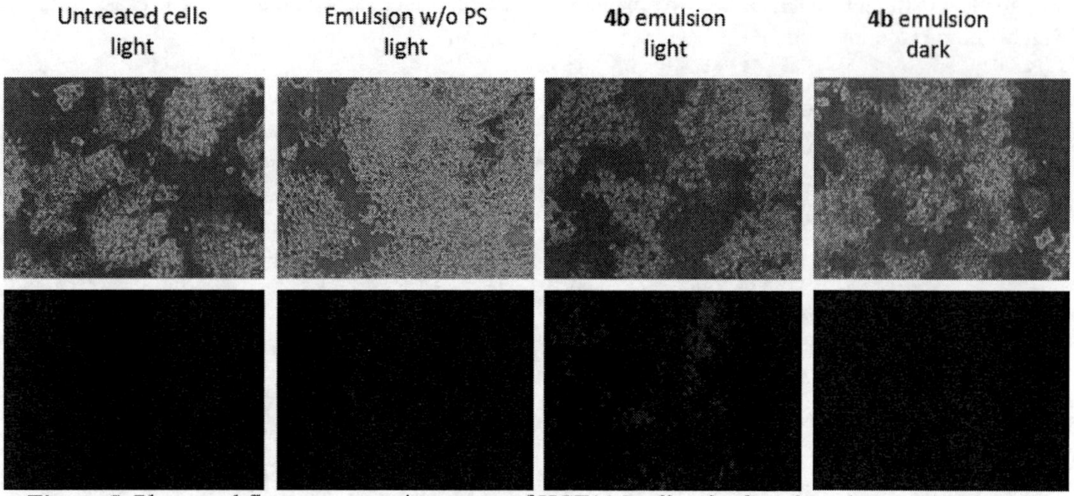

Figure 5. Phase and fluorescence microscopy of HCT116 cells, 1h after photodamage (633nm, 33 J/cm^2), propidium iodide staining.

PI-positive cells in which necrosis occurred due to photoexcitation **4b** of the emulsion was visualized by fluorescence microscopy 60 minutes after photoexcitation. We added PI to the cell buffer to the cells that accumulated the **4b** emulsion, and performed a photoexcitation of the **4b** PS with a 405 nm laser and a dose of 33 J / cm2 using a confocal microscope, recording the dynamics of PI entering the cell nucleus from the buffer. It was revealed that after photoexcitation the minimum time of PI entry is 50 minutes (Figure 6).

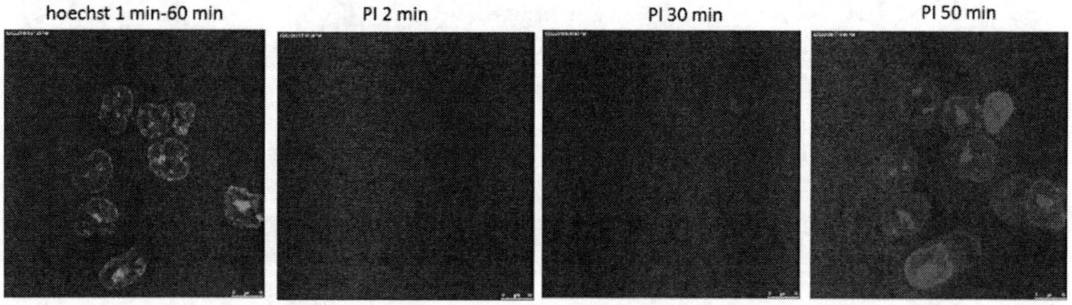

Figure 6. Nuclei staining of HCT116 cells that accumulated emulsion **4b**, (final concentration of PS 10 μM) and 2-50 minutes after photoexcitation (405nm, 33 J / cm^2). Green - hoechst Ex405Em415-485, Red - PI Ex488Em560-630.

Thus, photoinduced necrosis is the mechanism of action of fluoroporphyrin **4b** in emulsion form. An accumulation of emulsion **4b** in the cells and its cytoplasmic distribution takes place within 24 hours, peroxidation of membrane lipids during 2 minutes after photoexcitation at a dose of 33 J / cm2, and after 1 hour, cell necrosis. Necrosis develops irreversibly, as shown by staining with a propidium iodide - necrosis marker.

Conclusions

We synthesized fluorine-containing photosensitizers of the porphyrin class, soluble in fluorocarbons, for the photodynamic treatment of tumors. The introduction of fluorine substituents into the structure of the porphyrin cycle made it possible to achieve the solubility of photosensitizers in fluorocarbons, gas-transporting components of emulsions. The prepared fluorocarbon emulsions containing a fluorinated photosensitizer showed low dark cytotoxicity and submicromolar photoinduced cytotoxicity on human HCT116 adenocarcinoma cells. Emulsions with a photosensitizer accumulate in the cells within 24 hours and within 72 hours do not show a dark damaging effect. With photoexcitation of the photosensitizer in the emulsion, it was possible to achieve peroxidation of the cell membranes, including the cytoplasmic one, and to cause rapid photonecrosis, which develops within 1 h after photoexcitation. An emulsion without a photosensitizer is inert to cells for 72 hours both in the dark and after illumination. The data obtained make it possible to characterize fluorocarbon emulsions as promising compositions for placing fluorinated photosensitizers in them for photodynamic therapy of tumors.

The materials and methods used in the work are given in the Supplementary materials.

Acknowledgments

The reported sinthetical and biological study was funded by RFBR according to the research project № 18-315-00432. NMR studies, spectral characterization, elemental analysis were performed with the financial support from Ministry of Science and Higher Education of the Russian Federation using the equipment of Center for molecular composition studies of INEOS RAS.

References

[1] Scheer A, Kirsch M and Ferenz K B 2017 *J Nanosci Nanomed.* **1** 21–27
[2] Belyaeva E V, Markova A A, Kaluzhny D N, Sigan A L, Gervitz L L, Ataeva A N, Chkanikov N D and Shtil A A 2018 *Anticancer. Agents Med. Chem.* **18** 617–627
[3] Grancho J C P, Pereira M M, Miguel M D G, Gonsalves A M R and Burrows H D 2002 *Photochem. Photobiol.* **75** 249–256
[4] Day R A, Estabrook D A, Logan J K and Sletten E M 2017 *Chem. Commun.* **53** 13043–13046
[5] Yuhao Cheng, Hao Cheng, Chenhiao Jiang. Xuefeng Qiu, Kaikai Wang, Wei Huan, Ahu Yuan, Jinhui Wu and Yiqiao Hu 2015 *Nature communications* **6** 8785
[6] Wrobeln A, Laudien J, Groß-Heitfeld C, Linders J, Mayer C, Wilde B, Knoll T, Naglav D, Kirsch M and Ferenz K B 2017 *Eur. J. Pharm. Biopharm.* **115** 52–64
[7] Moisenovich M M, Ol'shevskaya V A, Rokitskaya Y I, Ramonova A A, Nikitina R G, Savchenko A N, Tatarskiy Jr. V V, Kaplan M A, Kalinin V N, Kotova E A, Uvarov O A, Agapov I I, Antonenko Yu N and Shtil A A 2010 *PLoS One* **5** e12717

Carbon-forming properties of oils influencing the service of engines and aircraft reducers

I S Melnikova[1], L V Kovba[1], D V Ratenko[1] and V S Dvoyeglazov[2].

[1]Federal State Unitary Enterprise State Research Institute of Civil Aviation, Moscow, Russia
[2]JSC "Ural Civil Aviation Plant", Moscow, Russian Federation

E-mail: melnikova_is@bk.ru; universal358@mail.ru; dvoeglazov@uwca.ru

Abstract. During the operation of gas turbine engines (GTE), failures are encountered due to structural, production, and operational defects. According to the damage of the gas turbine engine, the percentage of malfunctions in the lubrication and venting system is 2.9% [1]. This paper discusses the most common problematic issues in the operation of oil systems (OS) associated with the formation of solid decomposition products of oils on the surfaces of gas turbine engine parts. Laboratory studies of the creation of high-temperature carbon deposits on a heated aluminum surface were carried out by spraying the test oil on it for a specific time on the tilt plate instrument. This instrument was used to study NYCO TURBONIKOIL 98 aviation oil for aviation gas turbine engines (GTD-350, TV2-117, TV3-117, VK-2500) and main gearboxes (VR-2, VR-8A, VR-14, VR- 252) aircraft (Mi-2, Mi-8T, Mi-8MTV / AMT, Ka-32). Quantitative dependencies of the formation of carbon deposits on the oil temperature and the temperature of the aluminum plate are obtained, and a color template of the intensity of formation of solid high-temperature carbon deposits depending on the oil temperature and the temperature of the aluminum plate is presented.

1. Introduction

One of the most critical elements for ensuring reliable operation and engine life is a correctly selected oil, considering the conditions of its activity in the engine oil system. The oil is a lubricating element of the engine design that simultaneously provides heat dissipation from engine hot spots while undergoing thermal, mechanical, and chemical influences that affect the quality of the oil and its ability to perform its functions for a long time and reliably.

There are many different designs of engines. However, the construction of their lubrication systems is based on a single principle. It is providing forced lubrication under pressure of the most critical and loaded bearings, gear, and spline joints, provided that the oil is reused and its residence time is reduced to reduce oil heating, saturation its air, and reduce consumption.

Typically, the operating temperature range of lubricating oils is usually characterized by the temperature of the oil at the inlet and outlet of the engine at a steady state. According to the design bureaus (DB) and equipment manufacturers, the operating temperature range for most engines is in a narrow range of 80-130°C. However, the oil temperature at the engine outlet does not reflect the exact temperatures of local oil overheating in individual friction units and may exceed it by 10-60 and more degrees [2].

The oil experiences the highest temperature loads in a turbojet engine in contact with the bearings of the high-speed shafts of the compressor and turbine (the temperature of the compressor bearings

IOP Conf. Series: Materials Science and Engineering **848** (2020) 012055 doi:10.1088/1757-899X/848/1/012055

can reach 250°C, and turbines up to 250-350°C). In turboprop engines, specific loads also have a significant effect on oil, especially in the nodes of the propeller gearbox during intensive pumping in the system [3].

A generalization of operating experience shows that the main failures and malfunctions of the lubrication and venting system during operation are [4]:

1. Increased oil consumption and, as an external manifestation, smoke, and oil leak (Fig. 1a, b, c). Possible causes of this effect are a violation of the tightness of the system due to the appearance of cracks, destruction of pipelines, or weakening of their connection.

The consequences of a similar effect: there is a gradual decrease in pressure and an increase in oil temperature, further overheating of engine components, up to destruction from overheating of the most loaded units (bearings of the middle and rear rotor bearings, gears and bearings of the drive box) and, as a result, engine shutdown due to jamming of the rotor or disconnection of the kinematic chain of the drives.

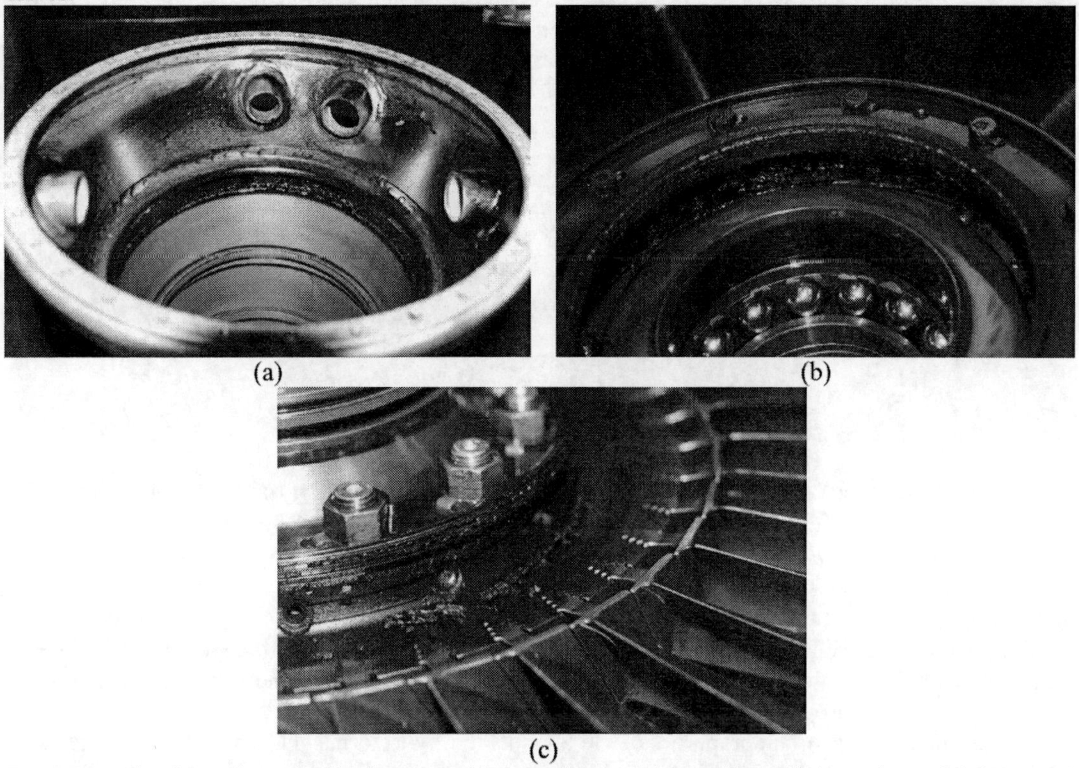

(a) (b) (c)

Figure 1. The formation of solid decomposition products of the working oil due to loss of tightness of the node: a) on the details of the 2nd support of the TV3-117 helicopter engine, b) and c) on the details of the labyrinth seal of the free turbine of the TV2-117 engine.

2. Supercharging of venting system with air or hot gas of high pressure.

Possible reasons for this effect are the destruction of the oil labyrinth seals of the bearing planes (Fig. 2a and 2b) of the rotor bearings; leaks of oil communications passing through the air or gas cavities of the high-pressure engine.

The possible consequences of this effect: the release of oil from the system through the safety valve of the oil tank or centrifuge. Further consequences are similar to those described above.

3. Disruption of oil circulation in the system.

Possible reasons for this effect:

- failure (destruction) of pressure head oil pumps.

Possible consequences of this effect: complete cessation of oil circulation, which is accompanied by a sharp drop in oil pressure to zero.

IOP Conf. Series: Materials Science and Engineering **848** (2020) 012055 doi:10.1088/1757-899X/848/1/012055

Failure of pumping oil pumps.

Possible consequences of this effect: stopping pumping oil from the cavities of the oil system, overfilling it with oil and ejecting oil from the system;

- destruction of pressure oil communications or blockage of nozzles (nozzles) leading oil to individual nodes.

As a result of the disturbance of circulation, possible consequences may be overheating and destruction of nodes, the flow of oil to which has stopped. With the continued operation, the engine may turn off due to the overheating of the most loaded units.

2. Experimental part

The oil system operability of the power plant mainly depends on correctly selected (for the real conditions of lubrication of the friction surfaces of the engine) oil. It provides lubricity and sufficient heat removal from its units and parts during the assigned resource. The cessation and reduction of the oil supply, regardless of the cause, leads to overheating of the engine mounts, the destruction of its bearings, jamming of the rotor of the gas turbine engine until the engine turns off.

(a) (b)

Figure 2. a) and b) Formation of solid decomposition products of the working oil on the contact seal details of the 1st support of the TV3-117 engine.

The temperature conditions deterioration of the work of rubbing parts leads to the use in the operation of oil that does not have a margin of quality in terms of thermo-oxidative stability and has a high ability to form carbon.

For the correct selection of oil, it is necessary to conduct tests not only in the volume of the internal standard (GOST) but also by qualification methods. At the same time, the practical experience of using oils on aircraft products should be taken into account.

One of the most important properties of the oil is its ability to form soot, which directly affects the ability of the oil to remove heat and prevent local overheating as follows from the materials presented.

Previously, the organization GOSNII GA conducted studies of the carbon-forming properties of oils [5, 6]. The positive experience has been gained in producing results. However, when studying the issue of carbon-forming properties of aviation oils, quantitative standards for the resulting carbon deposits were not established under various temperature conditions of oil on a metal surface, the appearance of carbon deposits and plates was not evaluated, and accordingly, a colour intensity template was not created. In addition, over time, there have been changes associated with the advent of new equipment with more loaded units, assemblies and with an increase in the assortment of various brands of oils on the market, which affected the actual operating conditions.

Therefore, we propose an improved methodology for assessing the carbon-forming properties of oils, taking into account actual operating conditions, using NYCO TURBONIKOIL 98 as the most widely used helicopter in civil aviation in this research.

The study of the formation of solid decomposition products of aviation oil TURBONIKOIL 98 from NYCO was chosen using the method from the list of sets of methods for qualifying oils of the Interdepartmental Commission for the admission to the production and use of fuels, oils, lubricants and special fluids under the State Standard of the Russian Federation “The tendency of oils to form high-temperature deposits "On the device" inclined plate" (GMK Decision No. 23 / 1-54 of 02/14/1979). The research was conducted on a device improved by automation.

On the “inclined plate” device, the ability of oil to form carbon is evaluated under conditions of short-term contact of the oil with a heated metal surface, which makes it possible to assess the possibility of the oil working in the lubrication system close to the actual engine operating conditions and to determine the mass amount of solid oil decomposition products at various working oil temperatures and washed parts. For this, the research conditions were expanded: the number of revolutions of the sprayer, the mode of spraying oil, the temperature of the oil, the temperature of the plate.

The essence of the method is to constantly spray heated oil onto a metal plate. The temperature of the plate is maintained at 250-350°C with a heating time of the plate for 6 hours.

As can be seen from the device diagram (Fig. 3), a roller with a metal comb of eight rows of teeth is located in the oil reservoir (2). The oil tank is situated at an angle of 25° to the horizontal. Oil in a volume of 300 cm^3 from the cylinder is poured into the tank. The teeth of the roller combs are partially submerged. During operation, the roller rotates at a constant speed of 1500 ± 50 rpm. The oil splashes onto the aluminium plate at the top (1). Part of the oil lingers on the plate; part flows back into the reservoir. The temperature of the plate is maintained by an electric heater (3) located on it. The test duration is 6 hours. The oil temperature during the test is 120-180°C, the temperature of the plate is 250-350°C. The set temperature is maintained with an accuracy of ± 5°C. The amount of oil soot is determined by the weight of carbon deposits formed on the aluminium plate.

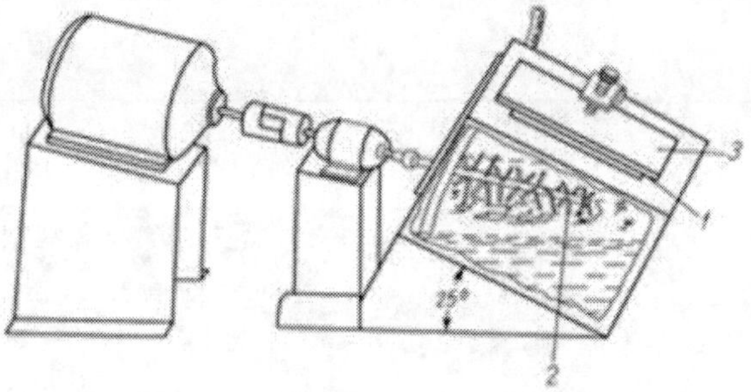

Figure 3. Diagram of a device for determining solid decomposition products of oils: 1 – aluminium plate, 2 – rotating oil spray, 3 – electric plate heater.

3. Results and discussion

Figure 4 showing the dependence of the mass of deposits on the plate temperature for different oil temperatures, equidistant within the experimental error. The curves presented in Fig. 4 show that the mass of deposits depends significantly on the temperature of the plate.

The presenting curves (Fig. 4) show that the intensity of carbon deposits on the parts of the bearing assemblies is not directly dependent on the oil temperature in the range from 120°C to 180°C. The amount of solid oil decomposition products formed increases with increasing temperature of the plate to 350°C, at oil temperatures of 120°C, 140°C, 160°C, 170°C to 320-330°C.

The heat capacity of oil vaporisation plays a decisive role in these conditions. So, at a constant intensity of oil spraying, an increase in thermal energy supplied to the plate, upon reaching a critical value, leads to the boiling of oil droplets on the plate and the beginning of the carbonation process. The formation of a significant amount of carbon deposits (70-90 mg under the conditions of this experiment) reduces the thermal conductivity of the plate. Therefore, the amount of heat transferred to the oil decreases, and the growth of deposits slows down or completely stops according to [7]. The mass of carbon deposits is reduced because the process of "burnout" of the oil begins at a temperature of 340°C and above.

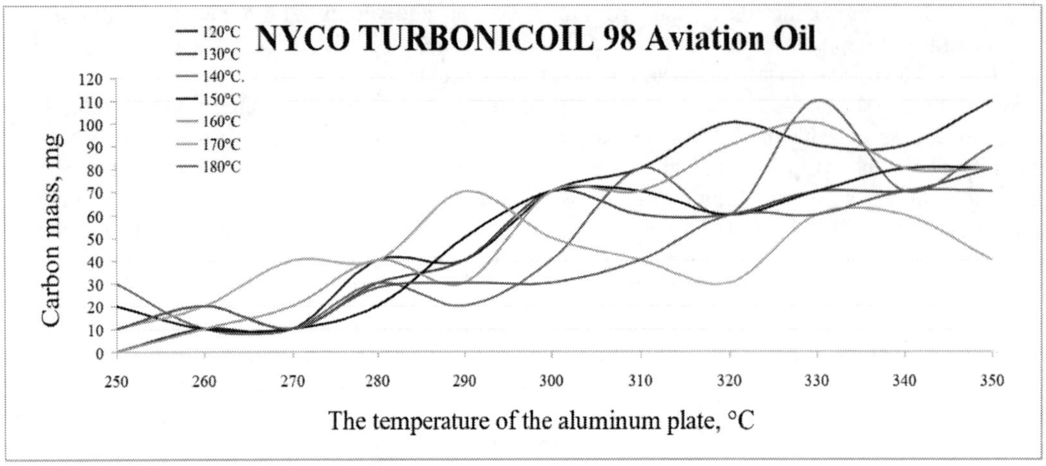

Figure 4. Dependence of the temperature of the aluminium plate and oil on the mass of solid decomposition products of oil (soot)

In the process of studying the dynamics of the formation of solid decomposition products of NYCO TURBONICOIL 98 aviation oil, a regular change in the appearance and physical properties of carbon deposits was established depending on various temperature conditions. In fig. 5 are photographs of carbon deposits obtained on aluminium plates.

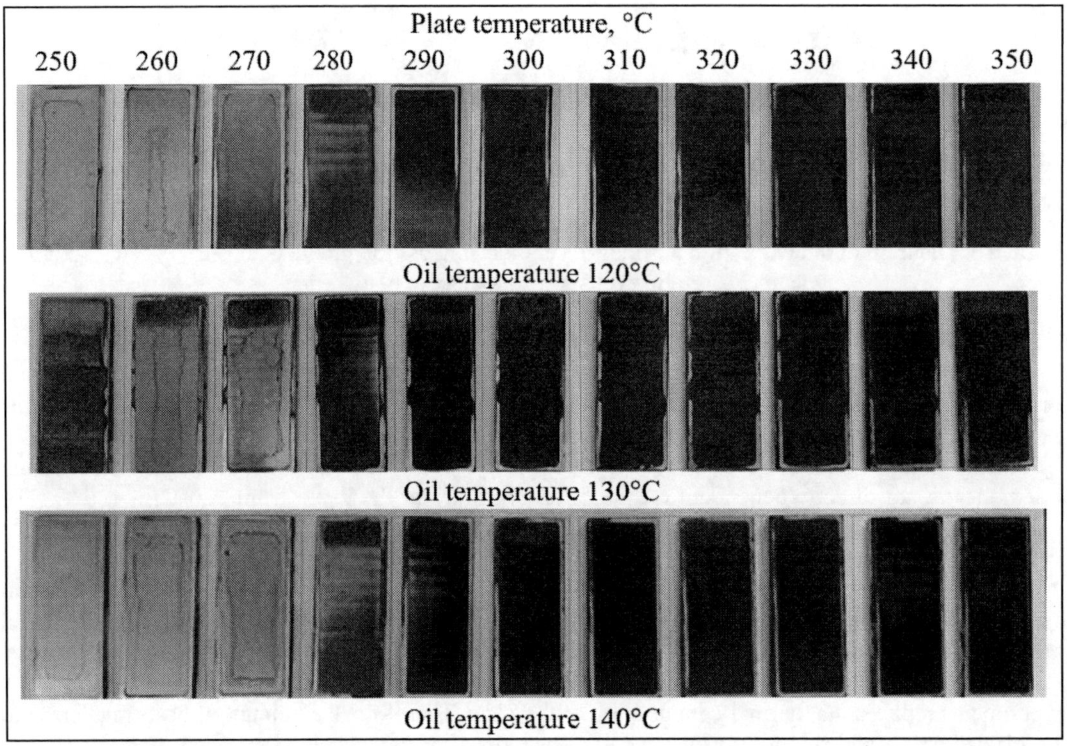

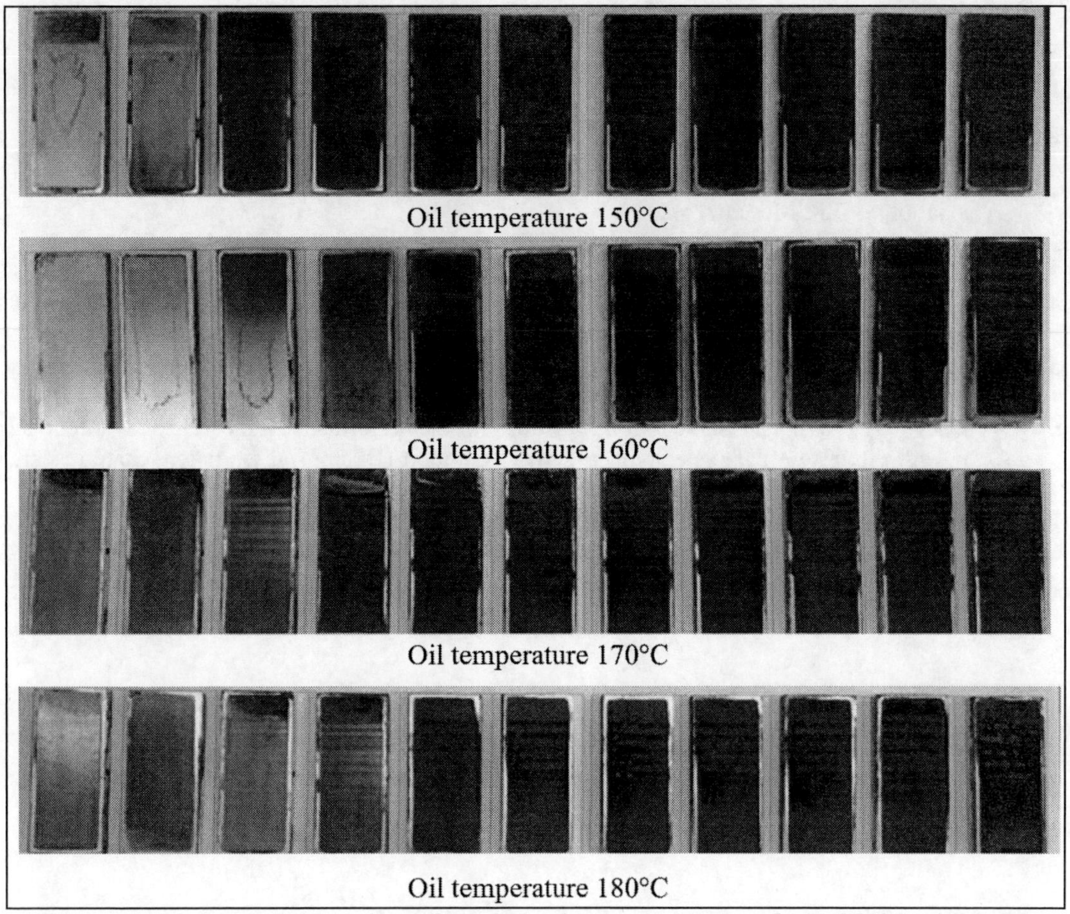

Figure 5. Pictures of aluminium plates heated at temperatures of 250-350°C with solid decomposition products NYCO TURBONIKOIL 98 oils in the temperature range 120-180°C

It was noted analysing the data obtained that at a plate temperature from 250°C to 270°C, a thin layer of deposits is formed in the form of varnish. It has a smooth, shiny surface of light yellow and brown colour. With increasing temperature of the plate from 280°C to 350°C, the nature of the deposits changes from an even thin layer to a matte one. With increasing temperature of the plate above 300°C, deposits become thin black soot and black with interspersed carbon particles, which give some surface roughness and the formation of a glossy black surface in the upper part of the plate.

4. Conclusion

It can be concluded based on the work carried out that the materials obtained will make it possible to predict the ability of the oil to perform its functions for a long time and reliably protecting the engine components and assemblies from overheating and, accordingly, to ensure reliable operation and engine life under various temperature conditions of its service.

In our opinion, it is advisable to continue the work on the accumulation of materials for a comparative assessment of the characteristics of carbonisation and the formation of solid decomposition products of various oils. The results obtained can be useful in the development of engine modes, the selection of new and modified oils with the introduction of unprincipled changes in the previously adopted production technology, for periodic verification of commissioned oils that have already been put into operation, as well as for the development of an active program of measures to eliminate failures.

IOP Conf. Series: Materials Science and Engineering **848** (2020) 012055 doi:10.1088/1757-899X/848/1/012055

References

[1] Sirotin N N 2002 Design and operation, damage and operability of gas turbine engines (Moscow: RIA "IM-INFORM") p 439
[2] Mikheichev P A 1982 The development of optimal service life of aviation oils in gas turbine engines of civil aviation aircraft (Doctoral dissertation, State research institute of civil aviation, USSR, Moscow) p 202
[3] Piskunov V A, Zrelov V N, Vasilenko V T, Litvinov A A, Chernova K S 1983 Directory. Chemotology in civil aviation (Moscow: Transport) p 248
[4] Alexandrov V G, Glazkov Yu A, Alexandrov A G, Sidrkin P N 1976 Monitoring the technical health of aircraft and helicopters. Directory (Moscow: Transport) p 360
[5] Kaverina N I, Lebedev O A, Kozlova N N, Shershebkov L A 1973 Trudy gosudarstvennogo nauchno-issledovatel'skogo instituta grazhdanskoy aviatsii 93 102
[6] Lebedev O A 1987 Establishment of limit values for the performance indicators of synthetic oils in order to ensure their operation in actual condition (Doctoral dissertation, State research institute of civil aviation, USSR, Moscow) p 171
[7] Papok K K, Piskunov V A, Yurenya P G 1971 Carbon deposits in jet engines (Moscow: Transport) p 112

Research of the intermetallics formation mechanism during the synthesis of functionally graded layered steel-aluminum compositions

R S Mikheev[1], I E Kalashnilov[2], L K Bolotova[2], A G Kolmakov[2]

[1] Faculty of Engineering Technology, Bauman Moscow State Technical University
5, 2-ya Baumanskaya str., Moscow, 105005, Russian Federation
[2] Laboratory of Strength and Plasticity of Metallic and Composite Materials and Nanomaterials, Baikov Institute of Metallurgy and Materials Science of Russian Academy of Sciences
49, Leninskiy prospect, Moscow, 119334, Russian Federation

E-mail: mikheev.roman@mail.ru

Abstract. The features of the formation and growth of intermetallic compounds in the production of functionally graded layered steel-aluminum compositions by the arc surfacing process have been revealed. The influence of the intermediate aluminum layer formed by the arc surfacing process on the diffusion zone formation mechanism has been shown.

1. Introduction

The friction unit elements (bearing shell, blocks of sliding bearings and others) responsible for the operating mode and resource of machines and mechanisms are mainly bimetallic structures [1-3]. The bimetallic structure consists of working antifriction layer and low-carbon steel substrate which restricts plastic deformation. However, the level of performance properties of existing antifriction materials from non-ferrous metal alloys (Al, Cu, Pb, Sn, etc.) has reached maximum values, and the potential for further improvement due to alloying or special processing has exhausted itself, as evidenced by the analysis of plain bearings emergency failure [4, 5]. Therefore the use for the working layer of new particle-reinforced composite materials (CM) based on aluminum is a promising and effective solution. The CM have both high performance properties and low specific weight, and also the possibility of multiple recovery by surface treatment [6, 7]. The use of fundamentally new functionally graded layered compositions with CM working layers will not only improve the level of performance properties, but also will reduce the friction units cost. However, the main obstacle limiting the wide using of such compositions is associated with the low adhesive strength between the coating and the substrate due to the formation of an iron-aluminum system intermetallic brittle layer at the interface [8]. Therefore, the study of the intermetallic layer formation mechanisms is important to overcome the obstacle when developing technologies for applying aluminum matrix composite coatings to steel substrates and their successful operation which has been the purpose of this work.

2. Materials and methods

Composite coatings were deposited on plates of size 150x200x2 mm made of high-quality low-carbon steel 20 (0,17-0,24 wt.% C; 0,17-0,37 wt.% Si; 0,35-0,65 wt.% Mn; ≤0,25 wt.% Cr; Fe - the rest according to GOST 1050), by an argon arc surfacing process, characterized by universality and efficiency. Preliminarily, an intermediate layer of A5 grade aluminum (0,2-0,35 wt.% Fe; 0,1-0,25 wt.% Si, ≤0,015 wt.% Cu; Al - the rest according to GOST 7871) had been formed on the steel substrate surface. According to previous studies the intermediate layer allows to limit interfacial interaction between the substrate material and the CM matrix [8]. The use of pure aluminum for the intermediate layer made it possible to take into account the influence on the formation mechanism and diffusion zone characteristics of alloying elements from only the filler metal composition.

A promising liquid phase method of arc surfacing - the process of "Cold Metal Transfer" (CMT), was used to obtain the aluminum intermediate layer. CMT process is characterized by minimal heat input and allows limiting the growth of intermetallic compounds. CMT technology was carried out on specialized welding equipment TransPuls Synergic 2700, Fronius using a A5 grade welding wire with a diameter of 1,2 mm. The CMT technological parameters (pulse current - 150 A; base current - 40 A; voltage - 18 V; surfacing speed – 0,5 m/min) ensured the formation of an intermediate layer up to 2 mm thick and were selected according to [9]. Previously the steel plate's surface was treated by hot dip galvanizing in order to improve its wetting with aluminum melt.

The presence or absence of high-strength, refractory reinforcing particles does not affect the formation and growth mechanisms of intermetallic compounds, and first of all determines the CM characteristics level. Therefore, more cheaper materials similar in chemical composition to CM matrix alloys in the form of OK Tigrod 4047 rods (≤0,6 wt.% Fe; 11-13 wt.% Si; ≤0,15 wt.% Mn; ≤0,05 wt.% Cu; Al - the rest according to AWS A5.10) with a diameter of 2,4 mm were used. The process of argon arc surfacing of the matrix alloy was carried out according to the scheme shown on Figure 1. Technological parameters of the process, the main of which are welding current (I) - 150-160 A; arc voltage (U) - 18-20 V; surfacing speed (V) - 11-13 m/h, were selected according to [7], and provided full remelting of the intermediate layer, i.e. contact of the melt with a solid substrate.

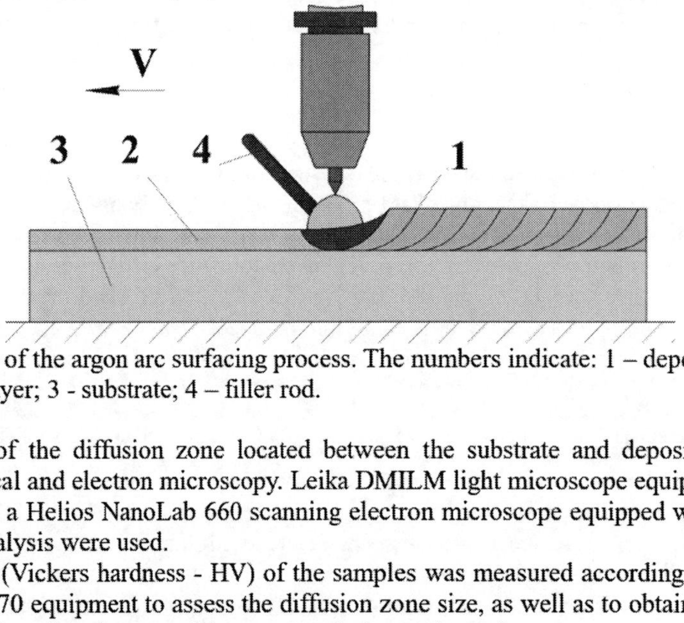

Figure 1. Scheme of the argon arc surfacing process. The numbers indicate: 1 – deposited layer; 2 - intermediate layer; 3 - substrate; 4 – filler rod.

The structure of the diffusion zone located between the substrate and deposited coatings was studied using optical and electron microscopy. Leika DMILM light microscope equipped with a digital camera, as well as a Helios NanoLab 660 scanning electron microscope equipped with an attachment for X-ray microanalysis were used.

Microhardness (Vickers hardness - HV) of the samples was measured according GOST 9450 with using a DuraScan 70 equipment to assess the diffusion zone size, as well as to obtain quantitative data on the characteristics of the intermetallic compounds that make up it.

The characterization of the functionally graded layered compositions properties was carried out by assessing the adhesive strength of the deposited coatings according to the results of peeling and shear tests according to [10]. These test schemes were chosen due to their wide using in industry for bimetallic structures of friction units and composite bimetallic materials manufacturing. The selected schemes were carried out using a tensile testing machine 2054 R-5. The load was applied stepwise until the samples were destroyed.

3. Results and discussion

During CMT process, the continuous intermetallic layer with a thickness of 4 to 10 μm (an average value of 8,2 μm) is formed at the interface between the steel substrate and the pure aluminum intermediate layer (Figure 2a). It should be noted that the longitudinal cracks dividing the intermetallic layer into two regions are observed along the entire length of the interface. The region located near the deposited layer has an average thickness of 3,6 μm and according to the X-ray microanalysis data consists of intermetallic compounds based on aluminum $FeAl_3$ and Fe_2Al_5 (Figure 2b). These phases have maximum hardness values among intermetallic compounds of the Fe-Al binary system [11].

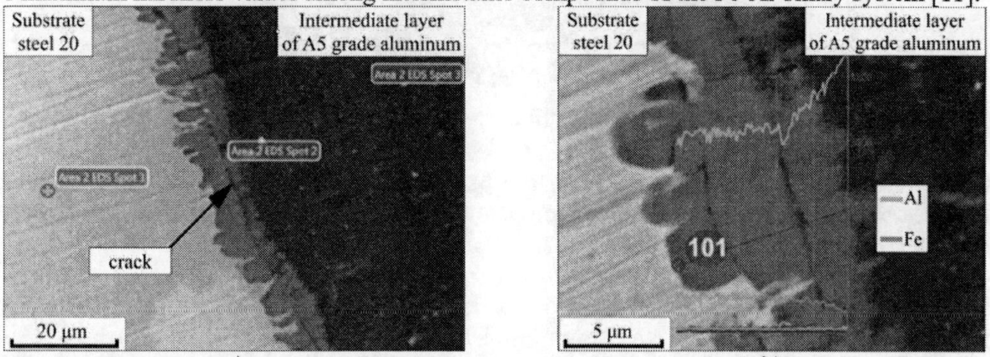

a) b)

Figure 2. Microstructure (a) and chemical elements distribution (b) in the intermetallic layer between the steel substrate and intermediate layer of A5 grade aluminum.

The formation of the $FeAl_3$ intermetallic phase occurs upon contact of the aluminum melt with solid steel due to the dissolution of iron in aluminum [8, 11, 12]. Subsequently, the Fe_2Al_5 compound is formed on the steel side as a result of chemical reaction:

$$FeAl_3 + Fe = Fe_2Al_5 \quad (1)$$

Iron-based intermetallic compounds (Fe_2Al_3, FeAl, Fe_3Al) are located in the region adjacent to the steel substrate. These intermetallics are characterized by the shape of tongue-like outgrowths in the direction of steel (Figure 2b). They form a continuous layer of variable thickness, the average value of which is 4,6 μm.

Different values of the average thickness of the intermetallic layer regions formed by compounds based on aluminum and iron, respectively 3,6 and 4,6 μm, are probably associated with a longer duration of the diffusion process of aluminum into iron compared with the opposite direction of diffusion of iron into aluminum. The unevenness of the intermetallic layer thickness is caused by the anisotropy of the intermetallic phase Fe_2Al_5 diffusion properties. The Fe_2Al_5 phase is characterized by the ability to pass aluminum atoms in the direction perpendicular to the interface. The maximum values of the intermetallic layer thickness on the steel substrate side are observed under the conditions of the Fe_2Al_5 crystal growth along the normal to the surface (Figure 3). The crystal growth deviation from the normal direction leads to a decrease in the growth rate until it is completely stopped.

The formation of intermetallic compounds is accompanied by a significant increase in the final phase volume, which reacts with volumes of steel and aluminum. The new intermetallic compounds layers have an effect on previously formed ones. Significant stresses arising as a result of such an action can lead to the appearance of longitudinal cracks in the intermetallic layer. It explains their presence in the obtained samples, and is consistent with the results of [11].

The thickness of the 4047 aluminum alloy coatings deposited by an argon arc welding process with full remelting of the intermediate layer is up to 5 mm. The main feature in this case is the interaction of the silicon-containing aluminum melt not with the surface of the steel substrate, but with the intermetallic layer existing at the steel-aluminum interface formed during the intermediate layer deposition.

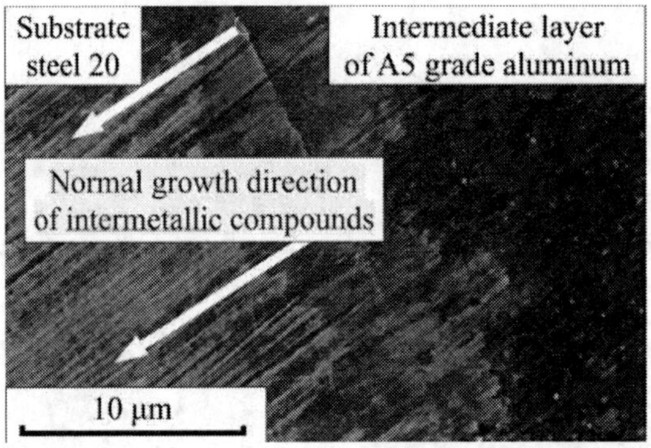

Figure 3. Scheme of the intermetallic compounds growth in the normal direction to the interface.

The significant changes occur at the interface between the obtained samples compared with the initial state. As a result two characteristic regions can be distinguished in them (Figure 4a):

- region I, resulting from the complete remelting of the aluminum intermediate layer;
- region II subjected to thermal influence during argon arc welding.

The intermetallic layer formed in region I, limited by the dimensions of the weld pool in contact with the solid steel, is continuous and uneven in thickness (Figure 4b). Its thickness values are in the range from 7 to 22 μm (average 18 μm). In region II, formed as a result of heating above the temperatures of intensive growth of the Fe-Al system intermetallic compounds and characterized by a length of up to 400 μm from the deposited metal, the thickness of the intermetallic layer is from 6 to 14 μm with an average value of 12 μm (Figure 4c).

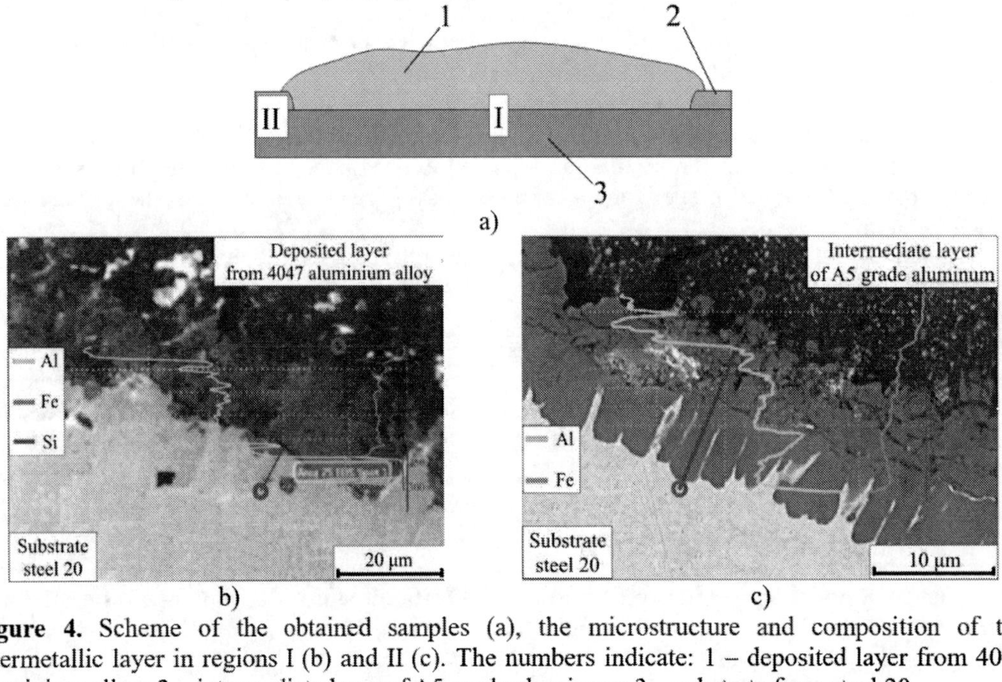

Figure 4. Scheme of the obtained samples (a), the microstructure and composition of the intermetallic layer in regions I (b) and II (c). The numbers indicate: 1 – deposited layer from 4047 aluminium alloy; 2 – intermediate layer of A5 grade aluminum; 3 – substrate from steel 20.

The complete remelting of the intermediate layer during surfacing leads to the partial dissolution of the Fe-Al binary system intermetallic layer formed during the aluminum intermediate layer deposition, as well as the development of silicon diffusion into it, as evidenced by the X-ray microanalysis results (Fig. 4, b). The maximum silicon concentration is achieved in the region of the diffusion zone located near the deposited aluminum alloy 4047 coating, and consisting of the $Al_{7,4}Fe_{1,8}Si$ intermetallic phase. The silicon amount in the diffusion layer composition decreases with further movement in the steel substrate direction. It indicates the appearance of the $Fe(Al, Si)_3$ compound. The formation of such ternary phases of the Fe–Al–Si system occurs due to the substitution of aluminum by silicon and occupation by it of structural vacancies in the initial Fe–Al binary intermetallic compounds (Fe_2Al_5, $FeAl_3$) [13-15]. It is important to note that the result of this is a decrease in the rate of aluminum and iron diffusion [15-17].

Full remelting of the Fe-Al binary system continuous intermetallic compounds layer located at the steel-aluminum interface is practically excluded during surfacing. The temperature of the weld pool which prevents direct exposure to a high-temperature electric arc does not exceed 1173 K according to [18]. This temperature is significantly lower than the intermetallic compounds melting temperature (Table 1).

Table 1. The melting temperature of the Fe-Al system intermetallic compounds of various stoichiometric composition [19]

Intermetallic compound	Intermetallic compound composition (in parts)	Melting temperature, K
FeAl	$Fe_{1/2}Al_{1/2}$	1583
$FcAl_2$	$Fe_{1/3}Al_{2/3}$	1365
$FeAl_3$	$Fe_{1/4}Al_{3/4}$	1430
Fe_2Al_5	$Fe_{2/7}Al_{5/7}$	1444
Fe_3Al	$Fe_{3/4}Al_{1/4}$	1825

Thus, the intermetallic layer formation mechanism during surfacing of coatings with full remelting of the intermediate aluminium layer obtained by the CMT process can be represented as follows (Figure 5):

1) at the first stage, the aluminum-silicon melt is wetted and spreads over the continuous intermetallic layer surface (Figure 5a). At the same time, the surface layer of intermetallic compounds having the $FeAl_3$ composition dissolves. It subsequently provides the possibility of physical contact of the melt with a layer of Fe_2Al_5 intermetallic compounds located near steel substrate (Figure 5b).

2) the second stage is characterized by the development of the diffusion process of silicon into the crystal lattice of Fe_2Al_5 intermetallic compounds which occurs by the vacancy mechanism. The silicon penetration depth achieved at this stage is almost comparable with the entire thickness of the intermetallic layer. As a result triple intermetallic compounds - θ-$Fe(Al, Si)_3$ with a melting point of 1373 K are formed (Figure 5c).

3) the third stage is associated with the formation on the site of the dissolved compound of the composition $FeAl_3$ of the intermetallic phase τ_5-$Al_{7,4}Fe_{1,8}Si$, which has a melting point of 1123 K (Fig. 5, c). At elevated temperatures (above 893 K), this compound is formed by a monovariant peritectic reaction [13, 14]:

$$L + \theta \rightarrow \tau_5, \quad (2)$$

The τ_5 phase crystallizes simultaneously with θ-$Fe(Al, Si)_3$ intermetallic compounds according the reaction. At temperatures less than 893 K the formation of the τ_5 phase occurs in accordance with the quasi peritectic reaction [12]:

$$L + \theta \rightarrow \tau_5 + (Al) \quad (3)$$

It is important to note that the formation of a continuous intermetallic layer of the τ_5-$Al_{7,4}Fe_{1,8}Si$ compound at the coating-substrate interface prevents the further growth of the θ-$Fe(Al, Si)_3$ phase.

4) the fourth stage is accompanied by an increase in the size of the τ_5-$Al_{7,4}Fe_{1,8}Si$ intermetallic compounds located near the deposited coating and crystallization of the aluminum-silicon melt (Figure 5d).

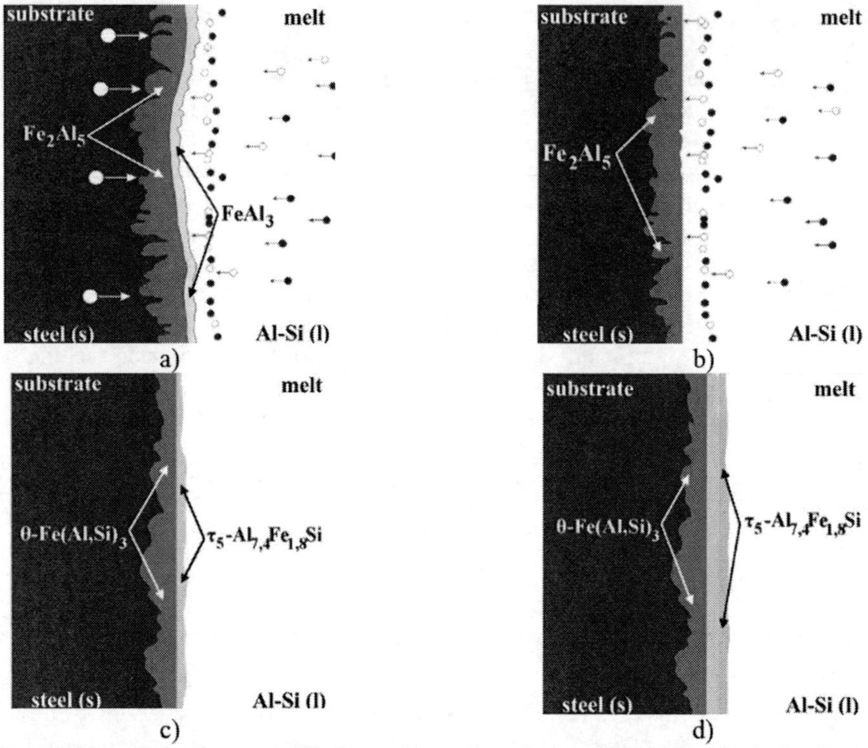

Figure 5. Scheme of the intermetallic layer formation during surfacing of an aluminum-silicon coating on steel with full remelting of the intermediate aluminum layer obtained by the CMT process. ● Atoms of Al and Si melt; ○ Atoms of Fe.

The phase composition of the intermetallic layer located in region II is characterized the similar structure to that formed when the intermediate layer of grade A5 aluminum is produced by the CMT process (compare Fig. 2, b and Fig. 5, b). The main differences are related to its thickness, as well as the presence of numerous cracks in it. The heating of the intermetallic layer located in region II during arc surfacing to temperatures exceeding the onset of intensive growth of intermetallic compounds leads to increasing of up to 50% (from 8 to 12 μm) of its average thickness compared to the initial state. The intermetallic layer located near the intermediate aluminum layer is characterized by many cracks. Their appearance is associated with high hardness and brittleness of the phases based on aluminum ($FeAl_3$ and Fe_2Al_5). The increase in the average thickness of the aluminum-based intermetallides interlayer reaches 66% (average thickness: 3,6 μm in the initial state, and 6 μm after arc surfacing). The other part of the intermetallic layer located near the steel substrate has a comparable average thickness (the maximum increase was 30% of the initial value: from 4,6 to 6 μm) and consists of iron-based Fe-Al binary intermetallic compounds (Fe_2Al_3, FeAl, Fe_3Al). The compounds formation and growth are associated with the diffusion process of aluminum into steel during arc surfacing process. These results confirm the earlier conclusion that in the solid phase, the diffusion process of iron into aluminum through the intermetallic layer has a higher speed compared to diffusion of aluminum into iron.

Tests for adhesive strength were carried out in order to assess the mechanical characteristics of the samples. The obtained values of adhesive strength did not exceed 15 and 25 MPa when tested for shear and peeling, respectively. Thus, both increase in the diffusion rate as a result of thermal action during the arc surfacing of an aluminum coating on steel with complete remelting of the intermediate layer of pure aluminum obtained preliminarily by the CMT process combined and uneven distribution of intermetallic compounds of various stoichiometric composition, lead to the thickness of the intermetallic layer increasing by up to 50% in comparison with the initial samples, as well as significant cracking mainly from the deposited layer side, whereby the adhesive strength values correspond to 25 MPa.

4. Conclusion

The mechanism for the formation and growth of intermetallic compounds upon contact of an aluminum-silicon melt with intermetallic compounds of the Al-Fe system during the process of arc surfacing of functionally graded layered steel-aluminum compositions with working coatings made of aluminum based CM is proposed.

It is shown that the CM matrix melt contacts with the Fe-Al system intermetallic layer located at the interface during arc surfacing on steel with a pre-deposited intermediate aluminum layer. The intermetallic layer partial dissolutes during the contact with CM matrix melt. The degree of its dissolution depends on temperature and time the existence of a weld pool. The dissolution rate of the initial intermetallic compounds is lower than the formation rate of new ternary Fe-Al-Si compounds. It leads to an increase from 8.2 to 18 μm in the intermetallic layer average thickness in the produced functionally graded layered steel-aluminum compositions, which can be recommended for use in friction units whose adhesive strength is 25 MPa.

References

[1] Yazykov A E, Kazanskiy V N and Buskunov R Sh 2010 *Bearings and Oil Systems: Tutorial* (Chelyabinsk: Cicero)
[2] Drokonov A M, Osipov A and Birykov A V 2009 *Turbine Bearings: Tutorial* (Bryansk: BGTU)
[3] Rubin M B and Bahareva V E 1987 *Bearings in Marine Technology: A Guide* (Leningrad:Sudostroenie)
[4] Gaziev R R, Zaharov N M, Babenko A F and Churagulova A G 2012 *Neftegazovoe Delo* **5** 236
[5] Pogodaev L I, Kuzmin A A and Lopyrev Y K 2015 *Reliability of Ship Equipment* (Saint-Petersburg: GUMRF)
[6] Chernyshova T A, Mikheev R S, Kalashnikov I E, Akimov I V and Kharlamov E I 2011 *Inorg. Mater. Appl. Res.* **2** 322-9
[7] Mikheev R S and Chernyshova T A 2008 *Zagotovitel'noye Proizvodstvo v Mashinostroyenii* **11** 44
[8] Kovalev V V, Mikheev R S and Kobernik N V 2016 *Vestnik MGTU. Seriya Mashinostroenie* **3** 134
[9] Zhang H T, Feng J C, He P, Zhang B B, Chen J M and Wang L 2009 *Mater. Sci. Eng. A* **1-2** 111
[10] Mikheev R S, Kobernik N V, Kovalev V V, Kalashnikov I E, Kolmakov A G, Bolotova L K and Kheifets M L 2019 *Svarka i Diagnostika* **2** 48
[11] Tanaka N, Nishimoto M, Kobayashi S and Takeyama M 2015 *Intermetallics* **7** 1
[12] Wang J, Lee P D, Hamilton R W, Li M and Allison J 2009 *Scr. Mater.* **7** 516
[13] Maitra T and Gupta S P 2003 *Mater. Char.* **49** 269
[14] Maitra T and Gupta S P 2003 *Mater. Char.* **49** 293
[15] Song J L, Lin S B, Yang C L and Fan C L 2009 *J. Alloys Compd.* **1** 217
[16] Dong H, Hu W, Duan Y, Wang X and Dong C 2012 *J. Mater. Process. Tech.* **2** 458
[17] Han W, Su X P, Yin F and Wang J 2010 *Trans. Mat. Heat Treat* **6** 28
[18] Szczepaniak A, Kostka A, Fan J and Raabe D 2012 *Adv. Eng. Mater.* **7** 464
[19] Ryabov V R 1984 *Welding of dissimilar metals and alloys* (Moscow: Mashinostroenie)

IOP Conf. Series: Materials Science and Engineering **848** (2020) 012057 doi:10.1088/1757-899X/848/1/012057

Effect of ions concentration in buffer solutions on the nucleation of hydroxyapatite surface on the octacalcium phosphate granules

P V Mikheeva, A Yu Teterina, I V Smirnov, A Yu Fedotov and V S Komlev

Institution of Russian Academy of Sciences A.A. Baikov Institute of Metallurgy and Material Science RAS, PhD in Materials Engineering, Leninskii pr. 49, Moscow, 117334 Russia

E-mail addresses: mikheevap7@gmail.com (P.V. Mikheeva)

Abstract. Octacalcium phosphate granules were subsequently immersed for different lengths of time from 1 day to 44 days at constant temperature 37±0.4°C in solutions, simulating the extracellular body fluids: Supersaturated Calcification Solution (SCS), Simulated Body Fluid (SBF), based on the ISO-10993 ''Biological Evaluation of Medical Devices'' standard. The role of buffer`s systems pH-value, time of the material exposure and calcium and phosphorus ionic concentrations were analysed with the main focus on kinetics of degradation, changes in the morphological characteristics and phase composition of octacalcium phosphate samples.

1. Introduction

The process of biological bones mineralization takes place in the intercellular body fluids. Hydroxyapatite (HA) is a major mineral component of bone tissue, but the mineralization starts from it`s precursor – octacalcium phosphate (OCP) that is why it is of special interest in bone regeneration due to it`s biocompatibility and biodegradability. This study was performed to examine in vitro stability of OCP granules. In the present investigation, the role of immersion OCP granules in static physiological solutions with a constant temperature of 37±1 ºC during 44 days was examined with the focus on the composition and morphologies changes after every chosen period.

Octacalcium phosphate ($Ca_8H_2(PO_4)_6 \cdot 5H_2O$) has the osteogenic capability and rate of resorption, close to speed of new bone tissue formation. However, the main direction in tissue development is creation bio-engineering structures with drugs or growth factors addition for directed bone healing, disease treatment and improvement implant`s osteoinduction and osteoconduction. Biomimetic deposition method suits the physiological temperature and pH requirements for infiltration biological agents [1]. Although this method was frequently used with hydroxyapatite (HA) [2,3] and many other materials [4], there is no previous report of using it on OCP granules.

The synthesis of pure OCP must be carried out under strict controlled conditions to prevent hydrolyzation OCP to HA [5]. Temperature, pH, solid-to-liquid ratio, Ca/P ratio, ion composition and concentration effect on OCP [6].

2. Materials and methods

In this work, octacalcium phosphate was prepared by chemical transformation method in the sodium acetate solution from α-TCP granules, which were synthesized by precipitation from solutions of calcium acetate (0.04 mol/L) and sodium hydrogen phosphate (0.04 mol/L). The pH value and

IOP Conf. Series: Materials Science and Engineering **848** (2020) 012057 doi:10.1088/1757-899X/848/1/012057

solution temperature varied in the physiological ranges of 5.0–6.5 and 37–90°C, respectively [7,8]. For this study, we used the samples from 1000 to 2000 μm.

Buffer solutions, which are SCS and SBF, were prepared by dissolving appropriate quantities of the chemicals in deionized water. Reagents were added one by one after completely dissolved previous reagent in 800 ml of water [9,10]. The chemical composition, pH values and ions concentration of the buffer solutions were determined before the experiment and detailed in **Table 1**.

The choice of solutions was based on physiological pH-value (≈ 7.4), temperature (37 °C), buffers system composition and ions concentration. According to previous research results, buffer solutions should have pH about 7.0 – 7.6 and Ca^{2+} (≈ 1.35 mM) concentration have to be lower than $(PO_4)^{3-}$ (≈ 2.0 mM) for launch of the OCP transformation to HA [11].

Table 1. Nominal concentrations of used buffer solutions and pH.

Solution	Ion	Na^+	K^+	Mg^{2+}	Ca^{2+}	Cl^-	HPO_4^{2-}	SO_4^{2-}	HCO_3^-	pH
SCS	Concentration (mM)	136.8	3.71	–	3.1	144.5	1.86	–	–	7.0
2xSBF		284.0	10.0	3.0	5.0	295.6	2.0	1.0	8.4	7.4

The experiment was carried out in closed systems: each sample was exposed to the solutions under static conditions at 37 °C ± 0.4 °C for determinate period of time: from 1 day up to 44 days. The samples one by one were carefully immersed in 50 ml volume of each solution for 1 g of solid material. After exposure, the samples were carefully washed off by distilled water and dried at the same temperature.

The Ca^{2+} and $(PO_4)^{3-}$ concentrations were determined by atomic emission spectroscopy (AES), solutions pH-value was measured using pH meter. Mass measurements of OCP granules were carried out before and after soaking in closed containers for comparison solubility rate and degradation behavior depending on time. Assessment of the degradation kinetics was carried out in the simulated extracellular body fluids according to Biological evaluation of medical devices - Part 14: Identification and quantification of degradation products from ceramics [12]. Complementary information about the surface morphology transformations during samples immersion in buffer systems was examined by scanning electron microscopy (SEM) Tescan VEGA II (Czech), previously covered with gold by Q150R Quorum Technologies (England). X-ray powder diffraction (XRD) studies were performed with a SHIMADZU XRD-6000 diffractometer (Japan) with CuKα radiation (λ = 1.5418 Å), 40 kV and 200 mA. The angular interval 2Θ analyzed was from 10 to 60, with 2°/min for the qualitative analysis. The phases were identified resorting to JSPDS. The Fourier transform Infrared spectroscopy *(FT-IR)* were recorded by Nikolet Avatar 330 spectrophotometer (USA) in the 4000–400 cm^{-1} wavelength region. The KBr pellet technique was used with 1 mg of powder in 50 mg of spectroscopic-grade KBr. Chemical analysis was performed by inductively coupled plasma atomic emission spectroscopy (AES) in an ULTIMA 2 instrument (France). Weighed portions of the studied materials were dissolved in 10 % hydrochloric acid. The main interest is concentrations changes of calcium and phosphate ions over time of immersion.

3. Results and discussion

Figure 1 shows the X-ray spectra of the OCP used, showing only OCP and HA phase diffraction lines. The presence of HA as an impurity in OCP associated with the method of obtaining and properties of the material.

IOP Conf. Series: Materials Science and Engineering **848** (2020) 012057 doi:10.1088/1757-899X/848/1/012057

The powder X-ray diffraction patterns of the samples from SCS solutions for each immersion duration are reported in Figure 2. The phase ratio shifts towards growing HA phase compared to OCP with increasing exposure time. This transformation is possible only in supersaturated solution, when Ca^{2+} and $(PO_4)^{3-}$ ions concentration is sufficient for sedimentation of HA on the surface layer. Ions concentration increases possible due to high solubility of OCP [13].

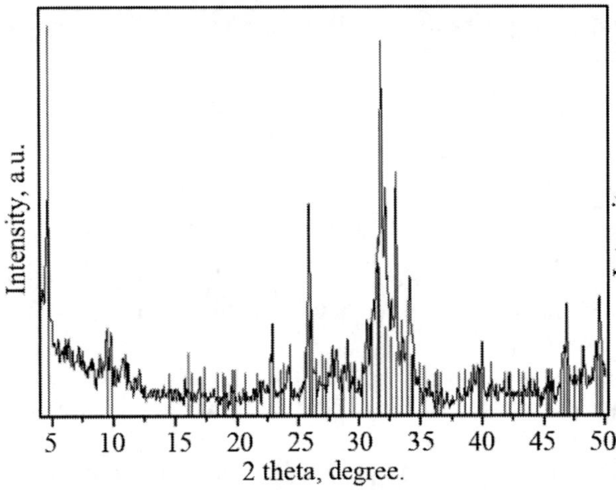

Figure 1. Powder X-ray diffraction patterns of the obtained OCP granules and XRD database card № 26-1056 (Calcium Hydrogen Phosphate Hydrate).

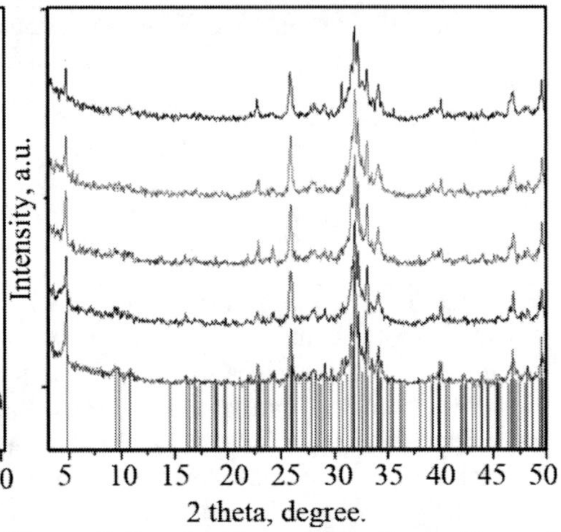

Figure 2. Powder X-ray diffraction patterns of the products obtained after OCP granules soaking in SCS solution for different periods: 1st, 3rd, 7th, 14th, 44th days of immersion.

In the SCS solution, granule composition ratio changes are noticeable in favour of an increase in the hydroxyapatite phase, which indicates the occurrence of recrystallization process at the phase boundaries. In the SBF buffer system, which is close to ionic composition of blood plasma material modification processes are similar with SCS solution. IR-spectroscopy results presented in Figure 3.

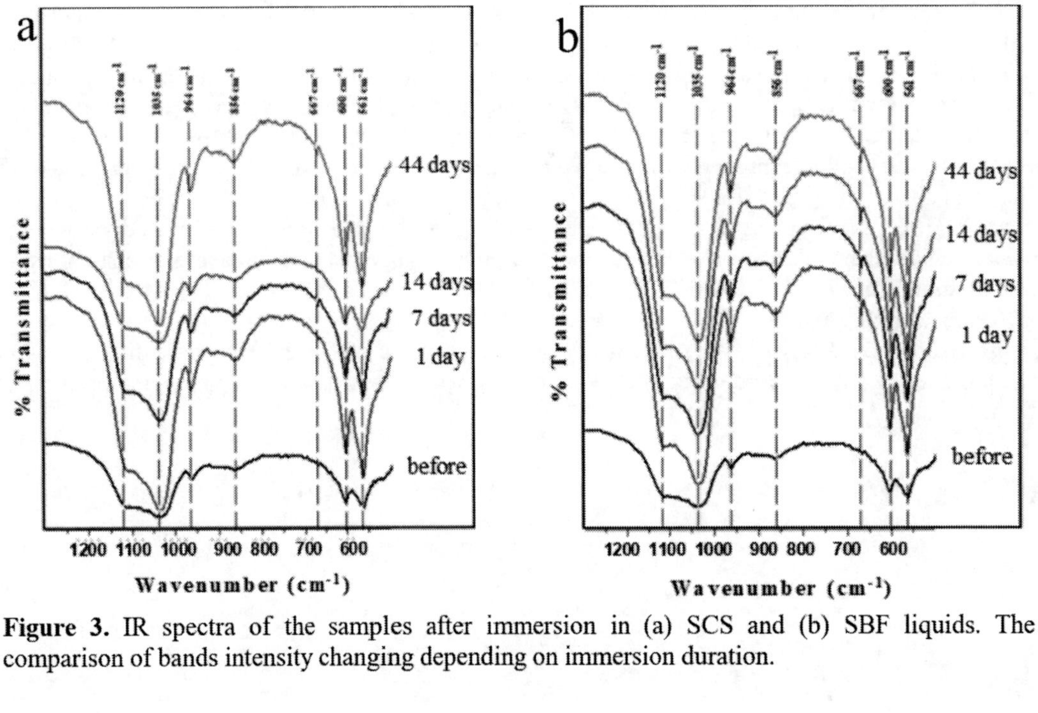

Figure 3. IR spectra of the samples after immersion in (a) SCS and (b) SBF liquids. The comparison of bands intensity changing depending on immersion duration.

revealed specific infrared spectra of octacalcium phosphate. The spectrum of sample exhibits a strong bands of the *v1* mode of $(HPO)_4^{2-}$ groups at 1020 cm^{-1} and the *v4* mode of the $(PO_4)^{2-}$ groups at

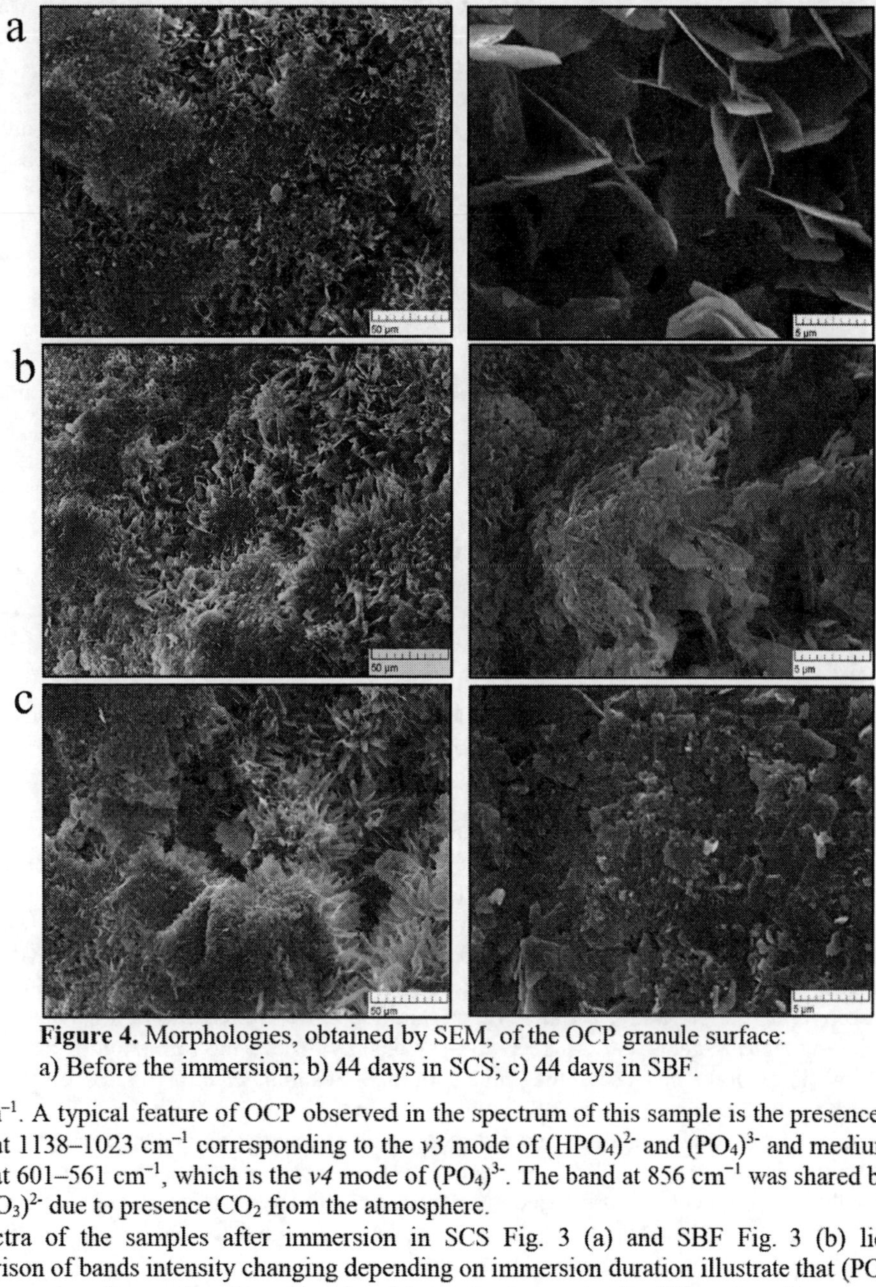

Figure 4. Morphologies, obtained by SEM, of the OCP granule surface:
a) Before the immersion; b) 44 days in SCS; c) 44 days in SBF.

601 cm^{-1}. A typical feature of OCP observed in the spectrum of this sample is the presence of intense bands at 1138–1023 cm^{-1} corresponding to the *v3* mode of $(HPO_4)^{2-}$ and $(PO_4)^{3-}$ and medium intensity bands at 601–561 cm^{-1}, which is the *v4* mode of $(PO_4)^{3-}$. The band at 856 cm^{-1} was shared by $(HPO_4)^{2-}$ and $(CO_3)^{2-}$ due to presence CO_2 from the atmosphere.

IR-spectra of the samples after immersion in SCS Fig. 3 (a) and SBF Fig. 3 (b) liquids. The comparison of bands intensity changing depending on immersion duration illustrate that $(PO_4)^{3-}$ groups become more polar as the absorption band intensity is greater in different soaking period, which indicates the phase changes in the material.

The crystal morphology of OCP granules and it`s appropriate size for the experiment requirements presented in Figure 4. Crystals of OCP samples is composed of needles, which are less than 1 µm thick and up to 10 µm long. The phase changes in OCP granules surface layer on Figure 4. in SCS (b) and SBF (c) solutions are well seen. The structure of crystals is heterogeneous. Most of the surface is similar to HA, but some parts are still looks like OCP. According to chemical analysis performed on

Figure 5, it is seen that maximum changes occur in the first two days of the experiment. A necessary sufficient concentration of the ions for recrystallization process is initially found in SCS. Ca and $(PO_4)^{3-}$ concentration in it decreased by more than 20 % of the original concentration after soaking for 2 days, because the recrystallization of OCP started immediately. By the 3rd day of the experiment ions exchange with the rest material is stabilized and tends to equilibrium.

In contrast, in a bipolar SBF solution there are not enough phosphorus ions. Therefore, the ion exchange between the surface layer of OCP granules and the solution leads to simultaneously degradation and recrystallization processes, as the calcium concentration rapidly decreases, while phosphorus ions gradually increase.

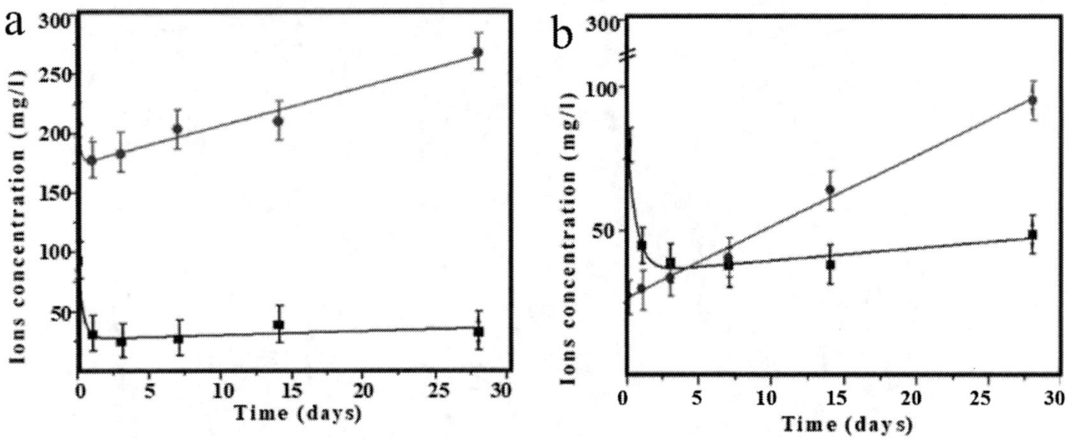

Figure 5. Ca^{2+} and $(PO_4)^{3-}$ concentrations in the (a) SCS and (b) SBF solutions as a function of soaking time.

4. Conclusions

The influence of immersion time in buffer solutions of various ionic composition, simulating the extracellular body fluids, on the kinetics of degradation, phase composition and morphological characteristics of calcium phosphate phases, including the microstructure and properties of material are reported. Interconnection between fluids ion concentration value and intensity XRD peaks of samples before and after immersion showed that recrystallization process on the surface layer is going faster in supersaturated solutions with sufficient concentration of Ca^{2+} and $(PO_4)^{3-}$ ions. Summing up all the research results, Supersaturated Calcification Solution buffer system has optimal characteristics for infiltration biological agents and drugs. The X-ray data evidenced the formation of the HA peaks on the OCP species with the increase of the immersion time. Scanning electron microscopy reveals the formation of heterogeneous and uneven distribution of spherical hydroxyapatite particles in SCS and SBF solutions by 44th day. The calcium and phosphor ionic concentration dependence from the OCP material exposure duration indicated that the reprecipitation process most quickly occurs in SCS. The results show that applicants can be infiltrated in material using buffer systems. Based on all the results presented above, it can be concluded that the SCS solution is optimal for the biomimetic deposition method for infiltration of various biogenic factors and drugs.

Acknowledgments

This work was supported by the Russian Science Foundation, project no.18-33-00955 mol_a «Surface engineering of mineral-polymer structures for highly effective infiltration of osteogenic factors». Authors also wishing to acknowledge assistance and encouragement from all the institute colleagues.

IOP Conf. Series: Materials Science and Engineering **848** (2020) 012057 doi:10.1088/1757-899X/848/1/012057

References

[1] F. Li, Q. Feng, F. Cui, H. Li, H. Schubert, *A simple biomimetic method for calcium phosphate coating, Surface and Coatings Technology*, 154(1), (2002), 88–93.

[2] Helebrant A, Jonasova L, Sanda L. *The influence of simulated body fluid composition on carbonated hydroxyapatite formation*, Ceram–Silikaty 46, (2002), 9–14.

[3] X. Lu, Y. Leng, *Theoretical analysis of calcium phosphate precipitation in simulated body fluid, Biomaterials* 26(10), (2005), 1097–1108.

[4] Y. Abe, T. Kokubo, T. Yamamuro, *Apatite coating on ceramics, metals and polymers utilizing a biological process. Journal of Materials Science: Materials in Medicine 1(4),* (1990), 233–238.

[5] A. Ramosa, G. Ybarrab, L. Pazosb, M. Parodib, L. Rodríguezb, M. Hernandez, E. Jesus, *Manufacture of nanosized apatite coatings on titanium with different surface treatments using a supersaturated calcification solution, Quimica Nova 39 (10),* (2016), 1159-1164.

[6] P. W. Brown, R. I. Martin, *An analysis of hydroxyapatite surface layer formation, J Physical Chemistry 103*, (1999), 1671–1675.

[7] A. Bigi, E. Boanini, R. Botter, S. Panzavolta, K. Rubini, *α-Tricalcium phosphate hydrolysis to octacalcium phosphate: effect of sodium polyacrylate, Biomaterials 23*, (2002), 1849–1854.

[8] V.S. Komlev, I.V. Fadeeva, A.S. Fomin, L. I. Shvorneva, D. Ferro, S. M. Barinov, *Synthesis of octacalcium phosphate by precipitation from solution, Doklady Chemistry 432(2)*, (2010), 178–182.

[9] P. Li, C. Ohtsuki, T. Kokubo, K. Nakanishi, N. Soga, *Apatite formation induced by silica gel in simulated body fluid, J Am Ceram Soc, 75*, (1992), 2094-2097.

[10] T. Kokubo, H. Takadama, *How useful is SBF in predicting in vivo bone bioactivity?, Biomaterials 27*, (2006), 2907 – 2915.

[11] P.T. Cheng, *Formation of octacalcium phosphate and subsequent transformation to hydroxyapatite at low supersaturation: a model for cartilage calcification, Calcif. Tissue Int. 40*, (1978), 339–343.

[12] *Biological evaluation of medical devices - Part 14: Identification and quantification of degradation products from ceramics* (ISO 10993-14:2001); German version EN ISO 10993-14:2009.

[13] M. Tung, N. Eidelman, B. Sieck, W. Brown, *Octacalcium phosphate solubility product from 4 to 37 ^{0} C, J Res Natl Bur Stand 93*, (1988), 613–24.

IOP Conf. Series: Materials Science and Engineering **848** (2020) 012058 doi:10.1088/1757-899X/848/1/012058

Study characteristics of failure multilayered steel materials in conditions at alternating symmetrical bending

A A Minakov, A I Plokhihk, D V Vlasova and K B Polikevich

Department of Material Science, Bauman Moscow State Technical University (BMSTU), 105005, 2nd Baumanskaya street, 5-1, Moscow, Russian Federation

E-mail: plokhikh@bmstu.ru

Abstract. The article presents a study of behaviour of multilayered steel materials in conditions of alternating symmetrical bending when heavy stress is applied at low-cycle fatigue conditions with the use of bending unbending test. The study showed different behaviour pattern when applying stress to two compositions of multilayered steel materials – W108+AISI 304 and W108+AISI 430. The fractographic analysis of fatigue rupture surfaces was used to explain the results obtained in studying the number of cycles prior to fracture, which showed that the mode of failure of the two compositions of multilayered steel materials was different.

1. Introduction

The modern level of mechanical engineering development calls for studies in the field of analysis and application of the new classes of structural materials. And among them it is possible to distinguish a class of multilayered steel materials with a multilayer (or lamellar) structure [1]. Lamellar structure is a multilayer composition including hundreds (or thousands) of layers with micron (submicron) thickness (Figure 1). Owing to laminarity of structure, a high complex of fatigue properties is formed in multilayered steel materials [2]. On the other hand, a laminar structure allows to obtain the anisotropy of fatigue properties. Therefore, the multilayered steel materials can develop their fatigue properties to the maximum extent when used in the parts of machines or constructions which operate at conditions of alternating symmetrical bending with normal stress [3]. With that, a short-term heavy stressing at low-cycle fatigue can be a limiting factor when using such materials at conditions of alternating symmetrical bending. Therefore, a study of behaviour and failure characteristics of multilayered steel materials in conditions at alternating symmetrical bending with heavy stressing at low-cycle fatigue is of high relevance.

IOP Conf. Series: Materials Science and Engineering **848** (2020) 012058 doi:10.1088/1757-899X/848/1/012058

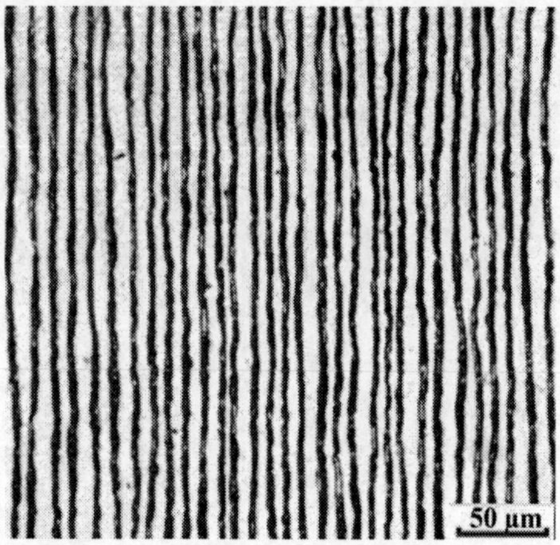

Figure 1. Lamellar structure in the multilayered steel material W108+AISI 304 after performance of the second process cycle.

2. Materials, methods and experiment

An experimental method of procedure was developed for the synthesis of multilayered steel materials, which consists of several process cycles each including the productive method of hot pack rolling [4]. Each process cycle is a series of same-type process operations following one after another:
1) measured cutting of parent steel sheets or finished steel sandwich;
2) preparation of billets (machining of their surfaces);
3) formation of multiple sandwich from billets;
4) hot rolling of multiple sandwich and generation of multilayered steel material sheet.

Rolled steel sheets of two steel grades with thickness of 0.5 mm were the source materials for the synthesis of multilayered steel materials. The following compositions based on carbon steel grade W108, austenitic stainless steel grade AISI 304 and ferritic stainless steel grade AISI 430 were taken for the purposes of this study: W108+AISI 304 and W108+AISI 430.

Only one process cycle of the experimental procedure method was used for the purposes of this study. At the start of the first process cycle the billets – measuring cards cut from the steel sheets of above-mentioned grades – were subjected to surface treatment. Next, a multiple sandwich was formed from the cards, consisting of 100 alternating steel grades, 50 pieces of each grade. The formed multiple sandwich was vacuumed and hot rolled at 1000 °C. As a result, after the first process cycle a multilayered steel material of two compositions was synthesized with the thickness of 2 mm, comprising 100 layers with the thickness of 20 μm each.

Specimens of type No. IV (GOST 25.502-79) were manufactured from the sheets of multilayered steel materials of compositions W108+AISI 304 and W108+AISI 430 for conduction of studies on the analysis of behaviour and failure characteristics of multilayered steel materials at alternating symmetrical bending when applying heavy stress at low-cycle fatigue conditions. For achievement of such type of bend under the given conditions, a device (Figure 2) was used, applicable in the bending unbending test (GOST 1579-93).

According to the method of experiment on this unit, first a test specimen 1 is placed between two drivers 2, which are fixed on the bending lever 3. From one side, a specimen is installed so that the fountain axis of bending lever is aligned with the central axis of the test specimen. From the other end, the specimen is placed between two mandrels 5. From the other side, a specimen is installed so that the mandrels are in one plane and parallel to each other and to the fountain axis of bending lever. Next, the test specimen is fixed by means of jaw details 7 and by means of grip 6. Loading of the test specimen is performed by turning the bending lever clockwise and back at 90° angle from initial position of the lever. Loading is performed until failure of the test specimen. Changing mandrels with different diameters, simultaneously count the number of cycles to failure.

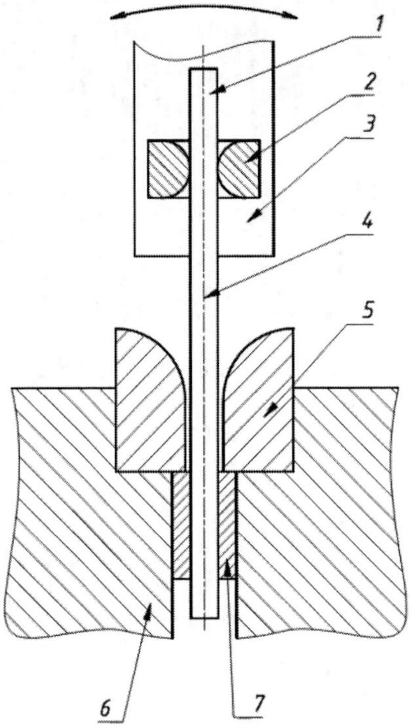

Figure 2. Layout of the device used in the bending unbending test (GOST 1579-93): 1 – test specimen; 2 – driver; 3 –bending lever; 4 – fountain axis of bending lever; 5 – mandrel; 6 –grip; 7 – jaw details.

The fractographic analysis of failure surfaces of the specimens of multilayered steel materials was carried out with the use of Vega Tescan 5130 scanning microscope at acceleration voltage of 20 kV.

3. Results and discussion

Results of the experiment (Figure 3) showed that by increasing the degree of loading by way of reducing the diameters of mandrels (from 30 mm, to 15 mm, and finally to 3.5 mm) caused the number of cycles to failure in W108+AISI 304 composition to decrease in 6 times (from 12 cycles to 2 cycles to failure). At the same time, on loading the composition W108+AISI 430 identically, it was discovered that when loaded, with the use of any diameter of mandrel, a specimen failed even before completion of one bend. I.e. the number of cycles to failure was zero for the composition W108+AISI 430.

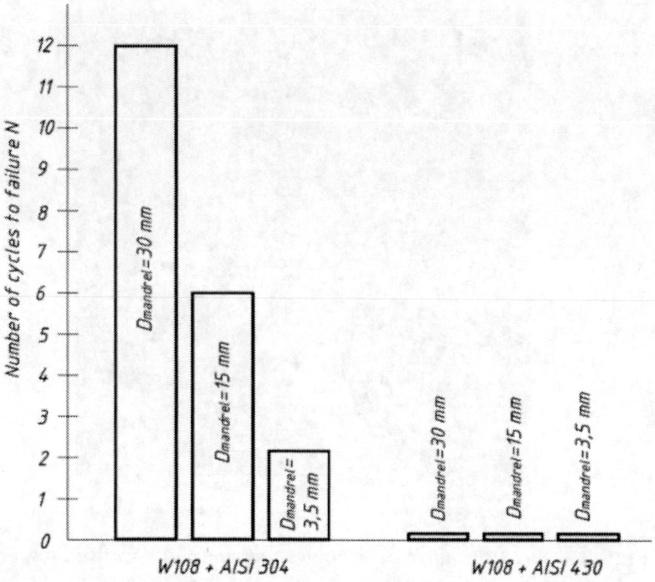

Figure 3. Results of the experiment on studying the behaviour of multilayered steel materials at alternating symmetrical bending with heavy stressing at low-cycle fatigue conditions.

The fractographic analysis showed that, at the fracture (Figure 4) of multilayered steel material of composition W108+AISI 304 in the layer of former W108 steel, it is possible to see the cleavage facets and a large number of dimples, initiating ductile failure around them. In the layer of former AISI 304 steel one can see the fatigue striations, as well as vertical exfoliation joints. The appearance of fatigue striations is explained by the fact that flat glide is typical for stainless steels under cyclic stressing [5]. A large amount of dimples is explained as follows. At the synthesis of multilayered steel material of this composition during the hot rolling both steels are in the same crystalline state (fcc lattice). Therefore, taking into account the processes of diffusion and considerable chemical affinity of carbon to chromium, carbon diffuses without difficulty through interlayer border to the layer of AISI 304 steel. As a result, a large amount of carbides is formed in the layer of AISI 304 steel, causing the appearance of a large number of micropores during stressing, which merge to form a large number of dimples.

Figure 4. Fractography of fractured surface in a multilayered steel material of compositionW108+AISI 304.

At the fracture (Figure 5) of multilayered steel material of composition W108+AISI 430 in the layer of former W108 steel, one can see cleavage cracks, exfoliation joints, lines of streamlet pattern. All this is indicative of the brittle mode of fracture in the layer of former W108 steel. In the layer of former AISI 430 steel one can see a small number of dispersed dimples and some cleavage cracks.

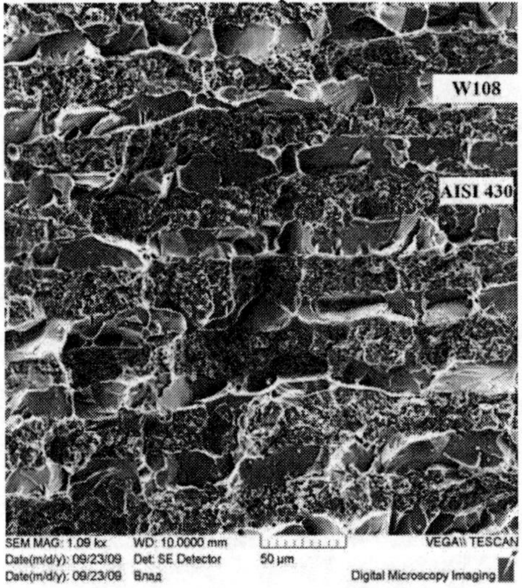

Figure 5. Fractography of fractured surface in a multilayered steel material of compositionW108+AISI 430

Brittle mode of failure is explained as follows. At the synthesis of multilayered steel material with composition of W108+AISI 430, the steels are in different crystalline state (W108 steel has fcc lattice, and AISI 430 steel has a stable bcc lattice). As the solubility of carbon is different in bcc and fcc lattices , the diffusion of carbon from fcc lattice to bcc lattice is inhibited. However at the synthesis of multilayered steel material it is gradually thinned. This is due to the fact that a process cycle consists of several passes during rolling. During rolling the actual temperature of the billet (Figure 6), due to gradual thinning, might become less than the temperature of direct diffusionless transformation. This will result in quenching of W108 steel layer on passing through the mill rollers, thus making the layer of W108 steel brittle when subjected to stress.

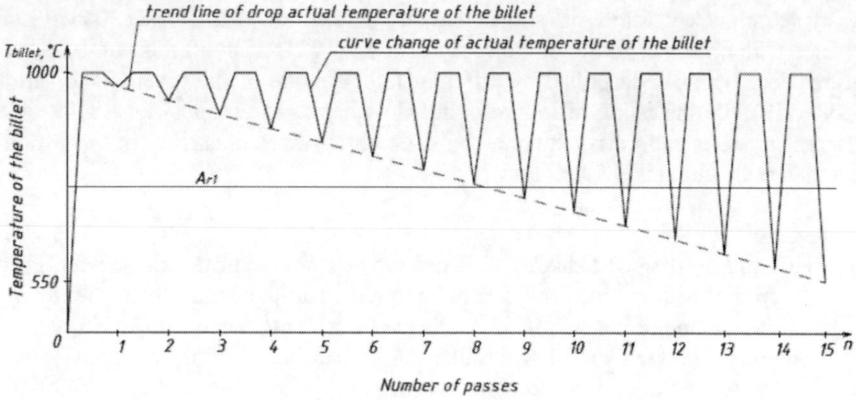

Figure 6. Flow chart of rolling

On the other hand, considerable chemical affinity of chromium to carbon should be taken into account, which will cause insignificant carbon diffusion and insignificant redistribution of chromium in AISI 430 steel. As a result the concentration of chromium in AISI 430 steel will be reduced [6], and the phase state of AISI 430 steel will change from α to α+γ according to the isothermal section of the ternary diagram of Fe-C-Cr phase equilibrium at 1000 °C (Figure 7), which will lead to the appearance of a small amount of carbides. Appearance of carbides will provoke the appearance of a small number of micropores, the coalescence of which will result in the appearance of a small quantity of dimples. Therefore the contribution of ductile failure to the general fracture mode will be much less than the contribution of brittle failure in the layer of W108 steel.

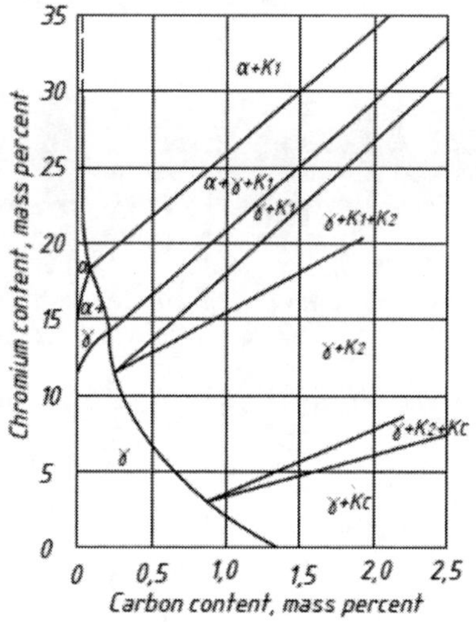

Figure 7. Isothermal section at 1000 °C of the Fe-C-Cr phase equilibrium diagram [6]

4. Conclusions

Consequently, the analysis of behaviour of multilayered steel materials with compositions W108+AISI 304 and W108+AISI 430 under conditions of alternating symmetrical bending with heavy stressing at low-cycle fatigue conditions has shown that the composition W108+AISI 304 withstands a large number of bends with unbends, while for the composition W108+AISI 430 this number was equal to zero. This different behaviour is described by different failure mode of the given compositions: in composition W108+AISI 304 the mode of failure is ductile whereas in composition W108+AISI 430 it is brittle. Different mode of failure is explained by different changes occurring in the structure of multilayered steel materials during their synthesis.

References

[1] Kolesnikov A G, Plokhikh A I, Komisarchuk Yu S and Mikhaltsevich I Yu 2010 A study of special features of formation of submicro- and nanosize structure in multilayer materials by the method of hot rolling *Metal Science and Heat Treatment* **52** N5-6 273-278

[2] Schmiedt A, Luecker L, Kolesnikov A, Plokhikh A I and Walther F 2017 Production- and microstructure-based fatigue assessment of metallic AISI 304/430 multilayer materials produced by hot pack rolling *Materialpruefung* **59** N2 123-129

[3] Minakov A A, Plokhikh A I , Schmiedt A and Walther F 2018 Comparative analysis of the fatigue characteristics of multilayer steel materials and nanolaminates *Mechanics, resource and diagnostics of materials and structures (MRDMS-2018): Proceedings of the 12th International Conference on Mechanics, Resource and Diagnostics of Materials and Structures* **2053** N1 040059

[4] Tabatchikova T I, Plokhikh A I, Yakovlev I L and Klyueva S Yu 2013 Structure and properties of a steel-based multilayer material produced by hot pack rolling *The Physics of Metals and Metallography* **114** N7 580-592

[5] *ASM Metals Handbook* (8th ed.) vol. 9 *Fractography and Atlas of Fractographs* American Society for Metals, Metals Park, OH 1974

[6] Houdremont E *Handbook of Special Steels* 3rd ed Springer: Berlin, Germany 1956

IOP Conf. Series: Materials Science and Engineering **848** (2020) 012059 doi:10.1088/1757-899X/848/1/012059

Hardness and fracture-toughness of hot-pressed LaB_6-TiB_2 ceramics

D D Nesmelov[1], E S Novoselov[1], A S Lysenkov[2], S V Vikhman[1] and S N Perevislov[3]

[1]Saint-Petersburg State Institute of Technology, Saint-Petersburg, Russia
[2]Baykov Metallurgy and Materials Institute (IMET RAS), Moscow, Russia
[3]Grebenschikov Institute of Silicate Chemistry (ISC RAS), Saint-Petersburg, Russia
E-mail: dnesmelov@yandex.ru

Abstract. Microstructural analysis and investigations of mechanical properties via indentation-techniques have been performed on a LaB_6 - 15 vol. % TiB_2 ceramics, obtained by the hot-pressing at T = 1900 °C and P = 30 MPa in argon gas. The relative density of hot-pressed material was up to 97% of theoretical. XRD, SEM and EDX analysis of ceramics were performed. The Vickers hardness (18.5 GPa) and fracture-toughness (3.9 MPa·m$^{1/2}$) were compared with the values, measured on a single-crystalline LaB_6 (Hv = 21 GPa, K_{Ic} = 2.4 MPa·m$^{1/2}$) and arc-melted eutectic composite LaB_6-TiB_2 (Hv = 23 GPa, K_{Ic} = 6.7 MPa·m$^{1/2}$).

1. Introduction

Traditionally LaB_6 is widely used as a thermionic emitter because of its low work function, high thermionic current density, and good stability at high temperatures [1-3]. During last decade LaB_6-based solid solutions became the object of interest because of their promising optical properties [4-7]. Another use of lanthanum hexaboride is to use it as a second phase to activate the sintering of ultrahigh-temperature ceramics based on refractory zirconium and hafnium diborides [7]. As a structural material lanthanum hexaboride exhibits relatively high strength, hardness and Young's modulus, which is due to strong covalent bonds within the B_6 octahedron. But insufficient fracture-toughness of single-crystalline LaB_6 (about 1.5-2.5 MPa·m$^{1/2}$) can be a critical factor in conditions of high mechanical loads. Near-full density polycrystalline LaB_6 demonstrates the same values of fracture-toughness. However, mechanical properties of LaB_6 can be enhanced by the strengthening with second phase, for example, MeB_2 (where Me = Ti, Zr, Hf, Cr, V, Nb etc.) [8-17]. All of the quasi-binary systems LaB_6-MeB_2 are eutectic-type [18-23], which allows obtaining dense ceramic materials via solid-state sintering at temperatures below $0.8T_{eut} \approx$ 2100-2200 °C. As shown in [24, 25], the creation of composite ceramic materials based on LaB_6-MeB_2 systems makes it possible to increase the fracture-toughness up to 3.7-4.5 MPa·m$^{1/2}$. This effect is associated with a crack deflection mechanism in hetero-phase structure. A significant decrease in sintering temperature of composite ceramics based on LaB_6-MeB_2 systems leads to the formation of more fine-grained structure in comparison with single-phase polycrystalline LaB_6. This effect is due to a two main factors: decreasing the sintering temperature caused by eutectic-type of interactions and lengthening of diffusion paths in the hetero-phase structure, which inhibits grain growth during sintering.

In the present study we investigated the microstructure, phase composition, hardness and fracture-toughness of LaB_6-TiB_2 ceramic composite, obtained by hot-pressing of submicron LaB_6 and TiB_2 powders. It was of considerable interest to investigate the Vickers hardness and fracture-toughness of

IOP Conf. Series: Materials Science and Engineering **848** (2020) 012059 doi:10.1088/1757-899X/848/1/012059

LaB_6-TiB_2 hot-pressed ceramic composite and to compare these results with similar characteristics of single-crystalline LaB_6 and arc-melted eutectic composite LaB_6-TiB_2, which were not quite correctly measured by us earlier in work [27].

2. Materials and Method

Commercial powders of LaB_6 (99.0 wt. %) and TiB_2 (99.0 wt. %) were used in this study. The average size of starting powders (the 50th percentile particle size, d_{50}) determined by laser diffraction was 12 and 15 μm respectively. Powders were mixed and milled on vibratory ball-mill in gasoline-solvent media with SiC balls (balls to material weight ratio B:M = 10:1) during 80 hours. Powders were mixed in ratio 85 vol. % LaB_6 – 15 vol. % TiB_2. The additional 2.3 mass. % SiC appeared in mixture due to the wear of SiC balls, which was determined by weighing the balls before and after grinding. After ball-milling, the average particle size decreased to d_{50} = 0.75 μm.

The vacuum-dried powder mixture was placed into graphite press-form with an internal diameter of 25 mm and sintered on a Thermal Technology HP20-3560-20 hot press at an isothermal holding temperature of 1900°C and a pressure of 30 MPa for 15 min in argon gas. After hot-pressing samples were polished for microstructural and micromechanical analysis.

Reference samples were obtained by the floating zone method (single-crystalline LaB_6) [27] and electric-arc melting (eutectic alloy LaB_6-TiB_2) [26].

The phase composition of ceramics was determined using the X-ray diffraction (XRD) analysis (Rigaku SmartLab 3, Cu Kα radiation, λ = 1.5406 Å) in the 2θ range of 10–90° with 0.01 ° step. The microstructure of samples was investigated using an optical microscope Meiji Techno 7200 and SEM Tescan Vega 3SBH. Analysis of elements concentration was carried out using Aztec X-Act analyzer (Oxford Instruments). The volume fractions of phases were calculated via the Thixomet-Lite system for image analysis using the fraction areas measuring from SEM-micrographs.

Indentation tests were carried out with the use of Vickers hardness tester PMT-3 under the load of 7.2 N. For calculation of *Hv* (Vickers hardness, GPa) we use equation (1):

$$Hv = 1854 \cdot \frac{P}{D^2}, \qquad (1),$$

where P – load, N; D – measured diagonal of the indentation, μm.

Fracture-toughness K_{Ic} was calculated according to the equation (2):

$$K_{IC} = 0{,}0889\sqrt{\frac{Hv \cdot P}{4l}}, \qquad (2)$$

where P – load, N; l – measured length of the indentation-induced crack from corner of indentation, μm.

At least 10 indentations were made, from which the mean values of the lengths of the diagonals and cracks were calculated.

3. Results and discussion

The relative density of hot-pressed LaB_6-TiB_2 samples, measured by the hydrostatical method, was up to 97% of the theoretical value. Such a high density, compared to polycrystalline LaB_6, obtained at the same conditions, is due to a eutectic-type of phase interactions in the LaB_6-TiB_2 system. The significant decrease in melting point is associated not only with liquid-phase sintering mechanism above the eutectic temperature but with activation of solid-state diffusion at the hetero-phase grain boundaries, which leads to a decrease in sintering temperature. As can be seen in the optical and SEM images of polished samples (Figure 1A and 1B) the structure of ceramics is two-phase: TiB_2 grains are distributed in LaB_6-matrix.

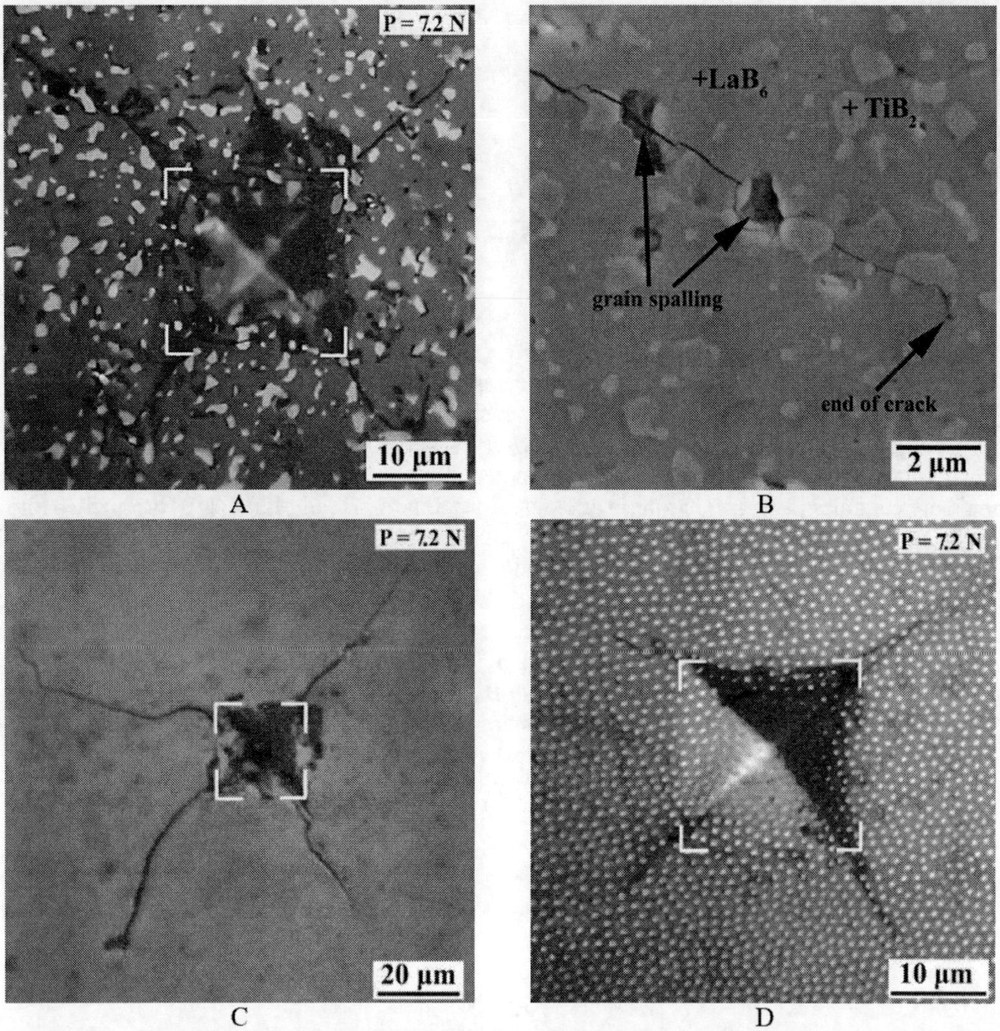

Figure 1. A – optical image of Vickers indentation in hot-pressed LaB_6-TiB_2 ceramics; **B** - SEM image of indentation-induced crack; optical images of Vickers indentations in: **C** - single-crystalline LaB_6; **D** - arc-melted LaB_6-TiB_2 eutectics.

Analyzing the morphology of the grains, we can note, that the main part of them has a complex shape, formed by association of several smaller particles during the hot-pressing. The median size of TiB_2 grains, calculated using statistical analysis of SEM micrographs, is about 1.8 µm, which is 2.5 times larger than the initial particle size. The phases area fractions calculation from SEM micrographs, as well as EDX-analysis, showed that the composition of ceramics, in addition to LaB_6 (84 mol %) and TiB_2 (14 mol %) includes 2 mol % of SiC. This result is in a good agreement with the value of SiC-balls wearing during the ball-milling, established by weighing the balls before and after milling.

The XRD-spectra of hot-pressed LaB_6-TiB_2 ceramics presented in Fig.2. There are distinct diffraction peaks of well-crystallized two main phases – cubic LaB_6 and hexagonal TiB_2. Also, the most intensive reflexes of hexagonal SiC-phase are clearly defined. The spectrum contains weak diffraction peaks of impurity phases; their intensity is close to the background intensity. They can be recognized as peaks of boron nitride and lanthanum nitride, which appeared due to the use of a protective coating of BN on a graphite mold.

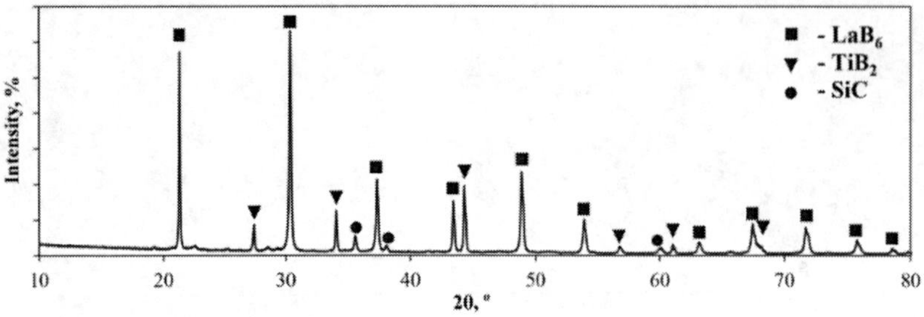

Figure 2. XRD patterns of hot-pressed LaB_6-TiB_2 ceramics

The average hardness of LaB_6-TiB_2 ceramic composite (Hv = 18.5 GPa at the load of 7.2 N) was significantly lower than the theoretical value calculated by the rule of mixtures (28.2 GPa, based on data from [28] and 25.9 GPa, based on data from [13]). This may be due not only to the effect of porosity in ceramics but also to the previously noted effect of microplastic deformation in heterophase composite ceramics [29]. The hardness of single-crystalline LaB_6 and LaB_6-TiB_2 arc-melted alloy was 21 GPa and 23 GPa, respectively.

The fracture toughness of LaB_6-TiB_2 ceramic composite was 3.9 $MPa{\cdot}m^{1/2}$, which is higher, than the fracture toughness of LaB_6 single crystal (3.1 $MPa{\cdot}m^{1/2}$) and quite lower, than the fracture-toughness of arc-melted LaB_6-TiB_2 alloy (6.7 $MPa{\cdot}m^{1/2}$).

All the experimental values of Vickers hardness and fracture-toughness of hot-pressed LaB_6-TiB_2, single-crystalline LaB_6 and arc-melted alloy LaB_6-TiB_2 presented at the Ashby plot on Fig. 3. as closed dots 1, 2 and 3, respectively. Open dots are the literature data from [8, 9, 13, 15]. Thus, three areas corresponding to: 1) polycrystalline materials obtained by hot pressing or spark-plasma sintering, 2) single crystals, and 3) directionally crystallized eutectics can be distinguished on the Ashby plot.

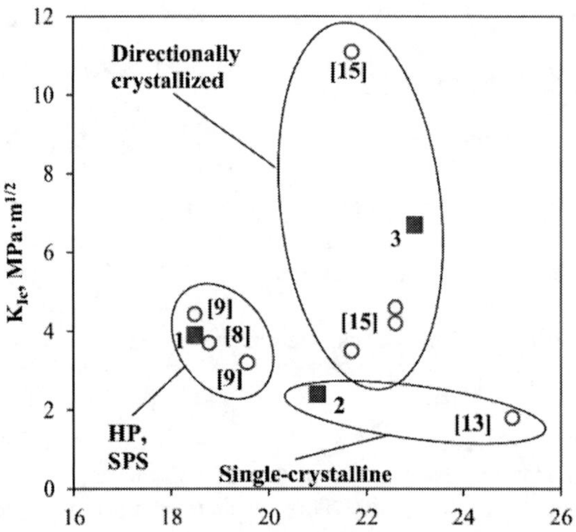

Figure 3. Ashby plot of LaB_6-MeB_2-based materials (Me = Ti, Zr).

The main mechanism for increasing the fracture-toughness of hot-pressed and arc-melted samples is caused by an increase in the length of the crack propagation path due to its deflection by TiB_2 grains (Figure 1B) or fibers (Figure 1D). In the case of hot-pressed LaB_6-TiB_2 ceramics, the crack propagation is mainly intergranular, but some TiB_2 grains cracked by the transgranular mechanism.

Also, we can see the TiB_2 grain spalling (Fig. 1D) when the reinforcing grains debonded from the matrix.

4. Conclusion

The LaB_6 - 15 vol. % TiB_2 ceramics with 97% density was obtained by the hot-pressing at T = 1900 °C and P = 30 MPa in argon gas. XRD, SEM and EDX analysis of ceramics were performed. The Vickers hardness (18.5 GPa) and fracture-toughness (3.9 $MPa \cdot m^{1/2}$) of hot-pressed ceramics were compared with the values, measured on a single-crystalline LaB_6 (Hv = 21 GPa, K_{Ic} = 2.4 $MPa \cdot m^{1/2}$) and arc-melted eutectic composite LaB_6-TiB_2 (Hv = 23 GPa, K_{Ic} = 6.7 $MPa \cdot m^{1/2}$). The reinforcement effect caused by the mechanism of crack deflection is most pronounced in the case of arc-melted LaB_6-TiB_2 alloy, and to a lesser extent observed in hot-pressed LaB_6-TiB_2 ceramics.

Acknowledgments

The reported study was funded by RFBR project No 18-33-20221. Authors are grateful to the Engineering Centre of Saint-Petersburg State Institute of Technology for the equipment provided.

References

[1] Lafferty J M 1951 *Journal of Applied Physics* **3** 299
[2] Taran A, Voronovich D, Oranskaya D, Filipov V, Podshyvalova O 2013 *Functional materials* **4** 485
[3] Paderno Y B, Taran A A, Voronovich D A, Paderno V N, Filipov V B 2008 *Functional materials* **1** 63
[4] Kowalczyk J M, Hadmack M R, Szarmes E B, Madey J M 2014 *International Journal of Thermophysics* **8** 1538
[5] Mattox T M, Coffman D K, Roh I, Sims C, Urban J J 2018 *Materials* **2** 226
[6] Mattox T, Urban J 2018 *Materials* **12** 2473
[7] Sani E, Mercatelli L, Meucci M, Zoli L, Sciti D 2017 *Scientific Reports* **1** 718
[8] Min G H, Gao R, Yu H S, Han J 2005 *Key Engineering Materials* **297** 1630
[9] Wang X., Zhang J X, Yang X Y, Hu K, Zhang J W 2017 *Advances in Applied Ceramics* **3** 132
[10] Bogomol I, Nishimura T, Nesterenko Y, Vasylkiv O, Sakka Y, Loboda P 2011 *Journal of Alloys and Compounds* **20** 6123
[11] Soloviova T O, Karasevska O P, Vleugels J, Loboda P I 2017 *Journal of Alloys and Compounds* **729** 749
[12] Soloviova T O, Karasevska O P, Loboda P I 2019 *Ceramics International* **7** 8677.
[13] Morozov V V, Mal'nev V I, Dub S N, Loboda P I, Kresanov V S 1984 *Izvestiya Akademii Nauk SSSR, Neorganicheskie Materialy* **8** 1421 in russian
[14] Bogomol I, Nishimura T, Vasylkiv O, Sakka Y, Loboda P 2010 *Journal of Alloys and Compounds* **1** 130.
[15] Deng H, Dickey E C, Paderno Y, Paderno V, Filippov V, Sayir A 2004 *Journal of materials science* **19** 5987.
[16] Volkova H, Filipov V, Podrezov Y. 2014 *Journal of the European Ceramic Society* **14** 3399.
[17] Loboda P I, Soloviova T O, Bogomol Y I, Remizov D O, Bilyi O I 2015 *Journal of Superhard Materials* **6** 394.
[18] Ordan'yan S S, Paderno Y, Khoroshilova I K 1983 *Soviet powder metallurgy and metal ceramics* **251** 946.
[19] Zakarian D, Kartuzov V, Kartuzov E, Khachatrian A, Sayir A 2011 *Journal of the European Ceramic Society* **7** 1305.
[20] Loboda P I, Kisla G P, Bogomol I I, Sysoev M A, Karasevskaya O P 2009 *Inorganic Materials* **3** 246.
[21] Ordan'yan S S, Paderno Y B, Nikolaeva E E, Khoroshilova I K 1984 *Poroshkovaya Metallurgiya*, 64-66 in russian

IOP Conf. Series: Materials Science and Engineering **848** (2020) 012059 doi:10.1088/1757-899X/848/1/012059

[22] Ordan'yan S S, Paderno Y B, Khoroshilova I K, Nikolaeva E E 1984 *Soviet powder metallurgy and metal ceramics* **2** 157.
[23] Ordan'yan S S, Nesmelov D D, Vikhman S V 2009 *Inorganic Materials* **7** 754.
[24] Ordan'yan S S, Rumyantsev V I, Nesmelov D D, Korablev D V 2012 *Refractories and Industrial Ceramics* **2** 108.
[25] Ordan'yan S S 1992 *Refractories and Industrial Ceramics* **9** 411.
[26] Nesmelov D D, Vikhman S V, Novoselov E S, Perevislov S N, Ordan'yan S S 2019 *IOP Conference Series: Materials Science and Engineering* **1** 012066.
[27] Otani S, Honma S, Ishizawa Y 1993 *Journal of alloys and compounds* **1-2** 286.
[28] Kosolapova T Y 1986 Properties, Production, and Application of Refractory Compounds (Moscow, Metallurgy)
[29] Ordanyan S S, Unrod V I 2005 *Refractories and Industrial Ceramics* **4** 276.

IOP Conf. Series: Materials Science and Engineering **848** (2020) 012060 doi:10.1088/1757-899X/848/1/012060

Reactive hot-pressed LaB_6-W_2B_5 ceramics

D D Nesmelov[1], E S Novoselov[1], D D Titov[2], S S Ordan'yan[1] and S N Perevislov[3]

[1]Saint-Petersburg State Institute of Technology (SPSIT), Saint-Petersburg, Russia
[2]Baykov Metallurgy and Materials Institute (IMET RAS), Moscow, Russia
[3]Grebenschikov Institute of Silicate Chemistry (ISC RAS), Saint-Petersburg, Russia

E-mail: dnesmelov@yandex.ru

Abstract. The heterophase LaB_6-W_2B_5 composite with LaB_6 content of 50 vol. % was fabricated by reactive hot pressing of the powder mixture of LaB_6, W and amorphous boron at 1800 °C and 30 MPa for 15 min. The structure and phase composition of the synthesis products were analyzed by SEM, EDX and XRD methods. The effect of the pressure-induced anisotropic W_2B_5 grains growth was observed. The elastic modulus of composites investigated by resonance frequency method, and the Vickers hardness via indentation method.

1. Introduction

Lanthanum hexaboride is one of the most widely used thermionic emitter material [1-3] and promising candidate for solar-heat absorption and other plasmonic resonance based applications [4-7]. High melting point of lanthanum hexaboride (2510 - 2715 °C) and strong covalent bonds B-B in the B_6 octahedron leads to the poor sinterability of polycrystalline LaB_6. An active increase in the diffusion mobility of atoms in the LaB_6 crystal lattice is observed only at temperatures above 0.8 of the melting point. So, the pressureless sintering temperature of polycrystalline LaB_6 reaches 2300 °C [8], which is quite high and leads to intensive grains growth. The most popular technics for activation of LaB_6 sintering with the use of externally applied pressure are the hot-pressing (HP) and spark-plasma sintering (SPS) [7, 9]. The densification of polycrystalline LaB_6 can be further enhanced by introducing second-phase additives, which interact with LaB_6 to form a eutectic [10-12]. The preparation of ceramic composites based on LaB_6-MeB_2 (Me = d-metals) and LaB_6-SiC eutectic systems was studied in most detail [13, 14].

Another way to decrease the sintering temperature is to provide the *in-situ* reactive sintering via hot pressing or SPS. Different starting mixtures were used by several researchers to obtain composites in LaB_6-based systems: La_2O_3-B_4C [15], La_2O_3-ZrO_2-B_4C [16], La_2O_3-ZrO_2-B_4C-C [17], Al-B_2O_3-Ti-LaB_6-TiB_2 [18], LaH_2-GdH_2-B [19], LaH_2-BaH_2-B [20] etc. In most cases, the reactive hot-pressing allows decreasing of sintering temperature and leads to the formation of a dense sintered body.

The aim of this work is to investigate the possibility of *in-situ* synthesizing LaB_6–W_2B_5 composite ceramics by reactive hot-pressing method from the LaB_6 + W + B starting mixture at a relatively low temperature of 1800 °C, and to analyze phase and chemical composition of LaB_6-W_2B_5 ceramics, using as a reference samples furnace-synthesized LaB_6-W_2B_5 powder mixtures. Solid-state synthesis of W_2B_5 usually carries out by the reaction of elemental tungsten and boron: $2W + 5B = W_2B_5$. The reaction is exothermic (ΔH = 176 kJ / mol), resulting in overheating of the mixture and rapid growth of large crystals. The LaB_6 particles role in our experiment was also to block the nucleating centers of W_2B_5 grains growth in order to reduce the rate of exothermic heat generation to inhibit crystal growth.

2. Materials and Method

Commercial powders of LaB_6 (99.0 wt. %) metallic tungsten (99.9 wt. %) and amorphous boron (99.0 wt. %) were used in this study. The average size of LaB_6, W and B_a starting powders (the 50th percentile particle size, d_{50}) determined by laser diffraction was 12, 60 and 1.1 μm respectively. Powders were mixed and milled on vibratory ball-mill in gasoline-solvent media with tungsten balls (balls to material weight ratio B:M = 10:1) for 20 hours. Powders were mixed in ratio of 26.5 wt. % LaB_6 – 64.1 wt. % W – 9.4 wt. % B_a for obtaining the LaB_6- 50 vol. % W_2B_5 composite. The addition of 7 wt. % W appeared in mixture due to the wear of W balls, which was determined by weighing the balls before and after ball-milling. This additional tungsten was taken into account when weighing the initial mixture. After the ball-milling, the average particle size decreased to d_{50} = 2.9 μm.

The vacuum-dried powder mixture was placed into graphite press-form with an internal diameter of 25 mm and sintered on a Thermal Technology HP20-3560-20 hot press at an isothermal holding temperature of 1800°C and a pressure of 30 MPa for 15 min in argon gas. After hot-pressing samples were polished for microstructural and micromechanical analysis.

As a reference sample, the furnace-synthesized LaB_6-W_2B_5 powder mixture was prepared from the same starting powders. The vacuum-dried powder mixture was compacted via cold pressing at 50 MPa to the 50 %-porosity green-body, which were placed into the graphite crucible and sintered at 1800 °C at 10^{-2} Pa for 1 hour in a SSHVA vacuum-furnace.

The phase composition of ceramics was determined using the X-ray diffraction (XRD) analysis (Rigaku SmartLab 3, Cu Kα radiation, λ = 1.5406 Å) in the 2θ range of 10–90° with 0.01 ° step. The microstructure of samples was investigated using an optical microscope (Meiji 7200 with Thixomet-Lite system for image analysis) and SEM Tescan Vega 3SBH. Analysis of elements concentration was carried out using Aztec X-Act analyzer (Oxford Instruments).

The elastic (Young's) modulus of ceramics was measured using the Zvuk-130 meter of natural oscillation frequencies implements the method of forced oscillations (resonance method) [21], based on the relationship of the propagation rate of acoustic waves, *Cl,* with the density, ρ, and Young's modulus, E:

$$Cl = \sqrt{E/\rho}, \qquad (1),$$

Indentation tests were carried out with the use of Vickers hardness tester PMT-3 under the load of 44 N. For calculation of *Hv* (Vickers hardness, GPa) we use equation (2):

$$Hv = 1854 \cdot \frac{P}{D^2}, \qquad (2),$$

where *P* – load, N; *D* – measured diagonal of the indentation, μm.

At least 10 indentations were made, from which the mean values of the lengths of the diagonals were calculated.

3. Results and discussion

Dense ceramics were obtained by hot-pressing, while about 25%-porosity samples were obtained after synthesis in a vacuum furnace at the same temperature. The relative density of hot-pressed LaB_6-W_2B_5 samples, measured by the hydrostatical method, was 92 % of the theoretical value. Taking into account the density of samples, obtained by both methods, the XRD measurements of hot-pressed ceramics were provided on the cross-sections, while the furnace-synthesized samples were crushed in a mortar and then analyzed by powder XRD method.

The XRD-spectra of initial powder mixture (blue curve), vacuum-furnace synthesized mixture (red curve) and hot-pressed LaB_6-W_2B_5 ceramics (green curve) presented in Figure 1.

The initial powder mixture spectra contain diffraction peaks of W metallic and LaB_6. The presence of boron is not detected because of its amorphous structure. Weak reflections of tungsten oxides are present above the background, but their intensity is so small that the scale of the illustration does not allow them to be distinguished. After the synthesis, there are distinct diffraction peaks of well-crystallized two main phases – cubic LaB_6 and hexagonal W_2B_5. The reflexes of impurity WB-phase are clearly defined only at the red curve – vacuum-furnace synthesized LaB_6-W_2B_5 spectra. There are no WB peaks in the spectra of hot-pressed LaB_6-W_2B_5. This fact is due to a relatively long holding time (1h) at a high temperature under vacuum, which leads to partial loss by vaporization and sublimation of reactants and results in chemical composition fluctuation. In contrast, during hot-pressing, the holding time is shorter and the evaporation of the components is prevented by the pressure of the argon atmosphere.

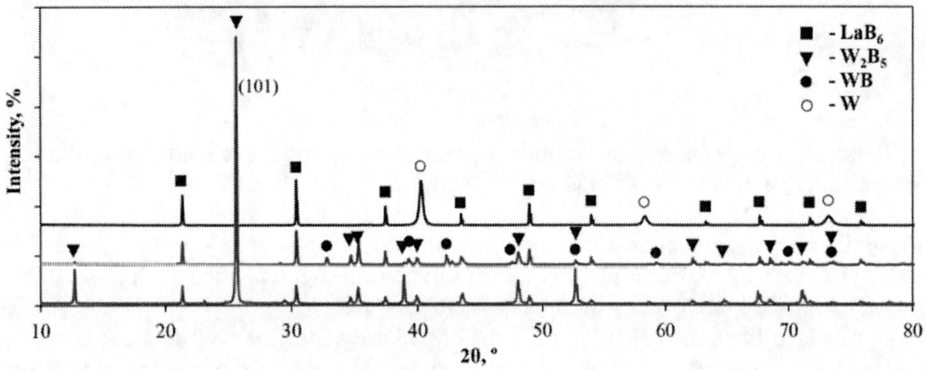

Figure 1. XRD patterns of LaB_6-W_2B_5 initial reaction mixture and synthesized composites

A disproportionately high diffraction peak of W_2B_5 at the $2\theta = 25.6°$ evidence a predominant reflection from (101) lattice planes. This indicates anisotropic W_2B_5 grain growth.

SEM-micrographs of hot-pressed samples (Figure 2A, B) demonstrates an uneven phase distribution in the structure of composites. Large-area white fields in Fig. 1A and B are a result of anisotropic W_2B_5 grains growth in a (101) plane, which is in a good agreement with XRD-spectra data.

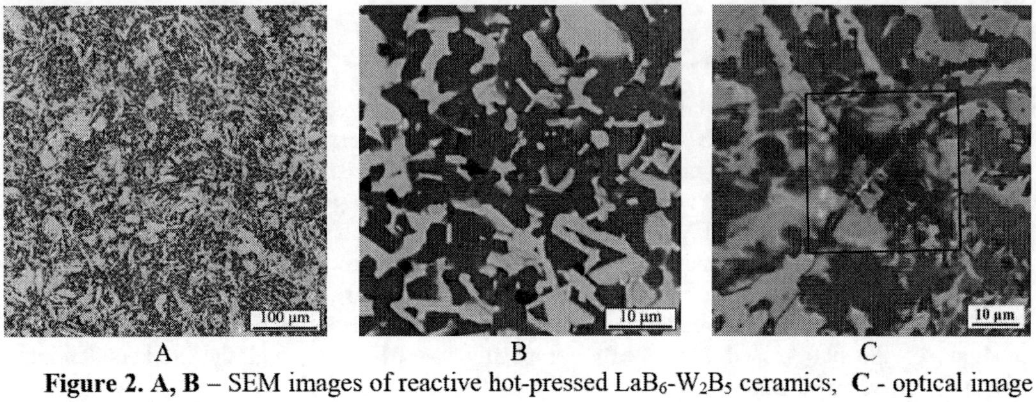

Figure 2. A, B – SEM images of reactive hot-pressed LaB_6-W_2B_5 ceramics; **C** - optical image of Vickers indentation

As can be seen from Fig. 1B, W_2B_5 grains have a plate-like morphology and different orientation, but predominantly orientation of the largest W_2B_5 grains is perpendicular to the pressure direction. The average size of the largest W_2B_5 particles along (101) plane is of several tens of μm, while the thickness of the plates is 1-5 μm. There were no plate-like grains observed in the reference samples of a vacuum-furnace synthesized mixture, so we can conclude anisotropic growth of W_2B_5 grains is caused by the external pressure effect (Figure 3).

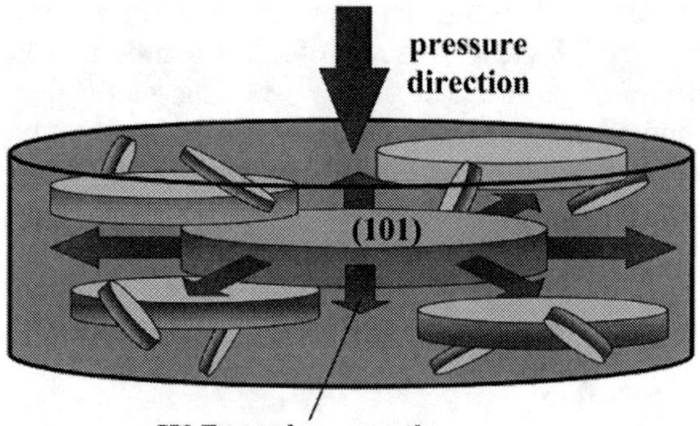

Figure 3. Scheme of W_2B_5 anisotropic grains growth under the load during the hot-pressing: W_2B_5 plate-like grains are distributed in a LaB_6 matrix

LaB_6 grains have a conventional morphology without any signs of anisotropic growth. The average LaB_6 grain size does not exceed 5 μm, but it is observed some agglomerates with a size of about 10-20 μm, caused by local inhomogeneity of initial powder mixture.

The phases area fractions calculation from SEM micrographs, as well as EDX-analysis, showed that the composition of ceramics is near the equivolume. The presence of a large number of plate-like W_2B_5 grains allows us to determine the structure of the equivolume composite as self-reinforced.

The average hardness of LaB_6-W_2B_5 ceramic composite was measured as Hv = 16 GPa at the load of 44 N. Such a low hardness, compared to the theoretical value calculated by the rule of mixtures (26.9 GPa), may be caused not only by the effect of porosity in ceramics but also by the previously noted effect of microplastic deformation in heterophase composite ceramics [22].

The average Young's modulus (610 GPa) also was decreased in comparison with theoretical value (632 GPa). This predictable decrease is due to the imperfection of the composite structure, primarily, the presence of porosity.

4. Conclusion

The heterophase LaB_6-W_2B_5 composite with LaB_6 content of 50 vol. % was fabricated by reactive hot pressing of the powder mixture of LaB_6, W and amorphous boron at 1800 °C and 30 MPa for 15 min and compared to the furnace-synthesized LaB_6-W_2B_5 powder mixture, prepared from the same starting powders at the same holding temperature. The structure and phase composition of the synthesis products were analyzed by SEM, EDX and XRD methods. The 92% relative density hot-pressed ceramics contained only two main phases – cubic LaB_6 and hexagonal W_2B_5. The impurity of WB-phase was detected in the vacuum-furnace synthesized LaB_6-W_2B_5.

The effect of the pressure-induced anisotropic W_2B_5 grains growth was observed. The elastic modulus of the hot-pressed LaB_6-W_2B_5 composite was 610 GPa, and the Vickers hardness via indentation method 16 GPa.

Acknowledgments
The reported study was funded by RFBR project No 18-33-20221. Authors are grateful to the Engineering Centre of Saint-Petersburg State Institute of Technology for the equipment provided.

IOP Conf. Series: Materials Science and Engineering **848** (2020) 012060 doi:10.1088/1757-899X/848/1/012060

References

[1] Lafferty J M 1951 *Journal of Applied Physics* **3** 299
[2] Taran A, Voronovich D, Oranskaya D, Filipov V, Podshyvalova O 2013 *Functional materials* **4** 485
[3] Paderno Y B, Taran A A, Voronovich D A, Paderno V N, Filipov V B 2008 *Functional materials* **1** 63
[4] Kowalczyk J M, Hadmack M R, Szarmes E B, Madey J M 2014 *International Journal of Thermophysics* **8** 1538
[5] Mattox T M, Coffman D K, Roh I, Sims C, Urban J J 2018 *Materials* **2** 226
[6] Mattox T, Urban J 2018 *Materials* **12** 2473
[7] Sani E, Mercatelli L, Meucci M, Zoli L, Sciti D 2017 *Scientific Reports* **1** 718
[8] Kauer E 1963 *Physics Letters* **3** 171
[9] Li-Hong B, Jiu-Xing Z, Shen-Lin Z, Yong-Feng W 2010 *Chinese Physics Letters* **10** 107901
[10] Ordan'yan S S, Paderno Y, Khoroshilova I K 1983 *Soviet powder metallurgy and metal ceramics* **251** 946.
[11] Loboda P I, Kisla G P, Bogomol I I, Sysoev M A, Karasevskaya O P 2009 *Inorganic Materials* **3** 246.
[12] Ordan'yan S S, Nesmelov D D, Vikhman S V 2009 *Inorganic Materials* **7** 754.
[13] Yang X, Wang X, Wang P, Hu K, Li Z, Zhang J 2017 *Journal of Alloys and Compounds* **704** 329
[14] Wang X, Zhang J X, Yang X Y, Hu K, Zhang J W 2017 *Advances in Applied Ceramics* **3** 132
[15] Sonber J K, Sairam K, Murthy T C, Nagaraj A, Subramanian C, Hubli R C 2014 *Journal of the European Ceramic Society* **5** 1155
[16] Min G H, Gao R, Yu H S, Han J 2005 *Key Engineering Materials* **297** 1630
[17] Gao R, Min G, Yu H, Zheng S Q, Lu Q, Han J, Wang W 2005 *Ceramics International* **1** 15
[18] Popov O, Loboda P, Klepko O, Solovjova T 2019 *Advances in Applied Ceramics* **4** 217
[19] Bao L H, Zhang J X, Zhang N, Li X N, Zhou S L 2012 *Physica Scripta* **3** 035710
[20] Zhou S L, Zhang J X, Bao L H, Yu X G, Hu Q L, Hu D Q 2014 *Journal of Alloys and Compounds* **611** 130
[21] Buravov A D, Vikhman S V, Kovarskaya E Z, Moskovenko I B, Nekrasova O K, Smirnova M A 2012 *Refractories and Industrial Ceramics* **6** 436
[22] Ordanyan S S, Unrod V I 2005 *Refractories and Industrial Ceramics* **4** 276.

IOP Conf. Series: Materials Science and Engineering **848** (2020) 012061 doi:10.1088/1757-899X/848/1/012061

Copper and cerium co-substituted hydroxyapatite: powders synthesis and sintering

Y O Nikitina, N V Petrakova, A A Egorov, D D Titov, A A Ashmarin, S M Barinov and V S Komlev

Baikov Institute of Metallurgy and Materials Science of the Russian Academy of Sciences (IMET RAS), Moscow, Leninskiy prospect 49, Russia
E-mail: nyo.94@mail.ru, petrakova.nv@mail.ru

Abstract. Copper- and cerium-substituted and copper-cerium co-substituted hydroxyapatite (HA) nanopowders were synthesized through aqueous precipitation method. The obtained materials were studied by XRD, SEM, BET absorption and dilatometry. The effects of the cation substitution for calcium ions in HA structure on the phase composition, lattice parameters, shrinkage and microstructure of the ceramics were studied and discussed. It was concluded that cerium ions incorporation in small amount (0,1 mol. %) contribute to a sintering process more than rather higher amount of copper ions (5 mol. %).

1. Introduction

Development of functionally-oriented biomaterials targeted for osteoplastic surgery devotes on bone reconstruction process and effective treatment. Eventually, the osteoconductive material for bone reconstruction has been the source of structural components while serves as matrix for new-bone formation. Various components could be involved into matrix material to design the specific function orientation. In proposed work we focused on development of the material for bone reconstruction that possessed antibacterial activity and luminescent properties. Bone defect treatment involves the extensive surgery that yields several complications such as postoperative infections, selection of systemic therapy and locale diagnostics. To find a way forward these problems we regard calcium hydroxyapatite (HA) ceramics with antibacterial and luminescent agents as functionally-oriented bone substitute. Being of great biocompatibility and bioactivity, HA ceramics provide high investigation value due to opportunities for improving mechanical and biological properties through the structure modifications. Cation and anion substitution in HA crystal lattice causes the micro-deformations and particle size decrease, rising of dissolution rate at the crystal-solution boundary. In the meantime, dopant type set the specific properties of biomaterial behavior *in vivo* [1-6].

Copper ions have been known as antimicrobial agent against a wide range of bacterial strains, particularly, *E. coli*, *St. aureus* [7-9]. Copper-substituted hydroxyapatites are also of interest in pigment synthesis due to copper ions availability in a range of different colors [10]. At the same time, cerium of number of rare earth elements have a significant ability to fluorescence and is used in fluorescent imaging [11-13]. Cerium ions possess bactericide effect against several gram-negative and gram-positive bacterial strains [14-17]. Cerium containing compounds are used in antibacterial and antitumor medicine and fluorescent visualization [18, 19]. HA structure allows the incorporation of wide range of ionic substitution from monovalent, divalent to trivalent and tetravalent ions [7, 15, 20-21]. Generally, the divalent copper ions stand into the Ca sites in HA structure and as a result the solid solution of substitution forms [22]. Trivalent Ce ions have greater charge then Ca ions and close values of electronegativity and radius (Ce^{3+}: $\chi=1.12, r=1.03$ Å and for Ca^{2+}: $\chi=1.0, r=1.0$ Å). Thus, the charge compensation can occur through cation vacancies and drifting of atoms in cation cites. On the other hand, copper ions have higher electronegativity and less radius than Ca (Cu^{2+}: $\chi=1.9, r=0.8$ Å) that could result in compensation in ionic sublattice of co-substituted HA structure.

Therefore, the bioactive functionally-oriented HA materials with complex of antibacterial and fluorescent properties is of great interest for developments of bone tissue engineering, cancer treatment and optical bioimaging diagnostic. Still, the Ca^{2+} co-substitution with Cu^{2+} and Ce^{3+} ions has not been studied before.

In this paper, we present the synthesis of copper- and cerium-substituted HA nanopowders in order to obtain antibacterial and luminescent properties. The effects of copper and cerium substitution for calcium ions in HA structure on the phase composition, lattice parameters, sintering process and microstructure of ceramics were studied and discussed.

2. Materials and methods

2.1. Materials and synthesis method

Pure HA, copper-substituted HA (Cu-HA), cerium-substituted HA (Ce-HA) and copper-cerium co-substituted HA (Cu-Ce-HA) powders were prepared by precipitation technique from aqueous solutions of calcium nitrate tetrahydrate ($Ca(NO_3)_2 \times 4H_2O$), copper nitrate trihydrate ($Cu(NO_3)_2 \times 3H_2O$) , cerium nitrate hexahydrate ($Ce(NO_3)_3 \times 6H_2O$) and ammonium hydrophosphate ($(NH_4)_2HPO_4$). Concentration of all used solutions was 0,5 M. All reagents used in the work were corresponded to the qualification of "analytical grade" and higher. Designation of compositions and calculated content of dopants are presented in table (1). Pure HA was prepared by slowly addition of $(NH_4)_2HPO_4$ solution to $Ca(NO_3)_2$ solution while mixing. Cu-HA, Ce-HA and Cu-Ce-HA were prepared the same way, only $(NH_4)_2HPO_4$ solution was added to the mixed solution of calcium nitrate and dopant nitrate. During the synthesis pH level was kept at 11 ± 0,5 by addition of ammonia solution. Precipitate was filtered, washed by water and dried at 80 °C during 24 h. Then the compact cakes were crushed to the powder using mortar and pestle and sieved through a sieve with cell size of 100 μm.

Table 1. Designation of compositions and calculated content of dopants

Sample	Cu^{2+}, mol. %	Ce^{3+}, mol. %	(Ca+Cu+Ce)/P
HA	0	0	1,67
Cu-HA	5	0	1,67
Ce-HA	0	0,1	1,67
Cu-Ce-HA	5	0,1	1,67

2.2. *Sample characterization*

Phase composition, lattice parameters and average crystallite size of obtained materials were estimated by X-ray diffraction analysis (XRD, Shimadzu) using CuKα radiation (average wavelength λ =1,54183 Å). Data were collected in 2θ range of 10–60 deg., step 2θ of 0.02 deg. and counting time of 2 s/step. X-ray analyses were performed on the powered samples calcinated at 1300 °C. The lattice parameters *a* and *c* (Å) for hexagonal HA structure, were calculated from the diffraction peaks *(002)* and *(211)*, respectively. The average crystallite size D (nm) of the powders was calculated from XRD data using the Scherrer's equation. The specific surface area of the obtained powers was measured by BET method (TriStar 3000, Micromeritics). Sintering of materials was evaluated by measuring continuous shrinkage of pressed samples using horizontal dilatometer (DIL 402C Netzsch) with corundum equipment. Cylindrical samples (pressure 50 MPa) with a diameter of 5 mm and height of 4 mm were pressed for sintering study. Measuring was performed in argon atmosphere at constant heating rate of 10 °C / min to the temperature of 1400 °C with. Microstructure of the ceramics was investigated by scanning electron microscope (SEM, Tescan Vega II SBU) with integrated Energy dispersive X-ray microanalyzer (EDX, INCA Energy).

3. Results

XRD analysis data of the synthesized and calcined powders indicated that all the materials consisted of the single crystalline phase of hydroxyapatite (Figure 1). Calculations of the lattice parameters revealed that the introduction of dopants into HA structure led to changes in crystal lattice (Table 2). Parameters *a* and *c* of Cu-HA (*a*=9,416 Å, *c*=6,884 Å) were less than for pure HA (*a*=9,425 Å, *c*=6.887 Å). This indicated the substitution of calcium with copper, since the ionic radius of copper (0.8 Å) was less than the ionic radius of calcium (1,06 Å). Parameters for Ce-HA (*a*=9,427 Å, *c*=6,886 Å) were insignificantly lowered than for pure HA, within error. Incorporation of cerium ions with larger radius (1,15 Å) in amount of 0,1 mol. % into the HA structure did not cause noticeable change of the HA crystal structure parameters. This could be explained by low amount of cerium and low method sensitivity, or by incomplete entry of the element into the structure. The co-substitution with copper and cerium ions of HA structure led to a decrease of parameter *a* and slightly increase of parameter *c* of the crystal lattice. This indicated an atoms redistribution in cationic/anionic HA sublattice.

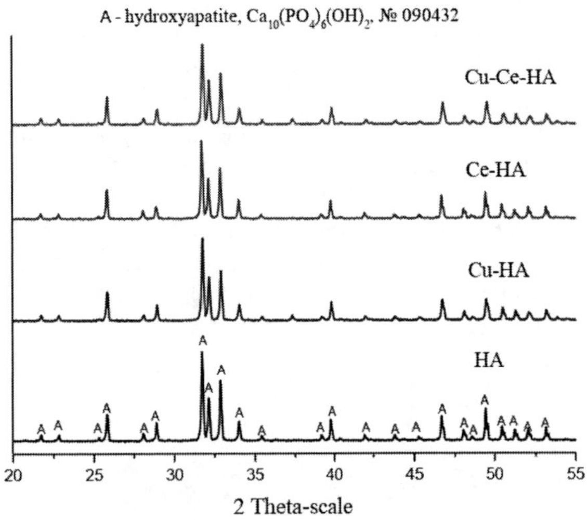

Table 2. Lattice parameters of substituted HA

Sample	a (Å)	c (Å)
HA	9,425	6,887
Cu-HA	9,416	6,884
Ce-HA	9,427	6,886
Cu-Ce-HA	9,406	6,888
$Ca_{10}(PO_4)_6(OH)_2$ 090432	9,418	6,884

Figure 1. XRDs of obtained powders of substituted HA.

The plot of linear shrinkage vs the temperature for a compacted samples of the pure and substituted HA powders are presented on Figure 2, a. Its derivative curves are given in Figure 2, b. All examined powders had a similar surface area in range of 120-140 m^2/g, i.e. an average particle size was about 13-16 nm in approximation of sphere particle shape. Shrinkage measuring of the materials revealed that a slight intrinsic expansion of the powder compacts occurred only for pure HA and Cu-HA at 210-550 °C. The initial sintering state, when the interparticle neck formed and growth, started at 570 °C for both HA and Cu-HA powders and at 500 °C for both Ce-HA and Cu-Ce-HA powders. The active sintering process with a classic linear curve trend occurred as follows: at above 880-1000 °C for both HA and Cu-HA samples ($\Delta l/l_0$=8,0÷22,6 %), at 810-1010 °C for Ce-HA sample ($\Delta l/l_0$=6,5÷22,6 %) and at 770-955 °C for Cu-Ce-HA sample ($\Delta l/l_0$=5.5÷19,5 %). During this intermediate stage the sintering rate is maximum. At the end of this stage for each sample the shrinkage curve had been gentler in slope corresponded to slowing down of the sintering stage. At 1300 °C, the traditional temperature of the end of HA sintering, shrinkage value was measured as follows: 26,6 % for HA, 24,7 % for Cu-HA and 25,1 % for both Ce-HA and Cu-Ce-HA. Based on dilatometry data it could be

IOP Conf. Series: Materials Science and Engineering **848** (2020) 012061 doi:10.1088/1757-899X/848/1/012061

noted that copper doping resulted in light decrease of HA sintering rate due to reducing of diffusion processes [23]. Cerium ions significantly decreased the sintering rate (Figure 3, b). While sintering is the mass transfer process, the sintering rate and other parameters depended on material's mass. So, if molar mass of Ce (M=140,12 g/mol) is much higher than for Ca (M=40,08 g/mol), its atom diffusion is restrained in compare with Ca-atoms. Co-substituted Cu-Ce-HA demonstrate the same sintering behavior as Ce-HA and low sintering rate. Also some widening took place in the plot of Cu-Ce-HA shrinkage rate vs the temperature pointed to an additional process of phase allocation or lattice deformation during thermal activation. So it could be conclude that cerium ions incorporation prevented mostly the sintering kinetic process of HA-ceramics.

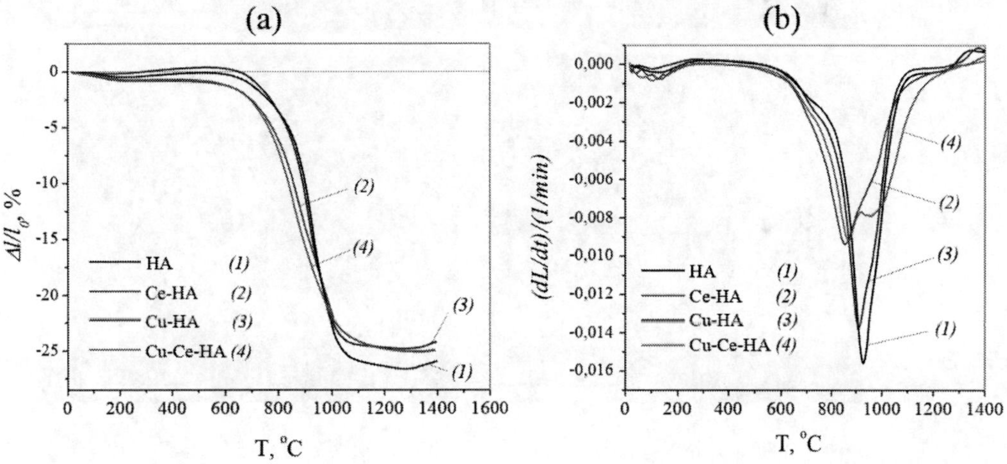

Figure 2. Linear shrinkage of obtained HAs vs the temperature (a) and shrinkage rate (b).

Figure 3 demonstrates the structure of spalling surface of ceramic substituted HAs samples. Pure HA ceramics consisted of crystals from 2 to 6 μm in size with strong boundaries (Figure 3, a). There was small amount of intercrystalline pores of average size of 0,5-1,0 μm. Copper incorporation into HA resulted in presence of irregularity of crystal shape and more friable surface. Also pores became bigger in size, number of pores has increased. The introduction of cerium into HA structure didn't significantly change the microstructure of HA ceramic. Cu-Ce-HA ceramics characterized by roughness of the crystal surface. Crystals had irregular shape and size from 2 to 5 μm. Intercrystalline pores are large and numerous.

Thus, the copper substitution of HA ceramics led to irregular crystal structure with rising of number of intercrystalline pores. Cerium introduction into HA structure didn't significantly affect the crystal structure, but inhibited sintering process and reduced the shrinkage of the materials.

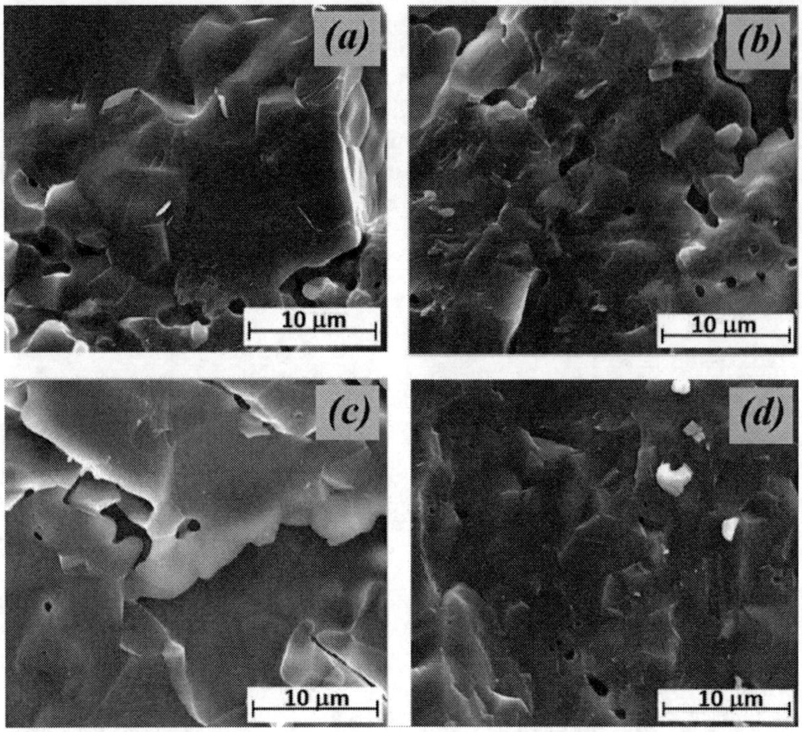

Figure 3. SEM-photographs of HA (a), Cu-HA (b), Ce-HA (c), Cu-Ce-HA (d).

4. Conclusion

In the work there were synthesized powders of Cu-HA, Ce-HA and Ce-Cu-HA with calculated copper content of 5 mol. %, cerium content of 0,1 mol. %, and Cu/Ce of 5/0.1 mol. %. The synthesis was performed through aqueous precipitation method. The obtained powders of cation substituted HAs were of average size of 13-16 nm. Materials were studied by XRD, SEM, BET absorption and dilatometry. It was found that the introduction of copper into HA structure reduced lattice parameters *a* and *c* due to smaller ion radius Cu than Ca. Also it led to irregular crystal structure with increase of number of intercrystalline pores in the HA ceramics. Incorporation of Ce ions with larger radius (1,15 Å) in amount of 0,1 mol. % into the HA structure did not change the crystal structure parameters. Ce didn't significantly affect the ceramics microstructure, but rather inhibited sintering process and reduced the shrinkage of the material. The co-substitution of HA with Cu and Ce ions led to a decrease of parameter *a* and slightly increase of parameter *c*, that caused for atoms redistribution in cationic/anionic HA sublattice. The sintering behavior of Cu-Ce-HA was similar to Ce-HA. Thus, small Ce amount (0,1 mol. %) contribute to a sintering process more than rather higher amount of copper ions (5 mol. %).

5. Acknowledgements

The reported study was funded by RFBR, project number 19-33-90235.

References

[1] Dorozhkin S V 2013 *J. Materials* **6** 3840
[2] Barinov S M and Komlev V S 2014 *Bioceramics based on calcium phosphates* (Moscow: Science) 201 p.
[3] Hench L.L. *J. Am. Ceram. Soc.* 1991 **74** 1487-1510

[4] Doremus R H 1992 *J. Mater. Sci.: Mater. Eng.* **27** 285
[5] Eliaz N and Metoki N 2017 *J. Materials* **10** 334
[6] Vallet-Regi M 2001 *J. Chem. Soc., Dalton Trans.* 97
[7] Fadeeva I V, Gafuov M R, Kiiaeva I A, Orlinskii S B, Kuznetsova L M, Filippov Y Y, Fomin A S, Davydova G A, Selezneva I I and Barinov S M 2017 J. *BioNanoScience* **7** 434
[8] Marques C F, Olhero S, Abrantes J C, Ferreira S, Vieira S and Ferreira J 2017 J. *Ceramics International* **43** 15719-15728
[9] Elfakhri S O 2014 *PhD Thesis* 240 p
[10] Kazin P E, Zykin MA, Tret'yakov Yu D and Jansen M 2008 *Russian Journal of Inorganic Chemistry* **53** (3) 362-366
[11] Chen F, Huang P, Zhu Y J, Wu J and Cui D-X 2012 J. *Biomaterials.* **33** 6447-6455
[12] Victor S P, Paul W, Vineeth V M, Komeri R, Jayabalan M and Sharma C P 2016 J. *Colloids Surf B Biointerfaces* **145** 539-547
[13] Okrushko E N, Seminko V V, Maksimchuk P O and Bespalova I I, Kononets N V, Viagin O G, Malyukin Y V 2017 J. *Low Temperature Physics* **43** 795-801
[14] Yingguang L, Zhuoru Y and Jiang C 2007 *J. Rare earths* **25** 452-456
[15] Ciobanu G Bargan A M and Luca C 2015 *Ceramics International* **41** 12192-12201
[16] Kaygusuz H, Torlak E, Akın-Evingür G, Özen I, Klitzing R and Erim F B 2017 International Journal of Biological Macromolecules **105** 1161-1165
[17] Rajeshkumar S and Naik P 2017 *Biotechnol Rep (Amst).* **17** 1-5
[18] Greenhalgh D G 2009 *J. Clin. Plast. Surg.* **36** 597-606
[19] Alpaslan E, Yazici H, Golshan N, Ziemer K and Webster T 2015 *J. ACS Biomater. Sci. Eng.* **1** 1096-1103
[20] Smirnov V V, Barinov S M, Smirnov S V, Krylov A I, Antonova O S, Goldberg M A, Obolkina T O, Konovalov A A and Leonov A V 2019 *Inorganic materials* **77** (7) 715-723
[21] Smirnov V V, Antonova O S, Smirnov S V, Goldberg M A, Komlev V S and Barinov S M 2017 *Inorganic materials* **53** (12) 1254-1260
[22] Zykin M A, Vasiliev A V., Trusov L A, Dinnebier R E, Jansen M and Kazin P E 2018 *Journal of Solid State Chemistry* **262** 38-43
[23] Nikitina Y O, Petrakova N V, Ashmarin A A, Titov D D, Shevtsov S V, Penkina TN, Kuvshinova E A, Barinov S M, Komlev V S and Sergeeva N S 2019 *Inorganic Materials* **55** (10) 1061-1067

IOP Conf. Series: Materials Science and Engineering **848** (2020) 012062 doi:10.1088/1757-899X/848/1/012062

Distribution of impurities in ceramics based on zirconium nitride obtained using the oxidative constructing approach

I A Kovalev[1,2], S V Shevtsov[1], K G Vorkachev[1], A A Ashmarin[1], A V Shokodko.[1,2], A I Ogarkov[1], N A Ovsyannikov[1], G P Kochanov[1,2], A V Tenishev[2], V N Kazakova[2], D P Shornikov[2], E A Shokodko[3], S S Strelnikova[2], A S Chernyavskii[1,2] and K A Solntsev[1,4]

[1]IMET RAS, Moscow, Russia
[2]National Research Nuclear University MEPhI (Moscow Engineering Physics Institute), Moscow, Russia
[3]Moscow State University of Civil Engineering (National Research University), Moscow, Russia
[4]Lomonosov Moscow State University, Moscow, Russia
E-mail:aogarkov@imet.ac.ru

Abstract. The localization of impurity phases in ceramics created by direct nitridation of zirconium rolled metal was established. For ceramics synthesized at temperatures above the melting point of the metal, the impurity phases are localized along grain boundaries and at macrodefects. In a composite heterostructure of the composition ZrN –(ZrN_x/*α*--solid solution of nitrogen in metal/ZrN_x) – ZrN, impurity phases are predominantly localized in an α-solid solution of nitrogen with their subsequent displacement to grain boundaries and macrodefects at the end of the nitridization process.

1. Introduction

Metal nitrides of the titanium subgroup have found application in various branches of technology: thin films in microelectronics [1–3], protective and hardening coatings [4–6], and functional ceramics [7–9]. Methods of creating a product from transition metal nitrides are based on classical technologies of molding nitrides of simple forms from powders with subsequent sintering; or on compaction of metal powders with nitriding of the manufactured compact. However, high requirements for the purity, granulometry and phase composition of metal and nitride powders suitable for compacting, as well as the complexity of the processes of forming thin-walled products and products of complex geometry, high sintering temperatures significantly limit the proposed range of ceramic products [10-12]. The oxidative engineering approach allows you to create ceramic products of various forms by direct single-stage oxidation of metal billets [13], including the nitridation process above the melting point of metal [14] due to the formation of a shell from a refractory nitride layer at the initial stage. In this case, the synthesis of ceramics passes through the stages of formation of $Zr_{solid\ solution}/ZrN_{1-x}/ZrN$ heterostructures [15–17].

The initial zirconium and nitrogen contain impurities, the composition and quantity of which correspond to the chemical qualification of the reagent, which can affect the composition and properties of the synthesized ceramics. The aim of this work is to study the distribution of impurity phases formed in the process of high temperature zirconium nitridation.

2. Experimental part

E-110 grade zirconium (TU 95.166-83) in the form of a 0.30 mm thick foil and a bar 4.0 mm in diameter were used as the starting material. Zirconium nitridation was carried out at a temperature of 2400 ° C in an atmosphere of nitrogen classification “ultrapure", 99.999% (table 1). During nitridation, the nitrogen pressure in the reactor was maintained at the level of $(1.5\pm0.1) \times 10^5$ Pa. The temperature of the sample was determined by a Luma Sense IMPAC ISR 50-LO optical pyrometer. Synthesized ZrN_x ceramics with different degrees of nitridation (x = 0.33-1), preserving the size of the original metal billet.

Table 1. The content of impurities in volume percent for Zr grade E-110 and gaseous nitrogen classification “ultrapure ", 99.999%.

E-110 grade zirconium	Impurities, %	Nb	Al	Be	C	Fe	Si	Ca	Cu
		0.9-1.1	0.008	0.003	0.02	0.05	0.02	0.03	0.005
		Mn	Ti	Pb	O_2	Hf	Ni	Cr	Mo
		0.002	0.005	0.005	0.1	0.05	0.02	0.02	0.005
Gaseous nitrogen classification “ultrapure ", 99.999%	Impurities, %	O_2	H_2	Hydrocarbons (CH_4)	-	-	-	-	
		$5\cdot10^{-4}$	$2\cdot10^{-4}$	$3\cdot10^{-4}$					

The phase composition was studied on an XRD-6000 X-ray diffractometer (Shimadzu) in monochromatized copper radiation. X-ray diffraction patterns were obtained from the surface layer (5–7 μm) of the synthesized samples and from powder samples prepared by grinding ceramic fragments.

The surface cleavage morphology was studied by scanning electron microscopy (SEM) (Carl Zeiss microscopes with energy dispersive microanalysis systems from Oxford Instruments: LEO 1420, with INCA Energy 300; LEO 430i with ISIS 300; CrossBeam 1540EsB, with INCA Energy 350XT).

3. Results and discussion

The diffraction patterns characterizing the initial structure of zirconium rolled metal and the synthesized zirconium nitride ceramic are shown in figure 1. It follows from this analysis that after the complete nitridation process, the surface and volume of the ceramic consist exclusively of stoichiometric zirconium nitride ZrN (diffraction patterns 2 and 3).

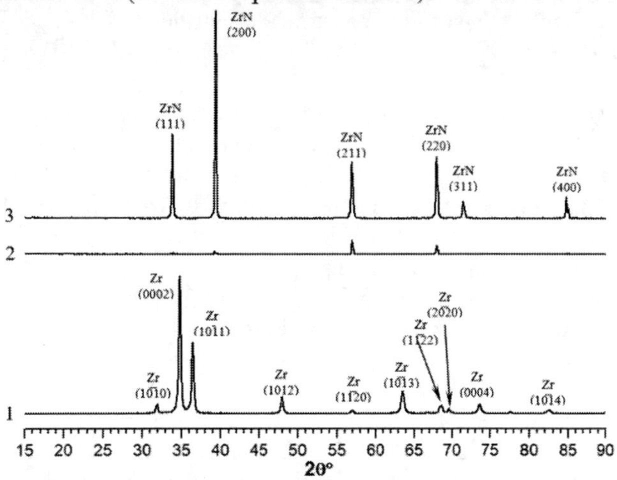

Figure 1. X-ray diffraction patterns of the studied samples: *1* - initial zirconium; *2* - surface of zirconium nitride; *3* - powder sample of zirconium nitride.

A typical image of the free surface of a completely nitridized zirconium sample is shown in Figure 2. Relief outlines of the grain boundaries appearing on the surface appear; randomly located pores with sizes up to several microns are present.

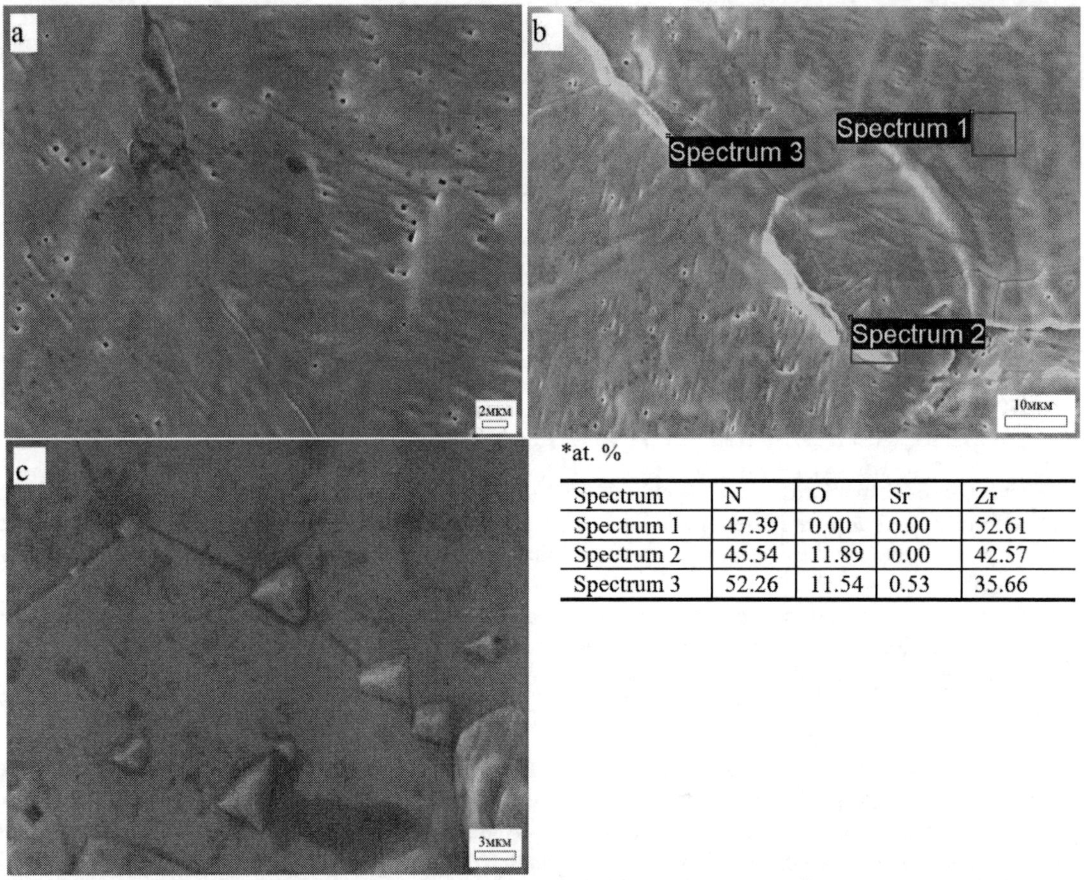

*at. %

Spectrum	N	O	Sr	Zr
Spectrum 1	47.39	0.00	0.00	52.61
Spectrum 2	45.54	11.89	0.00	42.57
Spectrum 3	52.26	11.54	0.53	35.66

Figure 2. A characteristic image of the free surface of nitride: without annealing (***a***), annealing for 1.5 hours (***b***) and 24 hours (***c***).

Figure 2b shows a typical view of the free surface of nitridized zirconium samples with an additional exposure (annealing) time in a nitrogen medium after completion of the main nitridization process. The microstructure of the surface did not undergo significant changes; however, the formation of local regions containing inclusions of a different elemental composition and located along grain-boundary regions was noted (Figure 2b, c). By the time of annealing at 24 hours, the impurity phase is formed crystals 1-3 microns in size.

This distribution of oxygen-containing phases differs from the phase distribution described in [18] for zirconium nitride annealed in the presence of oxygen.

A study of the transverse cleavage of a ceramic sample of zirconium nitride obtained by nitridation of a zirconium rod for 6 days indicates a structural heterogeneity of the obtained material. The cleaved surface morphology is represented by three different nitride layers (figure 3).

In the outer stoichiometric ZrN layer with a thickness of about 400 μm along the grain boundaries that make up the nitride layer, no defects in the form of pores and inclusions were detected. On a transverse cleavage of a layer of zirconium nitride below it (~ 1200 μm thick), which has a nitrogen deficiency of 2-12 at. % [15–17], grain-boundary segregation of the impurity phase was detected

IOP Conf. Series: Materials Science and Engineering **848** (2020) 012062 doi:10.1088/1757-899X/848/1/012062

(figure 4). The nature of the distribution of volume inclusions over the cleaved surface indicates intergranular destruction of nitride during cleavage, and local elemental analysis allows the impurity phase to be identified as zirconium oxide.

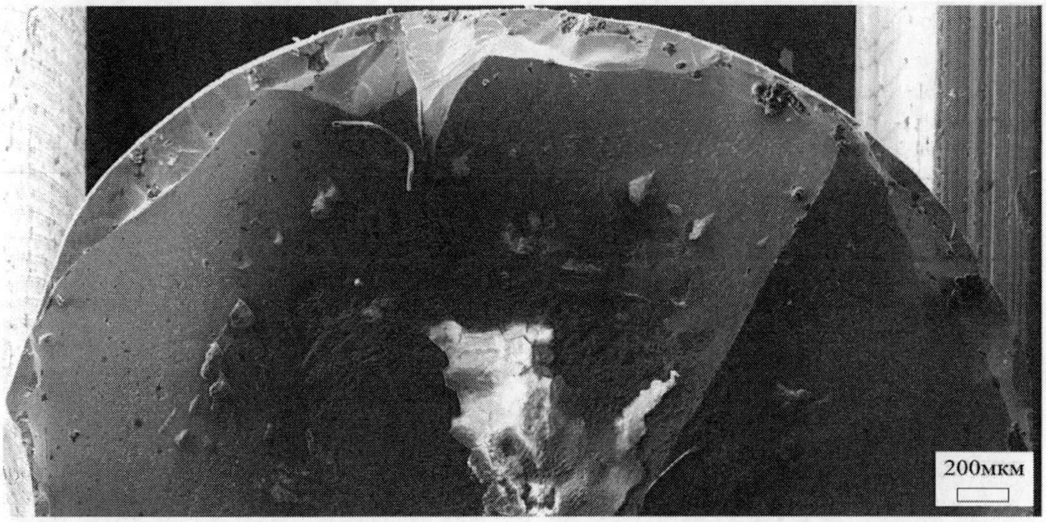

Figure 3. SEM image of a half transverse cleavage of a zirconium nitride sample obtained within 6 hours.

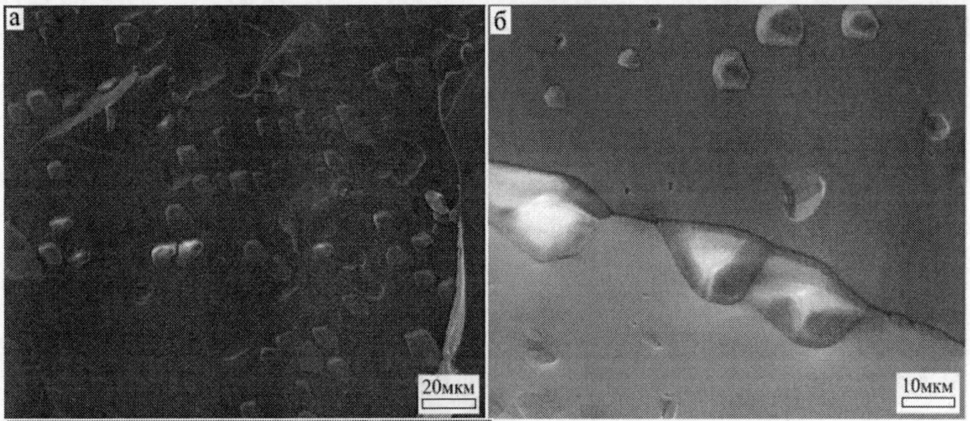

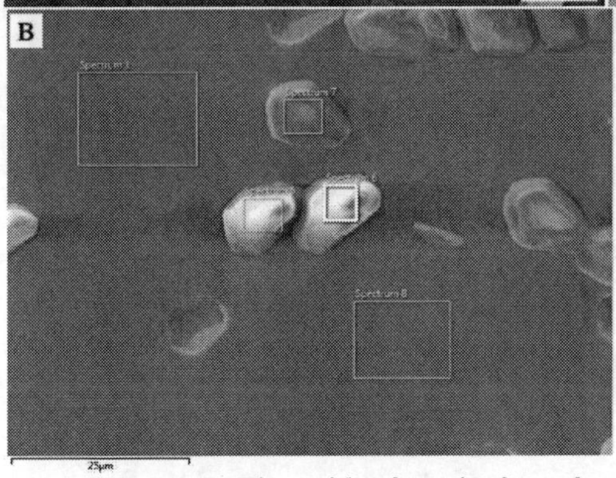

Spectrum	N	O	Zr
Spectrum 1	14.87	0.93	84.20
Spectrum 5	5.88	40.04	54.07
Spectrum 6	0.00	42.72	57.28
Spectrum 7	2.10	30.55	67.35
Spectrum 8	13.59	2.23	84.19

*mass. %

Figure 4. The oxide phase in the volume of non-stoichiometric zirconium nitride.

An inner ceramic layer is located in the central part of the sample with a diameter of ~ 2000 μm, a feature of which is the brooky pattern of the surface surface morphology (figure 5). Apparently, the hand-made relief is caused by amorphization of the nitride phase, or by the supersaturation of the impurity component (in this case, according to elemental analysis, with oxygen).

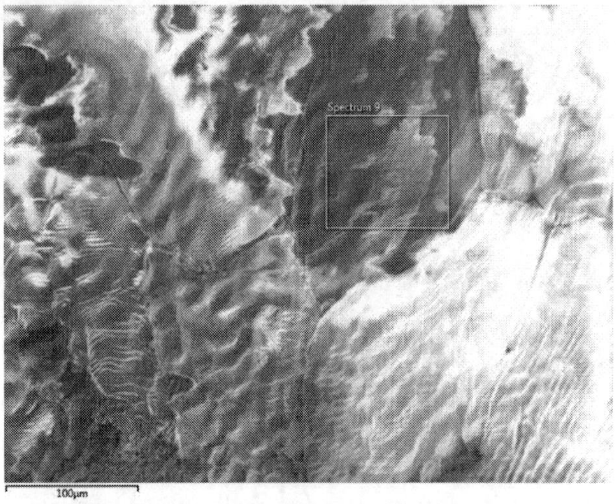

Spectrum	Spectrum 9
N	4.92
O	47.71
Zr	47.37
Total	100.00

*mass. %

Figure 5. SEM image of a brooky morphology pattern of the central part of the transverse cleavage of a ceramic sample after 6 days of nitridation.

A typical image of the transverse cleavage obtained for a sample of 300 μm foil is shown in figure 6. The internal volume of zirconium nitride exhibits a polycrystalline structure, the crystallite size reaches hundreds of microns and is limited in one direction by the thickness of the sample.

Figure 6. A typical image of a cleaved nitride sample obtained from a 300 μm zirconium foil for 120 min at 2400 °C.

A cavity partially occupied by phases based on impurity elements was found on one of the chips in the central part of the sample. The formation of cavities during the nitridation of titanium and zirconium was described in [17]. Ceramics synthesized by direct nitridation of a foil are characterized by the absence of a concentration of impurity phases over the entire cross section of the sample. This can be explained by the high degree and rate of nitridation. Since the complete nitridation of a zirconium foil 300 μm thick at 2400 ° C with the formation of stoichiometric ZrN nitride occurs in 30 minutes [16], the time is 120 minutes. It turns out to be enough to complete the process of recrystallization of the entire internal volume. This factor, as well as the low solubility of impurity elements in the nitride phase, close to the stoichiometric composition, caused their runoff and accumulation near the macrodefect (figure 7).

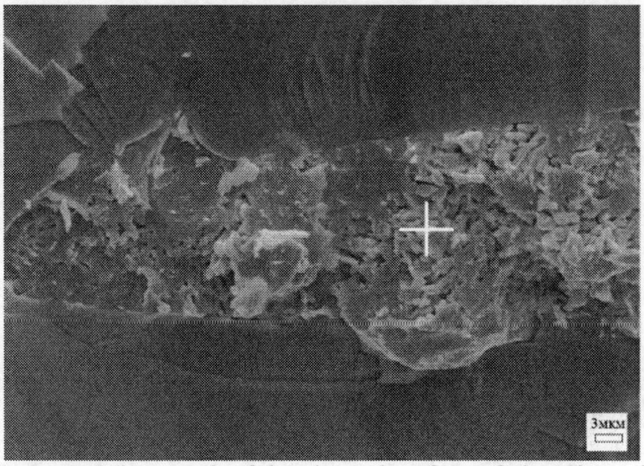

Element	At. %
N	83.96
Al	0.97
Si	1.10
Fe	3.57
Zr	3.82
Nb	6.58
Total	100.00

Figure 7. Micrograph of the cleaved surface of zirconium nitride obtained with pronounced localization of impurity elements near the macrodefect.

4. Conclusion

In the process of zirconium nitridation, impurities contained in the parent metal and nitrogen form independent phases. The localization of impurity phases is determined by the structure of nitride ceramics: the outer layer of stoichiometric nitride, below it is the region of non-stoichiometric nitride and in the center the region of the α-solid solution of nitrogen in the metal. At a temperature above the melting point of the metal, a liquid phase forms, which is retained in the bulk of the sample by a layer of pure refractory stoichiometric nitride. With the appearance of a melt inside the nitride shell, nitride formation occurs both in the liquid phase and on the inner surface of the outer nitride layer, which leads to an increase in the thickness of pure nitride and an increase in the concentration of impurities in the melt. This explains the increased concentration of impurities in the region of the α-solid solution. Upon completion of nitridation at a temperature $T > {}^{1}/_{3}T_{SL}$, a recrystallization process takes place in the ceramic with the formation of a coarse-grained nitride structure. This contributes to the additional displacement of impurity phases. Thus, to the impurities previously localized in the α-solid metal solution, phases displaced during recrystallization are added. All impurity phases are concentrated along the grain boundaries and in the regions of macrodefects of the structure.

Acknowledgments

The study was carried out as part of task No. 11.1957.2017/PCh on the implementation of the project on the theme “Development of an electric pulse technology for the consolidation of high-strength composite materials based on powders of refractory ceramics and boron carbide” (National Research Nuclear University MEPhI (Moscow Engineering Physics Institute), Moscow, Russia).

IOP Conf. Series: Materials Science and Engineering **848** (2020) 012062 doi:10.1088/1757-899X/848/1/012062

References

[1] Gillan E.G., Kaner R. B. Rapid Solid State Synthesis of Refractory Nitrides // Inorg. Chem. 1994. V. 33. No. 25. P. 5693-5700.
[2] Campbell S.A. The Science and Engineering of Microelectronic Fabrication. New York: Oxford University Press, 2001.
[3] Chauhan M., Gupta D.C. Structural, Electronic, Mechanical and Thermo Physical Properties of TMN (TM = Ti, Zr and Hf) under High Pressures: A First Principle Study // Int. J. Refractory Metals Hard Materials. 2014. V. 42. P. 77–90.
[4] Tsunekawa Y., Okumiya M., Kobayashi T. Thermal Spraying: Current Status and Future Trends // Proc. of the 14th Int. Thermal Spray Conf. Ibraki: High Temperature Society of Japan, 1995. P. 755.
[5] Andrievsky R.A., Spivak I.I. Strength of refractory compounds and materials based on them. Directory. Chelyabinsk: Metallurgy, 1989.368 p.
[6] Andrievsky R.A., Glezer A.M. Strength of nanostructures // Advances in Phys. sciences. 2009.Vol. 179. No. 4. P. 337–358.
[7] Jager D.A., Stover D., Schlump W. Thermal Spray: International Advances in Coating Technology // Proc. of Int. Thermal Spray Conf. Materials Park: ASM International, 1992. P. 69.
[8] Chen C.S., Liu C.P., Tsao C.Y.A., Yang H.G. Study of Mechanical Properties of PVD ZrN Films, Deposited under Positive and Negative Substrate Bias Conditions // Scr. Mater. 2004. V. 51. No. 7. P. 715–719.
[9] Yu H.Y., Li M.F., Kwong D.L. IEEE Trans. Electron Devices. 2004. V. 51. P. 609.
[10] Pierson H. O. Handbook of refractory carbides and nitrides: properties, characteristics, processing and applications. - William Andrew, 1996.
[11] Svetlov I. L. et al. High-temperature niobium composites hardened with niobium silicides // Journal of functional materials. - 2007. - T. 1. - No. 2 .-- P. 48.
[12] González-Gutiérrez J., Stringari G. B., Emri I. Powder injection molding of metal and ceramic parts // Some Critical Issues for Injection Molding. - InTech, 2012.
[13] Solntsev K.A., Buslaev Yu.A., Shustorovich E.M. Oxidative construction of thin-walled ceramics (OCTC) // Doklady Chemistry. 2001.V. 378. No. 4-6. P. 143–149.
[14] Kuznetsov K.B., Solntsev K.A., Chernyavsky A.S. A method of producing a refractory metal nitride, products from it, obtained by this method, and their application: Patent RU 2337058 C01B021 / 076 (01.2006).
[15] Shevtsov S.V., Ogarkov A.I., Kovalev I.A., Kuznetsov K.B., Prosvirnin D.V., Ashmarin A.A., Chernyavskii A.S., Solntsev K.A. Structural and phase transformations and hardness of ceramics produced by high-temperature zirconium nitriding // Russ. J. Inorg. Chem. 2016. V. 61. No. 12. P. 1573–1577.
[16] Kuznetsov K.B., Kovalev I.A., Zufman V.Yu., Ogarkov A.I., Shevtsov S.V., Ashmarin A.A., Chernyavskii A.S., Solntsev K.A. Kinetics of zirconium saturation with nitrogen during high-temperature nitridation // Inorg. Mater. 2016. V. 52. No. 6. P. 558–560.
[17] Shevtsov S.V., Ogarkov A.I., Kovalev I.A., Kuznetsov K.B., Ashmarin A.A., Chernyavskii A.S., Solntsev K.A. The nature of the structural inhomogeneity of ceramics produced by zirconium nitridation // Inorg. Mater. 2016. V. 52. No. 6. P. 600–603.
[18] Kovalev I.A., Shokodko A.V., Ogarkov A.I., Shevtsov S.V., Konovalov A.A., Kannykin S.V., Ashmarin A.A., Kolomiyets T.Yu., Chernyavskii A.S., Solntsev K.A. Phase transformations in the surface layer of compact on zirconium nitride based ceramics // Inorg. Mater. 2019. V. 55. No. 7, P. 724–729.

IOP Conf. Series: Materials Science and Engineering **848** (2020) 012063 doi:10.1088/1757-899X/848/1/012063

Hard radiation of reological explosion

I B Oparina and V V Shienok

Baikov Institute of metallurgy and materials Science, RAS, Moscow, Russian Federation
ibo@imet.ac.ru

Abstract. The paper presents the results of experimental research of hard radiation going with reological explosion of solid bodies at the Bridgman's anvils. The results of experiments gives evidence of electromagnetic character of radiation as in process of explosion, as after it. The energy of quanta of radiation, going with explosion was less, than some tens of keV and was less then some hundreds of keV in case of afterglow.

1. Introduction

Reological explosion is the specific kind of breaking of solid matter under pressure, combined with plastic deformation. It is rapid transformation of the energy of compression of the matter, on conditions that this energy is higher, than bond energy of this matter, to kinetic energy of components of this matter. The triggers of explosion are high-energy clusters, which are arising in the matter in the case of it's plastic deformation under the pressure. This phenomenon is known under different denominations as long as the investigations of behavior of solid bodies under high pressure and shear stresses were begun to carry out. The term «reological explosion» for this phenomenon was first introduced by Gorazdovski and Shienok in the year 1975 [1] and is using widely now.

Reological explosion have many features which are observing during experiments [2]. One of such features is hard radiation during the explosion and after it. Radiation during breaking of solids, such as heat and light radiation, is widely known phenomenon, but it don't attract an steadfast attention of investigators due to apparent simplicity of explanation of it's nature. Conventional explanation of said phenomenon consist of assumption, that it is heat radiation of incandescent breaking solids, which are heating due to transformation, through a mechanism of friction, as a rule, of mechanical energy of any kind to heat energy. Such explanation have good correlation with proportion between the energies of light quanta and mechanical energy per atom, which have a solid before breaking.

There is no explanation from this or any other generally accepted points of view of existence of hard radiation in process of breaking of solids with average energy of some electronvolts per atom. Hard radiation was observed in different cases of breaking and high-energy deformation of solids [3–7] and reological explosion is a one. Qualitative investigation of hard radiation of reological explosion was carried out in the works [8, 9]. Results of quantitative investigations are described here.

2. Experimental Procedure

IOP Conf. Series: Materials Science and Engineering **848** (2020) 012063 doi:10.1088/1757-899X/848/1/012063

The studies were carried out with the apparatus (figure 1) included one pair of Bridgman's anvils one of which were fixed at the ram of 100 tons hydraulic press and another at the shaft of mechanical reductor, which were mounted at this press. The reductor allowed to rotate said anvil around the axis, passed through the centers of both anvils. The angle of rotation was measured with accuracy of $\pm 4\times10^{-3}$ radian and was unlimited. Speed of rotation was constant with valuation about $1{,}8\times10^{-2}$ rad/s.

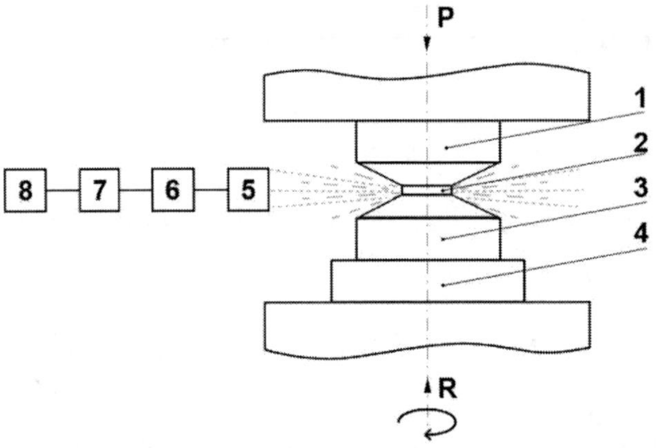

Figure 1. Scheme of the apparatus (1 – upper anvil, 2 – sample, 3 – lower anvil, 4 – shaft of redactor 5 – scintillation counter, 6 – amplifier-analyzer, 7 – counter of impulses, 8 – recorder).

The anvils had a form of truncated cones made of alloy steel forced up HRC60, or of cemented carbide with 8 % of cobalt. The working surfaces of anvils had a square of 2 – 3 cm^2 and were grinded for roughness of a few of microns.

Pressure was valuated as average force to square of working surface of anvils because it was not hydrostatic with maximum in the center. Moreover, gradient of pressure was changing during deformation. Thus exact measuring of pressure was useless. Deformation valued as the angle of rotation of anvil because it was different in different points of the sample and exact measuring was useless too. The samples exploded when pressure and deformation achieved some values, individual for each material. The objects of investigations were polycrystalline dielectrics. Most convenient object from the point of view of purposes of this work was chalk.

Measurements of radiation was carried out with scintillation counter, which included NaJ (doped Tm) crystal, guarded with Al–alloy shell, photomultiplier, amplifier-analyzer, counter of impulses and recorder. The crystal had volume 30 cm^3, working square 10 cm^2 and detected 80 % of ingressed electromagnetic quanta. Other kinds of particles don't excited scintillations in this crystal and because of this, there was not possible to know true the kind of primary radiation as electromagnetic radiation could be a result of interaction of other kinds of radiation with the shell or matter of crystal. The amplifier-analyzer allowed to adjust the range of energies of counted particles with required upper and lower bounds. The counter of impulses allowed to adjust duration of counting and interval between periods of counting. Minimal time in both cases was 0,5 s and was by a factor of 10^3 grater, than a time of explosion. This case strongly degraded signal-to-noise ratio, but it was necessary for catching the moment of explosion. Noise depended strongly on the width and limit values of the range of energies – the more the width, the more the noise and the lower bottom limit, the more the noise (effect of rising of upper limit was smallish). This case necessitated to adjust manifold the range of energies on the basis of getting of optimal signal-to-noise ratio. So, there was not possible to make conclusions about any characteristics of radiation, except of intensity in reference to noise and a value of energy in selected range. Recorder registered the results.

Expected intensity of radiation was low and it's energy was conjectural, so it were necessary meticulous measurements of native noise radiation, elimination of every possible obstacles and carrying out of a lot of pilot experiments.

The counter was mounted at shock damper, in the plane of projection of particles at the distance about 30 cm from exploding tablet and was protected with some layers of fabric from impacts of flying particles, what was necessary for preventing damage of the counter. No electric commutations were done in the moment of explosion and during 5 s before and after it, for preventing from breakthroughs. The radiation had a possibility to propagate free only in space angle of about 1,5 radian due to features of experimental apparatus. So, the counter can catch only about 10^{-3} of initial quantity of radiation only on geometric grounds. Moreover, the probable source of radiation was the tablet with relation of diameter to height from 10:1 to 20:1 and only a part of radiation, which passed through a lateral surface of a tablet, could be detected and other part was absorbed in the anvils. The quantity of radiation, which the counter could catch in such conditions, decreased by an order of magnitude.

Each explosion and results of measurements were very specific and so, any conclusions were possible only after analysis of great number of observations. Registration of any radiation in such conditions gave evidence of a great success in experiments or of a large-scale primary intensity of this radiation.

Measurements of afterglow of tablets after explosions was done too. The tablet was placed directly at the crystal of the counter as quickly, as possible and then the quantity of impulses were recorded. The series of some tens of measurements were than carried out with time of counting 1 s and dead time 0,5 s. The upper and the lower bounds of the range of energies were adjusted on the basis of getting minimal noise signal because of low intensity of measurable radiation. Noise was measured before and after counting the signals from the tablet. The noise in presence of source material with mass manifold greater, then a mass of tablet, was measured too. It was ascertained, that presence of any source material no changed the noise. The assessment of results was made with taking into consideration full quantity and average quantity per measurement of counted particles in series, maximal and average deviation from average quantity in series for each measurement.

3. **Results and Discussions**

There were gotten next results in such conditions. Immediate measurements of radiation which gone with explosion was carried out generally with chalk, as source material, due to maximal stability of characteristics of explosions. This measurements shown, that the exceeding of quantity of counted impulses over noise was changed in wide range from one experiment to another – from absence to manifold. Maximal exceeding (by the order of value) was observed in the range of energies from 1 – 5 to 20 – 30 keV, especially in the cases of very strong explosions. Increasing of the value of the lower bond leaded to decreasing of total amount of counted particles - as noise, as a value of exceeding from noise. Decreasing of this bond leaded to increasing of noise, which merged useful signals. Decreasing of the upper bond leaded to decreasing of total amount of counted particles - as noise, as a value of exceeding from noise. Increasing of the upper bond leaded to insignificant increasing of total amount of counted particles without effect on the exceeding of noise. This facts gives evidence, that the energies of majority of quanta of radiation, going with explosion were less than 20 – 30 keV with more or less surgeless distribution over the range. There was not possible to establish with confidence presence of quanta with higher energies and a value of lower bound of energies.

Measurements of afterglow were carried out with tablets of pure chalk, chalk, doped with materials, which could help to get maximum information about primary radiation in process of explosion and increase responsivity of the method due to possible nuclear reactions with atoms of this materials. Boron with effective interaction with neutrons and aluminum with short-lived isotopes was among such atoms. The best results was obtained for tablets of chalk, doped with boric acid in the range of energies 200 – 500 keV. The exceeding of average quantity of counted particles and average

IOP Conf. Series: Materials Science and Engineering **848** (2020) 012063 doi:10.1088/1757-899X/848/1/012063

deviation from average quantity over noise was about 50 % in this case and repeated well from one experiment to another. Maximum deviations from average were manifold in some cases, just when there was about 20 % for noise. The same results in this range, but with less exceeding, were observed for tablets of polymethylmethacrylate. The results were worse in other cases because other materials gave less quantity of particles in this range. The exceeding over noise was about some particles per second in this experiments. Reducing of lower bond of the range of energies for catching of the particles with lower energies leaded to increasing of noise and there was impossible to reveal reliably exceeding of noise, when noise was more, than some tens of impulses per second.

4. Conclusion

Some parameters of hard radiation, which is attending reological explosion, were detected in this work. The results of experiments gives evidence of electromagnetic character of radiation as in process of explosion, as after it. The energy of quanta of radiation, going with explosion was less, than some tens of keV and was less then some hundreds of keV in case of afterglow. Presence of boron enhanced intencivity of afterglow, what can point out involvement of neutrons in process of occurrence of radiation.

Acknowledgments
The work was carried out according to the state task No 075-00947-20-00

References

[1] Gorazdovskii T Y, Shienok V V 1975 *Russ. Trudi Moskovskogo Gidromeliorativnogo Instituta* **44** pp 113-116
[2] Oparina I.B., Schienok V.V. 2019 IOP Conf. Ser.: Mater. Sci. Eng. **525** 012065
[3] Shestakov A.I. 1975 *Russ. Voprosi sudostroeniya Ser. metallovedenie* **21** pp 97-103
[4] Shestakov A.I. 1975 *Russ. Voprosi sudostroeniya Ser. svarka* **19** pp 3-10
[5] Slabkii L.I, Odnovol L.A., Kozenko V.P. 1973 *Russ. Dokl. Ac. Sci.* **210** pp 319-320
[6] Bivin L.K., Viktorov V.V., Kulinich Y.V., Chursin A.S. 1982 *Russ. Izvestiya Ac. Sci.* **1** pp 183-186
[7] Sobolev G.A., Diomin V.M, Maibuk Y.Y., Kuznetzov O.L., Karanikolo V.F., Svechkariov Y.N. 1984 *Russ. Dokl. Ac. Sci.* **276** pp 583-585
[8] Gorazdovskii T Y 1967 *Russ. Journ. Exp. Theor. Phys. Lett.* **5** pp 78-82
[9] Gorazdovskii T Y, Yaroslavskii M A 1975 *Russ. Trudi Moskovskogo Gidromeliorativnogo Instituta* **44** pp 117-124

IOP Conf. Series: Materials Science and Engineering **848** (2020) 012064 doi:10.1088/1757-899X/848/1/012064

Scandium extraction from multicomponent systems by crystallization of complex sulfates

L A Pasechnik, I S Medyankina and S P Yatsenko

Institute of Solid State Chemistry, Ural Branch, Russian Academy of Sciences, 620990, Ekaterinburg, Russia

Corresponding author: pasechnik@ihim.uran.ru

Abstract. Red mud generated from alumina production is one of the most promising scandium resources. Leaching for concentrating scandium and other valuable metals is often performed with the use of acid solutions in specially selected conditions. Different precipitants can be employed for selective precipitation of scandium in the form of poorly soluble compounds. A new method for the crystallization of ammonium scandium sulfates from sulfuric solutions was carried out in this study. Depending on the ammonium cation concentration, precipitation is possible in the form of $NH_4Sc(SO_4)_2$ or $(NH_4)_3Sc(SO_4)_3$. The solubility values for $NH_4Sc(SO_4)_2$ and $(NH_4)_3Sc(SO_4)_3$ in water established by the isothermal saturation method were 9.02 and 12.89 g/l Sc_2O_3, respectively. Complete precipitation of the most promising gravimetric form of scandium represented by $NH_4Sc(SO_4)_2$ was achieved at 5-6 M H_2SO_4 and 0.5 M NH_4^+. The product of crystallization from sulfuric solution used for treating poor scandium concentrate produced from red mud is a mixture of polymorphs of the composition $NH_4Sc(SO_4)_2$. Co-precipitated impurities of Al, Fe, Ti, Zr etc. are separated after repeated dissolution and precipitation of ammonium scandium sulfate under analogous conditions. The recrystallization product after calcination at 1000°C contains 99% Sc_2O_3.

1. Introduction

About 70 million tons of red mud (RM) – wastes generated after extraction of alumina from bauxites in the Bayer process – are annually disposed into the environment worldwide. These wastes are universally recognized as a promising technogenic feedstock due to a high content of such components as iron, aluminum, titanium, calcium, and rare elements [1,2]. However, technologies for the utilization of red mud with obtaining of valuable products have not been introduced for different reasons including their low efficiency [3]. One of the most valuable components of RM is scandium classified, in spite of its high cost, as a critical metal of the future. Potential application of scandium in metallurgy, chemical and electronic industries is restrained by small volumes of its production in the world and a high price. Thus, the development of effective processes of scandium recovery from cheap secondary and technogenic raw materials is an urgent scientific challenge.

The scandium content in RM does not exceed 120-150 g/t Sc_2O_3, therefore scandium is recovered by selective extraction or by complete break-down of feedstock with subsequent separation of components [4]. Most often slime and other mineral raw materials are leached with acid solutions, and then scandium is extracted by hydrometallurgical methods – extraction, sorption and precipitation. Earlier we have developed carbonate recovery of scandium from RM pulp with the use of kiln gas emissions [5]. This technique allows mutual neutralization and involvement of two types of industrial waste in the processing – alkaline slime and thermal gases containing acid-forming oxides. A poor scandium concentrate with 2-5% Sc_2O_3 content was separated from the clarified carbonate solution. In this paper we propose selective recovery of scandium by crystallization of ammonium scandium sulfates from sulfuric solutions in the presence of impurities of the most abundant metals – Al, Fe, Ti, Zr etc. The structures of ammonium scandium sulfates of two compositions $NH_4Sc(SO_4)_2$ and $(NH_4)_3Sc(SO_4)_3$ were first studied by the authors in detail in the works where the conditions of existence of polymorphic modifications were reported [6,7]. These intermediate complex compounds extracted from sulfuric solutions can be calcined at temperatures 550-600°C to produce scandium sulfate and at 950-1000°C to obtain scandium oxide. The information about the solubility of ammonium scandium sulfates in aqueous solutions as a function of acidity was also studied to search for the optimal conditions of complete scandium recovery.

2. Methodology of crystallization from synthetic systems

The individual compounds $NH_4Sc(SO_4)_2$ and $(NH_4)_3Sc(SO_4)_3$ were preliminarily synthesized and certified as described earlier. The solubility of the complex sulfates was studied by the isothermal method by keeping the solutions having excessive concentration of the corresponding compound in hermetically sealed vessels at a temperature of 20±0.5°C with constant stirring until equilibrium was reached. The establishment of equilibrium was judged by the constant content of scandium in the solutions. To determine the solubility, it was assumed that the following equilibria are reached in the saturated solutions between the solid substance and the ions forming this substance:

$$NH_4Sc(SO_4)_2 \leftrightarrow Sc^{3+} + NH_4^+ + 2SO_4^{2-} \quad (1)$$

$$(NH_4)_3Sc(SO_4)_3 \leftrightarrow Sc^{3+} + 3NH_4^+ + 3SO_4^{2-} \quad (2)$$

The thermodynamic constants of these equilibria give the solubility products of the compounds. The solubility values calculated from these constants are presented in Table 1. The obtained values show that $NH_4Sc(SO_4)_2$ is less soluble and more promising for crystallization from aqueous solutions.

Table 1. Solubility of scandosulfates at 20±0.5°C, g/l Sc_2O_3

$NH_4Sc(SO_4)_2$	$(NH_4)_3Sc(SO_4)_3$
9.02	12.89

The synthetic solutions of scandium were crystallized to investigate the solubility of $NH_4Sc(SO_4)_2$ in the presence of 0.5M NH_4Cl at various acidity using the isothermal solubility method. The solutions with crystallized precipitates were held at 20±0.5°C while stirring for 120 h. The precipitates were filtered, washed with ethanol (95%), dried and weighed, and their composition was monitored. The content of Sc and H_2SO_4 in the solution was determined. The degree of scandium extraction (yield) η (%) in the form of complex scandium ammonium sulfate $NH_4Sc(SO_4)_2$ was estimated by Eq. (3):

$$\eta, \% = \frac{m_{Sc,s}}{m_{Sc,l}} \times 100\% , \quad (3)$$

where $m_{Sc,\,s}$ and $m_{Sc,\,l}$ indicate the quantity (mass) of Sc in precipitate of the complexity scandium-ammonium sulfate and in initial liquid, respectively (mg). Figure 1 shows the solubility in [Sc_2O_3], g/l of the cumulative yield (%) of $NH_4Sc(SO_4)_2$ phase with increasing concentration acid [H_2SO_4], M.

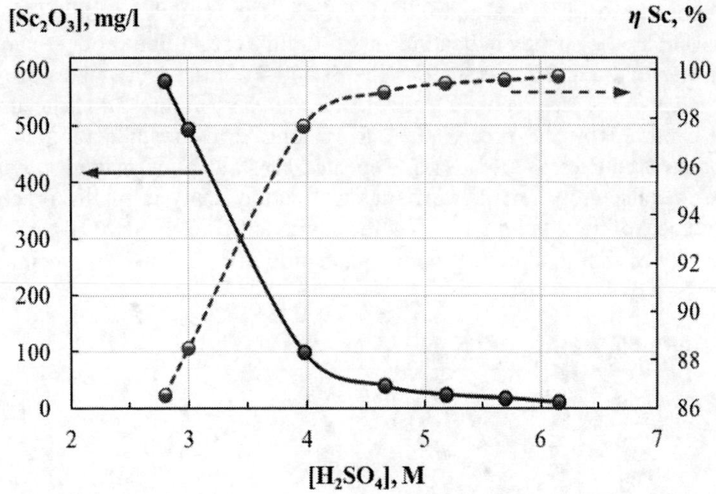

Figure 1. Solubility and yield of $NH_4Sc(SO_4)_2$ phase at various acidity.

As expected and shown in Figure 1, the solubility decreased linearly with increasing acidity from 2.5 to 4 M H_2SO_4, almost approaching a minimum at 5.5 M H_2SO_4. This means that a high recovery close to 99% of the stable phase of $NH_4Sc(SO_4)_2$ can be obtained at 5-6 M H_2SO_4 in the presence of 0.5 M NH_4Cl. The equilibrium solid phase in the selected acid concentration interval was the composition $NH_4Sc(SO_4)_2$ in the form of a mixture of two polymorphous modifications. The quantitative ratio of polymorphs in the precipitate depends on the intensity of mechanical impact in the processes of deposition, washing and preparation for analytical tests. The abrupt reduction of the solubility of the $NH_4Sc(SO_4)_2$ phase is due to the salting-out effects and the influence of the same ion in the solutions.

3. Crystallization of $NH_4Sc(SO_4)_2$ from Sc-strip liquor

The results of investigations of the model systems were worked through for the crystallization of scandium from the strip liquor produced by dissolution of poor scandium concentrate from RM. For carbonate recovery and primary concentration of scandium we used red mud from the Bogoslovskii alumina factory located in the Urals (Russia). The main elements contained in the employed red mud are found to be Fe, Al, Si, Ca, Na, Ti, as well as scandium and zirconium, as shown in Table 2. The poor scandium concentrate from RM contained zirconium, iron and titanium as the main impurities, see also Table 2.

Table 2. Compositional matrixes of samples, wt%

Product	Fe_2O_3	Al_2O_3	CaO	SiO_2	TiO_2	Na_2O	MgO	Sc_2O_3	ZrO_2
Red mud	43.00	14.10	9.85	9.77	4.52	3.58	1.60	0.012	0.064
I-Sc concentrate	1.47	0.52	1.50	0.25	2.26	2.00	0.50	3.99	20.82
$NH_4Sc(SO_4)_2$ phase	0.12	0.001	0.70	0.01	0.14	-	-	13.5	2.09

Ammonium scandium sulfate was extracted by crystallization from the strip liquor under optimal conditions with 98% yield Figure 2a shows the morphology of crystals obtained by crystallization from the strip liquor at 5-6 M H_2SO_4 and 0.5 M NH_4Cl. The crystals of the $NH_4Sc(SO_4)_2$ phase had regular hexagonal prismatic shapes indicating that a highly crystalline product can be obtained. The main microimpurities of Ca and Zr, as well as Fe, Si and Ti in the precipitates of $NH_4Sc(SO_4)_2$ were determined using ICP-AES (see Table 2) and EDX-analysis (Figure 2b). The impurities were co-precipitated and captured by the precipitate during long-term crystallization to provide complete recovery of scandium and decrease its losses. Repeated crystallization made it possible to completely separate impurity metals with the filtrate. X-ray diffraction analysis of the precipitate showed the simultaneous presence of two polymorphs of the composition $NH_4Sc(SO_4)_2$, which were detected earlier [6]. The purity of the final product upon calcination of $NH_4Sc(SO_4)_2$ precipitate at 1000°C was as much as 99.0% Sc_2O_3.

Figure 2. SEM image (a) and EDX spectra (b) for image area of $NH_4Sc(SO_4)_2$.

4. Conclusion
The solubility of the complex sulfates $NH_4Sc(SO_4)_2$ and $(NH_4)_3Sc(SO_4)_3$ and the conditions of selective precipitation from multi-components industrial liquors by crystallization have been substantiated by modeling the processes on synthetic systems. The solubility values of $NH_4Sc(SO_4)_2$ and $(NH_4)_3Sc(SO_4)_3$ in water were 9.02 and 12.89 g/l Sc_2O_3, respectively. A crystalline product corresponding to scandium sulfate $NH_4Sc(SO_4)_2$ was isolated by crystallization from the sulfuric strip liquor produced by dissolution of scandium concentrate from RM. The main impurities of Ca and Zr, as well as Fe, Si and Ti are co-precipitated or captured by the crystallizing precipitate and can be separated by repeated recrystallization. The technology of crystallization provided a very high yield of the solid product with very high crystallinity, which has a desirable impact on the filtration and washing efficiency.

Acknowledgments
We thank Vladimir Surikov for his assistance with analytical measurements. The work was carried out in accordance with the state assignment for the Institute of Solid State Chemistry of the Ural Brunch of Russian Academy of Sciences (theme No AAAA-A19-119031890028-0).

References
[1] Mishra B and Gostu S 2017 Materials sustainability for the environment: red-mud treatment. *Front. Chem. Sci. Eng.* **11** (3) 483–496.
[2] Zinoveev D V, Diubanov V G, Shutova A V et al. 2015 Recycling of red muds with the extraction of metals and special additions to cement *Russ. Metall.* **2015** (1) 19–21.
[3] Borra C R, Blanpain B, Pontikes Y et al. 2016 Comparative Analysis of Processes for Recovery of Rare Earths from Bauxite Residue *JOM* **68** (11) 2958–62
[4] Zhang N, Li H X, and Liu X M 2016 Recovery of scandium from bauxite residue—red mud: a

review *Rare Met.* **35** (12) 887-900.
[5] Pasechnik L A, Pigai I N, Skachkov V M, Yatsenko S P 2013 Extraction of Rare Elements from Residual Sludge of Alumina Production with the Use of Flue Gas of Sintering Kilns *Ecology and Industry of Russia*. 2013 (6) 36-38. (In Russ.)
[6] Pasechnik L A, Tyutyunnik A P, Enyashin A N et al. 2017 Synthesis and crystal structure of 3R and 1T' polytypes of $NH_4Sc(SO_4)_2$ *J. Solid State Chem.* **255** 50-60.
[7] Pasechnik L A, Tyutyunnik A P, Enyashin A N et al. 2018 Polymorphism and properties of ammonium scandium sulfate $(NH_4)_3Sc(SO_4)_3$: new intermediate compound in scandium production *CrystEngComm* **20** 3772-83.

IOP Conf. Series: Materials Science and Engineering **848** (2020) 012065 doi:10.1088/1757-899X/848/1/012065

Investigation of the influence LiF sintering additive on the properties of magnesium aluminate spinel ceramics

M O Senina, D O Lemeshev and M S Pedchenko

D Mendeleev University of Chemical Technology of Russia, Moscow, Russia
snnmarina@rambler.ru

Annotation. The relevance of developing a technology for creating transparent ceramics based on magnesium aluminate spinel is considered. The spinel precursor was obtained by the reverse coprecipitation method. The effect of the phase composition of spinel powder and sintering additive on the properties of ceramics is shown.

1. Introduction

More attention is focused on the issue of creating materials that have a whole range of functional properties. Polycrystalline materials with special properties, which have, for example, transparency in a certain spectral range are of interest. Such products should also have high mechanical properties in addition to transparency.

The technologies for creating optically transparent ceramics from aluminum oxynitride and polycrystalline aluminum oxide have been studied quite well.

Aforecited ceramic has high operational characteristics (Table 1), but its manufacturing technology is energy-consuming and expensive, because it requires the use of hot pressing processes (HP) and hot isostatic pressing (HIP) at high temperatures.

Table 1. Physical properties of certain ceramic materials [1].

Material	Strength, MPa	Elastic modulus, GPa	Density, g/cm^3	Melting temperature, K
AlON	300	323	3.68	2425
Sapphire	700	345-386	3.98	2300

Today, transparent polycrystalline ceramics based on magnesium aluminate spinel (AMS) is one of the most promising materials for the manufacture of transparent armor.

Magnesium aluminate spinel is a binary thermodynamically stable compound with a cubic crystalline structure.

It has a high level of light transmission in a wide range of radiation with a wavelength of 180 to 5000 nm, which includes 3 areas: ultraviolet (180-400 nm); visible (400-700 nm) and infrared (740-5000 nm). Spinel has a distinct transmission advantage over sapphire and AlON in the range of 4500–5500 nm, a region of particular importance for seeker and electro-optic imaging systems. Spinel also has high heat resistance, chemical resistance to aggressive environments, melting point above 2100 °C [2, 3]. In addition, spinel ceramics have a significant advantage over AlON due to the greater availability of the starting components and lower temperature of hot isostatic pressing [4].

IOP Conf. Series: Materials Science and Engineering **848** (2020) 012065 doi:10.1088/1757-899X/848/1/012065

In the synthesis of spinel with increasing temperature above 1000 °C, the composition of the spinel changes significantly. The so-called γ - non-stoichiometry arises: magnesium aluminate spinel is enriched with a rather significant excess of Al_2O_3 [5]. Aluminium oxide has good solubility in spinel with formation of a wide range of solid solutions from $MgO \cdot Al_2O_3$ to $MgO \cdot 5Al_2O_3$ [6].

Obtaining transparent ceramics is impossible if many factors are not observed. One of the requirements for creating a transparent ceramic material is its monophasity, the second phase leads to the absence of a high level of light transmission due to the difference in the refractive indices of different phases and light scattering. However, the authors [7] showed the preparation of transparent ceramics from magnesium aluminate spinel with an excess of aluminium oxide ($MgO \cdot nAl_2O_3$, n = 1.05–2.50). Its light transmission amounted to 84 %. It should be noted that in this work, hot isostatic pressing was used.

An alternative to HP and HIP processes is sintering in vacuum without applying of pressure, which is more energy efficient.

In this work, sintering of spinel ceramics in vacuum was studied. Sintering ceramics of magnesium aluminate spinel to an almost non-porous state is also necessary to obtain a transparent material. And it's impossible without using of sintering additives.

Lithium fluoride was chosen as such additive. At the initial stages of the process, it forms a melt and thereby intensifies liquid phase sintering, and when the temperature rises, it evaporates and enters into a defect formation reaction with magnesium aluminate spinel according to the following scheme:

$$3LiF \xrightarrow{MgAl_2O_4} Li'_{Mg} + 2Li''_{Al} + 3F^{\cdot}_{O} + V^{\cdot\cdot}_{O} \tag{1}$$

As a result, oxygen vacancies are formed. It helps to intensify diffusion mass transfer in the volume of material and therefore, accelerates sintering and allows to obtain a higher density of the product [8-11]. Also, the activation energy of the process is lowering, which leads to a decrease of sintering temperature.

The positive effect of LiF also lies in the fact that this additive does not allow volatilization of a large amount of magnesium oxide and violation of stoichiometry in spinel. According to the authors [12] slight heat evaporation of MgO is possible during the heat treatment of ceramics made of magnesium aluminate spinel. Magnesium oxide leaves the crystal lattice and this also leads to the formation of micropores, which are not completely healed during sintering, which leads to residual porosity and non-transparency of the ceramic [13]. The addition of LiF in this case makes it possible to "capture" magnesium oxide and return spinel to the crystal lattice, thereby preserving the stoichiometric composition of the compound.

2. Materials and Methods

The starting components were magnesium nitrate hexahydrate GOST 11088-75, aluminium nitrate nonahydrate GOST 3757-75, lithium fluoride TU 6-09-3529-78, ammonia aqueous solution GOST 24147-80.

Preliminary synthesis of $MgAl_2O_4$ powder was carried out by the method of reverse co-precipitation of magnesium and aluminum nitrates into a precipitator solution, followed by heat treatment. The LiF additive in an amount of from 0.5 to 4.5 wt. % was doped into the charge by wet method in a planetary mill for 15 minutes. Ceramic preforms were molded by semi-dry pressing at a pressure of 100 MPa. Firing was carried out in vacuum at a temperature of 1700 °C with exposure at a maximum temperature of 3 hours.

3. Results and discussion

The main phase of the powder obtained after calcination of the precursor is magnesium aluminate spinel (Figure 1).

IOP Conf. Series: Materials Science and Engineering **848** (2020) 012065 doi:10.1088/1757-899X/848/1/012065

The structure of the spinel powder is shown in Figure 2. The powder contains plate aggregates ranging in size from 1 to 10 μm. A powder with such particle size distribution can be used in synthesis of transparent ceramics.

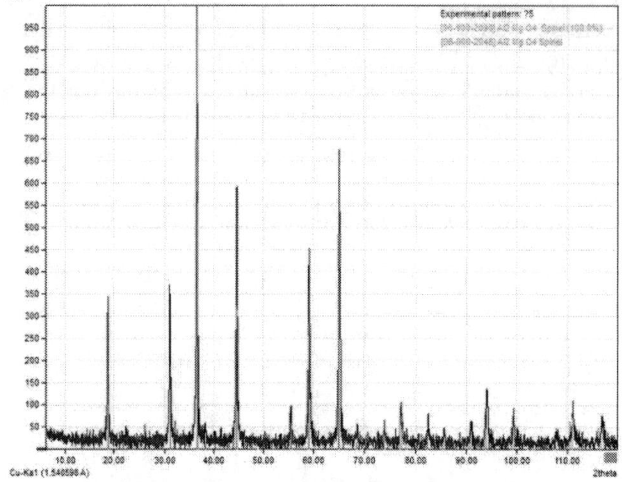

Figure 1. X-ray phase analysis of the obtained powder.

Figure 2. Microstructure photo of the obtained powder.

Figure 3 (a, b) shows the results of measuring the properties of ceramics.

According to the obtained data we can conclude that the optimal concentration of lithium fluoride at which the maximum compaction is achieved (open porosity - 4.9%, average density – 3.32 g/cm^3) is 2.5 wt. %. The similar nature of the curves can be explained by the fact that when the LiF content is less than 2.5 wt. % there is no significant compaction at the stage of sintering, and at the final stage of sintering, the remaining pores are not healed.

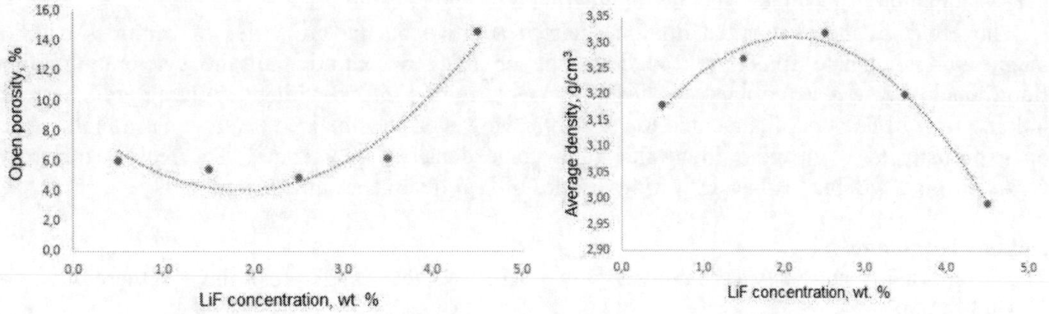

a) b)

Figure 3. Effect of LiF additive concentration on the properties of AMS ceramics, firing temperature 1700 °C.

The introduction of additives in excess of 2.5 wt. % promotes the formation of a large amount of melt. As a result, the melt during evaporation leaves a large number of pores, and the material does not sinter to a fully dense microstructure. Figure 4 shows the microstructure of a sample containing 2.5 wt. % LiF.

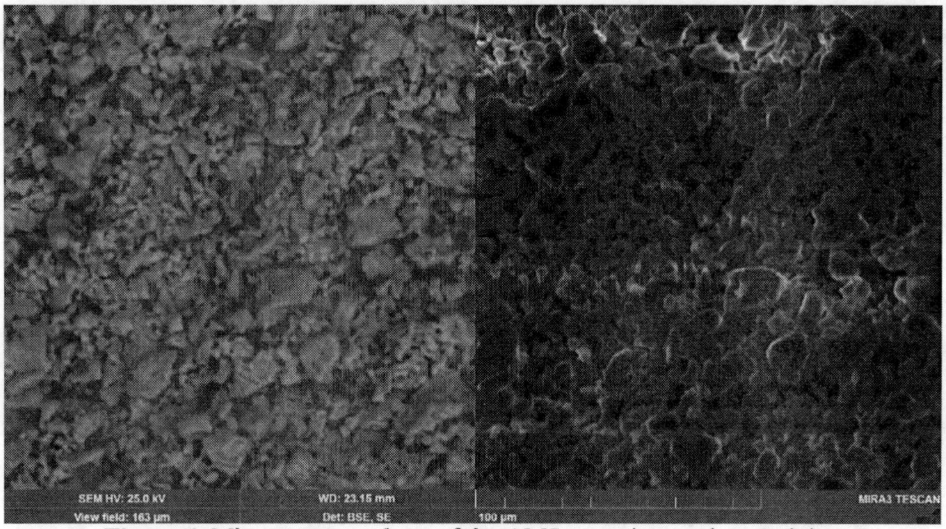

Figure 4. Microstructure photo of the AMS ceramic sample containing 2.5 wt. % LiF calcined at 1700 °C.

After sintering, the material is formed by grains with a structure close to cubic, up to 5 µm in size. Material does not sinter to a fully dense microstructure, the presence of pores is noticeable in the scanning electron microscopy image. It may adversely affecting the transparency of the product.

The resulting samples have high values of open porosity, which does not allow to achieve transparency of the product. In the future, it is advisable to control the synthesis parameters of the spinel precursor powder, as well as change the concentration of the sintering additive.

4. Conclusions

The method of heterophase coprecipitation is promising for obtaining a dispersed powder of magnesium aluminate spinel, suitable for creating transparent ceramics.

The effect of the evaporated lithium fluoride additive on the properties of ceramics based on magnesium aluminate spinel was studied. The optimal concentration of the evaporating lithium fluoride additive was determined at a firing temperature of 1750 °C without applying pressure with a holding time of 3 hours, it amounted to 2.5 wt. %. However, the obtained ceramic has a high value of open porosity (4.9 %) and a low value of average density (3.32 g/cm^3). To create a transparent ceramic, it is advisable to consider various modes of heat treatment of the material.

Acknowledgments
The reported study was funded by RFBR according to the research project № 18-33-00507.

References

[1] Suárez M, Fernández-Camacho A, Torrecillas R, Menéndez J L 2012 New Emerging Techniques p 527
[2] Pavlyukova L T, Lukin E S, Popova N A 2013 Proceedings of "The VII project contest of young scientists" p 28
[3] Vydrik G A, Solov'eva T V, Kharitinov F Y 1980 Transparent ceramic (Moskva: Energiya) p 97
[4] Bhatngara A 2011 Light ballistic materials: Front page from English (Moscow: Technosphere) p 392
[5] Kovtunenko P V 1997 Glass and ceramic 8 p 12
[6] Panda P C, Raj R 1986 Journal of the American Ceramic Society 69 p 365
[7] Hana D, Zhanga J, Liud P, Lia G, Ane L, Wanga S 2018 Ceramics International 44 p 3189
[8] Esposito L, Piancastellia A, Martelli S 2012 J Eur Ceram Soc p 1
[9] Rubat du Merac M. 2012 Ph.D. Thesis Colorado School of Mines Golden,CO and the Technical University Darmstadt Darmstadt Germany to be submitted December
[10] Johnson W C, Stein D, Rice R W 1974 Journal of the American Ceramic Society 57 p 342
[11] Reimanis I E, Kleebe H J 2007 Int J Mater Res 98 p 1273
[12] Esposito L, Piancastelli A, Miceli P, Martelli S 2015 Journal of the European Ceramic Society 35 p 651
[13] Mazzonia A D, Sainzb M A, Caballerob A, Aglietti E F 2002 Materials Chemistry and Physics 78 p 30

IOP Conf. Series: Materials Science and Engineering **848** (2020) 012066 doi:10.1088/1757-899X/848/1/012066

Thermal conductivity of SiC-B_4C materials obtained by reaction-sintering method

S N Perevislov[1], E S Motaylo[2], E S Novoselov[2] and D D Nesmelov[2]

[1] IV Grebenshchikov Institute of Silicate Chemistry of the Russian Academy of Sciences, St. Petersburg, Russia
[2]Saint-Petersburg State Institute of Technology, St. Petersburg, Russia
E-mail:perevislov@mail.ru

Abstract. In this research, temperature dependences of the thermal conductivity of reaction-bonded SiC – B_4C composite materials at the temperature range of 293 – 1400 K were investigated. The effect of B_4C concentration on thermal conductivity is shown. The thermal conductivity of SiC – B_4C materials were compared with the thermal conductivity of materials based on silicon carbide and boron carbide obtained by different methods.

1. Introduction
Materials based on silicon carbide are widely used in instrumental, metallurgical, ceramic, chemical, and electrical industries because of its unique properties. These materials have high hardness, low density, and outstanding chemical and erosion resistance. High melting points of covalent carbides make them promising for high-temperature applications. Thermal conductivity is one of the most important parameters for materials, which are operating at high temperatures and cyclic heating [1-4].

The thermal conductivity of silicon carbide is highly dependent on the preparation method. The room-temperature thermal conductivity ranges from about 30 - 40 W/(m·K) for recrystallized and liquid-phase sintered materials to 490 W/(m·K) for single-crystalline SiC [3].

Due to the high thermal conductivity, materials based on silicon carbide are used in the ferrous and non-ferrous metallurgy as lining blocks and bricks for blast furnaces, muffle, walk-through, rotary and shaft furnaces; as guide rails in blast furnaces and continuous furnaces; in the ceramic industry for receiving capsules, muffles which are used in the sintering of technical ceramics and porcelain; in radio engineering and electronics as a wave absorber, and as protective coatings for nuclear fuel particles, etc.

Boron carbide is widely used in nuclear reactors because of its high neutron absorption cross-section. Because of its high stability at high temperatures and interesting thermo-electrical properties, boron carbide is a promising semiconductor material, for instance, for high-temperature thermoelectric energy conversion. The thermal conductivity of B_4C is lover, then SiC, but it's quite high if related to most other high-temperature materials [2, 3, 5]. Thereby, the study of the thermal conductivity of composites based on SiC and B_4C is necessary for creating products with optimal combination of mechanical and thermophysical properties [6].

The aim of this research is studying the temperature dependence of thermal conductivity of SiC and B_4C- based materials obtained by the reaction-sintering method.

2. Analytical review

Heat transfer is carrying out by scattering of phonons by different ways, such as:

- scattering by another phonons (phonon-phonon scattering)
- scattering by crystal-lattice defects
- scattering by acceptor atoms

Single-crystalline SiC has high thermal conductivity, gradually decreasing with an increase in temperature (Table 1.)

Table 1. Dependence of thermal conductivity of pure single-crystalline SiC from values of temperature [3]

	Temperature, K							
	50	100	150	200	300	500	1000	1200
Thermal conductivity, W/(m·K)	5200	2800	1500	950	490	240	110	90

In [7] the temperature dependence of thermal conductivity of the single-crystalline silicon carbide 6H-SiC in the temperature interval 25÷2000 °C is presented. According to [7] thermal conductivity of monocrystalline silicon carbide complies with the regularity $\lambda = 61100/(T-115)$. At the room temperature value of thermal conductivity of the single-crystalline 6H-SiC is $\lambda = 334$ W/(m·K), at T=2300 K this value decreases to $\lambda = 28$ W/(m·K).

The thermal conductivity of different SiC polytypes studied in [8]. Values of thermal conductivity of cubic 3C-SiC, hexagonal 4H- SiC, and 6H-SiC are respectively $\lambda = 360$ W/(m·K) $\lambda = 370$ W/(m·K), and $\lambda = 490$ W/(m·K).

The grain boundaries, pores and interphase particles in polycrystalline materials are addition sources of phonon scattering. The thermal conductivity of real polycrystalline materials is determined by the concentration of impurities dissolved in the crystal lattice, the concentration of defects in the crystals, the size of the crystals, porosity and other phase components.

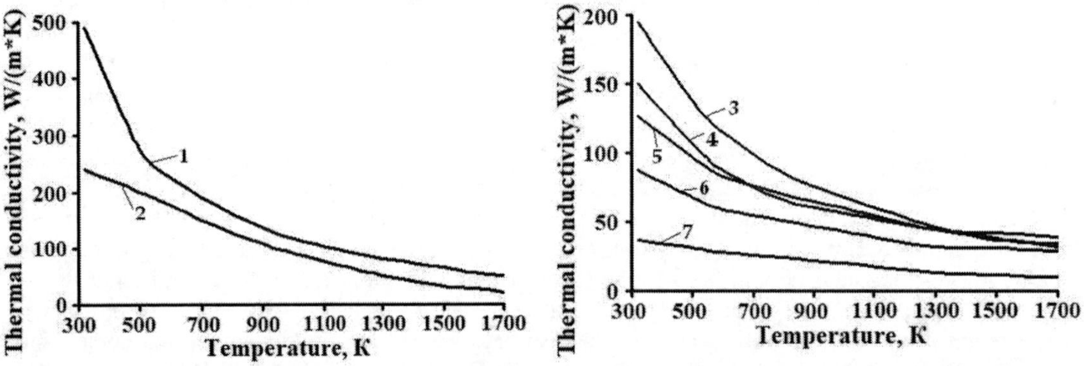

Figure. 1. Temperature dependence of thermal conductivity of SiC-materials obtained by different methods: 1 – MSiC; 2 – HPSiC; 3 – SiG; 4 – SiSiC; 5 – SSiC; 6 – LPSSiC; 7 – RSiC [3, 7 - 14]

The temperature dependences of the thermal conductivity of SiC materials obtained by various methods are shown in Figure 1. The thermal conductivity of a pure single crystal (MSiC) is marked by curve 1. The thermal conductivity decrease from $\lambda = 5200$ W/(m·K) to about 100 W/(m·K) during temperature increasing.

Hot-pressed materials with a low content of activating additive [9] has high value of thermal conductivity (HPSiC, curve 2). The values of thermal conductivity of SiC-materials obtained by solid-phase sintering method are lower (Figure 1b, curve 5). The material obtained by activated solid-phase sintering with an addition of 3% wt. of B_4C has a high level of mechanical properties and thermal conductivity $\lambda = 120$ W/(m·K) at room temperature [3].

Effect of oxide additives on thermal conductivity of SiC obtained by liquid phase sintering is determined by values of thermal conductivity of oxides [10-13]. Value of thermal conductivity of the hot-pressed SiC materials decrease to 5–11 W/(m·K) with the addition of 30 wt. % of the mixture of the $Al_2O_3 - Y_2O_3$ – CaO system [10]. On the contrary, thermal conductivity increases from 133 to 166 W/(m·K) with adding a mixture of oxides $Y_2O_3 - La_2O_3$ from 0.5 to 2 wt. % in the silicon carbide material, and with an increase of the additive above 4 wt. % the value of thermal conductivity begins to decrease [12].

The thermal conductivity of the reaction-sintered materials (Figure 1b, curve 4) depends on the percentage ratio of the polydisperse fractions in the initial silicon carbide powder, the concentration of SiC, C in the finally product, the ratio of primary particles (SiC^I) and phase of secondary silicon carbide (SiC^{II}) and the density of products [14]. Siliconized (curve 3) graphite has less functional parameters influencing the thermal conductivity, therefore the value of thermal conductivity is higher at room temperature, than at the silicon carbide obtained by the reaction-sintering method (Figure. 1b) [14].

The considered data shows, that materials obtained by the reaction-sintering method, hot-pressing method or chemical gas deposition without pores, with a high content of the main phase have higher values of thermal conductivity. The thermal conductivity of ceramic materials obtained by solid-phase and liquid-phase sintering is significantly lower (Figure 1b).

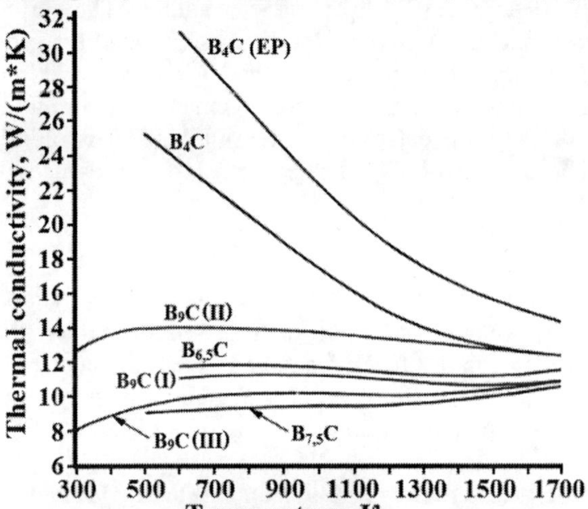

Figure 2. Temperature dependence of thermal conductivity of boron carbide materials [2, 15].

The temperature dependences of the thermal conductivity of various boron carbides (B_xC, where x = 4 ÷ 9) are shown in Fig. 2: hot-pressed B_4C obtained by sintering of powders B and C [2]; commercially available hot-pressed B_4C (Eagle-Picher) obtained by sintering of B_4C powders (EP samples); reaction-sintered B_4C [15] with a density of 2.38 g/cm^3 and a density of 1.85 g/cm^3. Samples B_9C I, II, and III were obtained by hot pressing under the load of 30 MPa, 55 MPa, and 62 MPa for 15 min. Materials $B_{6.5}C$ and $B_{7.5}C$ were obtained by hot pressing from elements with an excess of boron [2].

According to the data of [5], a noticeable increase in thermal conductivity is observed with an increase in the atomic concentration of carbon in the homogeneity range of B_4C. Thus, the thermal conductivity is influenced not only by the method of obtaining B_4C materials but also by its stoichiometric composition, in contrast to materials based on silicon carbide, which does not have a homogeneity range.

3.Materials and Method

Reaction-sintered materials SiC-B_4C were obtained by the impregnation porous preforms with liquid silicon. Preforms contain particles of primary silicon carbide (SiC^I), boron carbide (B_4C^I) and carbon black. The formation of secondary silicon carbide (SiC^{II}) occurs due to the dissolution of carbon and its transport through a silicon melt, and subsequent crystallization on the surface of primary SiC^I particles [16].

To obtain the close-packed preforms it is necessary to use powders of different particle size distribution. The experimental compositions presented in table 2. At the stage of mixing the powders into the material were introduced carbon in an amount of 10 wt. % in excess of 100 % of the starting components.

Table 2. Composition of materials studied in this research

№	Composition of components, wt. %		Carbon, wt. %
	SiC «M5»	B_4C «F180»	
1	50	50	10
2	40	60	10
3	30	70	10
4	-	100	10

The particle size of the powders of silicon carbide and boron carbide was determined by MasterSizer 3000 laser dispersion analyzer: SiC "M5" - $d_{0.5} = 3.7$ µm; B_4C "F180" - $d_{0.5} = 98.8$ µm.

Powders in the required ratio (Table 2) with an addition of 10% carbon black in excess of 100% were mixed in a ball mill. The mixture was plasticized, granulated and dried. The obtained powders were pressed in samples 25 mm in diameter × 5 mm high. The infiltration of green bodies by the silicon melt was carried out at 1600°C in a vacuum. The excess of silicon was removed from the surface of samples after sintering by sandblasting.

The density and porosity of the samples were determined by hydrostatic weighing. The elastic modulus was determined by the excitation of resonant vibrations, on the Sound-130 analyzer. The hardness and fracture-toughness coefficient were determined by introducing the Vickers pyramid into the material and measuring the size of the diagonal of the indent and the cracks emanating from it on the PMT-3M microhardness tester. Thermal conductivity was determined on an LFA 457 MicroFlash analyzer, at the temperature range from room temperature to 1400 K. The equipment was provided by the SPSIT Engineering Center.

4. Results and discussion

The porosity of investigated materials depends on the density of the green bodies, the fraction distribution of the initial SiC and B_4C particles, the amount of carbon in the starting mixture, the temperature, and duration of sintering of the samples, in contrast to the reaction-bonded silicon carbide material, usually sintering almost to a non-porous state ($P \leq 0.5$ vol.%) [16-19]. The relatively large size of particles of B_4C ($d_{0.5} = 90$ µm), used for mixtures preparing, limits solving of B_4C particles in a silicon melt.

The increasing of porosity from 1.0 to 1.6% with the increasing of B_4C content (table 3) is caused by the formation of the solid solution $B_{12}(C, Si, B)_3$. Materials with the highest content of SiC^{I} as well as SiC^{II} have highest density: $\rho = 2.86 \pm 0.02$ g/cm^3 (99% of theoretical).

The relationship between the density and mechanical properties of materials (Elastic modulus E, fracture toughness coefficient K_{1C}, Vickers hardness HV) are presented in table. 3.

Table 3. Mechanical properties of reaction-bonded materials SiC–B_4C

№ compound	$\rho \pm 0,02$, g/cm^3	$\Pi \pm 0,1$, %	$E \pm 10$, GPa	$K_{IC} \pm 0,1$, MPa·m$^{1/2}$	$HV \pm 0,2$, GPa
1	2,86	1,0	384	4,4	26,8
2	2,80	1,1	392	4,3	27,7
3	2,74	1,2	395	4,3	28,6
4	2,57	1,6	411	3,8	31,5

Mechanical properties depend on structural defects (pores, microcracks, etc.). The elastic modulus and Vickers hardness of the reaction-bonded SiC – B_4C materials rise with an increase in the content of B_4C, which have higher values of E or $H\mu$, than SiC. However, an increase in mechanical properties is limited by the formation of brittle phases Si and $B_{12}(B,C,Si)_3$ at high concentration of B_4C in the starting mixture. The highest fracture-toughness (K_{IC} = 4,4 ± 0,1, MPa·m$^{1/2}$) was observed at highest content of SiC in the initial mixture.

Experimental temperature dependences of thermal conductivity are decrease at all investigated temperature range (293-1400K) (fig.3). Composition 1 demonstrates decrease in thermal conductivity from 93 W/(m·K) at room temperature to 22 W/(m·K) at 1400 K. Relationship $\lambda = f(T)$ is similar to those of single- and polycrystalline SiC presented in Figure 1 and 2. This result is in good agreement with the data of [20]. In [20] the decreasing temperature dependence of SiC thermal conductivity is explained by the lattice component – phonon- phonon scattering. The most significant effect of a decrease in the thermal conductivity with increasing temperature observed in the range 273–800 K. The decrease in the thermal conductivity slows down with a further increase in temperature.

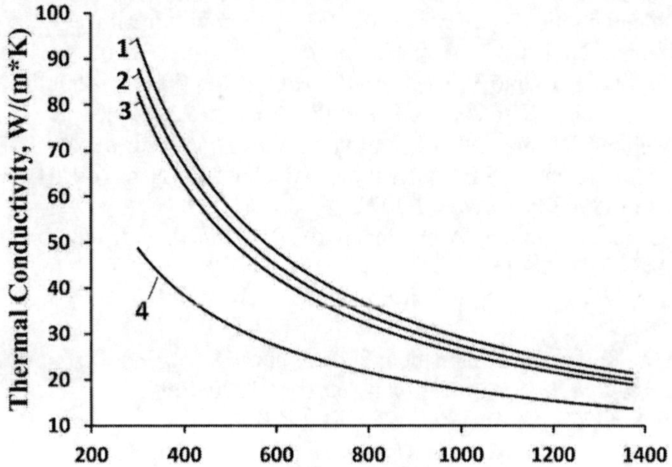

Figure 3. Experimental temperature dependences of thermal conductivity of reaction-sintering materials SiC–B_4C.

The concentration of boron carbide in composites predictably affects their thermal conductivity because of higher thermal conductivity of SiC compared to B_4C (Figure 1 and 3). Thermal conductivity gradually decreases over the entire temperature range with an increase in the content of B_4C in the initial mixture from 50 to 100 mass. %. At room temperature, the effect of B_4C concentration on the thermal conductivity of composites is most significant. An increase in the content of boron carbide in the initial mixture from 50 to 100 mass. % contributes to a decrease in thermal conductivity from 93 W/(m·K) (composition 1) to 46 W/(m·K) (composition 4). At high temperatures, the effect of B_4C concentration on thermal conductivity is less significant. At 1400 K, the thermal conductivity of these compositions decreases from 22 (50% B_4C) to 13 W/(m·K) (100% B_4C). The observed decrease of the thermal conductivity of SiC – B_4C composites with an increase in the concentration of B_4C is associated not only with the additive effect, but also with the influence of the porosity of materials. With an increase in the content of B_4C in the initial mixture from 50 to 100 mass. % porosity of reaction-sintered ceramics increases from 1 to 1.6%, which makes a certain contribution to the overall decrease of thermal conductivity. A similar effect of decreasing of thermal conductivity of reaction-sintered SiC ceramics was previously observed in [4] и [21].

5. Conclusion

The thermal conductivity of the reaction-bonded SiC – B_4C composite materials was studied in the temperature range from room temperature to 1400 K. The dependence is decreasing in the entire temperature range. The effect of a decrease of thermal conductivity with an increase of the concentration of boron carbide in the initial mixture was established. The effect of porosity on the value of thermal conductivity is noted. Composites with a maximum silicon carbide content have the highest thermal conductivity of $\lambda^{20} = 93$ W/(m·K).

Acknowledgments

This research was carried out with financial help of Russian Foundation for Basic Research 17-03-00863 A.

References

[1] S N Perevislov, A S Lysenkov, D D Titov, M V Tomkovich 2019 *Glass and Ceramics* **75** 400–407
[2] C Wood, D Emin, P E Gray 1985 *Physical Review B* **31(10)** 6811
[3] G G Gnesin Silicon carbide based materials 1977 (Moscow, Metallurgy) 216 p
[4] W Koji, N Hiromi, S Kimiyasu 2003 *J. Am. Ceram. Soc.* **10** 1812–1814
[5] I Gunjishima, T Akashi, T Goto 2001 *Materials Transactions* **7** 1445-1450
[6] I Gunjishima, T Akashi, T Goto 2002 *Materials transactions* **9** 2309-2315
[7] O Nilsson, H Mehling, R Horn 1997 *High Temperatures-High Pressures* **29** 73-79
[8] Yu Goldberg, M E Levinshtein, S L Rumyantsev, M S Shur 2001 New York 31-47 p
[9] D M Liu, B W Lin 1996 *Ceram. Int.* **22** 407–414
[10] M Khodaei, O Yaghobizadeh, A A Shahraki 2019 *Int. J. of Refractory Metals and Hard Materials* **78** 9-26
[11] M G Frolova, A V Leonov, Y F Kargin, A S Lysenkov 2018 *Inorganic Materials: Applied Research* **9** 675–678
[12] Y Zhou, K Hirao, K Watari, Y Yamauchi 2004 *J. of the European Ceramic Society* **2** 265-270
[13] Y K Seo 2017 *J. of the American Ceramic Society* **4** 1290-1294
[14] Y Takeda Y 1998 *Am.Ceram. Soc. Bull.* **67** 1961-1963
[15] M L Minges 1974 *Int. J. of Heat and Mass Transfer* **11** 1365-1382
[16] S N Perevislov, A S Lysenkov, D D Titov, M V Tomkovich 2019 *IOP Conference Series: Materials Science and Engineering* **525** 012074
[17] S N Perevislov, P V Shcherbak, M V Tomkovich 2018 *Refractories and Industrial Ceramics* **59** 32–36
[18] S N Perevislov, P V Shcherbak, M V Tomkovich 2018 *Refractories and Industrial Ceramics* **59** 179–183
[19] S N Perevislov, S.N. 2019 *Refractories and Industrial Ceramics* **60** 168–173
[20] I I Parfenova, Yu M Tairov, V F Tsvetkov 1990 *Physics and technique of semiconductors* **24** 258
[21] A S Lysenkov, K A Kim, D D Titov, M G Frolova 2018 *IOP Conference Series: J. of Physics* **1134** 012036

IOP Conf. Series: Materials Science and Engineering **848** (2020) 012067 doi:10.1088/1757-899X/848/1/012067

Physical and mechanical properties of composite materials in the $MoSi_2$-SiC-TiB_2 system

S N Perevislov[1], M A Markov[2], E S Motailo[3], S V Vikhman[3] and D D Titov[4]

[1]IV Grebenshchikov Institute of Silicate Chemistry of the Russian Academy of Sciences, St. Petersburg, Russia
[2]National Research Centre "Kurchatov Institute", Moscow, Russia
[3]Saint-Petersburg State Institute of Technology, St. Petersburg, Russia
[4]Baykov Institute of Metallurgy & Material Science RAS, Moscow, Russia
E-mail: perevislov@mail.ru

Abstract. In this work, we obtained ceramic materials based on $MoSi_2$, hardened by SiC and TiB_2 particles, which makes it possible to increase the heat resistance of the composite and increase the level of its physical and mechanical properties (at a relative density of more than 99,0 %, tensile strength in bending 520 ± 10 MPa). The shrinkage, density, porosity and weight loss of composite materials during sintering were studied. The value of the ultimate strength in bending of materials at the optimum sintering temperature is determined. The maximum values of physical and mechanical properties are achieved for a material composition of 50,0 vol. % $MoSi_2$ + 15,0 vol. % SiC + 35,0 vol. % TiB_2: $E = 471 \pm 10$ GPa; $\sigma_b = 520 \pm 10$ MPa, $K_{IC} = 4{,}4 \pm 0.1$ MPa • $m^{1/2}$; $HV = 22.1 \pm 0.1$ GPa obtained at a sintering temperature of 1950 ° C.

1. Introduction

Materials based on molybdenum disilicide retain stability and workability in an oxidizing environment at temperatures up to 1750 °C. Such materials are the main components of heaters, artillery gun liners, lining of the combustion chambers of jet engines, protective caps of high-speed flying vehicles, etc. [1, 2].

To solve modern production problems, it is often necessary to increase the level of mechanical properties of materials based on $MoSi_2$; for this, reinforcing components are introduced into the material - SiC_w, Si_3N_{4w} whiskers and SiC_f fibers, etc. [3, 4].

In [5], a method was proposed for hardening molybdenum disilicide by co-mixing the powder components of $MoSi_2$ with SiC particles, followed by hot pressing. Composite materials based on silicon carbide are known, containing silicon, $MoSi_2$ and Si, $NbSi_2$ and Si, which are synthesized during the impregnation of a chemically active melt [6].

A significant amount of work has been devoted to studies of the $MoSi_2$–SiC system; however, the possibilities of obtaining such ceramics by free sintering and the study of its mechanical properties have hardly been studied.

In [7–9], it was noted that materials of the $MoSi_2$–SiC–TiB_2(ZrB_2) system can be promising composite ceramics based on molybdenum disilicide. It was previously shown that SiC coexists with a large number of refractory compounds, among which carbides (TiC, ZrC, HfC, NbC, TaC) and borides (TiB_2, ZrB_2, HfB_2) of transition metals should be distinguished. Of greatest interest are the SiC – TiB_2 and SiC–ZrB_2 systems [10–13]. There is no information on the three-component state diagram of the $MoSi_2$–SiC–TiB_2 system; therefore, it is very promising to study the properties of both the two-component $MoSi_2$– SiC system and the system as a whole.

The aim of the work is the sintering of dense ceramic materials based on $MoSi_2$ by solid-phase sintering with the addition of reinforcing particles of SiC and TiB_2, the study of the structure and mechanical properties of the obtained composites.

2. Materials and Method

To obtain sintered composite materials of the $MoSi_2$–SiC, $MoSi_2$–SiC–TiB_2 systems, the following powders were used as initial components: $MoSi_2$, particle size $d_{0,5}$ = 6,5 μm; SiC, particle size $d_{0,5}$ = 0,8 μm; TiB_2, particle size $d_{0,5}$ = 14,3 μm. $MoSi_2$ and TiB_2 powders were additionally ground in a VML-1,5 vibration mill in isopropyl alcohol for 60 hours. The particle size of the powders after grinding was: $MoSi_2$ - $d_{0,5}$ = 0.,85 μm; TiB_2 - $d_{0,5}$ = 0,78 μm

The materials were prepared using standard ceramic technology [14–16], including mixing powders in a drum mixer in a gasoline environment with the simultaneous introduction of a temporary technological binder — 4 % wt. Paraffin. Over 100 % of charge components, molding of sample blanks measuring 6×6×50 mm and Ø25×8 mm on a hydraulic press at a pressure of 100 MPa. The samples were dried at a temperature of 120 °C for 3 h. and sintered in a vacuum oven at temperatures of 1800, 1850, 1900, and 1950 °C, in an Ar atmosphere, for 1 h. Shrinkage of materials was determined on a DIL 402c dilatometer. The compositions of the samples are presented in table. 1.

Table 1. The compositions of the initial components of the samples of the $MoSi_2$–SiC, $MoSi_2$–SiC–TiB_2 systems.

№	Content of components, vol. %			Theoretical density
	$MoSi_2$	SiC	TiB_2	$\rho_{theor.}$, g/cm^3
1	75	25	-	5,50
2	60	40	-	5,04
3	80	10	10	5,61
4	33,4	33,3	33,3	4,90
5	50	15	35	5,26
6	50	25	25	5,37

The particle size of the powders was determined using a Malvern Mastersizer 3000 laser dispersion analyzer. The density of the samples was determined by hydrostatic weighing. Porosity was measured using a Meiji Techno IM 7200 optical microscope and a Thixomet computer image analyzer. The elastic modulus was determined, using the excitation of resonant vibrations, on the ZVUK-130 unit. Strength in three-point bending was measured on a ShimadzuAG-300kNX unit. The hardness and coefficient of fracture toughness were determined by introducing the Vickers pyramid into the material and measuring the size of the diagonals of the indent and the cracks emanating from it on a TP-TR-1 macrohardness tester with a load of 2 kg. The microstructure was examined using a Meiji Techno IM 7200 optical microscope. The equipment was provided by the Engineering Center of St. Petersburg State Technological Institute.

3. Results and discussion

The optimal temperature of solid-phase sintering of materials was determined by analyzing the shrinkage curves by heating the samples to a temperature of 1900 °C in an Ar media (this temperature is lower than the eutectic). Previously, in [17], it was shown that the minimum temperature of the liquid phase formation (eutectic temperature) in the $MoSi_2$–SiC system is 1900 °C with a component ratio of 70 % vol. $MoSi_2$ + 30 % vol. SiC. The results of the shrinkage of the samples are presented in table. 2.

Shrinkage increases when a third component is introduced into the material. Maximum shrinkage is observed for materials of compositions 3–6 containing titanium diboride. The addition of TiB_2 and its uniform distribution in the material leads to the formation of a rigid refractory skeleton of SiC and TiB_2 particles, and molybdenum disilicide, as the least refractory component (T = 2030 °C) upon contact melting with SiC and TiB_2, acts as a source of the liquid phase that activates the process compaction of

IOP Conf. Series: Materials Science and Engineering **848** (2020) 012067 doi:10.1088/1757-899X/848/1/012067

materials (table. 2). When more refractory components (SiC - T = 2730 °C and TiB_2 - T = 3230 °C) are introduced into the material based on $MoSi_2$, the temperature to reach maximum shrinkage increases. The maximum shrinkage of 17,2 and 18,4 % is observed for materials of compositions 4 and 5 at a temperature of 1900 °C (Table 2).

Table 2. Physical properties of materials of the $MoSi_2$–SiC and $MoSi_2$–SiC–TiB_2 systems

№	Shrinkage Y (at $T_{sintering}$, °C), %	Density ρ_{rel} (at $T_{sintering}$, °C), %	Porosity Π (at $T_{sintering}$, °C), %	Mass loss Δm (at $T_{sintering}$, °C), %
1	8,8 (1800)	96,8 (1850)	3,1 (1850)	0,4 (1900)
2	9,8 (1880)	97,4 (1900)	2,5 (1900)	0,5 (1950)
3	12,5 (1820)	98,1 (1850)	2,0 (1850)	0,5 (1900)
4	17,2 (1900)	99,0 (1950)	1,0 (1950)	0,3 (1950)
5	18,4 (1900)	99,3 (1950)	0,8 (1950)	0,3 (1950)
6	16,8 (1890)	99,0 (1900)	0,9 (1900)	0,5 (1950)

The table 2 shows the values of density and porosity of sintered materials at the optimum temperature for each composition. Samples sintered at a temperature close to the temperature of maximum shrinkage do not reach the highest compaction, which is most likely due to the need for a slight overheating of the system. However, with significant overheating (more than 50 °C), increased porosity is observed, which is associated with the evaporation of oxides at temperatures above 1900°C (Table 2), present a thin layer on the starting particles, and at a higher temperature, loss of silicon by disilicide with the transition to Mo_5Si_3. The table 3 shows the mechanical properties of materials of the $MoSi_2$–SiC and $MoSi_2$–SiC–TiB_2 systems obtained under optimal technological sintering conditions, in accordance with the data given in Table 2.

Table 3. Mechanical properties of materials of the $MoSi_2$–SiC and $MoSi_2$–SiC–TiB_2 systems.

№	E±10, GPa	K_{IC}±0,1, MPa·m$^{1/2}$	HV±0,1, GPa
1	416	3,2	16,0
2	419	3,4	18,7
3	446	3,7	15,8
4	468	4,2	25,0
5	471	4,4	22,1
6	461	4,0	21,7

The elastic modulus of materials is affected by the presence of defects (microcracks, pores, etc.). Since the transition metal borides are characterized by a large value of the elastic modulus, with an increase in the TiB_2 content in the composition of the material, E increases (table. 3) due to the approximation of the density of the material to the theoretical. The values of the elastic modulus of compositions 1 and 2 depend on the quality of the obtained samples. For example, hot-pressed materials are characterized by E values close to theoretical.

An important characteristic of composite materials is crack resistance, which also depends on the density and structural defects (porosity) in the material. When moving from a two- to three-component system, it is possible to obtain a finer-grained structure of materials, since each additional component of the system is an inhibitor of grain growth of single- and two-phase materials, which reduces mass transfer and grain recrystallization during sintering. Therefore, compositions 3–6 are characterized by increased density (Table 2) and, as a consequence, higher levels of crack resistance coefficients (Table 3) and bending strength (Figure 1). The maximum value of the coefficient of crack resistance (4,4 ± 0,1 MPa · m$^{1/2}$) was achieved on a material with an equivalent volume content of components – 33,4 vol. % $MoSi_2$ + 33,3 vol. % SiC + 33,3 vol. % TiB_2.

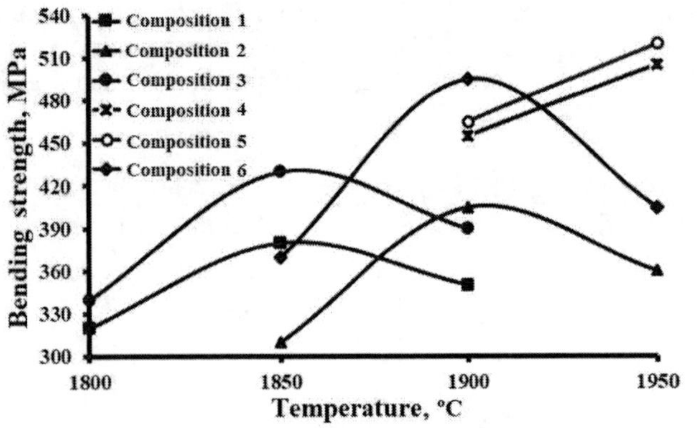

Figure 1. Dependences of flexural strength of composite materials of $MoSi_2$ - SiC and $MoSi_2$ - SiC - TiB_2 systems on sintering temperature.

The hardness of materials is defined as the additive value between the hardnesses of the phases contained in it, therefore, at a high titanium diboride content (Hμ = 33,7 GPa), the value of the HV composite increases. Cracks in the material propagate mainly by the intercrystalline mechanism, rounding grains, which, when using three-component systems, leads to an increase in the crack path and, as a result, to an increase in the coefficient of crack resistance (Table 3). An exception may be materials with a high content of molybdenum disilicide (compositions 1 and 3) having low hardness (Table 3), such materials are partially destroyed in accordance with the transcrystalline fracture mechanism [18, 19].

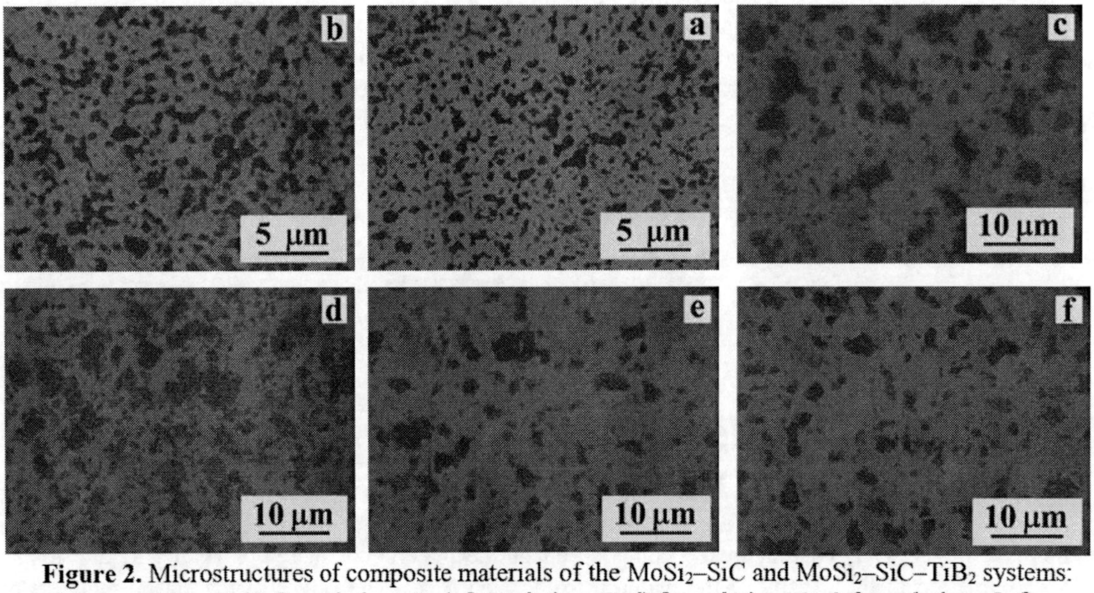

Figure 2. Microstructures of composite materials of the $MoSi_2$–SiC and $MoSi_2$–SiC–TiB_2 systems: a) compositions 1; b) formulations 2; c) formulations 3; d) formulations 4; e) formulations 5; f) formulations 6

Strength properties of sintered materials as well as elastic characteristics depend on the density determined by the optimal technological sintering conditions. High bending strength is achieved on materials of compositions 4–6 (Fig. 1) at a temperature that ensures partial melting of the components (based on $MoSi_2$) and, at the same time, their little evaporation. For compositions 4 and 5, most likely, with increasing temperature (more than 1950 °C), a sharp decrease in strength is observed, which is associated with the possible evaporation of silicon formed during partial metallization of the chemical bond in $MoSi_2$ with its conversion to Mo_5Si_3 and the liquid phase entering the material surface with the formation of pores in the volume of the material.

Figure 2 shows the microstructures of composite materials of the $MoSi_2$–SiC and $MoSi_2$–SiC–TiB_2 systems, where the light gray phase is $MoSi_2$, the dark phase is SiC, and the dark gray phase is TiB_2.

The high mechanical properties of the composite materials of the $MoSi_2$–SiC and $MoSi_2$–SiC–TiB_2 systems are ensured by the correct choice of sintering temperature, the use of fine powders and uniform distribution of the starting components over the volume of the material.

4. Conclusion

The microstructure was studied and the mechanical properties of composite materials of the $MoSi_2$–SiC and $MoSi_2$–SiC–TiB_2 systems obtained by solid phase sintering were studied. The effect of introducing the third component on the mechanical properties of ceramics is shown. When titanium diboride is introduced into the material, its modulus of elasticity, coefficient of crack resistance, and hardness increase. The bending strength of materials is determined by the selected mode of sintering of ceramics. The maximum values of mechanical properties are achieved on a material with a composition of 50,0 vol. % $MoSi_2$ + 15,0 vol. % SiC + 35,0 vol. % TiB_2: $E = 471 \pm 10$ GPa; $\sigma_b = 520 \pm 10$ MPa, $K_{IC} = 4{,}4 \pm 0{,}1$ MPa · $m^{1/2}$; $HV = 22{,}1 \pm 0{,}1$ GPa at a sintering temperature of 1950 °C.

Acknowledgments

This work was supported by a grant from the Russian Science Foundation (№ 19-73-10180).

References

[1] Daniel I M, Ishai O, Daniel I M, Daniel I 1994 Engineering mechanics of composite materials (New York: Oxford University Press) p. 256-256
[2] Lohfeld S, Schutze M, Bohm A, Guther V, Rix R, Scholl R 2005 *J. Materials and Corrosion* **56** 250–258
[3] Wang G, Jiang, W, Bai G, Wu L 2003 *J. of the American Ceramic Society* **86** 731-734
[4] Vasudevan A. K., Petrovic J. J. A comparative overview of molybdenum disilicide compo-sites // Materials Science and Engineering: A. - 1992. - V. 155. - No. 1-2. - P. 1-17.
[5] Maloney M J , Hecht R J 1992 *Materials Science and Engineering* **A155** 19-31
[6] Guo S Q, Nishimura T, Mizuguchi T, Kagawa Y 2008 *J. of the European Ceramic Society* **28** 1891-1898
[7] Ordan'yan S S, Rumyantsev V I, Nesmelov D D, Korablev D V 2012 *Refractories and Industrial Ceramics* **53** 108–111
[8] Ordan'yan S S, Nesmelov D D, Danilovich D P, Udalov Y P 2017 *Russian J. of Non-Ferrous Metals* **58** 545–551
[9] Markov M A, Ordan'yan S S, Vikhman S V, Perevislov S N 2019 *Refractories and Industrial Ceramics* **60** 385–388
[10] Zhang G J, Deng Z Y, Kondo N, Yang J F, Ohji T 2000 *J. of the American Ceramic Society* **83** 2330-2332
[11] Zhang S C, Hilmas G E, Fahrenholtz W 2008 *J. of the American Ceramic Society* **91** 26-32
[12] Zhao G, Huang C, Liu H, Zou B, Zhu H, Wang J 2014 *Ceramics International* **40** 2305-2313
[13] Zhao G, Huang C, Liu H, Zou B, Zhu H, Wang J 2014 *Int. J. of Refractory Metals and Hard Materials* **42** 36-41
[14] Frolova M G, Leonov A V, Kargin Y F, Lysenkov A S 2018 *Inorganic Materials: Applied Re-search* **9** 675–678
[15] Perevislov S N, Lysenkov A S, Titov D D, Tomkovich M V 2019 *Glass and Ceramics* **75** 400–407.
[16] Perevislov S N, Lysenkov A S, Titov D D, Tomkovich M V 2019 *IOP Conference Series: Materials Science and Engineering* **525** 012074.
[17] Carter D H, Petrovic J J, Honneu R E, Gibbs W S 1989 *Ceramic Engineering and Science Proceedings* 1121-1129.
[18] Lysenkov A S, Kim K A, Titov D D, Frolova M G 2018 *IOP Conference Series: J. of Physics* **1134** 012036
[19] Perevislov S N 2019 *Refractories and Industrial Ceramics* **60** 168–173

IOP Conf. Series: Materials Science and Engineering **848** (2020) 012068 doi:10.1088/1757-899X/848/1/012068

Sintering and physico-mechanical properties of materials based on silicon nitride nanoscale powders

S N Perevislov[1], O A Lukyanova[2]; A S Lysenkov[3]; K A Kim[3] and A B Vysotin[1]

[1]IV Grebenshchikov Institute of Silicate Chemistry of the Russian Academy of Sciences, St. Petersburg, Russia
[2]Belgorod State University, Belgorod, Russia
[3]Baykov Institute of Metallurgy & Material Science RAS, Moscow, Russia
E-mail: perevislov@mail.ru

Abstract. Obtained a high-density material (ρ = 3,28 g/cm^3, a porosity of 1,4 %) on the basis of silicon nitride powder by liquid phase sintering nanoscale. Displaying pressing pressure on the density of sintered materials the based silicon nitride. The process of pattern formation material during sintering. The main physico-mechanical characteristics: the density, porosity and grain size of the solid phase, the modulus of elasticity, flexural strength, fracture toughness ratio, Vickers hardness.

1. Introduction

Oxygen-free ceramics belongs to the brittle materials do not exhibit macroscopic ductility and fracture when loaded up to the dissociation temperature of the compounds. This phenomenon makes it possible to create a unique class of structural materials with high hardness, strength and stability of mechanical properties in a wide range of temperatures, exceeding significantly the properties of metal alloys [1].

Silicon nitride is widely used in industry due to its unique combination of physical and mechanical properties such as high hardness and mechanical strength, low thermal conductivity and good erosion properties that allows to apply the product of Si_3N_4 as a wear-resistant, heat-resistant and acid resistant materials, working in extreme conditions of high temperature and aggressive environments. Materials made of silicon nitride can be prepared by various methods: reaction sintering, activated sintering, chemical vapor deposition and hot, hot isostatic pressing. High properties are achieved at the materials produced by the last two methods, however, a large energy consumption, inability to obtain articles of complex shapes and low productivity limit the reception of materials by this method on an industrial scale. reaction sintering method is impossible to obtain dense materials with a porosity of less than 5 vol. %. Using the method of chemical precipitation from the gaseous phase leads to materials with nearly the theoretical density, but due to the technological difficulties their production is extremely difficult. The optimal method is the liquid-phase sintering with oxide dopant, which results in a material with high mechanical properties, with less energy consumption and processing cost [2]. For materials with a high level of mechanical properties necessary to use fine powders, what is feasible with the additional process step of grinding, however, to produce particles of less than 0,5 microns is extremely difficult. Using the silicon nitride nanoscale powders (n-Si_3N_4) allows to obtain dense materials with maximum mechanical characteristics, but complicates the intermediate preparation materials.

IOP Conf. Series: Materials Science and Engineering **848** (2020) 012068 doi:10.1088/1757-899X/848/1/012068

As the sintering additives are used as individual oxides of Al_2O_3, Y_2O_3, MgO and their eutectic mixture. The most common activating additive is a mixture of components Al_2O_3 and Y_2O_3 in a ratio of yttrium aluminum garnet [3–10]. Among the active energy-oxides can also include Li2O, BeO, MgO, CaO, Ce_2O, ZrO_2 and their double oxide mixture introduced into the materials on the based silicon nitride in the form of starting components and in the form of binary eutectic compounds.

The aim is the preparation and study of physico-mechanical properties of high density ceramic materials based on nano powder nitride silicon by liquid-phase sintering.

2. Materials and Method

We used a nanosized powder plasmochemical Si_3N_4 (OOO «Plazmotherm», Russia) with a particle size d = 50–80 nm and a silicon nitride obtained by CVC (ISMAN Sciences, Russia), ground in a planetary mill to a size $d_{0,5}$ = 1 μm. Nanosize powder has a high content of silicon and oxygen impurities – 0,8 and 1,7 wt. %, respectively. In the course of storage due to the high activity of n-Si_3N_4 adsorbs oxygen on the surface, forming a thin SiO_2 film, which during the sintering react with oxide additives to form a fusible silicate melts. The compositions of the materials investigated in the work shown in Table 1.

Table 1. The compositions of the ceramic materials of silicon nitride.

Number of composition	Content $5Al_2O_3{\cdot}3Y_2O_3$, wt. %	Content of nanosized Si_3N_4, wt. %	Contents micron Si_3N_4, wt. %
1	15	0	85
2	15	5	80
3	15	10	75
4	15	15	70
5	15	20	65
6	15	25	60
7	15	40	45
8	15	60	25
9	15	75	10
10	15	85	0

As a sintering additive, a mixture of Y_2O_3 and Al_2O_3 oxides (OOO «Plazmotherm», Russia), particle size d = 30–50 nm, in a ratio forming step of sintering yttrium aluminum garnet (YAG) [11]. A liquid phase-sintered materials based on silicon nitride ceramic prepared by techniques [12], the method comprising preparing the batch molding of the samples at different pressures (25; 50; 75; 100 MPa), high temperature drying and sintering.

To increase the bulk density of the n-Si_3N_4 powders and oxides repeatedly briquetted and granulated. The starting powders in a predetermined ratio (Table 1) were stirred in a drum mixer. The polyethylene glycol used as a temporary process binder. Sintering was performed at 1780 °C in N_2 atmosphere with isothermal exposure for 40 min.

density of the samples was determined by hydrostatic weighing. Porosity was measured using an optical microscope Meiji Techno IM 7200 and computer image analyzer. The elastic modulus was determined by excitation of the resonant vibrations on the installation Zvuk–130. Three-point bending strength was measured on installation ShimadzuAG–300kNX. Hardness and fracture ratio was determined by introducing into the material and measuring the Vickers pyramid diagonals print size and originating from cracks on its corners microhardness PMT–3M. The microstructure and grain size of the solids was investigated by the scanning electron microscope Tescan Vega 3. Equipment granted «Engineering Center» SPSIT.

3. Results and discussion

Obtaining ceramics of high quality with a maximum level of mechanical properties is possible while achieving high density (close to the theoretical). The sintering process may be controlled by selection of dispersion, particle morphology. For silicon nitride retardation characteristic of diffusion processes, as in the lattices covalent compounds suppressed processes diffusion-viscous flow responsible for the mass transfer and seal the solid-state sintering. Sintering can significantly intensify the introduction of additives, which at a certain temperature (below the sintering temperature of ceramics) form at the grain boundaries primary liquid phase component. To accelerate the seal material may also be reducing the size of the precursor particles, using nanoscale particles during sintering which reduces the diffusion path and increasing the mass transfer [13].

Sintering mechanism includes: a seal of the particles by the method of solid-phase sintering; education fusible silicates and glasses; redistribution Si_3N_4 particles in the presence of a silicate liquid phase; dissolving α-, β-Si_3N_4 in the silicate melt by the interface and their recrystallization from the melt in the form β-Si_3N_4. It plays an important role as the formation on the particles Si_3N_4 layer of a solid solution (phase Si–Al–O–N, known as «sialon») due to dissolution of nitride in the yttrium aluminum garnet, which is described by the formula $Si_{6\text{-}Z}Al_ZO_ZN_{8\text{-}Z}$ ($z = 0 \div 4{,}2$) [5]. During sintering Si_3N_4 with oxides based on aluminum are formed in the solid solutions Si_3N_4 Al_2O_3 (β_1- and X_1-phase), the latter being located in a bundle between the Si_3N_4 grains and β'-phase. β'-phase growth occurs by recrystallization of Si_3N_4 by the melt corresponding to the composition X_1-phase which is formed at a temperature over 1700 °C. The interaction of these elements leads to the formation of wurtzite structures (O-phase), apatite structure compounds (H-phase) and a solid solution of Y_2O_3 in the Si_2N_2O (Y-phase). Daylight α'→β' sialon to as α→β Si_3N_4 observed for the presence of liquid phase.

Thus, when the silicon nitride sintered with the oxides formed multiphase materials, whose structure is determined by the following factors: the composition and phase dispersion of Si_3N_4; type and oxide content, liquid phase sintering temperature and duration of high-temperature exposure [14].

Materials with a large number of n-Si_3N_4 achieve maximum density under the molding pressures of 50–75 MPa. Nano powder has a large specific surface, so the high-density materials prepared therefrom molding at lower pressures. According to [15] in the silicon nitride sintered α-phase Si_3N_4 completely transformed into β phase. This process is active in the temperature range 1100–1400 °C, when the density reaches only 60–65 % of the density of the sintered material. By increasing the sintering temperature goes into step increases densification rate, which is accompanied by a maximum shrink material.

Compressed to the optimum density of the samples have a high sintering density (97–98 % of theoretical for the latter two compounds) (Figure 1). Increasing the density of samples with increasing content of n-Si_3N_4 because nanoscale powders during sintering is achieved by high-density packing of the particles and increases the mass transfer of material from the surface to the grain boundaries. The radius of the particles is comparable with the diffusion layer thickness therefore activated sintering and is accompanied by volumetric shrinkage. When smaller n-Si_3N_4 content greater molding pressure necessary to obtain the best properties of materials. This trend is observed for formulations 1–3 samples for which the maximum density is achieved with a molding pressure of 100 MPa. Sample compositions 4 and 5 show the highest density value is already at a molding pressure of 75 MPa, wherein the further increase in pressure led to a decrease in density. With increasing concentration of n-Si_3N_4 (formulations 6–10) the maximum density is achieved with a molding pressure of 50 MPa. The highest density of the material at 75 % n-Si3N4 and 10 % micron-Si_3N_4 (Figure 1). Thus, it can be concluded that the nanoscale powder is not only increases the density of the sintered samples, but also intensifies the process of particle compaction.

An important parameter in the design of composite materials is porosity, which value is considerably affect the mechanical properties (see Table 2).

As seen from Table 2, by increasing the content of n-Si_3N_4 composed of the composite material porosity decreases, which is well supported by data reported for density (Figure 1). Table 2 also shows the dimensions of the grains of silicon nitride. The structure consists of elongated grains, and a reinforcing material imparting good mechanical properties to it (Figure 2). In the final stage of sintering the structure material is significantly increased content of elongated grains with clear-cut length to diameter ratio different 3÷5.

On the strength of the materials affected by the presence of structural defects (dislocations, impurity atoms pores and microcracks). The presence of the first two types of defects increases the overall strength of the material and reduces its last two. High speed cooling of the plasma during the synthesis of nanoscale silicon nitride particles results in a considerable concentration of lattice defects and, as a result, high activity of the powder. The appearance of significant amounts of edge and screw dislocations due to the kinetic characteristics of the process producing n-Si_3N_4. As shown in Table. 2, the total porosity of the materials based on n-Si_3N_4 does not exceed 2–3 %, resulting materials have a high strength level (Figure 3)

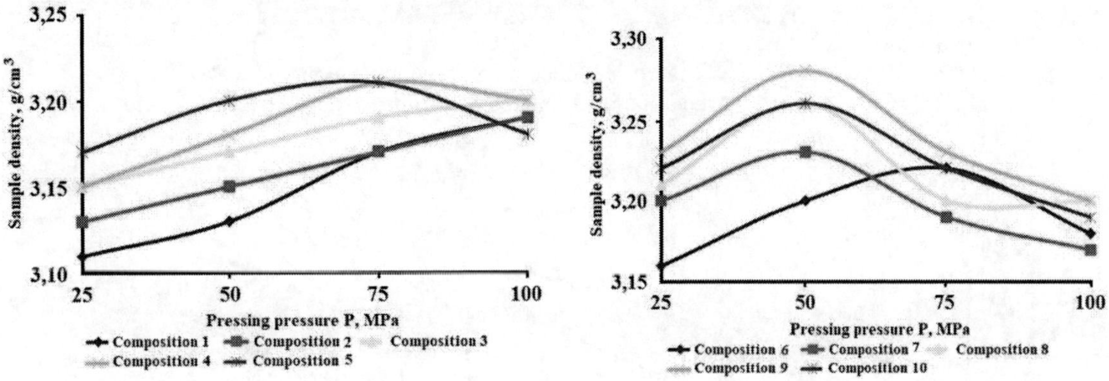

Figure 1. The dependence of the density of the sintered samples of different compositions by molding pressure.

Table 2. The values of porosity and grain size of materials based on silicon nitride.

№ composition	Porosity P (± 0,1),%	The pore size, microns		Grain size d_{max}, μm
		P_{max}	P_{min}	
1	3,5	13,2	1,2	5,1
2	3,3	12,7	1,4	4,7
3	3,1	10,4	1,2	4,8
4	2,5	11,9	1,1	4,4
5	2,6	10,6	0,8	3,9
6	2,1	11,3	1,1	4,0
7	2,0	9,4	1,2	3,1
8	1,8	8,8	1,0	2,9
9	1,4	7,5	0,8	1,8
10	1,1	3,1	0,5	0,3

The module of elasticity of materials is a function of the strength and also depends on the structural defects, i.e. the density and the bulk porosity. The paper [16] studied materials based on n-Si_3N_4 and concluded that by increasing the concentration of nano-sized powder of ceramics density increases.

Virtually all the compositions curves are extreme character with a maximum at the optimal properties of the molding pressure. A trend increase in the modulus of elasticity with increasing content of nano-sized Si_3N_4 powder composed of materials (Figure 4). Since the modulus depends directly on the density, the more denser materials at the maximum amount of n-Si_3N_4, showed a high level of properties (Figure 4).

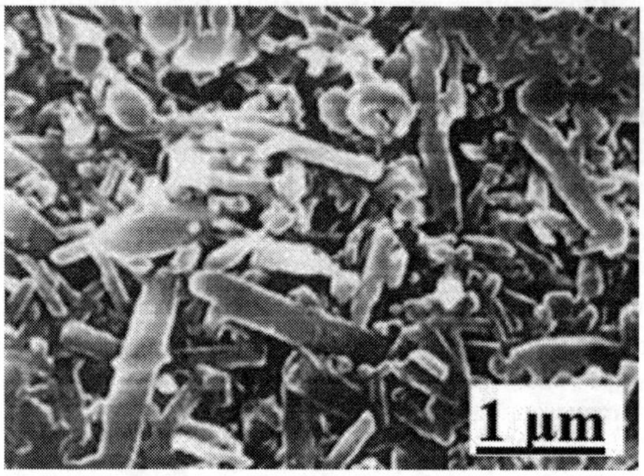

Figure 2. Etched microstructure of the section of the sintered Si_3N_4-material.

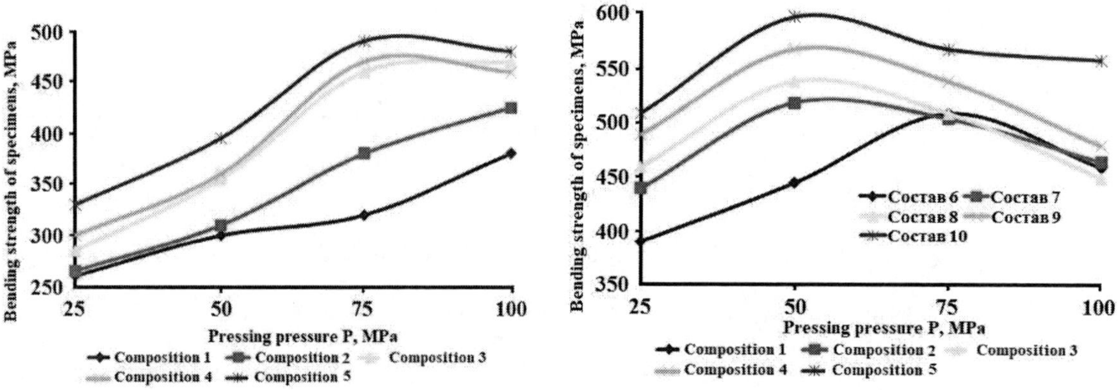

Figure 3. The dependence of the bending strength of samples of different compositions by molding pressure.

An important characteristic of composite materials is the ratio of fracture toughness (K_{IC}) (Fig. 5), which also depends on the density and on the presence of defects in the structure (porosity) in the bulk material. In [17] it is shown that by increasing the density of ceramics based on Si_3N_4 its K_{IC} values increase. Great influence on the mechanical properties of materials has materials based on silicon nitride nature intergranular phase. Since yttrium oxide additive increases the viscosity of the material, and alumina softens it somewhat. The value K_{IC} brittle ceramics can be enhanced by introducing the dispersed second phase inclusions. In [18, 19] has shown that when administered in the ceramics based on silicon nitride, a small amount of TiN increases its fracture toughness ratio of 1,5 times. According to [20, 21], with increasing concentration of n-Si_3N_4 volume percentage of the pores is reduced, thereby reducing the likelihood of local crack propagation and improving the value of K_{IC} (Figure 5). The materials are destroyed in accordance with the intergranular mechanism (destruction of grain bonded).

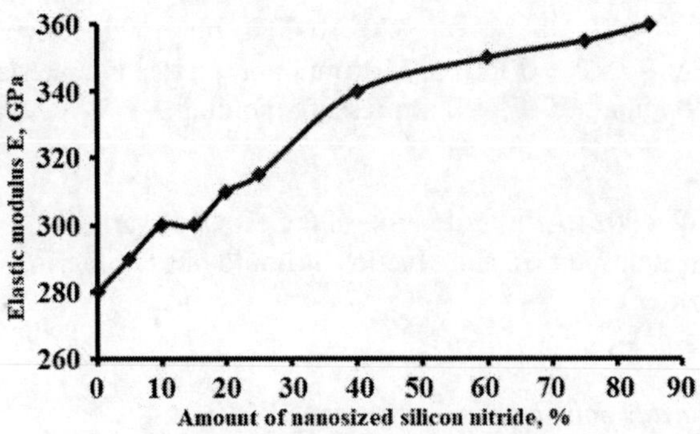

Figure 4. Dependence modulus of elasticity the material of the amount of n-Si_3N_4

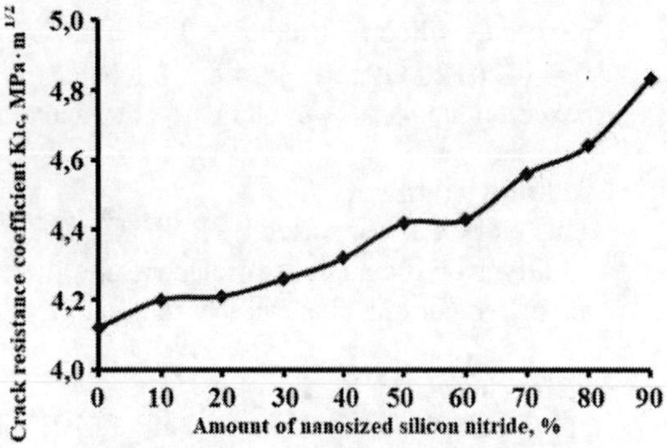

Figure 5. Dependence crack resistance the material of the amount of n-Si_3N_4

According to [22–24], the introduction of whiskers carbide, silicon nitride into the composition of a material based on silicon nitride, significantly increases its strength properties. Daylight $\alpha \rightarrow \beta$-Si_3N_4 grains elongated promotes crystallization, due to which the strength is increased.

The dependence of the compressive strength of the number of silicon nitride powder nano shown in Figure 6. Flexural strength and compressive hot-pressed materials may reach 900–1000 MPa and 1500–1700 MPa [25, 26] that 2,0–2,5 times higher than the strength values of the liquid-phase-sintered ceramics of micron powders. In this paper were obtained by liquid phase sintering materials based on n-Si_3N_4 with properties $\sigma_b = 600 \pm 15$ MPa; $\sigma = 1720 \pm 20$ MPa.

For test materials, the Vickers hardness is slightly lower than for ceramic based powders micrometer Si_3N_4 (Figure 7). Hardness heterophase materials is determined by the method of additive between the hardness phases within them. During sintering Si_3N_4 particles partly dissolve into the oxide bonded, forming a "Sialon" phase, reducing the overall hardness of the material.

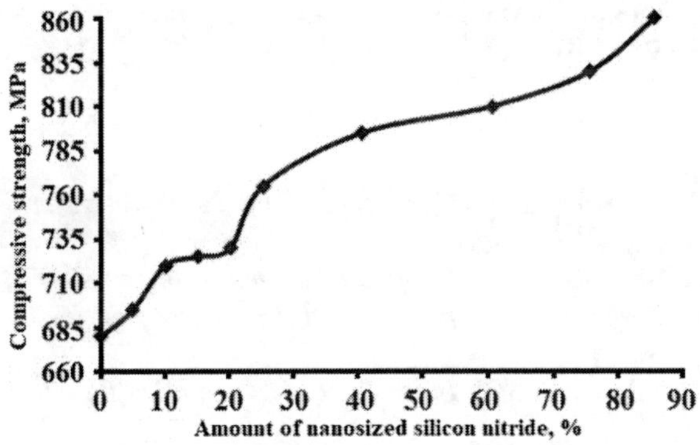

Figure 6. The dependence of the compressive strength materials from amount of n-Si_3N_4.

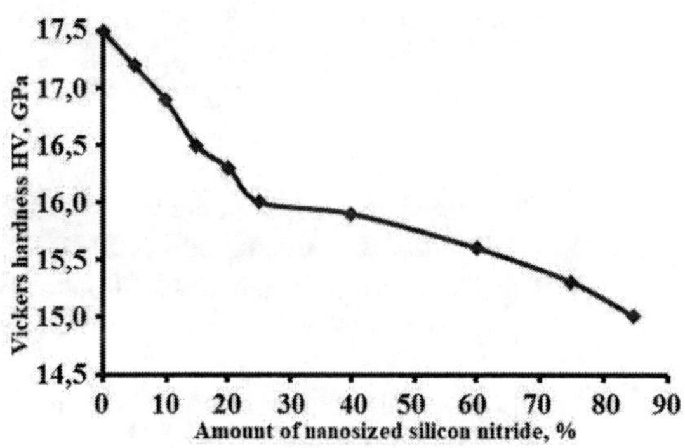

Figure 7. Dependence of Vickers hardness materials on the amount of n-Si_3N_4.

Thus, using a liquid-phase sintering of silicon nitride, the method was able to obtain materials with a sufficiently high level of mechanical properties that allow to recommend it for use in various fields of industry – in the production of sliding bearings, balls for rolling bearings, end seals for centrifugal pumps, of the cutting tool, abrasive nozzles, turbine blades and refractory products.

4. Conclusion

Obtained a high-density material ($\rho = 3{,}28$ g/cm^3) based nonorazmernogo silicon nitride powder by liquid phase sintering. The process of pattern formation material during sintering, as well as physical and mechanical characteristics. Maximum properties obtained at high material (85%) content of nanoscale silicon nitride: ρ = 3,20 ± 0,02 g/cm^3; P = 1,1±0,1%; E = 320±10 GPa; σb = 600±15 MPa; σ = 1720±20 GPa; K_{IC} = 4,8 ± 0,1 MPa·m$^{1/2}$; HV = 15,2 ± 0,1 GPa. Maximum properties in materials based on nanoscale powders are obtained at the optimal (50–75 MPa) pressure molding.

Acknowledgments

This work was supported by RNF (Grant № 18–79–00246 «Investigation of the effect of various additives on the complex dielectric and mechanical properties, and kinetics of liquid phase sintering advanced ceramic materials based on silicon nitride»).

References

[1] Han W, Li Y, Chen G, Yang Q 2017 *Materials Science and Engineering* **700** 19-24.
[2] Xu W, Yin Z, Yuan J, Wang Z 2017 *Materials Science and Engineering* **684** 127-134
[3] Markov MA, Ordan'yan SS, Vikhman SV, Perevislov SN 2019 *Refractories and Industrial Ceramics* **60** 385-388
[4] Dong H, Zhao Z, Wang C 2019 *Materials in Electronics* **30** 7590-7599
[5] Ordan'yan SS, Nesmelov DD, Danilovich DP, Udalov YP 2017 *Russian Journal of Non-Ferrous Metals* **58** 545-551
[6] Ahmad S, Ludwig T, Herrmann M, Mahmoud MM 2015 *J. of the European Ceramic Society* **35** 2261-2268
[7] Uludag A, Turan D 2015 *Int. J. of Materials: Mechanics and Manufacturing* **3** 105-109
[8] Kašiarová M, Tatarko P, Burik P, Dusza J 2014 *J. of the Europ. Ceram. Societyv* **34** 3301-3308
[9] Tatarko P, Kašiarová M, Dusza J, Šajgalík P 2013 *J. of the Europ. Ceram. Society* **33** 2259-2268
[10] Perevislov S N 2019 *Refractories and Industrial* **60** 168-173
[11] Tatami J, Komeya 2016 *Eng. Ceram.: Current Status and Future Prospects* 124-140
[12] Zhu XW, Sakka Y, Zhou Y, Hirao K, Itatani K 2014 *J. of the Europ. Ceram. Soc* **34** 2585-2589
[13] Naito M, Yokoyama T, Hosokawa K, Nogi K 2018 *Nanoparticle technology handbook* 720
[14] Perevislov SN, Lysenkov AS, Titov DD, Tomkovich MV 2019 *Glass and Ceramics* **75** 400-407
[15] Tatarko P, Kašiarová M, Chlup Z, Dusza J 2013 *Wear* **300** 155-162
[16] Gromov AA, Chukhlomina LN 2015 *John Wiley & Sons* 360
[17] Ordan'yan SS, Rumyantsev VI, Nesmelov DD, Korablev DV 2012 *Refractories and Industrial Ceramics* **53** 108-111
[18] Perevislov SN, Lysenkov AS, Titov DD, Tomkovich MV 2019 *IOP Conf. Ser.: Materials Sci. and Engineering* **525** 012074.
[19] Kandi KK, Thallapalli N, Chilakalapalli SPR 2015 *Int. J. of Applied Ceram. Tech.* **12** 909-920
[20] Lysenkov AS, Kim KA, Titov DD, Frolova MG 2018 *J. of Phys.: IOP Conference Series* **1134** 012036
[21] Frolova MG, Leonov AV, Kargin YF, Lysenkov AS 2018 *Inorg. Materials: Applied Research* **9** 675-678
[22] Perevislov SN 2013 *Glass and ceramics* **70** 265-268
[23] Lysenkov AS, Kim KA, Titov DD, Frolova MG 2018 *J. of Phys.: IOP Conference Series.* **1134** 012036
[24] Huang, S, Zhou W, Luo F, Wei P 2014 *Ceramics International* **40** 2785-2791.
[25] Kargin YF. Lysenkov AS. kim KA. Perevislov SN 2019 *IOP Conference Series: Materials Science and Engineering* **525** 012083
[26] Lukianova OA, Khmara AN, Perevislov SN, Kolesnikov DA 2019 *Ceramics International* **45** 9497-9501

Studying the properties of carbides in the system ZrC–HfC, TaC–ZrC and TaC–HfC

S N Perevislov[1], A B Vysotin[1], O Yu Shcherbakova[2]

[1] IV Grebenshchikov Institute of Silicate Chemistry of the Russian Academy of Sciences, St. Petersburg, Russia
[2]Federal STATE Autonomous educational institution of the Siberian Federal University, Krasnoyarsk, Russia
E-mail: perevislov@mail.ru

Abstract. Considers the chemistry of carbides of zirconium, hafnium and tantalum and their high temperature properties. By hot pressing the obtained dense materials in the systems ZrC–HfC, TaC–ZrC, and TaC–HfC analysed their microstructure, physical and mechanical properties.

1. Introduction

As is known in the systems ZrC–HfC, TaC–ZrC, and TaC–HfC can form complex carbides having a melting temperature of ~4000 °C, which exceeds the melting temperature of the individual carbides (T = 3532 °C (ZrC); T = 3900 °C (HfC); T = 3880 °C (TaC) [1]). To obtain dense bulk materials based on ZrC, HfC and TaC is possible only by hot pressing (HP) [2], hot isostatic pressing (HIP), spark plasma sintering, or pulse (SPS) [3, 4]. The study of patterns of synthesis and properties of single-phase complex carbides allows to obtain heat-resistant materials that can be used as main components for producing products used in missile and space technology.

In works [1, 5], the presence of anomalous physical properties of the carbide composition Ta_4HfC and materials in the systems ZrC–HfC and TaC–ZrC, but nhe most promising in the system TaC–HfC is a complex tantalum carbide-hafnium (Ta_4HfC_5), which has a record of a melting point over 4000 °C [6, 7].

Materials based on carbides and their compounds can be promising due to the unique properties themselves ZrC, HfC and TaCsuch as high hardness (~28,5 GPa (ZrC); ~20,0 GPa (HfC); ~19,0 GPa (TaC)) and the modulus of elasticity (412 GPa (ZrC); 425 GPa (HfC); 537 GPa (TaC)), heat resistance in a protective gas atmosphere to a temperature 2500 °C, high electrical conductivity etc., which allows the use of the materials and products on their basis when working in extreme conditions, as lining rocket engines, or the front edges of hypersonic aircraft and missiles and so on [8].

The aim of this work is to study the properties of synthesized single-phase connections of the complex carbides in the systems ZrC–HfC, TaC–ZrC and TaC–HfC by hot pressing, and the microstructure and physical-mechanical properties of obtained materials.

Features of the structure ZrC, HfC and TaC and high temperature properties of materials based on them.

Carbides ZrC, HfC and TaC have a face-centered a cubic lattice [9]. Hafnium carbide has exceptional resistance to high temperature oxidation, on its surface forms a stable oxide of hafnium (HfO_2), which serves as a barrier to oxygen diffusion into the bulk material, protecting it from further oxidation [10, 11].

The melting temperature of non-stoichiometric HfC increases with a deficit of carbon, until a maximum at the point of the congruent melting point – HfC0,82 (~ 45 at. % C) [12].

On the surface TaC formed oxide film, which protects the ceramics from oxidation at high temperatures (up to 1800 °C). Removal of oxygen impurity from the surface of the powder TaC is a thermochemical treatment at a temperature of 1000–1500 °C that leads to significant growth of the grains.

Tantalum carbide has relatively poor oxidation resistance and is completely oxidized at a temperature of ~ 850 °C after 5 h in an atmosphere of pure oxygen, forming the pentoxide of tantalum [13]. However, when TaC is used with minimal exposure to oxygen, for example, solid rocket motors with aluminum fuel, it is considered the most effective material for the ceramic veneer.

One of the possible ways of increasing the heat resistance of materials can serve as a synthesis of complex carbides and the introduction of more heat-resistant compounds, for example, adding $MoSi_2$ [14].

2. Materials and Method

The materials were prepared by standard ceramic technique [15–20]. For the synthesis used the commercially available powders of carbides with original size: ZrC (d0,5 = 30,2 µm); HfC (d0,5 = 28,6 µm); TaC (d0,5 = 44,1 kHz µm), which are separately milled in a planetary mill to a size d0,5 = 0,5 to 1,0 µm and, in appropriate ratio, mixed in a drum mixer in the environment of isopropyl alcohol for 20 h. Charge powders were dried and pelletized. Materials of size Ø25×5 mm were obtained from charge mixtures of powders by hot pressing (T = 1900 ° C, P = 30 MPa, t = 1 h), which were then further annealed at a temperature of 2000 ° C in a vacuum oven for 1 h in the atmosphere Ar. To determine the mechanical characteristics of sintered materials in the form of discs cut on samples with size 5×5×20 mm.

The density and porosity of the samples was determined by hydrostatic weighing. Phase composition and lattice parameters using the diffractometer Rigaku Smartlab 3. Ultimate strength in three-point bending was determined on a tensile testing machine ShimadzuAG-300kNX. The modulus of elasticity was determined on the installation of Sound-130. Examination of the microstructure was performed by scanning electron microscopy, determination of the elemental composition by the method of Microspectral analysis on the microscope Tescan Vega 3 SBH with the console EDX. The microstructure was analyzed using optical microscope Meiji Techno IM 7200. The equipment is provided by the Engineering center St. Petersburg State Technological Institute (Technical University).

3. Results and discussion

In work the method of X-ray analysis to determine the composition of hot-pressed ceramic carbides (Table 1), shows the change in the content of fixed carbon and the lattice parameter of the material before and after additional high temperature annealing. As can be seen from the results of the analysis, the content of bound carbon is markedly reduced, which affects the change in the composition of the sintered material. Thus, carbides of zirconium, hafnium and tantalum have changed the composition of $ZrC_{0,88}$, $HfC_{0,97}$ and $TaC_{0,98}$ at the $ZrC_{0,78}$, $HfC_{0,94}$ and $TaC_{0,93}$. It also indicates the change of lattice parameter (Table 1). Accordingly, we can assume that was a change in the stoichiometry of the complex carbides, for example, $(Zr0,6Hf_{0,4})C_{0,92}$ to $(Zr_{0,6}Hf_{0,4})C_{0,84}$ and $(Ta_{0,8}Hf_{0,2})C_{0,97}$ to $(Ta_{0,8}Hf_{0,2})C_{0,93}$ [21].

It is known that the maximum melting temperature is characteristic of nonstoichiometric carbides: $ZrC_{0,8\text{-}0,9}$ (T = 3410–3435 °C [22]); $HfC_{0,9}$(T = 3820 °C [23]); $TaC_{0,8}$ (T = 4000 °C [24]). Appreciable decarburization, which occur after the annealing of alloys ZrC–HfC, TaC–ZrC, and TaC–HfC gives reason to assume that the complex carbides are characterized by high melting temperature.

Table 1. Chemical composition and unit cell parameters carbide materials.

Alloy composition	Chemical composition, wt. %									Option cell a, Å	
			Bonded carbon								
	Total carbon	Carbon free	Before annealing	After annealing	Zr	Hf	Ta	N	O	Before annealing	After annealing
$ZrC_{0,88}$	11,2	<0,05	11,20	9,10	88,6	–	–	0,02	0,18	4,689	4,695
$HfC_{0,97}$	6,1	<0,05	6,00	5,80	–	93,8	–	0,01	0,09	4,640	4,639
$TaC_{0,98}$	6,1	<0,05	6,10	5,70	–	–	93,8	0,01	0,09	4,454	4,444
ZrHfC	10,5	1,80	8,70	7,50	43,2	46,0	–	0,23	0,07	4,663	4,664
Zr_3Hf_2C	9,9	0,24	9,66	7,70	52,0	37,6	–	0,17	0,33	4,670	4,669
Zr_4HfC	10,0	0,18	9,82	9,10	70,6	19,2	–	0,05	0,15	4,683	4,684
Ta_2ZrC	5,9	0,12	5,80	5,30	16,7	–	77,3	0,04	0,06	4,522	4,524
Ta_4ZrC	6,7	0,05	6,65	5,30	8,8	–	84,4	0,03	0,07	4,495	4,492
Ta_8ZrC	6,7	0,05	6,35	5,90	5,0	–	88,0	0,20	0,01	4,470	4,466
Ta_3HfC	5,3	0,13	5,17	5,05	–	22,0	72,1	0,43	0,17	4,486	4,483
Ta_4HfC	6,4	0,14	6,45	3,54	–	12,8	80,2	0,50	0,10	4,473	4,468
Ta_4HfC_5	15,4	0,09	15,3	10,8	–	11,4	72,7	0,40	0,10	4,453	4,458
Ta_5HfC	5,6	0,15	5,45	5,05	–	11,6	82,2	0,37	0,23	4,473	4,472

X-ray phase and microstructural analyses confirmed the formation in all materials, single-phase connection (Figure 1). The porosity of samples of complex carbides does not exceed 7,0–10,0 % (Table 2). Strength of materials maximum obtained for compositions located or close to equimolar ratio, however, the strength values are greatly influenced by porosity. The modulus of elasticity of materials is a function of strength and also depends on porosity (Table 2) as the concentration of carbon carbide in the complex increases its porosity and, consequently, reduced the level of mechanical characteristics (Table 2). By analogy with the nonstoichiometric compounds (ZrC0,8-0,9HfC0,9TaC0,8) having a high heat resistance, it can be assumed that the decrease of the concentration of carbon in the compounds, for example, from Ta3HfC to Ta5HfC also will increase the heat resistance of the material.

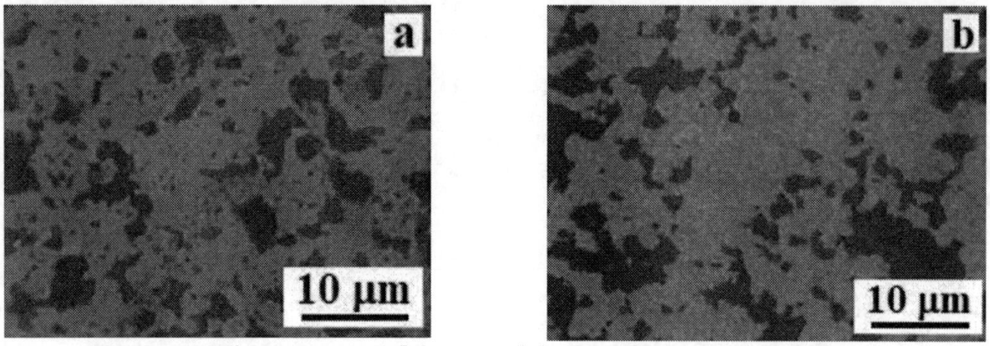

Figure 1. Microstructure of the materials Ta_4HfC_5 (a) and Ta_4HfC (b).

Table 2. Physical and mechanical properties of hot-pressed materials in systems ZrC–HfC, TaC–ZrC and TaC–HfC.

Composition carbides	Density, g/cm^3	Porosity, %	The modulus of elasticity, GPa	Bending strength, MPa
ZrHfC	8,82	6,8	358±5	418±8
Zr_3Hf_2C	8,19	8,1	410±8	384±12
Zr_4HfC	7,09	9,4	407±11	346±4
Ta_2ZrC	10,92	7,3	433±4	357±17
Ta_4ZrC	11,63	9,0	474±9	346±11
Ta_8ZrC	12,10	10,1	496±13	309±6
Ta_3HfC	12,56	8,8	418±5	394±13
Ta_4HfC	12,55	9,6	443±8	381±26
Ta_4HfC_5	12,77	6,6	459±6	384±11
Ta_5HfC	12,50	10,4	463±9	342±10

The most promising from the point of view of heat resistance, are system NbC–HfC, TaC–NbC, in which the synthesis and analysis of physical-mechanical and high temperature properties can be used to detect the formation of single-phase complex carbides, characterized by record melting temperature (over 4000 °C).

4. Conclusion

In the work by hot pressing of the synthesized single-phase complex carbides in the systems ZrC–HfC, TaC–ZrC and TaC–HfC, studied their chemical composition and determined the lattice parameters and physico-mechanical properties. Specificity of oxidation and explained the nature of high-temperature materials, the maximum temperature of the material melting in the system Zr–C corresponds to the composition ZrC0,8-0,9 (T = 3410–3435 °C) in the system Hf–C the composition of HfC0,9 (T = 3820 °C) and for the system Ta–C the maximum melting temperature is close to the composition of the material TaC0,8 (T = 4000 °C). The most promising system TaC–HfC are complex carbides of tantalum-hafnium composition Ta4HfC5 and Ta4HfC having a melting point over 4000 °C.

Acknowledgments

IOP Conf. Series: Materials Science and Engineering **848** (2020) 012069 doi:10.1088/1757-899X/848/1/012069

The work is executed at financial support of Russian Foundation for basic research (project № 18-53-18014\19 Bolga).

References

[1] S A Ghaffari, M A Faghihi-Sani, F Golestani-Fard, M Nojabayy 2013 *Int. J. of Refractory Metals and Hard Materials* **41** 180-184
[2] F Rezaei, M G Kakroudi, V Shahedifar, N P Vafa, M Golrokhsari 2017 *Ceramics International* **43** 3489–3494
[3] O Cedillos-Barraza, D Manara, K Boboridis, T Watkins 2016 *Scientific reports* **6** 37962
[4] L Feng, S H Lee, B L Yoon 2015 *Ceramics International* **41** 11637–11645.
[5] Q J Hong, A van de Walle 2015 *Physical Review B* **92** 020104
[6] O Gaballa, B A Cook, A M Russell 2013 *Int. J. of Refractory Metals and Hard Materials* **41** 293–299
[7] A A Lavrentyev, B V Gabrelian, V B Vorzhev, I Y Nikiforov 2008 J.*l of Alloys and Compounds* **462** 4–10
[8] V G Sevastyanov, E P Simonenko, N A Ignatov, Y S Ezhov 2011 *Russian J. of Inorganic Chemistry* **56** 661–672
[9] Q J Hong, A van de Walle 2015 *Physical Review B* **92** 020104
[10] O Cedillos-Barraza, S Grasso, N Al Nasiri, D D Jayaseelan 2016 *J. of the European Ceramic Society* **36** 1539–1548
[11] B C Schulz, B Wang, R A Morris, D Butts, G B Thompson 2013 *J.l of the European Ceramic Society* **33** 1219–1224
[12] W Shuo, Z Kan, A Tao, H Chaoquan 2015 *Applied Surface Science* **327** 68–76
[13] L Pienti, D Sciti, L Silvestroni, A Cecere, R Savino 2015 *J.l of the European Ceramic Society* **35** 1401–1411
[14] S A Ghaffari, M A Faghihi-Sani, F Golestani-Fard, S Ebrahimi 2013 *Ceramics International* **39** 1985–1989
[15] M A Markov, S S Ordan'yan, S V Vikhman, S N Perevislov 2019 *Refractories and Industrial Ceramics* **60** 385–388
[16] S N Perevislov 2019 *Refractories and Industrial Ceramics* **60** 168–173
[17] S N Perevislov, A S Lysenkov, D D Titov, M V Tomkovich 2019 *Glass and Ceramics* **75** 400–407
[18] S N Perevislov, A S Lysenkov, D D Titov, M V Tomkovich 2019 *IOP Conference Series: Materials Science and Engineering* **525** 012074
[19] M G Frolova, A V Leonov, Y F Kargin, A S Lysenkov 2018 *Inorganic Materials: Applied Research* **9** 675–678
[20] A S Lysenkov, K A Kim, D D Titov, M G Frolova 2018 *IOP Conference Series:J. of Physics* **1134** 012036
[21] S A Ghaffari, M A Faghihi-Sani, F Golestani-Fard, M Nojabayy 2013 *Int. J. of Refractory Metals and Hard Materials* **41** 180–184
[22] S Kim, I Szlufarska, D Morgan 2010 *J. of Applied Physics* **107** 053521
[23] F Réjasse, G Trolliard, O Rapaud, A Maitre, J David 2015 *RSC Advances* **5** 45341-45350
[24] T Dobatkina, N Bochvar 2010 Carbon–tantalum–zirconium *Refractory metal systems* 18-30 p

IOP Conf. Series: Materials Science and Engineering **848** (2020) 012070 doi:10.1088/1757-899X/848/1/012070

Structural changes after high-speed impact of tungsten powder with a steel target

E V Petrov[1], V S Trofimov[1], G R Saikova[1] and V O Kopytskiy[2]

[1] Merzhanov Institute of Structural Macrokinetics and Materials Science Russian Academy of Sciences, Academician Osipyan str., 8, Chernogolovka, Moscow Region, 142432, Russia

[2] Polzunov Altai State Technical University, Lenin ave., 46, Barnaul, Altai Region, 656038, Russia

Abstract. Shock-wave high-velocity impact of tungsten powder with a steel target was studied in the work. It is determined that shock wave treatment of the samples leads to a refinement, flattening and stretching of ferrite and perlite grains in the surface layer. The structural transition from plate perlite in the near-surface layer of the target to coarse-plate perlite in the bulk of the target, the microstructure of which does not differ from the microstructure of the initial sample, was detected in the structure of U8 steel after a particle flux. It is found that cellular supercooled austenite with tungsten carbide mesh along grain boundaries is formed in the structure of steel. The microhardness of the target in depth after exposure to tungsten particles is analyzed. It was shown that the microhardness distribution along the width of the sample is wave-like, and the microhardness decreases monotonically along the depth of the sample. The microhardness maximizes by 24% at a distance of 4 mm from the surface compared to the original microhardness.

1. Introduction

Study of structural transformations in metal materials exposed to a flux of high-energy microparticles is an important task of physical materials science, because the impact of the particles dispersed by the explosion energy, combined with shock-wave loading of detonation products lead to the improvement of physico-mechanical properties as the surface layer and throughout the volume of the obstacles. When a flux of different particles accelerated by explosion energy up to a velocity of 1–3 km/s interacts with a target surface, their material can penetrate into the target to a depth exceeding the size of initial particles by hundreds of times and affects the material structure of the sample. This effect was found studying the hardening of metals under shock-wave loading and this process is called the phenomenon of super deep penetration of particles [1, 2]. During the interaction of detonation products with powder particles in an explosive wave, these particles are entrained due to inelastic impact of molecules of detonation products with particles. The interaction of the flux of particles with the sample stops the most of them in the near-surface zone and results in the formation of a particle coating on its surface [3–7]. Shock-wave loading and impact of high-velocity particles provide improved physical and mechanical properties of the surface and the whole volume of the processed materials. When a high-velocity particle flux interacts with the surface of the target, the largest part of the particle flux forms a coating from these particles, and the thickness, structure, and phase composition of the formed layer are stable. Despite the fact that the nature of this effect has not yet been established [8–11], the method of the study closely simulates the impact of space debris with aircrafts in outer space.

IOP Conf. Series: Materials Science and Engineering **848** (2020) 012070 doi:10.1088/1757-899X/848/1/012070

The purpose of this work is to study the structural transformations in the steel target after high-velocity impact with tungsten powder.

2. Materials and Methods

Tool steel samples (U8 grade) with a diameter of 25 mm and a height of 30 mm were used as a steel target. The shape of the samples was chosen for the convenience of conducting experiments. Tungsten powder with a particle size of 10–16 μm was used. The particle size of the powder was determined on a Micro Sizer 201 laser particle analyzer.

The experimental arrangement is shown in Figure 1. The test sample (6) was placed into the directing tube (5), on top of which was a ring (3) with tungsten powder (4) with a bulk density of 5 g. A cut-plate (2) was installed on the ring in order for the detonation products to fly into the directing tube (5). Hexogen with a bulk density used as an explosive (1) was placed on the plate. An air gap was left between the tungsten powder and the explosive to provide a longer powder loading resulting in the formation of a flat shock-wave front [12]. The length-to-diameter ratio of the explosive was 2.5, which meets the condition for the generation of a stationary detonation wave [13] and the stable formation of a flux of particles. After detonation, the test sample was exposed to the flux of powder particles accelerated by shock wave and explosion products.

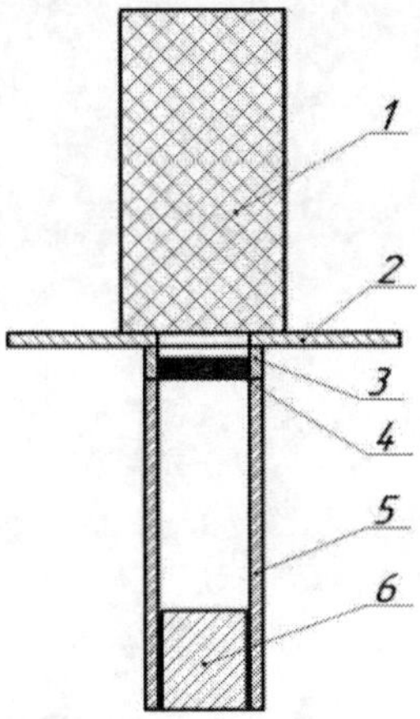

Figure 1. Experimental arrangement: (1) explosive charge; (2) cut-plate; (3) support ring; (4) tungsten powder; (5) directing tube; (6) obstacle (steel target).

Steel targets after high-speed processing were cut along the axis along the direction of loading by a stream of tungsten particles. Sample preparation included grinding and polishing operations for metallographic studies. The samples were studied by SEM using a Zeiss Ultra plus ultra-high-resolution scanning electron microscope and an Axiovert 200 MAT metallographic inverted microscope. The Vickers microhardness of the samples after interaction with a high-velocity flux of particles and in the initial state was measured using a PMT-3 hardness meter under a load of 100 g.

3. Results and Discussion

A metallographic analysis of the microstructure of samples treated with the high-velocity flux of tungsten particles showed that the shock-wave particle treatment of samples results in the grinding, flattening and extension of ferrite and perlite grains in the near-surface layer.

Figure 2a shows the microstructure of the initial steel sample (U8 grade) with the presence of lamellar perlite. A large number of the traces of particles are observed (black inclusions) at a distance of 2 mm from the sample surface treated with tungsten powder (figure 2b). Such traces are not observed for the sample treated only with a shock wave (experimental conditions are the same). This suggests that these traces are created by tungsten particles. Analysis of the microstructure of the near-surface zone shows the presence of a small amount of macrolamellar perlite. This is explained by the deformation of cementite plates under shock wave loading with penetrating particles with a pressure of more than 13 GPa and by the formation of defects remaining in the lattice after the passage of shock wave.

To obtain an estimate of the collision pressure of the flow of tungsten particles as a porous body with a steel barrier, the technique of matched impedances was used. In the calculation, the Hugoniot adiabat for tungsten particles was constructed according to the data of [14], where experimental data on the determination of wave velocities D in porous samples are presented. The calculations were carried out using the laws of conservation of thermodynamic parameters of shock compression. The calculations showed that the pressure in the incoming shock wave in steel is 34 GPa (the particle flux porosity m = 3.9). With porosity m = 4.5 pressure is 25 GPa. At a porosity of m = 8 (this value most closely matches the porosity of the tungsten particle flux during the experiment), it equals 10–13 GPa. The latter value is consistent with the results of [15], where the authors of the work obtained a pressure of 12 GPa.

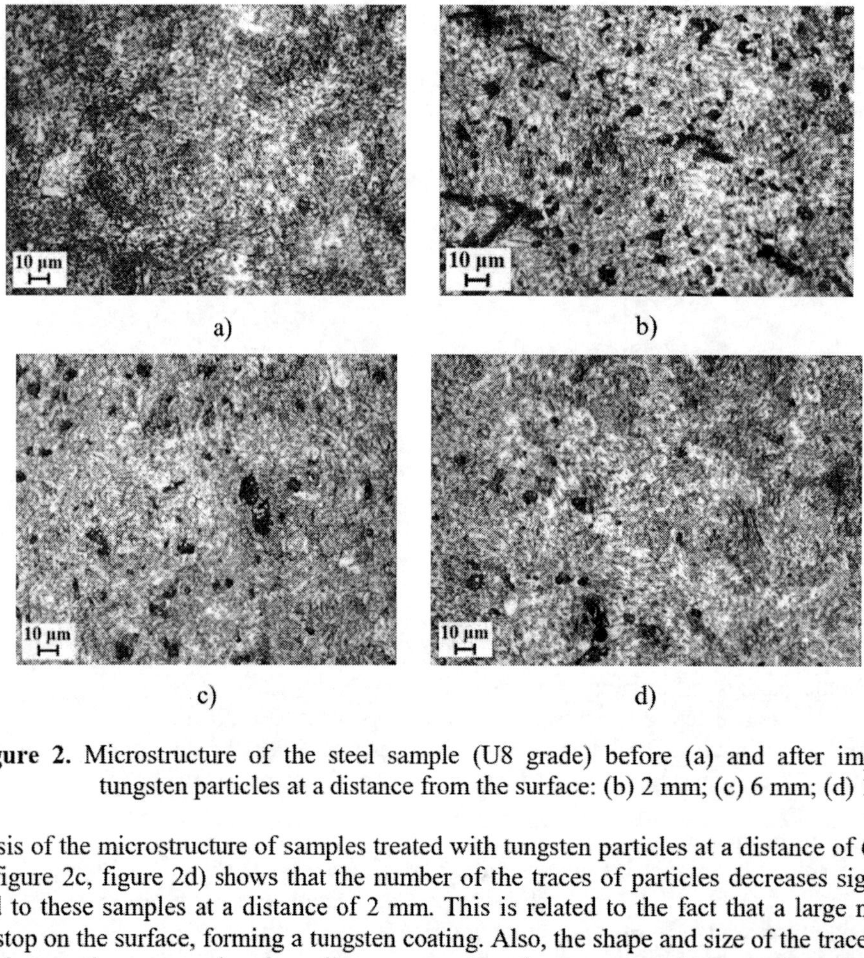

a) b)

c) d)

Figure 2. Microstructure of the steel sample (U8 grade) before (a) and after impact with tungsten particles at a distance from the surface: (b) 2 mm; (c) 6 mm; (d) 10 mm.

Analysis of the microstructure of samples treated with tungsten particles at a distance of 6 mm and 10 mm (figure 2c, figure 2d) shows that the number of the traces of particles decreases significantly compared to these samples at a distance of 2 mm. This is related to the fact that a large number of particles stop on the surface, forming a tungsten coating. Also, the shape and size of the traces of these samples change; they are oval and smaller, compared to the same samples at a distance of 2 mm. Analysis of the microstructure revealed that the content of the lamellar perlite increases in comparison with a distance of 2 mm. This is explained by the fact that at a distance of 6 mm and 10 mm from the surface, the effect of shock wave and powder particles on the structure of the sample material decreases, resulting in less heating compared to the surface layer.

Given the fact that the most tungsten particles stop on the surface during high-velocity impact with the sample, forming a coating, the near-surface area of the sample was studied. The SEM study of the interface between the sample and the coating after impact with a flux of tungsten particles revealed a transition zone (figure 3a) consisting of the particles of tungsten (white zone - spectrum 1), iron (gray zone - spectrum 4) and their intermetallic compound (light gray zone - spectrum 2, 3). The data of energy dispersive analysis are presented in Table 1. Analysis showed a high carbon content that was explained by the fact that the thin sections of the samples were polished using diamond paste.

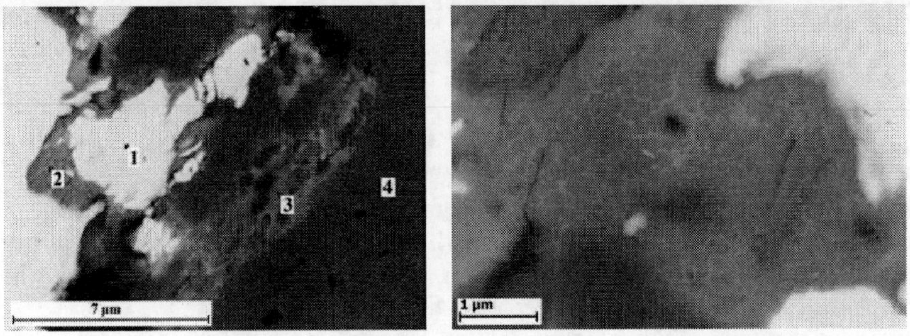

a) b)

Figure 3. Interface between the sample and the tungsten coating: (a) transition zone of the interface; (b) austenite cellular structure with tungsten carbide mesh.

Table 1. Energy dispersive analysis of the surface layer of the sample after impact with tungsten particles (weight %).

Spectrum	C	Fe	W
1	36.53	1.07	62.41
2	39.26	36.47	24.27
3	34.82	53.29	11.89
4	32.95	67.05	-----

Higher magnification of the transition zone (figure 3b) shows the presence of the cellular austenite structure with a tungsten carbide mesh along the grain boundaries of the intermetallic compound. That is, during impact of tungsten particles with steel, the temperature exceeds the temperature of the peritectoid reaction in the *Fe-W* system, resulting in the formation of high-carbon austenite, which is stabilized by dissolved tungsten during cooling, and a mesh cellular structure is formed [4].

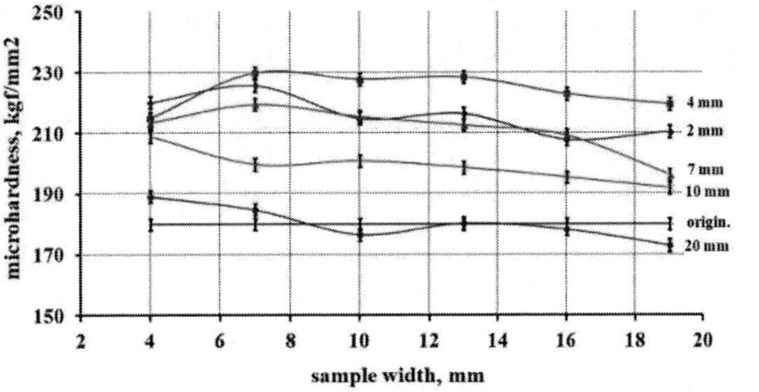

Figure 4. Distribution of microhardness over the width of the sample after impact with tungsten particles at different distances from the surface.

The study of microhardness showed (figure 4) that its distribution over the width of the sample was wave-shaped, and it monotonously decreased to the depth of the sample. The microhardness of the sample after exposure to tungsten particles at a depth of 20 mm is approximately equal to the initial microhardness of U8 steel, which was annealed to a temperature of 500 ° C and has a microhardness of 180 HV. It is seen that at a depth of 2 mm microhardness is less than at a depth of 4 mm. This is related to the fact that caverns are formed on the sample surface after impact of tungsten particles with the target. The average microhardness values are 215 HV at a depth of 2 mm, which is more than the microhardness values for the initial sample. The maximum average value of microhardness is 224 HV at a distance of 4 mm from the surface and varies in the range from 214 to 230 HV.

4. Conclusion

Studies of the microstructure of U8 steel samples treated with a high-speed stream of tungsten particles showed that the impact of particles in combination with shock-wave loading of detonation products is accompanied by a refinement of the perlite structure with a transition from lamellar perlite in the surface layer of the barrier to coarse plate perlite in the volume of the barrier, which does not differ from microstructures of the initial sample of steel U8.

It was revealed that when a flux of tungsten particles interacts with a target surface, the temperature in the reaction zone exceeds the temperature of the peritectoid reaction in the *Fe-W* system. As a result, cellular supercooled austenite with a tungsten carbide mesh along the grain boundaries is formed in the steel structure, which is identical to the structure of cast high-speed steel with complex carbide eutectic resembling ledeburite and located along the austenite grain boundaries.

A study of the microhardness of the target after treatment with a tungsten particle stream showed that the microhardness distribution along the width of the samples is wave-like, and the microhardness decreases monotonically along the depth of the samples. The study of microhardness of the target after treatment with a flux of tungsten particles showed the increase in microhardness compared to the initial microhardness of steel (U8 grade). The maximum increase in microhardness by 24% occurs at a distance of 4 mm from the processing surface in comparison with the initial microhardness.

Acknowledgments
The research was supported by the Russian Foundation for Basic Research, (project no. 19-08-00553).

References

[1] Usherenko S M 1998 *Ultra-deep penetration of particles into obstacles and the creation of composite materials* (Minsk: Research Institute of Pulse Processes Press) p 210.
[2] Usherenko S M 2002 *J. of Eng. Phys. and Thermophys.* **75** 753
[3] Usherenko Yu S, Usherenko S.M. and Yazdani J. 2017 *Procedia Engineering* **172** 1198
[4] Davydov S V and Petrov E V 2017 *J of Technical Phys.* **62** 1207
[5] Ganigin S Yu, Kalashnikov V V, Ibatullin I D, Murzin A Yu , Glazunova O Yu and Grigoryev A A 2013 *News of the Samara Scientific Center of the RAS* **15** (4-2) 339
[6] Korshunov L G, Zeldovich V I, Usherenko S M, Kheifets A E, Khomskaya I V, Chernenko N L and Frolova N Yu 2008 *Advanced Mater. Research.* **47–50** 423
[7] Aleksentseva S E and Krivchenko A L 2013 *Bulletin of SamGTU. Ser. Techn. Sci.* **2** 71
[8] Roman O V, Andilevko S K, Karpenko S S, Romanov G S and Shilkin V A 2002 *J. of Eng. Phys. and Thermophys.* **75** 997
[9] Chengzhi Qi and Jianjie Chen 2014 *J. of the Mechanical Behavior of Materials* **1-2** 21
[10] Makarov P V 2006 *Phys. Mesomechanics* **3** 61
[11] Kiselev S P and Kiselev V P 2002 *Inter. J. of Impact Engineering* **27** 135.
[12] Trofimov V S, Petrov E V and Alymov M.I. 2016 *Int. J. of Self-Propag. High-Temp. Synthesis* **2** 125
[13] Dremin A N, Savrov S D, Trofimov V S and Shvedov K K 1970 *Detonation waves in condensed matter* (Moscow: Science Press) p 164.
[14] Bakanova A A, Dudoladov I P and Sutulov Yu N 1974 *Prikl. Mekh. Tekh. Fiz.* **2** 117

[15] McQueen R G, Marsh S P, Tailor J W, Fritz J N and Carter W J 1973 *The equations of state of solids from shock wave studies, in High-Velocity Impact Phenomena* (Academic Press, New York) 293.

IOP Conf. Series: Materials Science and Engineering **848** (2020) 012071 doi:10.1088/1757-899X/848/1/012071

Oxidation and biodegradation of polymeric composites based on polylactide: structure and properties

M V Podzorova[1,2] and Yu V Tertyshnaya[2]

[1] Plekhanov Russian University of Economics, 36 Stremyanny per., Moscow, 117997, Russia
[2] Emanuel Institute of Biochemical Physics, Russian Academy of Sciences, 4 Kosygina str., Moscow, 119991, Russia

E-mail: mariapdz@mail.ru

Abstract. Polymer blends of polyethylene – polylactide of different composition were obtained and studied. Thermal degradation was investigated in the temperature range from 20 to 700 °C. It is shown that the polyethylene more resistant to the effects of temperature (peak of destruction - 414 °C) than polylactide (peak degradation – 364 °C). In blends with increasing polyethylene content in the matrix increases the maximum temperature of thermal degradation. The process of thermal oxidation carried out at temperatures of 80, 90, 110 °C and a pressure of 300 Torr. Binary polylactide–low-density polyethylene blends of various compositions were prepared, and their biodegradability in soil. The weight loss is higher upon incubation in laboratory soil compared to open soil.

1. Introduction

Polymer blends are systems, made through mixing of two or more polymers. In most cases polymer mixing is a process of mutual dispersion, possibly combined with further mutual solution of two or more liquids of higher viscoelasticity. Within such a system, inter-phase dispersability, morphology and adhesion largely depend upon interfacial energies that are of paramount importance for formation of a level of performance properties. In this connection, research of blends, their structures and properties is important for analysis of their performance.

A great number of scientific papers are dedicated to compositions based upon biodegradable polymers. The most popular biodegradable polymers: polylactide (PLA), polycaprolactone (PCL), polyhydroxyalcanoates (PHB, PHV) [1-7].

It is thought that PLA made with use of natural raw materials, such as corn, grain and other agricultural crops can partially replace polymers, being products of petrochemicals synthesis. PLA exhibits optimal mechanical properties, but shows a not high value of elongation per unit length and thermal stability [8].

That way, research works by various authors are mainly focused on improvement of PLA physical and mechanical properties. Extensive works were performed with purpose of manufacture of PLA blends with biodegradable, synthetic and secondary polymers. One of the benefits of PLA is its degradability under the influence of a number of various factors. Therefore, more extensive research is required to enable assessment of degradability of PLA and its compositions in a variety of conditions [9, 10].

IOP Conf. Series: Materials Science and Engineering **848** (2020) 012071 doi:10.1088/1757-899X/848/1/012071

2. Experimental

Blends of LDPE (TM 15803-020 from Neftekhimsevilen, OJSC, Russia) with PLA (4032D from Nature works, USA) in a wide range of compositions were studied. The PLA content in the polymer matrix was 20, 30, 50, 70, 80 wt% and pure polymers PLA and LDPE. There were also made three triple blends with 10, 20, 30 wt% of LDPEag (LDPE held in air, in an open vessel, at a temperature of 90°C for 250 hours, TM 15803-020 from Neftekhimsevilen OJSC, Russia) to neat LDPE respectively and 30 wt% of PLA[7].

The degree of crystallinity and melting temperature of the samples were studied with differential scanning calorimeter DSC 214 Polyma (Netzsch, Germany) at a heating rate of 10K/min and a sample weight of 5±0.1 mg. The experiment on thermal oxidation was performed in a static manometric installation at 80, 90, 110 °C and oxygen pressure of 300 Torr with the absorption of gaseous reaction products by solid KOH [7].

Thermogravimetric analysis using NETZSCH Tg 209 F1 Iris derivatograph (Germany) was carried out to study the thermal resistance of polymers. The heating rate was 20 °C / min, the analyzed temperature range was 25-700 °C.

The infrared spectra of the samples were recorded on the FT-IR spectrometer Perkin Elmer Spectrum 100 at T = 23 ± 2 °C in the wave number range $4600 \leqslant \nu \leqslant 450$ cm-1 in transmitted light [7]. For a complex study of the biodegradation of the materials, we performed a soil test on restored soil. The restored soil prepared according to GOST (State Standard) 9.060 simulates the real soil, leveling off the difference between various types of soils and ensuring high reproducibility of the results [11].

3. Results and discussion

As demonstrated in article [12], blend components are crystallizing polymers, and it makes the nature of their molecular level interaction more complex. According to numerous studies, the introduction of a high-molecular second component into a polymer matrix can ambiguously affect its molecular structure, material properties, and resistance to aggressive media.

Introduction of rigid-chain polymers into PE is, as a rule, accompanied by a lower elongation per unit length. In these PLA:LDPE compositions, a trend as follows is seen: higher PLA contents result in higher strength of the material. This is due to a rather high value of tensile strength of pure PLA, which ranges from 56 to 63 MPa.

In this paper, we studied the process of solid-phase oxidation of low-density polyethylene and polylactide blends at a temperature of 80, 90, and 110°C (figure 1).

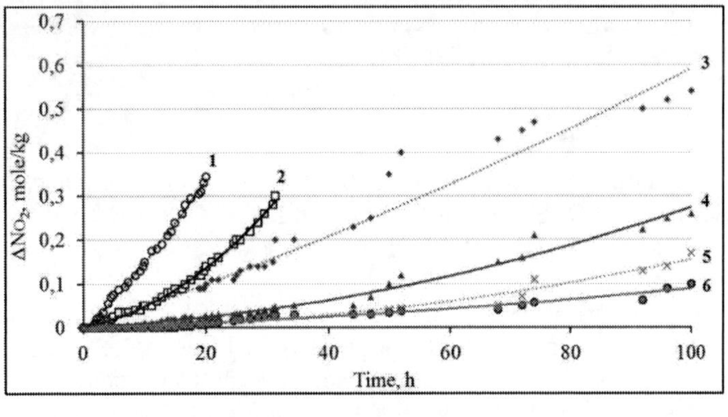

Figure 1. Oxygen uptake kinetic curves for various blends:
(1) 30PLA:40LDPE:30LDPEag and (2) 30 PLA:70LDPE at 110°C;
(3) 30PLA:40LDPE:30 LDPEag and (4) 30PLA:70LDPE at 90°C;
(5) 30PLA:40LDPE:30LDPEag and (6) 30PLA:70LDPE at 80°C.

IOP Conf. Series: Materials Science and Engineering **848** (2020) 012071 doi:10.1088/1757-899X/848/1/012071

The influence of temperature on the process of thermal oxidation is confirmed by the calculation of the oxidation rate of the compositions. The obtained dependence is shown in figure 2. It is noted that the oxidation rate increases with increasing temperature and content of LDPE in the blends.

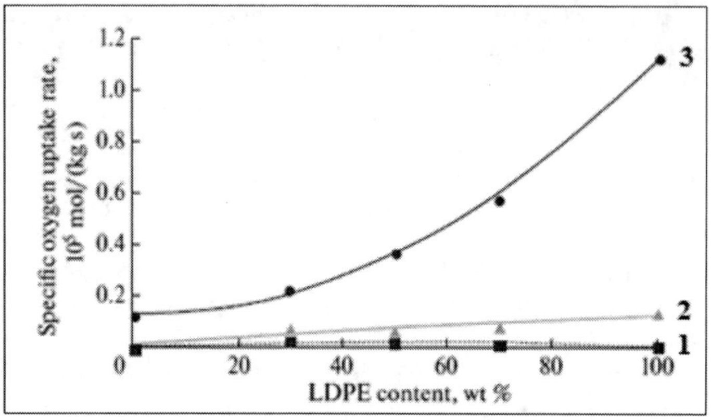

Figure 2. Dependence of the oxidation rate of PLA:LDPE blends on the content of LDPE in the blends at a temperature of (1) 80°C, (2) 90°C and (3) 110°C and an oxygen pressure of 300 Torr.

It should be noted that LDPEag oxidized faster than the original polyethylene, although according to the DSC melting point it is higher than the original LDPE. The above fact is confirmed by thermogravimetric analysis (TGA) (figure 3). It is shown that the polyethylene more resistant to the effects of temperature (peak of destruction - 414 °C) than polylactide (peak degradation – 364 °C).

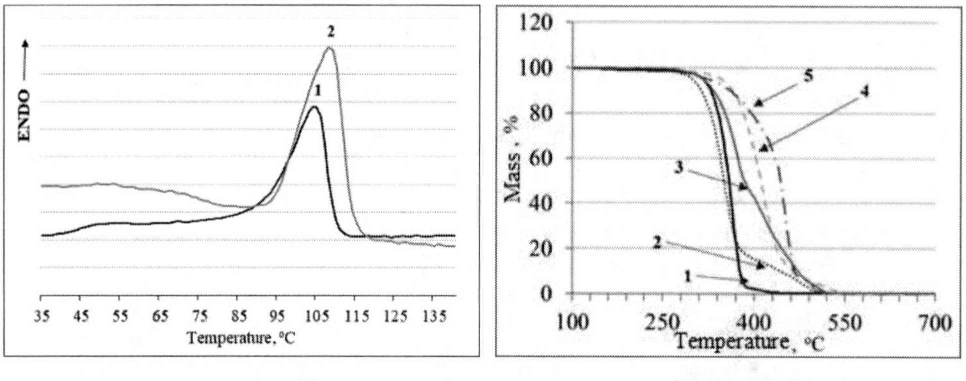

a *b*

Figure 3. DSC (a) melting thermograms: 1 – 100LDPE, 2 – 100LDPEag and TGA curves (b) of compositions PLA: LDPE: 1 – 100PLA, 2 – 70PLA:30 LDPE, 3 – 30PLA:70 LDPE, 4 – 100 LDPE, 5 – 100 LDPEag.

An important aspect of present-day polymer materials is their biodegradability under the influence of such a factor as soil microorganisms. In the paper [13], dedicated to research of PLA incubation in soils under normal conditions, it was found that it undergoes slow destruction, and degradation begins in several months. Other papers [14, 15] reveal that the number of PLA-decomposing microorganisms is considerably lower than that of those able to degrade PHB and PCL. On the other hand, PLA degradation can be accelerated by composting, where PLA hydrolysis leads to formation of smaller molecules (oligomers, dimers and monomers) in 45 – 60 days within a temperature range of 50–60 °C and then to formation of CO_2 and H_2O due to presence of microorganisms in compost. The experiments on degradation of the materials in soil allowed prediction of their behavior upon disposal in dumps. The main experiment was performed at constant temperature and moisture content (laboratory soil), and field tests were also performed for comparison. The weight loss Δm (%), changes in the specimen appearance, and intensity of the development of microorganisms on the specimen surface upon exposure to soil were used as the main parameters characterizing the degradation of the materials. After incubation of specimens in restored soil for 12 months, the largest weight loss, 18%, is observed for the 50 PLA:50 LDPE blend. For the other compositions, Δm is 5–10% [11].

In parallel to that laboratory test one conducted a 24-month outdoor research of PLA:LDPE binary mixtures. Figure 4 shows images of film samples of PLA:LDPE compositions upon exposure of materials to natural conditions within 24 months. The figure illustrates pores, cracks and blush. The latter is the most pronounced in samples of higher PLA contents, exposed to hydrolytic degradation.

After exposing the materials under natural environmental conditions, numerous mechanical damages are observed. It is important that some PLA:LDPE blends, e.g., those with 30, 50, and 100 wt. % polylactide content, undergo biofouling under field conditions more intensely compared to the materials subjected to laboratory tests. This fact can be attributed to higher concentration of microorganisms in soil on the test site. In laboratory soil, the concentration of microorganisms can decrease with time (figure 5). An average weight loss in the compositions upon natural tests comes to about 7 – 10% [11].

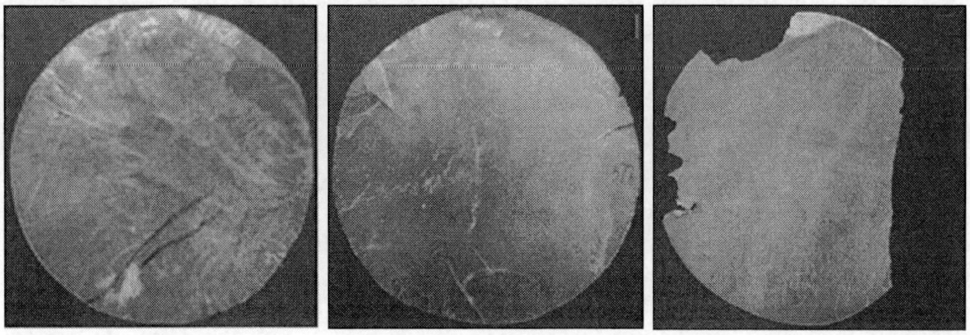

a *b* *c*

Figure 4. Photographs of specimens of PLA:LDPE blends. PLA content of the polymer matrix, wt %: (*a*) 30, (*b*) 70, (*c*) 100 after 24-month exposure to natural environmental conditions.

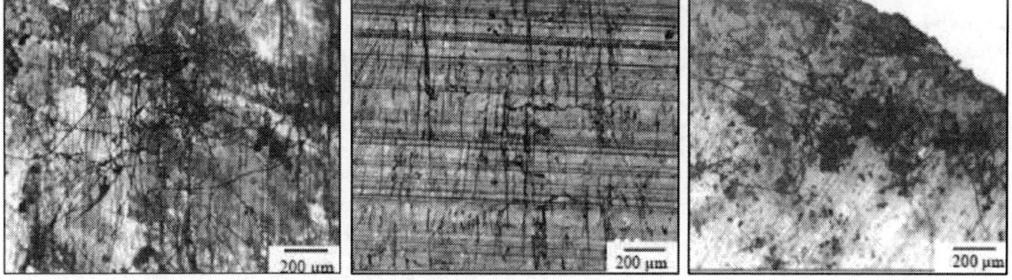

a *b* *c*

Figure 5. Photomicrographs of specimens of PLA–LDPE blends after exposure to soil under field conditions for 24 months. PLA content of the matrix, wt %: (a) 30, (b) 70, (c) 100.

To control the structure upon exposure in soils, one used IR spectroscopy and differential scanning calorimetry (DSC) methods. For PLA:LDPE samples subject to outdoor conditions within 24 months a ratio of crystal-phase optical densities 755 cm^{-1} to amorph-phase ones 870 cm^{-1} was calculated (D_{755}/D_{870}). An increase of this ratio is observed, which illustrates amorph phase destruction processes.

Research results of thermal and physical characteristics of the compositions upon outdoor exposure within 24 months are presented in Table 1, which shows that a degree of crystallinity is decreased by 5 – 15%, suggesting material destruction processes in progress.

Table 1. Thermal and physical characteristics of PLA:LDPE compositions after 24 months in soils.

Blends PLA:LDPE, wt.%	T_m, °C	T_m, °C after 24m soil	χ, %	χ, % after 24m soil
100:0	165	163	45	40
70:30	163:103	161:103	39:17	35:15
50:50	164:103	162:102	53:19	35:16
30:70	164:104	162:103	52:17	37:16
0:100	105	104	25	24

The results obtained enable to draw a conclusion that in a laboratory environment, subject to a constant temperature and contents of moisture, weight loss processes carry more actively than in soils. Analysis of the band intensity ratio (755 cm^{-1}/870 cm^{-1}) and DCS data reveals that destruction processes carry both in amorph- and crystalline-phases of PLA matrix.

4. Conclusion

The present study enables the conclusion that the process of oxygen uptake depends on the blend composition, i.e., on the content of PLA, as a component with a high degree of crystallinity. The addition of LDPEag, to the compositions containing 30 wt % PLA, accelerates the thermal oxidation of blends at all examined temperatures.

Upon incubation of the specimens in laboratory soil for 12 months, the weight loss is 5–10% (the largest weight loss, 18%, was observed for the 50:50 wt % blend), and after 24-month field tests the weight loss is 2–6% depending on the composition, with the largest weight loss observed for the 20 : 80 (about 11%) and 50:50 wt % (8%) blends. The weight loss in open soil is smaller than in the laboratory soil. This fact is associated with temperature fluctuations in the open soil and with the concomitant decreases in the microorganism activity. The most intense development of mycelial fungi in open soil was observed for the blends containing 30, 50, and 100 wt % polylactide.

Acknowledgments

The work was carried out using the scientific equipment of the Center of Shared Usage "New Materials and Technologies" of Emanuel Institute of Biochemical Physics and Joint Research Center of Plekhanov Russian University of Economics. This study was financed by a grant from the Plekhanov Russian University of Economics.

References

[1] Farah S, Anderson D G and Langer R 2016 *Adv. Drug Deliv. Rev.* **107** pp 367–392
[2] Peelman N, Ragaert P, Ragaert K, Meulenaer B D, Devlieghere F and Cardon L 2015 *J. Appl. Polym. Sci.* **132** pp 42305
[3] Ni J and Abe H 2015 *Polym.* **60** p 260
[4] Arrieta Ì P, Lopez J, Hernàndez A and Rayon E 2014 *Eur. Polym. J.* **50** p 255
[5] Tertyshnaya Yu V and Podzorova M V 2018 *Rus. J. Appl. Chem.* **91** p 417
[6] Zhang X, Espiritu M, Bilyk A and Kurniawan L 2008 *Polym. Degrad. Stab.* **93(10)** pp 1964–1970

[7] Podzorova M V, Tertyshnaya Yu V, Karpova S G and Popov A A 2019 *IOP Conf. Ser. Mater. Sci. Eng.* **525** 012043
[8] Bobel A, Lohfeld S, Shirazi R N and McHugh P E 2016 *Polym. Test.* **54** p 150
[9] Kim Y F, Choi C N, Kim Y D, Lee K Y and Lee M S 2004 *Fibers and Polym.* **5** pp 270-274
[10] Ploypetchara N, Suppakul P, Atong D and Pechyen C 2016 *Energy Procedia* **56** pp 201-210
[11] Tertyshnaya Yu V and Podzorova M V 2019 *Rus. J. Appl. Chem.* **92(6)** pp 737–744
[12] Tertyshnaya Yu V, Podzorova M V, Monakhova T V and Popov A A 2016 *Rus. J. Phys. Chem. B* **10(5)** pp. 825–829
[13] Ohkita T and Lee S H 2006 *J. Appl. Polym. Sci.* **100** pp 3009-3017
[14] Pranamuda H, Tokiwa Y and Tanaka H 1997 *Appl. Environ. Microbiol.* **63** pp 1637-1640
[15] Tokiwa Y and Calabia B P 2007 *J. Polym. Environ*. **15** pp. 259-267

IOP Conf. Series: Materials Science and Engineering **848** (2020) 012072 doi:10.1088/1757-899X/848/1/012072

Silver niobate doped lead-free perovskite KNN ceramics

E D Politova[1,2], G M Kaleva[1,2], N V Golubko[1], A V Mosunov[1], N V Sadovskaya[3], D A Kiselev[4,5], A M Kislyuk[4], T S Ilina[4], S Yu Stefanovich[1,6]

[1]L.Ya.Karpov Institute of Physical Chemistry; Vorontsovo pole, 10, Moscow 105064 Russia,
[2]Semenov Institute of Chemical Physics, Russian Academy of Sciences, Kosugina, 4, Moscow 119991 Russia,
[3]FSRC «Crystallography and Photonics» RAS, Leninskiy Prospekt 59, 119333, Moscow, Russia
[4]National University of Science and Technology "MISiS", Leninskii pr. 4, Moscow 119049 Russia,
[5]Fryazino branch of the Kotel'nikov Institute of Radioengineering and Electronics of Russian, Academy of Sciences, Vvedensky Square 1, Fryazino, Moscow region, 141190 Russia,
[6]Lomonosov Moscow State University, Leninskie gory 1, Moscow 119992 Russia

E-mail: politova@nifhi.ru

Abstract. Influence of $AgNbO_3$ doping on structure, microstructure, dielectric, ferroelectric and local piezoelectric properties of $(K_{0.5}Na_{0.5})NbO_3$ ceramics was studied. Decreased in the unit cell parameters correlated with ionic radii changes. High effective local d_{33} piezoelectric coefficient values (800 pm/V) were observed in compositions studied.

1. Introduction

Potassium-sodium niobate $(K,Na)NbO_3$ (KNN) perovskite ceramics are intensively studied last ten years as the most promising materials for replacement the toxic lead oxide containing materials [1 - 18]. The composition $(K_{0.5}Na_{0.5})NbO_3$ is close to the MPB between two orthorhombic phases and is characterized by two phase transitions near 470 K between ferroelectric phases and at 670 K between ferroelectric and paraelectric phases [10].

It should be noted that narrow sintering temperature interval for KNN-based ceramics comprises special problem, and difficulties in preparation of stoichiometric compositions due to the presence of high volatile potassium and sodium oxides [11 - 15]. Besides, the formation of oxygen vacancies in nonstoichiometric compounds leads to an increase in ionic conductivity of the samples [16-18].

Silver niobate $AgNbO_3$-based oxides showed great potential for use in power energy storage systems due to their high energy storage density. In [19], an investigation was performed analyzing the chemical structure and physical properties of $AgNbO_3$. The results confirmed that the dielectric anomaly at ~440 K originated from a phase transition between the non-centrosymmetric weakly polar and centrosymmetric non-polar phases. The weakly polar phase exhibiting antipolar ordered behavior was proposed to be ferrielectric one responsible for the strong field-induced double-like polarization hysteresis. The dielectric relaxor behavior at ~340 K correlates with the presence of local ferroelectric-type polar domains in a weakly polar/ferrielectric phase. The observed weak remnant polarization within double-like polarization hysteresis under strong field cycles was explained as originating from the polarization contributed by the residual strong field-induced ferroelectric phase. In this work, influence of $AgNbO_3$ additive on properties of KNN composition was studied.

2. Experimental

Ceramic samples $(1-x)(K_{0.5}Na_{0.5})NbO_3$ - $xAgNO_3$ with $x = 0.0 - 0.10$, $\Delta x=0.02$ were prepared by the sol-gel synthesis, annealed at T_1=673-773 K (4 h), T_2= 1073 K (6 h) and sintered at T_3= 1373-1393 K (2 h.

As initial reagents, tetrahydrate of sodium potassium tartrate $KNaC_4H_4O_6 \cdot 4H_2O$ ("pure"), silver nitrate $AgNO_3$ ("pure" containing 99.9 % of main substance) and niobium oxide Nb_2O_5 ("pure") were used.

Stoichiometric mixtures of salts and oxide were mixed in distilled water till their complete dissolving. Then the suspension was evaporated, annealed at 673-773 K (4 h), mixed in ethanol, pressed in tablets, annealed at 1073 K (6 h) and sintered at 1373-1393 K (2 h).

To improve sintering of ceramics the compositions were additionally modified by small amounts of overstoichiometric LiF additives (5 w. %) in order to improve the sintering of ceramics [16, 17].

Phase content and crystal structure parameters of ceramic samples prepared were characterized by the X-ray Diffraction method (DRON-3M, Cu-K_α radiation with wavelength λ=1.5405 A, in the 2 theta range of 5 – 70 degrees). Spontaneous polarization and phase transitions of the samples were studied by the Second Harmonic Generation (SHG) method (Nd:YAG laser, λ=1.064 μm in the reflection). Dielectric measurements of ceramics with fired silver electrodes were performed on heating with 10 K/min. and cooling in the temperature interval of 300 – 1000 K, and frequency range of 100 Hz – 1 MHz using Agilent 4284 A (1 V). Microstructure of the samples was examined using the Scanning Electron Microscopy (SEM) method (JEOL YSM-7401F with a JEOL JED-2300 energy dispersive X-ray spectrometer system).

Atomic Force Microscopy in Piezoresponse Force Mode (PFM) (MFP-3D, Asylum Research, USA) method with a Ti/Ir-coated conductive tip (Asyelec-02, Asylum Research, USA) with a radius of curvature ~ 28 nm was used for local piezoelectric properties measurements. The PFM out-of-plane images (mixed signal) were scanned in the PFM Dual AC Resonance Tracking (DART) mode at 1-2 V and a frequency of ~980 kHz.

Before PFM measurements, the samples were polished using polycrystalline diamond up to the root mean square roughness reached <10 nm. Effective piezoelectric coefficient d_{33} was calculated from PFM phase and PFM amplitude hysteresis loops using Igor Pro software version 6.37 (Asylum Research, USA). All measurements were repeated at least 4 times in the different locations of the same sample with strong PFM domains contrast.

3. Results and discussion

Pure ceramic samples with orthorhombic perovskite structure were prepared (Figures 1*a,b*). Increase in sintering temperature resulted in texturing of ceramics with increase in intensity of peaks at 23 grad. (Figure 1*b*).

Slight shift of the X-Ray diffraction peaks to higher angles points to a decrease in the unit cell parameters due to introduction of smaller Ag^+ cations into A-site positions of perovskite lattice (Figure 2*a*).

The SHG method measurements confirmed the polar nature of the samples prepared indicating to their ferroelectric properties (Figure 2*b*). With increasing Ag^+ content, slight increase in intensity of the SHG signal was observed at x=0.02 that pointed to an increase in spontaneous polarization of the doped samples, while further decreasing may be attributed to more intensive coloration of the doped samples.

IOP Conf. Series: Materials Science and Engineering **848** (2020) 012072 doi:10.1088/1757-899X/848/1/012072

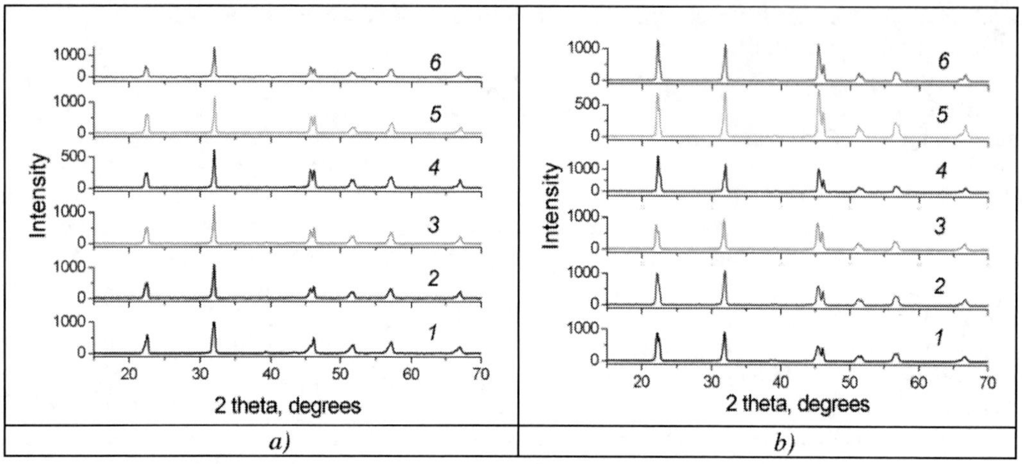

Figure 1. *a*) - X-Ray diffraction patterns of the samples (1-x)($K_{0.5}Na_{0.5}$)NbO_3 – $x$$AgNbO_3$ with x=0.0 (*1*), 0.02 (*2*), 0.04 (*3*), 0.06 (*4*), 0.08 (*5*), 0.10 (*6*) doped with 5 w. % LiF sintered at 1320 K (1 h). *b*) – sintered at 1320 K (5 min) → 900°C (4 h).

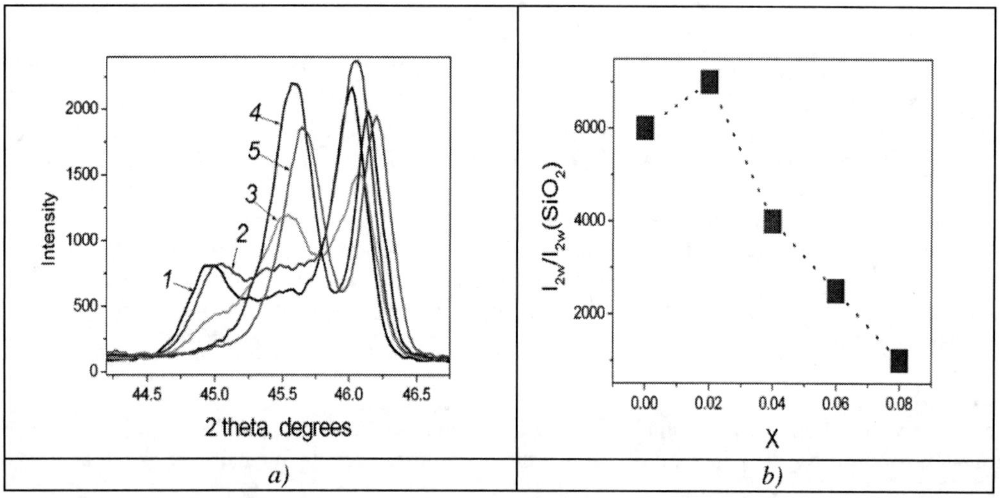

Figure 2. *a*) - Parts of the X-Ray diffraction patterns of the LiF doped samples (1-x)($K_{0.5}Na_{0.5}$)NbO_3 – $x$$AgNbO_3$ with x=0.0 (*1*), 0.02 (*2*), 0.04 (*3*), 0.06 (*4*), 0.10 (*5*). *b*) - Concentration dependence of the SHG signal q measured at the room temperature for the samples (1-x)($K_{0.5}Na_{0.5}$)NbO_3 – $x$$AgNO_3$ doped with 5 w. % LiF sintered at 1370 K (2 h).

Microstructure of the samples is sensitive to both substitutions and sintering conditions as well. With increasing sintering temperature enlargement of mean size of grains was observed, though mean size of grains slightly decreased with x increasing (Figure 3).

IOP Conf. Series: Materials Science and Engineering **848** (2020) 012072 doi:10.1088/1757-899X/848/1/012072

Figure 3. Microstructure of the samples with x=0.02 (*a*), 0.10 (*b*). Bars – 10 µm.

Dielectric properties of the samples were studied. Typical for the KNN-based compositions steps near ~ 350 - 400 K and maxima at ~ 600 – 700 K were revealed in the dielectric permittivity versus temperature curves (Figure 4). Slight decrease in temperature of the phase transition to paraelectric phase T_C, while an increase in the ferroelectric tetragonal to orthorhombic phase transition temperature were observed in the ceramic samples studied (Figure 5).

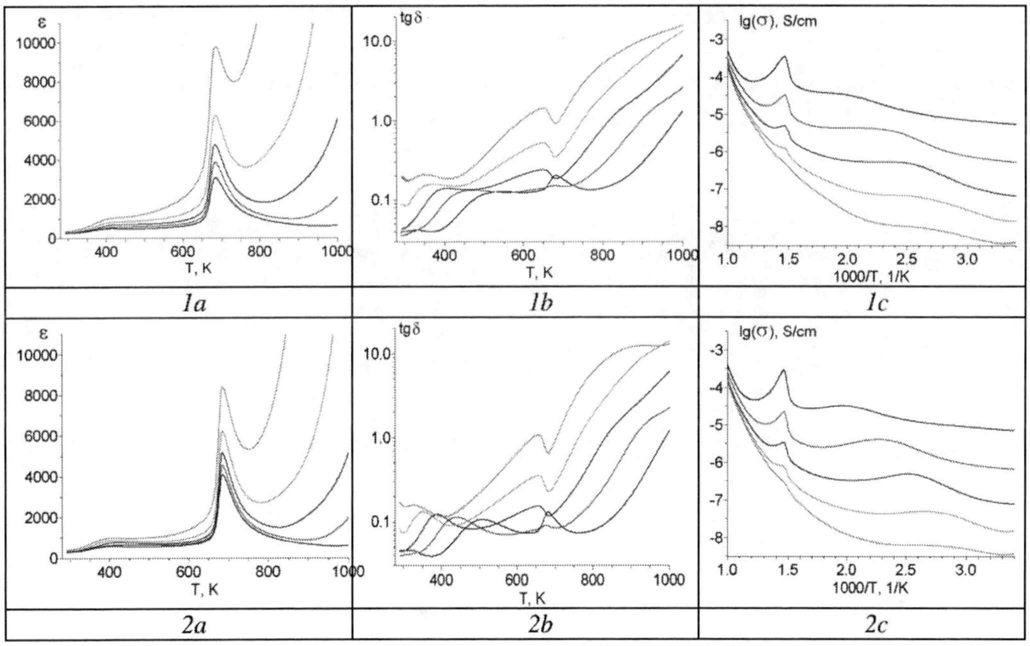

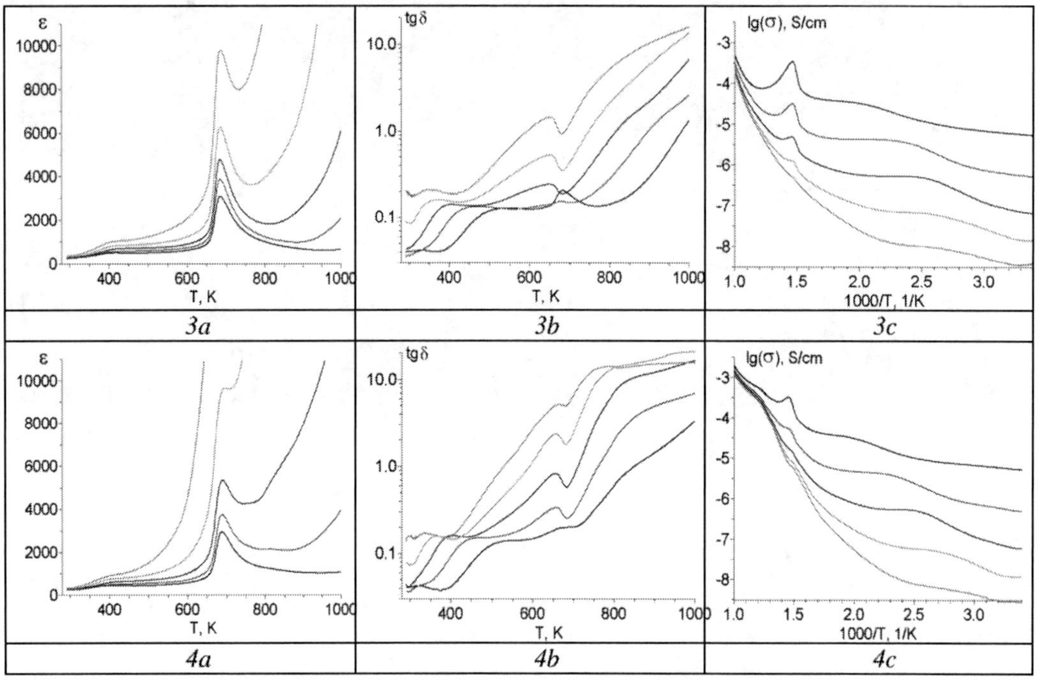

Figure 4. Temperature dependences of dielectric permittivity $\varepsilon(T)(a)$, dielectric loss $\tan\delta(T)(b)$ and $\lg\sigma(1000/T)(c)$ of the samples with x=0.0 (*1a-c*), 0.04(*2a-c*), 0.08(*3a-c*), 0.10(*4a-c*) sintered at T_3=1323 K (5 min) → T_4=1173 K (2 h) measured at frequencies f=100 Hz, 1, 10, 100 kHz, 1 MHz.

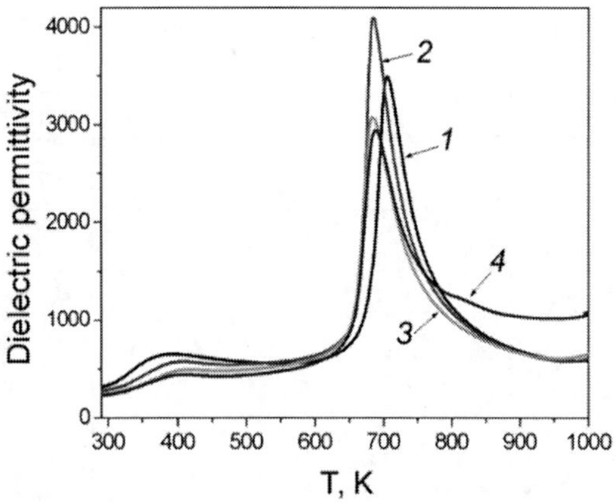

Figure 5. Temperature dependences of dielectric permittivity $\varepsilon(T)$ of the samples with x=0.0 (*1*), 0.04(*2*), 0.08(*3*), 0.10(*4*) sintered at T_3=1323 K (5 min) → T_4=1173 K (2 h) measured at frequency f = 1 MHz.

IOP Conf. Series: Materials Science and Engineering **848** (2020) 012072 doi:10.1088/1757-899X/848/1/012072

Using PFM method complex domain structure consisting of multiple domain patterns was found in investigated ceramics (Figure 6, top row). Simultaneous topography (not shown here) and out-of-plane PFM images confirmed that the imaged patterns were due to their piezoresponse [20 - 22]. To study the domain switching behavior of the Ag^{1+} ions doped KNN ceramics, 7.5×7.5 μm² positively polarized domains embedded in a 15×15 μm² equivalent negatively polarized domains were written ("box-in-box" structure). Figure 6 (middle row) presents an information of the writing voltage images, the dark and bright areas with applied –20V and +20V DC biases, respectively. Strong PFM contrast suggests complete switching process under poling for piezoelectric active grains. The domains created are stable in the time, and the PFM contrast of the written domains remained the same after 2 hours after the poling process (Figure 6, bottom row).

Local PFM hysteresis loops observed indicated ferroelectric polarization switching at nanoscale for the samples studied. An AC voltage (1-2 V) was superimposed onto a triangular square-stepping wave (f = 0.5 Hz, with writing and reading times 25 ms, and bias window up to ± 20 V) during the remnant piezoelectric hysteresis loops measurements.

Remnant hysteresis loops were received for separate grains (Figure 7*a*). All PFM hysteresis loops are asymmetric along the axis Y. According to Jesse et al., this offset may be mainly attributed to the electrostatic effect [20, 21]. Moreover, the local PFM hysteresis loops account for the polarization switching process in a nanoscale, just below the tip. Therefore, intrinsic polarization in the nanoregions may also contribute to the offset and asymmetry of the local piezoresponse hysteresis loops [22].

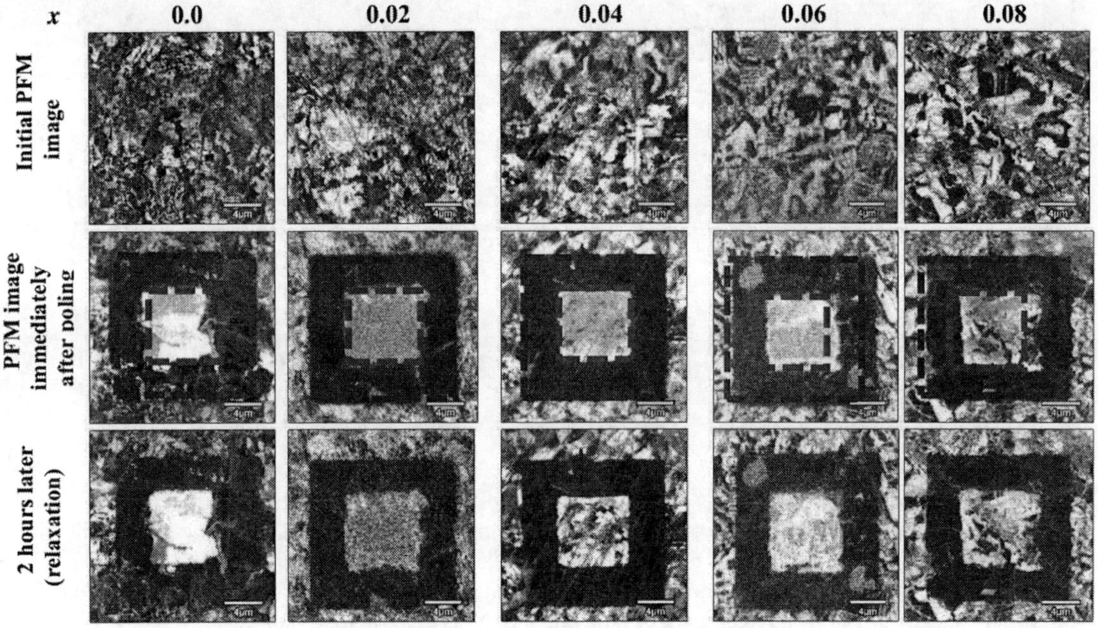

Figure 6. Initial domain structure and temporary changes of images of polarized ceramics with x=0.0, 0.02, 0.04, 0.06 and 0.08. Marked regions in the middle row present regions of 15×15 μm² poled with –20V DC bias and of 7.5×7.5 μm² poled with +20V DC bias.

IOP Conf. Series: Materials Science and Engineering **848** (2020) 012072 doi:10.1088/1757-899X/848/1/012072

The shifts along voltage axis of local PFM hysteresis loops indicate the presence of a biased-voltage due to the different contact between the top (PFM tip) moved electrode and surface of the investigated ceramic samples. The top electrode is the same (PFM tip coated Ti/Ir), so the shifts reflect the induced interface electrostatic potential step due to the built-in electric field at the interface.

Local PFM hysteresis loops revealed that remnant d_{33} piezoelectric coefficient reached the highest value of 800 pm/V for the initial KNN samples, decreased till 580 pm/V for the samples KNN-Ag with x=0.04, and then increased till 700 pm/V in the samples with x=0.08, with the saturation of loops reached at voltage value ~ 20 V (Figure 7*b*). Increase in the remnant d_{33} value at $x > 0.04$ may be explained by increasing texture of these ceramics (Figure 1*b*). The calculated value of the piezoelectric coefficient for the samples KNN-Ag is ≈ in 3-5 times larger than those previously observed for ceramic samples in the systems $(K_{0.5}Na_{0.5})NbO_3$ - $BaTiO_3$ [17] and $(K_{0.5}Na_{0.5})NbO_3$ – $Li(Nb,Mn)O_3$ [23].

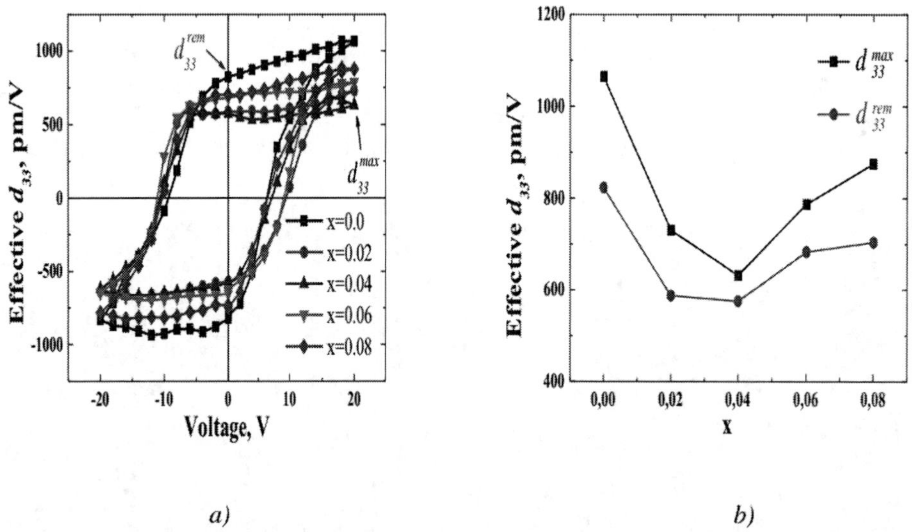

Figure 7. PFM hysteresis loops (*a*), concentration dependence of effective d_{33} values of the samples with x=0.0 – 0.08 calculated from PFM hysteresis loops (*b*).

4. Conclusions

Structure, microstructure, dielectric, ferroelectric and piezoelectric properties of the $(K_{0.5}Na_{0.5})NbO_3$ ceramics doped by Ag^{1+} ions were studied. Depending on composition and sintering conditions slight changes in the unit cell volume and temperatures of phase transitions were observed. High values of the remnant d_{33} piezoelectric coefficients ~ 600 – 800 pm/V were observed for ceramics studied. This confirms their prospects for development of new efficient piezoelectric materials.

Acknowledgment

The work was supported by the Russian Foundation for Basic Research (Project 18-03-00372).). PFM studies were performed at the Center for Shared Use "Material Science and Metallurgy" at the National University of Science and Technology "MISiS" and were supported by the Ministry of High Education and Science of the Russian Federation within the framework of the state task.

References

[1] Zhang S, Xia R, and Shrout R. Lead-Free Piezoelectric Ceramics: Alternatives for PZT? 2007 *J. Electroceram.* **19**, 251-7.

[2] Takenaka T, Nagata T, and Hiruma Y. Current developments and prospective of lead-free piezoelectric ceramics. 2008 *Jpn. J. Appl. Phys.* **47**, 3787-801.

[3] Damjanovich D, Klein N, Li J and Porokhonskyy V. What can be expected from lead-free piezoelectric materials? 2010 *Funct. Mater. Lett.* **3**, 5-13.

[4] Xiao D. Progresses and further considerations on the research of perovskite lead-free piezoelectric ceramics, 2011 *J. Adv. Dielectr.* **1**, 33-40.

[5] Coondoo I, Panwar N, Kholkin A. Lead-free piezoelectrics: Current status and perspectives, 2013 *J. Adv. Dielectr.* **3**, 1330002 (22 pages).

[6] Li J, Wang K, Zhu F, Cheng L, Yao F. (K,Na)NbO_3-Based Lead-Free Piezoceramics: Fundamental Aspects, Processing Technologies, and Remaining Challenges, 2013 *J. Am. Ceram. Soc.* **96**, 3677-96.

[7] Wu J, Xiao D, Zhu J. Potassium-sodium niobate lead-free piezoelectric materials: Past, present, and future of phase boundaries, 2015 *Chem. Rev.* **115**, 2559-95.

[8] Panda P and Sahoo B. PZT to Lead-Free Piezo Ceramics, 2015 Ferroelectrics 474, 128-43.

[9] Hong C, Kim H, Choi B, Han H, Son J, Ahn C, Jo W, Lead-free piezoceramics. Where to move on? 2016 *J. Materiomics* **2**, 1-24.

[10] Dai Y, Zhang X, and Chen K, Morphotropic phase boundary and electrical properties of $K_{1-x}Na_xNbO_3$ lead-free ceramics, 2009 *Appl. Phys. Lett.* **94**, 042905.

[11] Fang J, Wang X, Zuo R, Tian Z, Zhong C, Li L. Narrow sintering temperature window for (K,Na)NbO_3-based lead-free piezoceramics caused by compositional segregation, 2011 *Phys. Status Solidi A.* **208**, 791-4.

[12] Wang K, Li J. (K,Na)NbO_3-based lead-free piezoceramics: Phase transition, sintering and property enhancement, 2012 *Adv. Ceramics* **1**, 24-37.

[13] Zuo R, Roedel J, Chen R, Li L. Sintering and electrical properties of lead-free $Na_{0.5}K_{0.5}NbO_3$ piezoelectric ceramics, 2006 *J. Am. Ceram. Soc.* **89**, 2010-5.

[14] Tellier J, Malič B, Dkhil B, Jenko D, Cilensek J, Kosec M. Crystal structure and phase transitions of sodium potassium niobate perovskites, 2009 *Solid State Sci.* **11**, 320-4.

[15] Malič B, Koruza J, Hreščak J, Bernard J, Wang K, Fisher J, Benčan A. Sintering of lead-free piezoelectric sodium potassium niobate ceramics, 2015 *Materials* **12**, 8117-46.

[16] Politova E, Kaleva G, Golubko N, Mosunov A, Akinfiev V, Stefanovich S, Fortalnova E. Influence of NaCl/LiF Additives on Structure, Microstructure and Phase Transitions of $(K_{0.5}Na_{0.5})NbO_3$ Ceramics, 2015 *Ferroelectrics* **489**, 147-55.

[17] Politova E, Golubko N, Kaleva G, Mosunov A, Sadovskaya N, Stefanovich S, Kiselev D, Kislyuk A, Panda P. Processing and characterization of lead-free ceramics on the base of sodium-potassium niobate, 2018 *J. Adv. Diel.* **8**, 1850004.

[18] Ringgaard M, and Wurlitzer T. Lead-free piezoceramics based structure, microstructure and electrical properties of alkali niobates, 2005 *J. Eur . Ceram. Soc.* **25**, 2701-6.

[19] Tian Y, Jing L, Hu Q, Jin L, Yu K, Li J, Politova E, Stefanovich S, Xu Z, Wei X. Ferroelectric Transitions in Silver Niobate Ceramics, 2019 *Journal of Materials Chemistry C* **7**, 1028-1034; DOI: 10.1039/C8TC05451G

[20] Jesse S, Baddorf A, Kalinin S. Switching spectroscopy piezoresponse force microscopy of ferroelectric materials, 2006 *Appl. Phys. Lett.* **88**, 062908.

[21] Jesse S, Kalinin S, Proksch R, Baddorf A and Rodriguez B. The band excitation method in scanning probe microscopy for rapid mapping of energy dissipation on the nanoscale, 2007

Nanotechnology **18**, 435503 (8pp) doi:10.1088/0957-4484/18/43/435503.
[22] Trivedi H, Shvartsman V, Lupascu D, Medeiros M, Pullar R, Kholkin A, Zelenovskiy P, Sosnovskikh A, Shur V. Local manifestations of a static magnetoelectric effect in nanostructured $BaTiO_3$–$BaFe_{12}O_9$ composite multiferroics, 2015 *Nanoscale* **7**, 4489-96.
[23] Politova E, Golubko N, Kaleva G, Mosunov A, Sadovskaya N, Fortalnova E, Kiselev D, Ilina T, Kislyuk A, Stefanovich S, Panda P. Ferroelectric and local piezoelectric properties of modified KNN ceramics, 2019 *Integrated Ferroelectrics* **196** (1), 52-9.

IOP Conf. Series: Materials Science and Engineering **848** (2020) 012073 doi:10.1088/1757-899X/848/1/012073

Tunable parallel plate waveguide array based on VO_2 thin films

V I Polozov[1,2], S S Maklakov[1], S A Maklakov[1], V A Chistyaev[1], A A Politiko[1], K M Baskov[1], A D Mishin[1], D A Petrov[1] and V N Kisel[1]

[1] Institute for theoretical and applied electromagnetics RAS, 125412, Moscow, 13 Izhorskaya St., Russia
[2] Moscow Institute of Physics and Technology (State University), 141701, Moscow Region, Dolgoprudny, 9 Institutskiy per., Russia

Abstract. Vanadium dioxide (VO_2) is a material with a metal-insulator phase transition. We report a way of application of VO_2 thin films as functional elements of reconfigurable parallel plate waveguide array (PPWA). The PPWA presented here is designed to work in X, K_u and K bands. A transmission coefficient of the array with VO_2 elements can be altered using temperature by more than 10 dB across the entire experimentally studied frequency range. We demonstrate a method of design and manufacturing of the PPWA. The PPWA may be used as a basic component in more advanced reconfigurable devices.

1. Introduction

Parallel plate waveguide arrays are key components for electromagnetic beam steerers [1], lenses [2] and antennae [3]. The reconfigurability of those systems can only be achieved using mechanical methods. Mechanical switching methods can not be used in applications where the fast tuning of the device is required. Recent researches show that this issue can be resolved using functional metal oxides that have a metal-insulator transition property.

Among materials with the metal-insulator transition, vanadium dioxide (VO_2) thin films have proven to be the most promising, due to their large contrast in resistivity between two states. Switching between the states can be induced by heating the film to 60°C [4] or applying voltage bias [5]. Thanks to the metal-insulator transition, VO_2 thin films are reported as a switching element in reconfigurable RF slot antennae [6], RF coplanar switches [7] and tunable planar RF filters [8]. The typical size of the VO_2 switching elements utilized in those devices varies from 10×10 μm to 1×1 mm. In order to use VO_2 thin films as a switching element for the parallel plate waveguide array, a surface area of thin films should be in the region of square centimeters.

In our previous work [9] we presented the method of synthesis of the high quality 30 cm^2 VO_2 thin films with a 10^3 ratio of the sheet resistance in the insulator state to the resistance in the metal state. Metal-insulator transition attributes of the obtained thin films were uniform through the films' surface area. In the present work, we demonstrate a way of application of VO_2 thin films as walls of the tunable parallel plate waveguide array (PPWA). The frequency response of the device can be tuned by thermal actuation of the metal-insulator transition in VO_2 thin films. When VO_2 thin films switch from insulator to metallic state, the transmission coefficient of the PPWA drops by more than 10dB in a 6-24 GHz frequency range.

IOP Conf. Series: Materials Science and Engineering **848** (2020) 012073 doi:10.1088/1757-899X/848/1/012073

2. Model and numerical simulation

The shift of the PPWA transmission coefficient is related to the resistivity change of the waveguide walls material. While VO_2 is in the dielectric state, PPWA is transparent for RF electromagnetic waves. When VO_2 transits into the metallic state, it acts as a lossy conductor on the PPWA walls so that electromagnetic wave with vector ***E*** parallel to the walls attenuates. To estimate the shift of the PPWA transmission coefficient, a transmission coefficient of the single parallel plate waveguide (PPW) was analyzed.

Attenuation due to conductor loss α_c in a single parallel plate waveguide (figure 1) equals to [10]:

$$\alpha_c = 8.686 \cdot \frac{R_s}{Zd} \text{ dB/m} \tag{1}$$

where d is a distance between parallel plates, Z is an impedance of the medium in the waveguide and R_s is a surface resistance of the material which covers walls of the waveguide. According to (2), R_s depends on wall material resistivity ρ and skin depth δ_s, however, a skin depth of VO_2 thin films, obtained in this work exceeded their thickness across the entire frequency range. For example, at 12 GHz skin depth of VO_2 thin films in the metallic state, in which resistivity equals to $1.44 \cdot 10^{-5}$ Ohm·m, is $1.74 \cdot 10^{-5}$ m (calculated according to (3), where μ of VO_2 equals to μ_0 [11]). The thickness of VO_2 films used here was 120 nm. Therefore, instead of skin depth in (2), the thickness of the films should be taken. In that case, α_c will be 10 dB/cm. This means, that the transmission coefficient of PPW should decrease by 10 dB after switching from "opened" state (when VO_2 is insulator) to "closed" state (when VO_2 is a lossy conductor) for 1 cm length (l dimension on figure 1) PPW.

$$R_s = \frac{\rho}{\delta_s} \tag{2}$$

$$\delta_s = \sqrt{\frac{2\rho}{\omega\mu}} \tag{3}$$

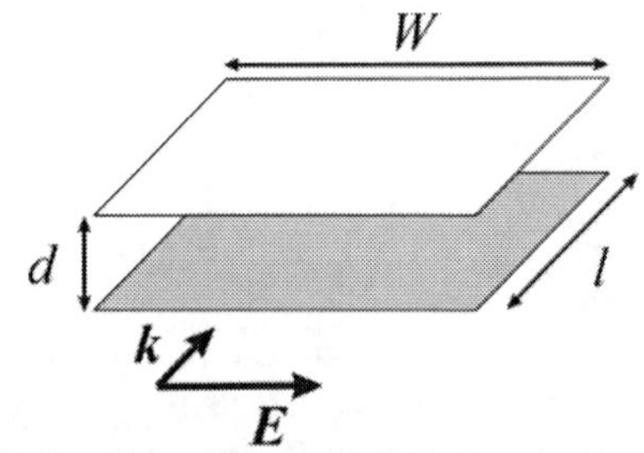

Figure 1. Schematic of the parallel plate waveguide, PPW.

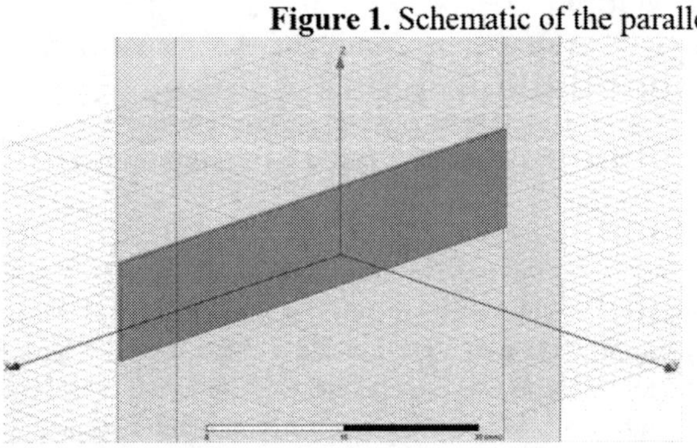

Figure 2. The model applied for the PPWA simulation in HFSS software.

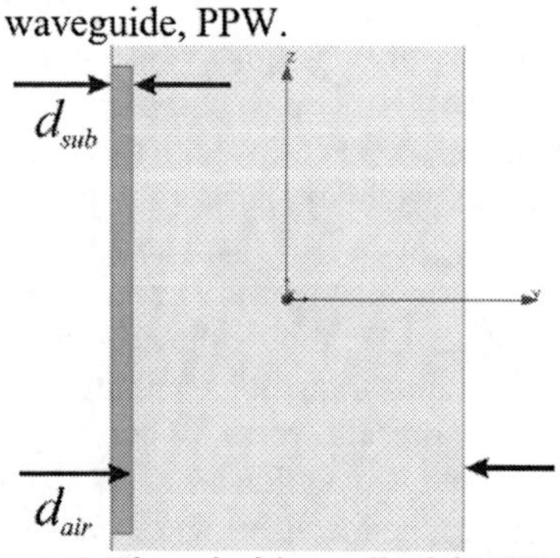

Figure 3. The primitive cell of the PPWA, a side-view.

From (1) it is evident that the bigger the distance between plates, the smaller α_c. Consequently, the usage of PPWA instead of single PPW would allow covering apertures with an area of several square centimeters without significantly extending length l to compensate for the decrease of α_c.

Numerical simulation was used to determine optimal dimensions of VO_2 elements used as PPWA plates and distance between them. PPWA with optimal dimensions should have had the highest transmission coefficient in "opened" state (when VO_2 is insulator) and the lowest transmission coefficient in "closed" state (when VO_2 is a lossy conductor). Numerical simulation was conducted employing HFSS electromagnetic simulation software.

A primitive cell of the PPWA was used for PPWA simulation (figure 2). The primitive cell consisted of a plate of polycrystalline Al_2O_3 substrate with dielectric permittivity $\varepsilon_{sub}=9.7$ and dielectric loss tangent $tan\ \delta=10^{-4}$ and a layer of air between two adjacent plates (figure 3). A thickness of Al_2O_3 substrate d_{sub} was fixed at 0.5 mm during the optimization process since the same substrate was used for thin film synthesis. Both variable parameters: d_{air} – a distance between adjacent plates, and length l of the pates were varied in 4-20 mm range. That range was chosen for reasons of ease of manufacturing PPWA from synthesized films. VO_2 layer was simulated by applying impedance boundary condition on the front side (figure 2) of the Al_2O_3 plate. In "opened" state value of impedance was $3\cdot10^5$ Ohm/□ and in "closed" state value of impedance was 120 Ohm/□, in accordance with measurements.

In order to simulate the interaction between the array of the primitive cells and plane electromagnetic wave, periodic boundary conditions were applied on the walls of the primitive cell. Periodic boundary conditions were applied on a pair of walls parallel to the YZ plane and on a pair of walls parallel to the XZ plane (figure 2). These conditions mean that the model represents a PPWA with an infinite W dimension consisted of an infinite amount of the primitive cells, translated along the Y-axis (figure 2). That model was simulated in the 6-24 GHz frequency range for waves polarized along the X-axis.

Using the simulation, an optimal d_{air} of 8.5 mm and l of 11 mm were found. Calculated spectra (figure 4) of the transmission coefficient of the PPWA in both states demonstrate, that the transmission can be tuned by more than 10 dB across the entire frequency range, as it was expected from the theoretical estimation.

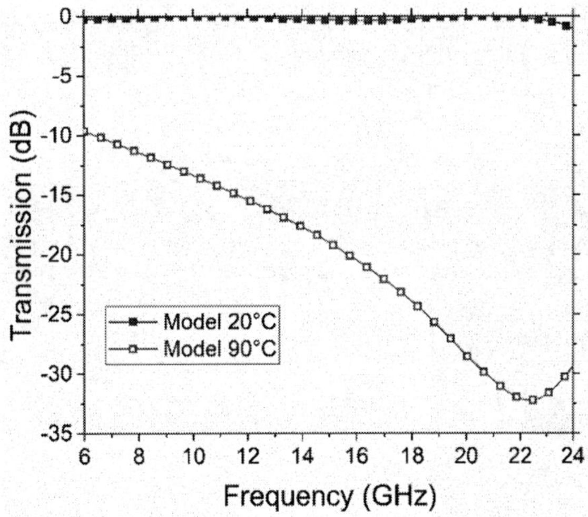

Figure 4. The modeled transmission spectrum of the PPWA.

3. Experimental

Vanadium dioxide thin films were deposited by means of ion-assisted RF magnetron sputtering of pure V (99.3%) target on polished polycrystalline Al_2O_3 substrate. The substrate dimensions were 48×60×0.5 mm. The deposition was preceded by Ar ion etching of the substrate for substrate cleaning. The sputtering was conducted in an argon-oxygen atmosphere with 13.5 vol.% of oxygen. The total pressure of the gas mixture was 1 mTorr. The substrate was kept at a temperature of 400°C and a negative voltage bias of -90V to enhance adatom mobility. Optimal sputtering parameters: oxygen concentration, substrate temperature, and voltage bias were determined in our previous work [9]. Deposition speed was 4 nm/minute and the thickness of the deposited VO_2 films was 120 nm. The thickness was obtained by an atomic force microscope (NT-MDT Solver P47H) in semi-contact mode (1 Hz), with a polycrystalline silicon ETALON© (NT-MDT) probe (tip radius curvature < 10 nm).

The magnetron sputtering was followed by post-deposition annealing in pure (99.999%) argon atmosphere in a tube-furnace. The duration of the annealing was 15 minutes and the temperature was 720°C. The argon flow rate during the annealing was 2000 sccm.

After the deposition, vanadium dioxide thin films were characterized by measuring sheet resistance vs temperature (R(T)) dependence, since it is a qualitative property of VO_2 thin films. R(T) dependence of the thin films was measured using the Solartron SI 1287 potentiostat in a two-probe configuration. To obtain sheet resistance of the VO_2 film, voltage bias of 1 V was applied to the sample and a current through the sample was recorded every 2 seconds while the sample was heated and cooled in a 20-90°C temperature range. The heating speed was 2°C per minute.

In order to form a parallel plate waveguide array, two thin films were cut to rectangle elements with length of 11 mm (l on figure 1) and width of 60 mm (W on figure 1). Those elements were installed in a holder made of PLA plastic (figure 3). Distance d (figure 1) between elements was 8 mm. 6 parallel plates were used to obtain an array with total dimensions of 60×40 mm, which can be measured in a present RF measuring bench.

The transmission and reflection coefficients of the sample at normal incidence of an electromagnetic wave in the frequency range from 6 to 24 GHz were measured using a methodology described in [12]. Ultra-wideband horn antennas with diaphragmatic lens and vector network analyzer R&S ZVA40 were used for the measurements.

The sample was placed (figure 4) in the aperture of a metal diaphragm positioned in the near field of the horn antennae to minimize the impact of the diffraction at the edges of both: the test sample the antennae. During the measurement of the reflection coefficient, a matched load was mounted in the area behind the diaphragm. Radio-absorbing material shaped like a three-dimensional pyramid-type block was used as the matched load. The matched load had a low reflection coefficient across the analyzed bandwidth (-50 dB). A time selection of clutter reflections with suppression of the Gibbs phenomenon was used in a digital signal processing for eliminating multiple reflections between the sample and the horn antenna [12].

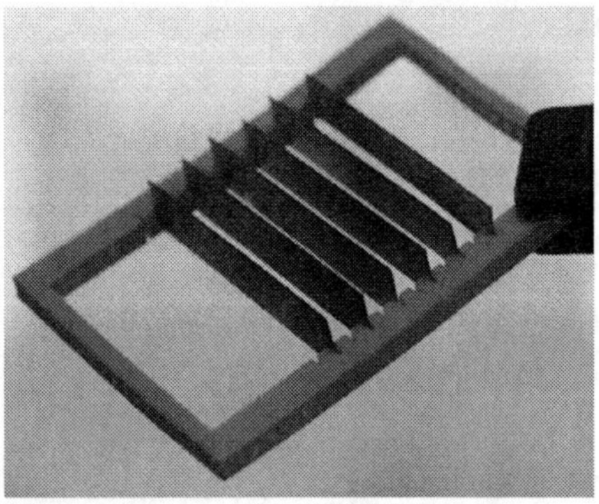

Figure 5. Assembled PPWA

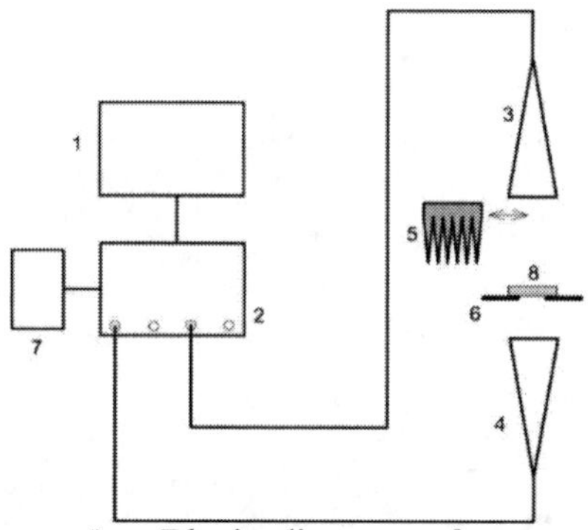

Figure 6. Block diagram of measuring bench. 1 – PC; 2 – R&S ZVA40 vector network analyzer; 3,4 – wideband lens horn antenna; 5 – pyramidal microwave absorber; 6 – metal diaphragm with a multi-layer radio absorbing material coating; 7 – rubidium frequency standard; 8 – sample.

The reflection and transmission coefficients of the sample were measured with a polarization of the electric field vector along the *W* dimension (figure 1) of the plates with VO_2. During the measurements, the sample was heated from 20°C to 90°C using a heat gun, and the temperature of the sample was measured using a pyrometer.

4. Experimental results and discussion

During metal-insulator transition in VO_2 thin films, used in PPWA, their resistance changed by 2600 times (figure 7).

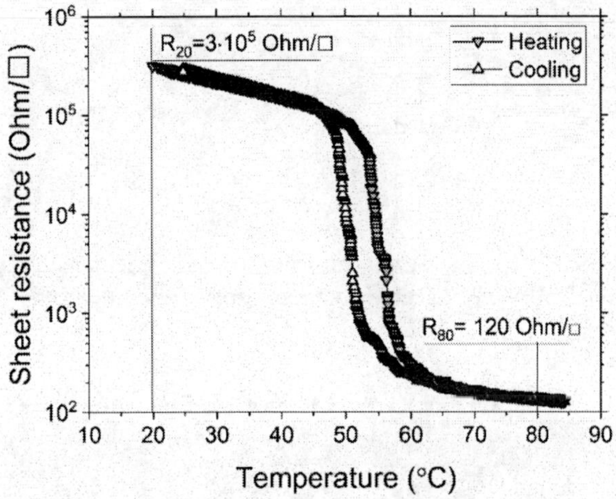

Figure 7. R(T) dependence of VO_2 thin films used in PPWA.

Numerical simulation of the frequency dependence of the transmission and reflection coefficients was in a good agreement with experimental results for "opened" state (figure 8 and figure 9). In the "closed" state the agreement was worse. There were two main sources of the discrepancy. The first source included discontinuities in the measurement bench and unprecise assembly of the PPWA. Dimensions of the manufactured PPWA were within 0.2 mm of the dimensions used for the numerical calculation. The second source included the use of the periodic boundary in the model, so certain dimensions of the modeled PPWA were infinite. Because of it, an influence of the manufactured PPWA finite size could not have been observed in the numerical simulation. The influence was the most significant in the "closed" state, where PPWA had metallic walls. That is why the simulated reflection coefficient spectrum had only one peak instead of four in the experimental spectrum (curves with solid symbols in figure 9). Measurement and simulation of the transition coefficient were also different in the "closed" state (curves with solid symbols in figure 8). The amplitude of the measured transition coefficient was higher because the manufactured PPWA did not entirely cover the aperture of the measurement bench. This discrepancy can be eliminated by using a model with finite dimensions. However, a simulation of such a model will require more computational power.

In the "opened" state of PPWA at 20°C the transmission coefficient of the PPWA was -0.3 dB on average in the analyzed frequency range. The loss of 0.3 dB was a result of the reflection from the upper faces of the PPWA walls and the attenuation due to dielectric loss in the Al_2O_3 substrate. The reflection coefficient spectrum of the PPWA in the "opened" state had an interference pattern (curves with open symbols in figure 9). The interference was evident on both: measured and simulated curves. The interference can be attributed to the distance between VO_2 plates.

Heating of the PPWA to 90°C induced metal-insulator transition in VO_2 thin films that resulted in the decrease of the PPWA transmission coefficient by more than 10 dB (figure 8). At the same time, the maximal value of the reflection coefficient increased from -11 dB in the "opened" state to -9 dB in the "closed" state, which means that the drop of the transmission coefficient was caused mainly by the growth of the attenuation due to conductive loss.

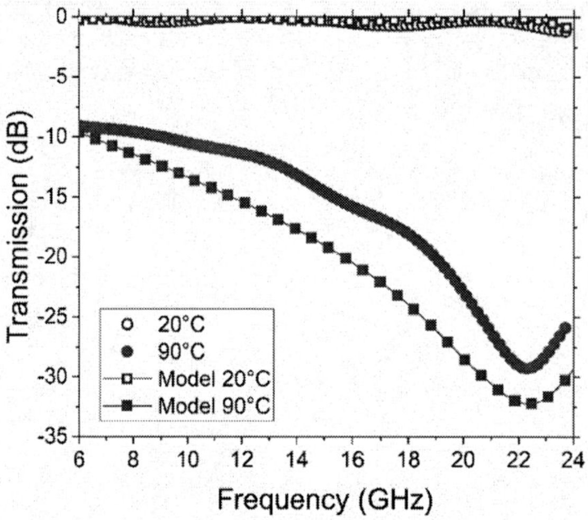

Figure 8. The measured transmission coefficient spectrum of the PPWA at selected temperatures in comparison with numerical simulation.

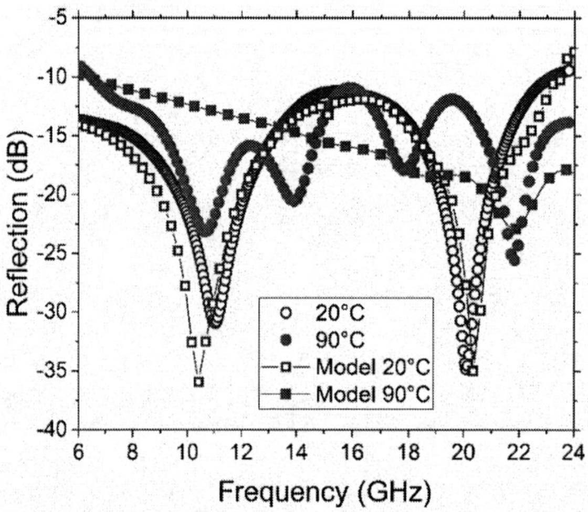

Figure 9. The measured reflection coefficient spectrum of the PPWA at selected temperatures in comparison with numerical simulation.

At 22.2 GHz the transmission coefficient in "closed" state had a minimum of -29 dB. The minimum is related to the length of the device (l dimension in figure 1). In the free space 22.2 GHz corresponds to 13.51 mm wavelength that is different from the length of the device, which was 11 mm. However, the primitive cell of the PPWA (figure 3) has effective dielectric constant ε_{eff} that increases the length of the PPWA for an electromagnetic wave by a factor of $\sqrt{\varepsilon_{eff}}$. The ε_{eff} of the layered medium consisted of air and Al_2O_3, can be calculated as follows [13]:

$$\varepsilon_{eff}=1+\frac{(\varepsilon_{sub}-1)d_{sub}}{(d_{sub}+d_{air})} \tag{4}$$

The dielectric constant of VO_2 was not taken into consideration in (4) since the thickness of VO_2 was 5000 times less than the thickness of the Al_2O_3 substrate. In the PPWA d_{sub} was equal to 0.5 mm and d_{air} was equal to 8 mm, so the ε_{eff} should be 1.48. Consequently, the effective length of the PPWA should be equal to $11\cdot\sqrt{1.48}$=13.39 mm, which is close to the resonant length of 13.51 mm obtained from the experimental data. It should be noted, that the reflection coefficient has the minimum at 22.2 GHz as well.

5. Conclusion

We report experimental fabrication and measurements of the VO_2-based parallel plate waveguide array with the tunable RF transmission coefficient. We present the method of synthesis of VO_2 thin films of sufficient size, method of numerical modeling, fabrication, and measurement of the PPWA. Fabricated PPWA changes transmission coefficient thought it from -0.3 dB to -16 dB on average in 6-24 GHz frequency range without significant growth of the reflection coefficient. To change the transmission coefficient of the PPWA, heating by hot air was used, which induced metal-insulator transition in VO_2. Proposed PPWA can be used as a reconfigurable element in various beam manipulating systems such as lenses and antennae.

Acknowledgments

References

1. Palm D, Dang Z and Rahm M 2019 Analysis and experimental investigation of a subwavelength phased parallel-plate waveguide array for manipulation of electromagnetic waves *Sci. Rep.* **9** 1–10
2. Lawrance J E, Christodoulou C G and Taha M R 2015 A High-Power Microwave Zoom Antenna with Metal-Plate Lenses *IEEE Trans. Antennas Propag.* **63** 3380–9
3. Eom H J 2001 Flanged Parallel-Plate Waveguide Array *Wave Scattering Theory: A Series Approach Based on the Fourier Transformation* (Berlin, Heidelberg: Springer Berlin Heidelberg) pp 35–55
4. Makarevich A M, Sadykov I I, Sharovarov D I, Amelichev V A, Adamenkov A A, Tsymbarenko D M, Plokhih A V., Esaulkov M N, Solyankin P M and Kaul A R 2015 Chemical synthesis of high quality epitaxial vanadium dioxide films with sharp electrical and optical switch properties *J. Mater. Chem. C* **3** 9197–205
5. Liao G M, Chen S, Fan L L, Chen Y L, Wang X Q, Ren H, Zhang Z M and Zou C W 2016 Dynamically tracking the joule heating effect on the voltage induced metal-insulator transition in VO2 crystal film *AIP Adv.* **6**
6. Anagnostou D E, Teeslink T S, Torres D and Sepulveda N 2016 Vanadium dioxide reconfigurable slot antenna *2016 IEEE Antennas and Propagation Society International Symposium, APSURSI 2016 - Proceedings* pp 1055–6
7. Ha S D, Zhou Y, Fisher C J, Ramanathan S and Treadway J P 2013 Electrical switching dynamics and broadband microwave characteristics of VO2 radio frequency devices *J. Appl. Phys.* **113**
8. Casu E A, Fernandez-Bolanos M, Krammer A, Schuler A, Ionescu A M, Fumarola A and Muller A A 2018 Vanadium Oxide Bandstop Tunable Filter for Ka Frequency Bands Based on a Novel Reconfigurable Spiral Shape Defected Ground Plane CPW *IEEE Access* **6** 12206–12
9. Maklakov S S, Polozov V I, Maklakov S A, Mishin A D, Ryzhikov I A, Trigub A L, Amelichev V A, Maslakov K I and Kisel V N 2018 Post-deposition annealing of thin RF magnetron sputter-deposited VO 2 films above the melting point *J. Alloys Compd.* **763** 558–69
10. Pozar D M 2011 *Microwave engineering* (Hoboken, NJ : Wiley) p 107
11. Mapleback B J, Nicholson K J, Taha M, Baum T C and Ghorbani K 2019 Complex Permittivity and Permeability of Vanadium Dioxide at Microwave Frequencies *IEEE Trans. Microw. Theory Tech.* **67** 2805–11
12. Semenenko V N, Chistyaev V A, Politiko A A and Baskov K M 2019 Ultrawide measurement

bench for measuring electromagnetic properties of materials in free space in a microwave range *Izmer. TEKHNIKA* 55–9

13. Perez M D 2012 *General effective medium model for the complex permittivity extraction with an open-ended coaxial probe in presence of a multilayer material under test* (Alma Mater Studiorum Università di Bologna)

IOP Conf. Series: Materials Science and Engineering **848** (2020) 012074 doi:10.1088/1757-899X/848/1/012074

The cobalt and nickel-containing polyethylene prepared by using metal-vapor synthesis

P V Pribytkov[1,2], A Yu Vasil'kov[3] and A V Naumkin[3,4]

[1] Department of Chemistry, M.V.Lomonosov Moscow State University, Moscow 119991, Russian Federation
[2] N. D. Zelinsky Institute of Organic Chemistry, Russian Academy of Sciences, Moscow 119991, Russian Federation
[3] A. N. Nesmeyanov Institute of Organoelement Compounds, Russian Academy of Sciences, Moscow 119991, Russian Federation
[4] Moscow Institute of Physics and Technology (State University), Dolgoprudny, 141701 Moscow region, Russian Federation

E-mail: pribytkov@general.chem.msu.ru, alexandervasilkov@yandex.ru

Abstract. Cobalt and nickel-containing low-pressure polyethylene samples (HDPE) were prepared by metal-vapor synthesis (MVS) of unstable bisarene complexes of Ni and Co type $(C_6H_5CH_3)_2M$ and their further decomposition at a temperature of 178 K in the volume of HDPE. The Faraday method shows that Ni and Co nanometer particles are stabilized by the polyethylene matrix up to T=370 K. When the Ni-HDPE material is heated above 420 K, irreversible processes of interaction of Ni clusters with the polyethylene matrix occur with the formation of a less magnetic phase. When the Co-HDPE material is heated above 420 K, Co clusters coalesce to form a more magnetic phase. At temperatures above 520 K, the magnetization of the Co/HDPE system begins to increase rapidly and approaches the value characteristic of massive cobalt. Changing the conditions of cryosynthesis significantly influences the properties of the resulting products. The sample of 2.5% Co-HDPE obtained at a temperature of 185 K is characterized by weak magnetic properties. Its magnetization at room temperature is lower than that of the samples with a similar metal content by 30-40 times. The XPS shows that nickel atoms are in the form of $Ni(OH)_2$.

1. Introduction

The production of hybrid materials based on polyethylene and metal nanometer particles has long attracted researchers. This is due to the fact that, firstly, the resulting hybrid materials, along with the characteristics inherent in polyethylene, such as high chemical resistance, dielectric properties, low sensitivity to shock (good shock absorber), extremely low adhesion, softening when heated (80 – 120° C) acquire additional valuable qualities possessed by metal nanometer particles and primarily their special magnetic and conductive properties. Secondly, polyethylene due to its chemical inertness can be considered as a convenient matrix for stabilizing highly active and unstable metal nanoparticles.

There are several main ways to produce hybrid materials based on polymers containing metal nanoparticles. It is possible to introduce directly metal nanoparticles or their oxides into the polymer matrix. In [1], a nanometer ZnO sol-gel powder was first obtained by the method of zinc acetate Zn $(CH_3COO)_2 \cdot H_2O$ using isopropanol and diethanolamine. Most of the particles had a size in the range of 40-75 nm. The resulting ZnO powder was then mixed with high density polyethylene (HDPE) powder and was driven through a two-roll mill heated to 140 °C, and then pressed in a hydraulic press at a pressure of 30 bar at a temperature of 160 °C. In the material, along with nanometer particles,

particles exceeding micron sizes were found in a small amount, which is due to the negative influence of heating and pressing in the process of manufacturing nanocomposites.

Introduction of metal nanoparticles into the polymer matrix is possible by reduction of various metal salts. For example, polyethylene fibers containing Ag nanoparticles were produced by chemically reducing $AgNO_3$ using ultrasonic irradiation. Ethylene glycol, ethanolamine (EA) and n-propanol were used as reducing agents. The smallest average size of silver nanoparticles was obtained using EA and amounted to 48 nm [2].

A common method is the destruction of metal-containing compounds in the polymer volume. An original version of this method was reported in [3], where elastic wave pulses were used to decompose iron sulfate, cobalt and nickel chlorides in a volume of high-pressure polyethylene (HPPE). The result was a HPPE containing Fe, Co, Ni particles with dimensions of 10-100 nm.

More often use thermolysis of metal-containing precursors, such as inorganic and organic salts and organometallic compounds [4]. For example, in [5, 6], a solution of Li acetate together with Mn, Co and Ni acetates in acetic acid was introduced into the molten polyethylene-hydrocarbon oil solution. HPPE was used as a polymer matrix. The average sizes of nanoparticles were: for the sample Li-Co-HPPE – 1,8±0,4, Li-Mn-HPPE – 2,0±0,3, Li-Ni-HPPE – 1,4±0,3 and Li-Ni-Mn-Co-HPPE – 2,3±0,4 nm.

It is of interest to use the method of the destruction of metal-containing compounds in the polymer volume under milder conditions, for example at temperatures not exceeding room temperature or even lower. This is possible with the use of organometallic compounds that are unstable at room temperature. In our work [7], a method of obtaining metal polymers by MVS of unstable bisarene complexes of nickel and cobalt type $(C_6H_5CH_3)_2M$ and their further decomposition at a temperature of 178 K in the volume of HDPE ($M = 2 \cdot 10^5$) was described.

2. Experimental setup and characterization techniques

A 5 liter MVS reactor [8-11] has a vacuumed. Cobalt or nickel shards (0.25-0.6 g) are evaporated in 10^{-2} Pa vacuum by a evaporator in the form of a tungsten spiral. Metal vapors are allowed to condense on the walls of the reactor cooled down to 77 K simultaneously with toluene (80-130 ml). Before the synthesis, toluene was distilled over sodium under Ar and further degassed in vacuum by several successive freeze-pump-thaw cycles. After the synthesis, the reactor walls are allowed to heat up and the metal-toluene ice is transferred right upon melting into jelly-like HDPE-toluene mixture. Metal-polymer composites have been prepared in two ways I and II (Fig. 1), the relevant conditions are presented in Table 1.

Table 1. Preparation conditions for metal-HDPE composites.

Sample	Metal	Amount of metal, wt %	Means of influence[5] (temperature, K)	Way
1, 2	Co	3.5, 4.6	MS (363)	I
3, 4	Ni	1.4, 4.3	MS (363)	I
5	Co	2.5	US (185)	II

[5] stirring of suspensions with a magnetic stirrer (MS) or under ultrasonication (US)

Magnetization measurements were carried in vacuum or in Ar by the Faraday method [12] in the temperature range of 77-573 K at magnetic field of 500-3200 Oe.

The X-ray photoelectron spectra were recorded using a XSAM-800 spectrometer (Kratos, UK) with Al Kα radiation in the fixed analyzer transmission mode. The spectra were measured with step size of 0.1 eV at ~20 °C and a pressure in the analytical chamber of $5 \cdot 10^{-8}$ Pa. The photoelectron spectra were approximated by Gauss function or the sum of Gauss functions, and the background caused by secondary electrons and photoelectrons that lost energy, was approximated by the straight line. The energy scale of spectrometer was calibrated according to the standard procedure taking into account the following binding energies: 932.7, 368.3 and 84.0 eV for Cu $2p_{3/2}$, Ag $3d_{5/2}$, and Au $4f_{7/2}$,

IOP Conf. Series: Materials Science and Engineering **848** (2020) 012074 doi:10.1088/1757-899X/848/1/012074

respectively. Quantification was performed using atomic sensitivity factors (ASF) included in the software of the spectrometer. Sample charging was corrected by referencing to the C-C/C-H peak in the C 1s spectrum (284.8 eV).

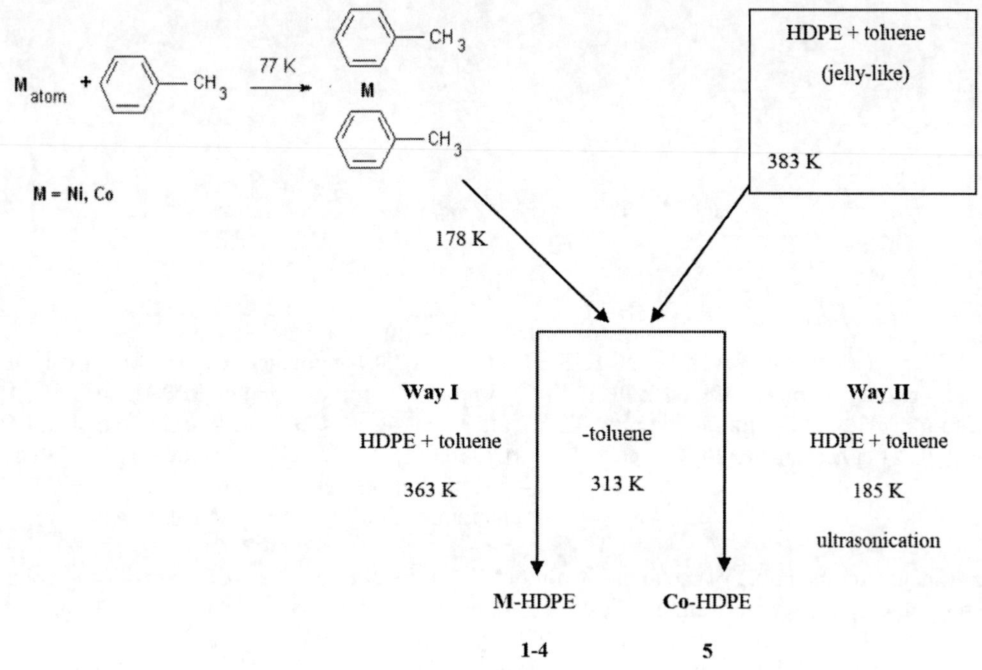

Figure 1. Scheme for preparation of Ni-HDPE and Co-HDPE hybrid materials.

3. Results and discussion

In both ways, at the initial stage of MVS, thermally labile bisarene complexes of Ni or Co are formed by interaction of metal vapors with toluene (Fig. 1). Resultant toluene solutions of these complexes just after the synthesis were transferred into the jelly-like HDPE-toluene mixtures prepared by boiling polyethylene in toluene under argon followed by cooling of the mixtures down to 363 and 185 K.

The synthetic way I initiates rapid thermal decomposition of the complexes in the bulk of the polymer at a temperature by far exceeding the decomposition point. Metal nanoparticles formed fill in cavities and defects present in the polymer and are immobilized therein due to sterical or other factors.

Effects of temperature on the properties of the composites were controlled by magnetostatic measurements. Thermomagnetic curves for Ni-HDPE hybrid materials (Fig. 2) demonstrate reversible behavior over 80-370 K, which suggests that Ni particles are stabilized by the polyethylene matrix at these temperatures.

An increase in the temperature of Ni-containing polyethylene to 420-430 K is accompanied by an irreversible decrease in magnetization and the appearance of new thermomagnetic curves, which also demonstrate reversible behavior over 300-420 K. This behavior of the samples may indicate that at temperatures above 370 K in the system there are irreversible processes of interaction of Ni clusters with the polyethylene matrix to form a less magnetic phase. This effect was observed for the Ni-nonadecan system [13].

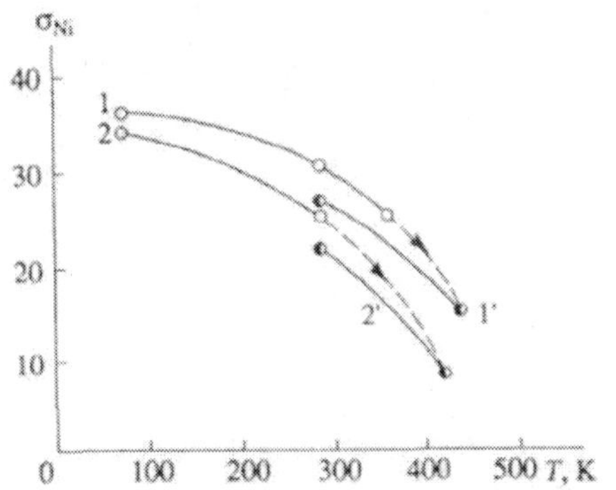

Figure 2. Thermomagnetic curves for Ni-HDPE samples: 1 - initial samples 1.4% Ni-HDPE; 1' - after heating at 430 K; 2 - initial samples 4.3% Ni-HDPE; 2' - after heating at 420 K.

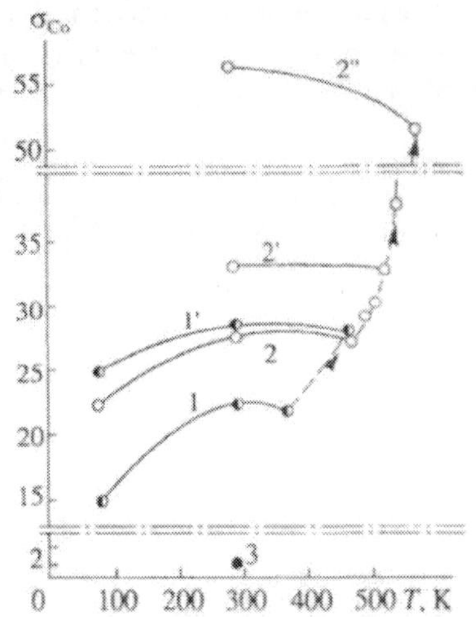

Figure 3. Thermomagnetic curves for Co-HDPE samples: 1 - initial samples 3.5% Co-HDPE; 1' - after heating at 470 K; 2 - initial samples 4.6% Co-HDPE; 2' and 2" - after heating at 530 and 586 K, respectively; 3 - samples 2.5% Co- HDPE maintained at a cryogenic temperature.

The influence of the nature of metal on the properties of metal polymers was studied using cobalt-containing polyethylene as an example. Co-HDPE samples manifest substantially different magnetic properties. Fig. 3 shows thermomagnetic curves of some representative Co-contaning samples. A 3.5% Co-HDPE composite exhibits a reversible thermomagnetic curve up to 370 K and demonstrates a 1.5-fold decrease in magnetization upon cooling to 80 K.

A 3.5% Co-HDPE hybrid materials is thermally stable to 370 K, which follows from the reversibility of the thermomagnetic curve (curve 1). When the temperature decreases to 80 K, a 1.5-fold decrease in magnetization is observed, apparently due to the anisotropy of the Co particle shape. An increase in the sample temperature to 470 K leads to an irreversible increase in magnetization and the appearance of a new, also reversible in this temperature range, thermomagnetic curve (curve 1'). This is due to the enlargement of Co particles, which, apparently, is due to a decrease in the stiffness of the matrix during the transition of polyethylene to the viscous state. At the same time, the anisotropy of the Co particle forms is still partially preserved, as indicated by the preserved decrease in magnetization when the temperature drops to 80 K. When the cobalt content in the polyethylene matrix increases, the initial magnetization value σ_{Co} increases (curve 2). The 4.6% Co-HDPE sample also demonstrates a decrease in magnetization when the temperature drops to 80 K, although not to the same extent as the 3.5% Co-HDPE sample. An increase in temperature to 470 K, as well as in the sample of 3.5% Co, leads to an irreversible increase in magnetization and the appearance of a new thermomagnetic curve (curve 2'). Further stepwise temperature increase to 585 K leads to the formation of a series of reversible thermomagnetic curves, each of which corresponds to the temperature dependence of the magnetization of cobalt particles, stable in these temperature ranges (curve 2"). At temperatures above 520 K, the magnetization of the Co-HDPE system begins to increase rapidly and approaches the value characteristic of massive cobalt.

Variations of cryosynthesis conditions strongly affect properties of the resultant products. A 2.5% Co-HDPE hybrid materials maintained at cryogenic temperatures (way II) is characterized by weak magnetic properties. Its magnetization at room temperature is lower than that of samples with a similar metal content synthesized via the route I by a factor of 30-40 (Fig. 3, spot 3). Furthermore it is characterized by a very weak ferromagnetic resonance signal.

The MVS synthesis of non-ferromagnetic Ni powders upon interaction of metal particles with alkanes has been described earlier. At the same time, cobalt blacks demonstrate totally different behavior [14]. Probably, the application of a lower temperature in the case of the toluene+HDPE system coupled with ultrasonication effectively encapsulate very small Co nanoparticles within the polymer shell.

Ni-HDPE hybrid materials was studied by XPS. XPS quantification based on ASF gives composition $C_{88.5}O_{9.1}Ni_{2.4}$. The Ni 2p spectrum consists of a spin-orbit doublet and two satellites (Fig. 4).

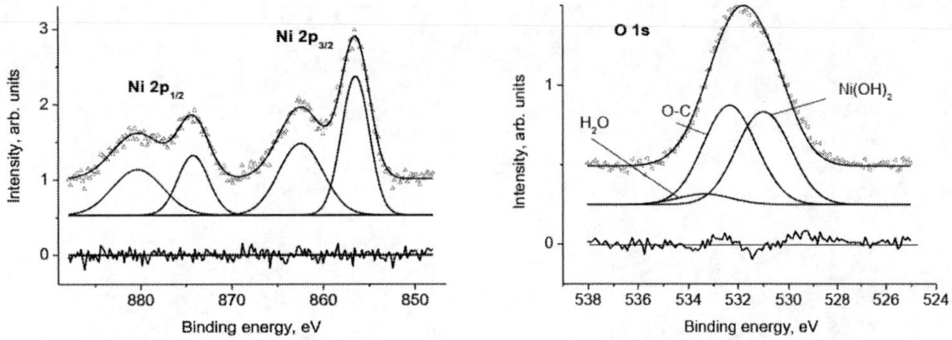

Figure 4. The Ni 2p and O 1s photoelectron spectra of Ni-HDPE samples.

The binding energies of the main Ni $2p_{3/2}$ and Ni $2p_{1/2}$ peaks are 856.7 and 874.4 eV, and the corresponding satellites are at 862.3 and 830.8 eV. The O 1s spectrum shows a wide peak at 531.4 eV, which can be equipped with three Gaussian peaks at 531.0, 532.3 and 533.3 eV with a width of 2.14 eV and relative intensities of 46:49:5. The peak at 531.0 eV represents the bound hydroxyl groups OH, and the peaks at 532.5 and 533.3 eV represents O-C bonds and water. Taking into account the characteristics of photoelectron spectra given in Table 2, such as binding energies (E_b) of the main and satellite peaks, satellite shifts ($\Delta E_1 = E_b$(Ni $2p_{3/2}$ sat) - E_b(Ni $2p_{3/2}$) and $\Delta E_2 = E_b$(Ni $2p_{1/2}$ sat) - E_b(Ni $2p_{1/2}$), spin-orbit splitting ($\Delta E_3 = E_b$(Ni $2p_{1/2}$) – E_b(Ni $2p_{3/2}$), an energy interval between Ni $2p_{3/2}$ and O 1s peaks ($\Delta E_4 = E_b$(Ni $2p_{3/2}$) – E_b(O 1s)) we assigned the Ni $2p_{3/2}$ spectrum to the $Ni(OH)_2$ phase [15-27].

In case of NiO, according to NIST XPS Database, the peak position of Ni $2p_{3/2}$ is in the range of 853.4 – 856.5 eV [28], while the position of O 1s is in the range 528.9-530.7 eV. It should be noted that for a reliable destination, we have collected only data containing satellite peak shifts.

To distinguish between $Ni(OH)_2$ and NiO we also compared the shape of Ni 2p line [29-31]. The Ni 2p binding energies of the Ni-HDPE samples compared to those in Table 2 show a positive shift relative to the original data. For the Ni/SiO_2 sample obtained by a similar method, the chemical shift was less by 0.8 eV [32, 33].

This indicates the different nature of the interaction of Ni particles with SiO_2 and polyethylene. A possible reason for this difference may be the presence of a large amount of O_2 in SiO_2 compared to polyethylene and the manifestation of differential charging. Another reason may be a large concentration of hydrogen in the polyethylene, which contributes to the predominant formation of $Ni(OH)_2$. The last hypothesis can be confirmed by the appearance of a signal at 282.8 eV in the C 1s spectrum, which can be attributed to Ni-C bond (Fig. 5).

The C 1s spectrum was provided with four states at 282.8, 285.0, 286.5 and 288.5 eV with Gaussian widths of 1.54, 1.81, 1.81 and 1.98 eV and relative intensities of 6:85:8:1, respectively (Fig. 5).

The peak at 285.0 eV represents the C-C/C-H bonds of polyethylene, while those at 265.5 and 288.5 eV represent C-O and C(O)O groups of oxidized carbon atoms. The peak at 282.8 eV is absent in the spectrum of the pristine polyethylene and may be assigned to formation of Ni-C bond and, therefore, indicates strong electronic interactions between Ni and C atoms.

Table 2. XPS characteristics of different Ni compounds: peak binding energies (E_b), satellite splitting ($\Delta E_{1,2}$), spin-orbit splitting (ΔE_3) and $\Delta E_4 = E_b(Ni\ 2p_{3/2}) - E_b(O\ 1s)$.

Sample	Ni $2p_{3/2}$	Ni $2p_{3/2}$ sat	ΔE_1	Ni $2p_{1/2}$	Ni $2p_{1/2}$ sat	ΔE_2	ΔE_3	O 1s	ΔE_4	Assignment for O 1s	Ref.
NiO	853.6 855.4	860.7	7.1					529.2			15
NiO	853.9 855.5	860.6	6.7					529.4			16
NiO	854.1 856.0	861.6	7.5					529.6			17
NiO	854.1 856.0	861.2	7.1					529.9			18
NiO	854.4 856.2	860.9	6.5					530.9			19
NiO	854.5 856.6	861.6 864.2	7.1 7.6	872.0 873.9	879.2 881.9	7.2 8.0	17.5	529.6 531.6	325.0	Ni-O O-C	20
NiO	855.22 857.13	861.42	6.2								21
α-$Ni(OH)_2$	855.5	861.2	5.7	873.1	879.8	6.7	17,6	531.0 532.4	324.5	$Ni(OH)_2$ H_2O, O-C	22
$Ni_2O_3 \cdot 6H_2O$	855.5	861.3	5.8	873.1	879.7	6.6	17,6	529.3 530.8	326.2	Ni_2O_3 $Ni(OH)_2$ O-C	23
γ-NiOOH	855.7	861.5	5.8	873.2	879.8	6.6	17.5	531.0 532.8	324.7	NiOOH H_2O, O-C	24
$Ni(OH)_2$	855.9	861.7	5.8	873.6	880.0	6.4	17.7	529.4 532.2	326.5	$Ni(OH)_2$ H_2O, O-C	25
Ni/SiO_2	855.9	861.5	5.6					531.5	324.4	Ni-O	8, 33, 34
β-$Ni(OH)_2$	856.2	861.8	5.6	873.8	880.1	6.3	17.6	531.0	325.2	$Ni(OH)_2$	26
$Ni(OH)_2$	856.5	862.4	5.9	874.1	880.6	6.5					27
$Ni(OH)_2$	856.7	862.3	5.6	874.4	880.8	6.4	17.7	531.0 534.5	325.7	$Ni(OH)_2$ O-C	this work

Note: ΔE1 = Eb(Ni 2p3/2 sat) - Eb(Ni 2p3/2), □E2 = Eb(Ni 2p1/2 sat) – Eb(Ni 2p1/2), □E3 = Eb(Ni 2p1/2) – Eb(Ni 2p3/2), □E4 = Eb(Ni 2p3/2) – Eb(O 1s). The energies are in eV.

This $Ni(OH)_2$ - C interface is also an important factor for reflecting the magnetic properties of Ni-polyethylene system. It should be noted that the C 1s photoelectrons are more sensitive to the surface that Ni 2p photoelectrons due to the difference in kinetic energies determining the inelastic mean free path of the electrons. The analysis of the XPS data for the Ni-HDPE hybrid materials showed that along with the metal oxidation process, the polymer matrix is destroyed and the Ni-C state is formed. The probability of these processes increases significantly with increasing temperature. This can lead to the formation of a non-magnetic metal-carbon phase, the influence of which seems to be evident in the study of magnetic properties.

Co-HDPE hybrid materials was also analyzed by XPS. No peaks characteristic of cobalt were observed in the survey spectrum of the sample. This indicates the presence of metal particles at a depth greater than the sampling depth of Co 2p photoelectrons.

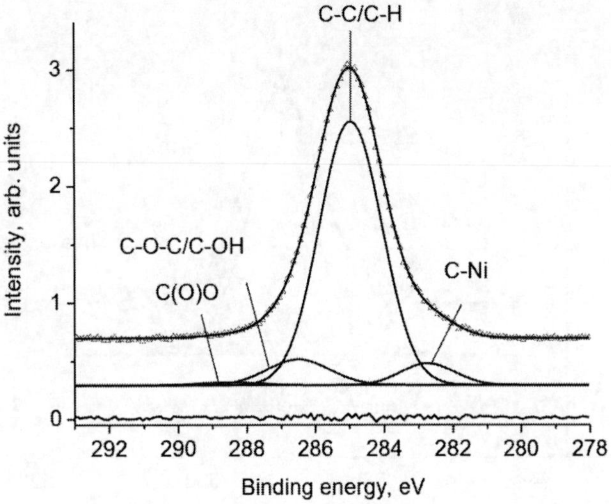

Figure 5. The C 1s photoelectron spectrum of Ni-HDPE samples.

The inelastic mean free path (λ) of Co 2p photoelectrons in pure Co is ~ 12 Å [35], while that in polyethylene is ~ 27 Å [36], the corresponding information depth 3 λ is ~ 81 Å. Since λ of the valence band and the C 1s photoelectrons is more than that of Co 2p photoelectrons they were analyzed also. The introduction of cobalt into polyethylene leads to a slight broadening of the C 1s peak, while significant changes in the spectrum of the valence band are observed (Figure 6). These changes reflect changes in the density of states near the Fermi level. Figure shows the spectra of C 1s and valence band of polyethylene and Co-HDPE.

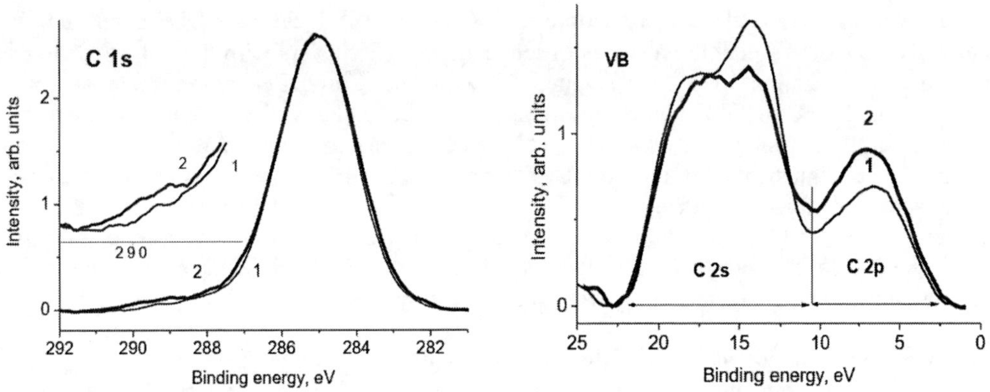

Figure 6. The C 1s and valence band (VB) spectra of HDPE (1) and Co-HDPE samples (2).

Figure 6 shows the C 1s and valence band spectra of polyethylene and Co-HDPE hybrid materials.

The C 1s spectrum of polyethylene can be provided by three Gaussian peaks at 285.0, 285.6 and 288.4 eV with a width of 1.98 eV and relative intensities of 93:4:3. This peaks are assigned to C-C/C-

H, C-O and C(O)O groups, respectively. The C 1s spectrum of Co-HDPE can be provided by three Gaussian peaks at 285.0, 285.6 and 288.7 eV with a width of 2.04 eV and relative intensities of 91:5:4.

Figure 7 shows the O 1s spectrum equipped with three peaks at 529.4, 532.4 and 534.5 eV with a width of 2.14 eV and relative intensities of 4:87:9.

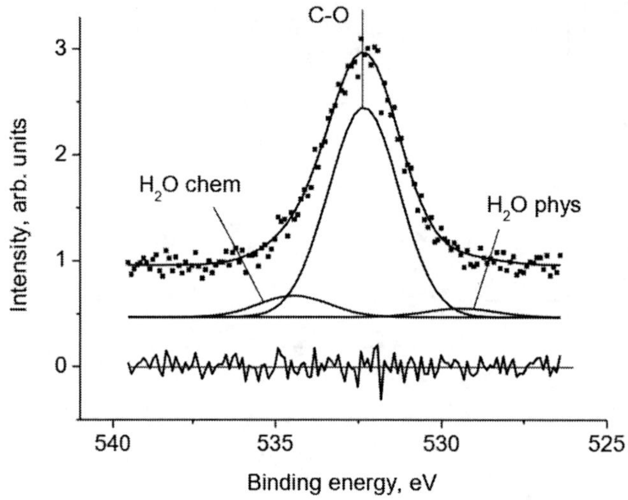

Figure 7. The O 1s spectrum of Co-HDPE samples.

The peak at 532.4 eV is for oxidized polyethylene carbon atoms, and the peak at 534.5 eV is for chemisorbed water. The peak at 529.4 eV can be attributed to physically absorbed water and/or Co-O bonds [23,29]. However, the latter assignment is less likely due to the small difference in the inelastic mean free paths of the Co 2p and O 1s electrons and great difference in ASF [30].

4. Conclusions

Ni and Co nanoparticles are stabilized by the HDPE matrix up to T=370 K. An XPS analysis of the Ni-HDPE hybrid materials showed formation of Ni-C bonds. When the Ni-HDPE nanocomposite is heated above 420 K, irreversible processes of interaction of Ni clusters with the polyethylene matrix occur with the formation of a less magnetic phase. The Co-HDPE hybrid materials behaves differently. When the Co-HDPE nanocomposite is heated above 420 K, the Co clusters fuse to form a more magnetic phase, which appears to be due to the coalescence of cobalt particles. At temperatures above 520 K, the magnetization of the Co-HDPE system begins to increase rapidly and approaches the value characteristic of massive cobalt.

The properties of nanocomposite materials with similar compositions are determined both by the synthesis conditions and post-synthesis factors. Thus, the 2.5% Co-HDPE sample obtained at a temperature of 185 K is characterized by weak magnetic properties.

This work was supported by the Ministry of Science and Higher Education of the Russian Federation and the Presidium of the Russian Academy of Sciences (Program "Nanostructures: physics, chemistry, biology, technology fundamentals", P13-2019-1).

References

[1] Mansour Sh A, Elsad R A and Izzulara M A 2016 *J. Sol-Gel Sci. Technol.* DOI 10.1007/s10971- 016-4109-x

IOP Conf. Series: Materials Science and Engineering **848** (2020) 012074 doi:10.1088/1757-899X/848/1/012074

[2] Abbasi A R, Kalantary H, Yousefi M, Ramazani A and Morsali A 2012 *Ultrasonics Sonochemistry* **19** 853

[3] Rakhimov R R, Jackson E M, Hwang J S, Prokof'ev A I, Alexandrov I A, Karmilov A Y and Aleksandrov A I 2004 *J. Appl. Phys.* **95** 7133

[4] Gubin S P, Yurkov G Yu and Kosobudsky I D 2005 *Int. J. Matet. Prod. Technol.* **23** 2

[5] Gubin S P, Spichkin Yu I, Koksharov Yu A, Yurkov G Yu, Kozinkin A V, Nedoseikina T I, Korobov M S and Tishin A M 2003 *J.Magnet. Magnet. Mater.* **265** 234

[6] Voronov V A and Gubin S P 2015 *Inorg. Mater.* **51** 1151

[7] Vasilkov A Y, Pribytkov P V, Fedorovskaya E A, Slinkin A A, Kogan A S and Sergeev V A 1993 *Doklady Akademii Nauk* **331** 179

[8] Vasil'kov A Yu, Migulin D A, Naumkin A V, Belyakova O A, Zubavichus Y V, Abramchuk S S, Maksimov Y V, Novichikhin S V and Muzafarov A M 2016 *Mendeleev Commun.* **26** 187

[9] Rubina M S, Kamitov E E, Zubavichus Ya V, Peters G S, Naumkin A V, Suzer S and Vasil'kov A Yu 2016 *Appl. Surf. Sci.* **366** 365

[10] Tsodikov M V, Ellert O G, Nikolaev S A, Arapova O V, Konstantinov G I, Bukhtenko O V and Vasil'kov A Yu 2017 *Chemical Engineering J.* **309** 628

[11] Rubina M S, Vasil'kov A Yu, Naumkin A V, Shtykova E V, Abramchuk S S, Alghuthaymi M A and Abd-Elsalam K 2017 *J. Nanostruc. Chem.* **7** 249

[12] Selwood P W 1975 *Chemisorption, Magnetisation* (NY: Acad. Press) p 164

[13] Sergeev V A, Slinkin A A, Fedorovskaya E A, Vasilkov A Y, Pribytkov P V and Lisichkin G V 1989 *Organometallic Chemistry in the USSR* **2** 1106

[14] Davis S C, Severson S J and Klabunde K J 1981 *J. Amer. Chem. Soc.* **103** 3024

[15] Lian K, Thorpe S J and Kirk D W 1992 *Electrochim. Acta* **37** 2029

[16] Venezia A M, Bertoncello R and Deganello G 1995 *Surf. Interface Anal.* **23** 239

[17] Dubé C E, Workie B, Kounaves S P, Robbat Jr A, Aksub M L and Davies G 1995 *J. Electrochem. Soc.* **142** 3357

[18] Bianchi C L, Cattania M G and Villa P 1993 *Appl. Surf. Sci.* **70** 211

[19] McIntyre N S, Chan T C and Chen C 1990 *Oxidation of Metals* **33** 457

[20] Mansour A N 1994 *Surf. Sci. Spectra* **3** 231

[21] Lian K K, Kirk D W and Thorpe S J 1995 *J. Electrochem. Soc.* **142** 3704

[22] Mansour A N and Melendres C A 1994 *Surf. Sci. Spectra* **3** 255

[23] Mansour A N and Melendres C A 1994 *Surf. Sci. Spectra* **3** 263

[24] Mansour A N and Melendres C A 1994 *Surf. Sci. Spectra* **3** 271

[25] Mansour A N and Melendres C A 1994 *Surf. Sci. Spectra* **3** 247

[26] Mansour A N 1994 *Surf. Sci. Spectra* **3** 239

[27] Chen X, Chen X, Zhang F, Yang Z. and Huang S 2013 *J. Power Sources* **243** 555

[28] Naumkin A V, Kraut-Vass A, Gaarenstroom S W and Powell C J 2012 NIST X-ray photoelectron spectroscopy database, version 4.1 *National Institute of Standards and Technology: Gaithersburg* http://srdata.nist.gov/xps

[29] Payne B P, Biesinger M C and McIntyre N S 2012 *J. Electron. Spectrosc. Relat. Phenom.* **185** 159

[30] Payne B P, Biesinger M C and McIntyre N S 2009 *J. Electron. Spectrosc. Relat. Phenom.* **175** 55

[31] Biesinger M C, Lau L W, Gerson A R and Smart R S C 2012 *Phys. Chem. Chem. Phys.* **14** 2434

[32] Naumkin A V, Vasil'kov A Yu, Volkov I O, Smirnov V V and Nikolaev S A 2007 *Inorg. Mater.* **43** 381

[33] Vasil'kov A Yu, Nikolaev S A, Smirnov V V, Naumkin A V, Volkov I O and Podshibikhin V L 2007 *Mendeleev Commun.* **17** 268

[34] Naumkin A V and Vasil'kov A Yu 2013 *Russ. Chem. Bull.* **62** 1

[35] Tanuma S, Powell C J and Penn D R 2011 *Surf. Interface Anal.* **43** 689

IOP Conf. Series: Materials Science and Engineering **848** (2020) 012074 doi:10.1088/1757-899X/848/1/012074

[36] Powell C J and Jablonski A 2000 NIST Electron Inelastic-Mean-Free-Path Database-Version 1.1 *National Institute of Standards and Technology, Gaithersburg, MD*

IOP Conf. Series: Materials Science and Engineering **848** (2020) 012075 doi:10.1088/1757-899X/848/1/012075

Surface modification TRIP \ TWIP steels

D V Prosvirnin, M D Larionov, A G Kolmakov, A V Larionova, M E Pruckov and S V Pivovarchik

IMET RAS, Moscow, Russia

Abstract. TRIP/ TWIP steels have high tensile strength, plasticity, and much higher yield stress then other structural steels. Currently class of TRIP/TWIP steels with austenite-martensite structure is used in highly loaded parts, such as helicopter propeller torsion bars. TRIP steels are most in demand in the automotive industry. However, this steel has a problem with fatigue strength, because martensite of deformation (product of austenite-martensite transformation induced by plastic deformation) is stress concentrator. Surface modification will increase fatigue strength and ductility, as well as high corrosion resistance in aggressive environments. It is supposed to evaluate the effect of a particular chemical element or chemical compound in the interaction with steels with a high austenite content and make first steps towards to technology of Surface modification TRIP \ TWIP steels.

Introduction

Metal matrix composites and steels with modified surface replace traditional metal materials because every new day industries require higher performance properties for production details with high relative strength properties. Materials that contain in composite cannot perform on the same level separately as composition of these materials, because interface (border region between phases) plays a huge role in properties of product.

In comparison with metal alloys, metal matrix composites and steels with modified surface can offer [1]:

- Increased static, dynamic and cyclic strength properties;
- Form stability;
- Heat resistance and high temperature strength;
- Corrosion resistance.

For creation of new advanced composite material, we supposed to use TWIP (Transformation Induced Plasticity) or TWIP (Twinning-Induced Plasticity) steels as matrix. In comparison to maraging or alloyed high-strength steels, TRIP/TWIP steels have higher strength, plasticity, and much higher yield stress (figure 1). In addition, these steels have high deformability and ability to absorb energy during impacts and crushing.

Currently class of TRIP/TWIP steels with austenite-martensite structure are used in highly loaded parts, such as helicopter propeller torsion bars. TRIP steels are most in demand in the automotive industry: their usage allows to optimize (reduce) the weight of products, calculate loaded car parts taking into account changes in material properties during collisions (important for ensuring passenger safety). Steel with transformation induced plasticity (TRIP) possess high strength and high plasticity: when material is plastically deformed austenite transforms into deformation martensite, and strain hardening, that leads to increase of strenght of the products, occurs [3-6]. Targeted kinetic transformation regulation in process of deformation will make it possible to obtain the required set of physic-mechanical characteristics.

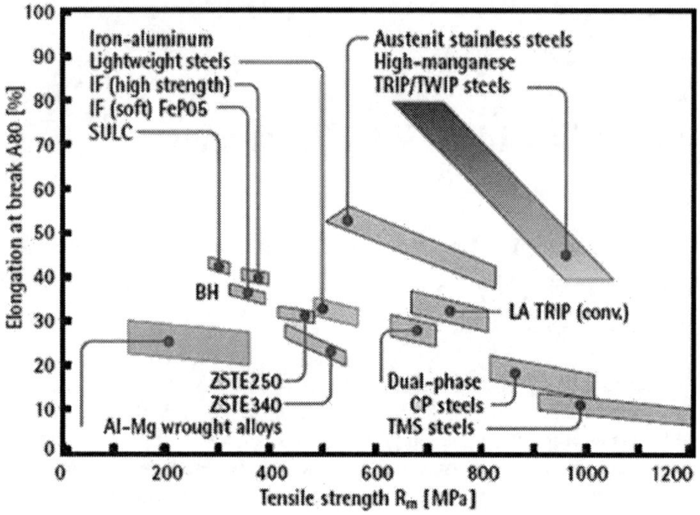

Figure 1. Comparative diagram of tensile strengths to elongation at break TRIP/TWIP and other steels [2].

In deformation process of TWIP steel dominant process is dislocation slip, but in the dynamic Hall-Petch effect, mechanical twins, the formation of stacking faults, also continuously form during deformation [7]. Since the formation of mechanical twins involves the creation of new orientations of crystals, twins gradually reduce the effective "mean free path" or effective slip distance of dislocations and increase the stress caused by dislocations leading to hardening [8]. The key to understanding the cause of formation of twins is to understand the effect of stacking fault energy (SFE): if SFE <20 mJ m^{-2}, deformation caused by the appearance of twins is more likely; if SFE 50> mJ m^{-2}, twinning is suppressed [8]. The manifestation of this or that effect is more dependent on the content of manganese Mn in the composition of the alloy and other alloying elements. Alloys with manganese content less than 15 wt. % are more prone to the TRIP effect, while with an increase manganese content in composition, appear the TWIP effect.

Such high properties provide steels with the TRIP / TWIP effect superiority over other metal materials common in automotive industry. Over the past decade, the percentage of these steels in the global automotive and aircraft industries has grown by 80%. However, high mechanical properties and impact characteristics of TRIP steels containing residual austenite depend on its diffusion-free transformation caused by deformation into high-carbon martensite. Formed deformation martensite will negatively affect fatigue properties of the material. The increased density of dislocations and a pronounced acicular structure have a strong effect on the rate of nucleation of microcracks. Also, due to the methods used for the production of these steels (large reduction ratios up to 80%), phase composition and, accordingly, physical and mechanical properties on the surface and in volume can vary. Surface micro- and macro-defects act as serious stress concentrators and especially negatively affect fatigue and corrosion properties, including wear resistance.

Traditional methods of heat and pressure treatment can significantly improve the set of mechanical characteristics. However, they leave a significantly defective surface with both micro- and macro-defects. By now, it has been theoretically established and experimentally confirmed that state of surface layers largely determines not only corrosion resistance of metallic materials, but also their mechanical (fatigue) properties. Due to the creation of thin (1-100 μm) modified layers on the surface of TRIP / TWIP steels, including usage of ion-vacuum technologies, it is possible to achieve an improvement in state (change in properties) of the surface and, as a result, simultaneously increase the fatigue strength characteristics. In this regard, the development of a composite based on these steels and metal coatings is especially relevant. This approach will ensure production of a material that, in comparison with the indicated steels, will have simultaneously increased strength and ductility, as well as high corrosion resistance in aggressive environments.

Approaches to solving the problem
One of main reasons for improving the characteristics of fatigue strength and ductility when modifying the surface of metallic materials is considered to be an increase in uniformity and a decrease in the localization of the processes of deformation and accumulation of defects in surface layers, as well as a slowdown process of nucleation surface microcracks (figure 2).

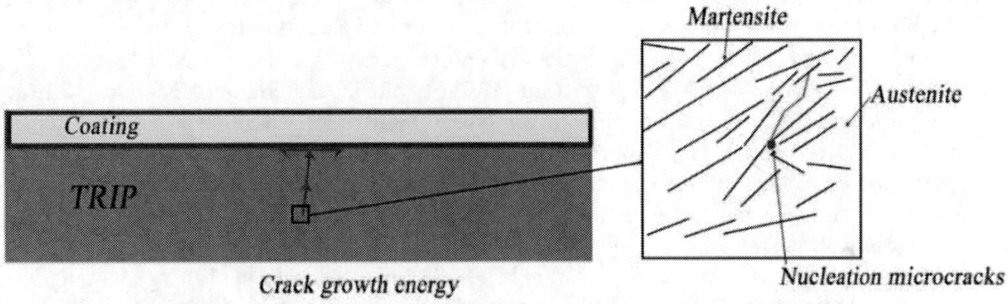

Figure 2. Crack nucleation and growth caused by martensite deformation.

An analysis of literature data cited in Russian sources on the effect of coating on the properties of TRIP / TWIP steels yielded not many results, since such steels themselves have not yet been sufficiently studied. So, in VIAM V.Ya. Belous dealt with the problem of preparing the surface of the VNS-9-Sh (Russian nomenclature) steel to create a composite material based on it [9], but this work had no further development. As part of a literature study, common works by Russian and German researchers was found (Joint Institute for Nuclear Research, Dubna; Institute for Metal Forming, Technical University of Freiberg Mining Academy, Freiberg, Germany) to study the mechanical properties of a composite material based on TRIP steel and ceramic from zirconium dioxide (Study of plastically deformed TRIP composites by neutron diffraction and small angle neutron scattering) [10, 11]. Other works related to such a composite material were also found in foreign sources [12-15], however, this composite is a material of a fundamentally different class, since in such a composite the content of ceramics can reach 50-70% of the total volume, which may complicate manufacturing technology and accordingly increase production costs.

It is supposed to evaluate the effect of a particular chemical element or chemical compound in the interaction with steels with a high austenite content. In particular, it is necessary to take into account not only the mechanical properties of the applied surfaces that meet the design requirements of composite materials, but also the thermodynamic interaction and phase equilibrium on interface.

First, using the finite element method, we should calculate the rate of crack nucleation and growth and analyze the effect of the applied coating using programs that simulate thermodynamic parameters.

We proposed to make on the surface of steel a layer with a thickness not more than 500 μm, the material selected at the stage of analysis that will increase the set of operational properties (fatigue life, ductility, corrosion resistance, etc.) at low additional cost in future mass production. The most promising materials for coating can be steels and alloys with higher plastic and anticorrosion properties compared to TRIP / TWIP steels, for example, austenitic stainless steels, aluminum alloys and alloys based on Ni, Cr and Co. Elimination of negative influence of surface defects of the base (matrix) material due to the applied protective and functional layer, leading to improved fatigue properties, can extend the service life of composite products. That will give a high economic effect due to the high cost of production of

TRIP \ TWIP steels, however, the influence of one or another element on the set of composite properties remains to be determined in future works.

Acknowledgments
This work was supported by a grant from RFBR № 20-08-00556 A. The literature research part of work was carried out according to the state assignment № 075-00947-20-00.

References

[1] NIKHILESH. CHAWLA. METAL MATRIX COMPOSITES. – Springer-Verlag New York, 2016.

[2] Knowledge M. Multi Phase Twinning-Induced Plasticity (TWIP) Steel., 2007

[3] Klimenko S.A., Kolmakov A.G., Lebedev V.Ya., Borodavko V.I. Inheritance and technological support of the operational properties of coated products // Bulletin of the National Academy of Sciences of Belarus. Series of physical and technical sciences, 2015. No. 3. P.31-40.

[4] Sirotinkin V.P., Terentyev V.F., Prosvirnin D.V., Shamray V.F., Slizov A.K. X-ray diffraction study of the change in the structure of thin-sheet TRIP steel VNS9-Ш after static tension // Metals. - 2015. - No. 2. - P. 61-67.

[5] Gvozdev A.E., Kolmakov A.G., Provotorov D.A., Minaev I.V., Sergeev N.N., and Tikhonova I.V. Grain size effect of austenite on the kinetics of pearlite transformation in low- andmedium-carbon low-alloy steels // Inorganic Materials: Applied Research, 2015. V.6, № 1. P.41-44.

[6] Nasakina E. O., Sevostyanov M.A., Dyomin K.Yu., Mikhaylova A.B., Goldberg M.A., Kolmakov A.G., Zabolotnyy V.T. The use of ion-atomic deposition in the fabrication of one-dimensional composites // Inorganic Materials: Applied Research, 2015. V. 7. №.6(4). P.293-297.

[7] Callister W. D. et al. Fundamentals of materials science and engineering. – London : Wiley, 2000.– V. 471660817.

[8] De Cooman B. C., Kwon O., Chin K. G. State-of-the-knowledge on TWIP steel //Materials Science and Technology. – 2012. – V. 28. – №. 5. – P. 513-527.

[9] Belous V. Ya. Et al. Corrosion resistance and surface preparation of cold-rolled steel tape VNS-9-Sh for the manufacture of metal-polymer composite material // Proceedings of VIAM. - 2015. - No. 11.

[10] Martin S. et al. Reinforcing Mechanism of Mg-PSZ Particles in Highly-Alloyed TRIP Steel //steel research international. – 2011. – V. 82. – №. 9. – P. 1133-1140.

[11] Biermann H. et al. Microstructure and compression strength of novel TRIP-steel/Mg-PSZ composites //Advanced Engineering Materials. – 2009. – T. 11. – №. 12. – C. 1000-1006.

[12] Hosseini S. M. K. et al. ANN model for prediction of the effects of composition and process parameters on tensile strength and percent elongation of Si–Mn TRIP steels //Materials Science and Engineering: A. – 2004. – V. 374. – №. 1-2. – P. 122-128.

[13] Gomez M. et al. Design of composition in (Al/Si)-alloyed TRIP steels //ISIJ international. – 2009.– V. 49. – №. 2. – P. 302-311.

[14] Maki J. et al. Galvanisability of silicon free CMnAl TRIP steels //Materials science and technology. – 2003. – V. 19. – №. 1. – P. 125-131.

[15] Baumgart C. et al. Comparative study of TRIP/TWIP assisted high density composite honeycomb structures under compressive load //Composite Structures. – 2016. – V. 136. – P. 297-304.

IOP Conf. Series: Materials Science and Engineering **848** (2020) 012076 doi:10.1088/1757-899X/848/1/012076

METALS RECOVERY FROM MOLYBDENITE CONCENTRATE BY ELECTROOXIDATION AND LEACHING

S N Rasulova, V P Guro, E T Safarov and X F Adinaev

Institute of general and inorganic chemistry of Uzbekistan Academy of sciences, Tashkent city, Uzbekistan.
Email: sitora_r91@mail.ru

Abstract. Mineral resources today are increasingly determined by the concept of "persistent". The reason for their difficult opening is in the composition of sulfide nature, leading to passivation of the surface and inhibition of hydrometallurgical processes. The way out in practice was found in oxidative firing and some other technologies. The work compares the results of leaching of molybdenite concentrate of Almalyk MMC JSC according to the schemes: the existing calcining-ammonia and non-calcining with nitric acid, hypochlorite opening and leaching. Non-calcining with hypochlorite opening and leaching scheme is not only ecologically cleaner in compare with pyro-metallurgy, but it relocates quantitatively to the solution not only molybdenum, rhenium, but gold and silver as well.

1. Introduction

Sulfide minerals are susceptible to surface passivation under oxidizing conditions [1]. Its overcoming is achieved through: 1 - oxidative roasting, 2 - in an autoclave, 3 - hydrometallurgical methods.

Firing is used in the processing of molybdenite concentrates (MoC) [2], containing Re [3]. Its disadvantages inherent in pyrometallurgy include the burden on the environment and high capital costs. MoC is oxidized in an autoclave with the addition of an HNO_3 catalyst [4]. An increase in temperature and pressure accelerates leaching in autoclave, but it is not feasible for low grade ores.

Hydrometallurgical processes are considered traditional, but also promising, due to the relative minimization of environmental threats. With the exception of bio-oxidation (carried out in H_2SO_4), sulfide minerals are opened in various solutions using oxidizing agents: chlorine [5], oxygen [6], HNO_3 with O_2 [7], salts of Fe (III) [8], ozone [9], sodium peroxodisulfate [10], bubbling air into a solution of HCl + $MgCl_2$ [11], reagent hypochlorite [12], hypochlorite generated in viva by electrolysis of 10% NaCl [13-16].

The mechanism of surface passivation of sulfides is extended to known minerals in the composition of the Mo-concentrate of Almalyk MMC JSC [17]. According to the technology existing there, it is granulated with kaolin (8-10% by weight), burned in a drum furnace with sublimation of rhenium oxide, ammonia leaching of the cinder [18]. To eliminate the effect of dilution of Mo caused by kaolin, an organic binder was proposed instead, with firing in a fluidized bed furnace [19-20]. Meanwhile, the hydrometallurgical processing of molybdenite concentrate remains an alternative to this technology [21]. One of its promising areas is hypochlorite electrochemical oxidation and leaching [22-23].

2. Methods

Samples of molybdenite concentrate (38% Mo in the MoS_2 phase) were used. The concentrations of Cu (II), Fe (III), Mo (VI), Re (VII) were determined photocolorimetrically (KFK-2) and spectrometrically (AAC Perkin-Elmer 3030B, ICP-Aligent 7500 ICP MS).

Microscopy of the samples was performed on SEM-EDS scanning electron microscopes: EVO-MA (Carl Zeiss, Germany, Oxford Instrum, UK, with a resolution of 3 nm) and JCM-6000 Neoscope II (Jeol, Japan, in the range of 10-60000 times and a resolution of up to 30 nm).

Kinetic measurements were performed with Mo-concentrate powders weighing up to 6 g, s : w = 1: 7 and 1:75, with stirring, 300 rpm, and a temperature of 20-70 ° C. Firing of Mo-concentrate granules (with kaolin or an organic binder) was carried out in furnaces: industrial drum and laboratory, fluidized bed. Hypochlorite leaching of MoC was carried out in solutions: 12% hypochlorite reagent (GOST 11086-76) and in a solution of 10% NaCl by electrooxidation, generating hypochlorite on the anode of the electrolyzer (Figure1).

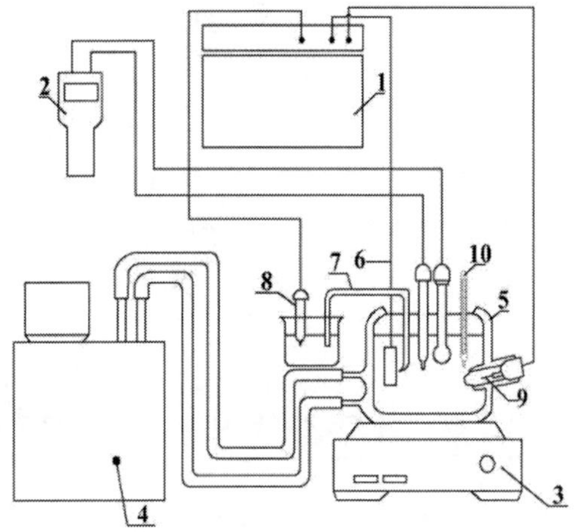

Figure 1. Scheme of electrooxidation of Mo-concentrate in a solution of 10% NaCl. *Designation:* 1-potentiostat PI-50-1, 2-pH meter, 3-magnetic stirrer, 4-thermostat, 5-cell, 6-anode, 7-capillary Luggin, 8-electrode silver chloride, 9-cathode, 10-thermometer.

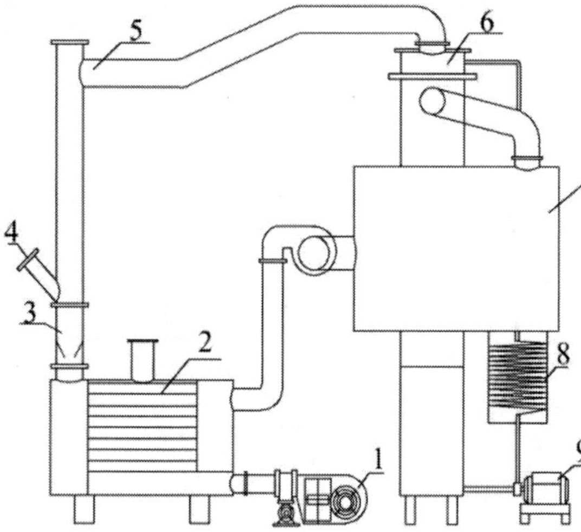

Figure 2. Scheme of the pilot installation of a fluidized bed furnace (FB) for firing MoC. *Designation:* 1-gas burner; 2-boiler; 3 - zone loading granules MoC; 4 - viewing window of the hearth of the firing zone; 5 - pipeline of gas products of firing; 6 - scrubber for absorption of rhenium oxide and SO_2; 7 - filter; 8 - water refrigerator, 9 - pump.

3. Results

Given the purpose of the work, it was of interest to compare the efficiency of the extraction of valuable metals from molybdenum concentrate under conditions of preliminary oxidative firing, changes in temperature, nature and concentration of the leaching agent, the process time (**Table 1**).

Table 1. Comparison of the degrees of extraction (in solutions) of metals from cinder and MoC (total initial raw materials). *Designation*: 1-12 - a cinder of firing MoC (570 °C): 1-3-3,7-9 in a fluidized bed furnace; 4-6,10-12 in a rotary kiln, in both - with O_2 access, 13-16 - MoC without firing, leaching: 13-15 - in ammonia after nitrate opening, 16-18 - opening and leaching at 12% hypochlorite, pH 11-12.

Sample, No.	Binder	Reagent leaching, %	T, °C	Leaching time, min	Extraction degree (%)				
					Cu	Mo	Re	Au	Ag
1- calcine	organic	NH_4OH		90	95.5	93.0	*97.0**	-	-
2- calcine		25%	20	180	96.8	94.0	*98.0**	-	-
3- calcine				270	96.9	97.6	*99.7**	-	-
4- calcine	kaolin		20	90	91.4	93.8	*99.5**	-	-
5- calcine				180	92.5	95.8	*99.3**	-	-
6- calcine				270	92.4	95.8	*99.8**	-	-
7- calcine	organic	NaClO	20	90	47.7	99.7	*99.5**	94.6	73.3
8- calcine		12%		180	55.3	99.8	*99.6**	95.3	79.4
9- calcine				270	64.5	99.8	*99.9**	96.5	99.0
10- calcine	Kaolin		20	90	66.4	99.6	*99.3**	92.3	72.0
11- calcine				180	68.4	99.8	*99.7**	93.5	77.3
12- calcine				270	98.2	99.9	*99.6**	95.1	97.5
13- MoC	None	HNO_3 56%	70	90	73.5	98.0	93.8	-	70.8
14- MoC		NH_4OH		180	75.5	98.5	94.0	-	72.0
15- MoC		25%		270	88.6	98.6	99.9	-	94.0
16- MoC	None	NaClO	40	90	83.5	98.1	99.9	93.8	81.0
17- MoC		12%		180	84.5	98.3	99.9	94.0	83.0
18- MoC				270	84.5	98.7	99.9	94.5	84.0

Note: * - extraction of rhenium - total: in the gas (in the form of Re_2O_7) and liquid phases.

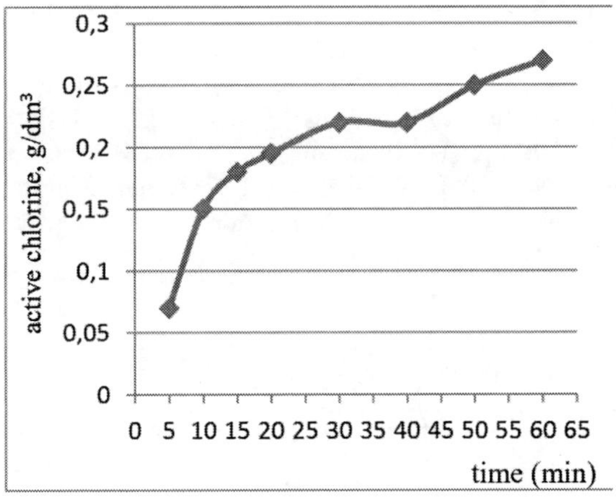

Figure 3. Change in the concentration of active chlorine in a 2-chamber membrane electrolyzer, galvanostatic mode I = 2.45 mA, 10 (mA/V), solution volume 75 cm^3. Over 60 min, the concentration of NaClO = 0.06% was achieved.

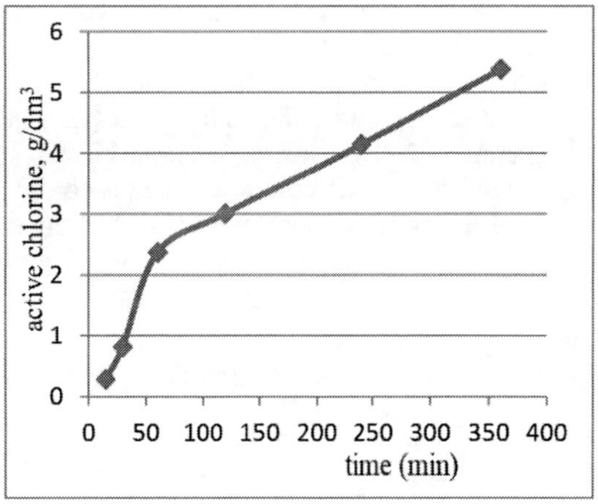

Figure 4. Change in the concentration of active chlorine in a 1-chamber electrolyzer, galvanostatic mode, I = 109.9 mA, 0.1 (A/V), the solution volume is 75 cm^3. Over 360 min, the concentration of NaClO = 1.13%.

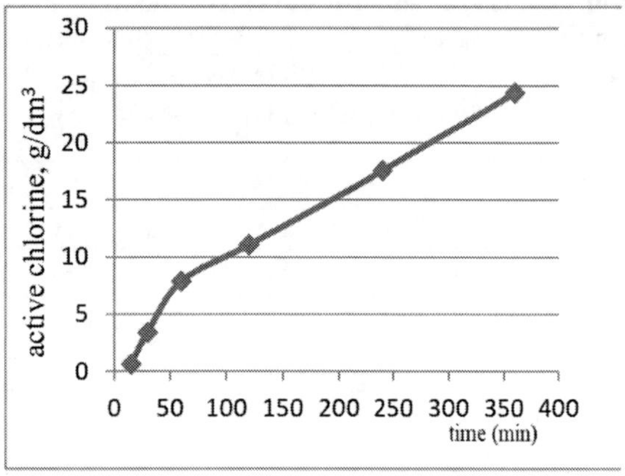

Figure 5. Change in the concentration of active chlorine in a 1-chamber electrolyzer, galvanostatic mode, I = 109.9 mA, 0.1 (A/V), solution volume 75 cm^3. Over 360 minutes, NaClO concentration = 5.12% was reached.

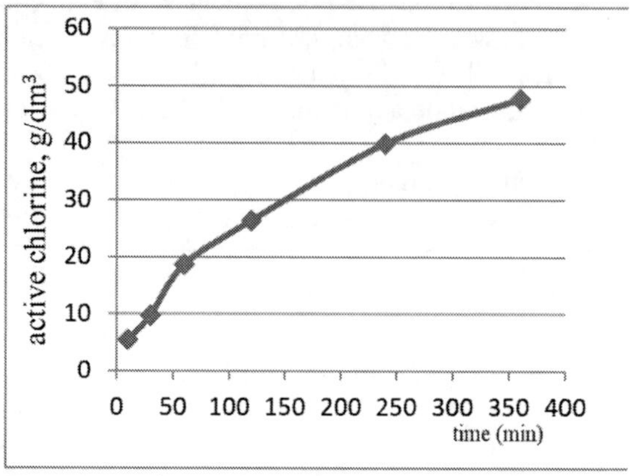

Figure 6. Change in the concentration of active chlorine in a 1-chamber electrolyzer, galvanostatic mode, I = 109.9 mA, 0.1 (A/V), solution volume 75 cm^3. Over 360 minutes, NaClO concentration = 10.0% was reached.

From Table 1 it follows: 1) from the cinder (samples 1-12), the extraction of Cu, Mo, Re - almost completely takes place in an ammonia medium in 90 minutes, in 12% hypochlorite, copper reaches the same level of extraction in only 270 minutes, Mo is already complete in 60 minutes; precious metals in ammonia are not extracted, in nitric acid, Au - is not extracted, Ag - partially; in hypochlorite: Au - completely in 90 minutes; Ag - completely in 270 min.

The material of the MoC binder granules did not affect the degree of metal extraction, with the exception of Au and Ag: the organic binder gave better results than kaolin. 2) from MoC: in nitric acid: Cu - completely in 270 minutes, Mo and Re — completely in 90 minutes, Ag - partially, Au - not recovered; in a hypochlorite solution: Cu - 84% in 270 minutes, Mo and Re - completely in 90 minutes, Au - in 94% to the maximum in 270 minutes, Ag - 84% in the same time.

The next task was to search for the generation modes at the anode of the hypochlorite electrolyzer in an electrolyte of 10% NaCl. Two types of synthesis were tested with pH control: in a two-chamber diaphragm electrolyzer with a volume of 90 cm^3, and in it in a 1-chamber (without membrane); electrodes - graphite, the ratio of the areas of the anode / cathode, varied: from 1:1 to 100: 1, temperature 20 °C, stirring on a magnetic stirrer, 300 rpmThe concentration of active chlorine generated in the cell was determined according to GOST 11086-76 (Figures 3-6). By increasing the ratio of the anode/cathode areas (from 1:1 to 100:1), the hypochlorite concentration in the 1-chamber electrolyzer was increased from 1.13% (Figure 4) to 5.12% (Figure 5) and 10.0 (Figure 6).

Table 2. The degree of Mo extraction by electrooxidation of MoC sulfides in the electrolyte (C NaClO = 1.03-10.0%) T: W = 1:7 (75 cm^3 of 10% NaCl electrolyte).

№	Time (hour)	Extraction rate of Mo [%] in the electrolyte with NaClO concentration:		
		1.13%	5.12%	10%
1	6	4	17	27
2	24	4.7	29.3	64.6
3	48	8.4	48.5	98.5

The next objective of the study was the leaching of MoC in 3 variants of electrooxidation: in medium 1.13; 5.12; 10.0% NaClO, which was achieved by adding MoC to the electrolytes to the ratio S:W = 1:7; continuous electrolysis, stirring suspensions on a magnetic stirrer, 300 rpm (Table 2).

IOP Conf. Series: Materials Science and Engineering **848** (2020) 012076 doi:10.1088/1757-899X/848/1/012076

The distribution of elements in MoC matrix, as the suspension of the powder in the sulfate-hypochlorite electrolyte is oxidized, occurs as follows (table 3).

Table 3. Change in the composition of samples of Mo-concentrate as they oxidatively leach in a sulfate-hypochlorite electrolyte, over time.

No. sample	Sulfide oxidation time, sec.	Content of elements in the solid phase, mass. %				
		Mo	S	Si	Cu	O
1	0	50.6	28.4	2.2	1.5	12.8
2	90	35.2	24.3	5.4	1.3	26.6
3	600	36.1	20.7	5.3	1.4	28.6
4	1800	4.9	9.5	16.1	0.0	42.3

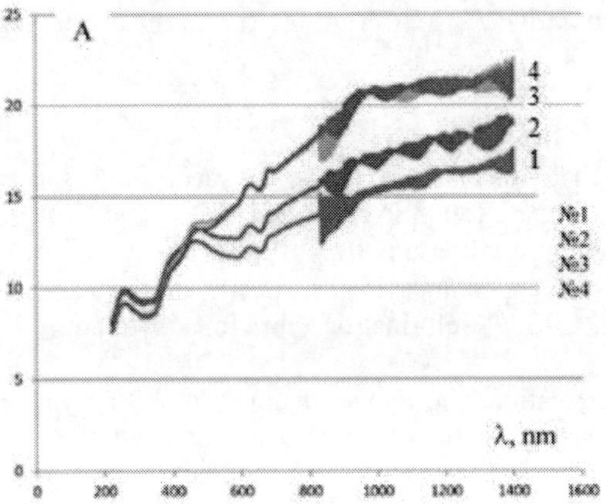

Figure 7. UV spectra of samples 1-4.

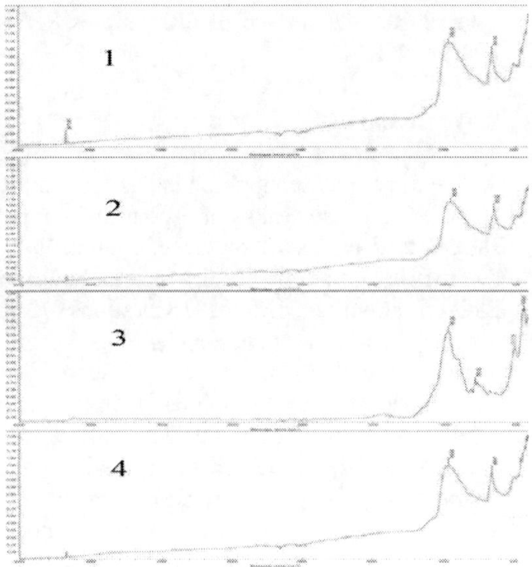

Figure 8. IR spectra of samples 1-4.

The designation in fig. 10-11: exposure of samples in a solution of sodium hypochlorite, sec: No. 1 - 0; No. 2 - 60; No. 3 - 600; No. 4 – 1800 (right).

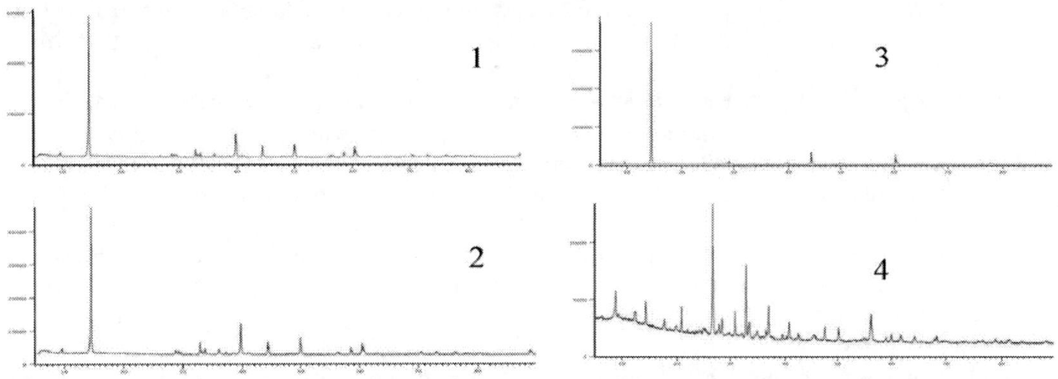

Figure 9. Radiographs (XRD Empyrean PANalytical) of samples No. 1-4. *Designation:* similar to figures 7-8.

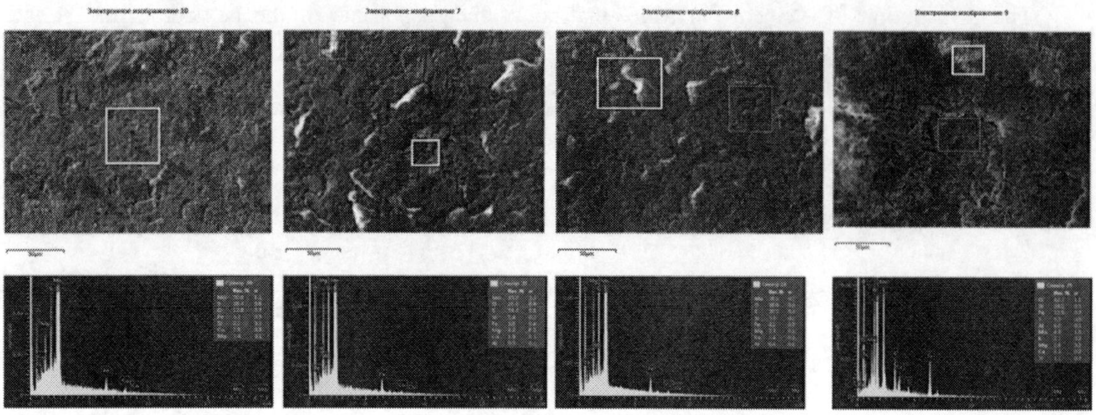

1 2 3 4

Figure 10. The surface of the samples SEM-EDS, brand EVO-MA (Carl Zeiss, Germany, Oxford Instrum, UK)

4. Discussion

The IR spectra of samples No. 1-3 do not differ from each other, in contrast to the spectrum of sample 4. The most intense vibration bands for samples No. 1-3 were found in the range of 966 and 664 cm^{-1} (Figure 7). These peaks correspond to symmetric stretching vibrations (S – S). Obviously, oxidation should lead to a shift of the vibrational bands and the appearance in the spectra of the bands of symmetric and asymmetric S–O stretching vibrations, O–S–O deformation vibrations, symmetric O–S–O vibrations, and $[MoO_4]$ 2 vibrations - [fourteen].

Despite the fact that samples No. 2–3 undergo oxidation, the above changes in the IR spectra associated with oxidation were observed only for sample No. 4, which was manifested in the appearance of new vibration bands in the range 981, 796, 517, 452 cm^{-1}. In addition, unlike sulfides, oxygen compounds of metals have different types of hydrogen bonds in the crystal structure and are characterized by wide vibration bands in the region of 3600 cm^{-1} and O–H bonds (1670 cm^{-1}), characteristic of stretching and deformation vibrations. These data confirm that in addition to metal sulfides, in sample No. 4 there are also components oxidized to sulfates or sulfites, molybdates.

At the same time, electronic reflection spectra are more sensitive to the ongoing oxidation of the samples, as indicated by a gradual shift of the reflection bands. Molybdenum sulfide compounds exhibit a d-d transition at 1000–588 nm, followed by charge transfer (from Mo to S) in the region from 588 to 333 nm. The spectra of oxidized samples No. 3-4 are almost identical and are characterized by a shift in the transition of charge transfer to the long-wavelength region.

The data of X-Ray phase analysis of samples No. 1-4 (Figure 9) also indicate significant changes in the structure and phase composition. In the diffraction patterns, in addition to peaks of molybdenite, pyrite peaks are observed (5-7%), impurities of quartz and natural silicates. From the diffraction patterns it is seen that sample No. 3 is characterized by a small number of diffraction peaks, in addition to peaks of molybdenite, which can serve as confirmation of the destruction of the crystal structure of the initial Mo-concentrate. For sample No. 4, growth of new crystalline phases was observed, which formed oxidation products in a heterogeneous system: "Mo-concentrate powder - Na-hypochlorite solution", during the interaction time of 1800 sec, at room temperature and atmospheric pressure.

A gradual "loss" of Mo and copper from the crystalline matrix of the material is observed, as a result of oxidative leaching of the sulfides of these metals from it, in the number of samples No. 1-No. 4 (Figure 10), as follows from the data of their surface control by scanning electron microscopy, with probe microanalysis of selected sites

Which is consistent with the previously obtained results on the leaching of chalcopyrite in the presence of an oxidizing agent [2]: the dissolution of sulfide minerals is presented in the sequence: - dissolution of the outer layer of the mineral at the beginning of the process leads to a short high speed of the process; - diffusion of Cu, Fe – ions from the bulk of the mineral, with their depletion of the surface of the solid phase; - enrichment of the mineral interface with sulfur, leading to a drop in the diffusion coefficients of metals; - continued thickening of the diffusion layer from the surface into the mineral.

From table 3 it follows that as the oxidative leaching of samples of samples No. 1-4 in a sulfate-hypochlorite electrolyte, over a period of 0-1800 seconds, the content of the following elements decreased: Mo - from 50.6 to 4.9%, Cu - from 1,5 to 0%, sulfur - from 28.4 to 9.5%, increased for the following elements: Si - from 2.2 to 16.1%, oxygen - from 12.8 to 42.3%. As follows from Figures 8-10 and Table 3, the time of oxidative leaching of MoC in 1800 sec was insufficient for 100% extraction of Mo into a solution and oxidation of sulfates to oxide-sulfate compounds. Nevertheless, the dynamics of the process are quite clearly demonstrated in the model of 4 samples of this system.

5. Conclusions

The results of leaching of molybdenite concentrate of JSC Almalyksky MMC are compared according to the schemes: the existing calcining-ammonia and non-calcining with nitric acid, hypochlorite opening and leaching. Preliminary oxidative firing was carried out both in the existing drum furnace of the enterprise (MoK granules with kaolin) and in the pilot fluidized bed furnace (granules with an organic binder, leveling the dilution of the cinder according to Mo). In addition to reagent hypochlorite leaching of MoC samples, MoC electrooxidation was used in a non-diaphragm electrolyzer charged initially with 10% NaCl, and the electrolysis time was up to 49 hours. In all cases, complete (> 97%) molybdenum recovery was achieved. In ammonia, Au Ag completely passed into a cake, and in hypochlorite - into solution. Nitric acid leaching led to an intermediate state: Au — in cake, Ag — partially in solution. By the example of oxidative leaching of samples of Mo-concentrate in a hypochlorite electrolyte, for up to 1800 sec, when studying the solid phase formed after reagent exposure, the destruction of sulfide compounds of molybdenum, copper, with the transition of most of them to a solution and the transformation of the remaining fraction from the sulfide to the oxide-sulfate phase, was established.

References

[1] S. Rasulova, V. Guro, M. Ibragimova, E. Safarov. Oxidation and passivation of sulfide ores in gold and molybdenum hydrometallurgy / Conference proceedings: Metal-2018, 27th International Conference on Metallurgy and Materials. May 23rd-25th 2018. Brno, Czech Republic, EU. P. 1442-1447.

[2] Eskandar Keshavarz Alamdari. Selective Leaching-Recovery of Re and Mo from Out-Gas Dust of Molybdenite Roasting Furnace. The Indian Institute of Metals - IIM 2016. DOI 10.1007 / s12666-016-1021-6

[3] Li G., Sun R., Peng Z., Zhou L., Zhang Y. (2016) Volatilization of Rhenium from Molybdenite Concentrate by Oxidative Roasting. In: Hwang JY. et al. (eds) 7th International Symposium on High-Temperature Metallurgical Processing. Springer, Cham. DOI https://doi.org/10.1007/978-3-319-48093-0_12

[4] K. M. Smirnov, N. A. Raspopov, Ya. M. Shneerson, A. Yu. Lapin, G. A. Bitkov, Yu. A. Men'shikov, P. N. Paskhin, and V. P. Kirichenko: Autoclave Leaching of Molybdenite Concentrates with Catalytic Additives of Nitric Acid: Russian Metallurgy (Metally), 2010, No. 4, pp. 11-19. DOI: 10.1134 / S0036029510070025.

[5] RF patent No. 2090637. The method of extraction of precious metals from ores. Gurov V.A .; Priorities: appl. : 1996-07-09; Publ.: September 20, 1997.

[6] RF patent No. 2117709. The method of oxidative autoclave leaching of polymetallic ferrosulfide materials containing precious metals. Korsunsky V.I., Timoshenko E.M., Naftal M.N. et al. 1997-10-13; publ.: 08.20.1998.

[7] RF patent No. 2105824. Method for hydrometallurgical extraction of metals from complex ores. Application Nr: 95116518/02; Date of submission: 07/26/1995; Date of publication: 02.27.1998; Author (s): Robert N. O'Brien [CA];

[8] RF patent No. 2087568 - Method for the extraction of precious metals from ores and materials.

IOP Conf. Series: Materials Science and Engineering **848** (2020) 012076 doi:10.1088/1757-899X/848/1/012076

Gurov V.A .; Priorities: appl.: 1996-07-09; publ. 08/20/1997.
[9] Allabergenov R.D., Akhmedov R.K. Possibilities of hydrometallurgy for processing copper sulfide raw materials // Uzb. chemical J., 2012.-No.5. - P. 49-51.
[10] LIU Zhi-xiong, YIN Zhou-lan, CHEN Yi-guang, XIONG Li-zhi. Leaching kinetics of molybdenum from Ni-Mo ore in sulfuric acid solution with sodium peroxodisulfate as oxidant. J. Cent. South Univ. (2015) 22: 874-879. DOI: 10.1007 / s11771-015-2596-5 /
[11] V. I. Lakshmanan, R. Sridhar, J. Chen, M. A. Halim. A Mixed-Chloride Atmospheric Leaching Process for the Recovery of Base Metals from Sulfide Materials. Trans Indian Inst Met (2017) 70 (2): 463-470 DOI 10.1007 / s12666-016-1004-7.
[12] Potashnikov Yu.M., Lutsik V.I., Lutsik V.A. The behavior of pyrite in solutions of sodium hypochlorite // News of universities. Non-ferrous metallurgy *in Russian). - 1987. - No. 3. - S. 45-48.
[13] Darjaa, T., Okabe, T. H., Waseda, Y., & Umetsu, Y. (1999). Electro-oxidation of Molybdenum Concentrate in Sodium Chloride Solution. Shigen-to-Sozai, (1999), V. 115 (10), Pp. 755-761. doi: 10.2473 / shigentosozai.115.755
[14] Darjaa, T., Okabe, T. H., Waseda, Y., umetsu, Y. (2000). Electro-Oxidation of Molybdenum Concentrate Using Bipolar Cell, and Recovery of Rhenium // Shigen-to-Sozai, (2000), V.116 (6), Pp.520-526. doi: 10.2473 / shigentosozai.116.520
[15] Darjaa, T., Okabe, T. H., Waseda, Y., Umetsu, Y. (2000). Recovery of Molybdenum from Sulfide Concentrate by Electro-Oxidation and Precipitation. Shigen-to-Sozai, (2000), V. 116 (3), Pp. 203–210. doi: 10.2473 / shigentosozai.116.203.
[16] CAO Zhan-fang, ZHONG Hong, JIANG Tao, LIU Guang-yi, WANG Shuai, XIA Liu-yin. Separation of rhenium from electric-oxidation leaching solution of molybdenite: J. Cent. South Univ. (2013) 20: 2103−2108 DOI: 10.1007 / s11771-013-1713-6.
[17] D Pugaev, M Nicol, G Senanayake. The mechanisms of the passivation of sulfide minerals in oxidative leaching processes. Proceedings of 6th Southern African Base Metals Conference, 2011.- P.39-48.
[18] V. P. Guro, F. M. Yusupov, M. A. Ibragimova, F. G. Rakhmatkarieva. The choice of optimal binder for molybdenite concentrate granulation. Tsvetnye Metally (Non-ferrous metals). 2016. No.2. Pp.68-73. DOI: http://dx.doi.org/10.17580/tsm.2016.02.11
[19] Guro V.P., Safarov Y.T., Ibragimova M.A. etc. A method of pelletizing sulfide molybdenite concentrates. UZ patent application for invention. Reg. No. IAP 20170198 dated 05/26/2017. Appl. – IGIC AN RUz.
[20] Safarov Y.T., Sharipov H.T., Guro V.P. Granulation of a molybdenite concentrate with an organic binder, increasing the yield of rhenium seven-oxide during firing // Proceedings of “RHENIUM, TUNGSTEN, MOLYBDENUM - 2016. "Institute GINCVETMET OJSC” conference; March 24-25, 2016, Moscow.- P.84-88.
[21] Rasulova Sitorabonu, Guro Vitaliy, Ibragimova Matluba, Safarov Edgorjon. Oxidation and passivation of sulfide ores in gold and molybdenum hydrometallurgy. Proceedings of the 22nd International Conference on Metallurgy and Materials. METAL 2018 - May 23rd - 25th 2018; Brno, Czech Republic, EU. Issued by Tanger, LTD, P.5.
[22] Cao, Z., Wang, M., Zhong, H., Chen, N., Xia, L., Fan, F., … Wang, S. (2015). Purification of bismuthinite concentrate by selective electro-oxidation of molybdenite. Hydrometallurgy, 154, 95–101. doi:10.1016/j.hydromet.2015.04.012
[23] DARJAA, T., OKABE, T. H., WASEDA, Y., & UMETSU, Y. (2000). Recovery of Molybdenum from Sulfide Concentrate by Electro-Oxidation and Precipitation. Shigen-to-Sozai, 116(3), 203–210. doi:10.2473/shigentosozai.116.203

IOP Conf. Series: Materials Science and Engineering **848** (2020) 012077 doi:10.1088/1757-899X/848/1/012077

Precision human bone prototypes manufactured by 3D printing for training and experimental studies

I D Romanov[1], A D Romanov[1], N A Kulinchenko[1], R O Gorbatov [2]and A A Klenin[2]

[1] Nizhny Novgorod State Technical University n.a. R.E. Alekseev, 603950, Nizhny Novgorod, Minin st. 24, Russia
[2] Institute of Traumatology and Orthopedics, Privolzhsky Research Medical University, 10/1 Minin and Pozharsky Square, Nizhny Novgorod, 603950, Russian Federation
E-mail: nil_st@nntu.ru

Abstract. The work shows the possibility of using additive technologies using 3D printing to build anatomical structures of a person using data of a computer-tomographic examination allows to create precision models taking into account external and internal structure of a biological object.

1. Introduction

The simulation of traumatological and surgical processes is currently performed by computer-aided design methods, which hold a prominent place in the planning and execution of operative interventions. Three-dimensional computer models enable to carry out a structural graphical analysis of the constitution of organs, to determine the patterns of their structure. The first experiment in this field was the American Visible Human Project launched in 1993 at the Colorado University, the USA. Later on other projects were implemented in different countries, namely: Voxel-Man (1998), Visible Korean Human (2001), Chinese Visible Human (2002), Ukrainian Visible Human Project (2005), etc. Nowadays quite a great number of atlases/ programs are available in the public domain. Many educational medical institutions have implemented complexes which enable to work based on the "medicine without patients" principle, that is, to create virtual operating rooms wherein both anatomy is studied and operations are simulated in the interactive mode.

There are actively used 3D-simulation technologies, in particular, CAD (computer-aided design) and CAM (computer-aided manufacturing). Figure 1.shows a knee bone model with guides and pins of a resection block manufactured for planning and performing the total knee arthroplasty. Figure 2.shows the practical application of this guide

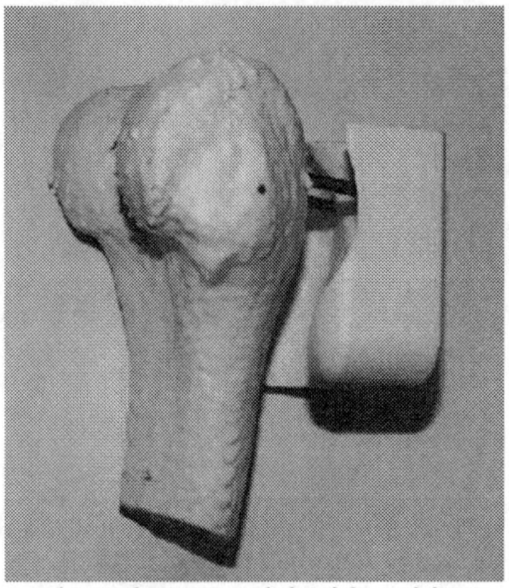

Figure 1. A knee bone model with guides and pins of a resection block.

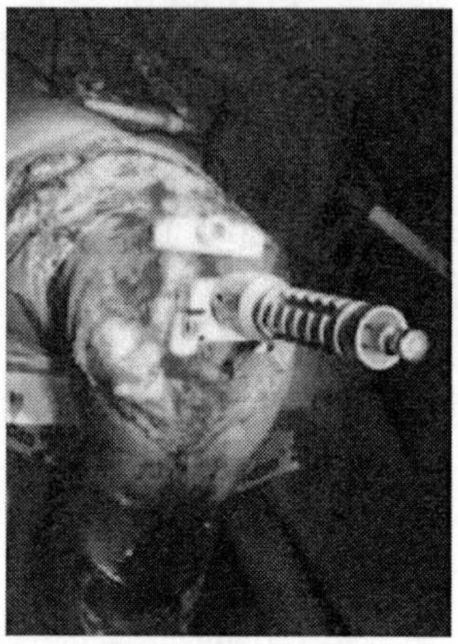

Figure 2. Intraoperative photos of the surgery stage [1].

The finite element method has become widespread for assessing the mechanical properties of the bone tissue [1-6]. Much attention is paid to the study of mechanisms in the bone tissue destruction during the normal life and under extreme loads, in particular, under the exposure to high-speed damaging elements [7]. However, the prediction of bone fractures is complicated by the uncertainty both as regards to mechanical bone properties and loads experienced by different skeleton areas. Furthermore, in reality due to many other factors in certain places of the volume the mechanical bone properties may to a considerable degree differ from average ones, however 3D models are "statistically average" and contain no pathologies whatsoever. In particular, Figure 3 shows an example of a bone joint tomogram.

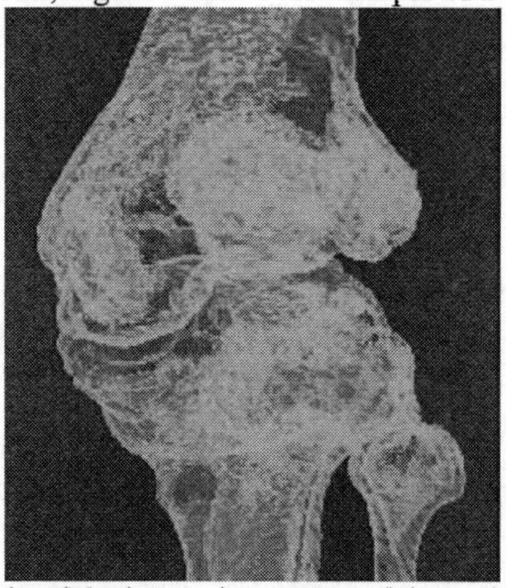

Figure 3. Example of the internal structure of the patient's knee bones.

Different components of the bone tissue have various strength characteristics and diverse destruction mechanisms under dynamic loads: from a brittle to ductile one. A cortical bone is more rigid than a spongy bone, so, it withstands larger stress but less relative deformation before rupture. The spongy bone does not break unless the relative deformation exceeds 75%. The cortical bone is destructed when the relative deformation exceeds 2%. Due to its porous structure the spongy bone has a greater ability to absorb energy. Thus, calculations of the stress-strain state even when using such a highly efficient numerical solution as the finite element method may cause an intolerable error.

Moreover, such predictive models do not correctly assess in some cases any possible consequences for the human body. Furthermore, it is known that the damage probability greatly depends on the choice of shoes both for athletes and in case of striking a mine.That is to say, in [8, 9] test results of the lower leg models of various systems (HIII (N) THOR-Lx (N) MiL-Lx (N)) in shoes (Figure 4) are compared. The development of the Frangible Surrogate Leg (FSL) model shown in Figure 5 was started by the Australia's Defense Science and Technology Organization (DSTO) in the early nineties as a part of its Human Surrogate Program but found the restricted use.

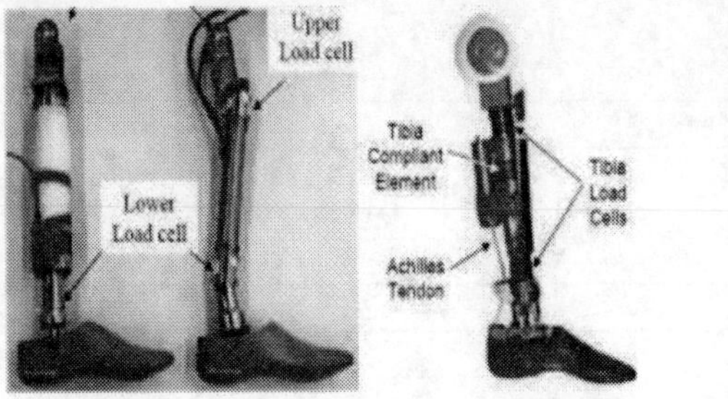

Figure 4. SA surrogate HIII THOR-Lx legs.

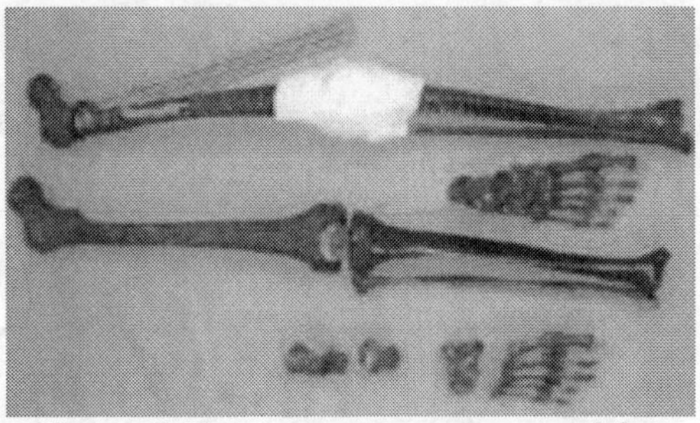

Figure 5. Bone system prototype used for damage analysis [6].

It should be noted that 61% of the body surface and more than 60% of all combat wounds fall on the limbs. Herewith as known the head and neck constitute about 12% of the body surface but they account for up to 20% of combat injuries and result in 47% of mortality. In [10, 11] the development of technique and technology for testing ballistic helmets is described and current approaches are presented to assessing the severity of damages in simulation of mine explosive injuries.

2. Practical Aspect

It is seen from literature data and material safety data sheets that based only on the "tensile strength"criterion the selection of a material as a bone tissue analogue for testing is not at all difficult. But in fact, everything is much more complicated, as some other factors such as loading speed, impact strength, hardness etc. also exert a significant impact.

As initial data for carrying out works the database of the human elemental bone system is used which has been created at the Privolzhsky Research Medical University or a new prototype is manufactured using computed tomography findings. Upon the completion of the computer-aided simulation using the "slicer" program a G-code is generated. The G-code is edited, if needed. Bone models are manufactured by 3D printers with two print-heads using plastics of different mechanical characteristics.

Figure 6 shows the test results of the bone analogue prototype. It should be noted that the sample in Figure 6.a has demonstrated the minimum impact strength of 0.095 MJ/m^2, while during tensile tests this sample (100% filling) had its strength equal to the strength of the material made therefrom.

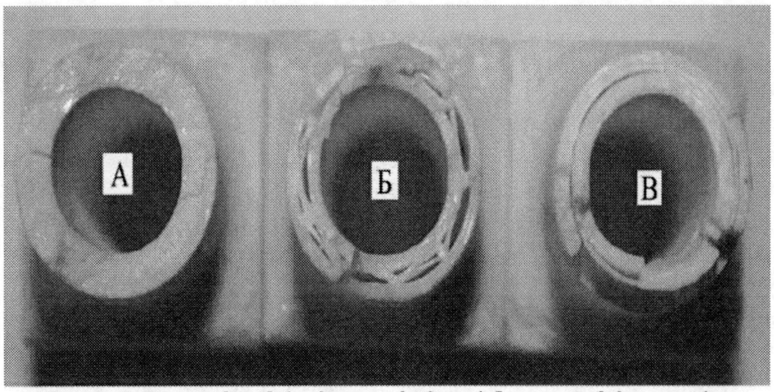

Figure 6. Example of the impact induced fracture of the samples.

It is herein to be noted that the anisotropy of elements manufactured under additive technologies and the part location on the table during 3D printing may crucially affect the experiment outcomes. Table 1 shows the outcomes of experiments in the impact resistance assessment for various design options.

Table 1. Outcomes of Experiments.

Sample number	**Mode of design**				**Impact strength (MJ/m^2)**
	Outside diameter, mm	**Inside diameter, mm**	**Number of layers**	**Filling degree**	
1	15	4,5	2	20	0,141
2	15	4,5	4	20	0,160
3	15	4,5	4	50	0,225
4	15	4,5	0	100	0,025
5	15	9,5	2	50	0,069
6	15	9,5	4	25	0,119
7	15	9,5	4	50	0,175
8	15	9,5	0	100	0,095

It is seen that the mechanical characteristics of the samples are substantially different. It enables to carry out different experiment with the same geometry of the bone model taking into account both the patient's age and diseases

It should be noted that printing with biodegradable materials such as PLA plastic may exert a significant impact on the result of the storage condition of the part after printing.

3. Conclusions

Calculations performed by the finite element method are undoubtedly quite an important area of development. The reproducibility of the results with reality thereat depends on the accuracy of conditions assigned for conducting mathematical simulation and the correctness of the mathematical model as a whole.

A physical experiment using bone models enables to obtain an actual pattern of destruction and in some cases to take into account any factors non-integrated into the mathematical model. Additive 3D printing technology for human anatomical structures using computed tomography findings enable to create precision prototypes taking into account the external and internal structure of a biological object. The creation of a test bench based on additive technologies with due consideration of the structure of the limbs and skull bones will improve the reliability of testing.

Moreover, the use of bone models makes it possible to bring the teaching of students to a new qualitative level, since the practical testing on the model of actions under conditions of the operation will reduce the probability of error.

References

[1] Gorbatov R.O., Malyshev E.E., Romanov A.D., Karyakin N.N. Total knee arthroplasty using virtual prototyping and additive manufacturing (2018) *Sovremennye Tehnologii v Medicine* Volume 10, Issue 3, Pp 146-152

[2] Lamadc W., Glombitza G., Demiris M.A. et al. Virtualle Operations planung in der Leberchirurgie (1999) *Chirurg* V. 70. P. 239-245.

[3] Ohkava M. et al. The role of three-dimensional computer tomography in the management of maxillofacial bone fractures (1997) *Acta Med. Okayama*. V. 51. № 4. P. 219-225

[4] Fedorov. V.D., Karmazanovsky G.G., Tsvirkun V.V. et al. Virtual surgery based on the use of spiral computed tomography (2003) *Surgery*. № 2. Pp 12-17

[5] Yamchikov O.N., Markov D.A., Abdullasyrov R.K., Athaniev D.V., Nenashev A.A. Computer modeling in pre-operational planning in the treatment of femoral fractures (2010) *Journal of Tambov University. Natural and Technical Sciences Series*. T. 15. Issue. 5. Pp 1508-1510

IOP Conf. Series: Materials Science and Engineering **848** (2020) 012077 doi:10.1088/1757-899X/848/1/012077

[6] Tahiri Y., Taylor J.A. In support of using computer-aided design and modeling for periorbital osteotomies (2018) *J. Craniofac. Surg*. Vol. 26, No 1. P. 100-103
[7] Belov N.N., Yugov N.T., Ishchenko A.N., Athaniev S.A.R., Habibullin M.V., Yugov A.A., Stukanov A.L. Mathematical modeling of bone tissue destruction at dynamic loading (2010) *Journal of the Tomsky State University. Mathematics and mechanics*. № 2. Page 28-37
[8] Thanyani Pandelani An investigation of the forces within the tibiae at typical blast loading rates – with different boots // Thesis presented for the Degree of MASTER OF SCIENCE in the Department of Mechanical Engineering in the special field of Impact Biomechanics University of Cape Town June 2014
[9] Bergeron, D. M. et al., 2001. Assessment of foot protection against anti-personnel landmine blast using a frangible surrogate leg. New Orleans, Canadian Centre for Mine Action Technologies.
[10] Chernyshov E.A., Romanov I.D., Romanov A.D., Gorbachev R.O. Modern approaches to the assessment of the severity of injuries in the simulation of mine-explosive injuries (2018) *Scientists of the note of Komsomolskiy-on-Amur State Technical University*. T. 1. No. 2 (34). Page 105-112.
[11] Chernyshov E.A., Romanov I.D., Romanov A.D., Gorbachev R.O. Development of technology and technology for testing ballistic helmets (2019) *Scientists notes of Komsomolsky-on-Amur State Technical University.* № 1 (37). Page 100-107.

IOP Conf. Series: Materials Science and Engineering **848** (2020) 012078 doi:10.1088/1757-899X/848/1/012078

Study of sintered aluminum materials with nanoparticles microadditions

L E Agureev[1], S V Savushkina[1, 2, 4], I N Laptev[1], B S Ivanov[1], A V Ivanov[1], V I Kostikov[3] and Zh V Eremeeva [3]
[1]Keldysh Research Center, Onezhskaya 8, Moscow 125438, Russia
[2]Moscow Aviation Institute, Volokolamskoe Shosse 4, Moscow 125993, Russia
[3]National University of Science and Technology «MISIS», Leninskii av. 4, Moscow 119049, Russia
E-mail: nanocentre@kerc.msk.ru

Abstract. Sintered aluminum materials with small additions (0.01-0.15 vol.%) of aluminum, zirconium and magnesium oxide nanoparticles were obtained by powder metallurgy. The microstructure of the samples and the effect of the nature and concentration of nanoparticles on the matrix grain size, density, microhardness, and corrosion resistance in 10% solutions of nitric and sulfuric acids were studied. The addition of nanoparticles led to matrix grains decrease, an increase in the density, microhardness, and corrosion resistance of sintered aluminum samples. The lowest value of grain size was shown by the sample with 0.1% vol. magnesium oxide. Sintered aluminum material with 0.05% vol. zirconium oxide nanoparticles had the highest density~ 2.67 g/cm^3. The highest Vickers microhardness at the level of 0.474 GPa was obtained for material with the addition of 0.1 vol.% Al_2O_3 nanoparticles. All samples showed significantly greater corrosion resistance in comparison with pure sintered aluminum. The samples with nanoparticles withstood 30 hours of testing, and pure aluminum dissolved in nitric acid in 15 hours, and in sulfuric acid in 10.

1. Introduction

Dispersion-strengthened metals and alloys are often used in areas of technology operating under conditions of high mechanical stress and corrosion. Composites "aluminum - ceramic particles" have a lower specific weight, an optimal ratio of strength and plasticity and sufficient corrosion resistance in combination with high operational mechanical characteristics [1-4]. Oxide particles are stable in the aluminum matrix and block the dislocations movement. As noted in [5, 6], the structure of an interfacial layer is much more complicated in comparison with intergrain boundaries, since neighboring crystallites can have not only different orientations, but also different structures. The structure of the interfacial layer is more complicated if the materials are not heterogeneous, include impurities, surface phases, voids, etc. Nanoparticles have high unrealized surface energy and affect the structure of the grain boundaries (interfacial layers) of the matrix. Even small additives (less than 0.2% vol.) can significantly affect the functional properties of the sintered material. The properties of the hardened metal depend on the uniform distribution of nanoparticles. The greatest hardening is associated with the provision of low aggregation of nanoparticles and their uniformly distribution over the matrix volume. This can be achieved by additions of nanomodifiers in low concentrations. In most works on aluminum and its alloys modification by nanoparticles, the concentration of additives

IOP Conf. Series: Materials Science and Engineering **848** (2020) 012078 doi:10.1088/1757-899X/848/1/012078

exceeds 1% vol [7-16]. However, in recent years, attention has been specifically growing to low concentrations of nanoparticles [1-4, 17-21].

Nanoparticles aggregates located at the grain boundaries of the metal matrix lead to an introduction of defects into the grains surface during a pressing process and penetrate into the metal. The nonequilibrium structure of the grain boundaries of disperse hardened materials is due to the presence of an excess dislocation density in the interfacial layers [22]. In [5] it is noted that defects distributed within the boundaries create internal stress fields that can impede the movement of dislocations in grains and their penetration into the boundaries. To make the deformation caused by external influence, it is necessary that the induced stress field, at which dislocations in the grain move, would be higher than the field σ_i, and also above the level of the short-range fields σ_0 created by dispersed particles, impurity atoms, etc. The condition for the development of intragrain deformation can be expressed as follows: $\sigma \geq \sigma_i + \sigma_0$.

In this work, the method of powder metallurgy was used to study the effect of small nanoparticles additions on the structure, density, microhardness, and corrosion resistance of sintered aluminum-based material.

2. Experimental setup and characterization techniques

An aluminum powder with average particle size 4 μm was used as a material matrix. The specific surface of the powder having a spherical particles was 0.58 m²/g. The oxide film thickness on the aluminum powder particles surfaces measured by X-ray photoelectron spectroscopy was ~ 5 nm. Nanoparticles Al_2O_3, ZrO_2, MgO were used as nanomodifiers (Table 1).

Table 1. Dimensional characteristics of nanoparticles.

Nanoparticles type	Average diameter, nm	Specific surface, m²/g
Al_2O_3	60	27
ZrO_2	50	32
MgO	45	40

Sintered materials were prepared in several stages: screening of the matrix powder (sieve - 60 μm); preparation of nanoparticles suspensions in alcohol; aluminum powder and nanoparticles mixing in ethanol at room temperature under the influence of ultrasound; drying the mixture in air; pressing the mixture at 400 MPa; free sintering in vacuum at 640 °C for 180 min. The structure of materials was studied using optical microscope Carl Zeiss and scanning electron microscope (SEM) Quanta 600. A density of the samples was determined by helium pycnometry using Ultrapycnometer 1200e. After measuring the pycnometric density of the samples, their closed porosity was calculated. The apparent density was measured by pycnometer after the coating of the samples by impervious to the gas varnish. The microhardness of the samples (Vickers method) was measured using Micromet 5114 instrument. Corrosion resistance was estimated by the rate of mass loss in 10% solutions of nitric and sulfuric acids.

3. Results and discussion

The average grain size of the sintered material without the addition of nanoparticles was ~ 7.1 μm. Figure 1 shows the microstructure of aluminum samples with the addition of zirconium oxide nanoparticles in various concentrations, obtained using SEM. Slightly elongated grains, as well as aggregates of zirconium oxide nanoparticles located at the grain boundaries are visible. Grains have various sizes. They are not abnormally large. The smallest average grain size (~ 4.2 μm) corresponds to the nanoparticles content equal to 0.01% vol., figure 1(a). In contrast to a material with 0.01% vol. ZrO_2 in the microstructure of the samples with 0.05% vol. zirconium oxide unevenness in the size distribution of grains is observed, figure 1(b). The SEM image shows large clusters of ZrO_2 nanoparticles located at the grain boundaries. The average grain size of aluminum was ~ 5.6 μm. The closed porosity of these samples was ~ 2.3%. SEM images of the material modified by 0.1% vol. ZrO_2

IOP Conf. Series: Materials Science and Engineering **848** (2020) 012078 doi:10.1088/1757-899X/848/1/012078

also show large clusters of nanoparticles located at the grain boundaries, figure 1 (c). The average grain size decreases to ~ 5.2 μm. The sample with 0.1% vol. nanoparticles has a closed porosity ~2.3%. In the Al + 0.15%vol. ZrO_2 material clusters of zirconium oxide nanoparticles are more uniformly distributed over grain boundaries, figure 1(d). Some clusters have a vermicular shape. This sample has a low closed porosity ~2.2%. The average grain size is ~ 5.4 μm. However, abnormally large grains with a size of 15–20 μm are observed. At a higher magnification, a large number of individual nanoparticles with a size of 30-40 nm located between the matrix grains is found.

Figure 2 shows the dependence of the grain size of aluminum on the content of various nanoparticles. An evaluation of the average grain size for all samples was carried out by calculating the number of crossings of grain boundaries by secants. The secants were applied to the SEM and optical microscope images. The secant must cross at least 10 grains, and at least 50 grains must be in full sight. The number of intersections was counted in the at least five sections. When determining the grain size, the average size L_a, mm was calculated by the formula: $L_a = L/N$, where L was the total length of the secants in same direction, reduced to the plane of the section; N was the total number of grains crossed by secant length L. The smallest grain size (~ 3.2 μm) was obtained for sintered material with additives of 0.1 vol.% MgO.

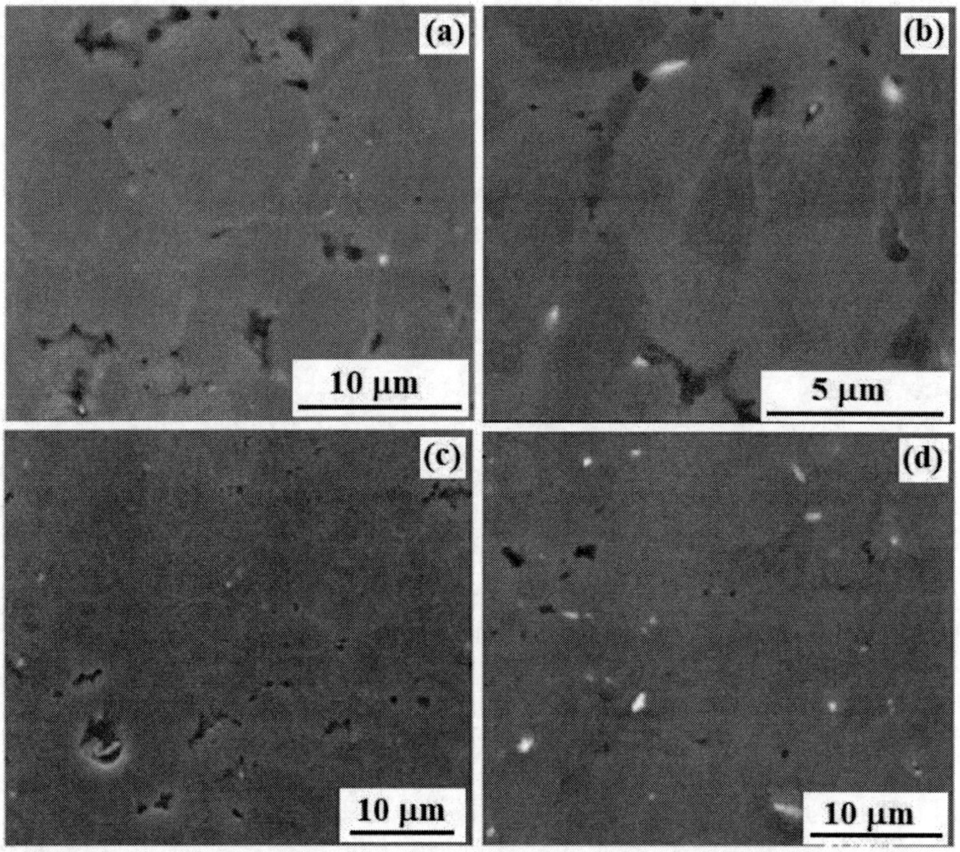

Figure 1. The microstructure of the sintered material with nanoparticles additions in the amount of: a) 0.01vol.%; b) 0.05 vol.%; c) 0.1 vol.%; d) 0.15 vol.%.

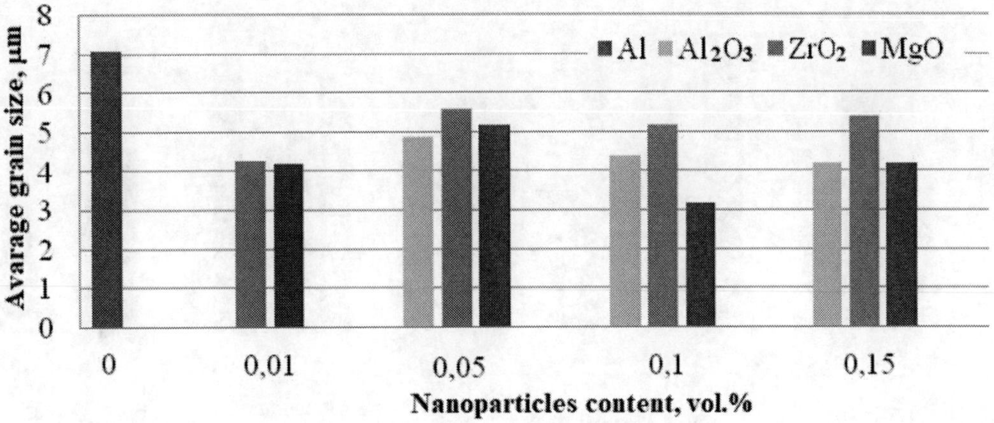

Figure 2. The average grain size of aluminum modified by different type and concentration of nanoparticles.

The density of aluminum samples containing 0.05, 0.1, and 0.15 vol.% alumina nanoparticles was measured before and after sintering. The final density increased with additive amount increase, although the samples with 0.05% vol. nanoparticles had the highest density before sintering, figure 3 (a). The total porosity of the samples after sintering for a concentration of 0.05% vol. Al_2O_3 nanoparticles was ~4.4%, for 0.1% vol. ~ 3.3%, and for 0.15% vol. ~ 4.0%. For sintered Al+ZrO_2 material, the sample modified by 0.05% vol. nanoparticles had the highest density (2.67 g/cm^3) after sintering, figure 3 (b). As in the previous case, before the sintering process the density was higher for the sample with 0.05% vol. Figure 3 (c) shows the dependence of the Al+MgO material density on the content of nanoparticles. An increase in the density of the samples up 2.62 g/cm^3 after sintering was observed for materials with 0.1 and 0.15% vol. The opposite effect was for concentrations of 0.01 and 0.05% vol. which is possibly associated with desorption of the dispersant or moisture at high temperature. Before to sintering the samples with 0.01% vol nanoparticles possessed the highest density.

Figure 4 shows the dependence of microhardness on the content of nanoparticles in sintered materials. The microhardness of sintered aluminum without additives was ~ 0.335 GPa. All types of nanoparticles additives led to an increase in the microhardness of aluminum. The sample with 0.1 vol.% Al_2O_3 nanoparticles showed the maximum value of 0.475 GPa. For material modified zirconium oxide, the microhardness was ~ 0.44 GPa for all samples. The least microhardness was shown by the samples with MgO additives.

All modified materials showed significantly greater corrosion resistance in solutions of nitric and sulfuric acids in comparison with pure sintered aluminum. The samples with nanoparticles withstood 30 hours of testing, and pure sintered aluminum dissolved in nitric acid in 15 hours, and in sulfuric acid in 10. The lowest mass loss rate of 1.1 mg/h was observed by the sample with 0.01 vol. % ZrO_2 additives, figure 5. In sulfuric acid, the material with 0.05 vol.% ZrO_2 nano-additives showed the best result (2.2 mg/h). The worst corrosion resistance in H_2SO_4 the sample with 0.15 vol.% nanoparticles showed. Perhaps this is due to the number and size of defects (cavities) introduced as a result of mixing the nanoparticles with aluminum powder. A nonlinear change in the corrosion resistance of aluminum sintered materials is possibly related to the shape of the interfacial layers around the nanoparticles. Nanoparticles concentrations of 0.01 and 0.05 vol.% lead to an optimal compaction level and a more correct distribution of additives in the matrix, which affects the diffusion of the oxidizing agent along the grain boundaries, preventing corrosion.

IOP Conf. Series: Materials Science and Engineering **848** (2020) 012078 doi:10.1088/1757-899X/848/1/012078

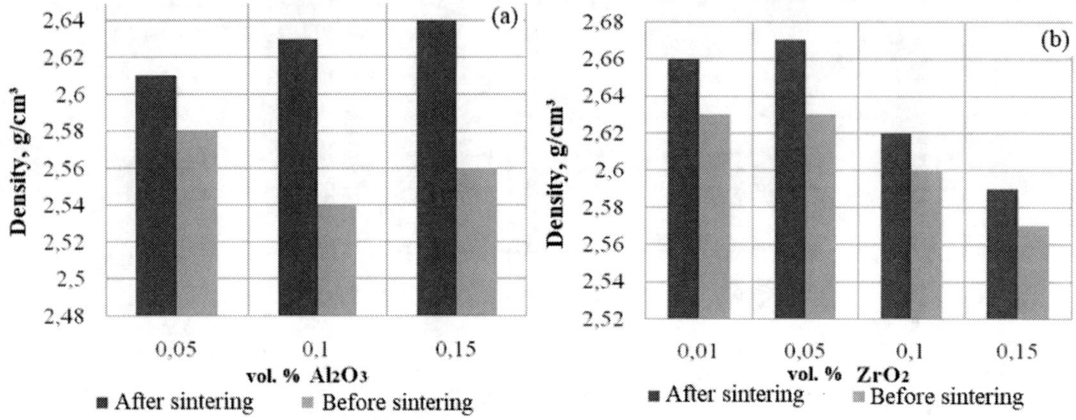

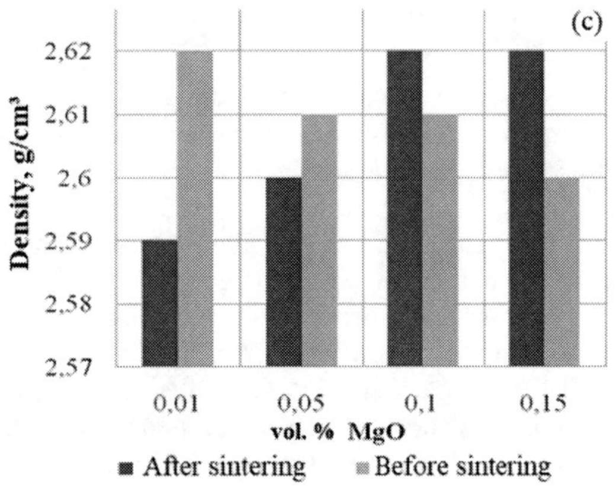

Figure 3. The dependence of aluminum samples density on the content of nanoparticles before and after sintering. a) Al_2O_3, b) ZrO_2, c) MgO.

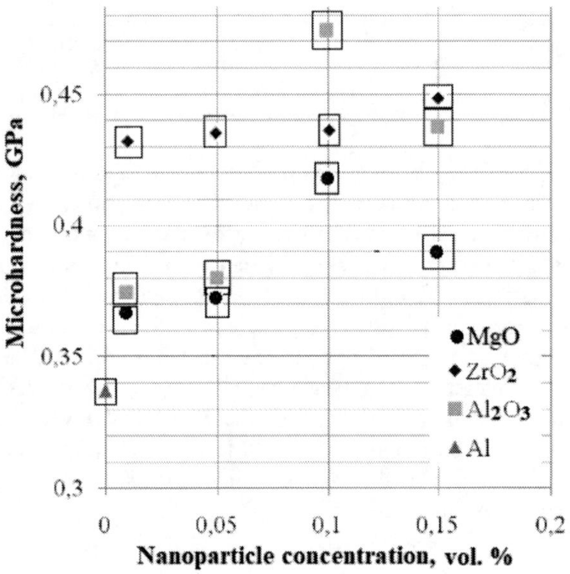

Figure 4. Dependence of aluminum materials microhardness on the nanoparticles content.

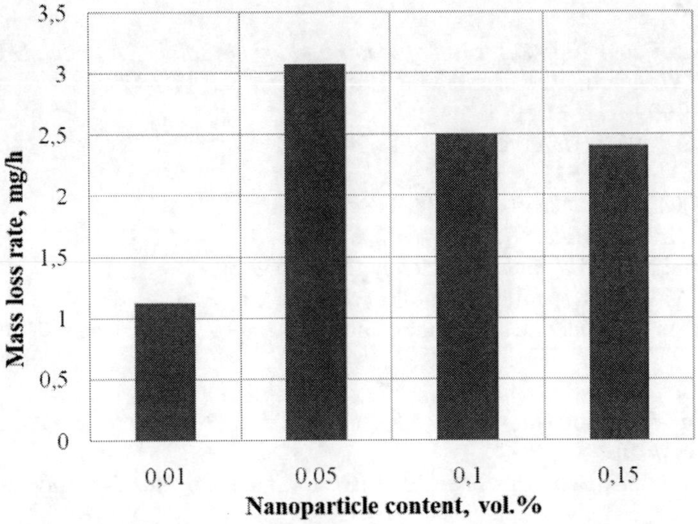

Figure 5. Dependence of the sintered material weight loss rate on the content of ZrO_2 nanoparticles tested in a solution of nitric acid.

Further improvement of microhardness, wear resistance and corrosion resistance of the sintered material can be achieved by metals alloying and applying or growing a functional surface coating, for example, from aluminum oxide [23-24].

4. Conclusions

The addition of nanoparticles led to matrix grains decrease, an increase in the density, microhardness, and corrosion resistance of sintered aluminum samples. According to the distribution of nanoparticles in the matrix, the best results were observed for aluminium materials with a minimum content of nanoparticles (Al_2O_3 - 0.05% vol., ZrO_2 - 0.01% vol., MgO - 0.01% vol.). It is associated with lower aggregation at these nanoparticles concentration and easier conditions for nanoparticles distribution inside the matrix during the mixing process. The average grain size for various aluminum composites was in the range of 3.2-5.6 μm. The lowest value corresponded to the sample with 0.1% vol. magnesium oxide and was 3.2 μm. Sintered aluminum material with 0.05% vol. zirconium oxide nanoparticles was closest to the theoretical density of aluminum and equal to 2.67 g/cm^3. The highest Vickers microhardness at the level of 0.474 GPa was obtained for the sample with the addition of 0.1 vol.% Al_2O_3 nanoparticles. In 10% solutions of nitric and sulfuric acids, the samples with 0.01% vol. zirconium oxide nanoparticles had the least rate of mass decrease equal to 1.1 mg/h. With an increase in the concentration of nanoparticles up 0.05% vol. the rate of weight loss increased to 3.1 mg/h, however, a further increase in nanoparticles to 0.1 and 0.15% vol. led to a decrease in corrosion rate to 2.5 and 2.4 mg/h, respectively. In sulfuric acid, the best result (2.2 mg/h) was showed by the material with 0.05 vol% zirconium oxide nanoparticles.

Acknowledgments
The study was supported by grant from the President of the Russian Federation No. MK-54.2019.8.

References

[1] Agureev L E, Kostikov V I, Yeremeyeva Z V, Barmin A A, Rizakhanov R N, Ivanov B S, Ashmarin A A, Laptev I N, Rudshtein R I 2016 *Inorg. Mater. Appl. Res.* **7** 507-10.
[2] Mironov V V, Agureev L E, Eremeeva Z V, Kostikov V I 2018 *Dokl. Phys. Chem.* **481** 2 110-13.
[3] Agureev L E, Savushkina S V, Laptev I N, Ivanov A V 2019 *J. Phys.: Conf. Ser.* **1396** 012001
[4] Kostikov V.I., Agureev L.E., Eremeeva Z.V. 2015 *J. Non-Fer. Met.* **56** 3 c. 325-28

IOP Conf. Series: Materials Science and Engineering **848** (2020) 012078 doi:10.1088/1757-899X/848/1/012078

[5] Chuvildeev V N 2004 *Nonequilibrium grain boundaries in metals. Theory and applications.* (Moscow: Fizmatlit)
[6] Bokshtein B S, Yaroslavtsev A B 2005 *Diffusion of atoms and ions in solids* (Moscow: MISiS)
[7] Hosking F M, Portillo F, Wunderlin R. Mehrabian R 1982 *J.Mater.Sci.* **17** 2 477-98
[8] Maruyama B, Hunt W P 1999 *JOM* **11** 59-61
[9] Srivatsan T S, Mattingly J 1993 *J. Mater. Sci.* **28** 611 - 20
[10] Pandey A B, Chawla N 1999 *JOM* **511** 11 69-72.
[11] Kang Y C, Chan S L I 2004 *Mater. Chem. Phys.* **85** 438-43
[12] Ma Z Y, Tjong S C, Li Y L 1997 *Mater. Sci. Eng., A* 225 125-34
[13] Mazahery A, Osfadshabani M 2011 *J. Compos. Mater.* **45** 24 2579-86.
[14] Zebarjad S M, Sajjadi S A, Vahid Karimi E Z 2008 *Res. Let. Mater. Sci.* 1-4
[15] Ansary Yar A, Montazerian M, Abdizadeh H, Baharvandi H R 2009 *J. Alloys Compd.* **484** 400-4
[16] Hemanth J 2009 *Mater. Sci. Eng., A* 507 110-13
[17] Sukhodaev P O, Redkin V E, Bogdanova T A, Kuznetsov V A 2017 *J. Siberian Federal University. Eng. Technol.* **10** 3 317-26
[18] Roudini G, Rasti Ghahfarokhi A, Behzadmehr A 2017 *IOP Conf. Series: Mat. Sci. Eng.* **201** 12-26
[19] Lurie S, Volkov-Bogorodskiy D, Solyaev Y, Rizahanov R, Agureev L 2016 *Comput. Mater. Sci.* **116** 62-73
[20] Kishore Babu N, Kallip K., Leparoux M., Al Ogab K A 2016 *Mater. Des.* **95** 534–44
[21] Kallip K, Kishore Babu N, Khaled A. Al Ogab K A, Kollo L, Maeder X, Arroyo Y, Leparoux M 2017 *J. Alloys Compd.* **714** 133-43
[22] Kosevich V M, Ievlev V M , Palatnik L S, Fedorenko A I 1980 *The structure of intergranular and interphase boundaries* (Moscow: Metallurgy)
[23] Agureev L et al. 2018 *Metals* **8** 6 459
[24] Savushkina S V, Agureev L E, Ashmarin A A, Ivanov B S , Apelfeld A V, Vinogradov A V 2017 *J. Surf. Invest.* **11** 6 1154–58

IOP Conf. Series: Materials Science and Engineering **848** (2020) 012079 doi:10.1088/1757-899X/848/1/012079

Research of thermoradiation properties and expansion of tantalum carbide at high temperatures

V N Senchenko[1], R S Belikov[1] and A V Tenishev[2]

[1] Joint Institute for High Temperatures of the Russian Academy of Sciences, 2 Bldg 13 Izhorskaya, Moscow, 125412, Russia

[2] National Research Nuclear University MEPhI (Moscow Engineering Physics Institute), 31 Kashirskoye Shosse, Moscow, 115409, Russia

pyrolab@ihed.ras.ru

Abstract. The article describes an experimental method for studying of the thermophysical properties of refractory carbides, in particular for studying thermal expansion and radiation spectra of the sample during millisecond pulse electrical heating. As a result emission spectra of tantalum carbide were obtained in the temperature range 2400-3000 K. Also the data obtained make it possible to determine the coefficient of thermal expansion and thermoradiation properties of high refractory carbides, in particular for tantalum carbide.

1. Introduction

Carbides of refractory metals are widely used in high-temperature engineering, in particular in the nuclear industry, as well as in aviation and space technology. At the same time, the thermophysical properties of these materials at the temperatures close to the melting point of about 4000 °C for tantalum carbide, such as thermal expansion coefficient (CTE) and emissivity, are not well investigated.

The preferred method for studying the thermophysical properties of refractory carbides at given temperatures is the method of pulse electrical heating. One of the advantages of this method is high measurement accuracy (up to 2 % for CTE measurements). In the case of tantalum carbide, this approach has proven itself in the study of materials made by both sintering and spraying [1].

2. Experimental setup and experimental procedure

The description of experimental setup and characteristics of its components are considered in detail, for example, in [2]. In this article, we restrict ourselves to providing the scheme of experimental setup for electrical pulse heating of refractory electrically conductive materials, see Figure 1.

Some advances were made to the diagnostics for measuring linear expansion of the material. A detailed description of the expansion measurement technique is given below.

The expansion measurement technique used in previous papers [2-3] was modified, which led to a significant increase in accuracy. The measurements of the relative expansion of the sample using the data for cross section of the specimen had an error of the order of 10 %. New approach allows measuring of the relative elongation of the heated sample with an accuracy of 1.5-2 %, which is better for almost the order of previous magnitude.

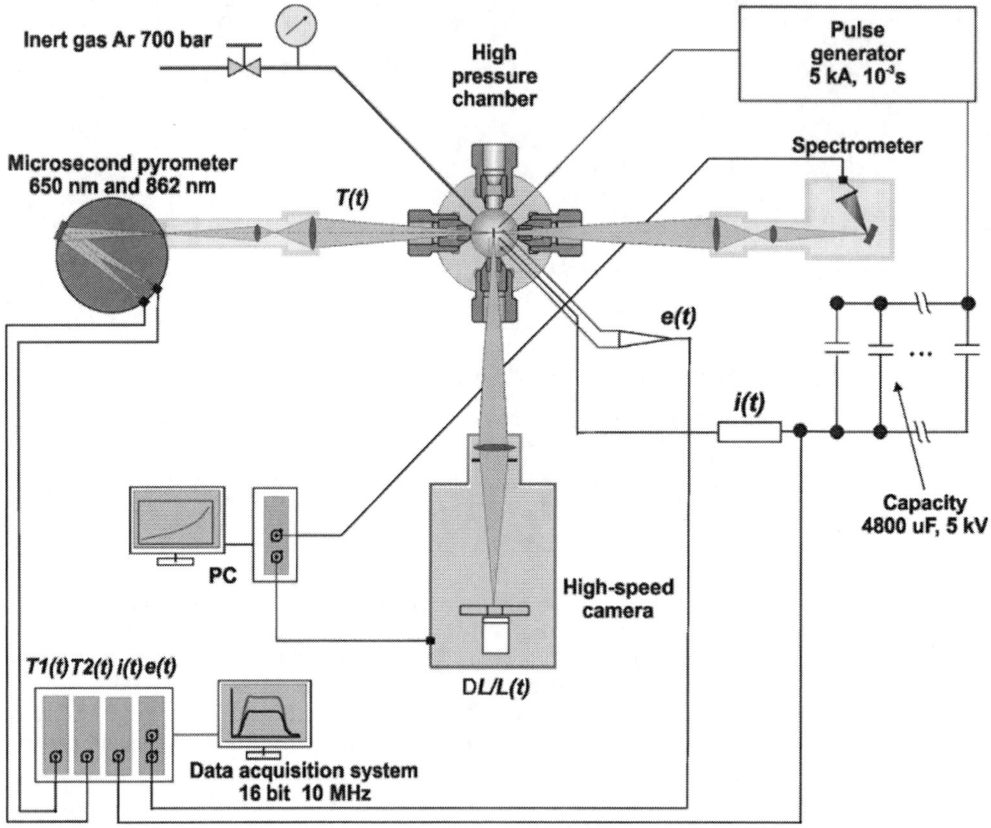

Figure 1. Scheme of the experimental setup.

The main principle of the method used is the measurement of the absolute linear elongation of a heated sample by determining the changing position of one of the collets with which the sample is attached, while the second collet remains stationary. To determine the position of the collet, a pointer is mounted on it, and a laser with a wavelength of 532 nm and a power of up to 0.5 W is used for illumination. The scheme of the experimental technique is presented in Figure 2.

It is important to note that due to the use of the backlight of the pointer allocated to the heated sample, the possibility of optical distortions due to heating of the dense gas layer near the sample (and a change in the refractive index of the gas) during heating is excluded. As was shown earlier [4], these distortions can lead to errors in the measurement of thermal expansion comparable with the magnitude of the expansion itself. Previously we used small protective glass plate for excluding this negative influence of refraction, but presence of the plate drastically (significantly) reduced number of good experiments [4]. Beside that, measurements errors were about 10 % for refractory metals.

Using new elongation measurement technique based on absolute measurement elongation of the heated sample (for the carbide samples, it can be of the order of 0.5 mm near melting temperature), allows effectively reduce the measurement errors in comparison with the method based on the analysis of thermal images of the cross section of the sample that was used in previous works [2-3]. In addition, the use of continuous laser illumination makes it possible to carry out measurements starting immediately from the moment the heating starts using the minimum possible exposure, which also leads to an increase in the number of shooting frames in a single experiment.

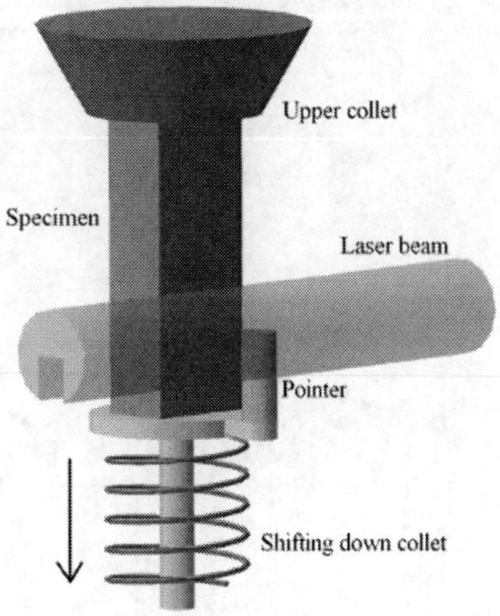

Figure 2. Linear expansion measurement technique.

3. Material and its preparation

For the experiments we used material obtained by sintering from stoichiometric tantalum carbide under pressure. The realized sintering parameters were: maximum temperature - 1950 °C, heating rate - 100 °C / min, pressure - 60 MPa, holding time - 15 minutes, matrix diameter - 15.4 mm.

The initial blanks had the size of ⌀ 15.2 x 1.1 mm and the density of 98 % of the theoretical density, the microhardness was 15.7 GPa. It should be noted that in order to avoid destruction of the material during rapid heating, it was important to achieve the maximum possible sample uniformity and material density close to theoretical. After the initial blanks were made, experimental specimens with the form of a parallelepiped with dimensions of about 1.1 x 1.5 x (9-15) mm were made using a low-speed saw.

Particle size for the initial powder was about 0.8-1.5 μm. Initial chemical composition of the material before the sintering is presented in table 1.

Table 1. Chemical composition of initial powder, weight %.

Total C	Free C	O	N	Al	Ca	Fe	Ti	Na	Nb	Si	S
6.23	0.1	0.12	0.025	0.004	0.003	0.034	0.002	0.005	0.08	0.005	0.005

To study the chemical composition of the material under question, an X-ray analysis of the phase composition and structural characteristics of the starting powder used for sintering of the samples was performed. The diffraction spectra were recorded to determine the phase composition of the samples and assess their structural state by a Bruker D8 Discover diffractometer using CuKα radiation and a LynxEye position-sensitive detector. The records were carried out with the increments of 0.01° along the Bragg angle 2θ and accumulation time of 1 s per detector strip, which in total gives about 170 s for each point in the angle. The phases were identified using the BrukerAXS DIFFRAC.EVA v.4.2 software and the ICDD PDF-2 international database; TOPAS was used to determine the structural characteristics. The X-ray reflection profile was adjusted using the data for the LaB6 reference sample (NIST SRM 660b).

In figure 3 the state diagram of the Ta-C system is depicted. As one can see, tantalum carbides have a wide homogeneity region, which amounts to 41.5–49.5 atomic percent carbon (corresponds to $TaC_{0.73}$ and $TaC_{0.96}$ compounds) [5]. Figure 4 shows obtained diffraction spectrum. The obtained values of the crystal lattice parameter indicate a slight deviation of the studied sample from the stoichiometric composition of TaC, for which it equals to 445.6 pm.

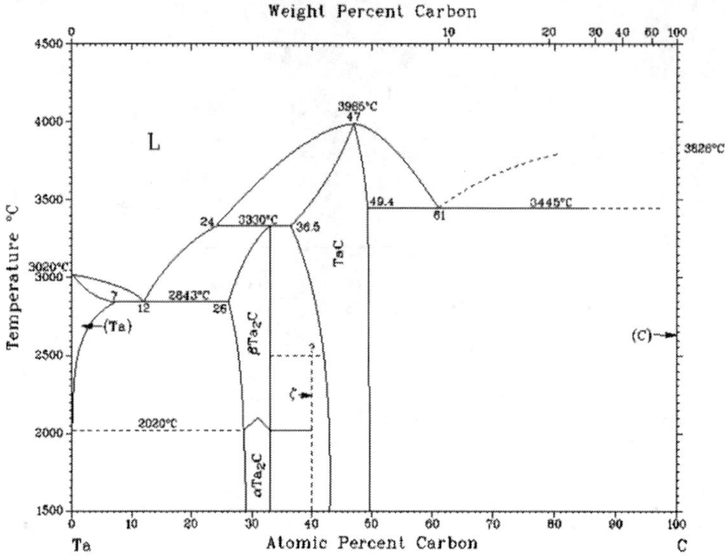

Figure 3. The state diagram of Ta-C systems [6].

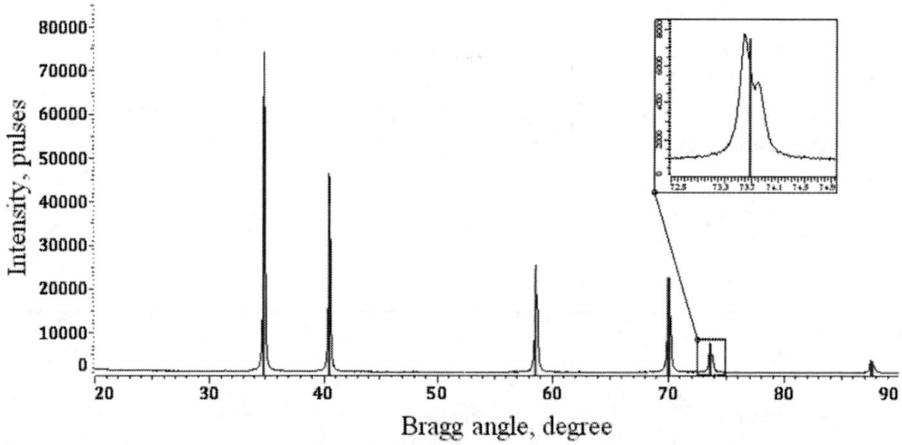

Figure 4. Diffraction spectrum of the TaC studied specimens.

4. Experimental results

A typical view of images captured by a high-speed camera is shown in figure 5.

The sequence of frames for clarity is located from left to right, left frames correspond to the initial stage of the experiment. Measurement of the position of the pointer relative to the vertical axis allows calculate the linear thermal expansion of the investigated material, and, accordingly, the CTE of this material.

During the experiment, emission spectra of tantalum carbide samples were obtained at high temperatures. One of the obtained spectra at a temperature close to 2400 K is shown in figure 6.

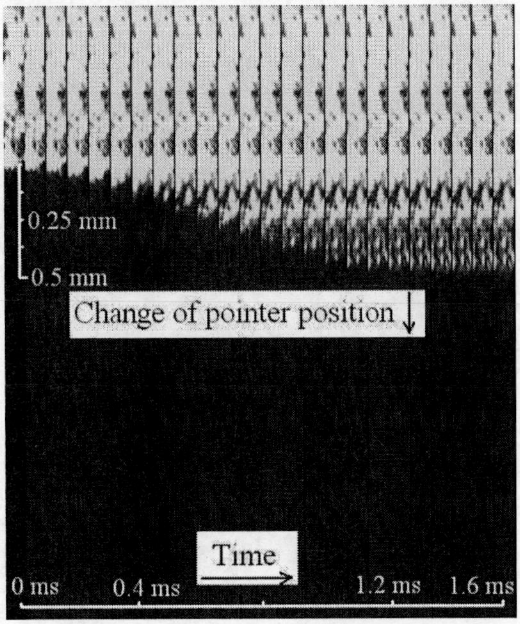

Figure 5. Typical strip photograph showing the pointer shifting while TaC specimen expands during heating.

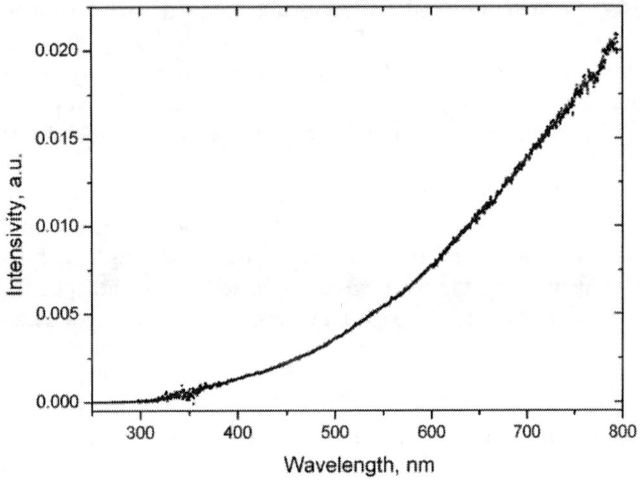

Figure 6. TaC emission spectrum at a temperature of 2437±25 K in the wavelength range of 250-800 nm.

The obtained spectra made it possible to calculate the true temperature of the samples by the method described in detail in [7]. For this, the obtained spectral dependences were recalculated in the so-called Wien coordinates. In this case, the slope coefficient of the straight line approximating the obtained values corresponds to the reciprocal temperature. A graph of one of the spectra obtained at a temperature of approximately 3000 K, plotted in these coordinates, can be seen in figure 7.

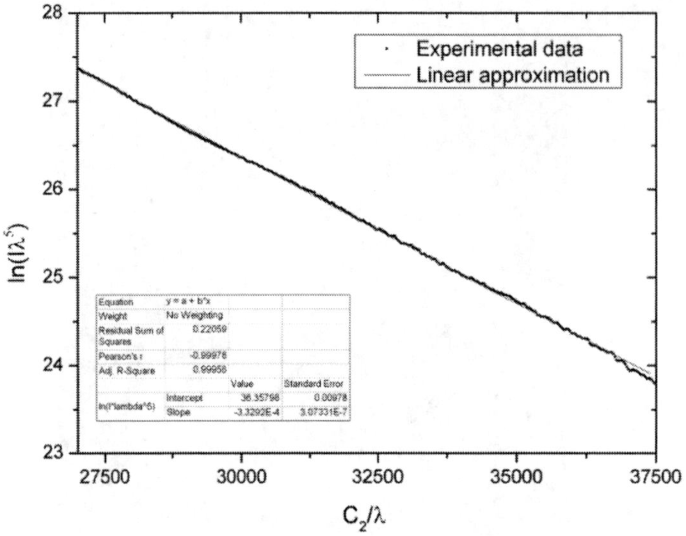

Figure 7. The TaC radiation spectrum at a temperature of 3003±30 K, plotted in the Wien coordinates.

5. Conclusion

The experiments performed demonstrated the applicability of the described method for measuring thermal elongation and radiation spectra of refractory materials during pulse electrical heating. As a result of the experiments, emission spectra of tantalum carbide were obtained in the temperature range 2400-3000 K. The data obtained make it possible to calculate the coefficient of thermal expansion and thermoradiation properties (such as spectral emissivity) of refractory carbides, in particular, tantalum carbide.

The data obtained may be of interest for high-temperature technology, since thermal expansion coefficients are an important parameter in the calculation of heat-shielding structures, and tantalum carbide (along with tantalum-hafnium carbide) is one of the promising materials for use in this field. In addition, with the help of CTE, the temperature dependence of the specific volume of tantalum carbide can be obtained, which is used to construct wide-range equations of state of substances at high temperatures.

Acknowledgments

This work was supported by a grant of Russian Foundation for Basic Research No. 18-08-00923.

References

[1] S V Onufriyev, A I Savvatimskiy, V I Yanchuk 2011 *Izmer. tekh.* **8** 49-52
[2] V N Senchenko and R S Belikov 2019 *J. Phys.: Conf. Ser.* **1147** 012011
[3] V N Senchenko and R S Belikov 2019 *J. Phys.: Conf. Ser.* **1370** 012034
[4] V N Senchenko *et al* 2016 *J. Phys.: Conf. Ser.* **774** 012020
[5] G V Samsonov 1974 Fizicheskoye materialovedeniye karbidov (Naukova dumka)
[6] Pierson H O 1996 Handbook of Refractory Carbides and Nitrides: Properties, Characteristics, Processing and Applications (Westwood, NJ: Noyes Publications)
[7] A N Magunov 2009 *Instrum. Exp. Tech.* **4** 5–28

IOP Conf. Series: Materials Science and Engineering **848** (2020) 012080 doi:10.1088/1757-899X/848/1/012080

Optimization potential of anaerobic biocatalytic processes using intracellular ATP concentration as the main criterion for decision making

O V Senko[1,2], O V Maslova[1] and E N Efremenko[1,2]*

[1] Faculty of Chemistry, Lomonosov Moscow State University, Leninskiye Gory 1/3, Moscow 119991, Russia

[2] Emanuel Institute of Biochemical Physics RAS, 4 Kosygina Str., Moscow 119334, Russia

*E-mail: elena_efremenko@list.ru

Abstract. The effective use of such an analytical indicator as the concentration of intracellular adenosine triphosphate (ATP) in the process of anaerobic transformation of sulfones was shown. The bioluminescent method for determining ATP evaluates to quickly evaluate the state of biocatalysts in such processes, to select the conditions for effective conducted of biotechnological processes and the functioning of cells, as well as media compositions used in the conversion of sulfur-containing compounds.

1. Introduction

Today, in connection with the presence of global environmental problems, research related to the environmentally friendly development of the chemical industry, the development of the best and most feasible technologies, in particular those that could improve the quality of hydrocarbon desulfurization, is relevant [1]. Low sulfur petrochemical feedstocks with sulfur content less than 10 ppm are promising for inclusion in the green schemes for the production of various compounds and materials [2-3]. A promising approach to sulfur recovery from oil content is currently the process of oxidative desulfurization of hydrocarbons [4]. During its implementation, after the stage of chemical oxidation of the feedstock, the oxidized forms of sulfur in the form of polar compounds, including sulfones, are separated from the feedstock by different methods, most often by extraction using different solvents [5]. The question of further rational use of sulfones is still open. In this case sulfones, obviously, can be considered as a source of raw materials for the production of sulfur, necessary for the production of various fertilizers, as well as new polymer materials and composites [6-8]. Today the volumes of "recovered" sulfur produced during the processing of oil and gas are approximately equal and in total make up more than 50% [9]. Organic hydrocarbon skeleton is also valuable in organic sulfones.

Recently, taking into account all mentioned factors, the process of biotransformation of sulfones into commercially valuable products (an inorganic sulfide and biogas) under the action of an anaerobic consortium artificially immobilized in cryogel poly(vinyl alcohol) (PVA) was proposed and successfully tested with model media [5]. It was revealed that the following factors have a significant influence on the productivity of biocatalysts and the implementation of this process: the initial concentration and chemical structure of the substrate (sulfone), and the nature of the solvent, in which organic sulfone is initially placed in the bioreactor. It should be noted that, depending on the chemical structure, the oxidized forms of sulfur in organic compounds, as well as solvents, can have an inhibitory effect on biocatalysts, but at the same time, under optimal conditions, complete biotransformation of sulfones into target products is possible [5]. During the implementation of such anaerobic processes, the duration of which is usually more than 24 hours [10], it is important to conduct a periodic assessment of the state of the biocatalyst in the reactor in order to, if necessary, timely adjust the process parameters and maximize its final productivity for the target products. Such monitoring is also necessary at the end of each working cycle, which allows us to draw conclusions about the possible reuse of biocatalysts.

The concentration of intracellular adenosine triphosphate (ATP) is a universal analytical signal that allows us to assess the viability and metabolic activity of microorganism cells [11-12]. This indicator allows to monitor the inhibitory effect of system components on the target activity of biocatalysts. Determining the value of this parameter to control the viability of biocatalysts during their storage and use in numerous aerobic processes in general has shown its high efficiency [10, 11, 13]. The concentration of intracellular ATP can also be proposed as a universal indicator of the metabolic activity of living cells during the development of methods for anaerobic biological destruction of organic sulfur-containing compounds. Since, when choosing a method for determining this parameter, one should take into account the fact that the ATP concentration in anaerobic cells is much lower as compared to aerobic cells. So, in this study, the ATP analysis was performed using the luciferin-luciferase bioluminescent method, which is characterized by high sensitivity and specificity of determination [14]. It is important to note other advantages of this method: possible usage of portable devices, simplicity of sample preparation, efficiency of measurements [15]. For now the method has been successfully adapted for the analysis of biocatalytic systems based on both free and immobilized cells of various microorganisms. By using this method it was repeatedly confirmed that the immobilization of cells in PVA cryogel makes them more tolerant to the presence of toxicants. Immobilized biocatalysts can be successfully used for biotransformation of renewable waste materials into various commercially significant products (organic acids, biofuels, etc) [16-18].

The aim of this work was to study the possible application of bioluminescent method for determining the concentration of ATP in the cells of an anaerobic consortium during the development and monitoring of the biotransformation of organic sulfones. As the organic sulfone, dibenzothiophenesulfone ($DBTO_2$) was chosen, which is one of the most commonly found components in various hydrocarbon feedstocks, for the purification of which from sulfur can be realized by the method of oxidative desulfurization with the separation of sulfones by extraction [19-20].

2. Materials and methods

2.1. Microorganisms

Anaerobic sludge sample was formed using the Bogatyr biopreparation (Moscow, Russia). Its characteristics were described previously [21].

The bacterial strains *Clostridium acetobutylicum* B1787 and *Desulfovibrio vulgaris* B4053 were obtained from the Russian National Collection of Industrial Microorganisms (www.genetika.ru) for introduction into the methane tank as additives to the anaerobic sludge. The *C. acetobutylicum* strain B1787 was cultivated in the following medium (g/L): glucose – 20; triptone – 10; yeast extract – 5 (pH 6.8). The *D. vulgaris.* B4053 was cultivated in the Postgate medium [14]. Cells were cultivated under anaerobic conditions at 30°C for 1 week. Cultivation of *C. acetobutylicum* and *D. vulgaris* cells was performed under anaerobic conditions in an argon atmosphere at 37°C for 20–24 h and 48 h, respectively.

2.2. *Methods*

2.2.1. Anaerobic fermentation

An artificial anaerobic consortium containing (wt.%): anaerobic sludge - 80%, *Clostridium acetobutylicum* B1787 - 10% and *Desulfovibrio vulgaris.* B4053 - 10% was used as a biocatalyst in the study. Its composition was previously optimized [2, 5, 22]. Each culture was separately immobilized in PVA cryogel [10, 17-18]. The granules of each biocatalyst were mixed in the required proportions and the resulting consortium was used in experiments. When assessing the energetic state of each type of cells in the consortium, the granules were selected by color and analyzed (Figure 1).

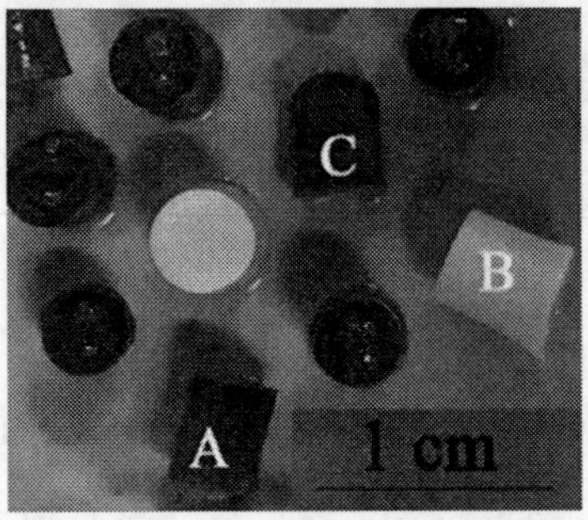

Figure 1. The used anaerobic consortia was a mixture of granules with different cells immobilized in PVA cryogel:

A – anaerobic sludge,

B – *C. acetobutylicum* B1787,

C – *D. vulgaris* B4053.

The initial inoculum concentration in batch reactors was 10% (v/v) for immobilized biomass as part of the sum biocatalyst. The anaerobic incubation was carried out at 32°C in all the experiments. Solution of 0.1 M phosphate buffer with the addition of 1 g/l glucose was used as a nutrient medium.

2.2.2. Determination of adenosine triphosphate (ATP) and sulfide concentration
The concentration of intracellular ATP in immobilized cells was determined by the bioluminescent luciferin–luciferase method using cell-containing granules immediately after sampling. For this purpose, granules were weighed (0.15 ± 0.05 g), transferred to dimethyl sulfoxide (1 mL) and allowed to stand at 25°C for 2 h to extract intracellular ATP [21, 22].

Sulfide ions in liquid phase was monitored spectrophotometrically at 660 nm using Shimadzu UV-1202 (Japan) [23].

2.2.3. Calculations
The data were shown as means of at least three independent experiments ± standard deviation (± SD). Statistical analysis was realized using SigmaPlot 12.5 (ver. 12.5, Systat Software Inc., San Jose, CA, USA).

The significant ($p \leq 0.05$) differences between obtained results were estimated by one-way analysis of variance (ANOVA).

3. Results and Discussion
To assess the feasibility of using the bioluminescent method for determining intracellular ATP in the process of anaerobic transformation of sulfones into inorganic sulfide, three model parts of experimental study were carried out to monitor the metabolic activity of the artificial anaerobic consortia.

3.1 The choice of the initial concentration of sulfone for transformation under the action of anaerobic consortium
Several initial $DBTO_2$ concentrations in the range from 0.5 mM to 2.5 mM initially dissolved in ethanol were tested in an anaerobic reactor (methane tank) for their effective conversion to sulfide. This range of sulfone concentrations was selected from the following considerations: the possibility of efficient transformation of $DBTO_2$ at a concentration of 0.15-0.45 mM under the influence of anaerobic sludge has already been established [5]. The choice of the upper limit of the tested concentration range (2.5 mM) was determined by the analysis of possible initial sulfur concentrations in real raw materials (vacuum gas oil ~ 14,800 ppm) [25] and the final concentrations

of oxidized forms of sulfur-containing compounds in ethanol extracts, taking into account their dilution with the extractant.

$DBTO_2$ at the indicated concentrations was introduced into a nutrient medium with an anaerobic biocatalyst, the mixture was exposed for 24 hours at 35 °C, and then the change in the concentration of intracellular ATP in the cells of the consortium was evaluated (Table 1).

Table 1. ATP concentration ($\times 10^{-12}$ mol/g biomass) of cells composing anaerobic artificial immobilized consortium (biocatalyst) in the process of $DBTO_2$ transformation (24h).

$DBTO_2$ concentration (mM)	Immobilized cells in consortium		
	Anaerobic sludge	*C. acetobutilycum*	*D. vulgaris*
0	57.2±2.2	66.4±3.1	17.3±0.8
0.5	56.2±2.1	64.2±2.6	15.8±0.8
1.0	54.7±2.1	63.4±2.4	18.1±0.8
1.5	56.1±2.9	61.1±2.1	16.7±0.8
2.0	53.8±2.7	54.3±2.2	16.8±0.8
2.5	54.1±2.8	51.3±2.4	17.7±0.8

Analysis of the results allows us to conclude that the most resistant to the toxic effects of $DBTO_2$ are immobilized cells *D. vulgaris*. At a concentration of 2.5 mM $DBTO_2$, all cells as components of the immobilized anaerobic consortium did not show a significant decrease in intracellular ATP, therefore, for this sulfone, it is acceptable as the initial concentration. It is obvious that in the transformation of other sulfones, as well as a mixture of oxidized forms of sulfur-containing compounds in the case of transformation of real raw materials, to select the initial concentration of the substrate or the degree of dilution, it is possible to conduct a similar experiment based on the analysis of changes in the concentration of ATP in cells.

3.2. The choice of extractant for the extraction of oxidized forms of sulfur-containing compounds from the organic phase

A number of solvents that mix well with water and are characterized by a different toxicity class have been tested as candidate extractants, the use of which is potentially possible for extraction of oxidized forms of organic sulfur-containing compounds (sulfones and sulfoxides) from the organic phase. To determine the effect of these potential extractants on the viability and metabolic activity of the cells inside the artificial anaerobic immobilized consortium, the consortium was exposed to presence of the studied substances (50 ml/L) (Table 2).

A slight positive change in the level of ATP in anaerobic sludge was observed as compared with other extractants in the presence of ethyl and isopropyl alcohols (Table 2). This was because of anaerobic sludge cells can use these substances as a substrate, as it is well known [25]. In the presence of water, changes in the concentration of intracellular ATP were within the measurement errors provided by the method. The concentration of intracellular ATP changed most significantly when acetonitrile, dimethylformamide and N-methylpyrrolidone were added to the media with anaerobic consortium.

Thus, it was found that analysis of changes in the concentration of intracellular ATP allows to make conclusions about the possibility and advisability of using one or another extractant for the implementation of the developed biocatalytic processes with the usage of anaerobic consortia, including those based on anaerobic sludge, to assess the energy status of cells, including immobilized cells used as biocatalysts for the conversion of organic sulfones to inorganic sulfide.

Table 2. The concentration of intracellular ATP ($\times$ 10^{-12} mol/g biomass) in the immobilized cells inside the artificially created anaerobic consortium, after their exposure for 24 hours in media with different extractants.

Extractant	Immobilized cells inside consortia		
	Anaerobic sludge	*C. acetobutilycum*	*D. vulgaris*
Acetonitrile	35.8±1.6	32.5±1.4	9.2±0.2
N-methylpyrrolidone	21.2±1.2	27.8±1.9	14.4±0.7
Dimethylformamide	32.4±2.1	25.4±1.1	11.8±0.4
Water	55.4±2.3	62.3±2.9	16.3±0.5
Isopropanol	57.4±2,5	64.9±2.0	15.8±0.6
Ethanol	64.3±2,9	64.7±2.8	16.2±0.5
Control[a]	56.1±2,3	65.6±3.0	17.3±0.8

[a] Control was based on 0.1 M potassium phosphate buffer (pH 7.2).

In this case, during the selection of extractants for the extraction of sulfones from the initial sulfur-containing media, from the point of view of the minimum inhibitory effect on the activity of biocatalysts, the use of ethanol, isopropanol or water turned out to be the most appropriate (Table 2).

3.3. Monitoring the state of the biocatalyst in the process of transformation of oxidized forms of sulfur-containing compounds

Initially dissolved in ethanol, $DBTO_2$ was introduced into a nutrient medium with an anaerobic consortium at a concentration of 2.5 mM, and then the concentration of intracellular ATP was monitored every 24 hours.

Under experimental conditions, the complete transformation of $DBTO_2$ into inorganic sulfide was achieved in 12 days. The level of ATP concentration in the cells as parts of sum anaerobic biocatalyst gradually decreased. At the end of the process, the residual ATP level in anaerobic sludge cells was 62 ± 3% of the initial level, in *C. acetobutilycum* and *D. vulgaris* cells it was 38.2 ± 1.9% and 70.1 ± 3.5% respectively. In this regard, it was decided to activate biocatalysts by exposing them for 24 hours in a nutrient medium of optimal composition. After activation, the concentration of ATP in cells increased to 90-95% of the initial level. After activation, the artificial biocatalyst was reused for biotransformation of the oxidized forms of sulfur-containing organic compounds. It was found that the 24-hour activation of the biocatalyst ensured its reuse in no less than 5 cycles of 12-15 days each, in which a 100% reduction of the oxidized form of sulfur in the form of 2.5 mM $DBTO_2$ to sulfide was observed.

Thus, monitoring the concentration of intracellular ATP allows to quickly respond to observed changes in the energy state of cells and the metabolic activity of the anaerobic biocatalyst used, as well as to evaluate the possible long-term and effective use of the biocatalyst under periodic conditions.

4. Conclusion

It was shown that the bioluminescent method for determining the concentration of intracellular ATP, as a way to control the energy state and metabolic activity of biocatalysts, can be effectively used in the development and implementation of new anaerobic biocatalytic processes.

The following parameters were selected for control within the process of anaerobic transformation of oxidized forms of sulfur-containing compounds using this method: allowable initial substrate ($DBTO_2$) concentration , extractant being non-toxic for the biocatalyst and suitable for extracting oxidized sulfur-containing compound from the organic phase for its transferring to the anaerobic reactor, the favorable mode of the biocatalyst use under periodic conditions (activation within 24 hours after each 12-15 days-cycle).

The results of the study seem to be useful to researchers involved in solving complex problems in the direction of the transition of existing production cycles to green chemical technologies based on the use of various biocatalysts in the form of whole cells.

Acknowledgment
The authors are sincerely grateful to the staff of the Department of Petroleum Chemistry and Organic Catalysis Faculty of Chemistry Lomonosov Moscow State University in the person of the head of the department Professor Anisimov A.V. for $DBTO_2$ provided for these studies.
This work was supported by the Russian Foundation for Basic Research (18-29-05064).

References

[1] Barthe P, Chaugny M, Roudier S and Delgado Sancho L 2015 Best available techniques (BAT) reference document for the refining of mineral oil and gas. *European Commission.* 754 p
[2] Maslova O, Senko O, Stepanov N and Efremenko E 2019 Perspective approaches with the use of biocatalysts for improving the processes of polyaspartic acid production from oil benzene fraction after oxidative desulfurization. *IOP Conf Ser Mater Sci Eng* **525(1)** 012037
[3] Rana M S 2017 Heavy oil refining processes and petrochemicals: A role of catalysis. *RAPSci* **2(1)** 555-580
[4] Koseoglu O R, Bourane A and Kressmann S 2018.*U.S. Patent No. 10,087,377.* Washington, DC: U.S. Patent and Trademark Office.
[5] Senko O, Maslova O, Gladchenko M, Gaydamaka S, Akopyan A, Lysenko S, Anisimov A and Efremenko E 2019 Prospective approach to the anaerobic bioconversion of benzo-and dibenzothiophene sulfones to sulfide. *Molecules* **24(9)** 1736
[6] Zhang Y, Glass R S, Char K and Pyun J 2019 Recent advances on the polymerization of elemental sulphur, inverse vulcanization and methods to functional chalcogenide hybrid inorganic/organic polymers (CHIPs) *Polym Chem* **10** 4078-4105
[7] Oishi S, Oi K, Kuwabara J, Omoda R, Aihara Y, Fukuda T, Takahashi T, Choi J-C, Watanabe M and Kanbara T 2019 Synthesis and characterization of sulfur-based polymers from elemental sulfur and algae oil. *ACS Appl Polym Mater* **1(5)** 1195-1202
[8] Gomez I, Leonet O, Alberto Blazquez J, Grande H J and Mecerreyes D 2018 Poly (anthraquinonyl sulfides): high capacity redox polymers for energy storage. *ACS Macro Letters* **7(4)** 419-424
[9] *Sulfur: history, technology, applications & industry.* Ed G Kutney (Toronto ChemTec Publishing) 2007. 260 p
[10] Senko O, Gladchenko M, Maslova O and Efremenko E 2019 Long-term storage and use of artificially immobilized anaerobic sludge as a powerful biocatalyst for conversion of various wastes including those containing xenobiotics to biogas. *Catalysts* **9(4)** 326
[11] Efremenko E N and Tatarinova N Y 2007 The effect of long-term preservation of bacterial cells immobilized in poly (vinyl alcohol) cryogel on their viability and biosynthesis of target metabolites *Microbiology* **76(3)** 336-341
[12] Efremenko E N, Azizov R, Raeva A and Abbasov V 2005 An approach to the rapid control of oil spill bioremediation by bioluminescent method of intracellular ATP determination *Int Biodeterior Biodegradation* **56(2)** 94-100

[13] Stepanov N, Senko O, Perminova I and Efremenko E 2019 A new approach to assess the effect of various humic compounds on the metabolic activity of cells participating in methanogenesis *Sustainability* **11** 3158

[14] Ismayilov I T, Stepanov N A, Efremenko E N and Abbasov V M 2015 Evaluation of biocidal properties of vegetable oil-based corrosion inhibitors using bioluminescent enzymatic method *Mosc Univ Chem Bull* **70(4)** 197–201

[15] Lomakina G Y, Modestova Y A and Ugarova N N 2015 Bioluminescence assay for cell viability *Biochemistry (Moscow)* **80(6)** 701–713

[16] Maslova O, Stepanov N, Senko O and Efremenko E 2019 Production of various organic acids from different renewable sources by immobilized cells in the regimes of separate hydrolysis and fermentation (SHF) and simultaneous saccharification and fermentation (SSF) *Bioresour Technol* **272** 1-9

[17] Efremenko E N, Nikolskaya A B, Lyagin I V, Senko O V, Makhlis T A, Stepanov N A, Maslova O.V. and Varfolomeev S D 2012 Production of biofuels from pretreated microalgae biomass by anaerobic fermentation with immobilized *Clostridium acetobutylicum* cells. *Bioresour Technol* **114** 342-348

[18] Efremenko E N, Stepanov N A, Nikolskaya A B, Senko O V, Spiricheva O V and Varfolomeev S D 2011 Biocatalysts based on immobilized cells of microorganisms in the production of bioethanol and biobutanol. *Catalysis in industry* **3(1)** 41-46

[19] Akopyan A V, Fedorov R A, Andreev B V, Tarakanova A V, Anisimov A V, Karakhanov E A 2018 Oxidative desulfurization of hydrocarbon feedstock *Rus J Appl Chem* **91(4)** 529-542

[20] Yang Y, Lv G, Deng L, Lu B, Li J, Zhang J and Du S 2017 Ultra-deep desulfurization of diesel fuel via selective adsorption over modified activated carbon assisted by pre-oxidation. *J Clean Prod* **161** 422-430

[21] Senko O, Maslova O, Gladchenko M, Gaydamaka S and Efremenko E 2019 Biogas production from biomass of microalgae *Chlorella vulgaris* in the presence of benzothiophene sulfone. *IOP Conf Ser Mater Sci Eng* **525(1)** 012089

[22] Stepanov N and Efremenko E 2017 Immobilised cells of *Pachysolen tannophilus* yeast for ethanol production from crude glycerol. *New Biotech* **34** 54–58

[23] Trukhina A I, Gladchenko M A and Kalyuzhnyi S V 2011 Optimizations of sulphide and organic modifications of the DEAMOX process. *Appl Biochem Microbiol* **47(9)** 841-845

[24] Akopyan A V, Plotnikov D A, Polikarpova P D, Kedalo A A, Egazar'yants S V, Anisimov A V and Karakhanov E A 2019 Deep purification of vacuum gas oil by the method of oxidative desulfurization. *Petrol Chem* **59(9)** 975-978

[25] *Immobilized cells: biocatalysts and processes*. Ed. E. Efremenko (Moscow RIOR) 2018, 499 p

IOP Conf. Series: Materials Science and Engineering **848** (2020) 012081 doi:10.1088/1757-899X/848/1/012081

Elementoxanealumoxanes – modifiers of silicon carbide ceramic composites components

G I Shcherbakova[1], P A Storozhenko[1], T L Apukhtina[1], M Kh Blokhina[1], D V Zhigalov[1], E A Novokovskaya[1], M S Varfolomeev[1,2], D D Titov[3] and A A Ashmarin[3]

[1]SSC RF "State Research Institute for Chemistry and Technology of Organoelement Compounds" (GNIIChTEOS), 105118, Moscow, Russia
[2]Moscow Aviation Institute (National Research University), 121552, Moscow, Russia
[3]Baikov Institute of Metallurgy and Material Science RAS, 119334, Moscow, Russia
E-mail: galina7479@mail.ru

Abstract. Pre-ceramic organoelementoxanealumoxane oligomers whose pyrolysis results in oxide ceramics of multicomponent composition: aluminum yttrium ($xY_2O_3 \cdot yAl_2O_3$), yttrium aluminum silicate ($xY_2O_3 \cdot yAl_2O_3 \cdot zSiO_2$), yttrium aluminum magnesium ($xY_2O_3 \cdot yAl_2O_3 \cdot zMgO$) ones, are modifiers of silicon carbide based composite materials that increase their resistance to high temperature oxidation.

1. Introduction
One of the most promising approaches to the development of advanced ceramic materials with tailor-made properties (high-strength, high-temperature and oxidation-resistant ceramic composites) is to use pre-ceramic organoelement oligomers and polymers as starting materials [1].

Silicon carbide composites (C/SiC_f и SiC/SiC_f) based on pre-ceramic polycarbosilanes (PCS) are priority for science intensive industries [2, 3]. However, such ceramic composites have a relatively low operating temperature (≈1300 °C) in an oxidizing environment, which significantly limits the application area of the material [1, 4].

It is known that the PCS modification with acetylacetonates $Y(acac)_3$ and $Al(acac)_3$ has a positive effect on the KD-Y and KD-SY fibers produced from YAlPCS, which have high strength and good thermal stability [5,6].

Oxides Al_2O_3, BeO, Y_2O_3, La_2O_3, CaO, MgO, ZrO_2 are most suitable sintering activators for SiC [7]. The use of additives of oxide composition reduces the sintering temperature of SiC and makes it possible to obtain a dense ceramic material, and the combination of oxides (Al_2O_3 and Y_2O_3, Al_2O_3 and MgO, etc.) provides higher stability of the components of the composite material [8, 9].

We previously showed that preceramic organoelementoxanealumoxanes synthesized at GNIIChTEOS are precursors of multicomponent oxide ceramics: $xY_2O_3 \cdot yAl_2O_3$; $xY_2O_3 \cdot yAl_2O_3 \cdot zSiO_2$; $xMgO \cdot yAl_2O_3$, $xY_2O_3 \cdot yAl_2O_3 \cdot zMgO$ and other compositions [10, 11].

2. Experimental details
Using a procedure described previously, we synthesized carbosilanes [1], ceramic-forming organoyttriumoxane alumoxanes [12], containing Al and Y polycarbosilanes (AlYPCS) [13], ceramic-forming organoyttriumoxane alumoxanesiloxanes [14], ceramic-forming organomagnesiumoxane alumoxanes [15], ceramic-forming organomagnesiumoxane yttriumoxanealumoxanes [16].

The main method for the production of oxygen-free silicon carbide fibers modified with Al, Y is the method of melt spinning of fiber-forming polymers to form polymer fibers that have low strength, subsequent thermochemical curing and pyrolysis at temperatures up to 1700 °C [13, 17].

The polymer fibers from them were pyrolyzed in a Nabertherm 50/500/11 tubular furnace under an argon atmosphere during heating to 1100 °C at a rate of 5 °C/min, followed by holding for 1 h and pyrolysis in an electric furnace "Tekhmashservice SVG 4.4.8" at temperatures up to 1700 °C [13, 17].

The organoelementoxane alumoxanes were heat-treated in a SNOL 12/16 resistance-heated electric furnace at 1500 °C.

The surface morphology and elemental composition of the polymer and ceramic fibers from the AlYPCS and the ceramics samples of organoelementoxane alumoxanes were assessed by scanning electron microscopy (SEM) on a Philips SEM 505 equipped with a Sapphire Si(Li) SEM10 energy dispersive detector and a Micro Capture SEM3.0M image capture system and on a Quanta 250.

The thermal analysis of organoyttriumoxane alumoxane siloxanes was carried out in the temperature range from room temperature to 1500 °C in an argon atmosphere in the DSC/TG mode on a NETZSCH STA 409 Luxx.

The ceramics produced from the organoyttriumoxane alumoxane siloxanes were characterized by XRD at room temperature on a Bruker D8 Advance diffractometer with monochromatized Cu$K\alpha 1$ radiation in reflection geometry. XRD intensities were measured by a Lynx eye position sensitive detector. The phases present were identified using ICDD PDF-2 and Inorganic Crystal Structure Data base (ICSD). Phase compositions were determined by the Rietveld profile analysis method.

Phase compositions ($xY_2O_3 \cdot yAl_2O_3 \cdot zMgO$) were determined by X-ray diffraction at room temperature on a Shimadzu XRD-6000 vertical X-ray diffractometer using monochromatized copper radiation with a wavelength $\lambda_{K\alpha cp} = (2\lambda_{K\alpha 1} + \lambda_{K\alpha 2})/3 = 1.54178$ [Å]. The crystalline and X-ray amorphous phases present were identified using ICDD PDF Release 2003 data.

Using a procedure described previously, we prepared film-forming agents and binders based on AlYPCS [13], organoyttriumoxane alumoxane siloxanes [18] and high-purity ceramics based on oxides of aluminum, yttrium and magnesium [19].

3. Results and discussion

The researches of GNIIChTEOS synthesized AlYPCS with the desired Al:Y molar ratio by cocondensation of raw PCS and organoyttriumoxane alumoxane oligomers, they are the base for polymer fibers [13].

It is demonstrated that the introduction of only 0.5 wt. % of organoyttriumoxane aluminoxane oligomers leads to the production of AlYPCS with good fiber-forming properties: softening temperature $T_1 = 210$ ° C, fiberization temperature $T_2 = 230$ ° C and dropping point $T_3 = 320$ ° C.

The pictures of AlYPCS fibers (figure 1: **a** – polymer, **b** – after curing and pyrolysis at 1100 °C, **c** – after pyrolysis at 1700 °C) and X-ray elemental microanalysis obtained by electron microscopy are presented below (figure 1).

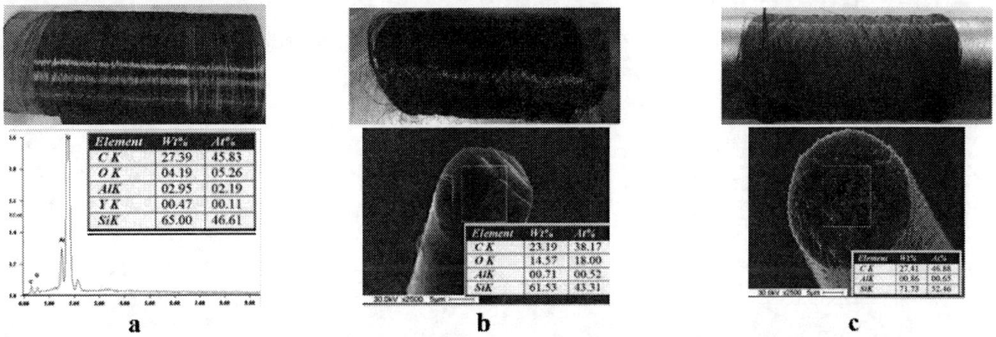

Figure 1. AlYPCS fibers (**a** – polymeric, **b** – after curing and pyrolysis at 1100 °C, **c** – after pyrolysis at 1700 °C) and X-ray elemental microanalysis.

IOP Conf. Series: Materials Science and Engineering **848** (2020) 012081 doi:10.1088/1757-899X/848/1/012081

Moreover, a solution of AlYPCS in toluene (concentration of 50 wt. %) was used as a binder, imparting plasticity and moldability to the ceramic paste. In addition, in the production of ceramics by cold molding from micropowders of silicon carbide SiC, it increases its oxidative stability. Ceramic samples after high-temperature sintering at 1500 °C in air, where: a – SiC sample {AlYPCS binder - 10 wt. % of yttriumoxanealumoxane), b – SiC sample {AlYPCS – 0.5 wt. % of yttriumoxanealumoxane), c – SiC sample (PCS binder) are presented below (figure 2).

a b c

Figure 2. Ceramic SiC samples on binders: a – AlYPCS - 10 wt % of yttriumoxanealumoxane, b – AlYPCS – 0.5 wt. % of yttriumoxanealumoxane, c – PCS.

Film-forming compositions based on organoyttriumoxane alumoxanesiloxanes were used to prepare protective high-temperature antioxidant coatings of Y_2O_3–Al_2O_3–SiO_2 composition on silicon carbide SiC fibers (figure 3) [1, 18].

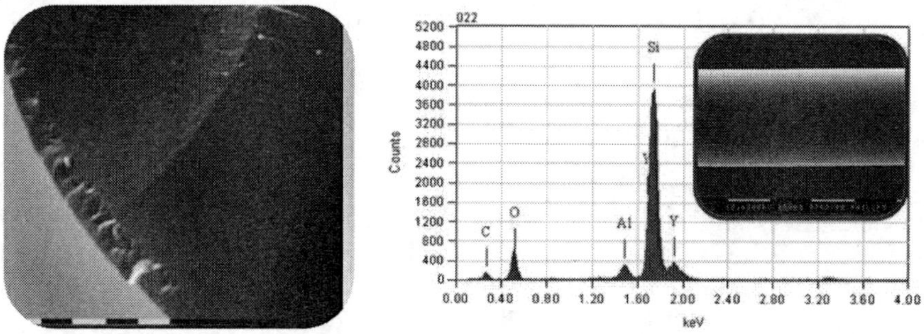

Figure 3. Protective coatings of Y_2O_3–Al_2O_3–SiO_2 composition on SiC fibers.

Protective coatings consist of crystalline phases with high melting points ($Y_2Si_2O_7$ – 1790 °C, $3Al_2O_3{\cdot}2SiO_2$ – 1870 °C) (figure 4), low heat conduction; temperature factor of linear expansion close to that of silicon carbide (38–$45{\cdot}10^{-7}$ K^{-1}); oxidation resistance at temperatures up to 1700 °C [18].

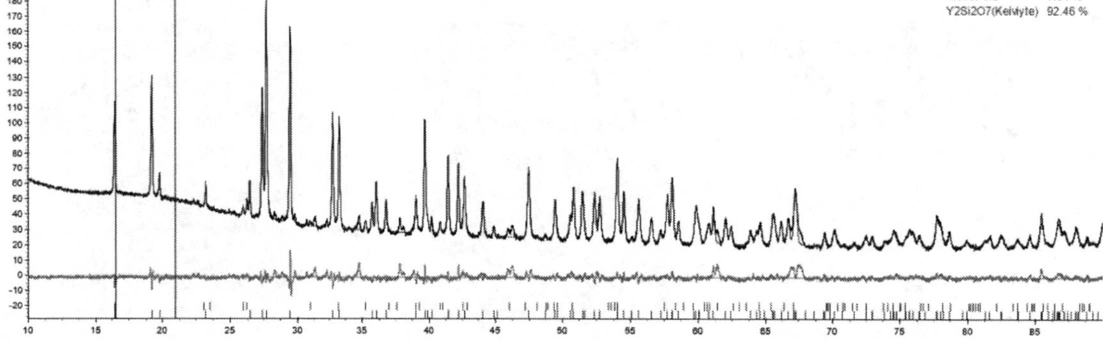

Figure 4. Diffractogram of organoyttriumoxane alumoxanesiloxane after pyrolysis at 1300 °C.

IOP Conf. Series: Materials Science and Engineering **848** (2020) 012081 doi:10.1088/1757-899X/848/1/012081

Organoyttriumoxane alumoxanesiloxanes up to 1500 °C (figure 5a) and the oligomer pyrolyzed at 700 °C to 1600 °C (figure 5b) were analyzed by DSC. The DSC results of pyrolyzed yttriumoxane alumoxanesiloxane at 700 ° C show (figure 5b) that the composition crystallizes first at a temperature of 1101 °C, then at 1365.7 and at 1403.8 °C. Melting of all crystalline phases is observed at 1418.3 °C, that is, glass formation occurs at this temperature, which confirms the possibility of the formation of a dense non-porous oxide coating.

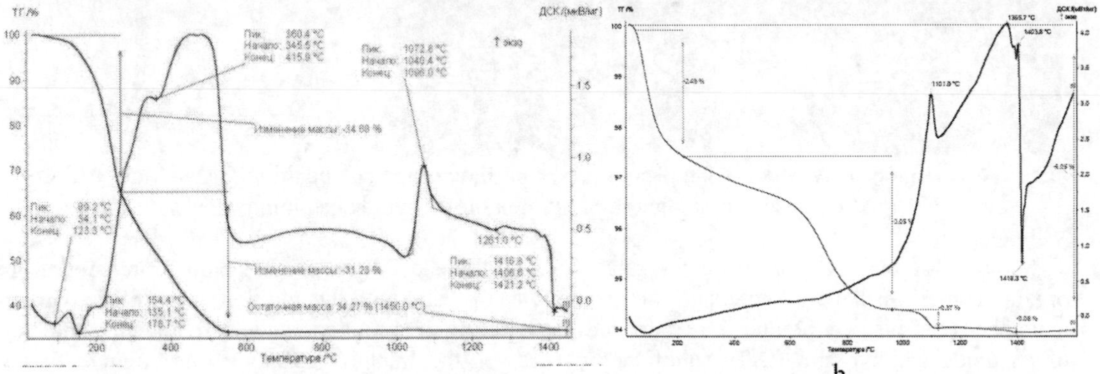

a b

Figure 5. A thermogram of organoyttriumoxane alumoxanesiloxane (a) and DSC results of yttriumoxanealumoxanesiloxane pyrolyzed at 700 °C (b).

The organomagnesiumoxane yttriumoxanealumoxanes have been synthesized at GNIIChTEOS [16], which are the organoelement precursors of oxide nanoceramics of the composition: $MgO-Al_2O_3-Y_2O_3$ (actual SiC sintering additives [8, 9]), This result opens up new prospects for the development of high-density modified ceramics based on silicon carbide [19].

The pyrolysis of organomagnesiumoxane yttriumoxanealumoxanes with a molar ratio of Al: Mg≈2 and Al: Y≈6 leads to the formation of ceramics of $MgAl_2O_4$ и $Y_3Al_5O_{12}$ composition according to the scheme:

$$[Mg(OR^*)O]_k \cdot [Al(OR)_l(OR^*)_g(OR^{**})_x(OH)_zO_y]_n \cdot [Y(OR^*)_s(OH)_tO_r]_m \xrightarrow{20\text{-}750^{\circ}C} \underset{\text{amorphous}}{kMgO \cdot nAl_2O_3 \cdot mY_2O_3} \xrightarrow{750\text{-}1500^{\circ}C} \underset{\text{spinel}}{MgAl_2O_4} \cdot \underset{\text{YAG}}{Y_3Al_5O_{12}}$$

The diffractogram, surface morphology and elemental composition of the ceramic sample follow (figure 6).

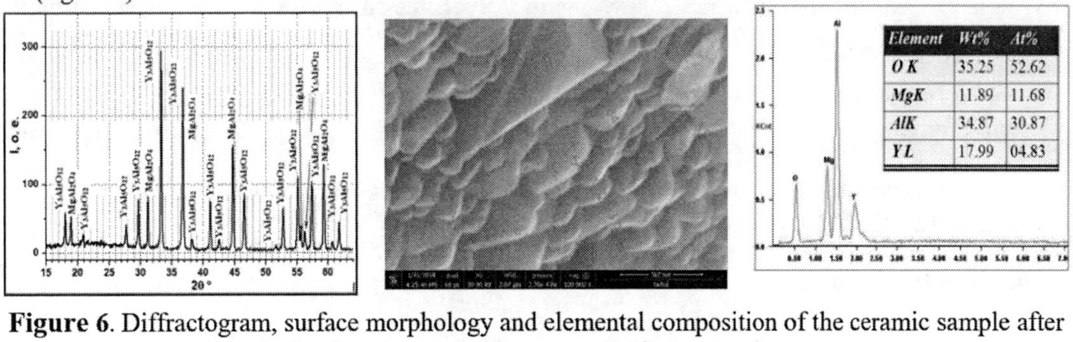

Figure 6. Diffractogram, surface morphology and elemental composition of the ceramic sample after pyrolysis of organomagnesiumoxane yttriumoxanealumoxane.

Dense SiC ceramics modified with magnesium, yttrium and aluminum oxides from micropowders of silicon carbide and a mixture of magnesium, yttrium and aluminum oxides (≈10 wt. %) with the addition of organomagnesiumoxane yttriumoxanealumoxane binder imparting plasticity and moldability to the ceramic paste was produced by high-temperature sintering of samples prepared by cold molding. Samples of ceramics with low porosity and high density, presented below were prepared. They are shown below (figure 7).

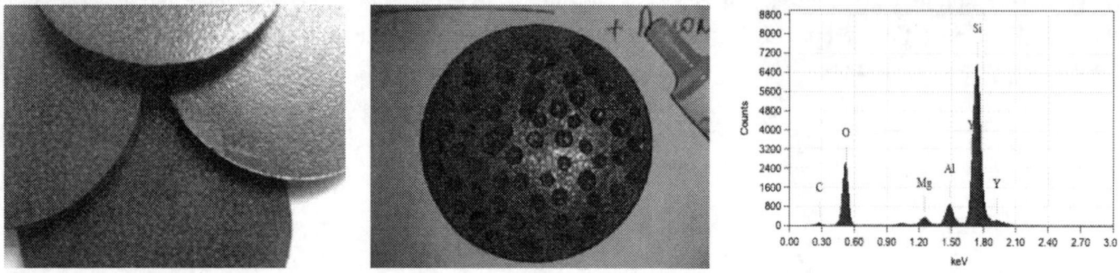

Figure 7. High density ceramic samples of mixed carbide-oxide composition: SiC + MgO + Y_2O_3 + Al_2O_3 based on organomagnesiumoxane yttriumoxanealumoxane.

Samples of the SiC composite with an oxide additive of eutectic composition prepared from organomagnesiumoxane yttriumoxanealumoxane (OMYA) and with the addition of multilayer carbon nanotubes (CNTs) were produced by spark plasma sintering. It is found that the combined introduction of an oxide additive and CNTs significantly increases the fracture resistance and strength of the composite [19]. Table 1 presents the test results of the samples for mechanical properties.

Table 1. Results of mechanical testing of samples

No	Material	Load P, g	Vickers hardness (HV), HPa	Fracture toughness coefficient C_{1C}, $MPa \cdot m^{1/2}$	Flexing strength σ_{fl} MPa
1	SiC	300	12.5	3.77	100
2	SiC + 10% CNTs	300	23.7	4.26	137
3	SiC + 0,5% OMYA	300	30.3	4.84	220

Conclusion

It is found that pre-ceramic organoelementoxanealumoxane oligomers, whose pyrolysis results in multicomponent oxide ceramics, are precursors for obtaining modified components (fibers, binders, coatings, powders) of silicon carbide ceramic composites with increased resistance to high-temperature oxidation, in particular:

- organoyttriumoxanealumoxanes – PCS modifiers; ceramic fibers AlYSiC were prepared on the base of AlYPCS, and AlYPCS binder was produced for press molding silicon carbide SiC micropowders;
- organoyttriumoxane alumoxanesiloxanes – precursors for film-forming compositions to prepare protective high-temperature antioxidant glass-ceramic ($Y_2Si_2O_7$ + $3Al_2O_3 \cdot 2SiO_2$) coatings on SiC-fibers;
- organomagnesiumoxane yttriumoxanealumoxanes – precursors of binding compositions $xY_2O_3 \cdot yAl_2O_3 \cdot zMgO$ to prepare high density ceramic samples of combined carbide-oxide composition and precursors of amorphous powder s – sintering additives of SiC-composite with higher fracture toughness and strength.

Acknowledgements

The work was financially supported by RFFI. Project 17-03-00331 A.

The work was financially supported by the Grant of President of the Russian Federation. Project MK-39.2019.3

IOP Conf. Series: Materials Science and Engineering **848** (2020) 012081 doi:10.1088/1757-899X/848/1/012081

References

[1] Abakumov G A et al 2018 *Russ. Chem. Rev.* **87** (5) 393–507
[2] Joo Y J, Khishigbayar K-E, Cho KY and Kim C J 2018 *Fibers and Polymers* **19** (9) 1806–12
[3] Mera G, Gallei M, Bernard S and Ionescu E 2015 *Nanomaterials*. **5** 468–540
[4] Flores O, Bordia R K, Nestler D, Krenkel W and Motz G 2014 *Adv. Eng. Mater.* **16** (6) 621–36
[5] Yang D, Song Y 2008 *Key Eng. Mater.* **368–372** 827–30
[6] Bai W, Jian K 2019 *IOP Conf. Ser.: Mater. Sci. Eng.* **2019** – 490 022057
[7] Ortiz A L, Cumbrera F L, Sánchez-Bajo F, Guiberteau F, Xu H, Padture N P 2000 *J. Am. Ceram. Soc.* **83** (9) 2282–6
[8] Johari M I A, Mustapha M, Mamat O, Kakooei S, Ginta T L, Ismadi M Z P 2016 *J. Eng. Appl. Sci.* **11** 13300–4
[9] Zharikov E V, Kapustin V V, Faikov P P, Popova N A, Barmin A A, Ivanov A V 2017 *IOP Conf. Ser.: Mater. Sci. Eng.* 2017 – **175**. 012065
[10] Shcherbakova G I et al. 2018 FNM2018. *IOP Conf. Ser.: Journal of Physics*: Conf. Series 2018 – **1134** 012054
[11] Shcherbakova G I, Storozhenko P A, Apukhtina T L, Zhigalov D V, Varfolomeev M S, Drachev A I, Ashmarin A A 2019 NMPT-4. *IOP Conf. Ser.: Mater. Sci. Eng.* 2019 – **525** 012057
[12] Shcherbakova G I et al 2012 *Inorg. Mater.* **48** (10) 1058–63
[13] Shcherbakova G I et al 2019 *RF Patent Application* 2019115647
[14] Shcherbakova G I et al 2014 *Inorg. Mater.* **50** (3) 306–13
[15] Shcherbakova G I, Storozhenko P A, Apukhtina T L, Varfolomeev M S, Kuznetsova M G, Drachev A I and Ashmarin A A 2017 *Polyhedron* **135** 144–52
[16] Shcherbakova G I et al 2019 *Inorg. Mater.* **55** (10) 1068–78
[17] Shcherbakova G I et al 2014 *Inorg. Mater.* **50** (4) 423–30
[18] Shcherbakova G I et al 2014 *Inorg. Mater.* **50** (6) 686–91
[19] Zabelina A A, Shcherbakova G I, Faikov P P, Zharikov E V 2019 *Ceramics International* (In Press CERI-D-19-06846)

IOP Conf. Series: Materials Science and Engineering **848** (2020) 012082 doi:10.1088/1757-899X/848/1/012082

Study of high-frequency magnetic properties of Fe-Ti-B films obtained by magnetron sputtering

E N Sheftel[1], E V Harin[1], V A Tedzhetov[1,*], G Sh Usmanova[1], S Y Bobrovskii[2], K N Rozanov[2], P A Zezyulina[2] and Ph V Kiryukhantsev-Korneev[3]

[1] A A Baikov Institute of Metallurgy and Materials Science, Russian Academy of Sciences, Moscow, Russian Federation
[2] Institute for Theoretical and Applied Electromagnetics, Russian Academy of Sciences, Moscow, Russian Federation
[3] National University of Science & Technology (MISIS), Moscow, Russian Federation
* corresponding author, e-mail: vtedzhetov@imet.ac.ru

Abstract. The Fe-Ti-B films were obtained by dc magnetron sputtering of the Fe + 15% TiB_2 and Fe + 30% TiB_2 targets. The chemical composition and structure were studied in the as-sputtered state and after vacuum annealing at 500°C. The static (saturation magnetization M_s and coercive field H_c) and rf magnetic properties (magnetic permeability μ', frequency of natural ferromagnetic resonance f_r and the frequency range in which μ' is kept) were determined.

1. Introduction

The main trends in modern magnetic microelectronics are miniaturization and performance, while ensuring efficient operation in the MHz and GHz frequency ranges of magnetic fields. Therefore, the development of new magnetic materials characterized by the properties that ensure the implementation of these trends is the most relevant task of materials science. The revolutionary changes in magnetic microelectronics that began in the mid-90s of the last century were largely initiated by experimental demonstration that nanocrystalline ferromagnets have unique magnetic properties [1]. This led to the development and the application of a number of Fe-based nanocrystalline alloys obtained by the quenching from the melt in the form of ribbons 10-50 μm thick [2].

At the same time, the scientific community showed interest in a new class of nanocrystalline soft magnetic alloys of Fe-Me-X systems (Me is one of the transition metals of IV group of the Periodic Table, X is one of the light elements N, C, O, B), obtained in the form of a film [3-5]. Particular interest in such films was shown in connection with the prospect of their use in miniature and high-performance audio- and video-equipment devices as magnetic cores of recording/playback heads [6]. For a new class of films, the physicochemical approach to the selection of chemical composition and structure formation conditions was first formulated in [7], which, as was shown later on the example of Fe-Zr-N films [8], provides a combination of high saturation magnetization M_s, low coercive field H_c and high hardness.

The approach consists in choosing a eutectic alloy in the Fe-MeX quasibinary system, producing films of this alloy in an amorphous or cluster state by magnetron sputtering, and a subsequent annealing, which forms a nanocrystalline structure in the films. The nanocrystalline structure should be represented by the main ferromagnetic Fe-based phase and the inclusions of the non-ferromagnetic thermodynamically-stable hard intercalation MeX phase (Fe/MeX nanocomposite structure).

The described approach was applied to soft-magnetic Fe-based bulk alloys obtained by casting. As a result, alloys were developed, including alloy (Fe - 9.5% wt. Si - 5.5% wt. Al) - TiB_2, characterized by a complex of magnetic (high saturation magnetization and magnetic permeability in the MHz frequency range) and mechanical properties (hardness, wear resistance), not achieved on other bulk soft-magnetic alloys [9-10].

Planar technologies, in particular magnetron deposition, make it possible to obtain films of controlled thickness and integrate them into multilayer structures of an electronic device in one technological process, which greatly simplifies the manufacturing technology of a device and reduces its cost. The structure of the films is controlled by the magnetron deposition parameters and/or the subsequent annealing. As applied to Fe-Ti-B films, studies were started by us relatively recently [11].

The present work is devoted to the study of static (saturation magnetization M_s and coercive field H_c) and rf magnetic properties (magnetic permeability μ', natural ferromagnetic resonance frequency f_r and frequency range in which μ' is kept) of films with a two-phase structure of the Fe-TiB_2 nanocomposite type.

2. Objects and methods

The Fe-Ti-B films were prepared by the dc magnetron sputtering of composite targets in Ar atmosphere at a pressure of 0.2-0.5 Pa (other parameters of the deposition process are given in Table 1). Two different composite targets with TiB_2 ceramic chips, uniformly distributed within the target erosion zone, covering 15 or 30% of the target surface (Figure 1), were used. The TiB_2 chips were produced by self-propagating high-temperature synthesis. The time of deposition of the films was from 8 to 25 min. The glass (glass slide 1.5 mm thick and cover glass 0.13 mm thick) and Ti substrates were used.

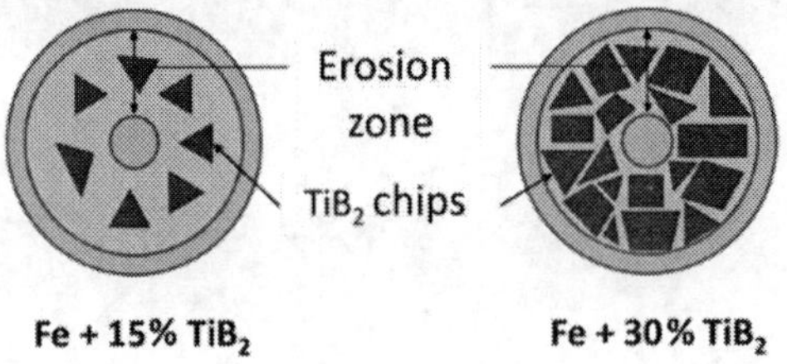

Figure 1. Scheme of the Fe + TiB_2 targets.

Table 1. The deposition parameters of the Fe-Ti-B films under study.

The area of TiB_2 on target, %	U, V	I, A	Ar pressure, Pa	Deposition time, min	Fe-Ti-B film thickness, μm
15	600	1.5	0.2	20	2.4
				15	1.9
				8	0.26
30			0.5	25	2.0
			0.4	20	1.57

The annealing of the films was performed at a 500°C temperature for 1 hour in a vacuum better than $2 \cdot 10^{-6}$ Torr. The annealing temperature was selected based on the previous studies [11].

The chemical composition, morphology, and film thickness were studied by EDX spectrometry using a Hitachi S3400N scanning electron microscope (SEM) with the Noran 7 Thermo Scientific attachment. The content of light elements (B, O, and N) was specified by the method of glow discharge optical emission spectroscopy (GDOES) using a Horiba Jobin Yvon Profiler-2 apparatus [12].

The phase-structural state of the films was studied by X-ray diffraction (XRD) on a Rigaku Ultima IV diffractometer equipped with a graphite monochromator in the Bragg-Bretano geometry using

CuKα-radiation. For quantitative phase analysis, a specialized software package [13] was used, that utilizes full-profile analysis according to the Rietveld method. The volume fractions of crystalline phases and their average grain sizes were determined. To estimate correctly the grain size, the physical broadening of the XRD peaks was analyzed taking into account not only the grain size D, but also the microstrains of the particles $<\varepsilon>$ (we will not further consider microstrains). To reduce the error in determining the interplane distances Δd, the lattice parameters a and c were calculated from the position of the centers of gravity of the high-angle peaks using the Voigt approximation.

The static magnetic properties of the films were studied on a vibrating magnetometer. The rf magnetic permeability of the studied films was measured using an HP 8720C vector network analyzer in the frequency range 0.1-2 GHz [14].

3. Results and discussion

3.1. Phase composition and structure of the Fe-Ti-B films under study

According to the chemical analysis, the films in the as-sputtered state contain Fe, Ti, B, and O (the latter in an amount of not more than 1.5 at.%). None of other uncontrolled impurities (N, C) were found in these films. The distribution profiles of the elements obtained by the GDOES method (Figure 2a) indicate a uniform distribution of the elements over the thickness of the films. The films are characterized by a dense pore-free structure (Figure 2b). The film thicknesses, measured over cross sections using SEM (Figure 2b), are given in Table 1.

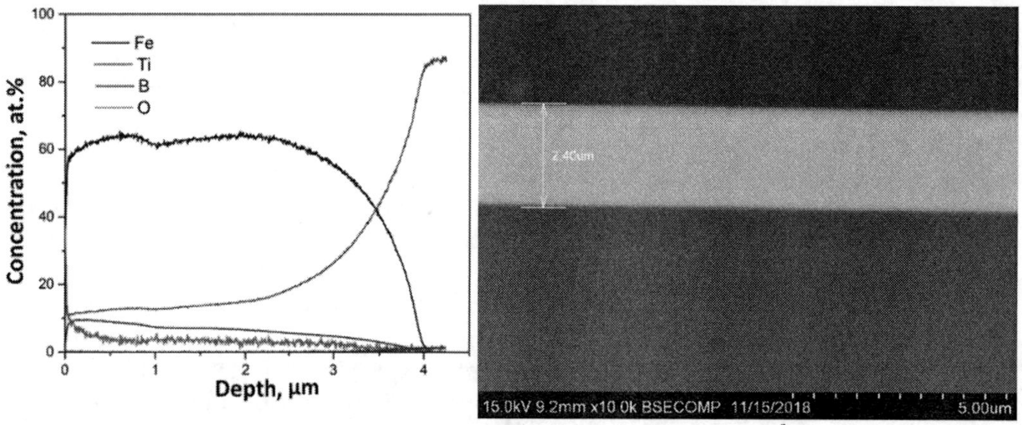

a b

Figure 2. Distribution of chemical elements by thickness (a, GDOES) and microphotograph of the cross section (b, SEM) of the $Fe_{81.5}Ti_{13.1}B_{3.9}O_{1.5}$ films on the Ti alloy and glass substrates, respectively.

According to the XRD results (Table 2), in the $Fe_{81.5}Ti_{13.1}B_{3.9}O_{1.5}$ films obtained by sputtering of the Fe + 15% TiB_2 target for 20 min, two-phase (bcc αFe(Ti) + hcp TiB_2, Figure 3) nanocrystalline (grain sizes of 3.5 and 3.6 nm, respectively) structure is formed during the growth process with a volume phase ratio of ~13:1. The bcc phase formed in these films upon sputtering is a solid solution, most likely Ti in αFe, as evidenced by the phase lattice parameter, which significantly exceeds that for pure αFe (2.866 Å). The assumption that only Ti, and not B or O, is involved in the formation of a solid solution is based on the fact that this element has significant solubility in αFe (up to 13.7 at.% at 882°C [15]). For B and O, the ultimate solubility does not exceed 0.001 and 0.1 at.%, respectively [15].

The $Fe_{70.4}Ti_{8.8}B_{20.2}O_{0.6}$ films, obtained by sputtering of the Fe + 30% TiB_2 target for 25 min, are characterized by a single-phase αFe-based amorphous (in terms of XRD) structure (Figure 3). Despite the presence of a significant amount of Ti and B, it is not possible to identify boride phases in these films.

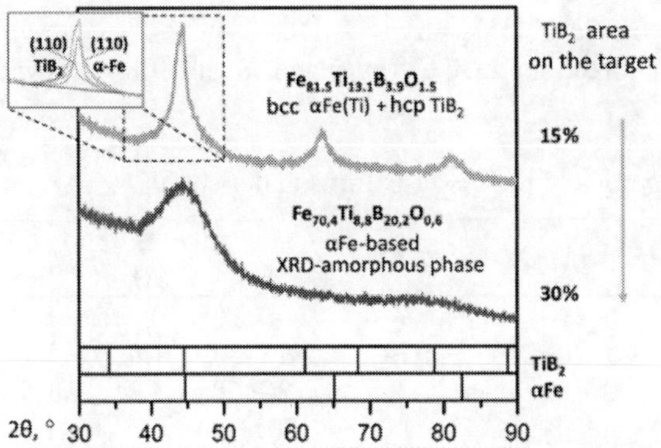

Figure 3. The XRD patterns of the Fe-Ti-B films in the as-sputtered state. The inset shows the decomposition of the asymmetric peak into two peaks corresponding to the different phases.

The annealing at 500°C of the $Fe_{81.5}Ti_{13.1}B_{3.9}O_{1.5}$ films leads to an increase of TiB_2 volume fraction up to 30 vol.% and to decrease of the bcc αFe(Ti) phase lattice parameter which becomes equal to that of αFe (2.866 Å). It means that depletion in Ti of the bcc solid solution phase carries out. Leaving the solid solution, Ti most likely interacts with B localized in the intergrain space (due to insolubility in αFe), which leads to an increase in the volume fraction of the TiB_2 phase. At the same time, the average grain size of the bcc phase increases up to 10 nm, while for the TiB_2 phase it practically does not change.

The amorphous (in terms of XRD) state, formed in the $Fe_{70.4}Ti_{8.8}B_{20.2}O_{0.6}$ films under sputtering, is saved after annealing at 500°C.

Table 2. The results of the XRD analysis of the Fe-Ti-B films under study.

Deposition parameters	Chem.composition of as-sput. films	Annealing	The results of the XRD analysis			
			Phase [vol.%]	*a*/*c*, Å	*D*, nm	<ε>, %
Fe+15% TiB₂, 20 min	$Fe_{81.5}Ti_{13.1}B_{3.9}O_{1.5}$	as-sputtered	bcc αFe(Ti) [93]	2.875	3.5	0.193
			hcp TiB₂ [7]	3.014/3.290	3.6	0.601
		500°C	bcc αFe [69]	2.866	10.1	0.506
			hcp TiB₂ [31]	3.030/3.292	3.7	0.149
Fe+30% TiB₂ 25 min	$Fe_{70.4}Ti_{8.8}B_{20.2}O_{0.6}$	as-sputt.	XRD-amorphous			
		500°C	XRD-amorphous			

3.2. Static and rf magnetic properties of the Fe-Ti-B films under study

For the films of the $Fe_{81.5}Ti_{13.1}B_{3.9}O_{1.5}$ composition obtained from the Fe + 15% TiB_2 target with a decrease in the deposition time from 20 to 8 minutes, the real magnetic permeability μ' increases from 4±2 to 130±50 (Table 3, Figure 4a). On Figure 4a, an uncharacteristic increase in the real magnetic permeability μ' is seen at frequencies above 1 GHz, when the imaginary part falls below zero. This may be due to the fact that the half-wave resonance in the measuring cell is superimposed on the ferromagnetic resonance of the sample, which is a purely measuring error. A decrease in the deposition time leads to a smaller film thickness; therefore, the substrate is less heated and the grain size is smaller. Table 3 also shows the frequency f, below which μ' is kept, and the frequency of the natural ferromagnetic resonance, f_r. The static magnetic properties of the films under study are given in Table 3. The magnetic permeability estimated from the hysteresis loop (Figure 5), $\mu = M_s/H_c$ [16], is consistent with μ' (Table 3).

The films of the $Fe_{70.4}Ti_{8.8}B_{20.2}O_{0.6}$ composition obtained from the Fe + 30% TiB_2 target, with a decrease in the deposition time from 25 to 20 minutes, also show an increase in μ' from 20±10 to 30±5 (Figure 4b). On Figure 4b, the imaginary permeability μ'' curve shows two resonances at frequencies of 0.7 and 0.9 GHz, which is caused by two anisotropy fields [17], and they are visible on the hysteresis loop: coercive field and perpendicular anisotropy (Figure 5b). For the same deposition time (20 min), the magnetic permeability μ' of the film from the Fe + 30% TiB_2 target is higher than from the Fe + 15% TiB_2 target (Table 3). Perhaps this is due to a single-phase XRD-amorphous structure in the first case. Measurements in two perpendicular directions in the film plane for all films showed very weak anisotropy (Figure 4).

Table 3. Static and rf magnetic properties of the Fe-Ti-B films under study

TiB_2 on target, %	Deposition time, min	Chemical composition, at.%	M_s, G	H_c, Oe	$\mu = M_s/H_c$	μ'	f, GHz	f_r, GHz
	20		–	–	–	4±2	0.8	1.2
15	15	$Fe_{81.5}Ti_{13.1}B_{3.9}O_{1.5}$	618±8	70±20	9±4	15±5	1.0	1.5
	8		–	–	–	130±50	1.0	1.5
30	25	$Fe_{70.4}Ti_{8.8}B_{20.2}O_{0.6}$	–	–	–	20±10	0.5	1.5
	20		230±10	7±2	33±6	30±5	0.6	1.0

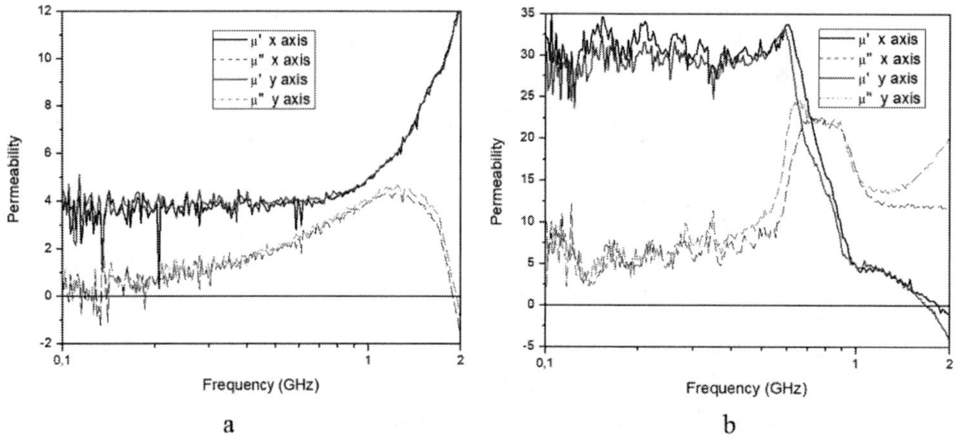

Figure 4. Frequency dependence of the magnetic permeability of the Fe-Ti-B films deposited from a target with 15 (a) and 30% (b) TiB_2 for 20 min. The real (μ') and imaginary (μ") magnetic permeabilities are measured in two perpendicular directions in the film plane.

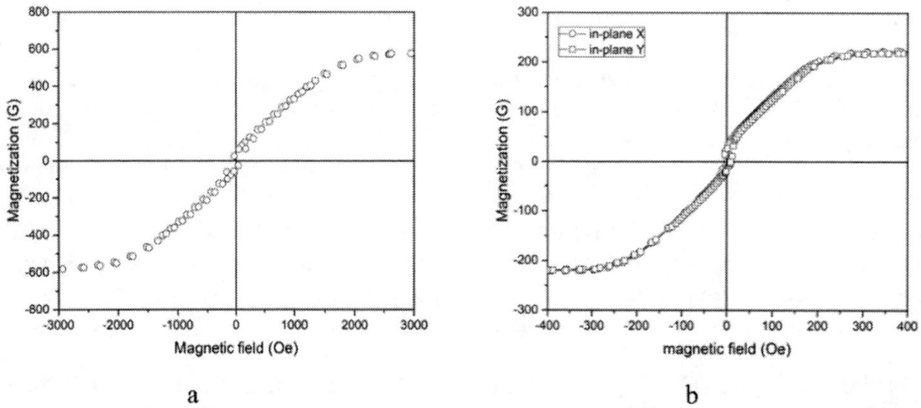

Figure 5. The hysteresis loops of the films deposited from a target with 15 (a) and 30% (b) TiB_2 for 15 and 20 minutes, respectively. The second loop is measured in two perpendicular directions in the film plane.

4. Summary

The Fe-Ti-B films with a thickness of 0.26 to 2.4 μm were obtained for 8-25 min by dc magnetron sputtering of the Fe + 15% TiB_2 and Fe + 30% TiB_2 targets.

According to the results of chemical and phase analysis, in the $Fe_{81.5}Ti_{13.1}B_{3.9}O_{1.5}$ films obtained by sputtering of the Fe + 15% TiB_2 target for 20 min in the process of growth, two-phase (bcc αFe(Ti) + hcp TiB_2) nanocrystalline (grain size about 4 nm) structure with a volume phase ratio of 93:7 is formed. The vacuum annealing at a temperature of 500°C led to an increase in the TiB_2 fraction up to 31%.

The $Fe_{70.4}Ti_{8.8}B_{20.2}O_{0.6}$ films obtained by sputtering of the Fe + 30% TiB_2 target for 25 min are characterized by a single-phase αFe-based XRD-amorphous structure. Despite the presence of a significant amount of Ti and B, it is not possible to identify boride phases in these films. The annealing at a temperature of 500°C did not lead to noticeable structural changes.

With a decrease in the deposition time of the Fe-Ti-B films (20, 15, 8 min and 25, 20 min for targets with 15 and 30% TiB_2, respectively), the real magnetic permeability μ' increases (from 4 to ~100 and from 20 to 30). For the same deposition time (20 min.), the magnetic permeability μ'of the film from the Fe + 30% TiB_2 target is higher than from the Fe + 15% TiB_2 target (30 and 4). The magnetic permeability estimated from the hysteresis loop, $\mu = M_s/H_c$, is consistent with μ'. The frequency f, below which the μ' value is kept, and the frequency of the natural ferromagnetic resonance f_r were determined. Measurements in two perpendicular directions in the film plane for all films showed very weak anisotropy.

Acknowledgements

This work was supported by RFBR (18-03-00502a "Creation of high-induction nanocrystalline Fe-based films for the microelectronics GHz devices").

References

[1] Yoshizawa Y, Oguma S and Yamauchi K 1988 *J. Appl. Phys.* **64** 6044
[2] McHenry M E and Laughlin D E 2000 *Acta Mater.* **48** 223
[3] Viala B, Minor M K and Barnard J A 1996 *J. Appl. Phys.* **80** 3941
[4] Chakraborty A, Mountfield K R, Bellesis G H, Lambeth D N and Kryder M H 1996 *J. Appl. Phys.* **80** 1012
[5] Nago K, Sakakima H and Ihara K 1992 *IEEE Transaction J. on Magn. in Japan.* **7** 119
[6] Rask M T and Longworth L L 1991 *US Patent* 5001589A
[7] Grigorovich V K and Sheftel' E N et al. 1995 *Report on the contract IMET-Philips* PLW-938018-D-WZ-86512
[8] Sheftel E N 2010 *Inorganic Materials: Applied Research* **1** 17
[9] Bannykh O A, Sheftel' E N, Grigorovich V K, Strug R E, Mkrtumov A, Polyukhova I R and Evdokimov A V 1992 *Russian patent* 4775860/02 [In Russian]
[10] Grigorovich V K, Sheftel' E N, Strug R E and Polyukhova I R 1993 *Izvestia Akademii nauk SSSR. Metally* No. 6, 173 [In Russian].
[11] Sheftel E N, Tedzhetov V A, Harin E V, Kiryukhantsev-Korneev F V and Usmanova G Sh 2016 *Physica Status Solidi C: Current topics in solid state physics* **13** 965
[12] Kiryukhantsev-Korneev Ph V 2012 *Protection of Metals and Physical Chemistry of Surfaces* **48** 585
[13] Shelekhov E V and Sviridova T A 2000 *Metal Science and Heat Treatment* **42** 309
[14] Starostenko S N, Rozanov K N, Osipov A V 2008 *J. Appl. Phys.* **103** 07E914
[15] Lyakishev N P (Ed.) *Phase diagrams of binary metallic systems* (Moscow: Mashinostroenie) 1997 Vol 2 p 586-591; 1999 Vol 3 Book 1 p 724-726 [In Russian]
[16] Dionne G F 2003 *IEEE Transactions on Magnetics* **39** 3121
[17] Zezyulina P A *et al.* 2016 *IEEE Magnetics Letters* **7** 3705804

Advantages of drawing and rolling metals with pulse current

O A Troitsky and V I Stashenko

Federal State Budget Office of Science
Institute of Mechanical Engineering. A.A. Blagonravova RAS
Russia, 101990, Moscow, M. Kharitonevsky per., 4
E-mail: oatroitsky@mail.ru and vis20-11@rambler.ru

Abstract. The results of studies on the drawing of copper and steel wires using pulsed current in the deformation zone are presented. The analysis of their structure and physicomechanical properties is given. As a result, the drawing forces of copper and steel wires are reduced by 30-35%, the electrical resistance is reduced by 18-20%, a more perfect axial texture appears, the ductility of the wires increases compared to warm drawing. In the steel wire, the α-phase content sharply decreases, which is predominantly released in the surface layer by several microns. We propose methods for the modernization of rolling equipment for technology using pulsed current. The efficiency of using a high-density pulsed current on a rolling mill for producing stainless steel from an initial billet of 2 mm thick and 100 mm wide 0.3 mm thick tape without intermediate annealing and oxide film is shown.
Keywords: wire drawing, pulsed electric current, structure, properties, modernization of mills

1. Introduction

The most important property of plastically deformed materials is used to obtain blanks and products. Form-changes of the blanks while obtaining the product are accompanied by complex structural changes, which requires the improvement of metal forming under pressure processes (MFP).

Many materials used in mechanical engineering have low technological ductility. Therefore, the urgent task is to extend the ductility interval, a targeted effect on the metal structure, on the value of internal stresses. Decisive in the formation of a particular structure and, therefore, the necessary properties of the metal is the rate of heating and cooling [1]. Great opportunities in solving these problems in MFP technologies are provided by the use of electric current in the zone of metal deformation. The action of an electric current provides fast and uniform heating of the workpiece in a wide temperature range, which allows combining the heating of the workpiece and its deformation, creating conditions for continuous, rational and high-performance production of products [2,3]. An increase in the heating rate significantly accelerates the diffusion movement of carbon atoms, improves the purification of ferrite and delays the coagulation process, preserving the lamellar form of cementite, which is a must condition for large total compressions, for example, for steel [1,3].

An additional reserve for increasing the ductility of deformable structural materials is the use of high-density pulsed current. The plasticizing effect of the pulsed current on the deformation of the metal (electroplastic effect - EPE) was established during tension and compression of zinc single crystals in liquid nitrogen [4]. The effect manifested itself in the form of jumps in plastic deformation under pulses of current [5]. It was confirmed by the work of [6,7] American scientists on polycrystalline titanium.

The force action of the current on the plastic deformation of the metal occurs mainly through two channels - through the electronic subsystem through the action of the "electronic wind" on the dislocation [8], and also through the lattice subsystem of the metal by creating vibration of the conductor due to the pinch action of the pulse current (pulsed current interaction with its own magnetic current field) [9] Recent studies have shown that under the action of current pulses, at the moments of the beginning of the leading and trailing edges of the pulse, significant jumps in acceleration (more than 1000 $м/с^2$) and surface displacements in the direction of the radius of the conductor occur [10], causing vibrations of the lattice sites similar to the effect caused by ultrasound on plastic deformation of metal.

For the implementation of EPE for any type of metal forming under pressure processes (MFP), the following conditions must be met:

- the workpiece must be under mechanical stresses above the yield strength $\sigma_т$;
- pulsed current of low voltage (20-25 V) and high density Jm must be supplied directly to the deformation zone;
- the shape of the pulses must be rectangular or trapezoidal;
- pulse duration τ should have values τ of about 10^{-4} seconds, do not exceed the 10^{-3} second time of normal deformation jumps so that significant concomitant heating does not occur;
- the amplitude current density J_m must be large, at least $10^5 \div 10^6$ A/$см^2$ so that conduction electrons acquire significant drift velocities v_e of the order of one meter per second ("electron wind" speed) in accordance with the formula $v_e = J_m/e\ n$, where e and n is the charge and concentration of free electrons and could transmit momenta of force and energy to dislocations and other mobile structural defects;
- pulse repetition rate F, defined by the formula $F = k\ v_{дв}/\Delta l$ where $v_{дв}$ - is the speed of movement of the workpiece, Δl - the length of the deformation zone and k is a factor of the order of two (two pulses per Δl) should be several hundred Hz to ensure that all sections of the workpiece are processed by the pulse current.
- The action of EPE during MFP gives the following results:
- metal resistance to deformation is reduced by 25-30%;
- increases the ductility of the metal during processing, as well as in the appearance of residual ductility by tens of percent;
- the degree of perfection of the axial texture of the wire during drawing increases, its electrical resistance decreases by 15-20%. The phase composition of the metal and structure are improved;
- during rolling, drawing and drawing of stainless steels, the austenitic-martensitic phase γ - α - transformation is almost completely suppressed, which makes the operations of expensive and energy-intensive austenizing annealing of workpieces unnecessary;
- reduces the likelihood of brittle fracture of the workpiece material during processing.

The aim of the work is the application of technology for electroplastic wire drawing and strip rolling, as well as ways to modernize machine tools for electroplastic deformation of metal are given.

2. Technology of electroplastic drawing of wire

The technology of electroplastic drawing (EPD) allows you to: increase the residual ductility of the wires and the work of their destruction, reduce the electrical resistance of the original wires by increasing the degree of perfection of the axial texture, reduce and ordering the number of remaining defects, including dislocations.

This allows you to get stronger, lighter, more flexible and reliable current-carrying wires with improved surface and contact properties. In addition, a large technical and economic effect is achieved by the abolition of intermediate annealing operations, which increases the productivity of mills.

The EPD technology is based on the supply of electric current (constant or pulsed) directly to the deformation zone of a metal wire during drawing, Figure 1 Schemes , Figure 1a) and Figure 1b) allow drawing with the passage of electric current through a section of wire, including a deformation zone, using carbide and diamond fibers, respectively.

Current is supplied to the drawing zone immediately before and after the die by means of rotating rollers or by means of sliding contacts. The entire EPD assembly is placed in a bath with a flowing emulsion, which guarantees the absence of thermal heating of the wire necessary for the effective action of EPE. A water-soap emulsion for copper, a triethanolamine emulsion and chlorine paraffin CP-7O for steels with unstable austenite were used as a cutting fluid. These lubricants cooled the die and the wire section subjected to current. The current supply to the moving wire was carried out using roller and brush contacts. A powder contacting device was also used [2].

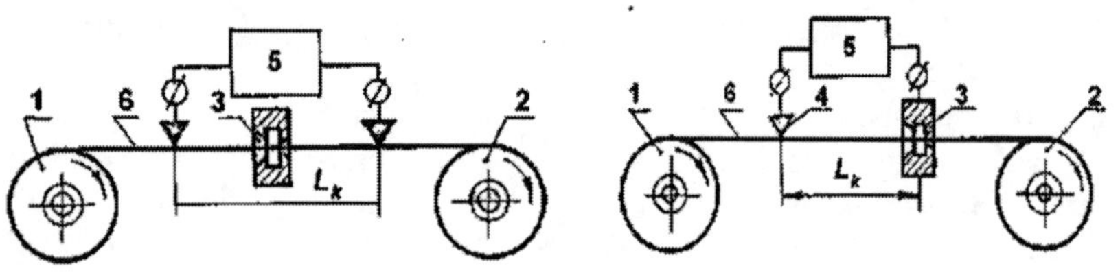

Figure 1. Schemes for supplying current to a moving wire during Drawing.
1 - transfer coil; 2-receiving coil; 3-wire; 4 - sliding contacts; 5 - current source; 6 - wire. l_k - current action section.

For most metals (copper, tungsten and molybdenum), the effect of reducing the drawing force is maximum when connecting the plus of the current source to the deformation zone, when the direction of the conduction electron current coincides with the direction of movement of the deformation zone. The current effect is maximum at small wire drawing speeds $u_в \leq 0,1$ м/c, and it has a smaller value at average drawing speeds $u_в$ = 0,3-0,5 м/c, with a further increase $u_в$, the electron drift velocity u_e exceeds the metal deformation rate $u_в$

The maximum allowable current density in copper wires is $J_m=10^3$ A/см^2. This corresponds to the electron drift velocity of 0.1 cm /s. Concomitant heating of billets, as a rule, does not exceed 100-200 ° C [7].

In semi-industrial experiments, pulsed current densities of (0,35-0,5)×10^6 A/см^2 corresponded to electron drift velocities u_e = 0,35-0,5 м/c. Thus, the applicability of EPE to intensify the process of metal drawing is undeniable.

The magnitude of the effect of the action of the current during EPD increases with increasing frequency, duration of current pulses, amplitude and average current density. During EPD on thin wires, the current power allocated in the drawing zone does not exceed 100 watts.

As a result of EPD of the stainless steel wire there is a sharp decrease in the content of the martensitic a-phase after 8-10 transitions on a wire with a diameter of 0.3 mm does not exceed 6-8%, and after the usual drawing 60-70%. The a-phase remaining after EPD the electron-hole emission is predominantly released at the surface of the wire, forming a jacket several microns thick.

The ductility of wire d (relative elongation) of metastable austenitic steel (MAS) as a result of EPD increases by 40%, which is more than the test results of samples after warm drawing (20-25%), and even more so after cold drawing (5-6%). The trip effect on the wire from the MAS, passed EPV, increases by more than two times at the pulsed current and one and a half times at constant current compared with warm drawing.

Improving the quality of the wire, reducing its cost is an urgent task. It becomes important to manufacture lightweight cables through the use of heavy copper substitutes with lighter and enough conductive materials, for example, aluminum and bimetals, which is primarily interested in the aerospace field of technology. In the case of cables, the ultimate goal should be to reduce the number of conductors and the weight of the cables.

The technology of electroplastic wire drawing was applied in the manufacture of cable. Compared to a standard cable, the new cable is almost twice as light, and in the future, the weight of the cable can be further reduced significantly (Figure 2).

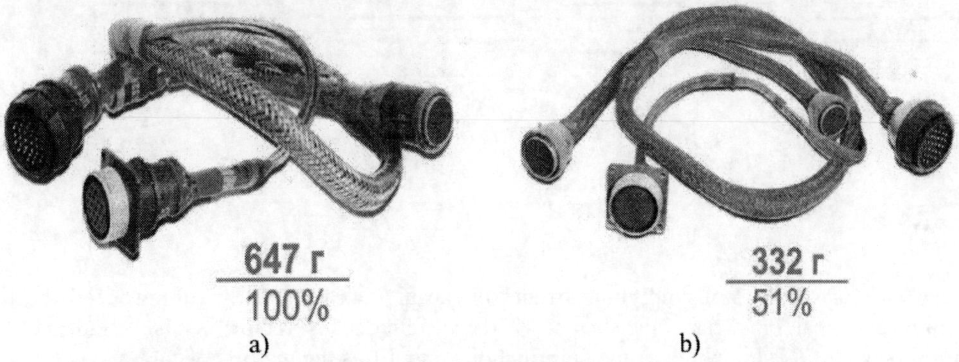

Figure 2. Typical existing cable a), new cable b) and prospective cable.

3. Methods of modernizing existing rolling mills for EPD technology

As an example, let's consider two main ways of applying current. The first method is to tear off one of the deformable shafts - the lower or upper part of the rolling stand from the mass of the mill, so that it is possible to supply the working current across the direction of movement of the workpiece according to the "from shaft to roll through the workpiece" scheme (Figure 3a) such a supply circuit current is most effective as an EPD method; it can be used when rolling sheet material and strips, as well as when flattening wire into tape on flattening mills; the accompanying thermal effect is minimal, not more than 100 ° C, and unit allowable reductions reach values of tens of percent. The method is not limited with respect to the resulting thickness of the rolled sheet or strip. Provided that the portion of the workpiece under current is minimized, the possibility of overheating is completely eliminated. If necessary, when dosed thermal preparation of the workpiece material is required, this method can combine electrical contact heating (ECH) and EPD, which as a result makes it possible to reduce the workpiece's resistance to deformation by more than 30-35%. The disadvantage of this method is the isolation of the deforming shafts from each other and from the rolling mill, i.e. modernization of the mill for EPD technology is not easy. However, this method is used to study EPD in laboratory conditions and on thin blanks.

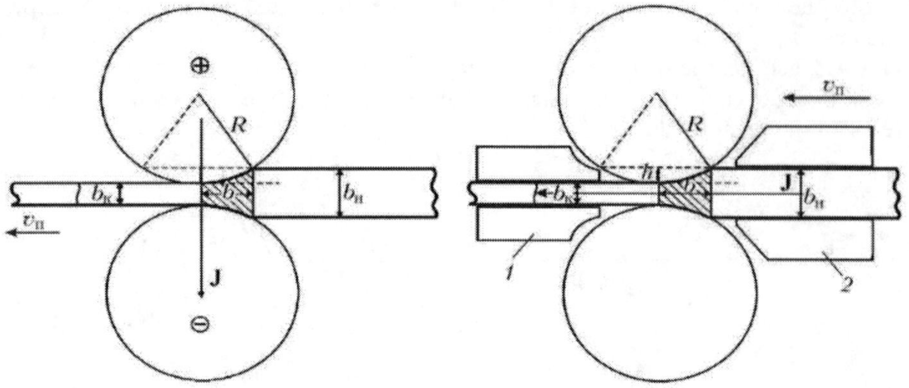

a) b)

Figure 3. a)- is a method of supplying current from a roll to a roll: v_n is the rolling speed; $b_н$, $b_к$ - the initial and final thickness of the strip; J - is the current density vector; R - is the radius of the deforming shaft; *B* is the width of the deformation zone; b)- is the method of supplying current with two sliding contacts before and after the ZD (1 and 2 sliding contacts).

The second method of supplying current using the contacts before and after the rolling zone has its advantages and disadvantages (Figure 3b). This method is the simplest, does not require major modernization of the mill, saves deforming shafts from destruction by electroerosion and allows you to combine EPD and ECH. It has no restrictions on the thickness of the workpieces compared to the first method. It is advisable to supply current from both sides of the working zone of the mill using sliding or roller contacts.

It is advisable to use this current supply circuit when introducing EPD technology into production for the following reasons. Deformation shafts are not included in the electrical circuit. At the same time, they process the section of the workpiece through which the working current passes and in which the EPD is realised. The resistance to deformation from the side of the workpiece is reduced by 30-35%. This leads to a decrease in wear of the deforming shafts and reduces the likelihood of a tack phenomenon in the operation of the mill. In this case, upgrading the mill to EPD technology is a low-cost and long-term operation of the mill.

For operation, it is necessary to provide a device for powerfully connecting current-carrying coaxial cables to the rolling contacts at the inlet and outlet of the mill, as well as to the part of the rolling stand isolated from the mass of the mill.

4. Rolling mill for stainless steel tape

It is advisable to use EPD technology only at low and medium rolling speeds (not more than 0,5 м/c), or high-speed EPD on very thin workpieces.

In the work, the pulse current modes for rolling stainless steel are established: the amplitude current density $J_m=10^4–10^5$ A/см^2, the pulse duration of $150–200\cdot10^{-6}$ c, the pulse repetition rate of 500-800 Hz. The method of supplying current with two sliding contacts was used (Figure 3 b).

During the testing of the mill, multiple strip rolling was performed from stainless steels of the 12X18H9 and 12X18H10T type from an initial thickness of 2 mm to a final thickness of 0.3 mm without intermediate annealing. At the same time, due to the rational choice of the pulsed current parameters, the appearance of an oxide film was prevented. The rolling speed was maintained in the range (0.3-0.5) m / s. The total electric power at maximum consumption was approximately 150 kW. The maximum amplitude value of the pulse current was 45 kA. Current density J_m in the steel strip in the deformation zone amounted to 300-800A/мм^2 depending on the current strip thickness.

IOP Conf. Series: Materials Science and Engineering 848 (2020) 012084 doi:10.1088/1757-899X/848/1/012084

In Figure 4, 5 show a view of a mill for producing stainless steel tape. In Figure 6 shows the results of measurements of physical and mechanical properties. The mill provided rolling of stainless steel tape 100 mm wide, from an initial thickness of 2.0 m to a final thickness of 0.3 mm with a rolling error of not more than 0.01.

The use of EPD technology in modern pulsed current mills opens up new possibilities for automatic regulation of the rolling process. By changing the amplitude and frequency of the pulses following the signals of the pressure sensors on the rolls, it is possible to strengthen or, on the contrary, weaken the electroplastic effect of the current. In this case, the gap between the rolls will remain unchanged and it will be possible to avoid an undesirable increase in the thickness of the workpiece (rolled) during the elastic squeezing of the rolls.

a)

b)

Figure 4. a) general view of the EPP mill; b) rolling of a steel strip

IOP Conf. Series: Materials Science and Engineering **848** (2020) 012084 doi:10.1088/1757-899X/848/1/012084

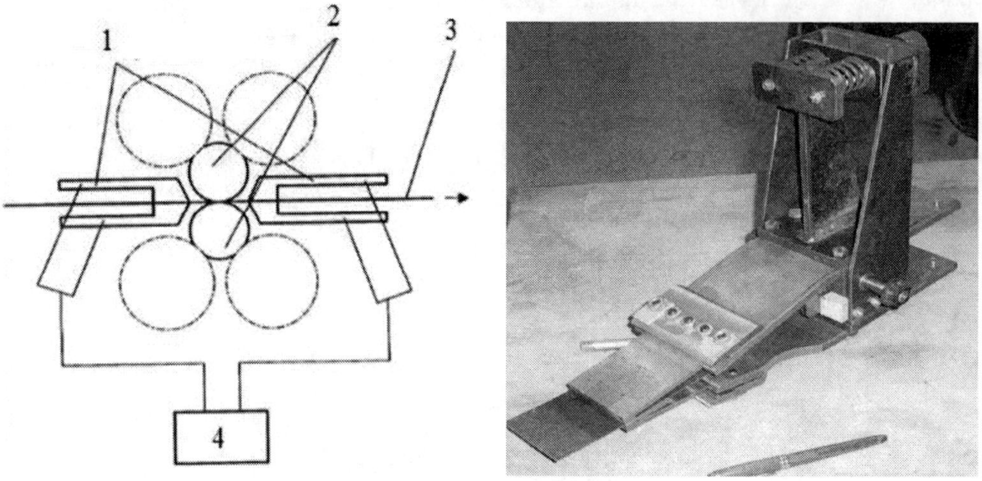

a) b)

Figure 5. a) circuit current to the deformation zone: 1-sliding contacts, 2-working rolls, 3-tape, 4-current source; b) type of sliding contacts

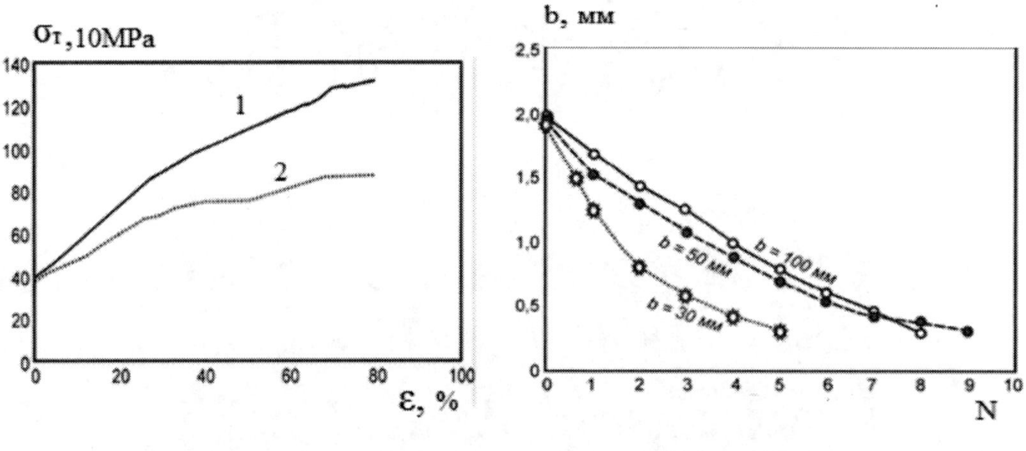

a) b)

Figure 6. a) change of yield strength σ_t for different relative deformation ε: 1-cold rolling; 2-electroplastic rolling; b) change of thickness b of steel strip on transitions N

References

[1] Kizin I N 1957 Phase transformations during accelerated heating of steel (Moscow: Metallurgiya) p 280
[2] Romanov D I 1975 Electrocontact heating of metals. (Moscow: Engineering) p 250
[3] Hasin G A et al. 1984 Electrothermal treatment and heat drawing of steel (Moscow: Metallurgiya) p149

IOP Conf. Series: Materials Science and Engineering **848** (2020) 012084 doi:10.1088/1757-899X/848/1/012084

[4] Troitsky O A 1969 *Electromechanical effect in metals* Letters in JETP **2** 18-22
[5] Stashenko V I, Troitsky O A 1982 *Influence of the shape and modes of the passage of current pulses on the plastic deformation of zinc crystals* DAN USSR **267** 638-640
[6] Conrad H, Sprecher A F 1978 *Dislocation in Solids*, edited by F.R.N. Ser. Metall. **12** 1063-1068
[7] Okazaki K, Kagawa M, Conrad H 1979 *A study of electroplastic effect in metals* Scripta Metall. **13** 473-477
[8] Kravchenko V I 1966 *Interaction of the directed electron flow with JETP dislocations* **51** 1676-1681
[9] Skvortsov O B, Stashenko V I, Troitsky O A 2018 *Elastic deformations of conductors under the influence of current pulses* Engineering and mechanical engineering **1** 33-49
[10] Stashenko V I., Skvortsov O B., Troitsky O A. 2019 *Excitation of shock vibration processes during electric pulse exposure to metals*. Conf. "Deformation and fracture of materials and nanomaterial" (Moscow. IMET RAS) 633-634

IOP Conf. Series: Materials Science and Engineering **848** (2020) 012085 doi:10.1088/1757-899X/848/1/012085

Microscopic research of amorphous alloys AlFeNiLa exposed to magnetic pulse processing

A A Viryus[1], T P Kaminskaya[2], M N Shipko[3]*, N D Bakhteeva[4], V V Korovushkin[5], A G Savchenko[5], M A Stepovich[6], E S Savchenko[5] and E V Todorova[4]

[1] Institute of Experimental Mineralogy, 4 Academician Ossipyan Street, Chernogolovka, Moscow District, 142432 Russia
[2] Lomonosov Moscow State University, Faculty of Physics, House 1, Building 2, Lenin Hills, 19991 GSP-1 Moscow
[3] Lenin Ivanovo State Power Engineering University, 34 Rabfakovskaya Street, Ivanovo, 153025 Russia
[4] Baikov Institute of Metallurgy and Material Science RAS, 49 Leninsky Ave., Moscow GSP-1 119991 Russia
[5] National Research Technological University "MISiS", 4 Leninsky Ave., Moscow, 119049 Russia
[6] Tsiolkovsky Kaluga State University, 26 Stepan Razin Street, Kaluga, 248023 Russia
E-mail: michael-1946@mail.ru

Abstract. The structure of the solid solution foil $Al_{85}Fe_7Ni_5La_3$ was studied, obtained by the spinning method, before and after magnetic pulses treatment. Comparison of the effects of pulses of weak magnetic field on the topology of the alloy surface allows us to conclude that when exposed to 30 pulses, the surface becomes more smooth. It was found that treatment with magnetic pulses significantly affects the magnetic characteristics of the alloy, which may allow this type of effect on amorphous alloys to be used in practice.

1. Introduction

The research for amorphous alloys has nowadays become noticeably interesting due to the increased possibility of using their unique features in space, aviation, and radio-electronic devices in practice. Unlike their crystalline substitutes, amorphous alloys enable obtaining a greater variety of features by changing their composition or due to various physicochemical effects. For instance, the short-range order of the atoms in the alloys, and therefore, their electromagnetic features can be influenced by these features [1, 2]. Previously [3-5], we studied the effect of influence of weak magnetic field pulses on the structure and properties of some electrotechnical materials. In the present work, similar studies were performed for amorphous $Al_{85}Fe_7Ni_5La_3$ alloys.

2. Research methodology

The amorphous $Al_{85}Fe_7Ni_5La_3$ alloy was studied in the form of a foil 20 μm thick, obtained by the spinning method, before and after magnetic pulse treatment (MPT). MPT was carried out by rectangular magnetic field pulses with a frequency of 10-20 Hz, intensity 10-100 kA/m. We studied the initial samples (before MPT) and samples subjected to 10, 20, and 30 magnetic pulses.

IOP Conf. Series: Materials Science and Engineering **848** (2020) 012085 doi:10.1088/1757-899X/848/1/012085

The structure and elemental composition of the samples were studied using a Tescan Vega II XMU scanning electron microscope (SEM) with an INCAx-sight energy dispersive X-ray spectrometer. Smaller details of the alloy surface were studied by atomic force microscopy (AFM) using a SMENA-A scanning probe microscope, Solver platform, NT-MDT, Russia, in semi-contact and contact modes at room temperature.

Magnetic parameters: residual magnetization, specific saturation magnetization, coercive force, shape of the magnetic hysteresis loop – were measured on a VSM-250 vibration magnetometer in a magnetic field of 20 kOe strength at room temperature.

3. Results

The surface topologies of the AlNiFeLa foil on the matte and shiny side differ dramatically; the shiny side is smooth and much more uniform – see figure 1-figure 5. Therefore, research was mainly carried out on the shiny side of alloys.

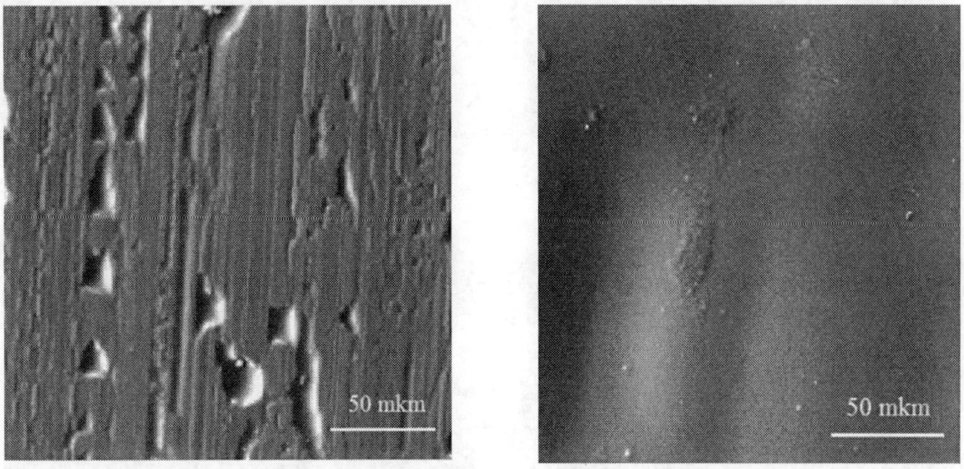

Figure 1. A micrograph of the matte (left) and shiny (right) side of the foil, obtained in SEM after exposure to the foil with 30 pulses of a magnetic field.

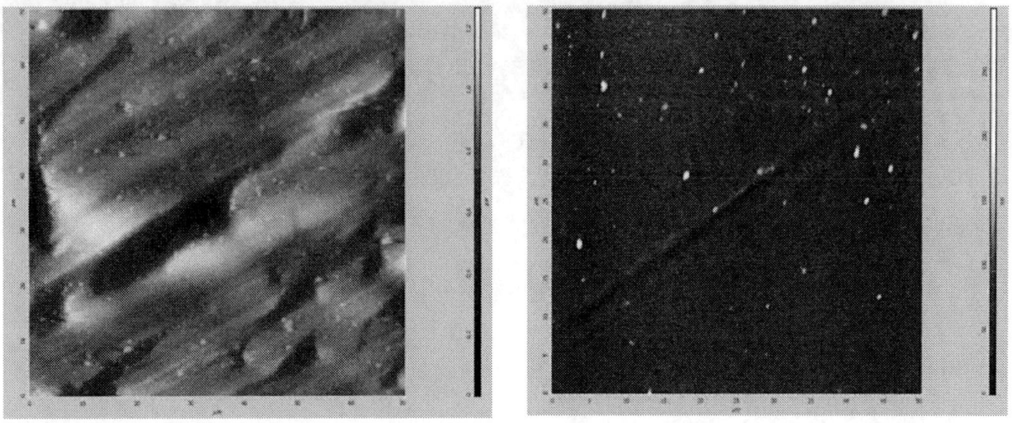

Figure 2. A micrograph of the matte (left) and shiny (right) side of the foil, obtained in AFM after exposure to the foil with 30 pulses of a magnetic field. The scanning area is 70 mkm×70 mkm.

IOP Conf. Series: Materials Science and Engineering **848** (2020) 012085 doi:10.1088/1757-899X/848/1/012085

Comparison of the surface topology of the shiny side of the initial sample and the sample after 30 pulses shows that when exposed to a pulsed magnetic field, the surface becomes clearly uniform and structured – see figure 1 and figure 2. The average size of the structural element of the surface is 40-60 nm. Comparison of the effects of 20 pulses and 30 pulses on the topology of the alloy surface allows us to conclude that when exposed to 30 pulses, the surface becomes more smooth and some of the pores are drawn out.

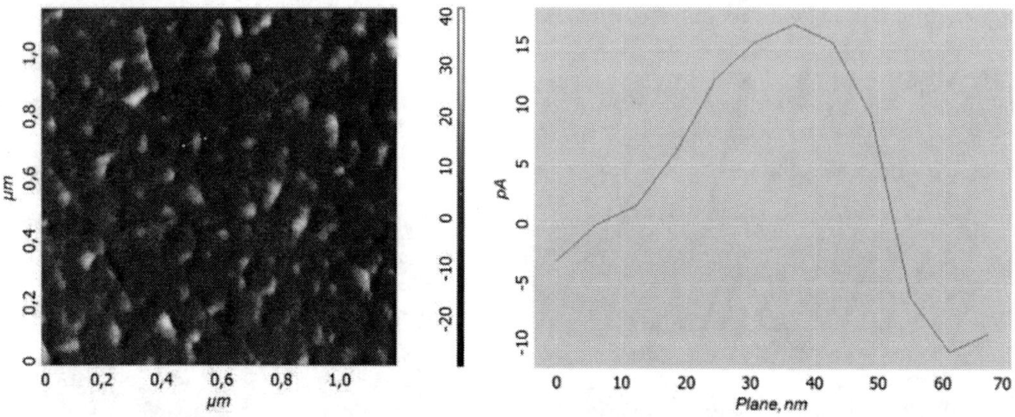

Figure 3. Micrograph (left) of the shiny side of the foil obtained in AFM, after exposure to the foil with 30 pulses of a magnetic field. The scanning area is 1.2 mkm×1.2 mkm. The linear profile (right) of the micrograph element marked on it with a straight line segment.

Figure 4. Three-dimensional micrograph of shiny side of the foil obtained in the AFM, after exposure to the foil with 30 pulses of a magnetic field. The scanning area is 2.5 mkm×2.5 mkm.

It was established that the MPT significantly affects the magnetic characteristics of the alloy. The hysteresis characteristics of the amorphous $Al_{85}Fe_7Ni_5La_3$ solid solution before and after the MPT: specific saturation magnetization in a field of 20 kOe, specific remanent magnetization, coercive force and hysteresis loop areas are given in the table, and the corresponding magnetic hysteresis loops obtained using a vibration magnetometer are shown in figure 5 [6].

Table 1. Hysteresis characteristics of the amorphous $Al_{85}Fe_7Ni_5La_3$ alloy before and after magnetic pulse treatment.

Number of pulses	Specific saturation magnetization (A·m²/кg)	Specific remanent magnetization (A·m²/кg)	Coercive force (kA/m)	Hysteresis loop areas (A·m²/кg)
0	3.6	1.4	110	300
10	2.3	0.5	34	110
30	22.7	4.4	9	215

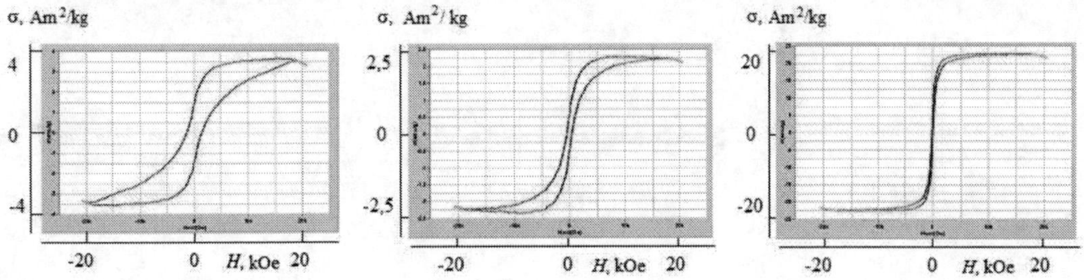

Figure 5. Magnetic hysteresis loops of the amorphous $Al_{85}Fe_7Ni_5La_3$ alloy obtained using a vibration magnetometer in the initial state (left) and after MPT: 10 (in the center) and 30 (right) by pulses.

4. Conclusions

The structure of the solid solution foil was studied, obtained by the spinning method, before and after magnetic pulses treatment. Comparison of the effects of pulses of magnetic field on the topology of the alloy surface allows us to conclude that when exposed to 30 pulses, the surface becomes more smooth. The average size of the structural element of the surface is 40-60 nm. It was found that treatment with magnetic pulses significantly affects the magnetic characteristics of the alloy, which may allow this type of effect on amorphous alloys to be used in practice.

Acknowledgments

The work was carried out on the State (Russian Federation) assignment No. 007-00129-18-00, as well as with partial financial support of the Russian Fund for Basic Research (project No. 19-03-00271), the RFBR and the Government of Ivanovo District (project No. 18-43-370012) and the RFBR and the Government of the Kaluga District (project No. 18-41-400001).

References

[1] Kekalo I B 2006 *Atomic structure of amorphous alloys and its evolution* (Publishing House "Study" MISiS, Moscow) 340 p. [in Russian]

[2] Glezer A M and Molotilov B V 1992 *Structure and mechanical properties of amorphous alloys* (Metallurgy, Moscow) 208 p. [in Russian]

[3] Shipko M N, Korovushkin V V, Kostishin V G, Isaev I M, Stepovich M A and Savchenko E S 2018 Effect of magnetic pulse processing on the structure and magnetic properties of ferrites *Bulletin of the Russian Academy of Sciences: Physics* **82** 203-7.

[4] Shipko M N, Tikhonov A I, Stepovich M A, Viryus A A, Kaminskaya T P, Korovushkin V V, Savchenko E S and Eremin I V 2018 Effect of Magneto-Pulse Processing on the microstructure and magnetic behavior of amorphous electrotechnical steel *Bulletin of the Russian Academy of Sciences: Physics* **82** 988-92.

[5] Viryus A.A., Shipko M.N., Stepovich M.A., Kaminskaya T.P., Korovushkin V.V., Tikhonov A.I. and Savchenko E.S. 2019 Analysis of the structure and composition of Fe_2O_3

IOP Conf. Series: Materials Science and Engineering **848** (2020) 012085 doi:10.1088/1757-899X/848/1/012085

oxides subjected to magnetic pulse treatment *Journal of Surface Investigation. X-ray, Synchrotron and Neutron Techniques* **13** 215-20.

[6] Shipko M N, Korovushkin V V, Savchenko A G, Stepovich M A, Bakhteeva N D, Savchenko E S and Todorova E V 2019 The effect of magnetic pulse treatment on the magnetic properties of amorphous solid solution $Al_{85}Fe_7Ni_5La_3$ *Applied Physic [Prikladnaya fizika]* **3** 81-5 [in Russian]

IOP Conf. Series: Materials Science and Engineering **848** (2020) 012086 doi:10.1088/1757-899X/848/1/012086

Novel materials for high-resolution three-dimensional printing using surface-selective laser sintering

M Syachina[1,2], S Minaeva[1], E Krumins[3], S Howdle[3] and V Popov[1]

[1]Institute of Photonic Technologies, Federal Research Centre "Crystallography and Photonics", RAS, Moscow, Russia
[2]Federal State Budgetary Scientific Institution "Research Centre for Medical Genetics", Moscow, Russia
[3]School of Chemistry, University of Nottingham, Nottingham, UK

E-mail: msyachina@crys.ras.ru

Abstract. A new approach for high-resolution three-dimensional printing using selective laser sintering is proposed. Novel ultrafine powder is synthesized and modified to face the requirements of the laser sintering process. The effect of the powder particle size and shape on the characteristics of the sintered structures is studied.

1. Introduction

Selective laser sintering is one of the most studied and established additive manufacturing (AM) technique based on the fusing of powdered material using laser radiation. This process involves fabrication of a three-dimensional (3D) object within a powder bed, which gradually indexes down as each layer is selectively fused according to a computer-aided design (CAD) model [1].

The only criteria when selecting the polymer material for SLS process is expected to be a powder form. However, the suitable material should also possess certain thermal properties and have absorption bands at the wavelength of the laser used. To overcome these limitations a modification of the conventional SLS – surface selective laser sintering (SSLS) – was proposed [2]. SSLS is based on layer-by-layer sintering of various polymer particles by heating absorbing nanoparticles (carbon or gold nanoparticles) or substances (water, various dyes) uniformly distributed over their surface. That makes fabrication of three-dimensional structures from any polymer particles possible using absorbing agent as a "glue".

Being a powder bed fusion technique, surface selective laser sintering allows to produce 3D objects of any shape and predetermined complex geometry as there is no need in the support structures during the build. The spatial resolution of the formed structures directly depends on the physical properties of the particles (e.g., particle size, size distribution and shape) as well as the laser radiation parameters (e.g., laser power, laser beam diameter and scan speed) [3].

Nowadays, the most widely used polymers in SLS technology are semi-crystalline thermoplastics, typically polyamides (PAs; nylon) 12 and 11 [4], - more than 95% of the total AM consumption. With the particle size from 40 to 80 μm and, thereby, spatial resolution of 200 μm PA12 meets the requirements of industrial applications. For the purpose of improvement of the spatial resolution of 3D structures significant research is conducted to produce new polymers available for laser sintering process as well as to modify current laser systems.

IOP Conf. Series: Materials Science and Engineering **848** (2020) 012086 doi:10.1088/1757-899X/848/1/012086

In this study, a unique combination of two promising approaches is proposed. On the one hand, the use of fundamentally new polymer powders for SLS consisting of micron sized spherical particles with practically monodisperse size distribution. On the other hand, the use of the earlier developed strategy of surface selective laser sintering to achieve a higher spatial resolution of three-dimensional objects formed.

2. Materials and methods

2.1. Materials
Commercially available polyamide (PA12, EOS, Germany) was used as a reference material. Poly(methyl methacrylate) (PMMA) particles were synthesized by the dispersion polymerization in supercritical carbon dioxide ($scCO_2$) as described in [5].

2.2. Experimental setup
Experiments in the formation of the layers and their further laser sintering were carried out on a specially developed setup shown in Fig. 1. The system consisted of a module for layer-by-layer application of ultrafine powders (Fig. 1 "1") and the module of the scanning focused laser beam (Fig. 1 "2-5").

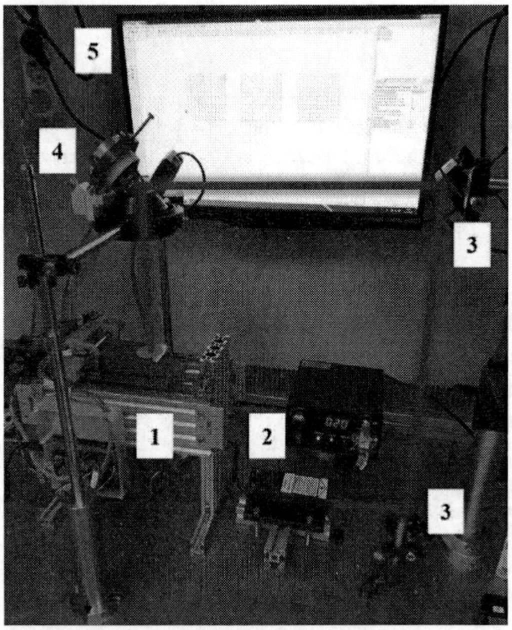

Figure 1. Experimental setup for SLS of polymeric powders:
1 – layer formation module;
2 – laser source with a wavelength of 405 nm;
3 – lens;
4 – galvanic scanner with f-theta objective;
5 – PC.

Basic building block of the layer formation module was made of stainless steel and contained two cylindrical chambers with pistons. Every piston was 30 mm in diameter equipped with an individual step drive providing maximum movement of 30 mm with a minimum pitch of 15 μm. The transfer of a new powder layer was carried out by a specialized device equipped with a removable polished hollow shaft (8 mm in diameter) which had a rotative levelling surface for the formation of a uniform and dense powder layer.

The main components of the scanning laser beam module included 405 nm diode laser (Fig. 1 "2"), model MDL-III-405 (Changchun New Industries Optoelectronics Tech. Co., Ltd, P.R. China), with the radiation power range from 10 to 90 mW and commercially available mirror galvanic scanner (Fig. 1 "4") (LScan XY, Ateko, Russia) operating in a wide range of wavelengths (from UV to near IR). The laser beam was focused on the surface of the powder with the system of lens (Fig. 1 "3") and f-theta SL-

266-100-160 objective (Ronar-Smith, Singapore) into a spot with a diameter of about 20 μm (at a level of 1 / e^2).

The movement of the pistons as well as the scanning laser beam was controlled with the developed in IPT RAS software installed on a personal computer (Fig. 1 "5").

2.3. Unsintered powder removal

One of the important and crucial stages of 3D printing of polymer structures by the SSLS method was the final procedure of removing powder particles that were not exposed to laser radiation. In the case of PMMA (the characteristic width of the sintered tracks is less than 100 μm) attempts to simply extract these particles with compressed air (as it is usually done for polyamide) led to the destruction of the resulting structures due to their insufficient mechanical strength.

In order to remove unsintered PMMA particles a method of washing out by processing the formed structures in an ultrasonic bath with a 0.25% aqueous solution of magnesium laureth sulfate (Industria Chimica Panzeri Srl, Italy) was proposed. After that, the samples were washed with distilled water and dried in air at a temperature of 20°C for 24 hours for further analysis.

2.4. Structural analyses

The initial powders as well as the structure of the resulting samples were examined without further processing with a Phenom ProX Scanning Electron Microscope (Phenom, The Netherlands).

3. Results and discussion

The appearance of the polymeric powder used in laser sintering process is presented in Fig. 2.

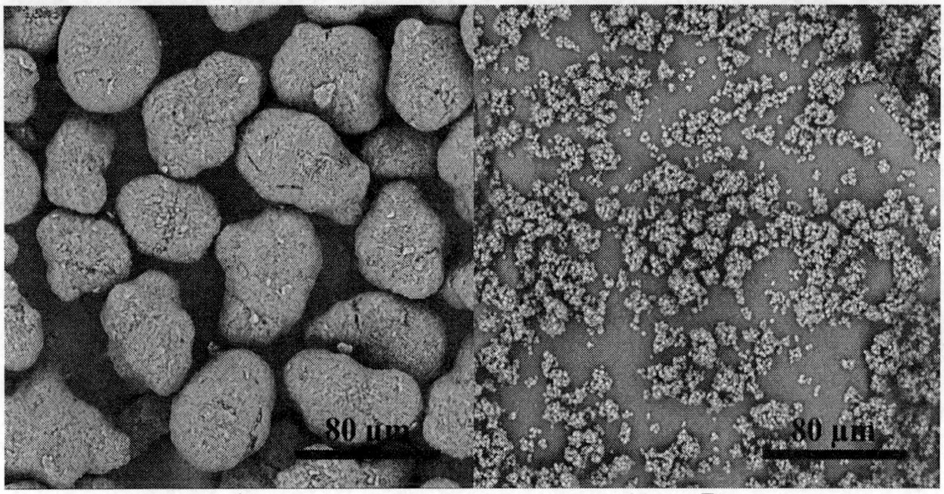

A B

Figure 2. Microphotographs of the raw powdered materials: A – polyamide 12, B – poly(methyl methacrylate).

PA12 had a characteristic particle size of 60 - 80 μm whereas the PMMA particles were produced from dispersion polymerization in supercritical carbon dioxide at ca. 3 μm particles. Because of the small size of the PMMA particles static forces predominate and the relatively large specific surface causes particle coagulation and formation of polymer agglomerates with a characteristic size of 20 μm.

To eliminate surface electrostatic effects in ultrafine polymer powder that could affect powder deposition levelling device and pistons of the developed layer formation module were electrically isolated from the rest of the experimental system and were able to be grounded or a constant potential was able to be applied to them in the range of ± 48 V. The resulting thicknesses of the powder layers were 150 μm and 40 μm for PA12 and PMMA, respectively, equivalent to two to three times the recommended average particle size. This ensures that most particles receive direct contact from the laser rather than relying on particle-to-particle conduction.

For high-resolution laser sintering process the 405 nm diode laser was used as it had a high quality monochromatic laser beam with the characteristic diameter of 20 μm. The fact that both PA12 and PMMA were virtually transparent to the wavelength used requires addition of an absorbent material. To do this we added a small amount of a disperse red dye to provide an absorption band at 405 nm and to carry out the SSLS process. The optimal laser parameters were laser power of 80 mW and scanning speed of 5 and 10 mm/s for PA12 and PMMA, respectively.

Typical structures obtained by the SSLS method from commercial PA12 and ultrafine PMMA powders under the action of continuous laser radiation at a wavelength of 405 nm are presented in Fig.3.

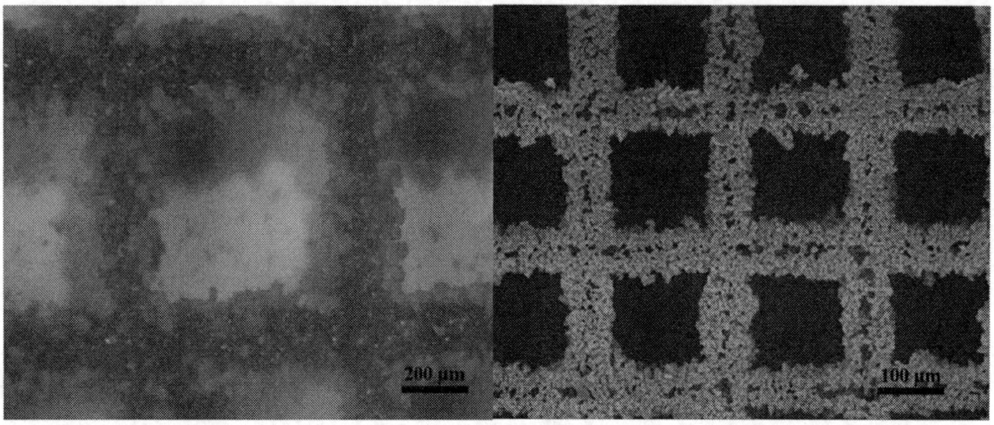

A B

Figure 3. Formed using surface selective laser sintering method structures from: A – polyamide 12, B – poly(methyl methacrylate).

Typical experimental width of the sintered tracks was 200-300 μm for PA12 powder and 80-90 μm – for PMMA powder which showed the significant improvement of the spatial resolution of the formed structures when using novel poly(methyl methacrylate) microparticles in comparison with commercially available polyamides.

4. Conclusion

Experimental system for forming homogeneous thin layers of ultrafine PMMA powders and their further laser sintering has been developed. Based on the used laser radiation source with the wavelength of 405 nm the optimization of powders optical characteristics has been conducted using a red dye as radiation absorbing agent for the surface selective laser sintering.

Thus, unique powder materials were developed for using in the process of three-dimensional printing based on surface selective laser sintering. There was a significant (almost threefold) increase in spatial resolution of formed polymer structures in comparison with traditional materials used.

5. Acknowledgments

This work was supported by the Ministry of Science and Higher Education within the State assignment of the Federal Research Centre "Crystallography and Photonics" RAS in part of the development of the SSLS method for the formation of polymer structures and by the Russian Foundation for Basic Research (grant No. 17-53-10014 KO_a) in part of the development of new materials for three-dimensional printing.

References

[1] Yap C, Chua C, Dong Z, Liu Z, Zhang D, Loh L and Sing S 2015 Review of selective laser melting: Materials and applications *Applied physics reviews* **2** 041101
[2] Antonov E, Bagratashvili V, Howdle S, Konovalov A, Popov V and Panchenko V 2006 Fabrication of polymer scaffolds for tissue engineering using surface selective laser sintering *Laser Physics* **16** 774-787
Antonov E, Bagratashvili V, Whitaker M, Barry J., Shakesheff K, Konovalov A, Popov V and Howdle S 2005 Three- dimensional bioactive and biodegradable scaffolds fabricated by surface- selective laser sintering *Advanced materials* **17** 327-30
[3] Yuan S, Bai J, Chua C, Wei J and Zhou K 2016 Material evaluation and process optimization of CNT-coated polymer powders for selective laser sintering *Polymers* **8** 370
[4] Koo J, Ortiz R, Ong B and Wu H 2017 Polymer nanocomposites for laser additive manufacturing *Laser Addit Manuf Mater Des Technol Appl* 205–35
[5] McAllister T, Farrand L and Howdle S 2016 Improved particle size control for the dispersion polymerization of methyl methacrylate in supercritical carbon dioxide *Macromolecular Chemistry and Physics* **217** 2294-2301

IOP Conf. Series: Materials Science and Engineering **848** (2020) 012087 doi:10.1088/1757-899X/848/1/012087

Mass-transfer kinetics, structure, and tribological properties of coatings deposited on steel in Ar or N_2+O_2 by electro-spark alloying using Cr_3C_2-NiAl electrodes

Ph V Kiryukhantsev-Korneev[1], A D Sytchenko[1*], V A Gorshkov[2] and E A Levashov[1]

[1] «National University of Science and Technology «MISiS», Leninsky prosp., 4, 119049 Moscow, Russia

[2] Merzhanov Institute of Structural Macrokinetics and Materials Science Russian Academy of Sciences, 142432 Chernogolovka, Russia

Abstract. Multicomponent coatings were obtained by electro-spark alloying using Cr_3C_2-NiAl electrodes, in air and in argon medium at direct and opposite polarity. Cr_3C_2-NiAl electrodes were manufactured by self-propagating high temperature synthesis. Mass transfer kinetics was determined by gravimetric method. The structure and composition of the coatings were evaluated by scanning electron microscopy and energy-dispersive analysis methods. Friction coefficient was measured using pin-on-disk test. Roughness parameters and wear tracks were studied by optical profilometry. It was found that the opposite polarity in combination with a protective environment is optimal in terms of growth rate, structure and tribological properties. The coatings obtained by the best regime had a relatively low roughness (Ra=4 μm), high wear resistance and a low friction coefficient at the level of 0.18.

1. Introduction

To improve the properties and increase the service life of important equipment components exposed to wear, high temperatures and aggressive environments, coatings based on chromium carbide, Cr_3C_2 [1-3], which has high mechanical characteristics and corrosion resistance, are used. The introduction of a NiCr metal binder into the coating composition reduces the coefficient of friction [4] and increases the resistance to oxidation at high temperatures [5]. Such coatings are also used to reduce abrasive and erosion wear at 850 ° C [6]. Increasing the heat resistance to temperatures of 1100 and 1200°C is achieved by using of NiAl binder [7]. The positive role of the NiAl was previously shown by the example of TiC-NiAl coatings [8]. It can be noted that NiAl is superior to NiCr in terms of corrosion resistance in KCl and KCl/K_2SO_4 at T = 600 °C. [9]. Cr_3C_2-based coatings are usually manufactured by thermal spraying [10-12]. A promising method to obtain these coatings is electro-spark alloying (ESA). The method has the following advantages: simplicity of equipment, high degree of use of the material, no strict requirements for surface preparation before deposition, high adhesion of the coating, low thermal impact on the substrate, satisfactory surface roughness, easy control of parameters and deposition conditions, composition control due to different media [13-16]. The aim of this work is to study the influence of the gas medium on the mass transfer kinetics and tribological properties of coatings obtained by ESA using the Cr_3C_2-NiAl electrode.

2. Materials and Methods

Electrodes 4×4×50 mm in size for ESA were cut from a workpiece obtained by self-propagating high-temperature synthesis (SHS) technology by SHS-casting method. The coatings were deposited onto

5140 steel discs Ø30×5 mm in size using the "Alier-Metal 303" machine with the following parameters: current of 120 A, voltage of 20 V, pulse duration of 20 μs, frequency of 640 Hz. Two types of polarity were used: DP - direct (cathode is a substrate) and OP - opposite (cathode is an electrode). The coatings were deposited in air or Ar at a normal pressure for 30 min. The mass transfer kinetics was studied by gravimetric method using analytical weights KERN 770. Morphology, elemental and phase composition of the coatings were estimated by scanning electron microscopy (SEM) and energy dispersion spectroscopy (EDS) on a Hitachi S-3400N microscope with the prefix Noran 7 Thermo. Tribological tests to determine the friction coefficient were carried out on the Tribometer CSM Instruments device according to the "pin-on-disk" scheme. The samples were tested in contact with a 6 mm diameter Al_2O_3 ball at room temperature with a linear velocity of 10 cm/s at a normal load of 1 N. An optical profiler "WYKO NT 110" was used to study the coating's surface roughness and wear tribo-tracks.

3. Results and Discussion

Mass transfer characteristics, such as a total weight gain of the coating (ΣΔC), total erosion of the electrode (ΣΔE), coating growth rate (V_c), and electrode erosion rate (V_e), are presented in Table 1.

Table 1. Mass transfer characteristics

Mode	Atmosphere	ΣΔC, mg	ΣΔE, mg	V_c, mg/min	V_e, mg/min
OP	Ar	0.34	0.45	10.3	13.6
OP	Air	0.31	1.06	9.3	31.7
DP	Ar	0.39	0.62	11.7	18.6
DP	Air	1.12	1.66	33.7	49.7

The obtained data indicate a weak influence of the medium on the growth characteristics of the coating in the opposite polarity mode. The total gain and growth rate of the coating were 0.31-0.34 mg and 9.3-10.3 mg/min, respectively. Carrying out the process in air, however, affected the characteristics of the erosion of the electrode, ΣΔE and V_e, increasing them almost twice, which can be attributed to the fact that the erosion is intensified by the use of active oxygen ions in the spark discharge plasma. The transition to direct polarity in the case of argon medium slightly increased the mass transfer characteristics (Table. 1). In the case of deposition in a protective environment in the DP mode, the growth rate of the coating and the erosion rate of the electrode were 11.7 and 18.6 mg/min, correspondingly. During the transition to the air medium, V_c and V_e increased 2.9 and 2.7 times, respectively. Other parameters also increased approximately three times. Apparently, in the air at DP there is a strong oxidation of the growing coating, which is confirmed by the data obtained with the help of EDS.

The curves describing mass transfer at ESD are very different (Figure 1). In all cases, when Ar was used, the coating mass increases smoothly, the electrode mass decreases, and the mass transfer kinetics curves have a character close to linear. During the process in the air in the OP mode from 0 to 9 minutes, an increase in mass was observed, after which, by 18 minutes of treatment, the mass of the sample dropped to the initial value. Further processing showed a decrease in the weight of the substrate, probably as a result of the reaction of oxygen with carbon and the formation of CO_2. The decrease in the substrate mass was accompanied by an increase in the rate of erosion of the electrode, which also indicates decarbonization. Similar processes occur during oxidation of carbon-containing ESA coatings during air annealing [17].

The coatings deposited at OP and DP had the same composition, thus Table 2 presents the elemental composition of samples obtained in the OP mode only. The results are averaged over the area of the EDS analysis.

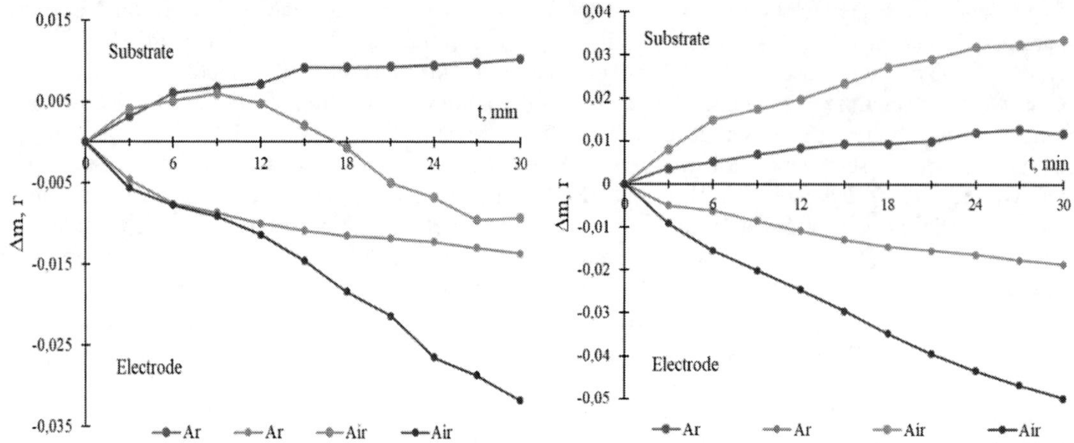

Figure 1. Results of gravimetric analysis of the weight gain of coatings and erosion of electrodes obtained in Ar and air in the OP (a) and DP (b) regimes

Table 2. Concentrations of elements for coatings obtained at OP, at. %

Atmosphere	C	Cr	Al	Ni	Fe	O
Ar	33.8	20.8	1.9	4.9	30.8	7.8
Air	27.4	14.5	1.2	1.5	27.4	28

For coatings obtained in Ar and air, the concentration of the substrate element (Fe) is observed at the level of 27.4-30.8 at.%. According to the content of the main elements, a significant difference was found between the samples obtained in Ar and in air. The concentration of C, Cr, Ni and Al for the coating obtained in air was lower by 19, 30, 70 and 37%, respectively. It can be explained by reaction of the coating material with oxygen during deposition, which is confirmed by the high concentration of oxygen of 28 at.%. Figure 2 shows SEM-images of the surface of coatings obtained by OP and DP in different gas media.

The samples have a structure typical for ESA coatings. Coatings have surface defects such as cracks, drops and splashes of molten material (Figure 2 (c, d)). On the surface of the sample obtained in argon at OP, small oxidation zones (FeO_x) are formed, corresponding to dark gray areas in micrographs. The total area of such areas does not exceed 5 % of the coatings area. Aluminum is present as irregular-shaped Al_2O_3 grains (black dots on micrographs). When ESA was performed in air in OP mode, the area with high oxygen content increases, and takes more than 50%. In this zone, there is the presence of iron and chromium oxides (dark gray areas). Areas with minimal oxygen content appear in the images as light gray areas. Note that Al_2O_3 particles are also found on the surface of the sample. Similar results on the influence of media on the structure of coatings were observed for coatings obtained in the DP mode. The roughness of the coatings deposited at OP in Ar and in air was 4.0 and 3.6 μm, respectively. When using DP, the roughness of the coatings obtained in Ar was equal to 3.6 μm, and when doped in air, Ra=4.6 μm. In the case of deposition in argon during the transition from OP to DP, a decrease in roughness is observed, as shown in [18].

The results of tribological tests are shown in Figure 3. The coating deposited in argon at OP had the lowest and most stable friction coefficient (f). The value of 0.22 was achieved at the first contact of the counter-body with the surface of the coating, after which at a distance of 3 to 50 meters the average coefficient of friction was 0.18.

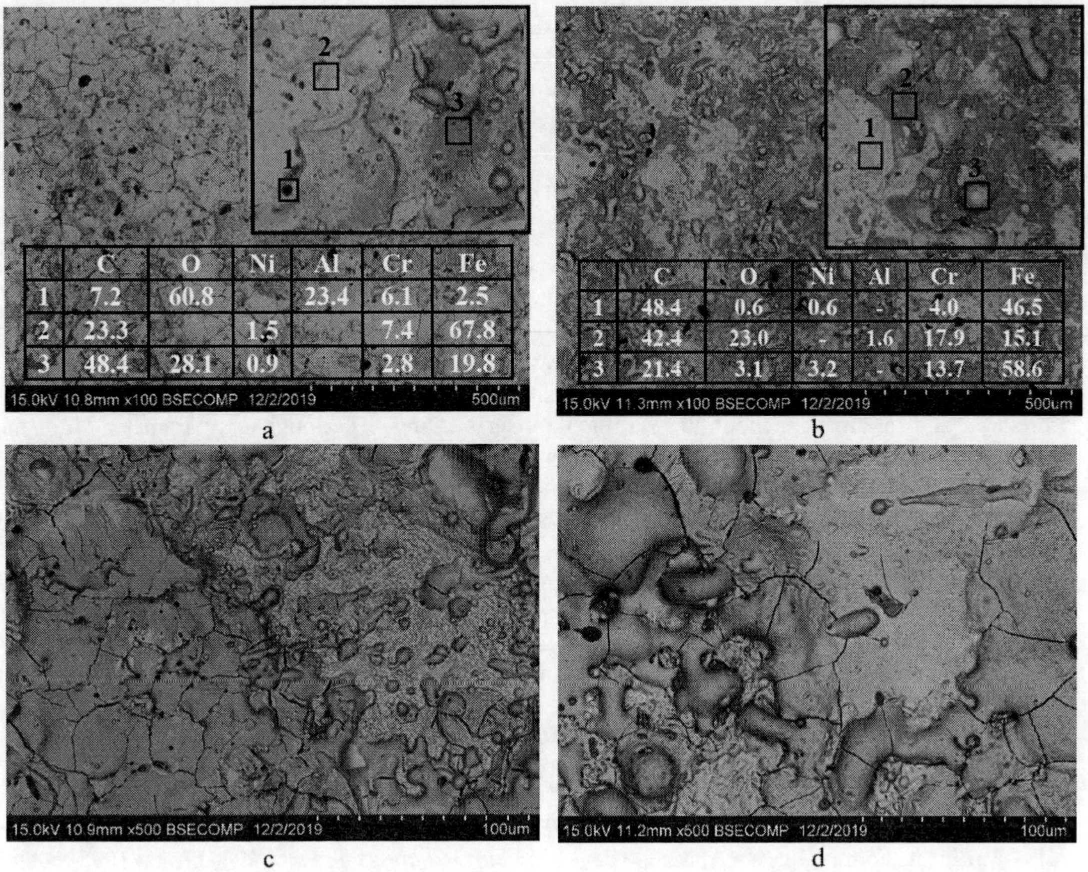

Figure 2. Top-view SEM-images of the coatings: a – OP in Ar, b - OP in N_2+O_2, c - DP in Ar, d - DP in N_2+O_2

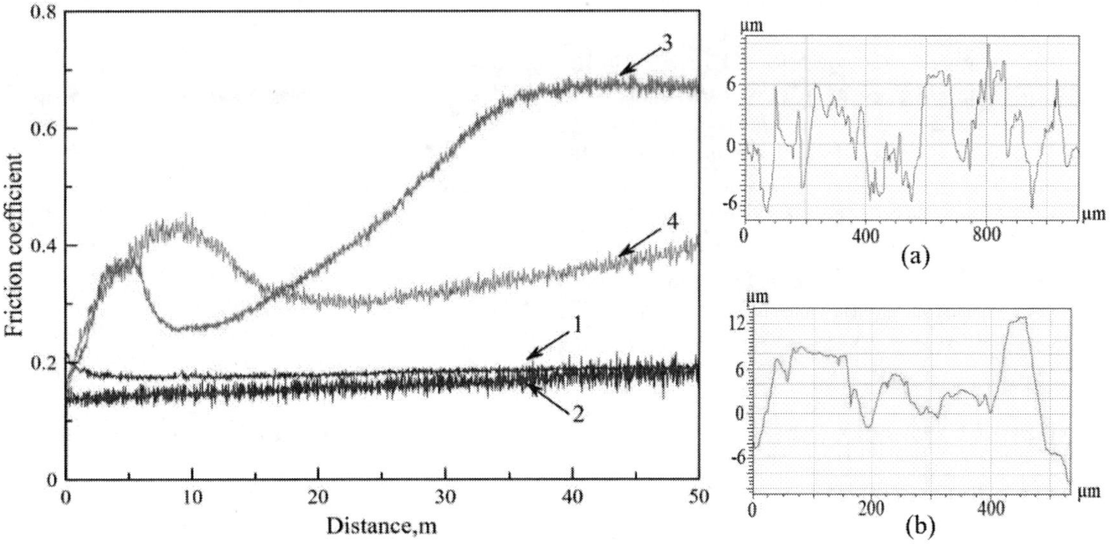

Figure 3. Friction coefficient of coatings obtained in Ar (1,3) and air (2,4) at OP (1,2) and DP (3,4) modes. Insert shows the 2D profiles of the wear tracks of samples obtained at DP in Ar (a) or air (b).

Close values were obtained for the coating deposited in the air in the OP mode. The friction coefficient of this sample at the beginning of the test had a value of 0.14, and throughout the distance it gradually increased to f=0.18. In the case of DP and Ar medium, f increased to 0.37 at a run length of 3 m, then the friction coefficient gradually increased and stabilized at 0.67. The coating obtained in the air in the DP mode also showed a jump of f at a distance of 20-50 m, the coefficient of friction gradually increased from 0.3 to 0.4. As shown on insert in Fig. 3 the wear of coatings is within the roughness and is difficult to estimate it.

4. Conclusion

Wear-resistant coatings were obtained by electro-spark alloying using Cr_3C_2-NiAl electrode. ESA was carried out in different gas media at direct and opposite polarity. According to SEM data, oxygen-containing zones were present on the surface of all samples, which in the case of argon environment were 5 %, and for air were about 50 % of the total surface area. The roughness parameter Ra of the coatings depended on the working medium and was 3.3-3.6 μm for samples obtained in Ar, and 5.4-5.6 μm for coatings manufactured in air. The friction coefficient of the coatings depended more on the polarity than on the gas medium. Samples deposited in the opposite polarity mode had the lowest coefficient of friction in the range of 0.14-0.20.

Acknowledgments

The work was financially supported by the Russian Science Foundation (Agreement No. 15-19-00203). The authors are grateful to N.V. Shvindina for the help with SEM study.

References

[1] Matthews S, James B, Hyland M 2013 *Corros. Sci.* **70** pp 203–211
[2] Sidhu H S, Sidhu B S, Prakash S 2006 *J. Mater. Process. Technol.* **171** pp 77–82
[3] Cekada M, Panjan P, Maček M, Šmíd P 2002 *Surf. Coatings Technol.* **151–152** pp 31–35
[4] J A Picas, Forn A, Igartua A, Mendoza G 2003 *Surf. Coatings Technol* **174–175** pp 1095–100
[5] Houdková S, Česánek Z, Smazalová E, Lukáč F 2018 *J. Therm. Spray Technol.* **26** pp 1720–31
[6] Janka L, Norpoth J, Trache R, Thiele S, Berger L-M 2016 *Surf. Coat. Technol.***291** 444–51
[7] Castello P, Stott F H, Gesmundo F 1998 *Oxidation of Metals* **49** pp 583–610
[8] Kiryukhantsev-Korneev Ph V, Sheveyko A N, Shvindina N V, Levashov E A and Shtansky D V 2018 *Ceram. Int.* **44** pp 7637–46
[9] Jafari R, Sadeghi E 2019 *Corros. Sci.* 108066
[10] Siao Ming Ang A, Howse H, Wade S A, Berndt C C 2016 *J. Therm. Spray Technol.* **25** pp 28–35
[11] Liu Q and et al 2019 *J. Alloys Compd.* **803** pp 730–41
[12] Berger L-M 2015 *Int. J. Refract. Met. Hard Mater.* **49** pp 350–364
[13] Kudryashov A E, Kiryukhantsev-Korneev Ph V, Petrzhik M I, Levashov E A 2019 *CIS Iron and Steel Review* **18** pp 46–51
[14] Johnson R N, Sheldon G L 1986 *J. Vac. Sci. Technol.* **4** pp 2740–46
[15] Peng H, Qian Y, Chang Z L and Wang R J 2007 *Sol. St. Phen.* **127** pp 325–30
[16] Shtansky D V, et al 2012 *Surf. Coatings Technol* **208** pp 14–23
[17] Kiryukhantsev-Korneev F V, Sytchenko A D, Kudryashov A E, Levashov E A 2018 *CIS Iron and Steel Review* **16** pp 57–62
[18] Kiryukhantsev-Korneeva Ph V, Sytchenko A D, Levashov E A 2019 *Russian Journal of Non-Ferrous Metals* **60** pp 662–72.

IOP Conf. Series: Materials Science and Engineering **848** (2020) 012088 doi:10.1088/1757-899X/848/1/012088

Reducing of holes effect on composite elements strength

E Kh Akhmedshin[1], A N Polilov[2] and N A Tatus[2,3]

[1]Moscow Polytechnic University, 38, Bolshaya Semyonovskaya str., Moscow, 107023, Russia
[2]Blagonravov Mechanical Engineering Research Institute of Russian Academy of Science, 4, Maly Kharitonyevsky Pereulok, Moscow, 101000, Russia
[3]Moscow State University of Civil Engineering (MGSU), 26, Yaroslavskoye Shosse, Moscow, Russia
fallenking74@gmail.com

Abstract. Reducing the effect of holes on the attachment points of composite elements is one of the key tasks from the time the composites appeared to the present. Since the composite material and the structure are inseparable, it is necessary to design the joints at the stage of manufacturing the structural element. The work is devoted to the experimental assessment of the effect of holes made by various technologies on the strength of composite specimen. A significant increase in the bearing capacity of composite flat specimen, holes in which are made by the separation of the fibers during the manufacturing. For fiberglass, the bearing capacity of a specimen with a drilled hole decreased by 4 times, and the specimen with fibers "flowing around" the hole by only 25% compared to a solid flat specimen.

Introduction

The manufacturing technology of composite elements is of primary importance. The composite does not exist outside the construction element. Subsequent machining carries a serious reduction in bearing capacity, so it should be avoided. Holes - a structural element, without which it is difficult to imagine the connection of parts. Cut fibers in composites are unacceptable, since it is the fibers carries the load. Making holes in composites by drilling breaks the integrity of the element and reduces the bearing capacity of not only a single composite element, but also the entire structure. Nature creates wood structures with many branches, but without cut fibers. (Figure 1). Moreover, the branch-trunk junction is so strong that when the ultimate loads are reached, the branch breaks, not the branch-trunk connection. A person has been working with wood materials throughout his existence, so it is intuitively clear that for fiber materials in the manufacture of holes it is better to expand the space between the fibers than to drill a hole. Why is this not done? - This is a question for technologists. Any, even the most insignificant change in technology leads to a partial or complete shutdown of production and is associated with serious costs - this may be the main problem.

An analysis of the literature shows that the authors are not alone in their research, they are engaged in similar work at the MSTU named after N.E.Bauman. Colleagues studied [1] the stresses in terms of fiber in the holes area, the coefficient of stress intensity near the hole was determined. Study was carried out with cross reinforcement ring specimens. An increase in bearing capacity specimens with a puncture hole compared to specimens in which holes were drilled by 44% is shown. Studies were also carried out [2] of the effect of various types holes on the bearing capacity of specimens with orthogonal reinforcement. When tensed, specimens with a puncture hole a showed a 12% increase in breaking strength compared to drilled.

IOP Conf. Series: Materials Science and Engineering **848** (2020) 012088 doi:10.1088/1757-899X/848/1/012088

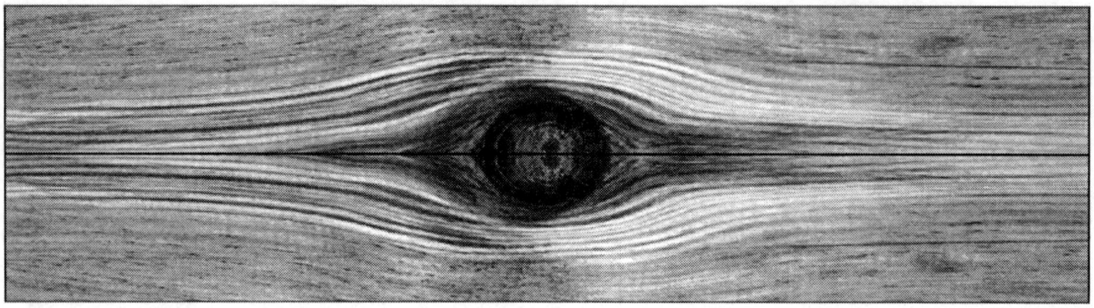

Figure 1. Pine knot "streamlined" by fibers.

Problem statement and experiment

That is, the decrease in the effect of holes made without breaking the fibers on the bearing capacity of specimens with cross reinforcement has been experimentally and theoretically proved [3]. The aim of this work is to investigate the effect of holes on the strength of unidirectional composites. It is clear that the effect of the holes on the bearing capacity of unidirectional specimens is higher than on the bearing capacity of specimens with cross reinforcement. The purpose of the work is to evaluate this effect. To achieve this goal, it was required to create an installation and work out the technology for manufacturing specimens with holes of various types. It was carrying out a series of experiments.

The installation for manufacturing composite specimens created at IMASH RAS (Figure 2) is a winding machine with a rotating mandrel for laying fibers. After winding a predetermined number of layers, the mandrel is placed in a vacuum bag for impregnation with binder. Glass fibers were wound with an elastic modulus of about 60 GPa, a tensile strength of about 2 GPa, and Etal-200M epoxy resin with Etal-1472 hardener acted as a binder.

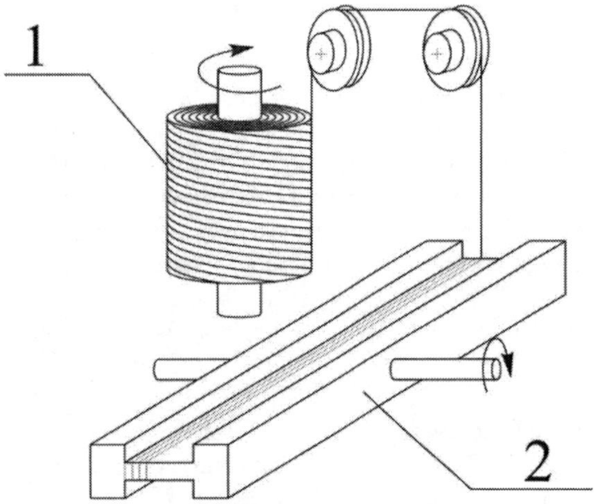

Figure 2. The scheme of manufacturing specimens: 1. - a coil with fiberglass thread, 2 mandrel.

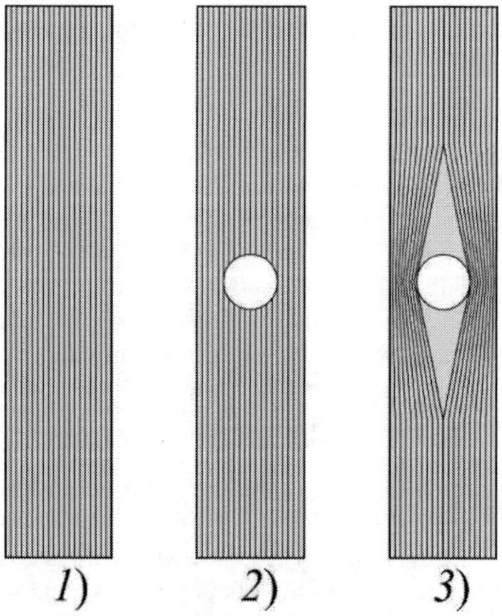

Figure 3. Types of specimens: 1) type 1, 2) type 2, 3) type 3.

After winding a predetermined number of layers, the mandrel is placed in a vacuum bag, where the specimen is impregnated with a binder, followed by curing at a temperature of 80° C. Specimens of three types were made: 1 - continuous (Figure 3, 1), 2 - with a hole drilled after curing (Figure 3, 2), 3 - with fibers "flowing around" the hole (Figure 3, 3).

Tensile tests of the specimens were carried out in self-tightening grippers. The specimen was brought to complete failure. The maximum load (bearing capacity) was measured. The data are presented in figure 4 and in table 1.

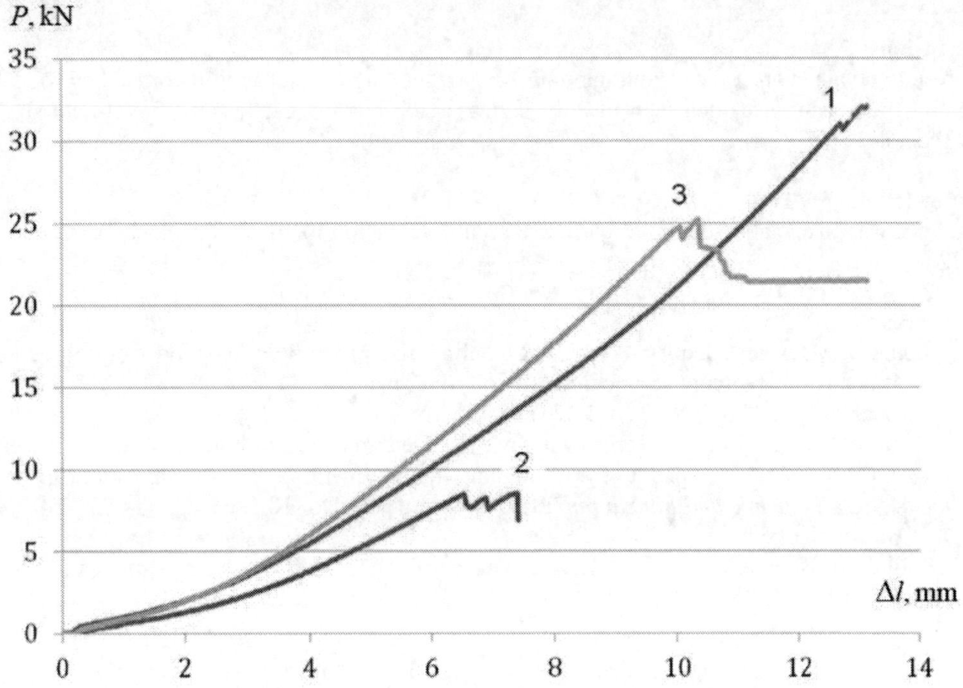

Figure 4. The load-displacement diagram for flat specimens of different types: 1 - type 1, 2 - type 2, 3 - type 3.

Table 1. The bearing capacity for flat specimens of different types.

Types of Flat Specimens	Type 1	Type 2	Type 3
Maximum load, N	32 090	8 520	25 200
Maximum load as a percentage of ultimate load for smooth specimens, %	100	27	79

The discussion of the results

Type 1 specimens, on which the manufacturing and testing technology was tested, showed variation in the maximum tensile load. The first few specimens broke at very low loads until the manufacturing technology was developed, which confirms the significant dependence of composites mechanical characteristics on the manufacturing technology. The bearing capacity of the specimens in the process of testing the technology increased from 8 to 32 kN.

Specimens of types 2 and 3 were made according to the technology developed in the manufacture of a specimen of type 1. The manufacturing quality of the specimens is quite high, despite the fact that they were manufactured not in production, but in laboratory conditions.

The destruction of type 2 specimens occurred according to the standard scheme for composites - the cut fibers practically do not carry loads and the specimen is destroyed by shear stresses, figure 4. The bearing capacity of such specimens decrease by 4 times compared with continuous type 1.

Type 3 specimens showed a decrease in bearing capacity of less than a quarter of the standard, which confirms the theory of the low impact on the bearing capacity of the holes around which the fibers are laid and explains the strength of the tree-trunk joints.

Conclusion

The results obtained are very encouraging and are in good agreement with theory. The production of holes by the puncture method significantly increases the bearing capacity of the structural element with the hole.

Acknowledgments

This work was financially supported by the Russian Foundation for Basic Research, grant No. 18-08-00372 A

References

[1] Komkov M A and Kolganov A V Modeling the process of forming holes in composite structures by piercing an uncured polymer fiber material *Bulletin of MSTU N E Bauman Ser. "Engineering"* 2007 No 3 p 33-47 (in Russian)

[2] Komkov M A, Bolotin Yu Z and Vasilyeva T V Determination of hole forming parameters in an uncured fabric composite by a piercing method with a pointed indenter. *Engineering Journal: Science and Innovation* 2017 No 9 (69) p 11 Doi 10.18698/2308-6033-2017-9-1678 (in Russian)

[3] Polilov A N and Tatus N A Biomechanics of the strength of fiber composites *Moscow: FIZMATLIT* 2018 328 p ISBN 978-5-9221-1760-9 (in Russian)

IOP Conf. Series: Materials Science and Engineering **848** (2020) 012089 doi:10.1088/1757-899X/848/1/012089

Modeling of paths and energy losses of high-energy ions in single-layered and multilayered materials

D I Tishkevich[1,2], S S Grabchikov[1], E A Grabchikova[1], D S Vasin[1], S B Lastovskiy[1], A S Yakushevich[1], D A Vinnik[2], T I Zubar[1,2], I V Kalagin[3], S V Mitrofanov[3], D V Yakimchuk[4] and A V Trukhanov[1,2]

[1]Laboratory of Magnetic Films Physics, Scientific-Practical Materials Research Centre of National Academy of Sciences of Belarus, P. Brovki str. 19, Minsk, 220072, Belarus

[2]Laboratory of Single Crystal Growth, South Ural State University, Lenin ave. 76, Chelyabinsk, 454080, Russia

[3]Nuclear Reactions Laboratory, Joint Institute of Nuclear Research, J. Curi str. 6, Dubna, 141980, Russia

[4]Cryogenic Research Department, Scientific-Practical Materials Research Centre of National Academy of Sciences of Belarus, P. Brovki str. 19, Minsk, 220072, Belarus

E-mail: dashachushkova@gmail.com

Abstract. Modern electronic systems and equipment used in aerospace and nuclear technology, as well as many scientific and medical devices, are used under the influence of a wide range of ionizing radiation (electrons, protons, heavy charged particles etc.). Protons or heavy charged particles exposure can lead to failures in the operation of spacecrafts electronic devices, which is associated with the radiation effects occurrence in an integrated circuits. One of the most effective ways to solve this problem is protection by radiation shields. Linear and mass paths of protons and Ar^+ ions in Al, Al_2O_3, Bi, and $W_{77.7}Cu_{22.3}$ composite shields were calculated using a SRIM software package. It is shown that the protection efficiency against high-energy ions by materials with large atomic charge values (*Z*) is higher from the position of linear ranges of particles, and lower from the position of mass ranges than materials with low *Z* values. The dependence of the threshold energy on the serial number of particles for Al, Bi and $W_{77.7}Cu_{22.3}$ composite shields is determined. The effect of the sequence in the arrangement and layer thicknesses in the Bi/Al/Al_2O_3 multilayered structures on the protection efficiency against high-energy ions was studied.

1. Introduction

During developing of new materials used in the elements and equipment of aerospace application, it is necessary to provide requirements for resistance to the effects of various types of ionizing radiation (IR). The influence of protons or heavy high-energy ions can lead to sudden failures in the operation of electronic devices and spacecraft (SC) units, which is associated with the occurrence of radiation effects in integrated circuits (ICs) (a single failure and burnout, snapping, etc.) [1]. The solution of this problem is a very complex and multi-level task. In practice, to increase the radiation reliability of electronic components, as a rule, are used: constructive-technological (circuitry) design methods, "majorization" (the creation of a second, third, etc. in an IC – back-up, which in case of failure of the first one will replace it) and constructive protection (constructive elements of the SC, specialized bodies, local protection) [1, 2].

Often, to provide radiation protection of the spacecraft's electronic components from the effects of electrons and protons of the Earth's radiation belts (ERB), plates or coatings in the form of local protection are used that are formed on individual elements of electronic equipment [3-7]. However, there is an opinion [8], according to which protection against exposure of heavy high-energy ions, (the term heavy charged particles (HCP) is often used in the scientific literature) by using radiation shields is ineffective, since flows of secondary particles and fragments of matter, created by the interaction of HCP with shields material, may have a more negative effect than primary radiation. At the same time, it is known [9] that the fluxes of protons and ERB electrons are relatively stable, and the IR spectra are determined for the main near-Earth orbits. The fluxes of solar cosmic rays (SCR) and galactic cosmic rays (GCR) are unstable factors and are probabilistic in nature. SCR fluxes mainly consist of low-energy electrons and protons ($E \sim 1$ keV) with a density of 10^8–10^9 cm^{-2}□s^{-1}, however, during solar flares, the energies can reach 10^8–10^9 eV [10]. Protons predominate in the composition of GCR fluxes ($E \leq 10^{12}$ eV), the remaining nuclei account for less than 10%. GCR fluxes are characterized by a relatively low density – up to 5 cm^{-2}□s^{-1}, but possess enormous energy up to 10^{20} eV [11]. According to the data of [12], the energy spectrum of particles in outer space can be characterized by the following values of particle fluxes per 1 mm^2 per year: ~ 30 particles with E = 10 GeV, ~ $7 \cdot 10^{-3}$ particles with E = 1 TeV, ~ $1 \cdot 10^{-6}$ particles with E = 1 PeV.

In near-Earth space, the GCR energy spectrum differs significantly from the GCR spectrum in interstellar space, which is associated with the influence of the solar wind (heliosphere) and the Earth's magnetic field. Moreover, the influence of these factors on particles with energies above $10Z$ GeV (Z is the charge of atomic nuclei) is insignificant. On the other hand, electronic components are usually located inside the spacecraft behind the outer body and constructive elements. Therefore, in near-Earth orbits, the energy and the number of particles acting on the electronic components will be lower than those given above, and the issue of protecting electronic components from high-energy ions through the use of radiation shields is on great practical interest. In this regard, the problem of studying the interaction of HCP with protective shields materials is very relevant. Experimental studies require expensive accelerators capable of accelerating ions to enormous energies, therefore, computer simulation of the ions transmission of a given energy in various materials (software packages SRIM, GEANT4, SuperMC etc.) are widely used. This paper presents calculations of the ranges and energy losses of high-energy ions in single-layered and multilayered materials using the SRIM software package.

2. Modeling technique

The linear ranges and energy losses of ions were calculated using SRIM – 2013.00 program [13], in which the model of continuous deceleration is used to simulate the transmission of HCPs in a substance. As high-energy impacts, protons, He^+, C^+, Ne^+, Ar^+, Fe^+, Kr^+ and Xe^+ ions with energies from 10 MeV to 27.4 GeV were used. All ions were directed perpendicular to the shields surface. The number of particles in the simulation was 10^3 ions. Linear, mass ranges and threshold energies (E_{tsh}) were fixed along the most probable path length of high-energy ions. Single-layered shields based on Al, Al_2O_3, Bi, $W_{77.7}Cu_{22.3}$ composite and Bi/Al/Al_2O_3 multilayered structures with different sequences and thicknesses of individual layers were as model objects. The shield thickness based on Al, Al_2O_3 and Bi was 1.4 mm, and the thickness of the $W_{77.7}Cu_{22.3}$ composite shields was 1.5 mm. The multilayered structures of the four types had the following structure:

– Bi/Al/Al_2O_3 and Al/Al_2O_3/Bi with partial layer thicknesses: Bi – 0.6 mm, Al – 0.2 mm, Al_2O_3 – 0.6 mm and total thickness 1.4 mm;

– Bi/Al/Al_2O_3/Bi/Al/Al_2O_3/Bi/Al/Al_2O_3 and Al/Al_2O_3/Bi/Al/Al_2O_3/Bi/Al/Al_2O_3/Bi with partial layer thicknesses: Bi – 0.2 mm, Al – 0.067 mm, Al_2O_3 – 0.2 mm and total thickness 1.4 mm.

3. Results and discussion

Linear range (R) – the distance traveled by a particle in a substance to a complete stop. With an increase in the charge of atomic nuclei in the shield from Al, Bi and $W_{77.7}Cu_{22.3}$ composite, linear ranges decrease. Accordingly, the energies required for the particles to pass through the shields of a given thickness (*d*) increase with Z rising. The increase in the ions linear ranges with a change in the shield material in the sequence $W_{77.7}Cu_{22.3}$, Bi and Al are primarily due to the lower density of the materials. According to the data of [2], the relative density (ρ) for Al = 2.7, Al_2O_3 = 3.9, Bi = 9.8 and the $W_{77.7}Cu_{22.3}$ composite 17.02 g/cm^3. Along with linear ranges, the quantity – mass range (R_M) $=R{\cdot}\rho$, its dimension g·cm^{-2} is often used.

The calculated linear and mass ranges of protons and Ar^+ ions in shields from Al, Al_2O_3, Bi and $W_{77.7}Cu_{22.3}$ composite are shown in Tables 1 and 2, respectively. It should be noted that the protection effectiveness against HCP by materials with large *Z* values is higher than materials with low *Z*, from the position of particle linear ranges (linear shield thickness). From the standpoint of mass ranges (mass shields thickness), materials with low *Z* have more effective protection than materials with large *Z* values. From the presented results it also follows that the values of the energies of Ar^+ ions at which $R \le d$ for shields from Al, Bi and $W_{77.7}Cu_{22.3}$ composite with thicknesses $d_{Al} = d_{Bi}$ = 1.4 mm and d_{WCu} = 1.5 mm are 2.0; 3.16 and 4.77 GeV, respectively.

Table 1. Proton linear and mass ranges in Al, Al_2O_3, Bi and $W_{77.7}Cu_{22.3}$ composite.

E, MeV		10	15	16	20	23.5	31
Al	*R*, cm	0.063	0.127	0.136	-	-	-
Z=13	R_M, g/cm^2	0.167	0.343	0.367	-	-	-
Al_2O_3	*R*, cm	0.039	0.080	-	0.133	-	-
Z=13;8	R_M, g/cm^2	0.153	0.313	-	0.518	-	-
Bi	*R*, cm	0.035	-	-	0.109	0.135	-
Z = 83	R_M, g/cm^2	0.345	-	-	1.068	1.323	-
$W_{77.7}Cu_{22.3}$	*R*, cm	0.019	0.036	-	0.058	-	0.116
Z=74;29	R_M, g/cm^2	0.279	0.541	-	0.870	-	1.752

Table 2. Ar^+ ions linear and mass ranges in Al, Al_2O_3, Bi and $W_{77.7}Cu_{22.3}$ composite.

E, GeV		1.0	1.4	2.0	2.6	3.16	4.77
Al	*R*, cm	0.046	0.077	0.139	-	-	-
Z=13	R_m, g/cm^2	0.123	0.207	0.375	-	-	-
Al_2O_3	*R*, cm	0.028	0.049	0.088	0.139	-	-
Z=13;8	R_m, g/cm^2	0.109	0.189	0.342	0.542	-	-
Bi	*R*, cm	0.023	-	0.067	0.098	0.139	-
Z = 83	R_m, g/cm^2	0.229	-	0.655	-	1.362	-
$W_{77.7}Cu_{22.3}$	*R*, cm	0.012	-	0.036	-	-	0.148
Z=74;29	R_m, g/cm^2	0.187	-	0.539	-	-	2.235

The results presented in Tables 1, 2 are usually described by the well-known Bethe-Bloch formula [8, 11], according to which the main energy losses of ions with *E*> 2 MeV/nucleon in a substance are associated with ionization losses. Based on the Bethe-Bloch formula, the ionization losses are proportional to the squared charge of the particle's core (Z_i) and proportional to the charge of the atomic nucleus of the material. Consequently, the HCP ranges are reduced in the case of heavier HCP and materials of the medium with *Z* large values.

The distribution profiles of high-energy C^+, Ne^+, Ar^+, Kr^+ and Xe^+ ions in the studied materials have a pronounced peak, the so-called Bragg peak (Figure 1) [14]. The Bragg peak nature is related to the nature of the distribution of absorbed energy along the mean free range in the substance. The main effect of energy absorption occurs at the end of the particle linear mean free range, and the cross section of this process grows with a drop in energy, as a result of which the particle loses the bulk of the energy before it stops. The proton distribution profiles have a completely different form (Figure 1 d, e, f) – the particle flux is scattered in different directions. The proton trajectory is significantly affected by multiple scattering effects. The deviation of the particle from the initial direction is proportional to the number of collisions, depending on the serial number and energy of the HCP.

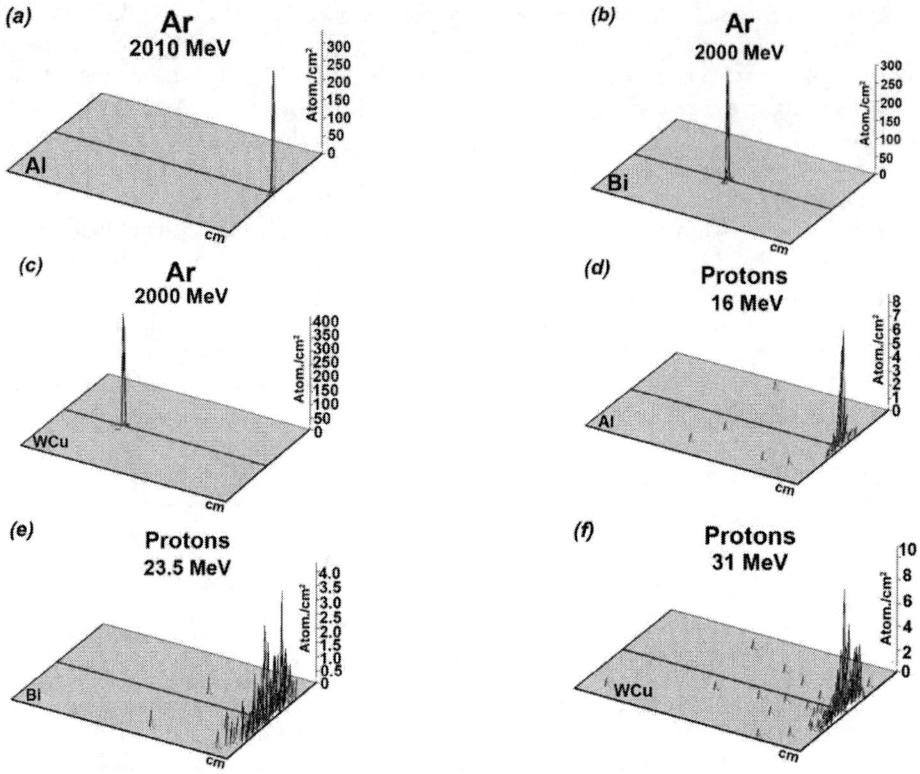

Figure 1. Ar^+ ions (*a, b, c*) and protons (*d, e, f*) histograms of the distribution in shields from Al (*a, d*), Bi (*b, e*) and $W_{77,7}Cu_{22,3}$ composite (*c, f*).

The obtained results are due to the interaction mechanisms of high-energy ions with matter and the associated losses of HCP energy (elastic collisions with substance atoms $E < 10^4$ eV, ionization losses at $E = 10^4$-10^8 eV and losses due to nuclear reactions at $E >$ 100 MeV/nucleon [11]). Figure 2 shows the ionization losses spectra (a) and nuclear reactions (b) during the transmission of protons with 22.5 MeV energy and Kr^+ ions with 9.23 GeV energy through 1.4 mm Bi shield. It can be seen that for protons and Ar^+ ions, the contribution to the energy losses from the ionization process is much higher (~ 10^2 - 10^3) than the contribution from nuclear reactions.

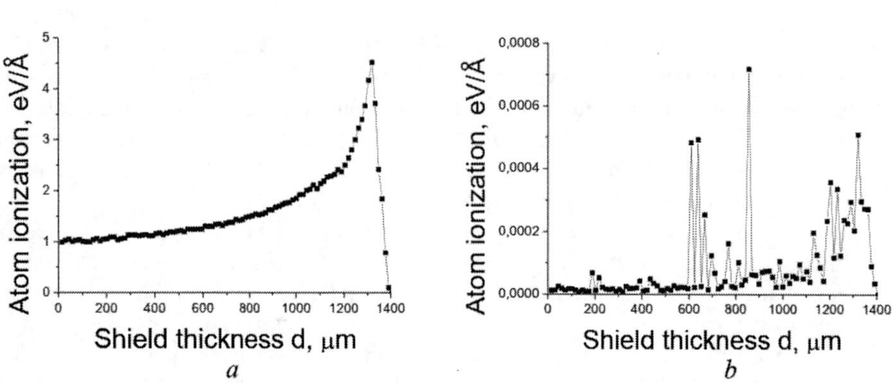

IOP Conf. Series: Materials Science and Engineering **848** (2020) 012089 doi:10.1088/1757-899X/848/1/012089

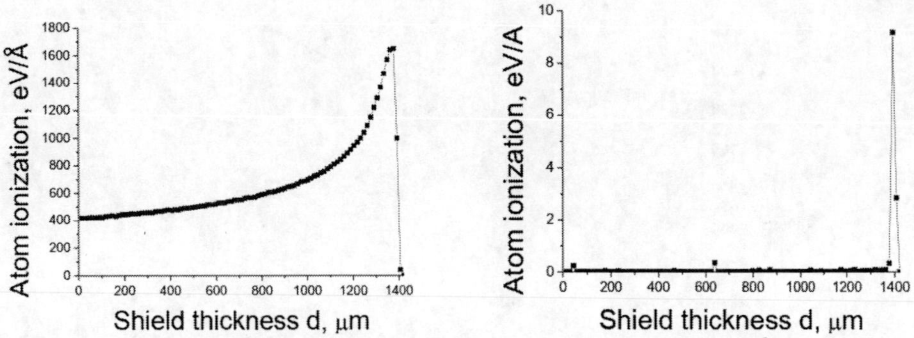

c *d*

Figure 2. The spectra of ionization losses (a, b) and nuclear reactions (b, d) during the protons transmission with 22.5 MeV (a, b) energy and Kr^+ ions with 9.23 GeV energy (b, d) through Bi shield with 1.4 mm thickness.

For radiation protection shields, of great interest is the parameter – the threshold ion energy (E_{tsh}), at which the linear range length is equal to the shield thickness. The calculations performed the high-energy protons, He^+, C^+, Ne^+, Ar^+, Fe^+, Kr^+ and Xe^+ ions ranges in Al, Bi and $W_{77,7}Cu_{22,3}$ composite shields and make it possible to plot the dependence of E_{tsh} on the HCP serial number (Figure 3). As can be seen from the above data, the shields based on the $W_{77,7}Cu_{22,3}$ composite exhibit the highest efficiency from the position of particle linear ranges. The threshold energies for this material with 1.5 mm thickness (d_m = 2.26 g/cm^2) when exposed to proton and He^+, C^+, Ne^+, Ar^+, Fe^+, Kr^+ and Xe^+ ions are 37; 480; 790; 1780; 4770; 7900; 14200 and 27000 MeV, respectively. The dependence of the threshold energy on the $W_{77,7}Cu_{22,3}$ composite linear thickness for Ar^+ ions is shown in Figure 4.

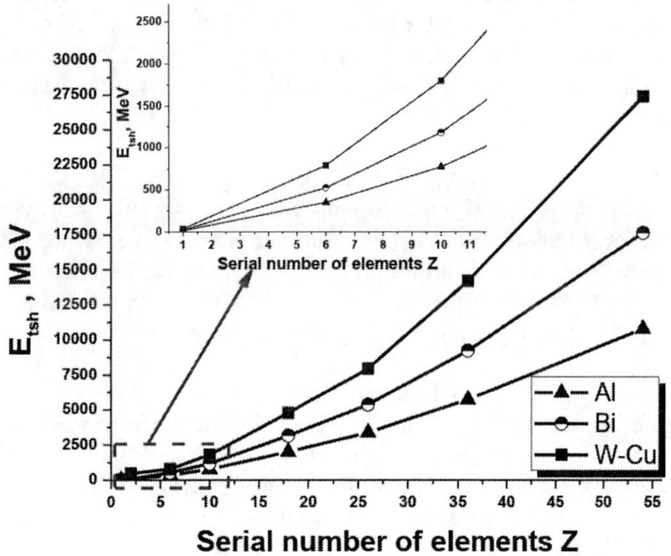

Figure 3. Threshold energy of shields based on Al, Bi and $W_{77,7}Cu_{22,3}$ composite with 1.5 mm thickness for protons, He^+, C^+, Ne^+, Ar^+, Fe^+, Kr^+ and Xe^+ ions.

IOP Conf. Series: Materials Science and Engineering **848** (2020) 012089 doi:10.1088/1757-899X/848/1/012089

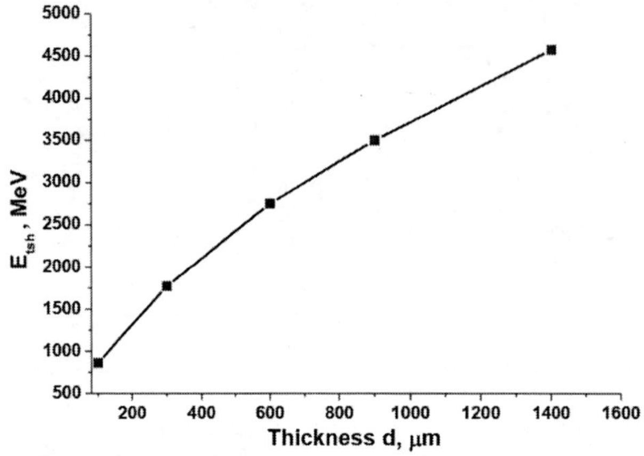

Figure 4. Threshold energy depending on the $W_{77,7}Cu_{22,3}$ composite thickness for Ar^+ ions.

The effect of multilayered structure in shields, namely, the sequence of arrangement and layers thickness, on the protection efficiency against high-energy ions, was studied on the $Bi/Al/Al_2O_3$ structures. Kr^+ ions ($Z = 36$) with an increased contribution of ionization processes to the total energy losses were selected as the acting particles. The calculated dependences of the linear ranges of Kr^+ ions on the number and layers sequence of arrangement of light and heavy elements in $Bi/Al/Al_2O_3$ multilayered structures are presented in Figure 5. For comparison, the same figure shows the simulation results for single-layered Al and Bi shields. From the linear particle ranges position, Al shields are the least effective, Bi shields are the most effective, and multilayered structures are located between them. In the three-layered $Bi/Al/Al_2O_3$ structures, the character of the dependences $R = R(E)$ is determined by the arrangement of individual layers sequence:

– in the case of the $Bi/Al/Al_2O_3$ structure, the $R = R(E)$ dependence up to 0.6 mm ($E \sim 5.4$ GeV) ranges coincides with a similar dependence for pure Bi, since this range corresponds to the thickness of the Bi layer of the multilayered structure. At energies $E > 5.4$ GeV, HCP penetrate into Al_2O_3 and Al layers, whose braking power is lower than that in Bi, as a result of which the efficiency of $Bi/Al/Al_2O_3$ shields at energies $E > 5.4$ GeV is lower than in single-layered Bi. The threshold energy for such structures is 8.1 GeV;

– in the case of the $Al/Al_2O_3/Bi$ structure, the starting point on the graph $R = R(E)$ corresponds to a 0.2 mm ($E = 2$ GeV) linear range and is equal to the thickness of the first Al layer. Therefore, this point coincides with that for single-layered Al. With a further increase in energy due to large losses in the Al_2O_3 layer, the curve $R = R(E)$ of the multilayered structure falls below the curve of single-layered Al. The third Bi layer reinforces this tendency and, with a 1.4 mm final thickness, the linear range approaches the range values for the $Bi/Al/Al_2O_3$ structure. The threshold energy for these structures is 7.9 GeV;

– in the case of the nine-layered $Bi/Al/Al_2O_3/Bi/Al/Al_2O_3/Bi/Al/Al_2O_3$ structure, a large number of layers and their small thicknesses neutralize the effects of different energy absorption by individual layers. It can be concluded that with an increase in layers number, the dependence of linear ranges on the ion energy approaches that shield of equivalent three-layered $Bi/Al/Al_2O_3$ material. Therefore, the dependence $R = R(E)$ for the nine-layered structure is located between the similar curves for the $Bi/Al/Al_2O_3$ and $Al/Al_2O_3/Bi$ structures.

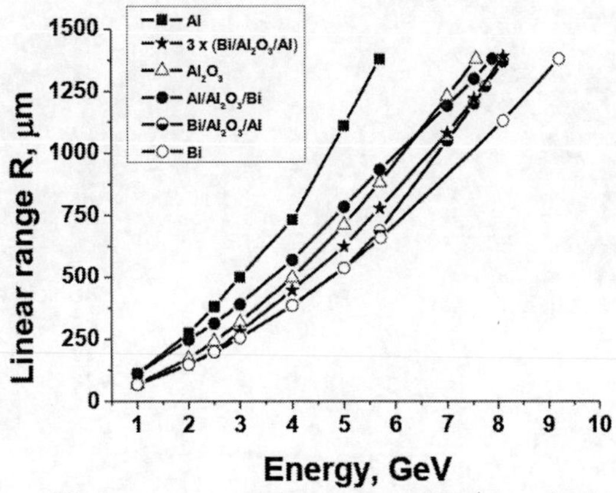

Figure 5. Dependences of linear ranges on the energy of Kr^+ ions in three-layered Bi/Al_2O_3/Al and Al/Al_2O_3/Bi structures; nine-layered structure Bi/Al_2O_3/Al/Bi/Al_2O_3/Al/Bi/Al_2O_3/Al; single-layered Al and Bi shields.

A value of R = 0.8 mm, which corresponds to the thickness of two Al and Al_2O_3 layers requires 5.2 GeV ion energy and a value of R = 1.4 mm requires 7.9 GeV an energy for the Al/Al_2O_3/Bi structures. Therefore, for the transmission through the Bi layer, the ions needed an additional energy of E = 2.7 GeV. For the Bi/Al/Al_2O_3 structure a value of R = 0.6 mm, which corresponds to the thickness of the Bi layer, requires an ion energy of about 5.4 GeV, and a value of R = 1.4 mm – 8.1 GeV. Therefore, for the transmission through Al and Al_2O_3 layers, an additional energy of 2.8 GeV was required. A different level of energy absorption by layers of a three-layered structure depending on their sequence is confirmed by ionization losses spectra calculated when protons with 20 MeV energy (a, b) and Kr^+ ions with 7.75 GeV energy (c, d) pass through multilayered Bi/Al/Al_2O_3 and Al_2O_3/Al/Bi structures with a 1.4 mm total thickness (Figure 6).

IOP Conf. Series: Materials Science and Engineering **848** (2020) 012089 doi:10.1088/1757-899X/848/1/012089

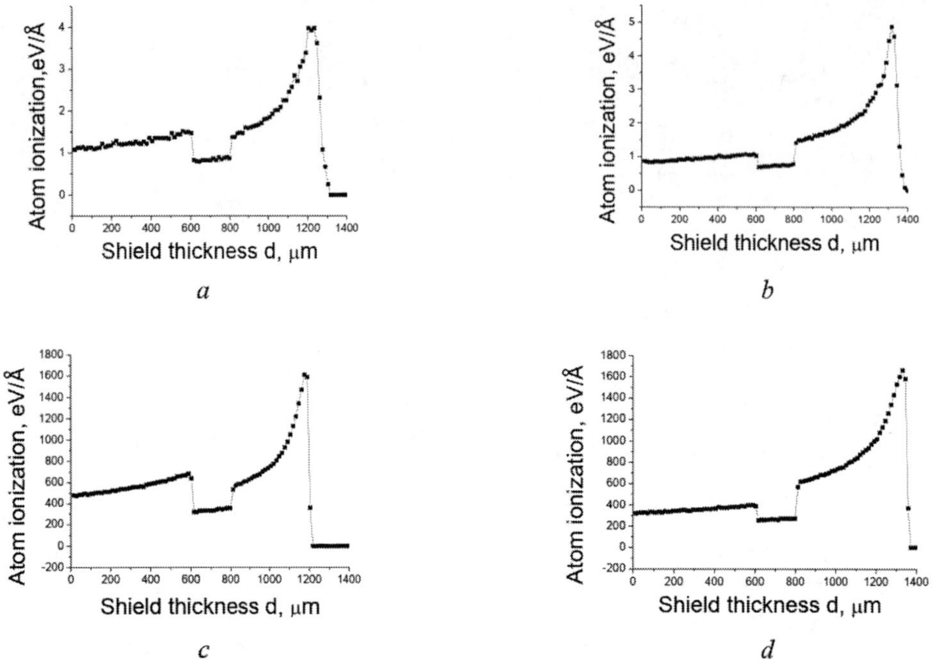

Figure 6. The ionization losses spectra for the protons transmissions with 20 MeV (*a, b*) energy and Kr^+ ions with 7.75 GeV (*c, d*) energy through the Bi/Al/Al_2O_3 (*a, c*) and Al_2O_3/Al/Bi (*b, d*) multilayered structures with 1.4 mm thickness.

From the data presented in Figure 6, it is clearly seen that the ionization losses for both protons and Kr^+ ions in the case of a multilayered structure, the first layer of which contains a heavy element (Bi), is higher than in the case when the first layer consists light elements (Al, Al_2O_3). It should also be noted that 20 MeV proton energy and the Kr^+ ion energy of 7.75 GeV are practically sufficient for high-energy particles to transmit through the Al_2O_3/Al/Bi multilayered structure, but not enough to pass through the Bi/Al/Al_2O_3 structure. Thus, in the three-layered Bi/Al/Al_2O_3 structure the character of the *R(E)* dependences and ionization energy losses are determined by the individual layers arrangement sequence, which can be explained by different levels of ionization losses in layers containing heavy and light elements. Therefore, structures in which the first layer contains a heavy element have more effective protective properties among multilayer shields with respect to monochromatic and high-energy ions. This is a Bi layer for the Bi/Al_2O_3/Al structure. An increase in the number of layers does not contribute to rise in the shielding efficiency of high-energy ions. Similar results were obtained in [15]. The calculation in the GEANT4 program of 6 MeV proton spectra when transmitting through W/Al two-layered shields with 50 μm individual layer thickness showed that the nature of the conversion of proton spectra depends on the material of first layer. The main transformation of the proton spectrum occurs in the heavy element layer, therefore, the changes in the relative fluxes of proton energy after passing through the W/Al and Al/W shields were 0.275 and 0.318 units, respectively. With an increase in the number of layers to 10, the relative energy flux of the transmitted ions approached 0.305, which corresponded to the W/Al alloy.

4. Conclusion

The linear and mass ranges of protons and Ar^+ ions in Al, Al_2O_3, Bi and $W_{77,7}Cu_{22,3}$ composite shields were calculated. It has been shown that the protection effectiveness against high-energy ions by materials with large values of the *Z* atomic nuclei charge is higher from the position of particles linear ranges and lowers from the position of mass ranges than materials with low *Z* values.

The dependence of the threshold energy on the HCP sequence number for shields based on Al, Bi and $W_{77,7}Cu_{22,3}$ composite was determined, according to which shields based on $W_{77,7}Cu_{22,3}$ composite have the highest efficiency from the standpoint of linear particle ranges. The threshold energies for this material with 1.5 mm thickness (2.55 g/gm^2) with exposure of protons and He^+, C^+, Ne^+, Ar^+, Fe^+, Kr^+ and Xe^+ are 37; 480; 790; 1780; 4770; 7900; 14200 and 27000 MeV, respectively.

The calculation of the ionization losses spectra for 20 MeV protons and 7.75 GeV Kr^+ ions during passing through multilayered Bi/Al/Al_2O_3 and Al/Al_2O_3/Bi structures showed that the braking of high-energy particles in the case when the first layer contains a heavy element higher than when the first layer contains a light element.

The effect of sequence in the arrangement and layer thickness in multilayered Bi/Al/Al_2O_3 structure on the protection effectiveness against HCP was studied. It is shown that the nature of the dependences *R(E)* and ion energy losses are determined by the sequence of individual layers arrangement, which is associated with the difference in the conversion of the spectra by the first layer material. More significant changes in the spectra occur in the layer with the heavy element, since the ionization losses in it are higher, and a higher level of energy flux attenuation corresponds to the case of heavy element/light element shield alternating layers.

Acknowledgement
The work was carried out with financial support of State program of scientific research "Physical Materials Science, New Materials and Technologies" subprogram "Nanomaterials and Nanotechnologies" (task 2.34), Belarusian Republican Foundation for Fundamental Research (BRFFR-JINR Grant No. Ф18Д-007), Grant of World Federation of Scientists (Geneva, Switzerland).

References
[1] Sexton F W 2003 Destructive single-event effects in semiconductor devices and ICs *IEEE Transactions on Nuclear Science* **50 (3)** 603.
[2] J H Heo, B J Ko, B J Chung 2009 A calculation of the cosmic aadiation dose of a semiconductor in a Geostationary orbit satellite depending on the shield thickness *Journal of the Korean Institute of Electrical and Electronic Material Engineers* **22 (6)** 476.
[3] RAD-COAT protection. Available at: http://www.spaceelectronics.com (accessed 18 December 2019).
[4] RAD-PAK protection. Available at: http:// www.maxwell/products/microelectronics (accessed 18 December 2019).
[5] WALOPACK protection. Available at: http:// www.3d-plus.com (accessed 18 December 2019).
[6] J P Spratt, B C Passenheim, R E Leadon, S Clark, D J Strobel 1997 Effectiveness of IC shielded packages against space radiation *IEEE Transactions on Nuclear Science* **44 (6)** 2018.
[7] T M Jordan, E G Stassinopoulos 1989 Effective radiation reduction in Space Station and missions in the magnetosphere *Advance Space Research* **9** 261.
[8] M Durante 2011 Physical basis of radiation protection in space travel *Reviews of modern physics* **83** 1245.
[9] Battison R, Fiandrini E, Sodano P 2002 Study of cosmic rays fluxes in low Earth orbit (LEO) observed with the AMS experiment (Madrid) p. 166.
[10] J P Spratt, C Conger, R Bredthauer , W Byers, R Groulx, R E Leadon, H Clark 2006 Proton damage effects in high performance P-channel CCDs *IEEE Transactions on Nuclear Science* **53 (2)** 423.
[11] J P Spratt, E A Burke, J C Pickel, R E Leadon 2001 Modeling high-energy heavy-ion damage in silicon *IEEE Transactions on Nuclear Science* **48 (6)** 2136.
[12] J P Spratt, G L Schnable, J D Standeven 1970 Impact of the radiation environment on integrated-circuit technology *IEEE Journal of Solid-State Circuits* **5(1)** 14.
[13] SRIM-2013. Available at: http:// www.srim.org/SRIM/SRIM-2013 (accessed 18 December 2019).
[14] M Yamaguchi, Yu Nagao, N Kawachi 2019 A simulation study on estimation of Bragg-peak shifts via machine learning using proton-beam images obtained by measurement of secondary

electron bremsstrahlung *IEEE Transactions on Radiation and Plasma Medical Sciences* DOI: 10.1109/TRPMS.2019.2928016.
[15] A B Aziz,F Rahman, M H Prodhan Comparison of lead, copper and aluminum as gamma radiation shielding material through experimental measurements and simulation using MCPN version 4c *International Journal of Contemporary Research and Review* **9 (8)** 20193.

Rheological properties of $Y_3Al_5O_{12}$ powder obtained by preceramic organoyttroxanealumoxanes

D D Titov[1], E A Gumennikova[1,2], K D Danilin[1,2], N V Petrakova[1], A S Lysenkov[1], M G Frolova[1], G I Shcherbakova[3], E A Novokovskaya[3], D O Lemeshev[2] and Yu F Kargin[1]
[1]A.A. Baikov Institute of Metallurgy and Materials Science, Russian Academy of Sciences,119334, Russia, Moscow, Leninskiy prospect, 49
[2]Mendeleev University of chemical technology of Russia, 125047, Russia, Moscow, Miusskaya sq., 9
[3]State Research Institute for Chemistry and Technology of Organoelement Compounds, 105118, Russia, Moscow, sh Entuziastov 38
e-mail: mitytitov@gmail.com

Abstract. Submicron powder $Y_3Al_5O_{12}$ (YAG) aluminium yttrium garnet was synthesized from an Al-Y-ceramic-forming precursor, according to the method developed in the State Research Institute for Chemistry and Technology of Organoelement Compounds (SRICTO). The advantage of the method is the possibility of obtaining close to the stoichiometric phase and obtaining high purity nanopowder. In this work, we studied the effect of the pyrolysis temperature of the polymer (700°, 800° and 900°C) and grinding time on the rheological properties of the powder. It was shown that after pyrolysis at 700° and 800°C, the powder has an amorphous structure, the ultimate strain, in this case, being 6.1 and 8.1% of the initial size. The compressibility modulus decreases with increasing temperature from 34.5 to 19.4 MPa. Due to the formation of YAG crystals after pyrolysis at 900°C, the compressibility of the powder increases and the compacts become denser.

1. Introduction

The need for modern technology, as well as the development and improvement of the production of ceramics from pure oxides, requires researchers to develop new methods for producing highly pure raw materials. One of the possible applications of such materials is the production of optically transparent ceramic materials, which in many physicomechanical properties can surpass opaque ceramics, and in some parameters single crystal materials. A special place among such materials is occupied by optically transparent ceramics from $Y_3Al_5O_{12}$ (YAG) yttrium aluminium garnet. YAG belongs to a group of minerals which have a complex cubic crystallographic structure. The YAG structure is particularly resistant to creep and is much less sensitive to orientation [1]. Ceramics based on it has a wide range of applications in various fields: nanophotonics, optoelectronics, aerospace engineering and other fields [2]. Yttrium aluminium garnet has high mechanical strength, excellent optical and thermal characteristics. Materials based on yttrium aluminium garnet are used for the manufacture of active elements of solid-state lasers in the near and middle IR ranges [3]. Due to the complex of unique properties of yttrium aluminium garnet and spinel, ceramics based on them are of great interest to industry.

IOP Conf. Series: Materials Science and Engineering **848** (2020) 012090 doi:10.1088/1757-899X/848/1/012090

In recent years, a new method for the synthesis of YAG from a ceramic-forming oligomer has been developed at the Research Institute for Chemistry and Technology of Organoelement Compounds (SRICTO). For the first time, hydrolytically stable in air soluble in organic solvents soluble in organic solvents keramo-forming organo-yttrioxanaluminoxoxanes, which are the organo-yttrioxanaluminoxoxanes, were the first oligomer to produce $Y_3Al_5O_{12}$ (YAG) [4]. The advantage of this method is the possibility of obtaining close to the stoichiometric phase, and obtaining high purity nano-powder (99.99 wt.%) [5], which is an essential criterion for optically transparent ceramics.

The work aim is to study the rheological properties of spinel and YAG powders. Namely, the study of the microstructure of the powders, the study of the influence of the pyrolysis temperature of the ceramic-forming oligomer and the grinding time on the rheological parameters of these powders during pressing.

2. Experimental part

2.1. Materials

The general scheme for the synthesis of ceramic-forming organoyttroxanealumoxanesiloxanes can be represented by the following equation [6]:

[(R**O)sY(OH)t Or]k·[Al(OR)l(OR*)x(OH)zOy]m·[(EtO)2SiO]p,

where k, m, p = 3–12; s + t + 2r = 3; l + x + 2y + z = 3; R = CnH2n + 1 with n = 2–4; R* = C(CH3)=CHC(O)OCnH2n + 1; and R** = C(CH3)=CHC(O)CH3.

The synthesized ceramic-forming oligomer was then subjected to pyrolysis for 2 hours in a muffle furnace at 700°, 800° and 900°C in air. Next, the resulting powder was crushed.

The grinding of the initial powder was carried out in a planetary mill for two hours at 250 rpm in a solution of isopropyl alcohol. The powder was ground before and after pyrolysis to compare rheological properties. After grinding, the powder was dried in an oven at 130 ° C until the alcohol was completely removed. The resulting spinel powder was characterized by particle size analysis, x-ray phase analysis, and using scanning electron microscopy.

Granulometric analysis of powders after pyrolysis was carried out on a Fritsch Particle Sizer 'analysette 22' laser diffraction analyzer. The dispersion was carried out in a liquid (laser wavelength 655 nm, incident beam diffraction 0.01–1000 μm, sample amount 0.1–2 cm^3, measured particle size of the powder 0.16–1160 μm); the Fraunhofer optical method was used for calculation. Grain composition was presented on a differential particle distribution curve.

The microstructure of spinel powders was studied on a Supra 50 VP high-resolution scanning electron microscope (LEO, Germany, 2003) with an INCA Energy + Oxford microanalysis system with an attachment for local X-ray spectral analysis. In both cases, the electron beam source is a cathode with field emission, the voltage at the accelerating electrode - up to 20 kV. Used magnifications - from × 45 to × 20,000.

2.2. Rheology

Rheological properties were tested on the mechanical testing machine (Instron 5581, Great Britain). The stain velocity was 1 mm/min. A sample of the test sample weighing 0.15 g was placed in a steel mould. At least 5 measurements were carried out for each sample, then the values were averaged (the average value of the standard deviation of deformation in the entire interval of the experiment was σ_{700}= 0.03% and σ_{800} =0.05% σ_{900} =0.16% for YAG powder after pyrolysis at 700°, 800° and 900°C, respectively), and calculations were carried out using the formulas below.

The deformation was calculated by the equation (1):

$$\varepsilon = \frac{\Delta h}{h0}$$

where Δh is the change in the height of the bulk layer over time; h0 is the initial height of the bulk layer. The compressibility modulus is numerically equal to the tangent of the linear section slope angle of the diagram strain against deformation, and is calculated by the equation (2):

$$G = \frac{\Delta p'}{\varepsilon *}$$

where Δp' is the limiting change in the linear section pressure value; ε * is the final value of the linear section deformation. The compressibility coefficient (k_c) characterises the reversible decrease in the sample height (volume) under the pressure action and is quantitatively determined by equation (3):

$$k_c = \frac{-1}{h_0} \cdot \frac{\Delta h'}{\Delta p'}$$

where Δh' and Δp' are the limiting values of the height and the linear section pressure variation, h_0 is the initial height of the backfill.

The physical meaning of the compressibility coefficient is characterisation the material ability to compact at the initial (linear) stage, during which the compaction intensity is maximal [7,8].

3. Discussion of results

Kinetic curves were obtained, representing the pressure versus time for YAG powder. These values were rearranged in the stress-strain curve. To study the effect of powder grinding on rheological properties, the obtained data were used to plot the strain dependences of the strain after grinding (figure 1). On the stress-strain diagram after grinding, the dependence of the rheological properties on the pyrolysis temperature can be traced. With an increase in the pyrolysis temperature, the limiting value of the strain increases from 6.1 to 9.9%; a linear section indicates this at the initial stage, where the stress obeys Hooke's law. Further, a deviation from linearity begins, and particle accommodation occurs on this site. Figure 1B shows that it is in this section that the curves for the powder spread after 700° and 800°C of pyrolysis. This difference is due to the partial crystallisation of YAG at 800°C.

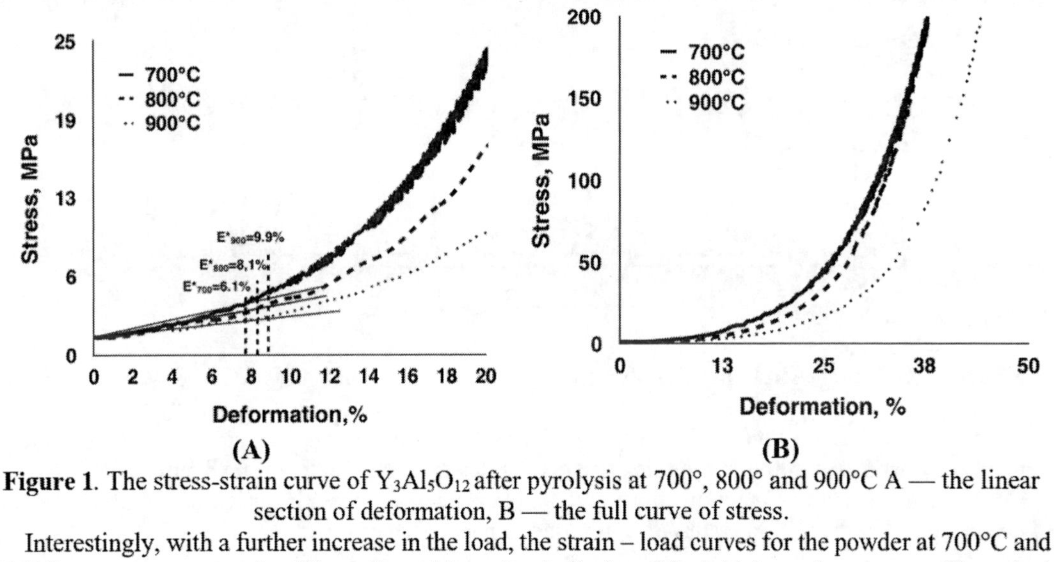

Figure 1. The stress-strain curve of $Y_3Al_5O_{12}$ after pyrolysis at 700°, 800° and 900°C A — the linear section of deformation, B — the full curve of stress.

Interestingly, with a further increase in the load, the strain – load curves for the powder at 700°C and 800°C converge at one point. We attribute this to the similarity of the initial powder, since, unlike spinel [7], complete crystallisation occurs after 1000°C. For this reason, the curve for the powder after 900°C at equal pressure has a large deformation. It means that the particles have a more complex shape, and when filling the mould, many voids are filled during the accommodation of the powder in the second section of the deformation – load curve.

The XRD analysis for spinel powder after pyrolysis at 700°C showed the presence of an amorphous phase (Figure 2A). A fully crystallised phase is observed only after pyrolysis at 1000°C (Figure 2B). After 1000°C, two phases were detected in the powder: YAG - 98 wt.%, YAM - 2 wt.%. Because the

IOP Conf. Series: Materials Science and Engineering **848** (2020) 012090 doi:10.1088/1757-899X/848/1/012090

non-crystallized phase remained in the powder after 700°C, the curve of the elastic deformation section for these powders is almost two times less than the powder for 900°C.

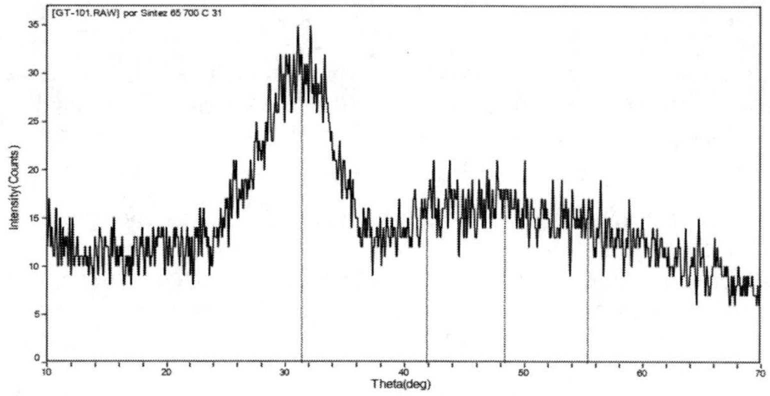

(A)

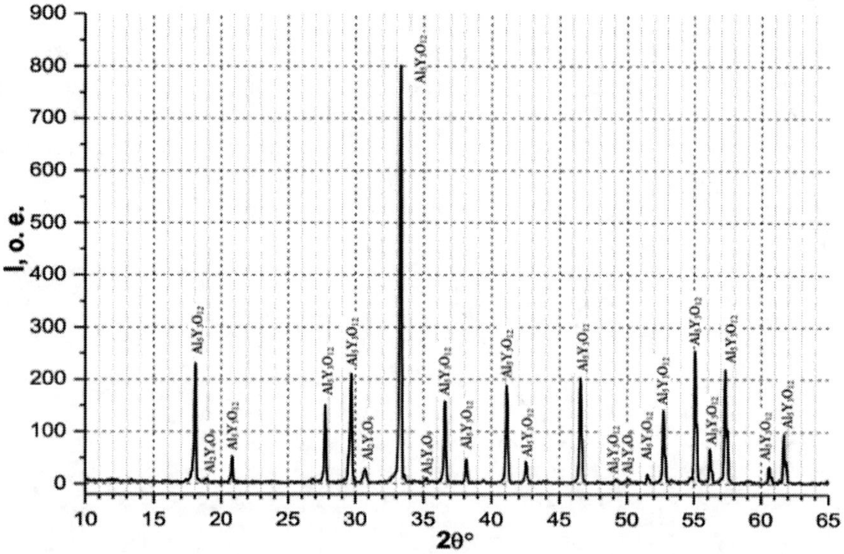

(B)

Figure 2. XRD of $Y_3Al_5O_{12}$ powder after pyrolysis at 700° and 1000°C (YAG - 98 wt.%, YAM – 2 wt.%).

The SEM analysis results confirm after heat treatment of the polymer in the air environment, there remain polymer flakes 100-120 nm thick and crystallites from 0.1 μm to 2 μm (Figure 3). Here are the SEM of the powder at two magnifications: overall plan and enlarged with particle size. In fig. 3B, D, F, it can be seen that the agglomerates are composed of nanoparticles.

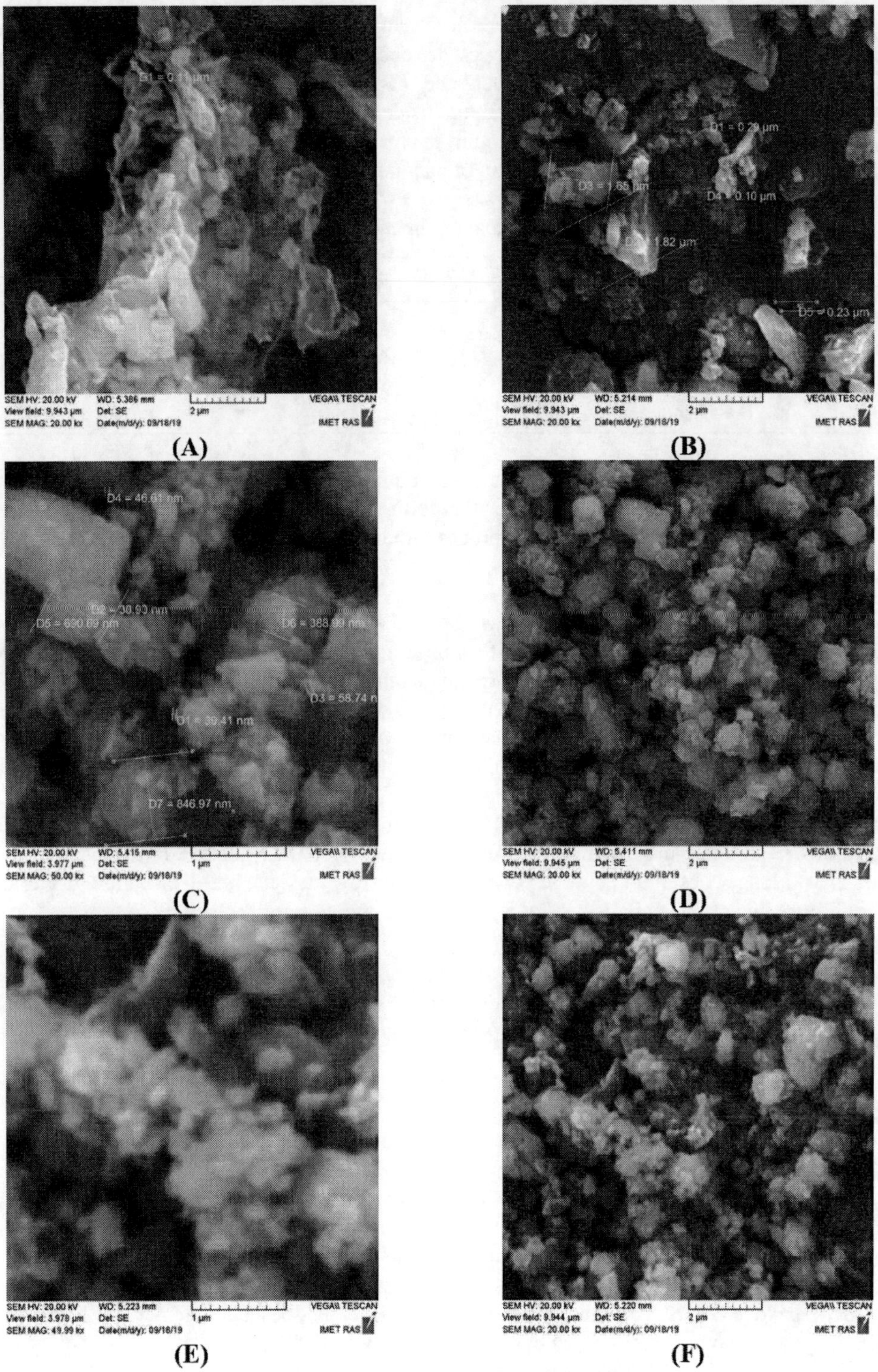

Figure 3. SEM analysis of $Y_3Al_5O_{12}$ powder after pyrolysis of A and B - 700°C, C and D - 800°C, E and F - 900°C.

The compressibility moduli for the powder after pyrolysis at 700 ° 800 ° and 900 ° C were calculated from the slope of the linear section when the powder obeys Hooke's law. According to Eq. 3, the value of the compressibility coefficient was calculated (Table 1). With an increase in the pyrolysis temperature, the value of the compressibility modulus decreases from 31.5 to 19.4 MPa and an increase in the ultimate strain value, due to the transition from an amorphous to crystalline material, the elasticity of the particles increases. Due to the formation of nanoparticles close to spherical, better compaction of the powder occurs.

Table 1. The rheological parameters of $Y_3Al_5O_{12}$ powder

Pyrolysis temperature	G, MPa	K_c, E^{-5}	ε^*, %
700°C	31.5	1.8	6.1
800°C	25.3	2.3	8.1
900°C	19.4	3.0	9.9

The data presented indicate the effect of grinding on the growth of the elastic modulus. This may be due to the destruction of particle agglomerates that form after pyrolysis. This once again proves that the particles created during the recrystallization process are brittle.

4. Conclusion

Yttrium aluminium garnet (YAG) powder was obtained after pyrolysis at 700°, 800°, and 900°C of the organottroxanealumoxanesiloxanes ceramic-forming oligomer. X-ray diffraction analysis showed that YAG has an amorphous structure up to a pyrolysis temperature of 900°C. In the powder after 700°C, the shape of thin flakes with a thickness of up to 100 nm remaining after burning of the polymer is preserved. An increase in the pyrolysis temperature to 900°C leads to the formation of nanosized YAG crystals agglomerated into particles of 1-2 μm. It was shown that an increase in the pyrolysis temperature from 700° to 900°C leads to a decrease in the compressibility modulus from 31.5 to 19.4 MPa and an increase in the ultimate strain value, due to the transition from an amorphous to crystalline material, the elasticity of the particles increases. At the maximum pyrolysis temperature in the powder, polymer residues are no longer observed, and the powder has a narrow particle size distribution, due to this, the powder deformation at 200 MPa is 45% at a pyrolysis temperature of 900°C.

Acknowledgments

The work was financially supported by the Grant of the President of the Russian Federation No. MK-39.2019.3. The methodical part of the work (X-ray and SEM) was carried out according to the state task No 075-00947-20-00.

References

[1] Bansal N P (Ed.) 2006. Handbook of ceramic composites 200 Springer Science & Business Media
[2] Senina M O, Lemeshev D O 2017 Successes in chemistry and chemical technology 31 182
[3] Aksenov D I, Zharikov E V, Faykov P P 2015 Advances in Chemistry and Chemical Technology 29 166 (in Russian)
[4] Shcherbakova G. I. et al. 2014 Inorganic Materials 50 636
[5] Storozhenko P A, Shcherbakova G I 2011 Inorganic Materials 47 167
[6] Shcherbakova G I, Storozhenko P A, Kutinova N B, et al. 2011 RF Patent 2 453 550
[7] E A Gumennikova et al 2019 J. Phys.: Conf. Ser. 1347 012062
[8] D D Titov et al 2019 IOP Conf. Ser.: Mater. Sci. Eng. 525 012077

IOP Conf. Series: Materials Science and Engineering **848** (2020) 012091 doi:10.1088/1757-899X/848/1/012091

Structure Of The $Al_{85}Ni_7Fe_4La_4$ Polycrystalline Alloy After Complex Impacts Of Deformation And Flash Lamp Annealing

N D Bakhteeva[1], E V Todorova[1] and S V Kannykin[2]

[1] Institution of Russian Academy of Sciences A.A. Baikov Institute of Metallurgy and Material Science RAS, Russia, Moscow, Leninsky av., 49
[2] Voronezh State University, Russia, Voronezh, Universitetskaya sq. 1
nbach@imet.ac.ru

Abstract. On polycrystalline alloy $Al_{85}Ni_7Fe_4La_4$ the effectiveness of complex processing was shown, including intensive plastic deformation (IPD) by shear under pressure and flash lamp annealing (FLA), to improve the strength characteristics of cast aluminum alloys.

1. Introduction

An effective way to strengthen aluminum-based alloys is their alloying with slightly soluble elements, which, reacting with aluminum and other alloying elements, form intermetallics of various types. The last have a high hardness, which increases the strength characteristics of the total alloy. Formation of intermetallic compounds of various types during crystallization of alloys along with hardening leads to the material embrittlement, which reduces the efficiency of its use in real structures. To eliminate this disadvantage, it is necessary to disperse the reinforcing phases in the alloy. The efficiency of using such alloys is achieved by optimal combination of alloying and processing.

The objective of this study was to optimize the processing modes of cast aluminum alloy with a high content of transition and rare earth metals to obtain the structure of a multiphase nanocomposite with a high complex of mechanical properties.

2. Material and techniques

Polycrystalline hardened samples of $Al_{85}Ni_7Fe_4La_4$ alloy were subjected to severe plastic deformation (SPD) by shear under pressure P=8 GPA on Bridgman anvils. The movable anvil rotation angle was $\varphi = 360°$. The samples obtained after IPD were subjected to flash lamp annealing (FLA) with fluence E = 10, 20 and 25 J/cm2 under the conditions of gas-discharge lamps emission.

Samples for metallographic studies were prepared by grinding-and-polishing machine MECAPOL (MODEL P 230, manufacturer - PRESI S. A., France) in accordance with the guidelines of the company PRESI S.A. for surface treatment of metallic materials.

For metallographic tests, polished samples were etched in reagents: nitric acid (4 ml) + hydrochloric acid (3 ml) + hydrofluoric acid (3 ml) + water (90 ml), etching time up to 20 sec. This reagent causes staining of the phases present in the alloy in different colors, which enables their identification.

The structure of the alloy $Al_{85}Ni_7Fe_4La_4$ after various types of processing and "color" etching was studied by metallography on an inverted light microscope company "Carl Zeiss" Axiovert 40 Mat with

IOP Conf. Series: Materials Science and Engineering **848** (2020) 012091 doi:10.1088/1757-899X/848/1/012091

digital high resolution camera Canon PowerShot A640. The images of the structure were obtained with magnifications from 200 to 1000 times in the modes of light and dark fields.

The phase composition of the alloy was studied by X-ray diffraction: DRON-3M diffractometer (Russia) using Cu-Ka - radiation (shooting mode: U = 38 kV, I = 20 mA, slit 1×2×0.5 + Soller slit system); ARL XTRA diffractometer, copper radiation, parallel beam geometry (parabolic mirror on the primary beam, thin - film collimator on the secondary beam). Phase analysis of the X-ray diffraction patterns was performed using PDF-2 data. The microhardness of the samples was measured on a 402 MVD – Wolpert Wilson Instruments microhardometer (Germany) at a load of 10 g and a loading time of 10 sec.

3. Results and its discussion

The literature review was held on the presence and types of intermetallic compounds, the allocation of which is possible in alloys based on aluminum with its content of more than 70 at. % alloyed with transition and rare earth metals. Such compounds - two, three and four – component were detected 16 [1].

Typical images of the alloy structure obtained by metallography are shown in Figure 1. Different color, contrast, morphology (shape, size) indicate the presence of four phases in the alloy, differing in composition. Light-field metallographic images of the tempered alloy structure (Figure 1a, b) supplemented with dark-field image (Figure 1c), which are formed by light scattered on the inhomogeneities of the structure, which gives an increase in contrast at the boundaries of phases, grains and twins. Dendritic crystallization of hypereutectic composition alloy begins with the selection of primary quaternary phase $Al_8Fe_{2-x}Ni_xLa$ with the formation of a large number of defects such as twins (Figure 1). When solid aluminum solution crystallizes in the form of round cells, excess alloying elements are pushed to the cell boundaries, where the eutectic structure, which is characterized by the simultaneous growth of two or more phases, is formed. By methods of metallography and X-ray diffraction it was found, that the structure of the alloy is four- component. It is shown that in the alloy, along with a crystalline solid solution of aluminum, there are three types of intermetallic compounds: Al_4La (rhombic system), $Al_8Fe_{2-x}Ni_xLa$ (rhombic system), $Al_9Ni_{2-x}Fe_x$ (monoclinic system) [2]. Since the investigated alloy contains a significant amount of iron (7 at.%), isolating of iron aluminide $Al_{13}Fe_4$ during crystallization is possible. On X-ray pictures the intensity peaks obtained by reflection from the specified phase are coincident with the maxima of reflection from other phases, which does not allow to state unambiguously that this phase is present in the alloy. Most intermetallic compounds have a size of more than one micron, some have a rack morphology (eg, lanthanum aluminide), which leads to embrittlement of the alloy after crystallization. The initial microhardness of the alloy is 216 HV.

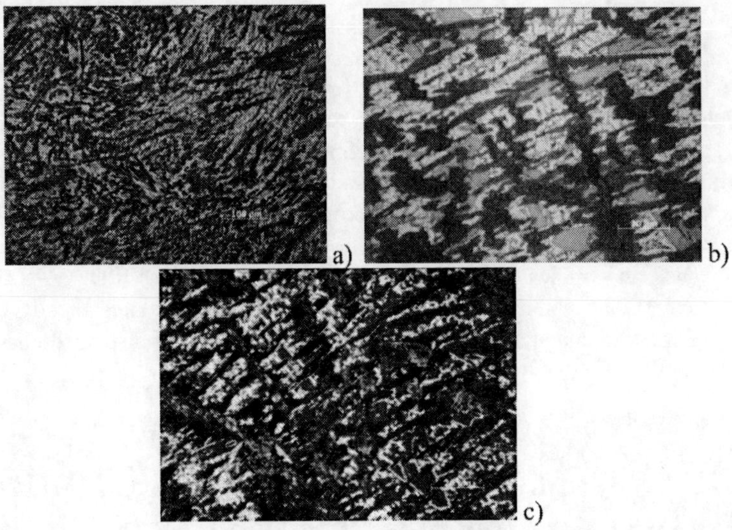

Figure 1. Microstructure of the alloy in initial state: light-field (a, b) and dark-field (c) images.

One of the known ways to optimize such a structure is deformation processing, which is used in the work on the scheme of intense plastic deformation by shear under pressure (IPD). After IPD $\varphi = 360°$, a "vortex" structure is formed (Fig. 2), accompanied by dispersion, spheroidization and partial dissolution of intermetallic compounds [3].

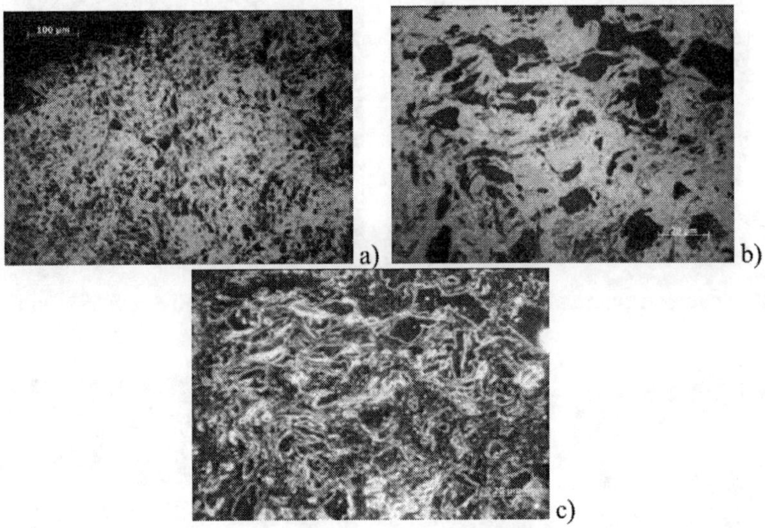

Figure 2. Alloy microstructure after IPD 8 GPA, $\varphi = 360°$: light-field (a, b) and dark-field (c) images.

The nanocrystalline structure, formed after IPD, characterized by a high concentration of defects, is thermodynamically unstable. Under high-energy influences, such as FLA, this structure undergoes changes: it develops a return and static recrystallization, and it is also possible to disintegrate the solid solution of aluminum with the precipitation of secondary intermetallides (Figure 3-5). The short-term effect of such a heat shock leads to increased dispersion of secondary precipitation without changing the phase composition, which is confirmed by the X-ray (Figure 6).

IOP Conf. Series: Materials Science and Engineering **848** (2020) 012091 doi:10.1088/1757-899X/848/1/012091

The broadening of peaks on diffractograms decreases, and their intensity increases. Secondary highly dispersed intermetallic discharges on dark-field images of the sample structure after complex exposure to IPD (φ= 360°) and FLA with a fluence of 20 J/cm^2 are particularly pronounced (Figure 4). By coloring on light-field metallographic images, these secretions are identified as intermetallic phase $Al_{11}La_3$. Thus, the development of two competing processes - return, leading to softening, and dispersion hardening - provides a non-monotonic dependence of the microhardness of the deformed polycrystalline alloy on the FLA fluence. The microhardness growth in the range of values E=10 ... 20 J/cm^2 indicates the predominant role of dispersion hardening. The maximum value of microhardness 569 HV was obtained on samples that underwent complex treatment of IPD (P = 8 GPa, φ = 360°) and FLA (E = 20 J/cm^2). The fluence increasing to 25 J/cm2 stimulates the development of static processes of softening and coagulation of secondary precipitations of dispersed phases, which leads to a decrease of the hardness of the alloy (Table 1).

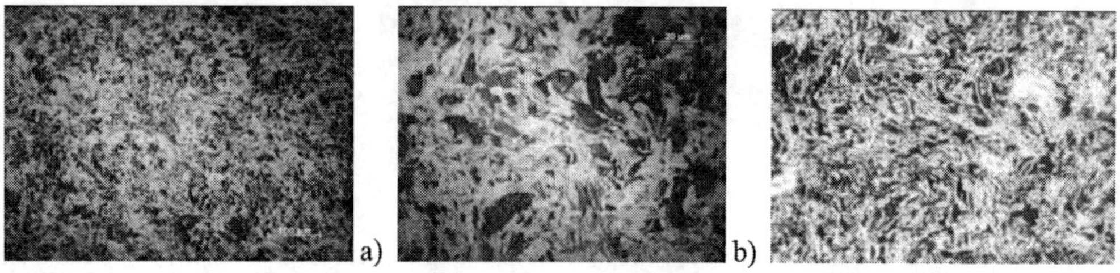

Figure 3. Alloy microstructure after IPD 8 GPA, φ = 360° followed by FLA 10 J/cm^2: light-field (a, b) and dark-field (c) images.

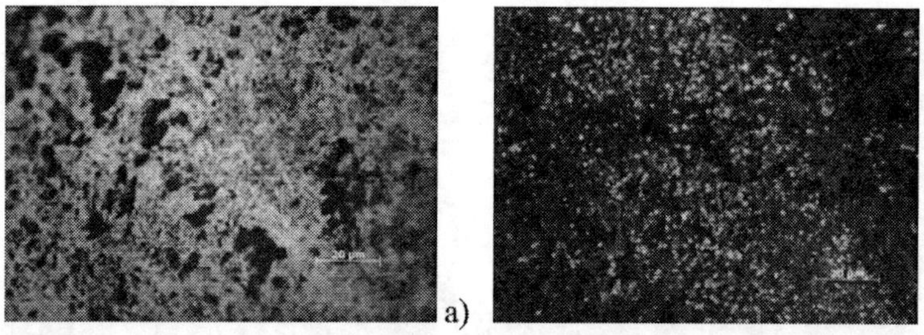

Figure 4. Alloy microstructure after IPD 8 GPA, φ = 360° followed by FLA 20 J/cm^2: light-field (a) and dark-field (b) images.

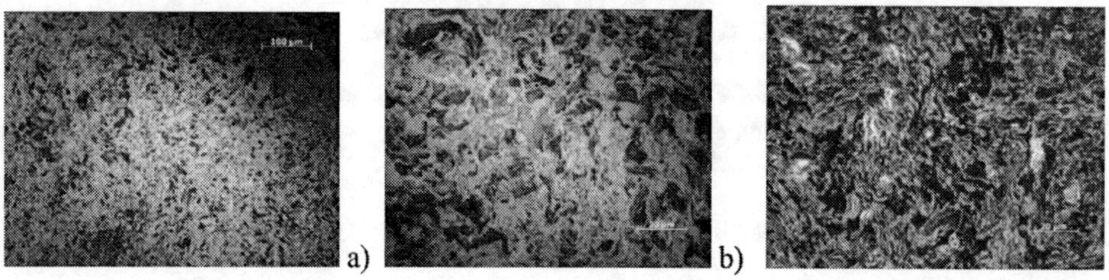

Figure 5. Alloy microstructure after IPD 8 GPA, φ = 360° followed by FLA 25 J/cm^2: light-field (a, b) and dark-field (c) images.

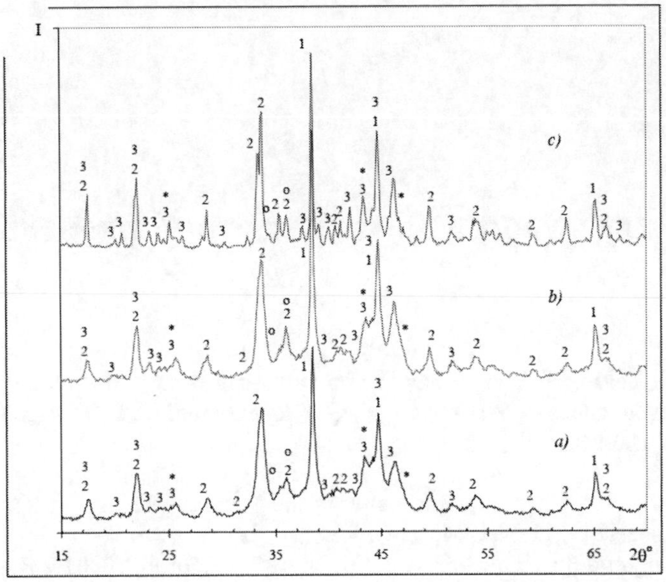

Figure 6. X-ray analisis of the alloy after IPD 8GPa, φ = 360° + FLA: 10 J/cm^2 (a), 20 J/cm^2 (b), 25 J/cm^2 (c); 1- Al, 2- $Al_{11}La_3$, 3 - $Al_9(Fe,Ni)_2$, * - Al_3Ni, o - $Al_8Fe_{2-x}Ni_xLa$.

Table 1. Microhardness of $Al_{85}Ni_7Fe_4La_4$ alloy in polycrystalline state after IPD of 8 GPa and subsequent FLA with energy doses of 10, 20 and 25 J/cm^2

	without FLA	**10 J/cm^2**	**20 J/cm^2**	**25 J/cm^2**
HV	492	522	569	533

4. Conclusions

In order to improve the strength characteristics, the complex and separate effects of deformation (IPD 8 GPA and the full angle of the Bridgeman anvil twist 360°) and pulsed photon irradiation using xenon lamps (FLA with a fluence from 10 to 25 J/cm^2) as an effective method of processing polycrystalline aluminum alloy (85 at.% Al) alloyed with Ni (7 at.%), Fe (4 at.%) and La (4 at.%). It was found that the complex treatment, including IPD + FLA (20 J/cm^2), provides a maximum increase in microhardness to 569 HV and introduces additional hardening compared to the microhardness of the hardened alloy (216 HV) and its microhardness after IPD without IFO (492 HV).

Acknowledgments

The investigations have been fulfilled according to the Federal order № 075-00947-20-00.

References

[1] Vasil'ev A.L., Bakhteeva N.D., Presniakov M.Yu., Lopatin S., Kolabylina N.N., Ivanova A.G., Todorova E.V. Microstructural Pecularities of Al-Rich Al-La-Ni-Fe Alloys 2019 *Metallurgical and Materials Transactions A* **50 (4)** 1995–2013

[2] Ivanova A.G., Bakhteeva N.D., Kolobylina N.N., Orechov A.S., Presnyakov M.Yu. and Todorova E.V. Microstructure of the Al-La-Ni-Fe system 2015 *Crystallography Reports* **60 (1)** 23-29

[3] Bakhteeva N.D., Todorova E.V., Sirotinkin V.P., Kolobylina N.N. and Vasil'ev A.L. Structure of polycrystalline Al-Ni-Fe-La alloys after quenching and plastic deformation by shear under pressure 2013 *Russian metallurgy (Metally)* **3** 206-216

IOP Conf. Series: Materials Science and Engineering **848** (2020) 012092 doi:10.1088/1757-899X/848/1/012092

Effect of rotary swaging on structure and properties of low-carbon steel

A A Tokar [1,5*], V A Lunev [1,5], A S Dolzhenko[2], O V Rybalchenko[1,5], M M Morozov[1], V S Yusupov[1], G V Rybalchenko[3], P D Odessky[4] and S V Dobatkin[1,5]

[1] A A Baikov Institute of Metallurgy and Materials Science of RAS, Leninsky prospect, 49, 119991, Moscow, Russia
[2] Belgorod State University, Pobeda st., 85, Belgorod, 308015, Russia
[3] P.N. Lebedev Physical Institute of RAS, Leninsky prospect, 53,119334 Moscow, Russia
[3] V A Kucherenko TSNIISK, 2nd Institutskaya, 6, 109428, Moscow, Russia
[4] National University of Science and Technology "MISIS", Leninsky prospect, 4, 119991, Moscow, Russia

*E-mail: tokarb2005@mail.ru

Abstract. The structural transformation and mechanical properties of a low-carbon 0.2% C steel and a 0.09% C–Mn–V steel in coarse- and ultrafine-grained (UFG) states were studied. The UFG structure with structural elements about 210-375 nm in size was obtained by rotary swaging (RS) and equal-channel angular pressing (ECAP). ECAP was used to compare the influence of the industrial deformation and SPD methods on the microstructure and properties of low-carbon steels. It was shown that an increase in true strain ratio and a decrease in final temperature of RS improve strength properties and reduce plasticity of these steels. In both steels after RS, σ_{UTS} within the range from 800 up to 920 MPa with elongation in the range of 15-17% at the final deformation temperature of 400 °C was obtained. It was found that the strength of 0.2% C steel after ECAP was on the same level with that steel after RS. The strength properties of low- carbon 0.09% C–Mn–V steel after ECAP were significantly higher than those after RS.

1. Introduction

Severe plastic deformation (SPD) methods lead to significant refinement of the microstructure and improve the strength and service properties of low-carbon steels [1,2]. The ultrafine-grained low-carbon steels were already obtained by such methods as equal-channel angular pressing (ECAP) [2-6] and high pressure torsion (HPT) [7-9]. However, these methods are hard to implement in industry. One of the industrial deformation schemes used, for example, in the automotive industry, is rotary swaging (RS) [10]. This method allows the molding of pipes, rods and wires to give the final shape to the deformed parts with minimal subsequent processing or without it. This technique was already approved for refinement of the microstructure of many alloys [11, 12].

Usually the unalloyed or low-alloyed low-carbon steels are subjected to severe plastic deformation in the initial ferritic-pearlitic state. The initial ferritic-pearlitic structure is convenient for SPD process since it does not need high deforming force. However, the obtained microstructure after deformation is

IOP Conf. Series: Materials Science and Engineering **848** (2020) 012092 doi:10.1088/1757-899X/848/1/012092

inhomogeneous and unstable during heating. On the other hand, the initial martensitic state provides a more dispersed microstructure and better strengthening than initially ferritic–pearlitic state [9,13]. Additionally, the formation of a fine carbide phase after quenching leads to increased thermal stability of such a microstructure. However, the deformation process of the initial martensitic structure requires increased deforming force. In our study we use high-temperature tempering of the steels after quenching. Such thermal treatment will secure uniform distribution of carbides after quenching and reduce the deforming force owing to high-temperature tempering.

The aim of present study is to evaluate the UFG microstructure that obtained by the RS method in the low-carbon steels and to clarify the difference between the microstructures processed by RS and SPD methods. The microstructure and thermal stability of strengthening of low-carbon 0.2% C steel and 0.09% C–Mn–V steels is studied and the effect of UFG structure of these steels on its mechanical behaviour is determined.

2. Experimental

The chemical compositions of the low-carbon 0.2% C and 0.09% C–Mn–V steels are shown in Table 1. The high-temperature treatment was performed by 2 regimes: (1) quenching at a temperature of 880°C (holding time of 1 h followed by water cooling) and tempering at a temperature of 600°C (holding time of 1 h) for the 0.2% C steel; (2) quenching at a temperature of 920°C (holding time of 1 h followed by water cooling) and tempering at 680°C (holding time of 1 h) for the 0.09% C–Mn–V steel. The microstructures of these steels in the initial state after quenching and high-temperature tempering are shown in figure 1.

Table 1. Chemical composition of steels.

Steels	Amount (wt.%)											
	C	Mn	Si	Ni	Mo	V	Cu	Cr	P	S	As	Fe
0.09% C–Mn–V	0.096	1.56	0.19	0.1	0.0027	0.05	0.13	0.12	0.02	0.014		balance
0.2% C	0.19	0.21	0.49	0.03			0.19	0.25			0.05	balance

RS was realized by a rotary swaging (RS) machine RKM 2129.02 (figure 2 a). Billets of 19.6 mm in diameter and 100 mm in length were deformed by RS machine with the frequency and magnitude of the impacts of the dies of 1920 min^{-1} and 3 mm, respectively. Eight impacts of dies per one revolution about the axis were applied to the billet [11, 12]. RS was carried out via two RS regimes with a stepwise decrease in the temperature of deformation within the range from 650 °C to 500 °C, or from 600 °C to 400 °C. The true stain increased from 0.6 to 2.3 during RS. The true stain (ε) was estimated by the formula:

$$\varepsilon = \ln(\mu),$$

where μ – is the total extrusion ratio equal to A_0/A_f, where A_0 and A_f are the initial and the final cross-sectional area of the billets, respectively [11, 12].

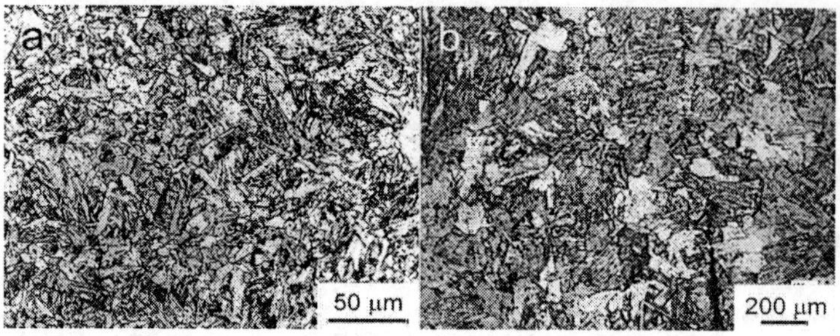

Figure 1. Light micrographs of the microstructure of 0.09% C – Mn – V (a) and 0.2% C (b) steels after quenching and high-temperature tempering.

ECAP (figure 2 b) of the specimens of 20 mm in diameter and 80 mm in length was carried out at a temperature of 400°C by route Bc (successive axial rotation of the billet by 90° after each pass). The intersection angle between the channels was 120°, and the numbers of passes N = 4 resulted in true strain of 3.6. The true strain applied to the billet per pass for the defined die geometry equals 0.9 (shear strain $\gamma = 1.5$) [14 ,15].

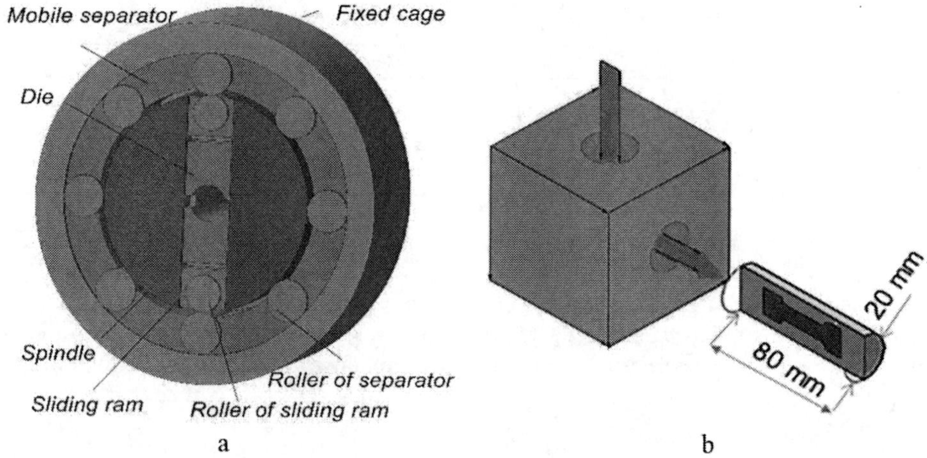

Figure 2 Schemes of RS (a) and ECAP (b).

RS of the 0.09% C–Mn–V steel was performed in two regimes with a stepwise decrease in the deformation temperature of each stage: regime1: 1 stage at 650°C ($\varepsilon = 0.6$) → 2 stage at 575°C ($\varepsilon = 1.2$) → 3 stage at 500°C ($\varepsilon = 2.3$); and regime 2: 1 stage at 600°C ($\varepsilon = 0.6$) → 2 stage at 500°C ($\varepsilon = 1.2$) → 3 stage at 400°C ($\varepsilon = 2.3$). RS of the 0.2% C steel was performed by the regime 2.

The microstructure was analysed in optical Olympus PME 3 microscope and in JEM- 1400 transmission electron microscope. The samples for the metallographic analysis were chemically etched with a solution of 3% nitric acid in C_2H_5OH. Thin foils for transmission electron microscopy (TEM) were mechanically ground to 90 μm and thinned to perforation electrolytically with a supersaturated solution of chromic anhydride in orthophosphoric acid, at a voltage of ~23 V.

The mechanical properties were determined using a tensile testing INSTRON 3380 machine with a load capacity of 100 kN. Microhardness was measured on a 402 MVD Wolpert Wilson instrument at a load of 100g.

IOP Conf. Series: Materials Science and Engineering **848** (2020) 012092 doi:10.1088/1757-899X/848/1/012092

3. Results and Discussion

The light micrographs of microstructure of the 0.09% C–Mn–V steel after RS are shown in two regimes with a decrease in the deformation temperature in figure 3. The microstructure of the 0.09% C–Mn–V steel after 1 stage of 1 RS regime with a true strain of 0.6 doesn't differ from the initial microstructure of this steel (figure 1, figure 3a). The 2nd stage of 1 RS regime led to the oriented initial grain microstructure (figure 3b). As we can see from the light micrograph of samples, microstructure after the final 3 stage of the RS is significantly oriented (figure 3c).

The TEM analysis revealed the formation of a submicrocrystalline grain-subgrain microstructure with an average grain/subgrain size of 325±30 nm (figure 4). The average size of the microstructural elements was determined from dark-field TEM images (figure 4b). During RS the refinement of microstructure occurred due to formation of the shear bands with a thickness of about 300 nm and the formation of subgrains inside of them. Point reflections on an annular electron diffraction patterns in the corresponding TEM micrographs (the inset of figure 4a), as well as a fringe contrast at the grain boundaries (figure 4a) proved the presence of high-angle boundaries, i.e. the grain microstructure.

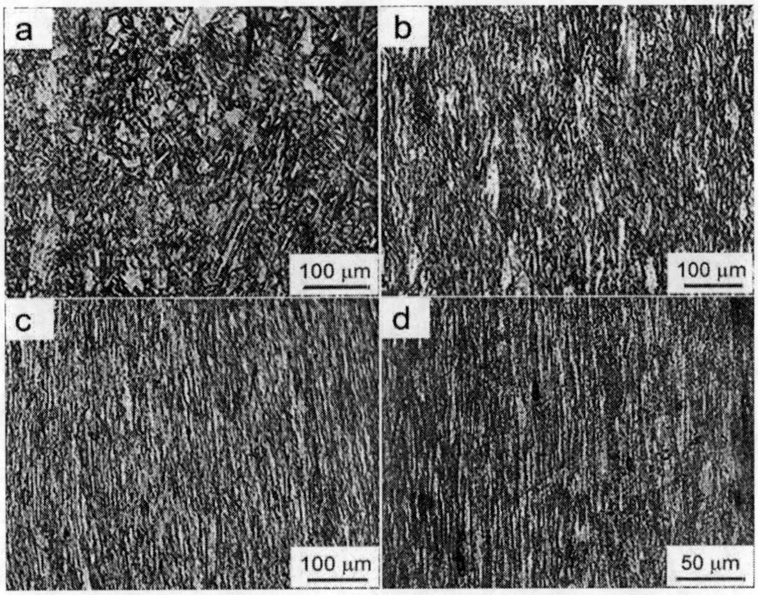

Figure 3. Light micrographs of the microstructure of 0.09% C–Mn–V steel after RS by 1 regime: stage 1 (a), stage 2 (b), stage 3 (c); 2 regime stage 3 (d)/

The metallographic analysis of the 0.09% C– Mn –V steel after the last stage of the second regime with a true strain of 2.3 revealed even more oriented microstructure comparing to that after 1 regime as a result of a lower final temperature (figure 3d).

IOP Conf. Series: Materials Science and Engineering **848** (2020) 012092 doi:10.1088/1757-899X/848/1/012092

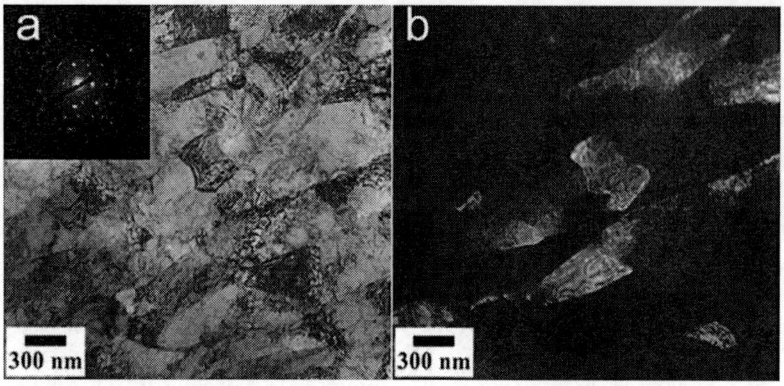

Figure 4. TEM micrographs and SAD pattern for a sample of 0.09% C–Mn–V steel after RS by 1 regime.

During RS by the second regime the refinement of 0.09% C–Mn–V steel microstructure also occurred due to the formation of the shear bands. The thickness of these shear bands is about 215±30 nm (figure 5). The TEM analysis revealed a small grain with an average size of 130 nm with a fringe contrast at the boundaries.

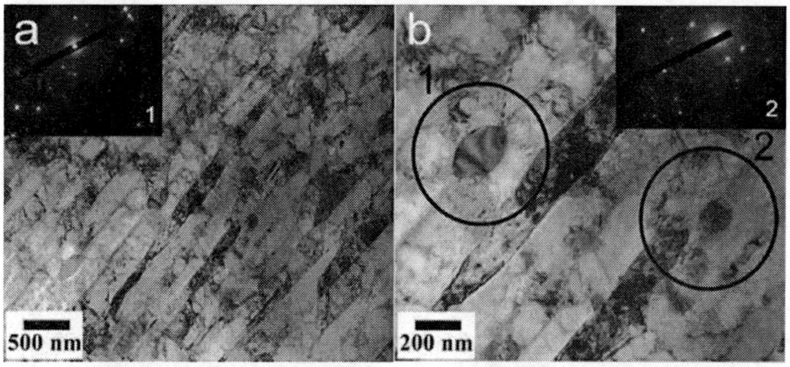

Figure 5. TEM micrographs and SAD patterns for a sample of 0.09% C–Mn–V steel after RS by 2 regime.

The TEM analysis of the 0.2% C steel after RS by the 2 regime also revealed a highly oriented structure (figure 6). Submicrocrystalline structure with an average grain size of 285 nm is seen in the TEM image.

Accordingly, the selected regimes of RS of the 0.2% C steel and the 0.09% C–Mn–V steel let to produce a predominantly UFG microstructure with grain sizes within the range of 285 - 375 nm. It should be noted that the subgrain structure of these steels was stabilized by the carbide phase precipitates.

IOP Conf. Series: Materials Science and Engineering **848** (2020) 012092 doi:10.1088/1757-899X/848/1/012092

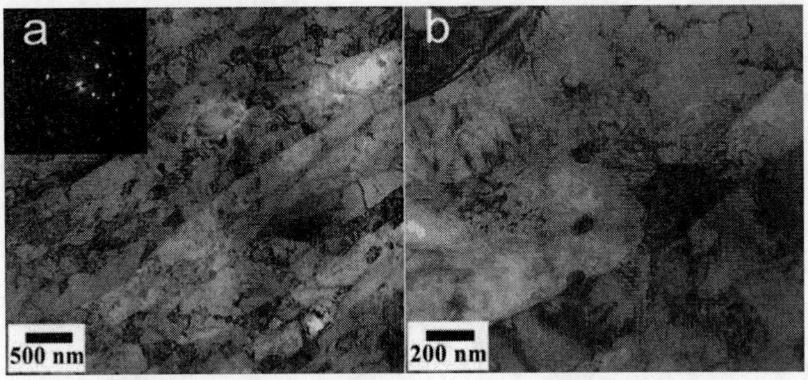

Figure 6. TEM micrographs and SAD pattern for a sample of 0.2% C steel after RS by 2 regime.

The TEM analysis of samples after ECAP doesn't reveal the formation of shear bands. The deformation subrgrains/grains have a quasi-equiaxial shape. The TEM micrographs of the samples after deformation by ECAP for a number of passes N = 4 showed a partially submicrocrystalline and, mostly, subgrain structure (figure 7). The high-angle misorientation of the boundaries is judged from the character of an annular electron diffraction patterns in the corresponding TEM micrographs (in the inset of figure 7) with individual point reflections, as well as a fringe contrast at the boundaries. The average size of the microstructure elements of the 0.2% C steel and the 0.09% C–Mn–V steel after ECAP is about 375± 30 nm and 209.4±19 nm, respectively. The grain refinement in the 0.2% C steel by ECAP was almost the same as that by RS. In contrast, the extent of refining the microstructure of the 0.09% C–Mn–V steel after ECAP was significantly higher than that after RS.

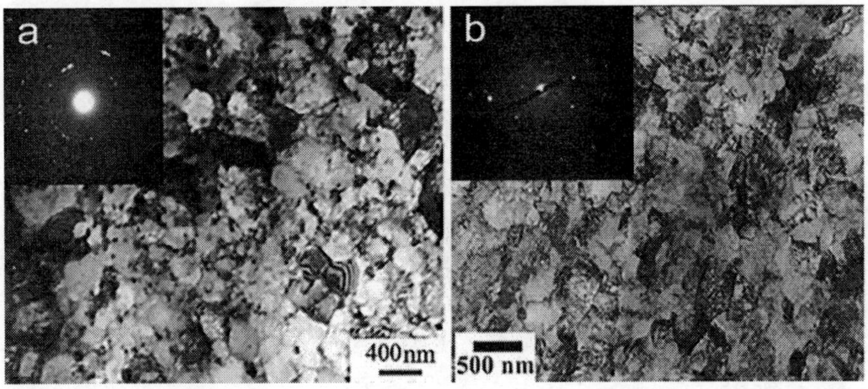

Figure 7. TEM micrographs and SAD patterns for the samples of 0.2% (a) and 0.09% C–Mn–V (b) steels after ECAP.

The thermal stability of strengthening of the 0.2% C steel and the 0.09% C–Mn–V steel after RS was studied according to the change in the microhardness during annealing. The values of microhardness increase with increasing the stage in each regime of RS (figure 8). After RS, the microhardness of the 0.09% C–Mn–V steel is higher as compared with that of 0.2% C steel (figure 8). The thermal stabilities of the 0.2% C steel and the 0.09% C–Mn–V steel after both regimes of RS are nearly the same. However, despite a significantly higher level of strengthening, the thermal stability of the samples after RS is lower than that after quenching and tempering.

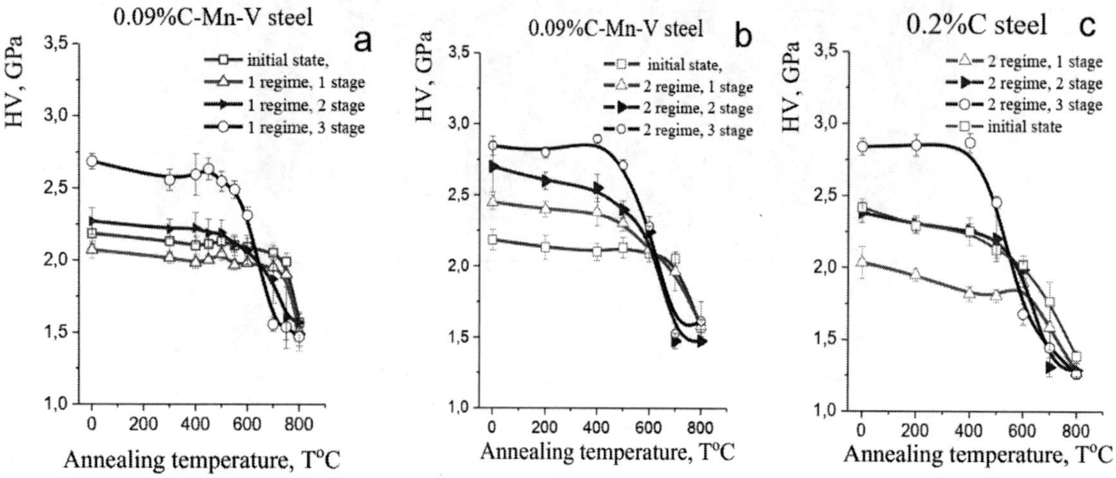

Figure 8. The thermal stability of 0.09% C–Mn–V (a, b) and 0.2% C (c) steels after RS by 1 regime (a) and 2 regime (b, c).

Comparing the strength properties of both steels deformed by the second regime, we can observe that RS at first stage with a true strain of 0.6 leads to a slightly higher σ_{UTS} of the 0.09% C–Mn–V steel and RS at the last stage with a true strain of 2.3 leads to a higher σ_{UTS} of the 0.2% C steel (figure 9, Table 2). It should be noted, however, that the values of σ_{UTS} of these steels are nearly the same. Consequently, σ_{UTS} in the range from 808 to 922 MPa was obtained for both steels at the final stage of RS with good elongation of $\delta = 16\text{-}17\%$.

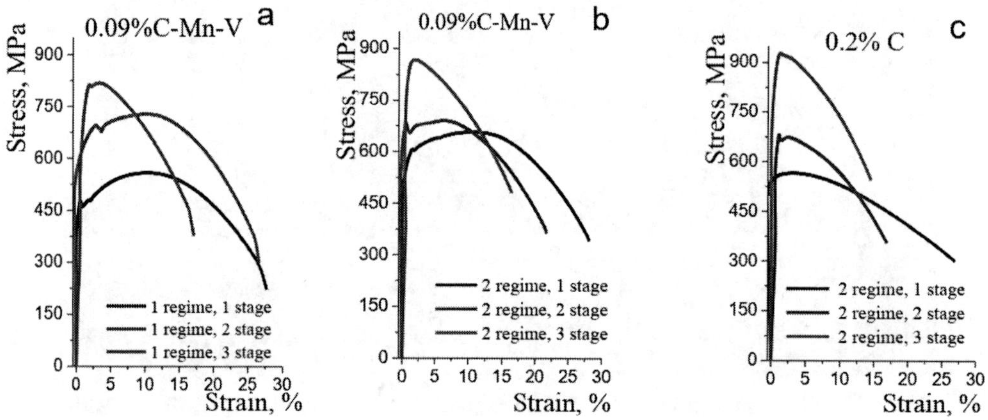

Figure 9. Mechanical properties of 0.09% C–Mn–V (a, b) and 0.2% C (c) steels after RS.

The strength properties of low- carbon 0.09% C–Mn–V steel after ECAP was significantly higher then those after RS (figure 10 a, Table 2). The ultimate strength of 0.2% C steel after ECAP was on the same level as that after RS (figure 10 b, Table 2). In contrast, the strength properties of low-carbon 0.09% C–Mn–V steel after ECAP are better then those after RS (figure 10), probably, because of the better structure refining and precipitation of fine special carbides during severe plastic deformation.

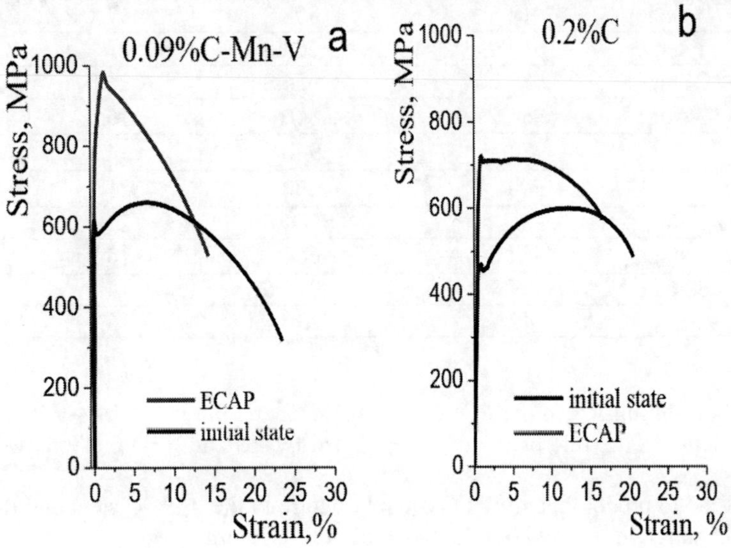

Figure 10. Mechanical properties of the 0.09% C–Mn–V steel (a) and the 0.2% C steel (b) after ECAP.

Table 2. Mechanical properties of 0.09% C–Mn–V and 0.2% C steels

Steels	Regime /stage,		σ_{UTS}, MPa	σ_{YS}, MPa	δ,%
0.09% C–Mn–V steel	Initial state		594	480.8	27
	RS	1 regime / 1 stage	558.4	460.4	28
		1 regime / 2 stage	727.8	678	26.6
		1 regime / 3 stage	818.5	808	17.1
		2 regime / 1 stage	653.9	603.8	28
		2 regime / 2 stage	689.6	654.2	21.8
		2 regime / 3 stage	866.7	846	16.5
	ECAP		984	940	14,3
0.2% C steel	Initial state		600	460	21
	RS	2 regime / 1 stage	565	504.6	27
		2 regime / 2 stage	672.7	664	16.9
		2 regime / 3 stage	927.4	922.6	14.5
	ECAP		742	730	16,4

4. Conclusions

The UFG microstructure evolution during RS in the low-carbon 0.2% C and 0.09% C–Mn–V steels was studied and the difference between the microstructures developed by RS and SPD methods was clarified.

1. The selected regimes of RS led to predominantly UFG microstructure in the 0.2% C steel and the 0.09% C–Mn–V steels with grain/subgrain sizes within the range of 285 - 375 nm.

2. The grain refinement in the 0.2% C by ECAP and RS was almost the same. However, refining the microstructure in the 0.09% C–Mn–V steel after ECAP was much pronounced than that after RS.

3. For both steels σ_{UTS} in the range from 800 to 920 MPa at admissible elongation of 16-17% were obtained at the final stage of RS.

4. The tensile strength of the low-carbon 0.2% C steel after ECAP does not exceed the tensile strength of this steel after RS at the final temperature. In contrast, the strength of the 0.09% C–Mn–V steel after ECAP significantly exceeds that after RS.

Acknowledgments

The work was carried out within the governmental task 075-00746-19-00 and supported by RFBR (grant 18-08-00321).

References

1. Valiev R Z, Zhilyaev A P, Langdon T G 2014 *Bulk nanostructured materials: Fundamentals and Applications* (New Jersey: TMS-Wiley).
2. Dobatkin S V 2000 *Investigations and Applications of Severe Plastic Deformation* (Dordrecht: Kluwer Academic) **3(80)** 13
3. Shin D H, Kim W-J, Choo W Y 1999 Scripta Mater .**41** 259
4. Park K-T, Shin D H 2002 Mater. Sci. Eng. A **334** 79
5. Shin D H, Kim W-J, Kim Y-S, and Park K-T 2000 *Acta Mater.* **48** 2247
6. Shin D H, Kim I, Kim J, and Park K T 2001 *Acta Mater.* **48** 1285
7. Astafurova E G, Dobatkin S V, Naidenkin E V et al. 2009 *Nanotechnologies in Russia* **4** 109
8. Degtyarev M V, Chashchukhina T I, Voronova L M et al. 2007 Acta Mater. **55** 6039
9. Dobatkin S V, Shagalina S V, Sleptsov O I, Krasilnikov N A 2006 *Russian Metallurgy* **5** 445
10. Radyuchenko Yu S 1962 *Rotary swaging. Processing of parts on rotary swaging crimping machines* (Moscow MASHGIZ) (in Russian).

11. Martynenko N S, Luk'yanova E A, Morozov M M et al. 2018 *Met. Sci. Heat Treat.* **60** (3-4) 253.
12. Martynenko N, Lukyanova E, Gorshenkov M et al. 2018 *Mater. Sci. Forum* **941** 808.
13. Sastry Sh M L, Dobatkin S V, Sidorova S V 2004 *Russian Metallurgy* **2** 129
14. Iwahashi Y, Horita Z, Nemoto M., Langdon T G 1997 *Acta Mater.* **45** 4733.
15. Iwahashi Y, Wang J, Horita Z, Nemoto M, Langdon T G 1996 *Scripta Mater.* **35 (2)** 143.

IOP Conf. Series: Materials Science and Engineering **848** (2020) 012093 doi:10.1088/1757-899X/848/1/012093

Influence of gallium oxide concentration on the properties of transparent ceramics based on magnesium aluminate spinel

A V Ulyanova, M O Senina and D O Lemeshev

D Mendeleev University of Chemical Technology of Russia, Moscow, Russia
alenkachim@mail.ru

Annotation. The article deals with the preparation precursor of magnesium aluminate spinel by the method of inverse co-precipitation, as well as the study of effect of firing temperature and the concentration of the gallium oxide addition on the optical and ceramic properties of the products is shown.

1. Introduction

To date, the production of optically transparent ceramics is being actively developed. This involves the use of ceramic products in the defense industry as helmet visors and armor windows of civil military equipment, as well as it is used to make lenses for lasers, night vision devices.

Currently, technologies for producing transparent armor based on aluminum oxynitride (AlON) and aluminum oxide, also known as sapphire, which have excellent functional properties, are being actively used. But this fabrication process of products from these materials requires large financial and energy costs due to the use of expensive equipment and high temperatures in the production process [1, 2].

In the second half of the last century, magnesium aluminate spinel $MgAl_2O_4$ (AMS) was synthesized, which in its properties is not inferior to AlON and sapphire, which are analogues. At the moment, transparent polycrystalline ceramics based on magnesium aluminate spinel are one of the most profitable materials for the production of transparent armor, due to the isotropic structure, which allows to achieve a high degree of light transmission up to 95 % in the visible area of the spectrum; low density (3.58 g/cm^3) compared to metals, while having excellent mechanical characteristics; also, production of the material is marked by simplicity in comparison with production by growing single crystals from a melt [3].

The purpose of the work was to study the influence of concentration and method of introduction of gallium oxide sintering additive, as well as the temperature of roasting on the properties of ceramics based on the magnesium aluminate spinel, obtained by the method of reverse heterophase co-precipitation.

2. Materials and Methods

The process diagram includes:

1. Synthesis of powder AMS.
2. Introduction of a sintering additive.
3. Introduction of temporary technological binder.
4. Dry pressing at 100 MPa.
5. Burning in the air at 1100 °C.
6. Burning in a vacuum at 1700 and 1750 ° C.

Highly dispersive powder of double magnesian-aluminum hydroxide of strictly stoichiometric composition, $MgAl_2(OH)_8$, which are precursors of the magnesium aluminate spinel, is synthesized by reverse heterophase co-precipitation, since this method does not require the use of special equipment and high temperatures, and is characterized by relatively short reaction time. Water nitrates of aluminum and magnesium ($Al(NO_3)_3 \cdot 9H_2O$ and $Mg(NO_3)_2 \cdot 6H_2O$), taken in a ratio of 50:50 mol % in terms of oxides, were used as the initial components for synthesis.

In order to determine the temperature of the precursor heat treatment, a differential-thermal analysis (DTA) was carried out, the results of which are presented in Figure 1.

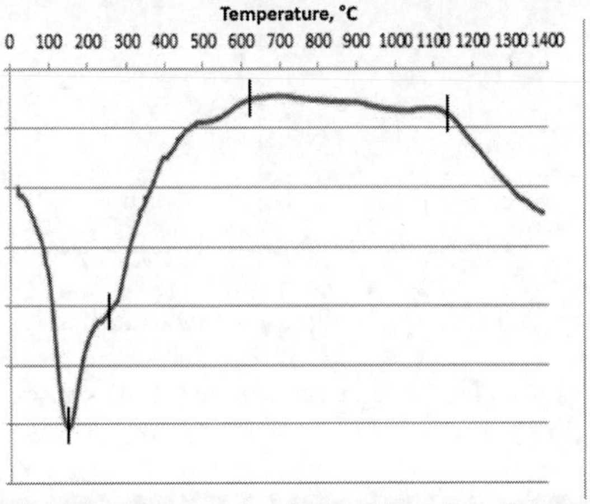

Figure 1. DTA of the spinel precursor obtained by reverse heterophase co-precipitation.

According to this differential-thermal data, we can assume that 1 peak (150 °C) refers to the removal of physically connected water; 2 peak (270 °C) refers to the removal of chemically bound water; 3 peak (610 °C) refers to the beginning of spinel formation; 4 peaks (1150 °C) – completion of spinel formation.

Based on the results of the differential-thermal analysis, the optimal temperature of the synthesis of the precursor powder for the production of the AMS phase equal to 1200 °C was chosen The choice of the synthesis method and the firing temperature is confirmed by the X-ray phase analysis data.

To reduce porosity, gallium oxide is selected as an additive, since it contributes to the formation of vacancies of solid solutions in both spinel undergrates, which improves sintering. The additive was introduced in the amount of 5 and 7 mol. %, in powder of spinel via wet method, with different methods of introduction: directly in the precursor and in the synthesized spinel. The semi-manufactured materials were formed by dry pressing at a pressing pressure of 100 MPa. The removal of temporary technological binder was carried out at a temperature of 1100 °C, and the burning of samples - at a temperature of 1700 and 1750 °C in a vacuum with a time of exposure of 3 hours. The material light transmission of the samples with different types of additive injection on the SHIMADZU UV-3600 spectrometer was measured.

3. Results and discussion

The stoichiometric composition of the magnesium aluminate spinel has the only crystal phase (Figure 2).

IOP Conf. Series: Materials Science and Engineering **848** (2020) 012093 doi:10.1088/1757-899X/848/1/012093

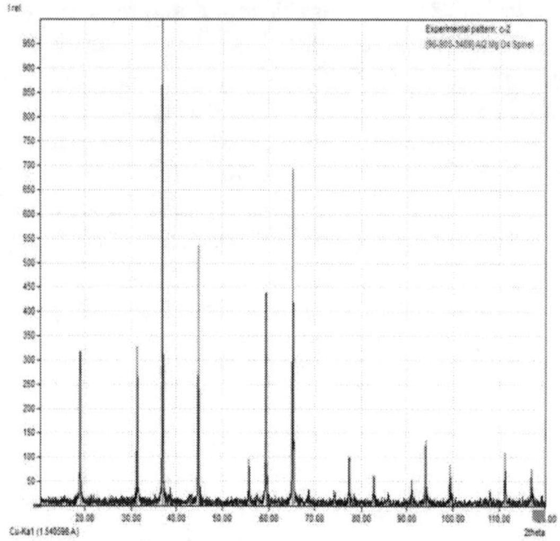

Figure 2. The $MgAl_2O_4$ difractogram obtained at 1200 °C.

Figure 3 (a, b) shows scanning electron microscopy (SEM) images of the precursor and spinel powder microstructures.

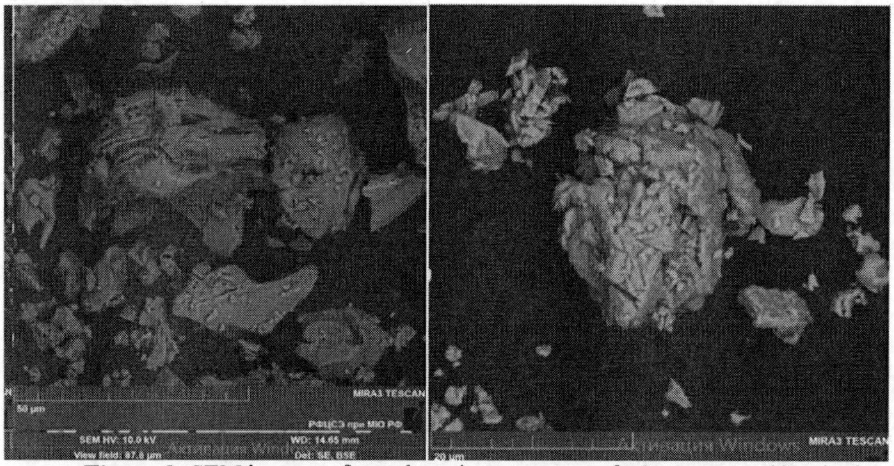

Figure 3. SEM images of powder microstructure of: a) precursor, b) spinel.

As can be seen in the figures, the precursor and spinel have a polyfraction composition with particles of different shape and size. As a result of this composition, pores can form in the material.

The X-ray phase analysis data (Figure 4) confirms the formation of a composition of gallium oxide and magnesium aluminate spinel of variable composition.

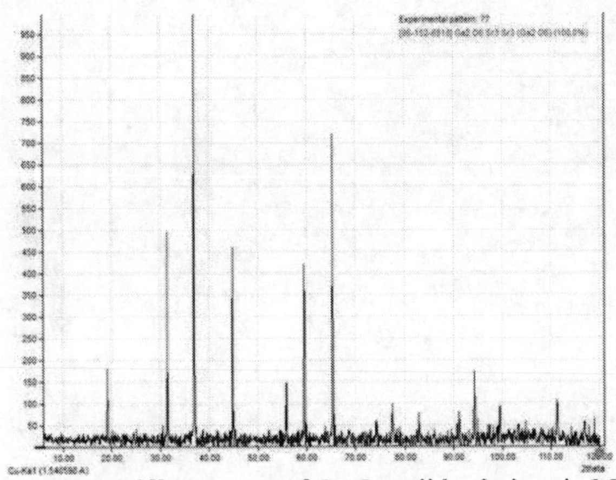

Figure 4. Diffractogram of Ga_2O_3 solid solutions in $MgAl_2O_4$.

The alloying addition introduced is not released in the form of an independant phase, which suggests the formation of a solid replacement solution in the lattice of the magnesium aluminate spinel.

The high dispersion of the producted powders provides their activity to sintering, which can lead to the pecking of particles and contribute to the process of aggregation already at the initial stages of pressing firing. It appears advisable to synthesizing of solid solutions at higher temperatures.

The following information about the properties of burned samples at 1700 °C was obtained during the studies.

Analysis of the graphs makes it possible to distinguish the best indicators of average density, open porosity at introduction of 7 mol. % of gallium oxide (Figure 5.)

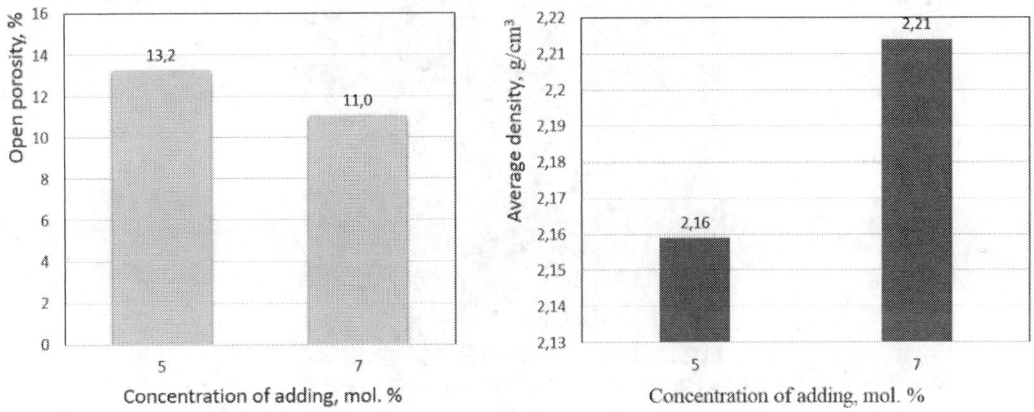

Figure 5. Properties of samples at a firing temperature of 1700 °C.

Than the properties of the samples produced by different methods of additive introducing with a concentration of 7 mol. % - into the precursor and into the synthesized spinel at a temperature of 1700 and 1750 °C were investigated. According to the results of the tests, it was revealed that the increase in temperature of firing leads to a significant improvement in the properties of ceramics. Thus, the best properties of the products were confirmed with temperature of firing (Figure 6).

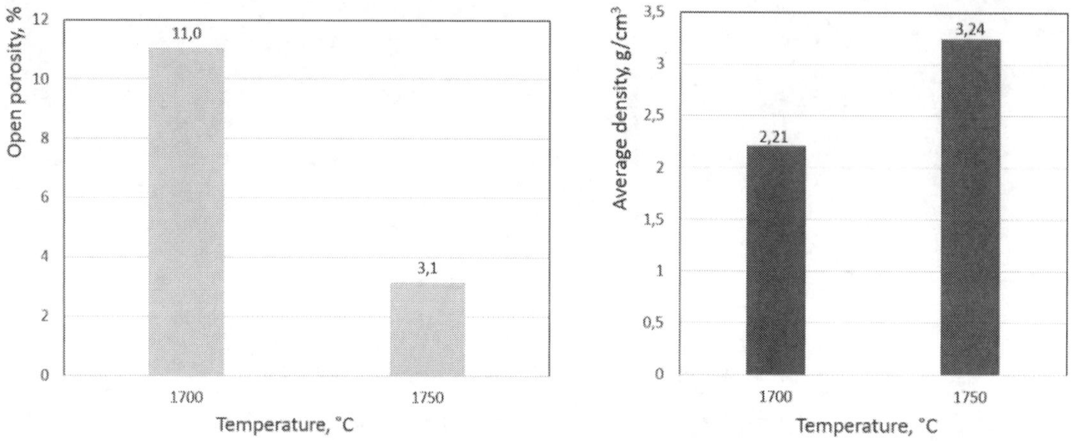

Figure 6. Comparison of the properties of samples with temperatures of firing of 1700 and 1750 °C.

Figure 7 shows the microstructure of the sample of 7 mol %. of gallium oxide burned at a temperature of 1750 °C. As a result of the sintering temperature increase, samples with a homogeneous structure were obtained, including a small number of intercrystalline and intra-crystalline pores.

Figure 7. Microstructure of ceramics from AMS with additive Ga_2O_3, $T_{firing} = 1750$ °C.

In Figure 8 a photograph of 7 mol % gallium oxide doped magnesium aluminate spinel ceramics is shown.

The light transmission graphs of ceramics are shown in Figure 9 (a, b).

Figure 8. Photograph of a ceramic sample of gallium oxide doped magnesium aluminate spinel.

The graphs show that the sample, obtained by the method of 7 mol. % additive introducing into the spinel has a higher degree of light transmission then 7 mol. % additive introduction into the precursor, temperature of firing is up to 1750 °C.

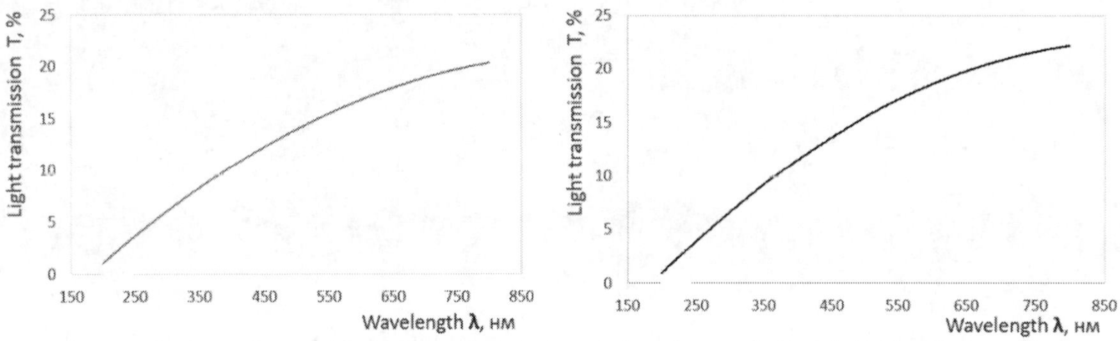

Figure 9. Dependence of light transmission on wavelength, sample 7 mol. % of gallium oxide added in a) precursor, b) spinel.

4. Conclusion

Thus, according to the experimental results, the following conclusions can be drawn:

1. The aluminum spinel is a promising material for the creation of optically transparent ceramics, since it has high optical, mechanical and inert properties with respect to aggressive environment, as well as it characterized by the isotropic structure, in which it is possible to achieve high light transmission, and the availability of raw materials.
2. One of the most relevant methods of obtaining spinels should be coprecipitation from aqueous solutions, since it does not require the use of special equipment and high temperatures, and also is characterized by a relatively short reaction time.
3. The best parameters of open porosity and average density are observed when adding 7 mol. % of gallium oxide after spinel synthesis. When adding an additive to the precursor, the quality of the sample is worse.
4. With the increase temperature of firing from 1700 to 1750 °C, the values of open porosity decrease, which indicates the beneficial effect of the firing temperature increase on the properties of ceramics from the aluminum magnesium spinel with the addition of gallium oxide to when it is inserted into the synthesized spinel.

Acknowledgments
The reported study was funded by RFBR according to the research project
№ 18-33-00507.

IOP Conf. Series: Materials Science and Engineering **848** (2020) 012093 doi:10.1088/1757-899X/848/1/012093

References

[1] Lukin E S 1996 Refractories and technical ceramics p 2
[2] Lukin E S, Popova N A, Glazachev V S, Pavlyukova L T, Kulikov N A 2015 Designs made of composite materials p 24
[3] Bhatngara A 2011 Light ballistic materials: Front page from English (Moscow: Technosphere) p 392

IOP Conf. Series: Materials Science and Engineering **848** (2020) 012094 doi:10.1088/1757-899X/848/1/012094

Effect of sintering temperature on magnetic hysteresis characteristics of powder alloy Fe-30Cr-8Co (wt.%)

A S Ustyukhin, V A Zelensky, I M Milyaev, A B Ankudinov and V S Shustov

Baikov Institute of Metallurgy & Materials Science (IMET RAS), Moscow, Russia

e-mail: fcbneo@yandex.ru

Abstract. In this work, we studied the effect of sintering temperature (1100 - 1400 °C) and thermomagnetic treatment (TMT) on the magnetic hysteresis characteristics of the Fe-30Cr-8 Co alloy, obtained by powder technology. It has been shown that at sintering temperatures of 1200 °C and below, it is possible to obtain materials with high magnetic characteristics that ensure their technical application. Properties exceed the characteristics of analogues obtained by casting technology. In the range 1200 - 1250 °C, the largest increase in the density of samples is observed with an increase in sintering temperature with a simultaneous increase in the values of magnetic characteristics. It was revealed that the total duration of heat treatment for 40-80 hours is sufficient for the formation of the necessary structure that determines the magnetic properties of the material. Long-term heat treatment (up to 160 hours) contributes to a higher level of magnetic properties.

1. Introduction

Hard magnetic materials are widely used as permanent magnets in various fields of technology, including power electronics (magnetrons, klystrons, travelling-wave lamps, etc.), electrical engineering (DC motors), instrumentation (magnetoelectric devices), and mechanical engineering (various types of separators), radio equipment (microphones and audio speakers), communications, medicine and. Etc [1-2]. Most classes of hard magnetic materials (cast alloys based on the Fe-Ni-Al-Co-Cu-Ti system, alloys based on the rare-earth compounds $SmCo_5$, Sm_2Co_{17}, $Nd_2Fe_{14}B$, magnetically hard barium and strontium ferrites) are characterized by relatively low mechanical (strength and plastic properties). However, in some cases of important technical application, magnetically hard material not only requires high magnetic hysteresis properties (residual induction B_r, coercive force H_c, maximum energy product $(BH)_{max}$) but also high strength ($\sigma_B \sim 600\text{-}800$ MPa).

Magnetic alloys of the Fe-Cr-Co system occupy a deserved place among hard magnetic materials in connection with their high mechanical properties in the presence of a sufficiently high level of magnetic hysteresis properties. The production of Fe-Cr-Co hard alloys is carried out in two ways - by casting and by powder metallurgy. The powder method has a number of additional advantages compared to casting technology: the ability to precisely control the chemical composition of alloys, an increased metal utilization rate in production, high productivity with large production volumes, and the possibility of economical manufacturing of piece lots of simple-shaped magnets. In this case, the magnetic hysteresis characteristics of powder magnets are not inferior to those of cast magnets [3-4].

The high vacuum sintering temperatures used in the production of magnetic alloys of the Fe-Cr-Co system cause evaporation of elemental components (mainly chromium) from the surface of the workpieces, which leads to a inhomogeneity of the chemical composition and deterioration of the material characteristics. Recent studies of powder Fe-Cr-Co alloys have shown that alloys with a low

IOP Conf. Series: Materials Science and Engineering **848** (2020) 012094 doi:10.1088/1757-899X/848/1/012094

cobalt content can be sintered at relatively low temperatures while maintaining high performance values [5].

In this work, we studied the effect of sintering temperature and thermomagnetic treatment on the magnetic hysteresis characteristics of the Fe-30Cr-8Co powder alloy. It has been shown that at sintering temperatures of 1200 °C and lower it is possible to obtain materials with high magnetic characteristics.

2. Materials and methods

Powder samples of the composition Fe-30Cr-8Co were made from industrial high-purity iron powder with a particle size of 10-20 μm, chromium powder with particles less than 50 μm in size and cobalt powder with particles less than 10 μm in size.

The powders were mixed in a C 2.0 turbulent mixer, pressing was performed on a KNUTH HP15 press in a detachable matrix with a diameter of 13.6 mm at a pressure of 600 MPa. The height of the compacts was about 20 mm. Sintering was carried out in a shaft furnace in a vacuum of no worse than 10^{-2} Pa with a holding time of 2.5 hours at maximum temperature. The density of the samples was determined by hydrostatic weighing. Quenching was carried out in a Nabertherm HT-16/16 high-temperature muffle furnace. Sintered powder samples were quenched from 1300 °C in water. Heat treatment was carried out in a muffle furnace and magnetic hysteresis properties were measured on a Permagraph L. hysteresograph. Photographs of the microstructure were obtained with an Olympus GX51 optical microscope. X-ray phase analysis was performed on a SHIMADZU XRD-6000 vertical x-ray diffractometer with monochromatized copper radiation.

3. Results and discussion

Green compacts had a relative density of about 80%. Sintering of the compacts was carried out in the temperature range 1100 - 1400 °C (sintering time - 150 min). The dependence of the relative density of sintered samples on sintering temperature is shown in Figure 1.

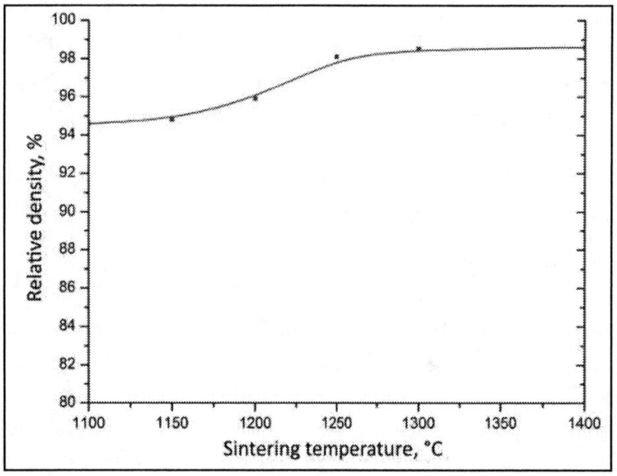

Figure 1. Dependence of the relative density of Fe-30Cr-8Co powder alloy samples on the sintering temperature.

Figure 1 shows that the alloy has a sufficiently high relative density (94.5%) even at low sintering temperatures of the order of 1100 - 1150 °C. The increase in density can be observed in the temperature range 1200 - 1250 °C, when the density reaches 98%. Heat treatment of the Fe-30Cr-8Co alloy samples, sintered at various temperatures, was carried out and magnetic properties were measured. The magnetic field during TMT was applied along the axis of easy magnetization, coinciding with the axis of the cylindrical sample. The results of measurement of the magnetic properties of the Fe-30Cr-8Co alloy samples, sintered at various temperatures are presented in table 1 and some hysteresis loops in Figure 2.

Table 1. Magnetic hysteresis properties of the Fe-30Cr-8Co alloy.

Alloy	Sintering temperature,°C	B_r, T	H_c, kA/m	$(BH)_{max}$, kJ/m^3
Fe-30Cr-8Co	1100	1,18 – 1,19	40,7 – 43,4	27,6 – 30,2
	1150	1,18 – 1,19	41,5 – 42,8	29,1 – 31,0
	1200	1,18 – 1,21	40,0 – 41,9	31,8 – 33,0
	1250	1,21 – 1,25	40,0 – 42,7	32,9 – 35,2
	1300	1,22 – 1,26	40,3 – 42,4	34,3 – 36,2
	1400	1,26 – 1,28	40,8 – 43,6	36,1 – 38,0

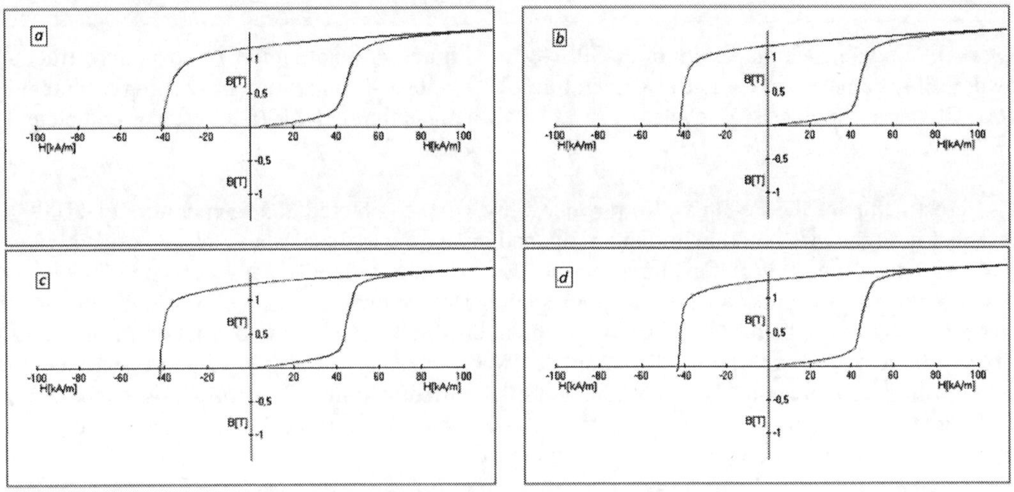

Figure 2. Hysteresis loops of Fe-30Cr-8Co samples, sintered at a temperature of 1100 (a), 1200 (b), 1300 (c) and 1400 °C (d) after a complete TMT cycle.

From the analysis of these results it follows that the magnetic properties of the powder alloy under all sintering conditions are at a high level, ensuring their technical application and exceed the characteristics of analogues obtained by casting technology. For example, the magnetic properties of alloy 28H10KA according to GOST 24897-81: B_r = 1.1 T, H_c = 38 kA / m, $(BH)_{max}$ = 26 kJ / m^3. It should also be noted that there is an increase in the magnetic characteristics in the temperature range of 1200 - 1250 °C. This fact is largely explained by an increase in the density of samples in the same temperature range. A further increase in sintering temperature up to 1400 °C gives a small increase in the residual induction B_r and maximum energy product $(BH)_{max}$ while the coercive force H_c remain at the same level.

It should be noted that the Fe-30Cr-8Co powder alloy has a high level of magnetic properties at low sintering temperatures of 1100 - 1150 °C. X-ray phase analysis of the samples sintered at a temperature of 1100 °C (Figure 3) shows the presence of only the α phase after a complete TMT cycle. A comparison of magnetic properties with the data of our previous studies about phase composition [5] shows that the absence of undesirable phases during sintering is an important factor for the formation of a high level of magnetic properties in Fe-Cr-Co powder alloys. In addition, sintering in the field of formation of a single-phase bcc α-solid solution also reduces the residual porosity in samples.

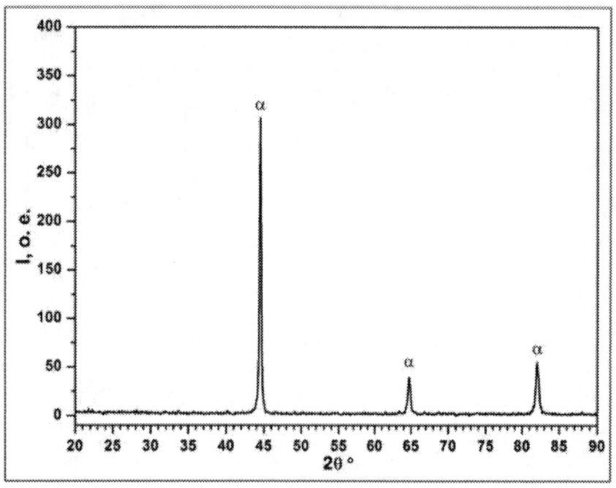

Figure 3. X-ray diffraction pattern of Fe-30Cr-8Co powder alloy sample, sintered at a temperature of 1100 °C after a complete TMT cycle.

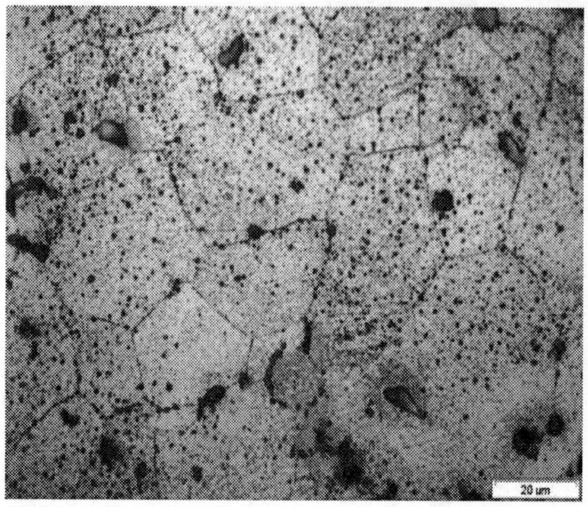

Figure 4. Photograph of the microstructure of Fe-30Cr-8Co powder alloy sample, sintered at a temperature of 1100 °C after a complete TMT cycle.

Microstructure of the Fe-30Cr-8Co powder alloy sample, sintered at a temperature of 1100 ° C, is presented in Figure 4. A mixture of aqua regia and glycerin was used as an etchant. Aqua regia is the main, recommended etchant for alloys of this system, however, when it is used without glycerin, structure blur often occurs. As experimental studies have shown, glycerin slows down the action of aqua regia and allows better identification of the microstructure. The microstructure of the Fe–30Cr–8Co powder alloy (Figure 4) after the complete TMT cycle consists mainly of grains with a diameter of 20–40 μm; dark areas in the photographs are identified as pores. The formation of the structure, responsible for the magnetic properties, occurs as a result of a certain heat treatment of the sintered sample. Heat treatment modes are presented in table 2.

Table 2. Heat treatment modes of a Fe-30Cr-8Co powder alloy, sintered at a temperature of 1400 °C.

Mode	Stage 1	Stage 2	Stage 3	Stage 4
1	670°C(10 min) + cooling to 500°C with V= 4 °C/h	-	-	-
2	670°C(10 min) + cooling to 500°C with V= 4 °C/h	+ annealing at 500°C(20 hours)	-	-
3	670°C(10 min) + cooling to 500°C with V= 4 °C/h	+ annealing at 500°C(20 hours)	+ cooling to 480°C(20 hours)	-
4	670°C(10 min) + cooling to 500°C with V= 4 °C/h	+ annealing at 500°C(20 hours)	+ cooling to 480°C(20 hours)	+ cooling to 460°C(20 hours)
5	670°C(10 min) + cooling to 500°C with V= 4 °C/h	+ annealing at 500°C(20 hours)	+ cooling to 480°C(20 hours)	+ cooling to 480°C(40 hours)
6	670°C(10 min) + cooling to 500°C with V= 4 °C/h	+ annealing at 500°C(20 hours)	+ cooling to 480°C(20 hours)	+ cooling to 480°C(80 hours)

Heat treatment consists of preliminary quenching from 1300 °C and the stage of TMT. As a result of TMT, a high-temperature α-solid solution decomposes into two isomorphic bcc α_1 and α_2 solid solutions [6]. This leads to the redistribution of components in the system and the formation of a structure that determines the magnetic properties. Heat treatment of the Fe-30Cr-8Co alloy samples, sintered at a temperature of 1400 °C, was carried out according to six modes (table 2) and magnetic properties were measured. Total TMT time increased from 40 hours (mode 1) to 160 hours (mode 6).

The dependences of H_c and $(BH)_{max}$ on the heat treatment time are presented in Figure 5. The residual induction B_r is independent of the TMT time and has a value of about 1,27 T.

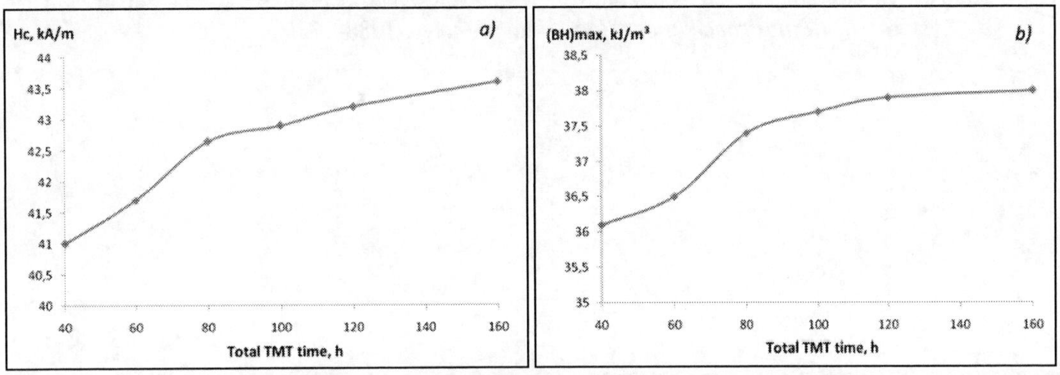

Figure 5. Dependence of H_c (a) and $(BH)_{max}$ (b) of Fe-30Cr-8Co alloys samples, sintered at a temperature of 1400 °C from the TMT time.

The coercive force H_c of the alloy (Figure 5a) grows with increasing heat treatment time. The dependence of the maximum energy product $(BH)_{max}$ on the heat treatment time (Figure 5b) has a similar form to the curve for the coercive force.

Samples subjected to TMT for 40-80 hours have the necessary level of magnetic hysteresis properties for technical use. However, if the consumer needs a higher level of magnetic properties of the products, then a long heat treatment (up to 160 hours) contributes to its production. From an economic point of view, heat treatment of alloys lasting less than 100 hours is sufficient.

4. Conclusion

The magnetic properties of the Fe-30Cr-8Co powder alloy, sintered at temperatures of 1100 - 1400 °C are at a high level ensuring their technical application. Properties exceed the characteristics of analogues obtained by casting technology.

In the temperature range 1200-1250 °C, the largest increase in the density of samples is observed with an increase in sintering temperature with a simultaneous increase in the values of magnetic characteristics.

TMT for 40-80 hours provides a sufficient level of magnetic hysteresis properties. Long-term heat treatment (up to 160 hours) contributes to a higher level of magnetic properties.

Acknowledgements
This study was supported by the Russian Foundation for Basic Research (project no. 18-03-00666).

References

[1] Artamonov E V, Libman M A, and Rudanovskii N N 2007 Magnetically Hard Materials for the Motors of Synchronous Hysteresis Electric Motors *Steel in Transl.* **37** 547-51

[2] S Jin, G Y Chin and Wonsiewicz B C 1980 A Low Cobalt Ternary Cr-Co-Fe Alloy for Telephone Receiver Magnet Use *IEEE Trans. on Magn.* **16** 3187-92

[3] Alymov M I, Ankudinov A B, Zelenskii V A, Milyaev I M, Yusupov V S and Vompe T A 2014 The effect of surfactant admixture during milling on pressing, sintering, and magnetic properties of FeCrCoMoW alloy *Inorg. Mat. Appl. Res.* **5** 530-5

[4] Shatsov A A, Ryaposov I V and Kozvonin V A 2017 Concertation-Inhomogeneous Hard Magnetic Alloys of the Fe – Cr – Co System with Elevated Content of Cobalt and Boron *Met. Sci. and Heat Treat.* **59** 45-9

[5] Ustyukhin A S, Ankudinov A B, Zelenskii V A, Milyaev I M, Ashmarin A A and Alymov M I 2018 Phase composition of powder hard magnetic Fe–30Cr–(8–24)Co alloys sintered under different temperature conditions *Dokl. Phys. Chem.* **482** 140-4

[6] Kaneko H, Homma M and Nakamura K 1971 New Ductile Permanent Magnet of Fe-Cr-Co System *AIP Conf. Proceedings. Magn. Magn. Mater.* 1088–92

IOP Conf. Series: Materials Science and Engineering **848** (2020) 012095 doi:10.1088/1757-899X/848/1/012095

Phosphors based on Ce:$(Pb,Gd)_3(Al,Ga)_5O_{12}$ epitaxial films: synthesis, optical properties, application

D A Vasil'ev[1], D A Spassky[2], V G Plotnichenko[3], V V Voronov[1], A V Khakhalin[4], A M Galstyan[1] and N V Vasil'eva[1]

[1] Prokhorov General Physics Institute of the Russian Academy of Sciences, Vavilov Str., 38, 119991 Moscow, Russia

[2] Skobeltsyn Institute of Nuclear Physics, Lomonosov Moscow State University, Leninskie Gory, 119991 Moscow, Russia

[3] Fiber Optics Research Center, Russian Academy of Sciences, Vavilov Str., 38, 119333 Moscow, Russia

[4] Physics Department, Lomonosov Moscow State University, Leninskie Gory, 119991 Moscow, Russia

E-mail: dimaphys@gmail.com

Abstract: Ce-doped $(Pb,Gd)_3(Al,Ga)_5O_{12}$ single crystalline garnet films were grown using liquid-phase epitaxy method from supercooled $PbO–B_2O_3$-based melt solutions on substrates from $Gd_3Ga_5O_{12}$ and $Gd_3Al_{2.26}Ga_{2.74}O_{12}$ single crystals. Optical absorption and photo- and cathodoluminescent properties of these epitaxial garnet films were studied. Ce-doped $(Pb,Gd)_3(Al,Ga)_5O_{12}$ garnet films can be used as a fast phosphor in the design of a PIF-01 electron-optical converter.

1. Introduction

Gadolinium - aluminum - gallium garnet doped with cerium ions (Ce:$Gd_3Al_2Ga_3O_{12}$, Ce:GAGG), is considered as a promising scintillation material among garnet crystals. Currently, 2-in. size Ce 1%:$Gd_3Al_{2.3}Ga_{2.7}O_{12}$ single crystal grown by Czochralski method shows a scintillation light yield approximately 58,000 photons/MeV [1]. The scintillation decay time (and its relative intensity) was 172 ns (88%) for the fast component in this crystal. The emission wavelength was 516 nm. Along with bulk single crystals, single crystalline films find their application as scintillation screens [2, 3]. Films can be grown by liquid-phase epitaxy (LPE) from supercooled melt solutions such as PbO - B_2O_3, Bi_2O_3 - B_2O_3 or BaO – B_2O_3 –BaF_2. The difference between epitaxial films and their bulk single crystal analogs is the presence of additional optical absorption, which is caused by impurity ions, which are introduced into the films from the melt solution and crucible material. For instance, the impurity ions in the epitaxial films grown from supercooled $PbO–B_2O_3$-based melt solutions are Pb^{2+}, Pb^{4+}, Pt^{4+} ions.

Cerium ions are the most common activator for scintillation detectors. The emission of Ce^{3+} arises due to an allowed 5d-4f electronic transition. As a result Ce doped scintillators are characterized by a high scintillation light yield and a fast decay time (tens of ns). The luminescence of the Ce^{3+} ion in the $Gd_3Ga_5O_{12}$ crystal at room temperature and atmospheric pressure is completely quenched due to the location of the $5d_1$ level of Ce^{3+} in the conduction band [4, 5], but under high pressure [6] or with the partial substitution of Ga with Al in the $Gd_3Ga_5O_{12}$ host Ce^{3+} emission can be observed [7].

In this work we report the results of the study of the influence of Ga/Al ratio variation on optical absorption, photo- and cathodoluminescent properties of epitaxial films Ce:$(Pb,Gd)_3(Al,Ga)_5O_{12}$, grown by LPE. Our main goal is to obtain samples with a maximum photoluminescence intensity of Ce^{3+} ions and to investigate the applicability of epitaxial films as a fast phosphor in the design of a PIF-01 electron-optical converter.

2. Experimental methods

The epitaxial films Ce:$(Pb,Gd)_3(Al,Ga)_5O_{12}$, were grown with standard isothermal LPE method. A schematic drawing of the set-up is given in the Figure 1. Before the growth of the films, the melt solution was homogenized in the platinum crucible for at least four hours.

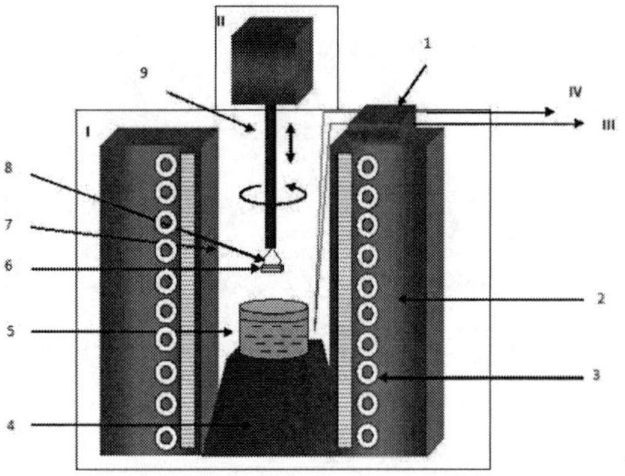

Figure 1. The set-up for LPE growth of films: I—heating furnace; II—rod displacement and rotation control; III—stabilization control and furnace temperature control by OMRON E5CN Digital Temperature Controller. IV- melt solution temperature control unit. 1 — thermocouples; 2 — thermal insulation; 3 — heating element; 4 — ceramic stand for crucible; 5 — platinum crucible; 6 — substrate; 7 — corundum pipe; 8 — platinum holder; 9 — ceramic rod.

The films were characterized by X-ray diffraction on a Bruker D8 Discover A25 Da Vinsi Design X-ray diffractometer (CuKα radiation). The quantitative chemical analysis of the grown films was performed with an electron-ion scanning microscope Quanta 3 D FEG. To simplify the spectroscopic studies, we did not remove the film from the back side of the substrate. Transmission spectra of the films were measured using Perkin Elmer Lambda 900 spectrophotometer in 250–550 nm wavelength range at room temperature. The optical density D was derived from the transmission using the formula $D = [\ln(T_s/T_{fsf})]$, where T_s is the transmission spectrum of the substrate and T_{fsf} is the transmission spectrum of the substrate with grown films on both sides. For the analysis of the absorption spectra of the films, we used the normalized optical density D/2h. It allows to compare the intensity of the absorption bands of the films with different thickness. Photoluminescence spectra of the films were measured at 300 K in the 400 - 700 nm region at E_{ex} = 165, 451 and 440 nm using specialized - spectroscopic set-ups. The set-ups allowed to perform the measurements of luminescence and luminescence excitation spectra in UV and VUV spectral regions. A 150 W Xe lamp combined with monochromator MDR-206 was used as the excitation source in UV spectral region. The luminescence was detected using Oriel MS257 spectrograph equipped with Marconi 30-11 CCD detector. A 150-W deuterium discharge lamp Hamamatsu L1835 was used as an excitation source in VUV spectral region. The excitation wavelength was selected using a McPherson Model 234/302 monochromator. The luminescence spectra were registered using a Shamrock 303i-B spectrograph equipped with an ANDOR iDUS 416 CCD camera. To measure the cathodoluminescence decay kinetics of cerium ions, we selected a sample of $Pb_{0.02}Ce_{0.05}Gd_{2.93}Al_{4.29}Ga_{0.71}O_{12}$ with a maximum thickness of 49.9 μm, on which an epitaxial film was ground on one side. This sample was glued using silver paste onto the inside of the fiber-optic disk of the screen assembly of the PIF-01 type tube having a silver-oxygen-cesium photocathode (S1). A YAP: Nd^{3+} laser, operating in the self-mode locking mode and generating a train of picosecond pulses at a wavelength of 1.08 μm, was used as a source of pulsed radiation at the input of the image intensifier tube.

3. Results and discussions

Ce-doped $(Pb,Gd)_3Al_xGa_{5-x}O_{12}$ single crystalline garnet films with 2 < x < 5 were grown using LPE method from supercooled $PbO–B_2O_3$-based melt solutions at concentrations $C(Gd_2O_3)$ = 0.2, 0.3, 0.4 and 0.5 mol %; $C(CeO_2)$ = 0.03, 0.2 and 0.3 mol % and $C(Al_2O_3)$ from 2 to 5 mol % in the mixture on single crystal garnet substrates: $Gd_3Ga_5O_{12}$ (GGG) with the (111) orientation and $Gd_3Al_{2.26}Ga_{2.74}O_{12}$ (GAGG) with the (320) orientation.

X-ray diffraction analysis showed that all films have a garnet type structure [8]. X-ray diffraction patterns obtained in the θ/2θ scan mode showed only strong 444 and 888 reflections from the films and the same weak reflections from its substrate, attenuated by the film. For example, for $Pb_{0.02}Ce_{0.06}Gd_{2.92}Al_{2.22}Ga_{2.78}O_{12}$ film grown from a melt solution at concentrations of gadolinium oxides of 0.2 mol %, cerium – of 0.2 mol % and aluminum – of 2.2 mol %, the unit-cell parameters of the substrate a_s == 12.3829 A and film a_f = 12.269 A were determined from the peak positions (the relative lattice mismatch $(a_s - a_f)/a_f$ is 0.9%). The unit-cell parameter of the substrate agrees with JCPDS Powder Diffraction File data for GGG (card no. 76_2290). It was found that the unit-cell parameter decreases with the increasing of film thickness.

The profiles of the asymmetric 880 and 12.60 reflections indicate that the films were single crystalline and grew epitaxially on its substrates. SEM images of the surface of $Ce:(Pb,Gd)_3Al_xGa_{5-x}O_{12}$ films at 0 ≤ x < 5 are shown in the Figure 2. The surface of the $Ce:(Pb,Gd)_3Ga_5O_{12}$ film is smooth. With the introduction of aluminum ions in the composition of the film lattice mismatch parameters of the film and substrate increases and the surface becomes rough, consisting of intergrowths of single crystalline grains oriented both in the direction of the [111] axis and in the plane of the substrate. With increasing of film thickness, surface smoothing occurs.

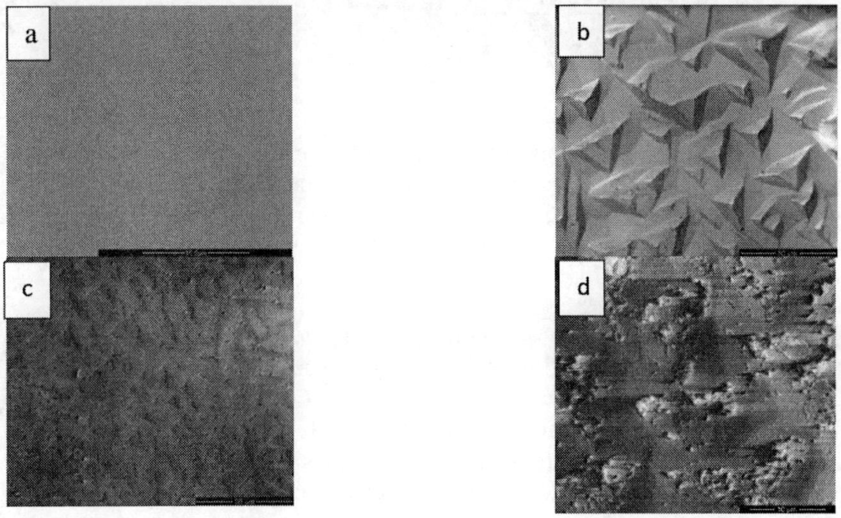

Figure 2. Electron images of the surface of films: a) - $Pb_{0.02}Ce_{0.03}Gd_{2.95}Ga_5O_{12}$ h = 3.6 μm (marker – 10 μm, magnification 5010ˣ); b) - $Pb_{0.03}Ce_{0.03}Gd_{2.94}Al_{2.79}Ga_{2.21}O_{12}$ h = 16.4 μm (marker – 50 μm, magnification 2000ˣ); c) - $Pb_{0.01}Ce_{0.05}Gd_{2.94}Al_{3.06}Ga_{1.94}O_{12}$ h = 33.8 μm, (marker – 50 μm, magnification 2000ˣ); d) – $Pb_{0.01}Ce_{0.02}Gd_{2.97}Al_{3.13}Ga_{1.87}O_{12}$ h = 43.3 μm, (marker – 10 μm, magnification 8000ˣ).

Study of the effect of the substitution of gallium ions with aluminum ions on optical absorption of cerium ions in the $Ce:(Pb,Gd)_3(Al,Ga)_5O_{12}$ films was performed by measuring the transmission spectra (Figure 3a). Two absorption bands in the range from 300 to 360 nm (level $5d_2$) and from 390 to 550 nm (level $5d_1$), corresponding to the parity allowed electronic transitions $4f(^2F_{5/2,7/2})$ - $5d_{1.2}$ Ce^{3+} ions were detected. These bands confirm the incorporation of cerium ions into the composition of the films.

Narrow absorption bands in the range from 240 to 313 nm, corresponding to electron transitions $^8S_{7/2} \rightarrow {}^6I_{15/2\text{-}7/2}$ Gd^{3+} ions in the transmission spectra of GGG substrates were also observed.

The normalized optical density spectra are presented in Figure 3b. The absorption band in the range from 250 to 300 nm, corresponding, according to [9], to the $^1S_0 \rightarrow {}^3P_1$ electronic transition in Pb^{2+} ($6s^2$) ions were observed in the absorption spectra of films with thickness ≤ 32 μm. Two other broad absorption bands correspond to the 4f - $5d_{1,2}$ Ce^{3+} electronic transition. It was found that with the increase of $C(Al_2O_3)$ from 2.1 to 5.0 mol % in the mixture at $C(CeO_2)$ = 0.2 mol % and $C(Gd_2O_3)$ = 0.2 mol %, the maxima of these two absorption bands shift in the following way. When the Al/Ga concentration ratio increases from 0.74 to 98.8 in the Ce:$(Pb,Gd)_3Al_xGa_{5\text{-}x}O_{12}$ films ($2.13 < x < 4.94$), the shift of 4f - $5d_1$ Ce^{3+} absorption band maximum to longer wavelengths was 30 nm (from 436 to 466 nm).

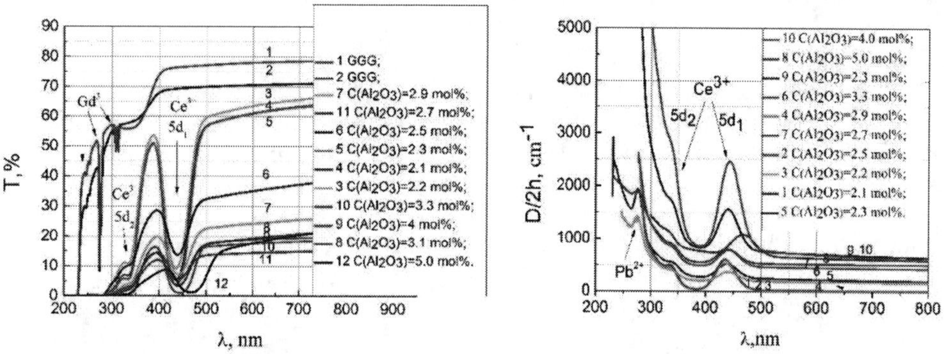

(a) (b)

Figure 3. a) - Transmission spectra of GGG substrates (curves 1 and 2) and samples of Ce:$(Pb,Gd)_3(Al,Ga)_5O_{12}$ epitaxial films grown from melt solutions with a concentration in the charge $C(CeO_2)$ = 0.2 mol %, $C(Gd_2O_3)$ = 0.2 mol % and various concentrations of $C(Al_2O_3)$ from 2.1 to 5.0 mol % indicated in the figure. b) - Spectral dependences of the normalized optical density D/2h on the wavelength of the same epitaxial films.

Analysis of the photoluminescence spectra under excitation at 451 (Figure 4a) and 440 nm (Figure 4b) Ce:$(Pb,Gd)_3(Al,Ga)_5O_{12}$ epitaxial films grown from melt solutions with $C(CeO_2)$ = 0.2 and 0.3 mol %, $C(Gd_2O_3)$ = 0.2 mol % in the mixture and at the concentration variation of $C(Al_2O_3)$ from 2.1 to 4.5 mol % showed that the films grown from melt solutions with $C(Al_2O_3)$ = 2.9, 3.1, 4.0 and 4.5 mol % in the mixture had the highest photoluminescence intensity. A blue shift of the emission band maximum from 555 to 550 nm was detected with a decrease in the Al^{3+} ion concentration from 3.14 to 2.13 in the film composition, which is consistent with the results obtained on single crystals of Ce:$Gd_3(Ga,Al)_5O_{12}$ (1 at. %) [10]. It was found that films grown from a melt solution with $C(Gd_2O_3)$ = 0.4 mol %, $C(CeO_2)$ = 0.2 mol % and $C(Al_2O_3)$ = 4.5 mol % had the maximum photoluminescence intensity of Ce^{3+} ions peaking at 555 and 532 nm under excitations at 440 (Figure 4b) and 165 nm (Figure 5a), respectively, at T = 300 K.

In the luminescence excitation spectra of $Pb_{0.01}Ce_{0.03}Gd_{2.96}Al_{3.14}Ga_{1.86}O_{12}$ film grown from melt solutions with $C(CeO_2)$ = 0.2 mol %, $C(Gd_2O_3)$ = 0.4 mol %, $C(Al_2O_3)$ = 4.5 mol % (1) four pronounced bands were observed. The bands at 448 nm and 343 nm are ascribed to electron transitions from 4f to $5d_1$ and $5d_2$ levels of Ce^{3+} ions. The band at 278 nm is a superposition of two bands related to $^1S_0 \rightarrow {}^3P_1$ and $^8S_{7/2} \rightarrow {}^6I$ in Pb^{2+} and Gd^{3+} ions, respectively. The latter indicates the energy transfer from Gd^{3+} and/or Pb^{2+} ions to Ce^{3+} ion. Non-elementary broad band peaking at 215 nm can be ascribed to the superposition of several bands connected with 4f - $5d_{3\text{-}5}$ transitions in Ce^{3+} and also with charge transfer transitions involving Ce^{4+} ions.

According to our previous study [11] the $Pb_{0.01}Ce_{0.03}Gd_{2.96}Al_{3.14}Ga_{1.86}O_{12}$ film shows the maximum value of the cathodoluminescence light yield of about 51,500 photons/MeV with fast decay times of the slow component of 61.0 ns (68%).

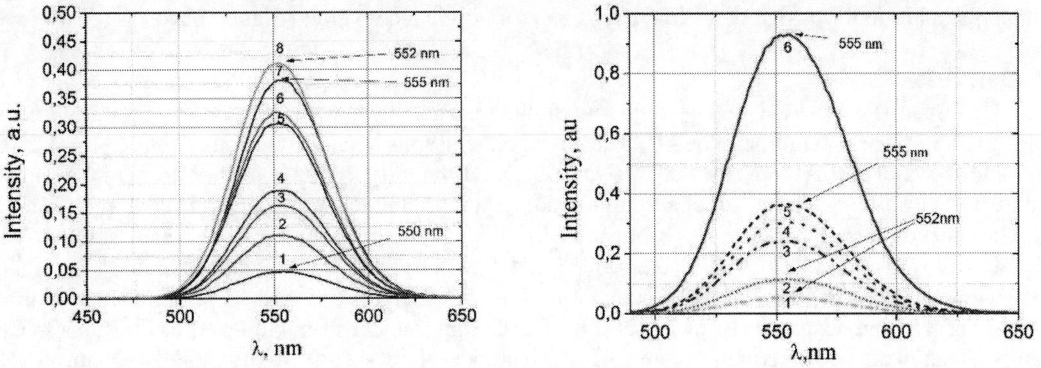

(a) (b)

Figure 4. a) - Photoluminescence spectra upon excitation at 451 nm of epitaxial films grown from melt solutions with $C(CeO_2)$ = 0.2 and 0.3 mol % and the concentration of $C(Al_2O_3)$ in the mixture: 2.1 mol % (1), 3.3 mol % (2), 2.7 mol % (3), 2.3 mol % (4), 4.5 mol % (5), 3.1 mol % (6), 4.0 mol % (7), 2.9 mol % (8). b) - Photoluminescence spectra upon excitation at 440 nm of epitaxial films grown from melt solutions with $C(Gd_2O_3)$ = 0.2 mol %, $C(CeO_2)$ = 0.2 mol %, $C(Al_2O_3)$ = 2.9, 3.1 mol % (1,2); $C(Gd_2O_3)$ = 0.4 mol %, $C(CeO_2)$ = 0.3 mol %,$C(Al_2O_3)$ = 4.5 mol% (3.4); $C(Gd_2O_3)$ = 0.3 mol %, $C(CeO_2)$ = 0.2 mol %, $C(Al_2O_3)$ = 4.5 mol % (5); $C(Gd_2O_3)$ = 0.4 mol %, $C(CeO_2)$ = 0.2 mol %, $C(Al_2O_3)$ = 4.5 mol % (6). Spectra were measured at room temperature.

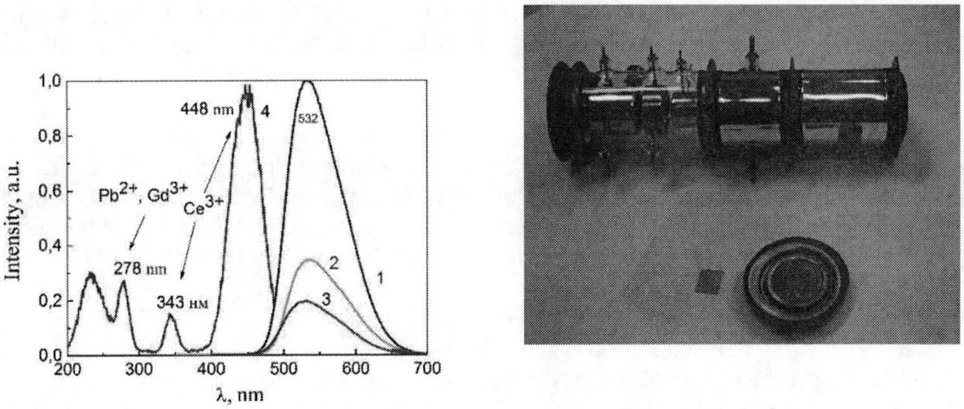

(a) (b)

Figure 5. a) - Photoluminescence spectra upon excitation at 165 nm of epitaxial films grown from melt solutions with $C(CeO_2)$ = 0.2 mol % and $C(Gd_2O_3)$ = 0.4 mol %, $C(Al_2O_3)$ = 4.5 mol % (1), $C(Gd_2O_3)$ = 0.3 mol %, $C(Al_2O_3)$ = 4.5 mol % (2), $C(Gd_2O_3)$ = 0.2 mol %, $C(Al_2O_3)$ = 4.0 mol % (3), T = 300 K and photoluminescence excitation spectrum of $Pb_{0.01}Ce_{0.03}Gd_{2.96}Al_{3.14}Ga_{1.86}O_{12}$ film grown from melt solutions with $C(CeO_2)$ = 0.2 mol %, $C(Gd_2O_3)$ = 0.4 mol %, $C(Al_2O_3)$ = 4.5 mol % (1) upon emission at 540 nm, T = 80 K. b) - Photography of an electron-optical converter of the PIF-01 type with a $Pb_{0.02}Ce_{0.05}Gd_{2.93}Al_{4.29}Ga_{0.71}O_{12}$ film phosphor.

An analysis of the cathodoluminescence decay kinetics of the films under excitation by a picosecond electron beam in the PIF-01 tube. The cathodoluminescence decay times of the $Pb_{0.02}Ce_{0.05}Gd_{2.93}Al_{4.29}Ga_{0.71}O_{12}$ film grown from a melt solution $C(Gd_2O_3)$ = 0.4 mol %, $C(CeO_2)$ = 0.3 mol % and $C(Al_2O_3)$ = 4.5 mol % were 22 ns (20%) for the fast component and 67 ns (80%) for the slow component [12]. Therefore, we conclude that the grown film with such parameters can be used as a fast phosphor in the design of a PIF-01 type electron-optical converter (Figure. 5b).

4. Conclusions

The Ce-doped $(Pb,Gd)_3Al_xGa_{5-x}O_{12}$ single crystalline garnet films with $2 < x < 5$ were grown using LPE method from supercooled $PbO–B_2O_3$-based melt solutions at concentrations $C(Gd_2O_3)$ = 0.2, 0.3, 0.4 and 0.5 mol %; $C(CeO_2)$ = 0.03, 0.2 and 0.3 mol % and $C(Al_2O_3)$ from 2 to 5 mol % in the mixture. The shift 4f-5d$_1$ Ce^{3+} of absorption band maxima to longer wavelengths was 30 nm when the Al/Ga concentration ratio increases from 0.74 to 98.8. The film grown from the melt solution with $C(Gd_2O_3)$ = 0.4 mol %, $C(CeO_2)$ = 0.2 mol % and $C(Al_2O_3)$ = 4.5 mol % demonstrated the highest photoluminescence intensity of Ce^{3+} ions at emission wavelengths 555 and 532 nm under excitations at 440 and 165 nm, respectively, at T = 300 K. The energy transfer from Gd^{3+} and/or Pb^{2+} ions to Ce^{3+} ion was shown. The maximum value of the cathodoluminescence light yield of about 51,500 photons/MeV at the decay time of the slow component of 61.0 ns (68%) was obtained for $Pb_{0.01}Ce_{0.03}Gd_{2.96}Al_{3.14}Ga_{1.86}O_{12}$ film. A film with such parameters can be used as a fast phosphor in the design of a PIF-01 type electron-optical converter.

Acknowledgments

This work was supported in part by the European Social Fund's Doctoral Studies and Internationalisation Programme DoRa. The work was performed using equipment purchased at the expense of the Moscow University Development Program.

References

[1] Kamada K, Kurosawa Sh, Prusa P, Nikl M, Kochurikhin V V, Endo T, Tsutumi K, Sato H, Yokota Y, Sugiyama K and Yoshikawa A 2014 Cz grown 2-in. size $Ce:Gd_3(Al,Ga)_5O_{12}$ single crystal; relationship between Al, Ga site occupancy and scintillation properties OPT MATER. **36** 1942-45.

[2] Kucera M and Prusa P 2017 LPE-Grown Thin-Film Nanocomposite, Ceramic and Thin Film Scintillators ed M Nikl (Singapore: Pan Stanford Publishing) chapter 5 pp 155-226.

[3] Douissard P-A, Cecilia A, Martin Th, Chevalier V, Couchaud M, Baumbach T, Dupre K, Kuehbacher M and Rack A 2010 A novel epitaxially grown LSO-based thin-film scintillator for micro-imaging using hard synchrotron radiation J SYNCHROTRON RADIAT. **17** 571-83.

[4] Doroshenko M E, Ivanov M A, Sigachev V B and Timoshechkin M I 1992 Influence of Ce^{3+} ions on the spectroscopic and lasing properties of $Y_3Al_5O_{12}:Ce^{3+}:Er^{3+}$and $Gd_3Ga_5O_{12}:Ce^{3+}:Er^{3+}$ crystals QUANTUM ELECTRON+ **22** 588-90.

[5] Kamińska A, Duzynska A, Berkowski M., Trushkin S and Suchocki A 2012 Pressure-induced luminescence of cerium-doped gadolinium gallium garnet crystal PHYS REV B. **85** 155111-21.

[6] Syvorotka II, Sugak D, Wierzbicka A, Wittlin A, Przybylińska H., Barzowska J., Barcz A, Berkowski M, Domagała J, Mahlik S, Grinberg M, Chong-Geng Ma, Brik MG, Kamińska A, Zytkiewicz ZR and Suchocki A 2015 Optical properties of pure and Ce^{3+} doped gadolinium gallium garnet crystals and epitaxial layers J LUMIN. **164** 31–7.

[7] Kamada K, Yanagida T, Endo T, Tsutumi K, Usuki Y, Nikl M, Fujimoto Y, Fukabori A and Yoshikawa A 2012 2 inch diameter single crystal growth and scintillation properties of $Ce:Gd_3Al_2Ga_3O_{12}$ *J CRYST GROWTH* **352** 88-90.

IOP Conf. Series: Materials Science and Engineering **848** (2020) 012095 doi:10.1088/1757-899X/848/1/012095

[8] Vasil'ev DA, Spassky DA, Voronov VV, Sokolov VO, Khakhalin AV, Vasil'eva NV and PlotnichenkoVG 2015 Effect of Al and Ce ion concentrations on the optical absorption and luminescence in $Gd_3(Al,Ga)_5O_{12}:Ce^{3+}$ epitaxial films *INORG MATER+* **51** 1008-16.
[9] Scott GB and Page JL 1977 Pb valence in iron garnets *J APPL PHYS* **48** 1342-49.
[10] Ogiegło JM, Katelnikovas A, Zych A, Justel Th, Meijerink A and Ronda CR 2013 Luminescence and luminescence quenching in $Gd_3(Ga,Al)_5O_{12}$ scintillators doped with Ce^{3+} *J PHYS CHEM-US* **117** 2479-84.
[11] Vasil'ev DA, Spassky DA, Omelkov SI, Vasil'eva NV, Khakhalin AV and Plotnichenko VG 2017 Study of the optical absorption and photoluminescence in $(Pb,Gd)_3(Al,Ga)_5O_{12}$:Ce epitaxial films grown from Pb-containing melt solutions *QUANTUM ELECTRON+* **47** 922-26.
[12] Vasil'ev DA, Vereshchagin KA, Vereshchagin AK, Spassky DA, Sokolov VO, Khakhalin AV, Vasily'eva NV, Galstyan AM and Plotnichenko VG 2015 Influence of Al ions on optical and kinetic properties of $(Pb,Gd)_3(Al, Ga)_5O_{12}$:Ce epitaxial films *APPL PHYS +* **4** 5-10.

IOP Conf. Series: Materials Science and Engineering **848** (2020) 012096 doi:10.1088/1757-899X/848/1/012096

Magnetic properties of the Fe-24%Cr-15%Co-3%Mo-1.5%Ti hard magnetic alloy in anisotropic and isotropic states

T A Vompe, I M Milyaev

A.A. Baikov Institute of Metallurgy and Materials Science, Russian Academy of Sciences, Leninskii pr. 49, 119334, Moscow, Russia

E-mail: tvompe@imet.ac.ru

Abstract. The magnetic properties of the Fe-24%Cr-15%Co-3%Mo-1.5%Ti (in wt. %) hard magnetic alloy in anisotropic and isotropic states after various heat treatments were studied. The alloy was prepared by the methods of powder metallurgy. The relative density was 98 %. The coefficient of thermal expansion of the alloy during heating and cooling was $17{,}4 \cdot 10^{-6}$ K^{-1} and $19{,}0 \cdot 10^{-6}$ K^{-1} in the temperature range of 100 °C – 1300 °C. The kinetics of σ-phase formation caused by an isothermal annealing at T = 750 °C was investigated by means of the hardness measurements. The best magnetic properties of the alloy in anisotropic state were obtained after thermomagnetic treatment with cooling rate 200 °C/h as B_r = 1.22 T, H_c = 42.2 kAm^{-1} and $(BH)_{max}$ = 31.2 kJm^{-3} and on the alloy in isotropic state with cooling rate 40 °C/h.

1. Introduction

Fe-Cr-Co alloys are well-known for their magnetic and high mechanical properties, ductility and corrosion resistance [1-2].

The addition of Mo and Ti to the ternary Fe-Cr-Co alloys as α-forming elements improves the magnetic properties [3-6]. They extend the α-phase region and prevent the σ-phase formation in the alloy, because σ-phase presence deteriorates properties of the alloy. It was found that the maximum σ-phase formation rate in Fe-Cr-Co alloys was at the temperature of 750 °C [7-8].

The magnetic properties of Fe-Cr-Co alloys are formed by the spinodal decomposition of high temperature α-phase into iron-rich phase α_1 and chromium-rich phase α_2. The magnetic properties are very sensitive to the condition of decomposition within the miscibility gap. It is important to study the phase transformation temperatures during various heat treatments to find the optimum conditions.

The typical heat treatment of the Fe-Cr-Co alloy includes solution treatment from α-phase region, high-temperature aging during spinodal decomposition and low temperature aging after it.

Isotropic and anistropic magnets can be produced from Fe-Cr-Co alloys depending on heat treatment [9-14]. Thermomagnetic treatment (TMT) allows to receive optimal magnetic properties of Fe-Cr-Co alloy. An anisotropic microstructure containing elongated α_1-phase particles is attained by the application of the magnetic field during the annealing process [15-17]. These particles are parallel to the applied field. Due to this heat treatment the residual induction B_r and coercive force H_c of the alloy. TMT also reduces time of receiving of the needed magnetic properties.

In this paper the magnetic properties of the Fe-24%Cr-15%Co-3%Mo-1.5%Ti (in wt. %) hard magnetic alloy in anisotropic and isotropic states after various heat treatments have been studied.

2. Materials and methods

2.1. Preparation

The alloy Fe-24%Cr-15%Co-3%Mo-1.5%Ti (in wt. %) was prepared by the methods of powder metallurgy. Metallic powders of iron, chrome, cobalt, molybdenum and titan of 99.9 wt. % purity have been used.

Powders were mixed in C 2.0 "Turbula" mixer (Russia) for 320 min. Single-ended dry pressing at 600 MPa in a zinc stearate lubricated hardened steel pressing die with internal diameter of 13.6 mm on a manual press KNUTH-130042 (Germany) was used to compact the samples.

The density of the samples was measured by hydrostatic weighing. Then compacted samples were sintered at the temperature of 1350 °C during 2.5 hours in a shaft furnace SshV-1,25/24-I (Russia) in 10^{-2} Pa vacuum. The relative density of sintered samples was 98 %.

2.2. Chemical composition

The chemical composition of the Fe-24%Cr-15%Co-3%Mo-1.5%Ti alloy was estimated by two different methods. The mean values are shown in Table 1.

Table 1. Chemical composition of the samples from Fe-24%Cr-15%Co-3%Mo-1.5%Ti (wt. %) alloy.

	Fe	*Cr*	*Co*	*Mo*	*Ti*
Initial	Bal.	24.0	15.0	3.0	1.5
XRF	Bal.	24.8	14.2	3.2	1.6
EDS	Bal.	22.5	14.0	3.5	1.1

Rigaku Ultima IV diffractometer (Japan) was taken to obtain the X-ray fluorescence (XRF) patterns. The scanning electron microscopy (SEM) TESCAN VEGA II SBU (Czech Republic) equipped with the energy dispersive X-ray spectroscopy (EDS) on large areas was also used. The oxidation of the alloy was detected by EDS method.

The results are in a good agreement with initial alloy composition due to the fact, that powder metallurgy process was successfully performed.

2.3. Characterization

The heat treatment of the samples of Ø 12 × 20 mm in size was carried out in the standard laboratory furnace equipped with armored electromagnet and laboratory muffle furnace (Russia). The temperature was kept and monitored by proportional–integral–derivative controller.

Magnetic properties (residual induction B_r, coercive force H_c and maximum energy product $(BH)_{max}$) were measured by hysteresisgraph Permagraph L EP-3 (Germany) at room temperature. The measurement error for coercive force H_c and residual induction B_r was 3 %, and it was 6 % for maximum energy product $(BH)_{max}$. The surface of sample was cleaned from slag using abrasive disk before measurements of the magnetic properties.

In order to study phase transformation temperature, the dilatometry experiments on dilatometer NETZCH DIL 402 C72G (Germany) was carried out in high purity argon atmosphere with the rate of the gaz 75 ml/min to prevent oxidation. The samples were heated to the temperature 1300 °C and cooled to room temperature with the rate of 10 °C/sec. The coefficient of thermal expansion (CTE) of the alloy was determined.

The X-ray diffraction (XRD) analyses were carried out using diffractometer Rigaku Ultima IV (Japan) with an x-ray tube with Cu Kα radiation. The XRD patterns were recorded in the 2θ range of 30–100° (step size of 0.05° and the time per step of 1 s).

The microstructure of the alloy was studied by optical metallography methods.

3. Results and discussions

3.1. Phase transformations and microstructures

The dilatometry curves for Fe-24%Cr-15%Co-3%Mo-1.5%Ti alloy sample are shown at the Figure 1. The average CTE during heating and cooling of the alloy in the temperature range 100 °C – 1300 °C was between 17,4*10^{-6} K^{-1} and 19,0 *10^{-6} K^{-1}.

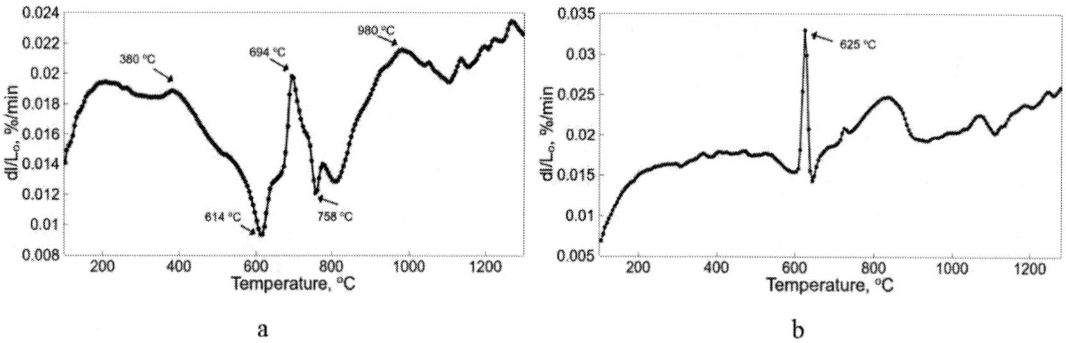

Figure 1. Dilatometry curves for the Fe-24%Cr-15%Co-3%Mo-1.5%Ti alloy during heating (a) and cooling (b).

According to the Fe-Cr and Fe-Co phase diagrams and obtained experimental data the phase transformation of $\alpha \rightarrow \alpha_1+\alpha_2$ would be expected at the temperature range 614 – 625 °C. There is the destabilization process of high-temperature α-phase because of the σ-phase presence before spinodal decomposition.

The microstructure of the solution treated sample for 15 min at 1300 °C followed by rapid quenching in water is shown at the Figure 2. The single α-phase can be observed.

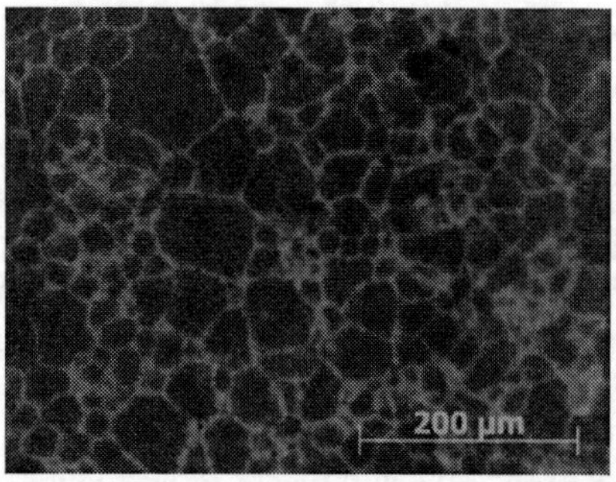

Figure 2. Microstructure of the Fe-24%Cr-15%Co-3%Mo-1.5%Ti alloy after quenching at 1300 °C during 15 minutes.

The kinetics of the σ-phase formation caused by an isothermal annealing at T = 750 °C was investigated by means of the hardness measurements. The hardness of the alloy increases from 32.3 to

49.3 HRC during 20 hours of annealing at 750°C (Figure 3). The microstructure of the Fe-24% Cr-15%Co-3%Mo-1.5%Ti alloy is shown at the Figure 4. The σ-phase occurs at grain boundaries.

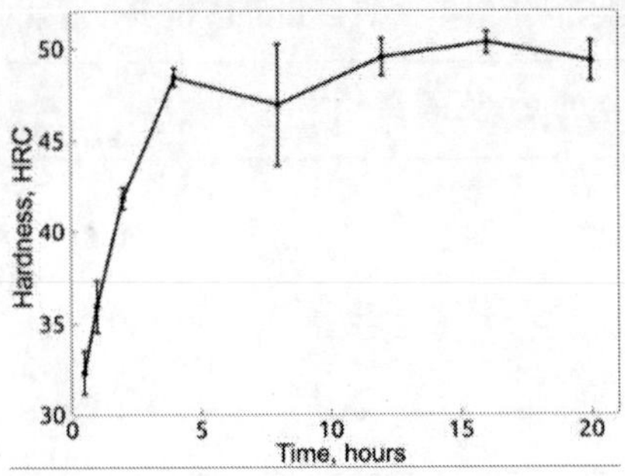

Figure 3. Hardness HRC of the Fe-24%Cr-15%Co-3%Mo-1.5%Ti alloy versus time of annealing at 750 °C.

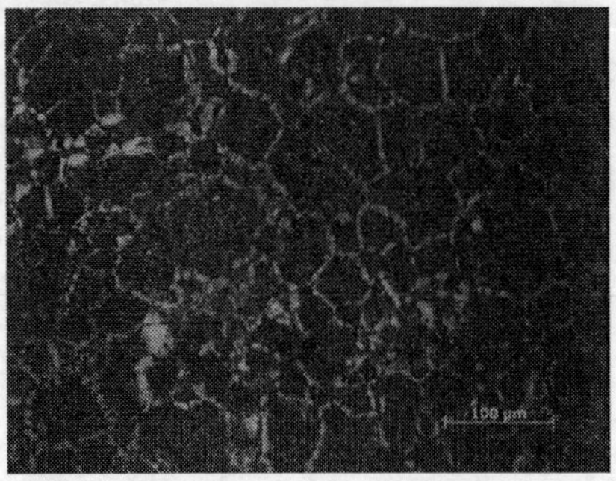

Figure 4. Microstructure of the Fe-24%Cr-15%Co-3%Mo-1.5%Ti alloy after annealing at 750 °C during 20 hours.

In accordance with XRD data the quantity of the σ-phase was about 5 vol. % after 20 hours of annealing (figure 6). The following lattice parameters have been calculated: α-phase – a = 0.2877 nm, σ-phase – a = 0.8835 nm, c = 0.4596 nm.

3.2. Magnetic properties

The heat treatment temperatures were chosen according to the phase transformation temperatures based on dilatometry curves and study of the σ-phase formation.

The samples were quenched in water from the temperature 1300 °C for 15 minutes, then heated to the temperature 710 °C which exceeded the temperature of spinodal decomposition of the α-solid solution, and cooled with a critical rate to the T = 680 °C. Then the samples were cooled with the cooling rate of v_1 from 40 to 250 °C/h to the temperature 600 °C under a magnetic field of 4 kOe or without it. From the temperature 600 °C to 500 °C the samples were cooled with a constant rate of 8 °C/h.

The coercive force of the Fe-24%Cr-15%Co-3%Mo-1.5%Ti alloy versus cooling rate v_1 during first step aging is shown at the Figure 5.

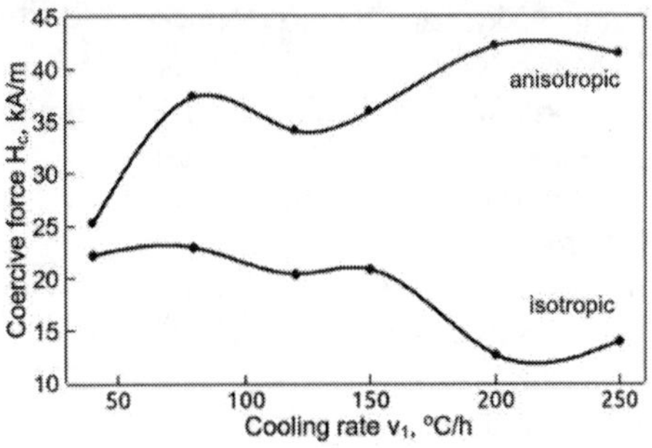

Figure 5. Variation of coercive force H_c of the Fe-24%Cr-15%Co-3%Mo-1.5%Ti alloy versus cooling rate v_1 during first step aging.

The coercive force increases after TMT (anisotropic state of the alloy) and decreases during heat treatment without magnetic field (isotropic state). The best magnetic properties were received after TMT with cooling rate v_1 = 200 °C/h.

Experimental data in table 2 show the best magnetic properties that were received on samples in different states.

Table 2. Magnetic properties of the Fe-24%Cr-15%Co-3%Mo-1.5%Ti alloy in different states.

State	*Heat treatment condition*		B_r, *T*	H_c, *kAm*$^{-1}$	$(BH)_{max}$, *kJm*$^{-3}$	$\eta = (BH)_{max}/(B_r*H_c)$
anisotropic	680 → 600 °C 200 °C/h with magnetic field	600 → 500 °C 8 °C/h no magnetic field	1.22	42.2	31.2	0.61
isotropic	680 → 600 °C 40 °C/h no magnetic field	600 → 500 °C 8 °C/h no magnetic field	0.56	22.2	3.5	0.28

The slower cooling rate, the better coercive force on the alloy in isotropic state is.

4. Conclusion

The hard magnetic alloy Fe-24%Cr-15%Co-3%Mo-1.5%Ti (wt. %) was prepared by the methods of powder metallurgy. The relative density was 98 %.

The coefficient of thermal expansion during heating and cooling of the alloy was 17,4*10^{-6} K^{-1} and 19,0 *10^{-6} K^{-1} in the temperature range of 100 °C – 1300 °C.

The kinetics of the σ-phase formation caused by an isothermal annealing at T = 750 °C was investigated in the alloy by means of the hardness measurements. The quantity of σ-phase in hard magnetic alloy Fe-24%Cr-15%Co-3%Mo-1.5%Ti (wt. %) was about 5 vol. %.

The best magnetic properties of the alloy in anisotropic state were obtained after thermomagnetic treatment with cooling rate 200 °C/h as B_r = 1.22 T, H_c = 42.2 kAm^{-1} and $(BH)_{max}$ = 31.2 kJm^{-3} and on the alloy in isotropic state with cooling rate 40 °C/h as B_r = 0.56 T, H_c = 22.2 kAm^{-1} and $(BH)_{max}$ = 3.5 kJm^{-3}.

Acknowledgments
The work was carried out according to the state assignment № 075-00947-20-00.

References

[1] Jin S and Chin G Y 1987 Fe-Cr-Co magnets (invited) *IEEE Trans. Magn.* **MAG-23** 3187–92
[2] Kaneko H, Homma M and Nakamura K 1971 New Ductile Permanent Magnet of Fe-Cr-Co System *AIP Conf. Proceedings. Magn. Magn. Mater.* 1088–92
[3] Chin T-S, Lee P, Chang C and Wu T 1984 Effect of alloying on magnetic properties of Fe-Cr-12 wt% Co permanent magnet alloys *J. Magn. Magn. Mater.* **42** 207–16
[4] Szymura S and Sojka L 1984 The effect of silicon and molybdenum on liquidus-solidus gaps in an Fe-Cr-Co alloy *J. Mater. Sci. Lett.* **3** 897–8
[5] Cherednichenko I V., Shubakov V S, Malinina R I, Perminov A S and Menushenkov V P 2010 Structure formation of the highly coercive state in Fe-Cr-Co-Mo alloys *Steel Transl.* **40** 93–7
[6] Ahmad Z, Ul Haq A, Husain S W, Ali A and Abbas T 2003 Influence of Ti content on magnetic and microstructural properties of Fe-28Cr-15Co-3.5Mo permanent magnets *J. Magn. Magn. Mater.* **257** 397–402
[7] Milyaev I M, Prutskov M E, Laisheva N V., Milyaev A I and Yusupov V S 2010 Kinetics of the σ-phase formation in Fe-Cr-Co hard magnetic alloys *Russ. Metall.* **2010** 1053–5
[8] Vompe T A, D'yakonova N P, Milyaev I M and Prutskov M E 2012 Kinetics of σ-phase formation in a strain aging hard magnetic Fe-33% Cr-12% Co-2% Cu alloy *Russ. Metall.* **2012**
[9] Sugimoto S, Honda J, Ohtani Y, Okada M and Homma M 1987 Improvements of the magnetic

properties of equiaxed Fe-Cr-Co-Mo hard magnets by two-step thermomagnetic treatment *IEEE Trans. Magn.* **MAG-23** 3193–5

[10] Chrost K and Klodas J 1989 The influence of heat treatment conditions in an external magnetic field on the structure and magnetic properties of the Fe-25Cr-12Co alloy *J. Magn. Magn. Mater.* **80** 359–66

[11] Kubota T, Member A, Wakui G and Itagaki M 1998 Hysteresis motor using magnetically anisotropic Fe-Cr-Co magnet *IEEE Trans. Magn.* **34** 3888–96

[12] Akbar S, Ahmad Z, Awan M S, Sarwar M N and Farooque M 2012 Single Step Heat Treatment Cycle for Development of Isotropic Fe-Cr-Co Magnets *Key Eng. Mater.* **510–511** 315–20

[13] Akbar S, Ahmad Z, Awan M S, Farooque M and Ali A 2012 Development of Fe-Cr-Co Permanent Magnets by Single Step Thermo-Magnetic Treatment *Key Eng. Mater.* **510–511** 507–12

[14] Milyaev I M, Yusupov V S, Ostanin S Y, Stelmashok S I, Milyaev A I and Laysheva N V. 2018 Magnetic Hysteresis and Mechanical Properties of Hard Magnetic Fe–27Cr–15Co–2Mo–Si–Ti–V Alloy *Inorg. Mater. Appl. Res.* **9** 523–9

[15] Livingston J D 1981 A review of coercivity mechanisms (invited) *J. Appl. Phys.* **52** 2544–8

[16] Jin S, Brasen D and Mahajan S 1982 Coercivity mechanisms in Fe-Cr-Co magnet alloys *J. Appl. Phys.* **53** 4300–3

[17] Milyaev I M, Milyaev A I and Yusupov V S 2009 Mechanism of formation of the high-coercivity state in nanostructured hard magnetic Fe-Cr-Co and Fe-Ni-Al-Co-Cu alloys *Russ. Metall.* **2009** 250–2

IOP Conf. Series: Materials Science and Engineering **848** (2020) 012097 doi:10.1088/1757-899X/848/1/012097

Preparation and researching the properties of organic aerogels based on epoxy resins

A E Voytik [1], G V Malkov [2], A T Kapasharov [2], E I Knerelman [2] and A Yu Kostin[2]

[1] Russia, Federal State Budgetary Educational Institution of Higher Education "Lomonosov Moscow State University. Lomonosov Moscow State University".
[2] Russia, Federal State Budgetary Institution of Science "Institute of Chemical Physics Problems of the Russian Academy of Sciences".
[1]Corresponding author E-mail: alexvoytik@gmail.com

Abstract. The aim of this work was to develop methods for the synthesis of organic gels based on epoxy resins and the production of their aerogels. In general, aerogels are obtained from gel samples that consist of a high-molecular spatial network and a low-molecular dispersion medium - solvent. Various drying methods are used to remove liquid from gel pores, the most reliable of which, allowing to obtain samples of aerogels with undamaged spatial structure, is drying in supercritical fluid medium. As a result of the work the aerogels of different density were obtained, the dependence of density on the solvent content in the samples was revealed. The obtained aerogels were characterized by density, studied by infrared spectroscopy to assess the degree of polymerization of the resin, as well as using the BET method to determine the specific surface area of the samples and obtain its dependence on the catalyst content and curing conditions.

1. Introduction

To date, much attention has been paid to obtaining, studying the properties and introducing into production aerogels - highly porous solid gels in which the liquid phase has been completely replaced by the gas phase. Due to the fact that in the pores of the aerogel is a gas rather than a liquid, these substances have an extremely low volume density (the lightest have a density of about 0,001 g/cm^3 [1]). They also demonstrate a number of properties unique to these materials, such as high strength, sorption capacity [2], low thermal and acoustic conductivity [3] etc.

The method of aerogel production consists of two stages: preparation of gels, whose pores are filled with solvent, and removal of liquid solvent with its replacement by gas. There are three main methods for drying gels in total: heating to boiling point of the solvent, freeze drying and supercritical drying. The first method is not suitable for the production of aerogels, as the transition of the solvent from the liquid phase to the gas phase due to the capillary effect and the huge surface tension (inversely proportional to the pore radius) breaks the gel structure.

The method of freeze drying (lyophilic drying) is based on sublimation of the solvent in cryogenic conditions in vacuum bypassing the liquid phase and, accordingly, the negative effect of the capillary effect. The result is a gel in whose pores the liquid was replaced by gas. However, this method is not suitable for drying, for example, hydrogels, where water acts as a solvent, because it expands when frozen, so that the structure of the gel can be destroyed. This is why the gels thus dried are released under the name of cryogels [1].

IOP Conf. Series: Materials Science and Engineering **848** (2020) 012097 doi:10.1088/1757-899X/848/1/012097

The method of supercritical drying, similar to the method of freeze drying, is based on the transfer of liquid to the gas without the formation of a mobile interface and, accordingly, capillary effects. The sample is placed in a reaction vessel and the pressure and temperature in it is raised to values above the critical for the solvent. The solvent passes from the liquid state to the supercritical state (supercritical fluid), in which the surface tension forces are no longer present - as a consequence, there is no capillary effect either. The pressure in the vessel is then changed so that the solvent is transferred from the supercritical state to the gaseous state and is removed from the vessel. This process results in an aerogel with an undamaged spatial structure [1]. Due to the fact that for organic solvents the critical values of pressure and temperature are rather high (for example, 4.7 MPa and 235.5 °C for acetone, 21.8 MPa and 374 °C for water), their achievement is very difficult, while for carbon dioxide the critical values are only 7.3 MPa and 35 °C. Therefore, supercritical drying is divided into "cold" using the supercritical fluid of carbon dioxide and "hot", where the critical pressure and temperature are reached for the solvent used [3].

The most common are aerogels obtained from various inorganic (SiO2, TiO2, Fe2O3, etc.) and carbon materials. Such aerogels are used as protective coatings for walls, ceilings and parts of equipment exposed to heavy wear and tear; as liquid and gas filters, and as space dust traps on the Stardast. In addition, aerogels can be obtained from organic materials (polyvinyl alcohol, cellulose, xanthan, polyisocyanates, etc.) [1, 4]. Some of these aerogels can be used in various biocompatible systems Others are used as additives in combined organic-inorganic aerogels to increase strength and plasticity [1]. The structure and properties of aerogels based on resorcinol-formaldehyde resins have been most well researched [5], which, for example, showed lower thermal conductivity than silica aerogels (0.012 W/m*K versus 0.016 W/m*K) [6], they are also promising for use as separators of gases and liquids, sorbents for contaminants, precursors for carbon nanoparticles and aerogels, which are considered as components of promising energy sources. [1]. At the same time, there is no data in the literature on the use of epoxy resins and binders based on them to produce organic aerogels of no less well-known and very widely used polymers forming a strong three-dimensional network [7, 8]. Epoxy resins are a large class of compounds containing oxirane cycles, capable of forming various 3D cross-linked structures under the action of various hardeners (polyamines, carboxylic acid anhydrides, etc.).

Therefore, the purpose of this work was to develop a methodology for obtaining gels and aerogels based on epoxy resins, to study their thermophysical properties, as well as to evaluate the specific surface area and determine their possible applications

2. Experimental

For the production of gels based on epoxy resin as reagents for the initial mixture took: epoxy diane resin ED-20 (based on DGEBA, the manufacturer of FKP "Zavod im. Sverdlov Plant", 22,8-23 wt.% of epoxy groups, Figure 1 (a)), curing catalyst 2,4,6-bis(dimethylamine methyl)phenol (alkophen, DMP-30, Figure 1 (b)), acetone (ch.p.), CO2 (ch.p.).

(a)

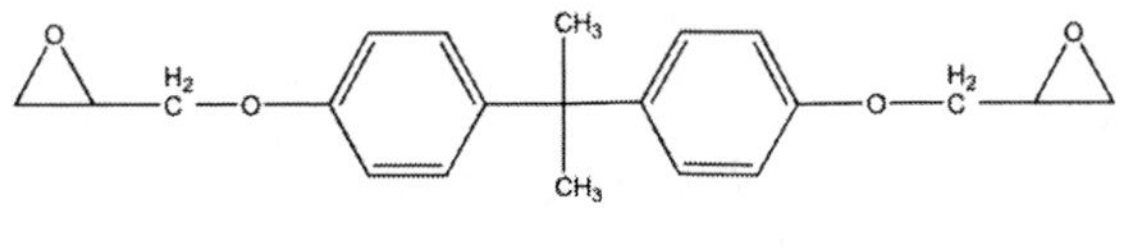

(b)

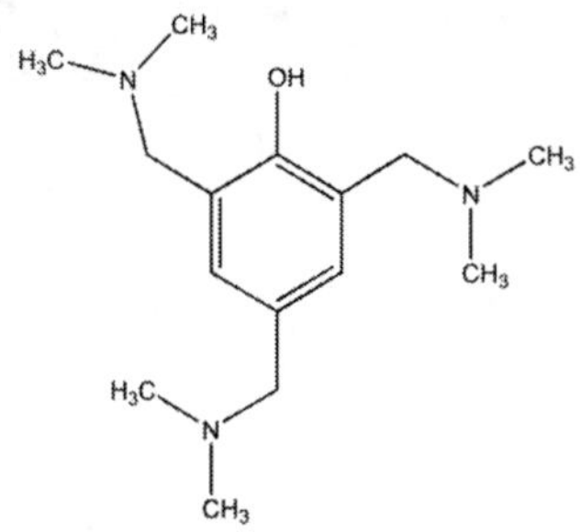

Figure 1. Structural formulas of initial compounds - ED-20 epoxy resin (a) and 2,4,6 bis(dimethylamine methyl)phenol (b).

2.1. Gel synthesis

IOP Conf. Series: Materials Science and Engineering **848** (2020) 012097 doi:10.1088/1757-899X/848/1/012097

In the process of preparation of the mixture to the resin attachment the calculated amount of acetone was added so that the total acetone content in the samples changed from 20 to 80 wt% (A-20, A-40, A-60 and A-80, respectively), and then a catalyst curing 1 wt%. After that the obtained mixtures were thermostatically controlled at 50°C for 5 in hermetically sealed polypropylene ampoules.

In the second series in ED-20 solutions with acetone concentration of 60 and 80 wt% varied catalyst concentration from 1 to 10 wt%: A-60 series (x = 1, 2, 3, 4, 5, 10 wt %) and A-80 series (x = 3, 5, 10 wt %). After that the received mixtures were thermostatically staticked at temperature 50°C during 15 days in hermetically sealed polypropylene ampoules.

Density of initial solutions at 23°C was determined by pycnometric method.

At the end of the resin curing process, the samples were washed in acetone for 1 week (acetone was changed daily). The samples were then dried in a CO_2 supercritical fluid atmosphere.

2.2. Supercritical drying

Drying was carried out on R250Sys Thar Instruments. The sample was placed in the reactor, which was then filled with liquid CO2 and heated to 35°C. The pressure in the vessel was gradually increased to 250 bar and it was kept in the vessel for 5 minutes, then reduced to 80 bar and then raised again to 250 bar. The cycle was repeated 10 times, after which the samples were taken out of the reaction vessel.

2.3. Characterization of aerogels

The bulk densities of the samples were calculated by their mass to volume ratio. Aerogels' shrinkage was estimated by the change of their geometric volume compared to the geometric volume of wet gels.

IR spectra of samples were recorded using the Alpha Bruker FTIR spectrometer (KBr tablets).

The specific surface area values were determined using a 5-point Brunauer, Emmett and Teller (BET) method at the relative pressure range of P/P0 = 0.05–0.25; the correlation coefficients of the corresponding linear regressions in the coordinates of the BET equation were not less than 0.9975. Pore volume distributions were determined according to the Barrett, Joyner and Halenda (BJH) method, by the analysis of a desorption branch of an adsorption-desorption isotherm (28 points). Prior to measurements, samples were flushed with a dry helium flow for 16 hours at 45 °C (Quadrasorb SI) or vacuum dried at room temperature (Autosorb-1 Quantachrome).

3. Results and discussing

The polymerization reaction in the mixture ED-20 - DMP-30 - acetone took place according to the scheme shown in Figure 2.

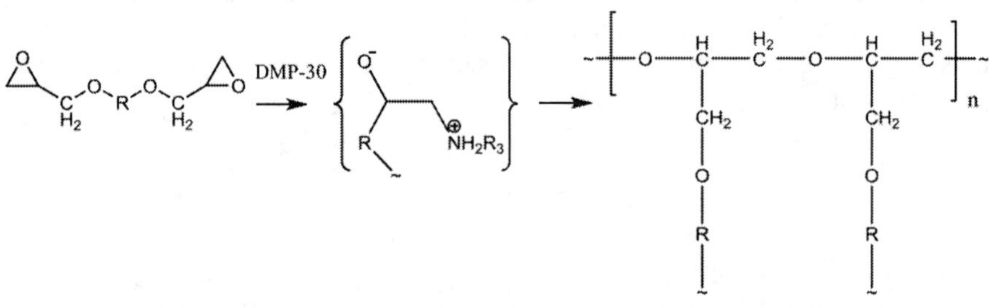

Figure 2. Schematic representation of epoxy resin polymerization process.

As a result of curing of ED-20 epoxy resin it was possible to obtain samples of cross-linked polymers in the form of gels at the concentrations of solvent 20, 40, 60 and 80 wt %, the photos of which before and after SC-drying are shown in Figure 3. Samples of A-20 and A-40, transparent solid and heavy cylinders, are gradually covered with cracks at long exposure times both without and after SC-drying (Figure 3 (a - c)). Samples of A-60 and A-80 are opaque, after SC drying they become noticeably lighter than the hardened resin and samples of A-20 and A-40. But A-60 occupies the whole volume corresponding to the initial solution, and A-80 occupies from 20 to 50 % of the initial volume.

It can be noted that the color of dried samples correlates with the concentration of the hardener - the more it is, the more saturated the color (Figure 3 (d - g, i - k)). It should also be noted that samples for a long time even after SC-drying, vacuuming and / or heating to 100°C retain a weak aroma of acetone, which disappears in the samples of series A-60 and A-80 only after prolonged (more than 100 hours) vacuuming at room temperature or heating above 130°C. As a result of the latter, samples of all series are significantly reduced in size and their density is close to that of the cured resin.

Figure 4 shows the type of dependencies of density of initial solutions and samples of obtained aerogels on acetone content in them. As expected, the density of the initial solutions changes smoothly. The density of samples with acetone content of 20 and 40 wt.% after SC drying is close to the density of cured resin, and with increasing acetone content it decreases sharply. At change of quantity of the curing catalyst in samples of A-60 and A-80 series the density practically does not change.

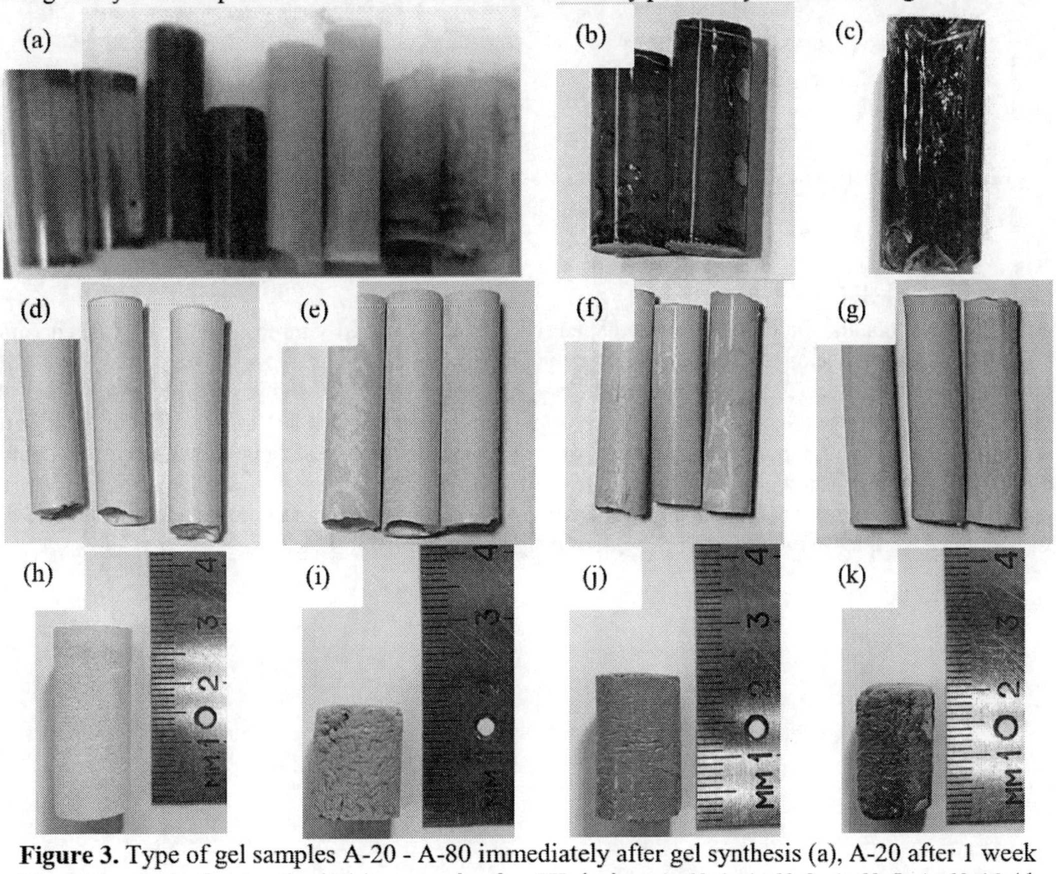

Figure 3. Type of gel samples A-20 - A-80 immediately after gel synthesis (a), A-20 after 1 week (b) and 1 month after synthesis (c), aerogels after SK drying: A-60-1, A-60-3, A-60-5, A-60-10 (d-g), and after trimming to measure the bulk density of A-60-1 (h), A-80-3, A-80-5, A-80-10 (i-k).

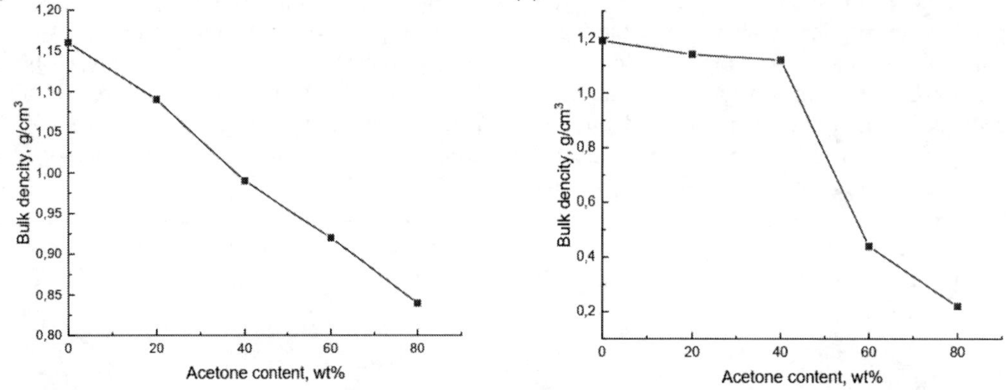

Figure 4. Dependence of the density of the stock solutions in acetone (a) and aerogels after SC drying (b) of the acetone content wt.%.

3.1. ИК-спектроскопический анализ

Analyzing the results obtained by infrared spectroscopy of SC-dried samples (Figure 6), one can draw conclusions about the close to quantitative course of the curing process of resin polymerization accompanied by the opening of epoxy groups. The infrared spectra shown in Figure 6(b) lack the characteristic absorption bands of the epoxy group in the region 971 and 915 cm-1, while in the infrared spectra of the ED-20 resin these bands are pronounced. In the spectra of samples with acetone content of 20 and 40 wt.%, shown in Figure 6 s, in the region 1710-1712 cm-1 there are absorption peaks corresponding to the fluctuations of the keto group C=0 acetone [9], from which we can conclude that the obtained samples even after drying contain a significant amount of acetone.

(a)

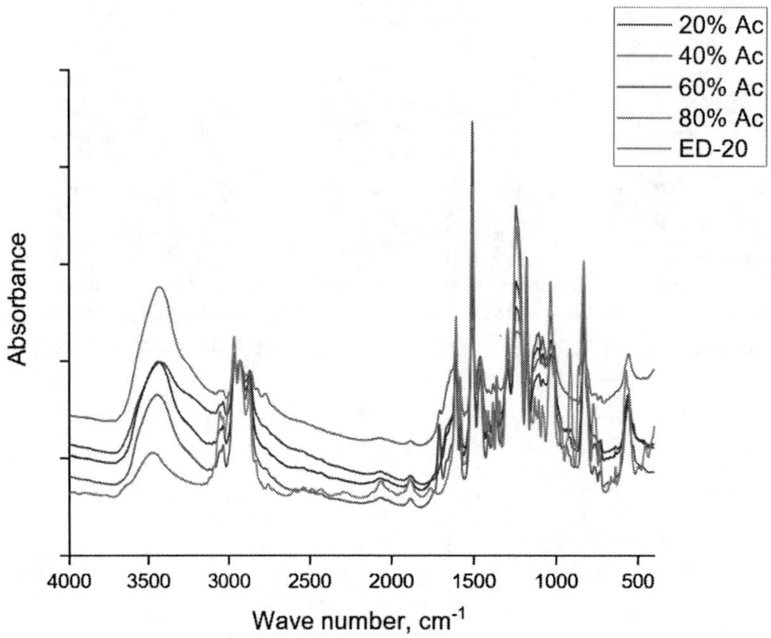

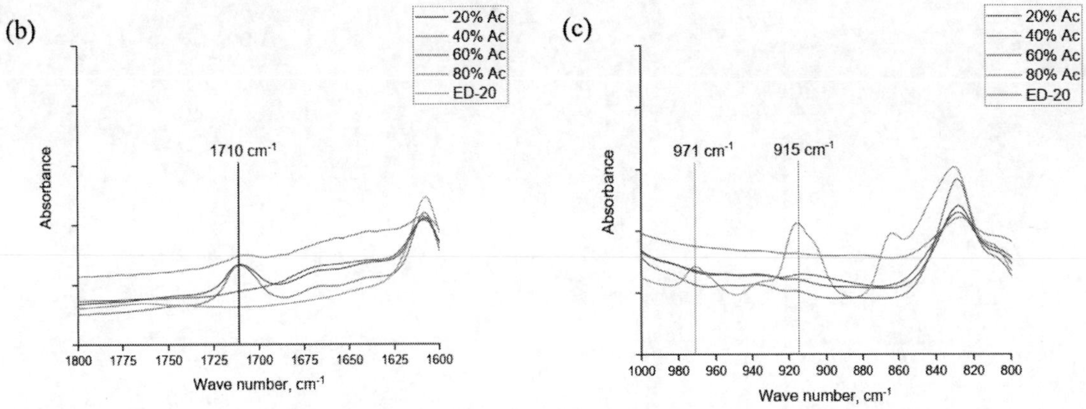

Figure 5. IR-spectra of the resin ED-20 and samples A-20 - A-80 after SC drying: general view (a), in the region of oscillations of groups C=O (b) and oxirane cycle (c).
The spectra are normalized by the characteristic band of aliphatic CH2 group oscillations.

3.2. Specific surface area estimation by Brunauer-Emmet-Teller (BET) method

The specific surface area was also determined for the samples by the BET method [9, 10, 11]. The research was carried out on two sorbetometers in order to evaluate the consistency of the results obtained with different methods of sample pre-treatment (vacuum drying and dry helium blowing). The results are presented in Figures 7 and Table 1.

Apparently, it is because of the fact that the number of catalysts in the samples of A-60-series differed, the values of specific surface area in them are reduced from 146 to ~7 m^2/g in proportion to the increase in DMP-30 concentration from 1 to 4 wt.%, and then virtually unchanged and remains equal to 6,4-7 m^2/g (Figure 7).

We believe that the reason for the reduction of the specific surface area of S may be a significant acceleration of resin polymerization with an increase in the catalyst content, as well as insufficient solubility of catalysis products in acetone, which is a complex with charge transfer, see Figure 1. The latter hypothesis is confirmed by the fact that when the order of application of the catalyst - addition to the resin and subsequent dissolution of the mixture in acetone, in the resulting mixture there is a flaky residue, which is not observed when the catalyst is added to drugs. It is also possible that gels with closed porosity are formed in samples with a large amount of catalyst, the area of which cannot be determined by the BET method.

Another factor affecting the size of the surface area is the duration of the curing process (Figure 7). Apparently, at a prolonged exposure at 50°C the polymer relaxing tightens pores or their estuaries as a result of nitrogen cannot penetrate into pores, adsorption can occur and the experimental value of the specific surface area is much lower than the real value.

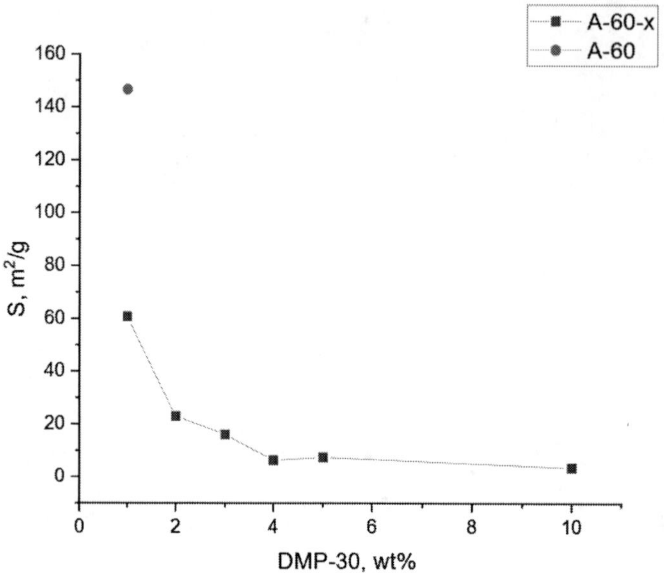

Figure 6. Dependence of specific surface area of samples on catalyst DMP-30 content.

Table 1. Dependence of density and specific surface of aerogel samples on the share of acetone in the initial mixture.

Sample	Share of acetone	Bulk density of acetone solution, g/cm³	Bulk density after SC-drying, g/cm³	Specific surface, m²/g	
				Autosorb-1 Quantachrome	Quadrasorb SI
A-60	0.603	0.920	0,481	146	-
A-60-1	0,605		0,512	75,1-65,9	57,3-60,8
A-60-2	0,608		0,478	-	22,8-23
A-60-3	0,562	0.919	0,565	-	16,1
A-60-4	0,616		0,487	-	6,4
A-60-5	0,621		0,491	-	7,4
A-60-10	0,641		0,598	-	3,6-6,7
A-80-3	0,806		0,256	2,64	<3
A-80-5	0,810	0.842	0,220	-	<3
A-80-10	0,820		0,212	3,8	<3

For the A-80-x series, it has been noticed that the introduction of a large number of catalysts makes it possible to obtain gels of a larger volume (as the number of catalysts increases, the height of the resulting array increases and the bulk density decreases slightly). This can be explained by the fact that more embryos are formed simultaneously in a unit of volume and the resulting gel particles do not have time to precipitate, stick together and fill a larger volume of the solution.

With the Barrett, Joyner and Halenda (BJH) method determined pore volume distributions for sample A-60-1 (Figure 7(b)). The sample contains only mesopores and macropores.

(b)

IOP Conf. Series: Materials Science and Engineering **848** (2020) 012097 doi:10.1088/1757-899X/848/1/012097

(a)

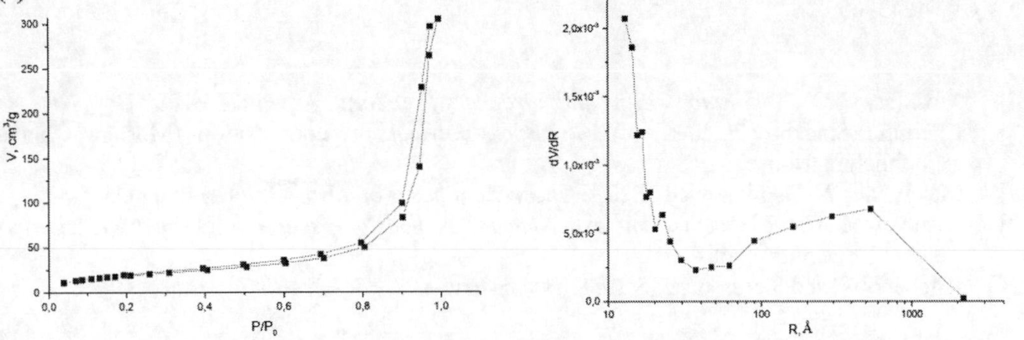

Figure 7. Isotherm of sorption-dessorption (a) and pore size distribution (b) for sample A-60-1.

The reason for the reduction of the specific surface area can be a significant acceleration of resin polymerization with increasing catalyst content. In this case, more resin comes into reaction and macromolecules are formed which do not have time to form a structure with a large number of pores. Another factor affecting the size of the surface area may be the duration of the curing process and the closing of pores by oligomeric molecules. As a result, nitrogen cannot be adsorbed in closed pores, and because of this, the experimental value of the specific surface area is much smaller than the real value.

A-60-1: total pore volume 0.47 cm^3/g, average pore radius 132 A. The pores are slotted, the distribution shows the size in the mesopores area.

A-80-3: specific surface S_{sp} = 2.64 m^2/g, total pore volume V_p = 0.026 cm^3/g, average pore radius r = 198 A.

A-80-10: specific surface S_{sp} = 3,8 m^2/g, total pore volume V_p = 0,018 cm^3/g, average pore radius r = 93 A.

4. Conclusion

In the course of this work it was possible to develop an experimental methodology for the synthesis of gels based on epoxy resins; the possibility of obtaining aerogel samples from them was shown. IR-spectroscopic analysis showed almost quantitative process of epoxy resin polymerization and presence of solvent in some samples after drying in the atmosphere of supercritical fluid. Primary data on the properties of epoxy resin-based aerogels were obtained. The BET method for gel and aerogel samples evaluated the specific surface area and specific compressive strength characteristics. It was shown that aerogel samples have rather low density (~0.5 g/cm3 - 0.2 g/cm3). The dependences of density and specific surface area values on the curing catalyst content and on the acetone content have been obtained.

Aknowledgements.
The authors thank Dr. S.A. Lermontov, Dr. E.R. Badamshina for their help and discussion of the results.

References

[1] Aegerter M, Leventis N and Koebel M 2011 *Aerogels Handbook* ed M A Aegerter and M Prassas
[2] Rao A, Hegde N and Hirashima H 2007 Absorption and desorption of organic liquids in elastic superhydrophobic silica aerogels *J. Colloid Interface Sci.* **305** pp 124–132.
[3] Wörmeyer K , Alnaief M and Smirnova I 2012 Amino functionalised silica-aerogels for CO_2-adsorption at low partial pressure *Adsorption* **18** pp 163–171.
[4] Hüsing N and Schubert U 1998 Aerogels - airy materials: chemistry, structure, and properties, *Angew. Chem. Int. Ed.* **37** pp 22–45.
[5] Pierre A and Pajonk G 2002 Chemistry of aerogels and their applications, *Chem. Rev.* **102** pp

IOP Conf. Series: Materials Science and Engineering **848** (2020) 012097 doi:10.1088/1757-899X/848/1/012097

4243–65.

[6] Moshinsky L Y 1995 *Epoxy resins and hardeners* (Tel-Aviv: Arcadia Press LTD)
[7] Chernin I , Smekhov F , Zherdev Y 1982 *Epoxy polymers and* compositions (Moscow: Chemustry Publiching House)
[8] Makarevich N, Bogdanovich N 2015 *Theoretical bases of adsorption* (Arkhangelsk)
[9] Ananyev M 2015 *Theoretical and experimental methods of research in chemistry* (Ekaterinburg: Ural Univercity Publisher)
[10] Plachenov T and Kolosentsev S 1988 *Porometry*

IOP Conf. Series: Materials Science and Engineering **848** (2020) 012098 doi:10.1088/1757-899X/848/1/012098

Ceria-based solid solutions for environmental application

I V Zagaynov[1*], I V Shelepin[1,2], S V Fedorov[1], A A Klimashin[1] and I V Kulbakin[1]

[1] Baikov Institute of Metallurgy and Materials Science
[2] Mendeleev University of Chemical Technology of Russia

*igorscience@gmail.com

Abstract. The work is focused on the investigation of the influence of metal oxide promoted ceria based materials with various dopants on the physicochemical, electrical and catalytic properties for CO oxidation in model reaction. The formation of stable solid solutions by doping gadolinium, bismuth, and third metal enhances the redox properties, oxygen vacancies, stability, and, of course, catalytic activity and electroconductivity.

1. Introduction
In the context of significant interest in energy efficient technologies and environmental protection, materials based on nanostructured ceria are promising for solving existing problems. The advantage of such materials is due to that CeO_2 has a large oxygen storage capacity (OSC) and high oxygen mobility, which can ensure its high catalytic activity and electrical conductivity Solid solutions based on ceria with the dopant of Bi have recently attracted attention to the production of both electrolytes and catalysts due to the special electron orbital structure to improve conversion efficiency [1-3]. The development of more stable systems based on GDC and bismuth oxide, using as a sintering additive without preservation of bismuth oxide or a second dopant with the preservation of the initial composition of the solid solution based on ceria, is proposed. Moreover, triple-doping or multi-doping in ceria has been studied and succeeded to improve the properties of ceria [4-6]. However, further investigations are still required on tri- or multi-doped ceria in order to develop new materials for future catalytic and SOFC applications. Thus, it is necessary to investigate systems, based on ceria solid solution with three dopants $Gd_{0.05}Bi_{0.15}M_{0.05}Ce_{0.75}O_2$ (M = Cu, Zr, Ni, Co, Mn) as a promising.

2. Materials and Method
$Ce(NO_3)_3{\cdot}6H_2O$, $Bi(NO_3)_3{\cdot}5H_2O$, $Gd(NO_3)_3{\cdot}6H_2O$ and $Cu(NO_3)_2{\cdot}3H_2O$ or $ZrO(NO_3)_2{\cdot}7H_2O$ or $Ni(NO_3)_2{\cdot}6H_2O$ or $Co(NO_3)_2{\cdot}6H_2O$ or $Mn(NO_3)_2{\cdot}4H_2O$ (Acros Organics) were used as initial salts. Appropriate amounts of salts were dissolved in concentrated nitric acid (68%) with the concentration of salts of 0.667 M. After the dissolution of salts, this mixture was added to distilled water, giving the concentration of 0.1 M and after stirring acetylacetone (Hacac) was added (Hacac/Σ(Me)=1). Then, the co-precipitation was carried out by the addition of 2.5 M KOH solution up to pH 11. Ultrasonic processing (35 kHz, 150 W) was used during all process at 30°C under stirring. The resulting precipitates were filtered, washed with distilled water-ethanol solution (H_2O/C_2H_5OH=9 vol.), dried at 150°C for 12 h, and calcined in static air by heating at a rate of 4°C/min from room temperature to 500°C and kept at 500°C for 1 h in the muffle furnace.

The as-obtained powders were pressed into pellets (with 5 wt.% binder made of 5% aqueous solution of polyvinyl alcohol) with 10 mm in diameter and 1 mm in thickness at 180 MPa. Then they were sintered at 750°C (T_{sint}) for 4 h in air with heating of 4°C/min. To fabricate symmetric cells for the impedance studies (Elins Z-350M impedance meter, the frequency range from 0.1 Hz to 1 MHz at the amplitude of AC signal of 30 mV), platinum paste was brushed onto both sides of the electrolyte pellets, and were dried at 150°C for 1 h and annealed at 750°C for 4 h in air. A platinum wire was used as current conductors.

The catalytic activity of the synthesized samples in the oxidation of CO was determined by the flow method at atmospheric pressure. The process was conducted in a U-shaped quartz reactor at the total flow rate of 60 ml/min within a temperature range of 20-500°C. The temperature was measured with a thermocouple placed in the center of the catalytic bed. The catalyst sample mass was 0.3 g. The model gas mixture has the following composition, vol.%: CO – 1; O_2 – 2; N_2 – balance. The concentrations of gases were determined by a gas chromatograph (Varian 450GC).

All powders and ceramics were characterized by XRD (Rigaku MiniFlex 600, Cu*Kα* radiation), SEM (TESCAN VEGA II SBU with INCA Energy 300 energy dispersive spectrometers), TG-DSC-MS (Netzsch STA449F3, the samples were heated to 1350°C at the rate of 10°C/min in air), TEM (Omega Leo-912AB), nitrogen adsorption-desorption method (TriStar 3000 Micromeritics).

3. Results and discussion

The stability and availability of solid solutions $Gd_{0.05}Bi_{0.15}M_{0.05}Ce_{0.75}O_2$ (M = Cu, Zr, Ni, Co, Mn) were evaluated. The endothermic peaks at 750-800°C can be attributed to the formation of other phases and at >1000°C with the evaporating process of Bi compounds.

Table 1. Some characteristics of initial powders and ceramics.

No	Sample	d_{XRD} (nm)	S_{BET} (m^2g^{-1})	T(100% conversion of CO, °C)	$\sigma_{600°C}$ ($*10^{-3}$, S^1cm^{-1})
0	$Gd_{0.05}Bi_{0.15}Ce_{0.8}O_2$	8	23	404	5.0
1	$Gd_{0.05}Bi_{0.15}Cu_{0.05}Ce_{0.75}O_2$	11	60	424	1.0
2	$Gd_{0.05}Bi_{0.15}Zr_{0.05}Ce_{0.75}O_2$	11	69	485	0.1
3	$Gd_{0.05}Bi_{0.15}Ni_{0.05}Ce_{0.75}O_2$	10	72	464	0.9
4	$Gd_{0.05}Bi_{0.15}Co_{0.05}Ce_{0.75}O_2$	12	62	441	2.4
5	$Gd_{0.05}Bi_{0.15}Mn_{0.05}Ce_{0.75}O_2$	8	95	395	8.3

According to XRD (Figure 1a), samples have single-phase with fluorite structure (Fm3m, JCPDS-34-0394). The XRD pattern of calcined powders at 500 and 1000°C (Fig. 1a) shows that the only single phase of the solid solution was formed, and no peaks of other phases were detected. When the spectrum is completely approximated, it is determined that it is also possible to form an amorphous copper oxide and cobalt oxide in the corresponding samples (rather in the form of very small particles). XRD pattern of calcined powders at 1000°C (Fig. 1a) shows that both phase of ceria solid solution and other phases were formed. In all samples, there is a phase of Bi_2O_3 (γ-Bi_2O_3, sillenite, cI66), which was not found in the system without a third dopant [7].

The shape of the adsorption-desorption curves for all samples (Figure 1b) corresponds to the IV type, which is typical for the formation of the mesoporous structure. For all samples, the shape of the hysteresis loop corresponds to the H2 type: apparently, individual or connected cylindrical and bottle-shaped pores may be present, which is also confirmed by the pore size distribution (Figure 1c).

The average crystallite size calculated using Scherrer formula was about 5-10 nm for all samples, which corresponds to the individual particles according to TEM (Figure 1d).

It is known that pure ceria is a poor oxide ion conductor ($\sigma_{600°C}$~10^{-5} S/cm), and the ionic conductivity can be significantly enhanced by increasing the oxygen vacancies, which can be created by the doping of ceria with about 20% concentration of dopant, because such systems achieve the highest conductivity, but a further increase in the dopant concentration results in decreasing one. The highest conductivity for intermediate temperature application was found in $Gd_{0.05}Bi_{0.15}Mn_{0.05}Ce_{0.75}O_2$ (Table 1). In addition, it was recently shown that the nature of manganese species in ceria may vary with the method of synthesis [8]. Moreover, after sintering Mn ions may be in solid solution with ceria, or as a segregated phase along the grains boundaries, or even on the surface of grains, therefore, the sintering mechanism of this system is a very complex phenomenon and requires further experiments.

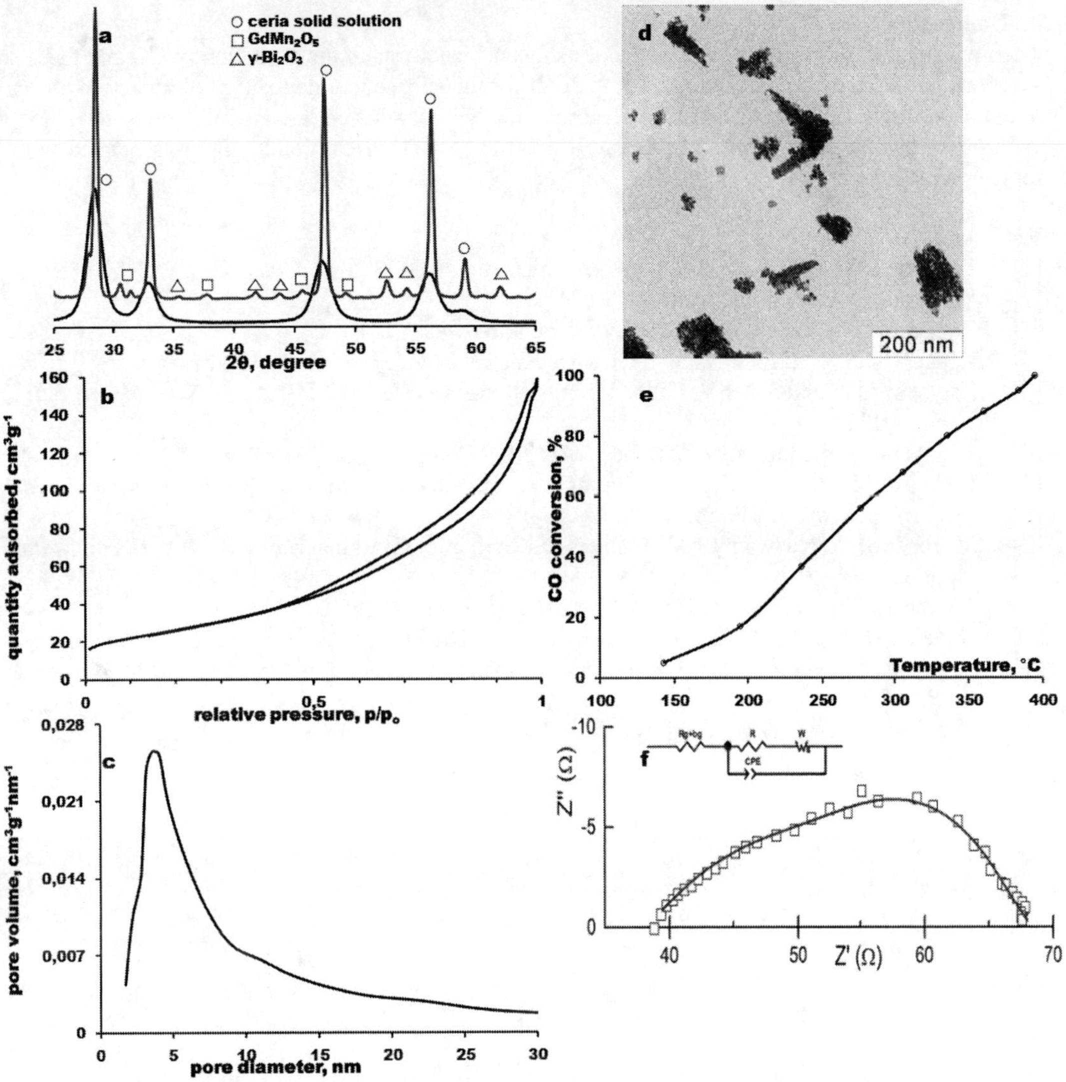

Figure 1. XRD patterns (a), adsorption-desorption isotherms (b), pore size distributions (c), TEM microphotos (d), catalytic activity in CO oxidation (e), and impedance spectrum at 600°C in air of of sample 5 ($Gd_{0.05}Bi_{0.15}Mn_{0.05}Ce_{0.75}O_2$).

The catalytic activity of systems was determined in the model oxidation reaction of CO. The conversion of carbon monoxide increases with the increase in temperature, and the curves have S-shaped type (Fig. 4). The 100% oxidation temperature of CO was following, °C: 404 (sample 0), 424 (sample 1), 485 (sample 2), 464 (sample 3), 441 (sample 4), 395 (sample 5). The full conversion temperature is higher than the original reference sample. However, the parameter $T_{50\%}$ is lower than that of the sample 0. As expected the modification by Cu or Mn [4] has a good effect on the surface, increasing the catalytic activity. The sample 5 ($Gd_{0.05}Bi_{0.15}Mn_{0.05}Ce_{0.75}O_2$) has the lowest oxidation temperature. The systems with the addition of zirconium had a conversion temperature higher by 100°C. When comparing the best catalyst with other systems based on ceria ($Zr_{0.12}Y_{0.08}Ce_{0.8}O_2$, $La_{0.12}Y_{0.08}Ce_{0.8}O_2$, etc. [9, 10]) under close test conditions, the catalytic activity of the developed system was higher (the temperature of complete oxidation of CO is lower by 50-100°C).

4. Conclusions

The investigation showed the expediency of using such catalysts in the oxidation of CO and these solid solutions could be used as the catalyst or support for the subsequent deposition of an active phase. As noted above, the catalytic activity and electrical conductivity for ceria based materials correlate with each other: sample $Gd_{0.05}Bi_{0.15}Mn_{0.05}Ce_{0.75}O_2$ had the best electrical conductivity associated with the ionic conductivity of oxygen [11, 12].

References

[1] Zagaynov IV, Fedorov SV, Konovalov AA, Antonova OS 2017 *Mater. Lett.* **203** 9
[2] Zagaynov IV 2017 *Appl. Nanosci.* **7** 871
[3] Zagaynov IV, Fedorov SV 2019 *Letters on Materials* **9** 424
[4] Zagaynov IV, Naumkin AV and Grigoriev YuV 2018 *Appl. Cat. B* **236** 171
[5] Zagaynov IV, Fedorov SV, Kulbakin IV, Klimashin AA 2019 *IOP Conf. Ser.: Mater. Sci. Eng.* **525** 012093
[6] Zagaynov IV, Fedorov SV, Goldberg MA 2019 *Process. Appl. Ceram.* **13** 244
[7] Prekajski M, Stojmenović M, Radojković A, Branković G, Oraon H, Subasri R, Matović B 2014 *J. Alloy. Compd.* **617** 563
[8] Murugan B, Ramaswamy AV, Srinivas D, Gopinath DS, Ramaswamy V 2005 *Chem. Mater.* **17** 3982
[9] Mukherjee D, Rao BG and Reddy BM 2016 *Appl. Cat. B* **197** 105
[10] Devaiah D, Tsuzuki T, Aniz CU and Reddy BM 2015 *Catal. Lett.* **145** 1206
[11] Zagaynov IV, Fedorov SV, Lyskov NV, Kulbakin IV, Antonova OS 2016 *Inorg. Mater.: Appl. Res.* **7** 756
[12] Zagaynov IV 2019 *Inorg. Mater.: Appl. Res.* **10** 42

IOP Conf. Series: Materials Science and Engineering **848** (2020) 012099 doi:10.1088/1757-899X/848/1/012099

Investigation of iron grains growth during solid-phase carbothermic reduction of red mud in the presence of iron metallic particles

A S Zakunov[1,2], P I Grudinsky[1], D V Zinoveev[1], A F Semenov[1], M O Panova[1], V G Dyubanov[1] and A L Petelin [2]

[1]A.A. Baikov Institute of Metallurgy and Material Science RAS, 49 Leninsky avenue, Moscow, 119334, Russia
[2]National University of Science and Technology "MISiS", 4 Leninsky avenue, Moscow, 119049, Russia
E-mail: az437@yandex.ru

Abstract – Red mud is a hazardous waste obtained from bauxite ore by the Bayer method. It has a high content of iron. In this study, the iron grain growth process during solid-phase carbothermic reduction of red mud in the presence of added iron metallic particles was investigated. Carbonyl iron, cast iron shots and metallurgical mill scale were used as the additives. The influence of amount and fraction of the additives on size of reduced iron grains obtained after reduction roasting of red mud at 1250-1350°C and 60-180 min duration was studied. It is showed that the best effect on size of iron grains has the addition of carbonyl iron and mill scale. In this way, besides red mud, some other iron-containing wastes can be recycled, e.g. mill scale.

1. Introduction

The production of alumina forms a waste called a bauxite residue. Another common name for this waste is red mud. According to the data of 2018, more than 4.5 billion tons of red mud have already been accumulated worldwide [1]. Moreover, the main problem is open storage of red mud in sludge fields. This waste has a high alkalinity and contains toxic compounds including salts of heavy metals, so it has a potential hazard for soil and groundwater. Also sludge field destruction can be a serious environmental disaster for nearby settlements [2].

Red mud has a high iron content, which can reach 45 % [3]. In addition to iron, red mud also contains aluminum, titanium, scandium and other rare-earth metals (REEs) that can be recovered [4–6].

A large number of studies were carried out on the topic of red mud recycling. The valuable elements from red mud can be recovered by hydrometallurgical and pyrometallurgical methods. Pyrometallurgical methods is generally used for recovery of iron at the temperatures over 1000 °C to obtain the slag [7,8] or the tailings [9,10] with high content of valuable elements. Hydrometallurgical methods can be used for recovery of aluminum, titanium and rare-earth metals by leaching using different mineral acids or alkalis [11]. NaOH solutions are applicable for selective recovery of aluminum and sodium from red mud because of low solubility of ferrous minerals in these solutions [12–14]. It is showed that more than 95% of Al and Fe from red mud can be extracted by HCl leaching [15,16].The best results for Ti [17] and Sc [18] selective extraction were achieved by H_2SO_4 leaching. The methods were proposed to prevent Fe and Si dissolution and to pass Sc into solution by HCl and H_2SO_4 leaching, respectively [6,19]. Some researchers studied the extraction of Sc and other REEs from red mud by different methods [20–22]. In [22], a comparative investigation of leaching efficiencies of different acids on solubility of Sc, Y, Nd, La, Ce and Dy showed the best results by application of HCl leaching.

High iron content makes possible to consider bauxite residue as a potential source for its extraction. Pyrometallurgical methods are the most applicable for iron recovery. Conventionally, they can be divided according to the applied approach. The first and most common method is reductive smelting at temperatures over 1450 °C [23,24]. As a result, valuable components are distributed between the iron ingot and slag. The second popular method is solid-phase reduction at temperatures between 950 and 1400 °C [25] with following magnetic separation of reduced iron [26]. The iron grains obtained are small in size and form intergrowths with an aluminosilicate phase [27] that decrease the efficiency of magnetic separation. To obtain high-quality iron concentrate it is necessary to find a way for enlargement of reduced iron grain and gangue-grain release after grinding. In the process of growth of reduced metal particles the formation of stable nuclei consumes most of energy. In this regard, we suggested that the addition of iron metallic particles to the charge for solid-phase carbothermic reduction can promote iron grain growth of the reduced iron grains. In this paper, the effect of addition of iron metallic particles on the growth of reduced iron particles was studied.

2. Materials and methods

2.1 Raw materials
Red mud of the Ural Aluminum Plant (Kamensk-Uralsky, Russian Federation) was the main raw material in the study. Long-flame coal containing 15 % of ash and 18 % of moisture was used as a reducing agent. Cast iron shot with a size of 0.3 mm, carbonyl iron and mill scale from AO "OMK-stal" (Vyksa, Russian Federation) were used as additives. Alkali components was preliminary leached from red mud by lime milk at 90°C for 3 hours [28]. Table 1 shows the composition of neutralized red mud.

Table 1. Chemical composition of red mud (wt %).

Fe_2O_3	SiO_2	Al_2O_3	TiO_2	CaO	MgO	MnO	Na_2O	P_2O_5	S	Sc	LOI
36.9	8.71	11.8	3.54	23.8	1.01	0.95	0.27	0.42	0.14	0.008	12.46

2.2 *Reduction roasting*
For the experiments red mud and the reducing agent were preliminarily ground and sieved up to the particle size of less than 0.2 mm. Red mud was dried at 110 °C for 4 hours. Cast-iron shot was 0.3 mm in size, the other additives were ground and sieved up to the required size of 0.1, 0.063 and 0.05 mm.

Red mud with the additives was mixed in a V-shell blender for 24 hours. Then the mixtures were pressed into tablets on hand-held hydraulic press Nordberg N3630A (China) with a pressure of 215 MPa. The weight of the tablets was 1 g, they were 17 mm in diameter and 2.5 mm in height.

The tablets were placed in a small corundum crucible of 62 mm in height and 45 mm in diameter. 10 tablets were mixed with charcoal so as to prevent their touching each other. The small crucible was filled by coal and then was turned over and placed in a large corundum crucible of 68 mm in height and 58 mm in diameter. The large crucible was half-filled with coal. This layout of the crucibles provides the reducing atmosphere inside.

The experiment was carried out in a muffle furnace at temperature range of 1250–1350 °C for 60–180 min. The temperature in the furnace was set using a platinum-rhodium thermocouple connected to an automatic programmable controller. The samples were placed in a preheated furnace. After roasting, the samples were quenched in liquid nitrogen to avoid reverse oxidation of reduced iron. After cooling, the roasted tablets were ground and sieved up to the particle size of less than 0.4 mm for further preparation of polished section and calculation of iron grain size

2.3 Calculation of the size of iron grains
The polished sections were investigated using a METAM LV-34 optical microscope (Russia) in reflected light with a magnification of x250. For each sample 16 microphotographs were combined

into one for further calculation. Full range of dimensional characteristics of all iron grains were obtained by Image Pro Plus software (USA). The Feret diameter was chosen as the main analyzed parameter which is the distance between two parallel straight lines tangent to the grain boundaries in a certain direction. The relative area of iron grains with mean Feret diameter in a certain range was taken as an argument in the obtained dependences.

3. Results

Figure 1 shows the influence of addition of cast iron shots on iron grains growth process during carbothermic reduction of red mud at 1200 °C.

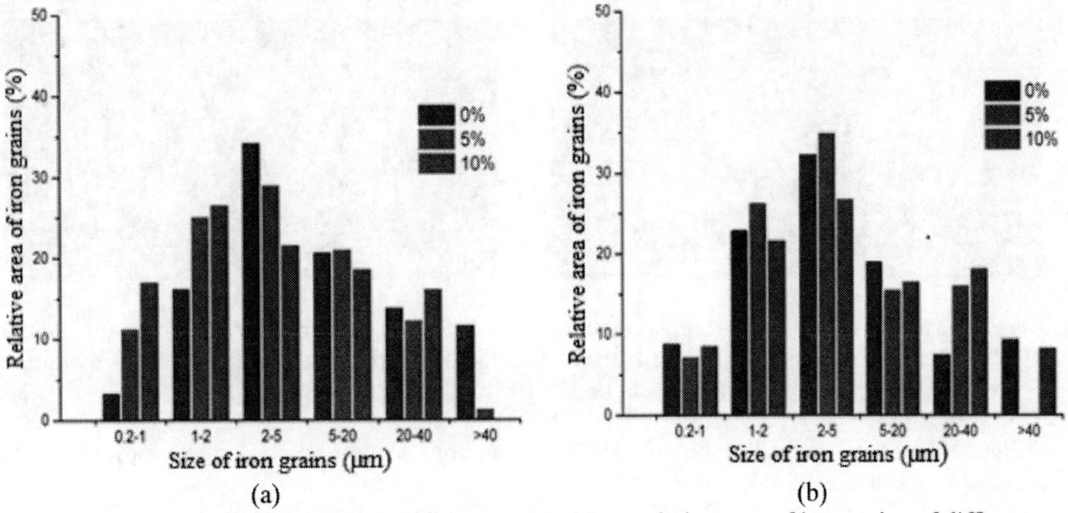

Figure 1. Effect of cast iron shots addition at 1200 °C on relative area of iron grains of different fractions after 60 min (a) and 120 min (b) of the roasting.

The addition of cast iron shots during carbothermic roasting of red mud at 60 minutes has no positive effect on size of iron grains (Figure 1(a)). Iron particles are mainly distributed between small fractions of 1-20 μm. Moreover, after the roasting there were no iron grains with size above 40 μm. An increase of roasting time from 60 to 120 min led to decreasing the amount of iron grains less 1 μm from 17 % to 9 % and to increasing the amount of iron grains above 40 μm from 0 % to 8 % at 10 % of cast iron shot addition (Figure 1(b)). Also amount of iron grains with size of 20-40 μm at 5 % and 10 % of cast iron addition was increased compared with the sample without additions. However, the influence of cast iron addition on iron grain growth is insignificant. It is likely due to the fact that a size of the added particles of cast iron shot was too large compared with a size of reduced iron grains. Figure 2 clearly demonstrates a lack of influence of cast-iron shot at 1200 °C .

IOP Conf. Series: Materials Science and Engineering **848** (2020) 012099 doi:10.1088/1757-899X/848/1/012099

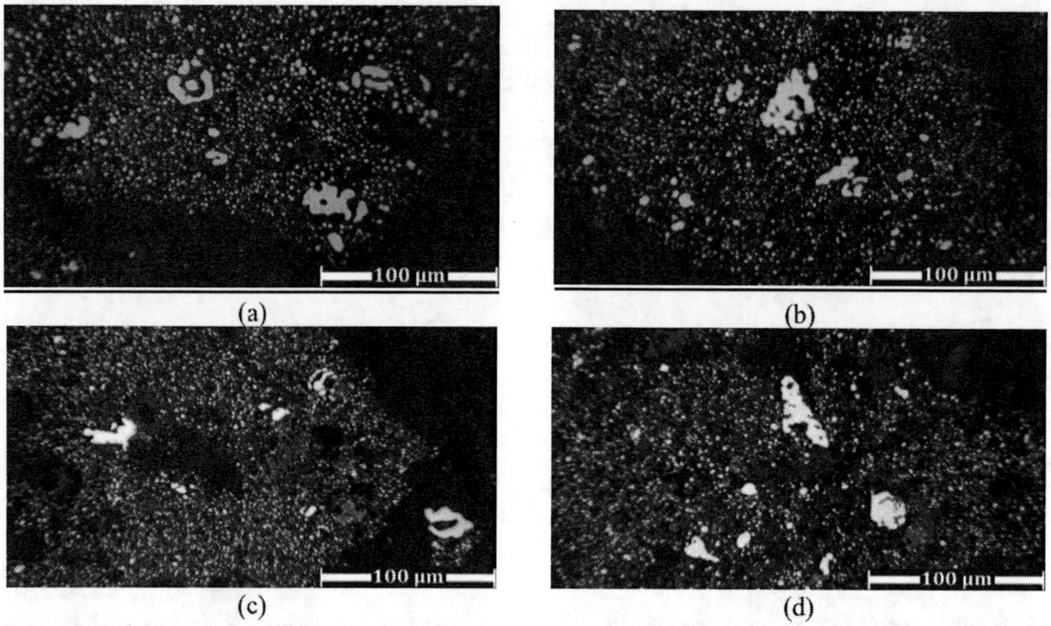

Figure 2. Microstructure of the samples after the roasting at 1200 °C with cast iron shots: (a) addition of 5 %, 60 min of roasting; (b) addition of 5 %, 120 min of roasting ; (c) addition of 10 %, 60 min of roasting; (d) addition of 10 %, 120 min of roasting.

Figure 3 shows the influence of cast iron addition on iron grain growth process during carbothermic reduction of red mud at 1250 °C.

IOP Conf. Series: Materials Science and Engineering **848** (2020) 012099 doi:10.1088/1757-899X/848/1/012099

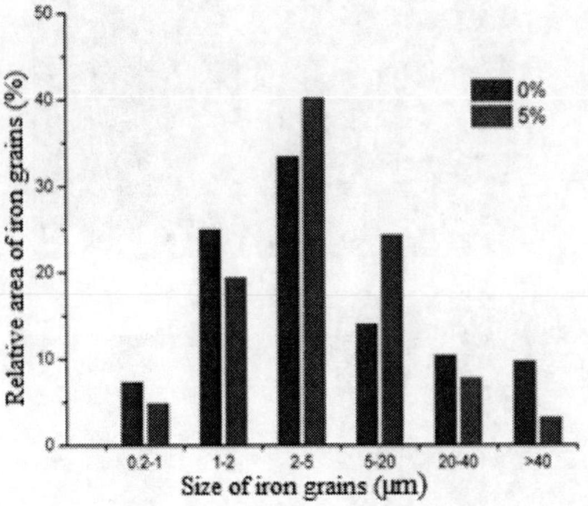

(a)

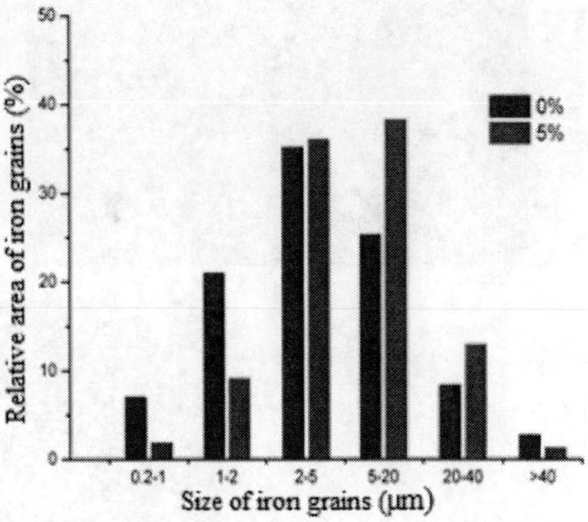

(b)

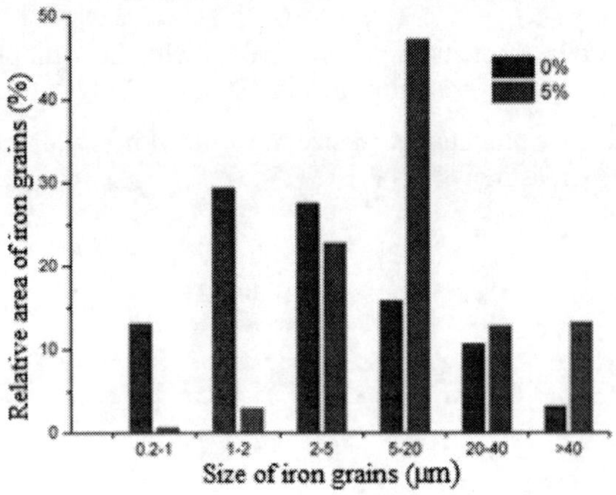

(c)

Figure 3. Effect of cast iron addition at 1250 °C on relative area of iron grains of different fractions after 60 min (a), 120 min (b) and 180 min (c) of the roasting.

An increase of roasting temperature from 1200 °C to 1250 °C led to increase the influence of cast iron shot additions on iron grain growth process (Figure 3). It is showed that there is appreciable growth of the iron grains with size of 5-20 μm at 60 and 120 min of the roasting. An increase of roasting time to 180 min led to increasing the amount of iron grains of 5-20 μm from 15% to 47% and to increasing the amount of iron grains above 40 μm from 3% to 12% compared with the sample without additions. It should be noted that the amount of iron grains less 2 μm was also considerably decreased by addition of cast iron shots.

Figure 4 shows microstructure of the samples after roasting at 1250 °C and indicates that cast iron shots locate separately from the reduced iron particles and have no effect on the iron growth process. It can be seen that cast iron shots save its original geometric shape and size. This fact was in all experiments with the addition of cast iron shots regardless of roasting duration and temperature.

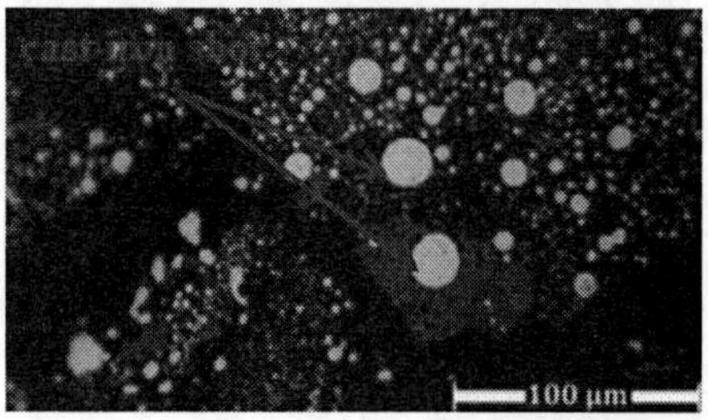

(a)

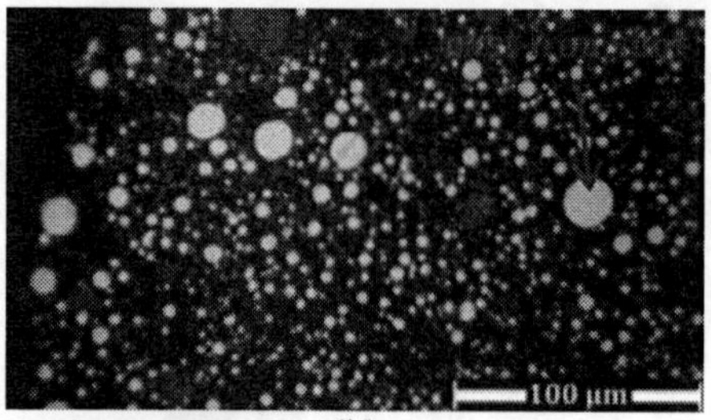

(b)

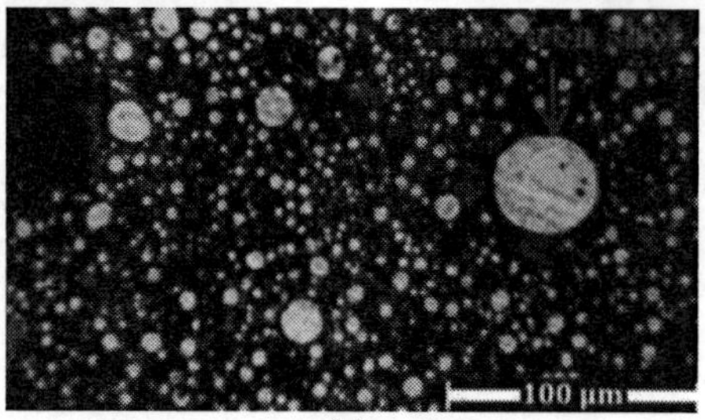

(c)

Figure 4. Microstructure of the roasted samples at 1250 °C with the addition of 5 % of cast-iron shots after 60 min (a), 120 min (b) and 180 min (c) of the roasting.

Figure 5 shows the influence of amount and size of carbonyl iron additions on size of iron grains at 1250 °C after 180 min of the roasting.

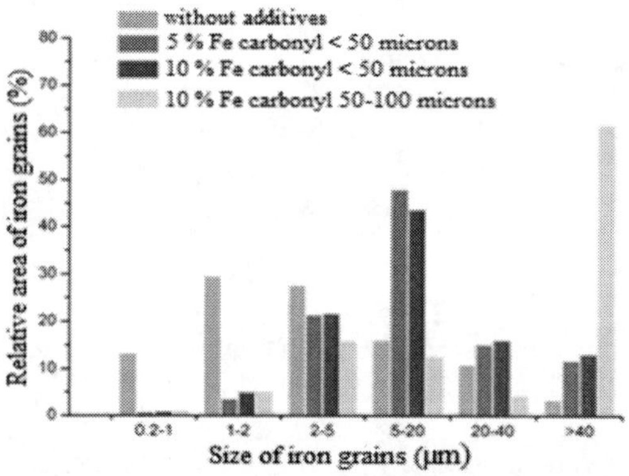

Figure 5. Effect of amount and size of carbonyl iron addition on relative area of iron grains of different fractions at 1250 °C after 180 min of the roasting.

The bar chart indicates that the main part of iron particles reduced without additives has size less than 5 μm. Taking this into account, the effect of carbonyl iron is clearly visible, namely a decrease of amount of the iron grains less 5 μm and a significant increase of amount of iron grains of 5-20 μm and 40 μm. The largest effect on iron grains growth was obtained with the addition of 10% carbonyl iron with particle size of 50-100 μm. It should be noted that size of the particles of carbonyl iron additions have a more significant effect on size of the reduced iron grains than their amount.

Figure 6 shows a comparison of the effect of carbonyl iron addition with different size on growth of reduced iron grains at 1350 °C after 60 min of the roasting.

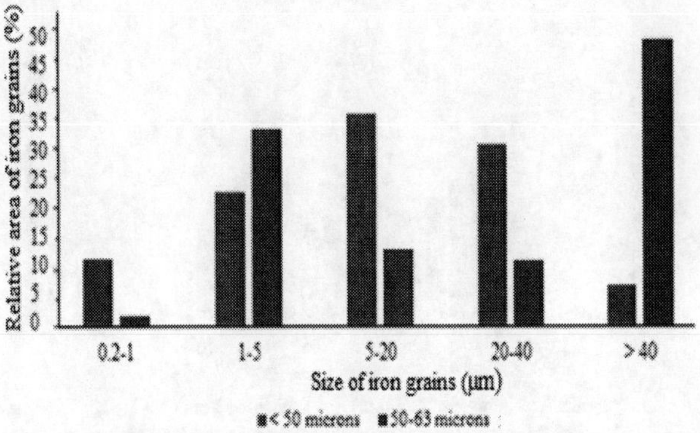

Figure 6. Influence of size of 5 % carbonyl iron addition on relative area of reduced iron grains of the different fractions at 1350 °C after 60 min of the roasting.

Figure 7 shows microphotographs of the samples obtained after the roasting with the addition of carbonyl iron at different temperatures. It can be clearly seen that the rise of temperature decreases an amount of small particles.

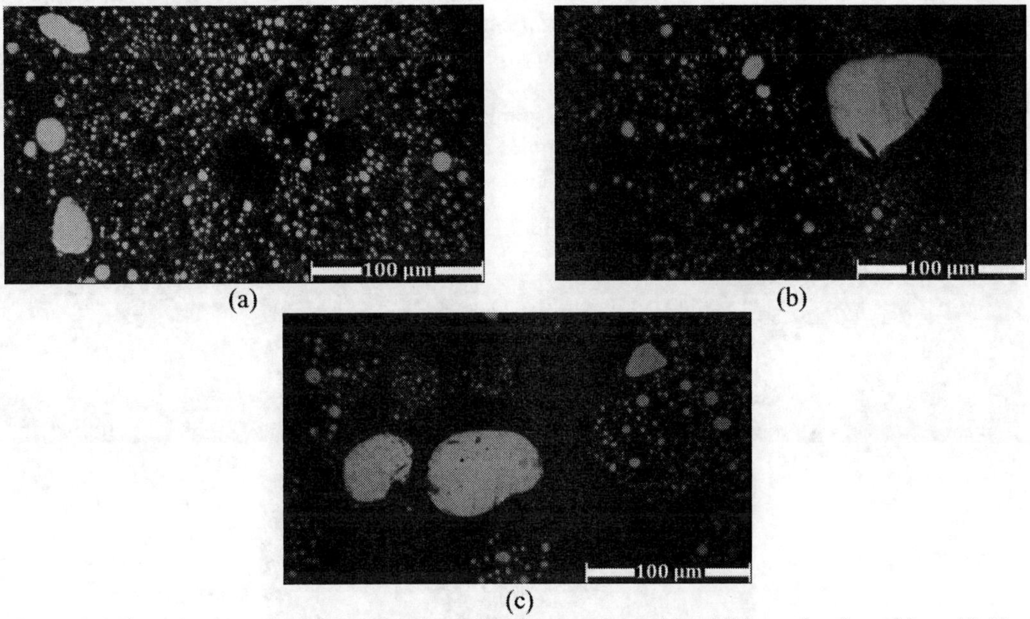

Figure 7. Microphotographs of the samples after the roasting with addition of carbonyl iron 10 % of addition 63-100 µm in size at 1250 °C 180 min (a), 10 % of addition 63-100 µm in size at 1300 °C 180 min (b) and 5 % of addition 50-63 µm in size at 1350 °C 60 min (c).

Figure 8 shows the fraction distribution of metallic iron particles after 180 min of the roasting at 1300 °C depending on the amount and size of mill scale particles.

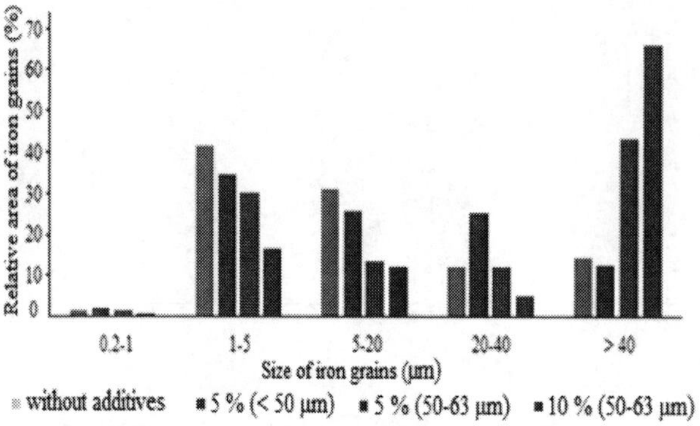

Figure 8. Effect of mill scale addition of various sizes on relative area of iron grains of different fractions at 1300 °C and after 180 min of roasting.

The bar chart indicates that the addition of mill scale has a positive effect on reduced iron enlargement in the same way with the addition of carbonyl iron. The addition of mill scale 50-63 μm in size led to the best results. The relative area of iron grains larger than 40 μm was 40 % and 65 % with the addition of 5 and 10 % of mill scale of this fraction, respectively. It is important to note that particles with a size of 1-5 μm occupy more significant area at the addition of 5 % mill scale compared with the addition of carbonyl iron of the same size. It also should be noted that the addition of 5 % mill scale with a particle size of less than 50 μm led to the lack of large particles; fractions of 1-5, 5-20, and 20-40 μm were the main part of the area of iron grains in this sample.

Figure 9 shows microphotographs of the samples after roasting with the addition of mill scale.

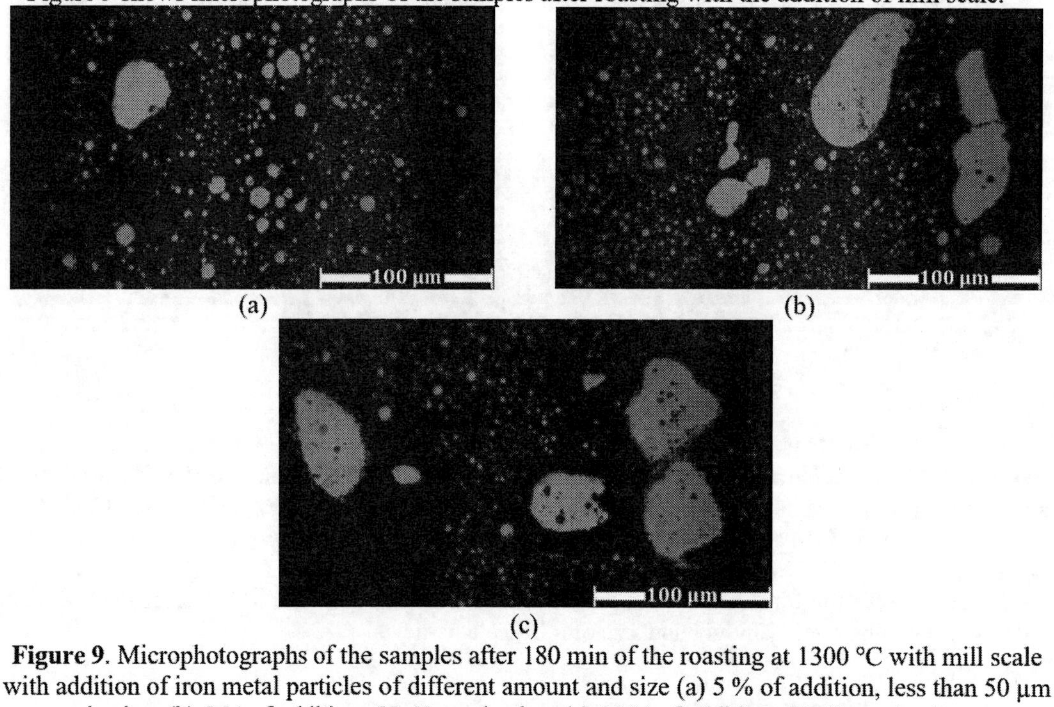

(a) (b)

(c)

Figure 9. Microphotographs of the samples after 180 min of the roasting at 1300 °C with mill scale with addition of iron metal particles of different amount and size (a) 5 % of addition, less than 50 μm in size; (b) 5 % of addition, 50-63 μm in size; (c) 10 % of addition, 50-63 μm in size.

As a generalization of the results, it should be noted that the addition of metallic iron particles into the charge for carbothermic reduction of red mud significantly increase a size of reduced iron grains if the particles have a size in the certain range. The added iron metallic particles should be slightly larger than the size of reduced iron grains obtained under similar conditions without the addition. Furthermore, it is interesting to note that a rise of the roasting temperature significantly increases the size of obtained iron grains. It is most likely due to the partial melting of slag components of the charge that facilitate diffusion of iron grains and promote their aggregation. Hence, the addition of fluxes which can reduce a melting point of red mud, e.g. alkali metal salts, can promote the migration of small grains of reduced iron to the added particles. Evidently, to improve the results it is preferable to use initial red mud in the process of roasting without neutralization stage. Probably, the geometric shape of the added metallic particles also has an impact on the grain growth. The irregular-shaped added particles with defects on the surface can grow easier than particles with regular shape. Figure 10 demonstrates a comparison of the surfaces of the added particles.

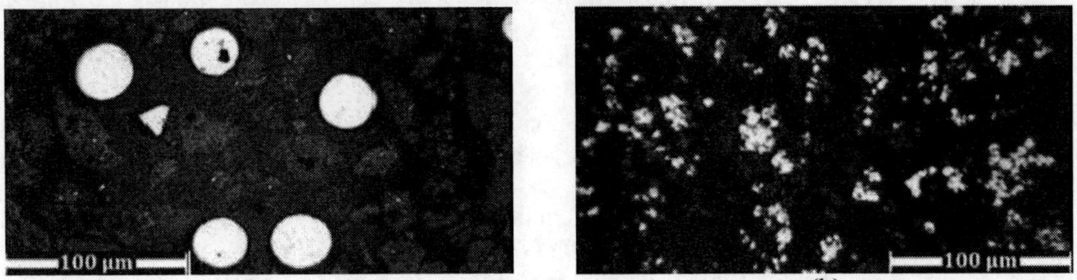

(a) (b)

Figure 10. Microphotographs of added particles of cast-iron shots (a) and carbonyl iron (b).

4. Conclusions

Thus, the study has found out a positive effect of addition of iron metallic particles during carbothermic reduction of red mud but the effect of such additives is limited. In this way, besides red mud, some other wastes can be recycled, e.g. mill scale.

To improve the process of iron grain growth during carbothermic reduction of red mud it is necessary to add iron metallic particles of slightly larger size than the size of reduced iron grains obtained under similar conditions without the additions. The optimal temperature range of the roasting is appeared to be between solidus and liquidus point with duration of more than 120 min.

Funding: The present study was funded by Program of Basic research of the Presidium of RAS №39 according to Research Project No. AAAA-A18-118031490064-3 (literature review) and RFBR according to Research Project No. 18–29–24186 (experimental work).

References

[1] Xue S guo, Wu Y jun, Li Y wei, Kong X feng, Zhu F, William H, Li X fei and Ye Y zhen 2019 Industrial wastes applications for alkalinity regulation in bauxite residue: A comprehensive review *Journal of Central South University* **26** pp 268–88

[2] Grenerczy G and Wegmüller U 2011 Persistent scatterer interferometry analysis of the embankment failure of a red mud reservoir using ENVISAT ASAR data *Natural Hazards* **59** pp 1047–53

[3] Gräfe M, Power G and Klauber C 2011 Bauxite residue issues: III. Alkalinity and associated chemistry *Hydrometallurgy* **108** pp 60–79

[4] Zhang N, Li H X and Liu X M 2016 Recovery of scandium from bauxite residue—red mud: a review *Rare Metals* **35** pp 887–900

[5] Loginova I V and Kyrchikov A V 2019 Complete Processing of the High-Iron Content Red Mud **946** pp 569–74

[6] Zhou K, Teng C, Zhang X, Peng C and Chen W 2018 Enhanced selective leaching of scandium

from red mud *Hydrometallurgy* **182** pp 57–63
[7] Pan J, Li S, Zhu D, Xu J and Chou J 2019 Migration of iron, aluminum and alkali metal within pre-reduced-smelting separation of bauxite residue *Minerals, Metals and Materials Series* pp 107–11
[8] Borra C R, Blanpain B, Pontikes Y, Binnemans K and Van Gerven T 2015 Smelting of Bauxite Residue (Red Mud) in View of Iron and Selective Rare Earths Recovery *Journal of Sustainable Metallurgy* **2** pp 28–37
[9] Gao F, Zhang J, Deng X, Wang K, He C, Li X and Wei Y 2019 Comprehensive Recovery of Iron and Aluminum from Ordinary Bayer Red Mud by Reductive Sintering–Magnetic Separation–Digesting Process *Jom* **71** pp 2936–43
[10] Deng B, Jiang T, Li G, Ye Q, Gu F, Rao M and Peng Z 2018 Effects of reductive roasting with sodium salts on leaching behavior of non-ferrous elements in bauxite ore residue *Minerals, Metals and Materials Series*
[11] Liu Z and Li H 2015 Hydrometallurgy Metallurgical process for valuable elements recovery from red mud — A review *Hydrometallurgy* **155** pp 29–43
[12] Bayca S U and Kisik H 2018 Optimization of leaching parameters of aluminum hydroxide extraction from bauxite waste using the taguchi method *Environmental Progress and Sustainable Energy* **37** pp 196–202
[13] Zhong L, Zhang Y and Zhang Y 2009 Extraction of alumina and sodium oxide from red mud by a mild hydro-chemical process *Journal of Hazardous Materials* **172** pp 1629–34
[14] Meher S N and Padhi B 2014 A novel method for extraction of alumina from red mud by divalent alkaline earth metal oxide and soda ash sinter process *International Journal of Environment and Waste Management* **13** pp 231–45
[15] Xie W, Zhang N, Li J, Zhou F, Ma X, Gu G and Zhang W 2017 Optimization of condition for extraction of aluminum and iron from red mud by hydrochloric acid leaching *Chinese Journal of Environmental Engineering* **11** pp 5677–82
[16] Alkan G, Schier C, Gronen L, Stopic S and Friedrich B 2017 A Mineralogical Assessment on Residues after Acidic Leaching of Bauxite Residue (Red Mud) for Titanium Recovery *Metals* **7** p 458
[17] Ghorbani A and Fakhariyan A 2013 Recovery of Al_2O_3 , Fe_2O_3 and TiO_2 from Bauxite Processing Waste (Red Mud) by Using Combination of Different Acids *J. Basic. Appl. Sci. Res*
[18] Lymperopoulou T, Georgiou P and Tsakanika L 2019 Optimizing Conditions for Scandium Extraction from Bauxite Residue Using Taguchi Methodology
[19] Alkan G, Yagmurlu B, Gronen L, Dittrich C, Ma Y, Stopic S and Friedrich B 2019 Selective silica gel free scandium extraction from Iron-depleted red mud slags by dry digestion *Hydrometallurgy* **185** pp 266–72
[20] Akcil A, Akhmadiyeva N, Abdulvaliyev R, Abhilash and Meshram P 2018 Overview On Extraction and Separation of Rare Earth Elements from Red Mud: Focus on Scandium *Mineral Processing and Extractive Metallurgy Review* **39** pp 145–51
[21] Borra C R, Blanpain B, Pontikes Y, Binnemans K and Van Gerven T 2019 Recovery of Rare Earths from Bauxite Residue (Red Mud) pp 343–56
[22] Borra C R, Blanpain B, Pontikes Y, Binnemans K and Van Gerven T 2016 Recovery of Rare Earths and Other Valuable Metals From Bauxite Residue (Red Mud): A Review *Journal of Sustainable Metallurgy* **2** pp 365–86
[23] Kaußen F and Friedrich B 2015 Reductive Smelting of Red Mud for Iron Recovery *Chemie-Ingenieur-Technik* **87** pp 1535–42
[24] Raspopov N A, Korneev V P, Averin V V, Lainer Y A, Zinoveev D V and Dyubanov V G 2013 Reduction of iron oxides during the pyrometallurgical processing of red mud *Russian Metallurgy (Metally)*
[25] Chun T, Li D, Di Z, Long H, Tang L, Li F and Li Y 2017 Recovery of iron from red mud by

hightemperature reduction of carbonbearing briquettes *Journal of the Southern African Institute of Mining and Metallurgy* **117** pp 361–4

[26] Zinoveev D V., Grudinskii P I, Dyubanov V G, Kovalenko L V. and Leont'ev L I 2018 Global recycling experience of red mud - a review. Part i: pyrometallurgical methods *Izvestiya Visshikh Uchebnykh Zavedenii. Chernaya Metallurgiya = Izvestiya. Ferrous Metallurgy* **61** pp 843–58

[27] Cardenia C, Balomenos E and Panias D 2019 Iron Recovery from Bauxite Residue Through Reductive Roasting and Wet Magnetic Separation *Journal of Sustainable Metallurgy* **5** pp 9–19

[28] Shiryaeva E V, Podgorodetskiy G S, Malyscheva T Y, Gorbunov V B, Zavodyaniy A V and Schapovalov A N 2014 Influence of low alkaline red mud on the properties and microstructure of the agglomerates from the charge materials JSC "Ural steel" *Izvestiya. Ferrous Metallurgy.* **57** pp 14–9

IOP Conf. Series: Materials Science and Engineering **848** (2020) 012100 doi:10.1088/1757-899X/848/1/012100

Study of active media based on microparticles of solid-state laser materials

O A Burdukova[1], V A Konyshkin[2], Yu L Kopylov[1,3], K V Lopukhin[1,3], V A Petukhov[1], Yu V Senatsky[1] and P G Zverev[2,4*]

[1] Lebedev Physics Institute of the Russian Academy of Sciences, Moscow 119991, Russia

[2] Prokhorov General Physics Institute of the Russian Academy of Sciences, Moscow 119991, Russia

[3] Kotel'nikov Institute of Radioengineering and Electronics of the Russian Academy of Sciences, 1 Vvedenskogo square, Fryazino, 141190, Russia

[4] Moscow Power Engineering Institute, 14 Krasnokazarmennaya street, Moscow, 111250, Russia

[*]Corresponding author: zverev@lst.gpi.ru

Abstract. Active media based on particles of solid-state laser materials distributed in liquid or gas may provide a new solution to the problem of heat removal from laser media. We propose active media based on particles of LiF crystal with F_2 color centers, as well as of Nd^{3+}-doped crystals and glasses. Mixtures of index-matched liquids with glass and crystal particles, methods of their preparation, and their optical characteristics are considered. The laser characteristics of the active media under excitation by nanosecond laser pulses are compared. A "slurry laser" based on a dense mixture of LiF:F_2 microparticles with an index-matched liquid is demonstrated. Lasing was obtained in small neodymium pentaphosphate crystals (NdP_5O_{14}) and unpolished bulk pieces of concentrated neodymium phosphate glass in immersion liquids.

1. Introduction

Active media based on particles of solid-state laser materials distributed in liquid or gas have been the object of research, since such an approach can provide a new solution to the problem of heat removal during laser operation [1-5]. Glass or crystal nano- and microparticles doped with Nd^{3+} or Yb^{3+} ions were considered as laser media [1-5]. Lasing was obtained in a dispersion of Nd_2O_3 nanoparticles (sizes ≤ 10 nm), scattering losses from such particles were negligible [3, 4]. However, nanoparticles may have some properties (emission lifetime, quantum yield) different from their bulk counterparts, which requires a surface modification of Nd_2O_3 nanoparticles to achieve laser operation [3, 4]. Particles of solid-state laser materials with dimensions over 1 μm are easier to manufacture and need no complicated modification, since their properties are identical to the properties of the original bulk single crystals or glasses. For such particles, index-matched liquids should be used to suppress Fresnel reflection and scattering. A "slurry laser" with an active medium in the form of a dense mixture of particles (sized 200÷1000 μm) of LiF crystal with F_2 color centers (CCs) and an immersion liquid was demonstrated in [6]. The possibility of developing such a laser based on particles doped with Nd^{3+} ions has been studied. The "slurry laser" concept may be attractive due to a potentially higher laser output energy from a dense pulp circulating through the laser module. In this paper, we present the results of the study of laser media based on particles of $LiF:F_2$, Nd:YLF, and NdP_5O_{14} crystals and Nd:glass. Compositions of immersion mixtures with glass and crystal particles, methods of their preparation, and their optical characteristics are considered; data on excitation of lasing in test samples are presented.

2. Fabrication of index-matched slurry-like active media

Slurry lasers can be based only on high-gain active media, because it is necessary to overcome the internal losses in the mixture of solid particles and a liquid due to incomplete immersion and existence of various scattering centers. LiF crystal with CCs [7-13] is an interesting material for preparing an active medium for slurry lasers. LiF has a high thermal conductivity (4 $W \cdot m^{-1}K^{-1}$) and can be used in cw and pulse-periodic lasers. The refractive indices of irradiated and non-irradiated LiF crystals are almost identical, namely, $n_{LiF} \approx 1.39$. This low value of the refractive index simplifies the search for an immersion liquid for LiF crystal. There are several types of LiF crystals with color centers, i.e., $LiF:F_2^-$, $LiF:F_2^+$, $LiF:F_2$, and some others [7-13]. One of the most promising materials is LiF with F_2 CCs. F_2 CCs have broad absorption (0.42÷0.47 μm) and emission (0.56÷0.84 μm) bands, as well as high emission and absorption cross sections, $\sigma \approx 7 \cdot 10^{-17}$ cm^2. Their characteristic feature is a very high concentration of centers, which results in a high amplification gain up to tens of cm^{-1} [7-9].

F_2 CCs with a concentration of 10^{18} cm^{-3} were produced in nominally pure oxygen-free samples of LiF crystals by irradiation at room temperature with γ-quanta from a Co^{60} source with a dose of $2 \cdot 10^7$ Rad. The ground $LiF:F_2$ powder was sequentially sieved through sieves with different mesh sizes to separate fractions with sizes from 200 to 1000 μm and larger. Several pieces of transparent non-irradiated LiF crystals were subjected to grinding and fractionation by the same procedure for comparison. Microphotograph of a non-irradiated LiF powder fraction obtained using a mortar with a pestle and then sieved through 200÷500 μm sieves is shown in figure 1a. The fractions of crushed material were placed together with an immersion liquid into rectangular glass cuvettes with dimensions of 24 × 40 × 8 mm^3 and a 3-mm gap between the walls (figure 1b).

As a rule, it is impossible to find a liquid with a refractive index that precisely coincides with the refractive index of solid particles and has the same dispersion characteristics in a given spectral region. However, it is possible to use a mixture of two liquids with higher and lower refractive indices. In this case, the refractive index of the mixture can be equal to the particle refractive index n_p, i.e.,

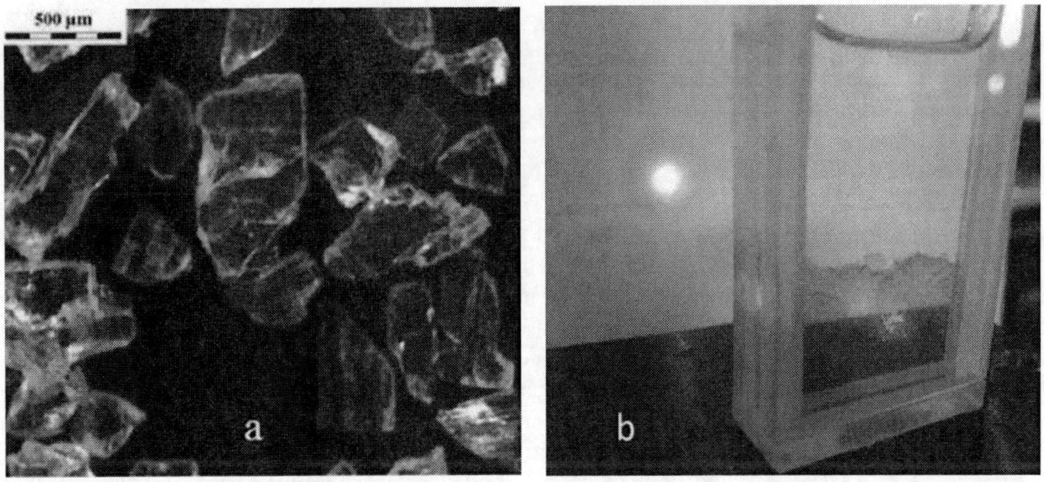

Figure 1. Microphotograph of the LiF crystal powder sieved through meshes (a); photo of a 3-mm-thick cuvette containing a slurry of 500÷1000 μm yellowish $LiF:F_2$ granules in an immersion liquid (heptane/vaseline oil) and the spot of a He-Ne laser beam passed through this slurry (b).

IOP Conf. Series: Materials Science and Engineering **848** (2020) 012100 doi:10.1088/1757-899X/848/1/012100

$$n_p = x \cdot n_1 + (1 - x) \cdot n_2, \tag{1}$$

Where n_1 and n_2 are the refractive indices of the first and second liquids, and x is the fraction of the first liquid. Table 1 shows refractive indices n of some laser crystals, glass, and liquids used in this work for the preparation of immersion mixtures. The refractive indices of materials at wavelengths λ of 650 nm and 1060 nm (with 4 decimal places) are given in accordance with the data of website [14]. For several materials, refractive indices are given for $\lambda = 589$ nm according to [15–19]. All liquids listed in Table 1 are transparent in the visible and near-IR spectral regions (400÷1000 nm).

Table 1. Refractive indices of laser materials and components of immersion liquids.

Material	Refractive index		
	$\lambda = 650$ nm	$\lambda = 589$ nm	$\lambda = 1060$ nm
Acetone	1.3574		1.3526
Ethanol	1.3600		1.3548
Heptane	1.3872		1.3827
LiF	1.3910		1.3866
Isobutyl alcohol	1.3913		1.3862
Ethylene glycol	1.4296		1.4232
Nd:YLiF	1.4524(o), 1.4750(e)		1.4480(o), 1.4702(e)
Glycerol	1.4701		1.4631
Toluene	1.4926		1.4813
Benzene	1.4915		1.4802
Vaseline oil		1.503	
Benzyl alcohol	1.536		
Anise oil	1.560		
Nd:glass			1.55
NdP_5O_{14}		1.60	
Cinnamaldehyde		1.62	

To prepare a slurry index-matched for the wavelength region near the expected operating wavelength of a given laser material, we used the laser setup shown in figure 2. A diode laser ($\lambda = 650$nm), a He-Ne laser ($\lambda = 633$ nm), and a Nd:YAG laser ($\lambda = 1064$ nm) with beam diameters of 1÷2 mm were used as reference light sources. To measure the power transmitted through a cuvette with a slurry, a power meter was placed behind an iris diaphragm near the focus of the lens with a focal length $f \approx 85$ mm. Liquids were added to the layer of powder (≈ 1 cm in height) in a cuvette using a syringe (dropwise, with stirring), so that a buffer layer of liquid was formed above the slurry. The optimum composition of the immersion liquid was achieved when the reference beam transmission through the slurry was maximum and the transmitted laser beam spot at a remote screen had the best quality. For LiF, the first liquid component with the higher index may be anise oil, vaseline oil, or toluene, and the second component with the lower index may be ethanol, acetone, or heptane. The vaseline oil/heptane and toluene/ethanol mixtures were used as immersion liquids for $LiF:F_2$ and LiF particles. At the optimal composition, the transmission of the He–Ne laser beam through the slurry of a fixed thickness depended on the size of particles. A decrease in the thickness of the layer of particles in the cuvette, as well as the use of grains of larger sizes (500÷1000 μm and larger), leads to a reduction of scattering in the medium. For 500÷1000-μm LiF or $LiF:F_2$ grains in a 3-

mm-thick cuvette, the fraction of transmitted laser radiation in the angle of 0.01 rad reaches 70%. For particles of smaller sizes (200÷500 μm and < 200 μm), the transmission decreases to 30÷60%.

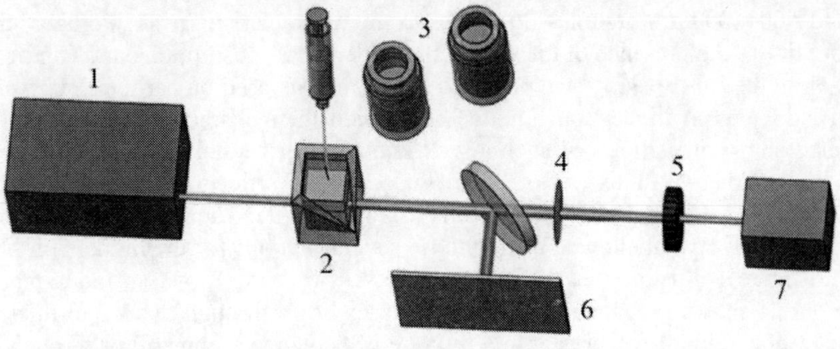

Figure 2. Scheme of the laser setup for preparation and testing of the index-matched slurry: (1) laser source; (2) cuvette with the slurry; (3) components for preparing immersion liquids; (4) lens; (5) diaphragm; (6) screen; (7) power meter.

All these observations correspond to the characteristics of the Christiansen filter (CF)—a well-known device in the form of a cuvette with glass or crystal powder in an immersion liquid, which is used to obtain spectral selection of radiation [20, 21]. CF is characterized by a peak of transmission at the operating wavelength. The transmission spectra of LiF and LiF:F_2 slurries with 500÷1000; 200÷500, and < 200-μm granules were measured on a Cary-5000 spectrophotometer (figure 3). The spectra of LiF:F_2 slurries in the region 300÷900 nm are similar to the spectrum of a LiF:F_2 bulk plate (figure 3a). These spectra exhibit the absorption band of F_2 centers near 450 nm and the absorption peak of the immersion liquid near 1200 nm.

Using the data on absorption for the slurry and the pure immersion liquid, it was possible to estimate the particle filling factor in the slurry to be ≈ 0.5. Figure 3(b) shows the spectrum of the slurry of pure LiF in a toluene/ethanol mixture with the transmission maximum near $\lambda = 570$ nm. This maximum can be shifted along the wavelength scale by changing the ratio of the immersion liquid components.

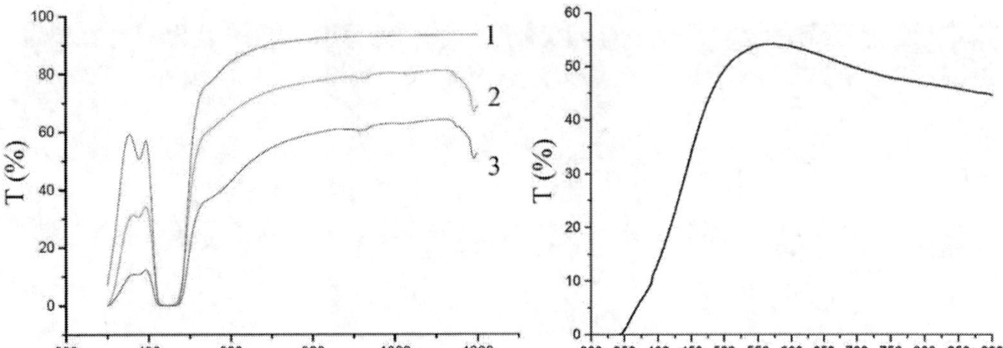

Figure 3. Transmission spectra of (a) (1) 2.5-mm-thick LiF:F_2 crystal plate and (2, 3) 3-mm-thick cuvettes with the LiF:F_2 slurry with (2) 500÷1000 and (3) 200÷500 μm granules in heptane/vaseline oil and (b) the 3-mm-thick cuvette with pure LiF granules (size < 200 μm) in toluene/ethanol mixture.

The calculations using data from [14] showed that the position of this maximum corresponds to the crossing point of the dispersion curves of LiF crystal and toluene/ethanol mixture, which testified that the spectrum in figure 3b is a CF spectrum. At the same time, the spectra of the LiF:F_2 slurry with 200÷500 and 500÷1000 μm granules do not have this characteristic peak. Perhaps, this feature was smoothed out due to the presence of the strong and wide LiF:F_2 absorption band (figure 3a).

In addition to the LiF:F_2 slurry, we prepared active media based on particles of crystals and glasses doped with Nd^{3+} ions in immersion liquids and studied their characteristics. The preparation of an index-matched slurry of particles of such a well-known laser crystal as Nd:YAG is not an easy task due to the high refractive index ($n \approx 1.82$) of this crystal. Particles of another widely used crystal, Nd:YLF, ($n \approx 1.45$) were obtained from a bulk Nd:YLF sample. The moderate value of the refractive index of the Nd:YLF crystal allowed us to choose several options for the immersion liquid. However, the photos in figure 4 illustrate the difficulties involved in trying to get an index-matched slurry for Nd:YLF. Figure 4 shows propagation of a He-Ne laser beam through the 3-mm-thick cuvette with a powder of isotropic (cubic) LiF crystal in a mixture of heptane/vaseline oil or through a powder from anisotropic Nd:YLF crystal in an immersion mixture of isobutyl and benzyl alcohols. Sizes of grains were 500–1000 μm in both cases. Comparison of the beam profiles on a remote screen indicates high scattering losses for the cuvette with Nd:YLF powder. The reason of these losses is that the refractive index of the liquid cannot be equal simultaneously to the ordinary and extraordinary indices of anisotropic Nd:YLF particles. Thus, high scattering losses (which may suppress lasing) are expected for slurry with particles of anisotropic materials.

Interesting materials for slurry lasers are highly concentrated Nd^{3+}-doped phosphate glass [15] and neodymium crystals, such as NdP_5O_{14} [18, 22, 23]. The concentration of Nd^{3+} ions in these media can exceed 10^{21} cm^{-3}, which results in the record laser gains (up to hundreds of cm^{-1} in NdP_5O_{14} [22]). The refractive indices of these materials are $n = 1.5 \div 1.6$, which allows using a number of liquids as immersion mixture components. Samples of NdP_5O_{14} crystal ($n \approx 1.60$) and LGSI-50 phosphate Nd:glass ($n \approx 1.55$; Nd^{3+} concentration $\approx 12\%$) were provided by IRE of the RAS. A powder with particle dimensions from 200 to 500 μm was obtained by crushing LGSI-50 glass bulk samples. For experiments with NdP_5O_{14} crystals, we used 0.2–0.4-mm thick plates with facets up to $2 \div 4 \times 1 \div 1.5$ mm taken directly from a growth crucible [23]. Such small crystals can, generally speaking, be considered as components of an active medium for a flow-through slurry-like laser. Because of this, we did not crush these small pieces of the NdP_5O_{14} crystal into powder.

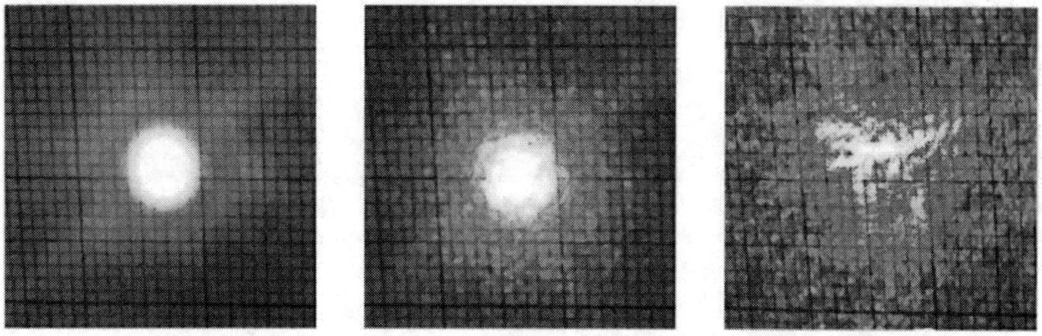

(a) (b) (c)

Figure 4. Images of a He-Ne laser beam on a screen at a distance of 1 m from the source in the cases (a) without a cuvette; (b) with a 3-mm thick cuvette with an index-matched slurry of LiF particles, and (c) with a 3-mm thick cuvette with a slurry of anisotropic Nd:YLF particles.

Figure 5 shows the spectra of LGSI-50 Nd:glass powder (a) and a 0.4-mm-thick NdP_5O_{14} plate (b) placed in a 1-mm-thick cuvette with immersion mixtures of cinnamaldehyde and benzyl alcohol. The immersion compositions for the spectral region near 1060 nm were prepared at a laser stand (figure 2) separately for each material. Both spectra show characteristic absorption bands of Nd^{3+} ions. The transmittance of the cuvettes near 1060 nm reached 80%. The transmission curve of LGSI-50 glass clearly shows an increase in scattering losses in the short-wavelength region (figure 5a). The absorption bands at 1200 and 1500 nm in the spectra correspond to the absorption by the immersion liquid.

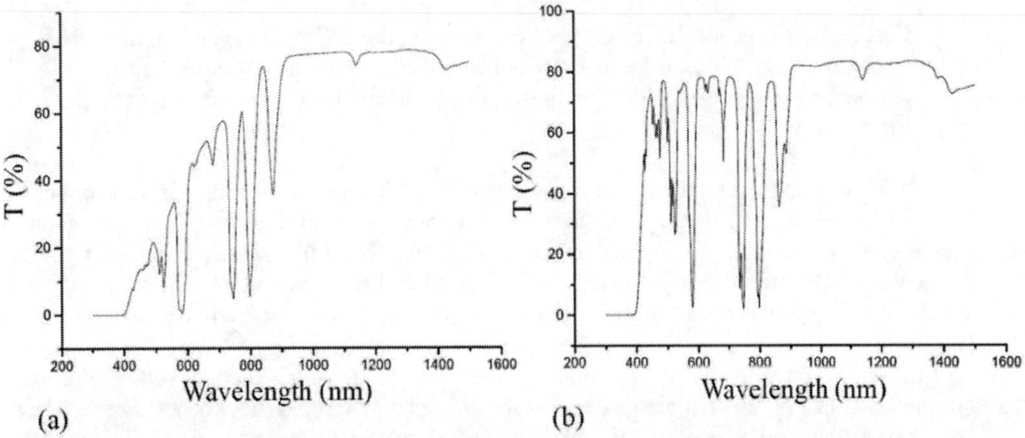

Figure 5. Transmission spectra of the cuvette (1-mm gap) with powder of LGSI-50 glass particles sized 200÷500 μm (a); and with a 0.4-mm thick NdP_5O_{14} crystal plate (b) in benzyl alcohol/cinnamaldehyde immersion mixtures.

3. Experiments on excitation of laser oscillation in active media

The experiments on lasing in samples were performed using a pulsed Nd:YAG laser LOTIS TII with conversion of 1060 nm radiation into the 2-nd and 3-rd harmonics. The scheme of the experiment on the LiF:F_2 slurry is shown in figure 6. A transversely pumped Coumarin 120 (C120) dye laser was used as a pump source. The cuvette with C120 dye solution in a resonator of length $L \approx 8$ mm was pumped by 12-ns pulses of the 3-rd harmonic (354 nm) of a pulsed LOTIS TII Nd:YAG laser. The energy in a single pump pulse was up to 8.7 mJ. This radiation was focused through the side wall of the cuvette into the dye solution. The C120 dye laser emitted 11-ns pulses with an energy up to 1.6 mJ at wavelength $\lambda_p \approx 446$ nm, which corresponds to the absorption band of F_2 centers (figure 3a). The pump beam diameter was 2 mm. The samples (LiF:F_2 bulk plates or cuvettes with a 2÷3-mm gap filled by LiF:F_2 slurry) were placed into a resonator ($L \approx 40$ mm) formed by two flat mirrors. The dye laser radiation was focused onto the samples by a lens with $f = 172$ mm along the resonator axis through the input mirror with a 90% transmission at λ_p and a high reflection (HR = 99%) in the red spectral region. The diameter of the pump spot at the samples was 0.5 mm, and the maximum pump intensity reached 60 MW·cm^{-2}. The pump radiation was almost completely (> 98%) absorbed in the LiF:F_2 media. Mirrors with transmission from $T = 1\%$ to $T = 23\%$ at a wavelength near 680 nm were used as an output coupler (OC). 10-ns pulses of red collimated radiation in the wavelength region near 680 nm were produced at the output of the resonator containing LiF:F_2 plates or cuvettes with LiF:F_2 granules.

The pulse energy from all types of LiF:F_2 active elements was measured using a calibrated silicon photodiode FD24 connected to a digital oscilloscope. Figure 7 compares the dependences of the output energy of the LiF:F_2 slurry lasers (for two compositions) and the LiF:F_2 single crystal laser on the pump energy. All measurements were carried out in the same resonator configuration with the OC transmission of 13%.

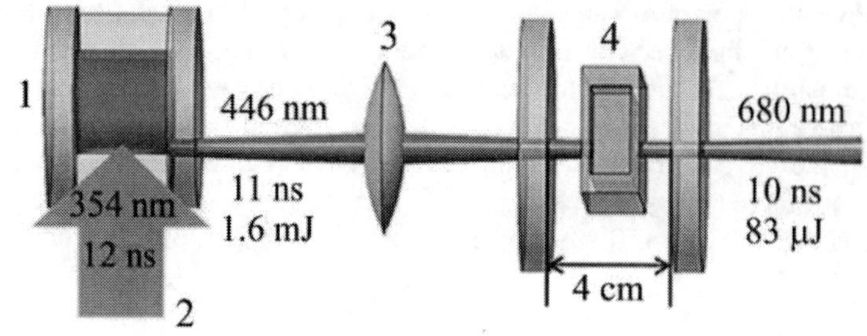

Figure 6. Scheme of the experiment on pumping the LiF:F_2 slurry active medium by a Coumarin 120 dye laser: (1) Coumarin 120 dye laser, (2) beam of the 3rd harmonic of a pulsed Nd:YAG laser; (3) focusing lens; (4) cuvette with the LiF:F_2 slurry.

From the graphs in figure 7, one can see that the efficiency of the slurry laser is only twice as low as that of the single crystal laser. The maximum output energy of the slurry laser was about 83 μJ at a pump energy of 1.34 mJ; the slope efficiency was 12.9%. The LiF:F_2 slurry laser beam divergence near the generation threshold was estimated to be 2 mrad and increased up to 5 mrad at higher pump energies. It was found that the lasing threshold is smaller and the laser output energy is higher for 500÷1000-μm granules as compared with 200÷500-μm granules. The output energy of the slurry laser with a 2-mm-thick cuvette was lower than in the case of a 3-mm-thick cuvettes. For particles < 200 μm in size, lasing was not achieved even for the OC with a reflectivity of 99%.

Experiments on excitation of media doped with Nd^{3+} ions were carried out both with powders (slurries) and with samples of the bulk material in an immersion liquid and without it. NdP_5O_{14} samples with dimensions of (2÷4) × (1÷1.5) × (0.2÷0.4) mm were used for lasing experiments. Despite the NdP_5O_{14} crystal anisotropy, a high gain in this crystal allowed us to expect lasing in these small samples even in conditions of incomplete immersion. The scheme of the experiment on the excitation of lasing in NdP_5O_{14} crystals is shown in figure 8. The active medium was pumped by pulses (20 ns, 5 mJ) of a pyrromethene 597 (PM597) dye laser at wavelength λ_p=581 nm, which coincides with the absorption peak of neodymium in NdP_5O_{14} crystals (figure 5b).

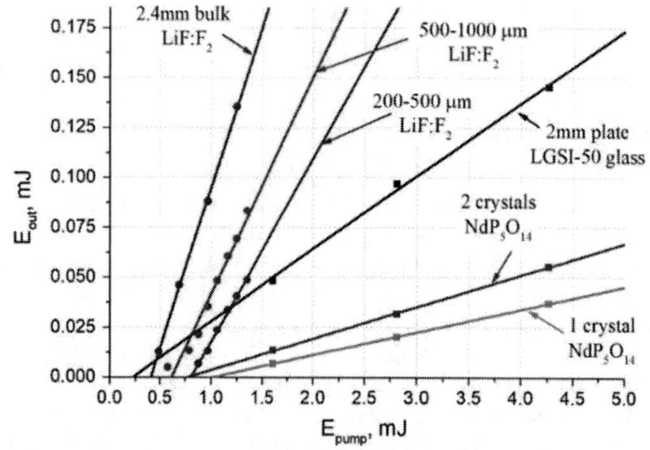

Figure 7. Dependences of the output energy on the pump energy for a laser based on a 2.4 mm thick LiF:F_2 bulk plate and on LiF:F_2 slurry in a 3 mm thick cuvette; for a laser based on 0.4-mm-thick NdP_5O_{14} crystals in a 1-mm-thick cuvette with an immersion liquid; and for a laser based on a 2-mm-thick LGSI-50 glass plate.

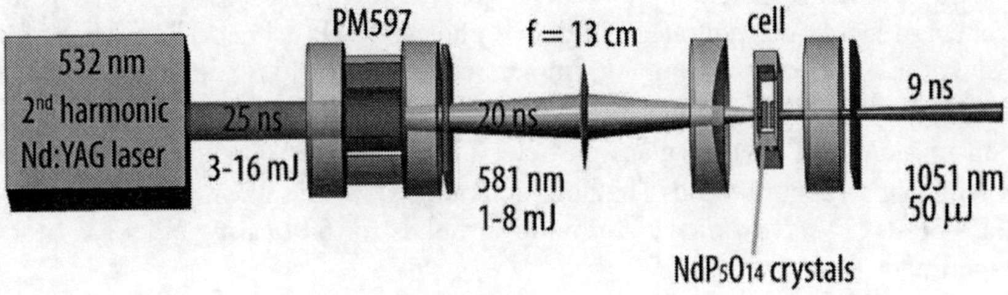

Figure 8. Scheme of the experiment on pumping of NdP_5O_{14} crystals in a 1-mm-thick cuvette with immersion liquid by a PM597 dye laser.

The dye laser radiation was focused on the cuvette with 1- or 2-mm gap by a 13-cm focus lens. A semiconfocal resonator with $L = 2.5$ cm was used: a HR spherical mirror with a radius of 5 cm and a flat output mirror with 1.7% transmission at the lasing wavelength, which is 1051 nm for NdP_5O_{14} crystal. The cuvette was located close to the flat mirror. The residual pump radiation passed through the cuvette and the output mirror was cut off from the generation by two IR filters. Figure 7 shows the dependences of the output energy on the pump energy for the lasers based on one and two 0.4-mm-thick NdP_5O_{14} samples in a cuvette with benzyl alcohol/cinnamaldehyde immersion; oscillograms of the pump and NdP_5O_{14} laser pulses are shown in figure 9.

In order to verify the lifetime of Nd^{3+} ions at the upper $^4F_{3/2}$ laser level in LGSI-50 glass, an experiment was carried out to measure this parameter. Luminescence in a polished LGSI-50 glass plate was excited by pulses of the 2-nd harmonic of a Nd:YAG laser. Radiation at the 1.06 μm $^4F_{3/2} \rightarrow {}^4I_{11/2}$ transition was recorded using a VM-25 monochromator. An FEU-62 photomultiplier tube and a Tektronix TDS-2022 digital storage oscilloscope recorded the luminescence dynamics. The obtained dependence is well described by an exponential function. The lifetime of the upper laser level for the LGSI-50 glass was found to be 45 μs. The experiments on lasing with LGSI-50 glass samples were carried out according to a scheme similar to the experimental setup for NdP_5O_{14} crystals (figure 8). The dependence of the output energy of a polished 2-mm-thick LGSI-50 glass plate (without immersion) on the pump energy is shown in figure 7. The energy of 10-ns pulses of the LGSI-50 glass laser reached 150 μJ. Pieces of LGSI-50 glass (0.8–2.0 mm thick) with different classes of surface finish were placed in cuvettes with a 2–3-mm gap filled with an cinnamaldehyde/benzyl alcohol immersion mixture. Laser pulses with an energy of up to 100 μJ were obtained in the same resonator configuration for all these samples in immersion, even for pieces with matted surfaces.

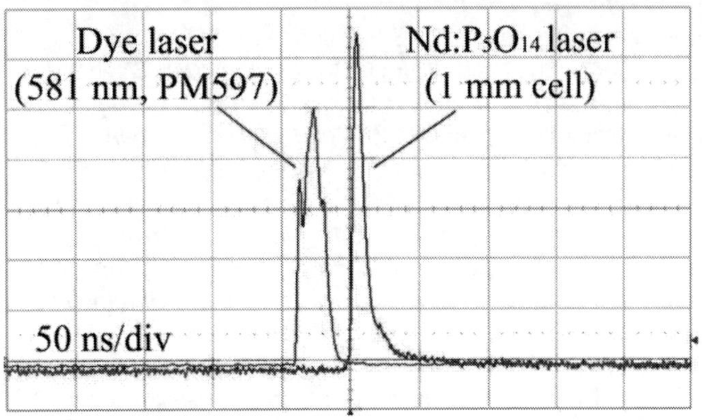

Figure 9. Oscillograms of the dye laser pulse and the pulse of the NdP_5O_{14} laser.

However, when cuvettes with LGSI-50 glass microparticles in immersion were installed in the resonator, the lasing threshold was not reached due to scattering losses. Similar results were obtained with particles of highly concentrated Nd-doped phosphate glass [15]. The source of losses in the glass powder slurry in addition to incomplete immersion can be various centers of scattering on the surface and inside the particles. Note that comparison of the transmittance of a 1÷3-mm-thick cuvettes with an immersion mixture and crushed glass particles ($T_{1.06}$ = 60÷80%) with the transmittance of the same cuvettes with polished glass beads (without activator) 0.7÷0.8 mm in diameter showed the advantage of the glass beads ($T_{1.06}$ = 90%). This provides grounds for fabricating Nd glass beads and conducting lasing experiments with them.

4. Conclusion

Active media based on particles of solid-state laser materials distributed in a liquid carrier may bring new solutions to the problem of heat removal in lasers due to active media circulation. The lasing characteristics of several active media based on particles of laser glass and crystals in an immersion liquid were measured upon pumping by nanosecond laser pulses in the gain switching regime. The technique of preparation and characterization of dense index-matched mixtures of solid-state microparticles and an immersion liquid (slurry) was tested for particles of LiF crystals with F_2 CCs, as well as for particles of laser glass and crystals doped with Nd^{3+} ions. The observed nanosecond pulses of LiF:F_2 emission near 680 nm in collimated beams from cuvettes with the slurry in a resonator and the dependences of the output energy on the pump energy give grounds for the conclusion that lasing from a dense active medium based on microparticles of a solid-state laser material in an immersion liquid (a slurry laser) was demonstrated. For cuvettes with Nd:glass powders in an immersion liquid installed in a resonator, the lasing threshold was not reached due to high scattering losses and insufficient gain. Lasing was obtained on unpolished samples of concentrated neodymium phosphate glass and on small neodymium pentaphosphate crystals in immersion liquids. The continuation of the study on the generation and amplification of cw and pulsed laser radiation in slurry-like active media seems reasonable. It is of interest to conduct experiments with such active media based on standardized particles, such as glass beads highly doped with Nd or Yb ions. Experiments with particles of LiF crystals with color centers should also be continued.

Acknowledgements
The authors thank B.N. Chichkov, B. I. Denker, E.A. Cheshev and A.A. Makovetsky for the help in organizing and conducting experiments. This study was supported by the Russian Foundation for Basic Research by the grant No. 18-02-00285 and partly by the grant No. 19-02-00344.

References

[1] Stefanov V 1989 *Infrared Phys.* **29** 381
[2] Perry M "Laser containing a slurry" US Patent, 2003/0161364 A1
[3] Tzuk Y, Goren Ch, Raanan D and Strum G 2012 *Opt. Letters* **37** 939
[4] Tzuk Y, Goren Ch, Strum G, Greenblatt J and Raanan D 2015 *Appl. Opt.* **54** 1157
[5] Yang X, Wang H, Yang Z and Xu X 2017 *Opt. Express* **25** 10574
[6] Burdukova O, Konyshkin V, Petukhov V, Senatsky Yu and Zverev P 2018 *Laser Phys. Lett.* **15** 095805
[7] Basiev T, Mirov S and Osiko V 1988 *IEEE J. Quantum Electron.* **24** 1052
[8] Mollenauer L F 1987 *Tunable Lasers* ed L F Mollenauer et al (New York: Springer) p 225
[9] Basiev T, Zverev P and Mirov S 2004 Color center lasers *Handbook of Laser Technology and Applications (Laser Design and Laser Systems* vol 2*)* ed C Webb et al (Bristol: Institute of Physics Publishing) ch B1.8 pp 499–522
[10] Basiev T T, Bykovsky N E, Konyushkin V A and Senatsky Yu V 2004 *Quantum Electron.* **34** 1138
[11] Men S, Liu Z, Cong Z, Rao H, Zhang S, Liu Y, Zverev P G, Konyushkin V A, Liu Z and Zhang X 2016 *Laser Phys.* **26** 025806

IOP Conf. Series: Materials Science and Engineering **848** (2020) 012100 doi:10.1088/1757-899X/848/1/012100

[12] Rao H, Cong Z, Zhang S, Liu Y, Men S, Liu Z, Zhang X, Zverev P G, Konyushkin V A, Yang H and Cheng W 2015 *Opt. Express* **23** 21884
[13] Baldacchini G, De Nicola E, Montereali R, Scacco A and Kalinov V 2000 *J. Phys. Chem. Solids* **61** 21
[14] http://refractiveindex.info/
[15] Denker B, Osiko V, Pashinin P and Prokhorov A 1981 *Soviet J. Quantum Electron.* **11** 289
[16] El'yashevich G, Kuryndin I and Rozova E 2017 *J. Opt. Technol.* **84** 481
[17] Heath H B 1981 *Source Book of Flavors: (AVI Sourcebook and Handbook Series* vol 2*)* (London: A Division of Albright & Wilson, Ltd.) p 221
[18] Ismailov I and Khalikov B 1978 *Soviet J. Quantum Electron.* **8** 537
[19] Cooper R 1983 *Appl. Optics* **22** 3070
[20] Hoffmann H-J and Steinhart R 1998 *Glastech. Ber. Glass Sci. Technol.* **71** 319.
[21] Exarhos G, Ferris K, Windisch C Jr, Bozlee B, Risser S and Swam S 2001 *Proc. SPIE* **4347** 432
[22] Weber H 1975 *Optical and Quantum Electron.* **7** 431
[23] Gaiduк M, Grigoryants V, Zhabotinsкy M, Makovetsky A and Tishchenкo R 1979 *Soviet J. Quantum Electron.* **9** 250

Ti-(15-25)Nb-5Ta ALLOY PLATE HARDNESS RESEARCH FOR MEDICAL APPLICATIONS

S V Konushkin[1], K V Sergienko[1], A S Baikin[1], A A Kolmakova[1], N V Berezina[1], A V Mikhailova[1], Y A Morozova[1], E P Balashov[2], A G Kolmakov[1] and M A Sevostyanov[1]

[1]A.A.Baikov Institute of Metallurgy and Materials Science, Russian Academy of Sciences
[2]State Scientific Institution "All-Russian Research Institute of Phytopathology", Russian Agricultural Academy",st. Institute, own 5, Bolshie Vyazemy, Odintsovo distr., Moscow reg., 143050, Russia

Abstract. This work is devoted to the study of the hardness of various compositions of TiNbTa alloys for medical applications in the hip joints. This TiNbTa alloy system has a shape memory effect and superelasticity. The paper describes the technology for producing plates 1 mm thick of TiNbTa alloys. Carrying out thermal and mechanical treatments of these alloys. The effect of Nb doping on the structure and hardness of TiNbTa alloys is shown.

1. Introduction

The desire to improve the quality of life of patients after injuries and extensive operations is the most important task facing physicians and materials scientists in connection with this, recent studies on the development of new nickel-free metallic medical alloys for the manufacture of implants and their parts [1-2], are widely used by research teams from many countries. Existing materials for restoration and replacement, in particular, fragments of bone tissue and joints have a number of significant drawbacks and require refinement, both in terms of corrosion resistance and biomechanical compliance with body tissues. At the moment, alloys such as stainless steel, Co-Cr-Mo, and Ti-6Al-4V alloys [3-4], which can withstand heavy loads and, unlike ceramic materials, are not brittle, are mainly used for the manufacture of an endoprosthesis [5]. However, there are several problems associated with the effectiveness of these implants, which lead to revision operations in 10-15 years [6-8]. It is known that the use of the alloys used is not optimal due to the high Young's modulus, which causes a discrepancy with the required mechanics for replacing bone tissue. Taking over most of the load on itself, the implant causes a decrease in the load on the bone, which in turn leads to a decrease in the density and strength of the bone, followed by the loss of strong adhesion of the bone and the implant, as well as a possible fracture of the bone. It is also known that Ti-6Al-4V alloy has a low tibological resistance caused by low adhesion of the surface oxide layer, which is prone to continuous fragmentation. One of the promising materials for the manufacture of bone implant parts and having the best range of properties in medical applications is the Ti-Nb-Ta alloy. The aim of the work was to study the hardness and microhardness of the surface of Ti-Nb-Ta alloys after rolling the alloy into sheets.

2. Materials and methods

IOP Conf. Series: Materials Science and Engineering **848** (2020) 012101 doi:10.1088/1757-899X/848/1/012101

Melting was carried out in a vacuum electric arc furnace with a copper water-cooled tray. As charge materials, iodide titanium, technically pure niobium and tantalum were used. The palletizing was carried out according to the principle of the gradient of the melting temperatures of the starting materials. Vacuum was carried out to a residual pressure of 10-2 mm RT. century, then the furnace chamber was filled with high purity argon to a pressure of 0.4 atm. To absorb impurities, at the first stage, a zirconium getter was melted in one of the holes of a copper water-cooled tray. Then, melt was made in other wells of the starting materials until a single ingot was formed. For uniform mixing of metals, at least 5 remelts were performed with ingots turning over after each remelting. The weight of the ingots is 20 g. Next, the ingots are remelted into one ingot weighing 180 grams with two remelts. For the final homogenization and elimination of the dendritic structure, homogenizing annealing in vacuum was carried out at a pressure of 10 ^ (- 5) mm. Hg. Art., within 12 hours at a temperature of 850 0C.

The primary deformation of cast billets with a thickness of 10 mm was carried out on a two-roll mill DUO-300 to a final billet thickness of 1 mm. The preforms were heated to a temperature of 600 ° C before deformation in a muffle furnace for 25 minutes.

As objects of study, we used 1 mm thick plates of the following compositions: Ti-15Nb-5Ta, Ti-20Nb-5Ta, Ti-25Nb-5Ta at.

Plates for further research were cut on an EDM machine. Samples for metallographic examination and hardness were prepared by sequential grinding after pressing on a Piatto diamond disk with a grain size of P120 for 5 minutes, P220 for 10 minutes, P600 for 15 minutes, on an Aka-Allegran-3 fine grinding diamond disk with DiaMaxx suspension Poly with a diamond particle size of 6 μm, on Akasel NAPAL velvet with a suspension of DiaMaxx Poly with a diamond particle size of 3 and 1 μm. The microstructure was detected by etching in a solution of hydrofluoric and nitric acids with distilled water in the ratio HF:2H2SO4:7H2O.

The investigations were carried out using a light microscope, which makes it possible to conduct observations using the bright and dark field methods, as well as in polarized light. A set of lenses allows you to get magnifications up to 2000.

Hardness measurements were carried out on an Instron Wolpert Wilson Instruments 930N universal hardness tester. In the work, the Vickers method was used according to GOST 2999-75, the essence of which is expressed in the indentation of a tetrahedral diamond pyramid with an angle of 136 ° between opposite faces into the prepared polished surface of the test sample, while the indentation force is 30 kgf and the load time is 10-15 seconds. After the indentation, the Vickers hardness is calculated from the size of the fingerprint after the indentation. For each sample, there were 5 measurement points.

3. Results

Images of the microstructure of the compositions Ti-15Nb-5Ta (Figure 1), Ti-20Nb-5Ta (Figure 2) and Ti-25Nb-5Ta (Figure 3) were obtained.

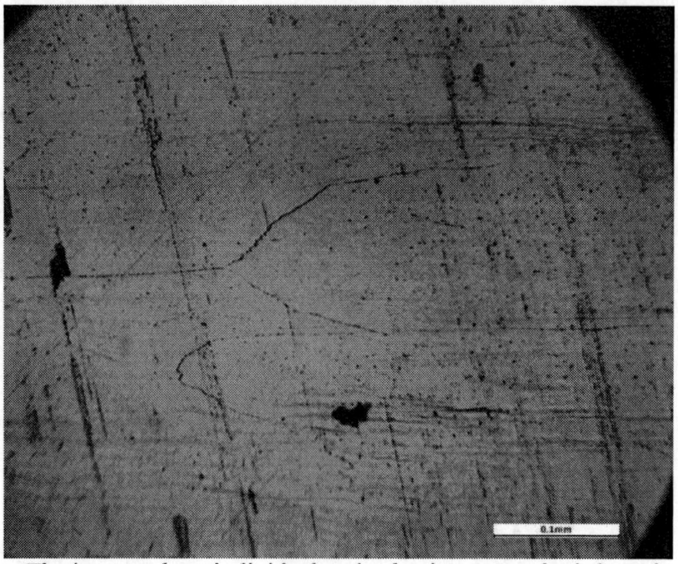

Figure 1. Ti-15Nb-5Ta.

The images show individual grains having a stretched shape in the rolling direction, while the aspect ratio of the grains is less than the degree of compression during rolling, which indicates partial recrystallization during heating before rolling. There are point inclusions of the alpha phase.

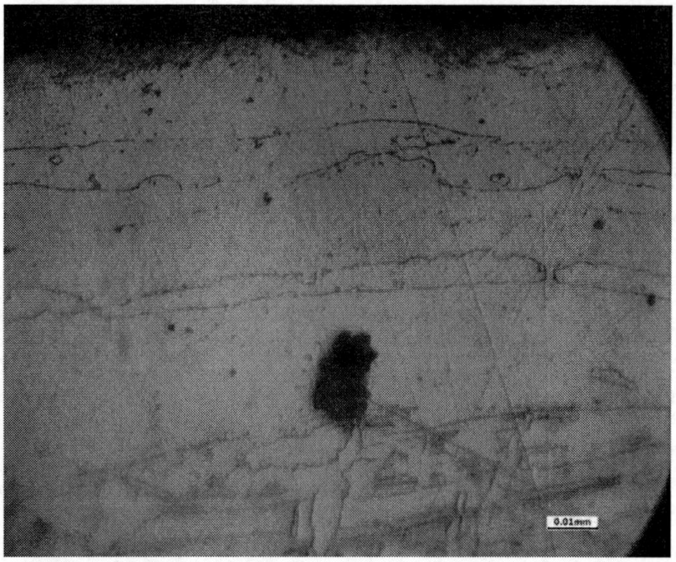

Figure. 2 Ti-20Nb-5Ta.

The images show individual grains having a very elongated shape in the direction of rolling, possibly a high voltage. There is also the formation of grains directed perpendicular to the rolling direction. There are point inclusions of the alpha phase.

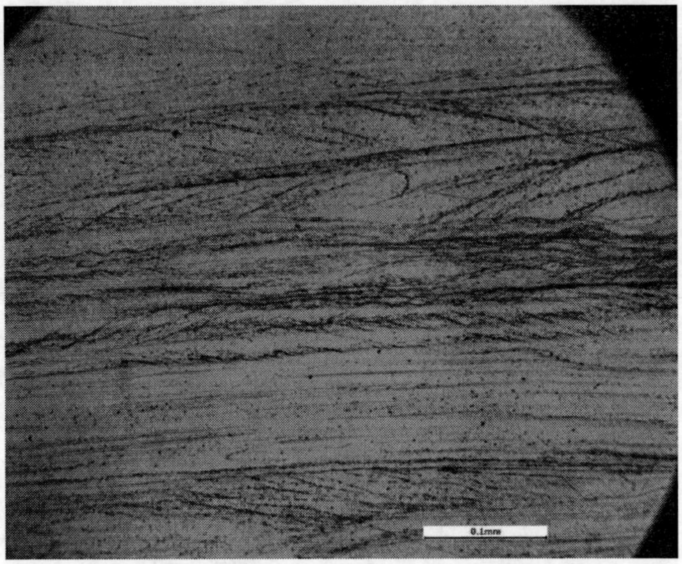

Figure 3. Ti-25Nb-5Ta.

The images show individual beta-phase grains representing the matrix, having a very elongated shape in the direction of rolling and dark bands of the alpha phase.

The hardness study was carried out on specimens prepared by polishing to a mirror shine, the temperature at which the studies were carried out was room temperature (23 °C), humidity 50%. Samples were not subjected to additional heat treatment except for those used for rolling. The obtained values for hardness are presented in table 1 and graph 1.

Table 1. The results of hardness of the alloy according to the method of Vickeras with a load of 30 kg.

Alloy composition	Greatness
Ti-15Nb-5Ta	173±2 HV
Ti-20Nb-5Ta	205±2 HV
Ti-25Nb-5Ta	175±2 HV

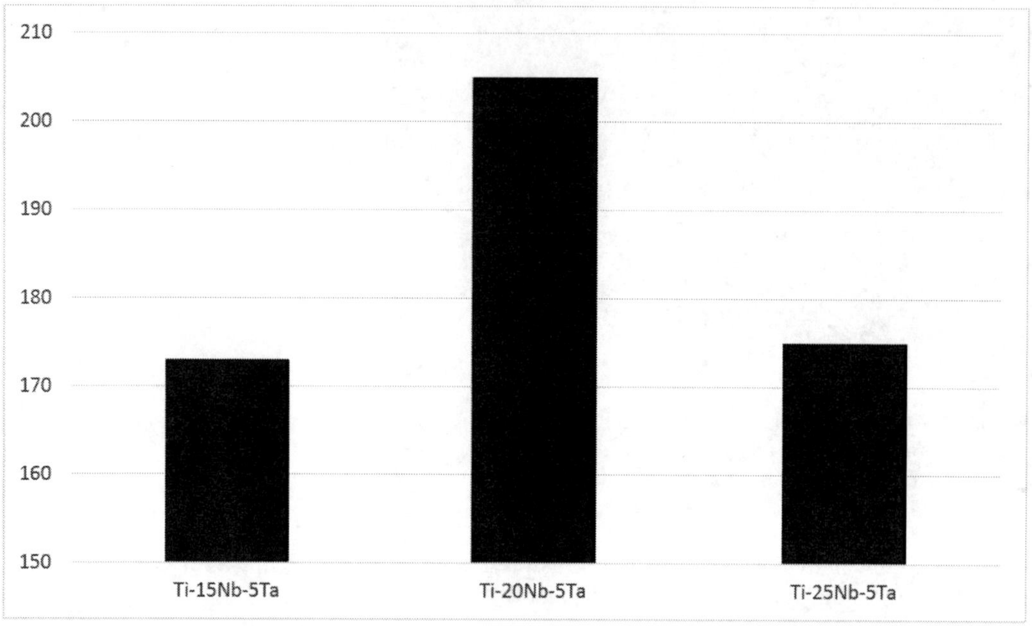

Graph 1. The hardness of the alloy according to the method of Vickeras with a load of 30 kg.

4. Conclusions

According to the data obtained, it can be noted that the hardness of the alloy is higher than the hardness of each initial component, so the hardness of titanium is 99 HV, niobium 135 HV and tantalum 85 HV. The highest hardness is possessed by samples of plates with the composition Ti-20Nb-5Ta, while the compositions Ti-15Nb-5Ta and Ti-25Nb-5Ta are comparable in hardness.

Acknowledgements

The work was supported by the Ministry of Education and Science of Russia (grant identifier RFMEFI60419X0242).

References

[1] M A Sevostyanov, A S Baikin, L A Shatova, E O Nasakina, A V Berezhnov, S V Gudkov, K V Sergienko, S V Konushkin, M I Baskakova and A G Kolmakov, 2018 *Biocompatibility of the Ti81Nb13Ta3Zr3 Alloy Doklady Chemistry* **482** 204-206 DOI: 10.1134/S0012500818090045

[2] M A Kaplan, M A Sevost'yanov, E O Nasakina, A S Baikin, K V Sergienko, S V Konushkin and A G Kolmakov 2018 Influence of the Surface Modification on the Mechanical Properties of NiTi (55.8 wt % Ni) Alloy *Wire for Medical Purposes Inorganic Materials: Applied Research* **9 (4)**, 751-756 DOI: 10.1134/S2075113318040159

[3] J V Giacchi, C N Morando, O Fornaro and H A Palacio 2010 Microstructural characterization of as-cast biocompatible Co-Cr-Mo alloys 2011 *Materials characterization* **62** 53-61 DOI: 10.1016/j.matchar.2010.10.011

[4] E Brandl, A Schoberth, and C Leyens 2012 Morphology, microstructure, and hardness of titanium (Ti-6Al-4V) blocks deposited by wire-feed additive layer manufacturing (ALM) *Materials Science and Engineering A* **532** 295-307 doi:10.1016/j.msea.2011.10.095

[5] E Askari, P Flores, D Dabirrahmani and R Appleyard Study of the friction-induced vibration and contact mechanics of artificial hip joints 2014 *Tribology International* **70** 1-10 doi.org/10.1016/j.triboint.2013.09.006

IOP Conf. Series: Materials Science and Engineering **848** (2020) 012101 doi:10.1088/1757-899X/848/1/012101

[6] J F Welter and V. M. Goldberg 2006 Bone grafts and bone substituents OKU *Hip and Knee Reconstruction 3. Rosemont : AAOS* 243-248

[7] I Yu Yezhov, A. A. Korytkin, S B Shchetinin, A V Shebashchev 2009 Experience of revision hip joint replacement in Nizhny Novgorod Research Institute of Nuclear *Surgery Endoprosthetics of large joints: materials scientific-practical. Conf.* 43-46

[8] N V Zagorodni, V I Makunin, A S Panteleeva, N G Zakharyan 2008 Analysis of revision endoprosthetics of the hip joint Osteosynthesis and endoprosthetics *materials of the Pirogov scientific-practical. conf.* 60-61

IOP Conf. Series: Materials Science and Engineering **848** (2020) 012102 doi:10.1088/1757-899X/848/1/012102

Corrosion resistance of nonnickel shape memory alloy

E O Nasakina[1], M A Sudarchikova[1], A M Tsareva[1], K V Sergiyenko[1], S V Konushkin[1], M A Kaplan[1], E P Balashov[2] and M A Sevost'yanov[1]

[1]Baikov Institute of Metallurgy and Material Sciences, 49 Leninsky Avenue, Moscow, Russia;
[2]State Scientific Institution "All-Russian Research Institute of Phytopathology", Russian Agricultural Academy",st. Institute, own 5, Bolshie Vyazemy, Odintsovo distr., Moscow reg., 143050, Russia.

Corresponding author's e-mail: nacakina@mail.ru

Abstract. Corrosion resistance of a Ti - (20-30) Nb - 5Zr shape memory alloy were studied. The structure and composition of the materials were determined using SEM and Auger electron spectrometry. Electro-chemical parameters and alloy dissolution in physiological modeling media were investigated. It has been shown that the alloys are quite corrosion-resistant: no dissolution and high Ebd potential were observed.

1. Introduction

Ni-free Ti-based alloys are currently being actively studied for its shape memory effect and superelasticity [1-6].

These alloys are considered potential candidates to replace nitinol, which is widely used in medicine, since the carcinogenic properties of nickel limit the subsequent medical use of this alloy [7-22]. The solution of this problem could be the creation of biocompatible and corrosion resistant surface layers [23]. However, in this case the question arises of the mutual adhesion of the components, especially in the case of variable dynamic loads. While obtaining the alloy only from non-toxic components does not imply this problem. It is only necessary to develop such an alloy, whose characteristics will be comparable with the mechanical properties of nitinol. Potentially mixtures based on the Ti-Nb-Zr system are useful in orthopedics and endovascular surgery.

The development of any material usually pursues the goal of obtaining a new facility with certain properties that are advantageous in terms of its intended use. In the modern world, the fields of science and technology, in connection with their rapid development, are increasingly intertwined, forming interdisciplinary directions, sometimes extremely difficult to separate from each other. Physical-chemical characteristics make it possible to evaluate the resistance of materials under the influence of an aggressive environment (water, oil, salt, alkali, acid, gas, fuel) during operation, including water and moisture absorption, chemical and heat resistance. In the case of metallic materials, this group can also combine chemical properties (oxidizability, solubility and corrosion resistance) and physical (color, specific gravity, density, fusibility, electrical conductivity, magnetic properties, thermal conductivity, and expandability when heated). As a result, a large-scale complex of operational properties is formed, the parameters of which will determine whether the alloy being developed is suitable for receiving minimally invasive medical implants (at the stage prior to biological research).

IOP Conf. Series: Materials Science and Engineering **848** (2020) 012102 doi:10.1088/1757-899X/848/1/012102

This work is devoted to a long-term study of the dissolution in static conditions of Ti–(20–30)Nb–5Zr shape memory alloys in the form of a thin wire and its electro-chemical parameters in modelling physiological media.

2. Materials and methods

The alloys of the Ti-Nb-Zr systems of the following compositions were studied: Ti-20Nb-5Zr, Ti-23Zr-5Zr, Ti-25Nb-5Zr, Ti-28Nb-5Zr, Ti-30Nb-5Zr (at.%). The alloys were smelted by the method of vacuum electric arc melting from pure charge materials. The ingots were annealed in vacuum at a temperature of 1000 °C for 12 hours and cooled with the furnace [6]. Then it was subsequently converted into bars and wire by rolling, forging and drawing. In order to improve the quality of the surface, wires in the initial state were subjected to successive grinding of the surface with emery paper of grit from 180 to 1000 and final processing with diamond paste to a mirror surface. In order to stabilize the structure and properties of the material, the wires were annealed at a temperature of 600 °C.

The morphology and layered elemental composition (including using transverse thin sections) of the surfaces of the materials were examined with a TESCAN VEGA II SBU scanning electron microscope (SEM) equipped with an INCA Energy dispersive energy analyzer, which also carried out fractographic studies of the samples, and electronic Auger spectrometer JAMP-9500F company JEOL in combination with ion etching when bombarded with argon at an angle of 30°.

The corrosion dissolution of the material was studied under static conditions by immersion in solutions of various acidity and composition. A neutral 0.9 mass % sodium chloride solution (NaCl, pH 6.31), artificial plasma (NaCl (92.3 mmol), NaHCO3 (26.3 mmol), K2HPO4 (0.9 mmol), KCl (2.7 mmol), NaH2PO4 (0.22 mM), CaCl2 (2.5 mM), MgSO4 • 7H2O (0.82 mm), Na2SO4 (1.48 mm), D-glucose C6H12O6 (1 g / l), pH 7.36) and standard buffer solution for reproducing at a given level of an acidic medium prepared from the corresponding mixture of the Merk company (potassium tetraoxalate KH3C4O8x2H2O, 0.05 M, pH 1.68) were used.

Samples of each type in the form of a 32.6 g wire were placed in polypropylene flat-bottomed flasks filled with 100 ml of the selected solution and kept for 30 days at 37 °C, making samples after 6, 13, 21, 30 and 60 days, respectively.

For the experiments, Ti-Nb-Zr alloy wires of 5 compositions were used in 3 states: after drawing (sample 1), after polishing (sample 2) and after annealing (sample 3). After a selected period of time, samples were taken from the solutions for analysis. The analysis was carried out on a sequential atomic emission spectrometer with induction plasma with the aim of using the ICP-AES method (inductively coupled plasma atomic emission spectrometry) for direct simultaneous determination of titanium, niobium, zirconium in solutions.

Electro-chemical corrosion indicators were investigated by the method of cyclic voltammetry (CVA) in a standard electrochemical cell using a universal potentiostat IPC-Pro. The potential sweep rate was 10 mV/s. Electrolyte was saline 0.9% NaCl at temperature 20 °C. Determination of corrosion parameters (stationary potential Ecorr, breakdown potential Ebd and repassivation potential Er) was carried out. The test samples were in the form of a wire with a diameter of 0.028 cm as the working electrodes. A glass graphite ring counter electrode was used as an auxiliary, and a saturated silver chloride silver electrode served as a reference electrode. The surface of the samples before conducting the experiments was treated with ethyl alcohol and washed with distilled water. Samples only after grinding the surface were investigated due to the need for surface homogeneity for analysis.

Scanning of CVA was started with a potential of -1.40 V and was carried out until a amperage of no more than 5 mA/cm^2 was achieved, based on the recommendations of the standard. Scanning in the opposite direction led to the value of the repassivation potential. The current density was measured in mA/cm^2.

3. Results and discussion

NMPT-5 IOP Publishing
IOP Conf. Series: Materials Science and Engineering **848** (2020) 012102 doi:10.1088/1757-899X/848/1/012102

During immersion tests, there was no metal yield from the solutions used, or was below the detection limit of the device (0.01 mg / l), which allows us to call the alloys highly corrosion-resistant. A slight dissolution, slowing down with time due to surface repassivation, was noted in an acidic solution, which is presumably connected with the titanium and tetraoxalate complex formation. There were no significant differences in the behavior of the alloys depending on the composition.

The results of electrochemical studies are shown in Figures 1-3.

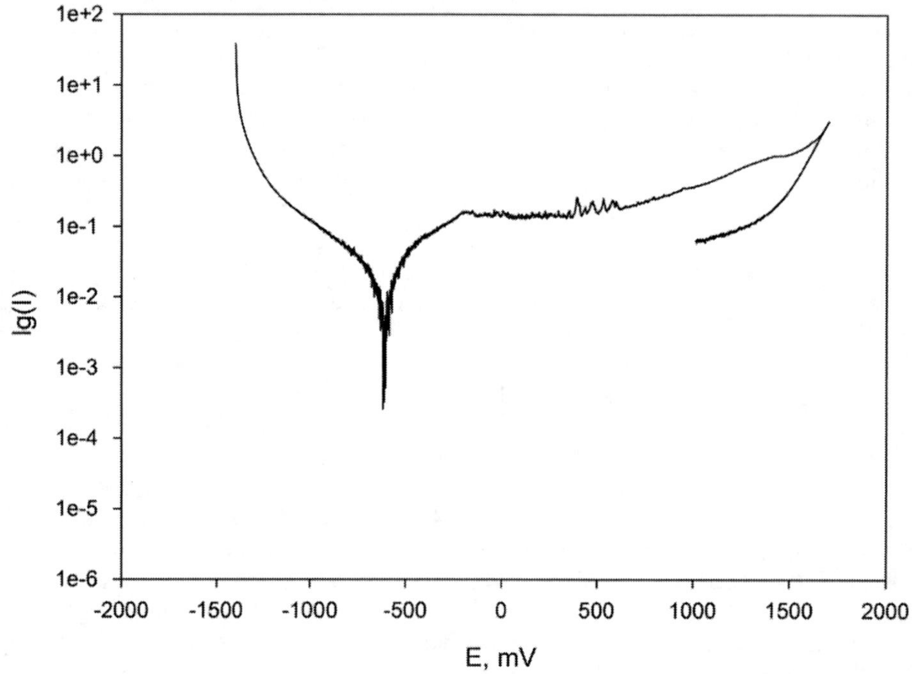

Figure 1. Cyclic voltammograms of Ti-25Nb-5Zr alloys in a 0.9% NaCl solution at a temperature of 20 °C.

IOP Conf. Series: Materials Science and Engineering **848** (2020) 012102 doi:10.1088/1757-899X/848/1/012102

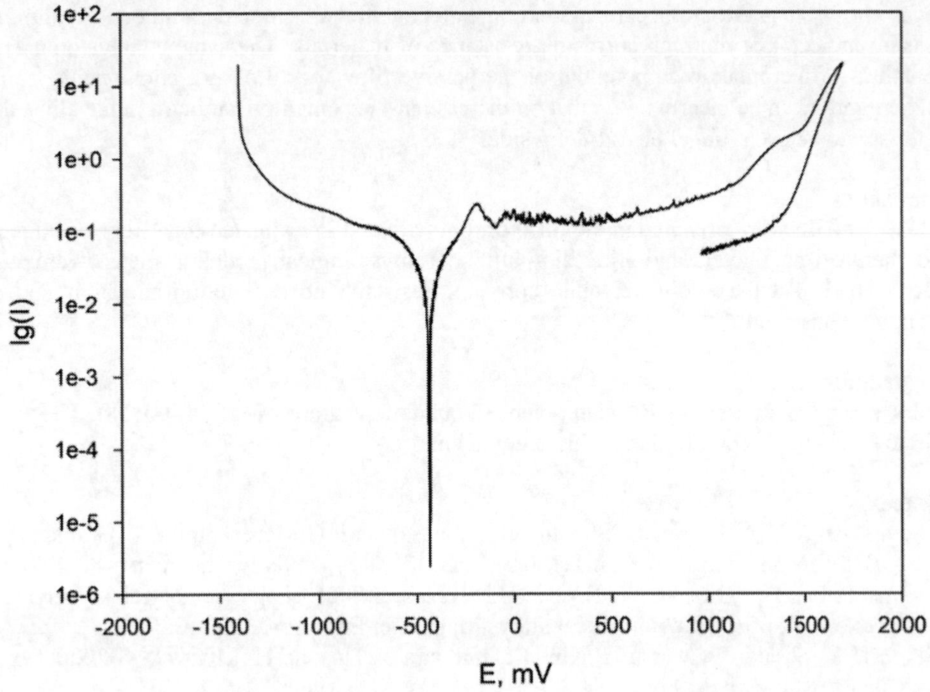

Figure 2. Cyclic voltammograms of Ti-20Nb-5Zr alloys in a 0.9% NaCl solution at a temperature of 20 °C.

Figure 3. Cyclic voltammograms of Ti-30Nb-5Zr alloys in a 0.9% NaCl solution at a temperature of 20 °C.

It was shown that the surface repassivation potential for all studied samples is within 1400 V, which is an indicator of the high corrosion resistance of materials. The same conclusion is supported by the values of the breakdown potential of the passive film, for all alloys, components from 500 V and higher, and it can be clearly noted that an increase in the content of niobium in the alloy shifts this characteristic to an even longer electro-long side.

4. Conclusions

Ti–(20–30)Nb–5Zr were prepared in the form of thin wire and investigated for its corrosion resistance. Electro-chemical parameters and alloy dissolution in physiological modeling media were studied. It has been shown that the alloys are quite corrosion-resistant: no dissolution and high Ebd and Ere potential were observed.

Acknowledgments

The work was supported by the Russian Science Foundation (grant № 18-79-00128). The authors are grateful to Kuznetsova O.G. for help with material analysis.

References

[1] Sevostyanov M A, Baikin A S, Shatova L A, Nasakina E O, Berezhnov A V, Gudkov S V, et al. 2018 Biocompatibility of the Ti81Nb13Ta3Zr3 Alloy Doklady Chemistry 482(1) 204–206.

[2] Kim J I, Kim H Y, Inamura T, Hosoda H,Miyazaki S 2005 Shape memory characteristics of Ti–22Nb–(2–8)Zr(at.%) biomedical alloys Mater. Sci.Eng. A 403 334–339.

[3] Kim H Y, Sasaki T, Okutsu K, Kim J I, Inamura T, Hosoda H, Miyazaki S 2006 Texture and shape memory behavior of Ti–22Nb–6Ta alloy Acta Mater 54 423–433

[4] Dubinskiy S M, Prokoshkin S D, Brailovski V, Inaekyan K, Korotitskiy A V, Filonov M R, Petrzhik M I 2011 Structure Formation During Thermomechanical Processing of Ti-Nb-(Zr, Ta) Alloys and the Manifestation of the Shape-Memory Effect PHYS MET METALLOGR 112(5) 529–542

[5] Sergienko K V, Sevost'yanov M A, Konushkin S V, Nasakina E O, Baikin A S, Shatova L A, Kolmakov A G 2018 Investigation of static properties of medical alloys Ti-(20-30)Nb-(10-13)Ta-5Zr *IOP Conf Ser: Mater Sci Eng* **347** 012049

[6] Konushkin S V, Baskakova M I, Leonov A V, Nasakina E O, Sudarchikova M A, Kolmakova A A, Bespamiatnova A, Sergiyenko K V, Leonova Yu O, Sevostyanov M A 2019 The structure of the alloy Ti - (20-30) Nb - 5Zr after smelting and homogenizing annealing *IOP Conf Ser: Mater Sci Eng* **525** 012060

[7] Davis J R. 2003 *Handbook of materials for medical devices* (ASM Int.) 350 p

[8] Nasakina E O, Sudarchikova M A, Sergienko K V, Konushkin S V, Sevost'yanov M A 2019 Ion Release and Surface Characterization of Nanostructured Nitinol during Long-Term Testing *Nanomaterials* **9(11)** 1569

[9] Petrini L, Migliavacca F 2011 Biomedical Applications of Shape Memory Alloys Journal of Metallurgy 2011 1-15

[10] Duerig T W, Melton K N, Wayman C M, Stockel D 1990 Engineering aspects of shape–memory alloys (Oxford: Butterworth Heinemann Ltd) pp 181 – 194

[11] Shabalovskaya S 1996 On the nature of the biocompatibility and medical applications of NiTi shape memory and superelastic alloys Bio Med Mater Eng 6 267 – 289

[12] Marjan Bahrami Nasab, Mohd Roshdi Hassan 2010 Metallic Biomaterials of Knee and Hip - A Review Trends in Biomaterials and Artificial Organs 24(1) 69 - 82

[13] Surdell D, Shaibani A, Bendok B, Eskandari M K 2007 Fracture of a Nitinol Carotid Artery Stent That Caused Restenosis J Vasc Interv Radiol 18(10) 1297–1299

[14] Bose A, Hartmann M, Henkes H A 2007 Novel, Self–Expanding Nitinol Stent in Medically Refractory Intracranial Atherosclerotic Stenosis: Wingspan Study Stroke 38 1531–1537

[15] Stoeckel D, Pelton A, Duerig T 2004 Self-expanding nitinol stents: material and design

IOP Conf. Series: Materials Science and Engineering **848** (2020) 012102 doi:10.1088/1757-899X/848/1/012102

considerations Eur Radiol 14 292-301

[16] Xiaoying Lu, Xiang Bao, Yan Huang, Yinghua Qu, Huiqin Lu, Zuhong Lu 2009 Mechanisms of cytotoxicity of nickel ions based on gene expression profiles Biomater 30 141–148

[17] Uo M, Watari F, Yokoyama A, Matsuno H, Kawasaki T 1999 Dissolution of nickel and tissue response observed by X-ray scanning analytical microscopy Biomater 20 747–755

[18] Wataha J, O'Dell N, Singh B, Ghazi M, Whitford G, Lockwood P 2001 Relating nickel-induced tissue inflammation to Ni release in vivo J. Biomed. Mater. Res. 58 537–544

[19] Balazic M, Kopac J 2007 Improvements of medical implants based on modern materials and new technologies J. Achiev. Mater. Manuf. Eng 25(2) 31-34

[20] Kim J H, Shin J H, Shin D H, Moon M W, Park K, Kim T H et al. 2011 Comparison of diamondlike carbon-coated nitinol stents with or without polyethylene glycol grafting and uncoated nitinol stents in a canine iliac artery mode Br J Radiol 84 210-215

[21] Tomić S, Rudolf R, Brunčko M, Anžel I, Savić V, Čolić M 2012 Response of monocyte-derived dendritic cells to rapidly solidified nickel-titanium ribbons with shape memory properties Eur. Cell. Mater. 23 58–81

[22] Chun-Che Shih, Shing-Jong Lin, Yuh-Lien Chen, Yea-Yang Su, Shiau-Ting Lai, Gaston J. Wu, Ching-Fai Kwok, Kwok-Hung Chung. 2000 The cytotoxicity of corrosion products of nitinol stent wire on cultured smooth muscle cells J. Biomed. Mater. Res. 52 395–403

[23] Sevost'yanov M A, Nasakina E O, Baikin A S, Sergienko K V, Konushkin S V, Kaplan M A et al. 2018 Biocompatibility of new materials based on nano-structured nitinol with titanium and tantalum composite surface layers: experimental analysis in vitro and in vivo *J. Mater. Sci. – Mater. Med.* **29** 33

IOP Conf. Series: Materials Science and Engineering **848** (2020) 012103 doi:10.1088/1757-899X/848/1/012103

Research of mechanical properties of composite material based on titanium nickelide with a titanium surface layer depending on magnetron sputtering time

E O Nasakina[1], M A Sudarchikova[1], A M Tsareva[1], K V Sergiyenko[1], S V Konushkin[1], M A Kaplan[1], A S Baikin[1], L A Shatova[2] and M A Sevost'yanov[1]

[1]Baikov Institute of Metallurgy and Material Sciences, 49 Leninsky Avenue, Moscow, Russia;

[2]Voronezh State Technical University, st. 20-letiya Oktyabrya, 84/4, Voronezh, Russia.

Corresponding author's e-mail: nacakina@mail.ru

Abstract. The regularities of the titanium surface and transitional monolayers formation using magnetron sputtering and ion bombardment depending on the deposition time, adhesion parameters and mechanical characteristics are investigated. Saturation regions of the surface and transition layers are noted. The transition layer provides high adhesion of the surface layer to the substrate. The application of a titanium surface layer to a nitinol reduces elongation and yield strength but increases tensile strength, and as the deposition time increases, the elongation decreases very slightly, and the yield strength and strength increase.

1. Introduction

In recent decades, composite materials have become more widespread and popular due to their much-increased capabilities compared to classical materials. The combination of several components in one object, forming a single, connected volume, and at the same time bearing each of their functional loads, allows not only to increase the characteristics of the future product but also to combine the properties of several traditional materials that were previously incompatible. Including layered composites, which are widely used in electronics, optics, medicine, in the creation of structural and other functional objects, can be effectively obtained using vacuum ion-plasma methods. One of such methods is magnetron sputtering, which contributes to the production of thin films and surface layers of a very different nature and thickness on the basis (substrates) of almost any composition and geometric configuration.

In this case, the parameters of the obtained layers are directly related to the process parameters, which can vary in a rather wide range: the time of sputtering, the relative position of the sputtered target from the material of the formed layer and the substrate onto which the layer is applied, in space, which determines the intensity of the resulting flow along with the power the process, the nature of the sprayed material and the condition of the surface on which it is applied, including determined by the phase composition of this surface. At the same time, the formation of new materials with a complex structure contributes to the multifunctionality of products made from them.

Nitinol exhibits biomechanical compatibility with living tissues, but also a controversial level of corrosion resistance and biocompatibility [1-16]. Titanium itself has high corrosion resistance and ductility, biocompatibility and a coefficient of thermal expansion close to nitinol, therefore it can be effectively used as a surface layer in a layered composition with nitinol. Moreover, these properties, among other things, depending on the structure of the material, which depends on the formation parameters.

The purpose of the work was to study the mechanical properties of such a composition with different lengths of the surface layer.

2. Materials and methods

As a basis (substrate) of the formed composite materials used wire diameter of 280 μm samples of titanium nickelide composition of 55.91 wt. % Ni - 44.03 wt.% Ti (hereinafter TiNi) after sequential grinding of the surface with sandpaper with a grain size of 180 to 1000 grit and diamond paste (6-1 μm) to improve the surface quality. To clean, activate and polish the surface of the substrate, argon ions were bombarded with discharge parameters Ue = 900 V, Ie = 80 mA - preliminary ion etching. Chemically pure titanium was used as the surface layer material, which forms a single phase regardless of the process parameters, which facilitates the identification of the results.

The creation of metal composite materials was carried out by forming surface layers using a direct current magnetron (~ 865 mA) at a voltage of ~ 400 V in an argon gas medium at a working and residual pressure of ~ 0.4 and $4x10^{-4}$ Pa, respectively, during the spraying time 5 -30 min at a spraying distance of 150 mm. The temperature on the surface of the substrates did not exceed 150 °C.

The morphology and layer-by-layer elemental composition (including using transverse sections) of the surface of the materials was studied using a TESCAN VEGA II SBU scanning electron microscope (SEM / SEM) equipped with an INCA Energy attachment for energy-dispersive analysis, and JEOL Auger electronic spectrometer JAMP-9500F combined with ion etching during argon bombardment at an angle of 30 °. The X-ray diffraction spectra of the coating samples were recorded on an UltimaIV X-ray diffractometer (Rigaku, Japan) with a vertical goniometer and a high-speed D/teX semiconductor detector in CuKα radiation according to the Bragg – Brentano method and oblique shooting with a fixed angle turning the x-ray tube. Phase analysis of coating samples was performed in the PDXL software package using the ICDD database.

The mechanical properties of the studied samples were determined under conditions of static tension on a mechanical test machine INSTRON 3382, with a test speed of not more than 2 mm/min. The system complies with all European standards. Five samples were tested per experimental point. The yield strength, tensile strength and elongation were determined. Cyclic tests were conducted.

3. Results and discussion

Composite materials were obtained: "oxide layer (the region at the very boundary of the solid with the surrounding gas medium, free from substrate elements, where the titanium content is not at a maximum, about 20 nm thick) – the surface layer of the deposited substance – a transition layer containing elements of both a surface layer and basis – the basis" (Figure 1). The general pattern of changes in the composition of the composites obtained in this work is approximately the same in-depth.

With an increase in deposition time (Figure 2), the thickness of the surface and transition layers increases nonlinearly. Saturation regions of the surface and transition layers are noted, after which the thickness practically does not change with changing conditions.

Externally, the surface layer repeats the morphology of the substrate. Even with a short spraying time (5 min), uniform films are formed that do not differ from those obtained with a longer time. According to previous studies, at first, the formation of the layer has an island character, and then a more uniform distribution of the deposited substance on the surface occurs, while the atomized atoms (which acquire additional energy when sputtering from the target) are constantly mixed with surface atoms (substrates or previously deposited), which causes formation of a transition layer. It can be concluded that with a layer thickness of about 300 nm or more, the islands are smoothed in this work.

IOP Conf. Series: Materials Science and Engineering **848** (2020) 012103 doi:10.1088/1757-899X/848/1/012103

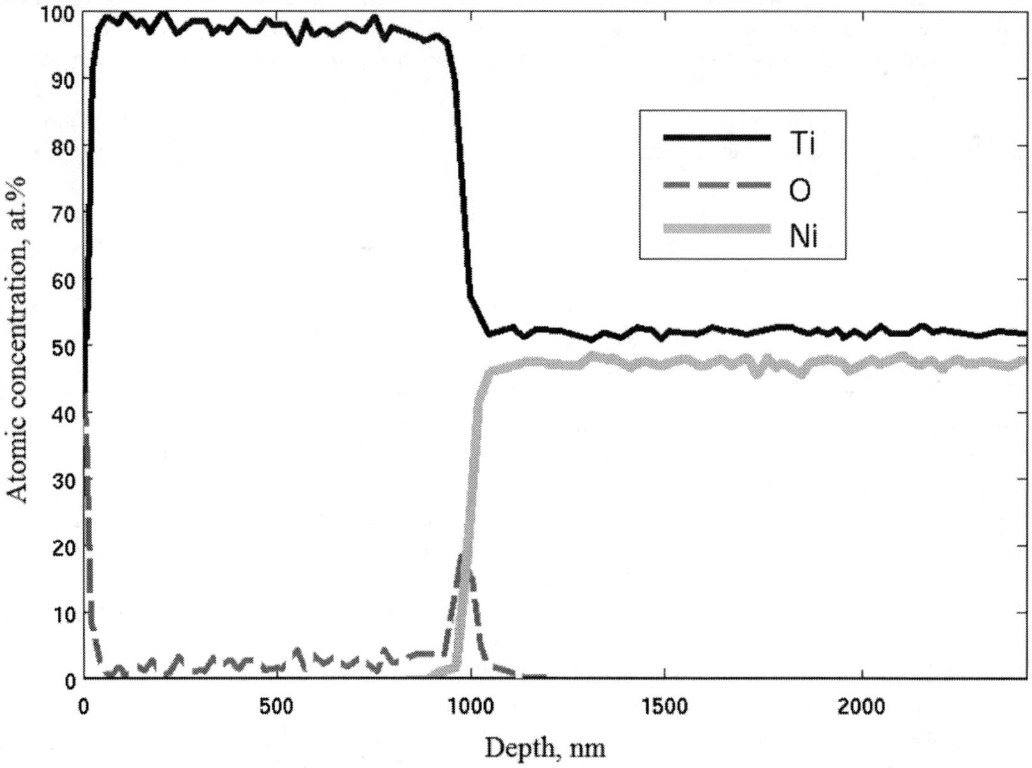

Figure 1. Layered composition surface morphology of a composite material with a titanium surface layer obtained in 30 min at a distance of 150 mm, a current of 865 mA and a voltage of 400 V, and a wire base made of nitinol.

X-ray diffraction patterns of samples with a titanium layer (Figure 3) show that the phase composition of the substrate does not change depending on the duration of the deposition of the layer. Regardless of the thickness of the layer, its phase composition also remains unchanged and is represented by beta-titanium, and does not correspond to the phase composition of the cast sputtering target.

Figure 4 shows a view of fractures of wire samples. It was noted that the layers had very high adhesion to the substrate and each other, breaking down as a single material and not peeling off except for the fracture zone. The destruction of all samples occurred almost identically. The fracture surface was oriented perpendicular to the tensile axis, representing a set of different-sized pores of a viscous fracture. Preliminary ion etching promotes better adhesion by promoting the formation of a uniform transition layer between the substrate and the deposited surface layer.

The results of the study of the mechanical properties of the composites are shown in Table 1. The application of a titanium surface layer to a titanium nickelide wire decreases the elongation and yield strength but increases the tensile strength, and as the deposition time increases, the elongation decreases very slightly, and the yield and strength limits increase.

According to the results of cyclic tests (Figure 5), we can conclude that delay and superelasticity are manifested in all samples. However, when applying a metal surface layer to a TiNi wire by magnetron sputtering, the flow stress decreases. Presumably, this occurs as a result of local heating of the surface layer and structural changes in the surface layers, including an increase in the initial fraction of martensite, which leads to a transition at lower loads. Also, the relaxation of residual stresses on the surface of the sample can occur.

IOP Conf. Series: Materials Science and Engineering **848** (2020) 012103 doi:10.1088/1757-899X/848/1/012103

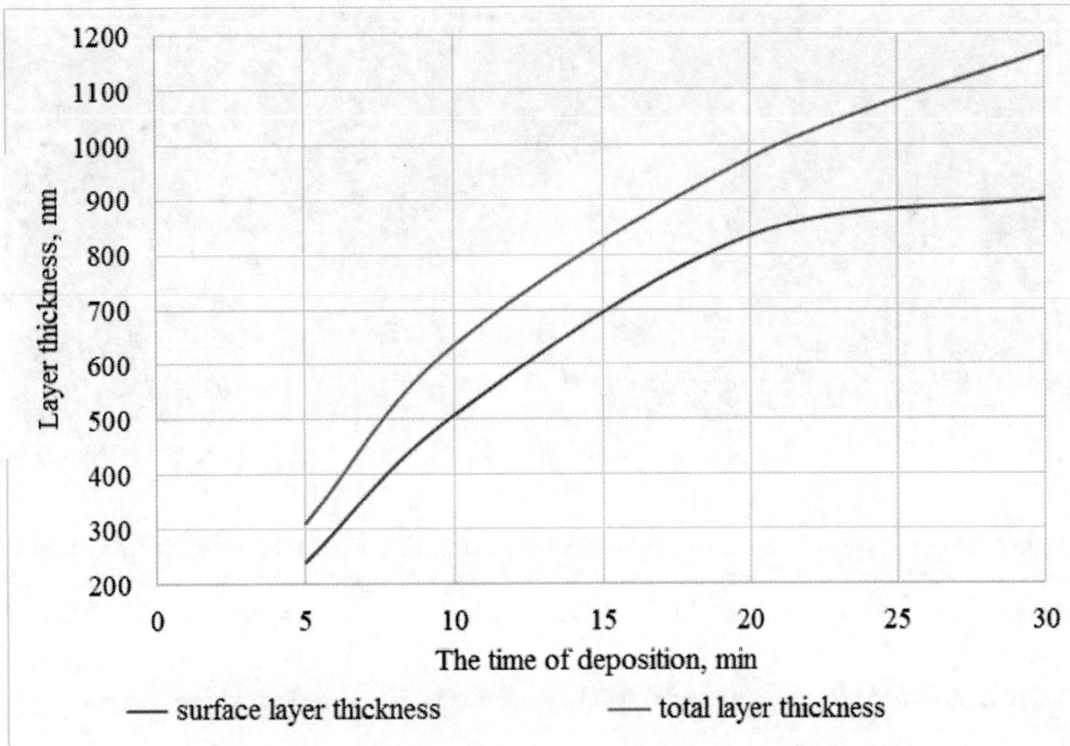

Figure 2. Dependence of the thicknesses of the surface and transition layers of a composite obtained in 30 min at a current of 865 mA and a voltage of 400 V, depending on the spraying distance.

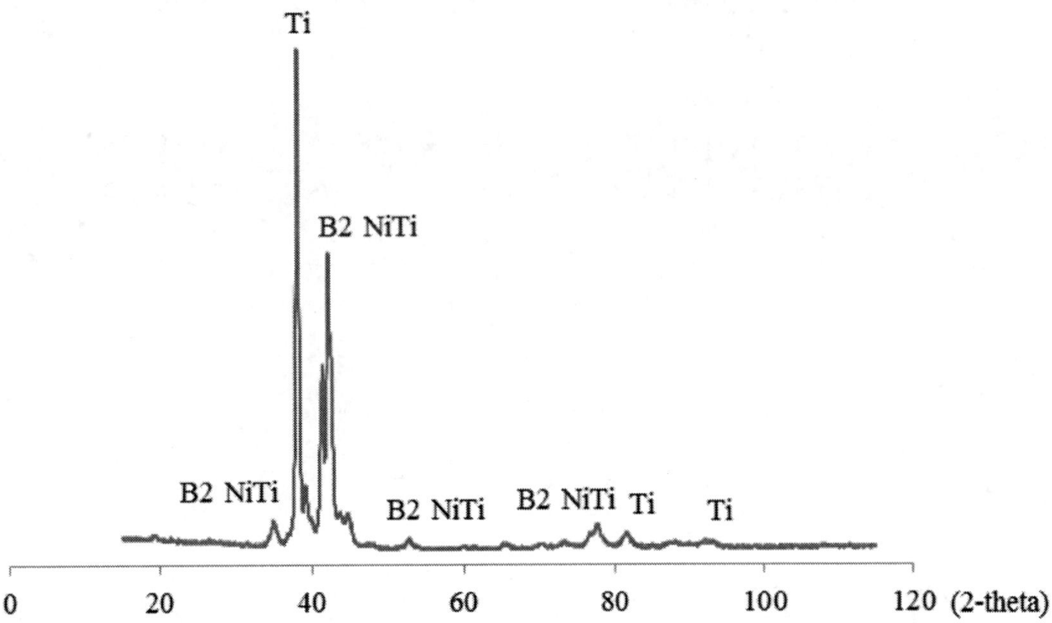

Figure 3. Composite material with a surface layer of titanium, obtained at a distance of 150 mm for 30 minutes.

IOP Conf. Series: Materials Science and Engineering **848** (2020) 012103 doi:10.1088/1757-899X/848/1/012103

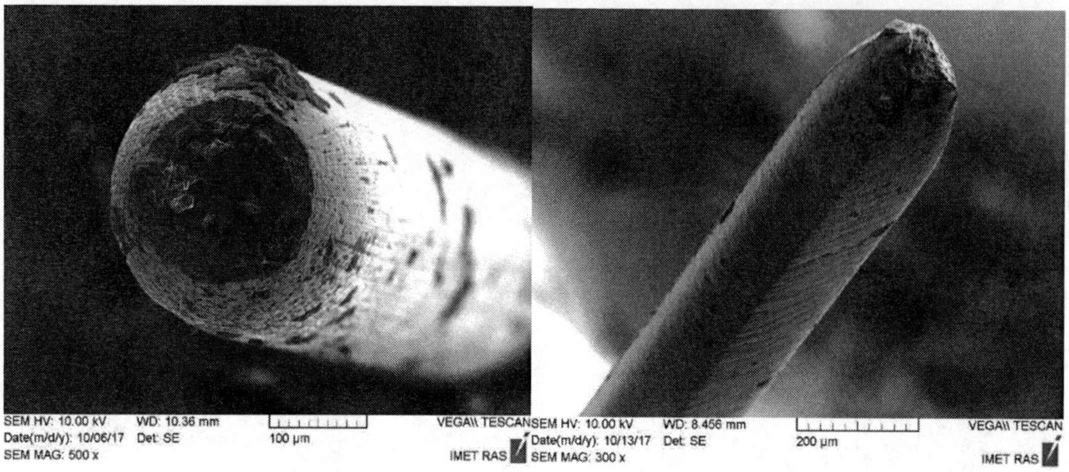

a b

Figure 4. SEM image of composite wires with a titanium (a) and tantalum (b) surface layers obtained in 30 min at a distance of 150 mm, a current of 865 mA and a voltage of 400 V, on a wire base of nitinol.

Table 1. Mechanical characteristics of a nitinol substrate and a composite with a titanium surface layer.

Sample	Elongation, %	Yield Strength (MPa)	Tensile strength (MPa)
As-received TiNi	7.57	661	1240
Polished TiNi	7.99	706	1242
Ti@NiTi 5 min.	6.64	447.00	1154.00
Ti@NiTi 10 min.	6.43	460.33	1452.00
Ti@NiTi 20 min.	6.29	480.00	1461.50
Ti@NiTi 30 min.	6.24	520.33	1475.6

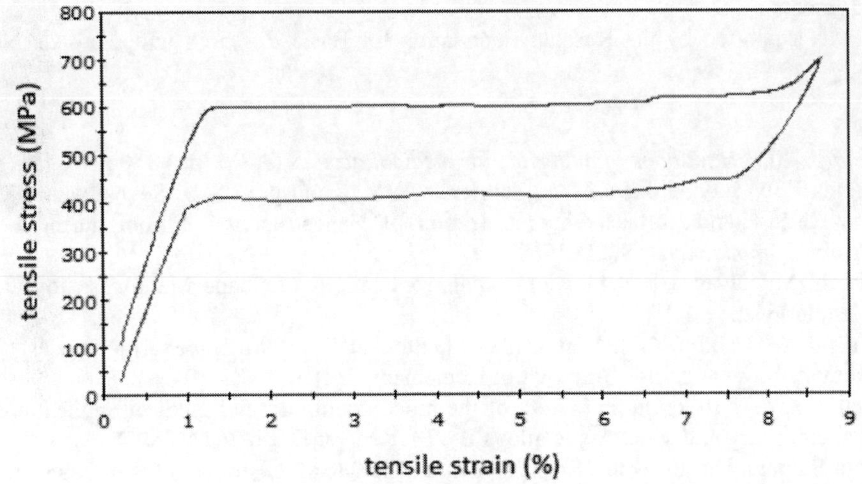

a

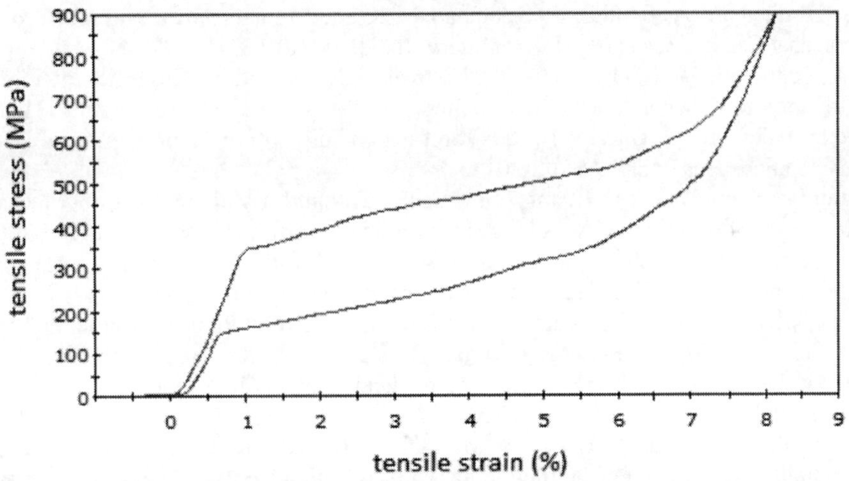

b

Figure 5. The results of cyclic mechanical tests for a) nitinol and b) a composite obtained in 30 minutes.

4. Conclusions

The regularities of the formation of surface and transitional monolayers of titanium using magnetron sputtering and ion bombardment depending on the time of deposition are investigated, and adhesion parameters and mechanical characteristics of all the materials obtained are determined.

The metal composite materials had the structure "oxide layer – the surface layer of the deposited substance – a transition layer containing elements of both a surface layer and basis – the basis". The surface morphology corresponds to the surface morphology of the substrate.

As the sputtering time increases, the thickness of the metal surface layer increases nonlinearly. Saturation regions of the surface and transition layers are noted, after which the thickness practically does not change with changing conditions. The transition layer provides high adhesion of the surface layer to the substrate.

The application of a titanium surface layer to a titanium nickelide wire reduces elongation and yield strength but increases tensile strength, and as the deposition time increases, the elongation decreases very slightly, and the yield strength and strength increase.

Acknowledgments
This work was supported by the Russian Foundation for Basic Research under project No. 19-08-00642 A.

References

[1] Davis J R. 2003 *Handbook of materials for medical devices* (ASM Int.) 350 p
[2] Nasakina E O, Sudarchikova M A, Sergienko K V, Konushkin S V, Sevost'yanov M A 2019 Ion Release and Surface Characterization of Nanostructured Nitinol during Long-Term Testing *Nanomaterials* **9(11)** 1569
[3] Petrini L, Migliavacca F 2011 Biomedical Applications of Shape Memory Alloys Journal of Metallurgy 2011 1-15
[4] Duerig T W, Melton K N, Wayman C M, Stockel D 1990 Engineering aspects of shape–memory alloys (Oxford: Butterworth Heinemann Ltd) pp 181 – 194
[5] Shabalovskaya S 1996 On the nature of the biocompatibility and medical applications of NiTi shape memory and superelastic alloys Bio Med Mater Eng 6 267 – 289
[6] Marjan Bahrami Nasab, Mohd Roshdi Hassan 2010 Metallic Biomaterials of Knee and Hip - A Review Trends in Biomaterials and Artificial Organs 24(1) 69 - 82
[7] Surdell D, Shaibani A, Bendok B, Eskandari M K 2007 Fracture of a Nitinol Carotid Artery Stent That Caused Restenosis J Vasc Interv Radiol 18(10) 1297–1299
[8] Bose A, Hartmann M, Henkes H A 2007 Novel, Self–Expanding Nitinol Stent in Medically Refractory Intracranial Atherosclerotic Stenosis: Wingspan Study Stroke 38 1531–1537
[9] Stoeckel D, Pelton A, Duerig T 2004 Self-expanding nitinol stents: material and design considerations Eur Radiol 14 292-301
[10] Xiaoying Lu, Xiang Bao, Yan Huang, Yinghua Qu, Huiqin Lu, Zuhong Lu 2009 Mechanisms of cytotoxicity of nickel ions based on gene expression profiles Biomater 30 141–148
[11] Uo M, Watari F, Yokoyama A, Matsuno H, Kawasaki T 1999 Dissolution of nickel and tissue response observed by X-ray scanning analytical microscopy Biomater 20 747–755
[12] Wataha J, O'Dell N, Singh B, Ghazi M, Whitford G, Lockwood P 2001 Relating nickel-induced tissue inflammation to Ni release in vivo J. Biomed. Mater. Res. 58 537–544
[13] Balazic M, Kopac J 2007 Improvements of medical implants based on modern materials and new technologies J. Achiev. Mater. Manuf. Eng 25(2) 31-34
[14] Kim J H, Shin J H, Shin D H, Moon M W, Park K, Kim T H et al. 2011 Comparison of diamondlike carbon-coated nitinol stents with or without polyethylene glycol grafting and uncoated nitinol stents in a canine iliac artery mode Br J Radiol 84 210-215
[15] Tomić S, Rudolf R, Brunčko M, Anžel I, Savić V, Čolić M 2012 Response of monocyte-derived dendritic cells to rapidly solidified nickel-titanium ribbons with shape memory properties Eur. Cell. Mater. 23 58–81
[16] Chun-Che Shih, Shing-Jong Lin, Yuh-Lien Chen, Yea-Yang Su, Shiau-Ting Lai, Gaston J. Wu, Ching-Fai Kwok, Kwok-Hung Chung. 2000 The cytotoxicity of corrosion products of nitinol stent wire on cultured smooth muscle cells J. Biomed. Mater. Res. 52 395–403

IOP Conf. Series: Materials Science and Engineering **848** (2020) 012104 doi:10.1088/1757-899X/848/1/012104

Study of the effect of the catalyst on the mechanical properties of RTV grade siloxane

A M Tsareva[1], E O Nasakina[1], M A Sudarchikova[1], E P Balashov[2], A S Baikin[1], L A Shatova[3] and M A Sevost'yanov[1]

[1]Baikov Institute of Metallurgy and Material Sciences, 49 Leninsky Avenue, Moscow, Russia;
[2]State Scientific Institution "All-Russian Research Institute of Phytopathology", Russian Agricultural Academy",st. Institute, own 5, Bolshie Vyazemy, Odintsovo distr., Moscow reg., 143050, Russia;
[3]Voronezh State Technical University, st. 20-letiya Oktyabrya, 84/4, Voronezh, Russia.

Corresponding author's e-mail: a.m.tsareva@yandex.ru

Abstract. A RTV grade siloxane film with a different percentage of catalyst was obtained. The mechanical properties of polymer films based on RTV grade siloxane are studied, as well as the effect of K-18 catalyst concentration on thickness, strength and ductility the properties of the obtained films was shown.

1. Introduction

Achievements of modern medicine are impossible without the fruitful work of scientists, engineers and doctors of various specialties. One of the important results of such a joint activity was the possibility of access to the organizers of various implant materials. Despite the achievements, the problem of the development and use of modern implantable surgical materials remains relevant.

Titanium alloys are widely used as materials for implants, stents, cava filters and other medical devices due to their high strength, low density and fracture toughness. However, from modern scientific research it is known that in direct contact with bones in a humoral environment, titanium alloys cannot form a bond with bone tissue and show obvious contact corrosion, which can lead to damage to the implant and its replicative replacement [1-2].

Polymers are complex chemicals that have found a wide range of applications. These compounds are also used in medicine. Due to their mechanical properties and biocompatibility, polymers have become indispensable materials for reconstructive surgery and prosthetics. The application of polymer films to the surface is one of the effective ways to solve the problems associated with corrosion resistance and biocompatibility. When used as barrier coatings on metals, siloxane provides adhesion to the surface of the metal coating [3-6].

Siloxanes are one of the most important groups of chemical compounds, which is associated with a number of their advantages. Siloxanes are generally considered non-toxic to humans and the environment. Many authors have proven that structure, degree of polymerization, molecular weight, lipophilicity or volatility affect the ability to cross the cell barrier, absorption into the body, migration into a living organism, the ability to accumulate, degradation and toxicity. Due to its unique physicochemical properties, siloxanes are indispensable in many areas of life. They are well tolerated and are an integral part of innovative treatments, healthcare and patient care [7-10]. However, various siloxanes, as well as their various physicochemical properties, require further research. Studies of effects on living organisms should relate to specific siloxanes, and not to the entire group. Siloxane is used as components or modifiers of organic resins or as the main binder component in protective coating compositions. Such coatings are highly resistant to elevated temperature, corrosion, abrasion, excellent biocompatibility and other quality characteristics that are obtained using silicone technology [11-12].

2. Materials and methods

The film-forming siloxane of the RTV brand is an excellent material for the manufacture of filled and unfilled adhesive coatings for stents, cava filters and other medical devices. The K-18 catalyst acted as a crosslinking agent for the RTV brand siloxane. Siloxane was mixed with 1-10% K-18 catalyst of the total mixture volume using automatic bags at a speed of 2.5 RPM second for 15-20 minutes. Then the resulting mixture was applied to a glass tray and evenly distributed using a glass rod. The resulting layer was dried in a fume hood for 24 hours. The film thickness is determined by the concentration of the catalyst in the solution and the volume of the deposited layer on the glass.

Strength tests of siloxane polymer films RTV were tested on a universal testing machine INSTRON 3382 with a loading speed of 10 mm / min. Samples of polymer films for testing were made according to GOST 14236-81, in the form of a double blade. The sample was fixed in the arms of the testing machine, which were uniformly tightened so that the sample did not slip during the test. Tests of polymer films with the definition of elongation, yield strength and tensile strength were carried out according to GOST 14236-81. Processing of the test results in determining the mechanical properties was carried out using the INSTRON Bluehill 2.0 software. The measurement error on the test machine is less than 1%.

5 samples were tested at the experimental point. The values of yield strength, tensile strength and elongation were determined.

3. Results and discussion

Table 1 presents the average results of mechanical tests of polymer films based on PTB grade siloxane with various catalyst contents.

Table 1. Mechanical properties of polymer films.

Siloxane volume, ml		The volume of catalyst,%	Relative elongation, %	Yield strength MPa	Tensile strength, MPa	Thickness, mm
1	50	1	42,52	0,044	0,125	0,710
2	50	2	55,93	0,042	0,247	0,710
3	50	3	34,09	0,046	0,155	0,718
4	50	4	29,03	0,045	0,128	0,715

5	50	5	34,61	0,040	0,132	0,743
6	50	6	39,09	0,035	0,142	0,750
7	50	7	46,01	0,052	0,204	0,820
8	50	8	41,97	0,047	0,206	0,737
9	50	9	43,49	0,045	0,221	0,735
10	50	10	36,19	0,059	0,200	0,720

Based on the results obtained, dependency graphs were constructed that can be used to track the trend of changes in mechanical characteristics (Figures 1-3).

Figure 1. The diagram of the dependence of tensile strength on the concentration of the catalyst K-18 in siloxane brand CRT.

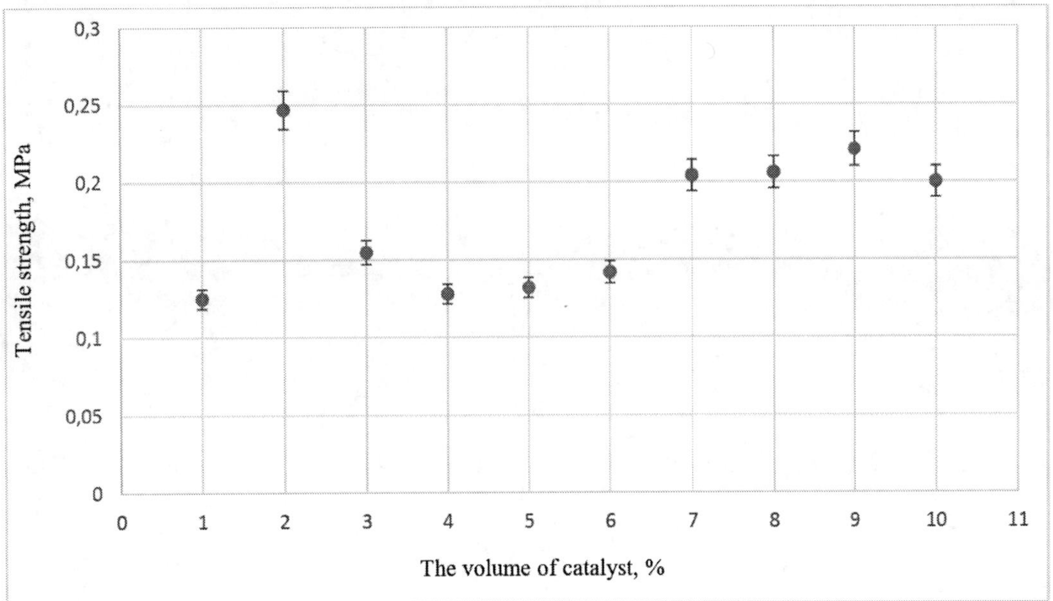

Figure 2. The diagram of the dependence of the yield strength on the concentration of the catalyst K-18 in siloxane brand CRT.

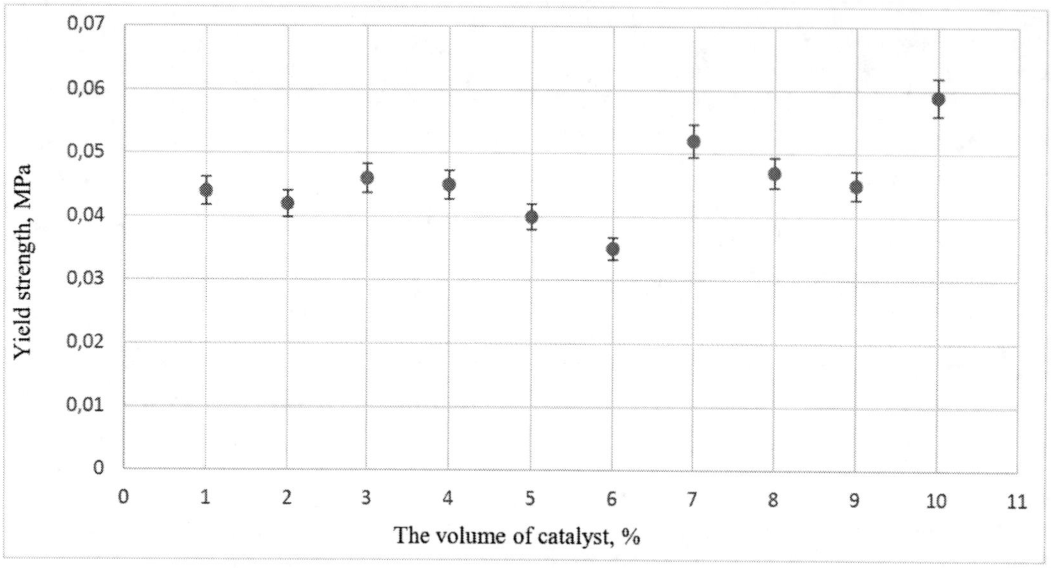

Figure 3. The dependence of the relative elongation on the concentration of catalyst K-18 in siloxane brand CRT.

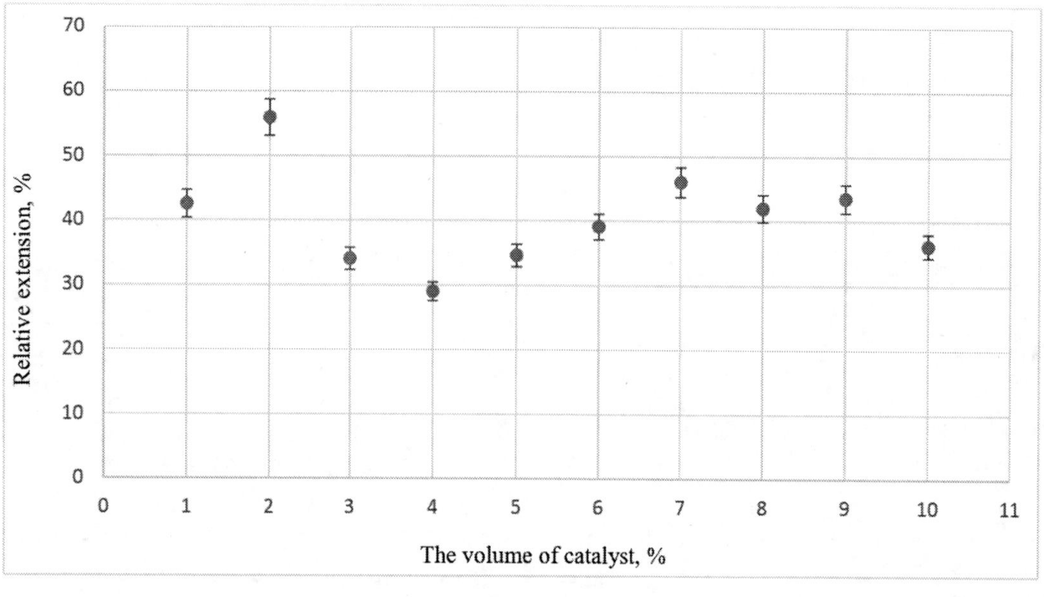

4. Conclusions

CRT grade siloxane with various catalyst contents was prepared in the form of a film. The mechanical properties of polymer films based on a siloxane type CRT with a K-18 catalyst were studied. We studied the effect of catalyst concentration on the mechanical properties of the polymer. A siloxane film sample with a catalyst content of 2% showed the best result. This sample demonstrates the following values: yield strength - 0.042 MPa, tensile strength - 0.247 MPa, relative elongation -

55.93%. However, it is worth noting that the percentage of K-18 catalyst does not significantly affect the final mechanical properties of CRT grade siloxane.

Acknowledgments
The work was supported by the Russian President Program for Young Scientists MK-1820.2020.8.

References

[1] M. Catauro, F. Bollino, F. Papale, R. Giovanardi, P. VeronesiCorrosion behavior and mechanical properties of bioactive sol-gel coatings on titanium implants, Mater. Sci. Eng. C, 43 (2014), pp. 375-382 (10/1)

[2] M.A.-H. Gepreel, M. NiinomiBiocompatibility of Ti-alloys for long-term implantation, J. Mech. Behav. Biomed. Mater., 20 (2013), pp. 407-415

[3] L.K. Wu, J.T. Zhang, J.M. Hu, J.Q. Zhang Improved corrosion performance of electrophoretic coatings by silane addition, Corros. Sci., 56 (2012), pp. 58-66

[4] M.Y. Jiang, L.K. Wu, J.M. Hu, J.Q. Zhang Silane-incorporated epoxy coatings on aluminum alloy (AA2024). Part 1: improved corrosion performance, Corros. Sci., 92 (2015), pp. 118-126

[5] Mojsiewicz-Pieńkowska K., Krenczkowska D. Evolution of consciousness of exposure tosiloxanes—review of publications, Chemosphere, Volume 191, January 2018, P. 204-217

[6] Nair B.P., Gangadharan Dh., Mohan N., Sumathi B., Nair Pr.D. Hybrid scaffold bearing polymer-siloxane Schiff base linkage for bone tissue engineering, Materials Science and Engineering: C, Volume 52, 1 July 2015, P. 333-342

[7] A. Franzen, C. Van Landingham, T. Greene, K. Plotzke, R. Gentry A global human health risk assessment for Decamethylcyclopentasiloxane (D5), Regul. Toxicol. Pharmacol., 74 (Suppl.) (2016), pp. 25-43, 10.1016/j.yrtph.2015.10.023

[8] A. Tiwari, M. Soucek (Eds.), Concise Encyclopedia of High Performance Silicones, Wiley-Scrivener Publishing, Beverly, MA (2014), pp. 243-251

[9] T.M. Tran, K.O. Abualnaja, A.G. Asimakopoulos, A. Covaci, B. Gevao, B. Johnson-Restrepo, et al., A survey of cyclic and linear siloxanes in indoor dust and their implications for human exposures in twelve countries, Environ. Int., 78 (2015), pp. 39-44, 10.1016/j.envint.2015.02.011

[10] Ch. Rücker, K. Kümmerer Environmental chemistry of organosiloxanes, Chem. Rev., 115 (1) (2015), pp. 466-524, 10.1021/cr500319v

[11] U.M. Eduok, M.M. Khaled Corrosion protection of non-alloyed AIAI 316L concrete steel metal grade in aqueous H2SO4: electroanalytical and surface analyses with Metiamide, Constr. Build. Mater., 68 (2014), pp. 285-290

[12] S.A. Umoren, U.M. Eduok Application of carbohydrate polymers as corrosion inhibitors for metal substrates in different media: a review, Carbohydr. Polym., 140 (2016), pp. 314-341

IOP Conf. Series: Materials Science and Engineering **848** (2020) 012105 doi:10.1088/1757-899X/848/1/012105

Study of the effect of the catalyst on the microstructure and film thickness of RTV siloxane

A M Tsareva[1], E O Nasakina[1], E P Balashov[2], A S Baikin[1], S V Konushkin[1] and M A Sevost'yanov[1]

[1]Baikov Institute of Metallurgy and Material Sciences, 49 Leninsky Avenue, Moscow, Russia;
[2]State Scientific Institution "All-Russian Research Institute of Phytopathology", Russian Agricultural Academy",st. Institute, own 5, Bolshie Vyazemy, Odintsovo distr., Moscow reg., 143050, Russia

Corresponding author's e-mail: a.m.tsareva@yandex.ru

Abstract. An RTV grade siloxane film with a different percentage of catalyst was obtained. The microstructure of the surface of polymer films based on PTB grade siloxane was studied and the effect of the K-18 catalyst concentration on the thickness and microstructure of the obtained films was shown.

1. Introduction

One of the common materials for medical products is metal. A stent is called an elastic metal structure in the form of a cylindrical frame placed in the lumen of hollow organs and blood vessels to expand the area narrowed by the pathological process [1]. For the manufacture of bare metal stents (BMS), stainless steel, alloys of cobalt and chromium, platinum, nickel, titanium, etc. are used [2]. The frequency of restenosis during implantation of stents of this type is 20–40% [3]. Antiproliferative-coated stents represent a significant progressive step in the fight against restenosis.

The development of stents with surfaces coated with polymeric materials is currently a promising area that can solve not only the problem of restenosis, but also increase the corrosion resistance and biocompatibility of a medical device [4-5].Polymers are complex chemicals that are widely used, including in medicine [6-7]. Siloxanes are one of the most important groups of chemical polymer compounds, which is associated with a number of their advantages. In many modern scientific studies, the authors proved that the structure, degree of polymerization, molecular weight, lipophilicity or volatility affect the ability to overcome the cell barrier, absorption into the body, migration into a living organism, the ability to accumulate, degradation and toxicity. Due to its unique physicochemical properties, siloxanes are indispensable in many areas of life. They have perfect biocompatibility, therefore they are an integral part of innovative methods of treatment, healthcare and patient care [8-15]. However, various siloxanes, as well as their various physicochemical properties, require further research. Studies of effects on living organisms should relate to specific siloxanes, and not to the entire polymer group of siloxanes. Siloxane is used as the main binder in protective coatings. Such coatings are resistant to elevated temperatures, corrosion, abrasion, excellent biocompatibility and other qualities that coatings acquire through the use of silicone technology [16-17].

IOP Conf. Series: Materials Science and Engineering **848** (2020) 012105 doi:10.1088/1757-899X/848/1/012105

2. Materials and methods

The film-forming siloxane of the RTV brand is an excellent material for the manufacture of filled and unfilled adhesive coatings for stents, cava filters and other medical devices. The K-18 catalyst acted as a crosslinking agent for the RTV brand siloxane. Siloxane was mixed with 1-10% K-18 catalyst of the total mixture volume using automatic bags at a speed of 2.5 RPM second for 15-20 minutes. Then the resulting mixture was applied to a glass tray and evenly distributed using a glass rod. The resulting layer was dried in a fume hood for 24 hours. The film thickness is determined by the concentration of the catalyst in the solution and the volume of the deposited layer on the glass.

The surface morphology of the obtained CRV siloxane films was studied using a Neophot-2 optical microscope manufactured by Carl Zeiss Jena. For research, lenses were used, which made it possible to obtain an increase of 10 and 40.

3. Results and discussion

Table 1 presents the average results of the obtained polymer films based on CRV grade siloxane with various catalyst contents. 5 samples were tested at each experimental point.

Table 1. The effect of the catalyst content on the thickness of the polymer film of the siloxane grade CRV.

Siloxane volume, ml		The volume of catalyst,%	Thickness, mm
1	50	1	0,710
2	50	2	0,710
3	50	3	0,718
4	50	4	0,715
5	50	5	0,743
6	50	6	0,750
7	50	7	0,820
8	50	8	0,737
9	50	9	0,735
10	50	10	0,720

Figures 1-2 show surface photographs of the obtained CRV siloxane films with various catalyst contents from 1 to 10%.

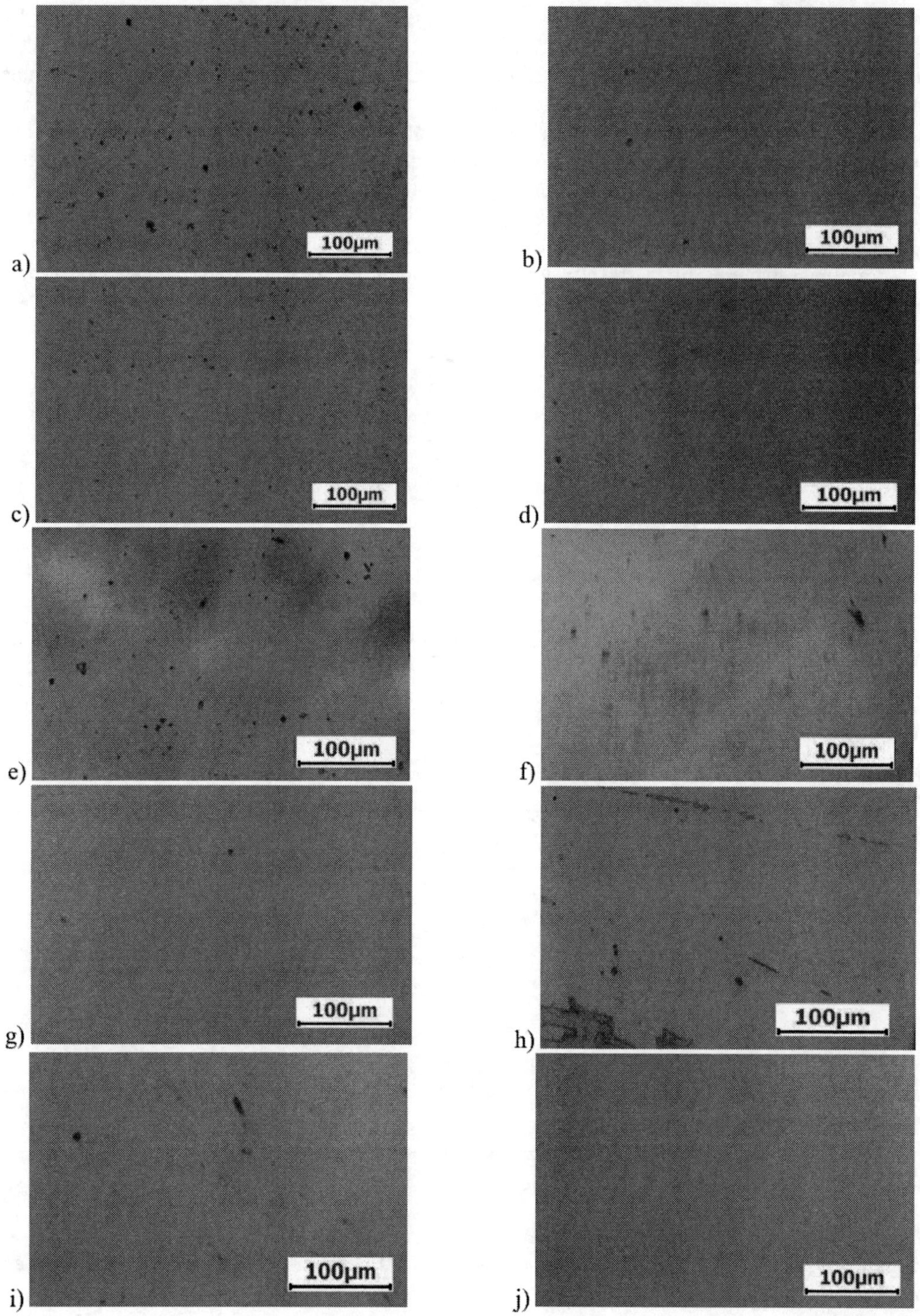

Figure 1. Images of surfaces of samples with different catalyst contents with an increase in the multiplicity of 10 a) 1%; b) 2%; c) 3%; d) 4%; e) 5%; f) 6%; g) 7%; h) 8%; i) 9%; j) 10%.

IOP Conf. Series: Materials Science and Engineering **848** (2020) 012105 doi:10.1088/1757-899X/848/1/012105

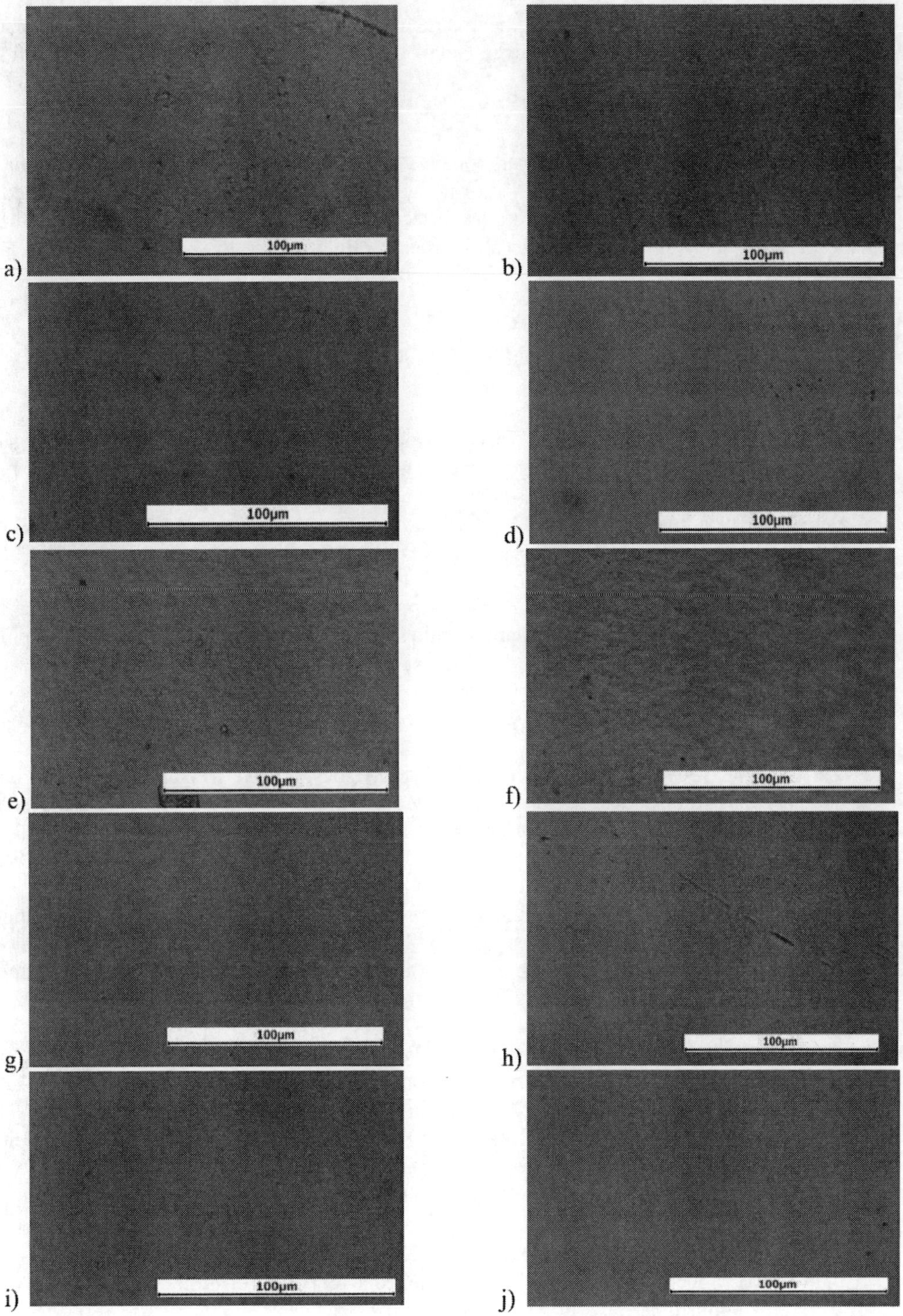

Figure 2. Images of surfaces of samples with different catalyst contents with an increase in the multiplicity of 40: a) 1%; b) 2%; c) 3%; d) 4%; e) 5%; f) 6%; g) 7%; h) 8%; i) 9%; j) 10%.

From the analysis of photographs of the microstructure, we can conclude that the surface is homogeneous, without any defects. However, small particles were found on the surface of some samples. Presumably it's dust.

4. Conclusions

CRT grade siloxane with various catalyst contents was prepared in the form of a film. The microstructures of polymer films based on a siloxane type CRT with a K-18 catalyst were studied. We studied the effect of catalyst concentration on the surface of a polymer film. It should be noted that the percentage of K-18 catalyst does not have a significant effect on the thickness of the CRT brand siloxane film and its structure.

Acknowledgments

Research was funded by the state assignment No. 075-00746-19-00.

References

[1] E.Yu. Zagorulko, A.A. Teslev, E.V. Flisyuk Modern technologies of covering coronary stents, releasing medicines (review), Development and registration of medicines, february № 1 (18), 2017, pp. 70-77
[2] Materials and Coatings for Medical Devices: Cardiovascular. – ASM International. 2009 450 p.
[3] I. Akin, H. Schneider, H. Ince. Second- and third-generation drug-eluting coronary stents: progress and safety // Herz. 2011 V. 36(3).P. 190–197.
[4] Lan Zhou, Guo-Hua Lv, Chao Ji, Si-ZeYang Application of plasma polymerized siloxane films for the corrosion protection of titanium alloy, Thin Solid Films, Volume 520, Issue 7, 31 January 2012, P. 2505-2509
[5] M. Catauro, F. Bollino, F. Papale, R. Giovanardi, P. Veronesi Corrosion behavior and mechanical properties of bioactive sol-gel coatings on titanium implants, Mater. Sci. Eng. C, 43 (2014), pp. 375-382 (10/1)
[6] P Zarrintaj, M Jouyandeh, M R Ganjali, B Sh Hadavand, M Mozafari, S S Sheiko, M Vatankhah-Varnoosfaderani, T J Gutiérrez, M R Saeb Thermo-sensitive polymers in medicine: A review, European Polymer Journal, Volume 117, August 2019, P. 402-423
[7] M.F.Maitz Applications of synthetic polymers in clinical medicine, Biosurface and Biotribology Volume 1, Issue 3, September 2015, P. 161-176
[8] L.K. Wu, J.T. Zhang, J.M. Hu, J.Q. Zhang Improved corrosion performance of electrophoretic coatings by silane addition, Corros. Sci., 56 (2012), pp. 58-66
[9] M.Y. Jiang, L.K. Wu, J.M. Hu, J.Q. Zhang Silane-incorporated epoxy coatings on aluminum alloy (AA2024). Part 1: improved corrosion performance, Corros. Sci., 92 (2015), pp. 118-126
[10] Mojsiewicz-Pieńkowska K., Krenczkowska D. Evolution of consciousness of exposure tosiloxanes—review of publications, Chemosphere, Volume 191, January 2018, P. 204-217
[11] Nair B.P., Gangadharan Dh., Mohan N., Sumathi B., Nair Pr.D. Hybrid scaffold bearing polymer-siloxane Schiff base linkage for bone tissue engineering, Materials Science and Engineering: C, Volume 52, 1 July 2015, P. 333-342
[12] A. Franzen, C. Van Landingham, T. Greene, K. Plotzke, R. Gentry A global human health risk assessment for Decamethylcyclopentasiloxane (D5), Regul. Toxicol. Pharmacol., 74 (Suppl.) (2016), pp. 25-43, 10.1016/j.yrtph.2015.10.023
[13] A. Tiwari, M. Soucek (Eds.), Concise Encyclopedia of High Performance Silicones, Wiley-Scrivener Publishing, Beverly, MA (2014), pp. 243-251
[14] T.M. Tran, K.O. Abualnaja, A.G. Asimakopoulos, A. Covaci, B. Gevao, B. Johnson-Restrepo, et al., A survey of cyclic and linear siloxanes in indoor dust and their implications for human exposures in twelve countries, Environ. Int., 78 (2015), pp. 39-44, 10.1016/j.envint.2015.02.011

[15] Ch. Rücker, K. Kümmerer Environmental chemistry of organosiloxanes, Chem. Rev., 115 (1) (2015), pp. 466-524, 10.1021/cr500319v
[16] U.M. Eduok, M.M. Khaled Corrosion protection of non-alloyed AIAI 316L concrete steel metal grade in aqueous H2SO4: electroanalytical and surface analyses with Metiamide, Constr. Build. Mater., 68 (2014), pp. 285-290
[17] S.A. Umoren, U.M. Eduok Application of carbohydrate polymers as corrosion inhibitors for metal substrates in different media: a review, Carbohydr. Polym., 140 (2016), pp. 314-341

IOP Conf. Series: Materials Science and Engineering **848** (2020) 012106 doi:10.1088/1757-899X/848/1/012106

The development of the high-strength corrosion resistant austenitic steels for the oil equipment shafts

S O Muradyan[1], O S Muradyan[2], M V Kostina[3] and V S Kostina[4]
[1] Researcher, Institute of metallurgy and material science by A. A. Baycov Russian academy of sciences, Moscow, Russia
[2] Head of steelmaking, Joint stock company "Izhevsk experimental mechanical plant", Izhevsk, Russia
[3] Head of laboratory, Institute of metallurgy and material science by A. A. Baycov Russian academy of sciences, Moscow, Russia
[4] Junior researcher, Institute of metallurgy and material science by A. A. Baycov Russian academy of sciences, Moscow, Russia

E-mail: muradianso@gmail.com

Abstract. The work proposes to use corrosion-resistant austenitic Fe-Cr-Mn-Ni-Mo-N steel with a nitrogen content of ~ 0,4-0,9% as a material for the shafts of the oil equipment. The steel was hardened by the solid-solution hardening. Previously, such materials in this area were not used.

1. Introduction

To manufacturing oil pump shafts, RPS shafts (reservoir pressure station) and other shafts, working in aggressive environment, such as:
- centrifugal pump installations (CPI);
- gas separators and water protection;
- reservoir pressure pumps

traditionally using nickel alloys monel and inkonel Alloy 925, Alloy 718, Alloy K-500 types. At first, these alloys was developed in USA to jet engine production. Unto the temperature 900°C this alloy differ by high percussive viscosity and strength, and at low temperatures (until -78°C) is not sensitive to incisions. All these alloys austenitic class with dispersive strengthening [1-5].

Using such nickel alloys to the oil equipment shafts explained by their high corrosion resistance and good strength indicators. This obtaining by high nickel containing and high cost elements alloying – molybdenum, copper, niobium, titanium. The shafts operating temperature lay in the interval from -50 to 180°C, and operating temperature interval of nickel alloys is excess for they.

Manufacturing shafts of these alloys is coupled with range of metallurgical problems. It is very difficult to provide a homogeneity of chemical compound at smelting at titanium and aluminum insertion, and exposure, which require to mixing, is adduct to high metal pollution by the products of their furnace lining interaction. For the same reason it is impossible to use own refuses in the melt charging. It is necessary to smelt alloys in vacuum. Otherwise atmosphere nitrogen will rough titanium and aluminum nitride formation, and, as a result, to plasticity and percussive viscosity loses.

Even greater difficulties causes by sensitivities of these alloys to slow cooling. Wherein it is flow an alloy dispersion hardening, accompanied by the volume reducing. A stresses arising from this lead to the many defects formation, up to the ingots or forging destruction. Therefore at all operating phases, up to 25 mm section they are necessary in hardening (cooling in water).

The heat treatment of alloys including the hardening and two-stage continuous tempering. To high strength obtaining it is necessary to provide a cold deformation additionally.

The main requirements to the materials – the high strength and corrosion resistance.

To the oil shaft pumps operating, working in aggressive environments, at present time using nickel corrosion resistance alloys Alloy 925, Alloy 718, Alloy K-500 types. The other way to obtain the necessary combination of high strength and corrosion properties is exist. It appeared relatively recently and till not widely used in Russia. This is alloying the corrosion resistant steel by nitrogen. Standard for the recommended materials for the oil and gas mining ISO 15156-3:2015 is recommend for working in this condition as well as tradition dispersion hardened nickel alloys the steels S-20910 (03Kh23G5N12M3 with nitrogen containing 0,2-0,4%), J95370 (03Kh25N18G2AM6D6V3 with nitrogen containing 0,7-0,8%), S32654 (03Kh25N23G4AM8D with nitrogen containing 0,45-0,55%).

Nitrogen steel from Fe-Cr-Mn-Ni-Mo-N system alloying with chromium containing ~25% and nitrogen ~0,7-0,9% mass. allows [6, 7]:

- to decrease in 8-10 times the product cost price, relatively to the inkonnels, and in 2-3 times relatively the steels from the standard ISO 15156-3:2015;
- to reach the more high category of strength – the liquidity limit up to 1800 MPa (alloy K-500 using for the shafts with the liquidity limit up to 1100 MPa, alloys 925 and 718 up to 1500 MPa, the steels from the standard ISO 15156-3:2015 can provide the strength category up to 1400 MPa because of not high enough nitrogen containing or the system of metal strengthening – nitrogen connected with a wolfram (the dispersion strengthening));
- to simplify the shafts operating technology (smelting without vacuum, exclude the hardening after each operation and the steel temper) [8-10].

2. Steels for the oil equipment shafts.

The job offers using as a material for the oil equipment shafts the corrosion resistant austenitic steel of alloying system Fe-Cr-Mn-Ni-Mo-N with nitrogen containing ~0,7-0,9%. Steel will strengthening by the hard-solution mechanism of strengthening from the impurity of implementation of nitrogen. Earlier such materials in this area was not applied. Nitrogen steels from the standard ISO 15156-3:2015 J95370 (03Kh25N18G2AM6D6V3 with nitrogen containing 0,7-0,8%), S32654 (03Kh25N23G4AM8D with nitrogen containing 0,45-0,55%) are dispersion-hardening and strengthening take a place due to allocation in them nitrides. Steel S-20910 (03Kh23G5N12M3 with nitrogen containing 0,2-0,4%) does not contain enough a number of nitrogen to obtain high strength categories and corrosion resistance, and high nickel containing are negatively affects on it cost.

In the work were developed and studied such models of the steel as in table 1 (numbers of lines 1-4). For comparison a mechanical properties of alloys Alloy K-500 and Alloy 925 are given. A typical austenitic microstructure of these steels after hot deformation is shown in Figure 1.

Steels 08Kh18AG10N2 and 05Kh20AG10N8M2F were tested to the resistance to the hydrogen sulfide cracking under stress (HSCS). Investigations were carried out by the standard NACE TM 0177-2005 method A (STO Gazprom 2-5.1-148-2007 (GOST9.901.1, GOST9.901.4). As a corrosion substance was a water solution 5%NaCl+0,5%CH_3COOH, saturated by H_2S, pH≤3,5, the temperature of solution - 24±2°C. Tests were carried out on the installation PN-16-4 at constant weight. The results of tests are showed in the table 2.

IOP Conf. Series: Materials Science and Engineering **848** (2020) 012106 doi:10.1088/1757-899X/848/1/012106

Table 1. Mechanical properties of the investigated models of steel.

№	A model of steel	Type and rate of deformation[a]	$\sigma_{0,2}$, MPa	σ_d, MPa	δ, %	Ψ, %	KCU
1	08Kh18AG10N2	H/d	560	850	45	56	18,3
		C/d 25%	1093	1143	22	59	14,4
		C/d 50%	1308	1359	18	55	12,4
2	05Kh20AG10N8M2F	H/d	730	1010	38	64	22,8
		C/d 22%	1230	1290	18	55	9,9
		C/d 52%	1510	1630	10	46	6,1
3	05Kh23AG11N5	H/d	950	1170	24	50	13,6
		C/d 20%	1290	1380	15	34	1,9
4	05Kh25AG10N5	H/d	893	1152	39	56	6,7
		C/d 22%	1380	1430	20	59	17,1
		C/d 54%	1790	1902	11	49	7,3
5	Alloy K-500	C/d+aging	1074	1412	31	53	19,3
6	Alloy 925	C/d+aging	1400	-	8	25	6

[a] H/d – hot deformation, C/d – cold deformation

Table 2. The results of tests of the steels 08Kh18AG10N2 and 05Kh20AG10N8M2F.

№	A model of steel	Type and rate of deformation	Load, MPa	Test result
1	08Kh18AG10N2	C/d 50%	0,5 from $\sigma_{0,2}$	Stand
2		C/d 50%	0,7 from $\sigma_{0,2}$	Not stand
3	05Kh20AG10N8M2F	H/d	0,7 from $\sigma_{0,2}$	Stand
4		C/d 22%	0,7 from $\sigma_{0,2}$	Stand
5		C/d 50%	0,7 from $\sigma_{0,2}$	Stand

3. Conclusions.

Analyzing data of tables 1, 2 and data from literary sources [11, 12] can conclude that nitrogen steels of alloying system Fe-Cr-Mn-Ni-Mo-N with chromium containing ~18-25% and nitrogen ~0,4-1,0 % mass, allows:

- in 8-10 times decreases cost price of product, comparison to inconells, and in 2-3 times comparison to the steels from the standard ISO 15156-3:2015;
- to get more higher category of strength (T18), i.e. limit of liquidity up to 1800 MPa. (For comparison: alloy K-500 using for shafts with limit of liquidity up to 1100 MPa, alloys Alloy 925 and 718 – up to 1400 MPa; steels from the standard ISO 15156-3:2015 can provide category of strength up to 1300 MPa due to unsufficient high nitrogen containing or fastening of nitrogen to nitrides);

Figure 1. Typical austenitic microstructure of steels from table 1 after hot deformation.

- to simplify a shafts operating technology (smelting without vacuum, exception after each operation hardening and tempering);
- using new steel will allow to open new directions of development and improvement of technics for a hydrocarbon production and transportation.

References

[1] Harris Z D, Agnew S R, Scully J R and Burns J T 2017 *ICF 2017 - 14th International Conference on Fracture* The effect of heat treatment on hydrogen environment-assisted cracking of nickel-based superalloys Vol 1 pp 366-367

[2] Marenych O, Kostryzhev A, Shen C, Pan Z, Li H and Van Duin S 1 January 2019 *Metals* Precipitation strengthening in Ni–Cu alloys fabricated using wire arc additive manufacturing technology Vol 9 Issue 1 № 105

[3] Sampath D, Akid R and Morana R 15 March 2018 *Engineering Fracture Mechanics* Estimation of crack initiation stress and local fracture toughness of Ni-alloys 945X (UNS N09946) and 718 (UNS N07718) under hydrogen environment via fracture surface topography analysis Vol 191 pp 324-343

[4] Sridhar N, Thodla R, Gui F, Cao L and Anderko A 13 March 2018 *Corrosion Engineering Science and Technology* Corrosion-resistant alloy testing and selection for oil and gas production Vol 53 pp 75-89

[5] Zhu Y L, Wang Z, Zheng W J, Song Z G and Feng H. November 2017 *Journal of Iron and Steel Research* Characteristics of high temperature flow behavior and constitutive equation of Monel K-500 alloy Vol 29 Issue 11 pp 933-938

[6] Pridancev M V, Talov N P and Levin F L 1969 *High strength austenitic steels* (Moscow: Metallurgy) p 248

IOP Conf. Series: Materials Science and Engineering **848** (2020) 012106 doi:10.1088/1757-899X/848/1/012106

[7] Lo K H, Shek C H and Lai J K L. 2009 *Materials Science and Engineering* Recent developments in stainless steels R 65 pp 39–104

[8] Pedrazzoli R and Speidel M O 1991 *Ergebnisse der Werkstoffforschung, Verlag Thubal-Kain* Korrosion und Spannungsrisskorrosion von stickstoffhaltigen Stahlen Schweiz Zurich pp103-121

[9] Speidel M O 2006 *Matwiss. u. Werkstoiftech* Nitrogen Containing Austenitic Stainless Steels 37 No 10 pp 875-880

[10] Pleva J 1991 *Ergebnisse der Werkstoffforschung, Verlag Thubal-Kain* Korrosionsfeste stickstofflegierte Stahle Schweiz Zurich pp 153-165

[11] Bela Leffler, Stainless steel and their properties, Outokumpu. http://www.outokumpu.com

[12] Uggowitzer P, Magdowski R and Speidel M O 1996 *ISIJ International* Nickel free high nitrogen austenitic steels Vol 36 №7 p 901-908

IOP Conf. Series: Materials Science and Engineering **848** (2020) 012107 doi:10.1088/1757-899X/848/1/012107

Study of the processes of three-dimensional printing of caprolactone copolymers with methylphosphate groups

A O Mariyanats[1]*, A V Mironov[1], I E Nifant'ev[2], A V Shlyakhtin[2] and V K Popov[1]

[1] Institute of Photon Technologies of Federal Scientific Research Centre "Crystallography and Photonics" Russian Academy of Sciences, Troitsk, Moscow, 108840, Russia.

[2] Lomonosov Moscow State University, Department of Chemistry, Moscow 119991, Russia.

*amariyanac@mail.ru

Abstract. Reconstruction of the damaged or missed organs or skeleton fragments is an ambitious target for regenerative medicine. Tissue engineering construction (TEC) comprising three-dimensional bioresorbable scaffold and living cells can become native tissue substitute. Scaffold should correspond to strict requirements. Generally, its physical-mechanical and biochemical characteristics should be comparable with tissue to be replaced. Besides, resorption rate of scaffold material should correlate with the tissue regeneration rate in each case. Thus, the effect of scaffold manufacturing methods on its properties is an important point in TEC design and fabrication. At present study, thermo-extrusion 3D printing of bioresorbable scaffolds based on polycaprolactone and its copolymer with polyethylene glycol and methylphosphate groups, synthesized by novel chemical route, was investigated. Scaffold surface morphology and its internal structures were studied by optical and scanning electron microscopy. The effect of thermal-induced degradation on polymers properties at various 3D printing processing parameters was analysed by gel-permeation chromatography, differential scanning calorimetry and mechanical testing. It was shown that optimizing experimental parameters the developed methods of thermo-extrusion 3D printing enable effective polymer scaffolds fabrication in accordance with their 3D computer model without significant changes of physicochemical properties of initial materials.

1. Introduction

Tissue engineering is a promising methodology that allows to create tissues and organs in vitro as well as to use biocompatible materials and cells to repair damaged or inherently abnormal tissues [1]. The combination of a bioresorbable scaffold, living cells and biologically active substances is called tissue-engineering construction (TEC) [2]. Scaffold is a three-dimensional substrate for cells, which should provide attachment, proliferation, neovascularization, possibility of nutrients penetration, induce tissue formation and even promote cell differentiation [3]. Thus, important aspects are both the choice of material for the scaffold and the methods of its manufacturing.

The scaffold material should be non-toxic and non-immunogenic. Its mechanical properties should be comparable with the characteristics of the native tissue. In addition, an important property of the scaffold material is biodegradability [4]. Besides, the resorption rate of the scaffold material should correlate with the rate of tissue regeneration in each case. Materials used for the scaffold fabrication can have either artificial or natural origin [5]. Synthetic polymers demonstrate promising features since their physicochemical and biochemical properties, and as a result, their mechanical properties and degradation rate, can be specially optimized for a particular application [6]. Nowadays, aliphatic polyesters are often chosen as the polymer base for TECs. Poly-ε-caprolactones and its copolymers are one of the main representatives of this class [7].

Nowadays, there is a huge variety of methods of polymer scaffolds fabrication for TECs. In particular, additive manufacturing (AM) technologies which based on layer-by-layer fabrication of solid copies of 3D computer models [8]. These methods are widely spread as they allow to produce complex custom-designed constructions of external shape and internal pore geometry. The main types of AM techniques currently used to fabricate polymer or composite structures are: different types of inkjet printing of polymer powder [9], fused deposition modeling (FDM) by molten polymer filament [10], hydrogel extrusion printing followed by chemical or UV solidifying of structures [11], laser stereolithography [12] and selective laser sintering [13]. Presumably, the most promising 3D printing method with thermoplastic polycaprolactones (and composites based on it) is a modified version of FDM. It allows using a wide range of thermoplastic polymer melts with high viscosity. Probability of thermal-induced degradation and initial molecular weight changes of thermolabile biocompatible polymers decreases in comparison with classical FDM [14]. One of the important issues is the effect of the thermo-extrusion printing process on physicochemical and mechanical properties of the initial materials and, accordingly, the spatial resolution and properties of the fabricated scaffolds.

Thus, the aims of the study were investigation of thermo-extrusion printing capability for biodegradable scaffolds made of new copolymers and analysis of the effect of printing process parameters on physical-chemical and mechanical properties of initial polymers and scaffolds made of them.

Materials and methods

1.1. *Materials*

Poly-ε-caprolactone (Mw = 39.4 kDa) as well as copolymer of caprolactone with polyethylene glycol and methylphosphate groups (Mw = 42.9 kDa) were synthesized by novel chemical route enabling polycaprolactone and its copolymer fabrication avoiding any presence of toxic tin octoate catalist. Briefly, Poly-ε-caprolactone (PCL) synthesis was carried out with polymerization of ε-caprolactone and BnO-MgBHT as a catalyst (in ratio 200:1). Synthesis of copolymer PEG 2000 1.2% - caprolactone 86.6% - methylphosphate 12.2% (PEG-PCL-MePh) was carried out with sequential polymerization of ε-caprolactone, methylphosphate, PEG and BnO-MgBHT as a catalyst (in ratio 28:12:1) [15].

1.2. *3D-printing*

The finding of optimal parameters of thermo-extrusion printing was carried out using our custom-designed and specially developed 3D printer (figure 1). It should be mentioned that in comparison with classical FDM printers the melting zone of our system extruder is limited to a small (ca. 10 mm3) volume located just before the nozzle. Thus, the negative effect of prolonged thermal stress on polymers can be significantly reduced. The process of thermo-extrusion 3D printing was carried out using following procedure. The grounded polymer powder is pressed and subsequently displaced into the heater zone (8) by the piston of the extruder (10). There powder turns into a melt which then is pushed through the nozzle (i.d. ca. 400 μm) of the extruder (7), forming a polymer filament. The working table (6) is moved by high precision stepping motors (3, 5), similar to the extruder which is moved by motor (9). That allows to lay down the extruded polymer fiber in accordance with 3D computer model. Optimal parameters of 3D printing by PCL and its copolymer melts were determined using this technique. In addition, testing scaffolds were fabricated to study the thermal-induced degradation of polymers caused by printing process, which in turn effect on the spatial resolution of obtained structures.

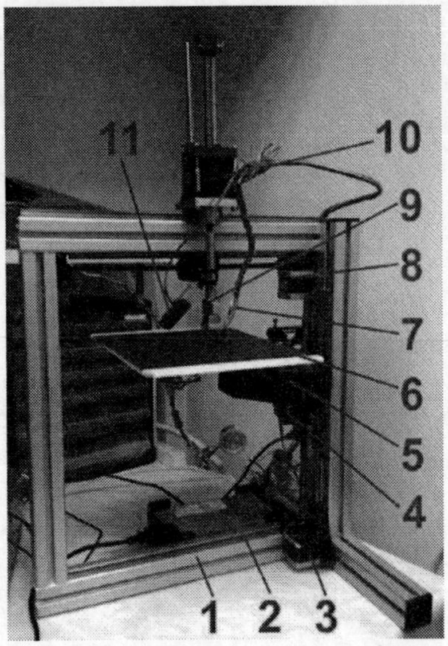

Figure 1. 3D thermo-extrusion printer. 1 – frame, 2 – power supply, 3 – stepper motor (Z-axis movement of the working table), 4 – Arduino MCU, 5 – stepper motor (Y-axis movement of the working table), 6 – working table , 7 – extruder nozzle, 8 – heater, 9 – stepper motor (X-axis movement of the extruder), 10 – extruder, 11 – stepper motor (movement of the extruder piston), 12 – cooling ventilator.

1.3. *Gel-permeation chromatography (GPC)*
The analysis of the molecular weight distribution of PCL and PEG-PCL-MePh (initial polymers and polymers subjected to thermal degradation during 3D printing) was conducted through gel-permeation chromatography on GPC/SEC "Stayer" (AO Akvilon, Russia) with the refractive index detector and column Phenogel 5 mm 10^4 A (300 7.80 mm 5 mm). Tetrahydrofuran was used as an eluent at 40°C and a flow rate of 1 mL/min. The recorded data was automatically processed by the program "MultiChrom 1.6". The chromatograph was calibrated with polystyrene standards.

1.4. *Optical and Scanning Electron Microscopy (SEM)*
Surface morphology, external shape and internal geometry of fabricated polymer scaffolds were observed by optical microscopy (Bresser Advance ICD, Germany) and scanning electron microscopy (LEO 1450, Carl Zeiss, Germany). For SEM characterization specimens were coated with a thin (~ 0.05 - 0.1 μm) gold film by sputtering at room temperature

1.5. *Differential scanning calorimetry (DSC)*
Thermal effects of phase transitions during the heating of polymers were investigated using a differential scanning calorimetry (Linseis, Germany). The polymer specimens (30 mg) were placed in alumina crucibles with a heating range of 20 to 300°C at a rate of 10°C/min in nitrogen atmosphere. The melting temperature and melting enthalpy of the PCL and its copolymer were determined from the first heating scan.

1.6. *Mechanical tests*
Mechanical properties of polymer specimens (made of PCL and PEG-PCL-MePh) were analyzed by using mechanical tester Shimadzu EZTest (Japan). 3-point bend test and tension test were carried out. Solid bar specimens (20*5*1 mm3) were fabricated by 3D printing for 3-point bend test. The rate of printing and of the polymer melt feeding were chosen to obtain homogeneous sample with good spatial resolution. For bend testing, a rectangular cross-section sample was put on steel supports. Then it was bent in the center between the supports at a constant speed. The experiment was carried out until sample destruction or a predetermined value of relative deformation/deflection consummation. The stress-strain curves were recorded at a bending speed of 1 mm/min, which is optimal for elastic modulus determination. Dogbone specimens (20*5*1 mm3) were fabricated by 3D printing for tension test. For tension testing, specimen was stretched along its main axis at a constant rate. The stress-strain curves

were recorded at a bending rate of 15 mm/min. Load, deflection and extension were measured constantly.

Results and discussion

Numerous experiments were carried out to determine optimal parameters (temperature of the polymer melt in the extruder nozzle zone, the distance between the nozzle and the working table surface, rate of the polymer melt extrusion, extruder scanning rate, etc.) of 3D printing with melts of PCL and its copolymer with PEG and methylphosphate groups.

Extruder motion rates (X and Y-axis), rate of the polymer melt extrusion and temperatures of 3D printing were chosen taking the following observations into account:

- the working temperature should be higher than the melting temperature of the polymer in order to ensure the formation of homogeneous polymer fiber, homogeneous in structure and diameter;
- the temperature of the polymer melt should be high enough so that the polymer fiber has steady adhesion to both the working table or the previously printed layer;
- the working temperature of printing should be minimal in order to avoid uncontrollable molecular weight distribution changes and in turn changes of physicochemical properties ;
- polymer fiber should cool down as fast as possible in order to avoid polymer melt overspreading and subsequent spatial resolution decrease.

In order to fulfill the last requirement, polymer fiber was cooled straight after extrusion by ventilator airflow. However, as it was demonstrated in our previous study [16], the complete solidification of the printed layer usually takes about 30 seconds, which corresponded to a cooling rate of about 20°C/min. Scaffolds (multilayer lattice discs) with diameter of 5 mm (figure 2) and 25 mm were fabricated by using our thermo-extrusion 3D printer.

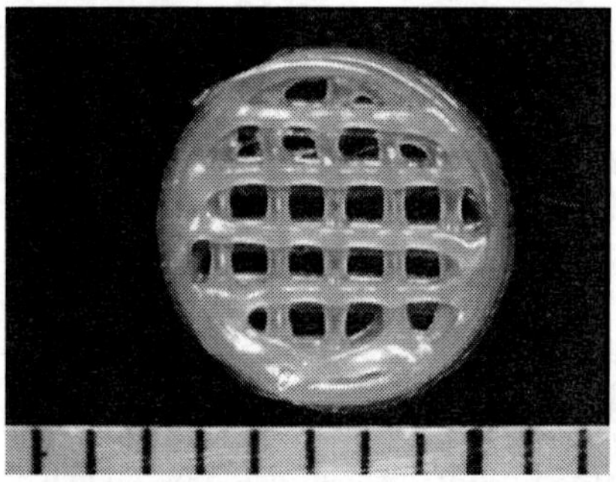

Figure 2. Optical image of PEG-PCL-MePh scaffold (5 mm diameter) fabricated by using thermo-extrusion 3D printer.

The scaffolds surface morphologies and their cross-section structures are shown on SEM images below (figure 3).

The melting temperatures of PCL (69°C) and its copolymer (63°C) were determined by DSC analysis. These temperatures are noticeably lower than the working temperatures of the developed 3D printing process (97°C and 120°C, respectively). However, in consequence of preliminary experiments, necessity of using heightened temperatures was determined. Optimal printing rate (6.5 mm/s for PCL and 4.8 mm/s for PEG-PCL-MePh) also was determined during experiments.

IOP Conf. Series: Materials Science and Engineering **848** (2020) 012107 doi:10.1088/1757-899X/848/1/012107

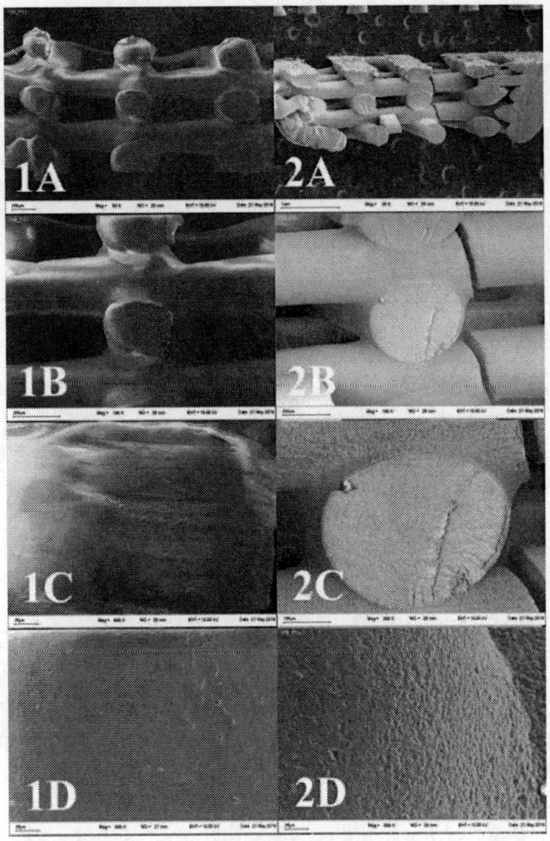

Figure 3. SEM images of polymer scaffolds fabricated by thermo-extrusion 3D printing. 1 – PCL, 2 – PEG-PCL-MePh. A, B - cross-section of scaffold, C - cross-section of polymer fiber, D – polymer fiber surface.

Fabricated polymer scaffolds were designed as a lattice discs with filling of about 40% (figure 2). They had 5 layers of polymer fibers with thickness of ca. 250-300 μm. 3D printing spatial resolution changed from 250 μm to 350-400 μm with PCL and PEG-PCL-MePh, respectively. Probably, this result relates to more apparent effect of the polymer macromolecules orientational relaxation. This effect arises during the process of polymer flow formation, when polymer melt is extruded through nozzle. Generally, scaffolds made of PCL and PEG-PCL-MePh both showed adequate spatial resolution (figure 3 - A1, A2). Overall PCL fiber had smooth and homogenous surface (figure 3 - D1). In some places faint spherulites with a diameter of 15 μm could be observed. Similarly spherulites with a diameter of 10 μm could be observed on the surface of PEG-PCL-MePh fiber (figure 3 – D2). Obtained specific microstructure related to fast heating of polymer powder and rapid extrusion of polymer melt. In this case, polymer powder particles only underwent partly melting in the area of nozzle.

Results of GPC analysis of an average molecular weight (M_w) and polydispersity index (D_m) of investigated polymers before and after thermo-extrusion 3D printing are shown at Table 1.

Table 1. Physicochemical characteristics of polymers before and after thermo-extrusion 3D printing.

	M_w, kDa	D_m	T_m, °C	ΔH, J□g-1
Initial PCL	73.3	1.62	69.3	-73.82
PCL after 3D printing	72.9	1.68	68.9	-51.36
Initial PEG-PCL-MePh	31.1	1.3	63.1	-47.46
PEG-PCL-MePh after 3D printing	30.1	1.35	65.9	-32.51

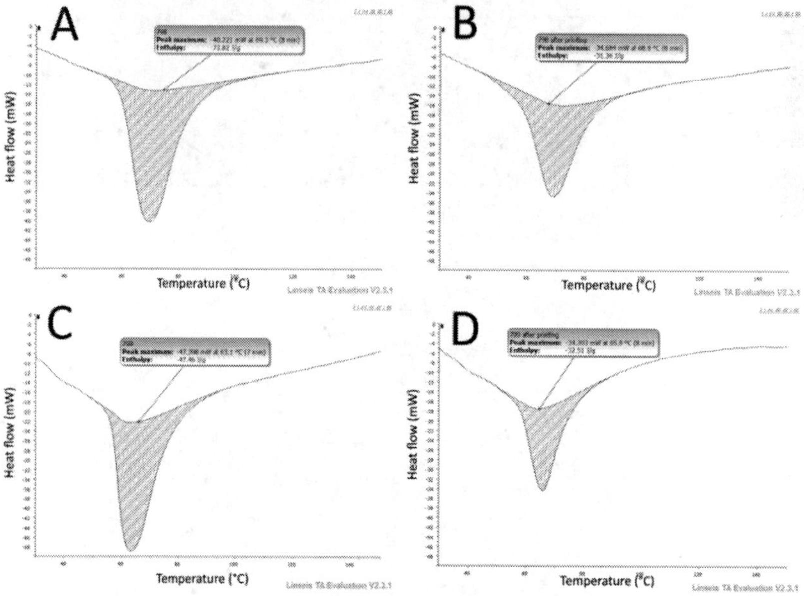

Figure 4. DSC thermograms of: PCL specimens before (a) and after (b) thermo-extrusion 3D printing process; PEG-PCL-MePh specimens before (c) and after (d) 3D printing process.

GPC analysis indicated that the average molecular weight loss percent for PCL specimens caused by thermal degradation (printing working temperature 97 °C) was only 0.6%. The average molecular weight loss percent for specimens made of PEG-PCL-MePh was 3.45% after thermo-extrusion printing (printing working temperature 120°C). We suppose that in the second case the weight loss was bigger due to higher working temperature.

To measure melting temperature (T_m) and melting enthalpy (ΔH) (see table 1) of investigated polymers DSC analysis was performed. Thermograms of polymer specimens before and after thermo-extrusion 3D printing were shown in figure 4. Since melting enthalpy is directly proportional to polymer crystallinity, then it can be concluded that for both polymers crystallinity decreases for about 30% after one-time heating.

The results of mechanical (tension) test (see figure 5) and 3-point bend test of polymer specimens are shown in Table 2.

Table 2. Mechanical characteristics of polymer specimens.

	Tensile strength, N□mm^{-2}	Break strain, %	Elastic modulus, MPa
PCL	26	842.5	300
PEG-PCL-MePh	18	455.4	195

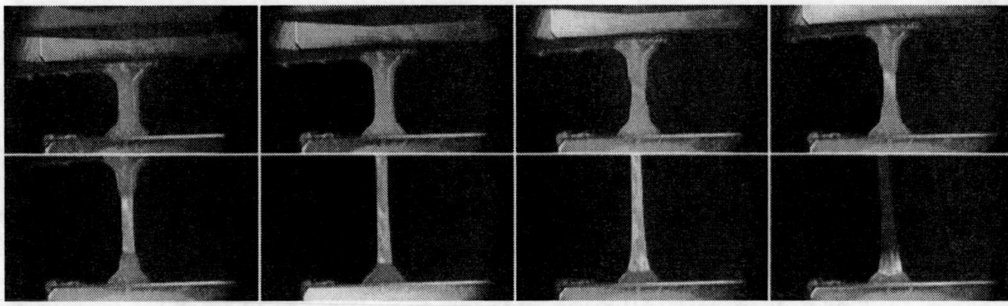

Figure 5. Tension test storyboard of PCL dogbone specimen.

Conclusions

A new thermo-extrusion 3D printing process of bioresorbable scaffolds based on polycaprolactone and its copolymer with polyethylene glycol and methylphosphate groups, synthesized by novel chemical route, was developed and studied.

Ultimately, the GPC analysis demonstrated that the average molecular weight loss of polymers used for 3D thermo-extrusion printing is insignificant (less than 3.5%). DSC analysis showed the polymer crystallinity decreased for 30% after one-time heating at working temperatures. The results of mechanical tests analysis indicated that PCL specimen are more elastic, but less mechanically strong than specimen made of caprolactone copolymer with polyethylene glycol and methylphosphate groups.

Based on characteristic evaluation of manufactured scaffolds, we showed that both polymers are promising for further application in TEC scaffold manufacturing. In summary, the developed methods of 3D printing enable effective fabrication of bioresorbable scaffolds in accordance with their 3D computer model without significant change of physicochemical properties of initial polymers.

Acknowledgments

This work was supported by the Ministry of Science and Higher Education within the State assignment FSRC «Crystallography and Photonics» RAS.

References

[1] Cui H, Nowicki M, Fisher J P and Zhang L G 2017 3D bioprinting for organ regeneration *Advanced healthcare materials* **6(1)** 1601118
Murphy S V and Atala A 2014 3D bioprinting of tissues and organs *Nature biotechnology* **32(8)** 773
[2] Hosseinkhani M, Mehrabani D, Karimfar M H, Bakhtiyari S, Manafi A and Shirazi R 2014 Tissue engineered scaffolds in regenerative medicine *World journal of plastic surgery* **3(1)** 3
[3] Hsu S H, Hung K C and Chen C W 2016 Biodegradable polymer scaffolds *Journal of Materials Chemistry B* **4(47)** 7493-505
[4] Hutmacher D W, Goh J C H and Teoh S H 2001 An introduction to biodegradable materials for tissue engineering applications *Annals-academy of medicine singapore* **30(2)** 183-91
[5] Yang E, Miao S, Zhong J, Zhang Z, Mills D K and Zhang L G 2018 Bio-based polymers for 3D printing of bioscaffolds *Polymer Reviews* **58(4)** 668-87
[6] Gunatillake P A and Adhikari R 2003 Biodegradable synthetic polymers for tissue engineering *Eur Cell Mater* **5(1)** 1-16
[7] Webb A R, Yang J and Ameer G A. 2004 Biodegradable polyester elastomers in tissue engineering *Expert opinion on biological therapy* **4(6)** 801-12
[8] Mota C, Puppi D, Chiellini F and Chiellini E 2015 Additive manufacturing techniques for the production of tissue engineering constructs *Journal of tissue engineering and regenerative medicine* **9(3)** 174-90
[9] Boland T, Xu T, Damon B and Cui X 2006 Application of inkjet printing to tissue engineering

Biotechnology Journal: Healthcare Nutrition Technology **1(9)** 910-17
[10] Zein I, Hutmacher D W, Tan K C and Teoh S H 2002 Fused deposition modeling of novel scaffold architectures for tissue engineering applications *Biomaterials* **23(4)** 1169-85
[11] Savelyev A G, Sochilina A V, Akasov R A, Mironov A V, Semchishen V A, Generalova A N ... and Popov V K 2018 Extrusion-based 3D printing of photocurable hydrogels in presence of flavine mononucleotide for tissue engineering *Sovremennye Tehnologii v Medicine* **10(1)**
[12] Melchels F P, Feijen J, and Grijpma D W 2010 A review on stereolithography and its applications in biomedical engineering *Biomaterials* **31(24)** 6121-30
[13] Williams J M, Adewunmi A, Schek R M, Flanagan C L, Krebsbach P H, Feinberg S E ... and Das S 2005 Bone tissue engineering using polycaprolactone scaffolds fabricated via selective laser sintering *Biomaterials* **26(23)** 4817-27
[14] Jang D S P, Jeng J Y, Dunaev A G, Krotova L I, Mironov A V, Mironova O A ... and Chen Y Y 2017 Thermal extrusion three-dimensional printing of copolymer polyethylene glycol and poly-ε-caprolactone matrix structures *Inorganic Materials: Applied Research* **8(2)** 286-91
[15] Nifant'ev I E, Shlyakhtin A V, Tavtorkin A N, Ivchenko P V, Borisov R S and Churakov A V 2016 Monomeric and dimeric magnesium mono-BHT complexes as effective ROP catalysts *Catalysis Communications* **87** 106-11
Nifant'ev I E, Shlyakhtin A V, Bagrov V V, Minyaev M E, Churakov A V, Karchevsky S G ... and Ivchenko P V 2017 Mono-BHT heteroleptic magnesium complexes: synthesis, molecular structure and catalytic behavior in the ring-opening polymerization of cyclic esters *Dalton Transactions* **46(36)** 12132-46
[16] Dunaev A, Mariyanac A, Mironov A, Mironova O, Popov V and Syachina M 2018 Phase transitions and structural formation of PEG-PCL-PEG copolymer in the processes of fused deposition 3D printing *IOP Conference Series: Materials Science and Engineering* **347(1)**

IOP Conf. Series: Materials Science and Engineering **848** (2020) 012108 doi:10.1088/1757-899X/848/1/012108

Study of products influence of rice waste liquid-phase catalytic oxidation on growth and plant development

A V Khvatov[1], S M Sevostyanov[2], P A Sakharov[1], E A Bocharnikova[2], N Ph Deeva[2], B K Son[2], D Yu Aladin[2], S M Lomakin[1], Yu K Lukanina[1], A Yu Zherebker[3], A S Kononikhin [1,3] and D V Demin[2]

[1]Emanuel Institute of Biochemical Physics, Russian Academy of Sciences, Moscow, Russia
e-mail hvatovanatoliy@gmail.com
[2]Russia, Institute of Basic Biological Problems Russian Academy of Sciences, Pushchino, Moscow state
e-mail nimedd@yandex.ru
[3]Skolkovo Institute of Science and Technology, Skolkovo, Moscow Region, Russia

Abstract. A method for producing growth stimulants using liquid-phase catalytic oxidation of rice waste was tested, with an assessment of the composition of the resulting solution, active silicon, and bioassay. Laboratory studies have shown that solutions from rice waste synthesized by liquid-phase catalytic oxidation are promising for the preparation of plant growth stimulants.

1. Introduction

Biostimulants are substances and/or microorganisms that, when processing plants, seeds or the rhizosphere, can enhance the growth and productivity of plants and improve crop quality by increasing the availability of nutrients for their absorption, as well as influencing the processes of metabolism and activation of the plant's protective system [1]. Biostimulants are usually used to treat directly plants or seeds in doses from a few grams to several kilograms per hectare or ton of seeds [2]. Modern biostimulants are mainly represented by humic/fulvic acids, vitamins, amino acids, microbial preparations, seaweed extracts, as well as inorganic substances [1]. Numerous studies indicate the ability of concentrated monosilicic acid to have a stimulating effect on the growth and development of plants [3, 4, 5].

The use of Si-containing compounds increased the yield of crops by 5-50% or more. Although the role of Si in plant physiology remains insufficiently studied, research results indicate an increase in plant viability under the influence of an element [6]. The positive effect of Si on plants is more pronounced under unfavorable growth conditions. The introduction of Si compounds allows one to control the transport system of plants, as well as the permeability of the plasma membrane in leaf cells [7]. Silicon-treated plants demonstrated enhanced chemical stability of DNA, RNA, and chlorophyll molecules, enhanced photosynthesis, and increased biosynthesis of plastid pigments [8, 3]. Under stress conditions, additional Si nutrition provided a decrease in the activity of oxidative degradation processes due to an increase in the content of specific and non-specific antioxidants [9]. It has been hypothesized that Si plays a signaling role in the response of plants to stress [4].

Since rice plants are silicophiles and are characterized by a high Si content (5–10%), it was suggested that it was possible to use the waste small pieces of rice grains remaining after threshing as a biostimulant.

We tested a method for producing growth stimulants using liquid-phase catalytic oxidation of rice waste, studied its composition and active silicon content, and evaluated the effect of watercress on the germination and germination of plants.

IOP Conf. Series: Materials Science and Engineering **848** (2020) 012108 doi:10.1088/1757-899X/848/1/012108

2. Materials and methods

2.1. Synthesis of Plant Growth Stimulator

For the synthesis of a plant growth stimulator, the method of liquid-phase catalytic oxidation by air of plant materials was used. Earlier, this method was used to obtain highly effective flame retardant compounds. Rice waste was used as plant material. During the oxidation process, a constant temperature of 60-70 °C was maintained. The oxygen pressure in the reactor was maintained at 3 atm. The end time of the oxidation reaction was determined by the concentration of free alkali by titration with 0.1 N HCl in solution up to 0.5 - 0.8 M. The reaction mixture was filtered on a nylon filter to separate from solid impurities. Thick cream-colored solutions were obtained having an alkaline medium - pH 12.3-12.7.

2.2. Mass spectrometric study

Bond Elut PPL cartridges (Agilent Technologies) (100 mg, 3 ml) were used to desalinate the oxidized samples. Samples were isolated according to the method developed for natural organic matter [2]. Briefly, 5 ml of the concentrated solution was passed through a 0.2 μm membrane filter followed by 5-fold dilution to avoid possible aggregation and gelation. The stained solutions were acidified with 1M HCl to pH 3, followed by passing through a PPL cartridge at ca. 5 ml * min-1. Then, the sorbed material was washed 3 times with 0.01 m HCl to remove most cations, after which the sorbent was thoroughly dried under vacuum. Elution of desalted samples was carried out with a 1.5% solution of ammonia in methanol. Methanol solutions were directly analyzed by FTICR MS.

All experiments were run on a FT MS Bruker Apex Ultra mass spectrometer equipped with a harmonized cell[4,5] (Bruker Daltonics), 7 T superconducting magnet, and electrospray ion source (ESI) in negative ionization mode. Before analysis, the test samples were diluted with methanol to a concentration of 100 mg/l. The solutions were injected into the ESI source using a microliter pump with a flow rate of 90 μl/h at a spray gas pressure of 138 kPa and a drying gas pressure of 103 kPa. To ensure rapid evaporation in the ionized droplets, the temperature of the source heater was maintained at 200 °C. Mass spectra were subjected to external and internal calibrations: the first was carried out using synthetic carboxylated polystyrene, and the second, using known residual peaks of fatty acids [6]. The resulting average mass determination accuracy was <0.2 ppm. Spectra were obtained with a data size of 4 megawords ESI(−) and 300 scans were accumulated for each spectrum. The resolution was 530,000 at m/z = 400. The FTICR MS data was processed using the open-source UltraMassExplorer browser application created by Leefmann et al. [8]. The calculated CHONS formulas were verified by applying a chemical filter characteristic of natural organic matter: (O/C ratio ≤ 1, $0.3 <$ H/C ratio ≤ 2.2, atomic restrictions ($C \leq 120$, $H \leq 200$, $0 < O \leq 60$, $N \leq 2$, $S \leq 1$) and mass accuracy <0.5 ppm). The assigned CHNOS formulas were further plotted on van Krevelen diagrams, which are projections on the axis of the H/C ratio and O/C ratio.

2.3. Plant Impact Assessment

Testing of the obtained preparations was carried out under the conditions of a short-term experiment on watercress seeds (Lepidium sativum L.) [10]. This test is quick, informative and technically simple. Watercress as a test object is convenient in that the effects of stressors and stimulants can be studied simultaneously on a large number of plants with a small area (Petri dishes, etc.).

Testing of drug solutions was carried out in a series of dilutions with distilled water 10^{-3}, 10^{-6}, 10^{-9}. Solutions in the amount of 10 ml were poured into Petri dishes, filter paper was placed, 30 cress seeds were distributed on it per option, including control. Petri dishes in those replicates were placed in a climatic chamber, where the temperature was maintained at +22 °C and illumination of 130 μmol of protons m^{-2} s^{-1}. After 120 hours, the number of seedlings was counted, the length of the roots and coleoptiles was measured. The average values for each option are calculated and the standard deviation is determined.

IOP Conf. Series: Materials Science and Engineering **848** (2020) 012108 doi:10.1088/1757-899X/848/1/012108

2.4. Determination of active silicon

To determine the content of monosilicic acid, a modified method of Mullin and Riley was used [11]. The modification consisted of replacing unstable naphthol with iron sulfate and changing the reaction time of monosilicic acid with molybdenum acid ammonium [12].

3. Results

3.1. Identification of synthesis products using mass spectrometry

Thousands of molecular formulas were allowed in the mass spectra of the studied samples, which is typical for products obtained by oxidative condensation [12]. In the samples, the dominance of N-containing compounds is noticeable, which is in good agreement with the peptide content in these samples (Figure 1).

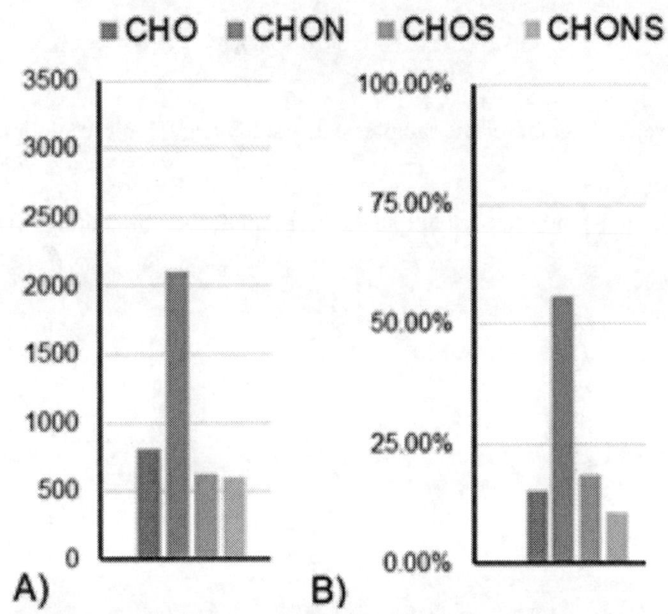

Figure 1. A) the amount and B) the relative content of different classes of compounds determined by high resolution mass spectrometry.

The traditional way to visualize data is the Van Krevelen diagram (Figure 2). From the data presented in the diagram, it can be seen that in the studied solutions obtained by liquid-phase catalytic oxidation, the distribution of compounds is characteristic of samples of biological origin. For such samples, the high density of compounds with H/C <1 is atypical [13]. Here, relatively reduced unsaturated compounds with O/C <0.5 and H/C> 1 are characterized by the highest density. It would also like to note the absence of aliphatic oxidized compounds in the products of oxidation.

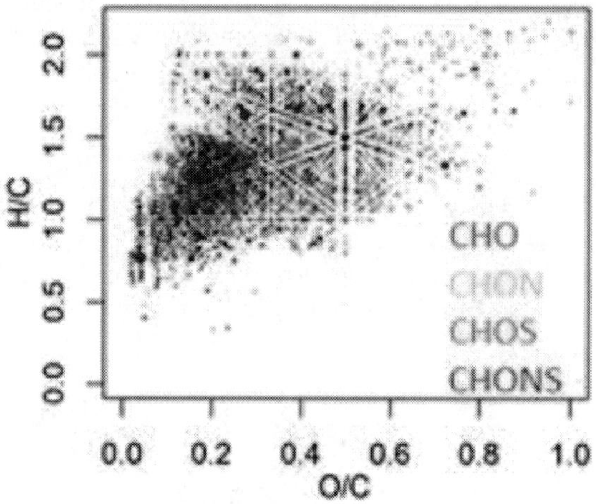

Figure 2. Van Krevelen diagram for the samples under study using high resolution mass spectrometry.

3.2. Testing on seedlings

It was found that diluted synthesized solutions had a pronounced stimulating effect on the growth of plant roots (Figure 3).

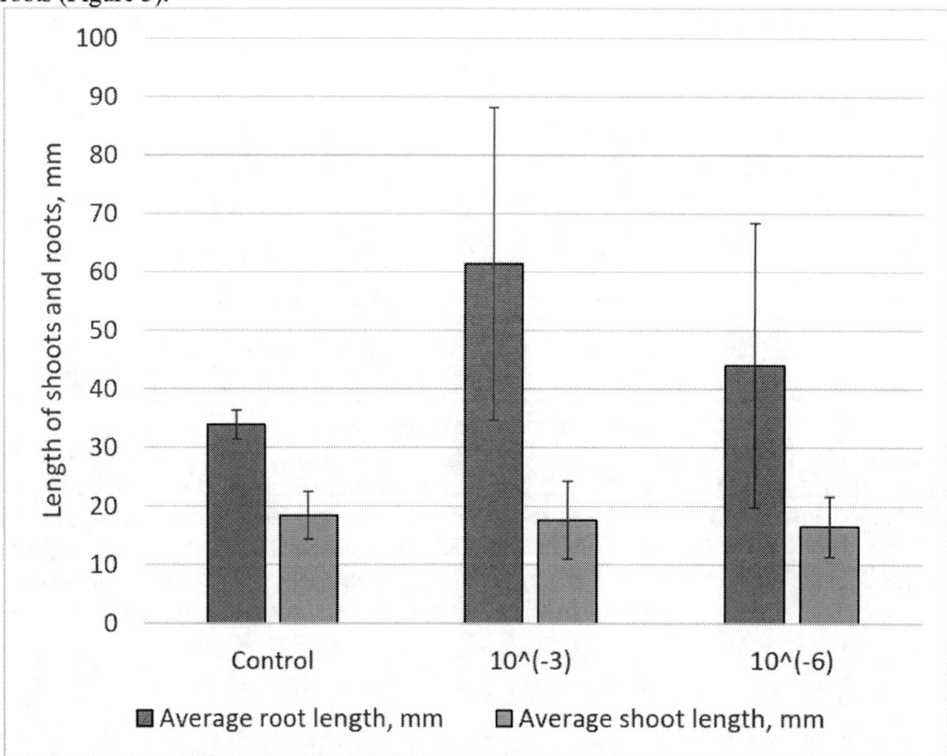

Figure 3. The effect of synthesized preparations on the length of the roots and coleoptiles of watercress.

Regarding the control (average 33.9 mm, taken as 100%), with 10^{-3} dilution, the root length was 61.4 mm, or 81% more than the control variant. When diluting 10^{-6}, the length of the roots of the plants was 44.1 mm, 30% more relative to the control. The effect on the length of the seedlings was not fixed, and with a dilution of 10^{-3} it was almost equal - 18.3 mm at 18.4 mm in the control. With a dilution of 10^{-9} relative to the control, the length of the seedlings was 10% less, but stronger roots and intensely colored shoots were visually recorded, that is, there were no signs of plant inhibition.

3.3. Determination of active silicon

The average value of the content of active Si (monosilicic acid) in the synthesized preparations was 335 ± 10 mg l^{-1}. In commercial Si-containing biostimulants, the concentration of monosilicic acid usually ranges from 100 to 500 mg l^{-1} Si. Thus, the preparation we obtained in terms of the content of active silicon can be qualified as a silicon plant growth stimulator.

4. Conclusion

Laboratory studies have shown that solutions of rice synthesized by liquid-phase catalytic oxidation are promising for the preparation of plant growth stimulants. Testing on watercress seedlings showed that the stimulating growth of plant roots is maximally manifested when diluted from 10^{-3}. Testing on other types of plants, in growing experiments, will reveal the optimal dose of the drug, as well as determine the most effective technology for the preparation of the drug.

References

[1] Brown P and Saa S 2015 Biostimulants in agriculture *Frontiers in Plant Science* **№6** pp 671

[2] Calvo P, Nelson L and Kloepper J W 2014 Agricultural uses of plant biostimulants *Plant and Soil* **383(1-2)** pp 3-41

[3] Soundararajan P, Manivannan A, Park Y G, Muneer S and Jeong B R 2015 Silicon alleviates salt stress by modulating antioxidant enzyme activities in Dianthus caryophyllus *Horticulture, Environment, and Biotechnology* **№56(2)** pp 233-239

[4] Fauteux F, Remus-Borel W, Menzies J G and Belanger R R 2005 Silicon and plant disease resistance against pathogenic fungi FEMS Microbiology Letters №249 (1) pp 1-6

[5] Maghsoudi K, Emam Y and Ashraf M 2016 Foliar application of silicon at different growth stages alters growth and yield of selected wheat cultivars Journal of Plant Nutrition №39(8) pp 1194-1203

[6] Luyckx M, Hausman J F, Lutts S and Guerriero G 2017 Silicon and plants: current knowledge and technological perspectives Frontiers in Plant Science №8 pp 411

[7] Matichenkov V V and Wei X 2019 Reduction Cd in soil-rice by Si Nova Science Publisher, Environmental research advances. New York pp 303

[8] Bocharnikova E A, Pakhnenko E P, Matychenkov V V and Matychenkov I V 2014 The effect of optimization of silicon nutrition on the stability of barley DNA Moscow University Soil Science Bulletin №69(2) pp 84-87

[9] Balakhnina T and Borkowska A 2013 Effects of silicon on plant resistance to environmental stresses International Agrophysics №27(2) pp 225-232

[10] Khvatov A V, Sakharov P A, Lomakin S M, Varfolomeev S D, Lukanina Y K, Minikh A A, Glinushkin A P, Seraya L G and Demin D.V 2019 IOP Conference Series: Materials Science and Engineering 525(1) 012103

[11] Varfolomeev S D, Lomakin S M, Sakharov P A and Khvatov A V 2019 Effective Chemical Methods of Combustion Control: New Threats and New Solutions Herald of the Russian Academy of Sciences №89(2) pp 151-156

[12] Zherebker A Y, Perminova I V, Konstantinov A I, Volikov A B, Kostyukevich Y I, Kononikhin A S and Nikolaev E N 2016 Extraction of Humic Substances from Fresh Waters on Solid-Phase Cartridges and Their Study by Fourier Transform Ion Cyclotron Resonance Mass Spectrometry J. Anal. Chem. 71 (4) pp 372–378

IOP Conf. Series: Materials Science and Engineering **848** (2020) 012108 doi:10.1088/1757-899X/848/1/012108

[13] Zherebker A, Turkova A V, Kostyukevich, Y, Kononikhin A, Zaitsev K V, Popov I A, Nikolaev E and Perminova I V 2017 Synthesis of Carboxylated Styrene Polymer for Internal Calibration of Fourier Transform Ion Cyclotron Resonance Mass-Spectrometry of Humic Substances Eur. J. Mass Spectrom. 23 (4) pp 156–161

[14] Leefmann T, Frickenhaus S and Koch B P 2019 UltraMassExplorer: A Browser-Based Application for the Evaluation of High-Resolution Mass Spectrometric Data. Rapid Commun. Mass Spectrom. 33 (2) pp 193–202

[15] Kuznetsov A. E. et al. 2010 Applied ecobiotechnology: a textbook M.: BINOM. Knowledge lab №2 pp 485

[16] Mullin J B and. Riley J P 1955 The colorimetric determination of silicate with special reference to sea and natural waters Analytica Chimica Acta. №12 pp 162–176

[17] Matichenkov V V, Bocharnikova E A and Ammosova Y M 1997 Method of determination of plant-available silicon in soils Agrochemistry №1 pp 76-80

[18] Zherebker A Y, Airapetyan D, Konstantinov A I, Kostyukevich Y I, Kononikhin A S, Popov I A, Zaitsev K V, Nikolaev E N and Perminova I V 2015 Synthesis of Model Humic Substances: A Mechanistic Study Using Controllable H/D Exchange and Fourier Transform Ion Cyclotron Resonance Mass Spectrometry Analyst 140 (13) pp 4708–4719

[19] Perminova I V, Shirshin E A, Konstantinov A I, Zherebker A, Lebedev V A, Dubinenkov I V, Kulikova N A, Nikolaev E N, Bulygina E and Holmes R M 2018 The Structural Arrangement and Relative Abundance of Aliphatic Units May Effect Long-Wave Absorbance of Natural Organic Matter as Revealed by 1 H NMR Spectroscopy Environ. Sci. Technol. 52 (21) pp 12526–12537

IOP Conf. Series: Materials Science and Engineering **848** (2020) 012109 doi:10.1088/1757-899X/848/1/012109

Synergetic flame retardant effect of bio-flame retardant based on oxidized wood in polyester's compositions

A V Khvatov[1], P A Sakharov[1], S M Lomakin[1], S D Varfolomeev[1], Yu K Lukanina[1], A A Minikh[2] and S M Frolov[3]

[1]Emanuel Institute of Biochemical Physics, Russian Academy of Sciences, Kosygin str., 4, Moscow, 19334 Russia
[2]Research and Development Center, Ltd. "NTC Tatneft, Ltd.", Nobel str., 5, Skolkovo, Moscow, 143025 Russia
[3]Semenov Institute of Chemical Physics Russian Academy of Sciences, Kosygin str., 4, Moscow, 119991 Russia
E-mail: hvatovanatoliy@gmail.com

Abstract. Synthesized bio-flame retardants from plant materials. A comparative analysis of the flammability indices of polyester resin samples containing complex double flame retardants and single flame retardants of bioantipyrene, aluminum hydroxide and ammonium polyphosphate showed a non-additive improvement in flame retardant compositions consisting of double flame retardant compositions.

Unsaturated polyester resins are polymer mixtures containing unsaturated groups and styrene, which acts as a crosslinking agent in the process of radical polymerization [1]. Unsaturated polyester resins can take two forms, depending on the polyester part: orthophthalic or isophthalic. Most unsaturated polyester resin formulations cure at room temperature. However, catalyst, accelerators, and temperature can be used to accelerate the reaction [2]. Additives, including flame retardants, can be easily incorporated into unsaturated polyesters at the processing stage. Good manufacturability, good mechanical properties, water and corrosion resistance and relatively low cost make them ideal for use (trains and ships) [3-5]. Unsaturated polyesters are highly flammable and emit large amounts of smoke during combustion [6].

The addition of halogens is very effective in reducing flammability and, despite environmental problems, it is still the main method by which low flammability of materials is achieved. Halogenated polyesters are effective only if they themselves are sources of ignition, otherwise they will burn together with other combustible components with the release of heat and toxic products. Also, halogen-containing resins have slightly worse mechanical properties compared to conventional polyesters [7].

The main reason for the high combustibility and increased heat release of polyester composites is that during their pyrolysis, 95 - 98% of combustible volatile substances are formed, and coke formation is practically absent [7].

The process of thermal decomposition of all unsaturated polyesters leads to the formation of various low molecular weight volatile substances, including CO, CO_2, methane, ethylene, propylene, butadiene, naphthalene, benzene, toluene, etc. [8 -10].

Significant disadvantages of polyesters under combustion conditions are their increased smoke formation due to the presence of styrene fragments in the polymer chain, as well as the formation of burning drops of liquid pyrolysis products. The accumulation effect can be reduced by using composite materials with a high content of inorganic inert flame retardants [11].

As a rule, aluminum hydroxide with the addition of phosphorus is used as flame retardant additives. It should be noted that magnesium hydroxide is not recommended for use in polyesters, since it accelerates their destruction. The combined use of aluminum hydroxide with ammonium polyphosphate is possible: in this case they form a composition with intumescent properties [12]. The

IOP Conf. Series: Materials Science and Engineering **848** (2020) 012109 doi:10.1088/1757-899X/848/1/012109

synergistic effect of reducing the combustibility of polymer composites with the combined use of ammonium polyphosphate and graphite was also previously shown [13]. The use of melamine and its salts as flame retardants is less effective compared to ammonium polyphosphate [14].

Some flame retardants, such as hydrated mineral fillers, nanoclay and other aluminosilicates are found in nature, and have low toxicity. Other inorganic flame retardants, such as ammonium polyphosphate, also have low toxicity, but decompose in the environment over time.

However, a global problem for many representatives of synthetic halogen- and phosphorus-containing flame retardants lies in their high toxicity and the potential for the formation of carcinogenic compounds in the process of destruction and combustion. Halogen-containing flame retardants, inhibiting the combustion process in the gas phase and reducing heat generation, contribute to the formation of products of incomplete combustion. In most cases, death from a fire occurs as a result of inhalation of toxic products of pyrolysis and combustion of materials, while halogen-containing flame retardants sharply increase the toxicity of combustion products.

Today, one of the solutions to the toxicity of bio-flame retardants (BFR) is the use of intumescent coke-forming systems [15, 16]. In particular, much attention is paid to developments in the field of creating modern environmentally friendly intumescent fire retardants of natural origin in the world.

In order to prepare a fire-retardant solution, 20 liters of hot water, 25 g of $CuSO_4 * 5H_2O$ and 2.8 kg of coniferous wood flour with a moisture content of about 5.5% are introduced into a steel sealed reactor with a volume of 0.04 m^3 heated to 60 °C . The mixture was stirred with a mechanical stirrer for 2 minutes at a stirring speed of 50 rpm, then 1.3 kg of NaOH was added. In addition, the stirring speed increases to 140 rpm, the reactor is sealed, and gaseous oxygen is supplied from above to the reactor at a rate of 0.2 l/min. During the reaction, a constant temperature of 60 °C is maintained. The oxygen pressure in the reactor during the entire oxidation process is maintained at 3 atm. The time of completion of the oxidation reaction is determined by the concentration of free alkali by titration with 0.1 N HCl in the solution to 0.5 - 0.8 M. The reaction mixture was filtered in order to separate from the solid impurities, the filtrate was dried at 60-70 °C. Finally the dried light brown product was crushed to the particle size less than 80 microns.

All the samples with BFR were prepared with the use of pristine polyester resin (RPE) with all components without flame retardant in accordance with the standard formulation using hardeners, a release agent and a UV stabilizer.

Further, complex compositions were prepared on the basis of the obtained mixtures (RPE) and various flame retardants, including flame retardant BFR, $Al(OH)_3$ and ammonium polyphosphate - (NH_4PO_3) n, where (n> 1000):

1. RPE - 100 mass parts (sample №1);
2. RPE -100 mass parts. / flame retardant BFR - 30 mass parts. (sample №2);
3. RPE -100 mass parts. / $Al(OH)_3$ - 30 mass parts. (sample №3);
4. RPE - 100 mass parts. / ammonium polyphosphate - 30 mass parts. (sample №4);
5. RPE - 100 mass parts. / flame retardant BFR - 20 mass parts. + $Al(OH)_3$ - 10 mass parts. (sample №5);
6. RPE - 100 mass parts. / flame retardant BFR - 10 mass parts. + $Al(OH)_3$ - 20 mass parts. (sample №6);
7. RPE - 100 mass parts. / flame retardant BFR - 20 mass parts. + ammonium polyphosphate 10 mass parts. (sample №7);
8. RPE - 100 mass parts. / flame retardant BFR - 10 mass parts. + ammonium polyphosphate - 20 mass parts. (sample №8).

The resulting mixture was stirred with a disk stirrer for 10 minutes at a speed of 800 rpm. Resulting composition was cured at 140-150 °C for 1 hour.

A comparative study of the combustibility characteristics of polyester compositions containing BFR flame retardant and aluminum trihydrate was the main task that was solved at the first stage. In this case, it was necessary to determine the optimal concentration of the introduced flame retardant A (as compared with $Al(OH)_3$) to create a low flammability polyester composition.

IOP Conf. Series: Materials Science and Engineering **848** (2020) 012109 doi:10.1088/1757-899X/848/1/012109

The flammability of samples of polyester resin compositions containing BFR flame retardant and aluminum trihydrate was studied using a device we developed — a mass calorimeter that allows one to determine the mass loss rate during burning of samples [RF Patent "Device for testing combustibility materials" No. 119115]. The principle of operation of the mass calorimeter is to determine the main parameters of the heat balance of the combustion process in an average fire, namely: the rate of heat generation and the rate of weight loss (ASTM 1354-92, ISO / DIS 13927).

In the general form, the scheme of heat balance for the cone (mass) calorimetry testing of samples is given in Figure 1.

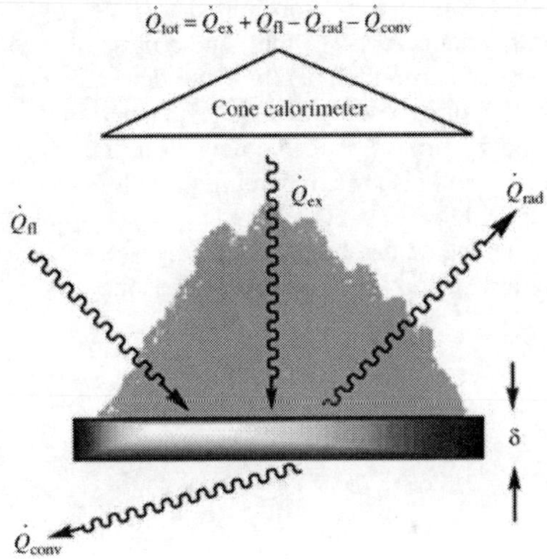

Figure 1. Heat balance scheme in cone calorimeter tests; δ is the sample thickness.

The total heat flux Q_{tot}, is composed of an incident heat flux from an external heater $Q_{ex.}$, a heat flux from the flame on the surface of the material $Q_{fl.}$, and heat losses due to convection $Q_{conv.}$ and radiation to the surroundings Q_{rad}.

A phenomenological equation relating the mass loss rate to the net heat flux during combustion may be represented by

$$\dot{m} = \frac{\dot{Q}_{tot.}}{L_g} + \frac{\dot{Q}_{fl.} - (\dot{Q}_{rad.} + \dot{Q}_{conv.})}{L_g} \quad (1)$$

where L_g (kJ/g) is the heat of gasification related to the total heat flux on the surface. In practice, Lg is determined from the slope of the plot of the mass loss rate as a linear function of external heat flux.

The basic equation relating the mass loss rate to the heat release rate during combustion is as follows:

$$\dot{Q}_{tot}(kW/m^2) = \chi \cdot \Delta H_{comb} \cdot \dot{m} \quad (2)$$

Here, χ is the combustion efficiency, ΔH_{comb} is the heat of complete combustion, and $\dot{m}$ is the rate of mass loss per unit surface.

If ΔH_{comb} is a constant value for given materials, i.e. since there are no gas-phase inhibition processes in the flame, the heat release rate linearly depends on the mass loss rate. In this case, the coefficient χ for the linear equation characterizing the combustion efficiency or the completeness of combustion directly depends on the amount and structure of the carbonaceous residue. It is known that the fire protection mechanism of materials containing intumescent and coke-forming flame retardants mainly describes reactions in the solid phase, while the effective values of the calorific value are independent of the introduced flame retardant. Therefore, the main indicator of combustibility - the rate of heat release in this case is directly proportional to the rate of mass loss *(dm/dt)*. Thus, the maximum mass loss rate (*dm/dt*) characterizes the combustibility of the material.

Under test conditions, samples placed on an electronic balance were subjected to a constant heat flux of 20 kW/m^2, providing a surface temperature of about 500 °C, from an infrared heater. After a certain time (induction period), ignition of the samples occurred. Data on the change in the mass of the sample at equal intervals (2 seconds) were transmitted to a computer. All studied samples had the same surface area (disks with a diameter of 15 mm) and a mass of 0.750 ± 0.200 g. By differentiating the time dependence of the mass loss on temperature, we determined the dependence of the mass loss rate on temperature (*dm/dt*). From the obtained data, the period of ignition induction was further determined as the time of the start of a noticeable deviation, the mass loss rate *dm/dt* proportional to the maximum heat release rate, and also the maximum mass loss rate satisfying the condition (*d2m/dt* = 0) when exposed to an external heat flux 20 kW/m^2.

Flammability tests were conducted for samples of a polyester resin containing BFR flame retardant, as well as $Al(OH)_3$ flame retardant and ammonium polyphosphate.

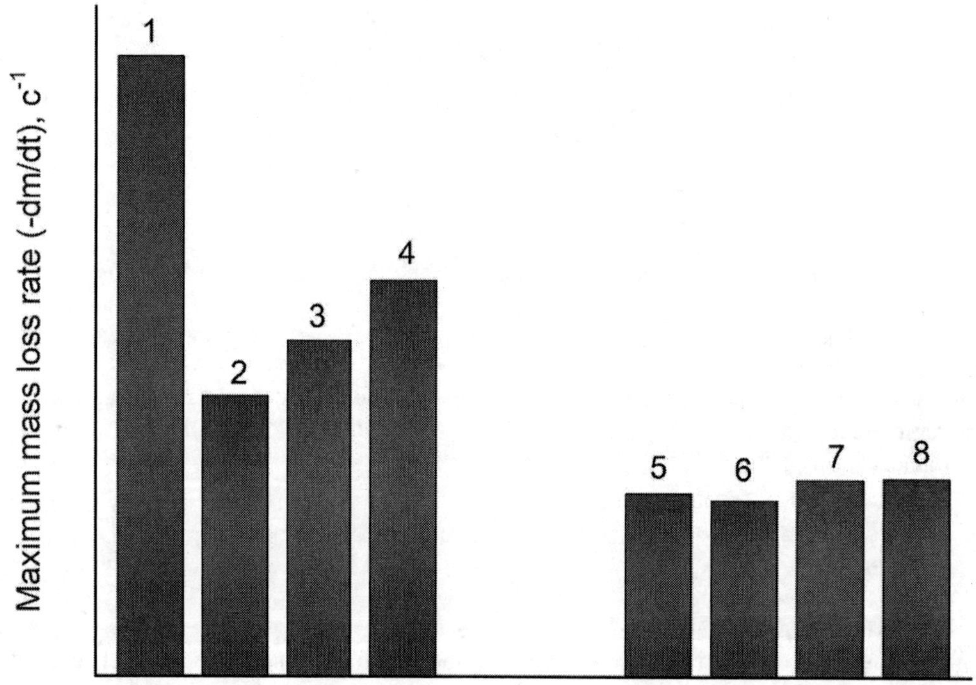

Figure 2. General diagram of the values of the maximum mass loss rate of samples of polyester compositions under conditions of combustion at an external heat flux of 20 kW/m^2.

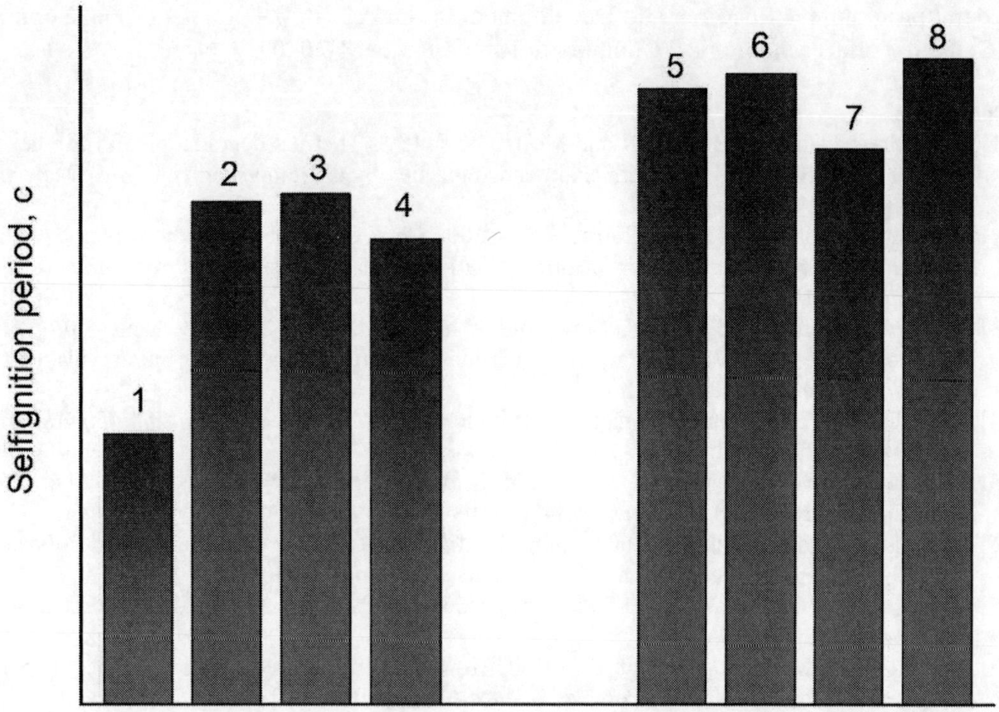

Figure 3. A general diagram of the values of the selfignition induction period of samples of polyester compositions under conditions of combustion at an external heat flux of 20 kW/m^2.

Comparison of the maximum mass loss rate of polyester compositions containing a flame retardant additive BFR at a concentration of 30 mass. p (sample №. 2) with similar compositions containing aluminum hydroxide and ammonium polyphosphate (samples № 3 and №. 4, respectively) shows a higher efficiency of the flame retardant additive BFR compared with aluminum hydroxide and ammonium polyphosphate in terms of the maximum mass loss rate.

The results of the flammability indicators of the compositions of the polyester resin containing a double flame retardant additive: a flame retardant additive BFR, aluminum trihydrate and ammonium polyphosphate (№ 5-8) clearly illustrate the synergism of the joint action of flame retardants.

A comparative analysis of the flammability indices of polyester resin samples containing complex double flame retardants and single flame retardants of aluminum hydroxide and ammonium polyphosphate showed a non-additive improvement in flame retardant compositions consisting of double flame retardant compositions. Moreover, the flammability indices of all flame retardant additives individually are worse than for binary flame retardant compositions, which is evidence of the synergistic effect manifested in systems containing flame retardant additives BFR, aluminum hydroxide or ammonium polyphosphate.

The introduction of BFR can significantly increase the fire retardant effect of traditional flame retardants, such as aluminum hydroxide and ammonium polyphosphate due to the combined thermochemical transformations during high-temperature pyrolysis and combustion.

This work was carried out with financial support of the Ministry of Education and Science of the Russian Federation on government contract № 05.607.21.0304 (RFMEFI60719X0304) "Development of effective ways of chemical inhibition of homogeneous and heterogeneous processes of burning and creation of the demonstration sample of technogenic waste detonation converter" under a special

IOP Conf. Series: Materials Science and Engineering **848** (2020) 012109 doi:10.1088/1757-899X/848/1/012109

federal programme "Research and Development in the Priority Directions of Development of a Scientific and Technological Complex of Russia for 2014-2020".

References

[1] Baudry A, Dufay J, Regneir N and Mortaigne B 1998 Thermal degradation and fire behaviour of unsaturated polyester with chain ends modified by dicyclopentadiene *Poly. Degrad. Stab.* **№61** pp 441-452

[2] Nazare S, Kandola B K and Horrocks A R 2006 Cone Calorimetric Study of Flame- Retardant Unsaturated Polyester Resin Incorporating Functionalised Nanoclays *Poly. Adv. Tech.* **№17** pp 294-303

[3] Baley C, Perrot Y, Davies P, Bourmaud A and Grohens Y 2006 Mechanical properties of composites based on low styrene emission polyester resins for marine applications *Appl. Comp. Mater.* **№13** pp 1-22

[4] Mathur V K 2006 Composite materials from local resources *Construction Building Mater.* **№20** pp 470-477

[5] Kootsookos A and Mouritz A P 2004 Seawater durability of glass-and carbon-polymer *Composites, Comp. Sci. Tech.* **№64** pp 1503-1511

[6] Weil E D and Levchik S V 2004 Commercial Flame Retardancy of Unsaturated Polyester and Vinyl Resins *Review Journal of Fire Sciences* **№22** pp 293-303

[7] Mouritz A P and Gibson A G 2006 Fire Properties of Polymer Composite Materials *Springer Netherlands* pp 394

[8] Bansal R J, Mittal J and Singh P 1989 Thermal stability and degradation studies of polyester resins *Journal of Applied Polymer Science* **№37** pp 1901-1908

[9] Gibson A G and Hume J 1995 Fire performance of composite panels for large marine structures *Plastics Rubber and Composites, Processing and Applications* **№23** pp 175-183

[10] Arii T, Ichihara S, Nakagawa H and Fujii N 1998 A kinetic study of the thermal decomposition of polyesters by controlled-rate thermogravimetry *Thermochimica Acta.* **№319** pp 139-149

[11] Knop S and Krieger W 2003 Flame retardant gelcoats on composite laminates *In: Proceedings of the 3rd Conference on Composites in FireNewcastle, UK* pp 9-10

[12] Hoerold S, Walz R and Zopes H P 2000 Halogen-free additives meet rail standards *Reinforced Plastics* pp 40-43

[13] Shih Y F, Wang Y T, Jeng R J and Wei K M 2004 Expandable graphite systems for phosphorus-containing unsaturated polyesters I Enhanced thermal properties and flame retardancy *Polymer Degradation and Stability* **№86** pp 339-348

[14] Ricciardi M R, Antonucci V, Zarrelli M and Giordano M 2012 Fire behavior and smoke emission of phosphate–based inorganic fire-retarded polyester resin *Fire Materials* **№36** pp 203-215

[15] Khvatov A V, Sakharov P A, Lomakin S M, Varfolomeev S D, Lukanina Y K, Minikh A A, Glinushkin A P, Seraya L G and Demin D.V 2019 IOP Conference Series: Materials Science and Engineering 525(1) 012103

[16] Varfolomeev S D, Lomakin S M, Sakharov P A and Khvatov A V 2019 Effective Chemical Methods of Combustion Control: New Threats and New Solutions *Herald of the Russian Academy of Sciences* **№89(2)** pp 151-156

IOP Conf. Series: Materials Science and Engineering **848** (2020) 012110 doi:10.1088/1757-899X/848/1/012110

Numerical solution of stress-strain state during electromechanical treatment of plasma sprayed coatings

A Yu Ivannikov

Baikov Institute of Metallurgy and Material Science, Russian Academy of Sciences, 49, Leninsky ave., 119334 Moscow, Russian Federation

ivannikov-a@mail.ru

Abstract. The principal aim of this study was to determine the distribution patterns of the stress-strain state at different times. The solution was performed using the finite difference method. The main results were presented depending on the time of action of the alternating electric current during the electromechanical treatment of the plasma sprayed coating.

1. Introduction

The electromechanical treatment is the modern method of the post-treatment of the plasma sprayed coating. This approach is based on the simultaneous resistive heating of the surface and the deformation of the plasma sprayed coating by the electrode. The plasma sprayed coating was densified after the EMT due to deformation and welding of the boundaries of the splats. The nanostructured crystal size was formed after the EMT because of high value of the cooling rates during the post-process; it was confirmed by the decrease of the crystal size and the increase of the microstrain [1, 2]. The cohesive strength of the plasma sprayed coatings increased after the EMT because of the resistive welding and the formation of the nanostructured crystal phases [3, 4]. The wear resistance of the plasma sprayed coatings increased, too. For this reason, the postprocessing of the plasma sprayed coatings by means of the concentrated energy fluxing during the EMT has great potential.

The numerical modeling of temperature, strain-stress state, etc. is the practical approach to determine the most effective modes of post-treatments. Also this approach was used for explanation of the structured transformation. For this reason, the papers [5-8] presented investigation process of the post-treatment of the coatings by means of the concentrated energy fluxing. The solution of this problem concerning stress and strain state is needed to ensure the determination of the effective modes of the post-treatment. By the way, the parameters of the strained condition of the two layers of the half-space, which are modelling the coatings and substrate, needs to be calculated.

The purpose of this study is to calculate strain-stress state and to determine the main parameters of the strained condition during the EMT process.

2. Method of the Solution

The solution of the thermo-force contact problem in the elastic-plastic formulation was performed. The high-temperature exposure and the intense contact pressure were also taken into account. The main boundary conditions were presented in the paper [9]. In this study, the solution was performed using the finite difference method.

An explicit finite-difference scheme was used for the approximate solution. The differential equation of equilibrium in the form of Dugamel-Neumann was approximated by a system of algebraic equations:

$$\left(1+\frac{1}{1-2\mu^n_{i,j,k}}\right)\cdot\frac{u^n_{i+1,j,k}+u^n_{i-1,j,k}-2\cdot u^n_{i,j,k}}{h_x^2}+$$
$$+\frac{u^n_{i,j,k+1}+u^n_{i,j,k-1}-2\cdot u^n_{i,j,k}}{h_y^2}+\frac{u^n_{i,j+1,k}+u^n_{i,j-1,k}-2\cdot u^n_{i,j,k}}{h_z^2}+$$
$$+\frac{1}{1-2\mu^n_{i,j,k}}\cdot\left[\frac{v^n_{i+1,j,k+1}-v^n_{i+1,j,k-1}-v^n_{i-1,j,k+1}+v^n_{i-1,j,k-1}}{4\cdot h_x\cdot h_y}+\right.$$
$$\left.+\frac{w^n_{i+1,j+1,k}-w^n_{i+1,j-1,k}-w^n_{i-1,j+1,k}+w^n_{i-1,j-1,k}}{4\cdot h_x\cdot h_z}\right]+$$
$$+\frac{2\cdot\left(1+\mu^n_{i,j,k}\right)}{1-2\mu^n_{i,j,k}}\cdot\left[\alpha^n_{i,j,k}\cdot\frac{\Delta T^n_{i+1,j,k}-\Delta T^n_{i-1,j,k}}{2\cdot h_x}+\Delta T^n_{i,j,k}\cdot\frac{\alpha^n_{i+1,j,k}-\alpha^n_{i-1,j,k}}{2\cdot h_x}\right]=0,$$

where all variables are assigned corresponding finite-difference indices: n – for the nodes of the time grid (n=0, 1,..., Nt; Nt – the number of steps in time); i, j, k – for the nodes of the spatial grid on the coordinate axes x, y, z: i=1, 2, ... , Ni; j=1, 2, ... , Nj; k=1, 2, ... , Nk (Ni, Nj, Nk – the number of points of the finite-difference grid in the direction of the x, y, z coordinate axes, respectively).

The deformation components were found after determining the components u, v, w:

$$\left.\begin{aligned}
\varepsilon_x &= \frac{\partial u}{\partial x}; & \gamma_{xy} &= \frac{\partial u}{\partial y}+\frac{\partial v}{\partial x};\\
\varepsilon_y &= \frac{\partial v}{\partial y}; & \gamma_{yz} &= \frac{\partial v}{\partial z}+\frac{\partial w}{\partial y};\\
\varepsilon_z &= \frac{\partial w}{\partial z}; & \gamma_{zx} &= \frac{\partial w}{\partial x}+\frac{\partial u}{\partial z}.
\end{aligned}\right\}$$

The stresses were calculated using physical ratios:

$$\left.\begin{aligned}
\sigma_x &= 2G\left[\frac{\partial u}{\partial x}+\frac{\mu}{1-2\mu}\cdot\theta-\frac{1+\mu}{1-2\mu}\cdot(\alpha_t\,\Delta T)\right];\\
\sigma_y &= 2G\left[\frac{\partial v}{\partial y}+\frac{\mu}{1-2\mu}\cdot\theta-\frac{1+\mu}{1-2\mu}\cdot(\alpha_t\,\Delta T)\right];\\
\sigma_z &= 2G\left[\frac{\partial w}{\partial z}+\frac{\mu}{1-2\mu}\cdot\theta-\frac{1+\mu}{1-2\mu}\cdot(\alpha_t\,\Delta T)\right];\\
\tau_{xy} &= G\left[\frac{\partial w}{\partial x}+\frac{\partial u}{\partial z}\right];\\
\tau_{yz} &= G\left[\frac{\partial w}{\partial x}+\frac{\partial u}{\partial z}\right];\\
\tau_{zx} &= G\left[\frac{\partial w}{\partial x}+\frac{\partial u}{\partial z}\right].
\end{aligned}\right\}$$

3. Results and Discussion

During the numerical simulation, the following results were obtained.

The main stresses were reallocated. The intensity of the main stress was increased. The characteristic zones of limit states in the coating were identified.

The analysis of the obtained results is presented below.

The presence of a non-stationary temperature field in the contact problem under consideration leads to a change in the precipitation of the half-space surface. The value of precipitation of the surface depends on the time and its value. Figure 1 shows the dependence of main stresses in the half-space under non-stationary thermo-force contact loading of a two-layer half-space. Analyzing this relationship, we note, that there is the relationship between the surface precipitation and the maximum temperature in the half-space. Thus, as the temperature increases, the precipitation decreases, and vice versa.

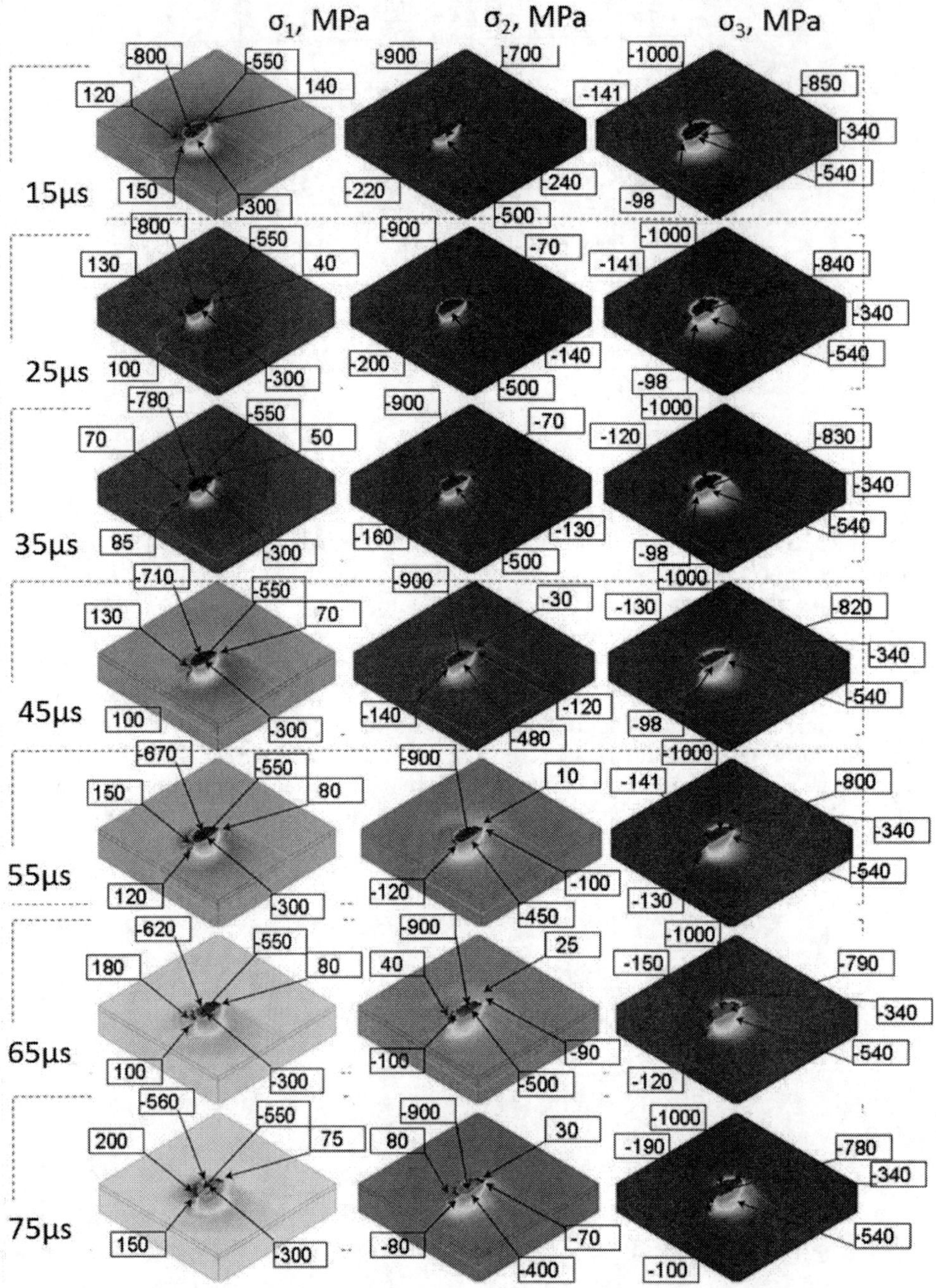

Figure 1. Distribution of main stresses in the half-space under non-stationary thermo-force contact loading of a two-layer half-space

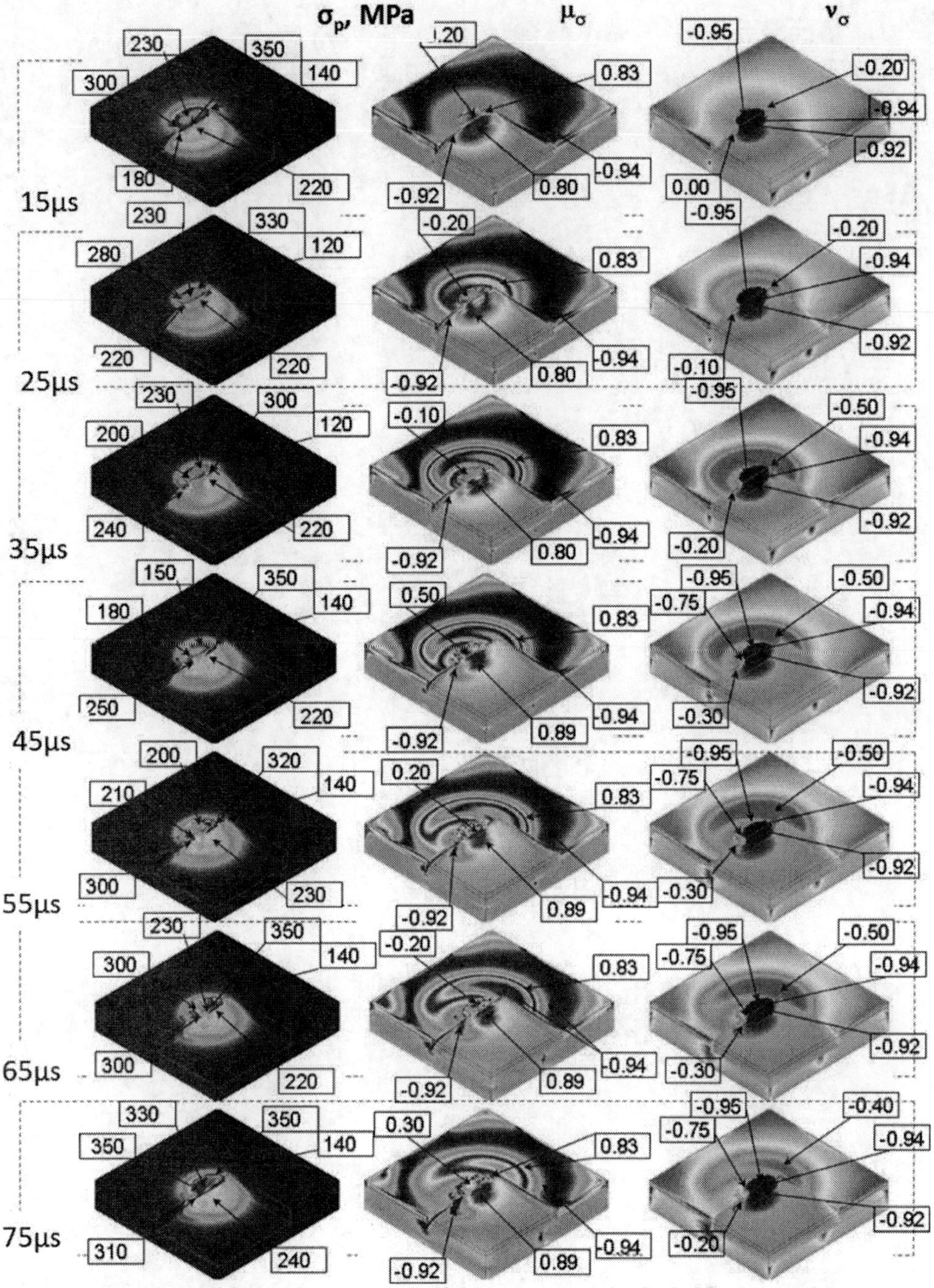

Figure 2. Stress intensities and parameters of the stress state in the half-space under non-stationary temperature-force contact loading of a two-layer half-space

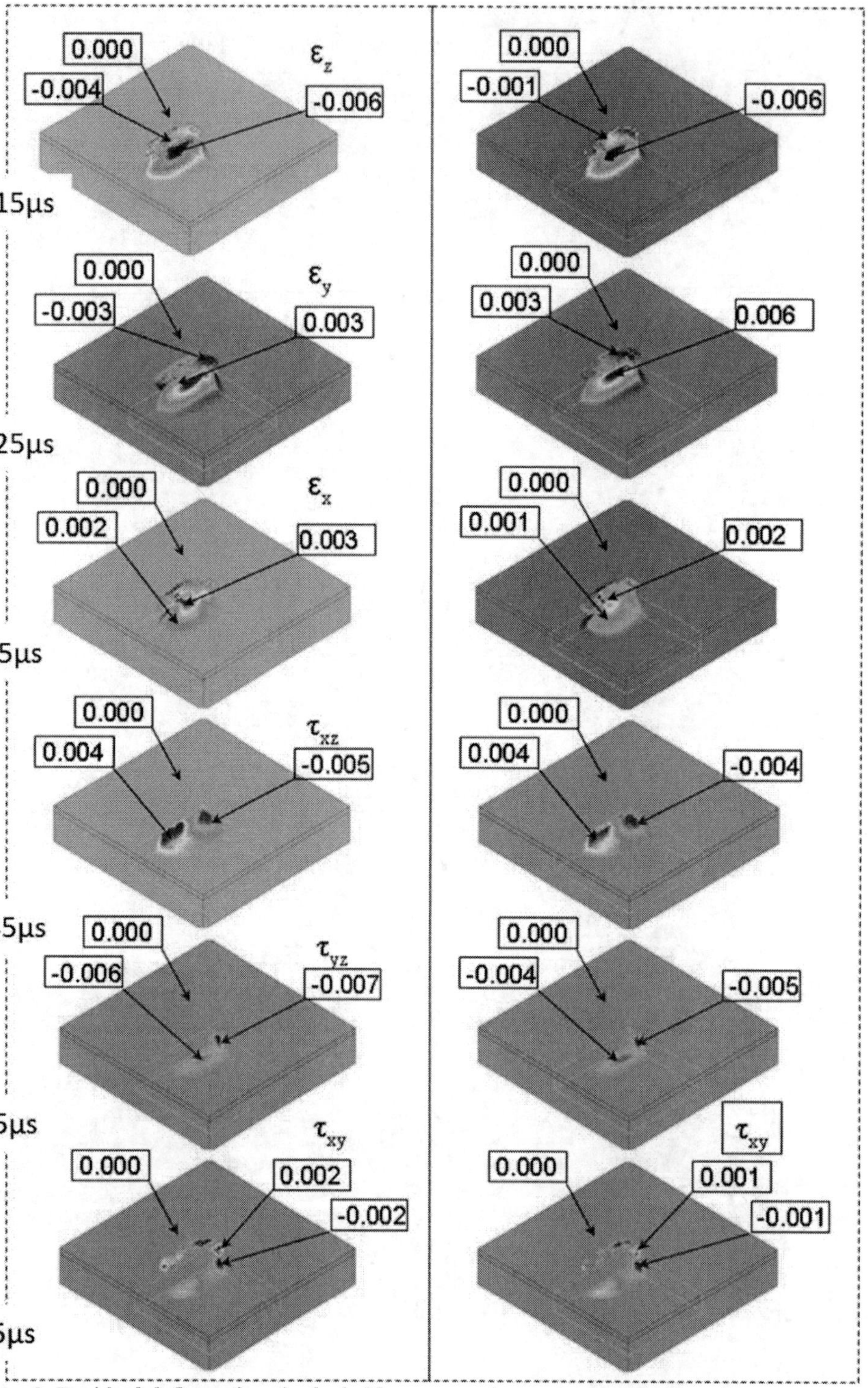

Figure 3. Residual deformations in the half-space: a – homogeneous, b – two-layer

Figure 2 shows the stress intensities and parameters of the stress state in the half-space under non-stationary temperature-force contact loading of a two-layer half-space. Analyzing this relationship, we note, that there is the relationship between the stress intensities, parameters of the stress state, and the maximum temperature in the half-space. Thus, as the temperature increases, the precipitation of the stress intensifies, parameters of the stress state increase, too.

Figure 3 shows residual deformations for homogeneous and two-layer half-space. Thus, in the case of a two-layer half-space, the value of residual deformation in the surface layer decreases (Fig. 3).

The obtained numerical results of residual deformations are confirmed by experimental results [10]. The investigation shows that the coating after the EMT had wavy structure. Due to the fact that part of the material was squeezed out along the roller's edges under pressure, the wavy structure was formed on the surface of the coating. The main geometric parameters of the wavy structure depend on the frequency of alternating electric current during the EMT.

The results of the numerical modeling obtained in this work allow us to determine the influence of the EMT modes on the stress-strain state in the coating. Therefore, an initial approach has been developed to determine the effective modes of the EMT of the plasma sprayed coating.

4. Conclusion

1. The solution of a non-stationary thermo-force contact problem is carried out. The change in the profile of the half-space surface depending on the size and the position of the temperature field during the EMT is shown.
2. The effect of the elastic properties on the stress-strain state of a two-layer half-space at a thermo-force contact during the EMT of the plasma sprayed coatings is examined.
3. It is determined that the developed model allows to choose the effective modes of the EMT of the plasma sprayed coatings.

The reported study was supported by the State target № 075-00947-20-00.

References

[1] A.Ivannikov, et al., 2019 J. Mater. Process. Technol., 266, 442-449
[2] D.Komlev, et al., 2013 Inorganic Materials: Applied Research.,4(3), 236-246
[3] V.Bagmutov, et al., 2009 Steel in Translation., 39(10), 870-877
[4] V.Bagmutov, et al., 2013 Steel in Translation., 43(6), 351-355
[5] V.Bagmutov, et al., 2007 Vestn. Voronezh. Gos. Tekhn. Univ. 3 (8), 135-140
[6] Zaguliaev, D., et al., 2019 J Mater Res Technol., 8(5), 3835-3842
[7] Hömberg, D., et al., 2016 Finite Elem. Anal. Des.,121 (15), 86-100
[8] C. Miehe, et al., 2015 Comput. Methods Appl. Mech. Eng., 294, 449-485
[9] V.Bagmutov, et al., 2017 PNRPU Mechanics Bulletin, (1), 233-250
[10] A.Ivannikov, et al., 2019 J. Therm. Spray Technol., 28(5), 883-892

IOP Conf. Series: Materials Science and Engineering **848** (2020) 012111 doi:10.1088/1757-899X/848/1/012111

Research and development of the inert gas atomization of the wire by means of arc spraying

A Yu Ivannikov, A A Kirsankin, T A Kalayda and M A Sevostyanov

Baikov Institute of Metallurgy and Material Science, Russian Academy of Sciences, 49, Leninsky ave., 119334 Moscow, Russian Federation

ivannikov-a@mail.ru

Abstract. The paper presents brief information on the theory and technology of electric arc wire spraying. The construction for spraying of four groups of wire materials is shown. Carbon and stainless steel, pure copper and titanium wires were atomized. A model of wire atomization during process of electric arc spraying is described.

1. Introduction

The method of electric arc spraying was originally developed by M. U. Schoop in 1910. But this process got commercial acceptance during 1960s. The technology of the electric arc spraying is the most economical among the methods of thermal spray processing. This is due to the high deposit efficiency of the material and the higher productivity of the spraying process. Traditionally, the electric arc spraying is used to protect large structures, such as bridge or tank surfaces. Currently, all major manufacturers of equipment for thermal spray processes have installations for electric arc spraying. For example, the world leaders Praxair S. T. Technology and Oerlikon Metco is used in a variety of situations from high-volume, low-cost applications to high-quality aircraft coatings that have replaced plasma.

Therefore, the theory and practice of electric arc spraying is discussed in detail in the fundamental works on thermal spray processes [1], as well as in specialized monographs [2]. It is known that the melting of wire metal is carried out by means of an electric arc. Liquid metal is sprayed with a gas stream (usually air). The flow of the droplets is formed. The droplets are transferred to the substrate and form a coating. The conditions of the dispersion process form a stream of narrow-fraction droplets. Therefore, the electric arc spray technology can be used to produce spherical powders of a narrow fractional composition.

In this paper, we consider a method of electric arc spraying to evaluate the possibility of obtaining spherical powders for the additive industry. The properties of powders have high influence on the results of additive manufacturing. Traditional atomization methods have known limitations. Therefore, it becomes relevant to study the process of atomization of wire materials using an electric arc spraying with an inert gas.

The purpose of this work is to develop the relevant equipment and to study the process of atomization of wire materials with different chemical compositions.

2. Materials and Method

Four types of materials were used for electric arc spraying (Table. 1). The selected materials are used in additive manufacturing.

IOP Conf. Series: Materials Science and Engineering **848** (2020) 012111 doi:10.1088/1757-899X/848/1/012111

The wire with a diameter of 0.8 mm was used for electric arc spraying of the low carbon steel and copper.

The spraying equipment is the twin-wire arc spray system C-16-CT(M) (GMBh "Thermal-Spray-Tec", Moscow, Russia) with the arc spray gun. Figure 1 shows open scheme of the arc spray gun. The maximum power of the invertor is 29 kW. The arc spray gun has cylindrical nozzle with diameter 4.5 mm. The spray gas is the argon. The wire feed mechanism has push/pull scheme.

Table 1. Materials for twin-wire arc spray atomization

Material	Chemical composition	Fractional composition	Application
Low carbon steel	0.9%C, 1.6% Mn, 0.8 %Si, Fe - base	<170	Mechanical design
Stainless steel	0.12%C, 18%Cr, 10 %Ni, 0.9% Ti, Fe - base	<180	Mechanical design in the corrosion conditions
Cupper	Cu	<200	Electrical products, heat transfer [3, 4]
Titanium	Ti	<250	Aircraft, medical applications

Figure 2 shows a schematic diagram of the spraying chamber for twin-wire electric arc spraying. The spraying is carried out in a closed chamber. There is a viewing window at the top of the chamber. A gas stream with liquid droplets and solidified particles passes through a cylindrical section, which is 5 m in length. The solidified powders are separated in cyclones. The purified gas is removed to the exhaust. In the first cyclone, large particles are collected. In the second cyclone, small particles are collected. To improve the classification of atomized powders, the number of cyclones can be increased.

IOP Conf. Series: Materials Science and Engineering **848** (2020) 012111 doi:10.1088/1757-899X/848/1/012111

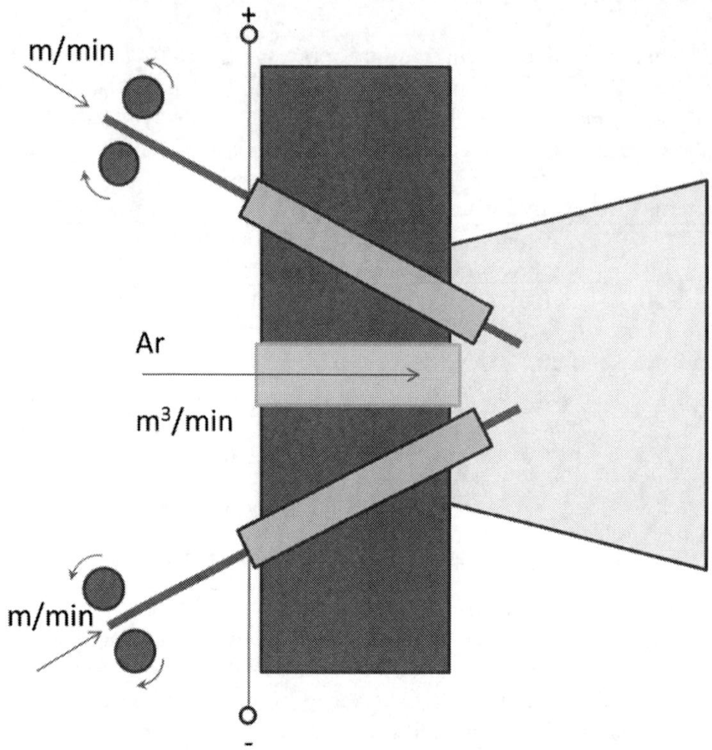

Figure 1. Open scheme of the arc spray gun.

3. Results and Discussion

A model of the electric arc spraying process is shown in Figure 3. There are three zones in the spray process. In the first zone the wire is melted by means of an electric arc. In this zone the material is in a liquid state and is drawn out by the gas flow. In the second zone liquid droplets are formed. Under the action of surface tension forces the droplets become spherical. In the third zone the drops are solidificated. The process of formation of liquid drops in the second zone is periodic. The voltage of the electric arc changes with the length of the liquid section in the first zone. Therefore, the voltage in the secondary circuit of the power supply has a variable value. Consequently, in the second and the third zones, there are periodic increase and decrease of the number of drops and solid particles, respectively (figure 4).

IOP Conf. Series: Materials Science and Engineering **848** (2020) 012111 doi:10.1088/1757-899X/848/1/012111

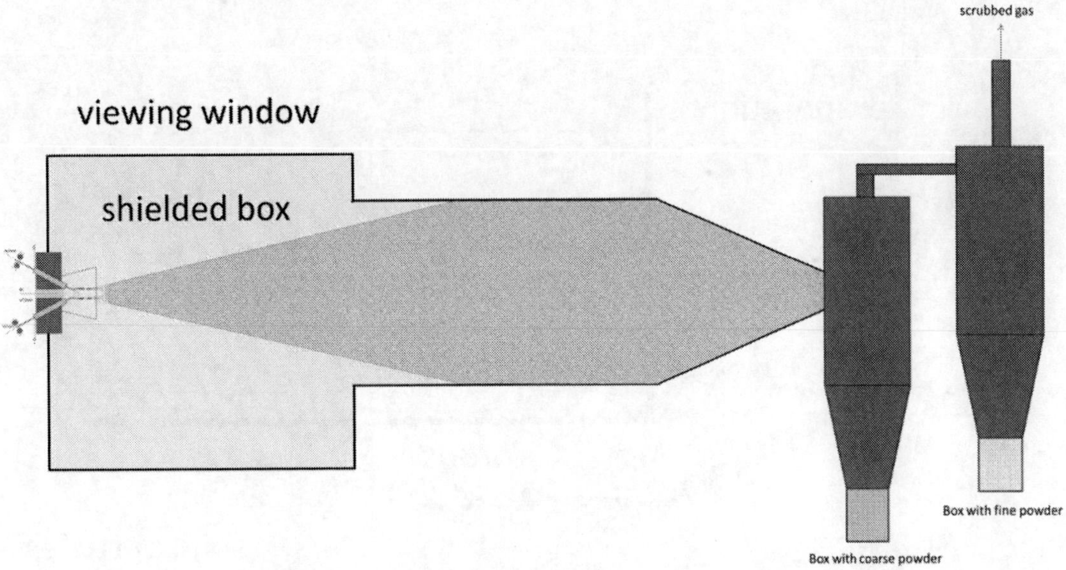

Figure 2. Schematic diagram of the spraying chamber.

The analysis of the effect of spray modes on the fractional composition showed that we had a normal distribution of particles as a result of electric arc spraying. Increasing the electric current and the wire feed speed increases the average fraction of the large powders. Increasing the voltage leads to the increase of the fine fraction. Increasing the gas flow rate reduces the size of the atomization powders.

Before spraying, the chamber was purged with argon. The additional vacuuming of the chamber was not performed. Therefore, during the spraying process, the powders were oxidized. For this reason, the chamber must be free of oxygen before spraying. Consequently, the design of the spray chamber requires modernization.

In addition, experiments have shown that using a smooth 4.5 mm nozzle leads to high gas consumption. Therefore, it is necessary to change the design of the spray gun.

IOP Conf. Series: Materials Science and Engineering 848 (2020) 012111 doi:10.1088/1757-899X/848/1/012111

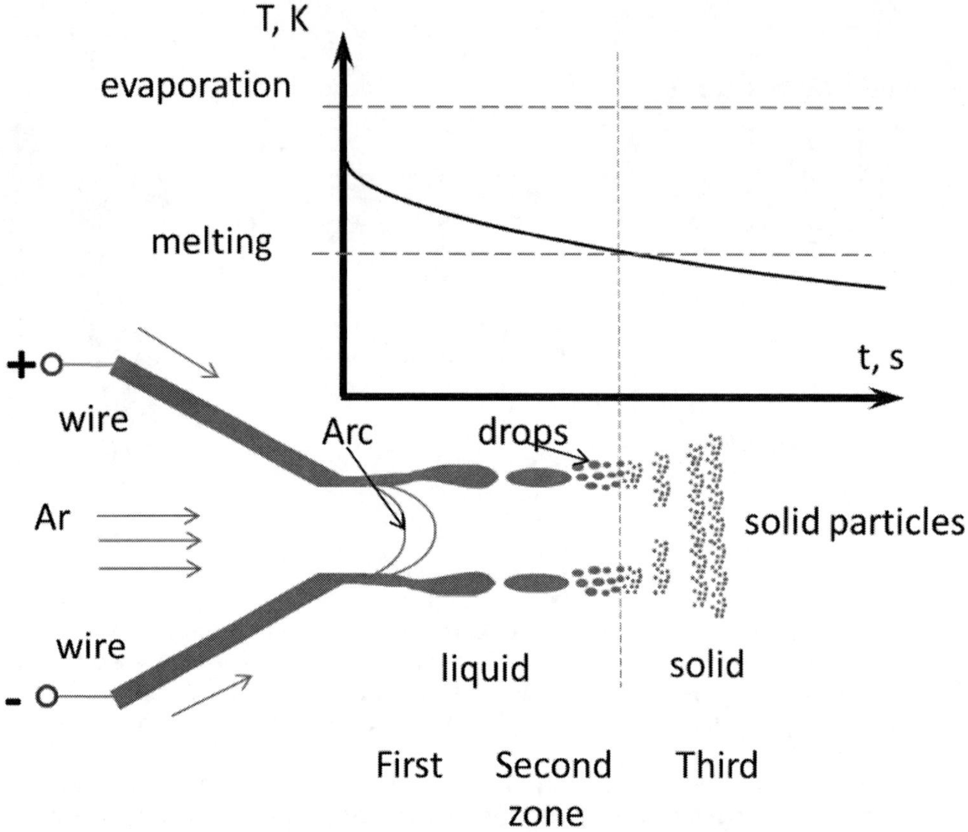

Figure 3. A model of the electric arc spraying process.

Figure 4. Sequence of video frames of the titanium twine-wire electrical arc spraying process.

4. Conclusion

The tests on wire atomization by means of the electric arc spraying were carried out. It is shown that the proposed process made it possible to obtain spherical powders of a narrow fractional composition from various electrically conductive materials.

A model of the electric arc spraying is proposed. The video footage confirming the variable frequency of the spraying process is presented.

The main directions of improving the spray technology and of selecting the optimal parameters for obtaining the required fractional composition of spherical powder are determined

IOP Conf. Series: Materials Science and Engineering **848** (2020) 012111 doi:10.1088/1757-899X/848/1/012111

The work was supported by the Ministry of Education and Science of Russia (grant identifier RFMEFI60419X0242).

References

[1] L. Pawlowski, 2008, The Science and Engineering of Thermal Spray Coatings, Wiley &Sons, Chichester, p. 627
[2] V. Boronenkov, et al., 2016, Fundamentals of Arc Spraying Physical and Chemical Regularities, Springer International Publishing, Switzerland, p. 256
[3] A. Pavlenko, et al., 2018 High Temperature, 56(3), 404-409
[4] A. Surtaev, et al., 2016 Technical Physics Letters42(4), 391-394

Formation of $Si_3Al_3O_3N_5$ oxonitride from mixtures of xerogels and silicon and aluminum nitrides in the nitrogen atmosphere

S N Ivicheva, A A Klimashin, N A Ovsyannikov, A S Lysenkov and Yu F Kargin

A.A. Baykov Institute of Metallurgy & Material Science RAS, Moscow, Russia

E-mail: ivitcheva@mail.ru

Abstract. To obtain silicon and aluminium oxonitrides, the transformations of xerogels and nitrides during firing in a nitrogen atmosphere were studied by thermal and x-ray phase analysis. It was established that the interaction between nitrides and xerogels in a nitrogen atmosphere with the formation of oxonitrides is due to reduction-nitriding reactions. The CO and hydrogen which are formed during the pyrolysis of adsorbed and chemisorbed alcohols (which are the products of the hydrolytic decomposition of silicon and aluminium alkoxides in xerogels) is a reducing agent for the finely dispersed oxides. The effect of the composition of the initial mixtures on the phase composition of the final product (SiAlON) and the morphological features of the powders obtained in one stage of firing in a nitrogen atmosphere is shown.

1. Introduction

The interest in aluminium and silicon alkoxides as components for the synthesis of sialons is associated primarily with the use of products of their hydrolysis to obtain pure (in chemical and phase composition) and active highly dispersed silicon and aluminium oxides and the possibility of their uniform distribution over the grain surface of the initial nitride powders. The oxides of aluminium and silicon obtained using alkoxotechnology differ in structural and textural characteristics from traditional crystalline and amorphous hydroxides species collected by coprecipitation, and by varying the hydrolysis conditions (pH, concentration, etc.) the main characteristics of the final product can be affected, including on its phase composition, distribution in the volume of the reaction mixture and dispersion. The use of freshly prepared sols makes it possible to evenly distribute the oxide component over the surface of nitride grains and introduce various dopants (for example, sintering or luminescent [1-4]) into the composition of ceramics. The result of hydrolysis and polycondensation of alcoholates on the surface of nitrides is the formation of highly dispersed active hydrated oxides (hydroxides). The polycondensation of intermediates can be controlled by adjusting the pH of the medium, concentration, solvent, and temperature. The phase transition temperatures of sol-gel technology products usually have different values compared to crystalline oxide mixtures and can contribute to the sintering of ceramic samples at lower temperatures, and themselves act as sintering additives. These and a number of other considerations led to the use of the products of the hydrolysis of silicon and aluminium alkoxides as an oxide component for the synthesis of sialons from silicon nitrides and aluminium and oxides and to study the possibility of synthesizing oxonitride aluminium silicates by nitriding with highly dispersed mixed xerogels of silicon and aluminium.

2. Materials and Method

A comparative analysis of the effect of the composition of the initial mixture on the phase composition of the final product and the morphological features of the sialons obtained under the same conditions (calcination temperature, nitrogen atmosphere) was carried out using reaction mixtures 1 (SiAlON-I): AlN + SiO_2 (sol); 2 (SiAlON-II): $2Si_3N_4 + 3Al_2O_3$ (sol); 3 (SiAlON-III): Si_3N_4 + AlN + Al_2O_3 (sol) and when using mixed sol 4 (SiAlON-IV): $2SiO_2$ (sol) + Al_2O_3 (sol), calculated on the same composition of the final product $Si_3Al_3O_3N_5$. The reaction mixtures were prepared by adding Si_3N_4 and AlN powders to neutral or acid sols based on silicon or aluminium alkoxides (tetraethoxysilane or aluminium sec-butoxide), simple or polyhydric alcohols, and organic and mineral acids.

It is known that a nitrogen molecule is powerful [5] and the dissociation of nitrogen molecules under experimental conditions (1600–1700°C) for nitriding the resulting xerogels is practically impossible. However, according to XRD data, for all reaction mixtures 1–4, phases of oxonitrides with a hexagonal crystal structure corresponding to β–$Si_3Al_3O_3N_5$ were obtained.

Figure 1 shows the X-ray diffraction pattern of SiAlON-IV samples obtained by nitriding at 1650°C a mixed Si-Al sol (not containing the initial nitrides), the phase composition of which is represented by β-SiAlON (main phase) mixed with Al_2O_3 (corundum). Thus, it is obvious that during the heat treatment of SiAlON-I - SiAlON-IV mixtures, the nitriding reaction of highly dispersed oxides proceeds.

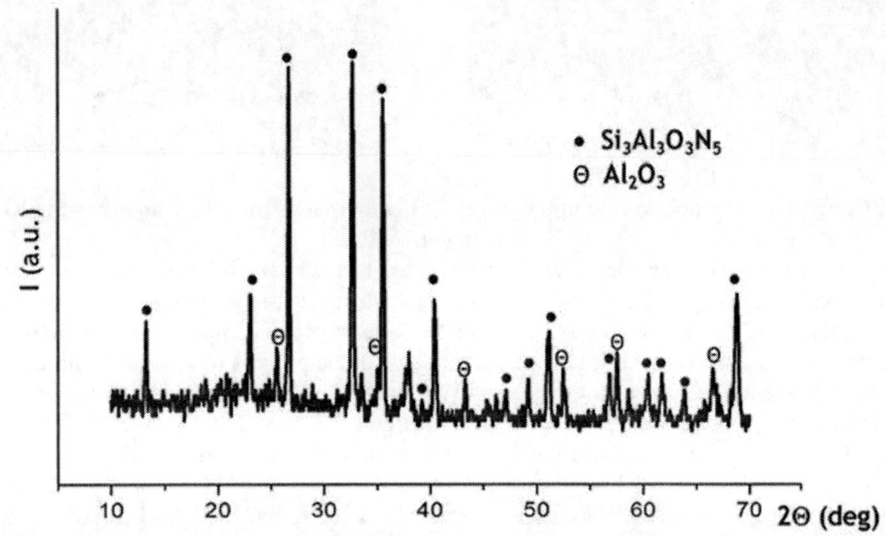

Figure 1. X-ray diffraction pattern of SiAlON-IV after firing at 1650°C in a nitrogen atmosphere

The micrograph (SEM) (Figure 2) shows the change in the morphology of the particles of the mixed sol of SiAlON-IV before and after firing at 1650°C in a nitrogen atmosphere. It can be seen that (Figure 2a) round-shaped nanoparticles ranging in size from 20 to 50 nm of the initial sol form agglomerates (Ssp. ~ 200 m^2/g) and, after firing at 1650°C, transform into strong intergrowths of faceted β – $Si_3Al_3O_3N_5$ crystals up to 15–20 µm long (Figure 2b).

The interaction of nitrides and xerogels in the synthesis of sialons from mixtures 1-3 (initial nitride powders with sols deposited on them) and nitriding conditions for mixed Si-Al xerogels were studied by thermogravimetric (DTA-TG) analysis of samples in a nitrogen atmosphere. To create the same synthesis conditions, nitrogen was supplied in a substantial excess with the same purge rate. The features of phase formation were studied taking into account thermal analysis data of the initial nitride powders with sols deposited on them after various processing steps: 1 - after polymerization and ageing of sols on crystalline nitrides at room temperature; 2 - after heat treatment (drying) at 300°C in air to reduce the amount of adsorbed reagents (water and hydrolysis products of Al and Si alkoxides) in the sintering reaction mixture; 3 - after heat treatment of the initial mixtures at 550°C in a nitrogen atmosphere. It has been established that different conditions of heat treatment of the initial mixtures after deposition of nitride particles and ageing of Si and Al xerogels on the particles can lead to the formation of the phases of different compositions. In particular, according to the XRD data, after preliminary drying in air at 300°C and subsequent firing of samples at 1650°C aluminosilicates (mullite, etc.) are formed, and not sialons, i.e., the nitriding process is not implemented. It can be assumed that preliminary heat treatment at 300°C in the air of initial mixtures of nitrides with deposited sols and mixed sols containing products of the hydrolysis of Si and Al alkoxides leads to hydrolytic oxidation and the removal of "active" adsorbed or chemisorbed radicals that activate the nitriding process.

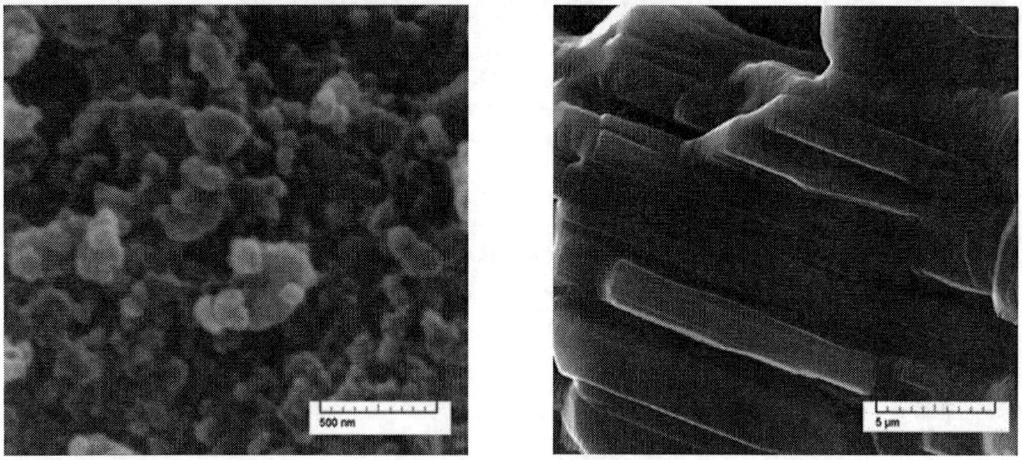

(a) (b)

Figure 2. SEM micrographs of the initial X-ray: (a) - amorphous mixed sol and (b) - SiAlON-IV after firing at 1650°C

According to DTA-TG analysis, it was found that the formation of β–$Si_3Al_3O_3N_5$ from mixtures 1–4 includes several stages: gel dehydration, pyrolysis in a nitrogen atmosphere of chemisorbed products of the hydrolysis of Si and Al alkoxides with the release of decomposition products of organic compounds, and the interaction of intermediate products in the process of nitriding with the formation of crystalline phases. Depending on the composition of the initial mixture an exothermic effect and a noticeable increase in mass are observed on the DTA and TG curves at 1100–1300°C indicating the initiation of the nitriding process and the formation of silicon-aluminium oxonitride. They are most pronounced in mixtures of SiAlON-I and SiAlON-III (in which used aluminium nitride). As an illustration in figure 3 DTA and TG curves of analysis of the initial SiAlON-I mixture in a nitrogen atmosphere are presented.

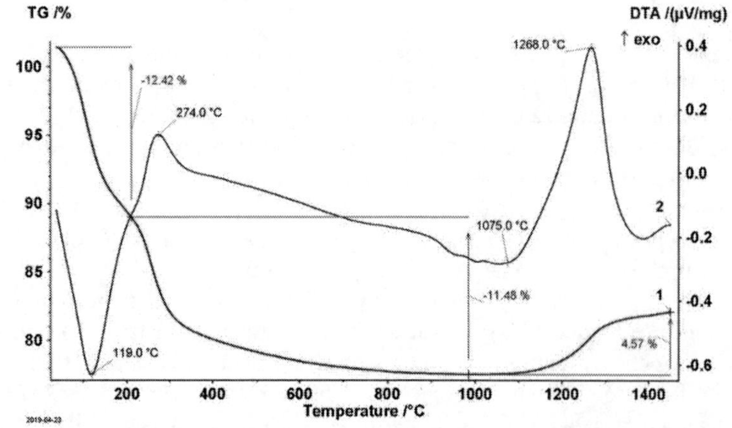

Figure 3. DTA and TG curves of the analysis of the initial AlN + SiO_2 mixture for SiAlON-I in a nitrogen atmosphere: 1 - mass change curve, 2 - DTA curve.

3. Results and discussion

For SiAlON-IV samples, there was no mass gain due to nitriding under TG experiment conditions (up to 1450°C), which may indicate that the nitriding process occurs at higher temperatures.

Changes that occur during heating of samples of mixtures 1–4 were studied using complex thermal analysis on an STA 409 Luxx Netzsch thermal analyzer, combined with a quadrupole mass spectrometer. Mass spectral (MS) analysis of the mixed Si-Al sol for SiAlON-IV was carried out in a nitrogen atmosphere.

During the MS analysis, a fairly high ionic current of hydrogen (H^{2+}) and carbon evolution (C^{+}) are recorded. Obviously, this is due to the dehydrogenation of the organic products of hydrolysis of Si and Al alkoxides that make up the gel. For example, alcohols (sec-butyl, isopropyl, ethyl, etc.) due to the catalytic action of the finely dispersed Al and (or) Si compounds in the gel upon heating. In [6], a hypothesis was put forward on the catalytic effect of alumina in relation to the high-temperature decomposition of alcohols, and hydrogen and aldehyde (which subsequently decomposes into simpler gases: ethylene, methane, CO (carbon monoxide) and hydrogen) are indicated as reaction products for ethanol.

According to TGA/MS in air and nitrogen, powder samples of all initial mixtures contain sorbed and chemically bound water, which are removed at 50–200°C and 200–300°C, respectively, as well as chemisorbed nitrate ions (when using acid sols), which are removed mainly at 200-300°C. It should be noted that the MS contains a high content of the organic component (C^{+}, CH^{4+}, $C_3H_8^{+}$ and other organic radicals with masses from 42 to 46), there remains a high ion current of hydrogen (H^{2+}), apparently associated with the catalytic dehydrogenation of organic alcohols and contributing to the reduction of a certain amount of SiO_2 to SiO. This assumption confirms that when the firing temperature is increased to 1700°C for SiAlON-IV according to XRD, in the samples, in addition to $Si_3Al_3O_3N_5$, there are other sialons - $Si_2Al_3O_7N$ (35-0023) and $Si_6Al_{10}O_{21}N_4$ (36-0832) [JCPDS-International Center for Diffraction Data. All rights reserved by PCPDFWIN v.2.4, 2003]. Their composition corresponds to a lower (relatively to $Si_3Al_3O_3N_5$) Si content, probably due to the volatility of SiO, which can be formed in the reducing environment of the experiment. There is a gas liberation (radicals with a mass of 28, which does not exclude the formation of carbon monoxide (CO^{+})), which is also characterized by reducing properties, according to the TGA/MS of the initial mixture for SiAlON-IV in the air at ~ 300°C and above.

4. Conclusion

Thus the sequence of chemical transformations for all the initial mixtures of SiAlON-I, II, III, IV is caused not only by the decomposition of hydrated oxides deposited on nitride particles but also by reduction processes involving radicals formed during the thermolysis of hydrated gels (obtained based on alkoxides aluminium and silicon) activating the nitriding process in the synthesis of sialons.

Acknowledgments
This work was supported by the Russian Foundation for Basic Research, project No. 20-03-00455a.

References

[1] S Ivicheva *et al* 2018 *IOP Conf. Ser.: Mater. Sci. Eng.* **347** 012046
[2] S Ivicheva *et al* 2019 *IOP Conf. Ser.: Mater. Sci. Eng.* **525** 012084
[3] Slesarev A I, Ivicheva S N, Lysenkov A S, Kargin Yu F, Solntsev K A, Ishenko A V, Yagodin V V, Shulgin B V 2017 *Problems of Spectroscopy and Spectrometry* **38** 3
[4] Yagodin V V, Ischenko A V, Ivicheva S N, Lysenkov A S, Ovsyannikov N A, Kargin Yu F, Baibalova G F 2018 *Problems of spectroscopy and spectrometry* **40** 87
[5] Knunyants I L 1988 *Chemical encyclopedia: Vol. 1* (Moscow: Publishing house "Soviet Encyclopedia") p 1220
[6] Ipatiew W. 1903 *J. Prakt. Chem.* **67** 420

Properties of composites SiC/SiC$_f$ obtained by hot pressing of SHS of silicon carbide powder

M G Frolova[1], A S Lysenkov[1], D D Titov[1], E I Istomina[2], S N Perevislov[3], K A Kim[1], V V Zakorzhevsky[4], S V Fedorov[1], Yu F Kargin[1]

[1]Baykov Institute of Metallurgy & Material Science RAS, Moscow, Russian Federation
[2]Federal Research Centre "Komi Science Centre of the Ural Branch of the Russian Academy of Sciences", Syktyvkar, Komi Republic, Russia
[3]Institute of Silicate Chemistry of Russian Academy of Sciences, Saint Petersburg, Russia
[4]Institute of structural Macrokinetics and problems of materials science. A. G. Merzhanova of the Russian Academy of Sciences (ISMAN), Chernogolovka, Moscow region, Russia

e-mail: frolovamarianna@bk.ru

Abstract. Dense ceramic samples based on SiC were obtained by hot pressing. We used the Use of silicon carbide powder as the initial components of SHS, due to the spherical shape of the particles, allows obtaining a denser workpiece. The reinforcing component in the form of silicon carbide fibers is obtained by siliconizing carbon fabric with SiO vapours. The phase composition of SiC fibers obtained by silicification by SiO vapour was determined. The properties of the obtained SiC / SiCf composites were studied: density 3.20 g/cm^3, strength 520 MPa (for monolithic ceramics 390 MPa), the value of the crack resistance coefficient 5.9 MPa·m$^{1/2}$.

1. Introduction

Modern technology, for its successful development and formation, is in dire need of materials capable of working in a wide range of operating temperatures and aggressive environments. High values of the physicochemical properties of silicon carbide ceramics make it widely applicable in various industries - rocket science, defense, chemical, metallurgy, oil industry, as well as cutting tools, heating elements, and abrasives, etc. However, brittleness is the major drawback of SiC ceramics and limits their uses as structural materials [1-3].

Monolithic ceramics low values of crack resistance can be improved with the help of reinforcing components in the form of fibers, fabrics and whiskers. The combination of ceramic matrix and reinforcing components allows to obtain ceramic materials composites with a high level of physico-chemical properties [4].

SiC/SiC$_f$ ceramic matrix composites (CMCs) are promising materials for replacing Ni-base super-alloys in high temperature applications as they exhibit as much as 1/3rd the density while maintaining sufficient strength and creep resistance to operate at significantly higher temperatures (i.e. >1000°C) [5].

The most successful for reinforcing silicon carbide ceramics are SiC fibers. The main manufacturers of SiC fibers are three main manufacturers found in the market are Nippon Carbon, COI Ceramics and Ube ind, producing the trade-marks Hi-Nicalon (Type-S), Sylramic (iBN) and Tyranno SA3 respectively, where the elemental composition of the precursors and the maximum production temperature varies [6]. However, the process of obtaining SiC fibers is quite time-consuming and expensive, and in Russia it is insufficiently mastered. In modern conditions, there is a need to develop an alternative technology for the production of fibers based on silicon carbide, capable of performing the function of reinforcing components, preserving the structure when exposed to high temperatures and aggressive environments.

2. Materials and Method

Thanks to the study [7], a method was implemented to create a reinforcing component based on silicon carbide by siliconizing carbon fabric with SiO vapour. For this, a siliconizing unit was created based on a vacuum furnace. Siliconization of carbon tissue (Figure 1) takes place in a high-temperature laboratory reactor of periodic action, placed in a vacuum furnace at temperatures up to 1400□C. The reactor consisted of two corundum crucibles nested in each other. Siliconizable material was wound around the inner crucible, and an equimolar mixture of Si and SiO_2 was placed on the bottom of the outer crucible as a source of SiO gas. This method allows to obtain a textile material based on silicon carbide, which can be used as a reinforcing component in the form of fabrics for layering, fibers and particles [4].

Figure 1. Carbon fabric.

The starting components in the research were silicon carbide powders, a sintering additive YAG, and SiC fibers obtained by siliconizing carbon fabric with SiO vapors. Nanosized powder particles of silicon carbide were obtained by the SHS method at ISMAN RAS, Chernogolovka. The synthesis of SiC powder was carried out under nitrogen pressure, and a temperature of 1900-1950°C since the thermal effect of the direct reaction of silicon with carbon is insufficient for the synthesis in self-sustaining combustion mode. The spherical shape of the silicon carbide powder SHS allows dense sintering of ceramic masses (Figure 2a).

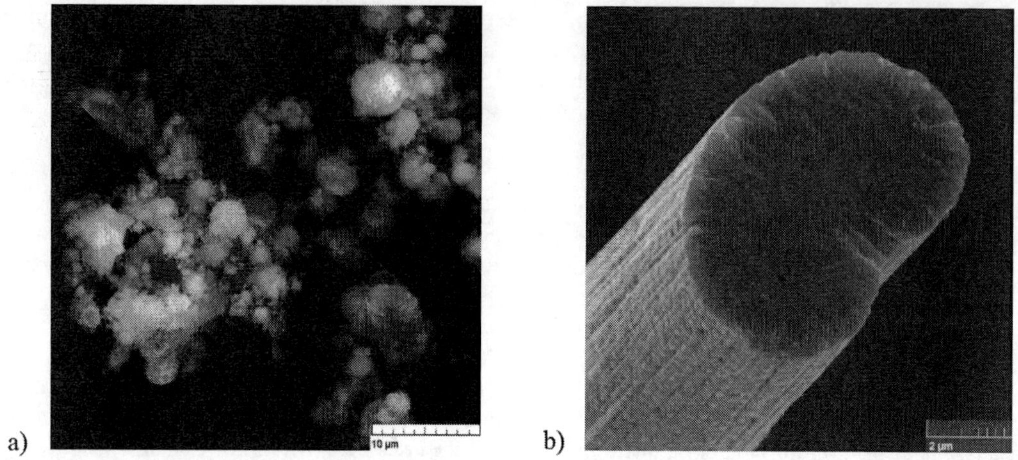

Figure 2. SEM a - silicon carbide powder obtained by SHS method, b - silicon carbide fiber obtained by siliconizing silicon fabric with SiO vapors.

Figure 2b shows a silicon carbide fiber obtained by siliconizing a carbon fabric with SiO vapour. The resulting fabric, consisting of continuous SiC fibers, the diameter of which is 5 µm, was mechanically crushed to a size of 1-2 mm. The fiber content in the samples ranged from 3 to 7 wt.%.

Due to the presence of a high proportion of Si-C covalent bond, the compaction of silicon carbide ceramics is difficult to realize without the participation of sintering additives. The most commonly used mixture of Y_2O_3 – Al_2O_3 oxides in a ratio of 3:5, which formed yttrium-aluminium garnet during sintering, was chosen as a sintering additive.

The starting components mixture was carried out in a planetary mill in the environment of isopropyl alcohol. The samples were formed by the method of one-sided semi-dry pressing. Sintering was carried out by hot pressing in argon at 1850°C for 30 minutes.

3. Results and discussion

In this paper, the microstructure and some physicochemical properties of hot-pressed silicon carbide reinforced silicon carbide materials obtained from SHS silicon carbide powders were studied (table 1). The SEM analysis of the obtained ceramic samples showed the absence of interaction between the matrix material and the reinforcing component (Fig. 3). The absence of interaction between the matrix material and the reinforcing component indicates that the resulting ceramics is a composite material and the implementation of ceramic reinforcement is carried out successfully.

Table 1. Mechanical properties of SiC/SiC_f ceramic matrix composites

SiC SHS, wt. %	SiC fibre, wt. %	Firing conditions	Density, g/sm³	Strength, MPa	K_{IC}, MPa·m$^{1/2}$
100	0	T=1850°C,	3,19	370±24	4,7
97	3	1 h,	3,20	455±22	5,2
95	5	Ar	3,20	471±22	5,7
93	7		3,20	520±26	5,9

IOP Conf. Series: Materials Science and Engineering **848** (2020) 012113 doi:10.1088/1757-899X/848/1/012113

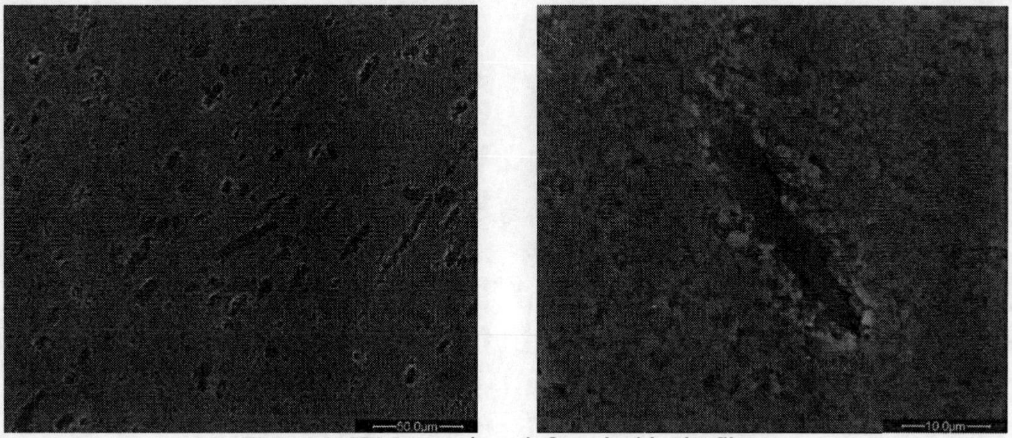

Figure 3. SEM ceramics reinforced with SiC fibers

The physicochemical properties of the resulting densely sintered composites were studied. The fibers concentration increases in SiC matrix leads to an increase in the mechanical properties of the composite. The maximum physicomechanical properties of ceramics based on silicon carbide, reinforced with SiC fibers, were the following. The samples density was obtained 3.20 g/cm^3, strength 520 MPa (for monolithic ceramics 390 MPa), the value of the crack resistance coefficient 5.9 MPa·m$^{1/2}$.

4. Conclusion.

The silicon carbide powder was used as the initial components of the SHS, due to the spherical shape of the particles. It made possible to obtain a denser workpiece. The use of silicon carbide fibers as a reinforcing material is a promising area. Reinforcing components tend to increase fracture toughness by redirecting the crack paths and diverting the strain energy, or minimizing its growth. Thus, reinforcing ceramics with fibers has a positive effect on the basic physicochemical properties of silicon carbide ceramics; namely, it makes it possible to increase the values of strength and fracture toughness coefficient. This is due to the level of physicomechanical properties of modern fibrous reinforcing fillers, and to the implementation of a more complex, in comparison with monolithic ceramics, mechanism of destruction of a ceramic composite, which is especially true for impact loads [1].

Acknowledgments
The work was carried out with the financial support of the program UMNIK № 14009GU/2019.

References

[1] Xie Y, Cheng L, Li L, Mei H, Zhang L 2013 *Journal of the European Ceramic Society* **33(10)** 1701
[2] Li M, Zhou X, Yang H, Du S, Huang Q 2018 *Scripta Materialia* **143** 149
[3] Patel M, Singh V, Singh S, Prasad V B 2018 *Materials Characterization* **135** 71
[4] M Frolova *et al* 2019 *IOP Conf. Ser.: Mater. Sci. Eng.* **525** 012085
[5] Bansal N P, Lamon J 2014 *Ceramic matrix composites: materials, modeling and technology* (New Jersey: John Wiley & Sons) p 712
[6] Bunsell A R, Piant A 2006 *Journal of Materials Science* **41(3)** 823
[7] E Istomina *et al* 2019 *IOP Conf. Ser.: Mater. Sci. Eng.* **525** 012059

IOP Conf. Series: Materials Science and Engineering **848** (2020) 012114 doi:10.1088/1757-899X/848/1/012114

Studies of the effectiveness of the chemical decomposition of polychlorinated biphenyls in alluvial meadow soils in situ under the influence of special physicochemical active compounds

D Yu Aladin, S M Sevostyanov, N Ph Deeva, D V Demin

Institute of Basic Biological Problems Russian Academy of Sciences, Pushchino, Moscow state, Russia
e-mail: aladindanila@gmail.com

Abstract. Polychlorinated biphenyls (PCB) are persistent organic pollutants used worldwide up 1990s. Although their use has been heavily restricted, PCB can be found in contaminated soils and sediments. The interaction of polychlorobiphenyls (PCBs) in soils with an alkaline amino acid reagent (NaL) and humic acids was studied. According to IR spectroscopy, intermolecular interaction of PCBs with NaL reagent was revealed. Chromatography and IR spectroscopy revealed the transformation and removal of chlorine in the aromatic ring of PCB macromolecules under the influence of an amino acid reagent. In the experiment, from 32 to 77% of PCBs undergo soil degradation, depending on the concentration of the reagent and the frequency of its introduction into soil samples.

1. Introduction

Among POPs, PCBs are among the most common. They were mass produced and used. PCBs belong to the class of aromatic compounds consisting of two benzene rings connected via the C — C internuclear bond and substituted by one to ten chlorine atoms in the ortho, meta, or para positions. There are 209 individual PCB congeners that differ in the number and position of chlorine atoms in the molecule, having the general formula C12H10-nCln, where n = 1-10. MPCs for PCBs, including the total toxic effect of all 209 congeners, in the soils are 0.06 mg / kg. Currently, about a third of PCBs produced have been released into the environment. The remaining two thirds are in a bound state in old electrical equipment and waste. PCBs are formed in the process of burning waste and all industrial processes using chlorine. PCBs are present in all components of the environment, and have been found in the tissues of animals living in untouched wild landscapes. [1,2,3,4,5] Removal of these xenobiotics is a global problem [6]. Cleaning soils contaminated with PCBs has been a challenge for decades [7]. The most common technologies used in the remediation of PCB-contaminated soils and technogenic formations are their burning or disposal in landfills, but these solutions are destructive and unstable. The chemical inertness of polyhalogenated dioxins and resistance to high temperatures only add to the problems when searching for approaches to their destruction.

Most technologies are still in their infancy and further research is needed on this issue. Field pilot experiments are important for evaluating the effectiveness of developed technologies. There is still no technology that is applicable for soil remediation from PCBs *in situ* [8]. Each technology is a unique case influenced by a number of physicochemical and biological factors and soil properties.

Of the physical factors, the most natural and economical way of degradation is solar radiation. However, most PCB congeners do not absorb electromagnetic waves with wavelengths of more than 300 nm, and therefore direct solar degradation is not very effective. In true solutions, especially in organic solvents, PCBs are relatively easily dechlorinated under UV irradiation. The efficiency of photodegradation depends on the level of chlorination of the congeners, the position of substituents in the PCB molecule (ortho> meta> vapor) and on the phase (liquid or gas) in which the PCBs are located [9]. Thermal irreversible degradation of PCBs occurs only in fairly harsh conditions - over 1200 C and at least 2 seconds [10]. Various methods of chemical destruction of PCBs have also been proposed, for example, by alkoxylation of technical mixtures of PCBs with sodium alcoholates, which leads to a decrease in the chlorine content and, thus, a decrease in their toxicity [11]. However, as a result of such a reaction, complex mixtures of products are formed containing dozens of compounds whose toxicity has not been studied [12]. The results of laboratory experiments on the treatment of PCB contaminated soil with NaBH4 reagent, with an exposure time of 3.5 to 30 hours, showed a decrease in the level of contamination from 2600 mg / kg to <0.2 mg / kg [13]. The final products of the interaction of PCBs and the reagent in this case were biphenyl, NaCl and boric acid. All of the above methods require the removal of contaminated soils, the availability of special equipment for further chemical treatment, which is quite time-consuming and dramatically increases the cost of work. Recently, biological methods of destruction of PCBs, based on the use of special strains — fungi and microorganisms [13, 14] have been of great interest. The biological destruction of PCBs occurs with the participation of enzyme systems of various types of microorganisms and can take place in two ways - aerobic (oxidative decomposition), and anaerobic (reductive dechlorination). These two processes proceed in different environmental conditions with the formation of specific metabolites. At the same time, bacterial dechlorination and degradation are not effective for PCB congeners with more than six chlorine atoms [15]. Microbiological methods have low efficiency and high cost, and the problem of controlling the possible mutation of microorganisms becomes very urgent, which makes the expediency of introducing them into natural environments very problematic. The use of the combined method of biological destruction is based on the introduction of sorbents (activated carbon, peat or other sorbents) into the soil that can absorb excess xenobiotics and a bacterial preparation of microorganisms-destructors. The sorbent provides a gradual release of the toxicant into the soil solution at concentrations non-toxic to the bacteria-destructors, which allows them to almost completely decompose the xenobiotic [16]. This method requires constant maintenance of the sorbent in a state of optimal moisture and aeration.

Therefore, all the above methods of detoxification of PCBs have one or another disadvantage. Successful soil restoration depends on the proper selection, design and regulation of remediation technology. Recent studies show that the use of a combined approach using the basics of biological and chemical destruction for soil remediation from PCBs is promising. In this regard, the search for cost-effective recovery technologies to remove these persistent organic pollutants from the environment is an urgent and sought-after problem for the whole world. The solution to this fundamental scientific problem can lead to the creation of a cost-effective technology. Therefore, the search for new effective methods for the rehabilitation of soils contaminated with organochlorine compounds is an urgent task. A promising area is a combination of chemical and biological methods.

2. Object and research methods

It is known that organochlorine compounds are able to interact with derivatives of primary and secondary amines, including simple and complex amino acids [17]. In this regard, we developed a method for the destruction of PCBs in soils using special amino acid-based reagents. As a source of amino acids, we used a reagent, which is an alkaline hydrolyzate obtained from waste leather and fur production, the main component of which is the sodium salts of amino acids (NaL).

The destruction of PCBs by an amino acid reagent was assessed on three samples of the surface humus horizons of alluvial soils from the basin of the Borovlyanka stream in Serpukhov. For all samples, some alkalization of the soil solution was noted (pH 7.60, 8.20, and 7.8), which is typical for urban soils. According to the granulometric composition, samples 1 and 3 belong to light loams, and sample 2 to medium loams. The humus content is 3.1, 4.2 and 3.2%, respectively.

A study on the destruction of PCBs with an amino acid reagent in substrates was evaluated by methods: by the content of PCBs by gas-liquid chromatography (GLC) using standards, and the determination of PCB congeners by GLC / LRMS (low resolution mass spectrometry) or GLC / HRMS (high resolution mass spectrometry).

From each moistened to 60-70% of the lowest moisture capacity of the sample, 3 samples of 500 g in terms of absolute dry weight were taken, placed in containers and processed with various doses of NaL. The treatment was carried out in two dose options: 17 ml of NaL solution (2 mol / L) and 34 ml of NaL solution (4 mol / L). The samples were mixed, left to undergo a reaction between NaL and PCB for 10 days, periodically mixed.

3.Results and discussion

The analyzes performed showed a several-fold decrease in PCB content in soil samples after NaL treatment (Table 1).

Table 1. Change in PCB content in soil samples after their chemical destruction of NaL (μg / kg).

Soil sample no.	The initial amount of PCB, μg / kg	The amount of PCBs after treatment with NaL (85 ml / kg), μg / kg	The amount of PCBs after treatment with NaL (170 ml / kg), μg / kg
1	85,6	20,7	21,5
2	107,6	26,2	20,4
3	320016	12872	5203

When treating the soil with a reagent in a dose of 17 ml of NaL solution (2 mol / L), the PCB content in the first and second samples decreased by 4.1 times, and in the third - by 24.9, respectively. When processing a double dose of reagent (170 ml / kg) for the first and second samples, approximately the same data were obtained. The same amount of undestructed PCB in the first and second samples at two doses of NaL may indicate that the residual PCB is tightly bound to the mineral part or humic substance of the soil. However, in the humus content, these soils vary greatly, therefore, it can be assumed that this amount of PCB is sorbed by the mineral part of the soil, as a result of which it is not accessible for chemical destruction during a single treatment.

In the third sample, PCB at a double dose of the reagent decreases by 61.5 times (by 98.4%). The bulk of PCBs is destroyed, however, in this case, the amount of reagent was not enough, since the soil remains significantly contaminated, and some PCBs are chemically accessible for destruction.

At the next stage, experiments on the destruction of PCBs were carried out in the field *in situ* in the most polluted area of the river basin Borovlyanka.

To determine the effect of amino acid salts on the destruction of PCBs in soils, experimental sites of 0.25 m2 were laid, to limit the area of which square tin boxes were used, which were driven into the soil to a depth of 20 cm, due to this they were isolated from the secondary side input of PCBs into upper horizon. The sides of the boxes, protruding above the soil, allowed for processing without spreading the reagent. Processing was carried out manually.

In the soils of the first square, which was treated with a minimum dose of 1.9 L of reagent, the PCB content decreased 2.95 times and amounted to 20359.9 μg / kg - 339 MPC. The ratio of congener groups did not change much, the sum of light fractions is 49.83% (Figure 2)

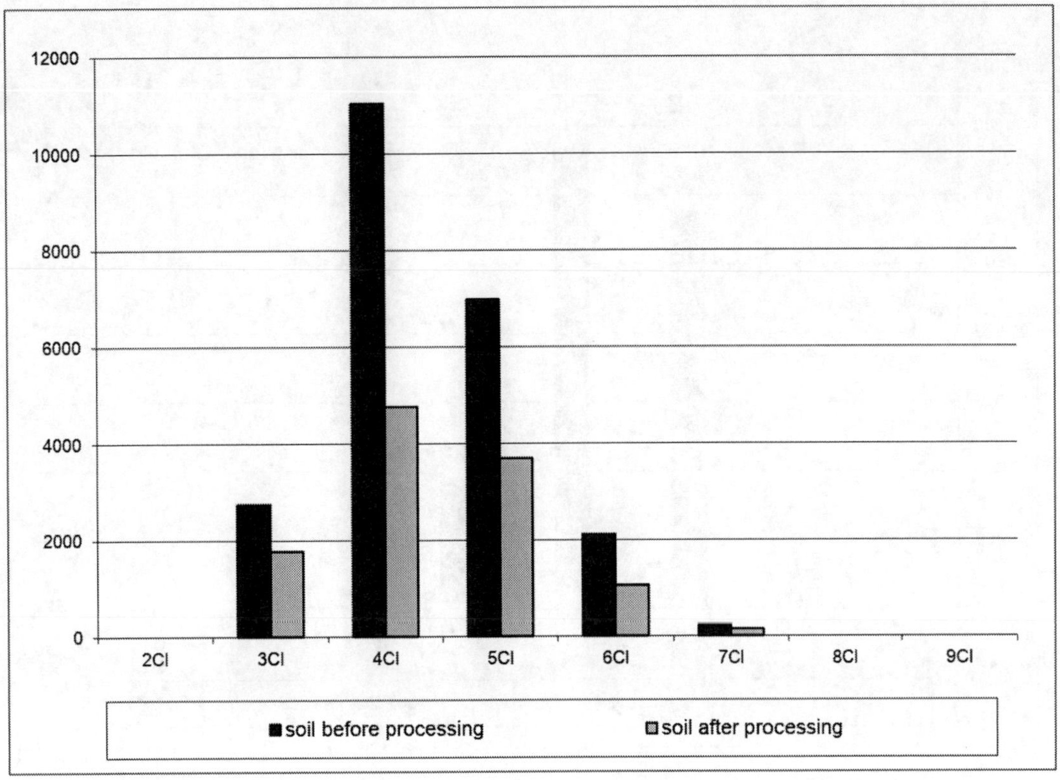

Figure 1. Concentration PCBs before processing by an aminoacidic reactant №1, mkg/kg.

In the soils of the 2nd square, where PCB was initially contained significantly less compared to other studied areas, with the introduction of the minimum dose of the reagent, the decrease in the level of contamination turned out to be the greatest - by 10.2 times, to 332.05 µg / kg, which is 5.5 MPC. The content of light fractions increased to 52% (Figure 2).

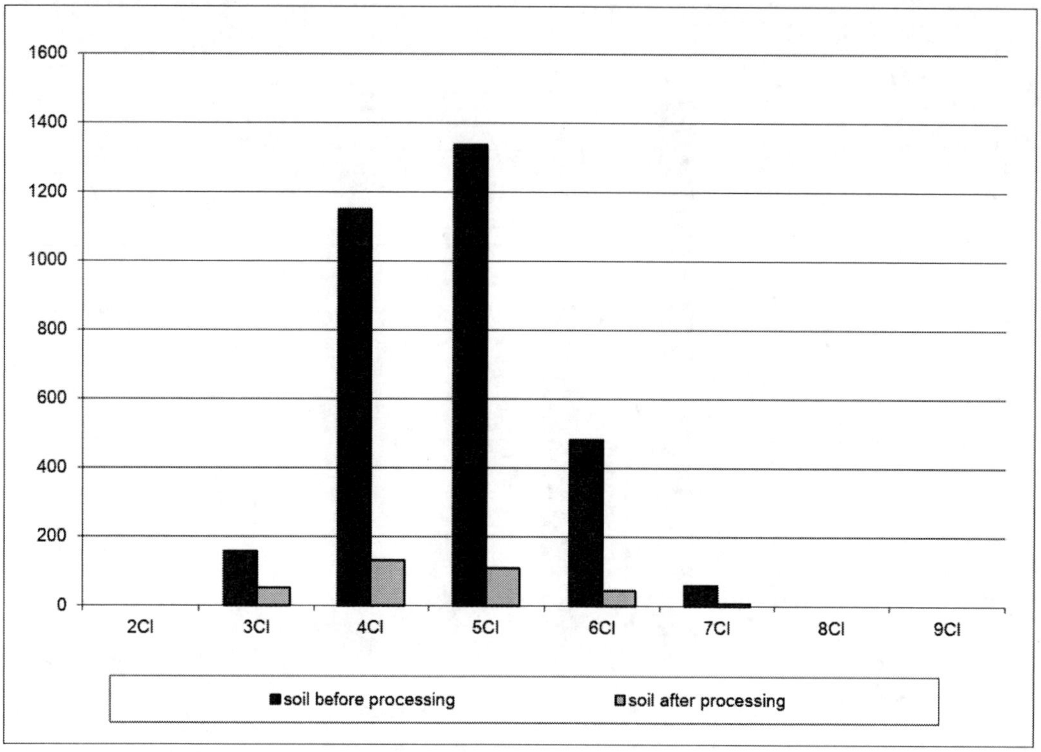

Figure 2. Concentration PCBs before processing by an aminoacidic reactant №2, mkg/kg.

In the soils of the third square, where the dose of the reagent was also minimal, the pollution level decreased 2.2 times - 11146 μg / kg (185.8 MPC) (Figure 3).

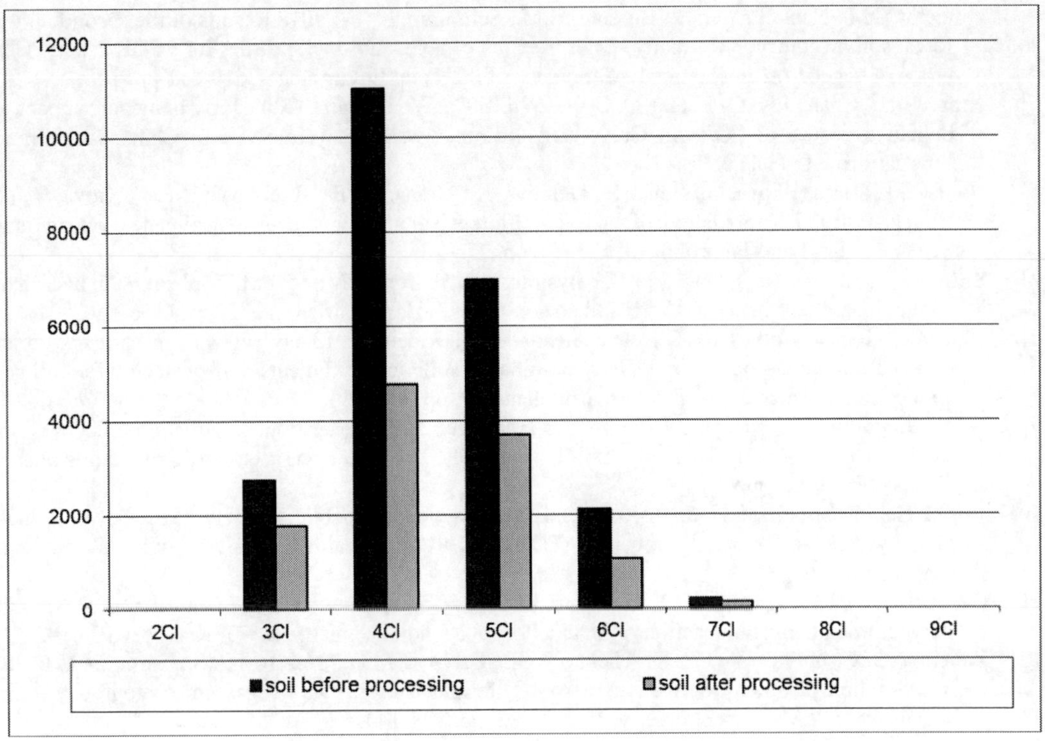

Figure 3. Concentration PCBs before processing by an aminoacidic reactant №3, mkg/kg.

Thus, after the introduction of various doses of the reagent, the level of PCB pollution decreased relative to the initial one: 1 square - by 75.8%, 2 - by 75.7%, and 3 square - by 98.4%.

4. Conclusions

1. Experiments on the detoxification of PCBs in samples from the upper soil horizons under laboratory conditions showed a several-fold decrease in the content of persistent organic compounds
2. The dose of the amino acid reagent 85 ml / kg (concentration 2 mol / l) leads to the destruction of the entire chemically available amount of PCBs for a given exposure time (10 days)
3. A change in the ratio of homologous groups of congeners in soils indicates the interaction of PCBs with the amino acid composition and the phased replacement of chlorine ions with an amino acid radical.

As a result, it can be emphasized that in experiments with field samples, positive results were obtained on the destruction of PCBs by amino acid salts. However, it is necessary to continue experimental studies to determine the doses of this reagent for soils with an extremely high degree of PCB contamination.

References

[1] Lupachev A.V., Deeva N.Ph., Aladin D.Yu., Sevostyanov S.M., Demin D.V. 2017 Analysis of pollution with oil products and organochlorines of soils in the vicinity of Russian Antarctic stations. Theoretic and applied ecology. № 2, pp. 20-24.

[2] S.G. Donaldson, J. Van Oostdam, C. Tikhonov, M. Feeley, B. Armstrong, P. Ayotte, et al. (2010) Environmental contaminants and human health in the Canadian Arctic. Sci Total Environ, 408 (22) (2010), pp. 5165-5234

[3] N.B. Hopf, M.A. Waters, AM. Ruder 2009 Cumulative exposure estimates for polychlorinated biphenyls using a job-exposure matrix. Chemosphere, Volume 76, Issue 2, Pages 185-193

IOP Conf. Series: Materials Science and Engineering **848** (2020) 012114 doi:10.1088/1757-899X/848/1/012114

[4] Pandelova, M., Henkelmann, B., Bussian, B.M., Schramm, K.-W. 2018 Results of the second national forest soil inventory in Germany - Interpretation of level and stock profiles for PCDD/F and PCB in terms of vegetation and humus type. Science of the Total Environment, 610-611 pp. 1-9

[5] Xing, G.H., Chan, J.K.Y., Leung, A.O.W., Wu, S.C., Wong, M.H. 2009 Environmental impact and human exposure to PCBs in Guiyu, an electronic waste recycling site in China. Environment International, 35 (1), pp. 76-82.

[6] Haluška, L., Barančíková, G., Baláž, Š., Dercová, K., Vrana, B., Paz-Weisshaar, M., Furčiová, E., (...), Bielek, P. 1995 Degradation of PCB in different soils by inoculated Alcaligenes xylosoxidans. Science of the Total Environment, 175 (3), pp. 275-285.

[7] S. Agarwal, S.R. Al Abed, D.D. 2009 Dionysiou A feasibility study on Pd/Mg application in historically contaminated sediments and PCB spiked substrates J. Hazard. Mater., 172, pp. 1156-1162

[8] Helena I. Gomes, Celia Dias-Ferreira, Alexandra B. Ribeiro 2013 Overview of in situ and ex situ remediation technologies for PCB-contaminated soils and sediments and obstacles for full-scale application. Science of the Total Environment 445, pp. 237-260

[9] G.G. Choudhry, J.A. van den Broecke, O. Hutzinger 1983 Formation of polychlorodibenzofurans (PCDFs) by the photolysis of polychlorobenzenes (PCBs) in aqueous acetonitrile containing phenols. Chemosphere, 12, pp. 487-492

[10] World Health Organization & International Programme on Chemical Safety 1992 Polychlorinated biphenyls, polychlorinated terphenyls (PCBs and PCTs : health and safety guide. World Health Organization, p.52

[11] Gorbunova T.I., Zapevalov A.Ya., Kirichenko V.E., Saloutin V.I., Chupakhin O.N. 2000 Polychlorodiphenyls in reactions with alcohols // Zh. adj. chemistry. V. 73. №4, pp. 610-614.

[12] Zabelina O.N., Yatluk Yu.G., Kirichenko V.E., Pervova M.G., Nazarov A.S., Saloutin V.I. 2005 Investigation of alkoxy derivatives of polychlorinated biphenyls by gas chromatography / mass spectrometry // Mass spectrometry. V. 2, No. 2. pp. 139-144.

[13] M. Aresta, P. Caramuscio, L. De Stefano, T. Pastore 2003 Solid state dehalogenation of PCBs in contaminated soil using NaBH4 Waste Manag., 23, pp. 315-319.

[14] P.A. Correa, L. Lin, C.L. Just, D. Hu, K.C. Hornbuckle, J.L. Schnoor, B.V. Aken 2010 The effects of individual PCB congeners on the soil bacterial community structure and the abundance of biphenyl dioxygenase genes. Environ. Int., 36, pp. 901-906.

[15] B.A. Mayes, G.L. Brown, F.J. Mondello, K.W. Holtzclaw, S.B. Hamilton, A.A. Ramsey 2002 Dermal absorption in rhesus monkeys of polychlorinated biphenyls from soil contaminated with Aroclor 1260. Regul Toxicol Pharmacol, 35, pp. 289-295

[16] M.S. Sokolov, O.A. Monastic, E.A. Pikushova 1994 Ecologization of plant protection; Pushchino: ONTI PNC RAS, 462 p.

[17] Demin D.V., Aladin D.Yu., Deeva N.F., Sevostyanov S.M. 2019 Structural - chemical studies of the destruction oforganochlorine compounds. IOP Conf. Series: Materials Science and Engineering, 525 (2019), 012105.

AUTHOR INDEX

Abdulmenova, E V 1
Adinaev, X F 499
Agureev, L E 512
Akentieva, N P 5
Akhmedshin, E Kh 567
Aladin, D Yu 705, 738
Aldoshin, S M 5
Aleksandrov, A A 13
Aleksandrov, V A 182
Alymov, M I 20
Andyyakova, A 188
Anikin, V N 97
Ankudinov, A B 609
Anokhin, D V 175, 316
Anuchkin, S N 23
Apostolova, M O 332
Apukhtina, T L 532
Apyari, V V 137, 152
Artem'ev, K V 244
Ashmarin, A A 406, 412, 532
Aslanli, A G 33
Avdeeva, L V 349
Averichev, O A 42
Averin, S I 20
Badamshina, E R 175, 316
Baikin, A S 662, 674, 681, 686
Bakhteeva, N D 552, 587
Balashov, E P 662, 668, 681, 686
Barinov, S M 406
Baskov, K M 477
Batueva, T 145
Bazhin, P M 49, 56
Belachenkov, I O 78
Belikov, R S 519
Belyaeva, E V 359
Benavides, V 332
Berezina, N V 662
Blokhina, M Kh 532
Bobrovskii, S Y 538
Bocharnikova, E A 705
Bogdanov, O 62
Bolotova, L K 375
Bolotskaya, A V 56
Bolshikh, A O 305, 311
Borisova, Yu 67
Borissevitch, I 261
Borshchev, O V 72
Burdukova, O A 652
Chaynikova, A S 78
Chepusov, A S 338
Chernikova, E A 278
Chernogorova, O P 332
Chernyavskii, A S 412
Chesnokov, M A 196
Chistyaev, V A 477
Chizhikov, A P 49, 56
Chkanikov, N D 359
Chunaev, D S 85
Churakova, A A 91
Danilin, K D 343, 581
Dashevskii, V Ya 13
Deeva, N Ph 705, 738
Demin, D V 705, 738
Dmitrienko, S G 137, 152
Dmitrieva, M V 359
Dobatkin, S V 592
Dokichev, V A 117
Dolzhenko, A S 592
Dremova, N H 5
Drozdova, E I 332
Drukarenko, N A 196
Dvoyeglazov, V S 368
Dyubanov, V G 641
Efremenko, E N 33, 525
Egorov, A A 406
Ekimov, E A 332
Eliseeva, E A 182
Eremeeva, Zh V 512
Eremin, S A 97
Ershov, A E 107
Fakhreeva, A V 117
Fedorov, S V 124, 637, 734
Fedoseev, A E 129
Fedoseeva, A E 129
Fedotov, A Yu 382
Frolov, S M 711
Frolova, M G 213, 343, 581, 734
Furletov, A A 137
Galstyan, A M 615
Garshev, A V 137, 152
Gizatullin, A R 5
Glukhov, L M 278
Golubko, N V 468
Gorbatov, R O 507
Gorbatsrvich, O B 278
Gorbunova, M A 175, 316

Gorbunova, M V ... 152
Gorbunova, M ... 145
Gorshkov, V A ... 562
Grabchikov, S S ... 571
Grabchikova, E A ... 571
Grachev, V P ... 238
Gridin, D V ... 343
Grudinsky, P I ... 641
Gryadunov, K I ... 163
Gumennikova, E A ... 213, 581
Guro, V P ... 499
Gusein-Zade, N G ... 244
Gvozdeva, O N ... 167
Harin, E V ... 538
Howdle, S ... 557
Ilina, T S ... 468
Imamutdinova, A M ... 175
Isaeva, I Yu ... 182
Istomina, E I ... 734
Ivannikov, A Yu ... 217, 250, 717, 724
Ivanov, A V ... 512
Ivanov, B S ... 512
Ivanova, D D ... 298
Ivanova, N ... 188
Ivicheva, S N ... 730
Kaibyshev, R O ... 129
Kalagin, I V ... 571
Kalaida, T A ... 217
Kalashnilov, I E ... 375
Kalayda, T A ... 724
Kaleva, G M ... 468
Kalinenko, A ... 67
Kamantsev, I S ... 196
Kambalin, I O ... 322
Kaminskaya, T P ... 552
Kanevskii, A G ... 13
Kannykin, S V ... 587
Kapasharov, A T ... 628
Kaplan, M A ... 217, 668, 674
Karasik, A Y ... 85
Kardashova, G D ... 200
Kargin, Yu F ... 213, 343, 581, 730, 734
Kazakov, A I ... 349
Kazakova, V N ... 412
Khakhalin, A V ... 615
Khomenko, N Yu ... 56
Khvatov, A V ... 705, 711
Kiiko, V M ... 205
Kim, K A ... 213, 343, 444, 734
Kirsankin, A A ... 217, 724
Kiryukhantsev-Korneev, Ph V ... 538, 562
Kisel, V N ... 477
Kiselev, D A ... 468
Kislyuk, A M ... 468
Klenin, A A ... 507
Klimashin, A A ... 637, 730
Knerelman, E I ... 628
Kochanov, G P ... 412
Kolesnikova, A M ... 97
Kolik, L V ... 244
Kolmachikhina, E B ... 222
Kolmakov, A G ... 375, 495, 662
Kolmakova, A A ... 662
Kolmykov, R ... 188
Kolobkova, E ... 62
Kolosova, O Yu ... 229
Komendant, R I ... 238
Komlev, V S ... 382, 406
Konchekov, E M ... 244
Konev, D V ... 288
Kononikhin, A S ... 705
Konstantinov, A S ... 56
Konushkin, S V ... 250, 662, 668, 674, 686
Konyshkin, V A ... 652
Konyushkin, V A ... 85
Kopylov, Yu L ... 652
Kopytskiy, V O ... 456
Korovushkin, V V ... 552
Kostikov, V I ... 512
Kostin, A Yu ... 628
Kostina, M V ... 255, 692
Kostina, V S ... 255, 692
Kostitsyna, E V ... 56
Kostyukov, A ... 261
Kovalev, I A ... 412
Kovaleva, V S ... 78
Kovba, L V ... 368
Kozlovskaya, E A ... 272
Krasovskiy, V G ... 278
Krumins, E ... 557
Kseneva, D S ... 272
Kudryashov, O Yu ... 97
Kulbakin, I V ... 124, 637
Kulikov, A V ... 288
Kulinchenko, N A ... 507
Kulkov, S N ... 1
Kurmaz, S V ... 288
Kurmaz, V A ... 288
Kurochkin, S A ... 238
Kuryakov, V N ... 298
Kuzmin, V ... 261
Kuznetsov, A V ... 196
Kuznetsova, O G ... 305, 311
Laptev, I N ... 512

Larionov, M D ... 495
Larionova, A V ... 495
Lastovskiy, S B ... 571
Lelecova, V A ... 316
Lemeshev, D O ... 213, 272, 343, 428, 581, 602
Leontiev, I A ... 97
Levashov, E A ... 562
Levin, A M ... 305, 311
Lobanov, A ... 188
Lodygina, V P ... 175
Lomakin, S M ... 705, 711
Lopukhin, K V ... 652
Lubyanoi, D A ... 322, 327
Lubyanoi, D D ... 322, 327
Lukanina, Yu K ... 705, 711
Lukina, I N ... 332
Lukyanova, O A ... 444
Lukyashin, K E ... 338
Lunev, V A ... 592
Lysenkov, A S ... 124, 213, 343, 395, 444, 581, 730, 734
Magadova, L A ... 117
Maklakov, S A ... 477
Maklakov, S S ... 477
Malkov, G V ... 316, 628
Mariyanats, A O ... 697
Markidonov, A V ... 322, 327
Markov, M A ... 439
Markova, A A ... 359
Maslova, O V ... 525
Medyankina, I S ... 423
Melnikova, I S ... 163, 368
Mestergazi, M ... 261
Mikhailova, A V ... 662
Mikheev, R S ... 375
Mikheeva, P V ... 382
Milyaev, I M ... 609, 622
Minaeva, S ... 557
Minakov, A A ... 388
Minikh, A A ... 711
Mironov, A V ... 697
Mishin, A D ... 477
Mitrofanov, S V ... 571
Mitschenko, D V ... 5
Modin, S Y ... 78
Mogucheva, A ... 67
Morenko, I V ... 182
Morozov, M M ... 592
Morozova, Y A ... 662
Mosunov, A V ... 468
Motailo, E S ... 439
Motaylo, E S ... 433
Muradyan, O S ... 692
Muradyan, S O ... 255, 692
Nakladov, A N ... 85
Nasakina, E O ... 250, 668, 674, 681, 686
Naumkin, A V ... 485
Nemchikov, M L ... 163
Nesmelov, D D ... 395, 401, 433
Nevolin, V M ... 322
Nieder, J B ... 5
Nifant'ev, I E ... 697
Nikitin, I S ... 129
Nikitina, Y O ... 406
Novokovskaya, E A ... 532, 581
Novoselov, E S ... 395, 401, 433
Odessky, P D ... 592
Ogarkov, A I ... 412
Oparina, I B ... 419
Ordan'yan, S S ... 401
Ostaeva, G Yu ... 182
Ovsyannikov, N A ... 412, 730
Palyulin, V ... 261
Panova, M O ... 641
Pasechnik, L A ... 423
Pedchenko, M S ... 428
Perepelitsina, E O ... 238
Perevislov, S N ... 213, 395, 401, 433, 439, 444, 451, 734
Petelin, A L ... 641
Petrakova, N V ... 343, 406, 581
Petrov, D A ... 477
Petrov, E V ... 456
Petukhov, V A ... 652
Pivovarchik, S V ... 495
Plokhihk, A I ... 388
Plokhikh, A I ... 343
Plotnichenko, V G ... 615
Podrugina, T ... 261
Podzorova, M V ... 462
Pogonin, V ... 261
Polikevich, K B ... 388
Polilov, A N ... 567
Politiko, A A ... 477
Politova, E D ... 468
Polozov, V I ... 477
Ponomarenko, S A ... 72
Popov, V K ... 697
Popov, V ... 557
Popova, A ... 188
Pribytkov, P V ... 485
Prichodchenko, T P ... 5
Prokhorov, D V ... 107
Prokopenko, D A ... 332
Prokopets, A D ... 56
Prosvirnin, D V ... 495

Pruckov, M E 495
Pulish, A V 244
Radchenko, A S 359
Radchenko, E 261
Rasulova, S N 499
Ratenko, D V 368
Romanov, A D 507
Romanov, I D 507
Rozanov, K N 538
Rybalchenko, G V 592
Rybalchenko, O V 592
Ryzhkova, E A 222
Sadovskaya, N V 468
Safarov, E T 499
Saikov, I V 20
Saikova, G R 456
Sakharov, P A 705, 711
Samodurov, I O 255
Sanina, N A 5
Saratovskikh, E A 349
Savchenko, A G 552
Savchenko, E S 552
Savchuk, O 5
Savushkina, S V 512
Sedenkov, P N 298
Semenov, A F 641
Senatsky, Yu V 652
Senchenko, V N 519
Senina, M O 272, 428, 602
Senko, O V 525
Sen', V D 288
Sergienko, K V 250, 662
Sergiyenko, K V 668, 674
Sevostyanov, M A 217, 250, 305, 311, 662, 724
Sevostyanov, S M 705, 738
Sevost'yanov, M A 668, 674, 681, 686
Shalin, A V 167
Sharapov, M G 255
Shatova, L A 250, 674, 681
Shcherbakova, G I 532, 581
Shcherbakova, O Yu 451
Shcherbakova, V A 349
Sheftel, E N 538
Shelepin, I V 637
Shevchenko, S Yu 327
Shevtsov, S V 412
Shienok, V V 419
Shipko, M N 552
Shkondina, N I 5
Shlyakhtin, A V 697
Shmykova, A 261
Shmytko, I M 107
Shokodko, A V 412
Shokodko, E A 412
Shornikov, D P 412
Shtil', A A 359
Shukhardin, D M 316
Shustov, V S 609
Sigan, A L 359
Silvestre, O 5
Skorotetcky, M S 72
Skvortsova, S V 167
Smirnov, I V 382
Smirnov, M A 217
Soldatov, A V 332
Solntsev, K A 412
Solomonov, V I 338
Son, B K 705
Spassky, D A 615
Spirikhin, L V 117
Starkova, D V 222
Stashenko, V I 544
Stefanovich, S Yu 468
Stepanov, S I 196
Stepovich, M A 552
Stepushin, A S 167
Stolbova, M O 343
Stolin, A M 42, 49, 56
Storozhenko, P A 532
Strelnikova, S S 412
Stroganova, T S 107
Sudarchikova, M A 668, 674, 681
Surin, N M 72
Svidchenko, E A 72
Syachina, M 557
Sytchenko, A D 562
Tatus, N A 567
Tedzhetov, V A 538
Telin, A G 117
Tenishev, A V 412, 519
Tertyshnaya, Yu V 462
Teterina, A Yu 382
Timoshenko, A N 163
Tishkevich, D I 571
Titov, D D 213, 343, 401, 406, 439, 532, 581, 734
Tkachenko, A N 298
Tkachev, E S 129
Todorova, E V 552, 587
Tokar, A A 592
Tolmacheva, V V 152
Torbov, V I 5
Trofimov, V S 456
Troitsky, O A 544
Troshin, P A 5

Trukhanov, A V ... 571
Tsareva, A M ... 668, 674, 681, 686
Tsybin, O I ... 305, 311
Ulyanova, A V ... 602
Usmanova, G Sh ... 538
Ustyukhin, A S ... 609
Vaganov, E V ... 238
Val'nev, V A ... 222
Varfolomeev, M S ... 532
Varfolomeev, S D ... 711
Vasil'ev, D A ... 615
Vasil'eva, N V ... 615
Vasil'kov, A Yu ... 485
Vasin, D S ... 571
Vikhman, S V ... 395, 439
Vinnik, D A ... 571
Viryus, A A ... 552
Vlasova, D V ... 388
Volkov, P A ... 137, 152
Voloshin, A I ... 117
Vompe, T A ... 622
Vorkachev, K G ... 412
Voronov, V V ... 615
Voytik, A E ... 628
Vysotin, A B ... 444, 451
Yakimchuk, D V ... 571
Yakushevich, A S ... 571
Yarullin, R N ... 349
Yashnov, Y M ... 97
Yatsenko, S P ... 423
Yusupov, V S ... 592
Yuzbekova, D ... 67
Zabelin, D A ... 78
Zagaynov, I V ... 637
Zakharov, Yu ... 188
Zakorzhevsky, V V ... 734
Zakunov, A S ... 641
Zaytsev, V D ... 137
Zelensky, V A ... 609
Zezyulina, P A ... 538
Zhdanova, M V ... 97
Zherebker, A Yu ... 705
Zhigalov, D V ... 532
Zhilenkov, A V ... 5
Zinoveev, D V ... 641
Zubar, T I ... 571
Zvarych, E B ... 327
Zverev, P G ... 85, 652